Mathematical Modeling of Food Processing

Contemporary Food Engineering

Series Editor

Professor Da-Wen Sun, Director

Food Refrigeration & Computerized Food Technology
National University of Ireland, Dublin
(University College Dublin)
Dublin, Ireland
http://www.ucd.ie/sun/

Contemporary Food
Engineering Series
Da-Wen Sun, Series Editor

Mathematical
Modeling of
Food Processing

Edited by Mohammed M. Farid

CRC Press
Taylor & Francis Group
Boca Raton London New York

CRC Press is an imprint of the
Taylor & Francis Group, an **informa** business

Published 2010 by CRC Press
Taylor & Francis Group
6000 Broken Sound Parkway NW, Suite 300
Boca Raton, FL 33487-2742

© 2010 by Taylor & Francis Group, LLC
CRC Press is an imprint of the Taylor & Francis Group, an informa business

First issued in paperback 2019

No claim to original U.S. Government works

ISBN 13: 978-0-367-45234-6 (pbk)
ISBN 13: 978-1-4200-5351-7 (hbk)

Library of Congress Cataloging-in-Publication Data

Mathematical modeling of food processing / editor, Mohammed M. Farid.
    p. cm. -- (Contemporary food engineering)
  Includes bibliographical references and index.
  ISBN 978-1-4200-5351-7 (hardcover : alk. paper)
  1. Food industry and trade--Mathematical models. I. Farid, Mohammed M.

TP370.9.M38M38 2010
664'.02015118--dc22   2009028738

Visit the Taylor & Francis Web site at
http://www.taylorandfrancis.com

and the CRC Press Web site at
http://www.crcpress.com

Contents

SECTION I Basic Principles

SECTION II Heat Transfer in Food Processing

SECTION III   Simultaneous Heat and Mass Transfer in Food Processing

SECTION IV   Low Temperature Thermal Processing

SECTION V   Non-Thermal Processing of Food

SECTION VI   Other Thermal Processing

SECTION VII   Mass and Momentum Transfer in Food Processing

SECTION VIII   Biofilm and Bioreactors

SECTION IX   Special Topics in Food Processing

Series Editor's Preface

CONTEMPORARY FOOD ENGINEERING SERIES

Food engineering is the multidisciplinary field of applied physical sciences combined with the knowledge of product properties. Food engineers provide technological knowledge essential to the cost-effective production and commercialization of food products and services. In particular, food engineers develop and design processes and equipment in order to convert raw agricultural materials and ingredients into safe, convenient, and nutritious consumer food products. However, food engineering topics are continuously undergoing changes to meet diverse consumer demands, and the subject is being rapidly developed to reflect the market's needs.

In the development of food engineering, one of the many challenges is to employ modern tools and knowledge, such as computational materials science and nano-technology, to develop new products and processes. Simultaneously, improving food quality, safety, and security remain critical issues in food engineering. New packaging materials and techniques are being developed to provide a higher level of protection for foods, and novel preservation technologies are emerging to enhance food security and defense. Additionally, process control and automation regularly appear among the top priorities identified in food engineering. Advanced monitoring and control systems are being developed to facilitate automation and flexible food manufacturing. Furthermore, energy savings and minimization of environmental problems continue to be important food engineering issues, and significant progress is being made in waste management, efficient utilization of energy, and the reduction of effluents and emissions in food products.

Consisting of edited books, the *Contemporary Food Engineering* book series attempts to address some of the recent developments in food engineering. Advances in classical unit operations in engineering applied to food manufacturing are covered, as well as such topics as progress in the transport and storage of liquid and solid foods; heating, chilling, and freezing of foods; mass transfer in foods; chemical and biochemical aspects of food engineering and the use of kinetic analysis; dehydration, thermal processing, nonthermal processing, extrusion, liquid food concentration, membrane processes and applications in food processing; shelf-life, electronic indicators in inventory management, and sustainable technologies in food processing; and packaging, cleaning, and sanitation. These books are intended for use by professional food scientists, academics researching food engineering problems, and graduate level students.

The editors of the books are leading engineers and scientists from many parts of the world. All the editors were asked to present their books in a manner that will address the market's needs and pinpoint the cutting edge technologies in food engineering. Furthermore, all contributions are written by internationally renowned experts who have both academic and professional credentials. All authors have attempted to provide critical, comprehensive, and readily accessible information on the art and science of a relevant topic in each chapter, with reference lists to be used by readers for further information. Therefore, each book can serve as an essential reference source for students and researchers at universities and research institutions.

Da-Wen Sun
Series Editor

Preface

Excellent texts and reference books are available on food processing; however, only a few of them deal with the mathematical analysis of food processing in a comprehensive way. I have taught this topic to both engineering and food science students and know the urgent need for such a reference for teaching and research.

This book is written for food scientists, food engineers, and chemical engineers. It is important at both the undergraduate and postgraduate teaching levels. However, some chapters deal with advanced topics, which make them more suitable for the postgraduate level. The book will also be of interest to scientists and engineers who are working in the food industry. It presents an elementary treatment of the principles of heat, mass, and momentum transfer in food processing, but it also contains material presented at a high level of mathematics. This makes the book of significant interest also to mathematicians who are interested in the mathematical modeling of food processing.

In the first six chapters of the book, the fundamentals of heat, mass and momentum transfer are introduced. Conduction, convective, and radiation heat transfer are analyzed with reference to food processing. Mass diffusion and convection and the principles of Newtonian and non-Newtonian flows of liquid food and suspensions are also introduced. The reader will also be familiarized with the topic of mass and energy balances in food processing, which will include heating, cooling, freezing, evaporation, etc. A comprehensive analysis of thermal, physical, and rheological properties of foods is introduced early in the book to assist the reader in understanding the various mathematical models presented in the following chapters. The sixth chapter introduces the fundamentals of computational fluid dynamics (CFD) while its applications are left to later chapters in the book.

The major part of the text deals with specialized topics in food processing. This includes *thermal processing* such as sterilization, pasteurization, evaporation, drying, baking, infrared heating, and frying; *low temperature processing* such as freezing, chilling, vacuum cooling, refrigeration, and crystallization; *nonthermal processing* such as high pressure, pulsed electric field, ozone treatment, membrane processing, and osmotic dehydration; *nonconventional thermal processing* such as microwave, radiofrequency, and ohmic heating; and *the analysis of biological and enzyme reactors*. The book also deals with the analysis of very *specialized topics* such as the use of artificial neural network, exergy analysis, process control, and Cleaning in Place (CIP) in industry. In all these chapters, emphasis is placed on the physical and mathematical analysis, which may vary from simple analytical or empirical methods to a comprehensive analysis using CFD. With the availability of high speed computers, and the advances in computational techniques, there is a growing trend in both education and research in food science and engineering to use mathematical modeling, and this requires a high level of computational skill.

This text presents for the first time, a comprehensive and advanced mathematical analysis of transport phenomena in food with 37 chapters written by world experts from industry, research centers, and academia.

Mohammed Farid
The University of Auckland
Auckland, New Zealand

Series Editor

Born in Southern China, Professor Da-Wen Sun is a world authority in food engineering research and education, he is a Member of the Royal Irish Academy, which is the highest academic honor in Ireland. His main research activities include cooling, drying, and refrigeration processes and systems, quality and safety of food products, bioprocess simulation and optimization, and computer vision technology. His innovative studies on vacuum cooling of cooked meats, pizza quality inspection by computer vision, and edible films for shelf-life extension of fruits and vegetables have been widely reported in national and international media. Results of his work have been published in over 200 peer-reviewed journal papers and in more than 200 conference papers.

He received a BSc with honors and an MSc in mechanical engineering, and a PhD in chemical engineering in China before working at various universities in Europe. He became the first Chinese national to be permanently employed at an Irish university when he was appointed college lecturer at the National University of Ireland, Dublin (University College Dublin) in 1995, and was then promoted to senior lecturer, associate professor, and full professor. Dr. Sun is now the professor of Food and Biosystems Engineering and the director of the Food Refrigeration and Computerized Food Technology Research Group at University College Dublin.

As a leading educator in food engineering, Professor Sun has significantly contributed to the field of food engineering. He has trained many PhD students, who have made their own contributions to the industry and academia. He has also given lectures on advances in food engineering on a regular basis at academic institutions internationally and delivered keynote speeches at international conferences. As a recognized authority on food engineering, he has been conferred adjunct/visiting/consulting professorships from ten of the top universities in China, including Shanghai Jiaotong University, Zhejiang University, Harbin Institute of Technology, China Agricultural University, South China University of Technology, and Jiangnan University. In recognition of his significant contributions to food engineering worldwide and for his outstanding leadership, the International Commission of Agricultural Engineering (CIGR) awarded him the CIGR Merit Award in 2000 and 2006, and the Institution of Mechanical Engineers (IMechE) based in the U.K. named him "Food Engineer of the Year 2004." In 2008 he was awarded the CIGR Recognition Award in honor of his distinguished achievements in the top one percent of agricultural engineering scientists in the world.

He is a fellow of the Institution of Agricultural Engineers and a fellow of Engineers Ireland (the Institution of Engineers of Ireland). He has also received numerous awards for teaching and research excellence, including the President's Research Fellowship, and twice received the President's Research Award of University College Dublin. He is a member of the CIGR executive board and honorary vice-president of CIGR, editor-in-chief of *Food and Bioprocess Technology—An International Journal* (Springer), series editor of the *"Contemporary Food Engineering"* book series (CRC Press/Taylor & Francis), former editor of the *Journal of Food Engineering* (Elsevier), and an editorial board member for the *Journal of Food Engineering* (Elsevier), the *Journal of Food Process Engineering* (Blackwell), the *Sensing and Instrumentation for Food Quality and Safety* (Springer), and the *Czech Journal of Food Sciences*. He is also a chartered engineer.

Editor

Professor Mohammed Mehdi Farid obtained his PhD and MSc in chemical engineering from the University of Swansea/Wales, U.K. and his BSc in chemical engineering from the University of Baghdad, Iraq. He founded the Department of Chemical Engineering at the University of Basrah in 1983. He has also worked as a full professor at the Jordan University of Science and Technology, the University Science Malaysia, and the University of Auckland, New Zealand, where he is currently working.

He is a fellow of the Institution of Chemical Engineers, London, and an active member of a number of international societies. He has published more than 220 papers in international journals and refereed international conferences, six books, nine chapters in books, and five patents.

Professor Farid is a member of the editorial board of several journals and has been a member of the International and Advisory Committee of a number of international conferences. He has received a number of international awards and was invited as a keynote speaker to a number of international conferences.

In the Department of Chemical & Materials Engineering, he is leading the research in food engineering and energy. His research is focused on the development of innovative food processing, which includes nonthermal processing (high pressure, pulsed electric field, and ultraviolet treatment of food), thermal processing (ohmic and microwave heating and cooking), analysis of food sterilization using computational fluid dynamics, supercritical CO_2, food dehydration, and other food-related topics, namely the production of biodiesel and bioethanol from waste products such as fat, oil, and cellulosic materials. He has developed a unified theory that describes most drying methods, such as freeze drying, steam drying, spray drying, as well as frying and freezing.

Contributors

Jasim Ahmed
Polymer Source Inc.
Dorval (Montreal)
Quebec, Canada

Sajid Alavi
Department of Grain Science and Industry
Kansas State University
Manhattan, Kansas

Brent A. Anderson
Research & Development
Mars, Inc.
Hackettstown, New Jersey

Konstantia Asteriadou
Department of Chemical Engineering
University of Birmingham
West Midlands, Birmingham, United Kingdom

L.A. Campañone
Centro de Investigación y Desarrollo en
    Criotecnología de Alimentos (CIDCA)
UNLP-CONICET

and

Departamento de Ingeniería Química
Facultad Ingeniería
Universidad Nacional de La Plata
La Plata, Argentina

Xiao Dong Chen
Department of Chemical Engineering
Monash University
Clayton Campus
Victoria, Australia

Donald J. Cleland
School of Engineering and Advanced
    Technology
Massey University
Palmerston North, New Zealand

Patrick J. Cullen
School of Food Science and
    Environmental Health
Dublin Institute of Technology
Dublin, Ireland

Jalal Dehghannya
Department of Bioresource Engineering
McGill University
Montreal, Quebec, Canada

Ibrahim Dincer
University of Ontario Institute of
    Technology (UOIT)
Oshawa, Ontario, Canada

Ferruh Erdoğdu
Department of Food Engineering
University of Mersin
Çiftlikköy-Mersin, Turkey

Mohammed M. Farid
Department of Chemical and Materials
    Engineering
The University of Auckland
Auckland, New Zealand

M. Alejandra García
Centro de Investigación y Desarrollo
    en Criotecnología de Alimentos
    (CIDCA)
UNLP-CONICET
Facultad de Ciencias Exactas
Universidad Nacional de La Plata
La Plata, Argentina

A.G. Abdul Ghani
Software Design Ltd.
Auckland, New Zealand

Daisuke Hamanaka
Laboratory of Postharvest Science
Kyushu University
Higashi-ku, Fukuoka, Japan

Tatiana Koutchma
Agriculture and Agri-Food Canada
Guelph Food Research Center
Guelph, Ontario, Canada

Prabhat Kumar
Department of Food, Bioprocessing, and
    Nutrition Sciences
North Carolina State University
Raleigh, North Carolina

Onrawee Laguerre
UMR Génie Industriel Alimentaire
    Cemagref-ENSIA-INAPG-INRA
Refrigerating Process Engineering
Cemagref, BP
Antony, France

Timothy A.G. Langrish
School of Chemical and Biomolecular
    Engineering
University of Sydney
Sydney, New South Wales, Australia

Roy Zhenhu Lee
Johnson Matthey Catalysts, Inc.
West Deptford, New Jersey

Fa-De Li
Department of Food Engineering
College of Mechanical and Electronic
    Engineering
Shandong Agricultural University
Tai'an City, Shandong Province, People's
    Republic of China

Alice A. Makardij
Orica Chemnet NZ
Tauranga, New Zealand

Gordon D. Mallinson
Department of Mechanical Engineering
The University of Auckland
Auckland, New Zealand

Chenchaiah Marella
Department of Agricultural and Biosystems
    Engineering
South Dakota State University
Brookings, South Dakota

Francesco Marra
Dipartimento di Ingegneria Chimicae
    Alimentare
Università di Salerno, via Ponte Don Melillo
Fisciano (SA), Italy

Gauri S. Mittal
School of Engineering
University of Guelph
Guelph, Ontario, Canada

Jean Moureh
Refrigerating Process Engineering
    Research Unit
Cemagref, BP
Antony, France

Kasiviswanathan Muthukumarappan
Department of Agricultural and Biosystems
    Engineering
South Dakota State University
Brookings, South Dakota

and

UCD School of Agriculture
Food Science and Veterinary Medicine
University College Dublin
Belfield, Dublin, Ireland

Michael Ngadi
Department of Bioresource Engineering
McGill University
Montreal, Quebec, Canada

Stuart E. Norris
Department of Mechanical Engineering
The University of Auckland
Auckland, New Zealand

Colm P. O'Donnell
UCD School of Agriculture
Food Science and Veterinary Medicine
University College Dublin
Dublin, Ireland

Quang Tuan Pham
School of Chemical Sciences and Engineering
University of New South Wales
Sydney, Australia

K.P. Sandeep
Department of Food, Bioprocessing, and
    Nutrition Sciences
North Carolina State University
Raleigh, North Carolina

Da-Wen Sun
Agriculture and Food Science Center
National University of Ireland
Dublin, Ireland

Paul Takhistov
Department of Food Science
Rutgers, The State University of
    New Jersey
New Brunswick, New Jersey

Fumihiko Tanaka
Laboratory of Postharvest Science
Kyushu University
Higashi-ku, Fukuoka, Japan

Nantawan Therdthai
Department of Product Development
Kasetsart University
Chatuchak, Bangkok, Thailand

Jianfeng Wang
School of Engineering and Advanced
    Technology
Massey University
Palmerston North, New Zealand

Lijun Wang
Biological Engineering Program
North Carolina Agricultural and Technical
    State University
Greensboro, North Carolina

Garth Wilson
Department of Chemical and Materials
    Engineering
The University of Auckland
Auckland, New Zealand

Vadim Yakovlev
Department of Mathematical Sciences
Worcester Polytechnic Institute
Worcester, Massachusetts

Zhengcai Ye
School of Chemical and Biomolecular
    Engineering
Georgia Institute of Technology
Atlanta, Georgia

Brent R. Young
Department of Chemical and Materials
    Engineering
The University of Auckland
Auckland, New Zealand

Noemi E. Zaritzky
Centro de Investigación y Desarrollo en
    Criotecnología de Alimentos (CIDCA)
UNLP-CONICET

and

Departamento de Ingeniería Química
Facultad Ingeniería
Universidad Nacional de La Plata
La Plata, Argentina

Lu Zhang
Department of Chemical and Materials
    Engineering
The University of Auckland
Auckland, New Zealand

Qixin Zhong
Department of Food Science and
    Technology
University of Tennessee
Knoxville, Tennessee

Weibiao Zhou
Department of Chemistry
National University of Singapore
Singapore

Section I

Basic Principles

1 Fundamentals of Momentum Transfer

Quang Tuan Pham
University of New South Wales

CONTENTS

1.1 BASIC CONCEPTS: STRESS, STRAIN, AND VISCOSITY

1.1.1 DEFINITION OF VISCOSITY

The study of fluid dynamics starts with the concept of viscosity. In contrast with solids which tend to keep their shapes, fluids are free to deform, but not entirely. A deforming fluid is subject to internal friction forces, arising from the interaction between its molecules as they move relative to each other.

Imagine two parallel solid plates with a thin layer of fluid of thickness Y sandwiched between them, the top plate being forced to slide in the x-direction at constant velocity V relative to the bottom plate (Figure 1.1). Due to molecular attraction, the fluid molecules next to the top plate will move at the same velocity V as the top plate, while those next to the bottom plate will remain stationary (the nonslip condition). A velocity gradient dv_x/dy will therefore exist in the fluid. In the simplest case, known as *Newtonian fluid*, this velocity gradient is constant and equal to

$$\frac{dv_x}{dy} = \frac{V}{\Delta y} \tag{1.1}$$

The force necessary to maintain the velocity of the top plate will obviously be proportional to the area of the plate. It is therefore, more appropriate to talk about the force per unit area, or *stress*, τ (measured in Pa or Nm^{-2}):

$$\tau = F/A \tag{1.2}$$

A larger stress will cause a larger velocity gradient. The viscosity μ of the fluid is defined as the ratio

$$\mu \equiv \left| \frac{\tau}{dv_x/dy} \right| \tag{1.3}$$

The SI unit for viscosity is Pa.s. The vertical bars indicate that the absolute value must be taken, i.e., viscosity is always positive. For so called *Newtonian fluids*, the viscosity is independent of the velocity gradient (although it may vary according to temperature or pressure).

Equation 1.3 can be written as

$$\tau = \gamma \frac{d(\rho v_x)}{dy} \tag{1.4}$$

where $\gamma \equiv \mu/\rho$ (measured in m^{2}s^{-1}) is called the kinematic viscosity. The stress τ is also called momentum flux since when you exert a force on an object you give it momentum. The term $d(\rho v_x)/dy$

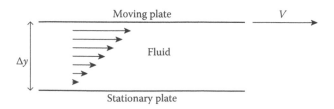

FIGURE 1.1   Sliding plate experiment.

represents a momentum gradient. The above equation shows that the momentum flux is proportional to the momentum gradient, just as the heat or mass fluxes are proportional to the temperature or concentration gradients. Therefore, γ is also known as the diffusivity of momentum (in analogy to thermal and mass diffusivities). In many situations, the diffusivities of heat, mass and momentum are of similar magnitude or closely related to each other.

It is stressed that Equation 1.3 holds for the particular case of one-dimensional flow only. Later we shall generalize this equation to three-dimensional flow.

1.1.2 VISCOSITY PREDICTION

1.1.2.1 Viscosity Prediction for Gases

The first model for the viscosity of gases is based on the kinetic theory. It assumes that gases are made up of rigid spherical molecules that move around randomly at an average speed that depends on temperature. There may be a mean underlying motion in addition to that thermal motion. The spheres interact with each other by direct contact only, bouncing off each other and any solid surface in a perfectly elastic manner (no energy loss). The average distance between collisions is called the mean free path and depends on the spheres' diameter and number of molecules per unit volume of space.

If one imagines a fast moving boat passing a stationary boat with passengers jumping back and forth from one to the other, then it can easily be seen that the passengers jumping from the moving boat will carry momentum with them and cause the stationary boat to start moving, while the passengers jumping from the stationary boat will cause the fast moving boat to slow down. This exchange of momentum can be interpreted as a force between the two boats, in effect a kind of "viscous friction." An observer who cannot see the passengers might conclude that there is some invisible force between the two boats.

Similarly, by considering the exchange of molecules between two layers of gas moving at different velocities (Figure 1.2), the resulting momentum exchange (or viscous stress) between these two layers can be calculated. The rate of molecular exchange depends on the density of the gas and the thermal velocity or the molecules, which in turn depends on temperature according to the kinetic theory of gases. Although the molecular movements are random, their effect can be averaged out. This approach results in the equation

$$\mu = \frac{2}{3\pi^{3/2}} \frac{\sqrt{\kappa_B m T}}{d^2} \tag{1.5}$$

where κ is the Boltzmann constant, m the molecular mass, T the temperature and d the diameter.

The above equation shows that the viscosity of gases is independent of pressure. This has been found to be true for pressures up to about 1 MPa and temperatures above the critical temperature. However, viscosity is found to increase faster than $\sqrt{T}$. A more accurate prediction equation for gas viscosity can be obtained from the Chapman–Enskog theory, which discards the assumptions of rigid

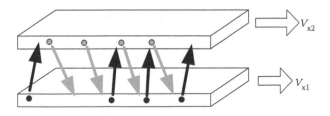

FIGURE 1.2   Illustration of the kinetic model of viscosity.

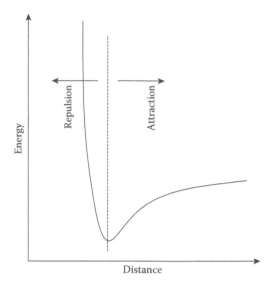

FIGURE 1.3 Lennard–Jones potential for molecular interaction.

spheres and treats the molecular interaction more realistically. This interaction can be expressed as an intermolecular potential energy field, whose gradient gives the intermolecular force, which may be repulsive or attractive depending on distance. Using the Lennard–Jones expression for the potential energy, represented in Figure 1.3, this approach leads to the following equation:

$$\mu = \frac{5}{16\pi^{1/2}}\frac{\sqrt{\kappa_B m T}}{\sigma^2 \Omega_\mu} \tag{1.6}$$

Compared with Equation 1.5, it can be seen that (apart from the constant factor) the rigid sphere diameter d has been replaced by the so-called collision diameter σ, and Ω_μ is a factor, called the *collision integral for viscosity*, which accounts for nonrigid sphere behavior.

1.1.2.2 Viscosity Prediction for Liquids

For liquids, the mechanism of viscosity is fundamentally different from that for gases. Here the molecules are constantly in contact with each other and for a molecule to move relative to the others it must pass an energy barrier, due to the necessity of squeezing past other molecules. No satisfactory predictive method is available at the moment, and empirical equations must be used. Also, in contrast to gases, the viscosity of liquids decrease with temperature according to the empirical equation

$$\mu = A \exp\left(\frac{B}{T}\right) \tag{1.7}$$

where A and B are empirical parameters, since the energy barrier is more easily passed at higher temperature.

1.1.2.3 Corresponding State Correlation

The behavior of widely different materials can be brought to the same basis by invoking the principle of corresponding states, according to which reduced properties of gases and liquids follow the same relationships with reduced pressure and temperature. *Reduced* means that the value is normalized by

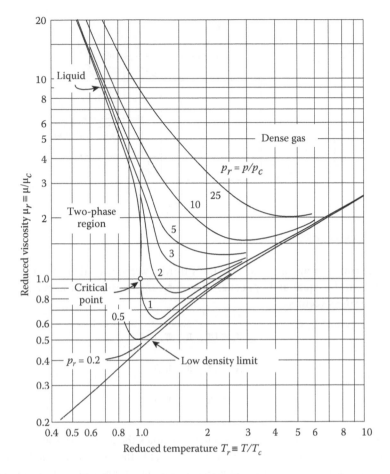

FIGURE 1.4 Plot of reduced viscosity as a function of reduced temperature and pressure. (Reproduced with permission from Hougen, O.A., Watson, K.M., and Ragatz, R.A., *Chemical Process Principles Charts*, Wiley, NY, 1960. Reproduced in Bird, R.B., Stewart, W.E. and Lightfoot, E.N., *Transport Phenomena*, Wiley, NY, 2002.)

dividing it by its value at the critical point. The reduced temperature and pressure can be interpreted as measurements of the deviation from ideal gas behavior, and hence of the interaction between molecules. Figure 1.4 plots reduced viscosity for many gases and liquids against reduced pressure and temperature.

1.1.2.4 Viscosity Prediction for Suspensions

Many liquid foods are suspensions, for which there is no fundamental way to calculate viscosity. Various empirical correlations have been proposed and reviewed in Ref. [2].

1.2   VISCOSITY EFFECTS IN THREE-DIMENSIONAL FLOW

1.2.1   THE STRESS TENSOR

Consider a very small cube of space in a fluid (Figure 1.5). Each of the six faces will experience a stress exerted by the surrounding fluid. For example, the face normal to the x-axis will experience a stress π_x. This stress is a vector and can be represented by $\pi_x = (\pi_{xx}, \pi_{xy}, \pi_{xz})^T$.

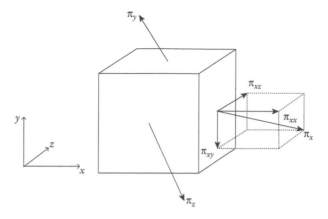

FIGURE 1.5   Stresses on a cube in a fluid.

As the cube shrinks toward a point, Newton's third law requires that the stresses on opposite faces of the cube will be equal and opposite. Therefore, the stresses on the cube can be described by the set of three stress vectors

$$\pi_x = (\pi_{xx}, \pi_{xy}, \pi_{xz})^T$$

$$\pi_y = (\pi_{yx}, \pi_{yy}, \pi_{yz})^T \qquad (1.8)$$

$$\pi_z = (\pi_{zx}, \pi_{zy}, \pi_{zz})^T$$

Note: in this chapter, each of the stress vector π_i denotes the stress experienced by the *positive* side of the cube (the side with the greater value of x, y or z, in other words the side facing toward $+\infty$). The stress experienced by its *negative* side is thus, $-\pi_i$. Some books, such as Bird et al. [2] use the opposite convention. Thus, $\pi_{xx} > 0$ implies a *tensile* stress on the face normal to x.

The matrix made up of the nine components of the three stress vectors is called the stress tensor, π:

$$\pi = \begin{bmatrix} \pi_{xx} & \pi_{yx} & \pi_{zx} \\ \pi_{xy} & \pi_{yy} & \pi_{zy} \\ \pi_{xz} & \pi_{yz} & \pi_{zz} \end{bmatrix} \qquad (1.9)$$

In the above tensor, the diagonal elements denote the normal stresses on the faces, while the non-diagonal elements denote the shear stresses. Each pair of shear stresses on opposite sides causes a torque. By considering the torque equilibrium on an infinitesimal element of fluid, we can show that $\pi_{xy} = \pi_{yx}$, i.e., the stress tensor is symmetric. Figure 1.6 shows how the stresses on the horizontal faces of the square create a torque, which must be counteracted by an equal an opposite torque due to the forces on the vertical faces.

The stress tensor is usually written as the sum of a viscous stress tensor τ and a mean normal stress tensor:

$$\pi = \tau + \frac{\pi_{xx} + \pi_{xx} + \pi_{zz}}{3} \mathbf{I} \qquad (1.10)$$

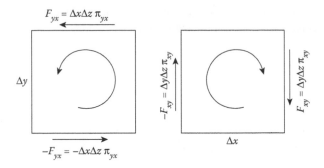

FIGURE 1.6 The torque by shear forces on y-faces, $F_{yx}\,\Delta y = \Delta x \Delta y \Delta z \pi_{yx}$ (left) must equal the torque exerted by shear forces on x-faces, $F_{xy}\Delta x = \Delta x \Delta y \Delta z \pi_{xy}$ (right), hence $\pi_{yx} = \pi_{xy}$.

The pressure p is defined as the negative of the average normal stresses $-(\pi_{xx} + \pi_{xx} = \pi_{zz})/3$ (the negative sign being due to stresses being defined in such a way that tensile stresses are positive and compressive stresses are negative), hence the above can also be written

$$\pi = \tau - p\mathbf{I} \tag{1.11}$$

The viscous stress tensor is also called the *deviatoric stress tensor* and is, of course, symmetric. It is to be noted that viscous stress can have a normal component (diagonal terms) as well as shear components.

1.2.2 THE VELOCITY GRADIENT TENSOR

In 3-D, the velocity is a vector $\mathbf{v} = [v_x, v_y, v_z]^T$. Let $\mathbf{v}$ be the velocity vector at position $\mathbf{x} = (x, y, z)^T$ and $\mathbf{v} + \delta \mathbf{v}$ be the velocity vector at a nearby point $\mathbf{x} + \delta \mathbf{x} = (x + \delta x, y + \delta y, z + \delta z)^T$. The velocity gradient is defined as

$$\lim_{\delta \mathbf{x} \to 0} \left(\frac{\delta \mathbf{v}}{\delta x}, \frac{\delta \mathbf{v}}{\delta y}, \frac{\delta \mathbf{v}}{\delta z} \right)^T = \left(\frac{d\mathbf{v}}{dx}, \frac{d\mathbf{v}}{dy}, \frac{d\mathbf{v}}{dz} \right)^T \tag{1.12}$$

$d\mathbf{v}/dx$ is the rate of change of $\mathbf{v}$ in the x-direction, etc. Since $\mathbf{v}$ is a vector, it may change in both magnitude and direction as we move from one point to another. Therefore $d\mathbf{v}/dx$, $d\mathbf{v}/dy$ and $d\mathbf{v}/dz$ are all vectors, i.e., the velocity gradient is a vector of vectors, or a *tensor*. Using the shorthand ∇ (the grad operator) to represent the vector of derivative operators, $[\nabla = \partial/\partial x, \partial/\partial y, \partial/\partial z]^T$ the velocity gradient can be written as

$$\nabla \mathbf{v} \equiv \begin{bmatrix} \dfrac{\partial v_x}{\partial x} & \dfrac{\partial v_y}{\partial x} & \dfrac{\partial v_z}{\partial x} \\[2mm] \dfrac{\partial v_x}{\partial y} & \dfrac{\partial v_y}{\partial y} & \dfrac{\partial v_z}{\partial y} \\[2mm] \dfrac{\partial v_x}{\partial z} & \dfrac{\partial v_y}{\partial z} & \dfrac{\partial v_z}{\partial z} \end{bmatrix} \tag{1.13}$$

We split the tensor into a symmetric component and an antisymmetric component, then separate out the mean diagonal term from the symmetric tensor:

$$
\nabla \mathbf{v} = \begin{bmatrix}
\dfrac{\partial v_x}{\partial x} - \dfrac{1}{3}e & \dfrac{1}{2}\left(\dfrac{\partial v_y}{\partial x} + \dfrac{\partial v_x}{\partial y}\right) & \dfrac{1}{2}\left(\dfrac{\partial v_z}{\partial x} + \dfrac{\partial v_x}{\partial z}\right) \\[2ex]
\dfrac{1}{2}\left(\dfrac{\partial v_y}{\partial x} + \dfrac{\partial v_x}{\partial y}\right) & \dfrac{\partial v_y}{\partial y} - \dfrac{1}{3}e & \dfrac{1}{2}\left(\dfrac{\partial v_z}{\partial y} + \dfrac{\partial v_y}{\partial z}\right) \\[2ex]
\dfrac{1}{2}\left(\dfrac{\partial v_z}{\partial x} + \dfrac{\partial v_x}{\partial z}\right) & \dfrac{1}{2}\left(\dfrac{\partial v_z}{\partial y} + \dfrac{\partial v_y}{\partial z}\right) & \dfrac{\partial v_z}{\partial z} - \dfrac{1}{3}e
\end{bmatrix}
+ \dfrac{1}{3}\begin{bmatrix} e & 0 & 0 \\ 0 & e & 0 \\ 0 & 0 & e \end{bmatrix}
$$

$$
+ \begin{bmatrix}
0 & \dfrac{1}{2}\left(\dfrac{\partial v_y}{\partial x} - \dfrac{\partial v_x}{\partial y}\right) & \dfrac{1}{2}\left(\dfrac{\partial v_z}{\partial x} - \dfrac{\partial v_x}{\partial z}\right) \\[2ex]
-\dfrac{1}{2}\left(\dfrac{\partial v_y}{\partial x} - \dfrac{\partial v_x}{\partial y}\right) & 0 & \dfrac{1}{2}\left(\dfrac{\partial v_z}{\partial y} - \dfrac{\partial v_y}{\partial z}\right) \\[2ex]
-\dfrac{1}{2}\left(\dfrac{\partial v_z}{\partial x} - \dfrac{\partial v_x}{\partial z}\right) & \dfrac{1}{2}\left(\dfrac{\partial v_z}{\partial y} - \dfrac{\partial v_y}{\partial z}\right) & 0
\end{bmatrix}
$$

$$(1.14)$$

where $e = (\partial v_x/\partial x) + (\partial v_y/\partial y) + (\partial v_z/\partial z)$ is a term that measures the dilation (expansion) rate. Let us give each of the tensors on the right a name: $\nabla \mathbf{v} = \mathbf{D} + \mathbf{E} + \mathbf{R}$. Each of these has a physical meaning, which we state here without proof:

- **D** measures the deformation rate, i.e., stretching and shear without volume change. It is, therefore, a generalization of the velocity gradient du/dy in the sliding plate experiment. We can therefore, expect that it gives rise to viscous stresses.
- **E** measures the expansion rate. We know that expansion or compression of a fluid gives rise to changes in pressure, but it may also cause *viscous* stresses due to the relative motion of the molecules.
- **R** (the antisymmetric component) measures the rotation rate. Solid body rotation does not cause viscous stresses because it does not cause molecules to slide past each other—therefore it can be dropped from the viscous stress calculation.

Total relative motion in an element of fluid arises from the addition of deformation, expansion and rotation (see Figure 1.7). Rotation does not cause any viscous stress, which comes mainly from deformation and (to a much smaller extent) from expansion.

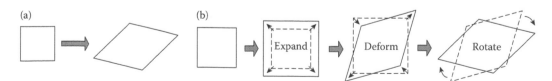

FIGURE 1.7 Any arbitrary small geometric transformation (a) can be decomposed into a sum of expansion, (b) deformation and rotation.

1.2.3   VISCOUS STRESS CORRELATION IN 3-D

From the above discussion we expect the viscous stresses to depend on the deformation tensor **D** and expansion tensor **E**. For a Newtonian fluid, this relationship is linear and we can express the viscous stress as

$$\tau = a\mathbf{D} + b\mathbf{E} \tag{1.15}$$

For the above equation to reduce to Equation 1.3 in the sliding plate experiment, where all terms except $\partial v_x/\partial y$ are zero, we must have $a = 2\mu$. We also put $b/3 = \kappa$ where κ is called the *bulk viscosity*. This leads to

$$\tau = \mu
\begin{bmatrix}
2\dfrac{\partial v_x}{\partial x} & \left(\dfrac{\partial v_y}{\partial x}+\dfrac{\partial v_x}{\partial y}\right) & \left(\dfrac{\partial v_z}{\partial x}+\dfrac{\partial v_x}{\partial z}\right) \\[2ex]
\left(\dfrac{\partial v_y}{\partial x}+\dfrac{\partial v_x}{\partial y}\right) & 2\dfrac{\partial v_y}{\partial y} & \left(\dfrac{\partial v_z}{\partial y}+\dfrac{\partial v_y}{\partial z}\right) \\[2ex]
\left(\dfrac{\partial v_z}{\partial x}+\dfrac{\partial v_x}{\partial z}\right) & \left(\dfrac{\partial v_z}{\partial y}+\dfrac{\partial v_y}{\partial z}\right) & 2\dfrac{\partial v_z}{\partial z}
\end{bmatrix}
-\left(\frac{2}{3}\mu-\kappa\right)
\begin{bmatrix}
e & 0 & 0 \\
0 & e & 0 \\
0 & 0 & e
\end{bmatrix} \tag{1.16}$$

or

$$\tau = \mu\left[\nabla\mathbf{v}+(\nabla\mathbf{v})^{\mathrm{T}}\right]-\left(\frac{2}{3}\mu-\kappa\right)e\mathbf{I} \tag{1.17}$$

For *incompressible* fluids ($e = 0$) the second term vanishes:

$$\tau = \mu\left[\nabla\mathbf{v}+(\nabla\mathbf{v})^{T}\right] \tag{1.18}$$

1.3   THE EQUATIONS OF CHANGE FOR FLUID FLOW

1.3.1   THE EQUATION OF CONTINUITY

The two fundamental transport equations in fluid dynamics are the continuity equation and the momentum equation. The continuity equation expresses conservation of mass in a fluid. It expresses the fact that the change in mass in a fixed volume of space is equal to the net flows across its surface. By considering the mass balance over a small cube it may be readily shown that, in Cartesian coordinates

$$\frac{\partial \rho}{\partial t} = -\frac{\partial(\rho v_x)}{\partial x} - \frac{\partial(\rho v_y)}{\partial y} - \frac{\partial(\rho v_z)}{\partial z} \tag{1.19}$$

or in tensor notation

$$\frac{\partial \rho}{\partial t} = -\nabla \cdot (\rho\mathbf{v}) \tag{1.20}$$

For incompressible fluids these are reduced to

$$\frac{\partial v_x}{\partial x} + \frac{\partial v_y}{\partial y} + \frac{\partial v_z}{\partial z} = 0 \tag{1.21}$$

12 Mathematical Modeling of Food Processing

or

$$\nabla \cdot \mathbf{v} = 0 \tag{1.22}$$

1.3.2 THE GENERAL MOMENTUM EQUATION

The momentum equation expresses conservation of momentum in a fixed volume of space. Momentum can be transported into or out of the volume in two ways: by convection (momentum carried by fluid entering or leaving the volume) or by molecular forces (pressure and viscous stresses). Momentum can also be created or lost within the volume because of body forces such as gravity. By carrying out the momentum balance over a small cube, it can be shown that in Cartesian coordinates the momentum balance in the i-direction is given by

$$\frac{\partial(\rho v_i)}{\partial t} = \left(\frac{\partial \tau_{xi}}{\partial x} + \frac{\partial \tau_{yi}}{\partial y} + \frac{\partial \tau_{zi}}{\partial z}\right) - \frac{\partial p}{\partial x_i} - \left(\frac{\partial \rho v_x v_i}{\partial x} + \frac{\partial \rho v_y v_i}{\partial y} + \frac{\partial \rho v_z v_i}{\partial z}\right) + \rho g_i \tag{1.23}$$

or in tensor notation

$$\frac{\partial(\rho \mathbf{v})}{\partial t} = \nabla \cdot \tau - \nabla p - \nabla \cdot (\rho \mathbf{vv}) + \rho \mathbf{g} \tag{1.24}$$

The symbol τ_{ji} means the i-th component of the viscous stress on the j-th face. In the above equations, the left hand side is the rate of change of momentum at a point in space. The terms on the right hand side express respectively the effects of viscous friction, pressure gradient, momentum convection, and body forces. It should be noted that the momentum equation is a vector equation, in effect a set of three equations, one for each coordinate. This is because momentum itself is a vector.

The above form of the momentum equation, where the left hand side describes the rate of momentum change in a fixed volume of space, is called the Eulerian form. An alternative form which may be more convenient to use in some cases is the Lagrangian form, which expresses the rate of momentum change of a chunk or packet of fluid moving in space, D/Dt. This is also called the *substantial derivative* and is related to the local or Eulerian rate of change by

$$\frac{D}{Dt} = \frac{\partial}{\partial t} + v_x \frac{\partial}{\partial x} + v_y \frac{\partial}{\partial y} + v_z \frac{\partial}{\partial z} \quad \text{(Cartesian coordinates)} \tag{1.25}$$

or

$$\frac{D}{Dt} = \frac{\partial}{\partial t} + \mathbf{v} \cdot \nabla \quad \text{(tensor notation)} \tag{1.26}$$

It can readily be shown that the Lagrangian form of the continuity equation is

$$\frac{D\rho}{Dt} = -\rho \nabla \cdot \mathbf{v} \tag{1.27}$$

while a somewhat more complicated proof, involving both the continuity and momentum equations, leads to the Lagrangian momentum equation:

$$\rho \frac{D\mathbf{v}}{Dt} = \nabla \cdot \tau - \nabla p + \rho \mathbf{g} \tag{1.28}$$

It should be noted that the convection term has disappeared from the equation, being incorporated in the *D/Dt* operator. This is because we are considering a given packet of fluid, without exchange of molecules with the outside. This packet accelerates or decelerates under the influence of viscous, pressure and body forces as shown on the right hand side, but there is no material flow through the surface of the packet.

1.3.3 THE NAVIER STOKES EQUATION

The momentum equation cannot be solved unless the viscous stresses are known. For incompressible Newtonian fluids, the viscous term can be calculated in terms of the velocity gradients, as shown earlier, and substituted into the momentum equation to give the famous Navier Stokes equation. Combination of Equations 1.27 and 1.28 give

$$\rho \frac{D\mathbf{v}}{Dt} = \nabla \cdot (\mu \nabla \mathbf{v}) - \nabla p + \rho \mathbf{g} \tag{1.29}$$

or

$$\rho \frac{\partial \mathbf{v}}{\partial t} = -\rho \mathbf{v} \cdot \nabla \mathbf{v} + \nabla \cdot (\mu \nabla \mathbf{v}) - \nabla p + \rho \mathbf{g} \tag{1.30}$$

If μ is constant then

$$\rho \frac{\partial \mathbf{v}}{\partial t} = -\rho \mathbf{v} \cdot \nabla \mathbf{v} + \mu \nabla^2 \mathbf{v} - \nabla p + \rho \mathbf{g} \tag{1.31}$$

The terms on the right of Equations 1.30 and 1.31 stand for the effects of convection, viscous forces, pressure gradient and body forces, respectively. In Cartesian coordinates, the symbol ∇^2 stand for

$$\nabla^2 \equiv \left(\frac{\partial^2}{\partial x^2} + \frac{\partial^2}{\partial y^2} + \frac{\partial^2}{\partial z^2} \right) \tag{1.32}$$

Again it should be remembered that there is a set of three equations to be solved, one for each component of **v**. In Cartesian coordinates the three equations are (for constant μ)

$$\rho \frac{\partial v_i}{\partial t} = -\rho \left(v_x \frac{\partial}{\partial x} + v_y \frac{\partial}{\partial y} + v_z \frac{\partial}{\partial z} \right) v_i + \mu \left(\frac{\partial^2}{\partial x^2} + \frac{\partial^2}{\partial y^2} + \frac{\partial^2}{\partial z^2} \right) v_i - \frac{\partial p}{\partial x_i} + \rho g_i \quad i = x, y, z \tag{1.33}$$

Although the assumptions of constant density and viscosity might seem restrictive, the Navier Stokes equation is often a good approximation in the case of small relative changes in temperature, pressure of viscosity. For small temperature changes the *Boussinesq approximation* is often used, where the density is treated as a constant but a buoyancy term due to temperature gradient $-\beta(T - \bar{T})\bar{\rho}\mathbf{g}$ is added to the body force term:

$$\bar{\rho} \frac{\partial \mathbf{v}}{\partial t} = -\bar{\rho} \mathbf{v} \cdot \nabla \mathbf{v} + \mu \nabla^2 \mathbf{v} - \nabla p + \bar{\rho} \mathbf{g} - \beta(T - \bar{T})\bar{\rho}\mathbf{g} \tag{1.34}$$

1.3.4 General Transport Equations in Fluids

The transport of heat and species in fluids involves convection and molecular diffusion, and therefore obey equations that are quite similar to the momentum equation. In fact, these equations are simpler since they are scalar, not vector, and do not involve the pressure gradient. The general transport equation for an incompressible fluid is, in Lagrangian form

$$\rho \frac{D\phi}{Dt} = \rho \nabla \cdot (\Gamma \nabla \phi) + S_\phi \qquad (1.35)$$

It expresses the fact that the rate of accumulation of ϕ in a given element of fluid is due in part to molecular diffusion and in part to some creation/destruction process (source term S_ϕ). Γ is a generalized diffusion coefficient. If the Eulerian form is used (rate of accumulation in a fixed volume of space) there will also be a convection term.

$$\rho \frac{\partial \phi}{\partial t} = -\rho \mathbf{v} \cdot \nabla \phi + \rho \nabla \cdot (\Gamma \nabla \phi) + S_\phi \qquad (1.36)$$

For heat transport, the equation becomes

$$\rho c_p \frac{DT}{Dt} = \rho c_p \nabla \cdot (\alpha \nabla T) + S_q \qquad (1.37)$$

where α is the thermal diffusivity. The volumetric heat source S_q (in Wm^{-3}) contain a heat term due to viscous dissipation, which can however be ignored in most cases (certainly in food processing applications), and may contain other sources such as chemical reaction, electrical or electromagnetic heat and heat of compression in high pressure processing of food products.

For species transport,

$$\rho \frac{D\omega_A}{Dt} = \rho \nabla \cdot (\Gamma_A \nabla \omega_A) + S_A \qquad (1.38)$$

where Γ_A is the diffusivity of A and S_A is the generation of species A (in $kg.m^{-3}s^{-1}$) due to chemical reactions.

1.4 DIMENSIONLESS GROUPS IN FLUID FLOW

The behavior of a system depends on the ratios between different competing influences, which can be represented by dimensionless groups or numbers. These dimensionless group appear when the governing differential equations are expressed in terms of dimensionless variables (ratios of the variables to the characteristic parameters of the problem). For example, in a flow problem with characteristic length L and characteristic velocity V, the Navier Stokes equation can be written in terms of dimensionless variables as

$$\frac{\partial \tilde{\mathbf{v}}}{\partial \tilde{t}} = -\tilde{\mathbf{v}} \cdot \tilde{\nabla} \tilde{\mathbf{v}} + \frac{1}{Re} \tilde{\nabla}^2 \tilde{\mathbf{v}} - \tilde{\nabla} \tilde{p} + \tilde{\mathbf{g}} \qquad (1.39)$$

The symbol Re is the well known Reynolds number, $Re = LV\rho/\mu$, which measures the ratio between inertial forces (momentum flux due to convection) and viscous forces. When the Reynolds

---

TABLE 1.1
Common Dimensionless Groups in Fluid Flow

Symbol	Formula	Name	Physical Interpretation
Fr	V^2/gL	Froude no.	Inertial force/gravity force
Gr	$g\beta\Delta TL^3/\gamma^2$	Grashoff no.	Buoyancy × inertial force/(viscous force)^2
Pe	RePr	Peclet no.	Convection/conduction
Pr	$c_p\mu/\lambda = \gamma/\alpha$	Prandtl no.	Momentum diffusivity/thermal diffusivity
Ra	GrPr	Rayleigh no.	See Gr
Re	$LV\rho/\mu$	Reynolds no.	Inertial force/viscous force
Sc	γ/D	Schmidt no.	Momentum diffusivity/species diffusivity
We	$\rho V^2 L/\sigma$	Weber no.	Inertial force/surface tension

---

number is very small and inertial forces can be neglected, we have creeping flow, described by dropping the momentum convection term $\rho\mathbf{v}\cdot\nabla\mathbf{v}$ from the Navier Stokes equation:

$$\mu\nabla^2\mathbf{v} - \nabla p + \rho\mathbf{g} = 0 \qquad (1.40)$$

As Re increases, the flow remains laminar up to a certain point or range. When this transition Reynolds number range is passed, the flow becomes turbulent, with the appearance of irregular and chaotic behavior, vortices and fluctuations in the flow pattern. For a given geometry, the transition Reynolds number will be the same, independent of scale, a fact that can be used in scaling up equipment.

Another class of dimensionless numbers characterise the fluid itself. An example is the Prandtl number, $Pr = c_p\mu/\lambda$, which is the ratio between momentum diffusivity and thermal diffusivity.

Several other dimensionless numbers are in current use to characterise fluid flow and associated phenomena. Those of greatest interest in food processing are listed in Table 1.1.

1.5 TURBULENCE MODELS

1.5.1 WHY TURBULENCE MUST BE MODELED

While analytical solutions have been found for many simple laminar flow problems, turbulent flows are far too complex and must almost invariably be solved by numerical methods or *computational fluid dynamics* (CFD). The equations of continuity and momentum, plus those describing the turbulence fields and any other variable of interest (such as concentration or pressure) are discretized in space using a finite difference, finite volume or finite element grid, to yield a set of ordinary differential equations that are solved by computer.

Turbulent flow is irregular and chaotic. Superimposed on the main average flow are eddies of many sizes, ranging from the scale of the flow geometry itself (diameter of conduit or obstruction) down to near-molecular size. The largest eddies are fed by energy from the main flow, e.g., by the velocity gradient near surfaces or buoyancy forces. They lose energy to smaller eddies, which lose energy to smaller eddies still, and so on until all the energy is dissipated as heat on the molecular scale.

Turbulence can thus be characterized by an *energy spectrum*, which describes the distribution of turbulent energy between eddies of various scales. Kolmogorov's spectrum law states that, in fully developed turbulent flow, the energy density is proportional to the wave number (the inverse of eddy size) to the power of −5/3. Two frequently mentioned parameters are the total turbulent energy, $k = (\overline{v_x'^2} + \overline{v_y'^2} + \overline{v_z'^2})/2$, and the length scale, l, which can be defined in various ways.

In principle, turbulent flow still obeys the Navier Stokes equation and can be calculated by direct solution of that equation. This approach, known as *direct numerical simulation* (DNS), requires huge computer resources for even the most elementary problem and thus is not used in practice. Instead some simplified model is used.

Turbulence modeling is a vast field. This chapter will gives only a brief overview. Interested readers should consult specialized texts [3]. Practical approaches usually rely on averaging out the turbulent fluctuating components of velocity in the momentum equation, to get a modified momentum equation (called the *Reynolds-averaged Navier Stokes* or RANS equation) in terms of the average velocities, with additional terms due to turbulence. An approach that is intermediate between DNS and RANS is the *large eddy simulation* (LES) model, where large eddies are solved rigorously while small eddies are solved by an approximate model. At this moment LES is still mainly a research tool.

In the RANS models, the velocity is expressed as the sum of a mean value and a turbulent fluctuating value

$$v_i = \bar{v}_i + v_i', \quad i = x, y, z \tag{1.41}$$

Substituting these into the (incompressible) continuity and Navier Stokes equations (Equations 1.21 and 1.33), then averaging over time (time-averaging the transient term on the left is problematic but let us just assume that it represents an underlying slow, non-turbulent change in the flow field) gives

$$\frac{\partial \bar{v}_x}{\partial x} + \frac{\partial \bar{v}_y}{\partial y} + \frac{\partial \bar{v}_z}{\partial z} = 0 \tag{1.42}$$

$$\rho \frac{D\bar{v}_i}{Dt} = -\rho \left[\frac{\partial}{\partial x}\left(\overline{v_x'v_i'} \right) + \frac{\partial}{\partial y}\left(\overline{v_y'v_i'} \right) + \frac{\partial}{\partial z}\left(\overline{v_z'v_i'} \right) \right] - \frac{\partial \bar{p}}{\partial x} + \mu \left(\frac{\partial^2}{\partial x^2} + \frac{\partial^2}{\partial y^2} + \frac{\partial^2}{\partial z^2} \right) \bar{v}_i + \rho g_i \tag{1.43}$$

These are just the standard continuity and Navier Stokes equations in terms of mean values of velocity and pressure, except that a new term $-\rho[(\partial/\partial x)(\overline{v_x'v_i'}) + (\partial/\partial y)(\overline{v_y'v_i'}) + (\partial/\partial z)(\overline{v_z'v_i'})]$ has been added to the second equation, which is due to the turbulent component of velocity. The terms $\rho\overline{v_i'v_j'}$, nine in all, can be interpreted as extra stress components due to turbulence and are termed the Reynolds stress tensor. Since the tensor is symmetric ($\overline{v_i'v_j'} = \overline{v_j'v_i'}$), there are six unknown terms due to turbulence. To solve for the flow, we need at least six new equations (that is, if we don't introduce yet more unknowns during the process). This is known as the closure problem.

Approaches to solve the closure problem can be classified into *Reynolds stress models* (RSM) and *effective viscosity models* (EVM). In the first, additional transport equations are introduced for each of the six Reynolds stress components $\rho\overline{v_i'v_j'}$.

$$\rho \frac{D\overline{v_i'v_j'}}{Dt} = \mu \left(\frac{\partial^2}{\partial x^2} + \frac{\partial^2}{\partial y^2} + \frac{\partial^2}{\partial z^2} \right) \overline{v_i'v_j'} + D_{T,ij} + S_{ij} \tag{1.44}$$

The first term on the right represents molecular diffusion, the second turbulent diffusion, and the third a source term. Calculating the latter two rigorously would introduce yet more unknowns, so they are usually modeled by some approximate equations.

In the EVM approach, the Reynolds stress is expressed as an additional viscous stress, dependent on the average velocity gradient, with the viscosity replaced by a turbulent or eddy viscosity μ_t. This is known as the Boussinesq assumption. The effective viscosity is then the sum of molecular and turbulent viscosities. The momentum equation now becomes, for incompressible Newtonian fluids:

$$\rho\frac{\partial \bar{v}_i}{\partial t} = -\rho\left(\bar{v}_x\frac{\partial}{\partial x} + \bar{v}_y\frac{\partial}{\partial y} + \bar{v}_z\frac{\partial}{\partial z}\right)\bar{v}_i - \frac{\partial p}{\partial x} + \left(\mu + \mu_t\right)\left(\frac{\partial^2}{\partial x^2} + \frac{\partial^2}{\partial y^2} + \frac{\partial^2}{\partial z^2}\right)\bar{v}_i + \rho g_i \qquad (1.45)$$

or

$$\rho\frac{\partial \mathbf{v}}{\partial t} = -\rho\mathbf{v}\cdot\nabla\mathbf{v} + \left(\mu + \mu_t\right)\nabla^2\mathbf{v} - \nabla p + \rho\mathbf{g} \qquad (1.46)$$

The problem now is to calculate μ_t. Several models have been proposed. We will look at some of them to give an idea of the "flavor" of the models.

1.5.2 Algebraic Effective Viscosity Models (EVM)

In algebraic models, the turbulent viscosity is calculated from an algebraic equation and hence there is no additional transport equation. The most well known is the mixing length model. For near-wall turbulence, the eddy viscosity is given by

$$\mu_t = \rho l^2 \left|\frac{\partial \bar{v}_x}{\partial y}\right| \qquad (1.47)$$

where l is the mixing length, which is assumed to be proportional to the distance from the wall. This model gives rise to the logarithmic *law of the wall* where velocity is a linear function of the logarithm of the distance from the wall. This relationship has been extensively verified and is often used as a boundary condition for more sophisticated turbulence models, to avoid the need to resolve the velocity profile near the wall, which would need a very fine CFD mesh. Instead, a coarse mesh can be used and the velocity at near-wall nodes is calculated from the law of the wall.

1.5.3 One-Equation Effective Viscosity Models (EVM)

In one-equation models, an additional transport equation is introduced to model the transport of some turbulence-related field variable. In the Spalart–Allmaras model [4] this field variable is the kinematic turbulent viscosity. The turbulent viscosity is calculated from the introduced field variable by an algebraic equation.

1.5.4 Two-Equation Effective Viscosity Models (EVM)

The most well known two-equation model is the k-ε model [5]. In this model, two field variables are used: the *turbulent kinetic energy k* and the *turbulent dissipation rate ε*. In an incompressible fluid they follow the transport equations

$$\rho\frac{Dk}{Dt} = \nabla\cdot\left[\left(\mu + \frac{\mu_t}{\sigma_k}\right)\nabla k\right] + G_k - \rho\varepsilon \qquad (1.48)$$

$$\rho\frac{D\varepsilon}{Dt} = \nabla\cdot\left[\left(\mu + \frac{\mu_t}{\sigma_\varepsilon}\right)\nabla\varepsilon\right] + C_{1\varepsilon}\frac{\varepsilon}{k}G_k - C_{2\varepsilon}\rho\frac{\varepsilon^2}{k} \qquad (1.49)$$

TABLE 1.2
Classification of Turbulence Models

Method	Typical Use
Direct numerical simulation (DNS)	Fundamental research only
Large eddy simulation (LES)	Very large computational resources
Reynolds average Navier Stokes (RANS):	Most practical problems
Reynolds stress models (RSM)	Highly anisotropic turbulence, swirling flows, secondary flows
Effective viscosity (EVM) or Boussinesq models:	
Algebraic models (e.g., mixing length)	Near wall region (well proven)
One-equation models (e.g., Spalart-Allmaras)	Relatively new and untested
Two-equation models (e.g., k-ε, k-ω)	Depends on version. Low anisotropy, nonswirling

and the turbulent viscosity is computed from

$$\mu_t = \rho C_\mu \frac{k^2}{\varepsilon} \tag{1.50}$$

In these equations, G_k is the generation of turbulent kinetic energy and can be calculated from the temperature gradient, while $C_{1\varepsilon}$, $C_{2\varepsilon}$, $C_{3\varepsilon}$, C_μ, σ_k, and σ_ε are empirical constants. If buoyancy is present there will be an additional turbulence source.

Several variants have been proposed for the k-ε model but their details will not be considered here. An alternative to the k-ε model is the k-ω model [6,7] where $\omega = \varepsilon/k$.

A summary classification of the turbulence models is shown in Table 1.2. It must be kept in mind that all the models mentioned above (except DNS) are approximate and their accuracy is by no means guaranteed. Ideally a model should be used only when it has been verified experimentally in a similar flow configuration.

1.6 NON-NEWTONIAN FLUIDS

1.6.1 GENERAL OBSERVATIONS

Up to now we have only considered Newtonian fluids, which has a constant viscosity. Many fluids—in particular polymers, which include most of those encountered in the food industry—exhibit non-Newtonian behavior, which cannot be characterized by a constant μ. Their long or complicated molecular shapes and complex intermolecular forces give rise to complex rheological behavior. Molecules may get entangled, form temporarily links that break up, or re-orient themselves under the influence of the velocity gradient.

The following phenomena have been experimentally observed in polymeric fluids:

- Recoil after cessation of flow: a liquid is pumped through a tube. When the pump is stopped, the fluid may pull back some distance, i.e., it shows signs of elasticity.
- Normal stress to velocity gradient: a rod is dipped into a container of fluid and rotated, causing velocity gradient in the horizontal plane. After a while, fluid starts to climb up the rod.
- Nonparabolic tube flow: the Navier Stokes equation predicts that laminar fluid flow in a circular tube should show a parabolic velocity profile. Instead, polymeric fluid will show a flatter-than-parabolic profile.
- Yield stress: Newtonian fluids cannot resist shear stress no matter how small. As soon as it is applied the fluid starts to deform. In contrast, some materials such as tooth paste behave as a solid and can resist shear stress up to a point, then start to flow like a liquid.

- Tubeless siphon: with Newtonian fluids a siphon will stop working when the higher tip of the siphon is lifted out of the liquid. With some polymeric fluids liquid will continue to rise from the surface to the tip and siphon out of the container, demonstrating solid-like cohesiveness.

It is obvious that these behaviors cannot be explained by the stress–strain relationship of Equation 1.18. Another relationship for stress is needed, for which one of two types of rheological models is commonly used: extended Newtonian models and viscoelastic models.

1.6.2 EXTENDED (GENERALIZED) NEWTONIAN MODELS

Extended Newtonian models are the simplest approach to modeling non-Newtonian behavior. They retain the same stress–strain relationship as Newtonian flow and simply assume that viscosity is some function of the velocity gradient. They do not attempt to model the time-dependent effects or elasticity. Thus, they are not suitable for use with unsteady state flows, but may provide a good enough description of steady state flows.

Defining the rate of strain tensor

$$\dot{\mathbf{S}} \equiv \nabla\mathbf{v} + \left(\nabla\mathbf{v}\right)^T \tag{1.51}$$

the 3-D strain-stress relationship for Newtonian fluids can be written

$$\tau = \mu\dot{\mathbf{S}} \tag{1.52}$$

For non-Newtonian fluids this equation is retained, but viscosity is a function of the magnitude of the rate of deformation tensor, $\dot{S} \equiv \sqrt{(\dot{\mathbf{S}}:\dot{\mathbf{S}})/2} = \sqrt{(\Sigma_{ij}\dot{S}_{ij}\dot{S}_{ij})/2}$:

$$\mu = \mu(\dot{S}) \tag{1.53}$$

Several empirical models have been proposed depending on the form of this relationship.

1.6.2.1 Power Law Models

$$\mu = \dot{S}^{n-1} \tag{1.54}$$

where n and m are constants specific to the fluid. In 1-D flow the power law reduces to

$$\mu = m\left(\frac{dv_x}{dy}\right)^{n-1} \tag{1.55}$$

or

$$\tau = m\left(\frac{dv_x}{dy}\right)^{n} \tag{1.56}$$

If $n > 1$ the fluid is called dilatant or shear-thickening, and if $n < 1$ it is called pseudoplastic or shear-thinning. For Newtonian fluids $n = 1$ and $\mu = m$.

1.6.2.2 Yield Stress Models

Some models assume that the material behaves as a solid until a threshold or yield stress τ_0 is exceeded, when it starts to flow as a fluid. A Bingham plastic material exhibits Newtonian behavior with respect to the excess stress:

$$\frac{dv_x}{dy} = 0, \quad \tau \leq \tau_0 \tag{1.57}$$

$$\frac{dv_x}{dy} = \frac{\tau - \tau_0}{\mu}, \quad \tau > \tau_0 \tag{1.58}$$

Similarly, a yield-dilatant material exhibits dilatant behavior with respect to the excess stress and a yield-pseudoplastic material exhibits pseudoplastic behavior when a threshold stress τ_0 is exceeded:

$$\frac{dv_x}{dy} = 0, \quad \tau \leq \tau_0 \tag{1.59}$$

$$\frac{dv_x}{dy} = \frac{\tau - \tau_0}{m\left(\dfrac{dv_x}{dy}\right)^{n-1}} \quad \text{or} \quad \tau = \tau_0 + m\left(\frac{dv_x}{dy}\right)^n, \tau > \tau_0 \tag{1.60}$$

The stress-rate of strain relationships for various extended Newtonian viscosity models are illustrated in Figure 1.8.

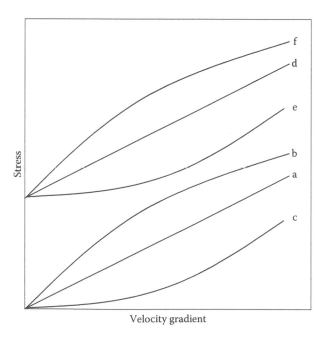

FIGURE 1.8 Extended Newtonian viscosity models: (a) Newtonian, (b) pseudoplastic, (c) dilatant, (d) Bingham plastic, (e) yield-dilatant, (f) yield-pseudoplastic.

1.6.3   TIME-DEPENDENT VISCOSITY MODELS

With some fluids there may be some structural rearrangement at the molecular level, causing the viscosity to change with time as shear is applied. *Thixotropic* fluids such as mayonnaise, clay suspensions and some paint and inks show decreasing viscosity with time. *Rheopectic* fluids such as bentonite sols, vanadium pentoxide sols and gypsum suspensions in water show increasing viscosity with time during shear.

1.6.4   VISCOELASTIC MODELS

1.6.4.1   Viscoelastic Behavior

Viscoelastic materials [8] exhibits both elasticity and viscous behaviors. While stress in a viscous fluid depends on the rate of strain, stress in an elastic solid depends on the strain itself. We shall first review the basic concepts of elastic behavior. Consider the 1-D situation of a rod subjected to a pulling force. Point A moves from x_A to x_A' and the displacement is denoted by $u_{xA} = x_A' - x_A$. Similarly point B is displaced by $u_{xB} = x_B' - x_B$ (Figure 1.9).

The difference in displacement $u_A - u_B$ causes strain. It can be expected that, the further apart A and B are, the greater the differential displacement. The ratio $(u_{xB} - u_{xA})/(x_B - x_A)$ is the (average) strain between A and B. Taking the limit as B approaches A, the strain in the x-direction at a point is defined as (du_x/dx) and can be seen as a displacement gradient.

Generalizing to 3-D, let the position of a point in a solid be denoted by $\mathbf{u} = (u_x, u_y, u_z)$. When force is applied, the solid may deform, each point in each moving by a different amount and the strain is defined as the displacement gradient tensor:

$$\nabla \mathbf{u} = \begin{bmatrix} \dfrac{\partial u_x}{\partial x} & \dfrac{\partial u_y}{\partial x} & \dfrac{\partial u_z}{\partial x} \\[2ex] \dfrac{\partial u_x}{\partial y} & \dfrac{\partial u_y}{\partial y} & \dfrac{\partial u_z}{\partial y} \\[2ex] \dfrac{\partial u_x}{\partial z} & \dfrac{\partial u_y}{\partial z} & \dfrac{\partial u_z}{\partial z} \end{bmatrix} \tag{1.61}$$

If the solid is moving at the same time as being strained, the rotational component of displacement has to be removed since it does not cause strain. This is done by taking the symmetric component of

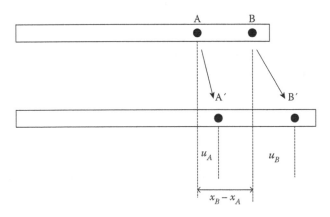

FIGURE 1.9   Displacements in a rod.

the displacement tensor and discarding the antisymmetric component, just as we did for the velocity gradient tensor. This gives the strain tensor:

$$\mathbf{S} = \nabla\mathbf{u} + (\nabla\mathbf{u})^T \tag{1.62}$$

Note that the time derivative of the strain tensor $\mathbf{S}$ is the rate of strain tensor $\dot{\mathbf{S}}$ of Equation 1.51:

$$\dot{\mathbf{S}} = \frac{\partial\mathbf{S}}{\partial t} \tag{1.63}$$

According to the linear elasticity model (Hooke's law) the shear (or deviatoric) stress tensor is then given by

$$\tau_{\text{elastic}} = K\mathbf{S} \tag{1.64}$$

or

$$\frac{\partial}{\partial t}\tau_{\text{elastic}} = K\dot{\mathbf{S}} \tag{1.65}$$

where K is the shear modulus of the material. Again note the parallel between Equations 1.52 and 1.64.

To summarize, in an elastic solid, stress is a function of the strain $\mathbf{S}$ while in a liquid it is a function of the rate of strain $\dot{\mathbf{S}}$. Viscoelasticity is a simultaneous happening or superposition of these two behaviors. Different models are proposed depending on how to superimpose the two behaviors. We will restrict ourselves to linear viscoelastic models, which do not involve higher powers of the stress or strain tensors.

1.6.4.2 Mechanistic Viscoelastic Models

Several mechanistic "spring and damper" models have been used to represent viscoelastic behavior.

a. *The Maxwell model*

The Maxwell model is described by the equation

$$\tau + \frac{\mu}{K}\frac{\partial}{\partial t}\tau = \mu\dot{\mathbf{S}} \tag{1.66}$$

The ratio $\mu/K = \theta$ is called the relaxation time: the more viscous and the less stiff the material, the longer it takes to settle down to equilibrium. For 1-D flow between parallel plates (Figure 1.1):

$$\tau + \theta\frac{\partial\tau}{\partial t} = \mu\frac{\partial v_x}{\partial y} \tag{1.67}$$

When the relaxation time is very small or the rate of strain very slow, the second term on the left vanishes and we obtain viscous behavior (Equation 1.3), while in the opposite case, elastic behavior is obtained. In between we have viscoelastic behavior. "Silly putty" is a material that behaves in such a way: it will bounce off the ground when dropped from a height, but you can squeeze it to make it flow like a viscous fluid. The Maxwell model can be illustrated by the spring-and-damper series arrangement of Figure 1.10, which illustrates the fact that the total strain is due partly to elastic stress and partly to viscous stress.

If a step change in strain is forced on a Maxwell element the resulting stress jumps (due to the spring) then relaxes exponentially. The stress step response is called the *relaxation modulus, G(t)*, and is of the form

$$G(t) = K \exp\left(-\frac{t}{\theta}\right) \quad \text{(see Figure 1.11)}$$

The stress jumps by an amount proportional to the stiffness K then decay with a characteristic time θ.

If a unit step stress is imposed, the strain response is called the *creep compliance* denoted by J. The creep compliance for a Maxwell viscoelastic material is shown in Figure 1.12.

b. *The generalized Maxwell model*

To improve goodness of fit with experimental data, a generalized Maxwell model may be used:

$$\tau = \sum_k \tau_k \tag{1.68}$$

where τ_k are the solutions to the equation

$$\tau_k + \theta_k \frac{\partial}{\partial t} \tau_k = \mu_k \dot{\mathbf{S}} \tag{1.69}$$

This can be represented by a parallel arrangement of several spring and damper units, each contributing a component of stress (Figure 1.13).

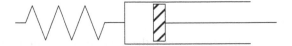

FIGURE 1.10   Spring and damper representation of the Maxwell model.

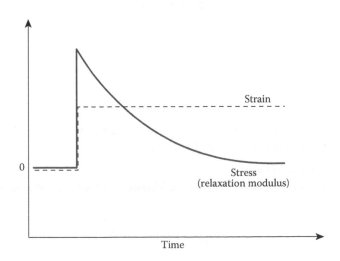

FIGURE 1.11   Response of Maxwell viscoelastic material to step change in strain (relaxation modulus).

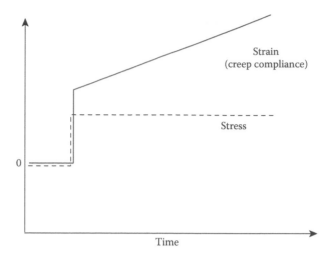

FIGURE 1.12   Response of Maxwell viscoelastic material to step change in stress (creep compliance).

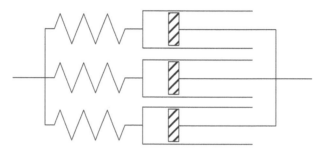

FIGURE 1.13   Generalized Maxwell model.

c. *The Kelvin–Voigt model*

The Kelvin–Voigt model superimposes strains rather than stresses as in the Maxwell model:

$$\tau = KS + \mu \dot{\mathbf{S}} \tag{1.70}$$

In 1-D:

$$\tau = K \frac{\partial u_x}{\partial y} + \mu \frac{\partial v_x}{\partial y} \tag{1.71}$$

The total stress is the sum of a viscous stress and an elastic stress. It can be represented by a parallel arrangement of spring and damper (Figure 1.14).

d. *The standard linear (Jeffrey) model*

The standard linear model is a combination of the Maxwell and Kelvin–Voigt models:

$$\tau + \theta_1 \frac{\partial}{\partial t} \tau = \mu \left(\dot{\mathbf{S}} + \theta_2 \frac{\partial}{\partial t} \dot{\mathbf{S}} \right) \tag{1.72}$$

It can be represented by a series-parallel combination of springs and damper (Figure 1.15).

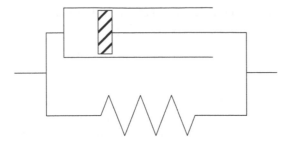

FIGURE 1.14   Kelvin–Voigt model.

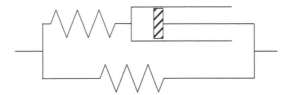

FIGURE 1.15   Representation of the standard linear model.

1.6.4.3   Response to Arbitrary Inputs: the Hereditary Integral

An arbitrary strain history can be divided into a series of infinitesimal steps $\delta S_{ij} = \dot{S}_{ij}(t)\delta t$, to which the response at time t_1 will be $\delta\tau_{ij} = G_{ij}(t_1 - t)\delta S_{ij} = G_{ij}(t_1 - t)\dot{S}_{ij}(t)dt$, provided that the behavior of the material is linear (the response being proportionate to the input). Using the principle of superposition, the response to these steps can be added together, giving the *hereditary integral*:

$$\tau_{ij}(t) = \int_0^t G_{ij}(t - \zeta)\dot{S}_{ij}(\zeta)d\zeta \tag{1.73}$$

where G_{ij} is the ij-th component of the 3-D relaxation modulus (response to unit step), which may be experimentally obtained or predicted from a mechanistic model such as Maxwell or Kelvin–Voigt. The above equation assumes that the system starts from a strain-free state at time 0. A similar approach can be used to calculate strain from an arbitrary stress history by using the response to a unit step change in stress (creep compliance) J:

$$S_{ij}(t) = \int_0^t J_{ij}(t - \zeta)\frac{\partial\tau_{ij}}{\partial\zeta}d\zeta \tag{1.74}$$

where J_{ij} is the ij-th component of the 3-D creep compliance.

1.6.4.4   Response to Small Oscillatory Input

The properties of viscoelastic materials is often experimentally investigated by dynamic mechanical analysis (DMA), where a sample is subjected to a small oscillatory force or displacement. By varying the temperature and measuring the stress–strain relationship, transitions in mechanical properties can be observed which gives clues to changes in the microstructure of the material. Consider a strain that varies sinusoidally with time with angular velocity ω:

$$u(t) = u_0 \sin \omega t \tag{1.75}$$

A purely elastic solid, represented by a spring, will respond with a time-dependent stress described by

$$\tau_{solid}(t) = Ku(t) = Ku_0 \sin \omega t \tag{1.76}$$

while a purely viscous liquid, represented by a damper, will respond with a stress

$$\tau_{\text{liquid}}(t) = \mu \frac{\partial u}{\partial t} = \mu \omega u_0 \cos \omega t \tag{1.77}$$

It can be seen that the stress response of the solid is in phase with the strain input while that of the liquid leads by $\pi/2$, since $\cos \omega t = \sin(\omega t + \pi/2)$. A viscoelastic material will display intermediate behavior: part of the stress response will be in phase with the strain and part will be out of phase by $\pi/2$. Its response to a unit sinusoidal strain will be

$$G(\omega) = \frac{\tau(t)}{u_0} = G'(\omega)\sin \omega t + G''(\omega)\cos \omega t \tag{1.78}$$

G' is known as the *storage modulus* since it relates to the elastic property of the material (ability to store energy), while G'' is called the *loss modulus* since it relates to its viscous (dissipative) property. In the Kelvin–Voigt model, where the total stress is obtained by adding the individual elastic and viscous stress responses, the sin term will be obtainable from Equation 1.76 and the cos term will be obtainable from Equation 1.77. Thus, G' will be independent of frequency and equal to the Young modulus k while G'' will be $\mu\omega$. However, this will not be true in the general case.

By trigonometric transformation the above equation can be written as

$$G(\omega) = \sqrt{G'^2 + G''^2} \cdot \sin(\omega t + \delta) \tag{1.79}$$

$$\tan \delta = \frac{G''(\omega)}{G'(\omega)} \tag{1.80}$$

Equation 1.79 shows that the stress leads the strain by an angle δ (Figure 1.16) whose tangent (Equation 1.80) is called the *loss tangent*.

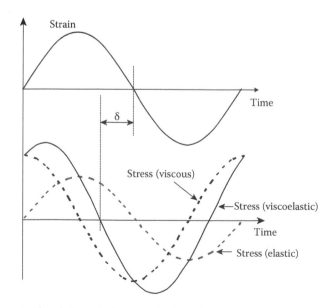

FIGURE 1.16   Stress–strain relationship for sinusoidal input.

A more compact notation can be used by defining a complex modulus $G^* = G' + iG''$, a complex strain $u^* = u_0 e^{i\omega t}$ and a complex stress $\tau^* = \tau_0 e^{i(\omega t + \delta)}$, where $i = \sqrt{-1}$. Using this notation the relationship between strain and stress can be written simply as

$$\tau^* = G^* u^* \tag{1.81}$$

where it is understood that the applied strain is the real part of u^* and the actual stress response is the real part of τ^*. The parameters G^*, G', G'' and $\tan \delta$ all depend on ω. Note that in the polymer literature, G^*, G' and G'' are used for shear deformation while the equivalent symbols for axial deformation are E^*, E' and E'' and for the general case M^*, M' and M'' [9].

ACKNOWLEDGMENT

The author would like to thank Professor Adesoji Adesina for his valuable comments on this chapter.

NOMENCLATURE

A	area, m^2
d	molecular diameter, m
D	deformation rate tensor, s^{-1}
$\dfrac{D}{Dt}$	substantial derivative, $\dfrac{D}{Dt} = \dfrac{\partial}{\partial t} + \mathbf{v} \bullet \nabla$, s^{-1}
e	volumetric expansion rate, $\dfrac{\partial v_x}{\partial x} + \dfrac{\partial v_y}{\partial y} + \dfrac{\partial v_z}{\partial z}$, s^{-1}
$\mathbf{E}$	expansion rate tensor, s^{-1}
F	force, N
$\mathbf{g}$	gravitational acceleration vector, ms^{-2}
$\tilde{\mathbf{g}}$	$L\mathbf{g}/V^2$, dimensionless gravitational acceleration
G	relaxation modulus, Pa
$\mathbf{I}$	identity matrix
J	creep compliance
k	specific turbulent kinetic energy $\dfrac{1}{2}\left(\overline{v_x'^2} + \overline{v_y'^2} + \overline{v_z'^2}\right)$, $m^2 s^{-2}$
K	shear modulus, Pa
l	mixing length, m
L	characteristic length, m
m	molecular mass, kg
n	viscosity index
p	pressure, Pa
$\tilde{p}$	$p/\rho V^2$, dimensionless pressure
$\mathbf{R}$	rotation rate tensor, s^{-1}
Re	Reynolds number
$\mathbf{S}$	strain tensor, $\nabla\mathbf{u} + (\nabla\mathbf{u})^T$
$\dot{\mathbf{S}}$	rate of strain tensor, $\nabla\mathbf{v} + (\nabla\mathbf{v})^T$, s^{-1}

$\dot{S}$	magnitude of the rate of deformation tensor, $\sqrt{\dfrac{1}{2}\sum_i \sum_j \dot{S}_{ij}\dot{S}_{ij}}$, s^{-1}
s_ϕ	source term for variable ϕ per unit volume
$\tilde{t}$	Vt/L, dimensionless time
T	temperature, K
$\mathbf{u}$	displacement vector, m
u_0	amplitude of sinusoidal strain
u_i	components of displacement vector, m
$\mathbf{v}$	velocity vector, ms^{-1}
$\tilde{\mathbf{v}}$	$\mathbf{v}/V$, dimensionless velocity
v'	turbulent velocity vector, ms^{-1}
v_i	components of velocity vector, m
v_i'	components of turbulent velocity vector, m
V	characteristic velocity, ms^{-1}
x, y, z	space coordinates, m
α	thermal diffusivity, $m^{-2}s^{-1}$
β	thermal expansion coefficient, K^{-1}
ε	turbulent dissipation rate, m^2s^{-3}
ϕ	a field variable
γ	kinematic viscosity μ/ρ, m^2s^{-1}
Γ	diffusivity, m^2s^{-1}
θ	relaxation time, s
κ	bulk viscosity, Pa.s
κ_B	Boltzmann constant, JK^{-1}
λ	thermal conductivity, $Wm^{-1}K^{-1}$
μ	viscosity, Pa.s
μ_t	turbulent viscosity, Pa.s
Ω_μ	collision integral for viscosity
π	total stress tensor, Pa
π_i	total stress on surface normal to i-th axis, Pa
π_{ij}	j-th component of total stress on surface normal to i-th axis, Pa
ρ	density, $kg\ m^{-3}$
σ	collision diameter, m
τ	viscous stress tensor, Pa
τ_i	viscous stress on surface normal to i-th axis, Pa
τ_{ij}	j-th component of viscous stress on surface normal to i-th axis, Pa
τ_0	yield stress, also amplitude of sinusoidal stress, Pa
ω	angular velocity of periodic motion
$\tilde{\nabla}$	$L\nabla$, dimensionless gradient operator

Subscripts

c	critical value
i, j	space coordinate index

REFERENCES

1. Hougen, O.A., Watson, K.M. and Ragatz, R.A. *Chemical Process Principles Charts*. 2nd Ed. Wiley, NY, 1960.
2. Bird, R.B., Stewart, W.E. and Lightfoot, E.N. *Transport Phenomena*. 2nd Ed. Wiley, NY, 2002.
3. Wilcox, D.C. *Turbulence Modeling for CFD*. DCW Industries Inc., La Canada, CA, 1998.
4. Spalart, P. and Allmaras, S. *A one-equation turbulence model for aerodynamic flows*. Technical Report AIAA-92-0439, American Institute of Aeronautics and Astronautics, Reston, VA, 1992.
5. Launder, B.E. and Spalding, D.B. *Lectures in Mathematical Models of Turbulence*. Academic Press, London, UK, 1972.
6. Kolmogorov, A.N. Equations of turbulent motion of an incompressible fluid. *Izv Akad Nauk SSR Ser Phys*, 6, 56, 1942.
7. Wilcox, D.C. Reassessment of the scale determining equation for advanced turbulence models. *AIAA J.*, 26, 1299, 1988.
8. Ferry, J.D. *Viscoelastic Properties of Polymers*. 3rd Ed. John Wiley & Sons, NY, 1980.
9. Kaye, A., Stepto, R.F.T., Work, W.J., Alemán, J.V. and Malkin, A.Ya. *Definition of Terms Relating to the Non-Ultimate Mechanical Properties of Polymers*. International Union of Pure and Applied Chemistry, Research Triangle Park, NC, 1997.

2 Rheological Properties of Food

Jasim Ahmed
Polymer Source Inc.

CONTENTS

2.1   INTRODUCTION

Rheology is the branch of science dealing with the flow and deformation of materials. It deals with the predictions of mechanical behavior based on the micro- or nanostructure of the material, e.g., the molecular size and architecture of food polymers in solution or particle size distribution in a solid suspension. Rheological instrumentation and rheological measurements have become essential tools in the analytical laboratory for characterizing component materials and finished products, monitoring process conditions, as well as predicting product performance and consumer acceptance. Rheological properties have been considered to provide fundamental insights on the structural organization of food. They play an important role in fluid flow and heat transfer processes. Rheologically speaking, food is complex materials varying from low viscous Newtonian fluid to high viscous non-Newtonian fluid or solid in nature that can be characterized by textural measurement. Today, rheological instrumentation and rheometry are accepted techniques to more fully characterize, understand, and provide control for food product development and quality assessment.

Rheological properties of solid foods have been measured commonly as a body and texture or consistency. Texture has been considered as one of the important quality attributes of food products. Besides eating quality, texture affects usage and handling properties of foods. Uniaxial compression tests are based on large destructive deformations which are especially important in determining fracture properties. Results of texture profile analysis (TPA) have been reported to correlate well with sensory assessments [1]. On the other hand, testing within the linear viscoelastic range (small deformations) can provide important data relating to structure down to the molecular level [2].

Fundamental rheological properties are independent of the instrument on which they are measured thus different instruments will provide the same results. This is an ideal concept and in fact we never obtain the same results using different set of instruments. However, the rheological measurement provides better understanding and more meaningful data over subjective measurements [3].

2.2   CLASSIFICATION OF MATERIALS

Rheological properties of materials measures the shear strain respond to the shear stress (or vice versa) in a particular situation. The parameters in any quantitative functional relation between the stress and the strain are the rheological properties of the material. A perfect solid (elastic) and fluid (viscous) behaviors represent two extreme responses of materials. An ideal solid deforms while a load is applied. The material comes back to its original configuration when the applied force is withdrawn. An ideal elastic material follows Hook's law where the resultant strain (γ) is directly proportional to the applied stress (τ):

$$\tau = G\gamma \tag{2.1}$$

Where G is the shear modulus. Hookean materials do not flow and are linearly elastic.

An ideal fluid deforms at constant rate while force is applied. The fluid cannot retain its original configuration upon release of the applied force. The flow of a simple viscous material is well represented by Newton's law where shear stress is directly proportional to the shear rate.

2.3 RHEOLOGY OF FLUID FOODS

A fluid may be considered as a matter composed of different layers. The fluid starts to move while a force acts upon it. The relative movement of one layer over another is due to shear force. The fluid exerts a resistance force opposite to the movement of a shear force and the resistance force is known as fluid viscosity.

In fluid flow, the shear stress (applied force per unit area) is commonly related to rate of shear (relative change of velocity divided by distance between two fluid layers). Newton observed that when the shear stress is increased, the shear rate also increases:

$$\tau \propto \dot{\gamma} \tag{2.2}$$

Where

$$\dot{\gamma} = \frac{du}{dy} \quad \text{and} \quad \tau = \frac{F}{A} \tag{2.3}$$

Equation 2.2 can be further written as:

$$\tau = \eta\dot{\gamma} \tag{2.4}$$

The proportionality constant η is known as coefficient of viscosity or simply viscosity of the fluid. The viscosity of fluid depends on the physico-chemical properties and temperature. The fluids, which follow the behavior described above is known as Newtonian fluids and most of the liquid foods fall into this category. Water, fruit juice, and milk are common examples of food that follow Newtonian fluids.

For low molecular weight liquids, the temperature dependence of viscosity follows a simple exponential relationship:

$$\eta = Ae^{-(E/RT)} \tag{2.5}$$

Where E is the activation energy for viscous flow and A is a constant. The flow of liquids can be explained by various molecular theories and Eyring's theory considering lattice structure [4].

By definition, Newtonian fluids have a straight line relationship between the shear stress and the shear rate with a zero intercept. Fluids which do not follow the rule are termed as non-Newtonian fluid and shear rate dependent (Figure 2.1). For non-Newtonian fluid, viscosity is commonly termed as apparent viscosity because of its dependency on shear rate. Most of food materials fall in this category. The deviation from Newtonian behavior lies on the composition of those foods and complexity in behavior during processing.

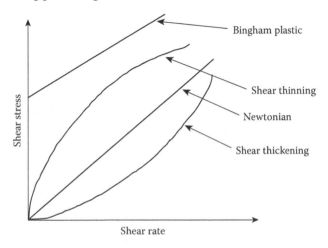

FIGURE 2.1 Rheograms for Newtonian and non-Newtonian food products.

Non-Newtonian foods can be divided into two categories namely time-dependent and time-independent. The former one depends only on shear rate at constant temperature whereas the later one depends on both shear rate and duration of shear. Time independent flow behavior further divided in two categories: shear thinning and shear thickening. Time dependent non-Newtonian fluid achieves a constant apparent viscosity while shear stress is applied for definite time period. These are also known as thixotropic materials and some starch suspensions are good examples of such behaviors.

2.3.1 SIMULATION OF STEADY FLOW RHEOLOGICAL DATA

A plot of shear-stress-shear rate can be mathematically modeled using various functional relationships. The simplest type of fluid behavior is described by the Herschel–Bulkley model:

$$\tau = \tau_o + K\dot{\gamma}^n \tag{2.6}$$

Where K is the consistency index, n is the flow behavior index and τ_o is the yield stress. The above equation can be applied for a wide range of food products and various steady flow rheological models (Newtonian, Bingham, power). It can be considered as a special case of the Herschel–Bulkley model [3]. For the Newtonian (Equation 2.4) and Bingham plastic models K becomes viscosity (η) and plastic viscosity (η_{pl}) (Equation 2.7), respectively. Many food materials like ketchup, margarine, cheese spread follow Bingham plastic behavior and are well described by the Bingham plastic model. The simplest form of the Herschley–Bulkley model is the power law where there is no yield term (Equation 2.8). There are two types of behavior commonly noticed for food materials. A decrease in shear stress with increasing shear rate is known as shear-thinning or pseudoplastic fluid ($0 < n < 1$). The opposite behavior (increase in shear stress with increase in shear rate) of fluid ($1 < n < \infty$) is termed shear-thickening or dilatent (Figure 2.1). Shear thinning results from the tendency of the applied force to disturb the long chains from their favored equilibrium conformation, causing elongation in the direction of shear [5]. Fruit puree behaved as shear thinning fluid and hot starch suspension is an example of shear-thickening food. The most important parameter of the Herschel–Bulkley model is the yield stress. The details of the yield stress, its measurement and applications are described separately for its important role in food rheology. Extensive compilations of magnitudes of flow model for various food products are available in the literature and, thus, those data are excluded from this chapter.

$$\tau = \tau_y + \eta_p\dot{\gamma} \tag{2.7}$$

$$\tau = K\dot{\gamma}^n \tag{2.8}$$

2.3.2 YIELD STRESS

There are strong interparticle interactions in concentrated solid–liquid suspensions that often exhibit plastic flow behavior and the presence of a yield stress [6]. The particles in flocculated suspensions can aggregate to form clusters which interact with one another to yield a continuous three-dimensional network structure extending throughout the entire volume [7,8]. The yield stress thus, measures the strength of the coherent network structure. The concentrated solid–liquid systems deform elastically with finite rigidity while small stress is applied and the material initiates flow like a viscous fluid when the applied stress exceeds the yield value. The yield stress is thus considered as a material property representing a transition between solid-like and liquid-like properties.

Yield stress is defined as the minimum shear stress required to initiate flow. The yield has industrial significance in process design and quality assessment of food materials. Yield stress has been strongly associated with consumer acceptance and structure retention of food products. An excessive high yield stress may result in unnecessarily high power consumption and hence high operating costs.

2.3.3 Yield Stress Measurement Techniques

Yield stress plays a significant role in food processing and it is, therefore, important that the yield stress be measured precisely. A strong interest in yield stress fluids has led to developments of a variety of experimental methods and techniques for measuring the yield stress property [9]. However, yield stress measurements are notoriously difficult to interpret. It has been observed that the results obtained from different methods with the same material, prepared and tested in the same laboratory varied significantly [10–13]. Such variability is indeed attributed to the differences in the principles employed by different techniques, the definition of the yield stress adopted and the time scale of the measurements involved [9,14,15]. The variable nature of yield stress measurements has led to a suggestion that an absolute yield stress is an elusive property and any agreement of results from different techniques is accidental [3,15].

There are both direct and indirect methods for the yield stress measurement. In the indirect method, the yield stress was obtained by (i) graphically or numerically extrapolating the shear stress-shear rate flow curve measured at low shear rates to zero shear rate [16], and (ii) fitting the shear stress-shear rate data using two nonlinear models for yield stress fluids, such as the Casson (Equations 2.9 and 2.10) and Herschel–Bulkley models (Equation 2.6).

$$\tau^{0.5} = K_{OC} + K_C \dot{\gamma}^{0.5} \tag{2.9}$$

$$\tau_{OC}{}^2 = K_{OC} \tag{2.10}$$

where K_{oc} is the intercept of the plot square root of shear stress against square root of shear rate. The Casson model fits well for chocolate, caramels, and other confectionaries.

It is rather difficult to obtain shear stress data at considerably low shear rate in traditional viscometers. In addition, instrumental defects like slip effect could lead to unreliable data at low shear rate. It has been observed for many food products that the values calculated from Casson model vary significantly from the values obtained by extrapolation technique. It is, therefore, recommended to measure the yield stress directly by independent and more reliable direct techniques [6]. However, some of the food products exhibited less slip effect and consequently less variation in the yield value obtained from different sources. Ahmed [17] compared between the direct yield stress value obtained from a rheometer and computed yield value from the Casson model (data obtained from rotational viscometer) and found there are insignificant differences between those values.

Various direct yield stress measurement techniques are available in the literature. These include: (i) vane technique [6], (ii) slotted plate technique [11], (iii) cylindrical penetrometer technique [18], (iv) inclined plane technique [19], (v) stress ramp technique [11], and (vi) creep technique. Details of each technique are available in the cited references. The vane technique has been considered as one of the best techniques for the direct measurement of the yield stress in highly concentrated food materials and the technique is described briefly.

2.3.3.1 The Vane Method

The vane method has been developed to eliminate slip effects commonly found in yield stress measurements with rotating cylinders. In addition, the vane method does not destroy product structure during sample loading because the original container may be used as the sample vessel [20]. The cylinder has been substituted by vane geometry (Figure 2.2). It consists of a small number of thin blades (usually two to eight) arranged at equal angles around a small cylindrical shaft. The geometry has another advantage that any disturbance caused by the introduction of the vane into the sample can be kept to a minimum. The geometry is well suited for fluids with significantly high thixotropy.

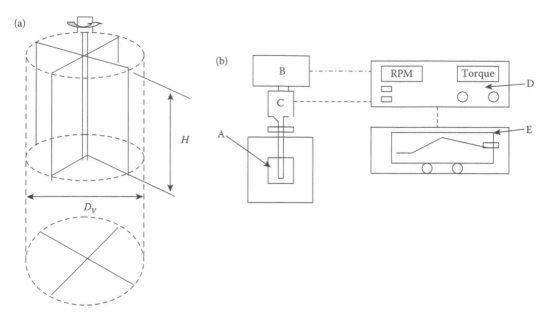

FIGURE 2.2 Schematic diagram of (a) vane and (b) vane apparatus. A: vane; B: motor; C: torsion head; D: instrument console; E: recorder. (Adapted from Nguyen, Q. D., and Boger, D. V., *J. Rheo.*, 27, 321, 1983.)

The vane method has been used to measure the true yield stress of studied materials under virtually static conditions. A vane test is carried out by gently introducing the vane spindle into a sample held in a container until the vane is fully immersed. The depth of the test sample and the diameter of the container should be at least twice of the length and diameter of the vane to minimize any effects caused by the rigid boundaries. The vane is rotated very slowly at a constant rotational speed, and the torsional moment required for maintaining the constant motion of the vane is measured as a function of time (or angle of rotation). For materials having a yield stress, the region of the suspension close to the edges of the vane blades would deform elastically while the vane rotates from the rest. Such linear behavior may be attributed to the mere stretching of the "network bonds" interconnecting the structural elements [21,22] (particles or aggregates or both). Finally, when all (or a majority of) the network bonds have been broken the network would collapse and microscopically the material may be said to yield. Furthermore, since hydrodynamic forces at extremely low shear are not strong enough to bring the separated structural elements close together for a reformation of the network bonds, the material would yield in an irreversible manner with "cracks" (visible or invisible) formed in a localized yield area [22]. This explains the existence of a maximum torque value followed by a rapid fall off in torque with time on the torque-time response.

Based on the above hypothesis it may be said that the presence of a distinctive peak on the torque-time curve obtained with the vane method is a true characteristic of suspensions with a yield stress and that the maximum torque can be related to the true yield stress.

The yield stress may be calculated from the maximum torque (M_o) and vane dimensions using the following relationship:

$$M_o = \left(\frac{\pi h d^2}{2} \right) \tau_o + 4\pi \int_0^{d/2} r^2 \left(\frac{2r}{d} \right)^m \tau_o dr \qquad (2.11)$$

where M_o is the peak torque (N.m), d is the vane diameter (m), h is the vane height (m), r is the vane radius (m), m is a constant (dimensionless), and τ_o is the yield stress (Pa).

Simplification of the above integral and solving for τ_o gives:

$$\tau_o = \left(\frac{2M_o}{\pi d^3}\right)\left(\frac{h}{d}+\frac{1}{m+3}\right)^{-1} \tag{2.12}$$

The parameter "m" is a measure of the shear stress over the ends (both top and bottom) of the vane. Assuming the end effects are small (i.e., $m = 0$) then the yield stress can be calculated as:

$$\tau_o = \left(\frac{2M_o}{\pi d^3}\right)\left(\frac{h}{d}+\frac{1}{3}\right)^{-1} \tag{2.13}$$

The yield stress can be determined from Equation 2.13 using selected dimension of equivalent diameters and heights of the vane. A plot of M_o against vane height (h) provides the yield stress directly from the slope ($\pi\tau_o d^2/2$).

2.3.4 TIME-DEPENDENT FLOW BEHAVIOR

A reliable measurement of time-dependent flow behavior of foods needs considerable care to avoid structure break down during measurcment. There are few mathematical models practiced to study such types of food.

2.3.4.1 Waltman Model

The Waltman model is used to study thoxytropic and nonthoxotropic food materials and the model is given below:

$$\tau = A - B\log t \tag{2.14}$$

where τ is the yield stress, t is the time of measurement, A (stress value after 1 s) and B are constants. A plot of τ against $\log t$ results a straight line. A negative slope indicates thixotropy and nonthixotropic nature is well represented by positive slope. The model has been used to represent time-dependent behavior of waxy corn starch dispersions [23].

2.3.4.2 Tiu and Boger Model

Tiu and Boger [24] studied the rheological behavior of mayonnaise with modification of Herschel-Bulkley model and introducing time-dependent structural parameter (λ) as shown below:

$$\tau = \lambda[\tau_o + K_H(\dot{\gamma})^{n_H}] \tag{2.15}$$

The detail descriptions are available in Tiu and Boger's publication [24].

2.3.5 VISCOSITY BASED RHEOLOGICAL MODELS

Most of the weak gels and hydrocolloids under steady shear conditions exhibit a strong shear-thinning behavior with similar types of flow curve. These complex materials, with an internal structure as a network gel, can be seen as soft solids. At moderate concentrations above a critical value (C^*) hydrocolloid solutions exhibit non-Newtonian behavior where their viscosity depends on the shear strain rate. To describe the structure of such type of fluids over a wide range of shears a model with at least four parameters is required. The Cross model is a good example of this type. Some of these

TABLE 2.1
Viscosity Based Flow Models

Model	Equation	Application
Sisko	$\eta = \eta_\infty + K\gamma^{n-1}$	Used for high shear rate range and when power law fails
Willamson	$\eta = \dfrac{\eta_o}{(1+K\dot{\gamma})^n}$	Used for low shear rate range
Cross	$\dfrac{\eta-\eta_\infty}{\eta_o-\eta_\infty} = \dfrac{1}{(1+K\dot{\gamma})m}$	Used for wide shear rate range
Ellis	$\dfrac{\eta-\eta_\infty}{\eta_o-\eta_\infty} = \dfrac{1}{(1+K\tau)^n}$	Used shear stress in place of shear rate
Carreau	$\dfrac{\eta-\eta_\infty}{\eta_o-\eta_\infty} = \dfrac{1}{[(1+K\dot{\gamma})^2]^{n/2}}$	Alternative to Cross model

TABLE 2.2
Carreau Model Parameters for Tamarind Paste

Temperature (°C)	η_o (Pa.s)	η_∞ (Pa.s)	K (s)	n (–)
10	447.5	3.82	625.2	0.60
30	275.5	0.26	578.0	0.81
50	21.96	0.12	34.38	0.84
70	13.15	0.05	20.43	0.86
90	18.28	6.07E-06	24.54	0.84

Source: From Ahmed, J., Ramaswamy, H.S., and Shashidhar, K.C., *Lebensmittel-wissenschaft und-technologie*, 40, 225, 2007. With permission.

mathematical models based on zero (η_0) and infinite (η_∞) shear rate viscosity are presented in Table 2.1. Applicability of Carreau model for juice concentrate [25] is presented in Table 2.2. The magnitudes of these parameters decrease significantly with temperature. An excessive high magnitude of zero shear rate viscosity was reported for protein-polysaccharide system [26].

2.4 VISCOELASTICITY

Most food materials deviate from Hook's law and exhibit viscous-like properties in addition to elastic characteristics. Viscoelastic materials are those for which the relationship between stress and strain depends on time. Specifically, viscoelasticity is a molecular rearrangement and is associated with the distortion of polymer chains from their equilibrium conformations through activated segment motion involving rotation about chemical bonds [27]. Following this rule, most of the food materials possess both viscous and elastic responses depending on the time scale of the experiment and its relation to a characteristic time of the material. These food materials are known as viscoelastic food. *Anelastic* solids represent a subset of viscoelastic materials: they have a unique equilibrium configuration and ultimately recover completely after removal of a transient load. Knowledge of the viscoelastic response of a material is based on measurement.

2.4.1 TIME

Time plays major role in characterization of viscoelastic materials. Time scale of experiment should be considered in relation to the "characteristic time" denoted by λ of materials (relaxation time). The characteristic times of the materials varies widely from zero to infinity for ideal elastic solid to viscous liquid. The characteristics time is related to time scale of observation by a dimensionless number known as Deborah number (De).

$$De = \frac{\lambda}{t} \qquad (2.16)$$

A high Deborah number (De $\gg$ 1) represents solid-like behavior and low values (De $\ll$ 1) exhibit liquid-like behavior with respect to the observation time. The viscoelastic property dominates while Deborah number is in the range of 1.

2.4.2 LINEAR VISCOELASTICITY

Linear viscoelasticity is the simplest type of viscoelastic behavior, in which viscoelastic behaviors are independent of the magnitudes of applied stress or strain [28]. Linear viscoelasticity of food materials is observed at very low stress–strain in which the structural breakdown is the minimum. Identifying a linear viscoelastic range is a challenge for many foods like dough, cheese, etc. Strain or strain rate dependent materials are known as nonlinear viscoelastic materials [29]. Linear viscoelasticity of food materials can be studied by either static or in dynamic method. In static tests, the sample is subjected to change in stress–strain being measured as a function of time. Oscillatory test is the most common dynamic method for viscoselasticity measurement.

2.4.3 STRESS RELAXATION

If elongation is ceased during the measurement of the stress–strain curve of a material, the force or stress decreases with time as the material approaches equilibrium or quasi equilibrium under the imposed strain. The observation and measurement of the phenomenon describes the stress-relaxation experiment. To determine the stress-relaxation of a material, the specimen is deformed at a given amount and the decrease in stress is recorded over a prolonged period of exposure at constant elevated temperature. At small strain, the stress–strain behavior is almost linear and can be represented by a time-dependent modulus of elasticity $G(t)$. Typical stress relaxation for Mozzarella cheese as reported by Ak and Gunasekaran [30] is presented in Figure 2.3, where $G(t)$ is plotted against time.

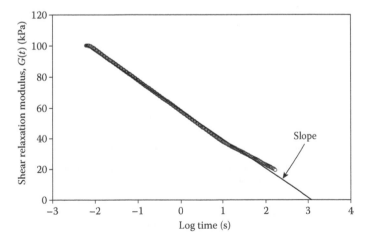

FIGURE 2.3 Stress relaxation spectrum for 14 days old Mozzarella cheese in shear at 20°C. (Adapted from Ak, M. M., and Gunasekaran, S., *J. Food Sci.*, 61, 566, 1996.)

2.4.4 CREEP

A creep test is carried out by applying a constant load to a tensile specimen at a constant temperature. Strain is then measured over a period of time. The stress that causes creep is usually less than the yield stress of a material. Creep tests are simple easier to apply, and more feasible for long period of time. It provides initial information on the viscoelastic behavior of food materials in contrast to metals where creep is undesirable because it results in a change of geometry of components and eventually in their failure.

Primary creep (Stage I, see Figure 2.4) is a period of decreasing creep rate. Primary creep is a period of primarily transient creep. During this period deformation takes place and the resistance to creep increases until it reaches Stage II. Secondary creep (Stage II) is a period of roughly constant creep rate. Stage II is referred to as steady state creep. Tertiary creep, Stage III, occurs when there is a reduction in cross sectional area due to necking or effective reduction in area due to internal void formation. The slope of the curve, identified in the Figure 2.5, is the strain rate during stage II or the creep rate of the material.

Creep and stress-relaxation are complementary aspects of plastic behavior and in many cases provide equivalent information of fundamental viscoelastic properties.

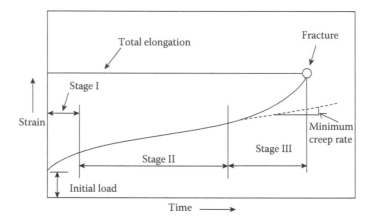

FIGURE 2.4   Creep test for a material.

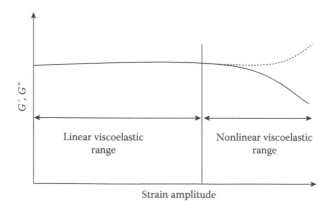

FIGURE 2.5   Strain sweep test to identify linear viscoelastic range.

2.4.5   DYNAMIC METHOD

Small amplitude oscillatory shear (SAOS) measurements afford the measurement of dynamic rheological functions without altering the internal network structure of materials tested [31] and are far more reliable than steady shear measurement [32]. The rheological changes among complex foods could be well represented by SAOS measurement where the strain is restricted to less than 5%.

The delayed response of a food to stress and strain affects its dynamic properties. If a simple harmonic stress of angular frequency (ω) is applied to a sample then the strain lags behind the stress by a phase angle whose tangent measures the internal friction. Dynamic rheological measurement in which sinusoidally oscillating stress or strain is applied to the sample is considered as the best method for characterizing viscoelastic foods. The technique allows to measure elastic (or complex) modulus of a material at different frequencies within a short time compared to other time consuming techniques. For food materials the test is carried out at a linear viscoelastic range and at a comparatively lower strain rate (less than 5%). With the commonly used dynamic test, food samples are usually subjected to a sinusoidal varying stress:

$$\tau = \tau_o + \sin(\omega t) \tag{2.17}$$

The resulting strain will be the sinusoidal response of the same frequency but out of phase by a phase angle δ and represented by the following equation:

$$\gamma = \gamma_o + \sin(\omega t + \delta) \tag{2.18}$$

where τ_o and γ_o are the amplitudes of the oscillation.

The complex modulus which is a complex number that can be derived from the following equations:

$$G^*(i\omega) = \frac{\tau_o}{\gamma_o}(\cos\delta + i\sin\delta) = G'(\omega) + iG''(\omega) \tag{2.19}$$

$$|G^*| = \sqrt{G'^2 + G''^2} \tag{2.20}$$

$$\tan\delta = \frac{G'}{G''} \tag{2.21}$$

where, G' is the storage or elastic modulus describing the amount of energy stored in the material and G'' is the loss or viscous modulus which represents loss in energy or viscous response. The imaginary unit i equals to $\sqrt{-1}$. The phase angle (δ) indicates how much the stress and the strain are out of phase. For a completely elastic material the phase angle is 0° and for a complete viscous fluid δ is 90°. The complex viscosity is another important parameter to characterize viscoelasticity and is given by the following equation:

$$\eta^* = \frac{G^*}{\omega} \tag{2.22}$$

2.4.5.1   Dynamic Measurements

Dynamic measurements are made under SAOS conditions, where the specimen disk is placed between two parallel plates and subjected to a sinusoidal oscillation. The experiments should be carried out in the linear viscoelastic range to keep the food structure intact. A strain or stress sweep is commonly carried out to determine the linear viscoelastic range of a specimen where strain or stress is varied over a selected range. Strain or stress sweep test is performed at a low frequency (e.g., 0.1 or 1 Hz) by increasing the amplitude of the imposed strain-stress. A sharp change in dynamic rheological parameters indicates the end of linearity (Figure 2.5). It is recommended to repeat the strain sweep test at the extremes of experimental variables since the linear viscoelastic region can vary with test frequency, temperature, sample age, and composition, etc. [30]. For precision, ω should be limited to three orders of magnitude, such as 0.1–10 Hz.

Scanning of both strain and frequency are possible and the results are capable of displaying changes in the dynamic rheological components G' and G''. In frequency sweep tests, G' and G'' are measured as a function of frequency at a constant temperature whereas temperature sweep test is carried out at constant frequency and G' and G'' are varied as function of temperature. Temperature sweep test detects the gel point (crossover G' and G'') of protein and starchy foods during heating/cooling. Time sweep test measures the change of G' and G'' as a function of time at constant temperature and frequency. This type of test provides gel stiffness or rigidity. For a specific food, magnitudes of G' and G'' are influenced by frequency, temperature, and strain. These viscoelastic functions have played significant role in protein, starch and complex food systems.

2.4.6 NONLINEAR VISCOELASTICITY

When deformation is rapid or large the linear viscoelastic theory is no longer valid. Relaxation modulus of food materials $G(t)$ depends not only on time but magnitude of deformation also. Many of the food operations deformation are larger like mixing and extrusion where viscoelasticity rule does not follow.

2.4.7 CONSTITUTIVE MODELS OF LINEAR VISCOELASTICITY

Viscoelastic food materials can be modeled in order to determine their stress–strain behaviors as well as their temporal dependencies. Various models are available to characterize viscoelastic food materials including the Maxwell, Kelvin–Voigt, and Burger models. Viscoelastic behavior is comprised of elastic and viscous components. Hookean elasticity is represented by a spring and Newtonian fluid by a dashpot. The viscoelastic behavior of a material can be described by a spring and dashpot model either in series Figure 2.6) or in parallel arrangements (Figure 2.7), respectively. These viscoelastic models are similar to electrical circuits. The elastic modulus of a spring is analogous to a circuit's resistance and the viscosity of a dashpot to a capacitor.

The two elements spring and dashpot are placed in series in Maxwell model (Figure 2.6). Maxwell element shows flow and elasticity on the application of stress. The spring elongates and the dashpot slowly yields under stress. On the removal of the stress the spring returns the original position while the dashpot does not. The strain is given by the following equation:

$$\frac{d\gamma}{dt} = \frac{1}{\eta}\sigma + \frac{1}{G}\frac{d\sigma}{dt} \tag{2.23}$$

where G is the modulus of elasticity and expressed as:

$$G(t) = \frac{\sigma(t)}{\gamma_o} \tag{2.24}$$

The relationship between creep and stress relaxation may be obtained by considering an experiment where a strain is obtained and further held by fixing the ends of the system. Equation 2.23 can then be solved by considering $(d\gamma/dt) = 0$:

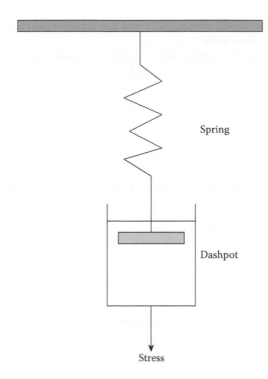

FIGURE 2.6   The spring-dashpot Maxwell model.

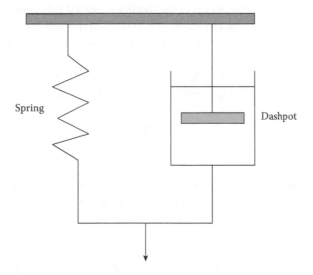

FIGURE 2.7   The spring-dashpot Kelvin model.

$$\sigma = \sigma_o \exp^{-(G/\eta)t} = \sigma_o \exp^{(-t/\tau_R)} \tag{2.25}$$

where the stress σ relaxes from its initial value σ_o exponentially as a function of time. The time η/G after which the stress reaches $1/e$ of its initial value is the relaxation time τ_R.

The relaxation behavior of food is not fully understood by considering a single exponent term and thus, it is desirable to include multiple exponential terms each with different relaxation time representing an array of Maxwell elements in parallel. At longer time ($t \gg \tau_R$), the Maxwell model predicts zero stress and the model cannot represent a truly viscoelastic solid which occasionally exhibit residual stress after relaxation.

The parallel combination of a spring and a dashpot is known as Kelvin model (or Voigt or Kelvin–Voigt model) (Figure 2.7). The spring and the dashpot are strained equally, but the total stress is the sum of the individual stresses. Such combination is used to explain the stress relaxation behaviors of viscoelastic materials. The equation for the strain is:

$$\eta \frac{d\gamma}{dt} + G\gamma = \sigma \tag{2.26}$$

If a stress is applied and removed after a time, the deformation time curve is represented by:

$$\gamma = \frac{\sigma}{G}(1 - \exp^{-(G/\eta)t}) = \frac{\sigma}{G}(1 - \exp^{-t/\tau}) \tag{2.27}$$

where τ is a retardation time. While the stress is removed the sample returns to its original position along the exponential curve:

$$\gamma = \gamma_o \exp^{-t/\tau} \tag{2.28}$$

The model is extremely good with modeling creep in materials, but with regards to relaxation the model is much less accurate.

The creep behaviors of whey protein (> 95% proteins) gels have been described by multicomponent mechanical models [33]. The creep curves of the thermally heated gels conformed to a six element mechanical model consisting of one Hookean, two Voigt (Kelvin) and one Newtonian component. The equation used is based on the following relationship between stress and strain:

$$\gamma(t) = \frac{\sigma}{G_H} + \sum \frac{\sigma}{G_{vi}[1 - \exp(-t/\tau_{ki})]} + \frac{\sigma}{\eta_n}t \tag{2.29}$$

$$\tau_{ki} = \frac{\eta_{vi}}{G_{vi}} \tag{2.30}$$

where $\gamma(t)$ is strain (dimensionless); σ is the stress; G_H is the elastic modulus of Hookean body; η_{vi} is the viscosity of Voigt body; τ_{ki} is the retardation time; η_n viscosity of Newtonian body and t is the time.

A typical creep curve and corresponding mechanical model as reported by Katsura et al. [33] is shown in Figure 2.8. A six element mechanical model described well the WPI gel. Although, a six-element mathematical model provides qualitative description of gel rheology upon applied stress, but provides less quantitative information concerning those interactions occurring at the molecular level [29]. Creep test provides the quantitative estimate if the response to the applied stress reached an equilibrium or equilibrium compliance ($J(e)$) [34]. Authors have used the following equation for creep compliance:

$$J(t) = \frac{e(t)}{S} = \frac{1}{E_H} + \sum \frac{1}{E_v(1 - \exp(-t/\tau_{ki})} + \frac{t}{\eta_n} \tag{2.31}$$

The creep compliance of gels at selected concentration decreased as concentration was increased. Since creep compliance is the reciprocal of the viscoelastic compliance and therefore the gels with

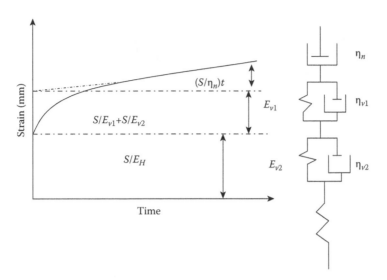

FIGURE 2.8   Typical creep curve of whey protein gel showing the corresponding mechanical model.

higher compliance exhibited lower viscoelastic constants and further indicated that whey protein gel became more solid in nature. A power type relationship was found between log of viscoelastic componenet (G or η) and concentration as shown below:

$$E(\text{or}\,\eta) = AC^a \tag{2.32}$$

The coefficients (A and a) are viscoelastic components and all mechanical components and a values ranged between 5.51 and 6.44.

2.5   COX–MERZ RULE

Some linear viscoelastic material functions measured from oscillatory testing can be related to steady shear behavior [3] by providing more reliable relationships especially at low frequencies and shear rate. Several empirical relations have been proposed that relate the viscometric functions to linear viscoelastic properties (LVP). Cox and Merz [35] found that the dynamic complex viscosity (η^*) and steady shear viscosity (η) of polymeric materials is nearly equal when the frequency (ω) and shear rate ($\dot{\gamma}$) are equal (Equation 2.33).

$$\left|\eta^*\right|(\omega) = \eta(\dot{\gamma})\big|_{\omega=\dot{\gamma}} \tag{2.33}$$

In the literature, there are examples of the Cox–Merz rule being valid for pure liquid and certain polymers [36]. This empirical correlation has been confirmed experimentally for several synthetic polymers, and for several solutions of random-coil polysaccharides. The Cox–Merz rule is fitted adequately for different food products including dilute starch suspensions, polysaccharide solution, concentrated dextran solution, locust bean gum, low methoxyl pectin, and apricot puree [37–41]. A typical applicability of Cox–Merz rule for selected fruit puree based baby foods is illustrated in Figure 2.9.

The values of η for heated cross-linked waxy maize starch dispersions were lower than those of η^* [23] and the empirical rule did not fit. It is reported earlier that the empirical rule does not fit well for structured fluids [42]. Therefore, the applicability of the Cox–Merz rule to polymeric structured fluid or complex food systems is debatable. Various modifications of the rule have been suggested, either by introducing "effective shear rate" or "shear stress equivalent inner shear rate" and testing

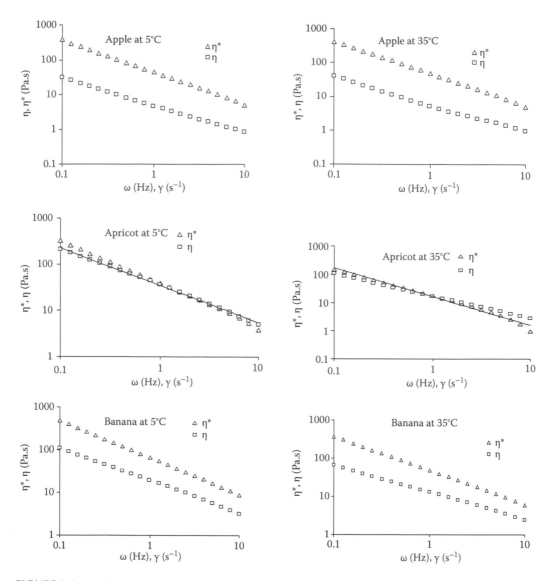

FIGURE 2.9 Applicability of Cox–Merz rule for selected fruit puree based baby foods.

them for various polymeric [43] and food materials (liquid and semisolid) with or without yield stress [31,32,44]. The rule can be applied to foods where oscillatory measurement is more convenient over steady shear measurement.

In many cases, it has been determined that the polymers follow the same general behavior as above when a shift factor is introduced (Equation 2.34) and the equation is known as generalized or extended Cox–Merz relation.

$$\left|\eta(\dot{\gamma})\right| = C\eta^*(\omega)\big|_{\omega=\dot{\gamma}} \qquad (2.34)$$

where C is the shift factor, which is determined experimentally. This is known as the extended or modified Cox–Merz rule.

The generalized or extended Cox–Merz relation has fitted well for various food products including tomato paste, wheat flour dough, and sweet potato puree [32,45–47]. With the introduction of the constant C (shift factor) in complex viscosity data the steady shear viscosity superimposed and fitted the Cox–Merz rule adequately. The constant C did not vary systematically with temperature. The nonfitting of the Cox–Merz rule for most of the food products is attributed to structural decay due to the extensive strain applied. Though applied strain is low in SAOS, it is sufficient enough in steady shear to break down structured inter- and intramolecular associations [31].

It is proposed that during oscillatory experiments the inner deformation increases due to the presence of rigid particles in suspensions [48]. The defect structure in a liquid crystal system changes over time in a steady-shear experiment until a steady-state viscosity and presumably a uniform defect structure is achieved at a given shear rate [49]. The Cox–Merz rule can only be applied if both steady shear and the dynamic oscillation experiments are probing a similar defect structure. This could be achieved by shearing the sample before dynamic oscillation measurement [50]. The difference in the studied samples is attributed to the shear characteristics and response of complex food structure during different types of shearing. The deviation has been explained by noncovalent interactions which determine the stability of the gel systems in literature [51].

The divergence from the Cox–Merz rule was, further, rectified by applying a nonlinear modification (Equation 2.33). A direct power shift factor (α) type relationship was introduced in addition to the constant C when power law parameters, especially the flow behavior index was in a narrow range [52].

$$\eta^* = C\eta^\alpha \big|_{\omega=\dot{\gamma}} \tag{2.35}$$

where C and α are constants to be determined experimentally. In most cases, both η^* versus ω and η versus $\dot{\gamma}$ can be approximated by a power-law, thus, when α equals 1, this relation reduces to the modified Cox–Merz rule.

Values for the multiplicative constant C and the power index α for pureed foods are presented in Table 2.3. The power indices of the baby foods ranged between 1.15 and 1.47 with an average value

TABLE 2.3
Power Model Coefficients of Equation 2.36 for Selected Fruit Puree Based Baby Food

Sample	Temperature (°C)	C	α
Apple	5	6.99	1.17
	20	5.87	1.17
	35	6.25	1.18
	50	5.91	1.15
	65	4.11	1.15
	80	4.64	1.20
Apricot	5	0.63	1.16
	20	0.54	1.20
	35	0.41	1.30
	50	0.64	1.29
	65	1.01	1.47
	80	1.26	1.44
Banana	5	2.22	1.15
	20	2.24	1.24
	35	1.91	1.26
	50	3.32	1.29
	65	11.8	1.33
	80	22.3	1.23

of 1.22 ± 0.06. Samples with α value close to unity show linear relationships between steady shear and complex viscosities. The C values indicate the difference between the complex and steady shear viscosity. Temperature has a significant effect on power model parameter, C. However, the parameter C varied nonsystematically as function of temperature. Most of the purees exhibited marginal change in α values with an exception for apricot puree where the change was significant at elevated temperature. This means that the different curves could be shifted to nearly belong to one band when the scale factor C is applied to them.

2.6   APPLICATION OF DYNAMIC RHEOLOGY TO FOOD

Rheometry is a necessary and powerful technique for explaining and predicting the quality of foods. Consequently, it has found use in a wide range of practical and scientific studies addressing the needs of processors, and, particularly, researchers. During the past 60 years, a variety of instruments based on various principles and techniques has been developed and applied to the study of food products. In this section the focus is given on the application of dynamic rheologial models to studied foods namely starch and protein gels and wheat protein dough. Development and research works in the area of rheology is tremendous and it is indeed difficult to put together. A brief overview is discussed in the following sections.

2.7   FOOD GEL RHEOLOGY

A gel is defined as a substantially diluted system which exhibits no flow [29]. Temperature affects chemical gelation in two major ways. High temperature accelerates the cross-linking reaction and, therefore, speeds up the gelation process. It also increases relaxation processes so that the characteristics time scales of rheology become reduced [53]. Cross-linking of polymers undergo a transition from liquid (sol) to solid (gel) state while the extent of cross-linking reaches a critical value at the gel point [53]. The leading molecular cluster and the corresponding longest relaxation time diverge to infinity for the ideal material exactly at gel point, the critical gel. The critical gel generally relaxes with a continuous power law relaxation time. There is much discussion as to whether a gel point can be defined dynamically by measurement at finite concentration of the eventually formed gel. It has been demonstrated for polymer gels that power law behavior is observed at the gel point for the shear modulii $G(t)$, $G'(\omega)$, and $G''(\omega)$ for permanently cross-linked chains [54].

2.7.1   STARCH GEL RHEOLOGY MEASUREMENT BY SAOS

Starch is a complex food system (polymers of α-D-glucose and partially crystalline polymer) where one would expect two phase transitions during heating in presence of excess water [55]. Starch granules absorb water resulting in swelling to several times of their original sizes to impart an enormous increase to the viscosity of the suspension and loss of crystallinity. The complete process is known as gelatinization. Gelatinization of starch paste involves changes in amylose and amylopectin [56]. Kinetics of starch gelatinization can be studied either in liquid water or steam in vitro using pure starch or in situ using whole grain. In the case of in vitro gelatinization, there is no physical barrier between starch granules and water molecules, and starch is readily accessible to water [57].

2.7.1.1   Effect of Frequency Sweep

Both G' and G'' increase with increase in frequency and G' values are found to be higher than G'' values over most of the frequency range. The frequency dependency of dynamic modulii indicating that quite fast molecular motions are experienced and there is no specific intermolecular interaction [58].

The viscoelasticity of the starch dispersions can be predicted by the calculated slopes of the linear regression of a power-type relationship (Equation 2.36) which is commonly used to predict the solid-like characteristics of polymers.

$$G' \quad \text{or} \quad G'' = A\omega^n \tag{2.36a}$$

Linear form:

$$\ln G' \quad \text{or} \quad \ln G'' = \ln A + n\ln\omega \tag{2.36b}$$

where n is the slope (dimensionless) and A is the intercept ($Pa.s^n$).

A true gel is characterized by a zero slope of the power law model while concentrated and viscous materials exhibit positive slopes [59]. For glutinous rice flour dispersion (4–8% dispersions), the slope ranged from 0.26 to 0.31 and from 0.34 to 0.38 for G' and G'', respectively indicating predominance of solid-like characteristics [60]. At higher concentration, rice starch shows solid-like characteristics and magnitudes of G' is found to be independent of frequency (slope < 0.07) [61]. Sweet potato starch also behaved as viscoelastic fluid with slight variation of dynamic modulii with frequency (slope < 0.1) [62]. G' values of corn and modified starches are independent of frequency (slope < 0.05) while G'' values are slightly frequency dependent with $G' > G''$ (0.1 < slope < 0.3) [63]. This type of spectrum is associated with weak-gel behavior [64].

2.7.1.2 Effect of Temperature Sweep

During temperature sweep biopolymers undergo sol-gel transition with substantial increase in dynamic modulii and a distinct gel point (crossover of G' and G''). Gelation is a critical phenomenon where the transition variable will be the connectivity of the physical and chemical bonds linking to the basic structure. For a better representation, only the G' as a function of temperature is considered since G' measures gel rigidity or gel strength during gelatinization. Based on studies by Ahmed et al. [61] for basmati rice starch dispersions (25% w/w), temperature sweep effect on G' at a constant heating rate of 2°C/min is shown in Figure 2.10. SAOS measurement was made with a controlled-stress rheometer using 60 mm parallel plate-plate geometry with a

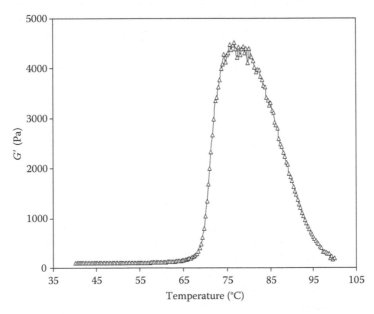

FIGURE 2.10 Temperature sweep of 25% basmati rice starch as function of elastic modulus. (From Ahmed, J., Ramaswamy, H. S., Ayad, A., and Alli, I., *Food Hydrocoll.*, 22, 278, 2008. With permission.)

gap width of 1000 μm. Starch sample was directly placed to the rheometer plate and heating was carried out *in situ* of the rheometer plate.

Starch dispersion initially showed a slow increase in G' in the temperature range of 40–60°C. The temperature was taken at the inflexion point when the G' and G'' values increase very quickly until reaching the maximum. A dramatic increase in G' was noticed between 60 and 82°C leading to sol-gel transformation of starch. The peak value of G' (termed as G'_{max}) was found at 76°C that represents gelatinization temperature for the starch dispersion. A considerable increase in G' of starch on heating is caused by formation of a three-dimensional gel network developed by leached out amylose and reinforced by strong interaction among the swollen starch particles [65,66]. A major reason for higher G'_{max} for basmati rice starch is the presence of significant amount of amylose content (24.6%). The gels of all high amylose content rice starches have been credited with higher G' indicating a well cross-linked network structures and increase in G' followed an exponential relationship with amylose content [67].

Compared to pure starch, the rice flour dispersion exhibited a constant G' values in the temperature range at the vicinity of peak temperature (gelatinization temperature). It is believed that above the melting temperature starches exhibited irreversible swelling and solubilization of leached out amylase content [67] results to a nearly constant G' value in those temperature ranges. The constant G' value of flour dispersion at wider temperature range could be attributed by resistance to amylose leaching in presence of lipid and protein. Minor constituents (lipids, phorphorus) also play important role in addition to concentration–temperature–time protocol on viscoelasticity of starch dispersions [68].

Heating above G'_{max}, the G' decreased appreciably, indicating break down of gel structure. The destruction of structure could be due to the "melting" of the crystalline regions remaining in the swollen starch granule or resulted from the disentanglement of the amylopectin molecules in the swollen particles that softens the particles [69]. The network collapse due to the loss of interaction between particles and network may be another reason for lowering G' value.

While working with corn starch, Yang and Rao [70] used a master curve to represent change of complex viscosity (η^*) with temperature which is very similar to what is shown in Figure 2.10. Authors have described the changes in terms of volume fraction of starch granules. The η^*–T rheogram has been divided into three steps namely pregelatinized, gelatinized and postgelatinized. At low temperature, the volume fractions of the granules were low and during heating it swelled by absorbing maximum amount of water. With further heating, the granules ruptured and disintegrated resulting in decreased volume fraction in the dispersion. The rupture of the granules happened as the amylose released and contributed to the viscosity in the continuous phase.

2.7.1.3 Starch Gel Kinetics by SAOS Measurement

Gelatinization kinetics of starch has been studied extensively by different techniques including differential scanning calorimetry (DSC), blue value technique, electrical conductance measurement, etc. It has been established that starch gelatinization follows first order reaction kinetics [57,71,72] irrespective of measuring technique. During starch gelatinization process, crystallites melt and both molecular and crystalline structures are disrupted [73]. Gelatinization occurs in nonequilibrium state and knowledge of reaction kinetics helps to understand reaction mechanism accurately.

SAOS measurements are a particularly useful method to study the gelatinization phenomenon, by monitoring the kinetics of network development, provided that the measurements are within viscoelastic limit [68]. Compared with DSC based starch gelatinization technique, much less is known about the rheological approach; nevertheless, a few studies on such kinetic approaches are available [61,74].

Changes in rheological characteristics of heated starch dispersions have been studied under isothermal and nonisothermal heating condition. Gelatinization kinetics of starch under isothermal heating follows first order reaction kinetics [75,76]. However, the isothermal heating has some limitations. The major limitation of isothermal heating is to attain the desirable temperature as most heating/cooling conditions are not instantaneous and thermal lag corrections are often applied to add the nonisothermal contribution to the total process. Kinetic handling of data under nonisothermal conditions would overcome those requirements and the methods have the advantage since the parameters can be estimated from a single experiment where temperature is varied over the range of interest, and samples are taken at various intervals. Parameters are estimated from a dynamic environment closer to commercial processing conditions, and the thermal lag problem is solved [77].

Dolan et al. [78] developed a model based on change of apparent viscosity with time during nonisothermal starch gelatinization. The details are available in the publication of Dolan and Steffe [79]. The model contains an exponential of time–temperature history and the Arrhenius equation to describe gelatinization phenomenon. The model can be presented at constant starch concentration and shear rate as:

$$\eta_{dim} = \frac{\eta_a - \eta_{aug}}{\eta_\infty - \eta_{aug}} = [1 - \exp(-k\psi)^n] \tag{2.37}$$

$$\psi = \int T(t) \exp\left(\frac{-\Delta E_g}{RT(t)}\right) dt \tag{2.38}$$

where η_a is apparent viscosity; η_{dim} is the dimensionless apparent viscosity; η_{aug} is the ungelatinized apparent viscosity; η_a is the highest magnitude of apparent visocity; ψ is the integral time–temperature profile; E_g is the activation energy during gelatinization; and R is the universal gas constant.

The above model was further verified by Yang and Rao [70] for corn starch where authors have used complex viscosity (η^*) instead of apparent viscosity.

Yamamoto et al. [80] studied alkali induced rice starch gelatinization kinetics using conventional flow models (power law and Newtonian). Authors have proposed a kinetic treatment involving a viscosity mixing rule of general power law type and found it to follow first order kinetics. The alkali gelatinization process of rice starch dispersion consisted of plural phases characterized by different values of a rate constant. Ahmed et al. [61] studied gelatinization kinetics of basmati rice starch under isothermal and nonisothermal heating condition. The isothermal heating was carried out in the vicinity of gelatinization temperature range (70–95°C) for 30 min. The starch dispersion (25% w/w) was nonisothermally heated from 30°C to 100°C at heating rate of 2°C/min. The change of G' as function of temperature and time has been modeled to obtain reaction order and process activation energy. The gelatinization temperature was initially detected from the DSC and the nonisothermal gelatinization kinetics was estimated between initial and the peak temperature of starch dispersion heating curve.

The nonisothermal kinetics is based on combination of the Arrhenius equation and time–temperature relationship [81]. The kinetic equation shown below (Equation 2.39) is in terms of rheological parameters (G' and dG') in stead of reactant concentration (C) and its change (dC) with time. The negative sign of conventional kinetic equation is replaced by positive sign due to increase in G' during heating (positive dG').

$$\ln\left(\frac{1}{G'^n}\frac{dG'}{dt}\right) = \ln k_o - \left(\frac{E_a}{R}\right)\left(\frac{1}{T}\right) \tag{2.39}$$

A multiple linear regression technique was employed with rice starch kinetic data set (at intervals of 10 s) to determine the order of the reaction (n) after changing the above equations into the following linear forms (Equation 2.40):

$$\ln\left(\frac{dG'}{dt}\right) = \ln k_o + n \ln G' - \left(\frac{E_a}{R}\right)\left(\frac{1}{T}\right) \tag{2.40}$$

The reaction order of the above equation was found to be 0.95 for 25% rice starch dispersion. The process activation energy was 42.2 kJ/mol. The reaction order was further verified by considering $n = 1$ and $n = 2$ in Equation 2.39 which confirmed better fit by first order model. An Arrhenius-type plot for starch gelatinization gave an activation energy (E_a) value of 32.2 kJ/mol.

2.7.1.4   Effect of Heating Time (Isothermal Heating)

Time sweep experiments provide insight whether temperature or time predominates in rice gelatinization process [82]. Isothermal heating behavior on gel rigidity of rice starch at selected temperatures (70–95°C) indicated that change of G' exhibited two different types of curves [61]. At the vicinity of gelatinization temperature ranges (70–75°C) G' decreased with an increase in temperature; however, the pattern was not followed systematically as function of time and temperature. At 70°C, a systematic increase in G' was found as function of time initially (up to 300 s), then exhibited constant values of G' between 300 and 1100 s and finally showed a decreasing trend above 1100 s. An initial increase in G' was also observed for first 500 s at 75°C followed by a decreasing trend while only a decreasing trend of G' was observed at 80 and 85°C, respectively from the beginning. These observations are close in agreement with DSC observations and clearly indicated that heating time has a pronounced effect on gel rigidity of rice starch in addition to temperature effect. A contrast temperature-time effect on G' exhibited by rice starch in the temperature range of 90–95°C. The G' of starch heated at 95°C for 15 min exhibited lower values than corresponding gel at 90°C. However, a sharp increase in G' was noticed at about 850 s. This unusual observation of G' could be contributed by melting of amylopectin crystallites above 90°C which enhanced gel rigidity significantly.

2.7.1.5   Effect of Concentration

The complex modulus ($G*$) and storage modulus (G') increase as a function of starch concentration. The gelatinization temperature decreases as function of starch concentration. A shift of gelatinization temperature is found for cassava starch from 57 to 62°C for dispersions of 15 and 2%, respectively [83]. Based on studies by Ahmed et al. [61], the effect of rice starch concentration on G' during temperature sweep at constant frequency (1 Hz) is illustrated in Figure 2.11. At lower starch concentration (10%) a very weak gel is formed while concentration at and above 20% significantly increase the gel rigidity. An increase in starch concentration from 10 to 30% increase gel rigidity substantially. As the starch concentration of the systems increased, the temperature of the sol to gel

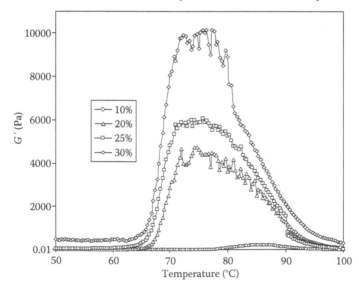

FIGURE 2.11   Concentration effect on gel rigidity of rice starch at heating rate of 2°C/min. (From Ahmed, J., Ramaswamy, H. S., Ayad, A., and Alli, I., *Food Hydrocoll.*, 22, 278, 2008. With permission.)

transformation increased [84]. It is obvious that a critical concentration (~20%) is required to obtain a moderate gel network (Figure 2.11). Although gel rigidity differs at the lower temperature range, the bulk of starch gelatinization occurs in the temperature range between 70 and 80°C except for 10% concentration. At higher temperature range (90–100°C) both 20 and 25% starch concentration superimposes with each other and exhibit almost similar gel rigidity. The strength of the starch gels, as measured by log G', was a linear function of starch concentration for wheat, maize, and other starches for a broad concentration range [84] however, a stronger concentration dependence as power law relationship is reported for rice starch gels (8–40%) [67]. It should be noted that no significant influence of the concentration on the temperature of the moduli overshoot was detected.

It is believed that rheology of starches depend on starch concentrations and on close packing concentration (C^*) [85,86]. At low concentration levels ($C < C^*$) the viscosity is determined in the first instance by the volume occupied by the swollen granules (termed as "dilute" regime), and, to a lesser extent, by the soluble fraction. However, in a concentrated regime ($C > C^*$), starch granules cannot swell to their equilibrium volume due to limited availability of water. The rheological characteristics of starch suspensions are then primarily determined by the particle rigidity of the swollen granules. In experimental practice though, C^* is not very well defined and quite a large transition domain between the concentrated and dilute regimes is observed [86].

The decrease in the G' and G'' values after the overshoot is attributed to the progressive weakening of the starch granules and dynamic acceleration of the gelled amylose matrix (released from granules during gelatinization) [83]. An empirical relationship was used to quantify this overshoot through the following expressions:

$$\Delta x = \frac{G'_{max} - G'_{80}}{G'_{max}} \tag{2.41}$$

$$\Delta x = \frac{\Delta G''_{max} - G''_{80}}{G''_{max}} \tag{2.42}$$

Where G'_{max} and G''_{max} are respectively, the overshoot values and G'_{80} and G''_{80} are the values for the moduli at 80°C.

2.7.1.6 Effect of Heating Rate

Heating rate affects G' and peak gelatinization temperature (T_d). Effect of heating rate on gel rigidity (G') and peak gelatinization temperature (T_d) of 25% rice starch as reported by Ahmed et al. [61] are

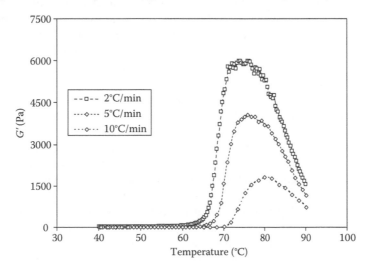

FIGURE 2.12 Effect of heating rate on gel rigidity of 25% basmati rice starch. (From Ahmed, J., Ramaswamy, H. S., Ayad, A., and Alli, I., *Food Hydrocoll.*, 22, 278, 2008. With permission.)

TABLE 2.4
Effect of Heating Rate on Peak Gelatinization Temperature of 10% Basmati Rice Starch Dispersion

Heating Rate (°C/min)	Peak Gelatinization Temperature (°C)
1	73.2
2	74.0
3	75.0
5	75.8
7	78.2
10	80

Source: From Ahmed, J., Ramaswamy, H. S., Ayad, A., and Alli, I., *Food Hydrocoll.*, 22, 278, 2008. With permission.

presented in Figure 2.12 and Table 2.4, respectively. The shape of the rheogram differs from one another while heating rate is varied. The T_d increased linearly from 73.2 to 80°C (about 9.3%) while heating rate was increased from 1 to 10°C/min. An increase in T_d for rice flour was earlier reported by Marshall [87] during increase in heating rate from 2 to 10°C/min. Different curves were obtained when complex viscosity of heated corn starch dispersions plotted as function of time at different heating rate (1.6–6°C/min) [70]. However, the rheograms superimposed with one another when plotted against temperature. The use of rapid heating rates may lead to temperature gradients within the sample, which may contribute to the broadening of the endotherm [88]. Firmer gel rigidity developed at slower heating rate could be attributed by combining effect of heating temperature and duration of heating.

2.7.1.7 Pasting Properties

The Brabender viscoamylograph (BVA) and rapid visco analyser (RVA) are used to study the rheological properties of starches. Conventionally, pasting properties of starch suspension are carried out by a heating-holding-cooling cycle. The sequence follows the steps: the sample holds at 50°C for 2 min, heat to 95°C in 9 min, hold at 95°C for 15 min, cool down to 50°C in the next 9 min and finally hold at 50°C for 10 min [86]. The BVA, in which measurements are made under nonlaminar flow conditions, and in addition, the starch paste is subjected to both thermal and mechanical treatment, thus making it difficult to relate true viscous behavior to only one of these parameters. There is thus an opportunity to extend the use of rheometers (in which thermal treatments are separated from shear effects) to investigate the rheology of gelatinized starch suspensions under well defined flow regimes and to compare these results with those obtained from BVA. An oscillatory probe rheometer is effective at measuring the viscosity of starch pastes and the viscoelastic properties of starch gels since when low shear strains are applied, the integrity of the gel is not disrupted during the testing [84,89].

Ahmed et al. [81] have used controlled stress rheometer to imitate viscoamylographic measurement steps to examine pasting properties of starch as followed by Vandeputte et al. [86]. The results are promising and provide better information with more precise control of the instrument. Authors have used complex viscosity (η^*) instead of Brabender viscosity to describe pasting properties. The starting gel point temperature was considered as the temperature where there was a crossover of G' and G''. The measurement of peak viscosity was based on the peak value of complex viscosity (η^*) which takes into account both dynamic modulii G' and G'' of rice starch during pasting. The difference between the peak and the minimum of η^* was considered as break down viscosity during heating ramp.

Pasting properties of 10% basmati rice starch dispersions is illustrated in Figure 2.13. The starting gel point temperature for starch suspensions was 65.9°C. The observed value was relatively higher than those of waxy starches (60–62°C) but similar to normal starches (64–77°C) [86]. The observed starting gel point was almost similar to the DSC onset gelatinization temperature (65.6°C). The viscosity of paste depends largely on the degree of gelatinization of the starch granules and the extent

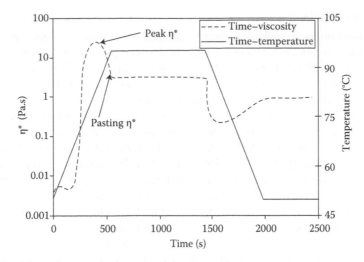

FIGURE 2.13 Pasting properties of 10% basmati rice starch dispersion. (From Ahmed, J., Ramaswamy, H. S., Ayad, A., and Alli, I., *Food Hydrocoll.*, 22, 278, 2008. With permission.)

of their molecular breakdown. The peak η^* obtained from the rheometer was 23.88 Pa.s (at 84.2°C during heating ramp). The η^* during pasting at 95°C was recorded as 2.90 Pa.s (Figure 2.13) which gradually decreased to 0.38 during holding periods (15 min). However, the final η^* value (during cooling) increased again from 0.87 to 1.31 during holding period (10 min) at 50°C. The breakdown paste complex viscosity was measured as $6.6.8 \times 10^{-2}$ Pa.s. Earlier reported values for rice starch ranged between 190 and 380 rapid visco analyzer unit (RVU) for normal rice starches.

2.7.1.8 Effect of High Pressure on Starch Rheology

High pressure treatment has been used to gelatinize starch dispersion and the resulted gel behaved differently from thermally treated gel. In addition, the functionality of starch changed dramatically leading to newer product development with desirable and novel texture with retention of sensory attributes. It is reported that the crystalline order of starches could be destroyed by means of high pressure treatment [90]. During high pressure treatment of rice starch dispersion a systematic sol-gel transformation was observed [91] because of gelatinization of starch and the gel behaved as a viscoelastic fluid. The G' of pressurized gel increased with applied pressure and rice concentration. A 15 min pressure treatment at 550 MPa was found sufficient to complete gelatinization of protein free isolated rice starch while the slurry required 650 MPa. The presence of proteins might have been responsible for the slower starch gelatinization in the rice slurry during pressure treatment.

2.7.2 PROTEIN GELATION

Protein gelation is an association or cross-linking of protein molecules to form a three-dimensional continuous network that traps and immobilizes water to form a rigid structure which is resistant to flow under pressure [92]. Protein denaturation is a prerequisite for an ordered gel formation. The effect of heat on protein system is particularly complicated since the protein exists in aggregates of varying size even without heat. The gelation properties of food proteins have been studied extensively [93–96].

Protein foods denture during thermal or pressure treatment and produce gel. The increase in G' values during gelation has been considered as the gel rigidity or stiffness for various protein foods [97]. The rheological characteristics (gel formation and viscoeleactic behavior) of proteins during heat treatment have been studied [93,98]. In the following section soy protein isolate (SPI) dispersion, the most studied food protein is discussed as one of the representative protein gels from rheological point of view.

2.7.2.1   Soybean Protein Gel

Among various plant proteins, soybean is the most studied for its gelation properties and health benefits. Soybean protein forms an excellent gel either heat and/or pressure treatment. Strain sweep and frequency sweep data of SPI dispersions indicated elastic nature of the dispersion at room temperature [99]. Elasticity increased rapidly while soy dispersions were heated [100]. The aqueous dispersion was changed to a dispersion consisting of particles embedded in heat-soluble protein sol matrix upon heat treatment, and further underwent a transition into a gel consisting of particles embedded in heat-formed protein gel matrix. The G' of undenatured SPI dispersions increased as a consequence of heating, but only when the temperature exceeded the denaturation point of the least stable soy fraction (7S fraction) [100,101]. A lower equilibrium G' was observed however upon heating of commercially processed SPI, with G' increasing upon cooling after heating [102]. The increase in G' during gelation has been considered as the stiffness or gel rigidity for protein foods [97]. The magnitude of G' is found to be proportional to protein denaturation [98]. At higher concentration ($> 26\%$), G' exhibits more frequency dependency than G' for soy protein gels [95].

During temperature sweep test, a consistently predominance of G' over G'' was observed for SPI dispersion and it continued as temperature was increased indicating protein gelation [100]. The G' was found to be significantly higher over G'' above 60°C. The mechanical response observed is characteristic of an entangled network of disordered polymer coils [29]. At low temperature range (<40°C), there is sufficient time for substantial chain disentanglement and rearrangement within the time-scale of the oscillation period. Hence, no significant dominance of elastic modulus over dissipative viscous flow of energy was found.

Effect of SPI concentration on gel rigidity of SPI is presented in Figure 2.14. It demonstrated that 5% SPI concentration could not achieve gelation during thermal scanning due to noninteraction among protein molecules resulted precipitation with scattered G' values. The G at the level of 10% increased with temperature and formed gel network due to hydrodynamic interactions between the protein molecules. Thus, 10% protein concentration has been considered as critical gel concentration. In concentrated solution like 15% SPI concentration, the hydrodynamic domains of the protein molecules come into contact and the interactions between the suspended particles play major role. Due to these particle interactions, concentrated protein dispersions exhibited non-Newtonian viscoelastic behavior. Further, increase in SPI concentration lowered the G' value. The decrease in G' at higher concentration could be attributed to the protein-protein interaction or the competition among active sites of protein molecules for hydration.

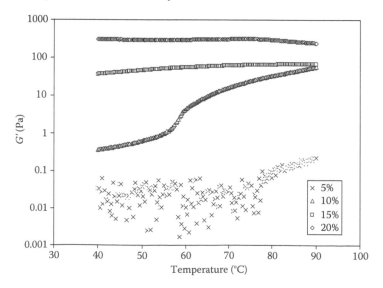

FIGURE 2.14   Effect of concentration on elastic modulus during temperature ramp of soy protein isolates.

TABLE 2.5
Regression Output of Equation $G' = k\omega^n$ for Two Modes of Heating of SPI Dispersion

Concentration (%)	Temperature Ramp to 90°C and Cooled to 20°C		Heated at 90°C for 30 Min and Cooled to 20°C	
	k	n	k	n
10	4.72	0.064	3.83	0.042
15	5.38	0.070	5.41	0.080
20	6.36	0.076	8.49	0.092

Effect of heating mode on gel characteristics of SPI dispersions were computed from power type equation (Equation 2.36) and presented in Table 2.5 [100]. The elastic modulus values were found to be relatively independent of frequency as the slope varied between 0.042 and 0.092. The slope of the power equation confirmed the viscoelasticity of heat treated SPI gels. The regression parameters increased as concentration was increased and consequently solid-like character increased.

Nonisothermal kinetics has been employed to investigate thermorheological behavior of proteins during gelation [100,103]. At appropriate concentration, protein dispersions produce three-dimensional network and, during the heating process, the material changes from a viscous liquid to a semi-solid finally to a hard gel. Gelation occurred at nonisothermal heating/cooling process at constant rate. Both G' and change of G' rate with time (dG'/dt) are found to be function of concentration during gelaion [103–105].

2.7.2.2 Nonisothermal Kinetic Studies

In thermal gelation, the protein unfolds followed by aggregation and cross-linking resulting gelation. Aggregation occur at relatively higher temperature (> 55°C) and is responsible for gel rigidity [106].

While working on nonisothermal heating of SPI dispersion (Equation 2.39) Ahmed et al. [100] reported that the reaction orders of the gelation were 1.16 and 1.45 for 10 and 15% SPI dispersions, respectively. The other kinetic parameters obtained from the equations are: $k_o = 1.16 \times 10^{-8}$ and 1.13×10^{-9} and $E_a = 27.28$ and 29.88 kJ/mol for 10 and 15% SPI dispersions, respectively. The reaction order was further verified and found to follow the second order reaction kinetics. It is reported that collision of two partially unfolded protein molecules (intermolecular aggregation) exhibited second order kinetics at higher temperature [103,107]. Among various kinetic models, nonisothermal has been accepted as the best technique since most of the gelation experimented in situ at constant heating/cooling rate and variation in G' measures structure development rate (dG'/dt) [103].

The activation energy (E_a) ranged between 129.84 and 35.52 kJ/mol for 10 and 15% SPI concentration, respectively. The significantly lower E_a at 15% SPI concentration indicated that the protein molecules were more actively participated in gel formation with less energy barrier to overcome compared to lower concentration (10%) where the energy barrier (E_a) were significantly higher. Comparing to fish protein denaturation [103]) the E_a for SPI denaturation is lower. The difference could be attributed to the types of protein (myofibrillar, glycinin, etc.) and its denaturation temperature ranges.

2.7.2.3 Isothermal Kinetic Studies

Isothermal heating of 15% SPI dispersions at selected temperature (70, 80, 85, and 90°C) for a time period of 30 min exhibited increase in G' values with shearing time except at 90°C where G' decreased with time. It was found that G'–t data fitted a second order reaction kinetics (Equation 2.43) adequately for temperature range of 70–85°C (Figure 2.15). A value of $n = 2$ has been frequently

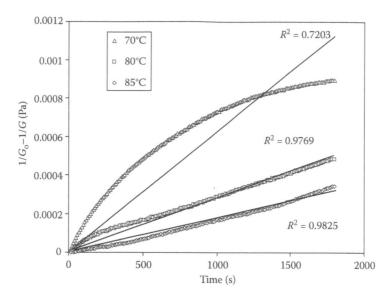

FIGURE 2.15   Second order isothermal gelation kinetics of 15% SPI dispersions at selected temperatures.

used for isothermal gelation following the work of Clark [108]. The slope of G' versus t decreased with increase in temperature and which is contradictory to earlier studies on whey protein isolates. This difference could be attributed by nature of protein and its thermal response during gelation. The process activation energy was found to be 74.01 kJ/mol. The magnitude of E_a in isothermal gelation process was significantly higher compared to nonisothermal gelation. The higher E_a during isothermal technique has been interpreted as the formation of a network by aggregation of unfolded protein molecules utilizing the available time period.

$$\frac{1}{G_o} - \frac{1}{G} = kt \tag{2.43}$$

2.7.2.4   Comparison between Isothermal and Nonisothermal Heating of Soy Protein Isolate (SPI) Gel

Viscoelastic behavior of thermally heated SPI gel made by two different heating techniques (nonisothermal heating 20–90°C at 1°C/min followed by cooling at 20°C and isothermal heating at 90°C for 30 min followed by cooling to 20°C) at selected concentrations (15 and 20%) was studied. Figure 2.16 illustrates the difference in gel rigidity between two different heat treatments. The isothermal heating technique resulted in stronger gels compared to nonisothermal heating. The difference in gel rigidity at 15% SPI concentration was not significantly ($P > 0.05$) higher while the difference was highly significant ($P < 0.05$) at protein concentration at and above 20%. The results suggested that protein gelation is the combined effect of concentration, temperature and heating time. The degree of gel rigidity of SPI dispersions was significantly more at constant temperature of 90°C for half an hour than that of nonisothermal heating from 20 to 90°C in 70 min. These maximum heating temperatures corresponded to the denaturation of glycinin.

The lower gel strengths of the gel obtained by nonisothermal heating, compared to gel from isothermal heating is likely due to insufficient time for proteins to denature and aggregate to form the network required for gelation. The denaturation temperature of SPI was 90.3°C as evidenced from DSC measurement; isothermal heating at 90°C resulted in protein denaturation and protein–protein interaction leading to high gel rigidity. Petruccelli and Anon [109] pointed out that denaturation of glycinin occurred at temperatures at above 85°C and heating rate exhibited changes in the

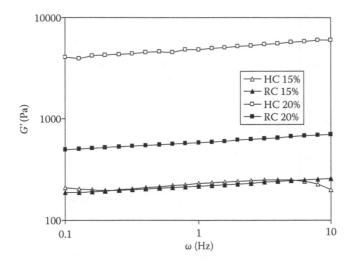

FIGURE 2.16 Effect of heating types on frequency sweep of SPI gel rigidity at selected concentration at 20°C (HC: isothermal holding at 90°C followed by cooling to 20°C; RC: ramp heating to 90°C followed by cooling to 20°C).

aggregation–dissociation state of the soy proteins above 85°C; these researchers suggested that longer heating period (30 min) favored protein aggregation regardless of temperature. Studies on thermal gelation of proteins show less strong or rigid gels during nonisothermal heating compared to isothermally heated gel [99,110].

2.7.3 Effect of High Pressure on Soy Protein Gel Rheology

Recently, high pressure processing of biological macromolecules has received tremendous research interest for the potential benefits in functionality leading to newer product development with desirable texture. These studies revealed high pressure affected protein foods differently: few cases high pressure assisted changes are insignificant (α-lactoalbumin, glycomacropeptides) [111,112], while others exhibited significant extent of modifications (secondary structure) [91]. Pressure treated gel of protein foods has industrial applications to produce newer product or analog products with minimal changes in quality and sensory properties. It is reported that pressure–time–temperature combination can be manipulated to manufacture desirable texture of soybean products [91,113].

Ahmed et al. [114] reported that commercial SPI dispersion was denatured completely and formed gel above 350 MPa. Application of various pressure levels (350–650 MPa) did not show any significant differences in G' of SPI dispersions and the changes were not systematic with pressure. However, 15–20% SPI concentrations exhibited excellent viscoelastic characteristics as supported by phase angle and complex viscosity data.

A plot of complex viscosity (η^*) versus angular frequency (ω) of 15% SPI dispersions at selected pressure levels (350–650 MPa for 15 min) is shown in Figure 2.17. It was found that there was no systematic change in η^* with pressure levels. The values of η^* significantly increased with concentration whereas decreased with angular frequency. Changes of η^* with concentration (C) at constant frequency (1 Hz) was followed power type relationship:

$$\eta = AC^n \tag{2.44}$$

The regression coefficients A and n of Equation 2.3 were varied between 5×10^{-8} and 1×10^{-6} for A and 6.99 and 8.13 for n.

Pressure induced gel showed less rigidity and viscoelasticity compared to thermally treated gel. Pressure treatment altered conformations of soy proteins duly supported by differential scanning

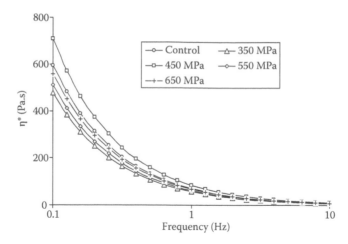

FIGURE 2.17   Effect of high pressure on visccoelasticity of 15% soy protein dispersion.

calorimetry and electrophoresis results. Pressure treated soft gel could be used to develop new food products in combination with other ingredients with known gel characteristics.

Wheat protein dough is one of the best examples of viscoelastic fluid and the viscoelastic properties of dough are briefly described in the following section due to its technical and industrial importance.

2.8   VISCOELASTIC PROPERTIES OF WHEAT PROTEIN DOUGH

Rheological properties of wheat flour dough are of primary importance because the flow and deformation behavior of the dough is central to the successful manufacture of bakery products. Rheologically, wheat flour dough is a composite system that comprises two dispersed filler phases (gas cells and starch granules) and a gluten matrix [115]. Gluten mainly consists of two major storage protein fractions namely gliadin (low molecular weight fraction) and glutenin (high molecular weight fraction) (about 65/45 w/w) [116] and they both contribute to rheological, structural and baking characteristics of wheat. The ratio of the rheological properties of the matrix to those of the fillers determines the behavior of the composite [117]. Wheat gluten proteins have unique viscoelastic properties and are capable of forming three-dimensional structures in aqueous systems.

Traditionally, wheat physical dough properties have been determined using dough mixing instruments such as the mixograph, or descriptive rheological methods such as the alveograph. Dough is a nonlinear viscoelastic fluid that is shear thinning, work hardening, and exhibits a small yield stress. The rheology of bread making is a vast and complicated subject and well reviewed by numerous authors. Dough is an example of complex rheology. Because of its complexity, and the fact that dough rheology changes during different stages, dough rheology mostly studied based on empirical measurements. The LVP of dough is related to dough mixing strength, baking characteristics and pasta quality [118,119], and that large deformation creep tests provide information on dough extensibility properties [120]. Measurement of fundamental of gluten and its building blocks, gliadin and glutenin, help to better understand the underlying macromolecular structures in gluten that are responsible for its unique physical properties. Dynamic oscillatory rheometers are powerful tools for examining the fundamental viscoelastic characteristics of dough [121].

Viscoelastic characteristics of wheat dough and gluten subfractions (gliadin or glutenin) during heat treatment have been studied extensively [122–124]. The thermal denaturation of dough was the result of rapid increase in the number of rheologically active cross-links in gluten [125]. Heating of gluten to 90°C leads to disulfide bond linked aggregates and conformational changes that affect mainly gliadins and low molecular weight albumin and globulin [126].

2.8.1 Frequency Sweep

Both G' and G'' of wheat protein dough are highly frequency dependent at 20°C and increase as frequency was increased. SAOS measurement of two commercial wheat isolates dough (Prolite 100 and 200) were reported by Ahmed et al. [127]. Prolite 100 dough demonstrated a more liquid-like behavior ($G'' > G'$) and Prolite 200 showed a more solid-like ($G' > G''$) characteristic within the frequency range studied. The differences between liquid-like and solid-like properties of dough were, however, not considerable and behaved like a true viscoelastic fluid. The viscoelasticity of the dough was further supported by the calculated slopes (0.48–0.57) of the power-type relationship (Equation 2.36). Contrary to this observation, Janssen et al. [128] observed a predominant liquid-like property (G'') of dough. The discrepancy could be attributed to the differences in moisture content (55%) and nature of protein subfractions.

2.8.2 Temperature Sweep

A large increase in G' for wheat protein dough was reported by Kokini et al. [129] while working on heating of gluten, gliadin or glutenin dough. The time–temperature controlled changes in the dynamic rheological parameters (G', G'', and δ) of two wheat protein isolates dough are illustrated in Figure 2.18. As the temperature increased, both G' and G'' decrease gradually, until a critical temperature was reached. The critical temperature is known as gel point (crossover of G' and G''). Gel point depends on the nature of protein and amount of gliadin/glutenin fraction. The gel points for Prolite 100 and Prolite 200 were 52.2 and 64.6°C, respectively. Cocero and Kokini [122] reported similar values for glutenin. After gel point, G' increased sharply as function of temperature. A higher G' value of Prolite 200 could be contributed to the high molecular weight glutenin. It is further believed that higher glutenin to gliadin ratio has significantly increased the dynamic modulii [128].

Prolite 200, however, behaved like a true viscoelastic fluid ($G' \approx G''$) at the beginning and gradually reached a peak at 90°C with considerable increase in its solid-like property. The dramatic increase in the G' was attributed to cross-linking/aggregation reactions happening among gliadin molecules resulting in the formation of a network structure [129]. The G'' of both doughs decreased while heating and reached a minimum value at 64.6 and 70.7°C, respectively. A further increase in temperature from 70 to 90°C, the G'' levelled and liquid-like property predominate. The magnitude

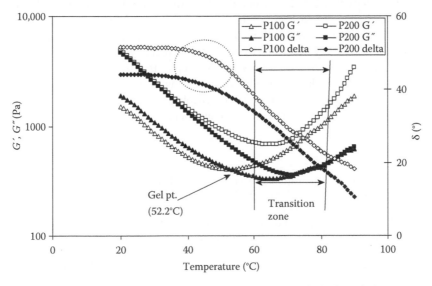

FIGURE 2.18 Effect of heating on mechanical properties of wheat protein isolates during temperature ramp. (From Ahmed, J., Ramaswamy, H. S., and Raghavan, V. G. S., *J. Cereal Sci.*, doi:10.1016/j.jcs.2007.05.013. With permission.)

of δ data clearly demonstrated a sharp transformation of wheat dough in the transition region. Phase angle data inside the circle were very much similar to onset temperature (Figure 2.17) of glass transition temperature.

2.8.3   EFFECT OF HEATING-COOLING ON DOUGH RHEOLOGY

Followed by isothermal heating at 90°C for 15 min, immediate cooling to 20°C dough sample exhibited strong mechanical property. Both G' and G'' increased linearly with frequency. The mechanical strength of dough samples dramatically changed after heat treatment. The regression slopes (Equation 2.38) decreased significantly from 0.57 to 0.11 [127]. The substantial decrease in the slope indicated a change in the entangled polymer flow to more rubbery structure during heating-cooling process.

Both δ and η^* values of dough samples are significantly been affected by heating-cooling operation. Frequency sweep demonstrated a reduction in the magnitude of δ after heating-cooling as compared to unheated sample suggesting increase in solid-like (rubbery) property. This is an evidence of permanently cross-linked density; however it is still a "weak viscoelastic gel" system [95]. Heat-cool process resulted in a very strong dependence of η^* on frequency as observed by Madeka and Kokini [123] for gliadin at higher temperature. Such behavior is common for highly cross-linked materials [107].

The large impact on G' during holding period of wheat protein dough at 90°C was attributed by thermal denaturation of proteins; completion of protein-protein aggregation reaction, and formation of highly cross-linked network. The network formation of gluten is well described in literature [130]. Higher temperature favors cross-linked of gliadin and cross-link through disulfide bonds of glutenins [131]. Protein fractions containing high molecular weight (HMW) glutenin appear to be less affected by heat whereas the extent of change was significant for the smaller polymers containing relatively higher amount of gliadins and subsequently affected rheological properties [132].

2.9   CONCLUSION

Rheology remains a key parameter for quality control of food products. With time, rheological measurement techniques have developed and newer information is available. Vane method has now been used to measure direct yield stress of food. Viscoelasticity of thermally treated foods including gels are truly characterized by SAOS measurement. Thermal starch gelatinization and protein gelation kinetics are adequately described by nonisothermal changes in elastic modulus as function of both temperature and time. Stress relaxation and creep tests provide structural behavior of food products. The structural break down of food during shearing over oscillatory measurement are testified by empirical Cox–Merz rule. On-line precise rheological measurements could provide better information on food microstructure and process equipment design.

REFERENCES

1. Zoon, P. The relation between instrumental and sensory evaluation of the rheological and fracture properties of cheese. *Bull. Int. Dairy Fed.* No. 268. IDF, Brussels, Belgium, 30, 1991.
2. Shoemaker, C. F. Instrumentation for the measurement of viscoelasticity. In: *Viscoelastic Properties of Foods,* Rao M. A. and Steffe J. F., Eds. Elsevier Applied Science, New York, NY, 233–246, 1992.
3. Steffe, J. F. *Rheological Methods in Food Process Engineering.* Freeman Press, East Lansing, MI, 1996.
4. Glasstone, S., Laidler, K. J., and Eyring, H. *The Theory of Rate Processes.* McGraw Hill Book Co. New York, NY, 1941.
5. Billmeyer (Jr), F. W. *Text Book of Polymer Science.* John Wiley and Sons, New York, NY, 1971.
6. Nguyen, Q. D., and Boger, D. V. Yield stress measurement for concentrated suspensions. *J. Rheo.,* 27, 321, 1983.

7. McDowell C. M., and Usher F. L. Viscocity and rigidity in suspension of fine particles. I. Aqueous solution. *Proc. R. Sot. London Ser. A,* 131, 409, 564, 1931.

8. Rehbinder, P. Formation of structures in disperse systems. *Pure Appl. Chem.,* 10, 337, 1965.

9. Nguyen, Q. D., and Boger D. V. Measuring the flow properties of yield stress fluids. *Annu. Rev. Fluid Mech.,* 24, 47, 1992.

10. James, A. E., Williams D. J. A. and Williams, P. R. Direct measurement of static yield properties of cohesive suspensions. *Rheol. Acta.,* 26, 437, 1987.

11. Zhu, L., Sun, N., Papadopoulos K., and De Kee D. A slotted plate device for measuring static yield stress. *J. Rheol.,* 45, 1105, 2001.

12. Uhlherr, P. H. T., Guo, J., Tiu, C., Zhang, X.-M., Zhou, J.Z.-Q., and Fang, T.-N. The shear-induced solid-liquid transition in yield stress materials with chemically different structures. *J .Non-Newt. Fluid Mech.,* 125, 101, 2005.

13. Nguyen, Q. D., Akroyd, T., De Kee1, D. C., and Zhu, L. Yield stress measurements in suspensions: an inter-laboratory study. *Korea-Australia Rheo. J.,* 18, 15, 2006.

14. Cheng, D.C.-H. Yield stress: a time-dependent property and how to measure it. *Rheol. Acta,* 25, 542, 1986.

15. Barnes, H. A. The yield stress-a review or 'παντα ρει'-everything flows? *J. Non-Newt. Fluid Mech.,* 81, 133, 1999.

16. Lang, E. R., and Rha, C. Determination of yield stress of hydrocolloid dispersions. *J Tex. Studies,* 12, 47, 1981.

17. Ahmed, J. Rheological behaviour and colour changes of ginger paste during storage. *Int. J. Food Sci. Technol.,* 39, 325, 2004.

18. Nguyen, Q. D., and Boger D.V. Direct yield stress measurement with the vane method. *J. Rheol.,* 29, 335, 1985.

19. Uhlherr, P. H. T., Guo, J., Fang, T.-N., and Tiu, C. Static measurement of yield stress using a cylindrical penetrometer. *Korea-Australia Rheo. J.,* 14, 17, 2002.

20. Coussot, P., and Boyer, S. Determination of yield stress fluid behaviour from inclined plane test. *Rheol. Acta,* 34, 534, 1995.

21. Briggs J. L., Steffe J. F., and Ustunol, Z. Vane method to evaluate the yield stress of frozen ice cream. *J. Dairy Sci.,* 79, 527, 1996.

22. Houwink R., and de Decker H. K. *Elasticity, Plasticity and Structure of Matter,* 3rd ed. Cambridge University Press, London, UK, 130, 1971.

23. Van den Temple, M. Rheology of plastic fat. *Rheol. Acta,* 1, 115, 1958.

24. Rao, M. A., Okechukwu, P. E., Da Silva, P. M. S., and Oliveira J. C. Rheological behavior of heated starch dispersions in excess water: role of starch granule. *Carbohyd. Poly.,* 33, 273, 1997.

25. Tiu, C., and Boger, D. V. Complete rheological characterization oftime-dependent food products. *J. Text. Studies,* 5, 329, 1974.

26. Ahmed, J., Ramaswamy, H. S., and Shashidhar, K. C. Rheological characteristics of tamarind (*Tamarindus indica* L.) juice concentrates. *Lebensmittel-wissenschaft und-technologie,* 40, 225, 2007.

27. Nunes, M. C., Batista, P., Raymundo, A., Alves, M. M., and Sousa, I. Vegetable proteins and milk puddings. *Coll. Surfaces B: Biointerfaces,* 31, 21, 2003.

28. Barnes, H. A., Hutton, J. F., and Walters K. *An Introduction to Rheology.* Elsevier, Amsterdam, The Netherlands, 1989.

29. Ferry, J. D. *Viscoelastic Properties of Polymer,* 3rd ed. John Wiley and Sons. New York, NY, 641, 1980.

30. Ak, M. M., and Gunasekaran, S. Dynamic rheological properties of Mozzarella cheese during refrigerated storage. *J. Food Sci.,* 61, 566, 1996.

31. Gunasekaran, S., and Ak, M. M. Dynamic oscillatory shear testing of foods—selected applications. *Tr. Food Sci. Technol.,* 11, 115, 2000.

32. Bistani, K. L., and Kokini, J. L. Comparisons of steady shear rheological properties and small amplitude dynamic viscoelastic properties of fluid food materials. *J. Tex. Studies,* 14, 113, 1983.

33. Katsuta, K., Rector, D., and Kinsella, J. E. Viscoelastic properties of whey protein gels: mechanical model and effects of protein concentration on creep. *J. Food. Sci.,* 55, 516, 1990.

34. Flory, P. *Principle of Polymer Chemistry.* Cornell University Press, New York, NY, Chapter 9, 1953.

35. Cox, W. P., and Merz, E. H. Correlation of dynamic and steady flow viscosities. *J. Polym. Sci.,* 28, 619, 1958.

36. Baird, D. G. Rheological properties of liquid crystalline solutions of poly-p-phenyleneterephthalamide in sulfuric acid. *J. Rheol.,* 24, 465, 1980.

37. Chamberlain, E. K., and Rao, M. A. Rheological properties of acid converted waxy maize starches. *Carbohydr. Polym.*, 40, 251, 2000.

38. da Silva, L. J. A., Goncalves, M. P., and Rao, M. A. Viscoelastic behavior of mixtures of locust bean gum and pectin dispersions. *J. Food Eng.*, 18, 211, 1993.

39. McCurdy, R. D., Goff, H. D., Stanley, D. W., and Stone, A. P. Rheological properties of dextran related to food applications. *Food Hydrocoll.*, 38, 609, 1994.

40. Oba, T., Higashimura, M., Iwasaki, T., Master, A. M., Steeneken, P. A. M., Robijn, G.W., and Sikkema, J. Viscoelastic properties of aqueous solutions of the phosphopolysaccharide viilian from *Lactococcus lactis* subsp *cremoris. Carbohydr. Polym.*, 39, 275, 2000.

41. Ahmed, J., and Ramaswamy, H. S. Dynamic and steady shear rheology of fruit puree based baby foods. *J. Food Sci. Technol.*, 44, 579, 2007.

42. Alhadithi, T. S. R., Barnes, H. A., and Walters, K. The relationship between the linear (oScillatory) and nonlinear (steady-state) flow properties of a series of polymer and colloidal systems. *Colloid Polym Sci.*, 270, 40, 1992.

43. Doraiswamy, D., Mujumdar, A. N., Tsao, I., Beris, A. N., Danforth, S. C., and Metzner, A. B. The Cox-Merz rule extended: a rheological model for concentrated suspensions and other materials with a yield stress. *J. Rheol.*, 35, 647, 1991.

44. Dus, S. J., and Kokini, J. L. Prediction of the nonlinear viscoelastic properties of a hard wheat flour dough using Bird-Carreau constitutive model. *J. Rheo.*, 34, 1069, 1990.

45. Berland, S., and Launay, B. Rheological properties of wheat flour doughs in steady and dynamic shear: Effect of water content and some additives. *Cereal Chem.*, 72, 48, 1995.

46. Ahmed, J., and Ramaswamy, H. S. Viscoelastic properties of sweet potato puree infant foods. *J. Food Eng.*, 74, 376, 2006.

47. Rao, M. A., and Cooley, H. J. Rheological behavior of tomato pastes in steady and dynamic shear. *J. Texture Studies*, 23, 415, 1992.

48. Gleissle, W., and Hochstein, B. Validity of the Cox-Merz rule for concentrated suspensions. *J. Rheo.*, 4, 897, 2003.

49. Grizzuti, N., Moldenaers, P., Mortier, M., and Mewis, I. On the time-dependency of the flow-induced dynamic moduli of a liquid crystalline hydroxypropylcellulose solution. *Rheol. Acta*, 32, 218, 1993.

50. Fujiwara, K., Masuda, T., and Takahashi, M. Rheological properties of a thermotropic liquid crystalline copolyester. *Nihon Reoroji Gakkaishi*, 19, 19, 1991.

51. Ross-Murphy, S. B. *Rheology of Biopolymer Solutions and Gels*. Blackie/Professional, London, UK, 1995.

52. Canet, W., Alvarez, M. D., Fernandez, C., and Luna, P. Comparisons of methods for measuring yield stresses in potato puree: effect of temperature and freezing. *J. Food Eng.*, 68, 143, 2005.

53. Izuka, A., and Winter, H. H. Temperature dependence of viscoelasticity of polycaprolactone critical gels. *Macromolecules*, 27, 6883, 1994.

54. Winter, H. H., and Chambon, F. Analysis of linear viscoelasticity of a crosslinking polymer at the gel point. *J. Rheol.*, 30, 367, 1986.

55. Maurice, T. J., Slade, L., Sirett, R. R., and Page, C. M. Polysaccharide-water interaction: thermal behavior of rice starch. In: *Properties of Water in Foods*, Simatos, D., and Multon, J. L., Eds. Martinus Nijhoff Publishers, Dortrecht, The Netherlands, 211–227, 1985.

56. Kim, C., Lee, S. P., and Yoo, B. Dynamic rheology of rice starch-galactomannan mixtures in the aging process. *Starch/Stärke*, 58, 35, 2006.

57. Turhan, M., and Gunasekaran, S. Kinetics of in situ and in vitro gelatinization of hard and soft wheat starches during cooking in water. *J. Food Eng.*, 52, 1, 2002.

58. Doublier, J. L., Launay, B., and Cuvelier, G. Viscoelastic properties of food gels. In: *Viscoelastic Properties of Foods*, Rao, M. A., and Steffe, J. F. Eds. Elsevier Applied Science, New York, NY, 371–434, 1992.

59. Ross-Murphy, S. B. Rheological methods. In: *Biophysical Methods in Food Research*, Chan, H. W. S. Ed. Blackwell Scientific Publications, London, UK, 1984.

60. Ring, S. G. Molecular interaction in aqueous solution of the starch polysaccharides: a review. *Food Hydrocoll.*, 1, 449. 1987.

61. Ahmed, J., Ramaswamy, H. S., Ayad, A., and Alli, I. Thermal and dynamic rheology of insoluble starch from basmati rice. *Food Hydrocoll.* 22, 278, 2008.

62. Fasina, O. O., Walter, W. M., Fleming, H. P., and Simunovic, N. Viscoelastic properties of restructured sweet potato puree. *Int. J. Food Sci. Technol.*, 38, 421, 2003.

63. Rosalina, I., and Bhattacharya, M. Dynamic rheological measurements and analysis of starch gels. *Carbohy. Polymers*, 48, 191, 2002.

64. Clark, A. H., and Ross-Murphy, S. B. Structural and mechanical properties of biopolymer gels. *Adv. Polymer Sci.*, 83, 57, 1987.

65. Eliasson, A. C. Vicoelastic behaviour during the gelatinization of starch I. Comparison of wheat, maize, potato and waxy-barley starches. *J. Tex. Studies*, 17, 253, 1986.

66. Hsu, S., Lu. S., and Huang, C. Viscoelastic changes of rice starch suspensions during gelatinization. *J. Food Sci.*, 65, 215, 2000.

67. Biliaderis, C. G., and Juliano, B. O. Thermal and mechanical properties of concentrated rice starch gels of varying composition. *Food Chem.*, 48, 243, 1993.

68. Biliaderis, C. G., Page, C. M., Maurice, T. J., and Juliano, B. O. Thermal characterization of rice starches: a polymeric approach to phase transitions of granular starch. *J. Agri. Food Chem.*, 34, 6, 1986.

69. Keetels, C. J. A. M., and van Vliet, T. Gelation and retrogradation of concentrated starch gels. In: *Gums and Stabilisers for the Food Industry,* Phillips, G. O., Williams, P, A., and Wedlock, D. J. Eds. IRL Press, Oxford, UK, 1994.

70. Yang, W. H., and Rao, M. A. Complex viscosity–temperature master curve of corn starch dispersion during gelation. *J. Food Process Eng.*, 21, 191, 1998.

71. Riva, M., Schiraldi, A., and Piazza, L. Characterization of rice cooking: isothermal differential scanning calorimetry investigations. *Thermochimica Acta,* 246, 317, 1994.

72. Ojeda, C. A., Tolaba, M. P., and Suarez, C. Modeling starch gelatinization kinetics of milled rice flour. *Cereal Chem.*, 77, 145, 2000.

73. Cooke, D., and Gidley M. J. Loss of crystalline and molecular order during starch gelatinization: origin of the enthalpic transition. *Carbohydr. Res.*, 227, 103, 1992.

74. Kubota, K., Hosokawa, Y., Suzuki, K., and Hosaka, H. Studies on the gelatinization rate of rice and potato starches. *J. Food Sci.*, 44, 1394, 1979.

75. Okechukwu P. E., Rao M. A., Ngoddy P. O., and McWatters K. H. Rheology of sol-gel thermal transition in cowpea flour and starch slurry. *J. Food Sci.*, 56, 1744, 1991.

76. Lund, D. Influence of time, moisture, ingredients, and processing conditions on starch gelatinization. *Crit. Rev. Food Sci. Nutr.*, 20, 249, 1984.

77. Cunha, L. M., and Oliveira, F. A. R. Optimal experimental design for estimating the kinetic parameters of processes described by the first-order Arrhenius model under linearly increasing temperature profiles. *J. Food Eng.,* 46, 53, 2000.

78. Dolan, K. D., Steffe, J. F., and Morgan, R. G. Back extrusion and simulation of viscosity development during starch gelatinization. *J. Food Process Eng.*, 11, 79, 1989.

79. Dolan, K. D., and Steffe, J. F. Modeling rheological behavior of gelatinizing starch solutions using mixer viscometry data. *J. Tex. Studies*, 21, 265, 1990.

80. Yamamoto, H., Makita., E., Oki, Y., and Otani, M. Flow characteristics and gelatinization kinetics of rice starch under strong alkali conditions. *Food Hydrocoll.,* 20, 9, 2006.

81. Rhim, J. W., Nunes, R. V., Jones, V. A., and Swartsel, K. R. Determinant of kinetic parameters using linearly increasing temperature. *J. Food Sci.,* 54, 446, 1989.

82. Hsu, S., Lu. S., and Huang, C. Viscoelastic changes of rice starch suspensions during gelatinization. *J. Food Sci.*, 65, 215, 2000.

83. Marques, P. T., Pérégo, C., Le Meins, J. F., Borsali, R., and Soldi, V. Study of gelatinization process and viscoelastic properties of cassava starch: effect of sodium hydroxide and ethylene glycol diacrylate as cross-linking agent. *Carbohydr. Polym.,* 66, 396, 2006.

84. Hansen, L. P., Hosek, R., Callan, M., and Jones, F. T. The development of high-protein rice flour for early childhood feeding. *Food Technol.*, 35, 38, 1981.

85. Steeneken, P. A. M. Rheological properties of aqueous suspensions of swollen starch granules. *Carbohydr. Polym.,* 11, 23, 1989.

86. Vandeputte, G. E., Derycke, V., Geeroms, J., and Delcour, J. A. Rice starches. II. Structural aspects provide insight into swelling and pasting properties. *J. Cereal Sci.*, 38, 53, 2003.

87. Marshall, W. E. Effect of degree of milling of brown rice and particle size of milled rice on starch gelatinization. *Cereal Chem.*, 69, 632, 1992.

88. Ozawa, T. Kinetic analysis of derivative curves in thermal analysis. *J. Therm. Anal.*, 2, 301, 1970.

89. Tsutsui, K., Katsuta, K., Matoba, T., Takemasa, M., and Nishinari, K. Effect of annealing temperature on gelatinization of rice starch suspension as studied by rheological and thermal measurements. *J. Agri. Food Chem.*, 53, 9056, 2005.

90. Knorr, D., Heinz, V., and Buckow, R., High pressure application for food biopolymers. *Biochimica et Biophysica Acta*, 1764, 619, 2006.
91. Ahmed, J., Ramaswamy, H. S., Ayad, A., Alli, I., and Alvarez, P. Effect of high-pressure treatment on rheological, thermal and structural changes of basmati rice flour slurry. *J. Cereal Sci.*, 46, 148, 2007.
92. Glicksman, M. *Food Hydrocolloids.*, CRC Press, Boca Raton, FL, Vol. I, 1982.
93. Kleef, F. S. M. Thermally induced protein gelation: gelation and rheological characterization of highly concentrated ovalbumin and soybean protein gels. *Biopolym.*, 25, 31, 1986.
94. Wang, C.-H., and Damodaran, S. Thermal gelation of globular proteins: influence of protein conformation on gel strength. *J. Agri. Food Chem.*, 39, 433, 1991.
95. Apichartsrangkoon, A., Bell, A. E., Ledward, D. A., and Schofield, J. D. Dynamic viscoelastic behavior of high-pressure-treated wheat gluten. *Cereal Chem.*, 76, 777, 1999.
96. Renkema, J. M. S., Gruppen, H., and van Vliet, T. The influence of pH and ionic strength on heat-induced formation and rheological properties of soy protein gels in relation to denaturation and their protein composition. *J. Agri. Food Chem.*, 50, 1569, 2002.
97. Clark, A. H., and Lee-Tuffnell, C. D. Gelation of globular proteins. In: *Functional Properties of Food Macromolecules,* Mitchell, J. R., and Ledward, D. A., Eds. Elsevier Applied Science Publishers, London, UK, 203, 1986.
98. Renkema, J. M. S., Knabben, J. H. M., and Van Vliet, T. Gel formation by beta-conglycinin and glycinin and their mixtures. *Food Hydrocoll., 15, 407,* 2001.
99. Hsu, S. Rheological studies on gelling behavior of soy protein isolates. *J. Food Sci.*, 64, 136, 1999.
100. Ahmed, J., Ramaswamy, H. S., and Alli, I. Thermorheological characteristics of soybean protein isolate. *J. Food Sci.,* 71, E158, 2006.
101. Owen, S., Tung, M., and Paulson, A. Thermorheological studies of food polymer dispersions. *J. Food Eng.*, 16, 39, 1992.
102. Chronakis, I. S. Network formation and viscoelastic properties of commercial soy protein dispersions: effect of heat treatment, pH and calcium ions. *Food Res. Int.*, 29, 123, 1996.
103. Yoon, W. B., Gunasekaran, S., and Park, J. W. Characterization of thermorheological behavior of Alaska pollock and pacific Whiting Surimi. *J. Food Sci.,* 69, E238, 2004.
104. Clark, A. H., and Ross-Murphy, S. B. Concentration dependence of gel modulus. *Br. Polymer J.,* 17, 164, 1985.
105. Renkema, J. M. S., and van Vliet, T. Concentration dependence of dynamic moduli of heat-induced soy protein gels. *Food Hydrocoll.,* 18, 483, 2004.
106. Niwa, E. Chemistry of surimi gelation. In: *Surimi Technology*, Lanier, T. C., and Lee, C. M., Eds. Marcel Dekker, New York, NY, 389, 1992.
107. Richardson, R. K., and Ross-Murphy, S. B. Mechanical properties of globular protein gels: 1. Incipient gelation behaviour. *Int. J. Biol. Macromol.*, 3, 315, 1981.
108. Clark, A. H. Gels and gelling. In: *Physical Chemistry of Foods*, Schwartzberg, H. G., and Hartel, R. W., Eds. Marcel Dekker Inc., New York, NY, 263, 1992.
109. Petruccelli, S., and Anon, M. C. Thermal aggregation of soy protein isolates. *J. Agri. Food Chem.*, 43, 3035, 1995.
110. Yoon, W. B., and Park, J. W. Development of linear heating rates using conventional baths and computer simulation. *J. Food Sci.,* 66, 132, 2001.
111. Ahmed, J., and Ramaswamy, H. S. Effect of high-hydrostatic pressure and temperature on rheological characteristics of glycomacropeptide. *J. Dairy Sci.*, 86, 1535, 2003.
112. Ahmed, J., and Ramaswamy, H. S. Effect of high-hydrostatic pressure and temperature on rheological characteristics of alpha-lactoalbumin. *Aus. J. Dairy Technol.*, 58, 233, 2003.
113. Apichartsrangkoon, A., Ledward, D. A., Bell, A. E., and Brennan, J. G. Physicochemical properties of high pressure treated wheat gluten. *Food Chem.*, 63, 215, 1998.
114. Ahmed, J., Ayad, A., Ramaswamy, H. S., Alli, I., and Shao, Y. Dynamic viscoelastic behavior of high pressure treated soybean protein isoate dispersions. *Int. J. Food Proper.*, 10, 397, 2007
115. Smith, J. R., Smith, T. L., and Tschoegl, N. W. Rheological properties of wheat flour doughs III. Dynamic shear modulus and its dependence on amplitude, frequency, and dough composition. *Rheol. Acta*, 9, 239, 1970.
116. Redl, A., Guilbert, S., and Morel, M. H. Heat and shear mediated polymerization of plasticized wheat gluten protein upon mixing. *J. Cereal Sci.*, 38, 105, 2003.
117. Nielsen, L. E. *Mechanical Properties of Polymers and Composites.* Marcel Decker, New York, NY, 1974.

118. Peressini, D., Edwards, N. M., Dexter, J. E., Mulvaney, S. J., Sensidoni, A., and Pollini, C. M. Rheological behaviour of durum wheat doughs and their relation to baking and pasta quality. In: *Proceedings of the Southern Europe Confernce on Rheology*, Calabria, Italy, September 7–11, 21, 1999.

119. Edwards, N. M., Peressini, D., Dexter, J. E., and Mulvaney, S. J. Viscoelastic properties of durum wheat and common wheat dough of different strengths. *Rheol. Acta.*, 40, 142, 2001.

120. Edwards, N. M., Dexter, J. E., Scanlon, M. G., and Cenkowski, S. Relationship of creep-recovery and dynamic oscillatory measurements to durum wheat physical dough properties. *Cereal Chem.*, 76, 638, 1999.

121. Petrofsky, E. A., and Hoseney, R. C. Oscillatory rheometry: the necessity for sample temperature correction. *J. Tex. Studies,* 27, 29, 1997.

122. Cocero, A. M., and Kokini, J. L. The study of the glass transition of glutenin using small amplitude oscillatory rheological measurements and differential scanning calorimetry. *J. Rheol.*, 35, 257, 1991.

123. Madeka, H., and Kokini, J. L. Changes in rheological properties of gliadin as a function of temperature and moisture: Development of a state diagram. *J. Food. Eng.*, 22, 241, 1994.

124. Savador, A., Sanz, T., and Fiszman, S. M. Dynamic rheological characteristics of wheat flour-water doughs. Effect of adding NaCl, sucrose and yeast. *Food Hydrocoll.*, 20, 780, 2006.

125. Schofield, J. D., Bottomley, R. C., Timms, M. F., and Booth, M. R. The effect of heat on wheat gluten and the involvement of sulfhydryl-disulphide interchange reactions. *J. Cereal Sci.*, 1, 241, 1983.

126. Guerrieri, N., Alberti, E., Lavelli, V., and Cerletti, P. Use of spectroscopic and fluorescence techniques to assess heat induced modifications of gluten. *Cereal Chem.*, 73, 368, 1996.

127. Ahmed, J., Ramaswamy, H. S., and Raghavan, V. G. S. Dynamic viscoelastic, calorimetric and dielectric characteristics of wheat protein isolates *J. Cereal Sci.*, doi:10.1016/j.jcs.2007.05.013.

128. Janssen, A. M., van Vliet, T., and Vereijken, J. M. Rheological behaviour of wheat glutens at small and large various isolation procedures. Comparison of two glutens differing in bread making potential. *J. Cereal Sci.*, 23, 19, 1996.

129. Kokini, J. L., Cocero, A. M., Madeka, H., and de Graaf, E. The development of state diagrams for cereal proteins. *Trends Food Sci. Technol.*, 5, 281, 1994.

130. Schofield, J. D., Bottomley, R. C., LeGrys, G. A., Timms, M. F., and Booth, M. R. Effects of heat on wheat gluten. In *Proceedings 2nd International Workshop on Gluten Proteins*, Gravcland, A., and Mooncn, J. H. E., Eds. TNO Institute for Cereals, Flour and Bread, Wageningen, The Netherlands, 81, 1984.

131. Slade, L., and Levine, H. Thermal analysis of starch. *Abstracts of Papers,* 1988 CRA Scientific Conference, Corn Refiners Association, Washington, DC, 169, 1988.

132. Stathopoulos, C. E., Tsiami, A. A., Dobraszczyk, B. J., and Schofield, J. D. Effect of heat on rheology of gluten fractions from flours with different bread-making quality. *J. Cereal Sci.*, 43, 322, 2006.

3 Fundamentals of Heat Transfer in Food Processing

Ferruh Erdoğdu
University of Mersin

CONTENTS

3.1   INTRODUCTION

Heat transfer is a significant transport phenomenon encountered in many unit operations in food processing. Among different heat transfer processes, cooling is applied to reduce enzymatic activity and rate of microbial growth. Pasteurization or sterilization is used for preservation and production of safe food. Freezing is needed for storage and preservation while thawing is required for immediate consumption or further processing of frozen commodities. Cooking (i.e., baking, roasting, grilling, and frying) is another operation that increases the eating quality of foods while attaining required safety. Cooking processes can also involve simultaneous heat and mass transfer where removal of moisture is obtained through the application of heat. In addition to these unit operations that have been applied in food processing for some time, emerging technologies, such as microwave, radio-frequency, infrared and ohmic heating, have also been used as an efficient method of heating.

It is important to understand the mechanism of heat transfer for the development of methodologies to model and further optimize a food processing operation. Therefore, the objective of this chapter is to introduce the fundamentals of heat transfer in food processing. For this purpose, following a brief introduction on heat transfer and thermal and physical properties of foods affecting the heat transfer processes, the boundary conditions and governing equations used in the different applications will be presented.

3.2 GENERAL INFORMATION ON HEAT TRANSFER IN FOOD PROCESSING

Heat transfer is the transfer of energy as a consequence of temperature differences between two objects. Hence, it travels from higher to lower temperature regions (or objects) by conduction, convection, or thermal radiation.

Conduction is considered to be the heat transfer mode when a temperature gradient exists between two solid objects or inside a single solid body. It is the transfer of energy from more energetic particles to the adjacent less energetic ones. If a temperature gradient occurs between fluids or between a fluid and a solid body, in other words if fluid motion is involved, then heat is transferred by convection. The faster the fluid motion and the higher the temperature difference, the higher the rate of convective heat transfer. Depending on the bulk of fluid flow, heat transfer may be controlled by free or forced convection. Free convection heat transfer is the result of buoyancy forces induced by density differences due to the effect of temperature. Forced convection heat transfer, on the other hand, occurs when the fluid motion is caused by an external force. Finally, based on the fact that any object with a temperature of above absolute zero radiates energy in the form of electromagnetic waves, thermal radiation is defined to be the electromagnetic radiation emitted by an object. With its distinctive feature, radiation propagates at the speed of light and does not need a medium for energy transfer unlike conduction and convection.

Temperature difference is the driving force for heat transfer, just as the concentration difference is the driving force for mass transfer and pressure difference for momentum transfer. For defining heat transfer modes, The First Law of Thermodynamics (conservation of energy) and Second Law of Thermodynamics (net increase in the entropy of a closed system leading to the result that heat flows from higher to lower temperature medium) with Fourier's law of conduction ($q'' = -k \cdot (\partial T / \partial x)$), Newton's law of cooling ($q'' = h \cdot (T_1 - T_2)$) and Plank's, Wien's displacement and Stefan–Boltzmann laws of radiation are all used to define heat transfer.

3.3 THERMAL AND PHYSICAL PROPERTIES OF FOODS

Thermal and physical properties of foods are important for modeling and optimization of heat transfer processes. Properties commonly used in mathematical modeling of heat transfer are thermal conductivity (k), specific heat (c_p) and density (ρ).

An accurate knowledge of these properties is essential for predicting temperature changes, process duration and energy consumption for processes involving heat transfer [1]. Among these properties, specific heat and density are significant in mass/energy balance analysis; and thermal conductivity is the key property in determining rate of thermal energy transfer within a material by conduction. Thermal diffusivity ($k/(\rho \cdot c_p)$), which combine the thermal properties is a key property in the analysis of transient heat transfer [2]. A general belief for thermal diffusivity is that the higher the thermal diffusivity the faster heat transfer rate. Recently, Palazoglu [3] presented a detailed discussion on the effect of thermal diffusivity on heat propagation in different heat transfer media. He concluded that the rate of heat penetration was a function of the combined effect of thermal diffusivity and heat transfer coefficient. Therefore, this issue was suggested to be taken into account in modeling and simulation studies. Specific heat, thermal conductivity, density, and thermal diffusivity are always used to model a conduction heat transfer process. Excellent reviews on these properties, for experimental and predictive methodologies are available in the literature [2,4,5]. For modeling freezing and thawing processes where a phase change, from liquid to solid in freezing and solid to liquid in thawing, is involved, enthalpy is used since specific heat becomes infinitely high at initial freezing or thawing point and it may not be possible to include the variations of specific heat in the model at this point.

When modeling of convection is desired, an additional property, viscosity of fluid product is needed to solve the momentum equations. Viscosity is the measure of resistance force exerted by fluid against shear force applied to the fluid. Rao [6] gives information on flow properties of liquid foods including a detailed discussion on rheological models applied to define viscosity changes with temperature, shear rate, and composition.

Additional properties are considered for modeling heat transfer in infrared, ohmic, microwave and radio frequency heating. In infrared heating, radiative properties of food products (emissivity, spectral transmittance, and reflectance) are required. These properties and their variation with temperature and wavelength have been reported by Ginzburg [7], Il'yasov and Krasnikov [8] and Dagerskog and Osterstrom [9]. Recently, Almeida et al. [10] gave a detailed review on measurement of radiative properties of foods in near- and mid-infrared region. As explained by Datta and Almeida [11], radiative properties are generally measured by having a source to emit electromagnetic waves and a detector to capture waves passing through or reflecting from the surface.

For ohmic heating, electrical conductivity (σ) of the product plays a significant role since heating rate is proportional to the strength of the applied electric field and electrical conductivity. Electrical conductivity is the ability to conduct an electrical current and ohmic heating relies on internal heat generation in the medium subjected to an electrical field. Due to the temperature dependence of electrical conductivity, more effective heating is accomplished through the product with higher temperature. In addition, depending upon the product composition, especially in the presence of ions (i.e., salts) electrical conductivity can be quite high. This brings the difficulty of uniform heating in multiphase products when electrical conductivity values are quite different. For this case, Salengke and Sastry [12,13] investigated effects of different electrical conductivities of particles in a static and mixed fluid medium. Methods for determination of electrical conductivity were also presented by Sastry [14].

In microwave and radio-frequency heating, dielectric properties of food products are taken into consideration. These are relative dielectric constant (ε) and relative dielectric loss (ε'') values where the term "loss" represents conversion of electrical energy into heat and the term "relative" shows the relative value with respect to vacuum. Dielectric properties of food products are affected by electromagnetic wave frequency, temperature, moisture content, and composition particularly by salt and fat contents [15]. Muira et al. [16], Venkatesh and Raghavan [17] and Datta et al. [18] provided a list of dielectric properties of different food products.

In addition to thermal and physical properties of food products, heat transfer coefficient, when the convection is present in the medium, is another parameter used in heat transfer studies. It depends on thermo-physical properties of medium, characteristics of food product (shape, dimensions, surface temperature and surface roughness), and characteristics of fluid flow (velocity and turbulence) [19].

There have been numerous expressions in the literature (different *Nusselt* number correlations as a function of *Reynolds*—for forced convection or *Grashof*—for natural convection and *Prandtl* numbers) to determine the value of convective heat transfer coefficient. A compilation for heat transfer coefficient data and empirical expressions can be found in Krokida et al. [20] and Zogzas et al. [21]. These specific expressions do not exist for irregular geometries. However, irregular shapes can be approximated by regular geometries enabling the use of existing correlations. It is best to determine heat transfer coefficient by experiments. Three approaches can be used for experimental calculation: quasi-steady state, transient (analytical and numerical solutions), and trial-and-error methods (analytical and numerical solutions). In the latter two methods, calculations are based on comparison of experimental time–temperature data with simulation results and might require extensive mathematical calculations [22].

3.4 GOVERNING EQUATIONS AND BOUNDARY CONDITIONS FOR HEAT TRANSFER

3.4.1 CONDUCTION AND CONVECTION

This section begins with an introduction of continuity, energy and momentum equations for thermal energy transport and proceeds with required equations and assumptions for conduction and convection heat transfer. By applying the law of conservation of mass, energy and Fourier's law on differential volume, the following equations are obtained in Cartesian coordinates:

Continuity equation:

$$\frac{\partial u_x}{\partial x} + \frac{\partial u_y}{\partial y} + \frac{\partial u_z}{\partial z} = 0 \tag{3.1}$$

Energy equation:

$$(\rho \cdot c_p) \cdot \left(\frac{\partial T}{\partial t} + u_x \cdot \frac{\partial T}{\partial x} + u_y \cdot \frac{\partial T}{\partial y} + u_z \cdot \frac{\partial T}{\partial z} \right) = k \cdot \left(\frac{\partial^2 T}{\partial x^2} + \frac{\partial^2 T}{\partial y^2} + \frac{\partial^2 T}{\partial z^2} \right) + Q \tag{3.2}$$

Momentum equation in x-direction:

$$\rho \cdot \left(\frac{\partial u_x}{\partial t} + u_x \cdot \frac{\partial u_x}{\partial x} + u_y \cdot \frac{\partial u_x}{\partial y} + u_z \cdot \frac{\partial u_x}{\partial z} \right)$$

$$= -\frac{\partial P}{\partial x} + \frac{\partial}{\partial x} \left(\mu \cdot \frac{\partial u_x}{\partial x} \right) + \frac{\partial}{\partial y} \left(\mu \cdot \frac{\partial u_x}{\partial y} \right) + \frac{\partial}{\partial z} \left(\mu \cdot \frac{\partial u_x}{\partial z} \right) + \rho \cdot g_x \tag{3.3}$$

Momentum equation in y-direction:

$$\rho \cdot \left(\frac{\partial u_y}{\partial t} + u_x \cdot \frac{\partial u_y}{\partial x} + u_y \cdot \frac{\partial u_y}{\partial y} + u_z \cdot \frac{\partial u_y}{\partial z} \right)$$

$$= -\frac{\partial P}{\partial y} + \frac{\partial}{\partial x} \left(\mu \cdot \frac{\partial u_y}{\partial x} \right) + \frac{\partial}{\partial y} \left(\mu \cdot \frac{\partial u_y}{\partial y} \right) + \frac{\partial}{\partial z} \left(\mu \cdot \frac{\partial u_y}{\partial z} \right) + \rho \cdot g_y \tag{3.4}$$

Momentum equation in z-direction:

$$\rho \cdot \left(\frac{\partial u_z}{\partial t} + u_x \cdot \frac{\partial u_z}{\partial x} + u_y \cdot \frac{\partial u_z}{\partial y} + u_z \cdot \frac{\partial u_z}{\partial z} \right)$$

$$= -\frac{\partial P}{\partial z} + \frac{\partial}{\partial x} \left(\mu \cdot \frac{\partial u_z}{\partial x} \right) + \frac{\partial}{\partial y} \left(\mu \cdot \frac{\partial u_z}{\partial y} \right) + \frac{\partial}{\partial z} \left(\mu \cdot \frac{\partial u_z}{\partial z} \right) + \rho \cdot g_z \tag{3.5}$$

Derivations for the given equations in Cartesian, cylindrical, and spherical coordinates can be found in heat transfer and fluid dynamics textbooks. The viscous dissipation term that appears on the right hand side of the given energy equation (Equation 3.2) is neglected since it is of significance only for high-speed flows where its magnitude can be comparable to the magnitude of conduction term. As noted by Datta [23], these equations are rarely solved with all the terms as given.

Some simplifications are first introduced to reduce the problem from three- to two- or one-dimension. Additionally, for the case of conduction heat transfer when there is no fluid motion or bulk flow involved (as in the case of porous media), the given equations are reduced to the following by neglecting viscosity and velocity terms:

$$(\rho \cdot c_p) \cdot \frac{\partial T}{\partial t} = k \cdot \left(\frac{\partial^2 T}{\partial x^2} + \frac{\partial^2 T}{\partial y^2} + \frac{\partial^2 T}{\partial z^2} \right) + Q \tag{3.6}$$

For one-dimensional conduction heat transfer (where heat transfer from other two dimensions or surfaces is negligible compared to the given principal dimension or surface area) with no heat generation involved, this equation leads to:

$$\frac{\partial T}{\partial t} = \frac{k}{\rho \cdot c_p} \cdot \left(\frac{\partial^2 T}{\partial x^2} \right) \tag{3.7}$$

In a more general format for one-dimensional heat transfer, the following equation can be used for slab ($n = 0$), cylinder ($n = 1$) and sphere ($n = 2$) geometries:

$$\frac{\partial T}{\partial t} = \frac{k}{\rho \cdot c_p} \cdot \frac{1}{x^n} \cdot \frac{\partial}{\partial x} \left(x^n \cdot \frac{\partial T}{\partial x} \right) \tag{3.8}$$

Equation 3.8 can then be written for cylindrical and spherical, respectively:

$$\frac{\partial T}{\partial t} = \frac{k}{\rho \cdot c_p} \cdot \left(\frac{1}{x} \cdot \frac{\partial T}{\partial x} + \frac{\partial^2 T}{\partial x^2} \right) \tag{3.9}$$

$$\frac{\partial T}{\partial t} = \frac{k}{\rho \cdot c_p} \cdot \left(\frac{2}{x} \cdot \frac{\partial T}{\partial x} + \frac{\partial^2 T}{\partial x^2} \right) \tag{3.10}$$

The simplest approach to solving Equation 3.8 is to assume a steady state condition where $\partial T / \partial t = 0$. This leads to the following:

$$\frac{\partial}{\partial x} \left(x^n \cdot \frac{\partial T}{\partial x} \right) = 0 \tag{3.11}$$

For this case, knowledge of two boundary conditions for one-dimensional heat transfer (since double integration in space variable is involved) is required. It is obvious that four conditions for two-dimensional and six conditions for three-dimensional steady state heat transfer problems are necessary. When steady state assumption does not hold, a condition that represents the integration in time (i.e., initial condition) must be used. The initial condition for unsteady state heat transfer problems is generally the continuous function of initial temperature distribution in the given object.

Boundary conditions across surfaces usually encountered in conduction heat transfer are prescribed surface temperature, prescribed heat flux and convection boundary (so-called third kind boundary condition):

- Prescribed surface temperature: surface temperature is constant or a function of space and/or time. This boundary condition is also known as Dirichlet condition and it is the easiest to work with since it is generally given as a constant temperature across the boundary.
- Prescribed heat flux: heat flux across the boundary is specified as constant or function of space and/or time. A common use of prescribed heat flux as a boundary condition in modeling heat transfer is its special condition where there is no heat flux across the surface (e.g., insulation), $\partial T / \partial x = 0$. In the solution of Equation 3.8, this special case is used for a boundary condition at the center due to symmetry.

- Convection boundary condition (third kind boundary condition): this boundary condition is expressed by the standard equation for convection heat transfer and is used where there is a fluid motion across the boundary:

$$q'' = -k \cdot \frac{\partial T}{\partial x}\bigg|_{\sigma} = h \cdot \left(T\big|_{\sigma} - T_{\infty} \right) \tag{3.12}$$

- As $h \to 0$, the third kind boundary condition tends to be the prescribed heat flux representing insulation condition where there is no heat transfer across the given boundary. On the other hand, it tends to be the prescribed surface temperature boundary condition as $h \to \infty$. In the case of an additional heat source, e.g., radiation, Equation 3.12 can be written as:

$$q'' = -k \cdot \frac{\partial T}{\partial x}\bigg|_{\sigma} = h \cdot \left(T\big|_{\sigma} - T_{\infty} \right) + \varepsilon \cdot \sigma \cdot \left(T\big|_{\sigma}^{4} - T_{\infty}^{4} \right) \tag{3.13}$$

- An additional boundary condition, called interface boundary condition between two solids, is the fourth boundary condition that can be applied in different problems. This is based on two medium (medium A and medium B) having a shared boundary where:

$$-k_A \cdot \frac{\partial T_A}{\partial x}\bigg|_{\sigma} = -k_B \cdot \frac{\partial T_B}{\partial x}\bigg|_{\sigma} \tag{3.14}$$

Exact solutions of temperature change for slab, cylinder and sphere geometries for unsteady state condition are called analytical solutions, and they can be, in addition to modeling purposes, used for determining thermal diffusivity and convective heat transfer coefficient [24]. Analytical solutions are generally obtained using convection boundary present across the surface and a symmetry condition at the center with a uniform initial temperature distribution by applying separation of variables solution methodology to Equation 3.8. The following equation represents one-dimensional analytical solution for slab, cylinder, and sphere:

$$\frac{T(x,t) - T_{\infty}}{T_i - T_{\infty}} = \sum_{n=1}^{\infty} \left[C_i(x) \exp\left(-\mu_n^2 \frac{\alpha \cdot t}{L^2} \right) \right] \tag{3.15}$$

where μ_n and $C_n(x)$ are given for these geometries, respectively.
   Slab:

$$N_{\mathrm{Bi}} = \mu \cdot \tan(\mu) \tag{3.16}$$

$$C_n(x) = \frac{2 \cdot \sin(\mu_n)}{\mu_n + \sin(\mu_n) \cdot \cos(\mu_n)} \cdot \cos\left(\mu_n \cdot \frac{x}{L} \right) \tag{3.17}$$

Cylinder:

$$N_{\mathrm{Bi}} = \mu \cdot \frac{J_1(\mu)}{J_0(\mu)} \tag{3.18}$$

$$C_n(x) = \frac{2 \cdot J_1(\mu_n)}{\mu_i \cdot \left[J_0^2(\mu_n) + J_1^2(\mu_n) \right]} \cdot J_0\left(\mu_n \cdot \frac{x}{L} \right) \tag{3.19}$$

Sphere:

$$N_{Bi} = 1 - \frac{\mu}{\tan(\mu)} \tag{3.20}$$

$$C_n(x) = \frac{2 \cdot [\sin(\mu_n) - \mu_n \cdot \cos(\mu_n)]}{\mu_n - \sin(\mu_n) \cdot \cos(\mu_n)} \cdot \frac{\sin\left(\mu_n \cdot \dfrac{x}{L}\right)}{\mu_n \cdot \dfrac{x}{L}} \tag{3.21}$$

where J_0 and J_1 are the first kind, zeroth and first order Bessel functions, L is half-thickness for slab and radius for cylinder and sphere, x is the distance from the center ($0 < x < L$), and N_{Bi} is Biot number.

Biot number is defined as the ratio of conduction resistance within the body to convection resistance across the boundary,

$$\left(N_{Bi} = \frac{h \cdot L}{k} \right).$$

$(\alpha \cdot t/L^2)$, on the other hand, is called Fourier number (N_{Fo}).

For center temperature of sphere, the given solution should be modified since

$$\lim_{x \to 0} C_n(x) = \frac{2 \cdot [\sin(\mu_n) - \mu_n \cdot \cos(\mu_n)]}{\mu_n - \sin(\mu_n) \cdot \cos(\mu_n)} \cdot \frac{\sin\left(\mu_n \cdot \dfrac{x}{L}\right)}{\mu_n \cdot \dfrac{x}{L}} = \frac{0}{0}.$$

Therefore, the coefficient C_n is reduced to

$$\lim_{x \to 0} C_n(x) = \frac{2 \cdot [\sin(\mu_n) - \mu_n \cdot \cos(\mu_n)]}{\mu_n - \sin(\mu_n) \cdot \cos(\mu_n)}$$

by applying the L'Hospital's rule.

To further simplify the given analytical solutions, the first term (C_1) of the given series is used when N_{Fo} is greater than 0.2. There have been discussions in the literature on limiting the value of N_{Fo}. McCabe et al. [25] reported that only the first term of the series is significant and all other terms can be neglected when N_{Fo} is greater than about 0.1. Kee et al. [26] stated that center solutions can be approximated by an exponential decay when N_{Fo} is greater than 0.15.

For the case of two- (e.g., finite cylinder) and three-dimensional (e.g., finite slab) geometries, given individual solutions are combined using the superimposition technique [27]:

$$\left[\frac{T - T_\infty}{T_i - T_\infty} \right]_{\substack{\text{finite} \\ \text{slab}}} = \left[\frac{T - T_\infty}{T_i - T_\infty} \right]_{\text{slab,depth}} \cdot \left[\frac{T - T_\infty}{T_i - T_\infty} \right]_{\text{slab,width}} \cdot \left[\frac{T - T_\infty}{T_i - T_\infty} \right]_{\text{slab,height}} \tag{3.22}$$

$$\left[\frac{T - T_\infty}{T_i - T_\infty} \right]_{\substack{\text{finite} \\ \text{cylinder}}} = \left[\frac{T - T_\infty}{T_i - T_\infty} \right]_{\text{cylinder}} \cdot \left[\frac{T - T_\infty}{T_i - T_\infty} \right]_{\text{slab}} \tag{3.23}$$

Even though the solutions are given for the third kind boundary condition, these solutions are easily adapted for the case of prescribed surface temperature condition where $h \to \infty$. This case, using Equation 3.12, can be shown as follows:

$$\lim_{h \to \infty} \frac{1}{h} \left(-k \cdot \frac{\partial T}{\partial x} \Big|_\sigma \right) = \left(T \big|_\sigma - T_\infty \right) \Rightarrow T \big|_\sigma = T_\infty \qquad (3.24)$$

The analytical solutions for slab, cylinder and sphere have also been reduced to relatively simple charts (Heisler charts) where the center temperature ratio is plotted as a function of N_{Fo} and N_{Bi}. These charts, in any heat transfer book, are given for N_{Fo} value between 0 and 100 and N_{Bi} value ranging from 0.01 to infinite. However, with extended computing and spreadsheet abilities, the best option would be to employ the infinite series solutions. The first step for this purpose is to determine the required number of roots in Equations 3.16, 3.18, and 3.20 to be as accurate as possible. A simple numerical procedure, i.e., bisection or Newton–Raphson method, can be used for this purpose [24]. Based on this, web based computer programs can be prepared, e.g., http://food.mersin.edu.tr/ferdogdu/Programs/AnalyticalSolutions/program.html.

Another solution in regular-shaped geometries is obtained when a semi-infinite body assumption is held for the heat transfer medium. A semi-infinite solid is defined as a body with infinite depth, width and length. For this geometry, heat transfer will be through one boundary to the end $(0 < x < \infty)$. A region of any thickness may be treated as a semi-infinite medium as long as temperature change, which started from one surface $(x = 0)$ with application of a certain boundary condition, does not penetrate to the other surface $(x = \infty)$ within a given time. It is generally accepted that semi-infinite medium assumption holds when the following criterion is satisfied [28]:

$$L \gg \sqrt{\alpha \cdot t} \qquad (3.25)$$

where L is the thickness of a given region.

Solution for semi-infinite medium approach, when the convection boundary is present with an initial uniform temperature distribution, is given by:

$$\frac{T(x,t) - T_i}{T_\infty - T_i} = \text{erfc}\left(\frac{x}{2 \cdot \sqrt{\alpha \cdot t}} \right) - \exp\left[\frac{h \cdot x}{k} + \left(\frac{h}{k} \right)^2 \cdot \alpha \cdot t \right] \cdot \text{erfc}\left[\frac{x}{2 \cdot \sqrt{\alpha \cdot t}} + \frac{h \cdot \sqrt{\alpha \cdot t}}{k} \right] \qquad (3.26)$$

where x is the distance from the surface, and "erfc" is the complementary error function $(\text{erfc}(y) = 1 - \text{erf}(y))$. If the convective heat transfer coefficient across the surface is assumed to be infinitely high, then Equation 3.26 is simplified to Equation 3.27 since $\lim_{y \to \infty} \text{erfc}(y) \to 0$:

$$\frac{T(x,t) - T_i}{T_s - T_i} = \text{erfc}\left(\frac{x}{2 \cdot \sqrt{\alpha \cdot t}} \right) \qquad \text{or}$$

$$\frac{T(x,t) - T_s}{T_i - T_s} = \text{erf}\left(\frac{x}{2 \cdot \sqrt{\alpha \cdot t}} \right) \qquad (3.27)$$

where T_s is the prescribed surface temperature.

When heat flux (q'') is present across the boundary of a semi-infinite medium, the following solution is obtained:

$$T(x,t) - T_i = \frac{q''}{k} \cdot \left[2 \cdot \sqrt{\frac{\alpha \cdot t}{\pi}} \cdot \exp\left(-\frac{x^2}{4 \cdot \alpha \cdot t} \right) - x \cdot \text{erf}\left(\frac{x}{2 \cdot \sqrt{\alpha \cdot t}} \right) \right] \qquad (3.28)$$

Under some special circumstances in conduction heat transfer, spatial temperature distribution within a solid during processing can be ignored. For this assumption to hold, thermal conductivity value must be high enough to enable uniform temperature distribution inside the body:

$$\lim_{k \to \infty} N_{Bi} = \lim_{k \to \infty} \left(\frac{h \cdot L}{k} \right) \to 0 \qquad (3.29)$$

In this condition, temperature only varies with time, and this condition is called lumped system methodology. This condition is possible when the internal resistance $((1/k) \cdot L)$ of the solid compared to the external resistance $(1/h)$ is smaller resulting in

$$N_{Bi} = \left(\frac{\dfrac{1}{k} \cdot L}{\dfrac{1}{h}} \right) \to 0$$

Governing equation for lumped analysis is given in Equation 3.30 after applying an energy balance:

$$\rho \cdot V \cdot c_p \cdot \frac{\partial T}{\partial t} = h \cdot A \cdot (T - T_\infty) \qquad (3.30)$$

For the solution, only initial condition is required. When this equation is solved by applying the uniform initial temperature condition $(T(t = 0) = T_i)$:

$$\frac{T - T_\infty}{T_i - T_\infty} = \exp \left[-\frac{h \cdot A}{\rho \cdot V \cdot c_p} \cdot t \right] = \exp \left[-N_{Bi} \cdot N_{Fo} \right] \qquad (3.31)$$

Validity of lumped system solution holds when $N_{Bi} \le 0.1$, characteristic dimension (L) for lumped system analysis is given by $L = (V/A)$. Lumped system analysis is commonly used to experimentally determine the convective heat transfer coefficient [22,24].

Solutions given above for different conditions are always restricted by the following assumptions:

- Solid object should conform to a regular geometry (i.e., slab, cylinder, or sphere) during the entire process (lumped system is an exception for this restriction).
- Initial temperature distribution should be uniform.
- Surrounding medium temperature should be constant.
- Thermal properties and physical dimensions of the solid should be constant during the process.

In nonuniform initial or medium temperature cases and variable thermal properties, solutions can be obtained using separation of variables technique, but the results will be complicated compared to the given cases.

Restrictions of analytical solution can be overcome by using numerical methods. Numerical solutions make discrete mathematical approximations of time and spatial variations and boundary conditions defined by partial differential equations. If numerical solutions are formulated and implemented correctly to reduce truncation errors, they are considered to be accurate and reliable. Numerical methods used in heat transfer analysis are finite difference and finite element methods.

With recent developments, computational fluid dynamics (CFD) packages have also been commonly used in heat transfer analysis bringing their own advantages with their extended abilities.

CFD is a powerful numerical tool that is becoming widely used to simulate many processes in the food industry [29]. Over the last two decades, there has been enormous development of commercial CFD codes to enhance their interaction with sophisticated modeling requirements [30–32].

In convection, it is difficult to obtain simple analytical solutions since simultaneous solution for given energy, momentum and continuity equations are required. Cai and Zhang [33] gave a detailed explanation for explicit analytical solutions for two-dimensional laminar natural convection along a vertical porous plate and between two vertical plates showing difficulties when compared to solutions for conduction. Complexity of these solutions obviously increases in three-dimensional cases and in cylindrical and spherical coordinates. Therefore, CFD solutions for convection heat transfer are generally preferred.

For cylindrical coordinates, continuity, energy, and momentum equations become:
Continuity equation:

$$\frac{1}{r} \cdot \frac{\partial}{\partial r}(r \cdot \rho \cdot u_r) + \frac{1}{r} \cdot \frac{\partial}{\partial \theta}(\rho \cdot u_\theta) + \frac{\partial}{\partial z}(\rho \cdot u_z) = 0 \tag{3.32}$$

Energy equation:

$$\frac{\partial T}{\partial t} + u_r \cdot \frac{\partial T}{\partial r} + \frac{u_\theta}{r} \cdot \frac{\partial T}{\partial \theta} + u_z \cdot \frac{\partial T}{\partial z} = \frac{k}{\rho \cdot c_p} \left[\frac{1}{r} \cdot \frac{\partial}{\partial r}\left(r \cdot \frac{\partial T}{\partial r} \right) + \frac{1}{r^2} \cdot \frac{\partial^2 T}{\partial \theta^2} + \frac{\partial^2 T}{\partial z^2} \right] \tag{3.33}$$

Momentum equation in the radial direction:

$$\rho \cdot \left(\frac{\partial u_r}{\partial t} + u_r \cdot \frac{\partial u_r}{\partial r} + \frac{u_\theta}{r} \cdot \frac{\partial u_r}{\partial \theta} - \frac{u_\theta^2}{r} + u_z \cdot \frac{\partial u_r}{\partial z} \right)$$

$$= -\frac{\partial P}{\partial r} + \mu \cdot \left[\frac{\partial}{\partial r}\left(\frac{1}{r} \cdot \frac{\partial}{\partial r}(r \cdot u_r) \right) + \frac{1}{r^2} \cdot \frac{\partial^2 u_r}{\partial \theta^2} - \frac{2}{r^2} \cdot \frac{\partial u_r}{\partial \theta} + \frac{\partial^2 u_r}{\partial z^2} \right] \tag{3.34}$$

Momentum equation in the angular direction:

$$\rho \cdot \left(\frac{\partial u_\theta}{\partial t} + u_r \cdot \frac{\partial u_\theta}{\partial r} + \frac{u_\theta}{r} \cdot \frac{\partial u_\theta}{\partial \theta} + \frac{u_r \cdot u_\theta}{r} + u_z \cdot \frac{\partial u_\theta}{\partial z} \right)$$

$$= -\frac{1}{r} \cdot \frac{\partial P}{\partial r} + \mu \cdot \left[\frac{\partial}{\partial r}\left(\frac{1}{r} \cdot \frac{\partial}{\partial r}(r \cdot u_\theta) \right) + \frac{1}{r^2} \cdot \frac{\partial^2 u_\theta}{\partial \theta^2} + \frac{2}{r^2} \cdot \frac{\partial u_\theta}{\partial \theta} + \frac{\partial^2 u_\theta}{\partial z^2} \right] \tag{3.35}$$

Momentum equation in the vertical direction:

$$\rho \cdot \left(\frac{\partial u_z}{\partial t} + u_r \cdot \frac{\partial u_z}{\partial r} + \frac{u_\theta}{r} \cdot \frac{\partial u_z}{\partial \theta} + u_z \cdot \frac{\partial u_z}{\partial z} \right) = -\frac{\partial P}{\partial z} + \mu \cdot \left[\frac{1}{r} \cdot \frac{\partial}{\partial r}\left(r \cdot \frac{\partial u_z}{\partial r} \right) + \frac{1}{r^2} \cdot \frac{\partial^2 u_z}{\partial \theta^2} + \frac{\partial^2 u_z}{\partial z^2} \right] + \rho \cdot g \tag{3.36}$$

For solution of these equations using CFD methodology, initial and boundary conditions are eventually required. In this case, however, boundary conditions regarding momentum equations are also needed in addition to temperature-related boundary conditions. Generally, no-slip condition at the walls was applied for this purpose. No-slip boundary condition states that velocity of fluid, which is in contact with a solid wall, is equal to the velocity of the wall. As the name implies there is no slip between the wall and the fluid itself. In other words, the fluid particles adjacent to the wall adheres to the wall and move with the same velocity. In conditions where the wall is motionless, the fluid adjacent to the wall has zero velocity [34]. Alternatively, wall velocity, free slip (zero shear stress) or a specified wall shear stress can also be set as a boundary condition [35]. In addition to the no-slip condition, pressure-related boundary conditions might also be applied to the inlet and outlet conditions depending upon the physical nature of the problem.

In addition to obtaining temperature changes, principles of convection heat transfer are also used in predicting convective heat transfer coefficient. This objective is accomplished by using different empirical equations. These equations are given as Nusselt (N_{Nu}) number as a function of Reynolds (N_{Re}) and Prandtl (N_{Pr}) numbers for forced convection and Grashof (N_{Gr}) and Prandtl (N_{Pr}) numbers for natural convection. A variety of these correlations is reported in any heat transfer textbook. A general format for these empirical equations is as follows for forced and natural convection, respectively:

$$N_{Nu} = \varphi \cdot N_{Re}^{a} \cdot N_{Pr}^{b} \tag{3.37}$$

$$N_{Nu} = \phi \cdot (N_{Gr} \cdot N_{Pr})^{c} \tag{3.38}$$

3.4.2 PHASE CHANGE HEAT TRANSFER PROBLEMS (FREEZING AND THAWING)

Freezing and thawing are examples of phase change heat transfer problems. Planck's equation and Pham's method are well known empirical equations used to predict freezing and thawing times of food products. Planck's equation describes only phase change period while ignoring pre- and post-freezing steps. Pham's method, on the other hand, relies on physical aspects of the process including prefreezing, phase change, and even further cooling after freezing is accomplished.

For more accurate determinations, numerical methods are preferred. For modeling of these processes where a phase change, from liquid to solid in freezing and solid to liquid in thawing, is included in the process, enthalpy is preferred since the specific heat becomes infinitely high at the freezing or thawing temperature. Mannapperuma and Singh [36] applied the enthalpy-formulation based finite difference solution to predict freezing and thawing times of food products with good agreement with experimental results. This method was based on rewriting the general Fourier equation into an enthalpy equation. If Equation 3.8 is re-written for this purpose including dependency of density, specific heat and thermal conductivity with temperature:

$$\rho(T) \cdot c_p(T) \cdot \frac{\partial T}{\partial t} = \frac{1}{x^n} \cdot \frac{\partial}{\partial x}\left(x^n \cdot k(T) \cdot \frac{\partial T}{\partial x} \right) \tag{3.39}$$

Then, using enthalpy as a product of specific heat and temperature, following general equation for enthalpy formulation is obtained:

$$\frac{\partial H}{\partial t} = \frac{1}{x^n} \cdot \frac{\partial}{\partial x}\left(x^n \cdot k(T) \cdot \frac{\partial T(H)}{\partial x} \right) \tag{3.40}$$

where H is the enthalpy in J/m³-K. The given equation can then be discretized for each volume element throughout a product with given initial and boundary conditions using a finite difference or finite element technique. This method was used by various authors, and the predictive equations used for obtaining thermo-physical properties and enthalpy in freezing-thawing temperature range were reported [1,37–39].

3.4.3 Heat Transfer with Internal Heat Generation

3.4.3.1 Microwave and Radio Frequency Heating

In Equation 3.2, the term Q (W/m³) refers to internal heat generation. A common source for internal heat generation is electromagnetic waves. Microwave-heating, for example, can be formulated as internal heat generation in the original energy equation. Microwaves are electromagnetic waves that enable volumetric heating of a product in an enclosed cavity. Fast and uniform heating are attractive features for potential applications of electromagnetic energy [40]. They have been used as a heat source since the 1940s and have been extensively applied in the food industry [41]. In fact, the food industry is the largest consumer of microwave energy where it is employed for cooking, thawing, drying, freeze-drying, pasteurization, sterilization, baking, heating, and re-heating of food products. Microwave energy penetrates food materials and produces volumetrically distributed heat source due to molecular friction resulting from dipolar rotation of polar solvents and from conductive migration of dissolved ions. Dipolar rotation is caused by variation of electrical and magnetic fields in the product [41]. Water, a major constituent of most food products, is the main source of microwave interactions due to its dipolar nature. Heat is generated throughout the material, leading to faster heating rates and shorter processing times compared to conventional heating. Using Maxwell equations, heat generation term Q is related to electric field E [42]:

$$Q = 2 \cdot \pi \cdot f \cdot \varepsilon \cdot \varepsilon'' \cdot E^2 \tag{3.41}$$

where ε is the relative dielectric constant, ε'' is the dielectric loss for the product, and f is the frequency. The governing equation for the electric field (E) is:

$$\nabla \left(\frac{\nabla \varepsilon}{\varepsilon} \cdot E \right) + \nabla^2 E + k^2 \cdot E = 0 \tag{3.42}$$

where k is the wave number. To avoid complex formulation of this methodology, Lambert's law, considering an exponential decay of microwave absorption within the product, can be used to determine internal heat generation assuming normal incident electric field to the material surface [43,44]:

$$Q(x) = Q_0 \exp(-2 \cdot \alpha \cdot x) \tag{3.43}$$

where Q is the power at a certain location inside the sample (W), Q_0 is the incident microwave power (W) at the surface, x is between 0 (surface) and maximum distance from the surface-penetration depth (m), and α is the attenuation factor (1/m):

$$\alpha = \frac{2 \cdot \pi}{\lambda} \cdot \sqrt{\frac{\varepsilon \cdot \left[\left(1 + \tan^2 \delta\right)^{0.5} - 1 \right]}{2}} \tag{3.44}$$

where $\tan(\delta)$, the loss tangent, is given by:

$$\varepsilon'' = \varepsilon \cdot \tan(\delta) \tag{3.45}$$

Based on Ayappa et al. [45], Lambert approximation is valid after a certain thickness of a given product. Based on this study, they concluded that application of Lambert's law or Maxwell equations represented similar numerical predictions for slabs 2.7 times thicker than the penetration depth.

For the solution of Equations 3.41 and 3.42 in the energy equation with continuity and momentum equations, it is best to use the CFD methodology. If flow is not included, the solution becomes rather easy. For example, for one-dimensional conduction heating, the problem can be formulated:

$$\frac{\partial T}{\partial t} = \alpha \cdot \frac{\partial^2 T}{\partial x^2} + \frac{Q}{\rho \cdot c_p} \tag{3.46}$$

where Q is the volumetric heat generation term (W/m³). This equation can be easily solved numerically with appropriate boundary conditions and Equations 3.41 and 3.42. As explained, dielectric properties of the product are required in addition to its thermal conductivity, density and specific heat. To include fluid flow, some assumptions might be needed in order to reduce the complexity of the problem and at the same time to retain the overall physics of the flow [46]. The energy equation should therefore be solved first for microwave energy distribution on the cross-sectional area of the product for a known set of dielectric properties. Then, it should be coupled with flow profile to provide theoretical temperature distributions.

Radio frequency heating of foods is also related to conversion of electromagnetic energy to heat within a food product. Radio frequency and microwave frequency bands are adjacent to each other in the electromagnetic spectrum with radio frequency having higher wave lengths. The expression for internal heat generation for radio frequency heating is expressed in a similar way to microwave heating.

3.4.3.2 Ohmic Heating

Ohmic heating is based on passing an electrical current through a food system where it serves as electrical resistance and internal heat is generated. Important factors in ohmic heating processes are applied electric field strength and electrical conductivity of the food product. In addition, product size, geometry, orientation, and composition play a significant role. Due to its temperature dependence of electrical conductivity, the higher the temperature the more effective ohmic heating on heat transfer rate through the product, highlighting difficulties of uniform heating in multiphase systems. Internal heat generation obtained in ohmic heating systems is a function of applied electrical field or voltage distribution which can be obtained by using Maxwell's equations or by combining Ohm's law and continuity equation for electrical current.

For a one-dimensional system, the equation to obtain voltage distribution can be written as:

$$\frac{\partial}{\partial x}\left(\sigma \cdot \frac{\partial V}{\partial x}\right) = 0 \tag{3.47}$$

where V is the voltage, and σ is the electrical conductivity (S/m). Magnitude of electric field strength (E, V/m) and the internal heat generation can then be obtained:

$$E = -\frac{\partial V}{\partial x} \tag{3.48}$$

$$Q = \sigma \cdot E^2 \tag{3.49}$$

At this point, the units should be used carefully. Since electrical conductivity is in S/m (S, Siemens is the inverse of electrical resistance unit, Ω), unit for internal heat generation becomes ($V \times I$) that can be converted to W by using the appropriate conversion factor. After obtaining the internal heat generation value as a function of electrical conductivity and applied electric field strength, the rest of the solution of this heat transfer problem would be to apply the required initial and boundary conditions.

3.4.3.3   High Pressure Processing

High pressure processing is mainly used in food processing as a preservation method to inactivate microorganisms and/or enzymes. During a high pressure process, temperature changes in food product and pressure medium might become different due to their composition and applied pressure. Applied pressure leads to volumetric heating as a result of increasing pressure's effect. The fluid motion generated, on the other hand, strongly affects spatial temperature distribution in a time-dependent manner. Solid foods treated under pressure are also influenced by free convection induced by the pressure medium.

Volumetric heating effect of high pressure can be modeled with internal heat generation theory. For this purpose, internal heat generation value (W/m³) can be obtained using the equation given by Rasanayagam et al. [47] to determine rise in temperature due to compression during high pressure processing:

$$\frac{\partial T}{\partial P} = \frac{\beta \cdot T}{\rho \cdot c_p} \tag{3.50}$$

where β is thermal expansion coefficient (1/K). This relationship is strictly applicable to small pressure changes only. Using this equation, heat source term $\left(Q = (1/V) \cdot m \cdot c_p \cdot (\partial T / \partial t)\right)$ in W/m³, as a function of pressure change, is obtained:

$$Q = \beta \cdot T \cdot \frac{\partial P}{\partial t} \tag{3.51}$$

Datta [22] also suggested the use of Equation 3.51 for heat source term.

3.4.4   Radiation and Infrared Heating

Radiation heat transfer is distinctly different from conduction and convection. It does not require the presence of a medium for heat transfer as it is mediated by electromagnetic waves at the speed of light. Basic laws of radiation are Plank's law, Wien's displacement law and Stefan–Boltzman law. The first two laws state that maximum amount of radiant energy can be emitted at a certain temperature and at a given wavelength bringing blackbody concept in radiation heat transfer $(\lambda \cdot T = 2897.6\,\mu m - K)$. The amount of heat emitted from a perfect radiator (black body) is determined using Stefan-Boltzmann law:

$$q'' = \sigma \cdot T^4 \tag{3.52}$$

where σ is the Stefan–Boltzmann constant (5.669×10^{-8} W/m²-K⁴), and T is temperature in K.

Since no object is a perfect emitter (blackbody), they are characterized by their efficacy of radiant emission called emissivity (ε). Based on this definition, Equation 3.52 takes the following form:

$$q'' = \sigma \cdot \varepsilon \cdot T^4 \tag{3.53}$$

The net radiation of heat energy exchange between two objects can then be written as:

$$q'' = \sigma \cdot \varepsilon \cdot \left(T_1^4 - T_2^4\right) \tag{3.54}$$

For computation of radiation exchange between any given two objects, an additional parameter is the view factor, a function of size and shape of the surfaces with their orientation against each other. Net radiation is proportional to view factor, and view factor is a quantity that is a fraction of radiation leaving one surface and intercepted by another. View factor relations and calculation methodologies can be found in any heat transfer textbook. When two surfaces are included in the analysis, their emissivity values definitely affect the quantity of heat transfer. For example, for two infinite parallel planes of 1 and 2, where plane 1 has the higher temperature, rate of heat transfer including view factor and emissivity values of the surfaces can be written as:

$$q = \frac{\sigma \cdot T_1^4 - \sigma \cdot T_2^4}{\dfrac{1-\varepsilon_1}{A_1 \cdot \varepsilon_1} + \dfrac{1}{A_1 \cdot F_{1 \to 2}} + \dfrac{1-\varepsilon_2}{A_2 \cdot \varepsilon_2}} \tag{3.55}$$

Assuming that parallel planes have equal surface areas and $F_{1 \to 2} = 1$ since radiation leaving plane 1 will be totally intercepted by plane 2, heat transfer rate then becomes:

$$q'' = \frac{q}{A} = \frac{\sigma \cdot (T_1^4 - T_2^4)}{\dfrac{1}{\varepsilon_1} + \dfrac{1}{\varepsilon_2} - 1} \tag{3.56}$$

Infrared energy, as a radiation heating process, is an excellent heat source for many food processing applications. Infrared region of electromagnetic spectrum is in the wavelength range of 0.75–1000 µm. In infrared heating processes, thermal energy is primarily absorbed by the surface of food products resulting in an increase in the surface temperature. Then, the rate of heat conduction to the interior becomes a function of thermo-physical properties of the product. Since food products have lower thermal conductivity, infrared heating is also preferred for surface pasteurization purposes. In designing infrared heating systems and determining surface temperature changes of food products, it is extremely important to know the fraction of infrared energy that has been absorbed by the surface.

A simple way of radiation modeling during infrared heating is to include as a boundary condition combined with convection (Equation 3.13). This inclusion of radiation as a boundary condition is a result of the fact that penetration depth of infrared radiation through the surface is negligible. If penetration depth is significant, then infrared heating must be included as a volumetric heat source term (Q, W/m³) where its effect decreased exponentially with increasing penetration depth:

$$Q = \frac{q_0''}{\delta} \cdot \exp\left(-\frac{x}{\delta}\right) \tag{3.57}$$

where δ is the penetration depth (m) and q_0'' is the heat flux (W/m²) at the surface:

$$q_0'' = -k \cdot \left.\frac{\partial T}{\partial x}\right|_{\text{surface}}$$

Ginzburg [7] and Datta and Almeida [11] reported the penetration depths of some food products as a function of wavelength of incident radiation.

3.4.5 SIMULTANEOUS HEAT AND MASS TRANSFER

Cooking food products (baking, frying, roasting, and grilling) is always coupled with moisture loss. This effect is included in solution methodologies with an additional term. This can be accomplished through the convective boundary condition:

$$h \cdot A \cdot \left(T\big|_\sigma - T_\infty\right) - \lambda \cdot \frac{\partial m}{\partial t} = -k \cdot A \cdot \left.\frac{\partial T}{\partial x}\right|_\sigma \tag{3.58}$$

where λ is the latent heat of evaporation (J/kg), A is the surface area (m²), and $\partial m/\partial t$ is the rate of moisture loss (kg/s) during the process. It is also possible to include a parameter in the boundary equation including mass transfer with diffusion coefficient instead of using moisture loss rate:

$$h \cdot \left(T\big|_\sigma - T_\infty \right) - \lambda \cdot D \cdot \frac{\partial C}{\partial x}\bigg|_\sigma = -k \cdot \frac{\partial T}{\partial x}\bigg|_\sigma \tag{3.59}$$

where D is the diffusion coefficient (m²/s) for water and $\dfrac{\partial C}{\partial x}\bigg|_\sigma$ is the concentration gradient at the surface.

The references by Farkas et al. [48,49] for frying, Singh et al. [50], and Towndsend et al. [51] for roasting, Zorrilla and Singh [52] for grilling, Broyart and Trystram [53], and Zhang and Datta [54] for baking and Meeso et al. [55] and Younsi et al. [56] for drying are suggested additional readings on simultaneous heat and mass transfer modeling.

3.5 MOVING BOUNDARY CONCEPT

In addition to freezing and thawing processes of foods, roasting, frying, grilling and baking are other examples of a moving boundary problem in food processing. Therefore, it is important to understand the fundamentals of transport phenomena that occurs across boundaries to develop the simulation models to improve these processes [57].

In these processes, after a certain time elapsed, two regions can be seen in the food product. For example, in freezing, frozen and unfrozen parts are seen with their distinctive thermal and physical properties (across the frozen front). In frying, crust formation on the outer surface forms with very different thermal and physical properties compared to the product's core region (across the evaporation front) right after the frying process is initiated. Similar observations are also seen in baking and roasting.

3.6 CONCLUSIONS

Thermal food processes involve one or more heat transfer mechanisms. In this chapter, a comprehensive review of the mathematical procedures for modeling conduction, convection, and radiation mechanisms of heat transfer with their applications in food processing operations was presented and significance of understanding and determining the heat transfer coefficient was covered. In addition, heat transfer with electromagnetic waves in different food processing applications was also discussed.

ACKNOWLEDGMENT

The author is greatly appreciative and thanks Dr. T. Koray Palazoğlu for proofreading this chapter.

NOMENCLATURE

A	Surface area	m²
c_p	Specific heat	J/kg-K
C	Concentration	kg/m³
		kg water/kg dry matter
d	Penetration depth	m
E	Electric field strength	V/m

f	Frequency	Hertz
g	Gravimetric acceleration	9.81 m/s²
h	Convective heat transfer coefficient	W/m²-K
H	Enthalpy	J/m³-K
I	Current	Ampere
k	Thermal conductivity	W/m-K
m	Mass	kg
q''	Heat flux	W/m²
L	Half-thickness of a slab or radius of a cylinder or sphere	m
	Characteristic dimension in lumped system analysis (V/A)	m
P	Pressure	Pa
Q	Power	W
Q	Volumetric heat generation term	W/m³
t	Time	s
T	Temperature	°C, K
u	Velocity component in the momentum equation	m/s
V	Volume	m³
	Voltage	V

Greek Letters

α	Thermal diffusivity	m²/s
α	Attenuation factor	1/m
β	Thermal expansion coefficient	1/K
δ	Penetration depth	m
δ	Loss tangent	
ε	Emissivity	
ε	Dielectric constant in vacuum	
ε''	Dielectric loss factor	
σ	Electrical conductivity	S/m
σ	Stefan-Boltzmann constant	5.669×10^{-8} W/m²-K⁴
λ	Latent heat of evaporation	J/kg
λ	Wavelength	μm
μ	Dynamic viscosity	kg/m-s
	Characteristic root of the Eqns. 16, 18 and 20	
ρ	Density	kg/m³

Subscripts

i	Initial
s	Surface
σ	Surface boundary
∞	Medium

REFERENCES

1. Fikiin, K.A. and Fikiin, A.G. Predictive equations for thermophysical properties and enthalpy during cooling and freezing of food materials. *J. Food Eng.*, 40, 1, 1999.
2. Heldman, D.R. Prediction models for thermophysical properties of foods. In *Food Processing Operations Modeling Design and Analysis*, Irudayaraj, J., Ed. Marcel Dekker Inc. New York, NY, 2002.

3. Palazoglu, T.K. Influence of convective heat transfer coefficient on the heating rate of materials with different thermal diffusivities. *J. Food Eng.*, 73, 290, 2006.
4. Nesvadba, P. Thermal properties of unfrozen foods. In *Engineering Properties of Foods*, Rao, M.A., Rizvi, S.S.H. and Datta, A.K. Eds. CRC Press, Boca Raton, FL, 2005.
5. ASHRAE. Thermal properties of foods. In *ASHRAE Handbook 2006 Refrigeration*, Owen, M.S., Ed. American Society of Heating, Refrigerating and Air-Conditioning Engineers, Inc. Atlanta, GA, 2006.
6. Rao, M.A. Rheological properties of fluid foods. In *Engineering Properties of Foods*, Rao, M.A., Rizvi, S.S.H. and Datta, A.K., Eds. CRC Press, Boca Raton, FL, 2005.
7. Ginzburg, A.S. *Application of Infra-red Radiation in Food Processing*. Leonard Hill Books. London, UK, 1969.
8. Il'yasov, S.G. and Krasnikov, V.V. *Physical Principles of Infrared Irradiation of Foods*. Hemi-sphere Publishing Corporation, New York, NY, 1991.
9. Dagerskog, M. and Osterstrom, L.,Infrared radiation for food processing I. a study of the fundamental properties of infrared radiation. *Lebensmittel-Wissenchauft Tech.*, 12, 237, 1979.
10. Almeida, M., Torrance, K.E., and Datta, A.K. Measurement of optical properties of foods in near- and mid-infrared radiation. *Int. J. Food Prop.*, 9, 651, 2006.
11. Datta, A.K. and Almeida, M. Properties relevant to infrared heating of foods. In *Engineering Properties of Foods*, Rao, M.A., Rizvi, S.S.H. and Datta, A.K., Eds. CRC Press, Boca Raton, FL, 2005.
12. Salengke, S. and Sastry, S.K. Experimental investigation of ohmic heating of solid-liquid mixtures under worst-case heating scenarios, *J. Food Eng.*, 83, 324, 2007.
13. Salengke, S. and Sastry, S.K. Models for ohmic heating of solid-liquid mixtures under worst-case heating scenarios. *J. Food Eng.*, 83, 337, 2007.
14. Sastry, S.K. Electrical conductivity of foods. In *Engineering Properties of Foods*, Rao, M.A., Rizvi, S.S.H. and Datta, A.K., Eds. CRC Press, Boca Raton, FL, 2005.
15. Tang, J. Dielectric properties of foods. In *The Microwave Processing of Foods*, Schubert, H. and Regier, M. Eds. CRC Press, Boca Rotan, FL, 2005.
16. Muira, N., Yagihara, S. and Mashimo, S. Microwave dielectric properties of solid and liquid foods investigated by time-domain reflectometry. *J. Food Sci.*, 68, 1396, 2003.
17. Venkatesh, M.S. and Raghavan, G.S.V. An overview of microwave processing and dielectric properties of agri-food materials. *Biosystems Eng.*, 88, 1, 2004.
18. Datta, A.K., Sumnu, G. and Raghavan, G.S.V. Dielectric properties of foods. In *Engineering Properties of Foods*, Rao, M.A., Rizvi, S.S.H. and Datta, A.K., Eds. CRC Press, Boca Raton, FL, 2005.
19. Rahman, S. *Food Properties Handbook*. CRC Press Inc. Boca Raton, FL, 1995.
20. Krokida, M.K, Zogzas, N.P. and Maroulis, Z.B. Heat transfer coefficient in food processing: compilation of literature data. *Int. J. Food Prop.*, 5, 435, 2002.
21. Zogzas, N.P. et al. Literature data of heat transfer coefficients in food processing. *Int. J. Food Prop.*, 5, 391, 2002.
22. Erdoğdu, F., Balaban, M.O. and Chau, K.V. Automation of heat transfer coefficient determination: development of a Windows-based software tool. *Food Tech. Turkey.* 10, 66, 1998.
23. Datta, A.K. Heat transfer. In *Handbook of Food and Bioprocess Modeling Techniques*, Sablani, S.S., Rahman, M.S., Datta, A.K. and Mujumdar, A.A., Eds. CRC Press, Boca Raton, FL, 2006.
24. Erdoğdu, F. Mathematical approaches for use of analytical solutions in experimental determination of heat and mass transfer parameters. *J. Food Eng.*, 68, 233, 2005.
25. McCabe, W.L., Smith, J.C. and Harriot, P. *Unit Operations in Chemical Engineering*, 4th ed. McGraw-Hill, Inc., New York, NY, 1987.
26. Kee, W.L., Ma, S. and Wilson, D.I. Thermal diffusivity measurements of petfood. *Int. J. Food Prop.*, 5, 145, 2002.
27. Newman, A.B. Heating and cooling rectangular and cylindrical solids. *Ind. Eng. Chem.*, 28, 545, 1936.
28. Hagen, K.D., *Heat Transfer with Applications*, Prentice-Hall, Inc. Upper Saddle River, NJ, 1999.
29. Norton, T. and Sun, D-W. Computational fluid dynamics (CFD)—an effective and efficient design and analysis tool for the food industry: a review. *Trends Food Sci. Tech.*, 77, 600, 2006.
30. Abdul Ghani, A.G. et al. Numerical simulation of natural convection heating of canned food by computational fluid dynamics. *J. Food Eng.*, 41, 55, 1999.
31. Abdul Ghani, A.G., Farid, M.M. and Chen, X.D. Numerical simulation of transient temperature and velocity profiles in a horizontal can during sterilization using computational fluid dynamics. *J. Food Eng.*, 51, 77, 2002.
32. Varma, M.N. and Kannan, A. CFD studies on natural convective heating of canned food in conical and cylindrical containers. *J. Food Eng.*, 77, 1024, 2006.

33. Cai, R. and Zhang, N. Explicit analytical solutions of 2-D laminar natural convection. *Int. J. Heat and Mass Transfer*, 26, 931, 2003.
34. Çengel, Y.A. and Cimbala, J.M. *Fluid Mechanics Fundamentals and Applications*. McGraw-Hill Companies, Inc., New York, NY, 2006.
35. Verboven, P., De Baerdemaeker, J. and Nicolai, B.M. Using computational fluid dynamics to optimize thermal processes. In *Improving the Thermal Processing of Foods*, Richardson, P., Ed. CRC Press, Boca Raton, FL, 2004.
36. Mannapperuma, J.D. and Singh, R.P. A computer-aided method for the prediction of properties and freezing/thawing times of foods. *J. Food Eng.*, 9, 275, 1989.
37. Miles, C. The thermophysical properties of frozen foods. In *Food Freezing: Today and Tomorrow*, Bald, W.B., Ed. Spring-Verlag, London, UK, 1991.
38. Reinick, A.C.R. Average and center temperature vs time evaluation for freezing and thawing rectangular foods. *J. Food Eng.*, 30, 299, 1996.
39. Schwartzberg, H.G. Effective heat capacities for the freezing and thawing of food. *J. Food Sci.*, 41, 152, 1976.
40. Zhong, Q., Sandeep, K.P. and Swartzel, K.R. Continuous flow radio frequency heating of particulate foods. *Innovative Food Sci. Emerging Tech.*, 5, 475, 2004.
41. Oliveira, M.E.C. and Franca, A.S. Microwave heating of foodstuffs. *J. Food Eng.*, 53, 347, 2002.
42. Prosetya, H. and Datta, A. Batch microwave heating of liquids: An experimental study. *J. Microwave Power Electromagnetic Energy*, 26, 215, 1991.
43. Campanone, L.A. and Zaritzky, N.E. Mathematical analysis of microwave heating process. *J. Food Eng.*, 69, 359, 2005.
44. Romano, V.R. et al. Modeling of microwave heating of foodstuff: study on the influence of sample dimensions with a FEM approach. *J. Food Eng.*, 71, 233, 2005.
45. Ayappa, K.G. et al. Microwave heating: an evaluation of power formulations. *Chem. Eng. Sci.*, 46, 1005, 1991.
46. Datta, A.K., Prosetya, H. and Hu, W. Mathematical modeling of batch heating of liquids in a microwave cavity. *J. Microwave Power Electromagnetic Energy*, 27, 101, 1992.
47. Rasanayagam, V. et al. Compression heating of selected fatty food materials during high-pressure processing. *J. Food Sci.*, 68, 254, 2003.
48. Farkas, B.E., Singh, R.P. and Rumsey, T.R. Modeling heat and mass transfer in immersion frying. I, model development. *J. Food Eng.*, 29, 211, 1996.
49. Farkas, B.E., Singh, R.P. and Rumsey, T.R. Modeling heat and mass transfer in immersion frying. II, model solution and verification. *J. Food Eng.*, 29, 226, 1996.
50. Singh, N., Akins, R.G. and Erickson, L.E. Modeling heat and mass transfer during the oven roasting of meat. *J. Food Proc. Eng.*, 7, 205, 1984.
51. Townsend, M.A. and Gupta, S. The roast: nonlinear modeling and simulation. *J. Food Proc. Eng.*, 11, 17, 1989.
52. Zorrilla, S.E. and Singh, R.P. Heat transfer in double-sided cooking of meat patties considering two-dimensional geometry and radial shrinkage. *J. Food Eng.*, 57, 57, 2003.
53. Broyart, B. and Trystram, G. Modelling heat and mass transfer during the continuous baking of biscuits. *J. Food Eng.*, 51, 47, 2002.
54. Zhang, J. and Datta, A.K. Mathematical modeling of bread baking process. *J. Food Eng.*, 75, 78, 2006.
55. Meeso, N. et al. Modeling of fat infrared irradiation in paddy drying process. *J. Food Eng.*, 78, 1248, 2007.
56. Younsi, R. et al. Computational modeling of heat and mass transfer during the high temperature heat treatment of wood. *App. Thermal Eng.*, 27, 1424, 2007.
57. Singh, R.P. Moving boundaries in food engineering. *Food Tech.*, 54(2), 44, 2000.

4 Mass Transfer Basics

Gauri S. Mittal
University of Guelph

CONTENTS

4.1 INTRODUCTION

Mass transfer in foods can occur in the form of liquid and gases in liquid and solid due to concentration gradient and convection. The chemical reaction can deplete or generate a component of the mass. Moisture transfer from food products to air occurs in drying when water vapor pressure in the food is higher than in the air. Food moisture content influences microbial, organoleptic, functional and structural qualities, enzymatic reactions, nonenzymatic browning, lipid oxidation, textural changes, and aroma retention of foods. Mass transport is required in the analysis of basic food processing operations such as drying, crystallization, humidification, distillation, evaporation, leaching, absorption, membrane separation, rehydration, mixing, extraction, and storage. In leaching, a solute is separated from solid matrix to a liquid solvent, while in extraction, a solute is separated from liquid mixture to a liquid solvent. In crystallization, a solute is transferred from a liquid phase (mother liquor) to a solid liquid interface. In many food processes mass transfer is accompanied by heat transfer such as drying, evaporation, and distillation.

Capillary or the liquid diffusion theory is generally applicable to the mass transfer in food materials. The capillary theory refers to the flow of liquid through the interstices and over the surface of a solid due to molecular attraction between the liquid and solid. The liquid diffusion is usually applicable in the later stages of drying and attributes water transfer to concentration gradients. For mass transfer controlled by capillary flow, the rate of mass transfer is inversely proportional to the food thickness, whereas for diffusion, the rate of mass transfer is inversely proportional to the square of the thickness.

4.2   MOLECULAR DIFFUSION

Molecular diffusion is the movement or transfer of individual molecules through a stationary fluid by means of random movements. The net diffusion is from high to low mass concentration. The sugar in a coffee cup will be mixed by molecular diffusion. However, it will be very slow and takes few hours to mix well in a whole cup if not mixed by turbulence. Thus, mass transfer by turbulence motion (convective mass transfer) is very rapid compared to molecular diffusion alone. Another example of molecular diffusion is the mixing of smoke from a point source into air. Again mixing will be faster if turbulence is applied along with molecular diffusion. The molecular diffusion in gases is many times faster than in liquids.

The diffusion of a binary mixture of components A and B is described by Fick's law:

$$J_A = -D_{AB} \frac{dC_A}{dz} \tag{4.1}$$

where J_A is the flux in gmol A/(m².s), C_A is the concentration of component A in gmol/m³, z is the distance in the direction of diffusion in m, and D_{AB} is the mass diffusivity of A in the binary mixture or simply the diffusion coefficient in m²/s.

The heat flux J_A is proportional to the concentration gradient. It is the flux of A relative to the molar average velocity v (m/s) of the whole fluid given by $v = x_A v_A + x_B v_B$. The total molar flux (N_A) of A relative to stationary coordinates is equal to the sum of diffusion flux of A plus the convective flux of A:

$$N_A = J_A + C_A v \tag{4.2}$$

The mass concentrations of binary systems (A and B) are expressed by various ways:

ρ_A = concentration of A (g of A/m³)
M_A = molecular weight of A (g/g mol)
$C_A = \rho_A/M_A$ = concentration of A (g mol/m³)
x_A = mole fraction of A
$C = C_A + C_B$ = total concentration (g mol A + B)/m³
$\rho = \rho_A + \rho_B$ = total concentration (g of A + B)/m³
$m_A = x_A M_A/(x_A M_A + x_B M_B) = \rho_A/\rho$ = mass fraction of A

$$m_A + m_B = 1 \tag{4.3}$$

$$M = x_A M_A + x_B M_B = \rho/C = (\text{g of A} + \text{B})/(\text{g mol A} + \text{B}) \tag{4.4}$$

4.2.1   STEADY STATE REACTION AND DIFFUSION

In dilute solutions, the flux in a solution can be predicted by the following equation if the solute is diffusing and reacting:

$$N_{AB} = -D_{AB} \frac{dC_A}{dz} \tag{4.5}$$

For first order equation, rate of generation or depletion is $-k\,C_A$, where k is reaction rate constant, and equation can be written as:

$$\frac{d^2C_A}{dz^2} - \frac{k}{D_{AB}} C_A = 0 \tag{4.6}$$

For unsteady state in a semi-infinite medium, the equation for reaction and diffusion can be written as:

$$\frac{\partial C_A}{\partial t} = D_{AB} \frac{\partial^2 C_A}{\partial x^2} - k\,C_A \tag{4.7}$$

These equations are for solid food in which convection mass flux is negligible. Moreover, the heat of biological reaction is usually small so it can be ignored.

The above equation is applicable where absorption at the surface of a stagnant fluid or solid occurs, and then unsteady state diffusion and reaction occurs in the fluid or solid [1].

4.2.2   MASS TRANSPORT BASICS

One dimensional mass transfer (Fick's law) is given after neglecting convective mass flux:

$$J = -D_m \rho \cdot \frac{dC}{dz} \tag{4.8}$$

For unsteady state:

$$\frac{\partial C}{\partial t} = \frac{\partial}{\partial z}\left(D_m \frac{\partial C}{\partial z}\right) \tag{4.9}$$

or when D_m is constant

$$\frac{\partial C}{\partial t} = D_m \frac{\partial^2 C}{\partial z^2} \tag{4.10}$$

where D_m = mass diffusivity, t = time, z = distance, C = mass concentration, ρ = density (kg/m^3), and J = rate of mass transfer or flux.

Mass transfer in three dimensional Cartesian co-ordinates is given by:

$$\frac{\partial C}{\partial t} = D_m \cdot \left[\frac{\partial^2 C}{\partial x^2} + \frac{\partial^2 C}{\partial y^2} + \frac{\partial^2 C}{\partial z^2}\right] \tag{4.11}$$

Mass transfer in cylindrical co-ordinates (r, θ, z) is given by:

$$\frac{\partial C}{\partial t} = D_m \left[\frac{\partial^2 C}{\partial r^2} + \frac{1}{r}\frac{\partial C}{\partial r} + \frac{1}{r^2}\frac{\partial^2 C}{\partial \theta^2} + \frac{\partial^2 C}{\partial z^2} \right]$$

(4.12)

Mass transfer in spherical co-ordinates (r, θ, φ) is given by:

$$\frac{\partial C}{\partial t} = D_m \left[\frac{\partial^2 C}{\partial r^2} + \frac{2}{r}\frac{\partial C}{\partial r} + \frac{1}{r^2}\frac{\partial^2 C}{\partial \theta^2} + \frac{\cot\theta}{r^2}\frac{\partial C}{\partial \theta} + \frac{1}{r^2 \sin^2\theta}\frac{\partial^2 C}{\partial \phi^2} \right]$$

(4.13)

4.2.2.1 Porous Materials

Most solid foods contain a mixture of air and water vapor in capillaries. Pores are known as capillaries when their diameter is less than 10^{-7} m, and such material is capillary-porous, others are known as porous. The mechanisms of liquid water transfer in porous solids are classified as:

Diffusion in solid phase, i.e., molecular diffusion, negligible
Liquid diffusion in pores
Surface diffusion of absorbed water
Capillary flow when capillaries are filled with water
Hydraulic flow in pores

4.2.3 BOUNDARY CONDITIONS IN MASS TRANSFER

Boundary conditions are needed in solving unsteady state mass transfer problems. Five commonly identified kinds are:

a. First kind: at $x = 0$, $p_s = p_a$, i.e., surface vapor pressure (p_s) = ambient vapor pressure (p_a) when h_m (mass transfer coefficient) is very small.
b. Second kind: fixed mass transfer rate at the surface, not so common.

$$J = -D_m \frac{\partial C}{\partial z}\bigg|_{z=0}$$

(4.14)

c. Third kind: convective mass transfer at the surface = diffusion at the surface, this is very common.

$$h_m(p_a - p_s) = -D_m \frac{\partial C}{\partial z}\bigg|_{z=0}$$

(4.15)

d. Fourth kind: this is not so common, $p_s = f(t)$.
e. Symmetry boundary condition: it is applied to a perfectly insulated surface or to an axis of symmetry.

$$\frac{\partial C}{\partial z}\bigg|_{z=0} = 0$$

(4.16)

The first kind is a special case of the third kind when $h_m \to \infty$, and symmetry condition is the third kind when $h_m = 0$.

For evaporative mass transfer:

$$J = h_m (p_a - p_s) \tag{4.17}$$

where p_a = vapor pressure in the ambient air (Pa), and p_s = vapor pressure in the boundary layer over the product surface (Pa).

The vapor pressure is taken from a psychrometric chart, or $p_a = p_{wa}$ RH and $p_s = p_{ws} a_w$, where RH is relative humidity, a_w is water activity of the food, p_{wa} is the vapor pressure of water at T_a, and p_{ws} is the vapor pressure of water at T_s.

4.2.4 MASS TRANSFER IN SOLIDS

Mass transfer in solids may or may not depend on its structure.

4.2.4.1 Diffusion Independent of Solid Structure

In this type of diffusion, fluid or solute diffusing dissolves in the solid forming homogeneous solution. Example of this is leaching of soybean oil from soybean-solvent solution. The total flux is given by:

$$N_A = -D_{AB} \frac{dC_A}{dz} \tag{4.18}$$

where D_{AB} is the diffusivity of A through solid B. This is generally measured experimentally, and is a function of temperature and concentration.

4.2.4.2 Diffusion Depending on Porous Solid Structure

Diffusion of liquids is affected by the sizes and types of pores or voids. The diffusion in this case is not in a straight path but in a random tortuous path and given by a factor τ, tortuosity. The effective diffusivity (D_{Ae}) is given by:

$$D_{Ae} = \frac{\varepsilon}{\tau^2} D_{AB} \tag{4.19}$$

where ε is the porosity.

4.3 MASS DIFFUSIVITY

An apparent (or effective) mass diffusion coefficient or diffusivity (D_a, D_e) describes the combined effects of the different mass transport mechanisms on the overall rate process. D_m ranged from 0.1(10^{-10}) m^2/s for gelatinized material to 50(10^{-10}) m^2/s for extruded products [2]. D_m in food can be determined by three methods: drying data analysis, sorption kinetics and permeability measurements. The permeability method is limited to food films. D_m from the drying data of foods can be calculated by three methods: from drying curve slope [3], by reducing the deviations between predicted and experimental moisture concentration data using computer based optimization techniques [4], and by the regular regime technique [5]. The regular regime method is used to determine concentration dependent D_m for foods where D_m decreases with decreasing moisture content below the critical moisture content. The first two methods gave similar results in materials when liquid diffusion predominated [6]. The slope method could provide quantitative information on mass transport of water and the type of moisture diffusion. Most of the reported D_m values were calculated using drying data. D_m is generally a function of temperature and moisture concentration.

The D_m is sometimes estimated from correlations such as the Wilke–Chang for small molecules or Stokes–Einstein form for macromolecules. These do not account for the effect of important parameters such as pH and ionic strength on the interactions between the solute and solvent. The drying constant (k) is related to D_m by $D_m = a^2 k/\pi^2$, where "a" is equivalent radius of the product [7,8]. The relationship between k and D_m for spherical shapes is: $k = \pi^2 D_m/(2.3\ R^2)$, where R is the radius of the sphere.

4.3.1 Activation Energy

Arrhenius equation is generally used to model the effect of temperature on rate constants and diffusivity, which is:

$$D_m = D_o \exp(-E/(R_g T_a))$$ (4.20)

Some times Eyring's absolute reaction rate theory is also used in place of Arrhenius equation:

$$D_m = K\ T_a/h \exp(\Delta s) \exp(-\Delta H/(R_g T_a))$$ (4.21)

where D_o is diffusivity at reference temperature or frequency factor, E = activation energy, Δs = entropy and ΔH = enthalpy of activation, R_g = gas constant, K = Boltzmann constant, h = Planck's constant, and T_a = absolute temperature. Table 4.1 provides values of E for some foods.

TABLE 4.1
Activation Energy (E) of the D_m for some Foods

Food	mc (db)	E (kJ/mol)	Reference
Apple	0.1–0.7	47.3–68.3	[5]
Apple tissue	0.1–0.7	51–83.2	[9]
Bread dough	0.3–0.65	52.7	[10]
Bread, white	0.1–0.7	50.9	[11]
Carrots	–	21	[12]
Corn	–	46	[12]
Fruits, air dried	0.1–1.0	52.3	[13]
Potato	0.1–1.0	52.3	[13]
Potato tissue	0.1–0.7	30–66	[9]
Rice (starchy endosperm)	0.34–0.13	28.5	[14]
Rice with bran	0.34–0.13	44.8	[14]
Sorghum (rewetted)	0.21–0.06	31.4	[15]
Starch gels	0.2–1.5	43.5	[3]
Starch/glucose gels	0.2–1.5	51.4	[3]
Wheat kernel	0.12–0.30	53.9–61.1	[16]

Source: Adapted from Mittal, G.S., *Food Review Int.*, 15(1), 19, 1999.
Note: mc = moisture content, dry basis (db).

4.3.2 MODELS AND DATA

Fick's model for 1-D flow was used for Cassava [17], $dm/dt = d/dx \, (D_a \, dm/dx)$ for slab shaped sample to calculate D_a as:

$$D_a = -4.567(10^{-4}) - 9.567(10^{-8})m + 9.867(10^{-8})e^m + 4.583(10^{-4})\exp(-1/T_a)3.717(10^{-6})\text{XR}$$

(4.22)

where m is moisture content (mc) dry basis (db) (fraction), XR is the relative humidity indicator, 0 if RH = 60%, and 1 if RH = 10%, and D_a is in m²/s. Other conditions were: air temperature (T) = 55, 65, 75°C, initial moisture content (m_o) = 2.14, 1.86, 1.71, 1.67, and air velocity (v) = 2.26 m/s. Liquid water diffusion was the most appropriate mechanism to describe the drying behavior of cassava.

The process conditions for potatoes were: RH = 30, 40, 50%, air velocity (v) = 1.6, 2.3, 3.1 m/s, initial mc = 5.3 db, and slab thickness = 4.5 mm. Fick's law was used to predict average drying time and internal moisture distribution. D_a values were between $1.596(10^{-10})$ and $8.487(10^{-10})$ m²/s [18]. The activation energy for moisture diffusion was 1543 kJ/kg. The models are given as:

$$\ln(D_a) = 11.782 - \frac{3326.45}{T_a}, \quad 40\% \text{ RH} \tag{4.23}$$

$$\ln(D_a) = 10.1 - \frac{2711.54}{T_a}, \quad 30\% \text{ RH} \tag{4.24}$$

Table 4.2 provides D_m values for typical fruits and vegetables.

Values of D_a using drying data for pasta at 40–85°C air temperature and 1.5–26% mc db were reported [29]: $D_a = 1.5(10^{-12})$ to $48(10^{-12})$ m²/s at 6% mc and 312 K, and 24% mc and 358 K, respectively. The model is given as:

$$D_a = A\exp\left(-\frac{B}{T_a}\right)[1 - \exp(-CM^D) + M^F] \tag{4.25}$$

where $A = 2.3920(10^5)$, $B = 3.1563(10^3)$, $C = 7.9082(10^0)$, $D = 1.5706(10^1)$, $F = 6.8589(10^{-1})$, and $\Delta H = B.R_g = 26$ kJ/mol, and M is mc db%.

The D_m for soybean was given as [30]:

$$D_m = 2.78(10^{-8}) \, [(0.0493 \exp(-0.59/(M_o-M_e)) + 0.0181 \, (M-M_e)) \exp(3137.6/T_a)] \tag{4.26}$$

where D_m is in m²/s, M is mc (db), M_o is initial moisture content, and M_e is equilibrium moisture content.

Table 4.3 provides D_e values for cereal products, and Table 4.4 for wheat components.

The D_m of emulsified finely comminuted meat varied from $0.403(10^{-10})$ to $2.121(10^{-10})$ m²/s [38]. The following Arrhenius model was fitted giving R^2 value of 0.86:

$$D_m = \exp(8.679 + 0.135 \, \text{FP} - 4341.5/T_a + 8.55 \, M) \tag{4.27}$$

where D_m is in m²/h, T_a is product temperature in K, FP is fat–protein ratio, and M is moisture concentration of the product in db.

The D_a of skim milk was determined [39] using drying data. The product density was 1470 kg/m³ and specific heat was 1790 J/(kg K). At 50°C, the models are given as:

$$D_{a,50} = \exp\left[\frac{-82.5 + 1700M}{1 + 79.61M}\right] \tag{4.28}$$

TABLE 4.2
D_m **for Fruits and Vegetables**

Food	D_m (m²/s)	Conditions	Method Used	Reference
Apple, red delicious	$D_a = 0.00275 \exp(12.97\,M-(4216\,M + 5267)/T_a)$	$0<M<1.3$ db, $288<T_a<318$ K	Sorption data, regular regime method	[19]
Apple, Granny Smith	$1.15(10^{-10})$, $2.6(10^{-10})$, $4.9(10^{-10})$	$76°C,<7$ db; $30°C,>15$ db; $30°C,<15$ db	–	[20]
Apple, freeze dried	$0.405(10^{-11})$	2.7 mm thick, 25°C	Sorption data	[21]
Potato	$1(10^{-10})$	<15% mc db, 65°C	Drying data	[22]
Potato slices, Chippewa	$3.1(10^{-10})$, $4.0(10^{-10})$, $4.8(10^{-10})$; $\Delta H = 13$ kJ/mol	49°C, 65°C, 80°C; <4 db	Drying data	[23]
Potato, raw and blanched	Raw: $2.25(10^{-10})\pm0.13$ (10^{-10}) Blanched: 2.48 $(10^{-10})\pm0.28$ (10^{-10})	60–80°C, 30%RH, 2 m/s air velocity	Drying and shrinkage data	[24]
Sugarbeet roots	$1.3(10^{-9})$, $0.4(10^{-9})$; $\Delta H = 29$ kJ/mol	80°C, 2.5–3.6 db; 40°C, 2.5–3.6 db	Drying data, initial drying rate	[25]
Tapioca roots	$0.9(10^{-9})$; $\Delta H = 23$ kJ/mol	97°C, 0.16–1.95 db	Drying data	[26]
Turnip, freeze dried	$0.761(10^{-11})$	2.5 mm thick, 25°C	Sorption data	[21]
Tomato	$D_a = 1.67(10^{-8})D_o \exp(-3024/T_a)$ $D_o = \exp(1.022\,v^{0.5} + 4.477)$	40–80°C air temp., $v = 0.4–1.8$ m/s.	Drying data	[27]
Walnut	$1.89(10^{-4})–14.64 (10^{-4})$ $5.69(10^{-4})–6.81(10^{-4})$	32–43°C, 20–48%RH, $v = 0.81$ m/s, 11.4–56.4%db to 4.1–15.5%db. 11–16°C, 70–91%RH, 3.7–8.1%db to 8–11.3%	Desorption data Sorption data	[28]
Raisin	$0.417(10^{-12})$	4.1 mm radius cylinders, 25°C	Sorption data	[21]

Source: Adapted from Mittal, G.S., *Food Review Int.*, 15(1), 19, 1999.

Notes: M = moisture content db, T_a = absolute temperature, K, v = air velocity, m/s, RH = relative humidity, %, db = dry basis, mc = moisture content, ΔH = enthalpy of activation, kJ/mol.

TABLE 4.3
D_e for some Cereal and Processed Cereal Products

Food	D_e (m²/s)	Conditions	Method Used	Reference
Biscuits	$9.35(10^{-10})$–$9.7(10^{-8})$	10–65% mc db, 20–100°C	Cooking data	[11]
Muffin	$8.4(10^{-10})$–$1.54(10^{-7})$	10–90% mc db, 20–100°C	Cooking data	[11]
Oatmeal cookie	$0.397(10^{-11})$	10.1 mm thick, 25°C	Sorption data	[21]
Drurm wheat pasta	$8(10^{-12})$–$8.9(10^{-11})$	13.6–27% mc db, 40–90°C	Drying data	[31]
Popcorn	k (h⁻¹) = 0.13 + 0.00203 exp(0.08T)–(0.0551–0.00235 T)/(M_o–M_f)	38–82°C air T, v = 0.25 m/s, T_{dp} = 8°C, M_o = initial mc, db; M_f = final mc, db	Drying data, 1 term series fitting	[32]
Sorghum	$1.4(10^{-11})$	40°C	Drying data	[33]
	$2.1(10^{-11})$	50°C		
Soybean	$D_e = 1.304(10^{-9})$ exp(−3437/T_a)	35–95°C	Drying data	[34]
Starch	$0.1(10^{10})$–$70(10^{10})$	A function of mc	Drying data	[6]
Starch gel	$1(10^{-14})$–$3.6(10^{-12})$	1–14% mc db, 25°C	–	[22]
Wheat flour	$0.386(10^{-11})$	3.6 mm thick, a_w = 0.11	Sorption data	[21]
	$0.32(10^{-10})$	8.3 mm thick, a_w = 0.75, 25°C		
Wheat, shredded	$0.553(10^{-11})$	2.9 mm thick, 25°C	Sorption data	[21]
White bread	$D_e = 2.8945$ exp(1.264 M–2.758 M^2 + 4.959 M^3–6117.4/T_a)	0.10<M<0.75 db, 293<T_a<373 K	Desorption data, regular regime theory	[11]

Source: Adapted from Mittal, G.S., *Food Review Int.*, 15(1), 19, 1999.

Notes: k = drying rate constant (1/h), a_w = water activity, v = air velocity (m/s), T_{dp} = dew point temperature (°C), mc = moisture content, db = dry basis, M_o = initial mc, M_e = equilibrium mc.

TABLE 4.4
D_m of Wheat and its Components

Food	D_m (10^{10} m²/s)	Conditions	Reference
Wheat (hard winter) in bulk	3.7	0.13 db mc	[35]
	3.3	0.20 db mc	
		20°C, drying data	
Wheat	6.9–2.8	12–30% mcdb, 20–80°C	[22]
Wheat, kernel	0.086	25°C, drying data, arbitrary shape,	[36]
		10 cm deep brass tube	
Wheat, kernel	0.069	21°C, drying data, sphere shape, three	[16]
		to four kernel deep	
Wheat	0.078–0.083	22.5°C, adsorption, slab shape,	[37]
		16 cm deep tube	

Source: Adapted from Mittal, G.S., *Food Review Int.*, 15(1), 19, 1999.
Notes: mc = moisture content, db = dry basis.

TABLE 4.5
Moisture Diffusivity Data for Food Products

Food	D_a (m²/s)	Conditions	Method Used	Reference
Milk, nonfat dry	0.213(10^{-10})	3.1 mm thick, 25°C	Sorption data	[21]
Sausage, pepperoni	5.7(10^{-11})	13.3% fat	Drying and cooking	[41]
	5.6(10^{-11})	17.4% fat	data	
	4.7(10^{-11})	25.1% fat		
		12°C, 0.19 db mc		
Beef, raw ground, freeze dried	0.307(10^{-10})	10.9 mm thick, 20°C	Sorption data	[21]
Dogfish	8.3(10^{-11})	–	Drying data	[42]
Sucrose	0.50(10^{-9})	In water at 25°C	Rotating disk method	[43]
Glucose	0.66(10^{-9})	In water at 25°C	Rotating disk method	[43]

Source: Adapted from Mittal, G.S., *Food Review Int.*, 15(1), 19, 1999.

$$D_{a,T_a} = D_{a,50} \exp\left[\frac{\Delta H}{R_g}\left(\frac{1}{T_a} - \frac{1}{323}\right)\right] \tag{4.29}$$

$$\Delta H = 1.39(10^5) \exp(-3.32\,M) \tag{4.30}$$

where M = mc (db), R_g = gas constant (J/(mol.K)), D_a in m²/s, and ΔH = activation energy (J/mol). Table 4.5 provides some more moisture diffusivity data for food products.

4.3.3 MOISTURE DIFFUSIVITY DURING FRYING

The D_a in rectangular muscle slices (65 × 25 × 3 mm) cut from chicken drum, during deep fat frying, was determined as a function of moisture (0.06–3.33 db) and temperature (120–180°C) [40]. The mean initial moisture content of the muscle samples was 3.33 db, and frying was conducted for

1920 s. D_a, determined using regular regime theory, ranged from $1.32(10^{-9})$ to $1.64(10^{-8})$ m²/s. D_a as a function of temperature (T_a, K) and moisture content (M, db) is given by:

$$D_a = 8.35(10^{-6}) \exp(-2930/T_a - 0.561\,M + 0.092\,M^2), \quad \text{where } D_a \text{ is in m}^2\text{/s}. \qquad (4.31)$$

The D_m for tortilla chips fried in fresh and used oil was $8.897(10^{-6}) \pm 1.773(10^{-7})$ m²/s and $6.851(10^{-6}) \pm 6.418(10^{-7})$ m²/s, respectively. The D_m values for meatball (beef, 4.7 cm diameter, 60 g mass) during baking and broiling processes were estimated by minimizing the root mean square of deviations between the observed and predicted moisture histories [44]. D_m values for baking and broiling were $3.9(10^{-8})$ and $2.5(10^{-8})$ m²/s, respectively. For frying meatballs in canola-based shortening at 160°C, D_m was $8.0(10^{-8})$ m²/s [45].

4.4 CONVECTIVE MASS TRANSFER

Convective mass transfer is due to bulk fluid motion created by turbulence. Molecular diffusion is still there but its contribution is very small. Fluid movement adjacent to a mass transfer surface is laminar or turbulent depending on flow conditions and geometry. There is an analogy between heat and mass transfer. Mass transfer coefficient (k_G or k_L or h_m) is needed to calculate convective mass transfer. The flux is given by $N_A = k_G\,(p_{AG} - p_{Ai})$ or $N_A = k_L\,(C_{Ai} - C_{AL})$, where subscript G is for gas, L is for liquid and i is for gas–liquid interface, and p is partial pressure of the gas. The typical dimensions for k_G is kmol/(N.s), and for k_L is m/s.

4.4.1 MASS TRANSFER COEFFICIENTS

The following dimensionless numbers are required for correlating mass transfer coefficients [1]:

$$\text{Reynolds number } (N_{Re}) = v\,L\,\rho/\mu \qquad (4.32)$$

$$\text{Schmidt number } (N_{Sc}) = \mu/(\rho\,D_{AB}) \qquad (4.33)$$

$$\text{Sherwood number } (N_{Sh}) = k_N\,\varphi_N\,L/D_{AB} \qquad (4.34)$$

$$\text{For equimolar counter-diffusion of liquids } N_{Sh} = k_L\,L/D_{AB} = k_x\,L/(c\,D_{AB}) \qquad (4.35)$$

$$\text{For diffusion of A through B (liquids) } N_{Sh} = k_L\,x_{BM}\,L/D_{AB} = k_x\,x_{BM}\,L/(c\,D_{AB}) \qquad (4.36)$$

$$\text{Peclet number } (N_{Pe}) = N_{Re}\,N_{Sc} = L\,v/D_{AB} \qquad (4.37)$$

$$\text{Stanton number } (N_{St}) = N_{Sh}/(N_{Re}\,N_{Sc}) = k_N\,\varphi_N/v \qquad (4.38)$$

$$\text{For equimolar counter-diffusion of liquids } N_{St} = k_L/v = k_x/G_M \qquad (4.39)$$

$$\text{For diffusion of A through B (liquids) } N_{St} = k_L\,x_{BM}\,/v = k_x\,x_{BM}\,L/G_M \qquad (4.40)$$

$$J \text{ factor } (J_D) = N_{St}\,(N_{Sc})^{2/3} \qquad (4.41)$$

$$\text{Conversion from } J_D \text{ to } N_{Sh} \text{ is: } N_{Sh} = J_D\,N_{Re}\,N_{Sc}^{1/3} \qquad (4.42)$$

where v = mass average velocity relative to stationary coordinates, L = length of plate in the direction of flow, ρ = density, μ = viscosity, $k_N = k_x = h_m = k_L$ = mass transfer coefficient, φ_N = bulk flow correlation factor = $(N_R - x_A)_{LM}/N_R$, $N_R = N_A/(N_A + N_B)$, LM = log mean, x_{BM} = log mean of x_{B1} and x_{B2}, c = total concentration, G_M = molar mass velocity, x_A or x_B = mole fraction of A or B.

Few correlations are given for some typical mass transfer processes [1]:

1. Turbulent flow in pipes or wetted-wall towers, when N_{Re} is above 2100:

$$J_D = 0.023 \, N_{Re}^{-0.17} \qquad (4.43)$$

2. Flow parallel to flat plates: For evaporation of liquids or mass transfer to gases, N_{Re} below 15000:

$$J_D = 0.664 \, N_{Re}^{-0.5} \qquad (4.44)$$

$$\text{For } N_{Re} \text{ above } 15{,}000\text{--}300{,}000: J_D = 0.036 \, N_{Re}^{-0.2} \qquad (4.45)$$

$$\text{For liquids (for } N_{Re} \text{ from } 3000\text{--}50{,}000): J_D = 0.99 \, N_{Re}^{-0.5} \qquad (4.46)$$

3. Flow past single sphere:
   (i) For liquids, $N_{Re} = 2\text{--}2000, N_{Sc} = 788\text{--}1680$:

$$N_{Sh} = 2 + 0.95 \, N_{Re}^{0.5} N_{Sc}^{1/3} \qquad (4.47)$$

   (ii) For gases, $N_{Re} = 1\text{--}48{,}000, N_{Sc} = 0.6\text{--}2.7$:

$$N_{Sh} = 2 + 0.552 \, N_{Re}^{0.53} N_{Sc}^{1/3} \qquad (4.48)$$

   (iii) For liquids, $N_{Re} = 2000\text{--}16{,}900$:

$$N_{Sh} = 0.347 \, N_{Re}^{0.62} N_{Sc}^{1/3} \qquad (4.49)$$

4. Flow past single cylinders:
   (i) For gases, $N_{Re} = D \, v \, \rho/\mu = 400\text{--}25000, N_{Sc} = 0.6\text{--}2.6$, where D = cylinder diameter, and v = average velocity in the duct ahead of the cylinder:

$$N_{Sh} = 0.281 \, N_{Re}^{0.6} N_{Sc}^{0.44} \qquad (4.50)$$

   (ii) For liquids, $N_{Re} = 750\text{--}12\,000, \quad N_{Sc} = 1000\text{--}3000$:

$$J_D = 0.281 \, N_{Re}^{-0.4} \qquad (4.51)$$

5. Flow through packed beds:
   (i) For gases in beds of spheres, $N_{Re} = D \, v \, \rho/\mu = 90\text{--}4000$, where D = sphere diameter, ε = porosity, and v = superficial mass average velocity in the tube without packing

$$J_D \, \varepsilon = 2.06 \, N_{Re}^{-0.575} \qquad (4.52)$$

   (ii) For liquids, $N_{Re} = D \, v \, \rho/\mu = 0.0016\text{--}55, N_{Sc} = 165\text{--}70600$,

$$J_D \varepsilon = 1.09 \, N_{Re}^{-2/3} \qquad (4.53)$$

   (iii) For liquids, $N_{Re} = 55\text{--}1500, N_{Sc} = 165\text{--}10690$,

$$J_D \varepsilon = 0.250 \, N_{Re}^{-0.31} \qquad (4.54)$$

6. Fluidized beds of spheres for gases and liquids, $N_{Re} = 20$–3000:

$$J_D = 0.010 + \frac{0.863}{N_{Re}^{0.58} - 0.483} \tag{4.55}$$

4.4.2 OVERALL MASS TRANSFER COEFFICIENT

Film mass transfer coefficients in gases and liquids are difficult to measure, hence an overall mass transfer coefficient (K_y or K_x) is considered as:

$$N_A = K_y (y_A - y_A^*) \quad \text{or} \quad N_A = K_x (x_A^* - x_A) \tag{4.56}$$

where $(y_A - y_A^*)$ or $(x_A^* - x_A)$ is overall driving force in gas or liquid phase respectively, and y_A^* is in equilibrium with x_A, and x_A^* is in equilibrium with y_A.

The total resistance is the sum of gas film resistance ($1/k_y$) and liquid film resistance (m'/k_x), where m' and m'' are Henry's constants given by:

$$p_A^* = m' C_{AL} \quad \text{or} \quad p_{AG} = m' C_A^* \quad \text{or} \quad p_{Ai} = m' C_{Ai} \tag{4.57}$$

Thus

$$\frac{1}{K_y} = \frac{1}{k_y} + \frac{m'}{k_x} \quad \text{and} \quad \frac{1}{K_x} = \frac{1}{m'' k_y} + \frac{1}{k_x} \quad \text{or} \quad 1/K_G = m'/K_L \tag{4.58}$$

4.5 MASS TRANSFER EXAMPLES

4.5.1 DIFFUSION AND CONVECTION OF IONS

Sodium and chloride ions transfer due to the movement of water in meat emulsions is taken as an example [46]. The preservation of meat by salt depends on salt reaching all parts of the meat, including the fatty tissues. For a given tissue, the rate of diffusion is largely governed by the concentration gradient and the temperature. Diffusion of salt into the tissue will cause liquid transport due to osmosis and changes in the protein structure [47]. Under the influence of either temperature or concentration gradients the ions effectively diffuse as ion pairs. Expressions for the flows of each ionic species can be obtained as linear functions of the driving forces operating in the system and with the restrictions that the solution as a whole is electrically neutral, and there is no microscopic charge separation.

If c_i is the concentration of ions of the i'th kind in the solution and v_i their velocity, the molar flow J_i per unit area per unit time across a reference frame fixed with respect to the solvent is given by expressions similar to Fick's law [48]:

$$J_+ = -DI_+ \operatorname{grad} c_+ \tag{4.59}$$

$$J_- = -DI_- \operatorname{grad} c_- \tag{4.60}$$

For one dimensional diffusion, these become

$$J_+ = -DI_+ \frac{\partial c_+}{\partial x} \tag{4.61}$$

$$J_- = -DI_- \frac{\partial c_-}{\partial x} \tag{4.62}$$

These equations express diffusion movement due to concentration difference. An additional term for bulk convection due to water movement is required to calculate the total movement of ions.

$$\text{Diffusion of ions} = -DI*dCI/dx*A \tag{4.63}$$

$$\text{Convection of ions} = CI*D_m*dC/dx*A \tag{4.64}$$

$$\text{Accumulation of ions} = dCI/dt*\Delta x*A \tag{4.65}$$

where CI = concentration of ions (mol/L), A = cross sectional area of the slab (m^2), t = time (h), c = moisture ratio or concentration of water (dimensionless), DI = diffusivity of ions (m^2/h), D_m = moisture diffusivity (m^2/h), and x = distance (m).

4.5.2   MASS TRANSFER BY DIFFUSION, CONVECTION, AND DEPLETION BY CHEMICAL REACTION IN AN INFINITE SLAB

In this section, mass transfer by diffusion and convection, and depletion by chemical reactions is modeled for an infinite slab, applied to hydrogen ion transfer and depletion in meat emulsions during cooking. The pH of meat is important because it affects meat color, bacterial growth, and water holding capacity (WHC). Acceleration of cured meat color development is encouraged by a decrease in tissue pH. Bacterial growth is considerably reduced if pH of meat is below 5.6 [49]. An increase in pH raises the WHC of meat products. Overall pH change as a consequence of heating has been investigated by various workers [50–52] for various meat products.

The movement of H_+ during meat emulsion processing is a case of convection with water, molecular diffusion and a chemical reaction. For these mechanisms the following mathematical equations are developed [53]:

$$\text{Initial conditions: } C_{IH}(x,0) = C_{IH0} \tag{4.66}$$

Mass balance: The general mass balance equation is given by:

Rate of mass in–rate of mass out–rate of mass depletion = rate of mass accumulation

where

Rate of mass in or out = due to diffusion and due to convection

$$\text{Diffusion of } H_+ = -D_{IH}*A*d(C_{IH})/dx \tag{4.67}$$

$$\text{Convection of } H_+ = C_{IH}*D_m*A*dC/dx \tag{4.68}$$

$$\text{Accumulation of ions} = d(C_{IH})/dt*\Delta x*A \tag{4.69}$$

$$\text{Rate of depletion} = -KR(C_{IH}-C_{IHE})*\Delta x*A, \text{ using first order chemical reaction} \tag{4.70}$$

where C_{IH} = concentration of hydrogen ion (mol/L), D_{IH} = diffusivity of H_+ (m^2/h), KR = reaction rate constant (1/h), C = concentration of moisture, Δx = incremental slice thickness, and C_{IHE} = equilibrium concentration of H_+ (mol/L).

The process models can be developed after putting these equations in the mass balance equations.

4.5.3   DEEP-FAT FRYING OF SPHERICAL FOOD PRODUCTS

Heat, moisture, and fat transfer both within and around the food, and the formation of a crust, is the basis of deep-fat frying. The fat not only acts as a heat transfer medium, but also enters or leaves the food product. Mittelman et al. [54] published two models showing that moisture diffusion during the frying of french fries was proportional to the square root of the frying time. They concluded that the models were only partially satisfactory due to over simplifications of physical conditions. Gamble et al. [55] used the same models to determine the drying rate during potato chip frying and achieved results in agreement with those of Mittelman et al. [54]. A simplification of the solution of the Fick's diffusion equation was used by Moreira et al. [56] and by Rice and Gamble [57], to predict water loss and fat uptake during frying of tortilla and potato chips, respectively. The one term exponential model was insufficient in describing fat profiles. The models were successful in predicting average moisture, fat, and temperature [58]. Guillaumin [59] attributed fat penetration to the formation of a crust during deep-fat frying of potato chips. Huang and Mittal [60] developed models describing heat and moisture transfer in beef meatballs during oven baking and boiling. The main limitation of these models was the neglection of fat transport. The example below describes the modeling of simultaneous fat, heat, and moisture transfer during deep-fat frying of a fatty product [45].

During frying, heat is transferred to the product surface by convection and to the geometric center of the product by conduction. It is assumed that during deep-fat frying, foods containing fat undergo two fat transfer periods: (i) the fat absorption period where fat diffuses from the surroundings into the product, and (ii) the fat desorption period where fat migrates from the product to the surroundings by capillary flow. Diffusion theory assumes that liquid moves through a solid body as a result of a concentration gradient, while capillary theory assumes that flow of liquid through the capillaries is caused by solid–liquid molecular attraction.

4.5.3.1 Fat Absorption Period

The following assumptions were made: (i) The product was homogeneous, isotropic, and spherical in geometry, (ii) the initial moisture and fat distribution in the product was uniform, (iii) the fat was mobilized by a concentration gradient, liquid fat diffusing from the surroundings into the product. Water was mobilized, also by a concentration gradient, and diffused in the liquid state throughout the product, followed by vaporization restricted to the product surface only, (iv) product shrinkage, and the effects of crust formation on physical properties were neglected, and (v) thermal, moisture and fat diffusivities were constant. Based on the above assumptions, the mathematical models characterizing one-dimensional moisture, fat and heat transfer in a sphere during the fat absorption period of the deep-fat frying could be represented as follows:

Moisture transfer

$$\frac{\partial m}{\partial t} = D_m \left[\frac{2}{r} \frac{\partial m}{\partial r} + \frac{\partial^2 m}{\partial r^2} \right] \tag{4.71}$$

Fat transfer

$$\frac{\partial \mathrm{mf}}{\partial t} = D_f \left[\frac{2}{r} \frac{\partial \mathrm{mf}}{\partial r} + \frac{\partial^2 \mathrm{mf}}{\partial r^2} \right] \tag{4.72}$$

Heat transfer

$$\frac{\partial T}{\partial t} = \alpha \left[\frac{2}{r} \frac{\partial T}{\partial r} + \frac{\partial^2 T}{\partial r^2} \right] \tag{4.73}$$

Initial conditions

Initially, moisture and fat were uniformly distributed throughout the product prior to frying, i.e.,

$$m(r,0) = m_0; \quad \mathrm{mf}\,(r,0) = \mathrm{mf}_0 \tag{4.74}$$

Boundary conditions

No temperature, moisture, or fat gradients exist at the product center. Therefore:

$$\frac{\partial T}{\partial r}\bigg|_{r=0} = 0; \quad \frac{\partial m}{\partial r}\bigg|_{r=0} = 0; \quad \frac{\partial \text{mf}}{\partial r}\bigg|_{r=0} = 0 \tag{4.75}$$

Heat supplied to the surface by convection is equal to the heat transferred to the center of the product by conduction, plus heat required to evaporate water (latent heat of vaporization, L_v), i.e.,

$$h(T_a - T_s) = k\frac{\partial T}{\partial r}\bigg|_{r=R} - D_m \rho_{dm} L_v \frac{\partial m}{\partial r}\bigg|_{r=R} \tag{4.76}$$

The boundary conditions expressing surface moisture and fat contents attaining instantaneous equilibrium with the fat medium are:

$$m(R,t) = m_e; \quad \text{mf}(R,t) = \text{mf}_e \tag{4.77}$$

where h = heat transfer coefficient, T_a = ambient temperature, T_s = surface temperature, k = thermal conductivity, r = radial increment, ρ_{dm} = density of dry matter, m = moisture content, dry basis, R = radius, L_v = latent heat of vaporization, mf = fat content, m_e = equilibrium moisture content, mf_e = equilibrium fat content, D_f = fat diffusivity, and α = thermal diffusivity.

4.5.3.2 Fat Desorption Period

The models used for moisture and heat transfer were the same as in the fat absorption period. The following additional assumption was made for fat transfer: The driving force for capillary flow is the difference between fat concentration and equilibrium fat concentration (i.e., a minimum fat concentration below which there is no transfer of fat). Based on the above assumptions, the mathematical model characterizing the desorption period of fat transfer could be represented as follows:

$$V\frac{\text{dmf}}{dt} = -\text{KL A(mf} - \text{mfep)} \tag{4.78}$$

Considering mass balance at the geometric centre: (mass in)–(mass out) = (mass accumulated), where mass in = 0; mass out = KL · $4\pi r_0^2(\text{mf}_0-\text{mfep})$; mass accumulated = $(\text{dmf}_0)/(dt)(4/3)\pi r_0^3$ Therefore:

$$\frac{\text{dmf}_0}{dt} = 3\frac{\text{KL}}{r_0}(\text{mf}_0 - \text{mfep}) \tag{4.79}$$

where KL = fat conductivity (m/s), mfep = equilibrium fat content, dry basis, and V = volume (m³).

4.5.4 Oxygen Balances for Gas–Liquid Transfer in a Bioreactor

In a continuous flow bioreactor, liquid stream (Li = liquid stream flow rate at inlet (m³/h); C_{Li} = oxygen concentration in liquid stream at inlet (kmol/m³)) enters and leave (Lo = liquid stream flow rate at outlet (m³/h); C_L = oxygen concentration in liquid stream at outlet and in the bioreactor (kmol/m³)) after mixing with gas stream in the bioreactor. The gas stream (Gi = gas stream flow rate at inlet (m³/h); C_{Gi} = oxygen concentration in gas stream at inlet (kmol/m³)) enters and leave (Go = gas stream flow rate at outlet (m³/h); C_G = oxygen concentration in gas stream at outlet and in the bioreactor (kmol/m^3)) after oxygen transfer and the rate of oxygen uptake by the cells in the bioreactor. The bioreactor is considered a well-mixed tank of volume containing gas (V_G = gas volume in the tank) and liquid (V_L = liquid volume in the tank) phases. The oxygen transfer in the bioreactor will involve accumulation, convective flow, interface transfer and biological oxygen uptake terms. For the gas phase the oxygen mass balance in the bioreactor is given by:

O_2 accumulation rate in the gas phase of bioreactor = O_2 flow rate in the inlet gas stream–O_2 flow out in the outlet gas stream–O_2 transfer rate from gas stream in the bioreactor

Or for the gas phase:

$$V_G \frac{dC_G}{dt} = \text{Gi}C_{\text{Gi}} - \text{Go}C_G - k_L a(C_L{}^* - C_L)V_L \tag{4.80}$$

where a = specific area for mass transfer or interfacial area per unit liquid volume (m^2/m^3) and k_L = mass transfer coefficient (m/s), $C_L{}^*$ = equilibrium solubility of O_2 corresponding to the gas phase concentration (C_G) and is given by Henry's law [61] as $C_G R_g T_a = H C_L{}^*$, where H = Henry's constant.

Similarly for O_2 transfer in liquid phase:

O_2 accumulation rate in the liquid phase of bioreactor = O_2 flow rate in the inlet liquid stream–O_2 flow rate out in the outlet liquid stream + O_2 transfer rate from gas stream in the bioreactor–O_2 consumption rate by cells

Or for the liquid phase:

$$V_L \frac{dC_L}{dt} = \text{Li}C_{\text{Li}} - \text{Lo}C_L + k_L a(C_L{}^* - C_L)V_L - q_{O2}XV_L \tag{4.81}$$

where X = biomass or cell concentration (kg/m^3) and q_{O2} = specific oxygen consumption rate by cells (kg O_2/kg cells).

REFERENCES

1. Geankoplis, C.J. *Mass Transfer Phenomena*. Ohio State University Bookstores, Columbus, OH, 1977, Chapters 2, 4, 6.
2. Mittal, G.S. Mass diffusivity of food products. *Food Review Int.*, 15(1), 19, 1999.
3. Saravacos, G.D. and Raouzeou, G.S. Diffusivity of moisture during air drying of starch gels. In *Engineering and Food*, Vol. 1. McKenna, B.M., Ed. Elsevier, New York, NY, 1984, p. 499.
4. Mittal, G.S. and Blaisdell, J.L. Moisture mobility in frankfurter during thermal processing: Analysis of moisture profile. *J. Food Proc. & Preservation*, 6(2), 111, 1982.
5. Schoeber, W.J.A.H. and Thijssen, H.A.C. A short cut method for the calculation of drying rates for slabs with concentration dependent diffusion coefficient. *AIChE Symp. Ser.*, 73(163), 12, 1977.
6. Karathanos, V.T., Villalobos, G. and Saravacos, G.D. Comparison of two methods of estimation of the effective moisture diffusivity for drying data. *J. Food Sci.* 55, 218, 1990.
7. Chittenden, D.H. and Hustrulid, A. Determining drying constants of shelled corn. *Trans. ASAE,* 9, 52, 1966.
8. Keey, R.B. *Drying Principles and Practices*. Pergamon Press, New York, NY, 1972, Chapter 5.
9. Luyben, K.Ch.A.M., Olieman, J.J. and Bruin, S. Concentration dependent diffusion coefficients derived from experimental drying curves. In *Proc. 2nd Int. Symp. on Drying, IDS'80*, Majumdar, A.S., Ed. Hemishere Publishing Co., New York, NY, 1980, 233.
10. Yoon, J. *Heat and Moisture Transfer during Bread Baking*. M.S. thesis, University of Minnesota, St. Paul, MN, 1985.
11. Tong, C.H. and Lund, D.B. Effective moisture diffusivity in porous materials as a function of temperature and moisture content. *Biotechnol. Prog.*, 6, 67, 1990.
12. Bimbenet, J.J., Daudin, J.D. and Wolf, E. Air drying kinetics of biological particles. In *Drying '85*, Toei, R. and Majumdar, A.S., Eds. Hemisphere Publishing Co., New York, NY, 1985, p. 178-185.

13. Saravacos, G.D. and Charm, S.E. A study of the mechanisms of fruits and vegetables dehydration. *Food Technol.*, 6, 78, 1962.
14. Steffe, J.F. and Singh, R.P. Moisture diffusivity in rice. *Trans. ASAE*, 23, 767, 1980.
15. Suarez, C., Viollaz, P.E. and Chirife, J. Diffusional analysis of air drying of grain sorghum. *J. Food Technol.*, 15, 523, 1980.
16. Becker, H.A. and Sallans, H.R. A study of internal moisture movement in the drying of the wheat kernel. *Cereal Chem.*, 32, 212, 1955.
17. Igbeka, J.C. Simulation of moisture movement during drying a starch food product—cassava. *J. Food Technol.*, 17, 27, 1982.
18. Yusheng, Z. and Poulsen, K.P. Diffusion in potato drying. *J. Food Eng.*, 7, 249, 1988.
19. Singh, R.K. and Lund, D.B. Mathematical modeling of heat and moisture transfer-related properties of intermediate moisture apples. *J. Food Proc. Preservation*, 8, 191, 1984.
20. Rotstein, E. and Cornish, A.R.H. Influence of cellular membrane permeability on drying behavior. *J. Food Sci.*, 43, 926, 1978.
21. Lomauro, C.J., Bakshi, A.S. and Labuza, T.P. Moisture transfer properties of dry and semi-moist foods. *J. Food Sci.*, 50, 397, 1985.
22. Chirife, J. Fundamentals of the drying mechanism during air dehydration of foods. In *Advances in Drying,* Vol 2. Mujumdar, A.S., Ed. Hemisphere Publishing Co., New York, NY, 1983, p. 73.
23. Hussain, A., Chen, C.S. and Clayton, J.T. Simultaneous heat and mass diffusion in biological materials. *J. Agric. Eng. Res.*, 18, 343, 1973.
24. Gekas, V. and Lamberg, I. Determination of diffusion coefficients in volume-changing systems—application in the case of potato drying. *J. Food Eng.*, 14, 317, 1991.
25. Vaccarezza, L.M., Lombardi, J.L. and Chirife, J. Kinetics of moisture movement during air drying of sugarbeet root. *J. Food Tehnol.*, 9, 317, 1974.
26. Chirife, J. Diffusional process in the drying of tapioca root. *J. Food Sci.*, 36, 327, 1971.
27. Hawlander, M.N.A., Uddin, M.S., Ho, J.C. and Teng, A.B.W. Drying characteristics of tomatoes. *J. Food Eng.*, 14, 259, 1991.
28. Alves-Filho, O., Rumsey, T. and Fortis, T. Apparent diffusivities for drying and rewetting of English walnuts, Paper No. 83-6012. ASAE conference. *Amer. Soc. Agric. Eng., St. Joseph, MI,* 1983.
29. Litchfield, J.B. and Okos, M.R. Moisture diffusivity in pasta during drying. *J. Food Eng.* 17, 117, 1992.
30. Sabbah, M.A., Meyer, G.E., Keener, H.M. and Roller, W.L. Simulation studies of reversed direction airflow drying method for soybean. *Trans. ASAE*, 22, 1162, 1979.
31. Andrieu, J. and Stamatopoulos, A. Duram wheat pasta drying kinetics. *Food Sci. Technol.*, 19, 448, 1986.
32. White, G.M., Ross, I.J. and Poneleit, C.G. Fully exposed drying of popcorn, *Trans. ASAE*, 24, 466, 1981.
33. Suarez, C., Viollaz, P.E. and Chirife, J. Kinetics of soyabean drying. In *Drying 80,* Vol 2. Mujumdar, A.S., Ed., Hemisphere Publishing Co., New York, NY, 1980, p. 251.
34. Misra, R.N. and Young, J.H. Numerical solution of simultaneous moisture diffusion and shrinkage during soybean drying. *Trans. ASAE*, 23, 1277, 1980.
35. Hayakawa, K.I. and Rossen, J.L. Isothermal moisture transfer in dehydrated food products during storage. *Food Sci. Technol.* 10, 217, 1977.
36. Becker, H.A. A study of diffusion in solids of arbitrary shape, with application to the drying of the wheat kernel. *J. Appl. Polymer Sci.*, 1, 212, 1959.
37. Pixon, S.W. and Griffiths, H.J. Diffusion of moisture through grain. *J. Stored Product Res.*, 7, 133, 1971.
38. Mittal, G.S. and Blaisdell, J.L. Heat and mass transfer properties of meat emulsion. *J. Food Sci. and Technol. (Leb.-Wiss. u.-Technol.),* 17, 94, 1984.
39. Ferrari, G., Meerdink, G. and Walstra, P. Drying kinetics for a single droplet of skim-milk. *J. Food Eng.* 10, 215, 1989.
40. Ngadi, M.O. and Correia, L.R. Moisture diffusivity in chicken drum muscle during deep-fat frying. *Can. Agric. Eng.* 37, 339, 1995.
41. Palumbo, S.A., Komanowsky, M., Metzger, V. and Smith, J.L. Kinetics of pepperoni drying *J. Food Sci.*, 42, 1029, 1977.
42. Jason, A.C. A study of evaporation and diffusion processes, in the drying of fish muscles. In *Fundamental Aspects of the Dehydration of Food Stuffs*, Fish, B.P., Ed. Soc. Chem. Ind., 1958, p. 143.
43. Ziegler, G.R., Benado, A.L. and Rizvi, S.S.H. Determination of mass diffusivity of simple sugars in water by the rotating disk method. *J. Food Sci.*, 52, 501, 1988.

44. Huang, E. and Mittal, G.S. Meatball cooking—modelling and simulation. *J. Food Eng.,* 24, 87, 1994.
45. Ateba, P. and Mittal G.S. Modelling the deep fat frying of beef meatballs. *Int. J. Food Sci. Technol.,* 29, 429, 1994.
46. Mittal, G.S., Blaisdell, J.L. and Herum, F.L. Diffusion dynamics of sodium and chloride ions: During cooking of meat emulsion slab. *J. Food Sci. Technol. (Leb.-Wiss. u.-Technol.),* 15, 275, 1982.
47. Wood, F.W. The diffusion of salt in pork muscle and fat tissue. *J. Sci. Food Agric.,* 17, 138, 1966.
48. Tyrell, H.J.V. *Diffusion and Heat Flow in Liquids.* Butterworths, London, UK, 1961.
49. Gibbons, N.E. and Rose, D. Effect of ante-mortem treatments of pigs on the quality of Wiltshire bacon. *Can. J. Res.,* 28F, 438, 1950.
50. Hamm, R. and Deatherage, F.E. Changes in hydration, solubility and protein changes of muscle protein during heating of meat. *Food Res.,* 25, 587, 1960.
51. Paul, P.C. The rabbit as a source of experimental material for meat studies. *Institute of Food Technologists Meeting,* Chicago, 1964.
52. Kauffman, R.G., Carpenter, Z.L., Bray, R.W. and Hoekstra, W.G. Biochemical properties of pork and their relationship to quality—pH of chilled, aged and cooked muscle tissue. *J. Food Sci.,* 29, 65, 1964.
53. Mittal G.S. and Blaisdell, J.L. Hydrogen ions profiles, apparent mobility, and disappearance during cooking of meat emulsion—slab geometry. *J. Food Process Eng.,* 6, 59, 1982.
54. Mittelman, N., Mizrahi Sh. and Berk Z. Heat and mass transfer in frying. In *Engineering and Food,* Vol. 1. Mckenna, B.M., Ed. Elsevier Applied Science, London, UK, 1984, p. 109.
55. Gamble, M.H., Rice, P. and Selman, J.D. Relationship between oil uptake and moisture loss during frying of potato slices. *Int. J. Food Sci. Technol.,* 22, 233, 1987.
56. Moreira, R.G., Palau, J., Castelle-Perez M.E., Sweat V.E. and Gomez, M.H. Moisture loss and oil absorption during deep-fat frying of tortilla chips, ASAE Conference Paper No. 91-6501. *Amer. Soc. Agric. Engr.,* St. Joseph, MI, 1991.
57. Rice, P. and Gamble M.H. Modelling moisture loss during potato slice frying. *Int. J. Food Sci. Technol.,* 24, 183, 1989.
58. Moreira, R.G., Palau, J. and Sweat, V.E. Thermal properties of tortilla chips during deep-fat frying, ASAE Conference Paper No. 92-6595. *Amer. Soc. Agric. Engr.,* St. Joseph, MI, 1992.
59. Guillaumin, R. Kinetics of fat penetration in food. In *Frying of Food, Principles, Changes and New Approaches,* Varela, G., Bender, A.E. and Morton, I.D., Eds. Ellis Horwood, Chichester, UK, 1988, p. 82.
60. Huang, E. and Mittal, G.S. Meatball cooking—modeling and simulation. *J. Food Engg.,* 24, 87, 1995.
61. Dunn, I.J., Heinzle, E., Ingham, J. and Prenosil, J.E. *Biological Reaction Engineering.* VCH, New York, NY, 1992, Chapter 5.

5 Mass and Energy Balances in Food Processing

Gauri S. Mittal
University of Guelph

CONTENTS

5.1 INTRODUCTION

Mass and energy balances are needed in the planning and designing of food processes, and engineering analysis. Many complex food processing operations can be simplified by using the principles of mass and energy balances. Many questions can be answered such as how much raw materials are required, how much finished products and waste are generated, and how much energy is needed or to be removed. Process analysis consists of mass and energy balances and the solution of these balances. In the designing of food manufacturing facilities, cooling and heating requirements, size of equipment, raw material requirement, based on the production rate of various products to be manufactured, are needed. These are calculated by mass and energy balances of various processes needed in the manufacturing of a product. Before mass and energy balances, various unit operations required to manufacture a product should be worked out. Mass and energy balances can be performed for a unit operation, for a process section, for any equipment or for entire manufacturing process. To solve complex problems, many mass and energy balances equations are needed.

5.2 MASS BALANCES

5.2.1 BASIC EQUATIONS

The equation for conservation of mass for a system can be written as Equation 5.1 when there is no chemical reaction:

Total mass or rate of mass, in = total mass or rate of mass, out
$$+ \text{ mass accumulated or rate of mass accumulation} \qquad (5.1)$$

Under steady state conditions, there should be no mass accumulation in the system, then Equation 5.1 will change to:

$$\text{Total mass or rate of mass, in} = \text{total mass or rate of mass, out} \qquad (5.2)$$

Equation 5.2 can be written for the total mass or for each component of a product. This can also be applied to each unit operation or a part of the process (combination of unit operations) or to a complete process. These equations are useful in analyzing complex food processing systems. After writing, mass balance equations, unknown mass or mass flow rate or concentration of a component in a food product can be computed by solving the simultaneous equations. The number of independent simultaneous equations should be equal to the number of unknowns.

5.2.2 COMPONENT MASS BALANCES

Mass balances can be performed on a complete process and on any subdivision of the process. The process section to be considered is determined by system boundary. The inlet and exit streams will cross the system boundary and these must balance with material generated or consumed within the boundary [1]. The proper choice of system boundaries can often greatly simplify difficult calculations. Selection of the proper subdivision of any process depends on insight of the problem structure. Proper judgment is required for this. The following general rules can be considered [1]:

   i. Include recycle streams within the system boundary if any
   ii. Take the boundary around the complete process
   iii. Select the boundaries to subdivide the process
   iv. Select the boundary around a stage so that number of unknowns is reduced

5.2.2.1 Component Mass Balances with no Chemical Change
The mass balance equation can be written as:

$$\text{Rate of mass accumulation of a component in a system}$$
$$= \text{Mass flow rate of the component into the system}$$
$$- \text{Mass flow rate of the component out the system} \qquad (5.3)$$

For component mass balance, under steady state, the combined mass flow rate of a component in all input streams (say m1, m2, m3) in a process is equal to the combined mass flow rates of the component at the outlet streams (say m4, m5). Hence x1.m1 + x2.m2 + x3.m3 = x4.m4 + x5.m5, where x is the mass fraction of the component in a stream to total mass flow rate (m), and the mass flow rate of the component is x.m. The component in a stream can be protein, fat, carbohydrate, minerals, vitamins, water, etc.

5.2.2.2 Component Mass Balances with Chemical or Biological Reaction
The mass balance equation can be written as:

$$\text{Rate of mass accumulation of a component in the system}$$
$$= \text{Mass flow rate of the component into the system}$$
$$- \text{Mass flow rate of the component out of the system}$$
$$+ \text{Rate of production of the component by the reaction} \qquad (5.4)$$

A negative sign is needed if the component is consumed in the reaction.

A differential mass balance is based on rates of mass in and out in continuous operations. While integral balance is based on the quantity of mass in batch and semibatch processes.

5.2.3 Chemical Composition

Compositions of mixtures and solutions of a process stream are usually represented in the following forms [2]:

i. Mole fraction of component A in a mixture of many components is defined as:

$$\text{Mole fraction of A} = \frac{\text{moles of A in a mixture}}{\text{total moles of all components in the mixture}} \quad \text{or} \quad X_A = \frac{n_A}{n_T} \tag{5.5}$$

It can be expressed in mole fraction percent by multiplying it by 100.

ii. Mass fraction of component A in a mixture is:

$$\text{Mass fraction of A} = \frac{\text{mass of A in a mixture}}{\text{total mass of all components in the mixture}} \quad \text{or} \quad X_A = \frac{m_A}{m_T} \tag{5.6}$$

It can be expressed as mass fraction percent by multiplying it by 100.

iii. Volume fraction of component A in a mixture is given by (usually for gases):

$$\text{Volume fraction of A} = \frac{\text{volume of A in a mixture}}{\text{total volume of all components in the mixture}} \tag{5.7}$$

It can also be expressed in volume fraction percent by multiplying it by 100.

Other less common ways of expressing concentration of a component in mixtures and solutions are:

iv. Mass per unit volume or density (kg/m^3)
v. Moles per unit volume (gmol/L)
vi. Parts per million (ppm) or part per billion (ppb)
vii. Molality (gmol/1000 g solvent)
viii. Normality (mole equivalent/L). For an acid or base, an equivalent g weight is solute weight that will produce or react with 1 gmol hydrogen ions.
ix. Degrees Brix (°Brix): with 20°C scale, each °Brix indicates 1 g sucrose per 10 g liquid.

Wet and dry basis moisture contents: wet basis (wb) moisture content (mc) is equal to mass fraction of water in a mixture or solution. However, dry basis (db) moisture content is the moisture content based on dry mass of the mixture or solution. Dry basis moisture content can be higher than 1.0 or 100%. The relationship between the two is given by:

$$\text{mcwb} = \frac{\text{mcdb}}{1+\text{mcdb}} \quad \text{and} \quad \text{mcdb} = \frac{\text{mcwb}}{1-\text{mcwb}} \tag{5.8}$$

5.2.4 Procedure for Mass Balances

The complex processes can be easily analyzed if the following procedure is followed carefully:

i. Choose the balance region in a system: first a process flow diagram should be drawn showing all relevant information such as all streams entering or leaving the system, quantitative information, mass flow rates, and mass compositions. The region should be selected in such that the variables are constant or change little within the selected region. After selecting the region, draw boundaries around the region. The region may be a reactor, a single cell, a reactor region, a single phase within a reactor, or a region within a cell. Some examples are given:

(a) Continuous stirred tank reactor (Figure 5.1). The boundary selected is shown by dotted lines:

$$\text{Total mass in the tank} = \rho V$$

$$\text{Mass of A in the tank} = C_A V$$

where ρ is sample density, V is reactor volume, and C_A is concentration of component A in the reactor.

(b) Tubular reactor: in this case the concentrations of the products and reactants vary along the reactor length w.r.t. time. However, under steady state conditions, concentrations at any location along the reactor length will be constant w.r.t. time. Under such conditions, the balance region should be selected sufficiently small so that the concentration of any component within the region can be assumed to be approximately uniform.

ii. Identify the mass flow streams flowing across the system boundaries. An example is shown in Figure 5.2 where mass input and output streams are due to convection and diffusion.
iii. All quantities of variables should be expressed in consistent units.
iv. Basis for the calculation should be selected. The mass balances calculations can be made simpler by correct choice of the basis. This selection will depend on judgments attained by experience as there is no hard and fast rule. Select as a basis, the stream flow for which most information is given. Generally it is the total amount of material fed or withdrawn in batch or semibatch processes, or the rate of material entering or leaving the continuous systems.
v. Reasonable assumptions of solving the problem should be stated clearly.

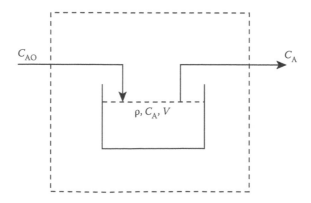

FIGURE 5.1 Continuous stirred tank reactor showing boundary, inputs, and outputs.

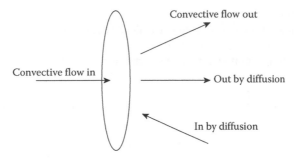

FIGURE 5.2 Mass flow streams across the system boundaries.

vi. System components involved in the reaction should be identified, which is required in the mass balances.

vii. Write the relevant mass balance, for example

$$\text{Accumulation} = \text{In} - \text{Out} + \text{Production}$$

viii. Express each balance term in mathematical form with measurable variables, such as

(a) Accumulation rate: taking the following symbols
C_i = concentration of component i (kg/m³ or kmol/m³)
M = mass (kg or mol)
p = total pressure (Pa)
p_i = partial pressure of component i (Pa)
t = process time (s, min, or h)
V = reactor volume (m³)
y_i = mole fraction of component i in gas phase

Thus, rate of mass accumulation of component i within the system is:

$$\frac{dM_i}{dt} = \frac{d(c_i V)}{dt} \tag{5.9}$$

For gas phase taking ideal gas law, $p_i V = n_i RT$, or

$$C_i = \frac{n_i}{V} = \frac{p_i}{RT} = \frac{y_i p}{RT} \tag{5.10}$$

$$\frac{dM_i}{dt} = \frac{d(c_i V)}{dt} = \frac{d}{dt}\left(\frac{p_i V}{RT}\right) = \frac{d}{dt}\left(\frac{y_i p V}{RT}\right) \tag{5.11}$$

Similarly, for total mass balance:

$$\frac{dM}{dt} = \frac{d\rho V}{dt} \tag{5.12}$$

(b) Convective flow:
     Convective mass flow rate = (volumetric flow rate) (density)
     For total mass balance, $dM/dt = F\rho$
     where F is volumetric flow rate, and ρ is density.
     Similarly for component mass flow: $dM_i/dt = F\,C_i$

(c) Transport of component by diffusion: Fick's law for molecular diffusion is

$$J_i = -D_i\frac{dC_i}{dZ} \tag{5.13}$$

A negative sign is needed as diffusive flow occurs in the direction of decreasing concentration.
     where

J_i = flux of any component i flowing across an interface (kg/(m².h), kmol/(m².h))
$D_i\,C_i/dZ$ = concentration gradient
D_i = molecular component diffusivity (m²/h)

An effective diffusivity value is used for diffusion in porous matrices and turbulent diffusion.

(d) Production or consumption of a component during reaction

Mass rate of production of component A = r_A
V = (reaction rate per unit volume) · (volume of the system)   (5.14)

r_A is positive when A is formed as product, and r_A is negative when reactant A is consumed.

5.2.5 Mass Balances with Various Streams

Generally food and bioprocesses involve recycle, bypass and purge streams (Figure 5.3). In such problems, several mass balances are needed to calculate all the unknowns. Separate total mass balances can be performed on each unit operation, as well as component mass balances can also be undertaken to generate many mass balance equations. In recycle processes a flow stream is returned to an earlier stage in the processing sequence. To solve such problems, the equations are set up with the recycle flows as unknowns. Recycled stream is used [3] in some processes to (i) increase the yield from the

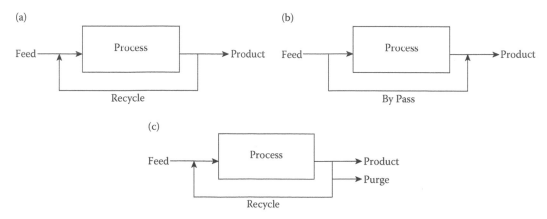

FIGURE 5.3 Various streams of material flow encountered in food processing (a) recycle, (b) bypass, and (c) purge streams.

process by recycling untreated materials, (ii) reduce the inlet concentration of a component by diluting the component, and (iii) reduce energy consumption by recycling high temperature exit streams. The *recycle ratio* is generally defined as the mass ratio of recycle to fresh feed. *Purge* is bleeding off a portion of a recycle stream to prevent built up of unwanted material. Bypass is used to control stream composition or temperature. In bypass, part of the flow stream is diverted around some unit operation.

5.2.6 STEADY STATE MASS BALANCES

Some examples are provided for steady state mass balances:

i. Total mass balance (see Figure 5.4)

$$I = W + C \tag{5.15}$$

ii. Component mass balance for each component separately Figure 5.5)
Concentration of a component "A" = mass of component A/total mass of mixture containing "A"

$$\text{Total mass balance: } W + Ad = D + Aw$$
$$\text{Component mass balance: } Ad + \text{water} = Aw \tag{5.16}$$

Example 5.1

A solution contains 1.8 kg of glucose (MW = 180) in 4.2 kg of water (MW = 18). Calculate the concentration of glucose in mass fraction, mole fraction, mass fraction percent and mole fraction percent.

Total mass of the solution = 1.8 + 4.2 = 6.0 kg
The mass fraction of glucose = 1.8/6.0 = 0.30
The mole fraction of glucose = ((1.8/180)/(1.8/180 + 4.2/18)) = 0.041
The mass percent = 0.30 × 100 = 30%
The mole percent = 0.041 × 100 = 4.1%

Example 5.2

4.5 kg water is added to 26.5 kg of dry flour. Calculate the moisture content of flour in wet and dry basis.

Moisture content, wet basis = 4.5/(4.5 + 26.5) = 0.145
Moisture content dry basis = 4.5/26.5 = 0.170

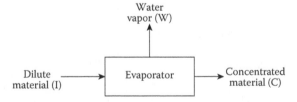

FIGURE 5.4 Example of an evaporator showing total mass balance.

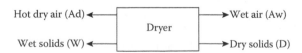

FIGURE 5.5 Example of a dryer to illustrate component mass balance.

Example 5.3

Diced carrots are dried in a dryer (Figure 5.6) from a moisture content of 81% wb to 15% wb using hot air with a humidity ratio (H) of 0.012 kg water/kg dry air at a mass flow rate of 400 times that of the dry carrots. Calculate the humidity ratio of the exit air.

Basis: Unknown flow rate of dry mass of carrots (C). Also the moisture content of the feed (F) on dry basis is 0.81/0.19 kg water per kg dry solids, hence F = 0.81/0.19 C

Water mass balance on the dryer provides:

Rate of water in = Rate of water out, or

0.81/0.19C + 400 C (0.012) = 0.15/0.85 C + 400 C H

C will be canceled, hence no need of writing overall mass balance. Therefore H = 0.022 kg water/kg dry air.

Example 5.4

Apple pieces are being dried as shown in Figure 5.7. Parallel flow of air and apples is used in the first dryer, and counter current flow in the second dryer. Part of the air leaving the second dryer is mixed with the air entering the first dryer. In the diagram, X is mass fraction of moisture in the solid; Y is kg of moisture per kg of dry air; and G is air flow rate in kg of dry air/kg of dry solids. Determine the quantity of air returned from the second to the first dryer, and the moisture content of the apples leaving the first dryer (A).

Consider 100 kg dry solids are entering at 1.

Mass of moisture at inlet (1) = 90/10*100 = 900 kg
Mass of moisture at outlet (3) = 15/85*100 = 17.65 kg

Moisture balance on the second dryer:

A(100) + 150 (100) (0.010) = 17.65 + 150 (100) (0.032)
A = 3.4765 kg water/kg dry solids

Moisture balance on the first dryer:

900 + 150 (100) (0.010) + B (100) (0.032) = (150 + B) (100) (0.042) + 3.4765 (100)
B = 72.35 kg dry air/kg solids

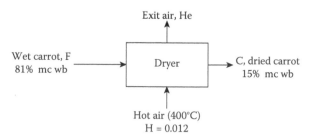

FIGURE 5.6 Diagram of the dryer for Example 5.3.

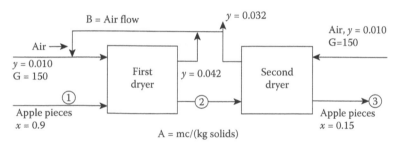

FIGURE 5.7 Two stage dryer for Example 5.4.

5.3 ENERGY BALANCES

In energy balances, an energy accounting system is set up to calculate the steam or cooling water amount required to maintain needed process temperatures. Energy balances can also be used to account for different energy forms utilized in a process. Energy balances assist in identifying locations where energy conservation can be applied, measuring the effectiveness of energy conservation, and planning energy conservation measures in a manufacturing operation. Energy balances will allow the calculation of the amount of heat required to be transferred in a process. This information is needed in process design so that proper heat transfer equipment are selected and proper sizes are calculated. Design of energy recovery operations also requires data from energy balances. Energy can input or exist in various forms such as heat, electrical energy, mechanical energy, and total energy is conserved.

The principle of the conservation of energy is explained by the first law of thermodynamics: "in any system the energy associated with matter entering the system plus the net heat added to the system is equal to the sum of the energy associated with matter leaving the system, net work done by the system, and the change in system internal energy," that is,

Energy accumulation in the system = Energy into the system – Energy out of the system   (5.17)

In Equation 5.17, heat and work are included in the total energy. The law of energy conservation further states that energy can neither be created nor destroyed. General form is written as:

$$\sum M(h + e_k + e_p)_{\text{in}} - \sum M(h + e_k + e_p)_{\text{out}} - Q + W_s = \Delta E \qquad (5.18)$$

where h = the enthalpy per unit mass, e_k = kinetic energy per unit mass, e_p = potential energy per unit mass, Q = heat removed or lost from the system, W_s = work done on the system, M = mass entering or leaving the system per unit time, and ΔE = total accumulation or change of system energy.

In food processing, most of the time we are interested in steady state conditions, at which the change in the system energy will be negligible. Also, the heat of biological reactions is small and can be ignored. Thus, for steady state conditions, the conservation of energy equation can be written as

Energy in = Energy out   (5.19)

The procedure for energy balance calculations is similar to mass balance calculations. There are some common assumptions applied in the energy balances as: (i) heat of mixing is generally neglected, (ii) evaporation of liquid may be considered negligible if the system is closed where negligible water is lost as vapor, (iii) heat losses from the system to the surrounding is often ignored if enough insulation is provided, (iv) sometimes shaft work can be neglected in the agitation of liquids in mixing, and (v) the system is considered homogeneous and well mixed [2].

The energy term may contain thermal, hydraulic, kinetic, or potential energy forms. In thermal processing of foods (heating, cooling, freezing, cooking, sterilization, pasteurization, etc.), the thermal energy is more pronounced, and other energy terms may be neglected under certain conditions. If they are not neglected, then they can be calculated using Bernoulli's equation. Two forms of thermal energy terms are generally encountered: (i) Sensible heat due to temperature change, and (ii) latent heat due to phase change. It can be for vaporization (from liquid to vapor), sublimation (solid to vapor) and freezing or fusion (liquid to solid). Steam tables are used for the total energy (or enthalpy) of the water. The enthalpy (h) contains heat energy and flow work and given by; $h = C_p (T - T_{\text{ref}})$.

Sensible heat is calculated by:

Sensible heat = (mass or mass flow rate).(specific heat, C_p) · (temperature change from a reference)

(5.20)

and the latent heat is computed by:

Latent heat = (mass or mass flow rate) · (latent heat per unit mass)   (5.21)

Approximate values of the latent heat of vaporization (h_{fg}) can be calculated using Trouton's rule stating that the "ratio" of molal latent heat of vaporization at atmospheric pressure (H_{fg}) to boiling point at atmospheric pressure (T_b) is a "constant"; i.e., H_{fg}/T_b = constant; for organic liquids, this constant is assumed to be 85 kJ/(mol K) [3]. However, values for latent heat of evaporation of most liquid are available in tabulated form in the literature [1,3].

5.4   PHASE DIAGRAM AND PROPERTY TABLES

5.4.1   PHASE DIAGRAM AND STEAM TABLES

The properties (specific volume, enthalpy, entropy, etc.,) of common substances (water, refrigerants), as a function of temperature and pressure, have been measured; and those for numerous pure substances have been cataloged in tabular forms. A complete set of tables is that for water, called the "steam tables." These tables provide the steam properties (enthalpy, specific volume, etc.), and are needed when calculating heat exchange involving foods and water or steam.

Water or any other pure substance can exist in three forms or phases: solid, liquid, and gaseous (vapor). A substance can exist in more than one phase at any time. Corresponding to any pressure is a temperature at which the liquid phase can coexist with the vapor phase. Such pressure and its corresponding temperature are called the "saturation pressure" and "saturation temperature," respectively (e.g., water at 101.325 kPa has a saturation temperature of 100°C; the saturation pressure corresponding to the saturation temperature of 100°C is, therefore, 101.325 kPa). For every pure substance definite relationships exist between temperature and their corresponding saturation pressures, not only for the liquid–vapor phases but also for the solid–liquid and solid–vapor phases.

Figure 5.8 shows the phase diagram ($P–T$) for a pure substance [4]. The following is the discussion on this phase diagram.

i. *Critical point*: the liquid–vapor saturation curve ends at the critical point, at which (a) the liquid and vapor phases are intermixed, and (b) there is no vaporization process.
ii. *Liquid–vapor-saturation (vaporization) curve*: between triple and critical points is liquid–vapor-saturation or vaporization curve. A certain amount of heat (latent heat or heat of vaporization) is required to vaporize a unit mass of a pure substance in the liquid phase. The heat of vaporization decreases as the critical point is approached. Two phases can coexist at any point on a saturation curve. When only liquid exists at the saturation pressure and temperature, the liquid is called "saturation liquid." When only vapor exists at the saturation pressure and temperature, then the vapor is saturated vapor. The temperature on the curve is boiling temperatures of water at various pressures.
iii. *Triple point*: the intersection of the three saturation curves (liquid–vapor, solid–vapor, and solid–liquid) is the triple point. All three phases of the pure substance exist in equilibrium at these conditions of pressure and temperature.
iv. *Solid–liquid saturation (fusion) curve*: solid and liquid phases coexist in equilibrium.
v. *Solid–vapor saturation (sublimation) curve*: solid and vapor phases coexist in equilibrium.

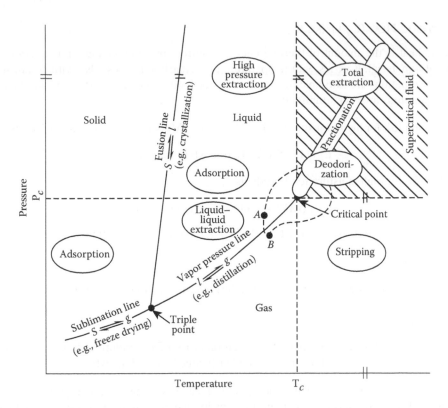

FIGURE 5.8 Phase diagram of a pure substance, water. (Adapted from Rizvi, S.S.H. and Mittal, G.S., *Experimental Methods in Food Engineering*, Van Nostrand Reinhold, New York, 1992.)

vi. *Subcooled or compressed liquid*: in this region, only the liquid phase exists. The temperature in the liquid phase is less than the saturation temperature for a given pressure, hence it is "subcooled liquid." The substance is also called "compressed liquid" because its pressure is greater than the saturation pressure at a given temperature.

vii. *Superheated vapor*: in this region, only the vapor phase exists. The temperature is greater than the saturation temperature for the existing pressure.

viii. *Quality (x)*: in the two phase regions, since pressure and temperature are dependent and thus another property is required to fix the state of the pure substance (e.g., on the vaporization curve), the part of the substance may be saturated liquid and part saturated vapor. The term "quality" is used for the vaporization curve, indicates what percent or fraction of the mixture is vapor. If $x = 0$, the substance is a saturated liquid, and if $x = 1$, all of the substance is a saturated vapor.

The properties of the liquid–vapor mixture are calculated by the following equations:

$$v = v_f + x \cdot v_{fg} \tag{5.22a}$$

$$h = h_f + x \cdot h_{fg} \tag{5.22b}$$

where v is specific volume (m³/kg); h is enthalpy (kJ/kg); subscript f stands for liquid, and subscript fg for the difference in vapor and liquid properties (e.g., $v_{fg} = v_g - v_f$). The h_{fg} is known as the latent heat of vaporization, and h_f is the sensible heat of the liquid phase, whereas h_g is the sum of sensible and latent heats.

5.4.2 Specific Heat (C_p, kJ/(kg.K))

The specific heat is the amount of thermal energy required for a unit change in temperature for a unit mass of a material. It varies with temperature particularly for gases, also with composition of foods. Some equations are available to calculate C_p for foods if composition is known. Siebel and Dickerson equations are over simplified.

i. Dickerson [5]: for high moisture food products i.e., "Moisture" content (mc) > 50% wb

$$C_p = 1.675 + 2.512 \text{ (mc wb in fraction)} \tag{5.23}$$

ii. Siebel [6]: for high moisture food products

$$C_p = 0.837 + 3.349 \text{ (mc wb in fraction)} \tag{5.24}$$

$$\text{For frozen foods: } C_p = 0.837 + 1.256 \text{ (mc wb in fraction)} \tag{5.25}$$

iii. Charm [7]:

$$C_p = 2.094 \text{ (fat fraction)} + 1.256 \text{ (solids not fat, fraction)} + 4.187 \text{ (mc wb)} \tag{5.26}$$

iv. Heldman [8]:

$$C_p = 1.424 \text{ (carbohydrate concentration)} + 1.549 \text{ (protein conc.)} \\ + 1.675 \text{ (fat conc.)} + 0.837 \text{ (ash conc.)} + 4.187 \text{ mc wb} \tag{5.27}$$

Tables 5.1 and 5.2 provide density and specific heat of various food components at various temperatures. These are useful to calculate the density and specific heat of foods at various temperatures and compositions.

TABLE 5.1
Effects of Temperature and Composition on Specific Heat and Density of Foods

Property	Component	Model	Standard Error	Standard Error (%)
		Models of Major Components of Foods		
ρ (kg/m³)	Protein	$\rho = 1.3299 \times 10^3 - 0.5184\,T$	39.9501	3.07
	Fat	$\rho = 9.2559 \times 10^3 - 0.41757\,T$	4.2254	0.47
	Carbohydrate	$\rho = 1.5591 \times 10^3 - 0.31046\,T$	93.1249	5.98
	Fiber	$\rho = 1.3115 \times 10^3 - 0.36589\,T$	8.2687	0.64
	Ash	$\rho = 2.4238 \times 10^3 - 0.28063\,T$	2.2315	0.09
C_p (kJ/(kg K))	Protein	$C_p = 2.0082 + 1.2089 \times 10^{-3}\,T - 1.3129 \times 10^{-6}\,T^2$	0.1147	5.57
	Fat	$C_p = 1.9842 + 1.4733 \times 10^{-3}\,T - 4.8008 \times 10^{-6}\,T^2$	0.0236	1.16
	Carbohydrate	$C_p = 1.5488 + 1.9265 \times 10^{-3}\,T - 5.9399 \times 10^{-6}\,T^2$	0.0986	5.96
	Fiber	$C_p = 1.8459 + 1.9306 \times 10^{-3}\,T - 4.6509 \times 10^{-6}\,T^2$	0.0293	1.66
	Ash	$C_p = 1.0926 + 1.8896 \times 10^{-3}\,T - 3.6817 \times 10^{-6}\,T^2$	0.0296	2.47

Source: Reproduced with permission from Choi, Y. and Okos, M.R., *Food Engineering and Process Applications*, Vol. 1, Le Maguer, M. and P. Jelen, Eds. Elsevier, New York, NY, 1987.
Note: T is in °C.

TABLE 5.2
Density and Specific Heats for Water and Ice as a Function of Temperature

	Model	Standard Error	Standard Error (%)
Water	$\rho_w = 9.9718 \times 10^2 + 3.1439 \times 10^{-3}\,T - 3.7574 \times 10^{-3}T^2$	2.1044	0.22
	$C_{pw1} = 4.0817 - 5.3062 \times 10^{-3}\,T + 9.9516 \times 10^{-4}\,T^2$	0.0988	2.15
	$C_{pw2} = 4.1762 - 9.0864 \times 10^{-5}\,T + 5.4731 \times 10^{-6}\,T^2$	0.0159	0.38
Ice	$\rho_i = 9.1689 \times 10^2 - 0.13071\,T$	0.5382	0.06
	$C_{pi} = 2.0623 + 6.0769 \times 10^{-3}\,T$	0.0014	0.07

Source: Reproduced with permission from Choi, Y. and Okos, M.R., *Food Engineering and Process Applications*, Vol. 1, Le Maguer, M. and P. Jelen, Eds. Elsevier, New York, NY, 1987.
C_{pw1} = for the temperature range of −40–0°C (frozen products).
C_{pw2} = for unfrozen products.
ρ = for the temperature range of 0 C_{pw1} = for the temperature range of −40–0°C (frozen products)

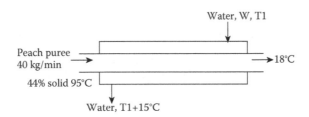

FIGURE 5.9　Heat exchanger set up for Example 5.5.

Example 5.5

Refer to Figure 5.9 and calculate the amount of water required to cool 40 kg/min of peach puree from 95 to 18°C. The peach puree contains 38% solids. The allowable increase in water temperature is 15°C while passing through the heat exchanger. Heat losses are negligible. The specific heat of water is 4187 J/(kg.K), and use Siebel equation to calculate specific heat of the puree.

Assuming T1 = inlet water temperature = 15°C, then exit water temperature = 15 + 15 = 30°C. For enthalpy calculations, a reference temperature of 0°C is considered. The specific heat of the puree is calculated using Siebel's equation:

C_p = 3349 (0.62) + 837.36 = 2913.7 J/(kg.K)
Energy balance is given by: Energy in = Energy out
Or W (15)(4187) + 40(95)(2913.7) = W (30)(4187) + 40(18)(2913.7)
Or 62805 W = 8974196 giving W = 142.9 kg/min

Example 5.6

The peach puree initially has a concentration of 7.2% total solids and is heated from 23 to 84°C by mixing steam. Refer to Figure 5.10 and calculate the concentration of total solids in the hot puree leaving the operation. The process is conducted at atmospheric pressure i.e., steam temperature of 100°C. The specific heat of solids is 2.10 kJ/(kg.K) and for water is 4.186 kJ/(kg.K).

Let us consider 100 kg of apple puree.

Total mass balance: 100 + A = B
Solids mass balance: (0.072) (100) = B(c)

or c = 7.2/B = 7.2/(100 + A)

FIGURE 5.10   Evaporator systematics for Example 5.6.

Energy balance:

100 (0.072) (23) (2.10) + 100 (1 − 0.072) (4.186) (23) + A (2675.5) = B(c) (2.10) (84) + B (1−c) (4.186) (84)

347.76 + 8934.60 + 2675.5 A = 0.072 (100) (176.40) + 351.62 (100 + A−7.2)

347.76 + 8934.60−1270.08−35162 + 2531.66 = 351.62 A−2675.5 A

or −24618.06 = −2323.88 A

A = 10.6 kg

B = 100 + A = 110.6 kg

c = 7.2/B = 7.2/110.6 = 0.065

REFERENCES

1. Sinnott, R.K. *Chemical Engineering, Vol. 6, Chemical Engineering Design.* Butterworth-Heinemann, Oxford, UK, 1998, Chapters 2, 3.
2. Doran, P.M. *Bioprocess Engineering Principles.* Academic Press, New York, NY, 1995, Chapter 2, 4–6.
3. Smith, P.G. *Introduction to Food Process Engineering.* Kluwer Academic/Plenum Publishers, New York, 2003, Chapter 4.
4. Rizvi, S.S.H. and Mittal, G.S. *Experimental Methods in Food Engineering.* Van Nostrand Reinhold, New York, NY, 1992, Chapter 2.
5. Dickerson, R.W. Jr. Thermal properties of foods. In *The Freezing Preservation of Foods*, Vol. 2. Tressler, D.K., Van Arsdel, W.B., and Copley, M.J., Eds. AVI Publishing Co., Westport, CT, 1969, 26.
6. Seibel, J.E. Specific heats of various products. *Ice Refrigeration*, 1892, 2, 256.
7. Charm, S.E. *Fundamentals of Food Engineering.* AVI Publishing Company, Westport, CT, 1971, Chapter 2.
8. Hedman, D.R. *Food Process Engineering.* AVI Publishing Co., Westport, CT, 1975, Chapter 3.
9. Choi, Y. and Okos, M.R. Effects of temperature and composition on thermal properties of foods. In *Food Engineering and Process Applications,* Vol. 1. Le Maguer, M. and Jelen, P., Eds. Elsevier, New York, NY, 1987, 93.

Section II

Heat Transfer in Food Processing

6 Fundamentals of Computational Fluid Dynamics

Gordon D. Mallinson and Stuart E. Norris
The University of Auckland

CONTENTS

6.1 INTRODUCTION

6.1.1 Fluid Dynamic Processes Typical of Food Engineering

Many processes associated with the production, transport, storage and processing of food rely on fluid motion, heat transfer and or thermodynamics. For example the production of milk products involves sterilization and separation processes that rely on heat transfer and multiphase fluid motion. Meat preservation and storage depends on freezing and refrigeration. Food preparation and cooking involves mixing and heat transfer by single or combined conduction, convection and radiation modes. It is therefore no surprise that the modeling of heat transfer, fluid mechanics and thermodynamics has become fundamental to understanding and improving many aspects of the food generation, storage and processing cycle.

Computational fluid dynamics (CFD) is a modeling technology that has been applied widely in the food engineering domain and a recent comprehensive review[1] attests to the variety of food engineering phenomena that are amenable to analysis using CFD. Many of the problems that have been studied are very complex and can easily stretch the capabilities of even the latest CFD technology and computational resources. There is a real need to apply CFD cautiously with the purposes of solving problems within the available resources and, more importantly, ensuring that the predictions have sufficient accuracy to be reliable and useful.

Although readily accessible via commercial packages or purpose written software, CFD is a complex technology. Best practice in its application relies on understanding the fundamental modeling processes and the numerical issues and restrictions that apply to almost any CFD simulation. The purpose of this discussion is to present some of the rigor that must be applied when using CFD.

The field of food engineering is very wide and there are many examples of physical processes that have been modeled by CFD and thermal modeling systems. This discussion does not attempt to describe these applications; instead the reader is referred to a recent comprehensive review by Sun.[2]

6.1.2 The Attraction of Computational Fluid Dynamics (CFD) and Issues Associated with its Use

CFD is often the method of choice for investigations of new processes or ways to improve existing processes. The mathematics of fluid motion, heat transfer and mass transfer are complex and CFD may be the only modeling technique that is available. CFD technology has developed to the level where packages provide comprehensive, easy to use interfaces that enable virtually anyone to use CFD. However, such users need to be cognisant of the principles of fluid mechanics and, where necessary, heat transfer and thermodynamics. They must also be acutely aware of principles of the numerical procedures that underpin CFD to ensure that the quality of the generated solutions is adequate. Unfortunately it can be very easy for users of even a well written and documented package to apply it to a problem under such conditions that the results it produces are meaningless, even though they "look plausible". It is important to realize that commercial CFD packages are written to "produce a solution for any situation" since the market place does not respond favorably to packages that do not produce solutions. This can lead to the generation of solutions for situations where there is no corresponding fluid flow configuration. A common example is the generation of a steady state solution for conditions for which a flow is, in reality, unsteady.

6.1.2.1 Philosophy of Application of Computational Fluid Dynamics (CFD)

CFD solutions must be validated. The application of CFD to practical problems must be made with caution and with due respect given to possible limitations. This consideration is especially important when, as is often the case, CFD is applied to "new" situations for which there are few or no independent data. The common sources of independent data are analytical solutions or experimental measurements. Of necessity analytical solutions represent simplified configurations of a problem and physical experiments are often very difficult to obtain with sufficient detail to validate CFD. However, as noted by Roache,[7] the purchase of a "really good code" does not remove the need for the user to do a validation study.

A CFD solution is a single instance of a numerical experiment. Whereas an analytical solution is usually expressed as functions of the relevant parameters a single CFD solution represents a single combination of those parameters. Many CFD solutions are required to generate data from which trends and functional relationships can be deduced. In this way CFD can be regarded in much the same way as a physical experiment and many CFD solutions will be required to produce a single correlation that represents the behavior of a process in the same way that a correlation derived from a set of physical experiments is generated. In both scenarios, careful use of nondimensional numbers can reduce the number of CFD calculations needed.

CFD is a computer intensive activity. A common misconception is that increasing the amount of computer resource will render even the most difficult problem tractable. Unfortunately, as will be discussed in Section 6.3.4.2, the rate at which computational requirements increase with increasing mesh resolution means that only small improvements may be achieved with quite substantial increases in computer resources. This is particularly true for 3D unsteady simulations where a halving of the effective mesh spacing requires at least 16 times the computational effort.

Verification and mesh sensitivity. In reality a compromise between mesh resolution and computational effort must usually be made. Many CFD simulations, especially those for industrial processes, are not made with sufficient accuracy to be confident that their predictions are correct. When using CFD, it is vital to estimate the effect of the compromise on solution accuracy using a process called verification and, if possible, error bounds for the output data. This process is called verification.

Verification and validation are essential cornerstones of CFD "best practice."

6.1.3 Philosophy of this Discussion

The equations of fluid mechanics, thermodynamics and heat and mass transfer are derived from conservation laws which are expressed in their fundamental forms as balances over control volumes. The common expressions of these laws are partial differential transport equations which for fluid motion are called the Navier–Stokes equations. Partial differential transport equations for energy, mass fractions of phases or chemical species represent additional processes relevant to food engineering.

The present discussion is made from the viewpoint that the conservation laws are fundamentally important for three reasons. (i) The partial differential equations and their approximations must faithfully represent the conservation laws at all levels of scale. (ii) The conservation laws are normally the mechanisms for presenting the qualitative interpretations of CFD solutions. (iii) Global conservation laws may, especially if simplifying assumptions are made, be evaluated analytically thereby providing approximate validation. This discussion therefore concentrates on conservation laws, their use in the derivation of the numerical models, and their value as quantitative interpretation metrics.

There is a body of CFD experience and expertise that now spans more than five decades. There are people wishing to use it in design or industrial contexts and they need to know the essential issues that must be addressed to obtain the best solutions that can be achieved in a given set of circumstances. A major objective of this discussion is to provide suitable guidance to new users of CFD on the principles of good practice.

6.2 THE FOUR ESSENTIAL STEPS OF A COMPUTATIONAL FLUID DYNAMICS (CFD) ANALYSIS

Before presenting the more specific details of CFD methodology, the general processes associated with numerical modeling are described and used to set the context for the remainder of the discussion. The steps presented below are generic and can be applied to any numerical modeling process.

TABLE 6.1
The Four Generic Steps Associated with a Numerical Solution

Generic Step	Methods	Considerations
1. Generation of governing equations	Physical modeling Conservation laws Scaling	Problem complexity Data availability Impact on numerical processes Modeling accuracy
2. Generation of approximate equations	Finite differences Finite volumes Finite elements Spectral methods Mesh generation Preprocessing	Consistency Convergence Discretization errors Impact on numerical processes Cost of solution
3. Solution of approximate equations	Direct solution Iterative methods Time sequence development	Stability Systematic errors Time discretization errors Termination criteria Efficiency
4. Interpretation and analysis of results	Computer graphics Postprocessing Analysis e.g., calculation of heat fluxes	Information reduction Comparison with external data, knowledge or common sense

Table 6.1 summarizes the four steps associated with the generation of a numerical solution or in this context a CFD model. Against each step the associated CFD methodology and relevant considerations are shown. The various generic terms appearing in the table are defined and discussed below.

6.2.1 Generation of Governing Equations

Numerical modeling is ultimately based on a set of governing equations, or underlying physical laws. In CFD, these equations will depend on modeling assumptions and methods of analysis that underpin the modeling of fluid mechanics, thermodynamics, and transport processes. Often scaling can be used to transform the equations into forms that are suited to particular combinations of the ranges of the relevant parameters and/or are dependent on a minimum number of parameters. There are two important considerations that affect the choice of governing equations (or the choice of options in CFD packages).

The complexity of the model's equations should be compatible with the data available for the problem. For example, it is of no use to allow for variation in properties such as thermal conductivity if suitable property data are not available.

The form of the governing equations will have a direct influence on the numerical methodology and its complexity. Simplifications may be required to allow a known numerical method to be applied.

For a new problem there may be several "iterations" through these two considerations.

6.2.2 Generation of Approximate or Algebraic Equations

The equations derived from the conservations laws are described in Section 6.4. In principle they can predict fluid motion very accurately, if only they could be solved. Approximations must be made, firstly to enable their solution using digital arithmetic and secondly to render complex phenomena such as turbulence amenable. Whatever approximations are involved, the objective is to generate sets of algebraic equations.

For CFD, the approximation step is difficult and is the core of much of the art or even folklore associated with its use. A subdivision strategy is used to replace continuum equations with discrete analogs. There are various methodologies such as finite differences, finite volumes, finite elements, boundary elements, or spectral methods. The first four use subdivision of geometric space to create the approximations whereas the last subdivides function space.

When geometric space is subdivided, the solution is represented by a set of point values distributed according to the approximation method and the design of a suitable mesh. When function space is subdivided the solution is represented by the linear combination of many functions. The discrete numerical values that represent the approximate solution are the coefficients of the linear combination. The five different methods are summarized below.

6.2.2.1   Finite Difference

Derivatives appearing in the governing equations are approximated by algebraic "difference" expressions which relate point, or nodal, values of the solution variables to each other. Application of these expressions at points distributed throughout the "solution domain" for the problem generate the required set of algebraic equations. This is the classical method and is a natural approach if the governing equations are expressed as partial differential equations.

6.2.2.2   Finite Volume

Algebraic equations are generated by applying the relevant conservation laws over small volumes or by integrating the governing equations over the same volumes. The former is now the dominant approach. Point values of the solution variables are distributed around the volumes according to the scheme used to approximate the conservation laws. In many cases the equations are the same as those arising from the finite difference approach. An advantage of this approach is that the approximate equations satisfy the conservation laws at the finite volume and domain levels whereas the finite difference based approximations have to be carefully constructed to achieve the same effect.

6.2.2.3   Finite Element

The solution domain is divided into a number of volumes or "elements" and the solution variables are approximated by functions (often polynomial) over each element. These functions are then "fitted" over the elements using a minimization technique that generates coefficients for the approximating functions. The coefficients are expressed in terms of nodal values of the solution variables and the equations for these coefficients can be combined to generate the required set of algebraic equations. The finite volume method is similar to the finite element method except that generally a minimization technique is not applied. In some cases a finite element method can yield the same set of algebraic equations as a finite volume or difference method.

6.2.2.4   Boundary Element

For certain kinds of flows, solutions of the governing equations can be expressed as integral equations, or sums of integral equations. Solutions are then found by fitting these integral expressions to the boundary conditions. Flows where viscosity can be neglected are particularly amenable to this method. A mesh can be created over the boundary and for each mesh element an integral term represents the contribution that part of the boundary makes to the flow. Summing the contributions of the effects of all these elements produces a solution for the whole flow field. These methods are called boundary element methods and are often used for predicting flows around aircraft and ships. The boundary element methods can be modified to account for the effects of the boundary layers where the influence of viscosity is concentrated.

6.2.2.5   Spectral

The solution functions are represented by sums of orthogonal "base" functions, (e.g., $\sin n\pi x$, $\cos n\pi y$ for a square solution domain.) Equations defining the coefficients for these sums are algebraic and can be used to generate a suitable solution set. In this method, the problem is expressed in terms of these coefficients rather than nodal values of the solution variables, but the end result is the same, a set of algebraic equations are solved. Spectral methods are not very well suited to irregularly shaped geometric domains.

6.2.3 Verification and Validation

How well the solution of the approximate equations represent the solution of the original governing equations is the issue that causes the most debate regarding the application of CFD to predict fluid motion. There are two processes that are used to assess how well a CFD model might represent reality.

6.2.3.1 Verification

Verification considers how well the numerical solutions represent those of the governing equations. Mesh convergence studies are part of this process and should be executed for every "new" problem. By obtaining solutions on three carefully chosen meshes, it is possible to determine the "order of convergence" of the solution of algebraic approximations to that of the continuous equations. It is also possible for some circumstances to estimate error bounds for the numerical solutions. A formal description of verification is presented in Section 6.3.4.3.

6.2.3.2 Validation

Validation considers how well numerical solutions represent reality. It really involves understanding both the errors associated with the approximations and the effects of simplifying assumptions made to set up the governing equations. For a the user of a CFD package the modeling assumptions are made via choices such as assuming the flow is incompressible, if buoyancy can be neglected, or selecting a particular turbulence model.

Validation is the process by which it is established that the correct equations (or models) are being solved. This involves using external data to compare with similar data produced by the numerical solutions. For example, does the lift estimate from an external flow analysis compare favorably with wind tunnel measurements? Or is a heat transfer prediction confirmed by experiment? Despite the confidence with which the major software packages are "sold," the process of validation still is essential when CFD is applied to any particular situation for which full and complete validation can not be inferred from previously published results.

6.2.4 Solution of the Approximate Equations

The methods by which the algebraic equations are solved are determined by advances in numerical analysis that have occurred over several decades and are generally of less interest to the user of a CFD package than are, or should be, the details associated with the formation of the equations, their approximations and the quality of the solutions that they produce. Iterative methods are used extensively and there are numerous forms. Once again, the food engineer will call on the advice of the package writer to decide on a particular method. Sometimes a certain amount of "tuning" may be required to optimize an iterative method. However modern methods for solving coupled sets of linear equations are usually very robust and very little tuning may be required.

Regardless of the kind of methodology actually used there are some generic considerations a user of CFD should be aware of.

6.2.4.1 Time Sequence Development

Many physical problems are such that a solution "evolves" with time. Numerically, this evolution can be approximated by a progression through a sequence of time steps. Algebraic equations are used to advance the solution variables over each step. This concept can be generalized by noting that any iterative method can be represented by a time sequence development. Hence the notion of time sequence development can be applied to the solution of steady state problems and well as unsteady problems.

6.2.4.2   Termination Criteria

An iteration sequence will, at some stage need to be terminated. This is an often neglected and/or misunderstood consideration. Many an iterative solution sequence has been stopped prematurely by a poor termination criterion.

6.2.4.3   Instability

Of the considerations which must be associated with this step, the most important in practice is that of stability (or instability). Often instability which results in the accumulation of round off errors to cause rapid growth of the solution variables until they are no longer reasonable representations of the solution. Constraints imposed to ensure stability are often critical in determining the viability of a numerical method. By the analogy described in Section 6.2.4.1, iterative methods are also affected by numerical instability.

6.2.5   Interpretation and Analysis of the Solutions

The entries in Table 6.1, for this step are obvious. The step has been included here to emphasize that obtaining a solution is not always the end of a CFD analysis. Numerical methods (especially transient solutions) can produce copious amounts of data which must be processed before useful interpretations can be made.

6.2.5.1   Estimation of Required Output Quantities

Estimations of averaged or integrated quantities are often the important outputs of a CFD model. For a food engineering problem, the estimation of heat fluxes and/or heat transfer coefficients can be an important postprocessing requirement. For example, separation processes can be represented by tracing the paths of elements of phases or species carried by the fluid. The accuracy of estimation of flow path lines is then paramount.

6.2.5.2   Visualization

Computer graphics techniques provide powerful interpretation aids, and there are several commercial packages available. However, there is no single strategy that can quickly reveal the important aspects of 3D or 4D CFD solutions[3] and there is a well established field of research dedicated to improving methods to visualize numerical predications of fluid motion. There are issues with some of the visualization methods that are commonly used. For example the construction of particle paths requires mass conservative interpolation to prevent false spirals and this condition is not often met. Apart from causing unrealistic artefacts,[4] particle tracking errors can be an issue when particle paths are used as part of post processing strategies to represent separation processes.

Often the human effort and computation associated with postprocessing can be difficult and time consuming. It may cost more to postprocess a solution than it does to generate it.

6.3   APPROXIMATIONS AND ERROR ESTIMATION

6.3.1   Approximation Metric

An approximation process has a metric that describes its "quality." This may be expressed in terms such as "resolution" or "mesh spacing" that indicate how the approximation can be adjusted. An alternative is to express the approximation in terms of its "cost." The number of mesh cells or mesh points is such a metric. The relationship between the two kinds of metrics is usually nonlinear. The assumption is made that increasing resolution, decreasing mesh spacing, or increasing the cost improves the quality of approximation.

The cost metric will be denoted here by N, where N could be the number of mesh points, finite volumes, finite elements, or spectral function coefficients. The selection of a method for generating the algebraic equations has a direct influence on the method which may be used for their solution. Moreover, the value of N will dictate how much computational effort or work is required to generate a solution. The value of N may even determine if a solution can even be generated at all on the available computer. The problem often reduces to one of being satisfied that the value of N dictated by computer cost restrictions is sufficient to satisfy consistency and convergence requirements.

6.3.2 ABSTRACTION OF THE GOVERNING EQUATIONS

The essential aspects of a numerical solution procedure can be illustrated by the following mathematical notation. The mathematical abstraction of a physical phenomenon such as a moving fluid can be represented by

$$\frac{\partial \varphi}{\partial t} + E(\varphi) = 0 \tag{6.1}$$

This is a very generalized notation where the function E represents an equation or set of equations in the most general sense. The variable φ similarly denotes a set of scalar or vector variables that may or may not be continuous. Equation 6.1 has been written as an evolution equation because fluid mechanics phenomena are unsteady. It can also be relevant and useful to consider the unsteady and steady aspects of the flow and solution processes separately.

The long term behavior of φ is determined by the limit of Equation 6.1 at infinite time, i.e.,

$$\lim_{t \to \infty} \left[\frac{\partial \varphi}{\partial t} + E(\varphi) \right] = 0 \tag{6.2}$$

If that limit exists in a way that $(\partial \varphi / \partial t) = 0$ then the system of equations that govern a steady state are

$$\lim_{t \to \infty} \left[E(\varphi) \right] = 0 \tag{6.3}$$

The reason that the equations are included in the limit is that for fluid dynamics modeling the equations themselves depend on the solution variables and hence may change with time. Equation 6.3 will be referred to as the steady state equation.

6.3.3 ABSTRACTION OF THE APPROXIMATION AND SOLUTION PROCESSES

6.3.3.1 Approximation

The approximation to Equation 6.1 can be represented by

$$\mathbf{L}_N(\boldsymbol{\varphi}'_N) = 0 \tag{6.4}$$

where the operator $\mathbf{L}_N(\boldsymbol{\varphi}'_N) = 0$ denotes a vector of approximate algebraic equations and $\boldsymbol{\varphi}'_N$ denotes a vector of values that approximate in some way the continuous function φ. $\mathbf{L}_N(\boldsymbol{\varphi}'_N) = 0$ is an evolution operator and incorporates the numerical approximation of the time derivatives. Assuming that Equation 6.4 has been correctly solved, $\boldsymbol{\varphi}'_N$ is the numerical solution that is assumed to be a good representation of φ. This notation will be used to formalize the following discussion of numerical accuracy.

Equation 6.4 represents the approximation to Equation 6.1. The question that arises is "What are the local errors caused by the use of only N points, N volumes, N elements, or N coefficients to represent the governing equations and solution variables? As CFD approximations rely on dividing geometrical or functional space into discrete pieces (elements, cell, coefficients, etc.) these local errors are loosely referred to as discretization errors. The effects of discretization errors are described by the concepts of consistency and convergence.

6.3.3.1.1   Consistency

As N is increased do the approximate equations correctly approach the governing equations? For example is the conservation of energy correctly represented by the approximate energy equation? Does the representation improve as N is increased? If this happens then the approximate equations are said to be "consistent". Using the notation of Equations 6.1 and 6.4 the concept of consistency can be described by

$$\lim_{N\to\infty}(\mathbf{L}_N(\boldsymbol{\varphi}'_N)) = \left[\frac{\partial\varphi}{\partial t} + E(\varphi)\right] \tag{6.5}$$

As Equation 6.5 implies, consistency embodies both the effects of the approximations to the equations and the resulting approximate solutions, bearing in mind that the coefficients in the approximate equations often depend on the solution variables.

6.3.3.1.2   Convergence

As N is increased does the numerical solution approach the solution to the governing equations? Convergence can be represented by

$$\lim_{N\to\infty}(\boldsymbol{\varphi}'_N : \mathbf{L}_N(\boldsymbol{\varphi}'_N) = 0) = \varphi \tag{6.6}$$

Equation 6.6 says that the $\boldsymbol{\varphi}'_N$ vector approaches φ as N increases, provided $\boldsymbol{\varphi}'_N$ is a solution of the approximate equations. The error in the solution can be written as

$$\boldsymbol{\varepsilon}_{\varphi,N} = \text{Diff}(\boldsymbol{\varphi}'_N, \varphi) \tag{6.7}$$

where Diff() is an operation that computes the difference between the vector of numerical estimates and φ. Remembering that the all the functions in Equation 6.7 are functions of time, this equation describes unsteady behavior of the difference between the numerical solution and the continuous solution to the governing equations.

Convergence is, in general, impossible to prove and may even be very difficult to demonstrate through numerical experimentation. Yet, this concept is fundamental to the application of CFD methods.

6.3.3.2   Solution of the Numerical Equations

6.3.3.2.1   Incomplete Solutions

In practice solution of the approximate equations is imperfect and will usually involve a decision regarding completion of an iterative or similar process. This can be denoted by

$$\mathbf{L}_N(\boldsymbol{\varphi}''_N(\mathbf{C})) = \boldsymbol{\varepsilon}_N(\mathbf{C}) \tag{6.8}$$

where $\mathbf{C}$ denotes a vector of solution completion parameters. The right hand side of Equation 6.8 represents the error arising from incomplete solution and is related to the "residuals" used more formally later in this discussion. It is assumed that the solution procedure is such that there is a termination point where no further reduction of the incompletion error can occur. This is an important point. The effects of finite arithmetic and numerical rounding mean that even if an "infinite number of iterations" or even a direct method are used there will still be a small but finite error in the numerical solution.

Equation 6.8 is a simplification of a rather complex situation. In reality, there will be termination criteria associated with iterative procedures for advancing through time steps. Hence for an unsteady problem there will be several, or even vectors, of termination criteria. This is why the termination criteria have been represented as a vector argument of the incomplete numerical solution.

Taking the incomplete solutions into account, Equation 6.6 can be expressed by

$$\lim_{N\to\infty}\left[\lim_{C\to C_f}(\boldsymbol{\varphi}_N''(\mathbf{C}):\mathbf{L}_N(\boldsymbol{\varphi}_N''(\mathbf{C}))=\boldsymbol{\varepsilon}_N(\mathbf{C}))\right]=\varphi \qquad (6.9)$$

where $\mathbf{C}_f$ is the vector of termination criteria that would generate exact solutions for Equation 6.4. The error in the solution can be written as

$$\boldsymbol{\varepsilon}_{\varphi,N}(\boldsymbol{\varepsilon}_N(\mathbf{C}))=\mathrm{Diff}(\boldsymbol{\varphi}_N''(\mathbf{C}),\varphi) \qquad (6.10)$$

to emphasize that the error does, in general, depend on the solution completion criteria.

6.3.3.2.2 Linearized Equations and Residuals

The concepts in Section 6.3.3.2.1 can be illustrated considering the situation when the approximate equations can be represented as a set of linear algebraic equations,

$$\mathbf{A}_N\boldsymbol{\varphi}_N'+\mathbf{B}_N=0 \qquad (6.11)$$

These equations can be normalized so that the diagonal coefficients in $\mathbf{A}$ are unity. This will be assumed to be done in this discussion.

At the termination of an iterative process the representation of the incomplete solution is

$$\mathbf{A}_N\boldsymbol{\varphi}''(\mathbf{C})+\mathbf{B}_N=\boldsymbol{\varepsilon}_N(\mathbf{C}) \qquad (6.12)$$

The vector $\boldsymbol{\varepsilon}_N(\mathbf{C})$ now represents the vector of residuals corresponding to the completion criteria. If $\mathbf{A}$ has been normalized, then each element in $\boldsymbol{\varepsilon}_N(\mathbf{C})$ represents the value that needs to be subtracted from the corresponding element of $\boldsymbol{\varphi}_N''(\mathbf{C})$ to locally satisfy Equation 6.11.

6.3.3.2.3 Iteration, Relaxation, and Time Steps

The use of the residual to correct the numerical solution vector, leads to the notion of a iterative process which can be described as

$$\boldsymbol{\varphi}_N''^{n+1}=\boldsymbol{\varphi}_N''^{n}+\boldsymbol{\varepsilon}_N^{n} \qquad (6.13)$$

where $\boldsymbol{\varphi}_N''^{n}$ is the vector of approximations to $\boldsymbol{\varphi}_N'$ after iteration n. The vector of residuals at that iteration is $\boldsymbol{\varepsilon}_N^{n}$. Note that the application of Equation 6.13 at a given iteration will not produce a set of zero residuals. In fact the simple process described by Equation 6.13 is prone to instabilities. One method to reduce instability is to introduce a relaxation factor, β.

$$\boldsymbol{\varphi}_N''^{n+1}=\boldsymbol{\varphi}_N''^{n}+\beta\boldsymbol{\varepsilon}_N^{n} \qquad (6.14)$$

Mostly the relaxation factor will be less than 1, corresponding to under relaxation. Sometimes over relaxation with $\beta > 1$ can accelerate the development of the solution.

If Equation 6.14 is manipulated,

$$\frac{\varphi_N''^{n+1} - \varphi_N''^{n}}{\beta} = \varepsilon_N^n \tag{6.15}$$

then β is seen to be similar to a time step. This means that there is a conceptual relationship between a sequence of iterations and a progression of time, with β being a relaxation parameter, or a time step accordingly. The residuals are a measure of the rate at which the solution is changing, which can be a useful interpretation of the residuals displayed by a CFD package during the solution process.

6.3.4 Convergence Behavior

6.3.4.1 Order of Approximation

The order of approximation is a measure of how the solution may be improved by increasing N. It is based on the concept of mesh refinement which means that it is usually described in terms of a Taylor's series analysis of finite difference equations. The concept can, however, be described more generally as is presented below.

Because the CFD approximate equations are generated by a discretization process, steps in spatial coordinates and time have physical meaning and their sizes impact directly on the approximation. It is convenient to treat the discretizations of space and time separately. The total number of values or degrees of freedom in a solution is the product of the number of spatial values, N_s and the number of time steps N_t.

$$N = N_s \times N_t \tag{6.16}$$

Let a spatial mesh size metric be denoted by h_s Then approximately

$$h_s = \frac{C}{\sqrt[d]{N_s}} \tag{6.17}$$

where C is an arbitrary constant and d is the number of spatial dimensions of the process being modeled.

The order of the approximation determines how the error in the solution changes with mesh size. If two different mesh sizes are considered, where $N_1 > N_2$ and $h_{s,1} < h_{s,2}$, the differences between the solutions produced by these meshes can be represented by

$$\frac{\left|\text{Diff}\left(\varphi'_{N_{s,1}}, \varphi\right)\right|}{\left|\text{Diff}\left(\varphi'_{N_{s,2}}, \varphi\right)\right|} = \left(\frac{h_{s,1}}{h_{s,2}}\right)^P \tag{6.18}$$

where p is the order of the approximation.

Using Equation 6.17 the change in the solution can be expressed in terms of the cost metric.

$$\frac{\left|\text{Diff}\left(\varphi'_{N_{s,1}}, \varphi\right)\right|}{\left|\text{Diff}\left(\varphi'_{N_{s,2}}, \varphi\right)\right|} = \left(\sqrt[d]{\frac{N_{s,2}}{N_{s,1}}}\right)^P = \left(\frac{N_{s,1}}{N_{s,2}}\right)^{-(p/d)} \tag{6.19}$$

For example for a 3D problem using a second order method,

$$\frac{\left|\text{Diff}\left(\varphi'_{N_{s,1}},\varphi\right)\right|}{\left|\text{Diff}\left(\varphi'_{N_{s,2}},\varphi\right)\right|} = \left(\frac{N_{s,1}}{N_{s,2}}\right)^{-(2/3)} \tag{6.20}$$

Turning this around the other way,

$$\frac{N_{s,1}}{N_{s,2}} = \left(\frac{\left|\text{Diff}\left(\varphi'_{N_{s,2}},\varphi\right)\right|}{\left|\text{Diff}\left(\varphi'_{N_{s,2}},\varphi\right)\right|}\right)^{-(d/p)} \tag{6.21}$$

6.3.4.2 Impact of Order of Approximation on Computational Cost

Consider the worst case of an unsteady problem. The number of time steps will be at least inversely proportional to the mesh spacing, however the effect of nonlinear coupling can reduce the time step and hence increase the number of steps.

$$\frac{N_{t,1}}{N_{t,2}} = \left(\frac{\left|\text{Diff}\left(\varphi'_{N_{s,1}},\varphi\right)\right|}{\left|\text{Diff}\left(\varphi'_{N_{s,2}},\varphi\right)\right|}\right)^{-r_t} \tag{6.22}$$

where $1 \le r_t \le 2$. This leads to

$$\frac{N_1}{N_2} = \left(\frac{\left|\text{Diff}\left(\varphi'_{N_{s,1}},\varphi\right)\right|}{\left|\text{Diff}\left(\varphi'_{N_{s,2}},\varphi\right)\right|}\right)^{-(r_t+(d/p))} \tag{6.23}$$

The computational effort will depend on the number of "values" to be computed, at best linearly, and at worst quadratically. Denoting computational effort by CE,

$$\frac{\text{CE}_1}{\text{CE}_2} = \left(\frac{\left|\text{Diff}\left(\varphi'_{N_{s,1}},\varphi\right)\right|}{\left|\text{Diff}\left(\varphi'_{N_{s,2}},\varphi\right)\right|}\right)^{-r_{CE}(r_t+(d/p))} \tag{6.24}$$

where $1 \le r_{CE} \le 2$.

Equation 6.24 can be used to estimate the computational cost of increasing the accuracy and samples are given in Table 6.2 for unsteady 3D models and Table 6.3 for steady 3D models. The estimates have used r_{cp} and r_t both equal to 1.1 which corresponds to the best performance of

---

TABLE 6.2

Estimates of Computational Power Factor and Times Estimated by Moores' Law to provide Improvement in the Accuracy of an Unsteady 3D CFD Simulation

Accuracy Improvement	First Order		Second Order		Fourth Order	
	CE Factor	Years	CE Factor	Years	CE Factor	Years
2	23	9	7.3	5.7	4.1	4
10	32,300	30	724	19	165	14.7

TABLE 6.3
Estimates of Computational Power Factor and Times Estimated by Moores' Law to provide Improvement in the Accuracy of a Steady 3D CFD Simulation

Accuracy Improvement	First Order		Second Order		Fourth Order	
	CE Factor	Years	CE Factor	Years	CE Factor	Years
2	9.85	6.6	3.14	3.3	1.77	1.65
10	2000	22	45	11	8	6

current solver technologies. Moore's law that computational power has doubled every two years has been used to estimate the number of years to wait until, all things being equal, the computational power will increase to meet the requirement. Of course computational demand can be met by increasing the number of processors, but the improvement is not equal to the number of processors. Speed up factors for CFD algorithms are often far from linear especially for large numbers of processors.

For the unsteady problem (Table 6.2), increasing the accuracy requirement by a factor of 10, can be expected to involve 32,300 times the computational effort for a first order method and 724 if the mesh convergence is quadratic, or one can wait 30 or 19 years, respectively for computational power to increase. This is a rather depressing result that indicates that major improvements in CFD technology are required if the complexity of problems that can be solved is to significantly increase.

For steady 3D problems the situation is substantially better. A gain of a factor of ten in accuracy requires only 45 times the computational effort if second order methods are used, or only eight times the effort if fourth order methods are used.

These estimates confirm history, that the major gains in the complexity of the problems that can be solved have been because of improvements in solver technologies, such as multigrid and coupled solvers, rather than simply increasing computer power.

6.3.4.3   Estimation of Convergence Errors

6.3.4.3.1   *Richardson Extrapolation*

Methods of error estimation are based on a process originally described by Richardson[5] and in subsequent texts.[6,7] It is based on the expansion of the solution as a Taylor series expressed in terms of the mesh spacing,

$$\varphi_N = \varphi + h_s \frac{\partial \varphi}{\partial h} + \frac{1}{2} h_s^2 \frac{\partial^2 \varphi}{\partial h^2} + O(h_s^3) \tag{6.25}$$

For a second order method,

$$\varphi_N = \varphi + \frac{1}{2} h_s^2 \frac{\partial^2 \varphi}{\partial h^2} + O(h_s^3) \tag{6.26}$$

For two meshes having different mesh spacing

$$\varphi_{N_1} = \varphi + \frac{1}{2} h_{s,1}^2 \frac{\partial^2 \varphi}{\partial h^2} + O(h_{s,1}^3) \qquad \varphi_{N_2} = \varphi + \frac{1}{2} h_{s,2}^2 \frac{\partial^2 \varphi}{\partial h^2} + O(h_{s,2}^3) \tag{6.27}$$

Eliminating the second derivative terms gives

$$\varphi = \frac{\varphi_{N_1} h_{s,2}^2 - \varphi_{N_2} h_{s,1}^2}{h_{s,2}^2 - h_{s,1}^2} + O\left(h_{s,1}^3, h_{s,1}^2 h_{s,2}^2\right) \tag{6.28}$$

Expressing the change in mesh spacing as a ratio

$$r_{2,1} = \frac{h_{s,2}}{h_{s,1}} \tag{6.29}$$

The estimate of the exact solution becomes

$$\varphi \cong \frac{\varphi_{N_1} r_{2,1}^2 - \varphi_{N_2}}{r_{2,1}^2 - 1} \tag{6.30}$$

For $r_{2,1} = 2$

$$\varphi \cong \frac{4}{3}\varphi_{N_1} - \frac{1}{3}\varphi_{N_2} \tag{6.31}$$

which is equivalent to the original expression derived by Richardson.[5]

6.3.4.3.2 The Issue of Unstructured Meshes

Richardson's original estimations were applied to second order methods that used central difference approximations on structured grids with uniform spacing. Extrapolation to nonuniform spacing and unstructured meshes is not strictly supported by a mathematical argument.[6] However, numerical experiments[8] have justified the use of approximate mesh norms to estimate the order of mesh convergence for numerical solutions on nonuniform grids. The procedure described in the following section uses a heuristic approach to obtain these estimates and has been extracted from editorial policy statements for the *Journal of Fluids Engineering*. Another good discussion of these concepts is the review by Roache.[7]

6.3.4.3.3 A Comment on Mesh Independence

Whatever the limitations, a study of mesh convergence this way is preferable to the often used alternative of seeking conditions that correspond to "mesh independence," a term that implies that a mesh has been refined to the point where further refinement will have no tangible effect. In reality this situation never occurs.

6.3.4.3.4 A Method of Estimating Mesh Convergence Behavior

The process is commenced by defining an average mesh cell size, using Equation 6.17

$$h_s = \frac{\text{Vol}}{\sqrt[d]{N_s}} \tag{6.32}$$

where we have replaced the arbitrary constant in Equation 6.17 by an estimate of the total volume. In practice the value of this constant does not influence the subsequent calculations. A suitable output variable is chosen for testing and is denoted here by φ. Particularly suitable are integral quantities such as lift and heat flow rates. Point values of variables are not good choices because they depend on interpolation to match values between different meshes. The problem is then solved using three different mesh sizes, where

$$h_{s,1} < h_{s,2} < h_{s,3} \tag{6.33}$$

The change in average mesh spacing is represented by the mesh ratios

$$r_{21} = \frac{h_2}{h_1} \qquad r_{23} = \frac{h_3}{h_2} \tag{6.34}$$

It is recommended that the mesh ratio between any two meshes is greater than 1.3. The following quantities can now be estimated.

$$\text{Convergence order:} \quad p = \frac{\left| \ln\left(\dfrac{\varphi_3 - \varphi_2}{\varphi_2 - \varphi_1} \right) + q(p) \right|}{\ln(r_{21})} \tag{6.35}$$

$$q(p) = \ln\left(\frac{r_{21}^p - s}{r_{32}^p - s} \right) \qquad s = sign\left(\frac{\varphi_3 - \varphi_2}{\varphi_2 - \varphi_1} \right) \tag{6.36}$$

$$\text{Estimated extrapolation:} \quad \varphi_{ext}^{21} = \frac{r_{21}^p \varphi_1 - \varphi_2}{r_{21}^p - 1} \tag{6.37}$$

$$e_a^{21} = \left| \frac{\varphi_1 - \varphi_2}{\varphi_1} \right| \tag{6.38}$$

$$e_{ext}^{21} = \left| \frac{\varphi_{est}^{12} - \varphi_2}{\varphi_{ext}^{12}} \right| \tag{6.39}$$

$$GCI_{fine}^{21} = \frac{1.25 e_a^{21}}{r_{21}^p - 1} \tag{6.40}$$

6.3.4.4 Example Problems

Two simple illustrative examples described, schematically in Figure 6.1, will be used throughout this discussion. They are problems that have become benchmarks for CFD and although simple, are sufficiently complex to illustrate some of the issues raised here.

The first example is that of a single phase fluid in an infinitely long box of square cross section. The two vertical walls are maintained at different uniform temperatures. Several benchmark solutions are available in the literature[9] and a typical solution is shown in Figure 6.2. It is usual to describe the fluid and the strength of the buoyancy forces arising from the applied temperature difference nondimensionally using the Rayleigh, Grashof, and Prandtl numbers, (Figure 6.2).

$$Gr = \frac{g\beta(T_h - T_c)L^3}{v^2} \qquad Pr = \frac{v}{\alpha} \qquad Ra = \frac{g\beta(T_h - T_c)L^3}{\alpha v} = Gr \cdot Pr \tag{6.41}$$

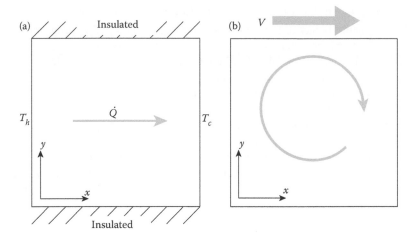

FIGURE 6.1 Examples: (a) Natural convection in a differentially heated cavity (b) Shear driven flow in a cavity with a sliding top.

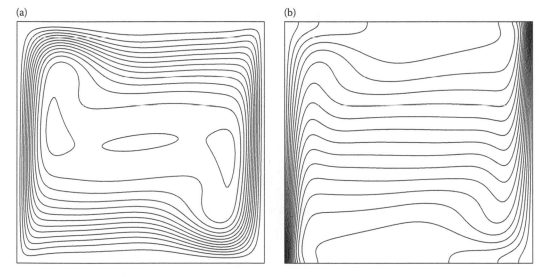

FIGURE 6.2 Differentially heated cavity with Ra = 10^6. (a) Stream lines (b) Contours of temperature.

An objective of a CFD analysis is to calculate the rate of heat transfer through the cavity as represented by the Nusselt number

$$Nu = \frac{qL}{k} \tag{6.42}$$

The second example is the flow induced in a square cross section cavity that has its top moving at constant velocity, and again good benchmark solutions are available.[10,11] A typical solution is shown in Figure 6.8. The strength of the driving motion of the lid is expressed as the Reynolds number,

$$Re = \frac{VL}{\nu} \tag{6.43}$$

6.3.4.4.1   Convergence Behavior for the Natural Convection Example

Estimates of convergence behavior are shown here for two different CFD programs. One (EHOA) used a structured grid and the other (ALE) an unstructured grid. Both methods used second order difference methods. The Nusselt number for the flow was calculated using either a surface integral of heat flux, or a volume integral[12] (see Section 6.4.6.5).

The calculated Nusselt numbers for the simulations are plotted in Figure 6.3. As the mesh is refined the four calculations converge to give a similar solution. As plotted the convergence shows an almost linear dependence on $1/N$ which for a 2D problem is proportional to h. This verifies that second order convergence has been achieved.

Note however that the extrapolated solution does differ between the four calculations (structured or unstructured mesh, surface or volume integrals). Error bounds for the extrapolated estimates are shown in Figure 6.4. Note that the extrapolated values, although different, agree with each other within the error bounds calculated using GCI.

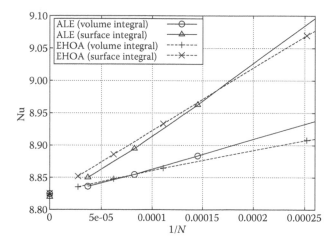

FIGURE 6.3   Convergence of the calculated Nusselt number with mesh refinement.

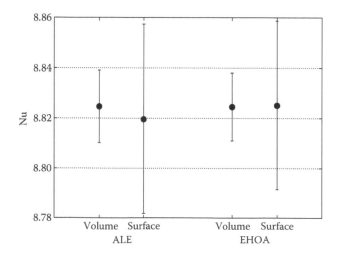

FIGURE 6.4   Extrapolated values with error bars.

6.3.4.4.2 Mesh Convergence of a Typical Food Engineering Industrial Problem

The efficiency of an open fronted display cabinet as used in supermarkets has been calculated using a 2D CFD model on a range of meshes. The convergence of the calculated efficiency is first order as is shown in Figure 6.5, and the calculation time in Figure 6.6. Note that extrapolation is only possible from the fine mesh solutions, and that extrapolation from course mesh solutions will give erroneous answers. This is generally true of all convergence calculations, and care should be taken that the solutions are in the region of monotonic convergence where the higher order terms of the Taylor series approximation are negligible. Note also that the computation time is approximately proportional to the number of points in the solution. This is typical behavior for the newer fully coupled/multigrid CFD solvers in modern commercial packages.

6.3.4.4.3 Mesh Convergence for the Cavity with a Sliding Top

The cavity with the sliding top problem is one of the standard problems used to test CFD algorithms (Figure 6.8). However, it does exhibit unexpected convergence behavior. The nondimensional power exerted by the sliding top can be estimated using

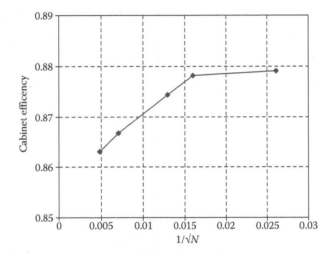

FIGURE 6.5 Convergence of calculated efficiency of open display cabinet.

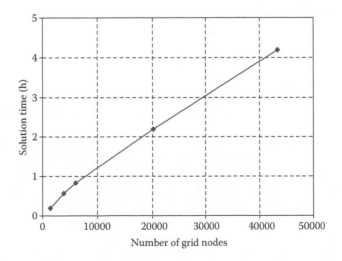

FIGURE 6.6 Solution time as function of mesh size for display cabinet calculation.

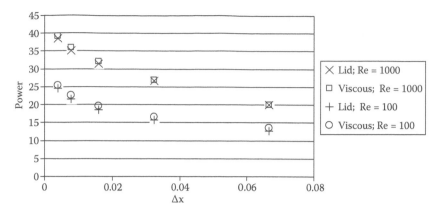

FIGURE 6.7 Variation of estimates of nondimensional power loss with mesh spacing in a cavity with a sliding top for Re = 100 and Re = 1000.

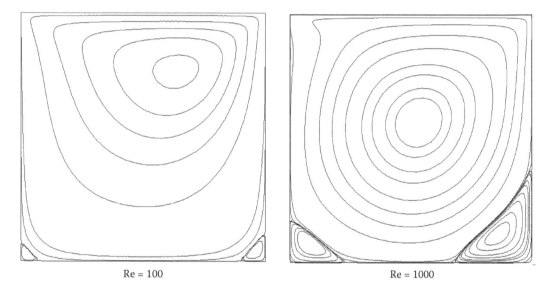

FIGURE 6.8 Flow in a cavity with a sliding top.

$$\dot{W}_{top} = -\frac{\int_0^1 u\frac{\partial u}{\partial y}\bigg|_{y=1} dx}{V^2} \qquad (6.44)$$

The Power can also be estimated by integrating the viscous dissipation over the cross section of the cavity.

The convergence as Δx decreases is poor and it is clear from that the estimates do not converge as $\Delta x \to 0$. The reason is, of course, that the problem as prescribed has discontinuities in the velocity boundary condition at the upper pair of corners. Each discontinuity is applied over a single cell which means that the problem being solved is one with a finite change in velocity over each corner cell, rather than one having a true discontinuity. As the mesh is changed the specification of these boundary conditions effectively changes. The estimate of power lost is sensitively affected by the way the discontinuity is represented, hence the poor convergence.

This example has been presented here to demonstrate that mesh convergence is not necessarily at the rate indicted by the order of the numerical approximations to the governing equations.

6.4 CONSERVATION LAWS, GOVERNING EQUATIONS AND FINITE VOLUME APPROXIMATIONS

The purpose of the discussion in this Section is to describe how conservation laws applied over control volumes can lead to the equations governing fluid motion and the relevant thermodynamics. As part of this presentation the finite volume method of generating approximate equations will be demonstrated. The approach is to use flow domain level conservation laws as the starting point in the processes used to generate the governing equations. This has been done to highlight the importance of these conservations laws which often are used to process the CFD solutions to provide interpretive data.

Conventional discussions[13,14] of the mathematical basis of CFD usually start with the equations of motion in partial differential equation form. They are derived from the laws of conservation of mass and energy and Newton's second law. Because a fluid is a moving medium, there are options regarding whether moving or fixed control volumes are used. Traditionally a volume moving with a fluid has been taken, which is especially relevant to Newton's law because it must be applied to a moving body or in this case an infinitesimal control volume containing a given fluid mass.

The perspective taken here and in recent texts on transport processes[15] is to start with stationary control volumes. An advantage of this approach is that the same discussion can be used to create numerical approximations, using the more common Lagrangian, or stationary, framework.

Another perspective is that there are many processes in food engineering for which the standard assumptions of fluid mechanics may not be relevant. Fluids may be non-Newtonian, stirring processes may involve complex body or internal forces that may not be conservative. Fluid properties may not be close to uniform. And there are a variety of heating sources. The conservation laws are presented initially in as much generality as possible to assist the reader to understand how these more complicated processes can be modeled.

6.4.1 CONSERVATION OF MASS

The conservation of mass in a stationary control volume that has no internal mass sources can be written as

$$\frac{\partial}{\partial t}\int_V \rho\, dv + \oint_A \rho \mathbf{v}.\mathbf{da} = 0 \tag{6.45}$$

The control volume illustrated in Figure 6.9 is of size V and contains fluid of spatially varying density ρ. An incremental element of area on the surface of the control volume is represented by the vector $\mathbf{da}$ which, by convention, is in the direction of the outward normal to the surface. The fluid moves through the boundary with velocity $\mathbf{v}$.

The first integration represents the rate of change of mass in the control volume. The second integration represents the rate of flow of mass through the control volume's boundaries.

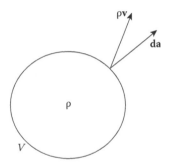

FIGURE 6.9   Conservation of mass in a control volume, V.

This equation is valid no matter how large or small the control volume may be. If the volume is vanishingly small the limiting form of Equation 6.45 is the partial differential equation that represents the conservation of mass; the so called continuity equation. If the control volume is very large it represents the conservation of mass in a large domain. At an intermediate level when the control volume is small but finite the equation can be represented approximately to become the basis of a numerical method.

It is very important to understand these three levels of representation are all expressions of the same law. This uniformity of concept is lost if one starts with the partial differential equations.

6.4.1.1 Large Control Volume

These principles can be illustrated in 2D Cartesian coordinates. Consider the 2D rectangular domain in Figure 6.10. Assuming that the density and the two velocity components are functions of x and y, Equation 6.45 becomes

$$\frac{\partial}{\partial t}\int_0^{L_y}\int_0^{L_x}\rho\,dx\,dy + \int_0^{L_y}(\rho u)_{x=L_x}\,dy - \int_0^{L_y}(\rho u)_{x=0}\,dy + \int_0^{L_x}(\rho v)_{y=L_y}\,dx - \int_0^{L_x}(\rho v)_{y=0}\,dx = 0 \qquad (6.46)$$

It is important to stress again that this is an exact expression, provided the integrations can be performed, and represents the global conservation of mass in the control volume. Equation 6.46 is valid for a control volume of any size.

6.4.1.2 Small Control Volume and a Finite Volume Approximation

If the control volume is very small we can make simplifying assumptions. Consider the control volume shown in Figure 6.11. The volume is sufficiently small that the density can be assumed to be represented by an average uniform density,

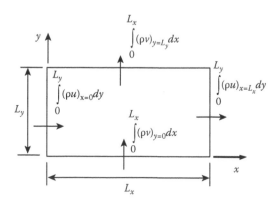

FIGURE 6.10   Stationary domain sized control volume representing the conservation of mass.

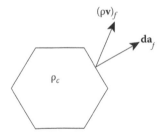

FIGURE 6.11   Conservation of mass in small but finite control volume.

$$\rho_c = \frac{1}{V} \int_V \rho dv \tag{6.47}$$

The boundary of the small volume can be represented by N plane surfaces, over each of which the momentum is constant. The mass conservation equation can be then be approximated by

$$V \frac{\partial \rho_c}{\partial t} + \sum_{f=1}^{N} (\rho \mathbf{v})_f \cdot \mathbf{da}_f = 0 \tag{6.48}$$

Equation 6.48 is a finite volume approximation to Equation 6.45. This is the kind of equation that is used to represent mass conservation in an unstructured grid constructed from arbitrary polyhedrons.

For a 2D rectangular control volume in Cartesian coordinates (Figure 6.12), Equations 6.46 or 6.48 become,

$$\frac{\partial}{\partial t} \int_y^{y+\Delta y} \int_x^{x+\Delta x} \rho dx dy + \int_y^{y+\Delta y} (\rho u)_{x=x+\Delta x} dy - \int_y^{y+\Delta y} (\rho u)_{x=0} dy + \int_x^{x+\Delta x} (\rho v)_{y=y+\Delta y} dx - \int_x^{x+\Delta x} (\rho v)_{y=y} dx = 0$$

$$\tag{6.49}$$

This can be replaced by

$$\frac{\partial \rho_c}{\partial t} \Delta x \Delta y + (\rho u)_e \Delta y - (\rho u)_w \Delta y + (\rho v)_n \Delta x - (\rho v)_s \Delta x = 0 \tag{6.50}$$

where

$$\rho_c = \frac{1}{\Delta x \Delta y} \int_y^{y+\Delta y} \int_x^{x+\Delta x} \rho dx dy \quad (\rho u)_w = \frac{1}{\Delta y} \int_y^{y+\Delta y} (\rho u)_{x=x} \quad (\rho u)_e = \frac{1}{\Delta y} \int_y^{y+\Delta y} (\rho u)_{x=x+\Delta x}$$

$$\tag{6.51}$$

$$(\rho v)_s = \frac{1}{\Delta x} \int_x^{x+\Delta x} (\rho v)_{y=y} \quad (\rho v)_n = \frac{1}{\Delta x} \int_x^{x+\Delta x} (\rho v)_{y=y+\Delta y}$$

As derived here the combination of Equations 6.50 and 6.51 is exact for the control volume. Equation 6.50 represents the conservation of mass. Equation 6.51 defines the face averaged mass fluxes used in Equation 6.50.

Equation 6.50 now contains an algebraic expression of the spatial mass balance for the cell. We have used a finite volume approach to generate this equation.

6.4.1.3 Infinitesimal Control Volume and a Partial Differential Equation

Equation 6.50 can be processed further by dividing by $\Delta x \Delta y$

$$\frac{\partial \rho_c}{\partial t} + \frac{(\rho u)_e - (\rho v)_w}{\Delta x} + \frac{(\rho v)_n - (\rho v)_s}{\Delta y} = 0 \tag{6.52}$$

If the control volume is allowed to shrink to infinitesimal size, the limiting form of Equation 6.52 is

$$\frac{\partial \rho}{\partial t} + \frac{\partial (\rho u)}{\partial x} + \frac{(\rho v)}{\partial y} = 0 \tag{6.53}$$

The equivalent vector form (which may be 3D) is

$$\frac{\partial \rho}{\partial t} + \nabla \cdot (\rho \mathbf{v}) = 0 \tag{6.54}$$

Equation 6.54 is often called the continuity equation and is, in fact, the starting point of a more traditional treatment.

6.4.1.3.1 Different Forms of the Continuity Equation

We can express Equation 6.54 in terms of the mass flux vector $\mathbf{m} = \rho \mathbf{v}$, which is also the momentum vector,

$$\frac{\partial \rho}{\partial t} + \nabla \cdot \mathbf{m} = 0 \tag{6.55}$$

This form of the mass conservation equation identifies $\mathbf{m}$ or $\rho \mathbf{v}$ as the mass transport vector.

Note that by comparing Equations 6.54 and 6.48 the following approximation for the divergence of momentum or in fact any vector follows.

$$\nabla \cdot \mathbf{m} \approx \frac{1}{V} \sum_{f=1}^{N} \mathbf{m}_f \cdot \mathbf{da}_f \tag{6.56}$$

Further insight can be obtained by processing Equation 6.54 using vector rules of differentiation,

$$\frac{\partial \rho}{\partial t} + \mathbf{v} \cdot \nabla \rho + \rho \nabla \cdot \mathbf{v} = 0 \tag{6.57}$$

The first two terms in Equation 6.57 represent what is called the material derivative which measures the rate of change of a quantity as it moves with the fluid (rather than being relative to a stationary frame of reference). Equation 6.57, becomes

$$\frac{D\rho}{Dt} + \rho \nabla \cdot \mathbf{v} = 0 \quad \text{where} \quad \frac{D}{Dt} \equiv \frac{\partial}{\partial t} + \mathbf{v} \cdot \nabla \tag{6.58}$$

Equation 6.58 demonstrated the use of the material derivative which represents, in a stationary coordinate system, the rate of change of a quantity as it moves with the fluid. It can be derived by applying control volume analysis to a moving element of fluid and using the Reynolds transport theorem to convert express the integration in terms of a stationary frame of reference.

6.4.2 MOMENTUM

The momentum equation can also be derived using the same control volume approach. For a stationary control volume, the net rate of generation of momentum in the control volume is equal to the sum of all forces acting on the control volume. This can be represented by

$$\frac{\partial}{\partial t} \int_V \mathbf{m} dv + \oint_A \mathbf{m}\mathbf{v} \cdot \mathbf{da} = \sum \mathbf{F} \tag{6.59}$$

where the sum is over all forces acting on the control volume (body forces or surface forces). The two terms on the left hand side represent the rate of increase of momentum in the control volume and the net flow of momentum leaving the control volume respectively. The vector **F** represents the various forces acting on the control volume. These may act on the control volume boundaries, or internally either on immersed bodies or directly on the fluid as a "body force."

As was the case with mass conservation, the conservation law expressed by Equation 6.59 can be applied at three levels of control volume size. At the whole domain level it can be used to calculate the macro momentum force balances of the flow. At the small but finite level it can be used to generate numerical approximations and at the infinitesimal level it can be used to generate partial differential equations.

At all levels the term **mv** needs to be interpreted. It is the convection of the vector **m** by the velocity **v**. For a 3D situation, each of the three components of **m** will be convected by each of the three components of **v**. Thus there are nine terms in the product **mv** and this is often called a tensor (although dyadic notation is used here). The simplest way of thinking about Equation 6.59 is to regard it as representing a set of equations for the components of **m** and **F**. For each of these components, the same analysis that was used for the mass conservation equation can be applied as will be demonstrated here for 2D flow.

6.4.2.1 Large Control Volume

At the domain level, if ρ in Equation 6.46 is replaced by m_x

$$\frac{\partial}{\partial t}\int_0^{L_y}\int_0^{L_x} m_x dx dy + \int_0^{L_y}(m_x u)_{x=L_x}\, dy - \int_0^{L_y}(m_x u)_{x=0}\, dy + \int_0^{L_x}(m_x v)_{y=L_y}\, dx - \int_0^{L_x}(m_x v)_{y=0}\, dx = F_x \quad (6.60)$$

and similarly for m_y

$$\frac{\partial}{\partial t}\int_0^{L_y}\int_0^{L_x} m_y dx dy + \int_0^{L_y}(m_y u)_{x=L_x}\, dy - \int_0^{L_y}(m_y u)_{x=0}\, dy + \int_0^{L_x}(m_y v)_{y=L_y}\, dx - \int_0^{L_x}(m_y v)_{y=0}\, dx = F_y \quad (6.61)$$

6.4.2.2 Small but Finite Control Volume

For a finite control volume the equation equivalent to Equation 6.48 is

$$V\frac{\partial \mathbf{m}_c}{\partial t} + \sum_{i=1}^{N}(\mathbf{vm})_i \cdot \mathbf{da}_i = \sum \mathbf{F} \quad (6.62)$$

For the 2D rectangular cell illustrated in Figure 6.13, Equation 6.62 reduces to

$$\frac{\partial m_{xc}}{\partial t} + \frac{(m_x u)_e - (m_x u)_w}{\Delta x} + \frac{(m_x v)_n - (m_x v)_s}{\Delta y} = \sum \frac{F_x}{\Delta x \Delta y}$$

$$(6.63)$$

$$\frac{\partial m_{yc}}{\partial t} + \frac{(m_y u)_e - (m_y u)_w}{\Delta x} + \frac{(m_y v)_n - (m_y v)_s}{\Delta y} = \sum \frac{F_y}{\Delta x \Delta y}$$

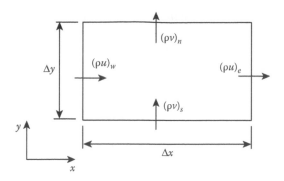

FIGURE 6.12   Finite rectangular control volume representing the conservation of mass.

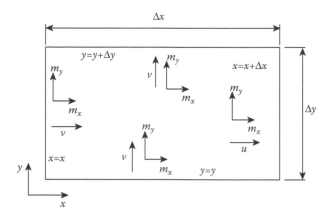

FIGURE 6.13   Momentum balance for a finite control volume.

6.4.2.3   Infinitesimal Control Volume

In the limit of a vanishingly small control volume

$$\frac{\partial m_x}{\partial t} + \frac{\partial (m_x u)}{\partial x} + \frac{(m_x v)}{\partial y} = f_x$$

$$\frac{\partial m_y}{\partial t} + \frac{\partial (m_y u)}{\partial x} + \frac{(m_y v)}{\partial y} = f_y$$

(6.64)

Equation 6.64 is the 2D component version of the more general vector equation

$$\frac{\partial \mathbf{m}}{\partial t} + \nabla \cdot (\mathbf{m}\mathbf{v}) = \rho \mathbf{f}$$

(6.65)

where $\mathbf{f}$ is the force per unit mass defined by

$$\mathbf{f} = \frac{1}{\rho} \lim_{V \to 0} \left(\frac{\sum \mathbf{F}}{V} \right)$$

(6.66)

Now

$$\frac{\partial \mathbf{m}}{\partial t} = \rho \frac{\partial \mathbf{v}}{\partial t} + \mathbf{v} \frac{\partial \rho}{\partial t}$$

$$= \rho \frac{\partial \mathbf{v}}{\partial t} - \mathbf{v} \nabla \cdot \mathbf{m} \quad \text{using the continuity equation}$$

(6.67)

Leading to

$$\rho \frac{\partial \mathbf{v}}{\partial t} + \mathbf{m} \nabla \cdot \mathbf{v} = \rho \mathbf{f} \tag{6.68}$$

or, using the more usual notation,

$$\rho \frac{\partial \mathbf{v}}{\partial t} + \rho \mathbf{v} \cdot \nabla \mathbf{v} = \rho \mathbf{f} \tag{6.69}$$

and finally

$$\frac{D \mathbf{v}}{Dt} = \mathbf{f} \tag{6.70}$$

This simple expression represents Newton's law applied to a moving infinitesimal element of fluid.

6.4.2.4   Force Intensity

The force per unit mass now needs to be evaluated. At the infinitesimal level the forces consist of the effects of body force such as gravity, magnetic, or rotational forces together with the internal stresses on the fluid. The body force is represented separately and the effects of internal stresses can be represented by the divergence of a tensor

$$\mathbf{f} = \mathbf{f}_b + \nabla \cdot \mathbf{\Pi} \tag{6.71}$$

The derivation of Equation 6.71 is described in many standard texts[15] on fluid mechanics or transport phenomenon. The stress tensor $\mathbf{\Pi}$ can be written as the sum of two terms, one representing the hydrostatic pressure, the other the stress resulting from fluid motion

$$\mathbf{\Pi} = -p\mathbf{I} + \mathbf{\sigma} \tag{6.72}$$

The contributors to these terms arising from the forces on an elemental 2D control volume are shown in Figure 6.14. Although not used directly here, understanding these contributions is important when formulating boundary conditions. Inserting Equation 6.71 into the momentum equation

$$\frac{\partial \mathbf{m}}{\partial t} + \nabla \cdot (\mathbf{mv}) = \rho \mathbf{f}_b + \nabla \cdot \mathbf{\Pi}$$

$$\Rightarrow \frac{\partial \mathbf{m}}{\partial t} + \nabla \cdot (\mathbf{mv} - \mathbf{\Pi}) = \rho \mathbf{f}_b$$

(6.73)

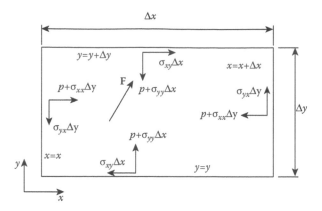

FIGURE 6.14 Forces on a small but finite control volume. The forces relate to the terms in Equation 6.72. The vector **F** is the total force on the control volume.

It is now convenient to introduce **M** as the momentum transport tensor defined by

$$\mathbf{M} = \mathbf{m}\mathbf{v} - \mathbf{\Pi}$$
$$= \mathbf{m}\mathbf{v} + p\mathbf{I} - \mathbf{\sigma} \tag{6.74}$$

In terms of **M**, the momentum equation is

$$\frac{\partial \mathbf{m}}{\partial t} + \nabla \cdot \mathbf{M} = \rho \mathbf{f}_b \tag{6.75}$$

It is the tensor **M** that characterizes much of a fluid and its flow behavior. More details of this approach can be found in Bird et al.[15] It needs to be clarified, however, that this discussion has used sign conventions that have been standard in the fluid dynamics literature. In contrast, Bird et al. define a tensor **τ** to represent the stresses due to fluid motion and that has the opposite sign to **σ** used here.

Note also that although the integral conservation laws have been used to derive the governing equations, it is still convenient to refer to the partial differential equations when describing the representation of physical phenomena. The partial differential equations are, of course, somewhat more compact than their integral counterparts.

6.4.3 ENERGY

The remaining law to be considered here is the conservation of energy. This can be expressed by

$$\frac{\partial}{\partial t} \int_V \rho e \, dv + \oint_A \rho e \mathbf{v} \cdot \mathbf{da} = \sum \dot{Q} - \sum \dot{W} \tag{6.76}$$

where e denotes the total internal energy per unit mass (or specific energy). Heat can flow into the control volume through the boundaries or as volumetric sources which means the summation includes volumetric and surface sources. Similarly work done by the control volume on its environment may occur at the boundaries or internally.

6.4.3.1 Total Energy

In this discussion, the total internal energy will be considered to be the sum of thermal and kinetic energies which are represented by the specific thermal and kinetic energies, respectively.

$$e = e_T + e_{KE} \tag{6.77}$$

where

$$e_{KE} = \frac{1}{2}|\mathbf{v}|^2 \qquad (6.78)$$

Note that this definition does not include potential energy as part of the total energy. This follows the conventions used in texts on the fundamentals of fluid mechanics[16] and transport processes.[15] As will be illustrated later, terms representing gravitational potential energy can be recovered from the body force work terms.

Contributions to the heat flow into the domain are the heat flux through the boundaries and volumetric heat generated within the domain

$$\sum \dot{Q} = -\oint_A \mathbf{q} \cdot \mathbf{da} + \int_V q''' dv \qquad (6.79)$$

Contributions to the work are the work against pressure at the boundaries, work done against viscous forces and work against the body force.

$$\sum \dot{W} = -\oint_A (\mathbf{v} \cdot \boldsymbol{\sigma}) \cdot \mathbf{da} + \oint_A \mathbf{v} p \cdot \mathbf{da} - \int_V \rho \mathbf{f}_b \cdot \mathbf{v} dv \qquad (6.80)$$

In general the body force may not be conservative. However, any arbitrary vector can be decomposed into conservative and nonconservative parts and the conservative part can be represented by a scalar potential,

$$\begin{aligned} \mathbf{f}_b &= \mathbf{f}_c + \mathbf{f}_{nc} \\ &= -\nabla\phi + \mathbf{f}_{nc} \end{aligned} \qquad (6.81)$$

This will be used later to extract the potential energy. For now the body force will be retained as the combination of both kinds of force.

Using these expressions for the heat and work rates produces the following form of the conservation of energy over a control volume.

$$\frac{\partial}{\partial t} \int_V \rho(e_T + e_{KE}) dv + \oint_A [(\rho e_T + \rho e_{KE} + p)\mathbf{v} - (\mathbf{v} \cdot \boldsymbol{\sigma}) + \mathbf{q}] \cdot \mathbf{da} = \int_V q''' dv + \int_V \mathbf{f}_b \cdot \mathbf{v} dv \qquad (6.82)$$

As was the case with the conservation of mass and momentum, this equation can be used at the domain level to calculate overall energy balances, at the finite cell level to generate finite volume approximations and at the infinitesimal level to generate partial differential equations. Accordingly the partial differential equation describing the conservation of total energy is

$$\frac{\partial \rho e}{\partial t} + \nabla \cdot [(\rho e + p)\mathbf{v} - (\mathbf{v} \cdot \boldsymbol{\sigma}) + \mathbf{q}] = q''' + \mathbf{f}_b \cdot \mathbf{v} \qquad (6.83)$$

The total energy transport vector can now be defined as

$$\mathbf{E} = (\rho e + p)\mathbf{v} - (\mathbf{v} \cdot \boldsymbol{\sigma}) + \mathbf{q} \qquad (6.84)$$

Leading to

$$\frac{\partial \rho e}{\partial t} + \nabla \cdot \mathbf{E} = q''' + \mathbf{f}_b \cdot \mathbf{v} \tag{6.85}$$

However to progress further it is usual to split this equation into mechanical energy and thermal energy transport equations. Note that these equations are referred to as transport rather than conservation equations to emphasize that the fundamental conservation law is that for total energy.

6.4.3.2 Mechanical Energy

The mechanical energy equation can be generated at the partial differential level by taking the dot product between the velocity and the momentum equation.

$$\mathbf{v} \cdot \frac{\partial \mathbf{m}}{\partial t} + \mathbf{v} \cdot (\nabla \cdot \mathbf{M}) = \rho \mathbf{v} \cdot \mathbf{f}_b \tag{6.86}$$

Substituting for $\mathbf{M}$ leads to

$$\mathbf{v} \cdot \frac{\partial \rho \mathbf{v}}{\partial t} + \mathbf{v} \cdot (\nabla \cdot (\rho \mathbf{v} \mathbf{v})) + \mathbf{v} \cdot \nabla p - \mathbf{v} \cdot (\nabla \cdot \boldsymbol{\sigma}) = \rho \mathbf{v} \cdot \mathbf{f}_b \tag{6.87}$$

Manipulation using vector and tensor identities leads to

$$\frac{\partial \rho e_{KE}}{\partial t} + \nabla \cdot \mathbf{E}_M = \rho \mathbf{v} \cdot \mathbf{f}_b + p \nabla \cdot \mathbf{v} - \Omega \tag{6.88}$$

where the mechanical energy transport vector is defined by

$$\mathbf{E}_M = [\rho e_{KE} + p] \mathbf{v} - \mathbf{v} \cdot \boldsymbol{\sigma} \tag{6.89}$$

and a generalized viscous dissipation term

$$\Omega = \nabla \cdot (\mathbf{v} \cdot \boldsymbol{\sigma}) - \mathbf{v} \cdot (\nabla \cdot \boldsymbol{\sigma}) \tag{6.90}$$

Equation 6.88 is complete because no assumptions have yet been made regarding the constitutive laws that define the properties of $\boldsymbol{\sigma}$ and the body force vector is still in its general form. Equation 6.88 can be integrated to produce a control volume equation

$$\frac{\partial}{\partial t} \int_V \rho e_{KE} dv + \oint_A \mathbf{E}_M \cdot \mathbf{da} = \int_V [\mathbf{f} \cdot \mathbf{v} + p \nabla \cdot \mathbf{v} - \Omega] dv \tag{6.91}$$

Equations 6.88 and 6.91 are respectively, the partial differential and integral transport equations for mechanical energy.

6.4.3.3 Thermal Energy

Equations 6.82 and 6.91 can be used to create a transport equation for thermal energy

$$\frac{\partial}{\partial t} \int_V \rho e_T dv + \oint_A \mathbf{E}_T \cdot \mathbf{da} = \int_V q''' dv - \int_V (p \nabla \cdot \mathbf{v} - \Omega) dv \tag{6.92}$$

where the thermal transport vector is

$$\mathbf{E}_T = \rho e_T \mathbf{v} + \mathbf{q} \tag{6.93}$$

The corresponding partial differential equation is

$$\frac{\partial \rho e_T}{\partial t} + \nabla \cdot \mathbf{E}_T = q''' - p\nabla \cdot \mathbf{v} + \Omega \tag{6.94}$$

It follows that

$$\mathbf{E} = \mathbf{E}_M + \mathbf{E}_T \tag{6.95}$$

6.4.3.4 Special Case of Conservative Steady Body Force and Negligible Viscous Dissipation

For the case when the body force is conservative and steady (e.g., the Earth's gravitational field) the total energy can be defined as

$$e = e_T + e_{\mathrm{KE}} + \phi \tag{6.96}$$

and the body force is set to zero in Equation 6.88 i.e.,

$$\frac{\partial}{\partial t}\rho(e_{\mathrm{KE}} + \varphi) + \nabla \cdot \mathbf{E}_{M\varphi} = p\nabla \cdot \mathbf{v} - \Omega \tag{6.97}$$

where

$$\mathbf{E}_M = [\rho(e_{\mathrm{KE}} + \varphi) + p]\mathbf{v} - \mathbf{v} \cdot \boldsymbol{\sigma} \tag{6.98}$$

Making the additional assumption that the terms representing the generation of heat from mechanical effects can be neglected, the terms $p\nabla \cdot \mathbf{v}$ which represents the reversible conversion between thermal and into kinetic energy, and Ω which represent irreversible conversion of mechanical energy into heat, can be removed from the energy equations which become

$$\frac{\partial}{\partial t}\rho(e_{\mathrm{KE}} + \varphi) + \nabla \cdot \mathbf{E}_{M\varphi} = 0 \tag{6.99}$$

and

$$\frac{\partial \rho e_T}{\partial t} + \nabla \cdot \mathbf{E}_T = q''' \tag{6.100}$$

In many situations Equation 6.100 is used as the thermal energy equation and is often the first equation that comes to mind at the "thermal energy equation."

6.4.4 Constitutive Equations

The previous sections have established that the conservation of mass, momentum and energy is characterized in each case by a transport vector. For convenience these are summarized in Table 6.4. To progress further, relationships are needed that enable the tensors or vectors that represent molecular

TABLE 6.4
Summary of Transport Vectors and Equations for Quantities used to Model Food Engineering

Transported Quantity	Transport Vector	Transport Equation
Mass ρ	$\mathbf{m} = \rho\mathbf{v}$	$\dfrac{\partial \rho}{\partial t} + \nabla \cdot \mathbf{m} = 0$
Momentum $\mathbf{m} = \rho\mathbf{v}$	$\mathbf{M} = \mathbf{mv} - \Pi$ $= \mathbf{mv} + p\mathbf{I} - \boldsymbol{\sigma}$	$\dfrac{\partial \mathbf{m}}{\partial t} + \nabla \cdot \mathbf{M} = \rho\mathbf{f}_b$
Total energy $\rho e : e = e_T + \dfrac{1}{2}\lvert\mathbf{v}\rvert^2$	$\mathbf{E} = (\rho e + p)\mathbf{v} - (\mathbf{v} \cdot \boldsymbol{\sigma}) + \mathbf{q}$	$\dfrac{\partial \rho e}{\partial t} + \nabla \cdot \mathbf{E} = q''' + \mathbf{f}_b \cdot \mathbf{v}$
Thermal energy ρe_T	$\mathbf{E}_T = \rho e_T \mathbf{v} + \mathbf{q}$	$\dfrac{\partial \rho e_T}{\partial t} + \nabla \cdot \mathbf{E}_T = q''' - p\nabla \cdot \mathbf{v} + \Omega$
Mechanical energy ρe_{KE}	$\mathbf{E}_M = [\rho e_{KE} + p]\,\mathbf{v} - \mathbf{v} \cdot \boldsymbol{\sigma}$	$\dfrac{\partial \rho e_{KE}}{\partial t} + \nabla \cdot \mathbf{E}_M = \rho\mathbf{v} \cdot \mathbf{f}_b + p\nabla \cdot \mathbf{v} - \Omega$
Mass of a component in a mixture	$\mathbf{N}_i = \rho\mathbf{v}m_i + \mathbf{j}_i$	$\dfrac{\partial \rho m_i}{\partial t} + \nabla \cdot \mathbf{N}_i = (\rho m_i)'''$
Entropy	$\mathbf{S} = \rho s\mathbf{v} + \dfrac{\mathbf{q}}{T}$	$\dfrac{\partial \rho s}{\partial t} + \nabla \cdot \mathbf{S} = s'''$

diffusion to be evaluated. Those of most interest here are the deviatoric stress tensor $\boldsymbol{\sigma}$ and the heat flux vector $\mathbf{q}$. Because it is the simpler situation, $\mathbf{q}$ will be considered first.

6.4.4.1 Thermal Energy

Fourier's law relates $\mathbf{q}$ to the temperature T,

$$\mathbf{q} = -k\nabla T \tag{6.101}$$

The thermal energy transport vector is,

$$\mathbf{E}_T = \rho e_T \mathbf{v} - k\nabla T \tag{6.102}$$

In order to produce an equation for temperature a relationship between e_T and temperature needs to be established. As discussed by Bird et al.[15] (p. 337) conventional understanding of this process involves the assumption of a Newtonian fluid and a restricted form of Equation 6.94 is,

$$\frac{\partial \rho C_V T}{\partial t} + \nabla \cdot \mathbf{E}_T = q''' - p\nabla \cdot \mathbf{v} + \Omega : \mathbf{E}_T = \rho C_V T\mathbf{v} - k\nabla T \tag{6.103}$$

For an incompressible liquid and negligible viscous dissipation.

$$\rho C_P \frac{\partial T}{\partial t} + \nabla \cdot \mathbf{E}_T = q''' \ : \ \mathbf{E}_T = \rho C_P T\mathbf{v} - k\nabla T \tag{6.104}$$

An alternative form of Equation 6.104 is

$$\rho C_P \left(\frac{\partial T}{\partial t} + \nabla \cdot (T\mathbf{v}) \right) = \nabla \cdot (k\nabla T) \tag{6.105}$$

In this context the important observation is that the divergence of the transport vector has produced a convection and a diffusion term, i.e.,

$$\nabla \cdot \mathbf{E}_T = \nabla \cdot (\rho C_p T \mathbf{v}) - \nabla \cdot (k \nabla T) \tag{6.106}$$

6.4.4.2 Momentum

In principle Equation 6.75 can model the momentum force balance in any situation, even turbulence. To reiterate, the momentum equation is

$$\frac{\partial \mathbf{m}}{\partial t} + \nabla \cdot \mathbf{M} = \rho \mathbf{f}_b \tag{6.107}$$

The divergence of the momentum transport tensor is

$$\nabla \cdot \mathbf{M} = \nabla \cdot (\mathbf{mv}) + \nabla p - \nabla \cdot \boldsymbol{\sigma} \tag{6.108}$$

To model turbulence the equation is simply modified to account for the fluctuating velocities and the effects of turbulence on the stress tensor. For now, laminar flow of Newtonian fluids will be considered and for these, the stress is related to the velocity gradients via the Stokes assumption.

What is essentially a linear relationship between the nine terms in the stress tensor and nine velocity gradients could involve 81 coefficients. However the number of independent coefficients can be shown to be two and that the stress tensor is symmetric. In vector notation the relationship can be written as

$$\boldsymbol{\sigma} = \mu \left(\nabla \mathbf{v} + (\nabla \mathbf{v})^{\mathrm{T}} - \frac{2}{3} \nabla \cdot \mathbf{vI} \right) + \kappa \nabla \cdot \mathbf{vI} \tag{6.109}$$

and the divergence of the momentum transport vector is

$$\nabla \cdot \mathbf{M} = \nabla \cdot (\mathbf{mv}) + \nabla p - \nabla \cdot \mu \left(\nabla \mathbf{v} + (\nabla \mathbf{v})^{\mathrm{T}} - \frac{2}{3} (\nabla \cdot \mathbf{v}) \mathbf{I} \right) - \nabla \cdot (\kappa (\nabla \cdot \mathbf{v}) \mathbf{I}) \tag{6.110}$$

It is not the intention to discuss the various forms that Equation 6.110 might take. There is an enormous number of options, depending on the type of flow. Using vector identities to process Equation 6.110,[17]

$$\nabla \cdot \mathbf{M} = \nabla \cdot (\mathbf{mv}) + \nabla p - \nabla \mu \cdot \left[\left(\nabla \mathbf{v} + (\nabla \mathbf{v})^{\mathrm{T}} \right) - \frac{2}{3} (\nabla \cdot \mathbf{v}) \mathbf{I} \right]$$

$$- \mu \left[\nabla^2 \mathbf{v} + \frac{1}{3} \nabla ((\nabla \cdot \mathbf{v}) \mathbf{I}) \right] + \nabla \kappa \cdot (\nabla \cdot \mathbf{v}) \mathbf{I} + \kappa \nabla ((\nabla \cdot \mathbf{v}) \mathbf{I}) \tag{6.111}$$

So that for incompressible flows

$$\nabla \cdot \mathbf{M} = \nabla \cdot (\mathbf{mv}) + \nabla p - \nabla \mu \cdot \left[\left(\nabla \mathbf{v} + (\nabla \mathbf{v})^{\mathrm{T}} \right) \right] - \mu \nabla^2 \mathbf{v} \tag{6.112}$$

If the viscosity is constant,

$$\nabla \cdot \mathbf{M} = \nabla \cdot (\mathbf{mv}) + \nabla p - \mu \nabla^2 \mathbf{v} \tag{6.113}$$

Inserting this result in Equation 6.107 produces

$$\frac{\partial \mathbf{m}}{\partial t} + \nabla \cdot (\mathbf{mv}) = \rho \mathbf{f}_b - \nabla p + \mu \nabla^2 \mathbf{v} \tag{6.114}$$

which is the very familiar form of the momentum equation for an incompressible fluid with uniform viscosity.

6.4.4.3 Non-Newtonian Fluids

Since many phenomena in food engineering involve non-Newtonian fluids it is necessary to consider how the constitutive equations are modified to accommodate departure from the Stokes model.

For example the generalized non-Newtonian models assume that the viscosity is dependent on the magnitude of rate of strain tensor,

$$\boldsymbol{\sigma} = \eta(\dot{\gamma})\left|\nabla\mathbf{v}+(\nabla\mathbf{v})^{\mathrm{T}}\right| \tag{6.115}$$

The function $\eta(\dot{\gamma})$ is estimated from physical measurements.

6.4.5 CONSERVATION EQUATIONS FOR OTHER QUANTITIES

This discussion has presented detailed descriptions of the equations governing the conservation of mass, momentum and energy, because they are all inexorably linked in the modeling of fluid motion.

There are other quantities that are of interest to a thermodynamicist or a food engineer. It will be assumed that the transport equations for such quantities can be written in the form

$$\frac{\partial\rho\varphi}{\partial t}+\nabla\cdot\Phi = \varphi''' \tag{6.116}$$

where φ is the specific quantity, Φ its transport vector and φ''' its volumetric rate of generation.

For example, the transport of a component i of a mixture can be represented by

$$\frac{\partial\rho m_i}{\partial t}+\nabla\cdot\mathbf{N}_i = (\rho m_i)''' \tag{6.117}$$

and the transport vector is

$$\mathbf{N}_i = \rho\mathbf{v}m_i + \mathbf{j}_i \tag{6.118}$$

where $\mathbf{j}_i$ is the mass diffusion vector for species i.

Similarly the transport of entropy can be written as

$$\frac{\partial\rho s}{\partial t}+\nabla\cdot\mathbf{S} = s''' \tag{6.119}$$

with the transport vector[4]

$$\mathbf{S} = \rho s\mathbf{v}+\frac{\mathbf{q}}{T} \tag{6.120}$$

6.4.6 BOUNDARY CONDITIONS AND GLOBAL CONSERVATION

The previous Section established a set of governing equations that used three levels of interpretation, the global domain level, the finite approximation level and the infinitesimal PDE level. The equations have been presented in their most general forms since these are the forms that are modeled by the most comprehensive packages and the user needs to be cognizant of the assumptions that are made with simpler models. The user also needs to be aware of what terms or effects are being turned on with various options.

The fluid motion and transport processes are, however, determined by more than just the governing equations; they are also dependent on conditions applied on the domain boundaries. Boundary conditions must be applied correctly and for every problem there is a correct number of boundary conditions that uniquely specify the problem. If too many boundary conditions are applied then the problem is over-specified. If there are too few, then the problem is under-specified.

Boundary conditions are inexorably related to the domain conservation laws and will be described here in that context.

6.4.6.1   Mathematical Boundary Conditions

The equations governing fluid motion are generally second order partial differential equations for scalar variables or scalar components of vectors. For an arbitrary scalar function φ the possible boundary conditions are specification of the value of one of: φ (Dirichlet), the gradient of φ (Neumann) or a linear combination of φ and its gradient (Robin). The mathematical descriptions of these three types of boundary conditions are

$$\begin{aligned} \varphi\,|_{boundary} &= b_{spec} &&\text{Dirichlet} \\ \nabla\varphi \cdot \hat{\mathbf{n}}|_{boundary} &= b_{spec} &&\text{Neumann} \\ (\lambda_1\varphi + \lambda_2\nabla\varphi \cdot \hat{\mathbf{n}})\,|_{boundary} &= b_{spec} &&\text{Robin} \end{aligned} \tag{6.121}$$

Specification of these boundary conditions is related to the domain level conservation laws as described in the following Sections.

6.4.6.2   Mass Conservation

The specification of mass flow though a boundary is via a Dirichlet boundary condition on the component of $\mathbf{m}$ normal to the boundary.

$$\mathbf{m} \cdot \hat{\mathbf{n}} = b_{spec} \tag{6.122}$$

The application of the mass conservation law by applying Equation 6.45 over a domain, such as that illustrated in Figure 6.10, implies that the mass flow cannot be specified independently over all sections of a boundary. This is considered for closed and open domains separately below.

6.4.6.2.1   Closed Domains

A closed domain has no mass flow through any section of its boundary. The two examples described in Figure 6.1 are closed domain flows. All boundaries are impervious which means that at any point on the boundary there is no component of velocity or mass flux normal to the boundary at that point.

6.4.6.2.2   Open Domains

Open domain have regions of their boundaries through which fluid can pass. The total mass flow into the domain is either zero or equal to the rate of increase of mass inside the volume. For most domains the rate of increase in mass is zero which means that there is no net mass flow into the domain. The implication of this statement is that if there are several regions of the boundary through which fluid can pass then the individual mass flows through these regions can not all be specified independently. One must be left unspecified and its net mass flow will be calculated as part of the solution process.

The simplest example is flow through a duct or pipe. The net mass flow through either the inlet or outlet can be specified. The flow through the other outlet will be constrained to be equal to that mass flow via application of the equation of continuity.

6.4.6.3   Momentum–Force–Pressure

The momentum transport equation is a vector equation which can be considered as a set of scalar equations for the components of momentum. Setting boundary conditions for these equations is complicated by the fact that there are restrictions on **m** implied by mass conservation and the boundary conditions can involve the transport of momentum or the application of tangential forces (shear) or normal forces (pressure). Although many boundary phenomena are such that setting an appropriate boundary condition is obvious there are others where the selection of the most appropriate boundary condition is difficult or even controversial.[18]

6.4.6.3.1   Tangential Components

Boundary conditions on the tangential components of **m** or **v** are usually easier to apply and will be discussed first. These boundary conditions often result from an assumption regarding shear stress which may be specified implicitly or directly. The implicit condition arises when the boundary is such that the fluid adjacent to the boundary moves at its velocity. This usually is associated with a *rigid* boundary. Often this is described as the boundary exerting shear by forcing the fluid to move with it. For example the boundary condition at the sliding surface in a lid driven cavity flow (Figure 6.1(b)) is the specification that the fluid adjacent to it moves at the boundaries velocity. The boundary exerts a shear force, but this force is calculated as part of the solution. This type of boundary condition is specified as Dirichlet conditions on the tangential components of **m** or **v**.

$$\mathbf{m} \cdot \hat{\mathbf{t}} = b_{\text{spec}} \quad \text{or} \quad \mathbf{v} \cdot \hat{\mathbf{t}} = b_{\text{spec}} \tag{6.123}$$

The specification may not necessarily be such that is uniform over the boundary. For example, a boundary that is rotating in its plane, e.g., the rotating top of a cylinder applies a condition on the normal component of vorticity,

$$(\nabla \times \mathbf{v}) \cdot \hat{\mathbf{t}} = b_{\text{spec}} \tag{6.124}$$

This boundary condition applies conditions on the tangential derivatives of the tangential components of **v**, which can be readily verified to be Dirichlet conditions for **v**.

The most common condition of this type is that the boundary condition is stationary and hence b_{spec} in Equation 6.123 is zero.

A direct specification of shear arises when a domain boundary is nonrigid. Usually the fluid at such a boundary is in contact with the same fluid (on the other side of the boundary) or another fluid which via its motion can exert shear. Hence the boundary condition is of the Neumann type,

$$\nabla(\mathbf{v} \cdot \hat{\mathbf{t}})\hat{\mathbf{n}}\,|_{\text{boundary}} = b_{\text{spec}} \tag{6.125}$$

where b_{spec} is proportional to the local value of shear stress. If the shear stress is zero then Equation 6.125 implies that the normal gradient of the tangential component of velocity is zero.

If there is no through flow through a boundary and the shear stress is zero then the boundary is called a symmetry boundary. A symmetry boundary has the useful property that there is no propagation of flow kinematics across the boundary and the flows on either side can be modeled separately.

6.4.6.3.2   Normal Components

As already mentioned in Section 6.4.6.2.2, a specification of the normal component of **m** or **v** at a boundary involves consideration of mass conservation. If the mass flux crossing the boundary can be specified then the resulting description of the variation of the normal velocity component over the boundary together with relevant specifications for the tangential components completes the specification. (Note that a swirling flow entering a domain requires specification of normal and tangential components of velocity at the inlet.)

When the satisfaction of mass conservation implies that the normal component of velocity cannot be directly specified then the typical practice is to specify the pressure at the boundary. As discussed in Section 6.4.7.1, this convention arises from experience with the Semi-Implicit Method for Pressure Linked Equations (SIMPLE) algorithm which effectively imposes a Neumann condition on the normal component of velocity at such a boundary. The default has been to apply a uniform pressure at an inlet or outlet when the normal velocity cannot be specified. However as discussed, for example, by Sani and Gresho[18] this is a simplification and alternatives such as specification of the normal gradient of the normal velocity component (as in the fully developed condition) may yield more satisfactory results.

Whatever boundary conditions are chosen at inlets and outlets, they are easy to apply only when the flow conditions are known to be relatively simple. If this cannot be assured then it is wise to provide extension of the domain to locations where the flow boundary conditions can be more easily specified. Failure to do this is a common weakness of CFD simulations.

6.4.6.4 Energy

The most common form of the energy equation is Equation 6.105. Relevant boundary conditions are:

$$
\begin{aligned}
T\,|_{\text{boundary}} &= T_{\text{spec}} & & \text{Dirichlet} \\
k\nabla T \cdot \hat{\mathbf{n}}\,|_{\text{boundary}} &= -\mathbf{q}_{\text{spec}} \cdot \hat{\mathbf{n}} & & \text{Neumann} \\
k\nabla T \cdot \hat{\mathbf{n}}\,|_{\text{boundary}} &= -h(T - T_\infty) & & \text{Robin}
\end{aligned}
\tag{6.126}
$$

where $\mathbf{q}$ represents heat flux and h a heat transfer coefficient.

For the natural convection example, the boundary conditions are

$$
\begin{aligned}
T_{x=0} &= T_{\text{hot}} & T_{x=L_x} &= T_{\text{cold}} \\
\frac{\partial T}{\partial y}\Big|_{y=0} &= 0 & \frac{\partial T}{\partial y}\Big|_{y=L_y} &= 0
\end{aligned}
\tag{6.127}
$$

If the Robin condition is used, the heat transfer coefficient h may represent quite complex or nonlinear processes, such as radiation, condensation, evaporation, or freezing.

6.4.6.5 Global or Domain Level Conservation

The discussion in Section 6.4.1 commenced with the conservation laws applied to a whole domain. It is essential that CFD solutions satisfy domain level conservation laws for two reasons. The first is for modeling consistency. The second reason is that often the CFD model has been constructed to predict behavior at the domain level. For example, the important information sought from a CFD analysis of air flow in an oven is the rate of heat transfer into the product, or the heat losses though the walls of the oven. The details of the movement of the air are of interest to explain how the heat transfer is occurring and may help toward designs to improve the oven's performance, but it is the overall heat transfer or efficiency of the oven that are paramount.

6.4.6.5.1 Global Conservation

The term global conservation is used here to refer to modeling consistency. Each finite volume, finite element or finite difference approximation represents a local application of a conservation law. If these are integrated or summed over a domain, then the result should represent domain level conservation and nothing else. There should not be terms resulting from discrepancies between the representation of fluxes between local control volumes sharing faces. Such terms result in sources or sinks of the conserved quantity.

For example, the central difference approximation to terms such as $u(\partial T/\partial x)$ which was used in early finite difference approximations for convection led to approximations that did not satisfy the global conservation requirement. Those based on $(\partial uT/\partial x)$ do. CFD methods using the latter approximations proved to be more stable as the Reynolds number was increased. A major advantage of the finite volume method for generating the approximation equations is that, by construction, it ensures that expressions used to calculate fluxes through a boundary are shared by the cells sharing that boundary, thereby satisfying the global conservation requirement.

6.4.6.5.2 Domain Level Conservation

Following from the discussion of global conservation, if all the algebraic equations that represent a CFD model are integrated or summed over the domain, the resulting equations represent conservation laws at the domain level. If these equations are used, the conservation laws should be satisfied to machine precision. If they don't there is something wrong with the model or its implementation. Most CFD packages report global conservation "balances." These should be inspected and used to check that the modeling is correct and that the solution has converged. Note that it is possible for residuals to decay to near zero, but for global conservation to be out of balance.

6.4.6.5.3 Estimation of Boundary Fluxes

The comments regarding global conservation lead to some interesting observations regarding the calculation of fluxes at a boundary. Consider the question of evaluating heat flux through a boundary, for example, on the $x = 0$ plane for the convection example in Figure 6.1(a). The local heat flux into the boundary is

$$q_{x=0} = \left|\mathbf{q}\right|_{x=0} = -k \left.\frac{\partial T}{\partial x}\right|_{x=0} \tag{6.128}$$

The total rate of heat transfer is

$$\dot{Q}_{x=0} = \int_0^{L_y} q\,dy = -\int_0^{L_y} k \left.\frac{\partial T}{\partial x}\right|_{x=0} dy \tag{6.129}$$

If the CFD algebraic equations are summed over the domain there will be an expression that represents the heat flow through this boundary. This expression will have, within it, expressions for the temperature gradient.

For a finite volume method, the expressions for the gradient may be first or second order estimates depending on how the finite volume equations have been set up at the boundaries. Finite volume methods normally have cell faces coincident with the boundary. This ensures that conservation at the cell level yields global conservation. If a flux boundary condition is applied, the face flux approximations may be modified for the boundary faces, or a "chimera" volume external to the domain is used. The integration will be a summation, corresponding to trapezoidal integration.

An alternative method for estimating the heat flow rate is to ensure that a second or higher order approximation is used to estimate the gradient and then use Simpson's rule to perform the integration.[9] This approach can produce a "more accurate" estimation of the heat transfer through the boundary. However, it may not be an estimate that satisfies conservation exactly.

In practice, the two different approaches should produce estimates that are within the error bounds produced by a mesh convergence analysis. If they are not, then this could be an indication that the mesh resolution is too coarse.

6.4.7 MOMENTUM PRESSURE EQUATIONS

The fluid dynamics equations are special in that the momentum and mass conservation equations form a set that have to be solved together. Traditionally each of the four equations (the three components of velocity and pressure) have been solved as a separate linear system, and then iterations have been performed to effect the non-linear coupling between them. The SIMPLE algorithm[19,20] and its variations have been the mainstay of this segregated solver approach, which has been commonly used in research codes and the older commercial packages (PHOENICS, CFX-4, Star-CFD, Fluent). Modern commercial solvers such as ANSYS-CFX, Fluent and Star-CCM + solve the whole set together as a single linear system in a fully coupled solver. However they still must iterate to solve for the nonlinear convective term for momentum.

This discussion will illustrate the ideas behind these methods for steady incompressible flow with constant properties. The mass conservation equation (6.55) reduces to

$$\nabla \cdot \mathbf{m} = 0 \tag{6.130}$$

For the purpose of illustrating the issues and solver methodologies the momentum equation,

$$\frac{\partial \mathbf{m}}{\partial t} + \nabla \cdot \mathbf{M} = \rho \mathbf{f}_b \tag{6.131}$$

will be written in the form

$$\frac{\partial \mathbf{m}}{\partial t} + \nabla \cdot \mathbf{M}'(\mathbf{v}) = -\nabla p + \rho \mathbf{f}_b \tag{6.132}$$

where

$$\begin{aligned} \mathbf{M}'(\mathbf{v}) &= \mathbf{M} - p\mathbf{I} \\ &= \mathbf{mv} - \mathbf{\sigma} \end{aligned} \tag{6.133}$$

and $\mathbf{M}'$ has been written as a function of $\mathbf{v}$ to emphasize that it depends on the velocity field. To put this notation in context with more conventional descriptions of the Navier–Stokes equations, for constant fluid properties

$$\mathbf{M}'(\mathbf{v}) = \mathbf{mv} - \mu \nabla \mathbf{v} \tag{6.134}$$

and Equation 6.131 becomes

$$\frac{\partial \mathbf{m}}{\partial t} + \nabla \cdot (\mathbf{mv}) = \rho \mathbf{f}_b - \nabla p + \mu \nabla^2 \mathbf{v} \tag{6.135}$$

The issue that the solver algorithms need to address is how to create equations for three velocity components and pressure from these two equations, and this process is described using Equations 6.130 and 6.132.

The approach that has been used for several decades is to "invent" an equation that can be used to couple pressure to the continuity equation.

One technique is that originally was developed by Chorin[21] introduces an artificial compressibility. As described by Chorin, the continuity equation for an incompressible fluid was represented by

$$\frac{\partial \rho}{\partial t} + \nabla \cdot \mathbf{v} = 0 \tag{6.136}$$

and the density was related to the pressure by an artificial equation of state,

$$p = \frac{\rho}{\delta} \tag{6.137}$$

Equations 6.131 and 6.136 are now a system that can be stepped through time to reach steady state.

A more popular technique has been to develop a pressure correction equation which is derived from the equation of continuity and this is that basis of the SIMPLE segregated solver.

6.4.7.1 The Semi-Implicit Method for Pressure Linked Equation (SIMPLE) Segregated Solver

The SIMPLE algorithm uses an equation of the form

$$\nabla^2 p' = C\nabla \cdot \mathbf{m}* \tag{6.138}$$

to compute a pressure correction to create a correction to a velocity or momentum field to make it divergence free. The velocity correction is

$$\mathbf{v}' = -C\nabla p' \tag{6.139}$$

The essential steps in the SIMPLE algorithm are

1. Given $\mathbf{v}^n$, p^n, the velocity and pressure at the nth time step, interpolate to get the cell face mass flow rates, $\dot{m}_f$.
2. Solve the discrete versions of the momentum equations to get a an approximation to $\mathbf{v}^{n+1}$. This approximation, denoted here by $\mathbf{v}*$ will not necessarily satisfy the continuity equation.
3. Interpolate to get the cell mass flow rates, $\dot{m}_f^*$ If the cell mass flow rates are mass conserving finish.
4. A pressure correction is then calculated using a discrete approximation to Equation 6.138
5. A correction to the velocity is then calculated using a discrete approximation to Equation 6.139.
6. New velocity and pressure fields are computed using

$$\mathbf{v}^{n+1} = \mathbf{v}^* + \mathbf{v}'$$
$$p^{n+1} = p^* + \alpha p' \tag{6.140}$$

   where α is a relaxation parameter.
7. Return to step 2 and repeat until the estimate for $\mathbf{v}^{n+1}$ and p^{n+1} converge.

The SIMPLE solver will be described in a little more detail using a mesh where the velocities and pressure are both defined at the cell centers.* A suitable conservation cell is illustrated schematically in Figure 6.15. The cell center is the point with index P. The surrounding N faces have indices f. Each adjacent cell has its center at a point with index c, f. Face values of variables are interpolated as required.

The derivation starts with the discrete form of the continuity equation

$$\sum_{f=1}^{N} \dot{m}_f = 0 \; : \; \dot{m}_f = \mathbf{m}_f \cdot \mathbf{da}_f \tag{6.141}$$

---

* Note that this differs from the original derivation where the pressures were at cell centers and the velocities were at cell faces, the so called staggered-mesh approach.

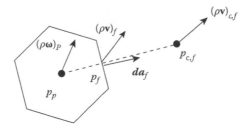

FIGURE 6.15 Schematic of a finite volume cell used to explain the SIMPLE algorithm.

and the discrete momentum equations which can be written as

$$V_p \left. \frac{\partial \mathbf{m}}{\partial t} \right|_{\text{Num}} + \sum_{f=1}^{N} \mathbf{M}'_f(\mathbf{v}_p, \mathbf{v}_{c,f}) \cdot \mathbf{da}_f = -\sum_{f=1}^{N} p_f \mathbf{n}_f da_f + V_p \mathbf{f}_p \qquad (6.142)$$

where the notation indicates that the momentum transport vector at each face depends on the velocity vector values stored at the mesh points and that the time derivative is represented by a numerical procedure which for the purposes of this discussion can be defined later. The form of the expression used to evaluate the transport vector at a face will depend on the approximation to the constitutive laws and the convection approximation used to interpolate the cell center velocities onto the faces. The face pressure is found by a linear interpolation between adjacent nodes. As discussed in Section 6.4.7.2, special care is required to calculate the face mass fluxes.

The discrete version of Equation 6.138 is

$$\sum_{f=1}^{N} \dot{m}_f = C \sum_{f=1}^{N} \left(p'_{c,f} - p'_p \right) da_f \qquad (6.143)$$

The set of equations for all cells can be solved to produce p' and then $\mathbf{v}''$. The relaxation factor used in the 6th step of the SIMPLE algorithm is necessary to reduce numerical instability. However the mass fluxes are not relaxed, ensuring that the mass flux fields are mass conserving.

Note that there are two flow fields in the solution: the momentum (conservation of momentum) equations are applied to the cell centered velocity field, whilst the continuity (conservation of mass) equation is applied to the cell face mass flux field. Therefore the cell centered velocity field $\mathbf{v}$ together with the pressure field p conserves momentum, but in general it will not be mass conserving. The cell face mass flux field $\dot{m}_f$ is mass conserving but will not conserve linear momentum.

The transient derivative is represented by an implicit time discretization (e.g., backward Euler or Crank–Nicolson) and is, as a consequence, stable. Because of the nonlinearities in the problem (the unknown face velocities) the solver must iterate on each time step to update the flow field.

For a steady state problem the sequence of iterates uses a steady state form of Equation 6.142 and the sequence continues until the state fields converge.

If the energy equation, or other scalars are required they can be solved after the flow field is calculated provided that they do not affect the flow field in any way. However, if they do affect the flow field (for example the turbulent eddy viscosity depends on the k and ε scalar fields, whilst buoyancy forces depend on T) they are solved between steps 6 and 7.

(a)   (b)

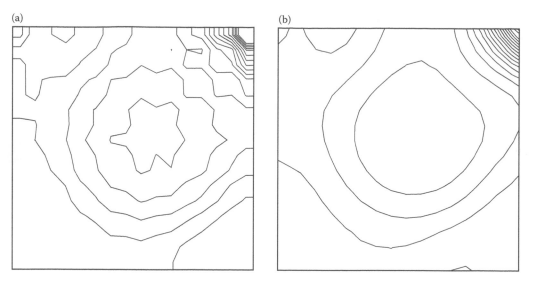

FIGURE 6.16   (a) Driven cavity pressure field calculated using linear and (b) Rhie–Chow velocity interpolation. Note the checker-boarding in the solution calculated using linear interpolation.

6.4.7.2 Mass Flux Interpolation

The matter of how the face mass fluxes are determined from the cell center velocity field has been glossed over. As a first approximation it might seem appropriate to calculate them using a linear interpolation,

$$\mathbf{m}_f = \frac{1}{2}\left(\mathbf{m}_{f,c} + \mathbf{m}_p\right) \tag{6.144}$$

As shown in Figure 6.16 this has been found to cause checker-boarding in the pressure field (i.e., oscillations in the pressure field of a wavelength twice the cell size). A common method to prevent this from happening is Rhie–Chow[22] interpolation, whereby the momentum equation is interpolated to the face using the gradient of the pressure normal to the face rather than the cell center value, using for example,

$$\mathbf{m}_{\text{Rhie–Chow}} = \frac{1}{2}[(\mathbf{m}_p + \mathbf{m}_{c,f}) + (C_p \nabla p_p + C_{c,f}\nabla p_{c,f}) - (C_p + C_{c,f})\nabla p_f] \tag{6.145}$$

6.4.7.3 Fully Coupled Solvers

Most CFD packages nowadays use fully coupled solvers, an example of which is given in Hutchinson et al.[23] This solver has less steps than SIMPLE, but is more complex to program and uses more memory. The algorithm is:

1. Guess an initial velocity and pressure field, $\mathbf{v}^*$ and p^*.
2. Interpolate to get the cell face mass flow rates $\dot{m}_f$.
3. Solve the momentum equations to get a correction to the velocity field $\mathbf{v}'$ and p' that conserves momentum, and the continuity equation so that these fields are mass conserving.
4. Update the velocity and pressure fields with $\mathbf{v} = \mathbf{v}^* + \mathbf{v}'$ and $p = p^* + p'$.
5. Return to step 2 and continue until there is negligible change in the mass flux field between iterations.

Compared with the SIMPLE scheme the solver is much more robust, in that mass conservation and momentum are enforced in the same step. The code must iterate to solve the nonlinearity of the mass flux terms, but it is no longer iterating to couple momentum and continuity.

Whilst the velocity and pressure fields are not relaxed, the iterative loop is implemented in the form of a false-transient method,[24] and so there is effectively relaxation through the size of the time step.

As with the SIMPLE scheme, scalars that affect the momentum equations must be solved inside the iterative loop, either in a separated linear solve after the solution of the velocity and pressure fields, or within the fully coupled linear system. Also, when solving to a truly transient solution the solver iterates at each time step to couple the non-linear terms. The transient discretization uses a first or second order backward Euler formulation.

6.4.8 Turbulence Models

The methods discussed to this point have been for laminar flow. Whilst some flows of interest to the food engineer may be laminar, a large number are turbulent, which must be accounted for in the CFD solution.

In theory there is no problem in modeling a turbulent flow with a laminar flow solver. The conservation equations described to this point still hold, and all that is needed is to solve for a transient flow using a mesh and time step small enough to resolve the finest features of the flow. Indeed, this approach is used in direct numerical simulation (DNS) which has found a niche in the investigation of the statistical properties of turbulent flows. The problem lies in the computational resources needed to solve a flow with even a moderate Reynolds number. For example, consider air ($\nu = 1.5 \times 10^{-5}$ m²/s) flowing past a 1 m square flat plate at a velocity of 3 m/s, giving a Reynolds number of $Re_x = 2 \times 10^5$. At the trailing edge the boundary layer has a thickness of $\delta \approx 40$ mm, but the smallest eddies in the flow are approximately 0.05 mm across. Therefore, to model a $0.1 \times 1 \times 1$ m domain requires approximately 10^{12} mesh points and so to store the u velocity field alone at a single time step would require 8 TeraBytes of storage. Although computing speeds and storage are increasing at a steady rate, it is estimated that it would take 50–100 years[25] at the current rate of growth before DNS could be applied to the majority of flows now modeled by CFD.

The approach adopted by the CFD community has been to model the turbulence, either across the entire turbulence spectrum (using the Reynolds averaged Navier–Stokes equations or RANS methods), or alternatively only modeling wavelengths less than the mesh spacing whilst simulating the larger structures using large eddy simulation, (LES). Whilst commercial CFD packages have started implementing LES, it remains a computationally expensive option, so the RANS methods are normally the option for the industrial user. The problem with this approach is that the solution is only as good as the turbulence model, and that the turbulence models are far from perfect.

6.4.8.1 Properties of Turbulent Flows

The essence of RANS methods is that the instantaneous turbulent flow is no longer calculated, but the average properties of the flow are. The problem is thereby reduced so that the mesh required is courser which further reduces the computational cost. The loss of information is not normally a disadvantage since the averaged properties are typically of more interest to the design engineer than the exact values.

The starting point of RANS modeling is the separation of flow variables into mean and fluctuating components. For example the instantaneous value of φ can be separated into its mean and fluctuating components

$$\varphi = \bar{\varphi} + \varphi' \tag{6.146}$$

where the mean is averaged over some interval Δt

$$\bar{\varphi} = \frac{1}{\Delta t} \int_{t_0}^{t_0 + \Delta t} \varphi \, dt \tag{6.147}$$

The choice of the time interval, Δt determines whether the mean value is unsteady or not. If a problem is known to be unsteady then Δt can be chosen to be much shorter than the time scale of the unsteady motion, but longer than the time scale of the turbulent fluctuations. In this way the resulting RANS equations can model the mean unsteady flow. Strictly the choice of Δt is then problem dependent. Because this discussion is illustrative, the assumption will be made that Δt is sufficiently large to encompass the turbulent fluctuations but short enough to allow the mean variables to represent unsteadiness.

The averaging process has the following rules that can be readily verified.

$$\overline{\varphi}' \quad \overline{\varphi^2} \neq 0 \quad \overline{\varphi'\psi'} \neq 0 \quad \overline{(\overline{\varphi}+\varphi')^2} = \overline{\varphi}^2 + \overline{\varphi'^2} \tag{6.148}$$

$$\overline{\overline{\varphi}} = \overline{\varphi} \quad \overline{\varphi+\psi} = \overline{\varphi} + \overline{\psi} \quad \overline{\overline{\varphi}\psi} = \overline{\varphi}\overline{\psi} \quad \overline{\frac{\partial \overline{\varphi}}{\partial x}} = \frac{\partial \overline{\varphi}}{\partial x} \tag{6.149}$$

6.4.8.2 Derivation of the Reynolds Averaged Navier–Stokes Equations (RANS) Equations for Steady Incompressible Flow

Separating the velocity and density into mean and fluctuating components leads to the following form of the conservation equation

$$\frac{\partial(\overline{\rho}+\rho')}{\partial t} + \nabla \cdot [(\overline{\rho}+\rho')(\overline{\mathbf{v}}+\mathbf{v}')] = 0 \tag{6.150}$$

Averaging this equation over the time interval gives

$$\frac{\partial(\overline{\rho})}{\partial t} + \nabla \cdot (\overline{\rho}\overline{\mathbf{v}} + \overline{\rho'\mathbf{v}'}) = 0 \tag{6.151}$$

For flows less than Mach 0.3, the effects of unsteadiness of the density can be ignored and for these flows Equation 6.151 can be written as Equation 6.152, where it is understood that ρ has no fluctuating component.

$$\nabla \cdot \rho \overline{\mathbf{v}} = 0 \tag{6.152}$$

The momentum equation can be treated in a similar manner. Equation 6.59 can be divided into its mean and fluctuating components

$$\frac{\partial}{\partial t} \int_V (\overline{\mathbf{m}} + \mathbf{m}')dv + \oint_A (\overline{\mathbf{m}} + \mathbf{m}')(\overline{\mathbf{v}} + \mathbf{v}') \cdot \mathbf{da} = \sum (\mathbf{F} + \mathbf{F}') \tag{6.153}$$

By applying the averaging rules Equations 6.148 and 6.149, and assuming that $\mathbf{F}$ does not involve products of fluctuating quantities, the above equation can be simplified to

$$\frac{\partial}{\partial t} \int_V \overline{\mathbf{m}}dv + \oint_A (\overline{\mathbf{m}\mathbf{v}} + \overline{\mathbf{m}'\mathbf{v}'}) \cdot \mathbf{da} = \sum \overline{\mathbf{F}} \tag{6.154}$$

The only additional term is the average of the product of the fluctuating components of $\mathbf{m}$ and $\mathbf{v}$. The corresponding partial differential equation is,

$$\frac{\partial \overline{\mathbf{m}}}{\partial t} + \nabla \cdot (\overline{\mathbf{m}\mathbf{v}} + \overline{\mathbf{m}'\mathbf{v}'}) = \rho \overline{\mathbf{f}} \tag{6.155}$$

By analogy with the discussion in Section 6.4.2.4, the time averaged form of Equation 6.75 is

$$\frac{\partial \overline{\mathbf{m}}}{\partial t} + \nabla \cdot \overline{\mathbf{M}} = \rho \mathbf{f}_b : \overline{\mathbf{M}} = \overline{\mathbf{mv}} + \overline{\mathbf{m'v'}} + \overline{p}\mathbf{I} - \overline{\sigma} \tag{6.156}$$

Equation 6.156 is the RANS version of the conservation of momentum. The term involving the fluctuating components represents additional stresses known as the Reynolds stresses.

6.4.8.3 Eddy Viscosity and the Boussinesq Approximation

Equation 6.156 suggests that time averaged effects of turbulence can be represented by a simple extension of the momentum equations to account for the Reynolds stresses. Herein lies their appeal, since the solvers developed for the NS equations can be reused to solve for the averaged flow properties. The one problem lies in the determination of the Reynolds stresses: a new equation has been derived but there are now more unknowns than equations.

To consider the options that are available, the momentum transport vector needs to be expressed in terms of the constitutive equations. For a Newtonian fluid, Equation 6.109 applies and

$$\overline{\mathbf{M}} = \overline{\mathbf{mv}} + \overline{\mathbf{m'v'}} + \overline{p}\mathbf{I} - \mu \left(\nabla \overline{\mathbf{v}} + (\nabla \overline{\mathbf{v}})^{\mathrm{T}} - \frac{2}{3} \nabla \cdot \overline{\mathbf{v}}\mathbf{I} \right) + \kappa \nabla \cdot \overline{\mathbf{v}}\mathbf{I} \tag{6.157}$$

where it has been assumed that fluctuations in the coefficients of viscosity can be ignored. If the flow is incompressible in the mean,

$$\overline{\mathbf{M}} = \overline{\mathbf{mv}} + \overline{\mathbf{m'v'}} + \overline{p}\mathbf{I} - \mu \left(\nabla \overline{\mathbf{v}} + (\nabla \overline{\mathbf{v}})^{\mathrm{T}} \right) \tag{6.158}$$

The various RANS turbulence models are characterized by how the Reynolds stress term is represented in a modified momentum transport vector. For example, in an eddy viscosity model the Reynolds stresses are assumed to produce a diffusion effect similar to that of viscosity. Hence the Reynolds stresses are represented by

$$\overline{\mathbf{m'v'}} - \frac{2}{3}\rho\mathbf{I}\left(\frac{1}{2}\overline{\mathbf{v}}^2\right) = -\mu_T \left(\nabla \overline{\mathbf{v}} + (\nabla \overline{\mathbf{v}})^{\mathrm{T}} \right) \tag{6.159}$$

or

$$\overline{\mathbf{m'v'}} - \frac{2}{3}\rho\mathbf{I}k = -\mu_T \left(\nabla \overline{\mathbf{v}} + (\nabla \overline{\mathbf{v}})^{\mathrm{T}} \right) \tag{6.160}$$

where k is the turbulent kinetic energy and μ_T is the eddy viscosity.

The resulting expression for the momentum transport vector is

$$\overline{\mathbf{M}} = \overline{\mathbf{mv}} + \left(\overline{p} + \frac{2}{3}\rho k \right)\mathbf{I} - (\mu + \mu_T)\left(\nabla \overline{\mathbf{v}} + (\nabla \overline{\mathbf{v}})^{\mathrm{T}} \right) \tag{6.161}$$

Equation 6.161 has two parameters, k and μ_T that must be estimated to complete the model. It is the basis of most RANS turbulence models, the differences between the models being in the method used to approximate μ_T. However, a little used class of turbulence models, the Reynolds Stress methods, do not make this approximation, but instead attempt to model the Reynolds Stresses directly through their own transport equations.

6.4.8.4 The k–ε Model

Possibly the most ubiquitous of turbulence models is the standard k-ε model derived in Jones and Launder,[26] but using the constants of Launder and Spalding.[27] It is described here to illustrate the process by which RANS turbulent models are generated. The model uses two transport equations, one for the specific kinetic energy of the turbulent fluctuations, k (or the *turbulent kinetic energy*) and the other for the rate that this energy is dissipated via viscosity, ε (or the *turbulent dissipation rate*).

An equation for k can be derived from the mechanical energy equation. The starting point is the expression Equation 6.89 for the mechanical energy transport vector with the constitutive law for an incompressible fluid

$$\mathbf{E}_M = \left[\rho \frac{1}{2}\mathbf{v}^2 + p\right]\mathbf{v} - \mathbf{v} \cdot \mu\left(\nabla \mathbf{v} + (\nabla \mathbf{v})^\mathrm{T}\right) \tag{6.162}$$

Substitution of mean and fluctuating properties and averaging leads to

$$\overline{\mathbf{E}}_M = \rho k \overline{\mathbf{v}} + \overline{p'\mathbf{v}'} + \frac{1}{2}\rho \overline{\mathbf{v}'\mathbf{v}'^2} - \mu \nabla k \tag{6.163}$$

The resulting equation for k is

$$\frac{\partial \rho k}{\partial t} + \nabla \cdot \overline{\mathbf{E}}_M = -\rho \overline{\mathbf{v}'(\mathbf{v}' \cdot \nabla)}\overline{\mathbf{v}} - \mu \overline{(\nabla \mathbf{v}')^2} \tag{6.164}$$

Various terms are replaced by conceptual relations

$$\overline{\mu(\nabla \mathbf{v}')^2} \equiv \rho\varepsilon$$
$$\varepsilon = \text{turbulent dissipation rate} \tag{6.165}$$

$$\overline{p'\mathbf{v}'} + \frac{1}{2}\rho\overline{\mathbf{v}'\mathbf{v}'^2} = \text{turbulent diffusion of turbulent kinetic energy}$$
$$\cong -\frac{\mu_T}{\mathrm{Pr}_k}\nabla k \tag{6.166}$$

$$\rho\overline{\mathbf{v}'}(\overline{\mathbf{v}'} \cdot \nabla)\overline{\mathbf{v}} = \text{Rate of production of turbulent kinetic energy}$$
$$\cong -\mu_T\left(\nabla\overline{\mathbf{v}} + (\nabla\overline{\mathbf{v}})^\mathrm{T}\right)\nabla\overline{\mathbf{v}} = -P_k \tag{6.167}$$

The modeled transport equation for turbulent kinetic energy is

$$\frac{\partial \rho k}{\partial t} + \nabla \cdot \overline{\mathbf{E}}_M = P_k - \rho\varepsilon \ : \ \overline{\mathbf{E}}_M = \rho k\overline{\mathbf{v}} - \left(\mu + \frac{\mu_T}{\mathrm{Pr}_k}\right)\nabla k \tag{6.168}$$

The transport equation for ε is found more through modeling than derivation, and is traditionally presented as received wisdom,

$$\frac{\partial \rho \varepsilon}{\partial t} + \nabla \cdot \left(\rho \varepsilon \bar{v} - \frac{\mu_T}{Pr_\varepsilon} \nabla \varepsilon \right) = C_{\varepsilon,1} P_k \frac{\varepsilon}{k} - \rho C_{\varepsilon,2} \frac{\varepsilon^2}{k}$$ (6.169)

The term in brackets can be interpreted as a transport vector for the turbulent dissipation rate. Pr_k and Pr_ε are turbulent Prandtl numbers, and $C_{\varepsilon,1}$ and $C_{\varepsilon,2}$ are empirical constants. The remaining parameter is the eddy viscosity. In the k-ε model this is expressed as

$$\mu_T = \rho C_\mu \frac{k^2}{\varepsilon}$$ (6.170)

The standard set of constants is

$$C_\mu = 0.09 \quad C_{\varepsilon,1} = 1.44 \quad C_{\varepsilon,2} = 1.92 \quad Pr_k = 1.0 \quad Pr_\varepsilon = 1.3$$ (6.171)

The k and ε equations are transport equations that can be solved using the methods developed for the momentum and energy equations, albeit with careful attention to the nonlinear source and sink terms.

The version of the k–ε model described above is what is known as the standard k-ε model,[27] often without reference. Many other turbulence models have been developed, see for example Wilcox,[28] Durbin and Petterson Reif,[29] Cebeci[30] and Hoffman and Chiang[25] for fuller introductions to the field.

6.5 APPROXIMATION PROCESSES

6.5.1 THE MESH AND ITS EFFECT

The mesh, as it is called, is fundamental to the approximation process. Increasing the number of mesh points, and so decreasing the element size (known as refinement) increases the resolution of the solution, allowing smaller structures in the flow to be resolved by the calculations. However there is a computational cost associated with refinement, and so solutions are a compromise between accuracy and solvability.

The creation of a mesh for a given problem is a complex process and mesh generation is a continuing and very comprehensive area of research. It is beyond the scope of this discussion to present little more than a brief statement of the general principles that must be considered when creating a CFD mesh.

The alignment of the mesh with the flow and the shape of the cells are important issues to consider in the generation of the mesh. The generation of a mesh that has elements that are aligned with the flow will lead to discretization errors of lower magnitude, with the numerical diffusion being minimized.[31] Similarly, high aspect ratio cells, or meshes that have sudden changes in mesh size, are sources of discretization error.

The mesh needs to be aligned with boundaries and have close spacing near surfaces. The alignment with the boundary is required to ensure that the mesh lines are aligned with the flow, but the close spacing near surfaces is to ensure that the fine structure in the boundary layers of the flow are adequately resolved. This is especially important in the case of modeling turbulent flow.

Current unstructured grids with inflation layers have to be carefully set up. It is important avoid having only a small number of layers in a boundary layer so that the effects of mesh refinement are dominated by the simple addition of mesh points in the boundary layer. This behavior can be observed when a mesh convergence study indicates a first order rate of convergence despite the method being of higher order.

6.5.2 CONVECTION AND ITS PROBLEMS

Much has been written in the CFD literature on the approximation of convection. Because the current discussion has concentrated on the fundamental equations rather than the discretization processes, the issue of approximation of convection has been left until this section where the influence of approximation strategies on CFD models and their solutions is considered as part of a general discussion of approximation processes.

Consider the approximation of the momentum equation

$$\frac{\partial \mathbf{m}}{\partial t} + \nabla \cdot \mathbf{M} = \rho \mathbf{f}_b \qquad (6.172)$$

Remembering that $\mathbf{M}$ is the momentum transport vector

$$\mathbf{M} = \mathbf{mv} + p\mathbf{I} - \sigma \qquad (6.173)$$

the transport of momentum is a balance between momentum generation via the pressure term, the dissipative effects of the fluid stresses represented by σ and convection of momentum by the fluid motions represented by the term $\nabla.(\mathbf{mv})$. Each component of $\mathbf{m}$ can be considered separately. For the transport of an arbitrary scalar φ, the convection of φ can be represented by $\varphi \mathbf{v}$ and the contribution to the conservation equation is $\nabla \cdot \varphi \mathbf{v}$.

The issues associated with the approximation of convection can be elucidated by considering 1D convection for a staggered mesh, using the notation in Figure 6.17. If the notation and conservation cell approach that led to Equation 6.52 are used, the approximation to $\nabla.\varphi \mathbf{v}$ is

$$\nabla \cdot \varphi \mathbf{v} \cong \frac{\varphi_e u_e - \varphi_w u_w}{\Delta x} \qquad (6.174)$$

The question that needs to be answered is what values are used for φ_e and φ_w, or u_e and u_w if they are not available? Perspectives and the presented error estimates change depending on whether values at only the mesh points are used or values at cell boundaries (i.e., finite volume approximations are used.)

The characteristics of various convection approximations are listed in Table 6.5. The list starts with the simple upwind and its variation, the conservative upwind approximation. Two common central difference approximations and the quadratic upstream interpolation for convective kinematics (QUICK)[32] scheme are also given.

The simple upwind difference scheme was introduced to finite difference schemes early in the development of CFD to overcome the limitations imposed by the onset of instability as the strength of the flow increased[33] although the use of one sided differences was known to mathematicians some decades earlier.[13] It is well documented that a formal Taylor series analysis indicates that the approximation is only first order correct. The leading term in the Taylor series expansion has the same form as terms that represent the effects of viscosity and, moreover, the error is proportional to the velocity. This means that the error associated with the first order upwind approximation introduces a dissipation process that is always sufficient to prevent numerical stability. Early studies[31–35]

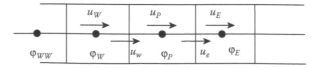

FIGURE 6.17 Approximation of convection for 1D finite volume.

TABLE 6.5
Characteristics of Convection Approximations for a Rectangular Grid

Approximation	φ_e	φ_w	u_e	u_w
Simple upwind $u > 0$	φ_P	φ_W	u_P	u_P
Conservative upwind	φ_P	φ_W	$\frac{1}{2}(u_P + u_E)$	$\frac{1}{2}(u_P + u_W)$
Central difference	$\frac{1}{2}(\varphi_P + \varphi_E)$	$\frac{1}{2}(\varphi_P + \varphi_W)$	u_P	u_P
Central difference self conservative	$\frac{1}{2}(\varphi_P + \varphi_E)$	$\frac{1}{2}(\varphi_P + \varphi_W)$	$\frac{1}{2}(u_P + u_E)$	$\frac{1}{2}(u_P + u_W)$
QUICK	$\frac{1}{8}(6\varphi_P + 3\varphi_E - \varphi_W)$	$\frac{1}{8}(6\varphi_W + 3\varphi_P - \varphi_{WW})$	$\frac{1}{2}(u_P + u_E)$	$\frac{1}{2}(u_P + u_W)$

showed that the dissipative effects were maximized when the flow was aligned at 45 degrees to the mesh lines and was sufficiently large to seriously compromise the accuracy of the CFD solutions.

The second upwind difference scheme or "donor cell" method[36] listed in the table is suitable for finite volume CFD methods and uses average values of the velocity to represent the velocity at cell faces. It is self conservative in that the momentum flux leaving one cell is identical that entering the adjacent cell. Applying a Taylor series analysis to predict the truncation error is subject to some argument, however it must be conceded that the approximation is first order and the primary affect of the error is to add false dissipation or diffusion.

The essential concept of an upwind approximation is to use information that is being carried to the mesh point at which the momentum transport is being approximated. This idea has physical validity since it represents the "transportive" nature of advection. Central difference approximations use data that are both up and downstream from the point under consideration and thus may not represent the "transportiveness" well. However the serious errors caused by the first order errors outweigh any advantage that their representation of transportiveness might have.

The QUICK scheme and its variations such as Ultra-QUICK[37] use upwind data and have second order convergence. They and higher order upwind schemes are preferred for the representation of convection.

The central difference approximations work well for viscous flows where convection is weak. In fact it can be shown that if the mesh Peclet number

$$Pe_{\Delta x} = \frac{|\mathbf{v}|\Delta x}{\alpha_D} \tag{6.175}$$

is less than two, central difference representations of convection will behave well. If this threshold is exceeded then the modeled flow will exhibit spurious oscillations. This observation led to the development of hybrid or power-law approximations that use a blended mix of central difference and first order upwind approximations to convection. These schemes use the second order central difference approximations for low Peclet or Reynolds numbers and then switch to the first order upwind when the mesh Peclet number is greater than two.

Today industry standard packages such as ANSYS-CFX offer their users second order approximations but still allow first order upwind approximation to be used. Although there are solutions, such as compact high order schemes, to the representation of the convection problem, they have not reached the commercial packages. This situation is exacerbated by the fact that often the approximation to the RANS k and ε transport equations uses first order upwinding only. The reader is referred to Leonard and Drummond's discussion[37] for indications of how seriously first order truncation errors can affect CFD solutions.

For the reasons discussed in this section it is imperative that a mesh refinement study is part of every CFD investigation of a problem.

6.6   PRINCIPLES OF GOOD PRACTICE

The purpose of this Section is to summarize the issues and recommendations highlighted in this Chapter and present them in the framework of the process of creating and using a CFD model of a food engineering process. The overarching ideas are:

- Keep in touch with the fundamentals, especially the conservation laws that underpin CFD technology
- Ensure that solutions are properly obtained making the best use of the methods offered by the CFD package that is being used
- Use mesh convergence analysis to estimate the order of convergence, to provide extrapolated results and error bounds.

6.6.1   STEPS INVOLVED IN SETTING UP AND CREATING A COMPUTATIONAL FLUID DYNAMICS (CFD) MODEL

6.6.1.1   Flow Domain and Physics

Understand the essential physics of the situation to be modeled. Model just what is required without adding superfluous complexity. Decide what essential physics need to be modeled. It may still be relevant to do an order of magnitude analysis to determine what effects need to be accounted for (although most packages are now "all embracing").

Determine what questions need to be answered by the model. Ensure that the model can produce the data to answer those questions.

Is the flow likely to be laminar turbulent? Is it steady or unsteady. Is it 2D or 3D? Is it axisymmetric? The answers to these questions lead to the selection of modeling options. If a turbulence model is being using make sure that appropriate boundary conditions can be set.

Ensure that the rest of the boundary conditions are set properly to reflect how the flow thermal and/or mass transfer conditions are applied. Make sure that the boundary conditions are consistent and are not over specified. Make sure inflow and outflow boundaries are sufficiently far from the region of interest to have minimal effect on the flow.

Define a consistent set of boundary conditions. This can be a difficult step.

Ensure that the physical properties are correct, and that there is dimensional consistency between properties and the geometry.

6.6.1.2   Approximation

What method is being used or is the most appropriate? The choice may be between finite difference approximation, finite volume or finite element methods. The bulk of the commercial packages in use today use finite volume methods. There are other specialized methods such as smoothed particle dynamics, lattice Boltzmann or boundary element methods that may be particularly applicable.

Decide what resolution needs to be captured. Determine the "best method" for creating the mesh. A lot will depend on the package being used and the complexity of the problem being solved.

6.6.1.2.1   Create Mesh

Make sure that important boundary shapes are properly represented. How critical is mesh smoothness? Decide if boundary layers are likely to occur and arrange meshing appropriately. Use inflation layers if available but do not have a sudden transition between long inflation layer cells and regular internal mesh elements.

Ensure that there are no sudden discontinuities in mesh resolution anywhere. An accepted guideline is that the expansion ratio between adjacent mesh cells should not exceed 1.2.

Ensure that there are no cells with small internal angles.

When modeling flow around a free standing object, make sure the far field boundaries are sufficiently remote so as not to affect the solution.

6.6.1.2.2   Set up the Computational Fluid Dynamics (CFD) Model

Define the flow domain, the properties of the fluid, the physical models and set the boundary conditions.

Test the boundary conditions and physical models on a coarse mesh solution with a large time step, to ensure that the model is correctly defined. It is a common occurrence for the naïve user to only spot errors in a model after running it for several days when a coarse mesh solution could have revealed the same errors much more quickly.

6.6.1.2.3   Choose the Approximation Accuracy

For example ensure that at least second order convection is chosen. Most packages have little better. In fact some use the term "high resolution" when the actual achieved approximation is struggling to reach second order.

Do not use first order upwind approximations unless absolutely necessary. In that case do not put any trust in the results.

6.6.1.3   Postprocessing and Visualization

Set postprocessing options so the output parameters are correctly defined. Devise ways to produce the data required from the analysis, such as forces, heat fluxes and flow rates. Are there additional methods such as particle tracking which will produce secondary data such as particle deposition on boundary walls, residence times, etc?

6.6.1.4   Solution Options

Set these to generate the required solution. This may have to be revisited as instability and slow rates of reduction of residuals occurs.

If a solution fails to converge it may be necessary to write out a solution file containing the solver residuals. This can be examined using visualization software to see if high values are restricted to particular regions. This may indicate that the mesh in these regions is of poor quality, or that the flow in that region is unsteady.

6.6.1.5   Verification and Validation

Once preliminary solutions have been obtained, perform a mesh refinement study and determine the order of convergence. Remember that mesh independence is an idealized "dream". Decide on a mesh suitable to production runs used to perform a parametric study.

Check that the solutions are physically plausible. For instance, does hot air rise? A common mistake is not understanding the sign convention used by a CFD package for **g** and the coefficient for thermal expansion and thereby getting the direction of the buoyancy force wrong.

It is acceptable to perform verification and validation for a set of representative solutions. It is important to be cautious when the flow, geometry, structure parameters or boundary conditions move away significantly from the representative configurations.

6.6.1.6   Production Solutions

Use the mesh selected in the previous step to produce production solutions.

6.6.1.7   Numerical Results

The final phase of a CFD analysis is the preparation of results. Invariably these will involve estimates of conservation balances. These should be done carefully, bearing in mind the comments made in Section 6.4.6.5. It may be possible to estimate the desired quantities in different ways. For example the force on a moving vane in a duct may be calculated by integrating the pressure and shear stresses on the vane. Alternatively a momentum balance over the whole domain can also be used to estimate the force exerted by the vane on the fluid. Both methods should give the same result, particularly when the correct global conservation equations are used. Using both methods can be an illuminating way to check the numerical consistency of the CFD model.

6.7   CONCLUSION

Food engineering CFD involves a wide range of applications, from the processing and preparation of products and ingredients to their transportation and storage. Because of the nature of many of the products the food engineer is likely to want to extend the modeling beyond the capabilities of an available package. Non-Newtonian fluids, phase changes, chemical reactions, particulate transport and multimode heat transfer processes are all part of the food engineering modeling domain and reviews of applications and relevant techniques have already been published.[2]

This discussion has, therefore sought to highlight CFD issues that lead to principles of best practice. For this reason, the discussion of mesh convergence and methods of error estimation were presented using an abstraction of the modeling and approximation processes to emphasize that this important aspect of CFD modeling is above the particular issues of meshing, numerical approximations and solution methodologies. An important part of this discussion was the presentation in Section 6.3.4.2 that estimated the high demands made by CFD on computational resources and that further advances in CFD technology are required to achieve substantial gains in the complexity of the problems that can it can address. Throwing more computational resource at a problem is no panacea for the limitations imposed by low orders of convergence.

The equations and numerical approximations were described in Section 6.4 from the perspective that the conservation laws are the primary drivers of the modeling process. This discussion uses the increasingly common notation of transport vectors[15] that can be embedded in integral or partial differential representations of the conservation laws. Coupling between the momentum and pressure fields was presented to provide the reader with an understanding of how important proper handing of these equations is and how advances in the associated numerical techniques have been responsible for the major improvements in CFD modeling solution speed and robustness. A description of the k-ε turbulent model provided an illustration of the modeling assumptions involved in RANS modeling. The discussion should not be seen as endorsing that particular model, it is up to the food engineer to research the most appropriate model for an application.

CFD is an attractive and useful technology that, if applied with cognizance of the limitations of some of its weaker aspects and a willingness to perform the appropriate mesh convergence studies, can be used very usefully for the design of new systems or the optimization of existing processes. Subject to the cautions discussed in this Chapter, CFD can be a valuable tool for the food engineer.

NOMENCLATURE

$\mathbf{a}$	Area vector
$\mathbf{A}_N$	Coefficient Matrix for a set of N algebraic equations
$\mathbf{B}_N$	Source Vector for a set of N algebraic equations
C_p	Specific heat at constant pressure
$\mathbf{C}$	Vector of termination criteria
d	Dimension
$\mathbf{da}$	Small area vector
Diff()	Difference between vectors of numerical estimates operator (equation (7))
e	Specific energy
e_{KE}	Specific kinetic energy
e_T	Specific thermal energy
$\mathbf{E}$	Total energy transport vector
$\mathbf{E}_T$	Thermal energy transport vector
$\mathbf{E}_M$	Mechanical energy transport vector
$E(\varphi)$	General set of equations (equation (1))
$\mathbf{f}$	Force per unit mass

$\mathbf{f}_b$	Body Force per unit mass
$\mathbf{F}$	Force
$\mathbf{g}$	Gravitation acceleration vector
Gr	Grashof number (equation (6.41))
h	Heat transfer coefficient, mesh interval
$\mathbf{j}$	Mass diffusion vector
k	Thermal conductivity, turbulent kinetic energy
L	Domain Length
$L_N(\varphi'_N)$	Approximation to $E(\varphi)$ based on N values
m_i	Mass fraction of species i.
$\dot{m}_f$	Mass flow rate through the face of an element
$\mathbf{m}$	Momentum vector ($\rho\mathbf{v}$)
$\mathbf{M}$	Momentum transport vector
N	Number of values or degrees of freedom
Nu	Nusselt number (equation (6.44))
p	Pressure, or order of approximation
Pr	Prandtl number (equation (6.42))
q'''	Volumetric rate of heat generation
$\mathbf{q}$	Heat flux vector
$\dot{Q}$	Heat flow rate into a control volume
r	computational load factor, mesh ratio
Ra	Rayleigh number (equation (6.42))
Re	Reynolds number (equation (6.46))
s	Specific entropy
$\mathbf{S}$	Entropy transport vector (equation (6.46))
t	Time
T	Temperature
u	x component of the velocity vector
v	y component of the velocity vector
$\mathbf{v}$	Velocity vector ($= u\mathbf{i} + v\mathbf{j} + w\mathbf{k}$)
V	Speed
Vol	Volume
w	z component of the velocity vector
$\dot{W}$	Rate of work done by a control volume on its environment
x	Cartesian coordinate
y	Cartesian coordinate
z	Cartesian coordinate
β	Relaxation factor
Δ	Small increment
ε	Turbulent dissipation rate
$\varepsilon_{\varphi.N}$	Solution error
$\varepsilon_N(\mathbf{C})$	Error associated with an incomplete solution (equation (6.8)), or Residual
ε_N^n	Error associated with an incomplete solution at iteration n based on N values
φ	Solution variable, or arbitrary function, or set of variables
φ'_N	Numerical solution vector based on N values
$\varphi''_N(\mathbf{C})$	Incomplete solution vector (equation (6.8))
φ''^n_N	Incomplete numerical solution vector at iteration n based on N values
ρ	Density
Π	Stress tensor
Ω	Generalised viscous dissipation term (equation (6.90))
σ	Deviatoric stress tensor

κ   Second coefficient of viscosity
μ   Dynamic viscosity
γ̇   Magnitude of the rate of strain tensor
η   Non-Newtonian viscosity

Suffixes

c   Cell Centre
CE   Computational Effort
e   East Direction
est   Estimated
f   Face index
i   i'th space coordinate
KE   Kinetic energy
M   Mechanical (energy)
Mφ   Mechanical with potential (energy)
s   Space or south direction
w   West direction
t   Time
top   Top
T   Thermal (energy)

REFERENCES

1. Norton, T. and Sun, D.-W. 2006. Computational fluid dynamics (CFD)—an effective and efficient design and analysis tool for the food industry: A review. *Trends in Food Science and Technology,* 17, 600–620.
2. Sun, D.-W. 2007. *Computational Fluid Dynamics in Food Processing.* Taylors & Francis, Boca Raton, FL.
3. Mallinson, G.D. 2008. CFD visualisation: challenges of complex 3D and 4D data fields. *International Journal of Computational Fluid Dynamics,* 22, 1, 49–59.
4. Mallinson, G.D. 2008. Vector lines and potentials for computational heat transfer visualisation. In de Vahl Davis, G. and Leonardi, E. (Eds.) *CHT'08 Advances in Computational Heat Transfer.* Begell House, Inc., Marrakech, Morocco.
5. Richardson, L.F. 1910. The approximate arithmetical solution of finite differences of physical problems involving differential equations with an application to the stresses in a masonry dam. *Transactions of the Royal Society London, Series A,* 210, 307–357.
6. Roache, P.J. 1998. *Verification and Validation in Computational Science and Engineering.* Hermosa Publishers, Albuquerque, NM.
7. Roache, P.J. 1997. Quantification of uncertainty in computational fluid dynamics. *Annual Review of Fluid Mechanics,* 29, 123–160.
8. Celik, I. and Karatekin, O. 1997. Numerical experiments on application of Richardson extrapolation with nonuniform grids. *ASME Journal of Fluid Engineering,* 119, 584–590.
9. de Vahl Davis, G. 1983. Natural convection of air in a square cavity: A bench mark numerical solution. *International Journal for Numerical Methods in Fluids,* 3, 249–264.
10. Ghia, U., Ghia, K.N. and Shin, C.T. 1998. High-Re solutions for incompressible flow using the Navier-Stokes equations and a multigrid method. *Journal of Computational Physics,* 8, 387–411.
11. Botella, O. and Peyret, R. 1998. Benchmark spectral results on the lid-driven cavity flow. *Computers and Fluids,* 27, 421–433.
12. Norris, S. and Mallinson, G. 2007. Volumetric methods for evaluating energy loss and heat transfer in cavity flows. *International Journal for Numerical Methods in Fluids,* 54, 12, 1407–1423.
13. Roache, P.J. 1972. *Computational Fluid Dynamics.* Hermosa Publishers, Albuquerque, NM.
14. Fletcher, C. 1991. *Computational Techniques for Fluid Dynamics.* Springer Verlag. Berlin, New York.
15. Bird, B.R., Stewart, W.E. and Lightfoot, E.N. 2006. *Transport Phenomena.* John Wiley & Sons, Inc. New York.
16. Aris, R. 1989. *Vectors, Tensors and the Basic Equations of Fluid Mechanics.* Dover Publications Inc. New York.

17. Owczarek, J.A. 1968. *Introduction to Fluid Mechanics*. International Textbook Company Scranton, PA.

18. Sani, R.L. and Gresho, P.M. 1994. Resume and remarks on the open boundary condition symposium. *International Journal for Numerical Methods in Fluids*, 18, 983–1008.

19. Patankar, S.V. and Spalding, D.B. 1972. A calculation procedure for heat, mass and momentum transfer in three-dimensional parabolic flows. *International Journal of Heat and Mass Transfer*, 15, 1787–1806.

20. Patankar, S. 1980. *Numerical Heat Transfer and Fluid Flow*. McGraw-Hill, NY.

21. Chorin, A.L. 1967. A numerical method for solving incompressible viscous flow problems. *Journal of Computational Physics*, 2, 12–26.

22. Rhie, C.M. and Chow, W.L. 1983. A numerical study of the turbulent flow past an isolated airfoil with trailing edge separation. *AIAA Journal*, 21, 1525–1532.

23. Hutchinson, B.R., Gaplin, P.F. and Raithby, G.D. 1988. Application of additive correction multigrid to the coupled fluid flow equations. *Numerical Heat Transfer*, 13, 133–147.

24. Mallinson, G.D. and de Vahl Davis, G. 1973. The method of the false transient for the solution of coupled elliptic equations. *Journal of Computational Physics*, 12, 4, 435–461.

25. Hoffmann, K.A. and Chiang, S.T. 2000. *Computational Fluid Dynamics*. Engineering Education System, Wichita, KS.

26. Jones, W.P. and Launder, B.E. 1972. The prediction of laminarization with a two-equation model of turbulence. *International Journal of Heat and Mass Transfer*, 15, 301–314.

27. Launder, B.E. and Spalding, D.B. 1974. The numerical computation of turbulent flows. *Computer Methods in Applied Mechanics and Engineering*, 3, 269–289.

28. Wilcox, D.C. 1993. *Turbulence Modelling for CFD*. DCW Industries La Canada, CA.

29. Durbin, P.A. and Petterson Reif, B.A. 2001. *Statistical Theory and Modelling for Turbulent Flows*. Wiley, NY.

30. Cebeci, T. 2004. *Analysis of Turbulent Boundary Layers*. Oxford, Elsevier, Amsterdam.

31. de Vahl Davis, G. and Mallinson, G.D. 1976. An evaluation of upwind and central difference approximations by a study of recirculating flow. *Computers and Fluids*, 4, 29–43.

32. Leonard, B.P. 1979. A stable and accurate convective modelling procedure based on quadratic upstream interpolation. *Computer Methods in Applied Mechanics and Engineering*, 19, 59–98.

33. Thoman, D.C. and Szewczyk, A.A. 1966. *Numerical Analysis of Time Dependent two Dimensional Flow of a Viscous, Incompressible Fluid over Stationary and Rotating Cylinders*. Heat Transfer and Fluid Mechanics Laboratory, Department of Mechanical Engineering, University of Notre Dame, IN.

34. Runchal, A.K. and Wolfshtein, M. 1969. Numerical integration procedure for the steady state Navier-Stokes equations. *Journal of Mechanical Engineering Science*, 11, 5, 445–453.

35. Wolfshtein, M. 1968. *Numerical Smearing in Onesided Difference Approximations to the Equations of Non-Viscous Flow*. Imperial College of Science and Technology, London, UK.

36. Gentry, R.A., Martin, R.E. and Daly, B.J. 1966. An Eulerian differencing method for unsteady compressible flow problems. *Journal of Computational Physics*, 1, 87–118.

37. Leonard, B.P. and Drummond, J.E. 1995. Why you should not use 'hybrid', 'power-law' related exponential schemes for convective modelling—there are much better alternatives *International Journal for Numerical Methods in Fluids*, 20, 421–442.

7 Computational Fluid Dynamics Analysis of Retort Thermal Sterilization in Pouches

A.G. Abdul Ghani
Software Design Ltd.

Mohammed M. Farid
The University of Auckland

CONTENTS

7.1 INTRODUCTION

Thermal sterilization is still the most commonly used technique for food preservation. Strict regulations and procedures are established by government agencies for thermal processing of low-acid canned foods because of widespread public health concerns about anaerobic *Clostridium botulinum,* a spore-forming microorganism that produces a toxin deadly to humans, even in very small amounts. Two different methods of thermal processing are known, the aseptic processing in which the food product is sterilized prior to packaging, and canning in which the product is packed and then sterilized [1]. In the design of thermal food process operations, the temperature in the slowest heating zone (SHZ) and the thermal center of the food during the process must be known. Traditionally, this temperature course is measured using thermocouples. There is growing interest in the use of mathematical models to predict food temperature during thermal treatment [2–6]. As will be explained later, it is difficult to measure the temperature at the SHZ because this is a region which keeps moving during the heating progress.

In heat transfer mechanism, canned foods were classified as either convection-heated, conduction-heated or both convection and conduction heated [7]. Most of the mathematical analyses have been carried out for conduction-heated products because simple analytical or numerical solutions are readily available. The analysis is acceptable for heating solid food, but not for liquid. Foods like canned tuna, thick syrups, purees, and concentrates are usually assumed to be heated by pure conduction. For these foods, the required processing time is generally determined by analytical or numerical solution of the heat conduction equation [8].

It was established experimentally that during heating, the convective circulating stream of the liquid food rises along the can wall and falls in the can center [9]. It was also established both experimentally [10] and theoretically [4] that the SHZ in a convection-heated food in a cylindrical can was a torroid continuously altering its location. Most of the numerical studies have been carried out for water-like liquid food products, with the assumption of constant viscosity [11]. The numerical predictions of the transient temperature and velocity profiles during natural convection heating of water have been well studied by Datta and Teixera [4,5]. The influence of natural convection heating during the sterilization process of sodium carboxyl-methyl cellulose (used as a model liquid food) has been studied [11–14] using computational fluid dynamics (CFD). The effect of sterilization on the bacteria inactivation during natural convection heating of viscous liquid in a cylindrical can was also studied by Ghani et al. [15].

The optimum sterilization should ensure that the SHZ is exposed to adequate heat treatment for a sufficient period of time to inactivate the microorganisms. The location of the SHZ for a convection-heated product is not as easily determined as for a conduction-heated product, and requires knowledge of the transient temperature and flow patterns during the sterilization process. In natural convection heating, the velocity in the momentum equations is coupled with the temperature in the energy equation because movement of fluid is solely due to buoyancy force. Because of this coupling, the energy equation needs to be solved simultaneously with the momentum equations.

Sterilization of food in cans has been well studied both experimentally and theoretically, but little work has been done on sterilization of food in pouches, which has been introduced to the market in recent years. Little information is known on the temperature distribution within the pouch during the sterilization process. The available knowledge on thermal sterilization of food in cans cannot be fully utilized to understand thermal sterilization of food in pouches due to the complicated geometry of the latter. Pouch analysis will require computer modeling in a three-dimensional domain. Bhowmik and Tandon [16] and Tandon and Bhowmik [17] developed a model to evaluate thermal processing of a two-dimensional pouch containing a conduction-heated solid or viscous liquid food.

The objective of the work presented in this chapter is to provide rigorous analysis of thermal sterilization of liquid food contained in a three-dimensional pouch and to predict transient temperature, velocity profiles and concentrations of bacteria and vitamin C profiles in the pouch as heating progresses. Also, the migration of the SHZ during this natural convection heating is simulated and analyzed. This investigation may be used to optimize industrial sterilization process with respect to sterilization temperature and time. As a result of CFD analysis and the model presented in this chapter, companies involved in the canning industry will be able to predict accurately the sterilization time required for any pouch containing any new food products. This optimization process will save both energy and time, which are of great value for the large production capacity needed in the canning industry.

Although CFD models have long been applied to different processing industries such as the aerospace, automotive and nuclear industries, it is only in recent years that they have been applied to food processing applications. Scott and Richardson [18] discussed the mathematical modeling techniques of CFD which can be used to predict the flow behavior of fluid food. Advances in computing speed and memory capacity of computers are allowing ever more accurate and rapid calculations to be performed, and a number of commercial software packages of practical use to the food industry have now become available, such as CFX, FLUENT, and PHOENICS, which is used in the simulations presented in this chapter. CFD models can be of great use in a variety of food engineering applications. It can be used for predicting mixing efficiency in specific mixer geometry, determining average residence times of turbulent flows through heat exchangers, predicting convection patterns in chillers or ovens, or determining flow patterns of airborne microorganisms in a clean-room factory environment.

The liquid food material used in the work presented here (carrot–orange soup) was one of the products of Heinz Watties Australasia, Hastings, New Zealand. The CFD code PHOENICS is used here. Saturated steam at 121°C was assumed to be the heating medium. The partial differential equations describing the conservation of mass, momentum and energy were solved numerically together with bacteria and vitamin concentrations using the finite volume method (FVM) of analysis.

In the work presented here, the following cases are analyzed:

1. Prediction of temperature distribution, velocity and concentration profiles of bacteria (*Clostridium botulinum*) and vitamin C for carrot–orange soup during in pouch sterilization. The migration of the SHZ during sterilization was also studied.
2. Simulation was also performed for the same pouch but with the assumption of pure conduction heating for the purpose of comparison.
3. Study the effect of the cooling period of the pouch on the sterilization process.

The results of the simulations show that the velocity of food particles in the pouch due to heating is very small due to the small height of the pouch and high viscosity of the soup analysed. In all simulations, the SHZ was found to migrate toward the bottom of the pouch into a region within 30–40% of the pouch height, closest to its deepest end. Sterilization time was found shorter compared to that for cans, which is attributed to the large surface area per unit volume of the pouch. The simulations also show the dependency of the relative bacteria (*Clostridium botulinum*) and Vitamin C concentration on both the temperature and velocity profiles.

7.2 BASIC MODEL EQUATIONS AND SOLUTION PROCEDURE

The computations were performed for a three-dimensional pouch with a width (W) of 0.12 m, height (H) of 0.04 m, and length (L) of 0.22 m. The pouch outer surface temperature (top, bottom and sides) was assumed to rise instantaneously and maintained at 121°C throughout the heating period. The effect of the retort come-up time was studied earlier and was found to be very small.

7.2.1 COMPUTATIONAL GRID AND GEOMETRY CONSTRUCTION

The boundary layer occurring at the heated walls and its thickness are very important parameters to the numerical convergence of the solution. Temperature and velocities have their largest variations in this region. To adequately resolve this boundary layer flow i.e., to keep the discretization error small, the mesh should be optimized and a large concentration of grid points is needed in this region. If the boundary layer is not resolved adequately, the underlying physics of the flow is lost and the simulation will be erroneous. On the other hand, in the rest of the domain where variations in temperature and velocity are small, the use of a fine mesh will lead to increase computation time without any significant increase in accuracy. Thus, a nonuniform grid system is needed to resolve the physics of the flow properly, which was used in all simulations.

Pouch volume was divided into 6000 cells: 20 in the x-direction, ten in the y-direction and 30 in the z-direction as shown in Figure 7.1. The natural convection heating of the soup used was simulated as follows:

(a)   (b)

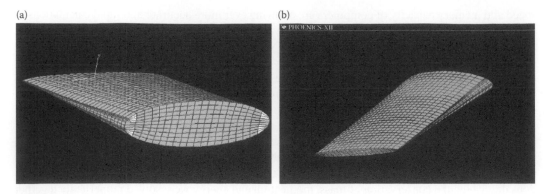

FIGURE 7.1 Pouch geometry and grid mesh showing (a) the widest end and (b) the narrowest end. (From Ghani, A. G., Farid, M. M., and Chen, X. D., *Proceedings of the Institution of Mechanical Engineers*, 217, 1–9, 2003. With permission.)

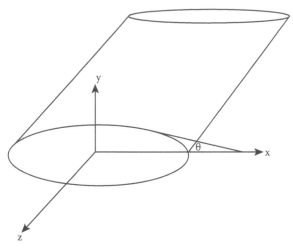

FIGURE 7.2 Geometry of the pouch. (From Ghani, A. G., Farid, M. M., Chen, X. D., and Richards, P., *Journal of Process Mechanical Engineering*, 215, Part E, 1–9, 2001. With permission.)

1. The total simulation time used for the sterilization of carrot–orange soup was 3000 s. It was divided into 30 time steps. It took 10 steps to achieve the first 200 s of heating, another ten steps for the next 800 s and ten steps for the remaining 2000 s.
2. The total simulation time for heating and cooling cycle during sterilization of carrot–orange soup was 4800 s. It took 60 time steps to achieve the total heating cycle (0–3600 s), and the remaining 20 time steps for the cooling cycle (3600–4800 s).

The simulations were conducted using the UNIX IBM RS6000 workstations at the University of Auckland. The solutions have been obtained using a variety of grid sizes and time steps as will be discussed later in this chapter. In the construction of pouch geometry, Equation 8.1 for ellipse was used for the construction of pouch grid in the *x*–*y* plane:

$$\left(\frac{x}{a}\right)^2 + \left(\frac{y}{b}\right)^2 = 1 \tag{7.1}$$

The height of the pouch, shown in Figure 7.2, can be written as:

$$y = \pm b \sqrt{1 - \left(\frac{x}{a}\right)^2} \tag{7.2}$$

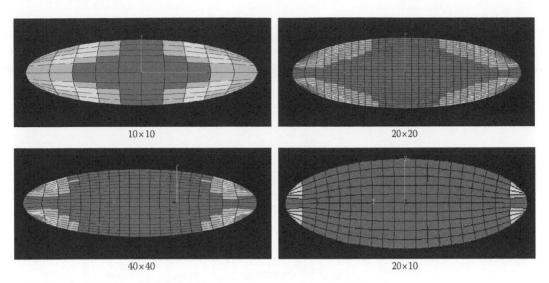

FIGURE 7.3 Different grid meshes used to test the cells of the pouch.

The three-dimensional grid was constructed using a series of two-dimensional grids in the $x - y$ plane with height varying with the z-coordinate as shown in Figure 7.2. The elliptical boundary created a distorted rectangle of cells. However, in order to minimize the distortion of grid cells in the corner, Equation 7.1 was rewritten in terms of θ and the discontinuity between x and y grid lines was placed at x, which can be written in terms of a, b, and θ as:

$$x = \sqrt{\frac{\left(\dfrac{a^4}{b^2}\right)\tan^2\theta}{1 + \left(\dfrac{a^2}{b^2}\right)\tan^2\theta}} \qquad (7.3)$$

where $\tan\theta$ is the gradient of the boundary at x, with $\theta = 45°$.

The key characteristic of the method is the immediate discretization of the integral equation of flow into the physical three-dimensional space (i.e., the computational domain covers the entire pouch, which was divided into a number of divisions in three dimensions). Details of the code can be found in the PHOENICS manuals, especially the Input Language (PIL) manual [19].

To construct the geometry of the pouch, body fitted coordinate (BFC) was used. For generating curvilinear grid within the subdomain command, the option of solving differential equations for the corner coordinates within the currently active domain was used. This option involves the solution of Laplace equations for the Cartesian coordinates of the cell corners. The finite-difference equations (FDE) solved for Cartesian coordinates were expressed in linearized form so that they can be solved by means of linear equation solvers.

In the simulations, a variety of grid sizes (Figure 7.3) and time steps were used. Through mesh refinement study, it is clear from Figure 7.3 that the optimum mesh possible was the one used (20×10), which is due to the dominant orthogonal cells (red cells) that improve the stability of the solution. The results obtained in this study showed that the solution is almost time-step independent and weakly dependant on grid size.

7.2.2 Convection and Temporal Discretization

An important consideration in CFD is the discretization of the convection terms in the finite volume equations. The accuracy and numerical stability of the solution depends on the numerical scheme used for these terms. The central issue is the specification of an appropriate relationship between the convected variables, stored at the cell center, and its value at each of the cell faces [20]. The convection discretization scheme used for all variables in our simulations is the hybrid-differencing scheme (HDS). The HDS of Spalding [21] used in PHOENICS switches the discretization of the convective terms between central differencing scheme (CDS) and upwind differencing scheme (UDS) according to the local cell Peclet number.

The cell Peclet number Pe (ratio of convection to diffusion) for bacteria and vitamins within the flow domain in the z-direction is:

$$Pe = \frac{u\Delta z}{\alpha} \tag{7.4}$$

where u and Δz are the typical velocity and typical distance of the cell in the z-direction, respectively. The thermal diffusivity α of the bacteria and the vitamins in the fluid is given by the Stockes–Einstein equation:

$$\alpha = \frac{k_T T}{6\pi\mu a} \tag{7.5}$$

where k_T is the reaction rate constant for particle at temperature T, μ is the apparent viscosity, and a is the radius of the particle.

The calculated cell Peclet numbers within the flow domain in this study are of the order of 10^4, and so the diffusion of bacteria and vitamins has been ignored. Within PHOENICS, the temporal discretizaion is fully explicit.

7.2.3 GOVERNING EQUATIONS AND BOUNDARY CONDITIONS

The partial differential equations governing natural convection motion in a pouch space are the Navier–Stokes equations in x, y, and z coordinates as shown below:

The continuity equation

$$\frac{\partial v_x}{\partial x} + \frac{\partial v_y}{\partial y} + \frac{\partial v_z}{\partial z} = 0 \tag{7.6}$$

The energy equation

$$\frac{\partial T}{\partial t} + v_x \frac{\partial T}{\partial x} + v_y \frac{\partial T}{\partial y} + v_z \frac{\partial T}{\partial z} = \frac{k}{\rho C_p}\left[\frac{\partial^2 T}{\partial x^2} + \frac{\partial^2 T}{\partial y^2} + \frac{\partial^2 T}{\partial z^2}\right] \tag{7.7}$$

The momentum equation in y-direction

$$\rho\left(\frac{\partial v_x}{\partial t} + v_x \frac{\partial v_x}{\partial x} + v_y \frac{\partial v_x}{\partial y} + v_z \frac{\partial v_x}{\partial z}\right) = -\frac{\partial P}{\partial x} + \frac{\partial}{\partial x}\left(\mu \frac{\partial v_x}{\partial x}\right) + \frac{\partial}{\partial y}\left(\mu \frac{\partial v_x}{\partial y}\right) + \frac{\partial}{\partial z}\left(\mu \frac{\partial v_x}{\partial z}\right) + \rho g_x \tag{7.8}$$

The momentum equation in x-direction

$$\rho\left(\frac{\partial v_y}{\partial t} + v_x \frac{\partial v_y}{\partial x} + v_y \frac{\partial v_y}{\partial y} + v_z \frac{\partial v_y}{\partial z}\right) = -\frac{\partial P}{\partial y} + \frac{\partial}{\partial x}\left(\mu \frac{\partial v_y}{\partial x}\right) + \frac{\partial}{\partial y}\left(\mu \frac{\partial v_y}{\partial y}\right) + \frac{\partial}{\partial z}\left(\mu \frac{\partial v_y}{\partial z}\right)$$
$$+ \rho_{ref} g_x (1 - \beta(T - T_{ref})) \tag{7.9}$$

The momentum equation in z-direction

$$\rho\left(\frac{\partial v_z}{\partial t} + v_x\frac{\partial v_z}{\partial x} + v_y\frac{\partial v_z}{\partial y} + v_z\frac{\partial v_z}{\partial z}\right) = -\frac{\partial P}{\partial z} + \frac{\partial}{\partial x}\left(\mu\frac{\partial v_z}{\partial x}\right) + \frac{\partial}{\partial y}\left(\mu\frac{\partial v_z}{\partial y}\right) + \frac{\partial}{\partial z}\left(\mu\frac{\partial v_z}{\partial z}\right) + \rho g_z \quad (7.10)$$

The boundary conditions used are: $T = T_w$, $u = 0$, $v = 0$, and $w = 0$ at top surface, bottom surface and side walls. The initial conditions used are: $T = T_{ref} = 40°C$, $u = 0$, $v = 0$, and $w = 0$.

For conduction dominated heating, Equations 7.7 through 7.10 are reduced to a single equation:

$$\frac{\partial T}{\partial t} = \frac{k}{\rho C_p}\left[\frac{\partial^2 T}{\partial x^2} + \frac{\partial^2 T}{\partial y^2} + \frac{\partial^2 T}{\partial z^2}\right] \quad (7.11)$$

7.2.4 PHYSICAL PROPERTIES

The properties of carrot–orange soup used in the simulations were: $\rho = 1026$ kg m^{-3}, $C_p = 3880$ Jkg^{-1} K^{-1}, and $k = 0.596$ Wm^{-1}K^{-1} [22,23]. The properties were calculated from the values reported for all food materials used in the soup using their mass fractions. In the simulation presented here, the viscosity was assumed function of temperature, following a second order polynomial. The viscosity of carrot–orange soup was measured at different temperatures and shear rates using Paar Physica Viscometer VT2 to obtain the constants of the polynomial. The values of the viscosities were taken at the extreme low shear rate, which closely simulate the situation in the pouch being sterilized. These values were 9.79 Pa.s, 5.14 Pa.s, and 2.82 Pa.s at 30°C, 50°C, and 70°C, respectively.

The variation of the density with temperature was governed by Boussinesq approximation (Equation 7.12).

$$\rho = \rho_{ref}[1 - \beta(T - T_{ref})] \quad (7.12)$$

where β is the thermal expansion coefficient of the liquid, T_{ref} and ρ_{ref} are the temperature and density at the reference condition [24]. The density is assumed constant in the governing equations except in the buoyancy term (Boussinesq approximation), where Equation 7.12 is used to describe its variation with temperature.

The magnitude of the Grashof number in the pouch during sterilization was in the range of 10^{-1}–10^1 (using maximum temperature difference and maximum viscosity). This magnitude of the Grashof number for the viscous liquid used in the simulations indicated that natural convection flow is laminar.

7.2.5 ASSUMPTIONS USED IN THE NUMERICAL SIMULATION

To simplify the problem, the following assumptions were made:

1. Heat generation due to viscous dissipation is negligible due to the use of high viscous liquid with very low velocities [25].
2. Boussinesq approximation is valid.
3. Specific heat (C_p), thermal conductivity (k), and volume expansion coefficient (β) are constants.
4. The assumption of no-slip condition at the inside wall of the pouch is valid.
5. The condensing steam maintains a constant temperature condition at the pouch outer surface.
6. The thermal boundary conditions are applied to liquid boundaries rather than to the outer boundaries of the pouch because of the low thermal resistance of the pouch wall.
7. The fluid is non-Newtonian, however due to the low shear rate it is considered as a Newtonian fluid.

7.3   HEATING AND COOLING CYCLE

The theoretical analysis for the cooling cycle in a three-dimensional pouch filled with carrot–orange soup based on FVM was also conducted for the three-dimensional pouch containing carrot–orange soup. The simulation covered both the heating and cooling cycles with a total time of 4800 s. The pouch used was the same as that used in other simulations. Governing equations, model parameters and boundary conditions were the same as those explained in Section 7.2. During the cooling cycle, these conditions changed and the temperature of the wall switched to 20°C (i.e., $T = T_w = 20$°C).

In the cooling cycle, a high temperature zone was found to develop in the core of the pouch and gradually migrate toward the widest end. The vertical location of this high temperature zone was about 60–70% of the pouch height, as described later.

7.4   BACTERIA INACTIVATION AND VITAMIN C DESTRUCTION

Bacteria (*Clostridium botulinum*) inactivation and vitamin C destruction were also studied and analyzed. Three-dimensional pouches filled with carrot–orange soup and heated by condensing steam at 121°C were tested.

The partial differential equations of continuity, momentum, and energy described earlier were solved together with that for bacteria and vitamin concentrations given below:

Mass balance for bacteria (concentration equation)

$$\frac{\partial C_{rb}}{\partial t} + v_x \frac{\partial C_{rb}}{\partial x} + v_y \frac{\partial C_{rb}}{\partial y} + v_z \frac{\partial C_{rb}}{\partial z} = -k_{Tb} C_{rb} \tag{7.13}$$

Mass balance for vitamins (concentration equation)

$$\frac{\partial C_{rv}}{\partial t} + v_x \frac{\partial C_{rv}}{\partial x} + v_y \frac{\partial C_{rv}}{\partial y} + v_z \frac{\partial C_{rv}}{\partial z} = -k_{Tv} C_{rv} \tag{7.14}$$

The boundary conditions used for bacteria inactivation and vitamins destruction on the pouch walls are:

$$\frac{\partial C_{rb}}{\partial x} = 0 \quad \text{and} \quad \frac{\partial C_{rv}}{\partial x} = 0 \tag{7.15}$$

$$\frac{\partial C_{rb}}{\partial y} = 0 \quad \text{and} \quad \frac{\partial C_{rv}}{\partial y} = 0 \tag{7.16}$$

$$\frac{\partial C_{rb}}{\partial z} = 0 \quad \text{and} \quad \frac{\partial C_{rv}}{\partial z} = 0 \tag{7.17}$$

Other boundary conditions for temperature and velocities are the same as those used in Section 7.2. The initial conditions used are: $T = T_{ref}$, $C_{rb} = 100\%$, $C_{rv} = 100\%$, $v_x = 0$, $v_y = 0$ and $v_z = 0$. The simplifications used in the simulation are identical as those used in Section 7.2.4. Furthermore, the effect of molecular diffusion of bacteria and vitamin C are neglected due to the dominating influence of convection motion. This assumption has been verified by Ghani et al. [26], which allows omitting the diffusion terms in Equations 7.13 and 7.14.

7.4.1   Kinetics of Bacteria Inactivation and Vitamin Destruction

The rate of bacteria inactivation and vitamin destruction is usually assumed to follow first order kinetics [27]. It is known that the reaction rate constants are functions of temperature and are usually described by Arrhenius equation:

TABLE 7.1

Kinetic Data for some Reaction Processes used in our Simulations as Reported by Fryer et al.

Property	D_{121} (min) Reported	D_{121} (min) used	E_a (kJmol^{-1}) Reported	E_a (kJmol^{-1}) used
For bacteria deactivation				
Clostridium botulinum	0.1–0.3	0.2	265–340	300
For vitamins destruction				
Ascorbic acid (C)	245	245	65–160	100

Source: Data from Fryer, J. P., Pyle, D. L., and Rielly, C. D., *Chemical Engineering for the Food Industry*. Blackie Academic and Professional, UK, 1997.

$$\text{For bacteria: } k_b = A_b \exp\left(\frac{-E_b}{R_g T}\right) \tag{7.18}$$

$$\text{For vitamins: } k_v = A_v \exp\left(\frac{-E_v}{R_g T}\right) \tag{7.19}$$

where, A_b and A_v are the reaction frequency factors, E_b, and E_v are the activation energies for the two processes respectively, R_g is the universal gas constant and T is the temperature. Accurate kinetic parameters such as reaction rate constants and activation energy are essential to predict quality changes during food processing [28]. In food process engineering, the decimal reduction time (D) is used more often. The relationship between the reaction rate constant and decimal reduction time is [29].

$$k_T = \frac{2.303}{D_T} \tag{7.20}$$

The decimal reduction times and activation energies of bacteria inactivation and destruction of vitamin C are shown in Table 7.1.

The reaction rate constants k_b and k_v at 121°C for bacteria and vitamin C are calculated from Equation 7.20 using the values of D reported in the literature. Equations 7.18 and 7.19 are then used to calculate the constants of Arrhenius equation A_b and A_v. These equations used to describe the kinetics of the biochemical changes are introduced to the existing software package using a FORTRAN code.

7.5 RESULTS OF SIMULATIONS

7.5.1 Temperature Distribution and Flow Profile

The objective here is to analyze the calculated temperature distribution inside a three-dimensional pouch filled with canned foods (carrot–orange soup) and sterilized by condensing steam heating at 121°C. The migration of the SHZ and the effect of the velocity profiles on its shape were also analyzed. The combined heating/cooling cycle was also presented and discussed.

Two cases of carrot–orange soup have been simulated and studied. The first was for the pouch heated with a convection dominating heating inside it, while the second was for the pouch assumed to be heated by conduction only. The results of temperature distribution and migration of the SHZ for both cases were compared to estimate the importance of natural convection heating in such a process.

Figures 7.4 through 7.6 show the temperature distribution at different x, y and z-planes in a pouch filled with carrot–orange soup at the end of heating (3000 s). The three figures combined together, provide a detailed picture of the location of the SHZ at the end of the heating period. Such detailed information can be obtained at any time during sterilization. Figure 7.5 clearly shows the settlement of the SHZ at about 30% of the pouch height. This figure shows that the SHZ is not a stationary region and its location is not at the geometric center of the pouches.

Figure 7.7 shows temperature distribution in the pouch at different periods of heating (60 s, 200 s, 300 s, 1000 s, 1800 s, and 2400 s). Initially, the content of the pouch is at uniform temperature. As heating progresses, the mode of heat transfer changes from conduction to convection. Figure 7.7a at $t = 60$ s shows a similar temperature profile as that in Figure 7.8a for conduction, indicating that the heating process is governed by conduction heating. At the later stage of heating (at t greater than 300 s), the isotherms are influenced by convection as may be seen from the comparison of the isotherms shown in Figure 7.7c and e for convection heating with those of Figure 7.8b and c for conduction heating only. At the end of heating (at $t = 3000$ s), the circulation of the flow almost ceased due to the low temperature differences, and temperature distributions become similar to conduction heating as discussed earlier.

Figure 7.7d shows the temperature distribution in the pouch after 1000 s of heating, this figure shows the presence of the SHZ at 30% height from the bottom, starting from the deepest end and extending to almost 60% of the pouch length. Within this region, Figure 7.7d shows the existence of a relatively hotter zone in the middle of the SHZ. These observations can be explained with reference to Figure 7.12 of the x-plane velocity vector, which shows a strong circulation. Although Figure 7.9 shows the existence of stagnant regions near both the deepest and narrowest end of the pouch, the SHZ appear only in the deepest end because conduction dominated heat transfer at the narrowest end of the pouch. In Figure 7.9, the maximum velocity is found near the deepest end, which raised the temperature of the location near this end.

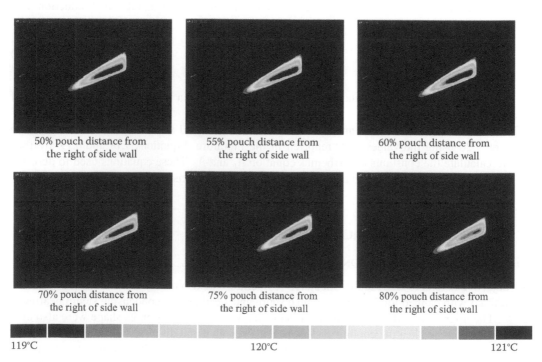

FIGURE 7.4 Temperature profiles at different x-planes in a pouch filled with carrot–orange soup.

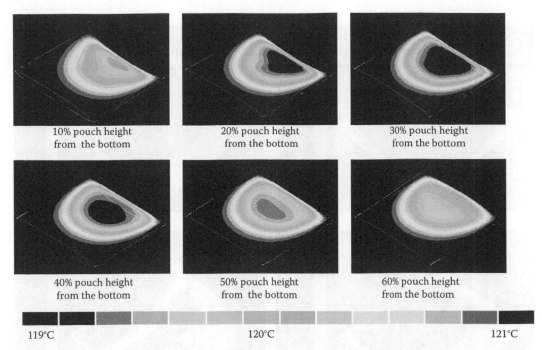

10% pouch height from the bottom	20% pouch height from the bottom	30% pouch height from the bottom
40% pouch height from the bottom	50% pouch height from the bottom	60% pouch height from the bottom

119°C   120°C   121°C

FIGURE 7.5   Temperature profiles at different *y*-planes in a pouch filled with carrot–orange soup and heated by condensing steam after 3000 s. (From Ghani, A. G., Farid, M. M., Chen, X. D., and Richards, P., *Journal of Process Mechanical Engineering*, 215, Part E, 1–9, 2001. With permission.)

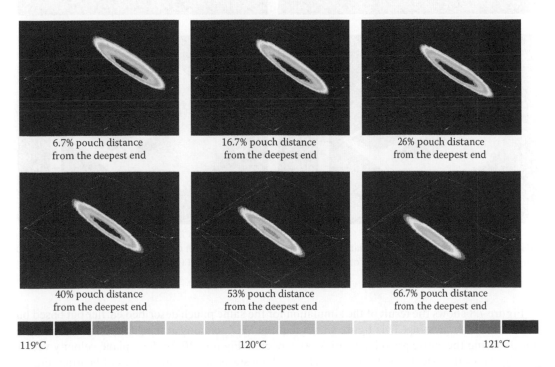

6.7% pouch distance from the deepest end	16.7% pouch distance from the deepest end	26% pouch distance from the deepest end
40% pouch distance from the deepest end	53% pouch distance from the deepest end	66.7% pouch distance from the deepest end

119°C   120°C   121°C

FIGURE 7.6   Temperature profiles at different *z*-planes in a pouch filled with carrot–orange soup and heated by condensing steam after 3000 s.

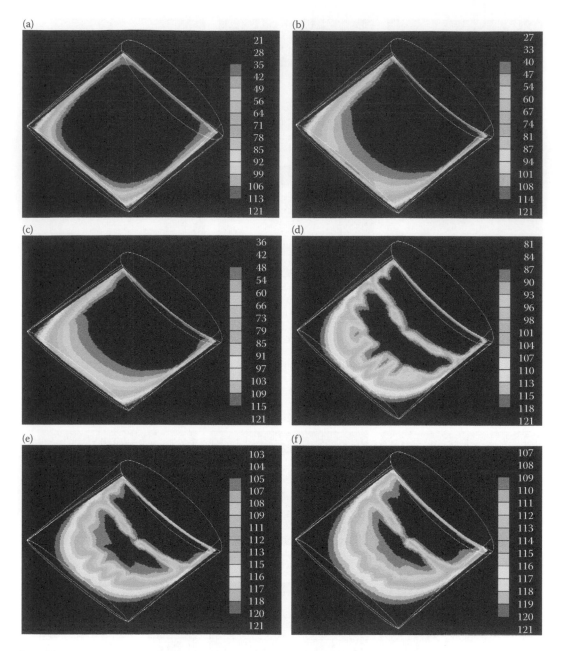

FIGURE 7.7 Temperature profile planes at 30% height from the bottom of pouch filled with carrot–orange soup and heated for different periods of (a) 60 s; (b) 200; (c) 300 s; (d) 1000 s; (e) 1800 s; (f) 3000 s.

Figure 7.8 shows the result of the simulation for the same pouch described earlier but based on pure conduction heating. This figure shows that the SHZ remains at the geometric center of the pouch during the entire period of heating, as expected. Figure 7.10 for the y-plane velocity vector clearly shows the effect of natural convection, which starts at the early stages of heating. Figure 7.10 also shows the axial velocity of the soup, which is found in the order of 10^{-2}–10^{-3} mm s^{-1}. The small velocity is due to limited buoyancy caused in the shallow pouch containing viscous soup. Convective circulation takes the form of a distorted torroid with flow rising up the pouch

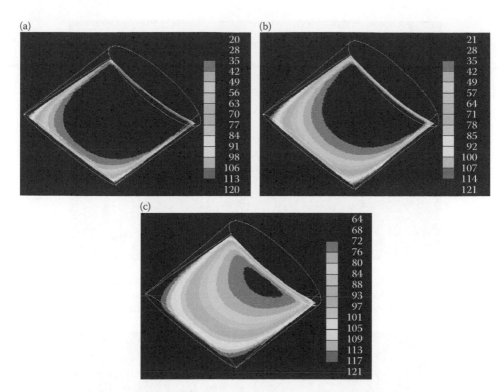

FIGURE 7.8   Temperature profile planes at 30% height from the bottom of pouch filled with carrot–orange soup and heated by conduction only for different periods of (a) 60 s; (b) 300 s; (c) 1800 s. (From Ghani, A. G., Farid, M. M., Chen, X. D., and Richards, P., *Journal of Process Mechanical Engineering*, 215, Part E, 1–9, 2001. With permission.)

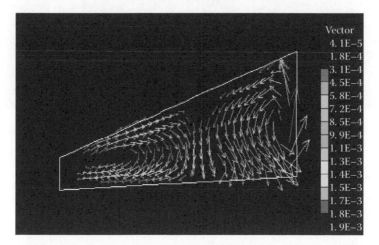

FIGURE 7.9   The center of *x*-plane velocity vector of carrot–orange soup in a pouch heated by condensing steam after 1000 s, showing the effect of natural convection. (From Ghani, A. G., Farid, M. M., Chen, X. D., and Richards, P., *Journal of Process Mechanical Engineering*, 215, Part E, 1–9, 2001. With permission.)

walls and descending in the center (Figure 7.11). At the same time of heating (at $t = 1000$ s), the value of the velocity vector in *x*-direction (Figure 7.9) is almost ten times higher than the value of the velocity in *z*-direction (Figure 7.11). This means that the velocity profiles in this direction has a clear dominant effect on the temperature distribution than those in *z*-direction as seen clearly from Figure 7.7d.

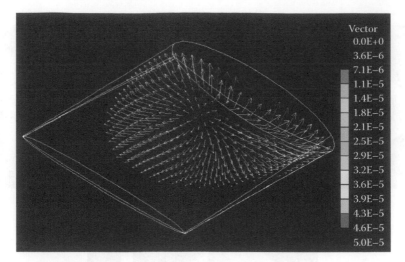

FIGURE 7.10   The *y*-plane velocity vector of carrot–orange soup in a pouch heated by condensing steam after 300 s, showing the effect of natural convection. (From Ghani, A. G., Farid, M. M., Chen, X. D., and Richards, P., *Journal of Process Mechanical Engineering*, 215, Part E, 1–9, 2001. With permission.)

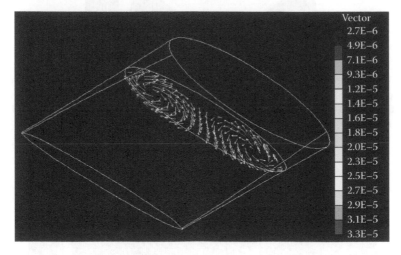

FIGURE 7.11   The *z*-plane velocity vector of carrot–orange soup in a pouch heated by condensing steam after 1000 s, showing the effect of natural convection.

7.5.2   HEATING AND COOLING CYCLE

It is necessary to analyze the cooling process following the heating process in food sterilization as the core of the pouch may stay hot for a significant length of time, which will influence the sterilization process. In this section, the results of the temperature distribution and the migration of the SHZ during heating and slowest cooling zone (SCZ), during cooling are compared and analyzed.

The results of the simulation are presented in Figures 7.12 and 7.13. Figure 7.12 shows that the location of the SCZ in the lower half of the pouch lies at the center of the pouch. At higher locations, the SCZ tends to move toward the widest end of the pouch, which is due to the effect of cooling from the upper surface of the pouch. Figure 7.12 also shows that the SCZ covers a wider area at a location close to about 70% of pouch height from the bottom unlike in heating, where the SHZ covered a wider area at location 30% of pouch height from bottom as shown clearly in Figure 7.5.

During cooling, the lowest half of the pouch will be less influenced by convection compared to the upper half of the pouch, which is due to the variation of the density with temperature. This

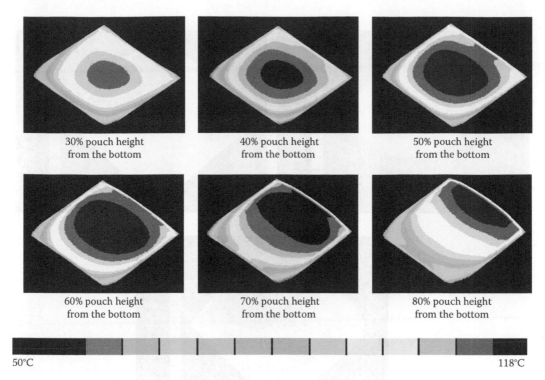

30% pouch height from the bottom

40% pouch height from the bottom

50% pouch height from the bottom

60% pouch height from the bottom

70% pouch height from the bottom

80% pouch height from the bottom

50°C   118°C

FIGURE 7.12   Temperature profiles at different y-planes in a pouch filled with carrot–orange soup after 600 s from the start of the cooling cycle. (From Ghani, A. G., Farid, M. M., and Chen, X. D., *Proceedings of the Institution of Mechanical Engineers*, 217, 1–9, 2003. With permission.)

is reflected on the temperature profiles shown in Figure 7.15, which shows conduction controlled heating at locations 30–40% of pouch height from bottom and convection control heating at other heights, especially at 70 and 80% of pouch height.

Observations of temperature profiles at different y-planes and for different periods during the heating cycle show different results compared to cooling. During heating, the lowest half of the pouch will be heated slower while the upper half of the pouch is heated slower in the case of cooling. This is why all of our results and profiles of temperature and velocity during the heating cycle are studied and presented in the lower part of the pouch, where the location of the SHZ and the effect of natural convection are dominant.

Figure 7.13 shows the temperature profile planes at 80% height from the bottom of pouch filled with carrot–orange soup at different periods of heating and cooling. Cooling with water at 20°C starts at $t = 3600$ s and continued until 4800 s. In this figure, the results for the temperature profiles during the cooling cycle show that the cooling process is influenced by convection as in the heating process.

7.5.3 VITAMINS DESTRUCTION

The effect of sterilization temperature on the rate of destruction of vitamin C is presented in Figures 7.14 through 7.16. Figure 7.14 shows the results of the simulation at times of 200 s, 1000 s, and 3000 s. At the early stage of heating ($t = 200$ s), heat transfer is solely controlled by conduction as shown by the comparison between the convection heating (Figure 7.7a) and conduction heating (Figure 7.8a) discussed earlier. The destruction of vitamin C in Figure 7.14a follows similar trends but with varying degrees.

As time progresses, heating is dominated by natural convection as shown in Figure 7.7d. Figure 7.14b shows the results of the simulation after relatively longer periods of 1000 s. This figure shows that vitamin C destruction occurs mostly at locations near the wall and especially at

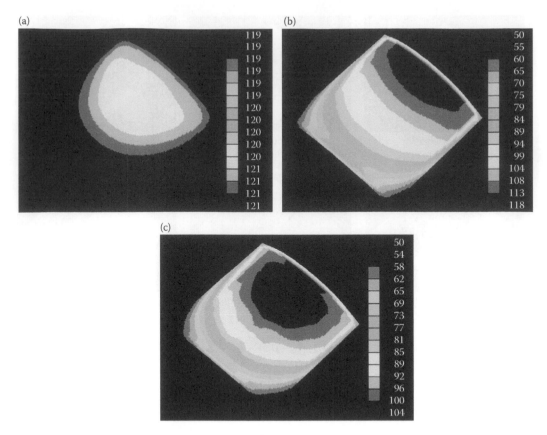

FIGURE 7.13 Temperature profile planes at 80% height from the bottom of pouch filled with carrot–orange soup after (a) 3000 s from the start of the heating cycle; (b) 600 s; (c) 900 s from the start of the cooling cycle. (From Ghani, A. G., Farid, M. M., and Chen, X. D., *Proceedings of the Institution of Mechanical Engineers*, 217, 1–9, 2003. With permission.)

the narrowest end of the pouch, which high temperature locations. The figures also show that the vitamins are concentrated at the SHZ and the destruction of vitamins has occurred mostly in other locations. The liquid and thus the vitamins carried with it are exposed to much less thermal treatment at the SHZ than the rest of the product. Figure 7.14b shows clearly the combined effects of the temperature distribution (Figure 7.7d) and velocity profile (Figures 7.9 through 7.11) on the shape and location of the high vitamin concentration zone (HVCZ). Figure 7.11 for the z-plane velocity vector at 8 cm from the widest end shows the effect of fluid circulation that leads to two stagnant zones, which influence the HVCZ shown by Figure 7.14b. These results lead to the conclusion that the vitamin profiles depend not only on the temperature distribution but also on the velocity profiles in the pouch. The locations of the high vitamin concentration zone occur almost within the stagnant zones, which belong to minimum liquid velocity.

Figure 7.15 shows the relative concentration profiles of vitamin C in the x-plane of the center of the pouch filled with carrot–orange soup at the end of heating (3000 s). This figure shows the locations of HVCZ at the two stagnant zones, which can be explained clearly in terms of x-plane velocity vector shown in Figure 7.9.

Figure 7.16 shows the locations of high vitamin C concentration zones at different z-planes in the pouch after 1000 s of heating. Figure 7.16a and b show that HVCZ occurs at the lower half of the pouch, which is due to the effect of natural convection currents such as that shown in Figure 7.11. At locations close to the narrowest end, Figure 7.16c and d show that the HVCZ stays at the mid-height of the plane. This is due to the conduction dominating heating in the locations at the narrowest end of the pouch.

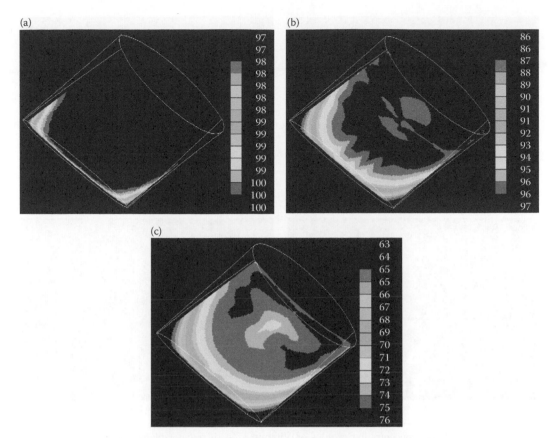

FIGURE 7.14 Relative concentration profiles of Vitamin C at 30% height from the bottom of pouch filled with carrot–orange soup and heated by condensing steam after periods of (a) 200 s; (b) 1000 s; (c) 3000 s.

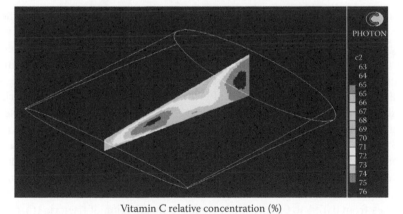

Vitamin C relative concentration (%)

FIGURE 7.15 Relative concentration profiles of Vitamin C at 50% of x-plane of pouch filled with carrot–orange soup and heated by condensing steam after 3000 s.

7.5.4 BACTERIA INACTIVATION

At the early stage of heating, the bacteria concentration profiles is almost identical to what is usually found in pure conduction heating, with live bacteria killed only at locations close to the wall of the pouch. At this early stage of heating, the bacteria concentration profile seems to be influenced mainly by temperature profile and not by flow pattern. This is evident from the higher rate of bacteria

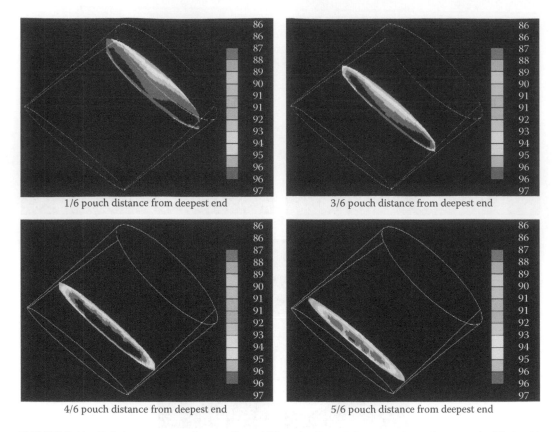

1/6 pouch distance from deepest end	3/6 pouch distance from deepest end
4/6 pouch distance from deepest end	5/6 pouch distance from deepest end

FIGURE 7.16   Relative concentration profiles of Vitamin C at different y-planes in a pouch filled with carrot–orange soup and heated by condensing steam after 1000 s.

inactivation at the narrowest end of the pouch where the temperature is higher, while the bulk of the pouch remains minimally affected.

As heating progresses (at $t = 1000$ s), the temperature profiles are strongly influenced by convection currents (Figure 7.7d) as described earlier. Figure 7.17 shows the results of the simulation (the relative bacteria concentration profiles of *Clostridium botulinum*) for a pouch filled with carrot–orange soup and heated by condensing steam after 1000 s. The temperature and bacteria concentration profiles shown in these figures (Figures 7.7d and 7.17) are very different from those observed at the beginning of heating. The SHZ keeps moving during heating and eventually stays at 30% of the bottom of the pouch. The liquid and thus the bacteria carried with it at these locations are exposed to much less thermal treatment than the rest of the product. This can be shown clearly from the relative bacteria concentration profile shown in Figure 7.17b at 30% pouch height from the bottom. After 1000 s of heating, the SHZ reaches 81°C (Figure 7.7d). In this zone SHZ, the relative bacteria concentration ranges from 0.07% to 0.95%, while complete inactivation of bacteria ($C_b/C_{bo} = 0$) occurs in most of the other locations (Figure 7.17).

Figure 7.17 also shows that bacteria deactivation is influenced significantly by both temperature (Figure 7.7d) and flow pattern (Figures 7.9, 7.10, and 7.11), unlike those observed at the early stage of heating. The highest concentration of bacteria shown in Figure 7.17 occurs at two locations. These locations belong to minimum liquid velocity and low temperature zones. Similar observations have been found at later stages of heating. At the end of heating, the SHZ remains at the same location as the HBCZ. It is only after 3000 s of heating that most bacteria have been inactivated. This is true only for the highly viscous liquid used in the simulation. A much shorter time would be required for the sterilization of liquid of low viscosity such as milk, orange juice, etc.

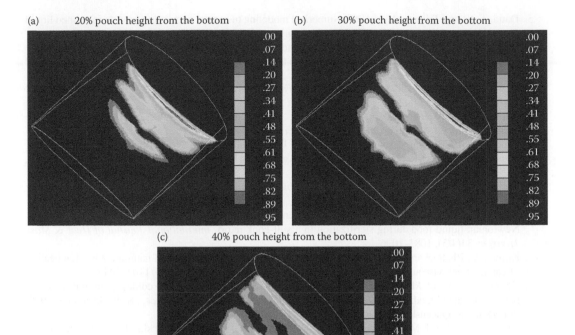

FIGURE 7.17   Relative concentration profiles of Clostridium botulinum at different y-planes in a pouch filled with carrot–orange soup and heated by condensing steam after 1000 s. (a) 20% pouch height from the bottom; (b) 30% pouch height from the bottom; (c) 40% pouch height from the bottom.

The relative bacteria and vitamin concentration profiles during natural convection heating are expected to depend on both temperature and flow pattern. Hence, the locations of low bacteria inactivation zone belong to low liquid velocity and low temperature zones. However, it was found from Figures 7.7d, 7.9, 7.11, 7.14b and 7.17 that, at the same time (at $t = 1000$ s), the relative bacteria concentration profile (Figure 7.17) seems to be influenced mainly by temperature profiles (Figure 7.7d). The influence of flow pattern (Figures 7.9 and 7.11) on vitamin C concentration profile (Figure 7.14b) is even stronger than those for the bacteria. This is because the low value of decimal reduction time of bacteria compared to that of vitamin C which makes bacteria inactivation more sensitive to temperature. This explains why the bacteria profile follows exactly the temperature profile.

REFERENCES

1. Barbosa-Canovas, G. V., Ma, L., and Barletta B. 1997. *Food Engineering Laboratory Manual*. Technomic Publishing Company, Inc, Lancaster, PA.
2. Teixeria, A. A., Dixon, J. R., Zahradnik, J. W., and Zinsmeister, G. E. 1969. Computer optimization of nutrient retention in thermal processing conduction heated foods. *Food Technology*, 23(6), 134–140.
3. Naveh, D., Kopelman, I. J., and Pflug, I. J. 1983. The finite element method in thermal processing of foods. *Journal of Food Science*, 48, 1086–1093.
4. Datta, A. K. and Teixeira, A. A. 1988. Numerically predicted transient temperature and velocity profiles during natural convection heating of canned liquid foods. *Journal of Food Science*, 53(1), 191–195.

5. Datta, A. K. and Teixeira, A. A. 1987. Numerical modeling of natural convection heating in canned liquid foods. *Transaction of American Society of Agricultural Engineers*, 30(5), 1542–1551.
6. Nicolai, B. M., Verboven, B., Scheerlinck, N., and De Baerdemaeker, J. 1998. Numerical analysis of the propagation of random parameter fluctuations in time and space during thermal food processes. *Journal of Food Engineering*, 38, 259–278.
7. Herson, A. C. and Hulland, E. D. 1980. *Canned Foods. Thermal Processing and Microbiology*. Churchill Livingstone, Edinburgh, UK.
8. Datta, A. K., Teixeira, A. A., and Manson, J. E. 1986. Computer based retort control logic for on-line correction process derivative. *Journal of Food Science*, 5, 480–483.
9. Hiddink, J. 1972. Natural convection heating of liquids with reference to sterilization of canned foods. Agriculture Research Report No 839. Center for Agricultural Publishing and Documentation, Wageningen, The Netherlands.
10. Nickerson, J. T. and Sinskey, A. J. 1972. *Microbiology of Foods and Food Processing*. Elsevier Publishing Co, New York, NY.
11. Kumar, A. and Bhattacharya, M. 1991. Transient temperature and velocity profiles in a canned non-Newtonian liquid food during sterilization in a still-cook retort. *International Journal of Heat & Mass Transfer*, 34(4/5), 1083–1096.
12. Kumar, A., Bhattacharya, M., and Blaylock, J. 1990. Numerical simulation of natural convection heating of canned thick viscous liquid food products. *Journal of Food Science*, 55(5), 1403–1411.
13. Ghani, A. G., Farid, M. M., and Chen, X. D. 1998. A CFD simulation of the coldest point during sterilization of canned food. *The 26th Australian Chemical Engineering Conference*, 28–30 September 1998, Port Douglas, Queensland, Australia, No. 358.
14. Ghani, A. G., Farid, M. M., Chen, X. D., and Richards, P. 1999. Numerical simulation of natural convection heating of canned food by computational fluid dynamics. *Journal of Food Engineering*, 41(1), 55–64.
15. Ghani, A. G., Farid, M. M., Chen, X. D., and Richards, P. 1999. An investigation of deactivation of bacteria in canned liquid food during sterilization using computational fluid dynamics (CFD). *Journal of Food Engineering*, 42, 207–214.
16. Bhowmik, S. R. and Tandon, S. 1987. A method of thermal process evaluation of conduction heated foods in retortable pouches. *Journal of food Science*, 52(1), 202–209.
17. Tandon, S. and Bhowmik, S. R. 1986. Evaluation of thermal processing of retortable pouches filled with conduction heated foods considering their actual shapes. *Journal of Food Science*, 51(3), 709–714.
18. Scott, G. and Richardson, P. 1997. The applications of computational fluid dynamics in the food industry. *Journal of Trends in Food Science & Technology*, 8, 119–124.
19. *PHOENICS Reference Manual*, Part A: PIL. Concentration Heat and Momentum Limited, TR 200 A, Bakery House, London, UK.
20. Malin, M. R. and Waterson, N. P. 1999. Schemes for convection discretization on PHOENICS. *The PHOENICS Journal* 12(2), 173–201.
21. Spalding, D. B. 1972. A novel finite-difference formulation for differential expressions involving both first and second derivatives. *International Journal of Numerical Methodology Engineering*, 4, 551–559.
22. Hayes, G. D. 1987. *Food Engineering Data Handbook*. John Wiley and Sons Inc., New York, NY.
23. Rahman, R. 1995. *Food Properties Handbook*. CRC Press, Inc., Boca Raton, FL.
24. Adrian, B. 1993. *Heat Transfer*. John Wiley and Sons Inc., New York, NY, 339–340.
25. Mills, A. F. 1995. *Basic Heat and Mass Transfer*. Richard Irwin, Inc. USA.
26. Ghani, A. G., Farid, M. M., Chen, X. D., and Richards, P. 2001. A computational fluid dynamics study on the effect of sterilization temperatures on bacteria deactivation and vitamin destruction. *Journal of Process Mechanical Engineering*, 215, Part E, 1–9.
27. Reuter. 1993. *Processing of Foods*. Technomic Publishing Company Inc.. Lancaster, PA.
28. Ilo, S. and Berghofer, E. 1999. Kinetics of color changes during extrusion cooking of maize gritz. *Journal of Food Engineering*, 39, 73–80.
29. Heldman, D. R. and Hartel R. W. 1997. *Principle Food Processing*. Chapman and Hall, New York, NY.
30. Fryer, J. P., Pyle, D. L., and Rielly, C. D. 1997. *Chemical Engineering for the Food Industry*. Blackie Academic and Professional, UK.
31. Ghani, A. G., Farid, M. M., and Chen, X. D. 2003. A computational and experimental study of heating and cooling cycles during thermal sterilization of liquid foods in pouched using CFD. *Proceedings of the Institution of Mechanical Engineers*, 217, 1–9.

8 Heat Exchangers

Prabhat Kumar and K.P. Sandeep
North Carolina State University

CONTENTS

8.1 INTRODUCTION

A heat exchanger is an equipment which transfers thermal energy between two or more fluids. Heat exchangers are used in various industries for a wide range of applications. Typical applications of heat exchangers include heating or cooling of a fluid and evaporation or condensation of single or multicomponent fluids. Common examples of heat exchangers are shell and tube heat exchangers, automobile radiators, condensers, evaporators, and cooling towers [1]. Heat exchangers can be classified into contact and noncontact types. There is direct physical contact between the product and heating or cooling system in the contact type of heat exchangers. In the noncontact type heat exchangers, the product is physically separated from the heating or cooling medium [2]. Steam injection and steam infusion are examples of contact heat exchangers. Some of the noncontact type heat exchangers are double tube, triple tube, multitube, plate, scraped surface, shell and tube, and helical coil heat exchangers. A particular type of heat exchanger needs to be selected for a given application. Selection of a heat exchanger depends on the characteristics of the product, potential changes in the product, cleanability, adaptability to regeneration, foot print, heat transfer coefficient, and cost.

8.2 CONVECTIVE HEAT TRANSFER COEFFICIENT

The term convection refers to the transfer of heat within liquids or through a solid–fluid interface. Convective heat transfer can be either free (natural) convection or forced convection. In free convection, motion of the fluid results from density changes whereas the fluid is forced to flow by pressure differences (by using pump or fan) in forced convection. The rate of heat transfer from a solid body to a fluid by convection at the solid surface is defined as follows [3]:

$$q = hA(T_w - T_f) \tag{8.1}$$

where q is the heat transfer rate (W), A is the area available for heat transfer (m^2), T_w is the temperature of the solid surface (K), T_f is the average or bulk temperature of the fluid (K), and h is the convective heat transfer coefficient (W.m^{-2}.K^{-1}).

Convective heat transfer coefficient (h) is not a property of the solid material, but it depends on the properties of the fluid (density, specific heat, viscosity, and thermal conductivity), velocity of the fluid, and geometry and roughness of the surface of the solid. A high value of h results in a high rate of heat transfer [2]. Determination of h is of great importance for designing heat exchangers and determining temperature distribution in fluids. The convective heat transfer coefficient is determined from empirical correlations. Dimensionless numbers such as the Reynolds, Nusselt and Prandtl numbers are used in correlations for computing the convective heat transfer coefficient. Reynolds number (N_{Re}), which is defined as the ratio of the inertial to the viscous forces is given by [3]:

$$N_{Re} = \frac{\rho v_{av} d_c}{\mu} \tag{8.2}$$

where ρ is the density of the fluid (kg.m^{-3}), v_{av} is the average fluid velocity (m.s^{-1}), d_c is the characteristic dimension (m), and μ is the viscosity of the fluid (Pa.s). For flow through a circular pipe, d_c is the diameter (D) of the pipe. Reynolds number provides an insight into the flow characteristics of a fluid. A Reynolds number of less than 2100 indicates laminar flow whereas a Reynolds number greater than 4000 indicates turbulent flow. Transition flow takes place when Reynolds number is between 2100 and 4000.

Nusselt number (N_{Nu}), which is the ratio of convective heat transfer to the conductive heat transfer is given by [2]:

$$N_{Nu} = \frac{h d_c}{k} \tag{8.3}$$

where h is the convective heat transfer coefficient (W.m^{-2}.K^{-1}), d_c is the characteristic dimension (m), and k is the thermal conductivity of the fluid (W.m^{-1}.K^{-1}).

Prandtl number (N_{Pr}), which is the ratio of momentum diffusivity to thermal diffusivity is given by [2]:

$$N_{Pr} = \frac{\mu c_p}{k} \tag{8.4}$$

where c_p is the specific heat of the fluid (J.kg^{-1}.K^{-1}). Prandtl number is an indicator of the relative thickness of the hydrodynamic boundary layer compared to the thermal boundary layer.

8.2.1 Forced Convection

For forced convection, N_{Nu} is a function of N_{Re} and N_{Pr}. For fully developed laminar flow in pipes with constant surface temperature, Nusselt number is given by [2]:

$$N_{Nu} = 3.66 \tag{8.5}$$

For fully developed laminar flow in pipes with constant surface heat flux, Nusselt number is given by [2]:

$$N_{Nu} = 4.36 \tag{8.6}$$

In Equations 8.5 and 8.6, thermal conductivity is obtained at the average fluid temperature and d_c is inside diameter of the pipe.

For both the entry region and fully developed region for laminar flow in pipes, Nusselt number is given by [2]:

$$N_{Nu} = 1.86 \left(N_{Re} \times N_{Pr} \times \frac{d_c}{L} \right)^{0.33} \left(\frac{\mu_b}{\mu_w} \right)^{0.14} \tag{8.7}$$

where L is the length of the pipe, d_c is the inside diameter of the pipe, and all fluid properties are obtained at the average fluid temperature except μ_w which is obtained at the surface temperature (T_w) of the wall.

For transition flow in pipes, Nusselt number is given by [2]:

$$N_{Nu} = \frac{\dfrac{f}{8}(N_{Re} - 1000)N_{Pr}}{1 + 12.7 \left(\dfrac{f}{8} \right)^{0.5} (N_{Pr}^{(2/3)} - 1)} \tag{8.8}$$

where friction factor, f, is obtained for smooth pipes using the following expression [2]:

$$f = \frac{1}{(0.790 \ln N_{Re} - 1.64)^2} \tag{8.9}$$

For turbulent flow in pipes, Nusselt number is given by [2]:

$$N_{Nu} = 0.023(N_{Re})^{0.8}(N_{Pr})^{0.33} \left(\frac{\mu_b}{\mu_w} \right)^{0.14} \tag{8.10}$$

For convection in noncircular ducts, an equivalent diameter, d_c is calculated as follows [2]:

$$d_c = \frac{4 \times A_s}{P_w} \tag{8.11}$$

where A_s is the free surface area and P_w is the wetted perimeter. For an annular pipe with inner diameter (ID) of the outer pipe as D_1 and outer diameter (OD) of inner pipe as D_2, d_c is determined as follows:

$$d_c = \frac{4 \times \pi \left(\dfrac{D_2^2}{4} - \dfrac{D_1^2}{4} \right)}{\pi(D_2 + D_1)} = D_2 - D_1 \tag{8.12}$$

For flow past immersed objects, heat transfer depends on the geometry of the object, orientation of the body, proximity to other objects, flow rate, and fluid properties. For a flow parallel to a flat plate, Nusselt number is given by [3]:

$$N_{Nu} = 0.664(N_{Re})^{0.5}(N_{Pr})^{1/3}, (N_{Re} < 3 \times 10^5; N_{Pr} > 0.7)$$

$$N_{Nu} = 0.0366(N_{Re})^{0.8}(N_{Pr})^{1/3}, (N_{Re} > 3 \times 10^5; N_{Pr} > 0.7)$$

(8.13)

where the characteristic dimension, d_c, is the length (L) of the plate.

For flow past a single sphere, Nusselt number is given by [2]:

$$N_{Nu} = 2 + 0.060(N_{Re})^{0.5}(N_{Pr})^{(1/3)}, (N_{Re} < 70,000; 0.6 < N_{Pr} < 400)$$

(8.14)

where the characteristic dimension, d_c, is the outside diameter of the sphere. In the above equations for flow past immersed objects, fluid properties are evaluated at the film temperature T_f given by [3]:

$$T_f = \frac{T_w + T_b}{2}$$

(8.15)

where T_w is the surface temperature of the wall and T_b is the bulk fluid temperature. For flow past a cylinder, bank of cylinders and packed beds, correlations have been provided by Geankoplis [3]. Correlations for Nusselt number arising from different forced convection situations can be found in the literature [4–6].

8.2.2   FREE CONVECTION

Free convection occurs when a solid surface comes in contact with fluid at different temperature. Density differences in the fluid arising from heating provide the buoyancy force required to move the fluid. Empirical correlations for Nusselt number during free convection are of the following form [2]:

$$N_{Nu} = a(N_{Ra})^m$$

(8.16)

where a and m are constants given in Table 8.1, and N_{Ra} is the Rayleigh number. Rayleigh number is the product of Grashof number (N_{Gr}) and Prandtl number (N_{Pr}). Grashof number (N_{Gr}), which is a ratio of the buoyancy forces to the viscous forces is defined as [2]:

$$N_{Gr} = \frac{d_c^3 \rho^2 g \beta \Delta T}{\mu^2}$$

(8.17)

where d_c is the characteristic dimension (m), ρ is the density (kg.m^{-3}), g is the acceleration due to gravity (9.8 m.s^{-2}), β is the coefficient of volumetric expansion (K^{-1}), ΔT is the temperature difference between the wall and surrounding fluid (K), and μ is the viscosity of the fluid (Pa.s). Correlations for Nusselt number arising from different free convection situations can be found in the literature [4–6].

TABLE 8.1
Constants a and m to Calculate Nusselt Number During Free Convection

Physical Geometry	N_{Ra}	a	m
Vertical planes and cylinders	$<10^4$	1.36	1/5
$[L < 1 \text{ m}]$	10^4-10^9	0.59	1/4
	$>10^9$	0.13	1/3
Horizontal cylinders	$<10^{-5}$	0.49	0
$[D < 0.20 \text{ m}]$	$10^{-5}-10^{-3}$	0.71	1/25
	$10^{-3}-1$	1.09	1/10
	$1-10^4$	1.09	1/5
	10^4-10^9	0.53	1/4
	$>10^9$	0.13	1/3
Horizontal plates	$10^5-2 \times 10^7$	0.54	1/4
Upper surface of heated plates or lower surface of cooled plates	$2 \times 10^7-3 \times 10^{10}$	0.14	1/3
Lower surface of heated plates or upper surface of cooled plates	10^5-10^{11}	0.58	1/5

Source: Adapted from Geankoplis, C.J., *Transport Processes and Unit Operations*, 3rd ed. Prentice Hall Inc., NJ, 1993.

8.3 OVERALL HEAT TRANSFER COEFFICIENT

Conductive and convective heat transfers occur simultaneously in many heating/cooling applications. For a hot fluid flowing in a pipe, heat transfer occurs first due to forced convection at the inside surface, then by conduction in the pipe wall, and finally by free convection at the outer surface of the pipe. Using the approach of thermal resistance values, the rate of heat transfer can be written as [2]:

$$q = \frac{T_i - T_o}{R_t} \tag{8.18}$$

where q is the rate of heat transfer (W), T_i is the temperature of hot fluid (K), T_o is the outside temperature (K), and R_t is the total thermal resistance. R_t is given by [2]:

$$R_t = R_i + R_k + R_o \tag{8.19}$$

where

$$R_i = \frac{1}{h_i A_i}$$

$$R_k = \frac{\ln\left(\dfrac{r_o}{r_i}\right)}{2\pi k L} \tag{8.20}$$

$$R_o = \frac{1}{h_o A_o}$$

The overall heat transfer coefficient (U) determines the rate of heat transfer in a heat exchanger. The rate of heat transfer using the overall heat transfer coefficient can be written as [2]:

$$q = \frac{T_i - T_o}{\dfrac{1}{U_i A_i}} \tag{8.21}$$

where U_i is the overall heat transfer coefficient based on inside area of the pipe (W.m^{-2}.K^{-1}) and A_i is the inside heat transfer area of the pipe (m^2). From the above equations, the expression for U_i can be written as:

$$\frac{1}{U_i A_i} = \frac{1}{h_i A_i} + \frac{\ln\left(\frac{r_o}{r_i}\right)}{2\pi k L} + \frac{1}{h_o A_o}$$

(8.22)

The overall heat transfer coefficient (U_o) based on the outside area (A_o) of the pipe is given by [2]:

$$\frac{1}{U_o A_o} = \frac{1}{h_i A_i} + \frac{\ln\left(\frac{r_o}{r_i}\right)}{2\pi k L} + \frac{1}{h_o A_o}$$

(8.23)

The rate of heat transfer using U_o and A_o can be written as [2]:

$$q = \frac{T_i - T_o}{\frac{1}{U_o A_o}}$$

(8.24)

To avoid using different areas for heat transfer (A_i and A_o) in a tubular heat exchanger, a log mean cross-sectional area (A_{lm}) is defined as:

$$A_{\text{lm}} = \frac{A_o - A_i}{\ln\left(\frac{A_o}{A_i}\right)} = \frac{2\pi r_o L - 2\pi r_i L}{\ln\left(\frac{2\pi r_o L}{2\pi r_i L}\right)} = \frac{r_o - r_i}{\ln\left(\frac{r_o}{r_i}\right)}$$

(8.25)

The rate of heat transfer using U and A_{lm} can be written as:

$$q = \frac{T_i - T_o}{\frac{1}{U A_{\text{lm}}}}$$

(8.26)

8.4   TYPES OF HEAT EXCHANGERS

Several types of heat exchangers are used for heating or cooling products. The choice of a heat exchanger is based on the characteristics of the product. Heat exchangers can be classified into contact and non-contact types. There is direct physical contact between the product and the heating or cooling system in the contact type of heat exchangers. In the noncontact type heat exchangers, the product is physically separated from the heating or cooling medium. Some of the common heat exchangers have been described below [2,7].

8.4.1   STEAM INJECTION

This is a contact type of heat exchanger used for homogeneous and high viscosity products and is particularly suited for shear sensitive products such as creams, desserts, and viscous sauces. The liquid product is heated by directly injecting steam into the product. The rapid heating by steam combined with rapid methods of cooling can yield a high quality product. However, this method is only suitable for liquids with no particles. Another disadvantage of this method is the reduced heat recovery of about 50% [7].

8.4.2 STEAM INFUSION

This is a contact type of heat exchanger used for homogeneous and high viscosity products and is particularly suited for shear sensitive products such as creams and desserts. This method, similar to steam injection, involves infusing a thin film of liquid product into an atmosphere of steam which provides rapid heating. The heated product with condensed steam is then released from the heating chamber. Water added to the product due to condensation of steam is sometimes desirable. Otherwise, the added water is flashed off by pumping the heated product into a vacuum cooling system.

8.4.3 DOUBLE TUBE HEAT EXCHANGER

The most common noncontact type of heat exchanger is the double tube heat exchanger. It consists of a tube located concentrically inside another tube of larger diameter. The two fluids flow in the inner and outer tube, respectively (Figure 8.1). Double tube heat exchangers can be classified according to the path of fluid flow through the heat exchanger. The three basic flow configurations are: cocurrent, counter-current, and cross-flow. In cocurrent heat exchangers, both the fluid streams enter simultaneously at one end and leave simultaneously at the other end as shown in Figure 8.1. In counter-current heat exchangers, both of the fluid streams flow in opposite directions. In cross-flow heat exchangers, one fluid flows through the heat transfer surface at right angles to the flow path of the other fluid [6]. When both fluid streams are simultaneously in cocurrent, counter-current and multipass cross-flow, then the arrangement is called mixed flow.

8.4.4 TRIPLE TUBE HEAT EXCHANGER

A slightly modified version of the double tube heat exchanger is the triple tube heat exchanger. It consists of three concentric tubes. Product flows in the inner annular space and the heating/cooling medium flows in the inner tube and outer annular space (Figure 8.2). Advantages of a triple tube heat

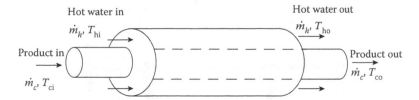

FIGURE 8.1 Schematic of a double tube heat exchanger for cocurrent flow configuration.

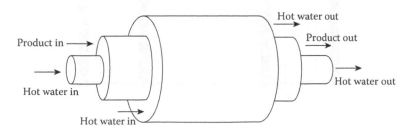

FIGURE 8.2 Schematic of a triple tube heat exchanger.

exchanger over a double tube heat exchanger include larger heat transfer area per unit length and higher overall heat transfer coefficients due to higher fluid velocities in the annular regions. Batmaz and Sandeep [8] developed an expression for effective overall heat transfer coefficient for a triple tube heat exchanger to facilitate comparison of a triple tube heat exchanger to an equivalent double tube heat exchanger.

8.4.5 MULTITUBE HEAT EXCHANGER

A multitube heat exchanger consists of four or more concentric tubes through which fluids flow. Advantage of using a multitube heat exchanger is increased surface area. However, higher pressure drop across the heat exchanger is one of the disadvantages for this type of heat exchanger.

8.4.6 PLATE HEAT EXCHANGER (PHE)

Plate heat exchangers (PHEs) are used for homogeneous and low viscosity (< 5 Pa.s) products (milk, juice and thin sauce) containing particle sizes up to approximately 3 mm. These heat exchangers consist of closely spaced parallel stainless steel plates pressed together in a frame. Gaskets, made of natural rubber or synthetic rubber, seal the plate edges and ports to prevent intermixing of fluids. There are three basic plate designs: flat, herringbone and modified herringbone. The flat plate design is used only to compare the performance of other plate designs. The herringbone design is the most popular design and has special patterns to increase turbulence in the product stream. The modified herringbone design is a herringbone design with an additional angle bend (to enhance turbulence) at the edges of the outside plate [9].

PHEs have widespread applications in various industries because it confers several advantages. PHEs are compact and thus require less floor space compared to traditional heat exchangers. Turbulence is achieved at lower Reynolds number (N_{Re} of 10–500) compared to other heat exchangers. PHEs provide rapid rate of heat transfer due to the large surface area for heat transfer and turbulent flow characteristics. A schematic of a plate heat exchanger is shown in Figure 8.3. The product is heated to the desired temperature in the heating section, the heated product then heats part of the incoming raw product in the regeneration section. Additional plates are required for regeneration, but it lowers the operating costs by decreasing the heating load in the heating section. Several studies have been done to develop correlations for Nusselt number in a (PHE) [9–14].

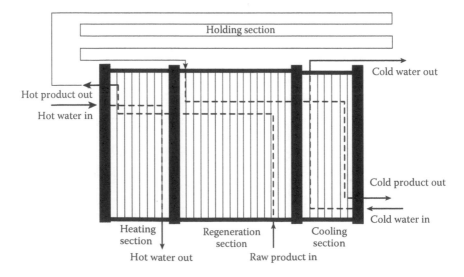

FIGURE 8.3 Schematic of a plate heat exchanger.

8.4.7 SCRAPED SURFACE HEAT EXCHANGER (SSHE)

Scraped surface heat exchangers (SSHEs) are used for viscous products (diced fruit preserves and soups) containing particles of sizes up to approximately 15 mm. Particle concentrations of up to 40% can be accommodated by these heat exchangers. These heat exchangers consist of a jacketed cylinder enclosing scraping blades on a rotating shaft. The rotating action of the blades prevents fouling on the heat exchanger surface and improves the rate of heat transfer and mixing. A summary of studies on heat transfer in (SSHEs) has been presented by Rao and Hartel [15].

8.4.8 SHELL AND TUBE HEAT EXCHANGER

Shell and tube heat exchanger is another common noncontact type of heat exchanger used in the food industry. It consists of a bundle of tubes connected in parallel and enclosed in a cylindrical shell. A schematic of a shell and tube heat exchanger is shown in Figure 8.4. One of the fluids flows inside the tubes while the other fluid flows over the tubes through the shell. To ensure the flow of shell-side fluid over all of the tubes, baffles are used in these heat exchangers. The segmented baffle is the most commonly used baffle and it causes the shell-side fluid to flow in a zigzag manner across the tube bundle. This enhances turbulence on the shell side of the heat exchanger and thus improves the rate of heat transfer. However, improperly designed segmented baffles cause low local heat transfer coefficients in flow stagnation regions where the baffles are attached to the shell wall. Baffles also result in excessive pressure drop by separating the flow at the edges of the baffles. Helical baffles are an alternative to segmented baffles as they offer the following advantages: improved shell-side heat transfer, lower pressure drop for a given shell-side flow rate, reduced shell-side fouling and prevention of flow induced vibrations. However, helical baffles are associated with fabrication and manufacturing difficulties [16].

8.4.9 HELICAL COIL (MULTICOIL) HEAT EXCHANGER

Helical coil or multicoil heat exchanger is a heat exchanger with two concentric coils in a vertical cylindrical enclosure (Figure 8.5). The coils can be used together or separately, either in series or parallel configuration. These heat exchangers offer several advantages such as compactness, enhanced mixing and improved rate of heating. Flow in a helical tube creates a secondary flow, normal to

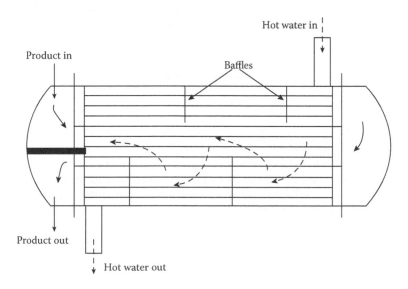

FIGURE 8.4 Schematic of a shell and tube heat exchanger.

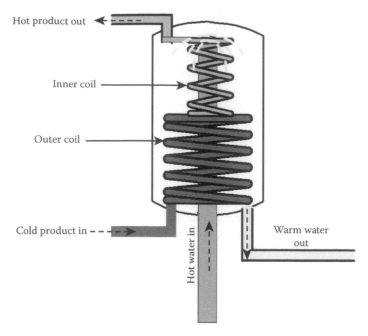

FIGURE 8.5 Schematic of a multicoil heat exchanger.

the direction of main flow due to imbalance of centrifugal forces between the fluid elements at the inner and outer radial location within this tube. Transition from laminar to turbulent flow in helical coiled tubes is gradual and fully developed turbulent flow occurs at Reynolds numbers as high as 6000–8000 [17]. Dean number (N_{De}), which governs the type of flow and rate of heat transfer in coiled tubes is given by [18]:

$$N_{De} = N_{Re}\sqrt{\frac{r}{R}} \tag{8.27}$$

where r is the radius of the tube (m) and R is the radius of the helical coil (m). The magnitude of dean number is a measure of the extent of secondary flow. Various studies have been conducted to determine correlations for Nusselt number in helical coil heat exchangers and a summary of those studies has been presented by Coronel and Sandeep [18].

8.5 PERFORMANCE ANALYSIS OF A HEAT EXCHANGER

8.5.1 LOGARITHMIC MEAN TEMPERATURE DIFFERENCE

The rate of heat transfer (q) is the quantity of primary interest in the analysis of a heat exchanger. Considering the cocurrent double tube heat exchanger shown in Figure 8.1, q may be determined by applying an energy balance to a differential area element dA in the hot and cold fluids. The decrease in temperature of the hot fluid is denoted by dT_h whereas the increase in temperature of the cold product is denoted by dT_c (it will decrease by dT_c for counter-current flow). The energy balance for adiabatic and steady state flow is given as [6]:

$$dq = -(\dot{m}c_p)_h dT_h = (\dot{m}c_p)_c dT_c$$

or   (8.28)

$$dq = -C_h dT_h = C_c dT_c$$

where subscript h represents the hot fluid, subscript c represents the cold fluid, $\dot{m}$ is the mass flow rate (kg.s^{-1}), c_p is the specific heat capacity (J.kg^{-1}.K^{-1}), C_h is the heat capacity rate for hot fluid (J.s^{-1}.K^{-1}), and C_c is the heat capacity rate for cold fluid (J.s^{-1}.K^{-1}).

The rate of heat transfer (dq) from the hot fluid to the cold fluid across the area dA can also be expressed in terms of overall heat transfer coefficient as [6]:

$$dq = UdA(T_h - T_c) \tag{8.29}$$

where T_h and T_c are the temperatures of the hot and cold fluids, respectively.

Equation 8.28 can also be written as:

$$dT_h - dT_c = d(T_h - T_c) = -dq\left(\frac{1}{C_c} + \frac{1}{C_h}\right) \tag{8.30}$$

By substituting the value of dq from Equation 8.29 into Equation 8.30, we obtain:

$$\frac{d(T_h - T_c)}{T_h - T_c} = -U\left(\frac{1}{C_c} + \frac{1}{C_h}\right)dA \tag{8.31}$$

Integration of Equation 8.31 over the entire length of the heat exchanger results in:

$$\ln\frac{T_{ho} - T_{co}}{T_{hi} - T_{ci}} = -UA\left(\frac{1}{C_c} + \frac{1}{C_h}\right) \tag{8.32}$$

where subscripts o and i refers to the outlet and inlet of the heat exchanger, respectively.

Equation 8.32 can be written as:

$$T_{ho} - T_{co} = (T_{hi} - T_{ci})\exp\left[-UA\left(\frac{1}{C_c} + \frac{1}{C_h}\right)\right] \tag{8.33}$$

Equation 8.33 for a counter-current arrangement in a heat exchanger becomes:

$$T_{ho} - T_{ci} = (T_{hi} - T_{co})\exp\left[UA\left(\frac{1}{C_c} - \frac{1}{C_h}\right)\right] \tag{8.34}$$

Substituting the value of C_c and C_h in Equation 8.32 in terms of q and solving for q yields:

$$q = UA\frac{(T_{ho} - T_{co}) - (T_{hi} - T_{ci})}{\ln\dfrac{T_{ho} - T_{co}}{T_{hi} - T_{ci}}}$$

$$\text{or} \tag{8.35}$$

$$q = UA\frac{\Delta T_2 - \Delta T_1}{\ln\dfrac{\Delta T_2}{\Delta T_1}}$$

where ΔT_1 is the temperature difference between the two fluids at the inlet of the heat exchanger and ΔT_2 is the temperature difference between the two fluids at the outlet of the heat exchanger.

Comparison of Equation 8.35 with Equation 8.26 reveals that appropriate average temperature difference between the hot and cold fluids over the entire length of heat exchanger can be written as:

$$\Delta T_{lm} = \frac{\Delta T_1 - \Delta T_2}{\ln \dfrac{\Delta T_1}{\Delta T_2}} \qquad (8.36)$$

where ΔT_{lm} is the log mean temperature difference (LMTD). The rate of heat transfer can be written as:

$$q = UA\Delta T_{lm} \qquad (8.37)$$

The rate of heat transfer in a tubular heat exchanger can be written by replacing area (A) in Equation 8.37 by log mean area (A_{lm}):

$$q = UA_{lm}\Delta T_{lm} \qquad (8.38)$$

A similar expression can be written for a counter-current heat exchanger. However, the inlet and outlet temperature differences are defined as $\Delta T_1 = (T_{hi} - T_{co})$ and $\Delta T_2 = (T_{ho} - T_{ci})$. When the heat capacity rates for the hot and cold fluid are equal, ΔT_1 and ΔT_2– become equal. LMTD becomes indeterminate for such a situation and either ΔT_1 or ΔT_2– can be used instead of LMTD.

The LMTD method developed above for single pass flow arrangements is not applicable for the analysis of cross-flow and multi-pass flow heat exchangers. Equation 8.38 can be used for multipass and cross-flow heat exchangers by multiplying ΔT_{lm} with a correction factor (F). F is dimensionless and it represents the degree of deviation of the true mean temperature difference from LMTD for a counter-current configuration. F is 1 for a counter-current heat exchanger and it is less than 1 for cross-flow and multipass heat exchangers. F depends on temperature effectiveness (P), heat capacity rate ratio (R), and the flow arrangement. P is a measure of the ratio of heat actually transferred to the heat which would be transferred if the outlet temperature of the cold fluid was raised to the inlet temperature of the hot fluid for a counter-current configuration. R is the ratio of heat capacity rate of the cold fluid to that of hot fluid. Thus, rate of heat transfer for cross-flow and multipass heat exchangers is given by [6]:

$$q = UAF\Delta T_{lm} \qquad (8.39)$$

where F can be determined from LMTD correction factor charts given in the literature [1,5,6].

8.5.2 ε-Number of Transfer Units (NTU) Method

During analysis of the rate of heat transfer in heat exchangers, trial and error method need to be applied for using the LMTD method in a situation when the inlet or outlet temperatures of the fluids are not known. The correct value of the LMTD should satisfy the requirement that the heat lost by the hot fluid in the heat exchanger is gained by the cold fluid. The method of the number of transfer units (NTU), based on the concept of heat exchanger effectiveness (ε), is used to avoid the trial and error method. Heat exchanger effectiveness (ε) is defined as the ratio of actual rate of heat transfer (q) to the maximum possible amount of heat transfer (q_{max}) if an infinite heat transfer area was available. q is given by [6]:

$$q = (\dot{m}c_p)_h(T_{hi} - T_{ho}) = (\dot{m}c_p)_c(T_{co} - T_{ci}) \qquad (8.40)$$

The fluid with minimum heat capacity rate (C_{min}) will undergo maximum temperature difference. The maximum possible amount of heat transfer (q_{max}) is given by [6]:

$$q_{max} = (\dot{m}c_p)_c(T_{hi} - T_{ci}), \quad \text{if } C_c < C_h$$

$$q_{max} = (\dot{m}c_p)_h(T_{hi} - T_{ci}), \quad \text{if } C_h < C_c \tag{8.41}$$

Heat exchanger effectiveness (ε) can therefore, be written as:

$$\varepsilon = \frac{C_h(T_{hi} - T_{ho})}{C_{min}(T_{hi} - T_{ci})} = \frac{C_c(T_{co} - T_{ci})}{C_{min}(T_{hi} - T_{ci})} \tag{8.42}$$

Equation 8.42 is valid for all heat exchanger flow configurations and the value of ε ranges from 0 to 1. NTU is defined as [6]:

$$NTU = \frac{UA}{C_{min}} = \frac{1}{C_{min}} \int_A U dA \tag{8.43}$$

Capacity rate ratio is defined as [6]:

$$C^* = \frac{C_{min}}{C_{max}} \tag{8.44}$$

Assuming $C_c < C_h$, Equation 8.33 can be written in terms of NTU and capacity rate ratio as:

$$T_{ho} - T_{co} = (T_{hi} - T_{ci})\exp[-NTU(1+C^*)] \tag{8.45}$$

Using Equation 8.42, T_{ho} and T_{co} in Equation 8.45 can be eliminated and the following expression is obtained for ε for a cocurrent arrangement in a double tube heat exchanger:

$$\varepsilon = \frac{1 - \exp[-NTU(1+C^*)]}{1+C^*} \tag{8.46}$$

A similar analysis may be applied to obtain the expression for ε for a counter-current arrangement in a double tube heat exchanger as:

$$\varepsilon = \frac{1 - \exp[-NTU(1-C^*)]}{1 - C^*\exp[-NTU(1-C^*)]} \tag{8.47}$$

Equation 8.47 yields an indeterminate value of ε for $C^* = 1$. For such situation ($C^* = 1$), the following equation is used to calculate ε for a counter-current arrangement in a double tube heat exchanger:

$$\varepsilon = \frac{NTU}{1+NTU} \tag{8.48}$$

It can be concluded from Equations 8.46 and 8.47 that heat exchanger effectiveness is a function of NTU, C^*, and flow configuration. Expressions for effectiveness of cross-flow and multipass heat exchangers can be found in the literature [1,5,6]. A counter-current heat exchanger arrangement has the highest value of ε for a given NTU and C^*. Thus, heat transfer is most efficient in a counter-current heat exchanger as compared to that in all other flow arrangements.

8.5.3 EXERGY ANALYSIS

Exergy is a term used in connection with the second law of thermodynamics to describe the irreversible losses that occur within any thermal cycle. Exergy is used to describe differences in energy quality. The exergy content of a system indicates its distance from thermodynamic equilibrium. Exergy (E) of a closed system at a specified state is given as [19]:

$$E = (e - e_0) + p_0(V - V_0) - T_0(S - S_0)$$ (8.49)

where e, V, and S denote the energy (sum of internal energy (u), kinetic energy (KE) and potential energy (PE)), volume and entropy of the system respectively and e_0, V_0 and S_0 are the respective values of energy, volume and entropy at dead state. Dead state is the state of a closed system where the system is in equilibrium with the environment. At the dead state, both the system and environment possess energy, but the value of exergy is zero because there is no spontaneous change within the system, environment or between them. The units of exergy are the same as that of energy. Exergy is the maximum theoretical work that could be done by a closed system and the environment if the closed system were to come into equilibrium with the environment. Thus, exergy is an attribute of the system and environment together. Exergy represents the quality of energy. Second law of thermodynamics states that not all of the heat energy can be converted to useful work. Exergy is the part of the heat energy which can be converted to work. Exergy is not conserved, but is destroyed by irreversibilities [19].

Change in exergy between two states of a closed system can be written using Equation 8.49 as [19]:

$$E_2 - E_1 = (e_2 - e_1) + p_0(V_2 - V_1) - T_0(S_2 - S_1)$$ (8.50)

where p_0 and T_0 are pressure and temperature of the environment respectively. Change in energy $(e_2 - e_1)$ and entropy $(S_2 - S_1)$ can be written as [19]:

$$e_2 - e_1 = \int_1^2 \delta Q - W$$

$$S_2 - S_1 = \int_1^2 \left(\frac{\delta Q}{T}\right)_b + \sigma$$ (8.51)

where W and Q represent work and heat transfers between the system and the surroundings respectively (J), T_b is the temperature of the system boundary (K) and σ accounts for internal irreversibilities. Thus, an exergy balance for a closed system can be written as [19]:

$$E_2 - E_1 = \int_1^2 \left(1 - \frac{T_0}{T_b}\right)\delta Q - [W - p_0(V_2 - V_1)] - T_0\sigma$$ (8.52)

Exergy change   Exergy transfers   Exergy destruction

Exergy rate balance for a control volume is given as [19]:

$$\frac{dE_{cv}}{dt} = \sum_j \left(1 - \frac{T_0}{T_b}\right)\dot{Q}_j - \left[\dot{W}_{cv} - p_0 \frac{dV_{cv}}{dt}\right] + \sum_i \dot{m}_i e_{fi} - \sum_e \dot{m}_e e_{fe} - \dot{E}_d \tag{8.53}$$

where $\dot{Q}_j$ represents the rate of change of heat transfer at the location of the boundary where the instantaneous temperature is T_j, $\dot{W}_{cv}$ is the rate of energy transfer by work other than flow work, $\dot{m}_i e_{fi}$ is the rate of exergy transfer due to mass flow and flow work at the inlet, $\dot{m}_e e_{fe}$ is the rate of exergy transfer due to mass flow and flow work at the exit, and $\dot{E}_d$ is the rate of exergy destruction due to irreversibilities within the control volume. Specific flow exergy (e_f), which accounts for exergy transfer due to both flow and work is given by [19]:

$$e_f = h - h_0 - T_0(s - s_0) + \frac{v^2}{2} + gz \tag{8.54}$$

where h and s represent specific enthalpy and entropy, respectively.

For a control volume at steady state, Equation 8.53 reduces to the following expression:

$$0 = \sum_j \left(1 - \frac{T_0}{T_b}\right)\dot{Q}_j - \dot{W}_{cv} + \sum_i \dot{m}_i e_{fi} - \sum_e \dot{m}_e e_{fe} - \dot{E}_d \tag{8.55}$$

The concept of exergy can be used for the performance analysis of a heat exchanger. For a cocurrent double tube heat exchanger, shown in Figure 8.1, that operates at steady state with no heat transfer with its surroundings, there is no work other than flow work ($\dot{W}_{cv} = 0$), and the temperature of both fluid streams above T_0. An exergy rate balance can be written as [19]:

$$\dot{m}_h(e_{fhi} - e_{fho}) = \dot{m}_c(e_{fco} - e_{fci}) + \dot{E}_d \tag{8.56}$$

where the term on the left side of Equation 8.56 accounts for the decrease in exergy of the hot stream, the first term on the right side accounts for the increase in exergy of the cold stream, and second term on the right side is the exergy destroyed. Exergetic heat exchanger efficiency (ε_e) can be written as [19]:

$$\varepsilon_e = \frac{\dot{m}_c(e_{fco} - e_{fci})}{\dot{m}_h(e_{fhi} - e_{fho})} = \left[1 - \frac{\dot{E}_d}{\dot{m}_h(e_{fhi} - e_{fho})}\right] \tag{8.57}$$

Exergetic efficiency is also known as the rational efficiency or second-law efficiency. Exergetic efficiency is an indication of the degree of thermodynamic perfection of the system. The higher the value of exergetic efficiency, the closer the heat exchanger operates to a reversible process [20]. Yilmaz et al. [20] have summarized all performance evaluation criteria for heat exchangers based on entropy and exergy. Wu et al. [21] introduced the concept of exergy transfer effectiveness (ε_e) using the analogy of heat exchanger effectiveness. Exergy transfer effectiveness (ε_{et}) was defined as [21]:

$$\varepsilon_{et} = \frac{\text{Actual exergy transfer rate}}{\text{Maximum possible exergy transfer rate}} \tag{8.58}$$

For a heat exchanger operating above environment temperature (T_0), the expression for ε_{et} is given by [21]:

$$\varepsilon_{et} = \frac{T_{co} - T_{ci} - T_0 \ln \dfrac{T_{co}}{T_{ci}}}{T_{hi} - T_{ci} - T_0 \ln \dfrac{T_{hi}}{T_{ci}}} \quad \text{for zero pressure drop}$$

(8.59)

$$\varepsilon_{et} = \frac{T_{co} - T_{ci} - T_0 \ln \dfrac{T_{co}}{T_{ci}} - \dfrac{\nu \Delta P}{c_{pc}}}{T_{hi} - T_{ci} - T_0 \ln \dfrac{T_{hi}}{T_{ci}}} \quad \text{for finite pressure drop}$$

where ν and c_{pc} are the specific volume and specific heat of the cold fluid, respectively and ΔP is the pressure drop along the length of the heat exchanger. Studies have been conducted for performance analysis of a heat exchanger based on the concept of exergy [21–23].

8.6   FOULING OF HEAT EXCHANGERS

Fouling of heat exchangers is defined as the accumulation of undesirable substances on the surface of a heat exchanger. Fouling reduces the effectiveness of a heat exchanger by decreasing the rate of heat transfer and increasing the pressure drop. The overall effect of fouling on heat transfer is usually represented by a fouling factor or fouling resistance (R_f). Overall heat transfer coefficient for a fouled surface (U_f) is defined as [6]:

$$\frac{1}{U_f} = \frac{1}{U_c} + R_{ft}$$

(8.60)

where U_c is the overall heat transfer coefficient for a clean surface and R_{ft} is the total fouling resistance given by [6]:

$$R_{ft} = \frac{A_o R_{fi}}{A_i} + R_{fo}$$

(8.61)

where R_{fi} and R_{fo} are the inside and outside surface fouling resistance, respectively. Fouling decreases the inside diameter and roughens the surface of a tube. This causes an increase in pressure drop.

Fouling of heat exchangers adds additional cost in the form of increased maintenance cost, downtime and energy consumption. Based on the principal processes causing fouling, fouling can be classified into five categories: particulate, crystallization, corrosion, biofouling and chemical reaction. Particulate fouling is caused by the accumulation of solid particles from the process stream onto the heat exchanger surface. Crystallization fouling is caused mainly by dissolved inorganic salts such as calcium and magnesium carbonates in the process stream. Corrosion fouling is caused by the reaction of a corrosive fluid with the heat exchanger surface. Biofouling is caused by the deposition and/or growth of biological materials on the heat exchanger surface. Chemical reaction fouling is caused due to chemical reactions taking place in the process stream [6].

In the design of heat exchangers, it is very important to predict the progression of fouling with time. Fouling as a function of time can be expressed as [6]:

$$\frac{dR_{ft}}{dt} = \phi_d - \phi_r$$

(8.62)

where ϕ_d and ϕ_r are the deposit and removal rates, respectively. Jun and Puri [24] have summarized different models of fouling for heat exchangers in dairy industries. Fouling can be controlled by using additives that act as fouling inhibitors. On-line or off-line surface cleaning can also be performed to control fouling [6].

8.7 FACTORS AFFECTING SELECTION OF A HEAT EXCHANGER

Many different types of heat exchangers have been described in Section 8.4. A particular type of heat exchanger needs to be selected for a given application. Selection of a heat exchanger depends on the characteristics of the product, potential changes in the product, cleanability, regeneration, heat transfer coefficient and cost. Characteristics of the product such as presence of particles, viscosity, and thermophysical properties should be taken into consideration when selecting a heat exchanger. PHEs are used for homogeneous and low viscosity products such as milk and juice whereas SSHEs are preferred for viscous products containing particulates. Cleanability of a heat exchanger depends on the extent of fouling. In applications involving moderate to severe fouling, either a shell and tube or a PHE is used. In a shell and tube heat exchanger, the tube side is generally selected for the fouling fluid because the tube side can be cleaned more easily. PHEs are selected where severe fouling occurs because disassembly, cleaning and reassembly of the plates can be done easily. Cost plays an important role in the selection of a heat exchanger. The total cost including capital, installation, operation and maintenance costs should be considered during selection of a heat exchanger [1].

8.8 HEAT TRANSFER AND PRESSURE DROP ANALYSIS IN HEAT EXCHANGERS

8.8.1 HEAT TRANSFER

The analysis of rate of heat transfer presented below is based on the configurations of a double tube heat exchanger shown in Figure 8.6. For a counter-current configuration, energy balance for the hot fluid over a small length δx can be written as [25]:

$$\dot{m}_h c_{\mathrm{ph}} T_h - \dot{m}_h c_{\mathrm{ph}} \left(T_h + \frac{dT_h}{dx} \delta x \right) - U \left(A \frac{\delta x}{L} \right) (T_h - T_c) = 0 \tag{8.63}$$

where $\dot{m}_h$ is the mass flow rate of the hot fluid (kg.s^{-1}), c_{ph} is the specific heat capacity of the hot fluid (J.kg^{-1}.K^{-1}), U is the overall heat transfer coefficient (W.m^{-2}.K^{-1}), A is the cross-sectional area of the tube, and L is the length of the heat exchanger.

Equation 8.63 can be rearranged as:

$$\frac{dT_h}{dx} = -\frac{UA}{\dot{m}_h c_h} \frac{1}{L} (T_h - T_c) = -\frac{\mathrm{NTU}_h}{L} (T_h - T_c) \tag{8.64}$$

A similar analysis for cold fluid results in the following equation:

$$\frac{dT_c}{dx} = -\frac{UA}{\dot{m}_c c_c} \frac{1}{L} (T_h - T_c) = -\frac{\mathrm{NTU}_c}{L} (T_h - T_c) \tag{8.65}$$

Equation 8.64 can also be written as [25]:

$$T_c = \frac{L}{\mathrm{NTU}_h} \frac{dT_h}{dx} + T_h \tag{8.66}$$

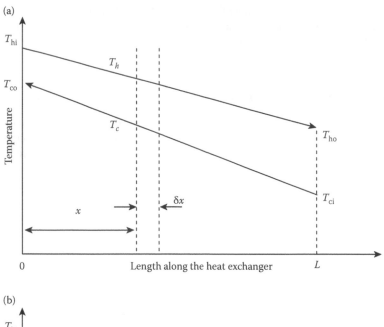

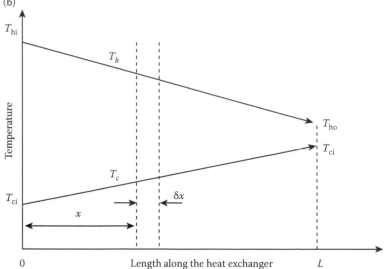

FIGURE 8.6 Temperature profile of fluids in a double tube heat exchanger: (a) counter-current (b) cocurrent.

Differentiating Equation 8.66 yields:

$$\frac{dT_c}{dx} = \frac{L}{\mathrm{NTU}_h}\frac{d^2T_h}{dx^2} + \frac{dT_h}{dx} \tag{8.67}$$

Substituting Equation 8.67 in Equation 8.65 results in the following equation:

$$\frac{d^2T_h}{dx^2} + \left[\frac{\mathrm{NTU}_h - \mathrm{NTU}_c}{L}\right]\frac{dT_h}{dx} = 0 \tag{8.68}$$

Similarly, for cold fluid:

$$\frac{d^2T_c}{dx^2} + \left[\frac{\text{NTU}_h - \text{NTU}_c}{L}\right]\frac{dT_c}{dx} = 0 \tag{8.69}$$

The solution to the second order differential Equation 8.68 is given by:

$$T_h = A_1 \exp\left(\frac{(\text{NTU}_h - \text{NTU}_c)x}{L}\right) + B_1 \tag{8.70}$$

With boundary conditions

$$x = 0 : T_h = T_{\text{hi}} \quad \text{and} \quad T_c = T_{\text{co}}$$
$$x = L : T_h = T_{\text{ho}} \quad \text{and} \quad T_c = T_{\text{ci}} \tag{8.71}$$

The following result is obtained for both fluids in counter-current configuration [25]:

$$\frac{T_{\text{hi}} - T_h}{T_{\text{hi}} - T_{\text{ho}}} = \frac{T_{\text{co}} - T_c}{T_{\text{co}} - T_{\text{ci}}} = \frac{1 - \exp\left[-(\text{NTU}_h - \text{NTU}_C)\dfrac{x}{L}\right]}{1 - \exp[-(\text{NTU}_h - \text{NTU}_C)]} \tag{8.72}$$

A similar analysis for cocurrent configuration yields the following result [25]:

$$\frac{T_{\text{hi}} - T_h}{T_{\text{hi}} - T_{\text{ho}}} = \frac{T_{\text{ci}} - T_c}{T_{\text{ci}} - T_{\text{co}}} = \frac{1 - \exp\left[-(\text{NTU}_h + \text{NTU}_C)\dfrac{x}{L}\right]}{1 - \exp[-(\text{NTU}_h + \text{NTU}_C)]} \tag{8.73}$$

8.8.2 PRESSURE DROP

Pressure drop due to fluid friction in a heat exchanger is an important parameter because it determines the power of the pump required to maintain the flow of fluids. The design and selection of a heat exchanger should optimize both heat transfer and pressure drop for an economical operation. Pressure drop (ΔP) in a tube of circular cross section is given by [6]:

$$\Delta P = 4\frac{L}{D}\left(\frac{\rho v_{\text{av}}^2}{2}\right)f \tag{8.74}$$

where L is the length of the tube (m), D is the inside diameter of the tube (m), v_{av} is the average velocity of the fluid (m.s^{-1}), ρ is the density of the fluid (kg.m^{-3}) and f is the Fanning friction factor. Fanning friction factor, which is the ratio between shear stress at the wall to the kinetic energy of the fluid per unit volume is given by [6]:

$$f = \frac{\tau_w}{\rho v_{\text{av}}^2/2} \tag{8.75}$$

For laminar flow conditions, f is given by [6]:

$$f = \frac{16}{N_{Re}} \tag{8.76}$$

For transition and turbulent flow conditions, f is determined from a chart known as Moody chart. The Moody chart represents friction factor as a function of Reynolds number for various magnitudes of relative roughness (e/D) of tube. Correlations for friction factor under turbulent flow conditions in a smooth tube are given by [6]:

$$f = 0.046\, N_{Re}^{-0.2} \quad \text{for} \quad 3 \times 10^4 < N_{Re} < 10^6$$

and   (8.77)

$$f = 0.079\, N_{Re}^{-0.25} \quad \text{for} \quad 4 \times 10^3 < N_{Re} < 10^5$$

Haaland [26] gave an explicit equation to estimate the friction factor for a rough tube under turbulent flow conditions as follows:

$$\frac{1}{\sqrt{f}} \approx -3.6 \log\left[\frac{6.9}{N_{Re}} + \left(\frac{e/D}{3.7}\right)^{1.11} \right] \tag{8.78}$$

where e is the surface roughness of the tube.

For flow in noncircular ducts, an equivalent diameter, d_c, as defined in Equation 8.11 is used for the diameter (D) to compute friction factor from the correlations for circular tubes. Srinivasan et al. [27] proposed the following correlations for friction factors under laminar flow conditions in terms of Dean number (N_{De}) to calculate pressure drop in helical coils:

$$\frac{f_c}{f_s} = \begin{cases} 1 & \text{for} \quad N_{De} < 30 \\ 0.419 N_{De}^{0.275} & \text{for} \quad 30 < N_{De} < 300 \\ 0.1125 N_{De}^{0.5} & \text{for} \quad N_{De} > 300 \end{cases} \tag{8.79}$$

where c and s represent curved and straight tube respectively. Srinivasan et al. [27] also proposed the following correlations for friction factor under turbulent flow conditions:

$$f_c \left(\frac{R}{r}\right)^{0.5} = 0.0084 \left[N_{Re} \left(\frac{R}{r}\right)^{-2} \right]^{0.2}$$

(8.80)

$$\text{for} \quad N_{Re} \left(\frac{R}{r}\right)^{-2} < 700 \quad \text{and} \quad 7 < \frac{R}{r} < 10^4$$

The total pressure drop in a heat exchanger is the sum of the pressure drops in straight tubes, bends, fittings, contractions, and expansions. The total pressure drop in a bend is the sum of the frictional loss due to the length and curvature of the bend and head loss due to excessive pressure drop in the downstream tube. The total pressure drop in a bend is given by [6]:

$$\Delta P = K \frac{\rho v_{av}^2}{2} \quad \text{where} \quad K = \frac{4 f_c L}{D_e} = \frac{4 f L}{D_e} + K^* \tag{8.81}$$

where f_c is the friction factor of the bend, f is the friction factor for a straight tube at Reynolds number similar to that of the bend, and K^* is the combined loss coefficient other than friction. All tube fittings such as elbows, tees, and valves contribute to frictional losses. The pressure drop associated with tube fittings is given by [2]:

$$\Delta P = C_{ff} \frac{\rho v_{av}^2}{2} \tag{8.82}$$

where C_{ff} is the loss coefficient and its typical value for various fittings can be found in the literature [2]. A sudden contraction in the cross-section of the tube causes pressure drop which can be computed as [2]:

$$\Delta P = C_{fc} \frac{\rho v_{av}^2}{2} \tag{8.83}$$

where v_{av} is the average velocity downstream and C_{fc} is contraction loss coefficient given by [2]:

$$C_{fc} = 0.4 \left[1.25 - \frac{A_2}{A_1} \right] \quad \text{when} \quad \frac{A_2}{A_1} < 0.715$$

$$C_{fc} = 0.75 \left[1 - \frac{A_2}{A_1} \right] \quad \text{when} \quad \frac{A_2}{A_1} > 0.715 \tag{8.84}$$

where A_1 is the upstream cross-sectional area and A_2 is the downstream cross-sectional area. Similarly, pressure drop due to an increase in the cross-sectional area of the tube is given by [2]:

$$\Delta P = C_{fe} \frac{\rho v_{av}^2}{2} \tag{8.85}$$

where the expansion loss coefficient, C_{fe}, is given by [2]:

$$C_{fe} = \left(1 - \frac{A_1}{A_2} \right)^2 \tag{8.86}$$

where A_1 is the upstream cross-sectional area and A_2 is the downstream cross-sectional area.

The pumping power requirement (W_p) is proportional to the total pressure drop and is given by [6]:

$$W_p = \frac{\dot{m} \Delta P}{\eta_p \rho} \tag{8.87}$$

where $\dot{m}$ is the mass flow rate of the fluid and η_p is the efficiency of the pump.

8.9　RECENT DEVELOPMENTS IN HEAT EXCHANGERS

One of the recent developments in heat exchangers has been to reduce the size (length and/or area) by enhancing the rate of heat transfer. This has been achieved by using a grooved surface or electro-hydrodynamic (EHD) enhancement. A grooved surface induces turbulence and increases the surface area available for heat transfer. EHD enhancement involves placing conductive plates around the heat exchanger and applying a voltage across the plates. The electric field generated between the EHD plates induces secondary flow and increases the rate of heat transfer in the heat exchanger. Improvement in PHEs has included the use of different types of fins and plates. Traditionally, segmented baffles have been used in shell and tube heat exchangers. However, segmented baffles result in low local heat transfer coefficients and excessive pressure drop. New geometries of baffles such as rods, grids, twisted tubes, and helical and angled baffles have been developed which allow longitudinal flow on the shell side. Another recent development in the area of heat exchangers has been to minimize fouling by better understanding fouling and integrating it as a part of heat exchanger design.

REFERENCES

1. Shah, R.K., Sekulic, D.P. *Fundamentals of Heat Exchanger Design*. John Wiley and Sons, Inc., NJ, 2003, pp. 1–941.
2. Singh, R.P., and Heldman, D.R. Heat transfer in food processing. In *Introduction to Food Engineering*, 3rd ed. Singh, R.P. and Heldman, D.R., Eds. Academic Press, Oxford, UK, 2001, pp. 63–331.
3. Geankoplis, C.J. *Transport Processes and Unit Operations*, 3rd ed. Prentice Hall Inc., NJ, 1993, chapter 4.
4. Rahman, S. *Food Properties Handbook*. CRC Press LLC, Boca Raton, FL, 1995, pp. 393–456.
5. Holman, J.P. *Heat Transfer*, 8th ed. McGraw-Hill, Inc., New York, NY, 1997, pp. 283–393.
6. Kakac, S., and Liu, H. *Heat Exchangers: Selection, Rating, and Thermal Design*, 2nd ed. CRC Press LLC, Boca Raton, FL, 2000, pp. 1–501.
7. Skudder, P.J. Ohmic heating. In *Aseptic Processing and Packaging of Particulate Foods*, Willhoft, E.M.A., Ed. Blackie Academic & Professional, London, UK, 1993, pp. 74–89.
8. Batmaz, E., and Sandeep, K.P. Calculation of overall heat transfer coefficients in a triple tube heat exchanger. *Heat and Mass Transfer*, 41, 271–279, 2004.
9. Talik, A.C., Swanson, L.W., Fletcher, L.S., and Anand, N.K. Heat transfer and pressure drop characteristics of a plate heat exchanger. *ASME/JSME Thermal Engineering Conference*, 4, 321–329, 1995.
10. Emerson, W.H. The thermal and hydrodynamic performance of a plate heat exchanger: I. flat plates. National Engineering Laboratories, Glasgow, UK, 283, 1967.
11. Buonopane, R.A., and Troupe, R.A. A study of the effects of internal rib and channel geometry in rectangular channels. *AIChE Journal*, 15(7), 585–596, 1969.
12. Cooper, A. Recover more heat with plate heat exchangers. *The Chemical Engineer*, 285, 280–285, 1974.
13. Focke, W.W., Zachariades, J., and Olivier, I. The effects of corrugation inclination angle on the thermodynamic performance of plate heat exchangers. *International Journal of Heat and Mass Transfer*, 28(8), 1469–1479, 1985.
14. Roetzal, W., Das, S.K., and Luo, X. Measurement of heat transfer coefficient in plate heat exchanger using a thermal oscillation technique. *International Journal of Heat and Mass Transfer*, 37(1), 325–331, 1994.
15. Rao, C.S., and Hartel, R.W. Scraped surface heat exchangers. *Critical Reviews in Food Science and Nutrition*, 46, 207–219, 2006.
16. Stehlik, P., Nemcansky, J., Kral, D., and Swanson, L.W. Comparison of correction factors for shell-and-tube heat exchangers with segmented or helical baffles. *Heat Transfer Engineering*, 15(1), 55–65, 1994.
17. Coronel, P., and Sandeep, K.P. Pressure drop and friction factor in helical heat exchangers under nonisothermal and turbulent flow conditions. *Journal of Food Process Engineering*, 26, 285–302, 2003a.
18. Coronel, P., and Sandeep, K.P. Flow dynamics and heat transfer in helical heat exchangers. In *Transport Phenomena in Food Processing*, Welti-Chanes, J., Velez-Ruiz, J.F., and Barbosa-Canovas, G.V., Eds. CRC Press LLC, Boca Raton, FL, 2003b, pp. 377–397.
19. Moran, M.J., and Shapiro, H.N. *Fundamentals of Engineering Thermodynamics*, 5th ed. John Wiley & Sons, NJ, 2003, chapter 7.

20. Yilmaz, M., Sara, O.N., and Karsli, S. Performance evaluation criteria for heat exchangers based on second law analysis. *Exergy, An International Journal*, 4, 278–294, 2001.
21. Wu, S.Y., Yuan, X.F., Li, Y.R., and Xiao, L. Exergy transfer effectiveness on heat exchanger for finite pressure drop. *Energy*, 32, 2110–2120, 2007.
22. Fang, Z., Larson, D.L., and Fleischmen, G. Exergy analysis of a milk processing system. *Transactions of the ASAE*, 38(6), 1825–1832, 1995.
23. Das, S.K., and Roetzel, W. Exergetic analysis of plate heat exchanger in presence of axial dispersion in fluid. *Cryogenics*, 35(1), 3–8, 1995.
24. Jun, S., and Puri, V.M. Fouling models for heat exchangers in dairy processing: a review. *Journal of Food Processing Engineering*, 28, 1–34, 2005
25. Smith, E.M. Thermal design of heat exchangers, A *Numerical Approach: Direct Sizing and Stepwise Rating*, John Wiley & Sons, Inc., NJ, 1997, pp. 51–90.
26. Haaland S. Simple and explicit formulas for the friction factor in turbulent pipe flow. *Journal of Fluids Engineering*, 105, 89–90, 1983.
27. Srinivasan, P.S., Nandapurkar, S.S., and Holland, F.A. Friction for coils. *Transactions of the Institution of Chemical Engineering*, 48, T156, 1970.

9 Thermal Phase Transitions in Food

Jasim Ahmed
Polymer Source Inc.

CONTENTS

9.1 INTRODUCTION

The stability of food remains a challenge to food processors. Newer development in the area of food processing enhances knowledge and better understanding of structure, mechanism, processing, storage, and quality of food products. Thermal transition is associated with heat treatment of food and thermo-analytical measurements are commonly used for qualitative and/or quantitative changes of biomaterials. For example, crystallization of ice-cream or state transitions (e.g., glass transition) of food is identified by thermal analysis using trace amount of sample.

Thermal analysis comprises of a group of techniques in which physical property of a substance is measured as a function of temperature, while the substance is subjected to a controlled temperature program. Among various thermal transitions, the interest in the glassy state of food has grown considerably, stimulated by the application of the concepts from polymer science and strongly supported by industrial demand [1]. The glass transition concept used in polymer science has proven useful in the understanding and control of the physico-chemical properties of food systems. The key parameter in the thermal transition of polymeric materials is the glass transition temperature or T_g, above which glassy polymers transform to the rubbery state. Numerous examples are available in the literature where T_g is related to the structural and textural properties, chemical reactions and microbiological activities in food systems. Thermal analysis is no longer only a research tool for analysts but it also offers a real-time monitoring technique in studying process-induced changes for the food technologists [2].

The glassy state of matter and the glass transition itself still remained unsolved problems in various disciplines of science and engineering. The glass transition is the most important property of amorphous materials both practically and theoretically since it involves a dramatic slowing down in the motion of chain segments that one can rarely observe in the static state. Material science provides a powerful theoretical format to understand the ongoing behaviors exhibited by amorphous materials. For food products and ingredients, the polymeric approach appears to be similar. The concept of glass transition was introduced to the food science and technology world more than 30 years ago as a controlling factor in the stability of low moisture food products [3]. The pioneering works of Slade and Levine [4,5] however, opened a new era in food science research. The concept is now found relevant to food processing to predict shelf-life, storage and packaging of amorphous food products. For example, many spray-dried dairy powders contain lactose in its amorphous state. Amorphous components are thermodynamically unstable and there is a tendency for them to crystallize at favorable conditions that finally leads to oxidation and product deterioration [6]. Furthermore, an amorphous component can lead to caking by absorbing moisture at higher temperature which is not desirable by food industry.

The basic difference in food application is the frequent heterogeneity in chemical composition and on the other the predominant role of water as a plasticizer [7]. The glass transition is a unique property of amorphous materials. The amorphous state is a solid state of materials that is different from crystalline state, whose molecular arrangement is highly ordered regular lattice. The amorphous solid is characterized by short-range molecular order similar to that in crystal although only a few molecular dimensions quantified as a few Angstroms [8,9]. The glass transition temperature (T_g) is used for designing and manufacturing a product that can work in a wider temperature range. The glass transition temperature varies widely for a specific food as it depends on many factors like sample preparation and size, heating/cooling rate, sample holding time, moisture content, etc. [10].

Most food products with reduced moisture content are partly or totally amorphous. Food products shelf-life, stability and desirable texture (crispiness, softness) are solely dependent on water content and composition and consequently on glass transition temperature of the product. Glass changes to a rubbery or liquid phase above glass transition temperature and then becomes soft and prone to physical and chemical changes. Processing operations like baking, air- and freeze-drying, extrusion, and flaking may pass through the glass transition domain. For many food products, the rates of a number of physical and chemical changes are considerably affected by the glass transition [5,11–14].

Crystallization, an important unit operation is extensively used in food processing and as a quality indicator. Depending on the product, the absence or presence of crystals, as well as their size and shape, are critical factors, for instance to the desired texture properties of confectioneries or ice-creams and for the free flowing characteristics and dissolution of powders [15]. The process may enhance the release of substances entrapped in the glass. Pure, crystalline solids have a characteristic melting point (T_m), the temperature at which the solid starts to melt and become a liquid. The disappearance of a polymer crystalline phase at the melting point occurs by changes in physical properties: discontinuous change in density, refractive index, heat capacity, transparency and other properties. The chains come out of their ordered arrangements and begin to move around freely. The transition between solid and liquid is very distinct for small samples of high purity where its melting point can be measured to $0.1^\circ C$. The melting point of polylactide for example, is about 115°C. The sharp melting point could identify a specific ingredient in a complex food system [10] and also indicate the presence of adulterant in a food sample [16]. It is possible to identify among various sugar moieties for example, glucose (m.p. = 150°C), fructose (m.p. = 103–105°C) and sucrose (m.p. = 185–186°C) by determining the melting point of a small sample. Melting of crystalline starch in the presence of water is the best known phenomenon in food processing known as gelatinization. The details of the thermal transitions associated with food have been discussed separately in this chapter.

9.2 PHASE CHANGE

Collection of molecules can exist in three physical states: solid, liquid and gas. In thermodynamics, "phase change" or "phase transition" is the transformation of a thermodynamic system from one phase to another. Phase change associates with a significant change in one or more physical properties, especially heat capacity, with a minor change in a thermodynamic variable like temperature.

The transformation of water (liquid phase) to ice (solid phase) termed as freezing is the best example of phase change. Similarly, vaporization of water is another example of phase transformation. Some common phase change operations are presented in Table 9.1.

In food materials, things are not as straightforward. Most food materials either decompose before they boil or cross-linked bio-polymers decompose before they melt. For many foods, the transition between the solid and liquid states is rather diffuse and difficult to trace. Amorphous food materials pass through a transition commonly known as glass transition (discussed later) where glassy state changes to rubbery state while food materials are being heated and crosses the glass transition temperature. Many food processing techniques involve phase change and phase separation including drying (hot air drying, spray drying and freeze drying), freezing and rapid cooling, grinding and extrusion [17]. Some common examples of food products that exhibit phase changes are milk powder, lactose, chocolate, ice-cream, starch, and various dried products. The wide existence of amorphous foods makes it important to understand the nature of amorphous state, its state transitions and the corresponding impact on food quality.

Phase transitions in food are classified as first-order and second-order. A *first-order transition* is characterized by changes in physical state of materials isothermally and latent heat is evolved. For example, melting and freezing are examples of such a transition. Second-order transitions do not have accompanying latent heat however; can be detected by significant variation in compressibility, heat capacity, thermal expansion coefficient, etc. Glass transition temperature is the example of second-order transition.

9.3 HEAT FLOW AND HEAT CAPACITY

Heat (Q) is an energy transferred from one body or system to another due to differences in temperature. Heat flows spontaneously from an object with a higher temperature to an object with a lower temperature. The transfer of heat from an object to another object with an equal or higher

TABLE 9.1
Phase Changes Associated with Food

Type of Phase Change	Terminology for the Change	Heat Movement During Phase Change
Liquid to solid	Freezing	Heat evolves from the liquid as it freezes
Liquid to gas	Evaporation, boiling	Heat goes into liquid
Solid to gas	Sublimation	Heat goes into the solid as it sublimates
Gas to liquid	Condensation	Heat leaves the gas as it condenses
Solid to liquid	Melting (gelatinization)	Heat goes into the solid as it melts

temperature can occur with the aid of an external source. Simply, heat flow tells us how amount of heat (q) flows per unit time (t).

$$\text{heat flow} = \frac{q}{t} \tag{9.1}$$

There is another term "heating rate" commonly used in calorimetry. It provides information on how fast we can heat an object. It is represented by:

$$\text{heating rate} = \frac{\text{increase in temperature}}{\text{time}} = \frac{\Delta T}{t} \tag{9.2}$$

The heat capacity of a substance is a measure of how well the substance retains heat. While a material is heated, it obviously causes an increase in the material's temperature. The heat capacity is an extensive variable since a large quantity of matter will have a proportionally large heat capacity. Mathematically, heat capacity is the ratio of an amount of heat increase to the corresponding increase in its temperature. Heat capacity is a ratio of heat flow to heating rate as shown below:

$$\frac{\text{heat flow}}{\text{heating rate}} = \frac{q/t}{\Delta T/t} = \frac{q}{\Delta T} = \text{heat capacity} \tag{9.3}$$

The heat capacity is not only of theoretical interest but has also considerable importance in practical application. Commonly, heat capacity is used to calculate enthalpy difference between two temperatures or form the basis for interpretation of differential scanning calorimeter (DSC) curves as described in the following section. Heat flow and heat capacity are measured by DSC in contrast to earlier expensive adiabatic calorimeter [18].

9.4   DIFFERENTIAL SCANNING CALORIMETRY

DSC is a convenient method for measuring thermal properties of a material [19]. This technique was developed initially as a tool to characterize thermal transitions in polymers but it is now frequently used in food and biopolymer research [20].

DSC is a technique for measuring the energy necessary to establish a nearly zero temperature difference between a substance and an inert reference material, since the two specimens are subjected to identical temperature regimes in an environment heated or cooled at controlled heating rate. In a DSC, an average temperature circuit measures and controls the temperature of the sample and reference holders to conform to predetermined time-temperature program. This temperature is recorded in one of the x–y axis. At the same time, a temperature difference circuit compares the temperature and reference holders and proportion power to the heater in each holder so that the temperatures remain same. While a sample undergoes thermal transition, the power of the two heaters is adjusted to maintain their temperature and a signal proportional to the power difference is plotted on the second axis. The area of the resulting curve represents the latent heat of transition.

A schematic diagram of heat flow with temperature and DSC heating unit are shown in Figures 9.1 and 9.2, respectively. The sample and reference are enclosed in the same furnace. Each pan is placed on top of a heater and interestingly, the two separate pans heat at the same rate with respect to each other with two separate heaters. The main assembly of the DSC cell is enclosed in a cylindrical, silver heating black which dissipates heat to the specimens via a constantan disc which is attached to the silver block. The heating rate is usually maintained at 5–10°C/min and the rate is precisely controlled by computer software. The temperature difference is finite only when heat is being evolved or absorbed due to exothermic or endothermic activity of the sample or when heat capacity of sample changes abruptly [18]. A plot of heat flow or heat capacity with changes in temperature is termed as a thermogram and typical DSC thermograms for a food sample is illustrated in Figure 9.3.

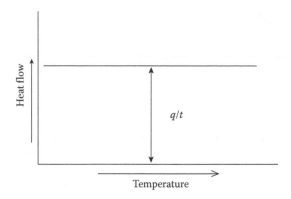

FIGURE 9.1   Heat flow as function of temperature.

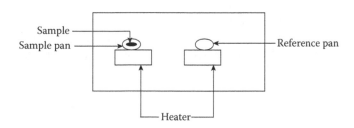

FIGURE 9.2   Components of a DSC heating unit.

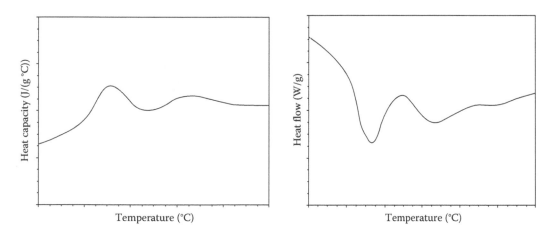

FIGURE 9.3   Typical DSC thermograms.

9.5   GLASS TRANSITION OF FOOD

9.5.1   GLASS TRANSITION

Glass transition temperature was considered as a reference temperature: below T_g, the food was expected to be stable; above this temperature, the difference between T_g and storage temperature T, i.e., $(T–T_g)$ was assumed to control the rate of physical, chemical and biological changes [7]. Moreover, the variations of mechanical and transport properties in the glass transition range could contribute to a better control of some food processing operations.

9.5.2 DEFINITION AND MOBILITY

The glass transition temperature (T_g) is defined as the temperature below which the physical properties of amorphous materials change in a manner similar to those of a solid phase (glassy state), and above which amorphous materials behave like liquids (rubbery state) (Figure 9.4). T_g is used for wholly or partially amorphous phases of materials. One property associated with the glassy state is a low volume coefficient of expansion that occurs as a result of changes in slope of the curve of volume versus temperature at T_g [18]. This behavior is shown by ε-caprolactone. In high temperature region, it shows rubbery behavior and below T_g at about −65°C it becomes a glass.

A material's molecular mobility becomes restricted in the glassy state (below T_g). Noncovalent bonds between polymer chains become weak compared to thermal motion above T_g, and the polymer becomes rubbery and exhibit elastic or plastic deformation without fracture. Above T_g, the specific volume of material increases to accommodate the increased motion of the wiggling chains. For materials, T_g is the mid-point of a temperature range in which they gradually become more viscous towards lower range of temperature and liquid changes to solid in higher range. The glass transition phenomenon is a kinetic and relaxation process associated with the so-called α relaxation of the material [15]. A given polymer sample does not have a unique value of T_g because the glass phase is not at equilibrium. The measured value of T_g will depend on the structure, molecular weight of the polymer, crystalline or cross linking, diluents, thermal history and age, on the measurement method and on the rate of heating or cooling.

9.5.3 MOBILITY ABOVE T_g

A theoretical interpretation of the temperature dependence of polymeric materials above glass transition is available based on free volume and viscosity [21].

Free volume can simply be considered as holes between rigid particles of polymers. In polymeric melts, the proportion of free volume remains as 30% of the total volume and the theory predicts that it collapses to about 3% at the glass transition temperature [22]. The fraction of free volume (f) may be written as:

$$f = f_g + (T - T_g)\Delta\alpha \qquad \text{when } T \geq T_g \qquad (9.4)$$

$$f = f_g \qquad \text{when } T < T_g \qquad (9.5)$$

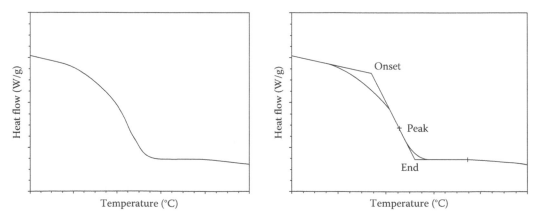

FIGURE 9.4 Typical glass transition temperature curves showing onset, peak, and end glass transition temperature.

The value of f and f_g remains constant below the glass transition temperature. The volume expansion coefficient (α) accounts for an increase in amplitude of molecular vibration with an increase in temperature. An increase in free volume is expected above T_g and can be estimated by expansion coefficient.

Willams, Landel, and Ferry [23] proposed that the log viscosity is a function of $1/f$ and varies linearly above T_g as shown below:

$$\ln\left(\frac{\eta}{\eta_g}\right) = \frac{1}{f} - \frac{1}{f_g}$$ (9.6)

Substitution into Equation 9.1 results:

$$\log\left(\frac{\eta}{\eta_g}\right) = -\frac{a(T - T_g)}{b + T - T_g}$$ (9.7)

The above equation is known as the Willams, Landel, and Ferry (WLF) equation. The equation was fitted with available data in the literature for polymeric materials to obtain numeric values of a and b. Those coefficients were also found to be valid for food materials [24]. Equation 9.3 can be rewritten as:

$$\log a_T = \frac{-17.44(T - T_g)}{51.6 + T - T_g}$$ (9.8)

The shift factor (a_T) is the ratio of the viscosity at T relative to that of T_g. The latter one is about 10^{14} Pa.s for many materials.

The WLF equation is valid over the temperature range from T_g to about $T_g + 100$ K.

While applying WLF equation for food materials, Roos and Karel [13] found that the relationship fitted well for amorphous sucrose and lactose with the "universal" constants. Soesanto and Williams [24] predicted the viscosity of fish protein hydrolyzate matrix at the onset of collapse during freeze-drying using the WLF equation. However, the validity of these universal constants has been challenged by Peleg [25] after experimenting with some polymers and amorphous sugars. The author found serious errors in the magnitude of viscosity when fitting the model over 20–30 K above T_g. Peleg [26] further found that the upward concavity of changes in a transitional region, which cannot be predicted by WLF or Arrhenius equation, can be described by another model with the structure of Fermi's function. These results inferred that though WLF equation is well fitted for various food systems, there is a need for careful examination before applying this equation universally for all complex food system.

9.5.4 Molecular Motion below T_g

Molecular mobility becomes restricted below glass transition temperature. However, molecular relaxation processes still take place at temperatures below T_g. Molecular motions still persist in the glassy state with lower amplitude and co-operatively than at T_g. Relaxation processes can be noticed in the glassy state with dynamical mechanical thermal analysis or dielectric spectroscopy; they also give rise to endothermic features on DSC curves. Sub-T_g relaxations are termed according to their relative position to the main relaxation-α. In addition to the main relaxation α, several secondary relaxations like β, γ, and δ are observed [27]. For biopolymers and low molecular weight sugars, secondary β relaxation has extensively been studied although its origin is still under investigation. β-relaxation are reported for various starches [28,29] and polysaccharides such as dextran, pullulan,

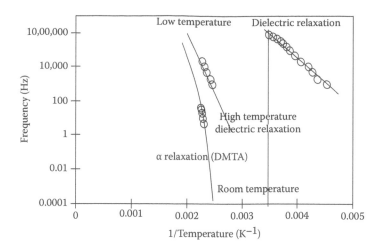

FIGURE 9.5 Relaxation map for dry bread. (Adapted from Roudaut, G., Simatos, D., Champion, D., Contreras-Lopez, E. and Le Meste, M., *Inno. Food Sci. Emer. Technol.*, 5, 127, 2004.)

etc. [30]. For polysaccharides, these relaxations are believed to be either lateral rotation (γ relaxation at low temperature) or local conformational changes of the main chain (β-relaxation closer to T_g) [30]. A secondary relaxation is observed in both dry white bread and in gelatinized dry starch but not in gluten in the temperature range of –60 and –40°C (depending on the frequency) with dielectric spectroscopy [28]. It inferred that secondary relaxation is associated with carbohydrate ingredients (sucrose and starch) of bread. Moreover, in starch-sucrose mixtures, the amplitude of the secondary relaxation exhibited sensitivity to sucrose concentration [28].

Roudaut et al. [28] presented secondary relaxations along with β-relaxation of cereal products in terms of mobility maps. A typical mobility map for dry crispy bread is illustrated in Figure 9.5. The map is a semilogarithmic plot between relaxation frequency and reciprocal values of absolute temperature. It can be used to characterize the different motions taking place in a material in a given temperature and frequency range. Different types of motions are shown in the mobility map: a first one observed at approximately –60°C and others at higher temperature that could be associated with the glass transition.

While a glassy material is placed between T_β and T_g, a microstructural evolution may occur and eventually, the system approaches the metastable equilibrium, with additional loss in enthalpy and volume [15]. This is termed as "physical aging," or "annealing." The more compact molecular organization and the strengthening of interactions result in changes in many physical properties: increased rigidity and brittleness, decreased dimensions and transport properties. Physical aging is expected to be of importance for the stability of low moisture products and is currently receiving a lot of attention. The rate of enthalpy relaxation was decreased as expected, upon addition of a substance with higher T_g, for instance dextran in sucrose [31,32]. In contrast, increasing the weight fraction of fructose in glucose-fructose mixtures also resulted in a decrease in the aging rate, although T_g was depressed [33].

9.6 MEASUREMENT OF GLASS TRANSITION TEMPERATURE

The measurement of glass transition temperature is an important part of phase transition studies. Several techniques are available namely DSC, dynamic thermal analysis (DTA), dynamic mechanical analysis (DMA) and dynamic mechanical thermal analysis (DMTA) for the determination of T_g and they are complementary depending on the nature of the studied material [7]. DMA measures glass transition in terms of the change in the coefficient of thermal expansion (CTE) as the polymer goes from glass to rubber state with the associated change in free molecular volume. A dynamic mechanical analyzer (DMA) measures the response of the materials to an oscillating deformation. However, the most common method used to determine glass transition is DSC that detects the change in heat capacity (i.e., a shift in the base line) occurring over the two periods. With modulated temperature DSC (MTDSC), a recent extension of conventional DSC using a modulated temperature input signal, information on the "amplitude of the complex heat capacity" is obtained, both in (quasi) isothermal and non-isothermal conditions. This complementary information, giving rise to a deconvolution of the (total) heat flow signal into "reversing" and "nonreversing" contributions, enables a more detailed study of complicated material systems [34]. In addition to these available methods, some other methods can be used as complementary techniques. In many cases, nuclear magnetic resonance (NMR), dielectric spectroscopy, and rheological measurement may help experimentally validate that the key factor in glass transition is the change in segmental mobility and relaxation behavior of materials. However, each method has its limitations, for example, in the sensitivity to certain material and the sample size used in the measurement.

Glass transition happens over a wide range of temperatures. Still, there is no consensus for the definition of the T_g point on a DSC curve. Commonly three points—onset (T_{gi}), midpoint (T_{gp}), end point (T_{gc}) have been chosen to define T_g. This is often a relatively narrow range of 10–20°C for amorphous sugars however a much larger range, for example 50°C, may be expected for the glass transition of polymers in foods. Within this temperature range, glass transition temperature can be referred to as the temperature initiating the onset of glass transition or the mid-point temperature of the change in specific heat capacity. It has become increasingly recognized that T_g should be characterized by at least two parameters indicating its onset or midpoint and the width of the transition [7].

9.7   GLASS TRANSITION OF WATER

Water is the major component of food products and is therefore better to understand glass transition behavior of water. Water can exist in more than one distinct amorphous form [35,36]. The conversion between different glass structures, the different routes producing glass structures and the relationship between the liquid and glass phases are under active debate [37]. The major concern is the identification of glass transition temperature at ambient pressure and the magnitude of the associated jump of specific heat.

DSC studies reported conflicting values of T_g. DSC scans detect glass transition in some cases where in some cases, it does not. An exothermal peak in the specific heat of properly annealed hyper-quenched water supports the estimate of a T_g value at –137°C [38], with a specific heat jump from 1.6 to 1.9 J/mol/K. The reported T_g value has been challenged by several researchers [39,40]. It has been argued [41] that the small peak measured in the work carried out by Johari et al. [38] is a pre-peak typical of annealed hyper-quenched samples preceding the true glass transition located at T_g = 165 K. Assigning T_g value of 108°C would explain some of the confusion related to glass transition in water [40]. The T_g at 108°C proposal can not be experimentally verified due to the homogeneous nucleation of the crystal phase at 123°C. A recent computer simulation study of the glass transition for water using the extended simple point charge potential claimed that T_g of water is above the accepted value of 147°C.

9.8   ANALYSIS OF GLASS TRANSITION TEMPERATURE
OF MISCIBLE COMPONENTS

The miscibility of binary polymer blends is commonly ascertained by measuring their glass transtion temperature(s). Two values of T_g indicate a two phase system and a single T_g between two polymers appears for miscible blends.

The general equation for a binary mixture assuming ΔC_{pi} independent of temperature is given by Couchman and Karasz [42] as:

$$\ln T_g = \frac{W_1 \Delta C_{p1} \ln T_{g1} + W_2 \Delta C_{p2} \ln T_{g2}}{W_1 \Delta C_{p1} + W_2 \Delta C_{p2}} \tag{9.9}$$

T_g is the glass transition temperature of the blend polymer, W_1 and W_2 are the weight fraction of two components and ΔC_{pi} is the difference in specific heat between the liquid and glassy states as T_{gi}.

The above equation can be modified with some assumptions. Assuming equal heat capacity ($\Delta C_{p1} = \Delta C_{p2}$), Equation 9.9 can be converted as:

$$\ln T_g = W_1 \ln T_{g1} + W_2 \ln T_{g2} \tag{9.10}$$

Again, if T_{g2}/T_{g1} is approximately one, then the logarithmic expansion of Equation 9.10 can be limited to the first term and the equation will be

$$T_g = W_1 T_{g1} + W_2 T_{g2} \tag{9.11}$$

Further, considering $\Delta C_{p1} = \Delta C_{p2}$, after rearrangement and expansion of the logarithmic term of Equation 9.10, Fox equation yields:

$$\frac{1}{T_g} = \frac{W_1}{T_{g1}} + \frac{W_2}{T_{g2}} \tag{9.12}$$

Instead of considering $\Delta C_{p1} = \Delta C_{p2}$ which is not a good approximation, we can consider $\Delta C_{p1}/\Delta C_{p2} = k$, a constant resulting in

$$\ln T_g = \frac{W_1 T_{g1} + k W_2 T_{g2}}{W_1 + k W_2} \tag{9.13}$$

Finally, with the first term of the expansion of the logarithmic term of Equation 9.9 and if T_{g2}/T_{g1} approaches to unity, therefore:

$$T_g = \frac{W_1 T_{g1} + k W_2 T_{g2}}{W_1 + k W_2} \tag{9.14}$$

where $k = \Delta C_{p2}/\Delta C_{p1}$; Equation 9.14 is known as the Gordon–Taylor equation.

Among food components, carbohydrates substantially affect the glass transition temperature of an amorphous dried food material. The glass transition temperature of common sugars like glucose, fructose, and sucrose are considerably low and therefore, these sugar moieties worked as plasticizer for multicomponent food systems. The influence of protein or fat does not affect T_g markedly [43–45].

Water plays a significant role in depression of T_g of food materials. The Gordon–Taylor equation was generally used for the prediction of water plasticization for binary systems. The equation fitted well to estimate the T_g of binary polymer system e.g., low and high molecular weight carbohydrates [46–48] and also for predicting water plasticization of lactose and lactose/protein mixtures, considering that all solid components contributing to the observed glass transition are miscible and formed a single phase [48]. The Couchmann and Karasz equation can be further extended and applied for multicomponent mixture systems such as water, glucose and sucrose [48,49].

In the Gordon–Taylor equation, the ratio of change in heat capacity has been replaced by a constant (k) for a binary mixture (water and solute) to be similar in form as Couchmann and Karasz equation. However, it is difficult to obtain exact values of ΔC_p experimentally especially with small molecules like water are involved. The ΔC_p reported values vary significantly depending upon the method of determination [48] and therefore, authenticity of experimentally determined k value is doubtful. Roos [50] suggested the following equation to predict k values for carbohydrates:

$$k = 0.0293T_g^\circ C + 3.61 \tag{9.15}$$

9.9 EFFECT OF MOLECULAR WEIGHT ON T_g OF FOOD POLYMER

For polymeric materials, glass transition temperature (T_g) is found to increase with number average molecular weight (M_n) in the absence of diluent [18] although it eventually attains a constant value irrespective of molecular weight. The trend has been followed adequately within a homologous family of food polymers (e.g., from the glucose monomer through maltose, maltotriose, and higher malto-oligosaccharides (e.g., maltodextrins) to the amylose and amylopectin high polymers of starch) [51,53] (Table 9.2). With a homologous family of food polymers like maltodextrins of different average molecular weights, the Fox and Flory [52] relationship has been used to determine the effect of molecular weight on T_g using the following relationship [14,53]:

$$T_g = T_{g\infty} - \frac{K_g}{M} \tag{9.16}$$

where T_g is the glass transition temperature of starch derivative, K_g a constant, $T_{g\infty}$ is the limiting T_g at infinite molecular weight (starch), and M is the molecular weight. $T_{g\infty}$ and K_g are material dependent.

This equation fits well for glucose polymers with three or more glucose chains [54]. A linear relationship was observed between T_g and dextrose equivalent (DE) over a wide DE ranges from 2 to 100 [55]. Those authors have advocated that the degree of hydrolysis of starch could be predicted from measured T_g using the following equation:

$$T_g = -1.4DE + 449.5 \tag{9.17}$$

TABLE 9.2
Change of T_g of Amorphous Food as a Function of Molecular Weight

Material	Molecular Weight	T_g	Reference
Fructose	180	5	[51]
Glucose	180	31	
Galactose	180	32	
Sucrose	342	62	[51]
Maltose	342	87	
Lactose	342	101	
Maltodextrin (DE 25)	720	121	[12–14]
Maltodextrin (DE 5)	3600	188	
Starch		243	Predicted

For food material with higher molecular weight, it is difficult to measure T_g which requires heating for a wider range of temperatures. Those food products commonly decompose at higher temperature and T_g can be predicted using equation. Slade and Levine [53] mentioned that although a general correlation between molecular weight and T_g of carbohydrates has been widely accepted, one should be aware that T_g can vary substantially even within a series of compounds of the same molecular weight due to their chemical structure. Microstructure of polymers plays a significant role in the T_g value where very little information is available on food matrix and its architecture.

9.10 CRYSTALLIZATION AND MELTING

9.10.1 CRYSTALLIZATION

Crystallization is the formation of solid particles within a homogenous phase. Crystallization from solution has industrial importance because of the variety of materials which are marketed in the crystalline form. Since crystallization is an ordering process, formation of crystal within a molten microdomain would generate a structure of "order-within-order" [56]. The objective of the crystallization process is two fold: firstly to obtain a crystal from an impure solution and secondly, to produce "purest-of-pure" from the mother liquor. In industrial practice, withdrawn crystal from mother liquor is known as magma [57].

The crystallanity of a material is characterized by specific volumes of the specimen (V), the pure crystals (V_c) and the completely amorphous material (V_a), as shown below:

$$w_c = \frac{V_a - V}{V_a - V_c} \qquad (9.18)$$

The relationship is valid only while the sample is free of voids. The crystalline specific volume is determined from x-ray unit cell dimensions, where V_a is obtained by extrapolation of the specific volumes as the polymer melts.

A crystal is a highly organized, nonliving matter and arranged in three dimensional arrays known as lattice. Many biopolymers and dried solid foods including sucrose and lactose show three dimensional ordered crystalline structure. Typical crystalline polymers are those whose molecules are chemically and geometrically regular in structure [18]. On the other hand, typical noncrystalline polymers are characterized by irregularity of structure and are commonly known as amorphous materials. Thermodynamically, the amorphous state is at a higher entropy level compared to crystals. The most direct evidence of the fact has been obtained from x-ray diffraction studies.

A crystalline material can be classified into three distinct categories (Table 9.3) based on degree of crystallinity and glass transition temperature [18]. An intermediate crystalline behavior is very common for both polymers and biopolymers with mechanical and engineering applications like tissue engineering and medical devices. Polymers with low crystallinity assume a glassy state and behave very similar to minimum cross-linked amorphous materials.

TABLE 9.3
Classification of Crystalline Materials

	Degree of Crystallinity		
T_g range	Low (5–10%)	Intermediate (20–60%)	High (70–90%)
Above T_g	Rubbery	Leathery, tough	Stiff, hard (brittle)
Below T_g	Glassy, brittle	Horn like, tough	Stiff, hard brittle

9.10.2 CRYSTAL FORMATION KINETICS

Most single crystals have the same general appearance and composed of thin, flat platelets about 100 A thick, commonly many microns in lateral dimensions [18]. The thickness increases by spiral growth of additional lamella from screw dislocation. In the melt, crystallization starts out from homogenous or heterogeneous nucleation followed by the growth of numerous lamellar crystals which aggregate into a super-structural identity called "spherulite" [58]. In the bulk melt, its concentration is the limiting factor that governs the movement of spherulites and finally crystal structure.

The size, shape, and regularity of crystals depend on their growth conditions such as: solvent, temperature, and growth rate. The growth and kinetics of crystal formation can be evaluated by the fixed cooling rate experiment in a DSC. Generally, samples are first annealed at temperature above melting (order-disorder transition), maintained for certain period of time and cooled at a specific cooling rate to record crystallization exotherm (Figure 9.6). The peak temperature of the exotherm is defined as the crystallization temperature (T_c). The area under the crystallization curve provides the latent energy of crystallization for the sample.

The development of crystallinity in a polymeric system is not instantaneous. A plot of specific volume as a function of time at temperature below the crystalline melting point is used for kinetic studies. The time required for crystallization process is infinite and therefore, it is customary to define rate of crystallization of a constant temperature as the inverse of the time needed to attain one-half of the total volume change [18]. The crystallization rate is a function of temperature and as the temperature is lowered, the rate increases reaching maximum, and decreased thereafter as mobility of molecules minimized and crystallization becomes diffusion controlled.

For low molecular weight polymer crystallization, Arvami equation can be fitted as shown below:

$$\ln\frac{(V_\infty - V_t)}{(V_\infty - V_0)} = -\left(\frac{1}{w_c}\right)kt^n \tag{9.19}$$

where V_∞, V_t, and V_0 are specific volumes at the times indicated in the subscripts, w_c is the fraction weight of materials crystallized, k is a constant describing the rate of crystallization, and n is an exponent that varies with the type of nucleation and growth rate. This equation however, has

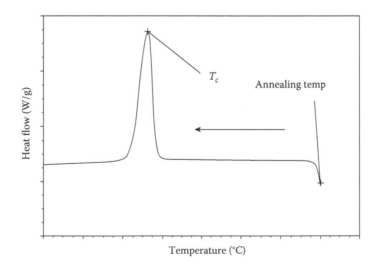

FIGURE 9.6 Thermogram showing annealing temperature and melt crystallization temperature.

limited application and fails to ascribe the shape of the curve while n ranges between two and four In fact, the crystallization continues for a longer period of time than the equation predicts. This indicates that crystallization occurs in two steps: first of which involves nucleation and fast formation of spherulite and the second is kinetically slow as it makes difficult improvement in crystal perfection [18].

The WLF equation can predict the crystallization rate of amorphous food products. The equation for crystallization kinetics is shown below:

$$\log\left(\frac{t}{t_g}\right) = \frac{-a(T - T_g)}{b + T - T_g} \tag{9.20}$$

where t is the time of crystallization, t_g the time to crystallize at glass transition temperature T_g, a and b are constants 17.44 and 51.6 K respectively, and T is temperature (K).

Many researchers [13,26,59] used the above equation to study the crystallization behavior of amorphous sugars including lactose, sucrose and sucrose/fructose.

9.10.3 Melting

The melting process is different from the glass transition phenomenon. Melting is a first-order transition that occurs only in crystalline polymers. Melting occurs while the polymer chains dissociate from their crystal structures and become a disordered liquid and the corresponding temperature of specific material melting is defined as melting temperature (T_m). Glass transition falls under the second-order time-temperature dependent transition (without change in latent heat). However, crystalline materials will have some amorphous portions that result in both T_g and T_m in some materials. For example, lactides possess both amorphous and crystallinity (Figure 9.7). It is important to remember that the chains that melt are not the chains that undergo the glass transition.

Measurements of the melting point of a solid can also provide information about the purity of the substance. DSC is most commonly used to detect melting temperature (T_m) of pure food materials. Pure, crystalline solids melt over a very narrow range of temperatures whereas mixtures melt over a broad temperature range. Mixtures also tend to melt at temperatures below the melting points of pure solids.

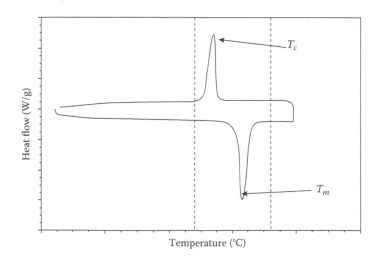

FIGURE 9.7 Thermogram of polylactide showing both crystallization and melting temperature.

For a homogenous molecular weight polymer, a statistical thermodynamic analysis based on a lattice model provides crystalline melting temperature by the following equation:

$$\frac{1}{T} - \frac{1}{T_m} = \frac{R}{\Delta H_m} \left(\frac{1}{xw_a} + \frac{1}{x - \zeta + 1} \right) \tag{9.21}$$

where w_a is the weight fraction of material that is amorphous at temperature T. The equilibrium melting point of pure polymer is obtained from the relationship: $T_m = \Delta H_m / \Delta S_m$. The degree of polymerization is x and ζ is the parameter that characterizes the crystallite size. The equation clearly indicates that for a uniform chain length polymer, melting happens over a finite temperature range.

For polymer-diluent systems, Florry Higgins equation can be used to predict melting temperature.

$$\frac{1}{T_m} - \frac{1}{T_m{}^o} = \frac{R}{\Delta H_m} \frac{V_2}{V_1} (v_1 - \chi_1 v_1{}^2) \tag{9.22}$$

Where T_m and $T_m{}^o$ are the melting points for polymer-diluent mixture and the pure polymer, respectively. V is the molar volume, v is the molar fraction and χ_1 is the polymer-diluent interaction parameter. Thus, melting point depression is directly related to the volume fraction of added diluent and its interaction with the polymer. The equation has wide applications in food biopolymers to predict melting point depression in polysaccharides and proteins.

9.10.4 Application of Crystallization and Melting in Food

Amorphous materials are meta-stable and therefore, can crystallize over time during storage. Crystallization is a phase transition occurring in freeze-dried and spray-dried amorphous lactose or other crystallizing carbohydrates. The rate of crystallization is a function of $(T - T_g)$, with increasing crystallization rates for higher temperatures [13,60]. Among food products, lactose is most studied. Numerous literatures are available on lactose crystallization and other dairy products during storage [43,44,61,62]. It is reported that amorphous lactose in skim milk powder crystallized and released water at 54% relative humidity (RH) at 20°C [62] where as Jouppila and Roos [44] found that freeze-dried lactose and lactose in milk powders crystallized at above 40% RH at 24°C. Nature of crystals depends on processing conditions. Scanning electron microscopy (SEM) observation of skim milk powder before and after spray-drying revealed tomahawk-like crystals in precrystallized (crystals formed before spray-drying) and needle-like crystals in postcrystallized (crystals formed after spray-drying) products [63]. Water sorption induces α-lactose monohydrate crystals formation in both spray-dried whole and instant skim milk powders [64].

Crystallization and melting temperatures of food products have been used to characterize and detect adulterant. Ferari et al. [16] studied isothermal crystallization for olive oil authentication. In general, any change in oil composition due to chemical or physical treatment affects in a typical way, the freezing heat flow curve. In other words, the nucleation and growth of the polymorphous crystalline fractions of triacyl glycerols are dependent on oil molecular composition. Crystallization and melting have been used to characterize composite foods like caramel by Ahmed et al. [10]. The melting behavior of caramels was mainly controlled by the nature of fats used in the formulation. The sharp melting point peak of one of the caramel samples has been attributed to the fractionated palm kernel oil-based fat in its formulation, while milk fat-based caramels exhibited mixed behavior. Caramels exhibit a broad endothermic peak with an onset and peak temperature. The crystal expected to depress the solubility of sucrose and the nature of fat. The presence of corn syrup also affects the nucleation and growth rate kinetics and may influence the relative shapes of the crystallization curves [65].

Melting and crystallization temperatures of honey samples collected from various sources have been characterized by Ahmed et al. [66]. A very wide and intense endothermic peak at 170–240°C corresponding to the melting of sugars caused by the presence of various sugars (mono-, di-, tri-, and oligosaccharides). The peak thermal transitions of fusion (melting temperature) of honey samples varied between 181.55 and 221.12°C. Some samples exhibit more than one temperature of fusion due to the presence of various sugar components. Since honey constitutes primarily of glucose and fructose, these two major sugar components contributed to the melting point/s of honey. Similar ranges of melting temperatures of honey have been reported by Cordella et al. [67].

The amorphous form of low molecular weight carbohydrates and protein hydrolyzates are very hygroscopic. The glass transition temperature is depressed while food samples absorb moisture from the surrounding and crystallization rate is accelerated. Bound and orderly molecular packing during crystallization cannot generally accommodate excess water in dehydrated powdery foods. It results in the loss of adsorbed water and absorption of this expels moisture to the surface of neighboring particles creating interparticulate liquid bridges resulting in caking [43,44] Surrounding particles, which absorb moisture will also be crystallized and the crystallization can proceed as a chain phenomenon. Crystallization process removes impurities including volatiles. Loss of diacetyl as a function of crystallization rate of lactose during storage has been reported by Senoussi et al. [60]. They found that when lactose was stored at 20°C above T_g, the amorphous product went through immediate crystallization and practically all diacetyl was lost after six days. Levi and Karel [59] also found increased rate of loss of a volatile 1-n-propanol in an amorphous sucrose system as a result of crystallization.

9.11   MELTING, GELATINIZATION, AND GLASS TRANSITION OF STARCH

Starch polysaccharides are among the first biopolymers to be studied and have been investigated for many years. There are two kinds of polysaccharides in starch, namely amylose which is essentially linear α-(1-4) linked D-glucose and amylopectin, which is highly branched myriad α-(1-6) linkage. Molecular weights are estimated at 10^5–10^6 and at 10^7–10^8 for amylose and amylopectin respectively. Both macromolecules are found in plant tissues in the form of granules. The granules contain crystalline and amorphous domains. The finding that the polysaccharides form double helices, which underlie the crystalline structures, is considered to be one of the pivotal events in the history of starch chemistry [68].

Thermal transitions of starch and gelatinization have been well reviewed in the literature and a large number of explanations of the molecular processes responsible for the gelatinization process are available in the literature [69]. However, as the knowledge base developed, many of these explanations have been superseded.

The origin of starch gelatinization enthalpic transition was thoroughly investigated by Cooke and Gidley [70] using DSC, NMR, and X-ray diffraction. The work reports that crystalline and helical orders are lost concurrently during gelatinization. However, the melting of crystalline and noncrystalline structures of starch is controversial. Most researchers' advocate that crystallite disruption occurred at a higher temperature by helix melting is inconsistent and there is no evidence to support that melting of noncrystalline helices occurred at a temperature lower than that of crystallite melting [71,72]. The other major finding of Cook and Gidley [70] was that the enthalpy of gelatinization primarily reflects the loss of molecular (double-helical) order.

The melting behavior of gelatinization describes that crystalline (or helical) zones in different granules have different stabilities [73]. While granules gelatinize, water migrates from one location to another and DSC peaks correspond to crystalline melting at different diluent levels are observed. It is reported that at significantly higher level of moisture content (>65%), each granule will absorb water without restriction and a single endothermic peak is observed on heating (Figure 9.8). If the water content falls below this level, there is competition by the granules for water absorption. In this situation, the least stable granules melt first, absorb water and consequently decrease the remainder

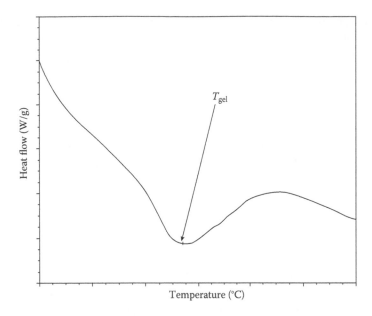

FIGURE 9.8 Typical starch gelatinization curve.

of the diluent. The latter particles melt at higher temperatures partly because they are more stable and partly because the effective volume fraction of diluent is reduced. The endotherm associated with the higher melting point fraction initially occurs in the form of a trailing shoulder on the first peak. As the water content of the sample is reduced further, fewer granules are able to gelatinize in an unrestricted water environment and as a consequence, the first endotherm decreases in size while the second increases and shifts to a higher temperature. If the volume fraction of water is reduced sufficiently, only the higher temperature endotherm is apparent. Changing the ratio of the two starches in the blend, it is observed that competition for water can be altered in a manner predicted by the melting explanation of gelatinization.

Glass transition studies in starch have largely concentrated on the use of DSC which has accompanied a debate over the role of glass transition in native starch, which is a highly nonequilibrium system. Biliaderis et al. [74] proposed a three-phase model to account for the observed thermal behavior. These phases are: (i) the amorphous regions, giving rise to a heat capacity change (ΔC_p) at the glass transition temperature (T_g), (ii) the crystalline regions, and (iii) the intercrystalline material. The intercrystalline material does not contribute to the ΔC_p observed at T_g as it is motionally restricted by its proximity to crystalline regions. An increase in heat capacity after starch gelatinization (swelling and melting of crystallites on heating in the presence of water) occurs due to glass transition happening immediately prior to gelatinization [5,74]. Zeleznak and Hoseney [75] advocated that the change of C_p happened due to granule swelling and the initiation of crystallite melting.

It is now well established that glass transition occurs in starch-water systems. It is indeed difficult to precisely determine the temperature with pure starch since thermal degradation of the dry polymer precludes measurement of T_g [76]. Extrapolation from measurements on starch-water systems indicates that T_g of pure starch is about 227°C [46]. The onset of crystallite melting depresses with water uptake following the same rule of polymer glass transition. T_g decreases by about 6°C per wt% water for the first 10% water addition [77]. Once the water content reaches about 22%, the starch is plasticized at room temperature. Further increase in water content markedly decreases T_g and eventually drops below the freezing point of the aqueous system. Benczedi et al. [78] studied the specific heat increment at the glass transition to estimate the transition temperature of samples containing up to 25 wt% water. Authors have found that T_g as a function of the weight fraction of water (w) in amorphous starch as measured by heat flow calorimetry and is represented by the following regression equation:

$$T_g(w_1) = 244.9 - 1565w_1 + 2640w_1{}^2 \tag{9.23}$$

9.12 STATE DIAGRAM OF FOOD

In food freezing system, fundamental factors that accompany ice formation are relatively complex and the freezing process has been dominated by heat and mass transfer properties and product compositions [79]. The formation of supersaturated state or glassy state resulting from freeze concentration has significant application to subzero temperature stability. Previously, freezing curve and glass transition line were used to construct the state diagram. Recently, attempts are being made to incorporate other structural changes with glass line, freezing curve and solubility line in the state diagram.

Simply, a state diagram represents a map of different states of food as a function of water or solids content and temperature [80]. The advantage of drawing a map is to help understand the complex changes while food water content and temperature are varied. It further identifies stability of food during storage and proper processing conditions. State diagram also serves as "stability map" for frozen and freeze dried foods [48,80].

The state diagram of food materials is constructed based on glass transitions, freezing curves and maximal freeze concentration condition in addition to pure component [81,82]. Numbers of microregions and new terminologies are being included in constructing the state diagram [83]. The terminologies (glass transition and melting temperature) used in frozen system and constructing the state diagram are different and probably to distinguish from sample measured at frozen condition. Franks et al. [84] used the symbol T_g' for the glass transition temperature found for a maximally freeze-concentrated solute phase in a frozen system. New terminologies related to state diagram of foods are well described by Rahman et al. [85]. The glass transition temperature and freezing point of pure components, unfreezable and freezable water can be measured by using DSC while initial freezing point and end point of freezing are available from cooling curve method [86,87]. Currently, the state diagrams of various foods are available in the literature [48,80,83,88–90].

One typical state diagram is illustrated in Figure 9.9 which has been adapted from Rahman [83]. It represents different states as a function of temperature and solids mass fraction. The freezing line (ABC) and solubility line (BD) are shown with respect to the glass transition line (EFS). The point F (X_s' and T_g') lower than T_m' (point C) is a characteristic transition (maximal-freeze-concentration condition) in the state diagram defined as the intersection of the vertical line from T_m' to the glass line EFS. The water content at point F or C is considered as the unfreezeable water ($1–X_s'$) based on unfrozen water mass that remained unfrozen even at significantly lower temperature. It takes into account both uncrystallized free water and bound water attached to the solid matrix. The point Q is termed as T_g'' and X_s'' as the intersection of the freezing curve to the glass line by maintaining similar curvature. Point R is known as T_g''' as the glass transition of the solids matrix in the frozen sample which is determined by DSC. This is due to the formation of same solid matrix associated with un-freezable water and transformation of all free water into ice even though the sample contains different level of total water before DSC scanning [85]. Various zones have been shown in the sate diagram and those zones are characterized by different behaviors. For example, the BDL line represents the melting line which is important for high temperature processing like frying, baking, roasting, and extrusion cooking.

Studies on state diagram of foods clearly predict stability and shelf-life of food during storage. The variation of stability however, below glass transition not following the rule indicating only glass transition temperature for developing the stability rule is inadequate. The stability of food should be considered from compositional and water content present in the food. Apart from water content of

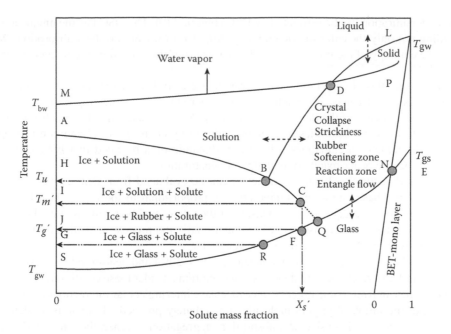

FIGURE 9.9 Typical state diagram of a food (Adapted from Rahman, M.S., *Trends in Food Sci. Technol.*, 17, 129, 2006. With permission.)

food, terms discussed above in relation to state diagram should be considered to better understand frozen food stability. The area has not been explored adequately and more research is needed to understand characteristic temperature of frozen food.

9.13 DENATURATION AND GLASS TRANSITION OF PROTEIN

Protein denaturation has been extensively studied by DSC where denaturation is indicated as an endothermic peak in thermogram [91]. Protein foods are denatured by various means (heat, pressure, pH, ionic strength). Heat is the common method for protein denaturation. Thermal denaturation of small globular proteins is generally considered reversible in high yield provided the reaction is carried out under conditions preventing aggregation and far from the isoelectric point [92]. This allows indirect thermodynamic evaluation of the process by applying equilibrium thermodynamics and assuming a two state model (Native to denature). Under these conditions, the equilibrium constant, K, of the process and simultaneously the standard enthalpy change, ΔH^o, from the van't Hoff equation can be estimated:

$$\Delta H^o = RT^2 d\ln(K) \tag{9.24}$$

The standard free energy change, ΔG°, and the standard change of entropy, ΔS°, may be then obtained from the following equations:

$$\Delta G^o = -RT\ln(K) \tag{9.25}$$

$$\Delta G^o = \Delta H^o - T\Delta S^o \tag{9.26}$$

Numerical estimates of the above thermodynamic parameters are of considerable significance in understanding the molecular aspects of the denaturation reaction [93].

Protein denaturation can be estimated by DSC (Figure 9.10). The DSC has the unique advantage of providing not only the calorimetric enthalpy, ΔH_{cal} (from the area of heat absorption), but also the effective enthalpy of the process (from the sharpness of the transition and using the van't Hoff equation) [91].

Thermal properties of compact globular proteins were studied by Privalov and Khechinashvili [91] by DSC using extremely dilute solutions (0.05–0.5%) under low heating rates (1°C/min). The temperature dependence of the denaturation enthalpy, ΔH, was determined from changes in AH at different conformational stabilities of proteins induced by changes in pH and considering small changes in the ionization enthalpies. The thermodynamic data revealed that pH change had no effect on the specific heat capacity or enthalpy of denaturation.

The protein denaturation kinetic parameters can be evaluated from DSC curves. The vertical displacement from base line at any temperature is proportional to the rate of heat flow into the sample (dH/dT), and consequently is a measure of the rate of reaction. Different methods are used to estimate the reaction rate constants [94–96] and finally the process activation energy may then be calculated from the Arrhenius plot of ln K versus $1/T$.

In practical DSC work, both protein concentration (5–20%) and heating rates (5–20°C/min) are quite high in order to resemble actual processing conditions. Under these conditions however, denaturation becomes an irreversible process since extensive intermolecular interactions are favored and aggregation of the unfolded protein molecules immediately proceeds. In contrast to denaturation which is connected with intensive heat absorption, aggregation is generally considered as an exothermic process. Therefore, it becomes more difficult to interpret ΔH_{cal} values quantitatively, since they represent the net product of a positive (denaturation) and a negative (aggregation) contributor. In addition to protein concentration, other parameters such as pH, ionic strength, tertiary and quaternary structure of the protein can also affect the observed AH_{cal} values by their influence on protein conformational stability [97–99].

Protein fractions of cereal proteins like wheat consist of several proteins and therefore, several denaturation endotherms are expected. However, DSC response to wheat proteins (gliadin and glutenin) is limited [100]. Most reported DSC data on gluten denaturation are contradictory. Ellison and Hegg [101] observed gluten denaturation endothermic peaks at 88 and 101°C with very small changes in enthalpy while Erdogdu et al. [102] did not observe any denaturation peak. Lacking a

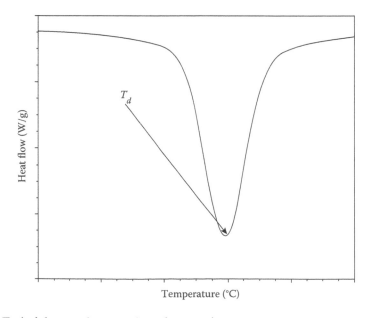

FIGURE 9.10 Typical denaturation curve for a plant protein.

tertiary-ordered structure, gluten cannot be put into a disordered form by heat, which explains the absence of DSC thermograms [103]. Recently Leon et al. [104] found gliadin denaturation temperature at 58°C whereas two endothermic peaks at 64 and 84°C for glutenins with high enthalpies. Those authors suggested that DSC protein denaturation peaks were a function of sample water content. Similar observation was recently reported by Ahmed et al. [105] for wheat protein isolates indicated distinct endothermic peaks in wider temperature range (50–130°C) at various moisture levels. While working on soy protein isolate, Ahmed et al. [106] found single denaturation peak for soy protein isolate at about 90°C corresponding to 11S fraction. However both 7S and 11S were detected from soy proteins [90].

Water has strong plasticizing effect on the glass transition of protein foods. Freeze-dried proteins should be stored below the glass transition temperature to ensure preservation of structure and activity of proteins or their long-term storage in the dry state [107]. The higher the T_g of the matrix, the better should be its stabilizing effect. It has been shown for various dried enzyme preparations that storage above T_g accelerated the loss of activity. The rate of glucose-6-phosphate dehydrogenase inactivation (in a glucose-sucrose matrix) was reported to conform to the WLF relation (for water content of 6–7%) [108]. The rate of inactivation of a pectinlyase preparation also greatly increased at temperature above T_g [109]. For globular proteins, the situation is less clear. For lysozyme, a marked change in heat capacity from 1.25 $Jg^{-1}K^{-1}$ to 1.55 $Jg^{-1}K^{-1}$ is observed with increasing water content from 7 to 20% w/w at room temperature [110].

As the heat capacity change is associated with the onset of enzyme activity and therefore, sufficient mobility for enzyme-catalyzed reaction, this transition can be likened to a glass transition. For a concentrated globular protein–water mixture the observed features in heat capacity as a function of temperature are relatively weak, span a comparatively wide temperature range and are difficult to attribute to a calorimetric glass transition [111]. It is believed that associated with the complex tertiary structure of proteins is a correspondingly complex dynamics [112], which cannot be adequately characterized through a single relaxation process. On the denaturation of globular proteins, with the consequent loss of tertiary structure, the calorimetric features associated with the glass transition become more evident and directly comparable with the behavior of flexible proteins [113,114].

Wheat gluten is a glassy, amorphous, and plasticizable polymer, and water has a strong plasticizing effect on the glass transition temperature (T_g) of flexible biopolymers like gluten [115]. For example, increasing the water content of a high molecular weight glutenin, from 3 to 12% w/w, depresses the calorimetric glass transition from 100 to 25°C [116]. Similar ranges of T_g (110–21°C) were obtained by Cocero and Kokini [117] while glutenin samples hydrated between 4% and 14% moisture content. Zhou and Labuza [118] studied the moisture-induced protein aggregation of whey protein powders (whey protein isolate, whey protein hydrolysates and beta-lactoglobulin) to elucidate the relationship of protein stability with respect to water content and glass transition. The heat capacity changes of WPI and BLG during glass transition were small (0.1–0.2 Jg^{-1} °C^{-1}), and the glass transition temperature (T_g) could not be detected for all samples. An increase in water content in the range of 7–16% caused a decrease in T_g from 119 down to 75°C for WPI, and a decrease from 93 to 47°C for WPH. For WPI and BLG, no protein aggregation was observed over the range of 0–85% RH, whereas for WPH, ~50% of proteins became insoluble after storage at 23°C and 85% RH or at 45°C and ≥ 73% RH, caused mainly by the formation of intermolecular disulfide bonds. This suggests that at increased water content, a decrease in the T_g of whey protein powders results in a dramatic increase in the mobility of protein molecules leading to protein aggregation in short-term storage.

9.14   CONCLUSIONS

Thermal analysis is a useful investigative tool for studying various heat-related phenomena in foods and their components by monitoring the associated changes in enthalpy. Glass transition plays an important role in determining stability and shelf-life of food products. Water acts as a plasticizer in food component/water system by depressing thermal transitions. Effect of water plasticization on thermal behavior of starch (gelatinization) and protein (denaturation) foods provide better understanding on stability and structure. The state diagram provides mobility-based information of binary or multi-component frozen food systems. More in-depth analysis and research are needed in the area however, and terms used to describe the state diagrams should be uniform. Studies on thermal analysis provide both thermodynamic and kinetic data that constitutes the main advantages of this technique. However, changes in enthalpy are nonspecific, the other complementary methods could be important to understand the physical nature of the phenomena observed.

REFERENCES

1. Slade, L. and Levine, H. Polymer science approach to water relationships in foods. In *Food Preservation by Moisture Control, Fundamentals and Applications,* Barbosa-Canovas, G.V. and Welti-Chanes. J., Eds. Technomic Publishing: New York, NY, 1995, 33.
2. Farkas, J. and Mohacsi-Farkas, C. Application of differential scanning calorimetry in food research and food quality assurance. *J. Therm. Anal.*, 47, 1787, 1996.
3. White, G.W. and Cakebread, S.H. The glassy state in certain sugar-containing food products. *J. Food Technol.*, 1, 73, 1966.
4. Slade, L. and Levine, H. Structural stability of intermediate moisture foods—a new understanding. In *Food Structure—Its Creation and Evaluation,* Mitchell, J.R. and Blanshard, J.M.V., Eds. Butterworths: London, UK, 1987, 115.
5. Slade, L. and Levine, H. Non-equilibrium behavior of small carbohydrate–water systems. *Pure Appl. Chem.*, 60, 1841, 1988.
6. Kim, M.N., Saltmarch, M., and Labuza, T.P. Non-enzymatic browning of hygroscopic whey powders in open versus sealed pouches. *J. Food Process. Preserv.*, 1981, 5, 49.
7. Champion, D., Le Meste, M., and Simatos, D. Toward an improved understanding of glass transition and relaxations in foods: molecular mobility in the glass transition range. *Trends Food Sci. Technol.*, 11, 41, 2000.
8. Hancock, B.C. and Zografi, G. Characteristics and significance of the amorphous state in pharmaceutical systems. *J. Pharm. Sci.*, 86, 1, 1997.
9. Simatos, D., Blond, G., and Martin, F. Influence of macromolecules on the glass transition of frozen system, In *Food Macromolecules and Colloids,* Dickinson E. and Lorient D., Eds. The Royal Society of Chemistry: Cambridge, UK, 1995, 519.
10. Ahmed, J., Ramaswamy, H.S., and Pandey, P.K. Dynamic rheological and thermal characteristics of caramels. *Lebensm-wiss u-technol*, 39, 216, 2006.
11. Roos, Y. and Karel, M. Differential scanning calorimetry study of phase transitions affecting the quality of dehydrated materials. *Biotechnol. Progr.*, 6, 159, 1990.
12. Roos, Y. and Karel, M. Phase transitions of mixtures of amorphous polysachharides and sugars. *Biotechnol. Progr.*, 7, 49, 1991.
13. Roos, Y. and Karel, M. Plasticizing effect of water on thermal behavior and crystallization of amorphous food models. *J. Food Sci.*, 56, 38, 1991.
14. Roos, Y. and Karel, M. Water and molecular weight effects on glass transitions in amorphous carbohydrates and carbohydrate solutions. *J. Food Sci.*, 56, 1676, 1991.
15. Le Meste M., Champion, D., Roudaut, G., Blond, G., and Simatos, D. Glass transition and food technology: A critical appraisal. *J. Food. Sci.,* 67, 2444, 2002.
16. Ferrari, C., Angiuli, M., Tombari, E., Righetti, M.C., Matteoli, E.G., and Salvetti M. Promoting calorimetry for olive oil authentication. *Thermochim Acta,* 459, 58, 2007.
17. Liu, Y, Bhandari, B, and Zhou, W. Study of glass transition and enthalpy relaxation of mixtures of amorphous sucrose and amorphous tapioca starch syrup solid by differential scanning calorimetry (DSC). *J Food Eng.,* 81, 599, 2007.
18. Billmeyer, F.W. (Jr). *Text Book of Polymer Science.* John Wiley & Sons: New York, NY, 1962.
19. Privalov, P.L. Three generations of scanning microcalorimeters for liquids. *Thermochim Acta,* 139, 257, 1989.
20. Nakamura, S., Todoki, M., Nakamura, K., and Kanetsuna, H. Thermal analysis of polymer samples by a round robin method. I. Reproducibility of melting, crystallization, and glass transition temperatures. *Thermochim Acta,* 136, 163, 1988.

21. Gibbs, J.H. and DiMarzio, E.A. Nature of the glass transition and the glassy state. *J. Chem. Phys.*, 28, 373, 1958.
22. Ferry, J.D. *Viscoelastic Properties of Polymers*, 3rd ed. John Wiley & Sons: New York, NY, 1980.
23. Williams, M.L., Landel, R.F., and Ferry, J.D. The temperature dependence of relaxation mechanisms in amorphous polymers and other glass forming liquids. *J. Am. Chem. Soc.*, 77, 3701, 1955.
24. Soesanto, T. and Williams, M.C. Volumetric interpretation of viscosity for concentrated and dilute sugar solutions. *J. Phys. Chem.*, 85, 3338, 1981.
25. Peleg, M. On the use of the WLF model in polymers and foods. *Crit. Rev. Food Sci. Nutr.*, 32, 59, 1992.
26. Peleg, M. On modeling changes in food and biosolids at and around their glass transition temperature range. *Crit. Rev. Food Sci. Nutr.*, 36, 49, 1996.
27. Roudaut, G., Simatos, D, Champion, D., Contreras-Lopez, E., and Le Meste, M. Molecular mobility around the glass transition temperature: a mini review. *Inno. Food Sci. Emer. Technol.*, 5, 127, 2004.
28. Roudaut, G., Maglione, M., van Duschotten, D., and Le Meste, M. Molecular mobility in glassy bread: a multi spectroscopic approach. *Cereal Chem.*, 1, 70, 1999.
29. Borde, B., Bizot, H., Vigier, G., and Buleon A. Calorimetric analysis of the structural relaxation in partially hydrated amorphous polysaccharides. I. Glass transition and fragility. *Carbohydr. Polym.*, 48, 83, 2002.
30. Montès, H., Mazeau, K., and Cavaillé, J.Y. The mechanical relaxation in amorphous cellulose. *J. Non-Cryst. Solids*, 235, 416, 1998.
31. Blond G. Mechanical properties of frozen model solutions. *J. Food Eng.*, 22, 253, 1994.
32. Shamblin, S.L. and Zografi, G. Enthalpy relaxation in binary amorphous mixtures containing sucrose. *Pharm. Res.*, 15, 1828, 1998.
33. Wungtanagorn, R. and Schmidt, S.J. Thermodynamic properties and kinetics of the physical aging of amorphous glucose, fructose and their mixture. *J. Thermal. Anal. Calorimetry*, 65, 9, 2001.
34. De Meuter, P., Rahier, H., and Van Mele, B. The use of modulated temperature differential scanning calorimetry for the characterisation of food systems. *Inter. J. Pharmaceutics*, 192, 77, 1999.
35. Jenniskens, P., Banham, S.F., Blake, D.F., McCoutra, M.R.S., Jenniskens, P. et al. Liquid water in the domain of cubic crystalline ice I_c *J. Chem. Phys.*, 107, 1232, 1997.
36. Debencdctti, P.G. Supercooled and glassy water. *J. Phys. Condens. Matter*, 15, R1669, 2003.
37. Giovambattista, N., Angell, C.A., Sciortino, F., and Stanley, H.E. Glass-transition temperature of water: A simulation study. *Phys. Rev. Lett.*, 93, 047801, 2004.
38. Johari, G.P., Hallbruker, A., and Mayer, E. The glass liquid transition of hyperquenched water. *Nature*, 330, 552, 1987.
39. Hallbrucker, A., Mayer, E., and Johari, G.P. Glass transition in pressure-amorphized hexagonal ice. A comparison with amorphous forms made from the vapor and liquid. *J. Phys. Chem.*, 93, 7751, 1989.
40. Ito, K., Moynihan, C.T., and Angell, C.A. Thermodynamic determination of fragility in liquids and a fragile-to-strong liquid transition in water. *Nature*, 398, 492, 1999.
41. Yue, Y. and Angell, C.A. Clarifying the glass-transition behaviour of water by comparison with hyper-quenched inorganic glasses. *Nature*, 19, 427, 717, 2004.
42. Couchman, P.R. and Karasz, F.E. A classical thermodynamic discussion of the effect of composition on glass-transition temperatures. *Macromolecules*, 11, 117, 1978.
43. Jouppila, K. and Roos, Y.H. Water sorption and time dependent phenomenon of milk powders. *J. Dairy Sci.*, 77, 1799, 1994.
44. Jouppila, K. and Roos, Y.H. Glass transitions and crystallization in milk powders. *J. Dairy Sci.*, 77, 2907, 1994.
45. Shimada, Y., Roos, Y., and Karel, M. Oxidation of methyl linoleate encapsulated in amorphous lactose-based food model. *J. Agri. Food Chem.*, 39, 637, 1991.
46. Orford, P.D., Parker, R., Ring, S.G., and Smith, A.C. Effect of water as a diluent on the glass transition behavior of malto-oligosaccharides, amylose and amylopectin. *Int. J. Biol. Macromol.*, 11, 91, 1989.
47. Aguilera, J.M., Levi, G., and Karel, M. Effect of water content of the glass transition and caking of fish protein hydrolyzates. *Biotechnol. Progr.*, 9, 651, 1993.
48. Roos, Y.H. Glass transition-related physicochemical changes in foods. *Food Technol.*, 49, 97, 1995.
49. Arvanitoyannis, I., Blanshard, J.M.V., Ablett, S., Izzard, M.J., and Lillford, P.J., Calorimetric study of the glass transition occurring in aqueous glucose: Fructose solutions. *J. Sci. Food Agric.*, 63, 177, 1993.
50. Roos, Y. Characterization of food polymers using state diagrams. *J. Food Eng.*, 24, 339, 1995.
51. Roos, Y. Melting and glass transitions of low molecular weight carbohydrates. *Carbohydr. Res.*, 238, 39, 1993.

52. Fox, T.G. and Flory, P.J. Second-order transition, temperatures and related properties of polystyrene. I influence of molecular weight. *J. Appl. Phys.*, 21, 581, 1950.
53. Slade, L. and Levine, H. Water and the glass transition dependence of the glass transition on composition and chemical structure: special implications for flour functionality in cookie baking. *J. Food Eng.*, 22, 143, 1994.
54. To, E.C. and Flink, J.M. Collapse, a structural transition in freeze dried carbohydrates II. Effect of solute composition. *J. Food Technol.*, 13, 567, 1978.
55. Busin, L., Buisson, P., and Bimbenet, J.J. Notion de transition vitreuse appliquee au sechage par pulverisation de solutions glucidiques. *Sciences des Aliments*, 16, 443, 1996.
56. Fairclough, J.P.A., Hamley, I.W., and Terrill, N.J. X-ray scattering in polymers and micelles. *Radiation Phy. Chem.*, 56, 159, 1999.
57. McCabe, W.I., Smith, J.C., and Harriot, P. *Unit Operations in Chemical Engineering*. McGraw Hill: New York, NY, 2001.
58. Chen, H.L., Hsiao, S.C., Lin, T.L., Yamauchi, K., Hasegawa, H., and Hashimoto, T. Microdomain-tailored crystallization kinetics of block copolymers. *Macromolecules*. 34, 671, 2001.
59. Levi, G. and Karel, M. The effect of phase transitions on release of n-propanol entrapped in carbohydrate glasses. *J. Food Eng.*, 24, 1, 1995.
60. Senoussi, A., Dumoulin, E.D., and Berk, Z. Retention of diacetyl in milk during spray-drying and storage. *J. Food Sci.*, 60, 894, 1995.
61. Jouppila, K., Kansikas, J., and Roos, Y.H. Glass transition, water placticization, and lactose crystallization in skim milk powder. *J. Dairy Sci.*, 80, 3152, 1997.
62. Lai, H.M. and Schmidt, S.J. Lactose crystallization in skim milk powder observed by hydrodynamic equilibria, scanning electron microscopy and ^{2}H nuclear magnetic resonance. *J. Food Sci.*, 55, 994, 1990.
63. Roetman, K. Crystalline lactose and the structure of spray-dried milk products as observed by scanning electron microscopy. *Neth. Milk Dairy J.*, 33, 1, 1979.
64 Saito. Z. Particle structure in spray-dried whole milk and instant skim milk powder related to lactose crystallization. *Food Microstruct.*, 4, 333, 1985.
65. Hartel, R.W. *Crystallization in Foods*. Aspen Publishers: Maryland, MD, 2001, 1.
66. Ahmed, J., Prabhu, T., Raghavan, G.S.V., and Ngadi, M.O. Physico-chemical, rheological and dielectric properties of Indian honey. *J. Food Eng.*, 79, 1207, 2007.
67. Cordella, Ch., Militao, J.S., Clement, M-C., and Cabrol-Bass, D. Honey characterization and adulteration detection by pattern recognition on HPAEC-PAD profiles.1. Honey floral species characterization. *J. Agri. Food Chem.*, 51, 3234, 2003.
68. Zobel, H.F. Starch granule structure. In: *Developments in Carbohydrate Chemistry*, Alexander, R.J. and Zobel, H.F., Eds. American Association of Cereal Chemists: St. Paul, MN, 1992.
69. Lelievre, J. Thermal analysis of carbohydrates. In: *Developments in Carbohydrate Chemistry*, Alexander, R.J. and Zobel, H.F., Eds. American Association of Cereal Chemists: St. Paul, MN, 1992.
70. Cooke, D. and Gidley, M.J. Loss of crystalline and molecular order during starch gelatinisation: origin of the enthalpic transition. *Carbohydr. Res.*, 227, 103, 1992.
71. Blanshard, J.M.V. Starch granule structure and function: a physiochemical approach. In: *Starch Properties and Potential*, Galliard, T., Ed. Wiley: Chichester, UK, 1987.
72. Tester, R.F. and Morrison, W.R. Swelling and gelatinization of cereal starches.1 Effects of amylopectin, amylase and lipids. *Cereal Chem.*, 67, 551, 1990.
73. Evans, I.D. and Haisman, D.R. The effect of solutes on the gelatinization temperature range of potato starch. *Starch/Staerke*, 34, 224, 1982.
74. Biliaderis, C.G., Page, C.M., Maurice, T.J., and Juliano, B.O. Thermal characterization of rice starches: a polymeric approach to phase transitions of granular starch. *J. Agric. Food Chem.*, 34, 6, 1986.
75. Zeleznak, K.J. and Hoseney, R.C. The glass transition in starch. *Cereal Chem.*, 64, 121, 1987.
76. Noel, T.R. and Ring, S.G. A study of the heat capacity of starch-water mixtures. *Carbohydr. Res.*, 227, 203, 1992.
77. Van Den Berg, C. In: *Concentration and Drying of Foods*, McCarthy, D., Ed. Elsevier Applied Science: London, UK, 1986.
78. Benczedi, D., Tomka, I., and Escher F. Thermodynamics of amorphous starch-water systems.1. Volume fluctuations. *Macromolecules*, 31, 3055, 1998.
79. Goff, H.D. and Sahagian, M.E. Freezing of dairy products. In *Freezing Effects on Food Quality*, Jeremiah, L.E., Ed. Marcel Dekker, Inc: New York, NY, 1996, 299.
80. Rahman, M.S. State diagram of date flesh using differential scanning calorimetry (DSC). *Int. J. Food Proper.*, 7, 407, 2004.

81. Sereno, A.M., Sa, M.M., and Figueiredo, A.M. Glass transitions and state diagrams for freeze-dried and osmotically dehydrated apple. In *Proc. 11th International Drying Symposium (IDS '98)*. Halkidiki, Greece, August 19–22, 1998.

82. Bai, Y., Rahman, M.S., Perera, C.O., Smith, B., and Melton, L.D. State diagram of apple slices: glass transition and freezing curves. *Food Res Int.*, 34, 89, 2001.

83. Rahman, M.S. State diagram of foods: its potential use in food processing and product stability. *Trends in Food Sci. Technol.*, 17, 129, 2006.

84. Franks, F., Asquith, M.H., Hammond, C.C., Skaer, H.B., and Echlin, P. Polymeric cryoprotectants in the preservation of biological ultrastructure. I. Low temperature states of aqueous solutions of hydrophilic polymers. *J. Microscopy*, 110, 223. 1977.

85. Rahman, M.S., Sablani, S.S., Al-Habsi, N., Al-Maskri, S., and Al-Belushi, R. State diagram of freeze-dried garlic powder by differential scanning calorimetry and cooling curve methods. *J. Food Sci.*, 70, E135, 2005.

86. Rahman, M.S. *Food Properties Handbook*. CRC Press: Boca Raton, FL, 1995.

87. Rahman, M.S., Guizani, N., Al-Khaseibi, M., Al-Hinai, S., Al-Maskri, S.S., and Al-Hamhami, K. Analysis of cooling curve to determine the end point of freezing. *Food Hydrocoll.*, 16, 653, 2002.

88. Kasapis, S., Rahman, M.S., Guizani, N., and Al-Aamri, M. State diagram of temperature vs date solids obtained from mature fruit. *J. Agric. Food Chem.*, 48, 3779, 2000.

89. Kantor, Z., Pitsi, G., and Thoen, J. Glass transition temperature of honey as a function of water content as determined by differential scanning calorimetry. *J. Agric. Food Chem.*, 47, 2327, 1999.

90. Morales, A. and Kokini, J.L. State diagrams of soy globulins. *J. Rheol.*, 43, 315, 1999.

91. Privalov, P.L. and Khechinashvili, N.N. A thermodynamic approach to the problem of stabilization of globular protein structure: a calorimetric study. *J. Mol. Biol.*, 86, 665, 1974.

92. Biliaderis, C.G. Differential scanning calorimetry in food research—a review. *Food Chem.*, 10, 239, 1983.

93. Tanford, C. Protein denaturation. *Adv. Protein Chem.*, 23, 121, 1968,

94. Borchardt, H.J. and Daniels, F. The application of differential thermal analysis to the study of reaction kinetics. *J. Am. Chem. Soc.*, 79, 41, 1957.

95. Kissinger, H.E. Reaction kinetics in differential thermal analysis. *Anal. Chem.*, 29, 1702, 1957.

96. Beech, G. Computation of kinetic parameters for differential enthalpimetric data. *J. Chem. Soc. A: Inorganic, Physical, Theoretical*, 1903, 1969.

97. Hermansson, A.M. Physico-chemical aspects of soy proteins structure formation. *J. Texture Studies*, 9, 33, 1978.

98. Hermansson, A.M. Aggregation and denaturation involved in gel formation. In: *Functionality and Protein Structure*, Pour-El, A. Ed. American Chemical Society: Washington, DC, No. 92, 1979.

99. Privalov, P.L. Stability of proteins. *Advan. Protein Chem.*, 33, 167, 1979.

100. Ma, C.Y. Thermal analysis of vegetable proteins and vegetable protein-based food products. In: *Thermal Analysis of Food*, Harwalkar, V.R., Ma, C.Y., Eds. Elsevier Science: New York, NY, 1990, 149.

101. Eliasson, A.C. and Hegg, P.O. Thermal stability of wheat gluten. *Cereal Chem.*, 57, 436, 1980.

102. Erdogdu, N., Czuchajowska, Z., and Pomeranz, Y. Wheat flour and defatted milk fractions characterized by differential scanning calorimetry (DSC). II. DSC of interaction products. *Cereal Chem.*, 722, 76, 1995.

103. Hoseney, R.C. and Rogers, D.E. The formation and properties of wheat flour doughs. *Crit. Rev. Food Sci. Nutri.*, 29, 73, 1990.

104. Le´on, A., Rosell, C.M., and Barber, C.B. A differential scanning calorimetry study of wheat proteins. *Euro. Food Res. Technol.*, 217, 13, 2003.

105. Ahmed, J., Ramaswamy, H.S., and Raghavan, G.S.V. Dynamic viscoelastic, calorimetric and dielectric characteristics of wheat protein isolates. *J. Cereal Sci.*, 47, 417, 2008.

106. Ahmed, J., Ramaswamy, H.S., and Alli, I. Thermorheological characteristics of soybean protein isolate. *J Food Sci.*, 71, E:158, 2006.

107. Franks, F. Thermomechanical properties of amorphous saccharides: their role in enhancing pharmaceutical product stability. *Biotech. Genetic Eng. Rev.*, 16, 281, 1999.

108. Sun, W.Q., Davidson, P., and Chan, H.S.O. Protein stability in the amorphous carbohydrate matrix: relevance to anhydrobiosis. *Biochim. Biophys. Acta.*, 1425, 245, 1998.

109. Taragano, V.M. and Pilosof, A.M.R. Calorimetric studies on dry pectinlyase preparations: impact of glass transition on inactivation kinetics. *Biotechnol. Prog.*, 17, 775, 2001.

110. Rupley, J.A. and Careri, G. Protein hydration and function. *Adv. Protein Chem.*, 41, 37, 1991.

111. Sartor, G., Mayer, E., and Johari, G.P. Calorimetric studies of the kinetic unfreezing of molecular motions in hydrated lysozyme, hemoglobin, and myoglobin. *Biophys. J.*, 66, 249, 1994.

112. McCammon, J.A. and Harvey, S.C. *Dynamics of Proteins and Nucleic Acids*. Cambridge University Press: Cambridge, UK, 1987.
113. Sochava, I.V. and Smirnova, O.I. Heat capacity of hydrated and dehydrated globular proteins - denaturation increment of heat capacity. *Mol. Biol.,* 27, 209, 1993.
114. Tsereteli, G.I., Belopolskaya, T.V., Grunina, N.A., and Vaveliouk, O.L. Calorimetric study of the glass transition process in humid proteins and DNA. *J. Therm. Anal. Calorim.,* 62, 89, 2000.
115. Hoseney, R.C., Zeleznak, K., and Lai, C.S. Wheat gluten as a glassy polymer. *Cereal Chem.,* 63, 285, 1986.
116. Noel, T.R., Parker, R., Ring, S.G., and Tatham. A.S., The glass transition behaviour of wheat gluten proteins. *Int. J. Biol. Macromol.,* 17, 81, 1995.
117. Cocero, A.M. and Kokini, J.L. The study of the glass transition and differential scanning calorimetry, *J. Rheol.* 35, 257, 1991.
118. Zhou, P. and Labuza, T.P. Effect of water content on glass transition and protein aggregation of whey protein powders during short-term storage. *Food Biophy,* 2, 108, 2007.

Section III

---

*Simultaneous Heat and Mass
Transfer in Food Processing*

10 Heat and Mass Transfer during Food Drying

Ibrahim Dincer
University of Ontario Institute of Technology (UOIT)

CONTENTS

10.1   INTRODUCTION

Drying is the process of thermally removing volatile substances (e.g., moisture) to yield a solid product. Mechanical methods for separating a liquid from a solid are not considered drying. When a wet solid is subjected to thermal drying, two processes occur simultaneously: (i) transfer of energy (mostly as heat) from the surrounding environment to evaporate the surface moisture, and (ii) transfer of internal moisture to the surface of the solid and its subsequent evaporation due to the first process.

The nature of the mechanism of moisture movement within the solid has received much attention in the literature. There appear to be four probable major modes of transfer: liquid movement caused by capillary forces, liquid diffusion resulting from concentration gradients, vapor diffusion due to partial pressure gradients, and diffusion in liquid layers adsorbed at solid interfaces. The mechanisms of capillarity and liquid diffusion have received the most detailed treatment. In general the former is most applicable to coarse granular materials and the latter to single-phase solids with colloidal or gel-like structures. In many cases it appears that the two mechanisms may be applicable to a single drying operation, i.e., capillarity accounting for the moisture movement in the early stages of drying while a diffusional mechanism applies at lower moisture contents [1].

The most important aspect of drying technology is the mathematical modeling of the drying processes and equipment. Its purpose is to allow design engineers to choose the most suitable operating conditions and then sizing the drying equipment and drying chamber accordingly to meet desired operating conditions. The principle of modeling for heat and mass transfer during drying of food products, which is based on solving a set of transient (time-dependent) heat and mass transfer equations, is important to adequately characterize the system and determine significant heat and mass transfer parameters for system analysis and design.

10.2   DRYING PERIODS

In a drying experiment, the moisture content data (dry basis) are usually measured versus drying time, as shown in Figure 10.1a. The curve in the figure represents the general case when a wet solid loses moisture first by evaporation from a saturated surface on the solid, followed in turn by a period of evaporation from a saturated surface of gradually decreasing area, and finally, when the latter evaporates within the solid. Figure 10.1b shows variation of drying rate versus moisture content (dry basis) of the solid. In addition, Figure 10.1c exhibits variation of drying rate versus drying time.

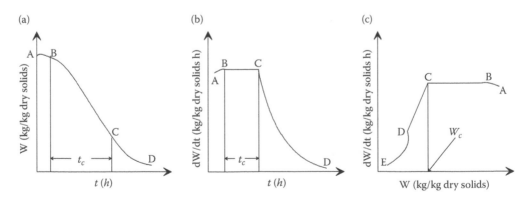

FIGURE 10.1   The drying periods for a solid. (a) Moisture content versus time. (b) Drying rate versus moisture content. (c) Drying rate versus time. (Drying curves for a wet material being dried at a constant temperature and relative humidity).

10.2.1 DRYING KINETICS

Drying kinetics is connected with the changes of average material moisture content and average temperature with time, contrary to drying dynamics which describes changes in temperature and moisture profiles throughout the drying body. Drying kinetics enables the amount of moisture evaporated, drying time, energy consumption, etc., to be calculated and determined to a considerable extent by the physico-chemical properties of the material. Nevertheless, the change of material moisture content and temperature is usually controlled by heat and moisture transfer between the body surface, the surroundings and the inside of the drying material.

The drying intensity which reflects the change in moisture content with time is influenced significantly by the parameters of the drying process, e.g., temperature, humidity (pressure), relative velocity of air or total pressure.

10.3 DRYING THEORY: BASIC HEAT AND MOISTURE TRANSFER ANALYSIS

During the drying of a moist product in heated air, the air supplies the necessary sensible and latent heat of evaporation to the moisture and also acts as a carrier gas for the removal of the water vapor formed from the vicinity of the evaporating surface. For the three stages given in Figure 10.1, we present basic aspects of heat and moisture transfer as follows:

Stage A–B

This stage represents a warming-up period of the solids and the solid surface conditions come into equilibrium with the drying air. It is often a negligible proportion of the overall drying cycle but in some cases it may be significant.

Stage B–C

It is the constant period of drying during which the rate of water removal per unit of drying surface is constant. The point C where the constant-rate period ends is termed as critical moisture content. In this period, moisture movement within the solid is rapid enough to keep a saturated condition at the surface and drying rate is controlled by the rate of heat transferred to the evaporating surface. During this period, the surface of the solid remains saturated with liquid water by virtue of the fact that movement of water within the solid to the surface takes place at a rate as great as the rate of evaporation from the surface. In fact, this period is highly affected by the heat and/or moisture transfer coefficients, the area exposed to drying medium, and the difference in temperature and relative humidity between the drying air and the moist surface of the solid. The rate of drying is dependent on the rate of heat transfer to the drying surface. The rate of moisture transfer can be expressed in the form of an equation as follows:

$$\frac{dW}{dt} = -h_m A(P_s - P_a) \tag{10.1}$$

where dw/dt is the drying rate, h_m is the moisture transfer coefficient, A is the drying surface area, P_s is the water vapor pressure at surface (i.e., vapor pressure of water at surface temperature), P_a is the partial pressure of water vapor in air.

Equation 10.1 can also be written in the following form:

$$\frac{dW}{dt} = -h_m A(H_s - H_a) \tag{10.2}$$

where H_s is the humidity at surface (i.e., saturation humidity of the air at surface temperature), H_a is the humidity of air.

Therefore, the rate of heat transfer to the drying surface can be expressed as

$$\frac{dQ}{dt} = hA(T_a - T_s) \tag{10.3}$$

where dQ/dt is the rate of heat transfer, h is the convection heat transfer coefficient during heating, A is the surface area, T_a is the dry-bulb temperature of air, T_s is the surface temperature of the material. Note that in the situation being considered here, i.e., convection heating only, T_s is the wet-bulb temperature of the air.

Since a state of equilibrium exists between the rate of heat transfer to the body and the rate of moisture transfer from it, these two rates can be related simply as follows:

$$\left(\frac{dW}{dt}\right)L = -\left(\frac{dQ}{dt}\right) \tag{10.4}$$

where L = latent heat of evaporation at T_s.

Combining Equations 10.3 and 10.4 we get

$$\frac{dW}{dt} = -\frac{hA}{L}(T_a - T_s) \tag{10.5}$$

If the drying rate is expressed in terms of the rate of change of moisture content W (dry-weight basis), Equation 10.5 can be written:

$$\frac{dW}{dt} = -\frac{hA_{ef}}{L}(T_a - T_s) \tag{10.6}$$

where A_{ef} is the effective drying surface per unit mass of dry solids.

For a tray of moist product of depth d, evaporating only from its upper surface, assuming no shrinkage during drying:

$$\frac{dW}{dt} = -\frac{h}{\rho_s Ld}(T_a - T_s) \tag{10.7}$$

where ρ_s = bulk density of the dry material.

Therefore, the drying time in the constant rate period can be obtained by the integration of Equation 10.7 as follows:

$$t_{CR} = -\frac{(W_i - W_c)\rho_s Ld}{h(T_a - T_s)} \tag{10.8}$$

where t_{CR} = constant rate drying time, W_i is the initial moisture content of solid, W_c is the moisture content at end of constant rate period.

The rate controlling factors during the constant rate period are therefore: (i) the drying surface area, (ii) the difference in temperature or humidity between air and drying surface, (iii) the heat or moisture transfer coefficients.

Note that in estimating drying rates, the use of heat transfer coefficients is considered to be more reliable than moisture transfer coefficients and suggest that for many cases, the heat transfer coefficient can be calculated from Nusselt–Reynolds (Nu–Re) correlations. Thus, the velocity of drying air, and system and product dimensions influence drying rates in the constant rate period. Alternative expressions for h are used where the air flow is not parallel to the drying surface or for through-flow situations. In addition, where heat is supplied to the material by radiation and/or conduction, in addition to convection, then an overall heat-transfer coefficient, taking this into account, must be substituted for h in Equation 10.7. Under these circumstances, the surface temperature during the constant rate period of drying remains constant, but at some value above the wet-bulb temperature of the air and below the boiling point of water.

Stage C–D

This drying period starts at the critical moisture content when the constant-rate period ends. From point C onward the surface temperature begins to rise and continues to do so as drying proceeds, approaching the dry-bulb temperature of the air as the material approaches dryness. When the initial moisture content is above critical moisture content, the entire drying process occurs under the constant-rate conditions. If it is below the critical moisture content, the entire drying process occurs in the falling-rate period. This period usually consists of two zones: (i) the zone of unsaturated surface drying and (ii) the zone in which internal moisture movement controls. Point E is the point at which all exposed surface becomes completely unsaturated and marks the start of that portion of the drying process during which the rate of internal moisture movement controls the drying rate. In Figure 10.1b, CE is defined as the first falling-rate period and DE is the second falling-rate period. In the falling rate periods, the rate of drying is mainly influenced by the rate of movement of moisture within the solid and the effects of external factors, in particular air velocity are reduced especially in the latter stage. Usually the falling rate periods represent the major proportion of the overall drying time.

For systems where a capillary flow mechanism applies, the rate of drying can often be expressed with reasonable accuracy by an equation of the type:

$$\frac{dW}{dt} = -h_{\mathrm{m}}(W - W_e) \tag{10.10}$$

extracting the moisture transfer coefficient as

$$h_{\mathrm{m}} = \frac{dW/dt}{(W_c - W_e)} \tag{10.11}$$

where dW/dt is the rate of drying at time t from the start of the falling rate period, W is the moisture content of the material at any time t, W_e is the equilibrium moisture content of material at air temperature and humidity.

After combining Equations 10.7, 10.10, and 10.11, the following can be obtained:

$$\frac{dW}{dt} = -\frac{h(T_a - T_s)}{\rho_s L d}\frac{(W_c - W_e)}{(W - W_e)} \tag{10.12}$$

Integrating the above equation under the following conditions, t from 0 to t and W from W_c to W, gives

$$t = \frac{\rho_s L d(W_c - W_e)}{h(T_a - T_s)}\ln\frac{(W_c - W_e)}{(W - W_e)} \tag{10.13}$$

Note that the above drying Equations 10.7, 10.8, and 10.12 are applicable when drying takes place from one side only. In cases where drying occurs from both surfaces, d will be taken as the half thickness.

10.4   MOIST PRODUCTS

Usually the products which are subjected to the drying process consist in general of the bone dry material (skeleton) and some amount of moisture, mainly in a liquid state. So-called "moist products" have different physical, chemical, structural, mechanical, biochemical and other properties which result from the properties of the skeleton and the states of water within it. Here, we list some key parameters of moisture products.

Although these parameters can significantly influence the drying process and hence determine the drying technique and technology, the most important in practice are the structural-mechanical properties, the type of moisture in the solids and the material-moisture bonding that will be discussed later.

10.4.1   PARAMETERS OF MOIST PRODUCTS

Here, we first present some basic concepts for a better understanding:

Moisture content. The moisture content of the material (X) can be defined in two ways:

$$\text{Dry basis: } X = \frac{m_m}{m_s} \tag{10.14}$$

where m_m is the mass of moisture (kg) and m_s is the mass of solid (dry) material (kg).

$$\text{Wet basis: } X = \frac{m_m}{m} = \frac{m_m}{m_m + m_s} \tag{10.15}$$

where m is the total mass of moist product (kg).

Sometimes, the moisture content is expressed as a percentage in the following forms:

$$X = \frac{X^*}{1 - X^*} \qquad \text{or} \qquad X^* = \frac{X}{1 + X} \tag{10.16}$$

where X is the dry basis content and X^* is the wet basis content.

In addition, the volumetric moisture content (X_V) is used:

$$X_V = \frac{V_{lm}}{V_s + V_{lm} + V_{vm}} \tag{10.17}$$

where V_{lm} is the volume of liquid moisture (m³), V_{vm} is the volume of vapor moisture (m³), and V_s is the volume of dry material (m³).

In this regard, the percentage saturation of moist product (Λ_V) becomes

$$\Lambda_V = \frac{V_{ml}}{V_{ml} + V_{mv}} \times 100 \tag{10.18}$$

One of the important drying aspects is the equilibrium moisture content, defined as follows. It is the moisture of a given material which is in equilibrium with the vapor contained in the drying agent under specific conditions of air temperature and humidity (so-called minimum hygroscopic moisture content). It changes with temperature and humidity of the surrounding air. However, at low drying temperatures (e.g., 15–50°C), the equilibrium moisture content may become independent of drying air temperature and zero at zero relative humidity. Equilibrium moisture content is strongly dependent on the nature of the solid. For nonporous (i.e., nonhygroscopic solids), it is essentially zero at all temperatures and humidities. For hygroscopic solids (e.g., wood, food, paper, soap, chemicals) it varies regularly over a wide range with temperature and relative humidity.

10.5 SOME LITERATURE DRYING MODELS

Table 10.1 presents some significant drying heat and moisture transfer models from the literature and exhibits their features. Below we summarize these models:

Model 1 describes the average outlet moisture content from a well-mixed fluidized bed dryer. Here $X(t)$ is a function describing the variation in moisture content X of a particle at constant bed temperature with time t as it dries and τ_B is the mean residence time of particles in the bed. The possible uses of this model includes an assessment of the process response to changes in operating

TABLE 10.1
A Summary of Some Drying Models Extracted from the Literature

Serial No	Model	Remarks	References
1.	$X_o = \dfrac{1}{\tau_B} \displaystyle\int_0^\infty X(t)\exp\left(\dfrac{-t}{\tau_B}\right)dt$	Well-mixed fluidized bed dryer	[2]
2.	$MR = \exp(-kt^n)$	Biological materials	[3,4]
3.	Concurrent flow drying model:		
	$G_G = G_{Go}$ (a)	Grain drying	[5]
	$G_S = G_{So}$ (b)		
	$\dfrac{dU_G}{dx} = \dfrac{f}{G_G}$ (c)		
	$\dfrac{dU_S}{dx} = \dfrac{-f}{G_S}$ (d)		
	$\dfrac{dT_G}{dx} = \dfrac{-q - fC_{PV}(T_G - T_S)}{G_G(C_{PG} + C_{PV}U_G)}$ (e)		
	$\dfrac{dT_S}{dx} = \dfrac{q - f\lambda}{G_S(C_{PS} + C_{PW}U_S)}$ (f)		
	Countercurrent flow drying model:		
	$\dfrac{\partial U_G}{\partial x} = \dfrac{f}{G_G}$ (g)		
	$\dfrac{\partial T_G}{\partial x} = \dfrac{-q - fC_{PV}(T_G - T_S)}{(C_{PG} + C_{PV}U_G)G_G}$ (h)		
	$\dfrac{\partial U_S}{\partial x} + v_S\dfrac{\partial U_S}{\partial x} = \dfrac{-f}{(1-\varepsilon)\rho_S}$ (i)		
	$\dfrac{\partial T_S}{\partial t} + v_S\dfrac{\partial T_S}{\partial x} = \dfrac{q - f\lambda}{(C_{PS} + C_{PW}U_S)(1-\varepsilon)\rho_S}$ (j)		
	where v_S stands for velocity of the solid, as given by:		
	$v_S = \dfrac{G_S}{(1-\varepsilon)\rho_S}$ (k)		

(Continued)

TABLE 10.1   (Continued)

Serial No	Model	Remarks	References
4.	$\dfrac{\partial M}{\partial t} = \nabla^2 K_{11}M + \nabla^2 K_{12}\theta + \nabla^2 K_{13}P$ (a)	Drying of porous media	[6]
	$\dfrac{\partial \theta}{\partial t} = \nabla^2 K_{21}M + \nabla^2 K_{22}\theta + \nabla^2 K_{23}P$ (b)		
	$\dfrac{\partial P}{\partial t} = \nabla^2 K_{31}M + \nabla^2 K_{32}\theta + \nabla^2 K_{33}P$ (c)		
5.	$\rho_o \dfrac{\partial X}{\partial t} = -W_{i,i}$ (a)		[7]
	where ρ_o is the mass density of dry body and capillary porous materials, W_i is the moisture flux proportional to the gradient of moisture potential μ		
	$W_i = -\Lambda_X \mu_{,i}$ (b)		
	$\mu_{,i} = \dfrac{\partial \mu}{\partial \varepsilon}\varepsilon_{,i} + \dfrac{\partial \mu}{\partial X}X_{,i} + \dfrac{\partial \mu}{\partial \chi}\chi, i$ (c)		
6.	$\rho_S + \rho_S \mathrm{div} V_S = 0$ (a)	First period of drying	[8]
	$\rho_S \dot{X} = -\mathrm{div} W_L + \overset{*}{X}_L$ (b)		
	$\rho_S X_V = \overset{*}{X}_V$ (c)		
	$\rho_S \dot{X}_A = 0$ (d)		
7.	$X = \dfrac{X_m CKa_W}{(1 - Ka_W)(1 - Ka_W + CKa_W)}$	GAB equation for moisture isotherms of foods	[9]
8.	$M_e = \dfrac{M_m.C.a_w}{(1 - a_w)\left[1 + (C-1).a_w\right]}$	BET model for equilibrium sorption isotherms	[10,11]
9.	$M_e = \dfrac{1}{B}.\left(Ln\dfrac{A}{R.T} - Ln(-Lna_w)\right)$		[12]
10.	$M_e = \dfrac{M_m.C.K.a_w}{(1 - K.a_w).\left[1 + (c-1).K.a_w\right]}$	GAB model for equilibrium sorption isotherms	[13,14]
11.	$M_e = \left(-\dfrac{A}{T.Lna_w}\right)^{\frac{1}{B}}$	Halsey model for equilibrium sorption isotherms	[15,16]
12.	$M_e = A.\left(\dfrac{a_w}{1 - a_w}\right)^B$	Henderson model for equilibrium sorption isotherms	[17,18]
13.	$M_e = \left(-\dfrac{Ln(1 - a_w)}{A.T}\right)^{1/B}.10^{-2}$	Oswin model for equilibrium sorption isotherms	[19,20]
14.	$M_e = A - B.Ln(1 - a_w)$	Smith model for equilibrium sorption isotherms	[21]
15.	$M_R = \dfrac{M_t - M_e}{M_i - M_e} = C \times \exp\left(-\dfrac{\pi^2 Dt}{R^2}\right)$	Diffusion model for large drying and re-wetting of grain	[22]
16.	$\left(\dfrac{\partial X}{\partial t}\right)_\xi = \left\{\dfrac{1}{\xi}\dfrac{\partial}{\partial \xi}\left(ID_{eff}(X,T)\xi\left(\dfrac{\partial X}{\partial \xi}\right)_t\right)\right\}_t$	Fickian diffusion in a date	[23]

TABLE 10.1 (Continued)

Serial No	Model	Remarks	References
16.	$$\left(\frac{\partial X}{\partial t}\right)_\xi = \left\{\frac{1}{\xi}\frac{\partial}{\partial\xi}\left(ID_{eff}(X,T)\xi\left(\frac{\partial X}{\partial\xi}\right)_t\right)\right\}_t$$ where $ID_{eff}(X,T) = \frac{ID(X,T)}{(1+\varepsilon X)^2}\left(\frac{r}{\xi}\right)^2$	Fickian diffusion in a date	[23]
17.	$$-\frac{dX_S}{dt} = k_M(X_S - X_{SE})$$ where $k_M = k_o d_p{}^{k1}T^{k2}V^{k3}a_w{}^{k4}$ for convective drying $k_M = k_o Q^{k1}P^{k2}$ for microwave drying	Mathematical model for drying kinetics	[24]
18.	$$X_{SE} = \frac{X_m CKa_w}{[(1-Ka_w)(1-(1-C)Ka_w)]}$$ where $C = C_o\exp(\Delta H_c/RT)$ $K = K_o\exp(\Delta H_k/RT)$	Mathematical model for equilibrium moisture content	[24]
19.	$$\frac{\partial M}{\partial\tau} = div(a_m\cdot divM + a_m\cdot\delta\cdot divT) \quad\text{(a)}$$ $$c\cdot\rho_0\cdot\frac{\partial T}{\partial\tau} = div(\lambda\cdot gradT)$$ $$+ k\cdot div\left(\frac{\partial M}{\partial\tau} + \delta\cdot gradT\right) \quad\text{(b)}$$	Model describing the heat and mass transport in capillary porous media	[25]

variables, a parameter sensitivity study and an evaluation of model controllability together with a simple optimization study.

Model 2 is widely used to describe the drying behavior of a variety of biological materials. The drying of most biological materials is a diffusion-controlled process. For the case of okra, the present model represents convective drying adequately.

Model 3 represents the concurrent and counter current flow drying of grains. The assumptions made were: (i) phases moves as plug flow; (ii) energy loss through the dryer walls is neglected; (iii) steady state operation. The advantage of this model employing parallel flow drying is that, it generally leads to more homogenous moisture and temperature distributions.

Model 4 describes the coupled heat and mass transfer through porous media, in which $M(x,t)$, $\theta(x,t)$ and P denotes the moisture content, temperature and pressure respectively in the material being dried. Since the total pressure gradient does not cause significant moisture flow within the product, the last equation (c) can be abandoned in the model and the terms $\nabla^2 K_{13}P$ and $\nabla^2 K_{23}P$ may be neglected as well. The above description is valid only in cases when the conductivities are constant properties.

Model 5 represents a thermo-mechanical drying of capillary-porous materials whose material constants depend on moisture content and temperature. The dried material is assumed to be an elastic capillary-porous body saturated with liquid. It can be applied to the problem of convective drying of a prismatic bar (two dimensional initial-boundary value problem).

Model 6 represents the mass balance equations for the first period of drying. Here, ρ_S is the mass density of the bone-dry body; X_L, X_V, X_A are the mass contents of liquid, vapor and dry air, all referred to the mass of the dry body; $X_L^* = -X_V^*$ are the liquid–vapor phase transition rates; v_S is the velocity of the porous solid; w_L is the flux of liquid. Dot over the symbol denotes the time derivative.

Models 7 through 14 are equations used to model equilibrium sorption isotherms, in which $M_e = f(a_W)$, where M_e is the equilibrium moisture content (decimal, dry basis) and a_W is the water activity of the product. The advantage of GAB equation over other equations is that, it describes temperature effect on the sorption isotherm by expressing the equation parameters with simple Arrhenius equations [13]. Parameter M_m represents moisture content when each sorption site contains only one water molecule (it is the monolayer moisture content of the product) [13].

Model 15 is a well-known diffusion model for large drying or re-wetting times, in which $C = 6/\prod^2$, M_R is the moisture ratio, M_t is the moisture content at any time (dry basis), M_e is the equilibrium moisture content (dry basis), M_i is the initial moisture content (dry basis), t is the drying time (h), D is the diffusion coefficient (m^2/h), R is the sphere radius (m). It is used to predict the re-wetting of single kernel grain with known diffusivity, radius, initial moisture content and equilibrium moisture content. The grain is assumed to be a spherical, homogenous, isotropic material drying or re-wetting under isothermal conditions.

Model 16 is based on heat and mass transfer equations, written in Eulerian coordinates. The term ID(X,T) is introduced as a parametric model of local moisture and temperature. It was found that ID is also a function of initial moisture content of the seeds. The temperature dependence of moisture diffusivity in foods has been verified by all researchers of this field and a general agreement to an Arrhenius-type relationship has been achieved.

Model 17 is an empirical model describing moisture transfer. It has the from of a general linear ordinary differential equation, in which right hand side contains an empirical mass transfer coefficient multiplied by the corresponding driving force. The mathematical models describing drying kinetics, rather than being strictly mechanistic, are often quasi-mechanistic and sometimes, mostly empirical. A complete description of the actual mechanisms involved, is usually not obtainable and would certainly be complex. Empirical models can be deduced from detailed mechanistic ones under certain assumptions, or it can be evaluated empirically, in the sense that they should account for the basic mechanisms in the process examined.

Model 18 is the form of GAB equation representing equilibrium moisture isotherms of sorption. It is the one that best represents the equilibrium moisture isotherms of sorption. The equilibrium material moisture content is strongly affected by temperature and humidity of surrounding air. The equilibrium moisture content is higher at lower air temperature and at higher humidity levels.

Model 19 describes the simultaneous heat and mass transfer through a capillary porous body. It was assumed that all the coefficients of heat and mass transfer in equations (a) and (b) are constant and that they can be used for the calculations by the phases of the process, small enough to satisfy the adopted linearization.

10.6   MOISTURE TRANSFER MODELING

Drying of moist products is a process of simultaneous heat and moisture transfer and can be considered an energy-intensive operation of some industrial significance. Drying is a process whereby moisture is vaporized and swept away from the surface, sometimes in vacuum but normally by means of a carrier fluid passing through or over the moist product. This process has found industrial applications in various forms ranging from wood drying in lumber industry to food drying in food industry. In the process, heat may be added to the product from an external source, by convection, conduction or radiation, or heat can be generated internally within the solid body by means of electric resistance. However, regardless of the mode of heating, moisture is always removed in a vapor phase.

In practice, there is an urgent need to develop new models and methods for determining the moisture transfer parameters in terms of moisture diffusivity and moisture transfer coefficient for products subject to drying. An accurate determination of such moisture transfer parameters is essential for an efficient moisture transfer analysis, leading to optimum energy use and operating conditions and hence, efficient and effective drying. Although there are a number of experimental and theoretical studies on the determination of drying profiles for various solid products, the number of models for moisture diffusivity and moisture transfer coefficient parameters is very few [e.g., 26–28].

10.6.1 Drying Process Parameters

These parameters are used in evaluating and representing a drying process. The dimensionless moisture content of a product being dried in any medium can be defined in terms of lag factor and drying coefficient as

$$\Phi = G \exp(-St) \tag{10.19}$$

Here, the drying coefficient (S) represents the drying capability of the product exposed to drying and lag factor (G) represents the magnitude of internal resistance to the moisture transfer from the product. The dimensionless moisture content values can be found using the experimental moisture content measurements from the following equation:

$$\Phi = \frac{W - W_e}{W_i - W_e} \tag{10.20}$$

10.6.2 Moisture Transfer Parameters

The moisture diffusivity model for IS, IC, and SP products developed by Dincer [26] becomes

$$D = \left(\frac{SY^2}{\mu_1^2} \right) \tag{10.21}$$

We had the correlations developed earlier for the corresponding characteristic root equations. Those were not that accurate and we now develop more useful correlations for the roots of characteristic equations with a correlation coefficient of 1.0. These correlations are then given below:

For IS

$$\mu_1 = -419.24G^4 + 2013.8G^3 - 3615.8G^2 + 2880.3G - 858.94 \quad \text{for } 0.1 \le \text{Bi} \le 100 \tag{10.22}$$

For IC

$$\mu_1 = -3.4775G^4 + 25.285G^3 - 68.43G^2 + 82.468G - 35.638 \quad \text{for } 0.1 \le \text{Bi} \le 100 \tag{10.23}$$

For SP

$$\mu_1 = -8.3256G^4 + 54.3842G^3 - 134.01G^2 + 145.83G - 58.124 \quad \text{for } 0.1 \le \text{Bi} \le 100 \tag{10.24}$$

Note that Equation 10.21 can easily be used to determine moisture diffusivity values for the plate, cylindrical and spherical products.

The equation determining the moisture transfer coefficient values can be written in the following form

$$k = \left(\frac{D}{Y} \right) (\text{Bi}) \tag{10.25}$$

where the Biot number (Bi) is determined using the following relation as derived and available elsewhere [26–28], based on the analogy between heat conduction and moisture diffusion within solids:

$$Bi = \frac{\mu_1}{\cot \mu_1} \qquad (10.26)$$

Therefore, the procedure used in determining the above drying process parameters is as follows:

- The experimental moisture content values against drying time taken from literature are nondimensionalized using Equation 10.20.
- The dimensionless moisture content distribution is regressed against the drying time in the exponential form of Equation 10.19 using the least square curve fitting method. Thus, the lag factor (G) and drying coefficient (S) are determined.
- The values of the characteristic root (μ_1) are estimated from Equations 10.22 through 10.24.
- The moisture diffusivity values are determined using Equation 10.21.
- The Biot number (Bi) is determined from Equation 10.26 using the known characteristic root (μ_1).
- Finally the moisture transfer coefficients are determined using Equation 10.25.

10.7 MODELING

In this chapter, simple models are proposed for determining the moisture transfer parameters in terms of moisture diffusivity and moisture transfer coefficient for infinite slabs (ISs) and infinite cylinders (ICs) subject to drying. The modeling is based on the most practical case, referring to $0.1 < Bi < 100$. Some new drying process parameters, i.e., drying coefficient and lag factor, as developed earlier [26–28], are incorporated into the present models which are then verified with actual drying data. The findings are in good agreement with data from the literature. We believe that these models will be beneficial to the drying industry. The main aim is to provide simple models for determining moisture transfer parameters in terms of moisture diffusivity and moisture transfer coefficient for ISs and cylinders subject to drying and to verify these models with an illustrative example using actual data.

Transient moisture diffusion process during drying of a solid product takes place as similar to the heat conduction process in such a solid product. So, the governing Fickian equation is exactly in the same form of the Fourier equation of heat transfer, in which temperature and thermal diffusivity are replaced with moisture and moisture diffusivity, respectively. The following assumptions are made:

- Infinite slab (IS) and infinite cylinder (IC) are the solid products subject to air drying process.
- Thermophysical properties of the solids and the drying medium are constant.
- The effect of heat transfer on moisture loss is negligible.
- There are finite internal and external resistances to moisture transfer in the solids (i.e., $0.1 < Bi < 100$).
- The moisture diffusion occurs in z direction in the slab and in r direction in the cylinder.

Under these conditions, the governing one-dimensional moisture diffusion equation in rectangular and cylindrical coordinates can be written as

$$D\frac{\partial^2 W}{\partial z^2} = \frac{\partial W}{\partial t} \qquad \text{for IS} \qquad \text{and} \qquad D\frac{\partial^2 W}{\partial r^2} + \frac{1}{r}\frac{\partial W}{\partial r} = \frac{\partial W}{\partial t} \qquad \text{for IC}$$

The solution of the above equation is extensively treated for the regular-shaped products earlier by the author [e.g., 26,27]. So, the dimensionless average moisture distribution for IS and IC becomes

$$\Phi = \sum_{n=1}^{\infty} A_n B_n \tag{10.27}$$

that can be simplified by ignoring the values of $(\mu^2 Fo)$ smaller than 1.2 (taking only the first term in Equation 10.27 into consideration), resulting in

$$\Phi = A_1 B_1 \tag{10.28}$$

where

$$\Phi = \frac{W - W_e}{W_i - W_e} \tag{10.29}$$

$$A_1 = \frac{2\sin\mu_1}{\mu_1 + \sin\mu_1 \cos\mu_1} \quad \text{for IS}$$

$$A_1 = \frac{2\,\mathrm{Bi}}{(\mu_1^2 + \mathrm{Bi}^2) + J_0(\mu_1)} \quad \text{for IC}$$

$$B_1 = \exp(-\mu_1^2\,\mathrm{Fo}) \quad \text{for both IS and IC.}$$

The equations for A_1 are already simplified earlier [26] to

$$A_1 = \exp\left[\frac{0.2533\,\mathrm{Bi}}{1.3 + \mathrm{Bi}}\right] \quad \text{for IS} \quad \text{and} \quad A_1 = \exp\left[\frac{0.5066\,\mathrm{Bi}}{1.7 + \mathrm{Bi}}\right] \quad \text{for IC} \tag{10.30}$$

The Biot and the Fourier numbers for moisture diffusion are defined as follows:

$$\mathrm{Bi} = \frac{kL}{D} \quad \text{for IS and} \quad \mathrm{Bi} = \frac{kR}{D} \quad \text{for IC} \tag{10.31}$$

$$\mathrm{Fo} = \frac{Dt}{L^2} \quad \text{for IS and} \quad \mathrm{Fo} = \frac{Dt}{R^2} \quad \text{for IC} \tag{10.32}$$

Due to the fact that cooling and drying have the same natural and exponentially decreasing trend (since in cooling we decrease the temperature to a certain temperature level and in drying we do it for moisture content), we establish the following equation for IS and IC products subject to drying, by introducing lag factor (G) and drying coefficient (S):

$$\Phi = G\exp(-St) \tag{10.33}$$

Both Equations 10.28 and 10.33 are in the same form and can be equated to each other by having $G = A_1$. Therefore, we can find the following moisture diffusivity models:

$$D = \frac{SL^2}{\mu_1^2} \quad \text{for IS and} \quad D = \frac{SR^2}{\mu_1^2} \quad \text{for IC} \tag{10.34}$$

Also, the expressions for the moisture transfer coefficients result in

$$k = \frac{\text{Bi}D}{L} \quad \text{for IS and} \quad k = \frac{\text{Bi}D}{R} \quad \text{for IC} \tag{10.35}$$

Example 10.1

Here, we briefly exhibit the use of the present moisture diffusivity and moisture transfer coefficient models and the verification of these models with actual data. So, we underline the procedure as follows:

- Measurement and/or calculation of average moisture content values, versus drying time in seconds, for the moist products being dried or to be dried.
- Nondimensionalization of moisture content values using Equation 10.29.
- Application of the regression analysis to the dimensionless moisture content distribution in exponential form, as in Equation 10.33, by means of the least-squares curve-fitting method.
- Determination of the lag factor (G) and drying coefficient (S) from the regression correlation, which can be defined as the most important drying process parameters. In fact, the lag factor indicates the magnitude of both the internal and external resistance to moisture transfer from the solid as a function of the Biot number. The drying coefficient shows the drying capability of the product.
- Determination of the Biot number from Equation 10.31 since $G = A_1$ is known.
- Determination of the value of μ_1 from the previously developed expressions as follows:
- $\mu_1 = \arctan(0.64\text{Bi} + 0.38)$ for IS, and $\mu_1 = [0.72\ln(6.79\text{Bi} + 1)]^{1/1.4}$ (for $0.1 < \text{Bi} < 10$) and $\mu_1 = [\ln(1.74\text{Bi} + 147.32)]^{1/1.2}$ (for $10 < \text{Bi} < 100$) for IC
- Determination of the moisture diffusivity and moisture transfer coefficient from Equations 10.34 and 10.35.

After listing the procedure above, we will verify the present models using actual moisture content values obtained from Lewicki et al. [29] for slab-shaped onion slice and Akiyama et al. [30] for a cylindrically-shaped starch powder. Table 10.2 shows thermal and physical data for these products.

The dimensionless average moisture content of each sample were taken from the above mentioned references and regressed in the exponential form, as in Equation 10.33. The experimental and regression profiles of dimensionless moisture content are shown in Figures 10.2 and 10.3, respectively. The following correlations are found through curve-fitting with high correlation coefficients ($r^2 > 0.99$):

$$\Phi = 1.207\exp(-0.0004t) \quad \text{for IS} \tag{10.36}$$

and

$$\Phi = 1.0082\exp(-0.0007t) \quad \text{for IC} \tag{10.37}$$

TABLE 10.2
Some Data for Slab and Cylindrical Products

Product	L (m)	R (m)	T_a (°C)	U_a (m/s)	Reference
Slab-shaped onion slice (IS)	1.5×10^{-3}	–	80	2	[29]
Cylindrically-shaped starch powder (IC)	–	5×10^{-3}	60	2	[30]

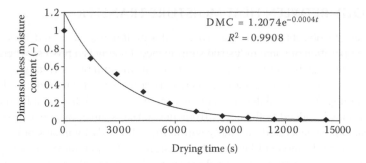

FIGURE 10.2 Dimensionless moisture content distributions for IS (♦, Data from Lewicki, P.P., Witrowa-Rajchert, D., and Nowak, D., *Drying Technology*, 16, 1&2, 59–81 and 83–100, 1998; —, regression).

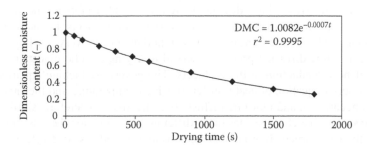

FIGURE 10.3 Dimensionless moisture content distributions for IC (♦, Data from Akiyama, T., Liu, H., and Hayakawa, K., *International Journal of Heat and Mass Transfer*, 40, 7, 1601–1609, 1997; —, regression).

In the above equations, the lag factors are 1.207 for IS and 1.0082 for IC and the drying coefficients as 0.0004 s^{-1} for IS and 0.0007 s^{-1} for IC. The necessary calculations are made as follows:

- The Biot numbers are calculated to be 3.73 for IS and 0.035 for IC.
- The values of μ_1 are calculated to be 1.28 for IS and 0.25 for IC.
- The moisture diffusivity values are determined as 5.45×10^{-10} m^2/s for IS and 3.04×10^{-7} m^2/s for IC.
- The moisture transfer coefficients are determined as 1.36×10^{-6} m/s for IS and 2.10×10^{-6} m/s for IC.

The moisture transfer parameters determined by the present approach are consistent with those from original literature sources. One should remember that these values are single values, representing the entire drying processes for the products, so that we can call these as effective moisture diffusivity and moisture transfer coefficient values. In numerous sources, such parameters are studied with respect to the temperature of air or the moisture content of the sample.

It can be seen in Figures 10.2 and 10.3 that experimental and regressed dimensionless moisture content profiles follow identical trends and decrease with increasing drying time. We note that the regressed dimensionless moisture content values at $t = 0$ are more than 1.0, leading to the lag factors, which indicate some finite internal and external resistances to the moisture transfer taking place during drying. Of course, these are accommodated in the case of Biot number, $0 < \text{Bi} < 100$.

In summary, in this illustrative example we studied some simple moisture transfer models for determining moisture diffusivity and moisture transfer coefficients during drying of solid products. The present models, involving drying process parameters, i.e., lag factor and drying coefficient, are verified with experimental data taken from the literature and the results are in good agreement with our experimental data.

10.8   ANALOGY BETWEEN HEAT MOISTURE TRANSFER

Drying of particulate objects, which is a process of simultaneous heat and mass transfer, is an energy-intensive operation of some industrial significance. Drying is a process whereby moisture is vaporized and swept away from the surface, sometimes in vacuum but normally by means of a carrier fluid passing through or over the moist object. This process has found industrial applications in various forms ranging from wood drying in lumber industry to food drying in food industry. In the process, the heat may be added to the object from an external source, by convection, conduction or radiation, or heat can be generated internally within the solid body by means of electric resistance. However, regardless of the mode of heating, the moisture is always removed in a vapor phase.

In practice, there is an urgent need to develop new models and methods for determining moisture transfer parameters in terms of moisture diffusivity and moisture transfer coefficient for the objects subject to drying. An accurate determination of such moisture transfer parameters is essential for an efficient moisture transfer analysis, leading to optimum energy use and operating conditions and hence, efficient and effective drying. Although there are a number of experimental and theoretical studies on the determination of drying profiles for various solid objects, the number of models for moisture diffusivity and moisture transfer coefficient parameters are very few [e.g., 26–28].

The transient moisture diffusion process observed in drying of solid objects is similar in form to the process of heat conduction in these objects. The governing Fickian equation is exactly in the form of the Fourier equation of heat transfer, in which temperature and thermal diffusivity are replaced with concentration and moisture diffusivity respectively. Therefore, similar to the case of unsteady heat transfer, one can consider three different situations for the unsteady moisture diffusion, namely the cases where Biot number takes values as follows: Bi ≤ 0.1, 0.1 < Bi < 100 and Bi > 100.

Since we deal with moisture transfer aspects using the analogy to heat conduction, in the first case where Bi ≤ 0.1 implies negligible internal resistance to the moisture diffusivity within the solid object. This may not be common in practice, and therefore may not be taken into account unless the product size is really small which will result negligible moisture transfer gradients within product. On the other hand, cases where Bi > 100 including negligible surface resistance to the moisture transfer at the solid object, are the most common situation, while cases where 0.1 < Bi < 100 including the finite internal and surface resistances to the moisture transfer, exist in practical applications. We therefore consider these two cases in our analysis.

The time-dependent heat and moisture transfer equations in one-dimensional rectangular, cylindrical and spherical coordinates for an IS, IC, and a sphere, respectively can be written in the following compact form:

$$\left(\frac{1}{y^m}\right)\left(\frac{\partial}{\partial y}\right)\left[y^m\left(\frac{\partial T}{\partial y}\right)\right]=\left(\frac{1}{a}\right)\left(\frac{\partial T}{\partial t}\right) \qquad \text{for heat transfer} \qquad (10.38)$$

$$\left(\frac{1}{y^m}\right)\left(\frac{\partial}{\partial y}\right)\left[y^m\left(\frac{\partial W}{\partial y}\right)\right]=\left(\frac{1}{D}\right)\left(\frac{\partial W}{\partial t}\right) \qquad \text{for moisture transfer} \qquad (10.39)$$

where $m = 0$, 1, and 2 for an IS, IC, and a sphere. $y = z$ for an IS, $y = r$ for IC and sphere. T stands for temperature (°C), W is the moisture content (kg/kg), a is the thermal diffusivity (m²/s), D is the moisture diffusivity (m²/s), and t is time (s).

To use experimental data in the mathematical model developed, the dimensionless temperature (θ) and dimensionless moisture content (Φ) can be defined as follows:

$$\theta = \frac{T - T_i}{T_a - T_i} \qquad \text{for heat transfer} \qquad (10.40)$$

$$\Phi = \frac{W - W_e}{W_i - W_e} \quad \text{for moisture transfer} \tag{10.41}$$

where subscripts a, e, and i stand for ambient, equilibrium and initial conditions.

10.8.1 DRYING PROCESS PARAMETERS

Two important drying process parameters are drying coefficient (S, 1/s) and lag factor (G, dimensionless unit). Drying coefficient shows the drying capability of the object or product and lag factor is an indication of internal resistances of object to the heat and/or moisture transfer during drying. These parameters are useful in evaluating and representing an entire drying process.

$$\Phi = G \exp(-St) \tag{10.42}$$

where t is the drying time (s).

10.8.2 THE DINCER NUMBER

A dimensionless number (so called Dincer number) which expresses the effect of flow velocity of heating or cooling of fluid on the heating or cooling coefficient of the objects with regular and irregular shapes is given as follows:

$$\text{Di} = \frac{U}{CL} \tag{10.43}$$

where U is the flow velocity of surrounding fluid (m/s), C is the heating or cooling coefficient (1/s), and L is the characteristic dimension (m) (e.g., radius for cylindrical and spherical objects and half thickness for slab ones).

Similarly, for a drying process:

$$\text{Di}_d = \frac{U}{SL} \tag{10.44}$$

which is the Dincer number for a drying process which represents a connecting link between the flow velocity of drying air and the drying rate of the object or product being dried. The use of these dimensionless numbers will be illustrated in the following section.

10.9 DRYING CORRELATIONS

In this section, we present various drying correlations as available elsewhere [31–34], for determining various moisture transfer parameters for use in the design and analysis of various practical drying applications, without going through costly experiments and complex mathematical analysis.

10.9.1 THE BIOT NUMBER–REYNOLDS NUMBER (Bi–Re) CORRELATIONS

Because of the thermophysical properties and velocity of the drying fluid, it is known to have a relationship between Biot number and Reynolds number (Re = 2UL/ν). Using experimental drying data available in the literature, the Bi–Re correlation was obtained for several kinds of food products being dried in an air flow (Figure 10.4), with the correlation coefficient as high as 0.72 (for further details, see Dincer et al. [31]):

$$\text{Bi} = 22.552\text{Re}^{-0.5897} \tag{10.45}$$

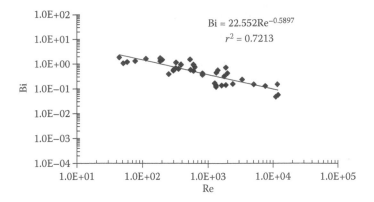

FIGURE 10.4 Bi–Re diagram for food products dried with air.

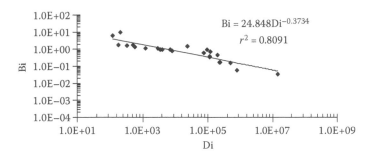

FIGURE 10.5 Bi–Di diagram for products dried with air.

10.9.2 THE BIOT NUMBER–DINCER NUMBER (Bi–Di) CORRELATION

In cooling, Dincer number expresses the effect of flow velocity of the cooling fluid on the cooling coefficient (cooling process parameter) for food products with regular or irregular shapes. Similarly in drying, Dincer number represents the effect of flow velocity of drying air on the drying coefficient (drying process parameter) for products of regular or irregular shapes. Using experimental drying data from literature works, we obtained the Bi–Di correlation for several products dried in air Figure 10.5), with the correlation coefficient of 0.8 as follows (for further details, see Dincer et al. [32]):

$$Bi = 24.848Di^{-0.3734} \qquad (10.46)$$

10.9.3 THE BIOT NUMBER–DRYING COEFFICIENT (Bi–S) CORRELATION

From Equation 10.26, we noticed that Biot number is a function of characteristic root (μ_1), which in turn is a function of lag factor (G). And since lag factor and drying coefficient are drying process parameters, there should be a relationship between Biot number and drying coefficient (S). Based on experimental data taken from various literature sources, the Bi-S correlation Figure 10.6) with correlation coefficient over 0.7 was obtained as follows (for further details, see Dincer et al. [33]):

$$Bi = 1.2388S^{0.3252} \qquad (10.47)$$

10.9.4 THE MOISTURE DIFFUSIVITY–DRYING COEFFICIENT (D–S) CORRELATION

The moisture diffusivity model, as developed by Dincer [26], is a function of drying coefficient, characteristic length of the product and characteristic root (μ_1). Based on experimental data taken

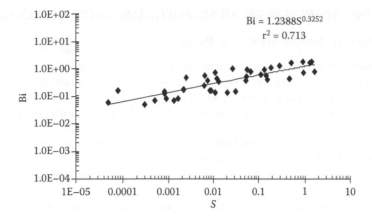

FIGURE 10.6 Bi-S diagram for products dried with air.

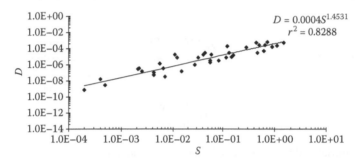

FIGURE 10.7 *D-S* diagram for products dried with air.

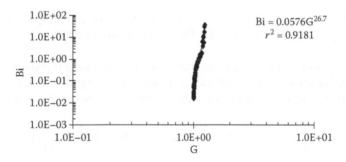

FIGURE 10.8 Bi-G diagram for products dried with air.

from literature, the following correlation was obtained between moisture diffusivity and drying coefficient (Figure 10.7) with a correlation coefficient over 0.8 as follows:

$$D = 0.0004S^{1.4531} \tag{10.48}$$

10.9.5 The Biot Number–Lag Factor (Bi–G) Correlation

As stated earlier, Biot number is a function of characteristic root (μ_1) which in turn is a function of lag factor (*G*). Using experimental data from various literature sources, a correlation (Figure 10.8) was developed between Biot number and lag factor with a correlation coefficient over 0.9 as follows (for further details, see Dincer et al. [34]):

$$Bi = 0.0576G^{26.7} \tag{10.49}$$

10.10   DRYING TIMES FOR REGULAR MULTIDIMENSIONAL PRODUCTS

10.10.1   DRYING OF INFINITE SOLID SLAB PRODUCTS

The transient moisture diffusion process during drying of a solid object takes place as similar to the heat conduction process in such a solid object. So, the governing Fickian equation has the same form of the Fourier equation of heat transfer, in which temperature and thermal diffusivity are replaced with moisture and moisture diffusivity respectively. The following assumptions are made:

- Thermophysical properties of the solid and the drying medium are constant.
- The effect of heat transfer on moisture loss is negligible.
- There are finite internal and external resistances to the moisture transfer within the moist product (i.e., 0.1 < Bi < 100).
- The moisture diffusion occurs in z direction (perpendicular to the slab surface) only.

Under these conditions, the governing one-dimensional moisture diffusion equation can be written as

$$D\frac{\partial^2 \Phi}{\partial z^2} = \frac{\partial \Phi}{\partial t} \tag{10.50}$$

where

$$\Phi = \frac{W - W_e}{W_i - W_e} \tag{10.51}$$

The following initial and boundary conditions are considered:

$$\Phi(z,0), \ (\partial\Phi(0,t)/\partial z) = 0, \quad \text{and} \quad -D(\partial\Phi(L,t)/\partial z) = h_m \Phi(L,t) \quad \text{for} \quad 0.1 \le \text{Bi} \le 100,$$

$$\Phi(L,t=0) \quad \text{for} \quad \text{Bi} > 100, \quad \text{and} \quad -h_m F\Phi(L,t) = \rho V c_p \partial(\Phi(L,t))/\partial t$$
$$\text{for} \quad \text{Bi} < 0.1 \quad (\text{F: Heat transfer surface area}).$$

where L is the half thickness of slab and the Biot number for moisture transfer is $\text{Bi} = h_m L/D$.

The solution of Equation 10.50 is extensively treated for the regular-shaped products earlier by the author [26–28]. So, the dimensionless average moisture distribution becomes

$$\Phi = \exp\left(\frac{-h_m t}{\sigma}\right) \quad \text{for the case where} \quad \text{Bi} < 0.1 \left[\sigma = \left(\frac{V}{F}\right) = L\right]. \tag{10.52}$$

and

$$\Phi = \sum_{n=1}^{\infty} A_n B_n \quad \text{for the case where} \quad \text{Bi} > 0.1 \tag{10.53}$$

Equation 10.53 can be simplified by ignoring the values of ($\mu^2 Fo$) smaller than 1.2 (taking only the first term into consideration), resulting in

$$\Phi = A_1 B_1 \tag{10.54}$$

where A_1 is given by

$$A_1 = \frac{2\sin\mu_1}{\mu_1 + \sin\mu_1 \cos\mu_1} \quad \text{for } 0.1 \le \text{Bi} \le 100 \tag{10.55}$$

and

$$A_1 = \frac{2}{\mu_1} \qquad \text{for Bi} > 100 \qquad\qquad (10.56)$$

and B_1 is

$$B_1 = \exp(-\mu_1^2 \, \text{Fo}) \qquad \text{for Bi} > 0.1 \qquad\qquad (10.57)$$

where the Fourier number is defined as $\text{Fo} = Dt/L^2$.

The corresponding characteristic equations are given as:

$$\cot(\mu_1) = \frac{1/\text{Bi}}{\mu_1} \qquad \text{for } 0.1 \le \text{Bi} \le 100 \qquad\qquad (10.58)$$

and

$$\mu_1 = \frac{\pi}{2} \qquad \text{for Bi} > 100. \qquad\qquad (10.59)$$

In one of the previous works [35] Equation 10.58 was simplified due to the complication in extracting the μ_1 values for practical purposes as:

$$\mu_1 = \arctan(0.64\,\text{Bi} + 0.38) \qquad \text{for } 0.1 \le \text{Bi} \le 100 \qquad\qquad (10.60)$$

Equation 10.55 can also be further simplified to

$$A_1 = \exp\left[\frac{0.2533\,\text{Bi}}{1.3 + \text{Bi}}\right] \qquad \text{for } 0.1 \le \text{Bi} \le 100 \qquad\qquad (10.61)$$

Due to the fact that drying has an exponentially decreasing trend, we establish the following equation for drying, by introducing lag factor (G) and drying coefficient (S):

$$\Phi = G\exp(-St) \qquad\qquad (10.62)$$

Both Equations 10.54 and 10.62 are in the same form and can be equated to each other by having $G = A_1$ and $S = \mu_1^2 D/L^2$. Therefore,

Moisture transfer coefficient:

$$h_m = \sigma S \qquad \text{for Bi} < 0.1 \qquad\qquad (10.63)$$

$$h_m = \frac{D}{L}\text{Bi} = \frac{D}{L}\left[\frac{1.3\ln G}{0.2533 - \ln G}\right] \qquad \text{for } 0.1 \le \text{Bi} \le 100 \qquad\qquad (10.64)$$

$$h_m = \frac{D}{L}\text{Bi} \qquad \text{for Bi} > 100 \qquad\qquad (10.65)$$

Moisture diffusivity:

$$D = \frac{SL^2}{\mu_1^2} \qquad\qquad (10.66)$$

where μ_1 is given in Equation 10.60.

10.10.2   DRYING TIME FOR INFINITE SOLID SLAB PRODUCT

Analytical drying time of a slab object can be obtained using Equations 10.52 and 10.54 as

$$t_{\text{slab}} = -\frac{\sigma}{h_m} \ln \Phi_c \qquad \text{for } \text{Bi} < 0.1 \tag{10.67}$$

$$t_{\text{slab}} = -\frac{L^2}{D\mu_1^2} \ln\left(\frac{\Phi_c}{A_1}\right) \qquad \text{for } \text{Bi} \ge 0.1 \tag{10.68}$$

where Φ_c is the dimensionless centerline moisture content.

On the other hand, using the experimental model given in Equation 10.62, the drying time can be related to the lag factor and drying coefficient as:

$$t_{\text{slab}} = -\frac{1}{S} \ln\left(\frac{\Phi_c}{G}\right) \tag{10.69}$$

10.10.3   DIMENSIONLESS DRYING TIME

Dimensionless time in the drying process can be defined as $\tau = Ut/(L)$, where U is the velocity of dry air over the product and L is the half thickness of the slab product, the characteristic dimension. Therefore, Equation 10.62 can be written as

$$\Phi = G \exp\left(-\frac{\tau}{\text{Di}}\right) \tag{10.70}$$

where Di is the Dincer number defined as Di $= U/(SL)$.

Figures 10.9a through d show the dimensionless moisture content variation with time for selected Di numbers. Bi number is fixed in each figure. Thus, the effect of Di number on the drying process can be observed. The lower Di number indicated faster drying. This is due to the fact that, by definition, the Di number is inversely proportional to the drying coefficient, S. As the Di number increases, the drying slows down as shown from these figures.

10.11   NUMERICAL ANALYSIS OF HEAT AND MOISTURE TRANSFER

In this section, numerical analysis of transient heat and moisture transfer during drying is presented in two-dimensional (2-D) form for three regular-shaped products, namely slab, cylindrical and spherical products. Some assumptions considered in the analysis are as follows:

   i. Thermophysical properties are constant.
   ii. Negligible shrinkage or deformation of objects during drying.
   iii. No heat generation inside the products.
   vi. Temperature of the drying air is constant.
   v. Variation of temperature and moisture is considered along "x" and "y" in slab, "r" and "z" in cylinder and "r" and "φ" in sphere.

10.11.1   THE 2-D SLAB PRODUCT

10.11.1.1   Analysis

The mathematical equations governing the drying process in 2-D cases with appropriate boundary conditions are given as:

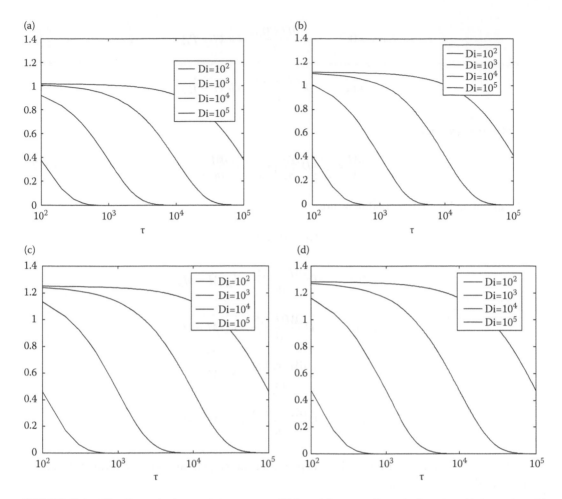

FIGURE 10.9 The dimensionless moisture content (Φ) in a slab versus dimensionless time (τ) for various Di numbers at (a) Bi = 0.1, (b) Bi = 1, (c) Bi = 10, and (d) Bi = 100, respectively.

- Heat transfer

$$\frac{1}{\alpha}\frac{\partial T}{\partial t} = \frac{\partial^2 T}{\partial x^2} + \frac{\partial^2 T}{\partial y^2} \tag{10.71}$$

The initial and boundary conditions are:

$$T(x,y,0) = T_i$$

$$\text{at } x = 0; \quad -k\frac{\partial T(0,y,t)}{\partial x} = h\ (T - T_d)$$

$$\text{at } x = 1; \quad -k\frac{\partial T(l,y,t)}{\partial x} = h(T - T_d)$$

$$\text{at } y = 0; \qquad -k\frac{\partial T(x,0,t)}{\partial y} = h\ (T - T_d)$$

$$\text{at } y = h; \qquad -k\frac{\partial T(x,h,t)}{\partial y} = h\ (T - T_d)$$

• Moisture transfer

$$\frac{\partial M}{\partial t} = \frac{\partial}{\partial x}\left(D\frac{\partial M}{\partial x}\right) + \frac{\partial}{\partial y}\left(D\frac{\partial M}{\partial y}\right) \tag{10.72}$$

The initial and boundary conditions are:

$$M(x,y,0) = M_i$$

$$\text{at } x = 0; \qquad -D\frac{\partial M(0,y,t)}{\partial x} = h_m(M - M_d)$$

$$\text{at } x = 1; \qquad -D\frac{\partial M(l,y,t)}{\partial x} = h_m(M - M_d)$$

$$\text{at } y = 0; \qquad -D\frac{\partial M(x,0,t)}{\partial y} = h_m(M - M_d)$$

$$\text{at } y = h; \qquad -D\frac{\partial M(x,h,t)}{\partial y} = h_m(M - M_d)$$

where D is the moisture diffusivity, whose dependence on temperature is of the form of Arrhenius equation [35]:

$$D = D_o \exp\left(\frac{-1119}{T}\right) \tag{10.73}$$

The temperature in the heat conduction equation and moisture content in the diffusion equation is nondimensionalized using the following equations

$$T^* = \frac{T - T_i}{T_d - T_i} \tag{10.74}$$

$$M^* = \frac{M - M_d}{M_i - M_d} \tag{10.75}$$

10.11.1.2 Solution Methodology

The solution of the above governing equations presented is difficult to obtain using analytical methods. Moreover, considerable assumptions have to be made in order to obtain a closed form solution with many inadequacies. Therefore, numerical methods of solution are used to solve them. The method used in the present study is the explicit finite difference approximation where the governing equations are first transformed into difference equations by dividing the domain of solution to a grid of points in the form of mesh and the derivatives are expressed along each mesh point referred to as a node. Knowing the dependent variable at each node initially, it is approximated for the next

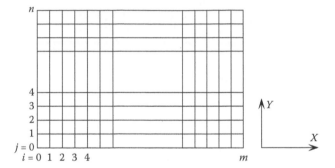

FIGURE 10.10 Numerical grid for a slab product.

time step and it continues until the final time step. The numerical grid of the solution domain is shown in Figure 10.10. It consists of two sets of perpendicular lines representing the x-direction and y-direction and the intersection of these lines constitute the nodes where the solution of the governing equations is obtained. The index i represents the mesh points in the X-direction starting with $i = 0$ being one boundary and ending at $i = m$, the other boundary while index j represents the mesh points in the Y-direction starting from $j = 0$. Thus, the finite difference representation of the mesh points will be as follows:

$$X_i = i\Delta X \qquad \text{for } i = 0,1,2...m$$

$$Y_j = j\Delta Y \qquad \text{for } j = 0,1,2...n$$

where ΔX and ΔY represents the grid sizes in the X- and Y-directions, respectively and the subscripts denote the location of the dependent under consideration, i.e., $T_{i,j}$ means the temperature at the i'th X-location and j'th Y-location.

The finite difference representation of the governing heat transfer equations can be written as:

$$T_{i,j,k+1} = r(\Delta Y)^2 T_{i+1,j,k} + [1 - 2r((\Delta X)^2 + (\Delta Y)^2)]T_{i,j,k} + r(\Delta Y)^2 T_{i-1,j,k}$$
$$+ r(\Delta X)^2 T_{i,j+1,k} + r(\Delta X)^2 T_{i,j-1,k} \tag{10.76}$$

where

$$r = \frac{\alpha \Delta t}{(\Delta X)^2 (\Delta Y)^2}$$

The initial and boundary conditions are:

$$T_{i,j,0} = T_i$$

$$\text{at } x = 0; \qquad T_{0,j,k} = \frac{T_{1,j,k} + C_1 \times T_d}{1 + C_1}$$

$$\text{at } x = 1; \qquad T_{m,j,k} = \frac{T_{m-1,j,k} + C_1 \times T_d}{1 + C_1}$$

$$\text{at } y = 0; \qquad T_{i,0,k} = \frac{T_{i,1,k} + C_2 \times T_d}{1 + C_2}$$

$$\text{at } y = h; \qquad T_{i,n,k} = \frac{T_{i,n-1,k} + C_2 \times T_d}{1 + C_2}$$

where

$$C_1 = \frac{h \times \Delta X}{k} \qquad \text{and} \qquad C_2 = \frac{h \times \Delta Y}{k}.$$

For mass transfer they can be written as follows:

$$M_{i,j,k+1} = r_d(\Delta Y)^2 M_{i+1,j,k} + [1 - 2r_d((\Delta X)^2 + (\Delta Y)^2)]M_{i,j,k} + r_d(\Delta Y)^2 M_{i-1,j,k}$$

$$+ r_d(\Delta X)^2 M_{i,j+1,k} + r_d(\Delta X)^2 M_{i,j-1,k} \tag{10.77}$$

where

$$r_d = \frac{D\Delta t}{(\Delta X)^2 (\Delta Y)^2}$$

The initial and boundary conditions are:

$$M_{i,j,0} = M_i$$

$$\text{at } x = 0; \qquad M_{0,j,k} = \frac{M_{1,j,k} + C_3 \times M_d}{1 + C_3}$$

$$\text{at } x = 1; \qquad M_{m,j,k} = \frac{M_{m-1,j,k} + C_3 \times M_d}{1 + C_3}.$$

$$\text{at } y = 0; \qquad M_{i,0,k} = \frac{M_{i,1,k} + C_4 \times M_d}{1 + C_4}$$

$$\text{at } y = h; \qquad M_{i,n,k} = \frac{M_{i,n-1,k} + C_4 \times M_d}{1 + C_4}$$

where

$$C_3 = \frac{h_m \times \Delta X}{D} \qquad \text{and} \qquad C_4 = \frac{h_m \times \Delta Y}{D}.$$

The above difference equations are solved to obtain temperature and moisture distributions inside the slab object at different time periods. The grid independent tests are conducted to ensure the grid independent results in the simulation. Stability analysis is performed in order to investigate the boundedness of the exact solution of the finite-difference equations using the von Neumann's method. The method introduces an initial line of errors as represented by a finite-Fourier series and applies in a theoretical sense to initial value problem. The stability criterions obtained for the above difference equations are:

$$\text{for heat transfer } \Delta t \leq \frac{(\Delta X)^2 (\Delta Y)^2}{2\alpha[(\Delta X)^2 + (\Delta Y)^2]} \qquad (10.78)$$

$$\text{for moisture transfer } \Delta t \leq \frac{(\Delta X)^2 (\Delta Y)^2}{2D[(\Delta X)^2 + (\Delta Y)^2]} \qquad (10.79)$$

Thus, the above criterions have to be satisfied in order to have converged solution.

10.11.1.3 Results and Discussion

This subsection presents the temperature and moisture distribution inside the slab as of rectangular slab of 0.03×0.02 m, respectively. The thermophysical properties and drying conditions used in the simulation are listed in Table 10.3. Figure 10.11 shows the temperature contours in the slab for different time periods, while Figure 10.12 shows the three-dimensional views of temperature distributions inside the slab. Temperature within the slab increases as the time period progresses. This is because of the higher temperature of the drying. The temperature contours in the slab present elliptic profiles due to the rectangular shape of the drying product.

In order to investigate the effect of heat transfer coefficient on temperature distribution inside the slab, five different heat transfer coefficients are considered ranging from 25 to 250 W/m²K. The variation of reduced center temperature and reduced surface temperature for different heat transfer coefficients are shown in Figures 10.13 and 10.14, respectively. The reduced center and surface temperature increases with the increase in heat transfer coefficient. The time required to reach the maximum limit (unity) becomes minimum for the maximum heat transfer coefficient. The optimal heat transfer coefficient in the drying application depends on the nature of drying, initial temperature of the product to be dried, type of product and its moisture content.

Furthermore, we note that the reduced center temperature gradient initially increases and then decreases with time for all heat transfer coefficients considered in the simulation. The initial rise is

---

TABLE 10.3
Thermophysical Properties and Drying Conditions Used in the Study

Product	Apple
k	0.219 W/mK (*)
ρ	856 kg/m³ (*)
c_p	851 J/kgK (*)
h	25 W/m²K
T_d	323 K
T_i	298 K
h_m	0.0001 m/s
M_i	5.25 kg/kg (db)
RH	0.53 kg/kg (db)

Source: Dincer, I., *Heat Transfer in Food Cooling Applications*, Taylor & Francis, Washington, DC, 1997.

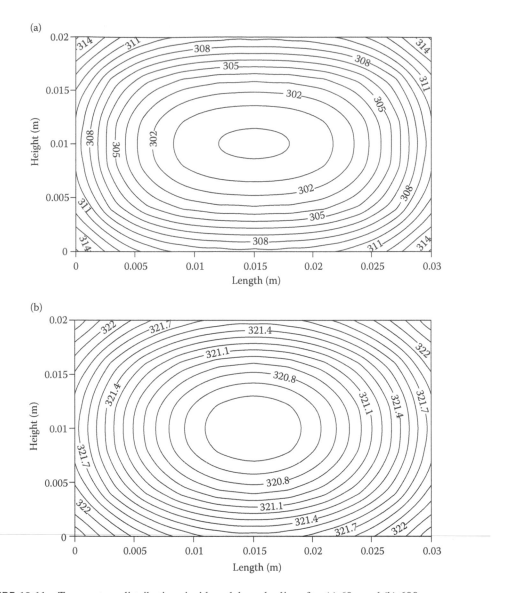

FIGURE 10.11 Temperature distributions inside a slab apple slice after (a) 60 s and (b) 600 s.

due to sudden increase of temperature after steady behavior of temperature for a certain period of time; also refer to as warming up of the solid. The reduced surface temperature gradient decreases with time and the maximum heat transfer coefficient has the highest gradient all the time and shows the highest heat transfer rate for the maximum heat transfer coefficient.

The moisture contours in a rectangular slab for different time periods is shown in Figure 10.15, while Figure 10.16 shows the three-dimensional views of moisture distribution in the slab. The moisture in the slab reduces as the time period progresses. Since the slab is rectangular, the resulting contours are in elliptic shape. The moisture distribution inside the slab is not uniform, therefore the time required to reduce the moisture content to half of its initial value varies at each location in the rectangular slab. Figure 10.17 shows variations of reduced center and surface moisture content with time. The reduced moisture content both at the surface and the center decrease as time progresses. Both profiles have the same trend but the reduction of moisture content is faster at the surface than that at the center. Furthermore, it exhibits the general trend in which rate of drying takes place in

(a)

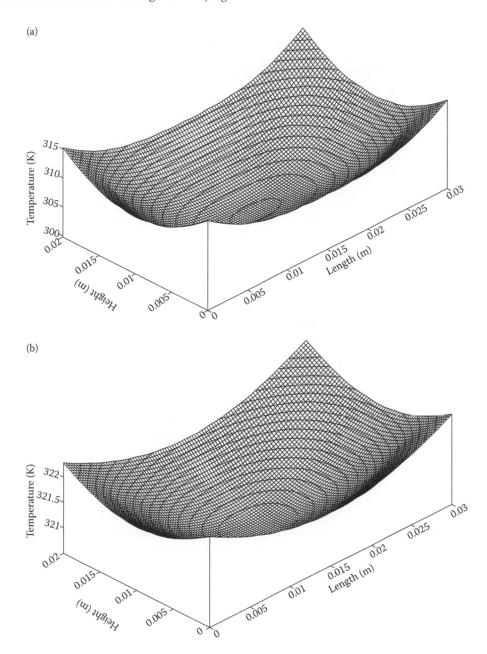

FIGURE 10.12 3-D plots showing temperature distribution inside a slab apple slice after (a) 60 s and (b) 600 s.

two periods. It is constant for some period of time representing the constant rate period and then starts decreasing as time progresses expressing the falling rate period.

10.11.2   THE 2-D CYLINDRICAL PRODUCT

10.11.2.1   Analysis

The mathematical equations governing the drying process in 2-D cylindrical object with appropriate boundary conditions are given as:

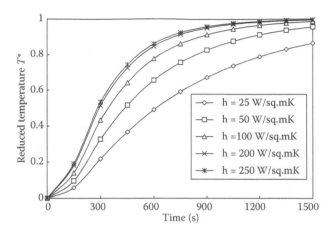

FIGURE 10.13 Variation of reduced center temperature distributions in a slab apple for different heat transfer coefficients.

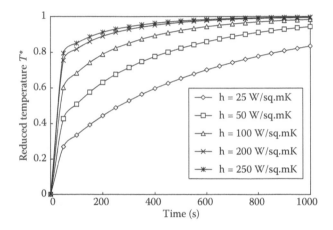

FIGURE 10.14 Variation of reduced surface temperatures in a slab apple for different heat transfer coefficients.

- Heat transfer

$$\frac{1}{\alpha}\frac{\partial T}{\partial t} = \frac{1}{r}\frac{\partial}{\partial r}\left(r\frac{\partial T}{\partial r}\right) + \frac{\partial^2 T}{\partial z^2} \tag{10.80}$$

The initial and boundary conditions are:

$$T(r,z,0) = T_i$$

$$\text{at } r = 0; \qquad \frac{\partial T(0,z,t)}{\partial r} = 0$$

$$\text{at } r = R; \qquad -k\frac{\partial T(R,y,t)}{\partial r} = h\ (T - T_d)$$

$$\text{at } z = 0; \qquad -k\frac{\partial T(r,0,t)}{\partial z} = h\ (T - T_d)$$

(a)

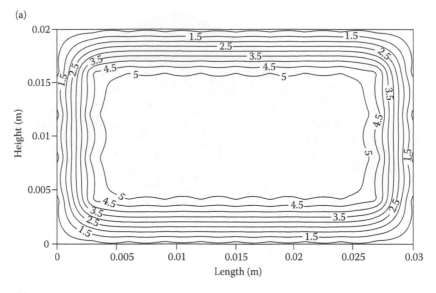

(b)

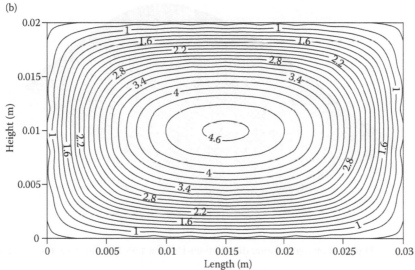

FIGURE 10.15 Moisture distributions inside a slab apple after (a) 60 s and (b) 600 s.

$$\text{at } z = L; \qquad -k\frac{\partial T(r,L,t)}{\partial z} = h\ (T - T_d)$$

- Moisture transfer

$$\frac{1}{D}\frac{\partial M}{\partial t} = \frac{1}{r}\frac{\partial}{\partial r}\left(r\frac{\partial M}{\partial r}\right) + \frac{\partial^2 M}{\partial z^2} \qquad (10.81)$$

The initial and boundary conditions are:

$$M(r,z,0) = M_i$$

$$\text{at } r = 0; \qquad \frac{\partial M(0,z,t)}{\partial r} = 0$$

(a)

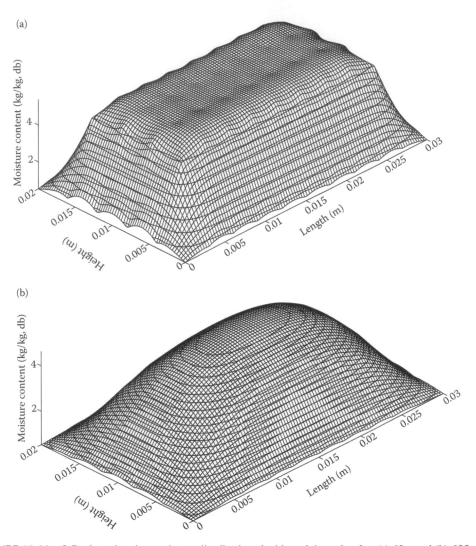

(b)

FIGURE 10.16   3-D plots showing moisture distributions inside a slab apple after (a) 60 s and (b) 600 s.

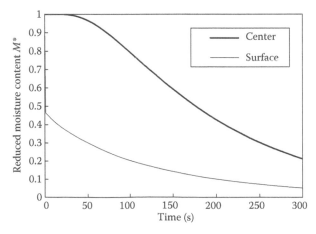

FIGURE 10.17   Variation of reduced center and surface moisture content distributions in a slab apple with respect to time.

$$\text{at } r = R; \qquad -D\frac{\partial M(R,y,t)}{\partial r} = h_m(M - M_d)$$

$$\text{at } z = 0; \qquad -D\frac{\partial M(r,0,t)}{\partial z} = h_m(M - M_d)$$

$$\text{at } z = L; \qquad -D\frac{\partial M(r,L,t)}{\partial z} = h_m(M - M_d)$$

10.11.2.2 Solution Methodology

The above governing equations are discretized using the explicit finite-difference method. Numerical grid of an axisymmetric cylindrical object is shown in Figure 10.18. The index i represents the mesh points in the Z-direction starting with $i = 0$ being one boundary and ending at $i = m$, the other boundary while index j represents the mesh points in the R-direction starting from $j = 0$. Thus, the finite difference representation of the mesh points will be as follows:

$$Z_i = i\Delta Z \qquad \text{for } i = 0,1,2\ldots m$$

$$R_j = j\Delta R \qquad \text{for } j = 0,1,2\ldots n$$

where ΔZ and ΔR represents the grid sizes in the Z- and R-directions, respectively and the subscripts denote the location of the dependent under consideration, i.e., $T_{i,j}$ means the temperature at the i'th Z-location and j'th R-location. Knowing the value of dependent variable at the initial time step, unknown values at next time steps are calculated using the finite difference equations. The finite difference representations of the governing equations can be written in the following form:

- Heat transfer

$$T_{i,j,k+1} = AT_{i+1,j,k} + (1 - 2[A + B])T_{i,j,k} + AT_{i-1,j,k} + B(1 + 0.5j)T_{i,j+1,k} + B(1 - 0.5j)T_{i,j-1,k} \qquad (10.82)$$

where

$$A = \frac{\alpha\Delta t}{(\Delta Z)^2} \qquad \text{and} \qquad B = \frac{\alpha\Delta t}{(\Delta R)^2}$$

The initial and boundary conditions can be written as

$$T_{i,j,0} = T_i$$

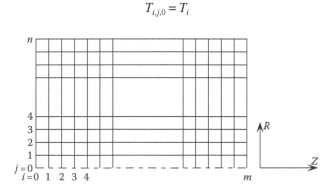

FIGURE 10.18 Numerical grid for an axisymmetric cylindrical object.

$$\text{at } r = 0; \qquad T_{i,0,k} = T_{i,1,k}$$

$$\text{at } r = R; \qquad T_{i,n,k} = \frac{T_{i,n-1,k} + C_5 \times T_d}{1 + C_5}$$

$$\text{at } z = 0; \qquad T_{0,j,k} = \frac{T_{1,j,k} + C_6 \times T_d}{1 + C_6}$$

$$\text{at } z = L; \qquad T_{m,j,k} = \frac{T_{m-1,j,k} + C_6 T_d}{1 + C_6}$$

where

$$C_5 = \frac{h \, \Delta R}{k} \qquad \text{and} \qquad C_6 = \frac{h \, \Delta Z}{k}.$$

- Moisture transfer

$$M_{i,j,k+1} = A_m M_{i+1,j,k} + (1 - 2[A_m + B_m]) M_{i,j,k} + A_m M_{i-1,j,k} + B_m (1 + 0.5j) M_{i,j+1,k}$$

$$+ B_m (1 - 0.5j) M_{i,j-1,k} \tag{10.83}$$

where

$$A_m = \frac{D\Delta t}{(\Delta Z)^2} \qquad \text{and} \qquad B_m = \frac{D\Delta t}{(\Delta R)^2}$$

The initial and boundary conditions can be written as

$$M_{i,j,0} = M_i$$

$$\text{at } r = 0; \qquad M_{i,0,k} = M_{i,1,k}$$

$$\text{at } r = R; \qquad M_{i,n,k} = \frac{M_{i,n-1,k} + C_7 \times M_d}{1 + C_7}$$

$$\text{at } z = 0; \qquad M_{0,j,k} = \frac{M_{1,j,k} + C_8 \times M_d}{1 + C_8}$$

$$\text{at } z = L; \qquad M_{m,j,k} = \frac{M_{m-1,j,k} + C_8 \times M_d}{1 + C_8}$$

where

$$C_7 = \frac{h_m \Delta R}{D} \qquad \text{and} \qquad C_8 = \frac{h_m \Delta Z}{D}.$$

The above difference equations are used to obtain temperature and moisture distributions inside the cylindrical object at different time periods. A stability analysis is performed in order to investigate the boundedness of the exact solution of the finite-difference equations. The stability criterions obtained for the above difference equations are:

$$\text{for heat transfer} \qquad \Delta t \leq \frac{(\Delta Z)^2 (\Delta R)^2}{2\alpha[(\Delta Z)^2 + (\Delta R)^2]} \qquad (10.84)$$

$$\text{for moisture transfer} \qquad \Delta t \leq \frac{(\Delta Z)^2 (\Delta R)^2}{2D[(\Delta Z)^2 + (\Delta R)^2]} \qquad (10.85)$$

Thus, the above conditions have to be satisfied in order to have converged solution.

10.11.2.3 Results and Discussion

In this subsection, temperature and moisture profiles obtained for the cylindrical object at different time periods are presented. The products considered in the simulation were *date* of diameter 0.0084 m and length 0.02 m and *banana* of diameter 0.02 m and length 0.03 m. The heat transfer coefficient was varied between 25 and 250 W/m²K. This case enables the investigation into the effect of heat transfer coefficient on temperature distribution. The thermophysical properties and the drying conditions used in the simulation of date drying are listed in Table 10.4. Figure 10.19 shows the temperature contours in the date cylinder for different time periods while Figure 10.20 shows the three-dimensional views of temperature distribution in the date cylinder. The temperature in the cylinder increases as the time period progresses. Moreover, temperature in the cylinder is nonuniform, maximum at the surfaces and minimum at the center.

The variations of reduced temperature at the center and at the surface for different heat transfer coefficients are shown in Figures 10.21 and 10.22. The temperature distributions increase with increasing heat transfer coefficient. The reduced center and surface temperatures attain the maximum limit (unity) for highest heat transfer coefficient considered here. The reduced surface temperature rises sharply during the first time step and then increases in a parabolic fashion. The rapid rise of temperature in the surface vicinity of the object is because of the internal energy gain in this region, which is due to convective heating of the surface.

TABLE 10.4
Thermophysical Properties and Drying Conditions Used in the Simulation of Date Drying

k	0.337 W/mK (*)
ρ	1319 kg/m³ (*)
c_p	$(0.837 + 1.256M) \times 1000$ J/kgK (*)
h	25–250 W/m²K
h_m	0.0001 m/s
T_i	298 K
T_d	323 K
M_i	3 kg/kg(db)
RH	0.42 kg/kg(db)

Source: Dincer, I., *Heat Transfer in Food Cooling Applications*, Taylor & Francis, Washington, DC, 1997.

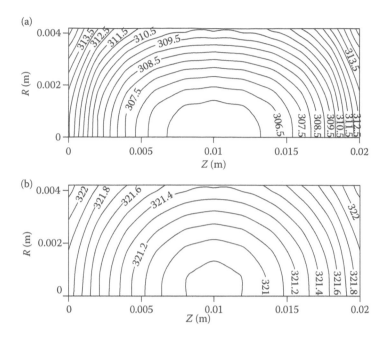

FIGURE 10.19   Temperature distributions inside a cylindrically-shap*ed d*ate after (a) 60s and (b) 300 s.

Furthermore, one should note that the reduced center temperature gradient initially increases for a certain period of time and decreases continuously after reaching a peak value. In this respect, the increase is due to considerably small change of temperature during early stages of heating and temperature increases rapidly due to gain in internal energy as the heating period progresses. The reduced surface temperature gradient decreases sharply at the initial time step and then becomes almost constant as the time period progresses. This is again due to the convective heating of the surface, which dominates over the conduction losses from the surface vicinity to the bulk of the solid object.

Figure 10.23 shows the moisture contours in a *date* cylinder for different time periods while Figure 10.24 exhibits a three-dimensional view of moisture distribution in a date cylinder. The moisture content inside the cylinder reduces as the time period increases. The reduction rate of moisture content in the surface region is higher compared to the interior of the object. Moreover, in the early heating period, moisture content reduces rapidly and as the heating progresses the rate of reduction of moisture content becomes less,. This is more pronounced in the surface vicinity. The rapid drop of moisture content in the early heating period is because of the high moisture gradient in this region, which in turn derives considerable diffusion rates from inside to the surface.

The variation of reduced center moisture content and reduced surface moisture content with time is shown in Figure 10.25. The reduction of moisture content is high in the early heating period and as the heating progresses, it reduces gradually until equilibrium moisture content is attained. The attainment of high moisture gradient in the early period is due to the diffusion process. The reduction of moisture content is more pronounced at the surface and as the heating progresses it becomes constant. One can further observe from the figure that drying takes place in two periods, first during the constant rate period in which rate of moisture reduction is constant which appears in the figure as straight line parallel to x-axis then the rate of moisture gradually reduces exhibiting falling-rate period.

Here, we can validate the model by comparing some experimental data in terms of temperature and moisture content distributions as taken from Simal et al. [37] with the numerical results obtained from the present model.

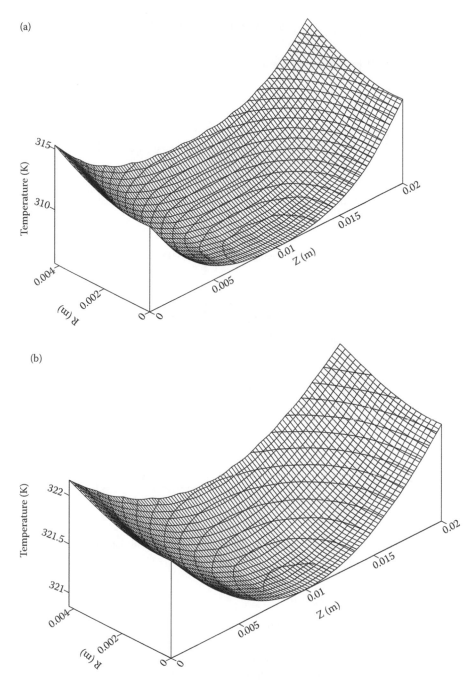

FIGURE 10.20 3-D plots showing temperature distribution inside a cylindrical date after (a) 60 s and (b) 300 s.

Figures 10.26 and 10.27 show some comparisons of center temperature and moisture distributions in a cylindrical product with the calculated and measured values taken from the literature. Experimental drying conditions and product are listed in Table 10.5. As shown in the respective figures, a good agreement can be found between the predicted and measured results. The mean percentage error between the measured and calculated values of temperature and moisture distribution is found to be ±0.76 and ±1.91, respectively.

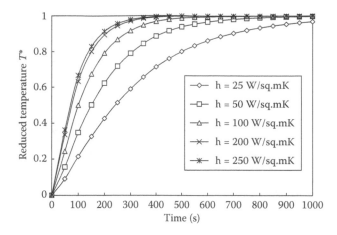

FIGURE 10.21 Variation of reduced center temperatures in a cylindrical date for different heat transfer coefficients.

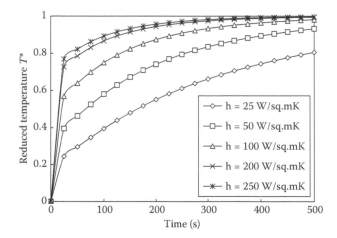

FIGURE 10.22 Variation of reduced surface temperatures in a cylindrical date for different heat transfer coefficients.

10.11.3 The 2-D Spherical Product

This subsection deals with temperature and moisture profiles in a spherical object during drying. The product considered in the simulation was potato of diameter 0.04 m. Thermophysical properties and the drying conditions used in the simulation are listed in Table 10.6.

Figures 10.28 and 10.29 show the variations of reduced center temperature and reduced surface temperature for different heat transfer coefficients. The temperature rises rapidly for all the heat transfer coefficients in the early heating period due to convective boundary condition at the surface. This is more pronounced in the surface vicinity of the object. Note that the rapid rise of temperature in the surface vicinity of the object is because of the internal heat gain in this region.

The variation of reduced temperature along the radius for different heating periods is shown in Figure 10.30. The reduced temperature increases all the time from the center to the surface of the sphere. As the heating period progresses, the reduced temperature gradient decreases until a stage is reached, in which heat gain due to convective boundary condition balances the heat transfer through conduction energy transport.

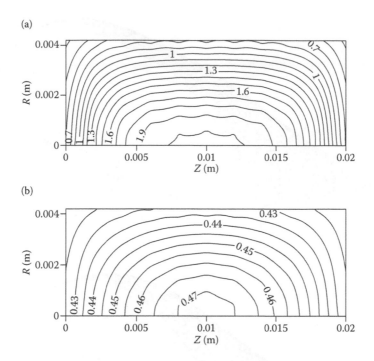

FIGURE 10.23 Moisture distributions inside a cylindrical date after (a) 60 s and (b) 300 s.

Figure 10.31 shows the reduced moisture content at the center and at the surface of the spherical object subjected to drying. The moisture content inside the sphere reduces as the time period progresses. In the early heating period, moisture content reduces rapidly and as heating progresses the rate of reduction of moisture content becomes less. This is more pronounced in the surface vicinity. The rapid drop of moisture content in the early heating period is due to the high moisture gradient in this region, which in turn derives considerable diffusion rates from bulk of the substrate to the surface. One can observe that during early heating period, the rate of drying is constant, thus exhibiting constant rate period. As the time period progresses, the rate of drying continuously decreases representing falling rate period. The variation of reduced moisture along radius at different heating period is shown in Figure 10.32. The reduced moisture content decreases both at the center and at the surface as the time period progresses.

The time required for the reduced moisture content to drop its half value at the center is about 1000 s. It is to be noted that the reduced moisture content is nonuniform in the sphere; consequently, half time requirement varies at each location.

Even though, we assumed the variation of reduced temperature and moisture in second dimension (coordinate) in our formulation, but the results obtained proves the fact that there is no variation in that direction due to convective heating at the surface. Thus, in a spherical product subjected to convective heating, the variation of temperature and moisture is only in the radial direction.

Here, we can validate the model by comparing some experimental data in terms of temperature and moisture content distributions as taken from McLaughlin and Magee [38] with the numerical results obtained from the present model. Thus, the comparison between the measured and calculated center temperature distribution and moisture content inside a spherical product are shown in Figures 10.33 and 10.34. A good agreement has been found between the predicted and measured results. The mean percentage error between the measured and calculated values of temperature and moisture distribution is found to be ±1.15 and ±2.26, respectively.

Thus in this section, the numerical solution for the temperature and moisture distribution inside the regular objects (i.e., slab, cylinder, and sphere), due to convective boundary conditions at the

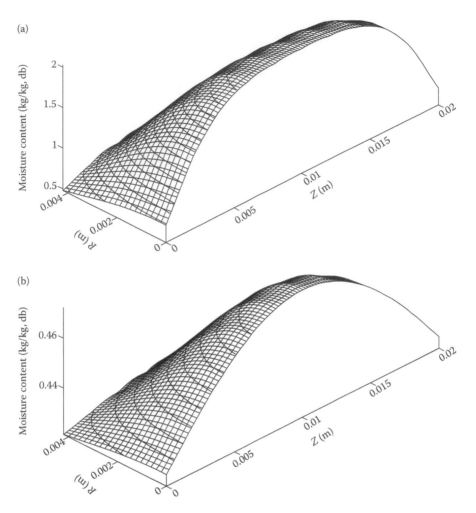

FIGURE 10.24 3-D plots showing moisture distribution inside a cylindrical date after (a) 60s and (b) 300s respectively.

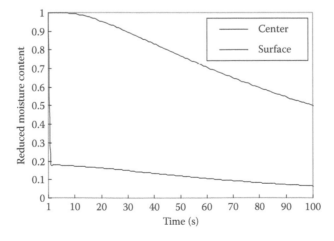

FIGURE 10.25 Variation of reduced center and surface moisture content distributions in a cylindrical date with time.

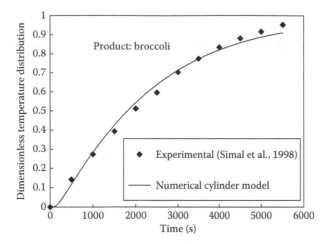

FIGURE 10.26 Measured and calculated center temperature distributions in a cylindrical broccoli.

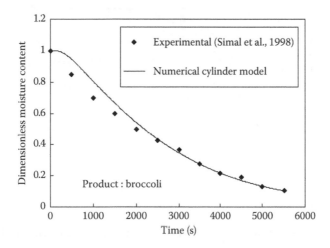

FIGURE 10.27 Measured and calculated center moisture distributions in a cylindrical broccoli.

TABLE 10.5
Thermophysical Properties and Drying Conditions used in the Study

Product	Cylindrical Broccoli
Size	Diameter 0.007 m and length 0.02 m
T_i	298 K
T_d	333 K
M_i	9.57 kg/kg (db)
RH	1.18 kg/kg (db)
k	$0.148 + 0.493 \times M_i$ (W/mK)*
ρ	2195.27 kg/m³ *
c_p	$(0.837 + 1.256 \times M_i) \times 1000$ (J/kgK)*
Reference	[37]

*Source: Dincer, I., *Heat Transfer in Food Cooling Applications*, Taylor & Francis, Washington, DC, 1997.

TABLE 10.6
Thermophysical Properties and Drying Conditions used in the Simulation

α	$1.31 \times 10^{-7}\,\text{m}^2/\text{s}$ (*)
h	25–250 W/m^{2}K
h_m	0.0001 m/s
T_i	296 K
T_d	333 K
M_i	5.25 kg/kg (db)
RH	0.42 kg/kg (db)

Source: Dincer, I., *Heat Transfer in Food Cooling Applications*, Taylor & Francis, Washington, DC, 1997.

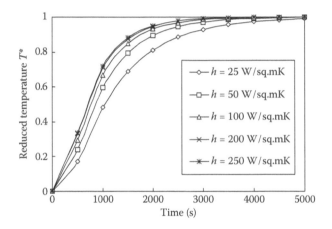

FIGURE 10.28 Variations of reduced center temperatures in a spherical potato for different heat transfer coefficients.

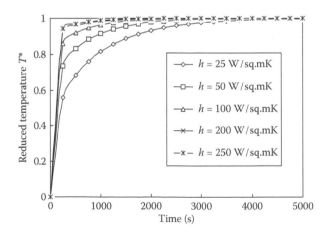

FIGURE 10.29 Variations of reduced surface temperatures in a spherical potato for different heat transfer coefficients.

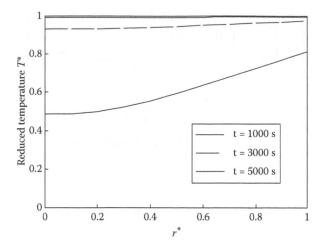

FIGURE 10.30 Variations of reduced temperature along radius for different drying periods.

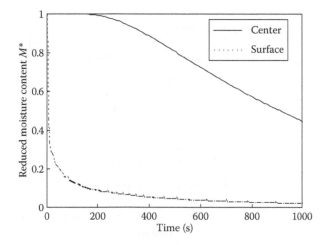

FIGURE 10.31 Reduced moisture distributions at the center and surface of a spherical product.

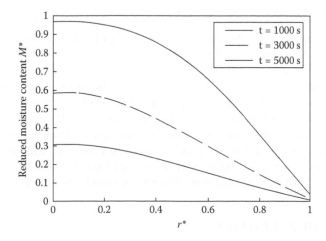

FIGURE 10.32 Variation of reduced moisture content along radius at different time periods.

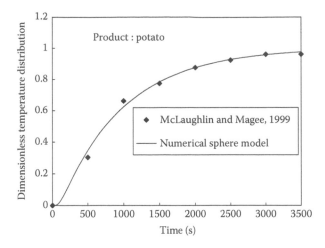

FIGURE 10.33 Comparison between the calculated and measured dimensionless temperature distribution in a spherical potato.

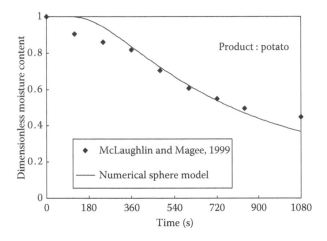

FIGURE 10.34 Comparison between the calculated and measured dimensionless moisture content in a spherical potato.

surface is presented for drying applications. It is found that the temperature rises rapidly in the early heating period and as the heating period progresses, the rise of temperature attains almost steady with advancing heating period. The moisture gradient is higher in the early heating period and as heating progresses, the moisture gradient remains almost steady. Furthermore, the effect of heat transfer coefficient on the temperature distribution inside the objects is also investigated. The temperature inside the objects increases as the heat transfer coefficient increases and the time required to reach the steady temperature is less for the maximum heat transfer coefficient. Moreover, validation of the results obtained from the present analysis is performed with the experimental data available in the literature. A fairly good agreement has been found between the calculated and measured values for the temperature and moisture distributions inside the objects.

10.12 CONCLUDING REMARKS

This book chapter has aimed to cover various topics of solids drying, such as fundamentals of food drying process and its periods, drying process parameters, moist products and their heat and moisture transfer behavior, moisture transfer models, analogy between heat conduction and moisture diffusion, practical drying correlations, and numerical heat and moisture transfer analysis for slab, cylindrical and spherical products with numerous illustrative examples and case studies. The emphasis is essentially put on the fundamental mechanisms and analysis methods during drying of food products. Model validation studies are also performed using drying data available in the literature.

NOMENCLATURE

a	specific heat and mass transfer surface (m^2-m^{-3})
A	drying rate (kg kg^{-1} s^{-1}); surface area (m^2); constant
a_m	moisture diffusion coefficient (m^2/s)
a_w	water activity
B	rate of solids temperature change (K s^{-1}); statistical parameter; constant
Bi	Biot number
c	mass quantity of heat (J/kg.K)
C	parameter of GAB equation; constant; cooling coefficient
Cp	specific heat (J-kg^{-1}-K^{-1})
C_{PG}	specific heat of dry air at constant pressure (kJ/kg.K)
C_{PS}	specific heat of dry solid at constant pressure (kJ/kg.K)
C_{PW}	specific heat of water at constant pressure (kJ/kg.K)
C_{PV}	specific heat of water vapor at constant pressure (kJ/kg.K)
D	moisture diffusivity, m^2/s
Di	Dincer number (U/SL)
e	energy content of gas per unit length (J/m)
E	energy flow rate for gas (W); shape factor
f	local mass transfer rate per unit volume of porous media (kg/s.m^3)
F	heat transfer surface area (m^2)
Fo	Fourier number
G	mass flow (kg-s^{-1}); mass velocity (kg/s.m^2); lag factor
h	convective heat transfer coefficient (W-m^{-2}-K^{-1}); height (m); energy content of solids per unit length (J/m)
h_m	moisture transfer coefficient (m/s)
H	enthalpy (J-kg^{-1}); energy flow rate for solids (W); humidity (%)
$J_0(\mu)$	Bessel function term
k	constant (kJ/m^3); moisture transfer coefficient, m/s
k_c	mass transfer coefficient (m-s^{-1})
K	parameter of GAB equation; constant
k_M	drying constant
L	half thickness (m); latent heat of vaporization (kJ/kg)
m	gas holdup per unit length (kg/m)
M	gas flow rate (kg/s); moisture content (%)
M_e	equilibrium moisture content (decimal dry basis)
MR	moisture ratio
p	solids holdup per unit length (kg/m)
P	solids flow rate (kg/s); pressure (Pa)
q	heat transfer rate per unit volume of porous media (kW/m^3)
Q	heat transfer (J or kJ)
r	radial coordinate
R	radius
S	drying coefficient (1/s)

t	drying time (s)
T	temperature (ºC, K); drying agent temperature (K)
U	moisture content in dry basis (kg/kg)
V	a volume in some place of dryer (m³)
v_S	velocity of the solid (m/s)
W	moisture content by weight
W_i	moisture flux (kg/m²s)
x	distance between a cross section of the dryer and the entrance of the grains (m)
X	moisture content (kg water-kg dry matter⁻¹); solids moisture content (kg kg⁻¹)
X^*_L	rate of liquid phase transition
X_m	monolayer moisture content (kg water/kg dry solids)
X_S	moisture content (kg/kg db)
X_{SE}	equilibrium material moisture content (kg/kg db)
Y	gas humidity (kg kg⁻¹)
z	distance (m); rectangular coordinate

Greek Symbols

δ	relative coefficient of thermal diffusion (kg/kgK)
ε	bed porosity (m⁻³ of gas phase-m⁻³ bed); void fraction of grain bed (m³/m³); coefficient phase transfer
λ	water latent heat of vaporization (kJ/kg); thermal conductivity (W/mK)
ρ	mass concentration (kg-m⁻³)
ρ_G	mass of dry air per unit volume of gas phase (kg/m³)
ρ_i	mass density of dry body (kg/m³)
ρ_0	oven-dry density (kg/m³)
ρ_S	bulk density of drying product (kg/m³)
τ_B	mean residence time of particles in the bed (s)
Φ	dimensionless moisture content
μ	transcendental equation root
β_1	ratio of second dimension to the characteristic length
β_2	ratio of third dimension to the characteristic length
σ	characteristic dimension (m)
θ	dimensionless temperature
τ	dimensionless time (Ut/L)

Subscripts

a	ambient; air
c	center
CR	constant rate
e	end; equilibrium
ef	effective
g	gas phase
G	gas property
i	initial; in solid-gas interface
lm	liquid moisture
n	nth value
o	outlet
s	in solid phase; solids; surface
S	solid property
v	volume

vm   vapor moisture
w   water
wv   water vapor
1   1st value

REFERENCES

1. Brennan, J.G., Butters, J.R., Cowell, N.D. and Lilly, A.E.V. 1976. *Food Engineering Operations.* Applied Science Publishers Limited, London, UK.
2. Langrish, T.A.G. and Harvey, A.C. 2000. A flowsheet model of a well-mixed fluidized bed dryer: applications in controllability assessment and optimization. *Drying Technology*, 18, 1&2, 185–198.
3. Sharaf-Eldeen, Y.I., Harndy, M.Y., Keener, H.M. and Blaisdell, J.L. 1979. Mathematical description of drying fully exposed grains, Paper No. 79-3034. ASAE, St Joseph, MI.
4. Byler, R.K., Anderson, C.R. and Brook, R.C. 1987. Statistical methods in thin-layer parboiled rice drying models. *Transactions of the ASAE,* 30, 2, 533–538.
5. Valenca, G.C. and Massarani, G. 2000. Grain drying in countercurrent and concurrent gas flow-Modeling, simulation and experimental tests. *Drying Technology*, 18, 1&2, 447–455.
6. Luikov, A.V. 1966. *Heat and Mass Transfer in Capillary-porous Bodies*, 1st ed. Pergamon Press, Oxford, UK.
7. Moreira, R., Figueiredo, A. and Sereno, A. 2000. Shrinkage of apple disks during drying by warm air convection and freeze drying. *Drying Technology*, 18, 1&2, 279–294.
8. Kowalski, S.J. and Strumillo, C.Z. 1997. Moisture transport in dried materials: boundary conditions. *Chemical Engineering Science,* 52, 7, 1141–1150.
9. Maroulis, Z.B., Tsami, E., Kouris, D.M. and Saravacos, G.D. 1988. Application of the GAB model to the moisture sorption isotherms for dried fruits. *Journal of Food Engineering*, 7, 63–78.
10. Adamson, A.W. 1963. *Physical Chemistry of Surfaces.* Wiley, New York, NY.
11. Brunauer, S., Emmett, P.H. and Teller, J. 1938. Adsorption of gases in multimolecular layers. *Journal of the American Chemical Society*, 60, 309–319.
12. Chung, D.S. and Pfost, H.B. 1967. Adsorption and desorption of water vapor by cereal grains and their products. *Transactions of the ASAE*, 28, 549–557.
13. Van den Berg, C. 1984. Description of water activity of foods for engineering purposes by means of the GAB model of sorption. In: McKenna, B.M., Editor. *Engineering and Food*, Vol. 1. Elsevier Applied Science, London, UK, 311–321.
14. Bizot, H. 1983. Using the 'G.A.B.' model to construct sorption isotherms. In: R. Jowitt, F. Escher, B. Hallström, H. Meffert, W. Spiess and G. Vos, Editors. *Physical Properties of Foods.* Applied Science Publishers, Essex, UK, 43–54.
15. Halsey, G. 1948. Physical adsorption on non-uniform surfaces. *Journal of Chemistry and Physics*, 16, 931–937.
16. Iglesias, H.A. and Chirife, J. 1978. An empirical equation for fitting water sorption isotherms of fruits and related products. *Canadian Institute of Food Science and Technology Journal*, 11, 12–15.
17. Henderson, S.M. 1952. A basic concept of equilibrium moisture. *Agricultural Engineering*, 33, 29–32.
18. Thomson, T.L., Peart, P.M. and Foster, G.H. 1968. Mathematical simulation of corn drying a new model. *Transactions of the ASAE*, 11, 582–586.
19. Oswin, C.R. 1946. The kinetics of package life. III. Isotherm. *Journal of the Society of Chemical Industry (London)*, 65, 419–421.
20. Chou, S.K., Hawlader, M.N.A. and Chua, K.J. 1997. Identification of the receding evaporation front in convective food drying. *Drying Technology*, 15, 5, 1353–1367
21. Smith, S. 1947. The sorption of water vapor by high polymers. *Journal of the American Chemical Society,* 69, 646–649.
22. Basunia, M.A. and Abe, T. 1998. Thin-layer drying characteristics of rough rice at low and high temperatures. *Drying Technology,* 16, 3–5, 579–595.
23. Kechaou, N. and Maalej, M. 2000. A simplified model for determination of moisture diffusivity of date from experimental drying curves. *Drying Technology*, 18, 4&5, 1109–1125.

24. Krokida, M.K., Karathanos, V.T. and Maroulis, Z.B. 2000. Compression analysis of dehydrated agricultural products. *Drying Technology*, 18, (1–2), pp. 395–408.

25. Dedic, A. 2000. Convective heat and mass transfer in moisture desorption of oak wood by introducing characteristic transfer coefficients. *Drying Technology*, 18, 7, 1617–1627

26. Dincer, I. and Dost, S. 1996. Determination of moisture diffusivities and moisture transfer coefficients for wooden slabs subject to drying. *Wood Science and Technology*, 30, 245–251.

27. Dincer, I. 1998. Moisture loss from wood products during drying. Part I: Moisture diffusivities and moisture transfer coefficients. *Energy Sources*, 20, 1, 67–75.

28. Dincer, I. and Dost, S. 1996. A new model for thermal diffusivities of geometrical objects subjected to cooling. *Applied Energy*, 51, 111–118.

29. Lewicki, P.P., Witrowa-Rajchert, D. and Nowak, D. 1998. Effect of drying mode on drying kinetics of onion. *Drying Technology*, 16, 1&2, 59–81 and 83–100.

30. Akiyama, T., Liu, H. and Hayakawa, K., 1997. Hygrostress multi-crack formation and propagation in cylindrical viscoelastic food undergoing heat and moisture transfer processes. *International Journal of Heat and Mass Transfer*, 40, 7, 1601–1609.

31. Dincer, I., Hussain, M.M., Sahin, A.Z. and Yilbas, B.S. 2002. Development of a new moisture transfer (Bi-Re) correlation for food drying applications. *International Journal of Heat and Mass Transfer*, 45, 8, 1749–1755.

32. Dincer, I. and Hussain, M.M. 2002. Development of a new Bi-Di correlation for solids drying. *International Journal of Heat and Mass Transfer*, 45, 15, 3065–3069.

33. Dincer, I., Hussain, M.M., Yilbas, B.S. and Sahin, A.Z. 2002. Development of a new drying correlation for practical applications. *International Journal of Energy Research*, 26, 3, 245–251.

34. Dincer, I. and Hussain, M.M. 2004. Development a new Biot number and lag factor correlation for drying applications. *International Journal of Heat and Mass Transfer*, 47, 4, 653–658.

35. Ruiz-Cabrera, M.A. et al. 1997. The effect of path diffusion on the effective diffusivity in carrot slabs. *Drying Technology*, 15, 1, 169–181

36. Dincer, I. 1997. *Heat Transfer in Food Cooling Applications*. Taylor & Francis, Washington, DC.

37. Simal, S., Rosselló, C., Berna, A. and Mulet, A. 1998. Drying of shrinking cylinder-shaped bodies. *Journal of Food Engineering*, 37, 426–435.

38. McLaughlin, C.P. and Magee, T.R.A. 1999. The effects of air temperature, sphere diameter, and puffing with CO_2 on the drying of potato spheres. *Drying Technology*, 17, 1&2, 119–136.

11 Mathematical Modeling Spray Dryers

Timothy A.G. Langrish
University of Sydney

CONTENTS

11.1 INTRODUCTION

The challenge of whether or not sufficient research had been done on spray drying was posed many years ago by Bahu,[1] who noted the need for mathematical modeling to understand the complex aspects of this type of equipment. In particular, the flow patterns in spray dryers are complex, and computational fluid dynamics (CFD) offers a prospect of addressing this basic aspect. A fundamental

301

challenge in scaling up spray dryers has been identified by Oakley.[2] The large number of parameters undergoing changes in spray drying including the chamber diameter, droplet diameters, atomizer dimensions and air velocity, makes dynamic scaling of the whole spray drying system virtually impossible. Again, this complexity points to the use of CFD as a scale-up tool.

CFD for spray drying almost inevitably involves the use of a turbulence model, since the chamber Reynolds number (cf. −2000+) is usually in the turbulent regime. Solving the fundamental challenges in turbulence modeling has also eluded Nobel-prize winning physicists. Even the advances in computer power are unlikely to address this challenge for at least another 20 years. This is not to say that progress in understanding turbulence models cannot be made, and this chapter will review some of this progress. The current status of turbulence modeling and CFD is somewhat of a contrast to the corresponding situation for stress modeling and finite element analysis (FEA), where a well-trained graduate can do sensible and rigorous work with a reasonable level of training. This situation cannot be emphasized too heavily. Industrialists tend to assume, or want to assume, that CFD can be treated automatically, like FEA, but it is unlikely to do so for the near future. In other words, CFD typically needs a higher level of understanding for appropriate application, usually involving postgraduate training.

While flow patterns are a basic aspect of spray dryer behavior, other basic aspects contribute to the complexity of spray drying. These other aspects involve heat and mass transfer, including basic drying kinetics, reaction engineering, particle technology, and process control. These aspects mean that there are a very wide range of applications and challenges in these applications, suggesting that significant future development may be expected.

Notwithstanding the value of CFD, other simpler and faster approaches may be fit for specific purposes, since Fletcher et al.[3] have shown that properly converged CFD simulations for the complex and transient flow patterns in spray dryers may take weeks to months to complete. These other approaches, including equilibrium-based mass and energy balances for well-mixed systems and parallel-flow simulations for tall-form spray dryers, will also be reviewed here.

11.2 SPRAY DRYER GEOMETRIES AND THE ROLE OF COMPUTATIONAL FLUID DYNAMICS (CFD)

Spray dryers can be divided into two basic types, short-form and tall-form designs Figure 11.1). Tall-form designs are characterized by height-to-diameter aspect ratios of greater than 5:1. This feature results in a significant plug-flow zone inside these dryers, as reported by Keey and Pham.[4] Short-form dryers, in which the height-to-diameter ratios are typically around 2:1, are most common for a variety of reasons. A significant one is the ease of accommodating the comparatively flat spray disk from a rotary atomizer (the most common industrial type of atomizer because of its flexibility) without giving excessive wall deposition. An implication of this design difference is the complexity of flow patterns in these dryers, as noted before, leading to the use of CFD for addressing the basic aspect of predicting the flow patterns.

The complexity of the flow patterns observed in short-form dryers is greater than that in tall-form dryers. Many short-form dryers have no plug-flow zone and a wide range of gas residence times. Arbitrary sequences of well-mixed and plug-flow stages together with bypasses to residence time distributions from helium-injection tracer measurements were the main features of the early attempts at modelling the gas flow patterns.[6,7] The complexity of the fitted sequences was significant (a 7-m diameter dryer, height 15 m, about six well-mixed stages in series;[6] a 6.7-m diameter dryer, height 24 m, five well-mixed stages with complicated bypass connections[7]). Both of these dryers were counter-current designs commonly used for the drying of detergents. At this stage, however, the mathematical modeling of spray dryer performance was limited to treating the equipment as parallel flow or well mixed and equilibrium-limited. These approaches still have some value today as computationally-modest methods for limited applications, where the complexity of the gas and particle flow patterns does not limit the equipment performance too significantly. Such applications include situations where the drying performance of the equipment is limited by equilibrium between the outlet gas and

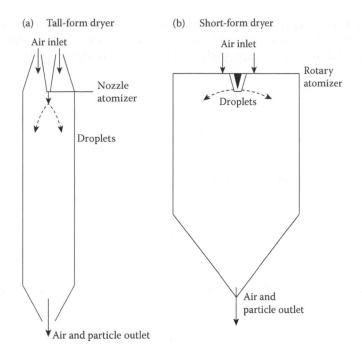

FIGURE 11.1 A classification of spray-dryer geometries. (From Langrish, T.A.G., and Fletcher, D.F., *Chemical Engineering and Processing*, 40, 345, 2001. With permission.)

the outlet solids.[8] Also included are situations where inertial impaction of particles with the walls dominates the deposition behavior, frequently in small-scale dryers.[9] The approaches both to parallel flow and to well-mixed, equilibrium-limited, simulations will now be described in more detail.

11.3 PARALLEL-FLOW DESIGN EQUATIONS

Tall-form dryers may be modeled, to a first approximation, by treating them as parallel-flow dryers. A clear description of equations for these situations is given in Truong et al.[10] and is summarized below. The approach, and the equations in it, was originally reported by Keey and Pham.

11.3.1 DROPLET TRAJECTORY EQUATIONS

The droplet trajectory equations are droplet axial, radial and tangential momentum balances. U_p and U_a represent the velocity (m s^{-1}) of the particles and the air, respectively. The axial distance from the atomizer is h (m). The subscripts x, r, and t represent the axial, radial and tangential components, respectively and h is the axial distance from the atomizer.

$$\frac{dU_{px}}{dh} = \left[\left(1 - \frac{\rho_a}{\rho_p} \right) g - \frac{3}{4} \frac{\rho_a C_D U_R (U_{px} - U_{ax})}{\rho_p d_p} \right] \frac{1}{U_{px}} \tag{11.1}$$

$$\frac{dU_{pr}}{dh} = \left[-\frac{3}{4} \frac{\rho_a C_D U_R (U_{pr} - U_{ar})}{\rho_p d_p} \right] \frac{1}{U_{px}} \tag{11.2}$$

$$\frac{dU_{pt}}{dh} = \left[-\frac{3}{4} \frac{\rho_a C_D U_R (U_{pt} - U_{at})}{\rho_p d_p} \right] \frac{1}{U_{px}} \tag{11.3}$$

Here ρ is the density (kg m^{-3}), d_p is the droplet diameter (m), U_R define is the relative velocity between the droplet and the air (m s^{-1}), and C_D is the drag coefficient. The subscripts a and p refer to the air and the particle or droplet, respectively. U_R and C_D are calculated as follows:[11]

$$U_R = \sqrt{\left(U_{px} - U_{ax}\right)^2 + \left(U_{pr} - U_{ar}\right)^2 + \left(U_{pt} - U_{at}\right)^2} \tag{11.4}$$

$$C_D = \frac{24}{Re_p}\left(1 + 0.15\,Re_p^{\,0.687}\right) \tag{11.5}$$

The particle Reynolds number is defined as:

$$Re_p = \frac{\rho_a U_R d_p}{\mu_a} \tag{11.6}$$

Here μ is air viscosity (kg m^{-1} s^{-1}).

The radial distance, r, of droplets as a function of axial distance from the atomizer is estimated as follows:

$$\frac{dr}{dh} = \frac{U_{pr}}{U_{px}} \tag{11.7}$$

11.3.2 DROPLET MASS BALANCE EQUATIONS

Based on the concept of a characteristic drying curve, the unsteady-state mass balance for the droplet can be stated as follows:[10]

$$\frac{dm_p}{dh} = -\xi \frac{A_p K_p}{U_{px}}\left(p_{vs} - p_{vb}\right) \tag{11.8}$$

Here m_p is the mass of the particle or droplet (kg), ξ is the relative drying rate, A_p is the droplet surface area (m^2), K_p is the mass-transfer coefficient (partial pressure based) (kg m^{-2} s^{-1} Pa^{-1}), p_{vs} is the partial pressure of the surface of the droplet (Pa), and p_{vb} is the partial pressure of water vapour in the bulk air (Pa). A_p can be calculated as follows:

$$A_p = \pi d_p^2 \tag{11.9}$$

The droplet diameter is expected to change due to shrinkage. The droplet diameter, d_p, is updated based on the assumption of balloon shrinkage without crust or skin formation. It has been suggested that free shrinkage of the sodium chloride droplets is a reasonable assumption.[12] Unlike milk droplets, sodium chloride droplets do not form a skin.

$$d_p = d_{pi}\left(\frac{\rho_{pi} - 1000}{\rho_p - 1000}\right)^{1/3} \tag{11.10}$$

$$\rho_p = \frac{1 + X}{1 + X\dfrac{\rho_s}{\rho_w}}\,\rho_s \tag{11.11}$$

Here, the variables that have not been defined previously are ρ_p, the particle density, ρ_w, the water density, d_{pi}, the initial droplet diameter (m), ρ_{pi}, the initial droplet density (kg m^{-3}), and X is the moisture content (dry-basis). The gas-phase mass-transfer coefficient is defined by the following equations:

$$K_p = \frac{M_w K_m}{M_a P} \tag{11.12}$$

$$K_m = \frac{\rho_a D_v \text{Sh}}{d_p} \tag{11.13}$$

Here, K_p is the mass-transfer coefficient (partial pressure based, in kg m^{-2} s^{-1} Pa^{-1}), M_w is the molecular weight of water (g mol^{-1}), M_a is the molecular weight of air (g mol^{-1}), K_m is the mass-transfer coefficient (kg m^{-2} s^{-1}), D_v is the diffusivity of water in air (m^2 s^{-1}), Sh is the Sherwood number and P is the total pressure (Pa). The diffusivity can be estimated from the equation:[13]

$$D_v = \frac{1.17564 \times 10^{-9} \times T_{abs}^{1.75} \times 101325}{P} \tag{11.14}$$

The Sherwood number is calculated from the equation:

$$\text{Sh} = 2.0 + 0.6\text{Re}_p^{0.5} \cdot \text{Sc}^{0.33} \tag{11.15}$$

The Schmidt (Sc) and Sherwood (Sh) numbers are defined below:

$$\text{Sc} = \frac{\mu_a}{\rho_a D_v}, \qquad \text{Sh} = \frac{K d_p}{D_v} \tag{11.16}$$

Here, T_{abs} is the absolute temperature of the droplet or particle (K) and Sc is the Schmidt number.

11.3.3 DROPLET HEAT BALANCE EQUATIONS

The unsteady-state heat balance for the droplet or particle is:[10]

$$\frac{dT_p}{dh} = \frac{\pi d_p k_a \text{Nu}(T_a - T_p) + \frac{dm_p}{dh} U_{px} H_{fg}}{m_s(C_{ps} + XC_{pw})U_{px}} \tag{11.17}$$

The Nusselt and Prandtl numbers are calculated from the equations:[14]

$$\text{Nu} = 2.0 + 0.6\text{Re}_p^{0.5} \text{Pr}^{0.33} \tag{11.18}$$

$$\text{Pr} = \frac{C_{pa}\mu_a}{k_a} \tag{11.19}$$

Other product and particle properties are calculated from the equations:

$$H_{fg} = 2.792 \times 10^6 - 160 T_{abs} - 3.43 T_{abs}^2 \tag{11.20}$$

$$m_s = \frac{X m_p}{100} \tag{11.21}$$

Here, T is the temperature, k_a is the thermal conductivity of humid air (W m^{-1} K^{-1}), Nu is the Nusselt number, H_{fg} is the latent heat of water evaporation (J kg^{-1}), m_s is the mass of solids in the droplet (kg), and C_p is the specific heat capacity (J kg^{-1} K^{-1}).

11.3.4 MASS AND ENERGY BALANCE EQUATIONS FOR DRYING MEDIUM

The mass-balance equation for the drying air is:

$$\frac{dY_b}{dh} = \frac{\sum_{\text{droplets}}\left(-\dfrac{dm_p}{dh}\right) n_{\text{droplets}}}{G} \tag{11.22}$$

Here, Y is the gas humidity (kg kg^{-1}), G is the mass flow rate of the dry air (kg s^{-1}), and n_{droplets} is the flow rate of droplets (number s^{-1}). The corresponding heat-balance equation for the drying air is:

$$\frac{dH_h}{dh} = -\frac{1}{G}\left(\sum_{\text{droplets}}\left(m_s\left(C_{\text{ps}} + XC_{\text{pw}}\right)\frac{dT_p}{dh}\right) - \frac{UA(T_a - T_{\text{amb}})}{L}\right) n_{\text{droplets}} \tag{11.23}$$

Here, H_h is the enthalpy of the humid air (J kg^{-1}), UA is the heat-transfer coefficient for heat loss from the dryer, T_{amb} is the ambient temperature, and L is the length of the spray-drying chamber.

11.4 HEAT AND MASS BALANCES FOR WELL-MIXED DESIGNS

Short-form dryers of a sufficient size that the outlet gas is close to being in equilibrium with the outlet particles[8] may be modeled, to a first approximation, by treating them as well-mixed dryers. Mass and energy balances may be used to solve the model.[15,16] Equilibrium between the gas and the solids means that the gas and solids outlet temperatures will be close ($T_{\text{So}} = T_{\text{Go}}$). Equilibrium also means that the gas humidity and gas and solids temperatures will affect the solids moisture content, as follows. This is not a limitation of the approach because it is easy to insert a fixed offset between the two temperatures into the approach. The approach is best described in the following set of exercises.

11.4.1 WORKED EXAMPLE: OVERALL AIM

The overall aim of this exercise is to use mass and energy balances to calculate the outlet temperature and moisture content in a dryer. If the dryer is well mixed, then these outlet conditions represent the conditions for the solids inside the equipment, so the stickiness or otherwise of the solids can be assessed.

11.4.2 ENERGY ENTERING THE DRYER

Energy enters the dryer mainly through the hot air, also through the liquid.

$$\text{Energy flow rate} = \text{enthalpy} \times \text{mass flow rate}$$

Enthalpy of air

$$H_a = C_{\text{Pa}}\left(T_a - T_{\text{ref}}\right) + Y\left[\lambda + C_{\text{Pv}}\left(T_a - T_{\text{ref}}\right)\right] \tag{11.24}$$

C_{Pa} = specific heat capacity of dry air $\approx$ 1 kJ kg^{-1} K^{-1}

T_a = air temperature, °C

T_{ref} = reference temperature, 0°C

Y = air humidity, kg water/kg dry air

λ = latent heat of vaporisation $\approx$ 2500 kJ kg^{-1}

C_{Pv} = specific heat capacity of pure water vapour $\approx$ 1.8 kJ kg^{-1} K^{-1}

Enthalpy of liquid water (same pattern for solids in water)

$$H_l = C_{Pl}(T_l - T_{ref}) \tag{11.25}$$

C_{Pl} = specific heat capacity of liquid water $\approx$ 4.2 kJ kg^{-1} K^{-1}

T_l = water temperature, °C

11.4.2.1 Example for Energy Entering the Dryer

In an industrial spray dryer, 2.5 kg s^{-1} liquid, 50% water, 50% milk solids (specific heat capacity 1.5 kJ kg^{-1} K^{-1}) at 60°C enters with 21 kg s^{-1} air having a temperature of 215°C and a humidity of 0.006 kg water/kg dry air. What is the total energy flow rate into the dryer?

Enthalpy of air

$$H_a = C_{Pa}(T_a - T_{ref}) + Y[\lambda + C_{Pv}(T_a - T_{ref})]$$

$$= 1\,\text{kJ kg}^{-1}\,\text{K}^{-1}\,(215°\text{C} - 0°\text{C})$$

$$+ 0.006\,\text{kg kg}^{-1}\,[2500\,\text{kJ kg}^{-1} + 1.8\,\text{kJ kg}^{-1}\,\text{K}^{-1}\,(215°\text{C} - 0°\text{C})]$$

$$= 232\,\text{kJ kg}^{-1}\,(\text{dry air})$$

Dry air flow rate = 21 kg s^{-1} × 1 kg dry air/1.006 kg total air (if the humidity is 0.006 kg water/kg dry air) = 20.9 kg s^{-1}

Energy flow rate with air = 232 kJ kg^{-1} × 20.9 kg s^{-1} = 4850 kW = 4.85 MW

Enthalpy of water

$$H_l = C_{Pl}(T_l - T_{ref}) = 4.2\,\text{kJ kg}^{-1}\,\text{K}^{-1}\,(60°\text{C} - 0°\text{C}) = 252\,\text{kJ kg}^{-1}$$

Water flow rate = 0.5 × 2.5 kg s^{-1} = 1.25 kg s^{-1}

Energy flow rate with water = 252 kJ kg^{-1} × 1.25 kg s^{-1} = 315 kW

Enthalpy of milk solids

$$H_s = C_{PP}(T_p - T_{ref}) = 1.5\,\text{kJ kg}^{-1}\,\text{K}^{-1}\,(60°\text{C} - 0°\text{C}) = 90\,\text{kJ kg}^{-1}$$

Milk solids flow rate = 0.5 × 2.5 kg s^{-1} = 1.25 kg s^{-1}

Energy flow rate with milk solids = 90 kJ kg^{-1} × 1.25 kg s^{-1} = 112.5 kW

Total energy flow rate entering = 4850 + 315 + 113 = 5280 kW

11.4.3 Energy Leaving the Dryer

Energy leaves the dryer mainly through the cooler, moister, air, also through the solids (which contain some moisture).

The outlet solids are close to be in equilibrium with the outlet gas, so the temperature of the gas and the solids may be assumed to be the same and the outlet moisture content of the solids can be assumed to be equal to the equilibrium moisture content of solids in contact with the outlet gas.

All the moisture that is evaporated from the solids is taken up by the gas, so a mass balance allows the outlet moisture content of the solids to be related to the outlet humidity of the gas.

11.4.3.1 Example for Energy Leaving the Dryer

Taking the industrial spray dryer example a step further, the unknowns are the outlet solids moisture content (X_o), the outlet solids temperature (T_{Po}), the outlet gas temperature (T_{Go}) and the outlet gas humidity (Y_o).

The inlet solids moisture content is known ($X_i = 1$ kg water/kg solids), as is the inlet gas humidity ($Y_i = 0.006$ kg water/kg dry gas). The dry solids flow rate in (F) is the milk solids flow rate of 1.25 kg s^{-1}, while the dry gas flow rate (G) has already been calculated as being equal to 20.9 kg s^{-1}. Hence the mass balance over the dryer gives the following equation:

$$G(Y_o - Y_i) = F(X_i - X_o)$$

$$Y_o = Y_i + \frac{F}{G}(X_i - X_o) \tag{11.26}$$

$$= 0.006 + \frac{1.25}{20.9}(1 - X_o) = 0.006 + \frac{(1 - X_o)}{16.7}$$

From the above example, the total energy flow rate entering the dryer is 5280 kW. The energy leaving the dryer is in the air and in the solids.

In the air

$$H_{Go} = C_{Pa}(T_{Go} - T_{ref}) + Y_o[\lambda + C_{Pv}(T_{Go} - T_{ref})]$$

$$= 1 \text{ kJ kg}^{-1} \text{ K}^{-1} (T_{Go} - 0°C)$$

$$+ Y_o [2500 \text{ kJ kg}^{-1} + 1.8 \text{ kJ kg}^{-1} \text{ K}^{-1} (T_{Go} - 0°C)] \tag{11.27}$$

$$= T_{Go} + Y_o [2500 + 1.8 T_{Go}] \text{ kJ kg}^{-1}$$

In the solids

$$H_{So} = C_{PP}(T_{po} - T_{ref}) + X_o C_{Pl}(T_{po} - T_{ref})$$

$$= 1.5 \text{ kJ kg}^{-1} \text{ K}^{-1} (T_{po} - 0°C) + X_o \, 4.2 \text{ kJ kg}^{-1} \text{ K}^{-1} (T_{po} - 0°C) \tag{11.28}$$

$$= 1.5 T_{po} + X_o \, 4.2 T_{po} \text{ kJ kg}^{-1}$$

The dry gas (20.9 kg s^{-1}) and dry solids (1.25 kg s^{-1}) flow rates are the same at the inlet and the outlet, so

$$5280 = 1.25[1.5 T_{So} + X_o \, 4.2 T_{So}] + 20.9[T_{Go} + Y_o (2500 + 1.8 T_{Go})] \tag{11.29}$$

The relative humidity (ψ) of the outlet gas (actual vapour pressure (p_v) divided by the saturation vapour pressure) needs to be calculated first, from the gas temperature (T_{Go}) and the gas humidity (Y_o).

$$\psi = \frac{p_v}{p_{vsat}} \qquad (11.30)$$

11.4.4 Explanation of Relative Humidity Calculation

The saturation vapour pressure (p_{vsat}) is the maximum vapour pressure at the outlet gas temperature (T_{Go}), and this vapour pressure may be calculated using the Antoine equation. For water, one version of the Antoine equation is:

$$p_{vsat}(Pa) = 133.3 \exp\left(18.3036 - \frac{3816.44}{T_{po}(^\circ C) + 229.02}\right) \qquad (11.31)$$

The actual vapour pressure (p_v) may be related to the outlet gas humidity (Y_o) by[15,16]

$$Y = 0.622 \frac{p_v}{P_{atm} - p_v} \qquad (11.32)$$

Why? Should a mixture of m_G kg of air and m_v kg of water vapour behave as an ideal gas, one has

$$p_G V = \frac{m_G}{M_G} RT \qquad (11.33)$$

$$p_v V = \frac{m_v}{M_v} RT \qquad (11.34)$$

in which p_G and p_v are the partial pressure of air and water, respectively, M_G and M_v are the molecular weights of air and water, respectively, R is the gas constant and T is the absolute temperature. These pressures may be added together to give the total pressure, P_{atm}:

$$p_G + p_v = P_{atm}$$
$$p_G = P_{atm} - p_v \qquad (11.35)$$

Backsubstituting into Equation 11.33 gives:

$$(P_{atm} - p_v)V = \frac{m_G}{M_G} RT \qquad (11.36)$$

$$p_v V = \frac{m_v}{M_v} RT \qquad (11.37)$$

The ratio of m_v to m_G is the gas humidity, Y_o, so

$$Y_o = \frac{M_v}{M_G} \frac{p_v}{P_{atm} - p_v} \qquad (11.38)$$

as required.

Rearranging Equation 11.32 to calculate the actual vapour pressure (p_v) from the gas humidity (Y_o) gives:

$$p_v = \frac{(Y_o/0.622)P_{atm}}{[1+(Y_o/0.622)]} \tag{11.39}$$

Equations 11.30, 11.31 and 11.39 give the relative humidity (ψ) from the gas temperature (T_{Go}) and the gas humidity (Y_o). This relative humidity, together with the gas and solids temperatures ($T_{So} = T_{Go}$), is used to estimate the equilibrium moisture content (X_{emc}), which is an estimate of the outlet moisture content (X_o). There are different equations for the equilibrium moisture content for different materials. For skim milk powder, one sorption isotherm is:[17]

$$X_o = X_{emc} = 0.1499\exp\left[-2.306\times10^{-3}(T_{So}+273.15)\cdot\ln\left(\frac{1}{\psi}\right)\right] \tag{11.40}$$

11.4.5 SOLUTION OF EQUATIONS

Hence the equations to be solved are:

$$Y_o = 0.006 + \frac{(1-X_o)}{16.7} \tag{11.26}$$

$$X_o = 0.1499\exp\left[-2.306\times10^{-3}(T+273.15)\cdot\ln\left(\frac{1}{\psi}\right)\right] \tag{11.40}$$

$$\psi = \frac{p_v}{p_{vsat}} \tag{11.30}$$

$$p_{vsat} = 133.3\exp\left(18.3036 - \frac{3816.44}{T+229.02}\right) \tag{11.31}$$

$$p_v = \frac{(Y_o/0.622)P_{atm}}{[1+(Y_o/0.622)]} \tag{11.39}$$

$$1.25[1.5T + X_o\ 4.2\ T] + 20.9[T + Y_o\ (2500 + 1.8\ T)] = 5280 \tag{11.29}$$

In these equations, the six unknown variables are X_o, Y_o, ψ, T, p_v, and p_{vsat}, and there are six equations, so one way to solve this set of equations is to use the following iterative procedure.

1. Guess X_o, say 0 kg kg^{-1}.
2. Use Equation 11.26 to calculate Y_o, here 0.06588 kg kg^{-1}.
3. Use Equation 11.29 to calculate T, here 72.9°C.
4. Use Equation 11.39 to calculate p_v, here 9704 Pa.
5. Use Equation 11.31 to calculate p_{vsat}, here 38401 Pa.
6. Use Equation 11.30 to calculate ψ, here 0.2527.
7. Use Equation 11.40 to calculate a new value of X_o, here 0.0500 kg kg^{-1}. Then, return to step 1 until this iterative procedure converges.

Repeating this iterative procedure converges quickly to an outlet moisture content of 0.0410 kg kg^{-1} and an outlet gas and solids temperature of 77.6°C.

The sticky-point curve can be represented by the following equation:[18]

$$T_{st} = \frac{[(1-X_o)T_{glact} + X_o T_{gwater}]}{(1+7.48X_o)} + 23.3 \qquad (11.41)$$

where T_{st} is the sticky-point temperature (°C), T_{glact} is the glass transition temperature of lactose (101°C) and T_{gwater} is the glass transition temperature of water (−137°C).

Hence, at the outlet moisture content of 0.0410 kg kg^{-1}, the sticky-point temperature is given by

$$T_{st} = \frac{[(1-X_o)T_{glact} + X_o T_{gwater}]}{(1+7.48X_o)} + 23.3$$
$$= \frac{[(1-0.0410)101 + (0.0410) - 137]}{(1+7.48 \cdot 0.0410)} + 23.3 = 93.1°C \qquad (11.42)$$

Since the sticky-point temperature is above the current outlet temperature, the skim milk is not in the sticky-point region and is unlikely to cause wall deposition problems.

11.5 COMPUTATIONAL FLUID DYNAMICS (CFD)

Reay[19] identified some weaknesses in parallel-flow modeling approaches with adjustments (such as axial dispersion), which require accurate measurements using helium injection and flow visualisation equipment to be performed in existing dryers, and as such are unsuitable for designing new drying chambers. The data may be fitted equally well by a variety of zone sequences (well-mixed, plug flow, bypass) may fit the data equally well, and it is then difficult to decide which one is most appropriate. The effects of varying chamber geometry or operating parameters, which are likely to have significant effects on the flow patterns, cannot be assessed using these empirical techniques. In turn, these significant effects on the flow patterns will affect the product moisture content and wall deposition rates in the dryers.

The need to use more fundamental approaches for predicting the flow patterns was recognized implicitly by Crowe[20] in his use of CFD as a technique. The central features of these techniques have not changed, including the use of discrete approximations to the time-averaged conservation equations and the application of a concept known as the particle source-in-cell or discrete droplet model. This concept is essentially a matter of treating the gas as a continuum (Eulerian) and the spray as a tracked phase in a Lagrangian manner. The coupling between the spray and gas frequently occurs both ways (two-way gas-particle coupling). The axial, radial and tangential components of the gas velocities are initially calculated by neglecting the influence of the spray. Tracking of a large number of droplets is then carried out through the gas inside the dryer. The range of droplet sizes leaving the atomizer is chosen so that the sum of the flow rates of each droplet size equals the total liquid flow rate. Following this calculation, the heat, mass and momentum transfer rates from the droplets to the gas phase are estimated, and these rates are feed back to the calculation of the components of the gas velocities again.

This trend toward the use of CFD has been made possible by the development of powerful workstations at accessible cost. However, many issues remain significant, including the selection of the turbulence model.

11.5.1 TURBULENCE MODELING AND TRANSIENT FLOW PATTERNS

As noted by Bradshaw et al.,[21] *"No current turbulence model gives results of good engineering accuracy for the full range of flows tested."* This means that no final determination can be made regarding the "best" or the only approach to turbulence modeling.

Oakley et al.[22] found that the predictions of the air flow patterns were sensitive to the values of the turbulence parameters selected at the annular air inlet. The experimental measurements of these parameters are not normally available and so either they must be treated as fitting parameters (which would reduce the predictive power of the model for design purposes) or they must be obtained from a separate numerical simulation of the air inlet. In a subsequent paper, Oakley and Bahu[23] reported the results of a modified numerical simulation which incorporated a separate simulation of the air inlet (as they had previously recommended) and a differential Reynolds stress model for turbulence, a model that might normally be considered appropriate for strongly swirling conditions. Even though an improvement was noted in the agreement between the model predictions and the experimental measurements, the use of the complex differential Reynolds stress model involved substantially greater computational effort than the previous k–ε model.

This led Livesley et al.[24] to revert to the use of the k–ε model when predicting particle sizes and mean axial gas and particle velocities inside industrial spray dryers concerning solutions and slurries. In the dryers studied in their work, which ranged from 0.7 m diameter by 1.4 m high to 5.4 m diameter by 10 m high, satisfactory agreement was found between the predictions of the numerical simulation and the experimental measurements. Thus, in spite of the limitations of the k–ε model for modeling turbulence in spray dryers, this model currently represents an acceptable compromise between accuracy and computational effort in many situations. Subsequently, workers such as Huang et al.[25] have also used the k–ε model for these reasons. However, some workers, such as Bayly et al.,[26] advocate the use of the differential Reynolds stress model for better predicting the shape of the velocity profile (a Rankine vortex) at the exit of a tall-form dryer than the k–ε model (as in Harvie et al.[27]).

Virtually all the above studies have been for steady-state simulations. Prior to the recent work on transient flow behavior in spray dryers, the origin of the unsteady flow behavior in spray dryers was poorly understood. The motivation for this work came from careful observations and hot-wire measurements of the flow patterns inside the dryers,[28] which showed some coherent behavior among the superficially chaotic flow patterns in this equipment. There were also indications from early numerical simulations using CFD, assuming steady axi-symmetric flow, that the flow patterns were not steady or axi-symmetric, but gave predicted oscillations (Figure 11.2) that suggested the presence of a three-dimensional core precessing around the central axis of the dryer.[22] For transient simulations, the situation regarding the choice of turbulence models may be subtly different. When assessing the ability of CFD simulations to predict oscillations in the flow behavior in spray dryers, Guo et al.[29] assessed the suitability of turbulence models for this application in the following way. They simplified a spray dryer down to a key geometrical feature, namely a sudden expansion (at the inlet of the dryer). They took this simplification even further, by reducing the three-dimensionality

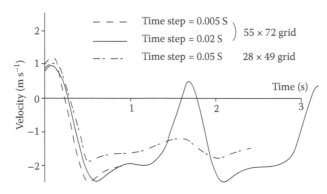

FIGURE 11.2 Predictions of time-dependent flows in a pilot-scale spray dryer. (From Oakley, D.E., Bahu, R.E., and Reay, D., *Proceedings of the 6th International Drying Symposium (IDS '88)*, Versailles, France, ed. M. Roques, OP 373, 1988.)

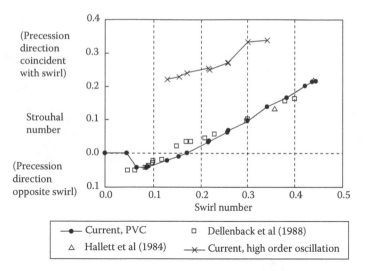

FIGURE 11.3 Variation of Strouhal number with swirl number (expansion ratio 1.96, Re = 10^5). (From Guo, B., Langrish, T.A.G., and Fletcher, D.F., *AIAA Journal*, 39, 96, 2001. With permission.)

of the spray dryer down to two dimensions. A key benefit of this two-dimensional situation is that a considerable body of experimental data is available on two-dimensional sudden expansions from the area of slab casting. For a wide range of data sets, Guo et al.[29] came to the rather surprising conclusion that the k–ε model outperformed the differential Reynolds stress model for predicting the regularity and frequency of the oscillations.

In three dimensions, some data are available from work on swirl-stabilized combustion systems, for low to medium swirl regimes. Here, Guo et al.[30] have found good agreement between experiment and simulation for the oscillation frequencies with the k–ε model (Figure 11.3). The images in Figure 11.4 show the helical nature and the increasing complexity as the swirl number increases, with the precession direction indicated relative to the swirling direction.

Bearing the additional geometrical complexity of a spray dryer compared with a sudden expansion in mind, numerical simulations (using CFX5) of the flow inside a 0.29 m diameter, 0.5 m tall, hydraulic model of an industrial spray dryer were performed by LeBarbier et al.[31] The precession of the central jet of inlet air may be connected to the occurrence of wall deposits inside this equipment. Experiments showed that the flow was strongly time-dependent, with two characteristic frequencies that depended on the angle of the inlet swirl vanes over a range of angles from 0 to 40°. The main frequency of precession of the central jet occurred over a time scale of a few seconds, and image analysis allowed the characteristic frequencies of the precession to be quantitatively assessed.

The simulations with CFX5 predicted the same flow behavior as observed during the experiments, showing strongly time-dependent flow behavior for all the swirl vane angles. Greater swirl caused more expansion of the central jet, with the vortex breaking down at a swirl vane angle of 40° both in the simulations and the experiments. An example of the simulation results obtained for the three different swirl vane angles for an arbitrary time of 1.5 sec after the start of the transient simulation is shown in Figure 11.5 for a vertical slice through the central axis of the chamber. This figure shows how the flow changes with the swirl angle. For the simulation without swirl, the central jet was very narrow and moved slowly. For 25° swirl, the central jet moved more vigorously. Moreover, the profile of the velocities at the entrance of the outlet pipe was different, and the central jet did not flow directly to the outlet pipe. The simulations with the 40° swirl angle showed a flow that changed completely in comparison with the simulation done without swirl (see Figure 11.6). The profile of the velocities at the entrance of the outlet pipe showed a large curvature (see Figure 11.6a). This phenomenon already existed for 25° swirl, but was much more pronounced for 40° swirl. The central jet moved so strongly that it moved out of the visualisation plane, as shown in Figure 11.6b.

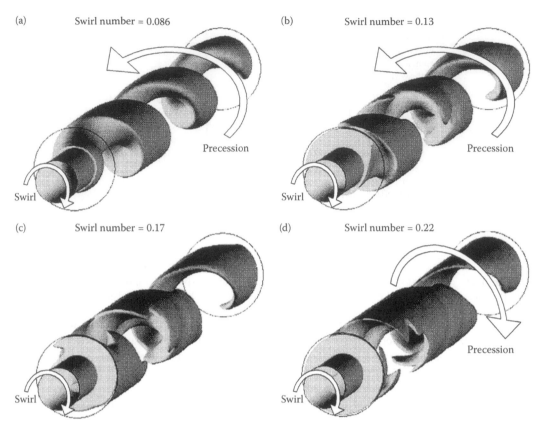

FIGURE 11.4 Iso-surfaces of velocity for different swirl numbers (expansion ratio 1.96, Re = 10^5). (From Guo, B., Langrish, T.A.G., and Fletcher, D.F., *AIAA Journal,* 39, 96, 2001. With permission.)

The simulated and observed flow behavior was also quantitatively compared. The Strouhal numbers found by the simulations and by the experiments are listed in Table 11.1. The results obtained by the simulations and by the experiments are in the same range, and for no swirl and a swirl angle of 40° are within the experimental error, with the difference at a swirl angle of 25° being only slightly outside the error bounds. It appears that the agreement is better for no swirl, and this may correspond to the known limitation of the k–ε turbulence model for highly swirling flows. Nevertheless, the agreement is very reasonable, and confirms the good agreement achieved in the simulations of Guo et al.[30] for the data of Dellenback et al.[32] with the k–ε model for a small (two fold) expansion ratio over a range of swirl numbers up to vortex breakdown. The underlying reason for this good agreement, in spite of the known limitations of the k–ε model, appears to be that the flow instabilities that lead to precession occur (within the chamber) very close to the expansion (here into the drying chamber). Hence the inability of the k–ε model to predict the size of the recirculation zones in spray dryers[22] or to predict the decay of swirl in a pipe is not too serious for this practical purpose. This in turn suggests that the k–ε turbulence model is suitable for estimating the frequencies of precession in spray dryers, where the precession may be linked to the occurrence of wall deposits, since precession deflects the central jet of air and particles toward the dryer walls.

Subsequent work (unpublished) has confirmed that the time dependent precession persists when particles are present in the flow, although the frequencies change by over an order of magnitude when particles are introduced. Following on from the work on sudden expansions, suggestions for mitigating the problem may be given. The effect of swirl varies with different expansion ratios. A slight swirl (e.g., $S = 0.2$ for an expansion ratio of five between the inlet tube and the main chamber

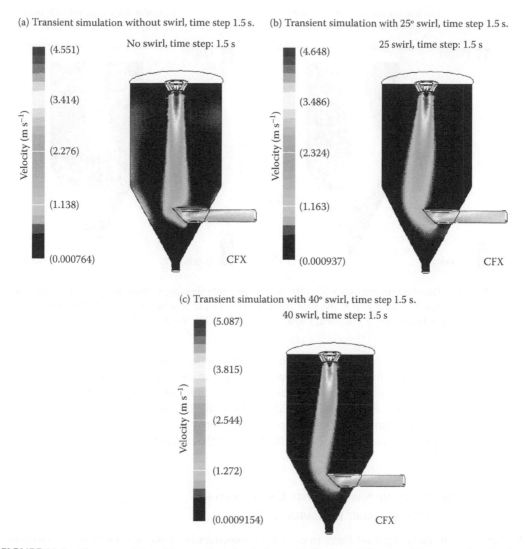

(a) Transient simulation without swirl, time step 1.5 s. (b) Transient simulation with 25° swirl, time step 1.5 s.

(c) Transient simulation with 40° swirl, time step 1.5 s.

FIGURE 11.5 Time step: 1.5 s of the transient simulations with the three different swirl vane angles showing vertical slices through the central axis of the chamber. The time step is the time from the start of the simulation, here showing corresponding times for simulations at different swirl angles. (From LeBarbier, C. et al., *Transactions of the Institution of Chemical Engineers*, 79, 260, 2001. With permission.)

TABLE 11.1

Comparison between the Frequencies of the Flow Patterns for Three Swirl Vane Angles Found by the Simulations and by the Experiments for a Real Spray Drying Geometry

Swirl Vane Angle (°)	Strouhal Number (Simulations)	Strouhal Number (Experiments)	Difference (%)
0	0.11 ± 0.02	0.10 ± 0.01	10
25	0.17 ± 0.02	0.13 ± 0.01	30
40	0.19 ± 0.03	0.22 ± 0.01	15

Source: From LeBarbier, C. et al., *Transactions of the Institution of Chemical Engineers*, 79, 260, 2001. With permission.

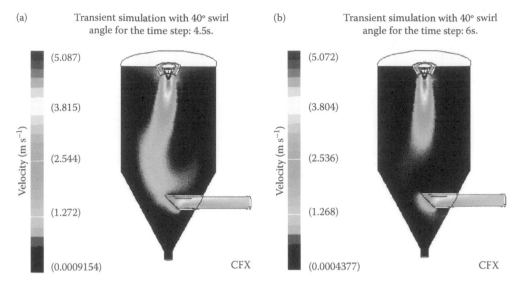

(a) Transient simulation with 40° swirl angle for the time step: 4.5s.

(b) Transient simulation with 40° swirl angle for the time step: 6s.

FIGURE 11. 6 Transient simulations with a 40° swirl vane angle showing vertical slices through the central axis of the chamber. The time step is the time from the start of the simulation. (From LeBarbier, C. et al., *Transactions of the Institution of Chemical Engineers,* 79, 260, 2001. With permission.)

of the expansion) tends to suppress the oscillations for a larger expansion, but may cause a stable flow to become unsteady for a small expansion. For an expansion ratio of about five, no steady symmetric flow pattern appears to be achievable. However, inlet swirl with an intensity below the cross-over swirl number ($S \approx 0.23$) may be used to stabilize the flow to a certain degree, since (1) the flapping oscillation is suppressed; (2) the precession frequency is decreased; and (3) the amplitude of the precessing vortices is reduced. These effects may make particle deposition more predictable, and easier to address.

11.5.2 A REVIEW OF THE NAVIER–STOKES EQUATIONS FOR COMPUTATIONAL FLUID DYNAMICS (CFD)

The problem of analysing fluid flows in complex geometries by numerical methods is commonly encountered by engineers. This science is known under the name of the CFD. Recently, the development of CFD software has permitted the analysis of more complex problems. Initially, this review will list the equations governing the conservation of mass and momentum. Subsequently the key elements of CFD will be explained and finally attention will be given to the turbulence modeling.

11.5.2.1 Equations for Fluid Flow

The fundamental equation is the conservation of mass:[33]

$$\frac{D\rho}{Dt} + \rho * \nabla \cdot \vec{v} = 0 \tag{11.43}$$

where D/Dt D/Dt is the substantial derivative. This equation is commonly known as the continuity equation. For incompressible flow, this reduces to:[33]

$$\nabla \cdot \vec{v} = 0 \tag{11.44}$$

For an incompressible, Newtonian fluid, the conservation of momentum is expressed by the Navier–Stokes equations, which in vector forms are[33]

$$\rho * \frac{D\vec{v}}{Dt} = -\vec{\nabla}p + \rho\vec{g} + \nabla \cdot (\mu\nabla\vec{v}) \tag{11.45}$$

In Cartesian coordinates, the x-component of Equation 11.45 is (with μ constant):[33]

$$\rho * \frac{Du}{Dt} = -\frac{\partial p}{\partial x} + \rho g_x + \mu\left(\frac{\partial^2 u}{\partial x^2} + \frac{\partial^2 u}{\partial y^2} + \frac{\partial^2 u}{\partial z^2}\right) \tag{11.46}$$

In the case of isothermal flow, no further equation is needed. An equation of conservation of energy is required when there is heat transfer, combustion or when the fluid is compressible. The total energy is given by:[33]

$$E = e + \frac{1}{2}v^2 + gz \tag{11.47}$$

where e is the internal energy. The conservation of energy is given by:[33]

$$\frac{\partial \rho E}{\partial t} + \nabla \cdot (\rho E \vec{v}) + \nabla \cdot (p\vec{v}) + \nabla \cdot \vec{q} = 0 \tag{11.48}$$

If the flow is laminar, then these equations are sufficient to solve the problem. However in case of turbulent flow, additional equations have to be used to model the Reynolds stresses, which arise in the momentum and energy equations. The turbulence modeling will be discussed later. All these differential equations can be written in the general form:[33]

$$\frac{\partial}{\partial t}(\rho\phi) + \nabla \cdot (\rho\vec{v}\phi) = \nabla \cdot (\Gamma_\phi \nabla_\phi) + S_\phi \tag{11.49}$$

The first term represents the transient effect: it is the unsteady term. The second term is the convection term and represents the convective transport of ϕ (1 = conservation of mass; u = axial component of velocity, etc). The third term is the diffusion term, with Γ_ϕ being the diffusion coefficient. The final term is known as the source term. The aim of CFD is to solve these equations. CFD has three key elements:[33]

- Grid generation
- Algorithm development
- Equations of motion

The grid generation and the algorithm development are very precise and well established. A brief overview will be given next. The turbulence modeling is the most uncertain part of CFD and will be discussed later.

11.5.2.2 Grid Generation and Algorithm Development

In order to solve the equations in an identified domain, a grid must be generated. Two main types of grids must be distinguished: structured grids and unstructured grids. In a structured grid, a cell is determined in regards to its neighbours. In an unstructured grid, no such relationships exist to determine which volumes are next to a given volume. The advantage of unstructured grids is their flexibility. The most commonly used volume is tetrahedral. In most CFD packages, a ready made system for unstructured grid generation is available. The steps in the generation of a grid are listed below:[33]

1. Define the different solids of the geometry.
2. Identify the key part of the geometry and apply constraints on them.
3. Define the surfaces to apply the boundary conditions.
4. Choose a compromise between a fine grid and an excessive number of cells.
5. Generate the grid.
6. Check the quality of the grid.

When the geometry is complex, the best way to mesh it is often to begin with a simple geometry and successively add more parts. The quality of the grid is essential for the quality of the solution. The Navier–Stokes equations are usually solved by using a finite volume method. In this method, the partial differential equations governing the flow of fluid (see Equation 11.49) are converted into a set of coupled algebraic equations with Gauss's theorem. A detailed explanation can be found in Fletcher.[33] These coupled algebraic equations are solved on each cell of the grid, over which all fluid variables are assumed to be constant. An iterative method must be used to solve the equations. At each step of the iteration a matrix equation, $Ax = b$ (representing the coupled algebraic equations, where A is a matrix, x the solution vector and b a vector), must be solved. Different methods can be used, including Newtonian methods, iterative methods or acceleration techniques.[33] The solution is considered to be converged when the residual, $r = Ax-b$, calculated with one of the previous methods, is as small as required. The norm of the residual is the square root of the sum of the squares of the elements of r divided by the number of elements. The user must set a criterion for convergence. When the residual is smaller than the criterion for convergence, the iterations are stopped and the solution is considered to be converged.

11.5.2.3 Turbulence Models and Equations

According to Hinze,[34] turbulence is defined as follows: "Turbulent fluid motion is an irregular condition of flow in which the various quantities show a random variation with time and space coordinates so that statistically distinct average values can be discerned." The phenomenon of turbulence is very complex, and the Navier-Stokes equations cannot be solved exactly by numerical methods for turbulent flows, as they can be in the case of laminar flows. In a turbulent flow, the range of length scales is very large, and no computer can solve the problem using the spatial resolution required to resolve all the turbulent length scales. The most commonly used approach is to perform turbulence modeling by a process of averaging. The Reynolds averaging can be either time averaging (stationary turbulence), spatial averaging (homogeneous turbulence), or ensemble averaging. In most problems the turbulence is inhomogeneous, in which case time averaging is used.

The velocity can be written as:

$$u = \bar{u} + u' \tag{11.50}$$

where $\bar{u}$ is the time averaged part and u' is the fluctuating part. The averaged part is:

$$\bar{u} = \frac{1}{T} \int_{t}^{t+T} u\,dt \tag{11.51}$$

The Reynolds-averaged version of the Navier–Stokes equations then become:

$$\frac{\partial \bar{u}}{\partial t} + \bar{u}\frac{\partial \bar{u}}{\partial x} + \bar{v}\frac{\partial \bar{u}}{\partial y} + \bar{w}\frac{\partial \bar{u}}{\partial z} = -\frac{1}{\rho}\frac{\partial \bar{p}}{\partial x} + \frac{1}{\rho}\frac{\partial}{\partial x}\left(\mu\frac{\partial \bar{u}}{\partial x} - \overline{\rho u'^2}\right)$$

$$+\frac{1}{\rho}\frac{\partial}{\partial y}\left(\mu\frac{\partial \bar{u}}{\partial y} - \overline{\rho u'v'}\right) + \frac{1}{\rho}\frac{\partial}{\partial z}\left(\mu\frac{\partial \bar{u}}{\partial z} - \overline{\rho u'w'}\right) \tag{11.52}$$

The terms of the form $\overline{\rho u'v'}$ are known as the Reynolds stresses (correlations of fluctuating velocity components). The overbar represents time averaging.

The averaging produces more degrees of freedom, so the problem is not closed. The aim of turbulence modeling is to model the Reynolds stresses in order to solve the problem. Hence turbulence modeling is known as solving the turbulence closure problem. Four common classes of turbulence models are:

1. Algebraic (zero-equation) models
2. One-equation models
3. Two-equation models
4. Stress transport models

The three first models use the Boussinesq eddy-viscosity hypotheses, which relate the Reynolds stresses to the turbulent shear stress, as follows:

$$-\overline{\rho u'v'} = \mu_t \frac{\partial \bar{u}}{\partial y} \tag{11.53}$$

where μt is called the turbulent or eddy viscosity.

11.5.2.3.1 Algebraic Models

As described in Fletcher,[33] Prandtl suggested the mixing length concept, which is a typical zero-equation model. If a particle of fluid is moved a distance l_m before it mixes with the fluid surrounding it, and the typical turbulence velocity is u_t, then by analogy to the molecular theory of gases.

$$\mu_t = \rho l_m |u_t| \tag{11.54}$$

If the velocity is defined in the following way,

$$u_t = u_{y+l_m} - u_y = l_m \frac{du}{dy} \tag{11.55}$$

Then the eddy viscosity will be:

$$\mu_t = \rho l_m{}^2 \frac{du}{dy} \tag{11.56}$$

The mixing length is estimated close to the wall:

$$l_m = \kappa\, y \tag{11.57}$$

where κ is the von Karman constant, with a typical value of 0.41, so

$$\frac{du}{dy} = \frac{u_t}{\kappa y} \tag{11.58}$$

Hence

$$\overline{u^+} = \frac{1}{\kappa}\ln y^+ + \text{constant}, \quad \text{with} \quad u^+ = \frac{u}{u_t} \tag{11.59}$$

This model is very useful for calculating the flow near the walls. Nevertheless it is incomplete for calculating complex flows, since it does not account for convection or diffusion of turbulence, and so inaccurately describes the mixing in complex flows.

11.5.2.3.2   One Equation Models

Turbulence energy equation models have been developed to incorporate non-local and flow history effects in the concept of an eddy viscosity. The turbulent kinetic energy, k, has been defined via:[33]

$$k = \frac{1}{2}(\overline{u'^2} + \overline{v'^2} + \overline{w'^2}) \tag{11.60}$$

A transport equation can be written for the turbulence kinetic energy by using the Reynolds average.[33]

$$\rho\frac{\partial k}{\partial t} + \rho\bar{v}\frac{\partial k}{\partial y} = -\overline{\rho u'v'}\frac{\partial \bar{u}}{\partial y} - \rho\varepsilon + \frac{\partial}{\partial y}\left(\mu\frac{\partial k}{\partial y} - \frac{1}{2}\rho\overline{u'^2v'} - \overline{p'v}\right) \tag{11.61}$$

The quantity ε is the dissipation per unit mass and is defined by:[33]

$$\varepsilon = \nu\overline{\left(\frac{\partial u'}{\partial z}\right)^2} \tag{11.62}$$

To close the mathematical description of this model, it is necessary to specify $\overline{\rho u'v'}$. Prandtl ref established arguments for each term in the equation. In general, one-equation models have a few of the disadvantages as well as most of the advantages of the mixing-length model. To improve the model, transport effects for the turbulence length scale must be included.

11.5.2.3.3   Two Equation Models

These models calculate not only the turbulence kinetic energy (k) but also the turbulence length scale l.[33]

The k–w model

The kinetic energy of the turbulence (k) is the first parameter, and the dissipation rate per unit turbulence kinetic energy (w) is the second parameter. The most tested version of the model has been presented by Wilcox[35] and is presented below.

Eddy viscosity

$$\mu_t = \rho\frac{k}{w} \tag{11.63}$$

Turbulence kinetic energy

$$\rho\frac{\partial k}{\partial t}+\rho\bar{v}\frac{\partial k}{\partial y}=-\overline{\rho u'v'}\frac{\partial \bar{u}}{\partial y}-\beta^*\rho k\omega+\frac{\partial}{\partial y}\left((\mu+\sigma^*\mu_t)\frac{\partial k}{\partial y}\right) \tag{11.64}$$

Specific energy dissipation rate

$$\rho\frac{\partial \omega}{\partial t}+\rho\bar{v}\frac{\partial \omega}{\partial y}=-\alpha\frac{\omega}{k}\overline{\rho u'v'}\frac{\partial \bar{u}}{\partial y}-\beta\rho\omega^2+\frac{\partial}{\partial y}\left((\mu+\sigma\mu_t)\frac{\partial \varepsilon}{\partial y}\right) \tag{11.65}$$

The closure coefficients are:

$$\alpha=5/9,\quad \beta=3/40,\quad \beta^*=0.09,\quad \sigma=0.5,\quad \sigma^*=0.5 \tag{11.66}$$

Auxiliary relations include

$$\varepsilon=\beta^*\omega k \text{ and } l=\frac{k^{1/2}}{\omega} \tag{11.67}$$

The k–ε model

This is the most commonly used two equation model.[33] The equations governing kinetic energy of the turbulence (k) and the energy dissipation rate (ε) are obtained by a combination of manipulation of the fundamental equations, to construct an equation for the transport of these quantities, followed by approximation to remove the unknown terms that the process introduces. The standard k–ε model may be described as follows:

Eddy viscosity

$$\mu_t=\rho C_\mu\frac{k^2}{\varepsilon} \tag{11.68}$$

Turbulence kinetic energy

$$\rho\frac{\partial k}{\partial t}+\rho\bar{v}\frac{\partial k}{\partial y}=-\overline{\rho uv}\frac{\partial \bar{u}}{\partial y}-\rho\varepsilon+\frac{\partial}{\partial y}\left((\mu+\frac{\mu_t}{\sigma_k})\frac{\partial k}{\partial y}\right) \tag{11.69}$$

Specific energy dissipation rate

$$\rho\frac{\partial \omega}{\partial t}+\rho\bar{v}\frac{\partial \varepsilon}{\partial y}=-C_{\varepsilon1}\frac{\varepsilon}{k}\overline{\rho u'v'}\frac{\partial \bar{u}}{\partial y}-C_{\varepsilon2}\rho\frac{\varepsilon^2}{k}+\frac{\partial}{\partial y}\left(\left(\mu+\frac{\mu_t}{\sigma_\varepsilon}\right)\frac{\partial \varepsilon}{\partial y}\right) \tag{11.70}$$

Closure coefficients include:

$$C_{\varepsilon1}=1.44,\quad C_{\varepsilon2}=1.92,\quad C_\mu=0.09,\quad \sigma_k=1.0,\quad \sigma_\varepsilon=1.3 \tag{11.71}$$

Auxiliary relations include:

$$\omega=\frac{\varepsilon}{C_\mu k}\quad \text{and}\quad l=\frac{C_\mu k^{3/2}}{\varepsilon} \tag{11.72}$$

The above model is for a high Reynolds number flow (Re > 20000) and does not apply in the region close to the walls. A special treatment near the walls has to be applied.

11.5.2.3.4   Stress Transport Models

These models directly solve the Reynolds stresses themselves. In the differential Reynolds stress model (DSM), partial differential equations are solved for each of the components for the Reynolds stresses and the energy dissipation rate. Further details may be found in Fletcher.[33]

11.5.2.3.5   Comparison between the Models

The k–ε model has the advantage of being computationally cheaper to use. There are only two equations added to the mass and momentum equations to solve the k–ε model, compared with seven for the DSM, making the k–ε model easier to converge than the DSM. On the other hand, DSM allows better predictions of anisotopic flows. In Fletcher et al.,[36] these two models were compared in a steady state simulation of flow and combustion in an entrained flow biomass gasifier. The predicted flows were very different, but the characteristic quantities, such as the exit temperature and the gas composition, were virtually the same, and since the k–ε model is computationally less costly, it would be preferred here. However, in Fletcher et al.,[37] the steady-state simulation of a rotary swirl cyclone could not be performed by the k–ε model, because the false decay of swirl completely changed the flow field. The comparison with the experimental observations pointed to the use of DSM in this high swirl case. For steady-state simulations, the choice of the model depends on which characteristics are modeled. Nevertheless, no generalisations can be made and the use of each model must be checked for each case.

In transient simulations, Guo et al.[29] compared the two turbulence models in a two dimensional slice of a submerged entry nozzle. The two-equation k–ε model reproduced the regular period of oscillation observed experimentally but predicted a decay of the vortices that was too rapid. The DSM produced a complex oscillation with a cycle to cycle variation in both amplitude and period. The average frequency was lower than predicted by the k–ε model, and the k–ε model matched the experimentally measured frequencies much better than DSM. For transient flows, this suggests that the k–ε model is therefore, preferred.

The shear stress transport (SST) model combines the best features of the k–ε model in the bulk of the flow and the k–w model near the walls.[35,38] This model has been used successfully by Langrish et al.[39] for simulating the experimentally measured transient flow patterns for air only in a pilot-scale spray dryer.

In the review by Fletcher et al.,[3] several important practical points emerge. On a normal personal computer, with 1 GHz Pentium CPU, run times may vary from a few hours for an air flow only type simulation to weeks for a fully coupled gas-particle transient simulation that is properly converged. The computational time for the software is therefore not quite real time—in other words, the software simulates the situations more slowly than they actually run. FLUENT and CFX are currently two of the most popular software packages, although STAR-CD and other CFD packages may also be suitable.

11.6   IMPLICATIONS OF UNDERSTANDING FLOW PATTERNS IN SPRAY DRYERS

What this work on flow patterns means is that we are now in a position to make more meaningful inroads into classical problems of agglomeration, aroma loss, wall deposition and thermal degradation in spray dryers.

11.6.1   Agglomeration

Various challenges that have been encountered in the process of developing validated Lagrangian and Eulerian models for simulating particle agglomeration within a spray dryer have been reported by Nijdam et al.[40] These have included the challenges of accurately measuring droplet coalescence rates within a spray, and modeling properly the gas-droplet and droplet-droplet turbulence interactions. The relative versatility and ease of implementation of the Lagrangian model compared with the Eulerian model has been demonstrated, as has the accuracy of both models for predicting turbulent dispersion of droplets and the turbulent flow-field within a simple jet system. The Lagrangian and Eulerian predictions are consistent with each other, which implies that the numerical aspects of each simulation are handled properly, suggesting that either approach can be used with confidence for future spray modeling. As with the work of Verdurmen et al.,[41,42] a clear preference was made for the Lagrangian approach, since Eulerian approaches currently involve representing each particle size fraction as an additional solids phase. Each phase requires another set of equations. It is clear that considerable research must be done in the area of particle turbulence modeling and accurate measurement of particle agglomeration rates before any CFD tool can be employed to accurately predict particle agglomeration within a spray dryer.

11.6.2 AROMA LOSS AND THERMAL DEGRADATION

Classical work on the loss of volatile aroma components in spray drying has been reviewed by Kerkhof,[43] including the work of Coumans et al.[44] Typically, the system is simplified to a ternary one, with water, a trace aroma component and solids. Both a straight aroma diffusion coefficient (with the driving force for one component of the aroma flux being the gradient in aroma concentration) and a cross-diffusion coefficient (with the driving force for another component of the aroma flux being the gradient in moisture concentration) are important for the transport of the aroma component. The aroma diffusion coefficients are low compared with those for water at low moisture contents, creating effectively semi-permeable dry layers and reducing the rate of aroma transport significantly.

Thermal degradation is important in the drying of heat sensitive products, such as milk and vitamin C. Important influences on the extent of thermal degradation include the temperature-time history and the range of residence times, with a large range of residence times giving a range of moisture contents for any product. Thermal degradation is often modeled using first-order reaction kinetics,[43] with the degradation rate constant depending on the temperature and moisture content of the material. The flow patterns in the dryer determine the history of the material temperature and moisture content through the dryer. All of the works reviewed by Kerkhof were carried out before the fluid flow patterns were understood in great detail, and the predictions were carried out using first approximations to the particle and fluid flow patterns. In some cases (including a simulation of the thermal degradation of skim milk particles), there was no comparison with experiments. In the case of the thermal degradation of the milk particles, the droplet diameter was predicted to have a significant influence on the final milk activity, due to the effect of diameter on residence time. Using the assumption that the gas and the particles moved through a spray dryer in cocurrent plug flow, Rulkens and Thijssen[45] found that the experimentally measured aroma retention, using methanol, n-propanol and n-pentanol as aroma model components, was around 70% of that predicted using the model assumptions. The uncertainty about the flow pattern meant that it was unclear how much of this discrepancy was due to this aspect (flow pattern) of the simulation. Another reason for the discrepancy was uncertainty about the particle morphology (whether the particles were hollow or solid). The model predicted that the aroma retention should increase with increasing solids content, due to the lower aroma diffusion coefficient, and this trend was also found in the experiments. However, this trend is almost self-evident from the fundamental kinetics assumed in the model, which has been largely constructed to mirror the experimental trends anyway.

11.6.3 WALL DEPOSITION

Particles build up due to the adhesion of particles to initially clean walls of spray dryers. Subsequent layers of particles become attached to this initial layer (cohesion). At the same time, Masuda and

Matsusaka[46] have pointed out that particles may be removed from the wall deposits by the shear stress created from the gas flow past the wall, so that a dynamic equilibrium is established between newly attached particles and detaching layers. Abbott[47] indicates that wall deposition may pose a potential fire risk and compromise hygiene requirements, as well as reducing product quality and yield. Such hazards include ignition of explosible dust clouds, dust deposits, bulk powder deposits and flammable vapour.

The interaction of substantial wall deposition with the residence time distribution was noted by Kieviet,[48] who concluded that the time taken to slide down the conical wall to the outlet was the most important factor in determining residence times with high wall deposition rates. Chen et al.[49] suggested that a combination of modified near wall airflow patterns and inlet temperature distribution could be used to significantly reduce wall deposition in industrial-scale spray drying of milk in New Zealand. Such changes are likely to be substantially less costly than making major modifications to the design of an existing dryer.

Particle stickiness is a key issue in wall deposition for spray dryers, determining the fate of particles that hit the walls. Whether or not particles hit the wall is a function of the fluid flow patterns in the equipment, and this is an area in which CFD can offer guidance, as suggested by Masters.[50] CFD allows the particle trajectories to be predicted, so that it is possible to estimate whether the particles hit the walls and the temperatures and moisture contents of such particles. However, whether the particles bounce off the walls, adhere to them, or cohere to other particles that are already adhering to the walls, is closely related to whether the particles are sticky or not.

Foodstuffs containing sugars and acids undergo a transition between a glassy and a more sticky (rubbery) state above a certain temperature.[51] One arrangement for detecting this transition mechanically is a stirred laboratory-scale beaker arrangement. Papadakis and Bahu[52] found this method to give reproducible results. The glass transition may be important both for cohesion of particles to one another in agglomeration (the situation that was strictly involved in this test) and adhesion of particles to walls. The glass transition temperature may also be measured in a differential scanning calorimeter, by detecting the change in specific heat capacity associated with the transition from the amorphous to the rubbery phase. Bhandari et al.[51] have found that stickiness can begin 10–20 K above the glass transition temperature, since the critical viscosity associated with stickiness does not necessarily occur exactly at the phase transition point. They gave formulae to predict the glass transition temperature for sugars and acids as a function of the moisture content. Ozmen and Langrish[18] also noted a good agreement between glass transition and sticky-point temperatures (with a slight offset) for skim milk powder.

The effect of different wall materials on wall deposition rates has been found by Ozmen and Langrish[53] to be insignificant in a pilot-scale spray dryer and by Murti et al.[54] to be insignificant when measured using a particle "gun." On the other hand, Kota and Langrish[55] found that nylon plates gave a significantly different deposition flux to stainless steel plates under some conditions, so there is some uncertainty in this area. Bhandari et al.[51] and Langrish et al.[56] have noted the significant effects of particle or droplet material on wall deposition behavior, particularly glass transition effects in sugary materials. Kim et al.[57] and Nijdam and Langrish[58] reported the importance of fat, which is sticky and also preferentially migrates to the surface of particles. Whey proteins, which are less sticky than many sugars and fats and which also preferentially migrate to the surfaces of particles, have also been shown by Adhikari et al.[59] to decrease wall deposition rates.

Electrostatic effects have been found to have surprisingly small effects on wall deposition rates. A preliminary study by Chen et al.[49] found that either charging or earthing plates had no effect on the amount of deposit build up per unit area of plate. Ozmen and Langrish[53] found that the average deposition flux did not change significantly when the spray dryer was earthed or not earthed. The lack of influence of electrostatic change of deposition in spray dryers may be attributed to the short range nature of electrostatic forces.

The role of CFD is emphasized by the work of Kota and Langrish,[60] who found that conventional pipe correlations for particle deposition cannot match the deposition behavior found in spray dryers.

The work of Hanus and Langrish[9] suggests that future work is needed to better understand the physical processes behind wall deposition

11.7 CONCLUSIONS

There have been numerous developments in the modeling of spray-dryer behavior over the past 20 years, which are meeting the challenges of simulating the highly transient and three-dimensional complex flow patterns inside the equipment. Simulations are being used for investigating methods to reduce the degree of wall deposition and of thermal degradation for particles by modifying the air flow patterns in the chamber through small changes in the air inlet geometry. Challenges include building particle drying kinetics and reaction processes, as well as agglomeration behavior, into these simulations. The numerical simulations are proving to be valuable supplements to pilot-scale testing, enabling more extensive and accurate optimisation to be carried out than hitherto possible.

ACKNOWLEDGMENTS

Acknowledgments and grateful thanks are due to Dr. M.N. Haque, CSIRO Minerals, for reviewing this chapter and for his helpful and constructive criticism and feedback, which are greatly appreciated.

NOMENCLATURE

A_p	droplet/particle surface area (m^2)
C_D	drag coefficient (−)
C_P	specific heat capacity ($J\ kg^{-1}\ K^{-1}$)
C_{Pa}	specific heat capacity of air ($J\ kg^{-1}\ K^{-1}$)
C_{Pl}	specific heat capacity of liquid water ($J\ kg^{-1}\ K^{-1}$)
C_{PP}	specific heat capacity of particles ($J\ kg^{-1}\ K^{-1}$)
C_{Pv}	specific heat capacity of pure water vapour ($J\ kg^{-1}\ K^{-1}$)
d_p	droplet diameter (m)
d_{pi}	initial droplet diameter (m)
D_v	diffusivity of water in air ($m^2\ s^{-1}$)
E	total energy (J)
e	internal energy (J)
F	dry solids flow rate ($kg\ s^{-1}$)
G	mass flow rate of the dry air ($kg\ s^{-1}$)
$\vec{g}$	acceleration due to gravity ($m\ s^{-2}$)
h	height (m)
H_{fg}	latent heat of water evaporation ($J\ kg^{-1}$)
H_h	enthalpy of humid air ($J\ kg^{-1}$)
k	turbulent kinetic energy per unit mass ($m^2 s^{-2}$)
k_a	thermal conductivity of humid air ($W\ m^{-1}\ K^{-1}$)
K_m	mass-transfer coefficient ($kg\ m^{-2}\ s^{-1}$)
K_p	mass-transfer coefficient (partial pressure based) ($kg\ m^{-2}\ s^{-1}\ Pa^{-1}$)
L	length of the spray-drying chamber (m)
l	turbulence length scale (m)
l_m	mixing length (m)
M_a	molecular weight of air ($g\ mol^{-1}$)
m_p	mass of a particle or droplet/particle (kg)
m_s	mass of solids in the droplet/particle (kg)
M_w	molecular weight of water ($kg\ mol^{-1}$)

n_{droplets}	flow rate of droplets/particles (number s^{-1})
Nu	Nusselt number (–)
P	total pressure (Pa)
P_{atm}	atmospheric pressure (Pa)
p_{vb}	partial pressure of water vapour in the bulk air (Pa)
p_{vs}	partial pressure of water vapour at the surface of the droplet (Pa)
p_v	actual vapour pressure (Pa)
p_{vsat}	saturation vapour pressure (Pa)
R	Universal Gas Constant (J mol^{-1} K^{-1})
r	radial distance of droplets/particles as a function of axial distance from the atomiser (m)
Re_p	particle Reynolds number (–)
S	swirl number (–)
Sc	Schmidt number (–)
Sh	Sherwood number (–)
t	time (s)
T	temperature (K)
T_{abs}	absolute temperature of the droplet or particle (K)
T_{amb}	ambient temperature (K)
T_l	temperature of liquid water (K)
T_{glact}	glass transition temperature of lactose (K)
T_{gwater}	glass transition temperature of water (K)
T_{Go}	gas outlet temperature (K)
T_{Po}	outlet solids temperature (K)
T_{st}	sticky-point temperature (°C)
UA	product of heat-transfer coefficient and external wall area for heat loss from the dryer (W K^{-1})
U_a	velocity of air (m s^{-1})
U_{ax}	velocity of air in the axial direction (m s^{-1})
U_{ar}	velocity of air in the radial direction (m s^{-1})
U_{at}	velocity of air in the tangential direction (m s^{-1})
U_p	velocity of droplet/particle (m s^{-1})
U_{px}	velocity of droplet/particle in the axial direction (m s^{-1})
U_{pr}	velocity of droplet/particle in the radial direction (m s^{-1})
U_{pt}	velocity of droplet/particle in the tangential direction (m s^{-1})
U_R	relative velocity between droplet/particle and the air (m s^{-1})
u	turbulent air velocity (m s^{-1})
$\bar{u}$	time averaged part of overall velocity (m s^{-1})
u'	fluctuating part of overall velocity (m s^{-1})
u_t	turbulence velocity (m s^{-1})
V	volume of air (m^3)
w	the dissipation rate per unit turbulence kinetic energy (s^{-1})
X	moisture content (kg kg^{-1})
X_{emc}	equilibrium moisture content (kg kg^{-1})
X_i	inlet moisture content (kg kg^{-1})
X_o	solids moisture content (kg kg^{-1})
Y	gas humidity (kg kg^{-1})
Y_i	inlet gas humidity (kg kg^{-1})
Y_o	outlet gas humidity (kg kg^{-1})

Greek

ε	dissipation rate of turbulence kinetic energy per unit mass (m^2 s^{-3})
ϕ	any variable in Equation 49, representing conservation of mass or momentum
Γ_ϕ	diffusion coefficient (m^2 s^{-1})
κ	von Karman constant (–)
λ	latent heat of vaporisation (kJ kg^{-1})
μ	air viscosity (kg m^{-1} s^{-1})
μ	turbulent or eddy viscosity (kg m^{-1} s^{-1})
ρ	density (kg m^{-3})
ρ_a	density of air (kg m^{-3})
ρ_p	density of a particle or droplet (kg m^{-3})
ρ_{pi}	initial droplet density (kg m^{-3})
ρ_s	density of solids (kg m^{-3})
ρ_w	density of water (kg m^{-3})
$\overline{pu'v'}$	Reynolds stresses (kg m^{-1} s^{-2})
ω	turbulence time scale (s)
ξ	relative drying rate (–)
ψ	relative humidity (–)

REFERENCES

1. Bahu, R.E. Spray drying—maturity or opportunities? In *Drying ' 92, Proceedings of the 8th International Drying Symposium (IDS '92)*, A.S. Mujumdar, I. Filkova, Eds. Elsevier, Amsterdam, Vol. A, 74, 1992.
2. Oakley, D.E. Scale-up of spray dryers with the aid of Computational Fluid Dynamics. *Drying Technology*, 12, 217, 1994.
3. Fletcher, D.F. et al. What is important in the simulation of spray dryer performance and how do current CFD models perform? *Applied Mathematical Modelling*, 30, 1281, 2006.
4. Keey, R.B., and Pham, Q.T. Behaviour of spray dryers with nozzle atomizers. *Chemical Engineering*, 516, 1976.
5. Langrish, T.A.G., and Fletcher, D.F. Spray drying of food flavours and applications of CFD in spray dryers. *Chemical Engineering and Processing*, 40, 345, 2001.
6. Place, G., Ridgway, K., and Danckwerts, P.V. Investigation of air-flow in a spray-drier by tracer and model techniques. *Transactions of the Institution of Chemical Engineers*, 37, 268, 1959.
7. Paris, J.R. et al. Modelling of the air flow pattern in a countercurrent spray-drying tower. *Industrial Engineering Chemistry Process Design and Development*, 10, 157, 1971.
8. Ozmen, L., and Langrish, T.A.G. A study of the limitations to spray dryer outlet performance. *Drying Technology*, 21, 895, 2003.
9. Hanus, M.J., and Langrish, T.A.G. Re-entrainment of wall deposits from a laboratory-scale spray dryer. *Asia-Pacific Journal of Chemical Engineering*, 2, 90, 2007.
10. Truong, V., Bhandari, B.R., and Howes, T. Optimization of co-current spray drying process of sugar-rich foods. Part I Moisture and glass transition temperature profile during drying. *Journal of Food Engineering*, 71, 55, 2005.
11. Rhodes, M. *Introduction to Particle Technology*. John Wiley & Sons, New York, 2–4, 1998.
12. Langrish, T.A.G., and Zbicinski, I. The effect of air inlet geometry and spray cone angle on the wall deposition rate in spray dryers. *Transactions of the Institution of Chemical Engineers*, 72, 420, 1994.
13. Perry, R.H., Green, D.W., and Malony, J.O. *Perry's Chemical Engineers' Handbook*, 7th Edition. McGraw-Hill, New York, 2–166, 1997.
14. Ranz, W.E., and Marshall, W.R. Evaporation from drops. *Chemical Engineering Progress*, 148, 141, 1952.
15. Keey, R.B. *Introduction to Industrial Drying Operations*. Pergamon, Oxford, UK, 15–99, 1978.
16. Strumillo, C., and Kudra, T. *Drying: Principles, Application and Design*. Gordon and Breach, New York, NY, 45–54, 1986.

17. Kockel, T.K. et al. An experimental study of the desorption equilibrium for skim milk powder at elevated temperatures. *Journal of Food Engineering*, 51, 291, 2002.
18. Ozmen, L., and Langrish, T.A.G. Comparison of glass transition temperature and stick-point temperature for skim milk powder. *Drying Technology*, 20, 1177, 2002.
19. Reay, D., Fluid flow, residence time simulation and energy efficiency in industrial dryers. In *Proceedings of the 6th International Drying Symposium (IDS '88), Versailles, France*, M. Roques, Ed., KL 1, ENSIC–INPL, Nancy, France, 1988.
20. Crowe, C.T. Modelling spray-air contact in spray-drying systems. In *Advances in Drying*, Mujumdar, A.S. Ed., Volume 1. Hemisphere, New York, NY, 63, 1980.
21. Bradshaw, P., Launder, B.E., and Lumley, J.L. Collaborative testing of turbulence models. *Journal of Fluids Engineering, Transactions of the ASME*, 118, 243, 1996.
22. Oakley, D.E., Bahu, R.E., and Reay, D. The aerodynamics of cocurrent spray dryers. In *Proceedings of the 6th International Drying Symposium (IDS '88)*, Versailles, France, M. Roques, Ed. OP 373, ENSIC-INPL, Narcy, France, 1988.
23. Oakley, D.E., and Bahu, R.E. Spray/gas mixing behaviour within spray dryers. In *Drying '91*, A.S. Mujumdar and I. Filkova, Eds. Elsevier, Amsterdam, 303, 1991.
24. Livesley, D.M. et al. Development and validation of a Computational Model for spray-gas mixing in spray dryers. In *Drying ' 92, Proceedings of the 8th International Drying Symposium (IDS '92)*, A.S. Mujumdar, I. Filkova, Eds. Elsevier, Amsterdam, Vol. A, 407, 1992.
25. Huang, L., Kumar, K., and Mujumdar, A.S. A parametric study of the gas flow patterns and drying performance of co-current spray dryer: results of a Computational Fluid Dynamics study. *Drying Technology*, 21, 957, 2003.
26. Bayly, A. et al. Airflow patterns in a counter-current spray drying tower—simulation and measurement. In *Drying 2004, Proceedings of the 14th International Drying Symposium (IDS 2004), Sao Paulo, Brazil*, M.A. Silva, S. Rocha, Eds. State University of Campinas, Campinas-SP, Brazil, Vol. B, 775, 2004.
27. Harvie, D.J.E., Langrish, T.A.G., and Fletcher, D.F. A Computational Fluid Dynamics study of a tall-form spray dryer. *Transactions of the Institution of Chemical Engineers*, 80, 163, 2002.
28. Langrish, T.A.G. et al. Time-dependent flow patterns in spray dryers. *Transactions of the Institution of Chemical Engineers*, 71 (A), 355, 1993.
29. Guo, B., Langrish, T.A.G., and Fletcher, D.F. An assessment of turbulence models applied to the simulation of a two-dimensional jet. *Applied Mathematical Modelling*, 25, 635, 2001.
30. Guo, B., Langrish, T.A.G., and Fletcher, D.F. Simulation of turbulent swirl flow in an axisymmetric sudden expansion. *AIAA Journal*, 39, 96, 2001.
31. LeBarbier, C. et al. Experimental measurement and numerical simulation of the effect of swirl on flow stability in spray dryers. *Transactions of the Institution of Chemical Engineers*, 79, 260, 2001.
32. Dellenback, P.S., Metzger, D.E., and Neitzel, G.P. Measurement in turbulent swirling flow through an abrupt axisymmetric expansion. *AIAA Journal*, 26, 669, 1988.
33. Fletcher, C.A.J. *Computational Techniques for Fluid Dynamics*. Springer Verlag, Berlin, Heidelberg, Vol. 2, 1, 47, 81, 333, 1991.
34. Hinze, J.O. *Turbulence*. McGraw-Hill, New York, NY, 2, 1975.
35. Wilcox, D.C. Reassessment of the scale determining equation for advanced turbulence models. *AIAA Journal*, 26, 1299, 1988.
36. Fletcher, D.F. et al. A CFD based combustion model of an entrained flow biomass gasifier. *Applied Mathematical Modelling*, 24, 165, 1999.
37. Fletcher, D.F. et al. Mathematical modelling of a rotary swirl cyclone. *Chemical Engineering Communications*, 161, 65, 1997.
38. Wilcox, D.C. *Turbulence Modelling for CFD*. DCW Industries, Inc, CA, 84, 1994.
39. Langrish, T.A.G., Williams, J., and Fletcher, D.F. Simulation of the effects of inlet swirl on gas flow patterns in a pilot-scale spray dryer. *Transactions of the Institution of Chemical Engineers*, 82, 821, 2004.
40. Nijdam, J.J. et al. Challenges of simulating droplet coalescence within a spray. *Drying Technology*, 22, 1463, 2004.
41. Verdurmen, R.E.M. et al. Simulation of agglomeration in spray drying installations: the EDECAD project. *Drying Technology*, 22, 1403, 2004.
42. Verdurmen, R.E.M. et al. Agglomeration in spray drying installations (the EDECAD project): Stickiness measurements and simulation results. *Drying Technology*, 24, 721, 2006.
43. Kerkhof, P.J.A.M. The role of theoretical and mathematical modelling in scale-up. *Drying Technology*, 12, 1, 1994.

44. Coumans, W.J., Kerkhof, P.J.A.M., and Bruin, S. Theoretical and practical aspects of aroma retention in spray drying and freeze drying. *Drying Technology*, 12, 99, 1994.
45. Rulkens, W.H., and Thijssen, H.A.C. Numerical solution of diffusion equations with strongly variable diffusion coefficients. Calculation of flavour loss in drying food liquids. *Transactions of the Institution of Chemical Engineers*, 47, T292, 1969.
46. Masuda, H., and Matsusaka, S. Particle deposition and reentrainment. In *Powder Technology Handbook*, 2nd ed. Gotoh, K., Masuda, H., Higashitani, K. Eds. Marcel Dekker, New York, NY, 143, 1997.
47. Abbott, J.A. Ed. *Prevention of Fires and Explosions in Dryers—A User Guide*, 2nd ed. Institution of Chemical Engineers, Warwickshire, UK, 30, 1990.
48. Kieviet, F.G. Modelling quality in spray drying. Ph.D. Thesis, T.U. Eindhoven, The Netherlands, 72, 1997.
49. Chen, X.D., Lake, R., and Jebson, S. Study of milk powder deposition on a large industrial dryer. *Transactions of the Institution of Chemical Engineers*, 71, 180, 1993.
50. Masters, K. Deposit-free spray drying: dream or reality? In *Proceedings of the 10th International Drying Symposium (IDS '96), Drying '96, Krakow, Poland*, C. Strumillo and A.S. Mujumdar, Eds. Lodz Technical University, Lodz, Poland, Vol. A, 52, 1996.
51. Bhandari, B.R., Datta, N., and Howes, T. Problems associated with spray drying of sugar-rich foods. *Drying Technology*, 15, 671, 1997.
52. Papadakis, S.E., and Bahu, R.E. The sticky issues of drying. *Drying Technology*, 10, 817, 1992.
53. Ozmen, L., and Langrish, T.A.G. An experimental investigation of the wall deposition of milk powder in a pilot-scale spray dryer. *Drying Technology*, 21, 1253, 2003.
54. Murti, R.A. et al. Controlling SMP stickiness by changing the wall material: feasible or not? In *Proceedings of Chemeca 2006, Auckland, New Zealand*, B. Young, Ed. CD ROM, Paper 209, 2006.
55. Kota, K., and Langrish, T.A.G. Fluxes and patterns of wall deposits for skim milk in a pilot-scale spray dryer. *Drying Technology*, 24, 993, 2006.
56. Langrish, T.A.G., Chan, W.C., and Kota, K. Comparison of maltodextrin and skim milk deposition rates in a pilot-scale spray dryer. *Powder Technology*, 179, 84, 2007.
57. Kim, E., Chen, X.D., and Pearce, D. On the mechanisms of surface formation and the surface compositions of industrial milk powders. *Drying Technology*, 21, 265, 2003.
58. Nijdam, J.J., and Langrish, T.A.G. The effect of surface composition on the functional properties of milk powders. *Journal of Food Engineering*, 77, 919, 2006.
59. Adhikari, B. et al. Effect of surface tension and viscosity on the surface stickiness of carbohydrate and protein solutions. *Journal of Food Engineering*, 79, 1136, 2007.
60. Kota, K., and Langrish, T.A.G. Prediction of wall deposition behaviour in a pilot-scale spray dryer using deposition correlations for pipe flows. *Journal of Zhejiang University Science A*, 8, 301, 2007.

12 Modeling of Heat and Mass Transfer during Deep Frying Process

L.A. Campañone, M. Alejandra García, and Noemi E. Zaritzky
CIDCA-Centro de Investigación y Desarrollo en
Criotecnología de Alimentos (CONICET-UNLP)

CONTENTS

12.1   INTRODUCTION

The frying process is considered one of the oldest cooking methods. It is used to cook food in oil and to confer unique textures and tastes; the organoleptic characteristics depend on the frying conditions [1,2]. Frying is usually done by the immersion of the product in oil at a temperature of 150–200°C, where a simultaneous heat and mass transfer takes place [2,3].

Heat conduction occurs in the core of solid food and is strongly influenced by the physical properties of the food that change continuously during the frying process [2]. Simultaneously, heat transfer by convection takes place between the oil and the surface of the food. When the formation of bubbles begins, the heat transfer is accelerated because bubbles contribute to the turbulence in the frying medium. However, the formation of foam in the oil produces a significant decrease in the heat transfer rate [2]. Convection heat transfer decreases as the amount of vapor decreases due to the depletion of water in the core of the food. The rate of heat transfer toward the food core is influenced by the thermal properties and viscosity of the frying medium and the agitation conditions. During the production of bubbles on the surface, forced convection becomes the main regimen.

Water transfer is produced by surface boiling, and, when evaporation decreases, diffusion of water from the food core toward the surface is the dominant physical phenomenon. Water leaves the product as bubbles of vapor while it migrates internally by means of different mechanisms [2–4]. After an initial time, when surface moisture is evaporated, a dehydrated zone begins to form on the surface of the food. Under these conditions, the surface temperature becomes closer to that of the frying medium while moisture is strongly reduced, reaching values close to the bound water value. Thus, an important gradient of moisture between the core and the surface is established. In starchy products, a water competition is found between the moisture necessary to gelatinize the starch and the moisture that is leaving the product.

The mechanism of oil penetration is a subject of controversy [5–7]. Absorption of oil on the surface of a fried product occurs when samples are removed from the frying medium; the oil that remains on the piece surface enters the product [3,4,7–13]. According to these authors, oil does not invade the product itself and oil uptake (OU) during frying is negligible. The conditions at which products are removed from the frying oil seem decisive for the uptake of oil; this would be related to the adhesion of oil to the surface and how it is drained.

Many factors affect OU in deep-fat frying such as oil quality, frying temperature, residence time, product shape and size, product composition (initial moisture, and protein content), pore structure (porosity and pore size distribution), and prefrying treatments (drying, blanching, surface coating). Interfacial tension was also reported by Pinthus and Saguy [5] to have a significant influence on OU after deep-fat frying. Many researchers have suggested that fat absorption is primarily a postfrying phenomena [10,11,14–16], thus, fat transfer occurs in food products during the cooling stage. Moreira and Barrufet [9] reported that 80% of the total oil content was absorbed in tortilla chips during the cooling period. They explained that as the product temperature decreases, the interfacial tension between gas and oil increases, raising the capillary pressure. This sucks the surface oil into the porous medium, thus increasing the final oil content.

Ufheil and Escher [8] also suggested that the absorption theory was based on surface phenomena, which involve the equilibrium between adhesion and the drainage of oil during cooling. Fat absorption in chips due to surface adherence was the result of steam condensation, as it was also suggested by Rice and Gamble [17]. OU depends on structural changes during the process; differences in the starting food microstructure can be expected to be important in the evolution of the characteristics of the end product. Meat products are mainly an arrangement of protein fibers, which may relax and denature upon heating [18] while flour based products, such as doughnuts and tortillas, have a different microstructure.

Several models have been developed for heat, moisture, and fat transfer during the frying of foods [10,19,20]. Ateba and Mittal [21] developed a model for heat, moisture, and fat transfer in the deep-fat frying of beef meatballs. Rice and Gamble [17] used a simplification of the Fick's equation to predict moisture loss and oil absorption during potato frying. Farkas, Singh, and Rumsey [22,23] developed a model for frying considering heat and moisture transfer and Singh [24] treated the crust as a moving boundary. Farid and coworkers [25,26] have made an important contribution to the moving boundary analysis that can be applied to describe a large number of processes such as frying, melting, solidification, microwave thawing, spray-drying, and freeze-drying.

The large volume of fried foods that are produced and its influence on the consumption of lipids enhances the study of the frying process because of its strong economical and nutritional impacts. From an economical point of view, the higher oil content in fried food products increases production costs [4]. However, the reduction of lipid content in fried foods is required mainly owing to its relation with obesity and coronary diseases. An alternative to reduce OU in fried foods is the use of edible films or coatings.

The application of hydrocolloid coatings allows the reduction of the oil content of deep-fat fried products due to their lipid barrier properties; the most widely studied are gellan gum and cellulose derivatives [7,14,15,27]. Cellulose derivatives, including methylcellulose (MC) and hydroxypropylmethylcellulose (HPMC), exhibit thermogelation. When suspensions are heated they form a gel that reverts below the gelation temperature, and the original suspension viscosity is recovered [28]. These cellulose derivatives reduce oil absorption through film formation at temperatures above their gelation point, or they reinforce the natural barrier properties of starch and proteins, especially when they are added in dry form [29].

With regard to the modeling of deep-fat frying of coated foods, Mallikarjunan et al. [30] and Huse et al. [31] demonstrated the effectiveness of various edible coatings in reducing oil absorption in starchy products. Williams and Mittal [14,15], working on the frying of foods coated with gellan gum, determined the effectiveness of edible films to reduce fat absorption in cereal products, and developed a mathematical model for the process.

In the present chapter the following objectives are proposed:

1. To model heat and moisture transfer during the deep-fat frying of food by solving numerically the partial differential equations of heat conduction and mass transfer within the food, which allows the prediction of temperature profiles and moisture concentrations, as a function of operating conditions
2. To validate experimentally the mathematical model with regard to the temperature profiles and the water losses from the food product
3. To relate OU measurements after the frying process with the effects of frying time on sample microstructural changes as well as on the growth of the dehydrated zone
4. To analyze the performance of applying an edible coating based on MC on a food model dough system by measuring moisture content and OU in the deep-frying process
5. To analyze quality attributes such as texture and color and to observe, using scanning electron microscopy (SEM), the surface microstructures of coated and uncoated fried products at different processing times

12.2 MATHEMATICAL MODEL

12.2.1 MATHEMATICAL MODEL OF HEAT AND MOISTURE TRANSFER DURING THE FRYING PROCESS

Deep-fat frying is a complex process that involves simultaneous heat and mass transfers. The process induces a variety of physicochemical changes in both the food and the frying medium.

A mathematical model of the frying process is proposed based on the numerical solution of the unsteady state heat and mass transfer partial differential equations. The frying process is modeled for disc-shaped food. Discs are considered as infinite slabs and unidirectional; heat and mass transfer equations are solved for $0 \leq x \leq L$ (L = the half-thickness of the sample). In the proposed model, different stages are considered.

Stage 1: This stage represents the initial heating of the product, in which heat and mass transfer between the product and the frying oil occurs. The model is based on the assumption of a uniform initial temperature distribution in the sample (T_{ini}).

To get the temperature profiles, the following differential equation, based on microscopic energy balance, is solved:

$$\rho c_p \frac{\partial T}{\partial t} = \nabla(k \nabla T) \tag{12.1}$$

Thermal properties correspond initially to the nonfried product and change with temperature and moisture content.

The initial and boundary conditions are

$$t = 0 \quad T = T_{ini} \quad 0 \leq x \leq L \tag{12.2}$$

$$x = 0 \quad \frac{\partial T}{\partial x} = 0 \quad t > 0 \tag{12.3}$$

$$x = L \quad -k\frac{\partial T}{\partial x} = h_1(T - T_{oil}) + L_{vap} m_{vap,1} \quad t > 0 \tag{12.4}$$

where $m_{vap,1}$ is the water flux from the product and h_1 is the convective heat transfer coefficient. The heat transfer mechanism is considered to be governed by natural convection.

The value of the heat transfer coefficient (h_1) for natural convection can be calculated for horizontal plates using two different equations [32,33]:

$$Nu = \frac{0.108\, Gr^{4/11}\, Pr^{9/33}}{(1 + 0.44\, Pr^{2/3})^{4/11}} \tag{12.5}$$

$$Nu = 0.55\,(Gr.Pr)^{1/4} \tag{12.6}$$

In order to evaluate water profiles, the microscopic mass balance is solved:

$$\frac{\partial C_w}{\partial t} = \nabla(D_w \nabla C_w) \tag{12.7}$$

The following initial and boundary conditions are used:

$$t = 0 \quad C_w = C_{w,ini} \quad 0 \leq x \leq L \tag{12.8}$$

$$x = 0 \quad \frac{\partial C_w}{\partial x} = 0 \quad t > 0 \tag{12.9}$$

$$x = L \quad -D_w\frac{\partial C_w}{\partial x} = m_{vap,1} \quad t > 0 \tag{12.10}$$

$m_{vap,1}$ is calculated from Equation 12.10 with $C_w = 0$ at the sample/oil interface, because it is considered that water evaporates and immediately leaves the frying medium.

For each time interval the model calculates the average value of moisture content as follows:

$$C_{w,ave} = \frac{\int_0^L C_w dV}{V} \tag{12.11}$$

The second stage starts when the surface temperature of the product reaches that of the water vaporization.

Stage 2: Water boiling is produced on the surface and a dehydrated layer is formed in the external zone of the product, having different thermo-physical and transport properties from those of the wet core (Table 12.1). Several authors [11–13,19,22,23] used the denomination of crust for the dehydrated zone. In the present model the crust is considered to be the portion of the dehydrated zone containing oil.

When water on the surface is no longer available, vaporization occurs in a front that moves toward the inner zone of the product. During this phase of the frying process, the vapor that is released from the product surface impedes oil penetration into the product [2,3,6].

Microscopic energy balances given by Equation 12.1 are solved for both (core and dehydrated) zones to obtain temperature profiles, considering the corresponding thermal properties.

The initial temperature profile in this stage is given by the final profile obtained at the end of Stage 1. The following boundary conditions are applied:

$$x = 0 \quad \frac{\partial T}{\partial x} = 0 \quad t > 0 \tag{12.12}$$

TABLE 12.1
Core, Dehydrated Zone, and Oil Properties Used in the Model

Property	Core Zone	Dehydrated Zone	Sunflower Oil
Initial water content	0.42	–	–
Density (kg/m³)	623[d]	579[a]	876[c]
Thermal conductivity (W/(mK))	0.6[d]	0.05[a]	0.6[c]
Specific heat (J/kg°C)	2800[d]	2310[a]	2033[c]
Viscosity (Pa.s)			2.04×10^{-3c}
Porosity	–	0.42[b]	–
Tortuosity	–	8.8[b]	–
Diffusion coefficient of vapor in air (dehydrated zone) (m²/s)		$\dfrac{5.68\ 10^{-9}(T+273.16)^{1.5}}{1.789 - 2.13\ 10^{-3}(T+273.16)}$ [e]	
Moisture diffusion coefficient in dough (core zone) (m²/s)		2×10^{-9f}	

Sources:

[a] Rask C., *J. Food Eng.*, 9, 167–193, 1989.
[b] This work.
[c] Moreira, R.G., Castell-Perez M.E. and Barrufet, M.A., *Deep Fat Frying: Fundamentals and Applications.* Aspen Publishers Inc., Gaithersburg, MD, 1999.
[d] Zanoni, B., Peri, C. and Gianotti, R., *J. Food Eng.*, 26, 497–510, 1995.
[e] Bird, R.B., Stewart, W.E., and Lightfoot, E.N., *Transport Phenomena.* John Wiley and Sons, New York, NY, 1976.
[f] Tong, C.H. and Lund, D.B., *Biotechnol. Progress*, 6, 67–75, 1990.

$$x = x_1 \quad T = T_{\text{vap}} \quad t > 0 \tag{12.13}$$

$$x = L \quad -k_d \frac{\partial T}{\partial x} = h_2(T - T_{\text{oil}}) \quad t > 0 \tag{12.14}$$

with x_1 = the position of the vaporization front (measured from the center of the slab).

Besides the thickness of the dehydrated zone (d_Z), measured from the sample/oil interface, can be calculated as $d_Z = L\text{-}x_1$.

Values of the heat transfer coefficient (h_2) higher than 250 W/m²K are found in the literature for the whole frying process. Costa et al. [34] reported a value of $h_2 = 650 \pm 7$ W/m²K for potato chips fried at 180°C; Dagerskog and Sorenfors [35] reported values of h_2 ranging between 300 and 700 W/m²K for meat hamburgers fried at 190°C; Moreira et al. [36] obtained values of 285 and 273 W/m²K for fresh and used soy oil, respectively. These high heat transfer coefficients allow us to assume a constant temperature at the product–oil interface.

Moisture content decreases due to water vaporization; vapor diffuses through the dehydrated zone, which is considered to be a porous medium.

The following mass balance is solved in the dehydrated zone:

$$m_{\text{vap},2} = -D_v \frac{\partial C_v}{\partial x} \tag{12.15}$$

D_v is calculated through the following relationship:

$$D_v = D_{\text{va}} \frac{\varepsilon}{\tau} \tag{12.16}$$

The following relationships are considered:

$$x = x_1 \quad C_v = C_{v,\text{equi}} \tag{12.17}$$

$$x = L \quad C_v = 0 \tag{12.18}$$

where $C_{v,\text{equi}}$ is the equilibrium vapor concentration at T_{vap}; it is calculated using the Clausius–Clapeyron equation, assuming ideal gas behavior for the water vapor.

Water vapor concentration at the product–oil interface is assumed to be negligible according to Equation 12.18.

To calculate the position of the vaporization front as a function of time, the following equation is proposed:

$$-C_{w,\text{ave}} \frac{dx_1}{dt} = -D_v \frac{\partial C_v}{\partial x} \tag{12.19}$$

where $C_{w,\text{ave}}$ is the average value of water content (WC) in the core calculated at the end of Stage 1, because it represents the volumetric water distribution inside the food.

Equation 12.19 considers that the amount of water available for vaporization in the core is the same as the humidity travelling in the dehydrated zone; therefore,

$$\frac{dx_1}{dt} = \frac{D_v}{C_{w,\text{ave}}} \frac{\partial C_v}{\partial x} \tag{12.20}$$

When the vaporization front reaches the center of the fried product, the dehydrated food system increases its temperature until the process is completed. Water transfer is absent and the moisture content corresponds to the bound water. To estimate temperature profiles the microscopic energy balance (Equation 12.1) with the boundary conditions (Equations 12.3 and 12.14) is solved.

Stage 3: The final step corresponds to the cooling of the fried product when it is removed from the hot oil medium. The energy microscopic balance (Equation 12.1) is solved with the boundary condition (Equation 12.3) and considering a natural convection heat transfer coefficient at the food–air interface, as follows:

$$x = L \quad -k_d \frac{\partial T}{\partial x} = h_3 (T - T_a) \quad t > 0 \tag{12.21}$$

h_3 can be calculated using the following equation [35]:

$$h = 1.32 \left(\frac{\Delta T}{H} \right)^{0.25} \text{ for horizontal plates with } H = \text{diameter (m)} \tag{12.22}$$

with ΔT = positive temperature difference between the surface of the fried food and the surrounding air (K).

In this step the OU is produced.

The mathematical model is solved for the different stages. The microscopic energy and mass balances lead to a system of coupled nonlinear partial differential equations.

12.2.2 Numerical Solution

The mathematical model is solved for the different frying stages. The microscopic balances of energy and mass lead to a system of coupled nonlinear partial differential equations. The system is solved by an implicit finite-differences method (Crank–Nicolson centered method) that provides an acceptable predictive quality, being numerically stable when compared to explicit schemes [38].

Stage 1: In this stage the following equations are used:

$$\frac{\partial U}{\partial t} = \frac{U_i^{n+1} - U_i^n}{\Delta t} \tag{12.23}$$

$$\frac{\partial U}{\partial x} = \frac{U_{i+1}^{n+1} - U_{i-1}^{n+1} + U_{i+1}^n - U_{i-1}^n}{4\Delta x} \tag{12.24}$$

$$\frac{\partial^2 U}{\partial x^2} = \frac{U_{i+1}^{n+1} + U_{i-1}^{n+1} - 2U_i^{n+1} + U_{i+1}^n + U_{i-1}^n - U_i^n}{2\Delta x^2} \tag{12.25}$$

where U corresponds to temperature or WC, i denotes node position, n is the time interval, Δx is the space increment, and Δt is the time step, such that $x = i\Delta x$ and $t = n\Delta t$, with $i = 0$ (center), b (border); n stands for time t, while $(n + 1)$ corresponds to time $(t + \Delta t)$. Assuming that thermal conductivity and the diffusion coefficient have a negligible change between times n and $(n + 1)$ Δt, one gets

$$\frac{\partial V}{\partial x} = \frac{V_{i+1}^n - V_{i-1}^n}{2\Delta x} \tag{12.26}$$

where V is the thermal conductivity or diffusion coefficient.

Finite differences (Equations 12.23 through 12.26) are replaced into Equation 12.1 to obtain the general equation to estimate temperatures, valid for $0 < i < b$:

$$T_{i+1}^{n+1}\left(-\frac{k_i^n}{2\Delta x^2} - \frac{k_{i+1}^n - k_{i-1}^n}{8\Delta x^2}\right) + T_i^{n+1}\left(\frac{\rho_i^n Cp_i^n}{\Delta t} + \frac{k_i^n}{\Delta x^2}\right) + T_{i-1}^{n+1}\left(\frac{-k_i^n}{2\Delta x^2} + \frac{k_{i+1}^n - k_{i-1}^n}{8\Delta x^2}\right)$$

$$= T_{i+1}^n\left(\frac{k_i^n}{2\Delta x^2} + \frac{k_{i+1}^n - k_{i-1}^n}{8\Delta x^2}\right) + T_i^n\left(\frac{\rho_i^n Cp_i^n}{\Delta t} - \frac{k_i^n}{\Delta x^2}\right) + T_{i-1}^n\left(\frac{k_i^n}{2\Delta x^2} - \frac{k_{i+1}^n - k_{i-1}^n}{8\Delta x^2}\right) \quad (12.27)$$

For the center point ($i = 0$), using the boundary condition Equation 12.3, the following is obtained:

$$T_0^{n+1}\left(\frac{\rho_o^n Cp_o^n}{\Delta t} + \frac{k_o^n}{\Delta x^2}\right) + T_1^{n+1}\left(\frac{-k_o^n}{\Delta x^2}\right) = T_1^n\left(\frac{k_o^n}{\Delta x^2}\right) + T_o^n\left(\frac{\rho_o^n Cp_o^n}{\Delta t} - \frac{k_o^n}{\Delta x^2}\right) \quad (12.28)$$

For the food surface ($i = b$), the corresponding equation is obtained by means of the boundary condition Equation 12.4:

$$T_i^{n+1}\left(\frac{\rho_i^n Cp_i^n}{\Delta t} + \frac{k_i^n}{\Delta x^2} + \frac{2h_1\Delta x}{k_i^n}\left(\frac{k_i^n}{2\Delta x^2} + \frac{k_i^n - k_{i-1}^n}{8\Delta x^2}\right)\right) + T_{i-1}^{n+1}\left(\frac{-k_i^n}{\Delta x^2}\right)$$

$$= T_i^n\left(\frac{\rho_i^n Cp_i^n}{\Delta t} - \frac{k_i^n}{\Delta x^2} + \frac{2h_1\Delta x}{k_i^n}\left(-\frac{k_i^n}{2\Delta x^2} - \frac{k_i^n - k_{i-1}^n}{8\Delta x^2}\right)\right) + T_{i-1}^n\left(\frac{k_i^n}{\Delta x^2}\right)$$

$$+ \frac{2h_1\Delta x T_{\text{oil}}}{k_i^n}\left(\frac{k_i^n}{\Delta x^2} + \frac{k_i^n - k_{i-1}^n}{4\Delta x^2}\right) + \frac{\Delta x L_{\text{vap}} D_{\text{wi}}^n (C_{\text{wi-1}}^n + C_{\text{wi-1}}^{n+1})}{k_i^n}\left(\frac{k_i^n}{\Delta x^2} + \frac{k_i^n - k_{i-1}^n}{4\Delta x^2}\right) \quad (12.29)$$

Similarly for water concentration the following general equation is obtained (valid for $0 < i < b$):

$$C_{\text{wi+1}}^{n+1}\left(\frac{-D_{\text{wi}}^n}{2\Delta x^2} - \frac{(D_{\text{wi+1}}^n - D_{\text{wi-1}}^n)}{8\Delta x^2}\right) + C_{\text{wi}}^{n+1}\left(\frac{1}{\Delta t} + \frac{D_{\text{wi}}^n}{\Delta x^2}\right) + C_{\text{wi-1}}^{n+1}\left(\frac{-D_{\text{wi}}^n}{2\Delta x^2} + \frac{(D_{\text{wi+1}}^n - D_{\text{wi-1}}^n)}{8\Delta x^2}\right)$$

$$= C_{\text{wi+1}}^n\left(\frac{D_{\text{wi}}^n}{2\Delta x^2} + \frac{(D_{\text{wi+1}}^n - D_{\text{wi-1}}^n)}{8\Delta x^2}\right) + C_{\text{wi}}^n\left(\frac{1}{\Delta t} - \frac{D_{\text{wi}}^n}{\Delta x^2}\right) + C_{\text{wi-1}}^n\left(\frac{D_{\text{wi}}^n}{2\Delta x^2} - \frac{(D_{\text{wi+1}}^n - D_{\text{wi-1}}^n)}{8\Delta x^2}\right) \quad (12.30)$$

For the center of the food, $i = 0$, using the boundary condition Equation 12.9, the following equation is obtained:

$$C_{\text{wo}}^{n+1}\left(\frac{1}{\Delta t} + \frac{D_{\text{wo}}^n}{\Delta x^2}\right) + C_{w1}^{n+1}\left(\frac{-D_{\text{wo}}^n}{\Delta x^2}\right) = C_{w1}^n\left(\frac{D_{\text{wo}}^n}{\Delta x^2}\right) + C_{\text{wo}}^n\left(\frac{1}{\Delta t} - \frac{D_{\text{wo}}^n}{\Delta x^2}\right) \quad (12.31)$$

In the border point, $i = b$ the humidity content is considered, to be zero (Equation 12.10).

The general temperature and WC equations and the particular expressions for the border and center lead to a system of algebraic equations that is solved by the Thomas algorithm, using a tridiagonal matrix of coefficients.

Stage 2: A variable grid is applied to overcome the deformation of the original fixed grid due to the shift of the vaporization front. The domain has been discretized as shown in Figure 12.1, where 0 indicates the food center, m is the moving front (the point m moves with time) and b is the surface. The scheme used in the present work is similar to that developed previously [39]. In this work, the numerical model uses three different spatial increments:

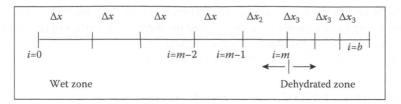

FIGURE 12.1 Scheme of the variable grid employed in the numerical solution (Stage 2).

Δx: constant spatial increment of the humidity zone.

Δx_2: variable spatial increment (corresponds to the first increment within the humidity zone, and withdraws according to the advance of the moving front).

Δx_3: variable spatial increment, in the dehydrated area.

In the core, Equation 12.27 is used to calculate the internal temperature, valid for $0 < i < m-2$; for the center point Equation 12.28 is applied.

In the point $(m-1)$ a discontinuity is produced because space increments toward the core are fixed in a Δx value; however, near the moving front, the space interval Δx_2 is variable (see Figure 12.1). In this case, the finite difference equations (Equations 12.23 through 12.26) are replaced by appropriate expressions that include the change of spatial increments [37]. Thus, the equation corresponding to $i = m-1$ is obtained:

$$T_f\left(-\frac{2\Delta x k_i^n}{(\Delta x + \Delta x_2)(\Delta x^2 + \Delta x_2^2)} - \frac{k_{i+1}^n - k_{i-1}^n}{2(\Delta x + \Delta x_2)^2}\right) + T_i^{n+1}\left(\frac{\rho_i^n Cp_i^n}{\Delta t} + \frac{2k_i^n}{\Delta x^2 + \Delta x_2^2}\right)$$

$$+ T_{i-1}^{n+1}\left(\frac{-2\Delta x_2 k_i^n}{(\Delta x + \Delta x_2)(\Delta x^2 + \Delta x_2^2)} + \frac{k_{i+1}^n - k_{i-1}^n}{2(\Delta x + \Delta x_2)^2}\right) = T_{\text{vap}}\left(\frac{2\Delta x k_i^n}{(\Delta x + \Delta x_2)(\Delta x^2 + \Delta x_2^2)}\right)$$

$$+ \frac{k_{i+1}^n - k_{i-1}^n}{2(\Delta x + \Delta x_2)^2}\right) + T_i^n\left(\frac{\rho_i^n Cp_i^n}{\Delta t} - \frac{2k_i^n}{\Delta x^2 + \Delta x_2^2}\right) + T_{i-1}^n\left(\frac{2\Delta x_2 k_i^n}{(\Delta x + \Delta x_2)(\Delta x^2 + \Delta x_2^2)} - \frac{k_{i+1}^n - k_{i-1}^n}{2(\Delta x + \Delta x_2)^2}\right)$$

$$(12.32)$$

To solve the energy balance in the dehydrated zone, the partial derivatives should be replaced by the finite differences equations 12.23 through 12.26. Thus, the general equation for the intermediate points $(m < i < b)$ is obtained similar to Equation 12.27 with the thermal properties of the dehydrated product. For the food surface $(i = b)$, the corresponding equation is obtained considering the boundary condition given by Equation 12.14. The expression is equivalent to Equation 12.29 without the surface vaporization term.

The set of equations allows us to calculate temperature profiles as a function of frying times. The equation system is solved by the Thomas algorithm. In this stage, the model uses 16 nodes in the core and 4 nodes in the dehydrated zone. The number of nodes in each zone changes as long as the vaporization front progresses. At the end of this stage, when the moving front reaches the center of the product, all the nodes are located in the dehydrated zone (20 nodes).

At each time step, the position of the moving front can be calculated by Equation 12.20 as follows:

$$x_1^{n+1} = x_1^n - \frac{D_v^{n+1}}{C_{w,\text{ave}}}\frac{C_{v,\text{equi}}}{(L - x_1^n)} \tag{12.33}$$

Stage 3: In the cooling period, Equations 12.27 through 12.29 are applied to evaluate temperature profiles, considering the air–product interaction and the thermal properties of the dehydrated product, using a space increment Δx_3 (the final value of Stage 2).

The model used employs 20 nodes and an interval of time of 0.1 s.

12.3 VALIDATION OF THE MATHEMATICAL MODEL

12.3.1 DESCRIPTION OF THE MODEL FOOD SYSTEM AND DETERMINATION OF FRYING CONDITIONS

To validate the mathematical model a dough system was used, prepared with 200 g of refined wheat flour (Molinos Río de La Plata, Argentina) and approximately 110 ml distilled water. From the obtained dough, discs of 60 mm in diameter and 7 mm thick were cut and immediately utilized. For each experiment, six discs were fried. The mathematical model was also validated on dough discs 5 mm thick.

Sensory characterization was performed to select time–temperature frying conditions. In all cases, panelists could not distinguish between the coated and the uncoated samples. Sensory analysis (color, flavor, texture and overall appearance) determined that 12 min at 160±0.5°C is the best frying condition for dough discs. This is the time required to complete starch gelatinization in dough discs (confirmed by microscopy observation) and consequently fried samples could be considered cooked.

12.3.2 FRYING EXPERIMENTS

The samples were fried in a controlled temperature deep-fat fryer (Yelmo, Argentina) filled with 1.5 l of commercial sunflower oil (AGD, Córdoba, Argentina). Oil composition was 99.93% lipids, with 25.71% mono-unsaturated and 64.29% poly-unsaturated fatty acids. The used oil was replaced with fresh oil after four frying batches; in each batch up to four dough discs were fried, maintaining the same product–oil volume ratio in all experiments.

Different constant frying temperatures were tested to select the working frying conditions according to sample characteristics; these temperatures ranged between 150±0.5°C and 170±0.5°C. Frying times between 5 and 15 min were analyzed and an initial temperature of 20°C was considered. The optimum time-temperature frying conditions were determined by a nontrained sensory panel of six members; the panelists judged color, flavor, texture, and overall appearance of the samples as described in previous work [40].

12.3.3 APPLICATION OF A HYDROCOLLOID COATING TO REDUCE OIL UPTAKE (OU) DURING FRYING

In order to analyze the effect of hydrocolloid coating to reduce OU in fried products, coating formulations were prepared using 1% (w/w) aqueous solution of MC (A4M, Methocel, COLORCON S.A., Argentina), and 0.75 % (w/w) sorbitol (Merck, USA) as plasticizer [40].

Dough samples were dipped in the coating suspensions for 30 s and immediately fried at the selected conditions, as previously described.

12.3.4 PHYSICAL PROPERTIES AND EXPERIMENTAL DATA INTRODUCED IN THE MATHEMATICAL MODEL

Table 12.1 shows the physical properties of the raw and fried dough systems fed to the mathematical model and the properties of the used oil.

The natural convection heat transfer coefficient for horizontal plates was calculated using Equation 12.5 and 12.6. The following dimensionless numbers were calculated:

$$Gr = \frac{L^3 \rho_{oil}^2 \, g \beta \Delta T}{\mu^2} \qquad (12.34)$$

$$Pr = \frac{cp\mu}{k} \qquad (12.35)$$

By replacing the properties shown in Table 12.1 the following values are obtained: $Gr = 5 \times 10^5$ and $Pr = 24.7$. Using Equation 12.5 a value of $Nu = 17.58$ is determined leading to $h_1 = 52.75$ W/m²K, with Equation 12.6, values of $Nu = 14.62$ and $h_1 = 43.86$ W/m²K are obtained. Therefore an approximate value of $h_1 = 50$ W/m²K is adopted.

It must be emphasized that the mathematical model is only fed with properties obtained from the literature and that the tortuosity factor is the unique adjusted parameter in the water vapor effective diffusion coefficient. The value used in this work is 8.8, similar to that reported for several food and nonfood products [41,42].

12.3.5 Heat Transfer Validation: Thermal Histories of Coated and Uncoated Samples during the Frying Process

Temperatures of coated and uncoated samples as a function of frying time have been measured using copper-constantan thermocouples (Omega). Temperature measurements in different internal points of the sample were performed by introducing the thermocouple carefully through the disc from the border toward the center. Temperatures were measured and recorded each 5 s using type T thermocouples linked to a data acquisition and control system, Keithley KDAC Series 500 (Orlando, FL). Figure 12.2a and b shows the measured values (symbols) at different selected points. The core temperature remains constant due to water evaporation while the temperature in the dehydrated zone increases up to the end of the process (15 min). A similar behavior of the thermal histories was reported by Costa et al. [34], working on sliced potatoes fried in sunflower oil at 140–180°C.

A cooling period can be observed when the product is removed from the fryer.

Also in Figure 12.2, simulated temperatures using the proposed model, with samples 5 or 7 mm thick are shown (lines) and a good agreement with the experimental thermal histories is observed.

Experimental thermal histories of coated and uncoated samples do not differ significantly ($P > 0.05$). These results can be explained by taking into account the fact that the average coating thickness determined by SEM is approximately 10 μm [40,43], leading to a negligible heat transfer resistance

12.3.6 Water Transfer Validation

WC was determined by measuring the weight loss of fried products upon drying in an oven at 110°C until reaching a constant weight [40]. At different frying times the relative variation of water retention % (WR) in the coated product relative to the uncoated one is calculated as follows:

$$WR = \left(\frac{WC_{coated}}{WC_{uncoated}} - 1 \right) \times 100 \qquad (12.36)$$

where WC is the water content of the samples (dry basis). For each frying time condition, results were obtained using all the samples from at least two different batches. The equilibrium WC is defined as the humidity reached at long frying times (1080 s).

Experimental values of WC for coated and uncoated samples are shown in Table 12.2; non-significant differences ($P > 0.05$) are detected between coated and uncoated discs. This could be attributed to the poor water vapor barrier of MC films [44–46]. The equilibrium WC values, which

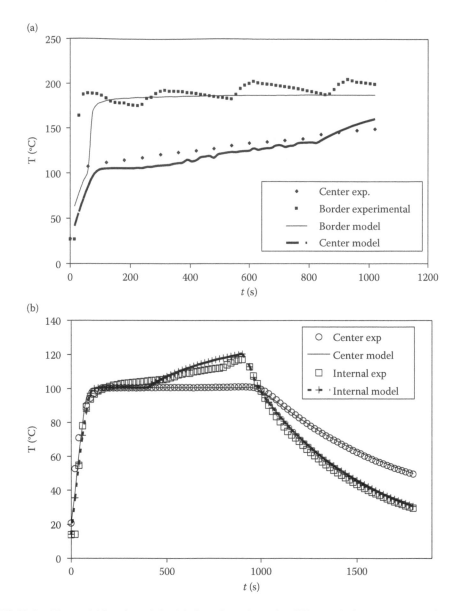

FIGURE 12.2 Thermal histories of the fried product along the different frying stages: experimental data (symbols) and numerical simulations (lines). (a) sample thickness = 5 mm; position of the thermocouples: center and border. (b) Sample thickness = 7 mm; positions of the thermocouples: center and internal point (1.6 mm from the border).

correspond to bound water, are 0.157 g water/g dry solid and 0.1695 g water/g dry solid for uncoated and coated samples, respectively.

The presence of the coating does not change the moisture content of the samples during frying because the hydrophilic MC coating becomes a negligible barrier to water transfer.

Simulated and experimental water concentrations versus frying times are shown in Figure 12.3a and b for samples of 5 or 7 mm thickness and a good agreement is achieved. Figure 12.4a shows the predicted position of the vaporization front as a function of frying time. At 15 min of process, the vaporization front reaches the center of the product, and the humidity corresponds to the bound water.

TABLE 12.2
Lipid and Water Contents of Uncoated and Coated Dough Discs as a Function of Frying Time

Time (s)	OU of Uncoated Samples (g oil/g dry solid)	OU of Coated Samples (g oil/g dry solid)	WC of Uncoated Samples (g water/g dry solid)	WC of Coated Samples (g water/g dry solid)	WR, Water Retention Relative Variation (%)	OUR (%)
0	0.00	0.00	0.721 ± 0.043[a]	0.786 ± 0.059	9.09	–
180	0.053 ± 0.010	0.045 ± 0.010	0.492 ± 0.017	0.477 ± 0.013	–3.06	14.07
360	0.058 ± 0.016	0.049 ± 0.001	0.356 ± 0.013	0.352 ± 0.012	–1.05	15.36
540	0.073 ± 0.008	0.054 ± 0.001	0.300 ± 0.013	0.296 ± 0.007	–1.34	26.36
720	0.074 ± 0.016	0.053 ± 0.006	0.253 ± 0.037	0.200 ± 0.006	–21.12	29.07
900	0.082 ± 0.017	0.062 ± 0.013	0.142 ± 0.016	0.159 ± 0.016	11.43	24.62
1080	0.089 ± 0.005	0.063 ± 0.005	0.157 ± 0.005	0.170 ± 0.004	8.03	29.91

[a] Value ± standard deviation.

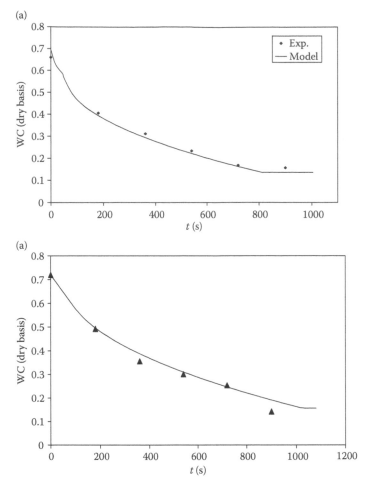

FIGURE 12.3   Water content along the different frying stages. Experimental data (symbols) and numerical simulations (lines). Sample thickness: (a) 5 mm; (b) 7 mm.

WC correlates linearly with the vaporization position x_1 that corresponds to the thickness of the humidity core (Figure 12.4b). Considering $d_Z = L - x_1$, the following equation is obtained:

$$\mathrm{WC} = 0.14893 + 158.4(L - d_Z) \quad r^2 = 0.9966 \tag{12.37}$$

12.3.7   TEMPERATURE AND WATER CONTENT (WC) PREDICTIONS USING THE MATHEMATICAL MODEL

Once the mathematical model is validated, it can be used to predict temperature and WC profiles as shown in Figure 12.5a and b at different frying times. As can be observed, the sample temperature of the dehydrated zone increases during the frying process reaching values higher than the equilibrium water vaporization temperature.

12.3.8   OIL UPTAKE (OU) OF COATED AND UNCOATED PRODUCTS

12.3.8.1   Experimental Results

The OU of fried products was determined by measuring the lipid content of dried samples using a combined technique of successive batch and semicontinuous Soxhlet extractions. The first batch

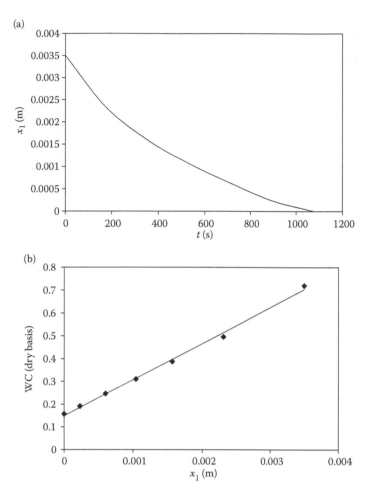

FIGURE 12.4   (a) Position of the vaporization front as a function of frying time. (b) Water content of fried dough discs as a function of the position of the vaporization front.

extraction was performed with petroleum ether: ethylic ether (1:1) followed by a Soxhlet extraction with the same mixture and another with n-hexane. Oil uptake relative (OUR) variation % in the coated product relative to the uncoated one is calculated as follows:

$$OUR = \left(1 - \frac{OU_{coated}}{OU_{uncoated}}\right) \times 100 \qquad (12.38)$$

For each frying time condition, results were obtained by using all the samples from at least two different batches.

The OU values of uncoated dough discs at each frying time are shown in Table 12.2. The final lipid content, at long frying times for these samples is 0.0894 g oil/g dry solid. The OU values of coated dough discs at each frying time are also shown in Table 12.2. As can be observed, MC coating reduces the lipid content of dough discs significantly ($P < .05$); the OU is 30% lower than that corresponding to uncoated samples (Table 12.2). Similarly, Williams and Mittal [14] reported that MC coatings reduced the lipid content of potato spheres by 34.5% compared to the control samples. This result was attributed to the presence of hydrocolloid films acting as lipid barriers, particularly MC, due to its thermal gelation properties. The final lipid content value, at long frying

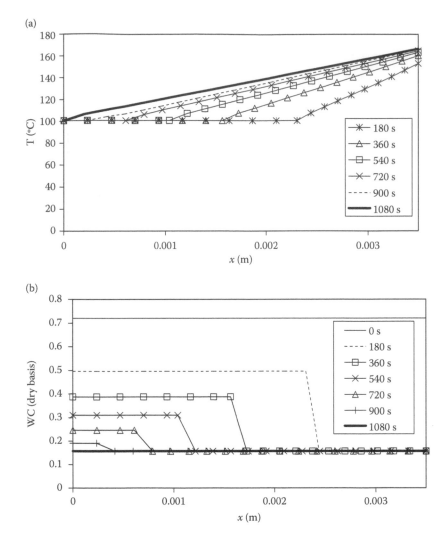

FIGURE 12.5 Simulated profiles obtained by the proposed mathematical model. (a) Temperature and (b) water content of a fried dough disc.

times, for coated samples is 0.0626 g oil/g dry solid; thus the ratio between the lipid content (db) of uncoated and coated samples is 1.4 at 1080 s, verifying that the MC coating acts as an effective oil barrier.

12.3.8.2 Microstructure Analysis of the Fried Product

SEM techniques and environmental scanning electron microscopy (ESEM) were used to observe the structure of the fried dough. SEMs of coated and uncoated samples were performed with a JEOL JSMP 100 SEM (Japan). The coated pieces were mounted on bronze stubs using double-sided tape and then coated with a layer of gold (40–50 nm). All samples were examined using an accelerating voltage of 5 kV.

The microstructure of the fried samples was also analyzed by using an Electroscan ESEM 2010.

The effects of frying time and coating application on the structure were studied. Figure 12.6 shows that the MC coating becomes dehydrated during the frying process and remains attached to the surface of the product, explaining the lower lipid content of the coated product. The thickness of

 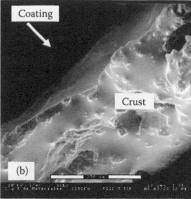

FIGURE 12.6 Cross-section micrographs of a fried dough disc coated with 1% methylcellulose plasticized with 0.75% sorbitol. (a) SEM, magnification 100 μm between marks; (b) ESEM.

the coating was measured on the micrographs, obtaining values ranging between 9 and 24 μm after 12 min of frying.

Figure 12.6 also shows the integrity of the MC layer and the good adhesion of this coating to the fried product. The addition of a plasticizer (sorbitol) to the MC coatings is necessary to achieve coating integrity [40]. The formation of a uniform coating on the surface of the sample is essential to limit mass transfer during frying [31]. Sorbitol addition improves the barrier properties of coatings by decreasing oil content compared to coated samples without a plasticizer. Similarly, Rayner et al. [47] reported that the performance of a soy protein film applied to dough discs increased through the addition of glycerol as a plasticizer, reducing the fat uptake by the food.

The coating does not prevent the formation of a dehydrated zone on the surface of the dough.

The release of water vapor leads to the formation of blisters at the outer surface of the crust that are smaller in size and number in the coated dough than in the uncoated one. In the first period of frying, the MC coating loses its water, the dough keeps its humidity, and the starch gelatinizes with a higher WC than in uncoated dough, leading to a more compact network (Figure 12.7a and b).

During frying in both coated and uncoated systems, the core is progressively dehydrated and starch gelatinization is completed in 12 min. The dehydrated zone shows holes due to the water vapor release; the size of the holes increases through the frying process (Figure 12.7c and d).

12.3.8.3 Relationship between Oil Uptake (OU) and Microstructure Changes during the Frying Process

OU depends on the structural changes that occur during the process; differences in the starting food microstructure can be expected to be important determinants in the evolution of the characteristics of the end product. Microstructural changes are produced during the frying time; the dehydrated zone increases, allowing the oil retained by the surface to penetrate into the pores formed by water evaporation.

The mathematical model allows us to estimate the thickness of the dehydrated zone ($d_Z = L-x_1$). Figure 12.8 shows OU versus frying time for both coated and uncoated samples. Simultaneously, the same figure shows the growth of the dehydrated zone with frying time. The predicted d_Z (dehydrated zone thickness) curve as a function of time is the same for coated and uncoated samples, according to the proposed mathematical model.

A simple equation is proposed to interpret the experimental results:

$$OU = a\,(1 - e^{-bt})$$

(12.39)

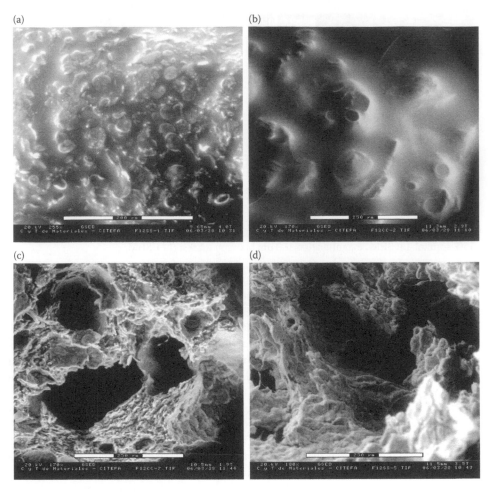

FIGURE 12.7 ESEM micrographs of a dough disc fried during 12 min: (a) Surface of the uncoated sample; (b) surface of the coated sample; (c) dehydrated zone of the coated sample; (d) dehydrated zone of the uncoated sample.

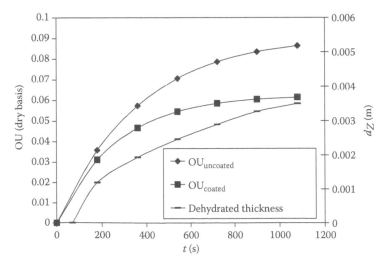

FIGURE 12.8 Oil uptake of uncoated and coated samples and dehydrated zone thickness (d_Z) as a function of frying time.

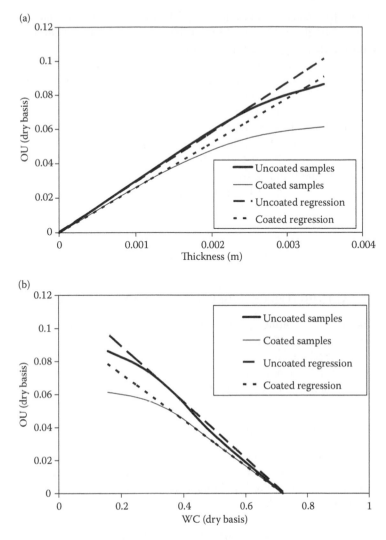

FIGURE 12.9 Oil uptake of uncoated and coated samples as a function of (a) dehydrated zone thickness and (b) the water content of the fried samples.

TABLE 12.3
Oil Absorption Parameters of Eq. (12.39)

Sample	Parameter A (g oil/g dry solid)	Parameter B (1/s)	Correlation Coefficient (r^2)
Uncoated	0.091	2.76×10^{-3}	0.999
Coated	0.062	3.81×10^{-3}	0.997

where a is the oil concentration at long times and b is the coefficient that takes into account the structural changes as a function of frying time. The parameters of Equation 12.39 are estimated by a nonlinear regression showing high values of the correlation coefficient r^2 (Table 12.3).

The results shown in Figure 12.8 allow us to correlate OU with the thickness of the dehydrated zone (d_Z), as shown in Figure 12.9a. A linear behavior of OU versus dehydrated zone thickness is maintained up to OU values of 0.031 g/g dry solid and 0.071 g/g dry solid for both coated and uncoated samples, respectively.

The values of OU for coated samples are lower than those of the uncoated ones. For low d_Z values, the oil retained by the surface could be incorporated into the dehydrated zone, when the sample is removed from the frying medium (linear relationship Figure 12.9a). When the dehydrated zone is large, a deviation from the linear behavior is observed. This could be attributed to the fact that the amount of oil retained at the sample surface is limited, being the oil surface wetting the property related to the interfacial tension governing these phenomena [6].

The presence of an MC coating with thermal gelation properties modifies the surface wetting and also becomes a mechanical barrier leading to a decrease in the OU of the coated samples.

The OU can also be correlated with the WC, showing a negative slope. At high water contents (WC) the results are linearly correlated; however deviations are observed at WC = 0.3 and 0.35 (dry basis) for uncoated and coated samples, respectively. Besides considering that WC is a function of d_Z (Equation 12.37) linear relationships are held for the initial frying periods (Figure 12.9b).

Following the OU criteria introduced by Pinthus and Saguy [5], the ratio between OU and WC is obtained from Table 12.2, being 0.17 for uncoated and 0.14 for coated samples; these results agree with the findings reported by Moyano and Pedreschi [48].

12.4 QUALITY ATTRIBUTES OF FRIED COATED AND UNCOATED SAMPLES

To evaluate whether the MC coating application affected the quality attributes, color, and texture parameters of coated and uncoated fried dough samples are analyzed.

SYSTAT software (SYSTAT, Inc., Evanston, IL) version 10.0 was used for all statistical analysis. Analysis of variance (ANOVA), nonlinear regression analysis, and Fisher LSD mean comparison tests were applied. The significance levels used were 0.05 and 0.01.

12.4.1 COLORIMETRIC MEASUREMENTS

Assays were carried out with a Minolta colorimeter CR 300 Series (Japan) calibrated with a standard ($Y = 93.2$, $x = 0.3133$, $y = 0.3192$). The CIELab scale was used, lightness (L^*) and chromaticity parameters a^* (red–green) and b^* (yellow–blue) were measured. L^*, C^* (chroma), H° (hue), and color differences (ΔE) were also calculated as

$$\Delta E = \sqrt{(\Delta L^*)^2 + (\Delta a^*)^2 + (\Delta b^*)^2} \tag{12.40}$$

where:

$$\Delta L^* = L_0^* - L_t^*, \ \Delta a^* = a_0^* - a_t^*, \ \Delta b^* = b_0^* - b_t^*$$

with L_t^*, a_t^*, b_t^* being the color parameter values of samples fried at different frying times. L_0^*, a_0^*, b_0^* were selected as the color parameters of samples fried for 3 min at 160°C and not the color of the raw sample in order to analyze the effect of frying time.

Samples were analyzed in triplicates, recording four measurements for each sample.

Nonsignificant ($P > 0.05$) differences were observed in the analyzed color parameters of coated and uncoated dough discs. The chromaticity parameter b^* increases significantly ($P < 0.05$) with frying time, while lightness L^* and chromaticity parameter a^* are independent of frying time. Besides, chroma parameter and color differences (ΔE) increase and hue decreases significantly ($P < 0.05$) as a function of frying time for $t < 720$ s Figure 12.10a). During the overcooking of the samples, all parameters remain constant.

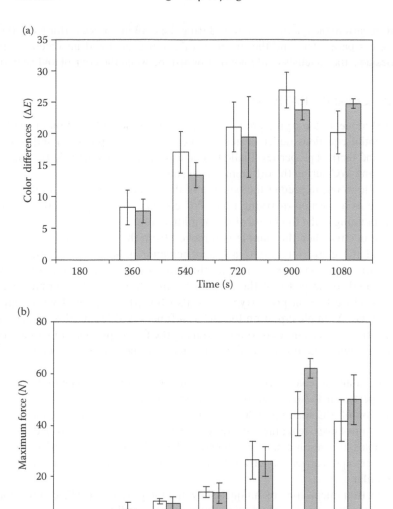

FIGURE 12.10 Quality attributes of coated (filled bars) and uncoated (empty bars) fried dough discs as a function of frying time. (a) Color differences; (b) firmness.

12.4.2 TEXTURE ANALYSIS

Breaking forces of the samples were measured by puncture tests using a texture analyzer TA.XT2i–*Stable Micro Systems* (Haslemere, Surrey, UK) with a 5 kg cell. The samples were punctured with a cylindrical plunger (2 mm diameter) at 0.5 mm/s. Maximum force at rupture was determined from the force–deformation curves. At least 10 samples were measured for each assay. Samples were allowed to reach room temperature before the tests.

During the frying process (for $t < 720$ s) similar values of the maximum force were obtained for coated and uncoated fried samples. In all cases, the maximum force to puncture the samples increased as a function of frying time, due the formation of a dehydrated zone (Figure 12.10b). At longer frying times, when the discs were considered overcooked, the highest maximum forces for coated samples were observed.

Thus, with regard to the quality attributes during the cooking process (for $t < 720$ s) differences between the color parameters and the firmness values of coated and uncoated samples were not significant; besides, the panelists could not distinguish between the control and coated samples.

12.5   FINAL CONSIDERATIONS

A mathematical model of the frying process based on the numerical solution of the heat and mass transfer differential equations under unsteady state conditions is proposed and solved using measured physical and thermal properties. It allows one to simulate satisfactorily the experimental data of temperature and WC during the different frying stages.

The model allows one to predict the position of the vaporization front and the thickness of the dehydrated zone as a function of frying time. WC correlates linearly with the vaporization front position corresponding to the thickness of the humidity core.

The OU that occurs when the sample is removed from the frying medium is correlated with the thickness of the dehydrated zone; a linear behavior is held for the initial frying period. However, deviations are observed when the thickness of the dehydrated zone increases. This could be attributed to the fact that the amount of oil retained at the sample surface is determined by the surface tension property. OU is also linearly correlated with water loss at the initial frying stage. A simple equation for OU as a function of frying time is proposed, considering the microstructural changes developed during the frying process, in which the dehydrated zone increases allowing the oil retained by the surface to penetrate into the pores left by water evaporation.

The presence of an MC coating reduces the OU due to the thermal gelation behavior, modifying the wetting properties and also becoming a mechanical barrier to the oil.

The application of MC coatings reduces significantly ($P < 0.05$) the oil content of dough discs with regard to control ones, reaching a decrease of 30%. However, the coating does not modify the WC of the samples; this result can be attributed to the hydrophilic characteristics of the films that lead to poor water vapor barrier properties; besides the thermal histories of the coated and uncoated samples are similar.

Scanning electron microscopy techniques show the integrity of the MC coating and good adhesion to the food product after its dehydration during the deep oil frying process.

NOMENCLATURE

A	Area, (m^2)
b	Index indicating surface position
Cp	Specific heat, (J/(kg °C))
C_v	Concentration of water vapor in the dehydrated zone, (kg/m^3)
C_w	Concentration of water in the dough, (kg/m^3)
d	Distance from surface, (m)
D_v	Effective diffusion coefficient of vapor in the dehydrated food, (m^2/s)
D_W	Effective diffusion coefficient of water in the food matrix, (m^2/s)
d_Z	Thickness of the dehydrated zone, (m)
g	Universal gravitational constant (m^2/s)
h	Heat transfer coefficient, (W/(m^2 K))
i	Position index
k	Thermal conductivity, (W/(m K))
L	Half thickness of the sample, (m)
L_{vap}	Latent heat of vaporization of water, (J / kg)

L*	Lightness
n	Time index
OU	Oil uptake in dry basis, (g oil/g dry solid)
OUR	Oil uptake relative variation, (%)
t	Time, s (or min)
T	Temperature, (°C)
U	Generic variable: temperature or concentration in finite differences equations
V	Generic property: thermal conductivity or diffusivity in finite differences equations
Vol	Volume of the sample, (cm^3)
WR	Water retention relative variation, (%)
WC	Water content of the samples in dry basis, (g water/g dry solid)
x	Spatial coordinate, (m)
x_1	Position of the vaporization front, (m)

Greek symbols

α	Thermal diffusivity, (m^2/ s)
β	Volumetric coefficient of expansion of the fluid (1/ K)
ε	Porosity
τ	Tortuosity factor
ρ	Density, (kg/m^3)
(Δa*)	Difference of the chromaticity parameter a*
(Δb*)	Difference of the chromaticity parameter b*
(ΔE)	Color difference (ΔE)
Δx	Spatial increment, m
Δt	Time increment, s

Dimensionless numbers

Gr	Grashof number, $Gr = \dfrac{L^3 \rho^2\, g\beta\Delta T}{\mu^2}$
Nu	Nusselt number, $Nu = \dfrac{hL}{k}$
Pr	Prandtl number, $Pr = \dfrac{cp\mu}{k}$

Subscripts

a	Air
ave	Average
equi	Equilibrium
d	Dehydrated
f	Fictitious
ini	Initial
w	Water
oil	Oil
v	Vapor
va	Vapor-air interface
vap	Vaporization

REFERENCES

1. Shaw, R. and Lukes, A.C. Reducing the oil content of potato chips by controlling their temperature after frying. USDA, ARS-73-58. US Department of Agriculture, Washington, DC, 1968.
2. Singh, R.P. Heat and mass transfer in foods during frying. *Food Technol.*, 49 (4), 134–137, 1995.
3. Aguilera, J.M. and Hernández, H.G. Oil absorption during frying of frozen parfried potatoes. *J. Food Sci.*, 65 (3), 476–479, 2000.
4. Aguilera, J.M. Fritura de alimentos. In J.M. Aguilera, *Temas de tecnología de alimentos*. México, D.F, Programa Iberoamericano CYTED, Instituto Politécnico Nacional, vol. 1, 185–214, 1997.
5. Pinthus, E.J. and Saguy, I.S. Initial interfacial tension and oil uptake by deep-fat fried foods. *J. Food Sci.*, 59 (4), 804–807, 1994.
6. Kassama, L.S. *Pore development in food during deep-fat frying.* Department of Bioresource Engineering, Macdonald Campus of McGill University, Quebec, Canada, 2003.
7. Mellema, M. Mechanism and reduction of fat uptake in deep-fat fried foods. *Trends in Food Sci. Technol.*, 14, 364–373, 2003.
8. Ufheil, G. and Escher, F. Dynamics of oil uptake during deep-fat frying of potato slices. *LWT,* 29 (7), 640–644, 1996.
9. Moreira, R.G. and Barrufet, M.A. A new approach to describe oil absorption in fried foods: A simulation study. *J. Food Eng.*, 31, 485–498, 1998.
10. Yamsaengsung, R. and Moreira, R.G. Modeling the transport phenomena and structural changes during deep-fat frying. Part I Model development. *J. Food Eng.*, 53, 1–10, 2002.
11. Bouchon, P., Aguilera, J.M. and Pyle, D.L. Structure oil-absorption relationships during deep-fat frying. *J. Food Sci.*, 68, (9), 2711–2716, 2003.
12. Bouchon, P. and Pyle, D.L. Modelling oil absorption during post-frying cooling, I: Model development. *Trans IChemE, Part C, Food and Bioprod. Proc.*, 83 (C4), 1–9, 2005.
13. Bouchon, P. and Pyle, D.L. Modelling oil absorption during post-frying cooling, II: Solution of the mathematical model, model testing and simulations. *Trans IChem, Part C, Food and Bioprod. Proc.*, 83 (C4), 1–12, 2005.
14. Williams, R. and Mittal, G.S. Water and fat transfer properties of polysaccharide films on fried pastry mix. *LWT,* 32, 440–445, 1999.
15. Williams, R. and Mittal, G.S. Low-fat fried foods with edible coatings: modeling and simulation. *J. Food Sci.*, 64 (2), 317–322, 1999.
16. Perkins, E.G. and Erikson, M.D. *Deep Frying: Chemistry, Nutrition, and Practical Applications.* Champaign, IL: AOCS Press, 1996.
17. Rice, P. and Gamble, M.H. Modeling moisture loss during potato slice frying. *Int. J. Food Sci. Technol.*, 24, 183, 1989.
18. Aguilera, J. M. and Stanley, D.W. *Microstructural Principles of Food Processing and Engineering.* Gaithersburg, MD: Aspen Publishers Inc., 1999.
19. Ni, H. and Datta, A.K. Moisture, oil and energy transport during deep-fat frying of food materials. *Trans IChemE, Part C, Food Bioprod Proc.*, 77, 194–204, 1999.
20. Yamsaengsung, R. and Moreira, R.G. Modeling the transport phenomena and structural changes during deep-fat frying. Part II Model solution and validation. *J Food Eng.*, 53, 11–25, 2002.
21. Ateba, P. and Mittal, G.S. Modelling the deep-fat frying of beef meatballs. *Int. J Food Sci. Technol.*, 29, 429–440, 1994.
22. Farkas, B.E., Singh, R.P. and Rumsey, T.R. Modeling heat and mass transfer in immersion frying. I. Model development. *J. Food Eng.*, 29, 211–226, 1996.
23. Farkas, B.E., Singh, R.P. and Rumsey, T.R. Modeling heat and mass transfer in immersion frying. II. Model solution and verification. *J. Food Eng.*, 29, 227–248, 1996.
24. Singh, R.P. Moving boundaries in food engineering. *Food Technol.*, 54 (2), 44–53, 2000.
25. Farid, M. The moving boundary problems from melting and freezing to drying and frying of food. *Chem.l Eng. Proc.*, 41, 1–10, 2002.
26. Farid, M.M. and Chen, X.D. The analysis of heat and mass transfer during frying of food using a moving boundary solution procedure. *Heat and Mass Transfer*, 34, 69–77, 1998.
27. Balasubramaniam, V.M., Mallikarjunan, P. and Chinnan, M.S. Heat and mass transfer during deep-fat frying of chicken nuggets coated with edible film: Influence of initial fat content. *CoFE 1995,* 103–106, 1995.
28. Grover, J.A. Methylcellulose and its derivates. In R.L. Whistler and J.N. Be Miller, *Industrial Gums* (pp. 475–504). San Diego, CA: Academic Press, 1993.

29. Meyers, M.A. Functionality of hydrocolloids in batter coating systems. In K.Kulp and R. Loewe, *Batters and Breadings in Food Processing* (pp. 17–142), St. Paul, MN: American Association for Cereal Chemists, 1990.
30. Mallikarjunan, P., Chinnan, M.S., Balasubramaniam, V.M. and Phillips, R.D. Edible coatings for deep-fat frying of starchy products. *LWT*, 30, 709–714, 1997.
31. Huse, H.L., Mallikarjunan, P., Chinnan, M.S., Hung, Y.C. and Phillips, R.D. Edible coatings for reducing oil uptake in production of akara (deep-fat frying of cowpea paste). *J. Food Proc. Pres.*, 22, 155–165, 1998.
32. Coulson J.M. and Richardson J.F. *Chemical Engineering,* Vol.1. London, UK: Pergamon Press, 176, 1956.
33. Geankoplis, C. *Transport Processes and Unit Operations,* Third Edition. Englewood Cliffs, NJ: Prentice Hall, 1993.
34. Costa, R.M., Oliveira, F.A.R., Delaney, O. and Gekas, V. Analysis of heat transfer coefficient during potato frying. *J. Food Eng.*, 39, 293–299, 1998.
35. Dagerskog, M. and Sorenfor, P. A comparison between four different methods of frying meat patties: Heat transfer, yield, and crust formation. *LWT,* 11, 306–311, 1978.
36. Moreira, R.G., Palau, J., Sweat, V.E. and Sun, X. Thermal and physical properties of tortilla chips as a function of frying time. *J. Food Proc. Pres.*, 19, 175–189, 1995.
37. Mc Adams W.H. *Heat Transmission,* Third Edition. New York, NY: McGraw Hill Book Co., 1954.
38. Forsythe, G.E. and Wasow, W.R. *Finite Difference Methods for Partial Differential Equations.* New York, NY: John Wiley and Sons, 1960.
39. Campañone, L.A, Salvadori, V.O. and Mascheroni, R.H. Weight loss during freezing and storage of unpackaged foods. *J. Food Eng.*, 47, 69–79, 2001.
40. García, M.A., Ferrero, C., Bértola, N, Martino, M. and Zaritzky, N. Edible coatings from cellulose derivatives to reduce oil uptake in fried products. *Innov. Food Sci. Emerg. Technol.*, 3, 391–397, 2002.
41. Harper J.C. Transport properties of gases in porous media at reduced pressures with reference to freeze-drying. *AICHE J.*, 298–302, 1962.
42. Moldrup, P., Olcsen, T., Komatsu, T., Schjonning, P. and Rolston, D.E. Tortuosity, diffusivity and permeability in the soil liquid and gaseous phases. *Soil Sci. Soc. Am. J.*, 613–623, 2001.
43. García, M.A., Ferrero, C., Bértola, N, Martino, M. and Zaritzky, N. Methylcellulose coatings reduce oil uptake in fried products. *Food Sci. Technol. Int.,*. 10 (5), 339–346, 2004.
44. Donhowe, I.G. and Fennema, O.R. The effects of plasticizers on crystallinity, permeability, and mechanical properties of methylcellulose films. *J. Food Proc. Pres.*, 17, 247–257, 1993.
45. Krochta, J.M. and De Mulder-Johnston, C. Edible and biodegradable polymers films: Challenges and opportunities. *Food Technol.*, 51 (2), 61–74, 1997.
46. Debeaufort, F. and Voilley, A. Methylcellulose-based edible films and coatings: 2. Mechanical and thermal properties as a function of plasticizer content. *J. Agric. Food Chem.*, 45, 685–689, 1997.
47. Rayner, M., Ciolfi, V., Maves, B., Stedman, P. and Mittal, G.S. Development and application of soy-protein films to reduce fat intake in deep-fried foods. *J. Sci. Food Agric.*, 80, 777–782, 2000.
48. Moyano, P.C. and Pedreschi, F., Kinetics of oil uptake during frying of potato slices: Effect of pre-treatments. *LWT,* 39 (3), 285–291, 2006.
49. Rask C. Thermal properties of dough and bakery products: a review of published data. *J. Food Eng.*, 9, 167–193, 1989.
50. Moreira, R.G., Castell-Perez M.E. and Barrufet, M.A. *Deep Fat Frying: Fundamentals and Applications.* Gaithersburg, MD: Aspen Publishers Inc., 1999.
51. Zanoni, B., Peri, C. and Gianotti, R. Determination of the thermal diffusivity of bread as a function of porosity. *J. Food Eng.*, 26, 497–510, 1995.
52. Bird, R.B., Stewart, W.E. and Lightfoot, E.N. *Transport Phenomena.* New York, NY: John Wiley and Sons, 1976.
53. Tong, C.H. and Lund, D.B. Effective moisture diffusivity in porous materials as a function of temperature and moisture content. *Biotechnol. Progress*, 6, 67–75, 1990.

13 Baking Process: Mathematical Modeling and Analysis

Weibiao Zhou
National University of Singapore

CONTENTS

13.1   INTRODUCTION

Bakery products, particularly bread with its long history, are a constant, daily element in the diet of most people. Most recent archaeological discovery indicated that baking might be dated back as early as 21,000 B.C. [1], when people knew wheat and learnt to mix wheat grain meal with water and bake it on stones heated by fire, thereafter the first flat bread was made. Leaven bread and its baking oven can be traced back to ancient Egyptians around 3000 B.C. Nowadays, varieties of bread can be found in supermarkets and bakeries. As consumers pay more attention to a healthy diet including traditional and organic food, bakeries using traditional ingredients and techniques such as sourdough, wholemeal and multigrain bread have become increasingly popular.

For bakery products, the baking process is a key step in which raw dough is transformed into light, porous, readily digestible and flavorful product under the influence of heat. As bread represents the largest category of bakery products, this chapter focuses on bread-baking. However, the mathematical models and analysis approaches presented in this chapter are generic in nature and can easily be extended to baking of other bakery products.

During bread-baking, the most apparent phenomenal changes are volume expansion, crust formation, inactivation of yeast and enzymatic activities, protein coagulation, partial gelatinization of starch in dough and moisture loss [2,3]. From the production viewpoint, it is highly desirable to establish optimum baking conditions to produce bread with lowest moisture loss and consistent quality attributes including texture, color, and flavor. This requires an in-depth understanding of heat and mass transfer phenomena both in an oven chamber and within dough pieces.

On the dough side, bread-baking can be taken as a multiphase flow problem. Heat flows from the bread surface toward the center. Liquid water, evaporating at the surface, diffuses from the bread center toward the surface. Meanwhile, with thermal expansion and evaporation of carbon dioxide and water, gas cells are formed inside the bread. Formation of these gas cells provides a pathway for water vapor to diffuse from hot area to cool area where it is condensed back to liquid. As a result, liquid water content in the center tends to increase in the early stage of baking. This evaporation–condensation mechanism also explains the fact that heat transfer in bread during baking is much faster than that described by conduction alone. The simultaneous heat, liquid water and water vapor transfers together with evaporation-condensation yield a model consisting of three partial differential equations augmented by algebraic equations. However, numerically solving this model albeit for a one-dimensional case proves to be an issue. This will be demonstrated in Section 13.2.

On the oven side, the challenge is to establish the temperature profile for a piece of dough/bread during its residence period inside the oven. In an industrial baking process, dough is conveyed continuously into the oven chamber as a first-in-first-out system. Key process parameters including conveyor speed, oven load, air temperature and air flow pattern, all play important roles in determining the product temperature profile and therefore, the final product quality. This requires establishing a model that covers the temperature distribution in the entire oven chamber. Computational fluid dynamics (CFD) modeling technique may be the only effective tool in dealing with the complexity of a continuous baking process. Both two-dimensional and three-dimensional CFD models have been established. Oven operating parameters were studied. Dynamic response of the traveling tin temperature profiles could be predicted in accordance with a change in the oven load. The modeled tin temperature profiles showed good agreement with the measured tin temperature profiles from the actual industrial baking process. Positioning of the controller sensors has also been investigated through a sensitivity study based on the CFD models. Applications of the CFD models have been extended to designing process controllers for the oven. These results will be reviewed in Section 13.3.

13.2 MODELING OF HEAT AND MASS TRANSFER WITHIN DOUGH DURING BAKING

13.2.1 GENERAL CONSIDERATIONS

Heat transfer in dough is a combination of conduction/radiation from band or tins to the dough surface, conduction in the continuous liquid/solid phase of the dough, and evaporation–condensation in the gas phase of the dough. De Vries et al. [4] described a four-step mechanism for heat transport: (1) water evaporates at the warmer side of a gas cell that absorbs latent heat of vaporization; (2) water vapor then immigrates through the gas phase; (3) when meeting the cooler side of the gas cell, water vapor condenses and becomes water; (4) finally heat and water are transported by conduction and diffusion through the gluten gel to the warmer side of the next cell. Water diffusion mechanism becomes important because dough tends to be a poor conductor that limits heat transfer via conduction.

Diffusion, with evaporation and condensation has been assumed to be the mass transfer mechanisms inside dough [5,6]. The transport of water is driven by the gradients in water content. Thorvaldsson and Skjoldebrand [7] found that at the center of a loaf, the measured water content decreased until the center temperature was at 70±5°C because of volume expansion. Then there was a change in the measured water content, which began to increase. To reduce the partial water vapor pressure due to temperature gradient, water moved toward the loaf center and the surface by evaporation and condensation. As a result, the rise in crumb temperature was accelerated.

In the model by Zanoni et al. [8], an evaporation front inside the dough was assumed to be always at 100°C. This evaporation front progressively advanced toward the center as bread's temperature increased. Crust was formed in the bread portion above the evaporation front. With similar parameters, Zanoni et al. [9] developed a two-dimensional axi-symmetric heat diffusion model. The phenomena were described separately for the upper and lower parts (crust and crumb). The upper part (crust) temperature was determined by equations including heat supply by convection, conductive heat transfer toward the inside, and convective mass transfer toward the outside. The lower part (crumb) temperature was determined by Fourier's law. In addition to the Cartesian co-ordinate models, a one-dimensional cylindrical co-ordinate model was also established by De Vries et al. [4].

13.2.2 SEQUENTIAL MODEL

The internal evaporation-condensation mechanism well explains the fact that heat transfer in bread during baking is much faster than that described by conduction alone in dough/bread. It also supports the observation that there is an increase in water content at the center of bread during the early stage of baking, rather than a monotonous decrease resulting from liquid water diffusion and surface evaporation only. Therefore, the best model for bread-baking should be a multiphase flow model which consists of three partial differential equations for the simultaneous heat transfer, liquid water diffusion and water vapor diffusion respectively, along with two algebraic equations describing water evaporation and condensation in gas cells. Saturated conditions can be assumed for gas cells in the model.

Thorvaldsson and Janestad [6] used this multiphase flow model to describe a one-dimensional case where bread of 12 cm × 12 cm × 2 cm was baked directly inside an oven, as shown in Figure 13.1. Baking tin was absent. As the first two dimensions of the bread are significantly larger than the third dimension, the bread can be taken as a thin slab where only one-dimensional heat and mass transfer are significant along the 2 cm thickness. The system was symmetric. The model can be written as follows:

$$\frac{\partial T}{\partial t} = \frac{1}{\rho c_p} \frac{\partial}{\partial x}\left(k \frac{\partial T}{\partial x}\right) + \frac{\lambda}{c_p} \frac{\partial W}{\partial t}, \quad 0 < x < x_L/2, \quad t > 0. \tag{13.1}$$

$$\frac{\partial V}{\partial t} = \frac{\partial}{\partial x}\left(D_V \frac{\partial V}{\partial x}\right), \quad 0 < x < x_L/2, \quad t > 0. \tag{13.2}$$

$$\frac{\partial W}{\partial t} = \frac{\partial}{\partial x}\left(D_W \frac{\partial W}{\partial x}\right), \quad 0 < x < x_L/2, \quad t > 0. \tag{13.3}$$

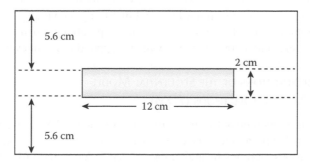

FIGURE 13.1 Bread-baking as an approximate one-dimensional system. (From Zhou, W., *International Journal of Computational Fluid Dynamics*, 19, 73, 2005. With permission.)

with the following boundary and initial conditions:

$$-k\left[\frac{\partial T}{\partial x}\right]_{x=0} = h_r\left(T_r - T_s\right) + h_c\left(T_{\mathrm{air}} - T_s\right) - \lambda\rho D_w\left[\frac{\partial W}{\partial x}\right]_{x=0}, \quad \left[\frac{\partial T}{\partial x}\right]_{x=x_L/2} = 0, \quad t > 0. \tag{13.4}$$

$$T(x,0) = T_0(x), \quad 0 \le x \le x_L/2.$$

$$\left[\frac{\partial V}{\partial x}\right]_{x=0} = h_V\left(V(0,t) - V_{\mathrm{air}}\right), \quad \left[\frac{\partial V}{\partial x}\right]_{x=x_L/2} = 0, \quad t > 0.$$

$$V(x,0) = V_0(x), \quad 0 \le x \le x_L/2. \tag{13.5}$$

$$\left[\frac{\partial W}{\partial x}\right]_{x=0} = h_W\left(W(0,t) - W_{\mathrm{air}}\right), \quad \left[\frac{\partial W}{\partial x}\right]_{x=x_L/2} = 0, \quad t > 0.$$

$$W(x,0) = W_0(x), \quad 0 \le x \le x_L/2. \tag{13.6}$$

where $T(x,t)$ is temperature in K, x is space co-ordinate in m, t is time in s. $V(x,t)$ is water vapor content in (kg water)/(kg product), D_V is water vapor diffusivity in m^2/s. ρ is apparent density in kg/m^3, c_p is specific heat in J/kgK, k is thermal conductivity in W/mK, λ is the latent heat of evaporation of water in J/kg, $W(x,t)$ is liquid water content in (kg water)/(kg product), x_L is the thickness of the bread slab. h_r and h_c are the heat transfer coefficients due to radiation and convection respectively, in W/m^2K. T_r and T_{air} are radiation source temperature and surrounding oven air temperature respectively, in K. $T_s = T(0,t)$ is bread surface temperature in K. D_W is liquid water diffusivity in m^2/s. h_V and h_W, in 1/m, are the mass transfer coefficients of water vapor and liquid water at the bread surface, respectively. V_{air} and W_{air} are the water vapor content and liquid water content of the oven air, respectively. T_0, V_0, and W_0 are the initial temperature, initial water vapor content and initial liquid water content of the bread, respectively.

Equations 13.1 through 13.3 are augmented with an algebraic equation which describes the relationship between water vapor content V and liquid water content W under saturation condition. The physical parameters in Equations 13.1 through 13.6 are not constants and they depend on the local temperature $T(x,t)$ and moisture content $W(x,t)$.

Implementation of this model may be described as "sequential." At every numerical iteration step, Equation 13.1 is solved first, followed by the algebraic equation applied to adjust W and V based on the new temperature according to saturation condition, which decides the amount of water evaporation. Next, Equation 13.2 is solved to calculate water vapor content V, followed by applying the algebraic equation again to adjust V and W according to saturation condition, which accounts for water condensation after water vapor diffusion. Finally, Equation 13.4 is solved to calculate W. This procedure is repeated at each time step until the end of baking. Obviously, it is critical to employ a stable and reliable numerical scheme to solve the three parabolic partial differential equations together with the augmented algebraic equation which is applied twice in every iteration.

13.2.3 NUMERICAL SOLUTIONS OF THE SEQUENTIAL MODEL BY FINITE DIFFERENCE METHODS (FDM)

Zhou [10] applied the theta method of the finite difference methods (FDM) to solve the above system. There were three algorithm parameters including theta θ, time step Δt and space step Δx. Different values of θ yielded various methods including the explicit Euler method ($\theta = 1$), the implicit Euler method ($\theta = 0$) and the Crank–Nicolson method ($\theta = 1/2$). The method of fictitious boundaries was used to handle the boundary conditions which were all Neumann and Robin types.

Figure 13.2 shows a typical temperature and moisture profiles within bread solved from the above-described model by the θ method. To evaluate the performance of various numerical schemes and their sensitivity to the algorithm parameters, quantities at several critical points in the temperature and moisture profiles were taken to characterize the profiles. The first of these is the time when bread temperature reaches 100°C. During baking, crust starts to form at this temperature when all liquid water evaporates. The second characteristic value is the time when liquid water content at the center of bread reaches its peak. As stated earlier, water content at the bread center was observed to increase during the early baking stage, and among the various models the internal evaporation-condensation model might be the one that best describes this phenomenon. The third characteristic value is the

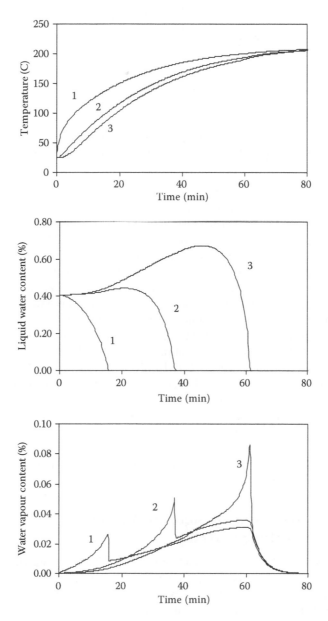

FIGURE 13.2 Typical temperature and moisture profiles. 1: surface; 2: halfway; 3: center. (From Zhou, W., *International Journal of Computational Fluid Dynamics*, 19, 73, 2005. With permission.)

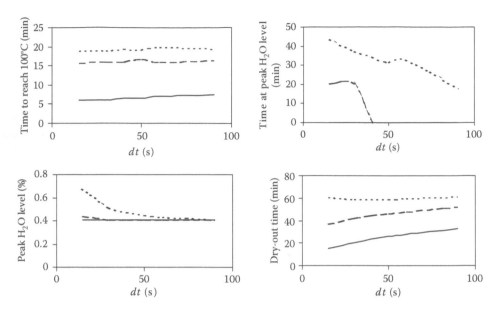

FIGURE 13.3 Sensitivity of backward FDM to time step, $\Delta x = 0.125$ cm. Solid lines: surface; dashed lines: halfway; dotted lines: center. (From Zhou, W., *International Journal of Computational Fluid Dynamics*, 19, 73, 2005. With permission.)

level of peaked liquid water content at the center. This value, though higher than the initial moisture content of the dough, should be reasonably restrained due to the small diffusivity of water vapor in bread. The fourth characteristic value is the time when water content at the center of bread becomes zero. This is the time when the whole bread turns into crust.

It was found that among various θ values between 0 and 1, only $\theta = 0$ produced converged results. All other values of θ produced erroneous temperature and moisture profiles at all time step and space step examined [10]. This means that only the implicit Euler method can be used for this system. All other methods including the Crank–Nicolson method, failed.

With the implicit Euler method, diverged temperature and moisture profiles were also produced when time step $\Delta t \leq 5$ s. Meanwhile, the initial increase in liquid water profile at the bread center vanished when Δt was sufficiently large (≥ 120 s). This means that with a large Δt the unique feature described by the internal evaporation-condensation model was lost in the corresponding numerical solution. Therefore, both small Δt and large Δt produced inaccurate or useless results. Figure 13.3 shows the sensitivity of the numerical solution to Δt in the range of 15–90 s by which reasonable results were produced. The numerical solution was very sensitive to the value of time step Δt. Meanwhile, for a given Δt within this range, it was evident that different values of space step Δx in the range of 0.0125–0.25 cm produced almost the same results. In other words, the numerical solution was not sensitive to the space step Δx.

13.2.4 Numerical Solutions of the Sequential Model by Finite Elements Methods (FEM)

Finite element methods (FEM) were also applied to solving the one-dimensional system by Zhou [10]. The two types of shape functions adopted in the FEM schemes were the linear pyramid shape functions and the high order Hermite cubic shape functions. Both uniform and nonuniform meshes were applied. It was found that FEM with pyramid shape functions under a uniform mesh produced almost the same result as the implicit Euler FDM. All characteristic values under all time steps and space steps were the same, except in one or two cases where small differences were negligible.

Furthermore, reasonable results were produced only when Δt was in the range of 15–90 s. Similar to the implicit Euler method, the numerical solution by FEM was not sensitive to the space step Δx, but highly sensitive to the time step Δt. Adopting a nonuniform mesh in FEM only reduced the already small difference between the results by different space steps. It changed neither the narrow workable range of Δt, nor the high sensitivity of the numerical solution to Δt.

FEM with the Hermite cubic shape functions under a uniform mesh produced comparable results to other schemes. The workable range of Δt remained the same and the sensitivity of the numerical solution to the time step remained high. However, unlike the other schemes, the numerical solution by this scheme was also sensitive to the space step. Adopting a nonuniform mesh in this scheme reduced the sensitivity to the space step. However, both the high sensitivity to Δt and the workable range of Δt were unchanged.

Obviously, there is a need to continuously search for a better numerical method for solving the sequential evaporation-condensation model with the possibility that the model itself needs to be revised. Such a method should be capable of producing stable results within a finite range of the algorithm parameters. Consistent characteristic values of the temperature and moisture profiles should be produced by the method.

13.2.5 REACTION-DIFFUSION MODEL

Recently, Huang et al. [11] carried out a mathematical analysis of the sequential model described by Equations 13.1 through 13.6 augmented by the algebraic equation for saturation relationship between liquid water and water vapor. It was revealed that there were two types of instabilities; one was instability associated with a particular numerical scheme and the other was diffusive instability associated with the model. The sequential model and its implementation method as described in the above lead to both numerical instability and diffusive instability.

It was suggested by Huang et al. [11] that the following reaction-diffusion model should be adopted:

$$\frac{\partial T}{\partial t} = \frac{1}{\rho c_p} \frac{\partial}{\partial x}\left(k \frac{\partial T}{\partial x}\right) + \lambda \Gamma, \quad 0 < x < x_L/2, \quad t > 0. \tag{13.7}$$

$$\frac{\partial V}{\partial t} = \frac{\partial}{\partial x}\left(D_V \frac{\partial V}{\partial x}\right) - \frac{\Gamma}{\rho}, \quad 0 < x < x_L/2, \quad t > 0. \tag{13.8}$$

$$\frac{\partial W}{\partial t} = \frac{\partial}{\partial x}\left(D_W \frac{\partial W}{\partial x}\right) + \frac{\Gamma}{\rho}, \quad 0 < x < x_L/2, \quad t > 0. \tag{13.9}$$

where Γ is the rate of phase change (kg/m³s) and given by the modified Hertz–Knudsen equation as follows:

$$\Gamma = E(1-\phi)\sqrt{\frac{M}{2\pi R}} \frac{P_v - c P_s}{\sqrt{T}} \tag{13.10}$$

where E is the rate of condensation/evaporation, ϕ is the porosity of the bread, M is the molecular weight of water, and R is the universal gas constant, P_v and P_s are the vapor pressure and saturation pressure respectively, and c is a constant for phase change. The boundary and initial conditions remain as those described by Equations 13.4 through 13.6.

Combining numerical tests and linear stability analysis, it was demonstrated that the reaction-diffusion model did not lead to numerical instability when sufficiently small time step size was used. However, diffusive instability is an intrinsic feature of the model and as a result the two-phase region where vapor and liquid water coexist may become unstable. Clearly, new model structure with intrinsic stability needs to be developed.

13.3 MODELING AND ANALYSIS OF AN INDUSTRIAL BAKING OVEN

13.3.1 HEAT TRANSFER IN BAKING OVENS

In a baking oven chamber, molecules of air, water vapor or combustion gases circulate throughout the oven and transfer heat by convection until they contact solid surfaces such as the baking tin, band and conveyor, etc. Then heat transfer mode is changed to conduction. Radiant heat comes from all hot metal parts in the oven and may also come from the burner flames. It travels straight until hitting an object that is opaque to the radiation. Heat radiation is responsive to changes in the absorptive capacity of dough. For example, color changes influence the progression of baking by increasing the absorption of infrared rays. An increase in the absorptive capacity for infrared rays, though not apparent visually, is an almost invariable concomitant of the visible change. As a result, there is a tendency for color changes to accelerate after the first browning appears. Such a tendency might be either good or bad depending on the desired characteristics in the final products. Therefore, radiation tends to cause localized temperature differentials of an exposed surface particularly the darkened area, whereas convection tends to even out temperature gradients [12].

In industrial bread making, dough is delivered continuously from a prover to an oven. As a first-in-first-out system, each piece of dough is baked continuously from the first position to the last position inside the oven. Ideally every piece of dough should obtain the same amount of heat from its traveling period in the oven. As dough moves deep into the oven, the degree of baking increases.

In the oven chamber, combination of heat transfer modes depends on types of product and oven. For example, radiation was found to be the most important heat transfer mode to bake sandwich bread [13,14]. However, conduction played the most important role in baking Indian flat bread [15]. Major parameters determining heat distribution in an oven chamber include heat supply [12], air-flow pattern [13,16–18], humidity and composition of gas in the chamber [13], oven load, and baking time.

Heat absorption by dough is huge at the beginning of baking, because of significant differences between oven air temperature and initial dough temperature. Dough absorbs heat to activate some thermal mechanisms and increases its temperature. Therefore, the part of an oven that contains cold dough would require higher heat transfer rate to supply enough energy for heating the dough. When dough continuously travels deep into the oven, the degree of baking increases and temperature differences between the dough and the oven air are reduced; as such, heat flux on the dough surface is decreased. Meanwhile, half-baked dough/bread requires less energy to increase its temperature due to a decrease in its specific heat. This means that as baking progresses, less energy is required for baking the dough/bread. Consequently, each part of the oven requires different amounts of energy depending on oven load (number of loaves in the oven) and state of baking. Therdthai et al. [19] demonstrated the existence of an optimum temperature profile that produced bread of the best quality.

When the oven is full of dough/bread (i.e., 100% oven load) moving at the same speed, energy is balanced between the heat supply from the burners and the heat loss to dough/bread and the environment. After that, the oven temperature in each part is at steady state. In industry, it is normally controlled at a preset level to ensure enough heat supply to compensate for heat loss. However, the operation of an oven is crucial for producing high quality products. To be able to manipulate oven conditions to achieve optimum temperature profiles, the relationship among the various parameters and quantities needs to be established. CFD modeling may be the only method to effectively solve such a complicated problem.

CFD is a numerical technique for solving partial differential equations. The main characteristic of this technique is the immediate discretization of the equation of flow into the physically two-dimensional or three-dimensional space. The solution domain is divided into a number of cells known as control volume [20]. The transport equations describing the conservation of mass, momentum and energy as fluid flows are used to solve the fluid dynamics problems, supplemented by additional algebraic equations such as the equation of state and the constitutive equation [21]. A summary on the applications of the CFD approach to optimize various aspects of baking oven design to improve oven efficiency and product quality can be found in Wong et al. [22]. In the following section, we will demonstrate the effectiveness of CFD modeling in studying an industrial traveling-tray bread-baking oven.

13.3.2 TRAVELING-TRAY BAKING OVEN

Figure 13.4 shows a schematic diagram of an industrial traveling-tray baking oven (Baker Perkins Pty Ltd). Dimensions of this continuous oven are 16.50 m in length, 3.65 m in width, and 3.75 m in height. The oven can hold 50 rows of trays at any one time, which convey from the front to the back of the oven and then making a U-turn to the front on a lower track. Each row contains 12 trays (0.55 m × 0.12 m × 0.28 m). Each tray consists of four tins (0.12 m × 0.12 m × 0.28 m) and air gaps in

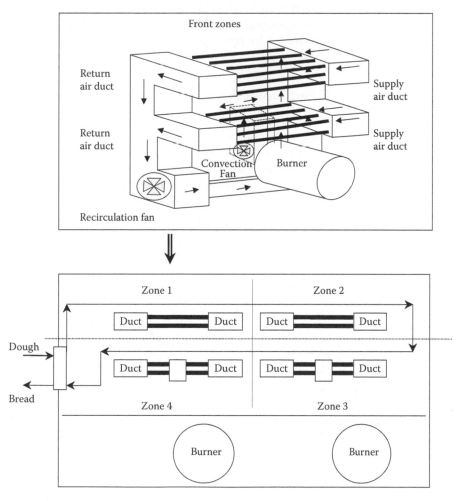

FIGURE 13.4 Schematic diagram of the industrial traveling-tray baking oven. (From Therdthai, N., Zhou, W., and Adamczak, T., *Journal of Food Engineering*, 60, 211, 2003. With permission.)

between. The entire oven is divided into two parts: front part and back part. Each part occupies half the oven chamber and has a pair of heating ducts that regulate the temperatures in that part. Both the front and back parts can be further divided into two zones, therefore each zone occupies one fourth of the oven chamber and takes one-fourth of the total baking time. For the heating system, there are two burners located below the lower tray conveyer. Hot air generated from these two burners is circulated inside the four split ducts (supply and return air ducts) and heats up their walls. Dampers are used to regulate flow rate from the bottom to the top of the supply air ducts. In this indirect heating system, heat is transferred from the duct surfaces to air in the oven chamber by convection and the hot air in the chamber bakes bread. In the oven chamber, there are also two convection fans located around the middle of the bottom heating ducts (tubes). These two fans help circulate airflow which increases heat transfer rate in the oven chamber.

Quality of bread baked in such an oven depends on the temperature profile of the tins which is the combined result of all baking parameters. Therdthai et al. [19] developed mathematical models that could predict the final crumb temperature, crust color and total weight loss using the tin temperature profile. Based on the models, by formulating and solving a constrained optimization problem, the optimal tin temperature profile was found to be at 115°C, 130°C, 156°C, and 176°C in zone 1, zone 2, zone 3, and zone 4, respectively, for a total baking time of 27.4 minutes. Optimal tin temperature profiles for shorter baking times were also presented. All these optimum temperature profiles aimed to minimize weight loss during baking with complete gelatinization and an acceptable crust color.

In order to investigate how oven operating condition affected the tin temperature profile and under what condition the optimal temperature profile could be achieved, CFD models were developed in both two-dimensional and three-dimensional domains for the studied oven.

13.3.3 Two-Dimensional Computational Fluid Dynamics (CFD) Models

For two-dimensional CFD models, a structured grid system was used with a steady state assumption. The model was capable of providing vital information on temperature profiles and airflow patterns inside the oven when energy supply and heat loss were in balance [23]. A total of nine different oven operating conditions were studied, including normal operating condition currently employed in industrial production. The simulation results clearly showed the influence of duct temperatures and convection fan volumes, as well as their interactions. Figure 13.5 presents the temperature profiles under five of the nine studied conditions. Comparing the optimal temperature profile to

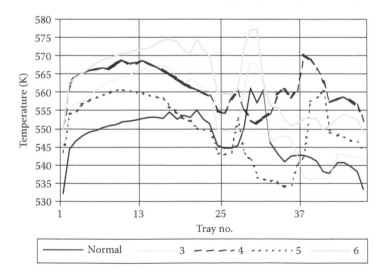

FIGURE 13.5 Variation of temperature profile due to changes in oven operating condition. (From Therdthai, N., Zhou, W., and Adamczak, T., *Journal of Food Engineer*ing, 60, 211, 2003. With permission.)

the temperature profile under normal oven operating condition, temperature in zone 3 should be reduced while temperature in zone 4 should be increased. However, directly increasing the temperature in zone 4 by raising the relevant duct air temperature would also increase the temperature in zone 1, which resulted in an adverse effect on bread quality. From the CFD results however, it was clear that the temperature in zone 4 could be increased without drastically affecting the temperature in zone 1 by manipulating fan volume.

Positioning of the controller sensors in the oven was also studied by the two-dimensional CFD modeling. The sensitivity of the controller at various locations was evaluated by establishing the correlation between sensor temperature and tin temperature under a wide range of oven operating conditions. The optimal position for placing the controller sensor was determined without considering the effect of oven load change [23].

13.3.4 THREE-DIMENSIONAL COMPUTATIONAL FLUID DYNAMICS (CFD) MODELS

Dynamic responses could not be studied using the above two-dimensional CFD model due to its steady state assumption. Employing moving grid, a three-dimensional CFD model was further developed with a transient state assumption by Therdthai et al. [24] and Zhou and Therdthai [25]. This model was capable of simulating the actual movement of dough traveling into the oven; therefore the effect of an oven load change on heat and mass transfer in the oven chamber could be studied. The model can be described by the following equations:

Mass conservation equation (Continuity equation):

$$\frac{\partial \rho}{\partial t} + \nabla \cdot (\rho \vec{v}) = 0 \tag{13.11}$$

Momentum conservation equation (Navier–Stokes equation):

$$\frac{\partial}{\partial t}(\rho \vec{v}) + \nabla \cdot (\rho \vec{v} \vec{v}) = -\nabla P - \nabla \cdot \overline{\overline{\tau}} + \rho \vec{f} \tag{13.12}$$

Energy conservation equation:

$$\frac{\partial}{\partial t}(\rho E) + \nabla \cdot (\rho E \vec{v}) = -\rho \dot{q} + \nabla \cdot (k \nabla T) - \nabla \cdot (P \vec{v}) - \nabla \cdot (\overline{\overline{\tau}} \vec{v}) + \rho \vec{f} \vec{v} \tag{13.13}$$

where $\vec{v}$ is velocity vector in m/s, P is static pressure in Pa, $\overline{\overline{\tau}}$ is viscoelastic stress tensor in Pa, $\vec{f}$ is body force vector per unit mass in N/kg, $E = e + (1/2)\vec{v}^2$ is total energy per unit mass in J/kg, e is internal energy per unit mass in J/kg, $\dot{q}$ is heating rate per unit mass in W/kg. Details of the boundary and initial conditions for the above model can be found elsewhere [24,25].

Using CFD simulations, the velocity profile at a position between the traveling trays was investigated under normal oven operating condition. The modeled profile was compared to the actual profile which was measured from the industrial baking process using an in-line anemometer developed by Therdthai et al. [26]. There was good agreement between the modeled profile and the measured profile except in the area around the U-turn. This could be due to the simplification of the oven configuration in the model where the U-turn movement was omitted due to limited capability of the software in grid deformation. The correlation coefficient and mean square error (MSE) between the modeled values and measured values were 0.6105 and 0.0337, respectively.

Wong et al. [27] extended the study to overcome the limitation of the CFD model in terms of the moving grid. In their new model, transient simulation of the continuous movement of dough/bread in the oven was achieved using a sliding mesh technique [28] in a two-dimensional domain. The U-turn movement of bread was successfully simulated by dividing the solution domain into two parts, then flipping and aligning them along the traveling tracks.

Using the three-dimensional CFD model in Therdthai et al. [24], surface temperature profiles were investigated when the oven was full, again under normal oven operating condition. The correlation coefficients between the modeled and measured profiles were 0.9132, 0.9065 and 0.9096 for the top, bottom and side surface temperature profiles, respectively. The average weighted temperature profile was predicted to be 116°C, 130°C, 172°C and 170°C for zone 1, zone 2, zone 3 and zone 4, respectively. The corresponding quality attributes were then estimated using these predicted average weighted temperature profiles through the mathematical models in Therdthai et al. [19]. The predicted values of the quality attributes were in good agreement with their actual values measured from the industrial baking process. A relative error of 2.86%, 0.56%, 2.24%, 0.96% and 1.52% was found in predicting the weight loss, crumb temperature, top crust color, side crust color and bottom crust color respectively [25].

Using the three-dimensional CFD model, changes in airflow pattern inside the baking chamber during the start-up of a continuous baking process were also investigated, as shown in Figure 13.6. They were partly due to the continuing tray movement which directly forced the airflow to change.

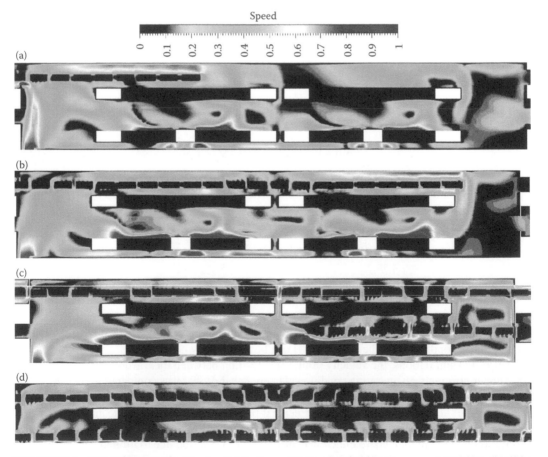

FIGURE 13.6 Top to bottom: air velocity distribution patterns when baking time was at (a) 240 s, (b) 600 s, (c) 960 s and (d) 1440 s, respectively.

Furthermore, each step of the tray movement increased the oven load unless the oven was already full. Oven load affected the corresponding temperature gradient, which in turn changed the convective flow therefore the overall airflow in the oven chamber. The temperature profiles of the later tins showed a small decrease compared to those of the earlier tins. The amount of temperature decrease was reduced as oven load approached its maximum. Consequently, variations in product quality were evident during this period with crust color being most noticeable. This result was consistent with the actual baking observations in industry, not only during a start-up period but also when a production gap was found in the oven chamber. With the simulation results, the CFD model may be further used to modify the existing control system so that it can respond to changes in oven load accordingly.

13.3.5 CONTROL SYSTEM DESIGN BASED ON COMPUTATIONAL FLUID DYNAMICS (CFD) MODEL

Wong et al. [29] explored applying a two-dimensional CFD model to process control design for the traveling-tray baking oven. A feedback control system was incorporated into the CFD model through user-defined functions (UDF). UDF was used to monitor temperature at specific positions in the oven and to define the thermal conditions of burner walls according to the control algorithm. A feedback control system with multiple decoupled PI controllers was designed and evaluated. The controller performed satisfactorily in response to disturbances and set-point changes. Such a control system outperforms traditional controller design methods. Firstly, evaluation of the impact of the controller output is not just limited to particular parameters or particular sensor points in the process. All information on the fluid flow (velocity, temperature, pressure, etc) is made available by the simulation tool. In addition, information is available for any position in the whole process. With the establishment of this method, users can gain a better understanding of the process which subsequently is beneficial to the controller design.

With the establishment of the new process control system, the need for a preheating step in typical industrial operations was re-evaluated. It was found that, under the control system, the elimination of the initial preheating to 550 K would not significantly affect the dough/bread top surface temperature profile across all baking zones. Elimination of the preheating stage will reduce the start-up period and save energy while producing the same quality of bread.

13.3.6 TOWARD OPTIMAL TEMPERATURE PROFILE

By integrating the three-dimensional CFD model in [24] and the mathematical models in [19], the normal oven operating condition would yield an average weighted temperature profile of $116^\circ C$, $130^\circ C$, $171^\circ C$, and $170^\circ C$ for zone 1, zone 2, zone 3 and zone 4, respectively. The corresponding weight loss in bread was predicted to be 9.35% after 25 minutes baking.

To minimize weight loss without changing baking time, the duct temperatures and/or the airflow volume need to be adjusted. The optimum average weighted temperature profile for a total baking time of 25 minutes was found to be at $106^\circ C$, $130^\circ C$, $166^\circ C$, and $179^\circ C$ for zone 1, zone 2, zone 3 and zone 4, respectively [19]. To drive the current profile toward the optimal profile, the average weighted tin temperatures in zone 1 and zone 3 should be reduced whereas the average weighted tin temperature in zone 4 should be increased. From the simulation results by the two-dimensional CFD model, adjusting airflow pattern inside the oven chamber would be the most efficient way to manipulate baking conditions. However, higher airflow volume would also increase baking temperature therefore heat supply to the oven should be reduced concurrently.

Simulations were carried out aiming at obtaining optimum temperature profile by varying airflow volume and duct temperatures, respectively. The simulation results revealed that to produce a temperature profile close to its optimum, heat supply should be reduced by 3.54%, 3.33%, and 3.54% in zone 1, zone 3, and zone 4, respectively. However, the flow volume through the convection fan in zone 3 needed to be doubled. By operating the oven at this condition, the weight loss was predicted to be 7.95%, a significant reduction of 1.4% from the current level of 9.35%. Crust colors on the top, bottom and side of the loaf were all within the acceptable range. The bread was also guaranteed to be completely baked despite decreased heat supply, indicated by its predicted final internal temperature of 98.7°C [25] and complete starch gelatinization inside the bread [30]. The completeness of starch gelatinization under the new operating condition was proven by integrating the three-dimensional CFD model in [19] with a kinetic model for starch gelatinization [30].

13.3.7 PHYSICAL PROPERTIES OF FOOD MATERIALS IN COMPUTATIONAL FLUID DYNAMICS (CFD) MODELING

There is an important issue associated with developing CFD models in general. That is, how to choose values of physical properties involved? For food materials during processing, the physical properties, physical structure and even composition of the materials change during the process. This leads to difficulty in correctly setting up material properties in a CFD model [31]. Many researchers used standard material blocks with known physical properties provided in CFD software which may compromise the quality of models.

Wong et al. [31] highlighted the importance of carefully selecting physical properties in CFD modeling. Through mathematical models, it was demonstrated that settings in some of the physical properties could significantly affect the simulated temperature profiles. Therefore, care should be taken when setting up a CFD model so as to minimize the error generated from the setting itself.

13.4 CONCLUSIONS

Recent development in mathematical modeling and analysis of bread-baking process has been reviewed here. For heat and mass transfers within dough pieces and those inside an oven chamber during baking, different approaches have been adopted in mathematical modeling. Within a piece of dough, multiphase flow model based on the internal evaporation-condensation mechanism have been developed to describe the simultaneous heat, water vapor, and liquid water transfers. The resulting partial differential equations were augmented by algebraic equations. However, numerical solving of the model proved to be an issue. Various finite difference methods (FDM) and finite element methods were used but the results were sensitive to the time step parameter. Adopting the reaction-diffusion model could eliminate the numerical instability but the intrinsic diffusion instability remains as an issue. Improved numerical methods or models with new structure and intrinsic stability need to be developed.

On the oven side, there are many factors affecting heat distribution in an oven chamber. For a continuous industry oven in particular, it has been proven that it is the temperature profile a piece of dough/bread experiences throughout the oven that determines the final product quality. Through mathematical models, this profile can be optimized for any specified total baking time. CFD modeling has been used to study the complicated interactions among various oven operating parameters. Both two-dimensional and three-dimensional CFD models were developed. These models provided much insight on heat distribution in the oven, changes brought by heat supply and oven load, variations in product quality due to these changes, etc. The CFD model also provided guidance on how to modify oven operating conditions to achieve optimal temperature profile that produced bread of best quality. Furthermore, better process control systems could be designed based on CFD models. Future research efforts are expected to focus on utilizing CFD models beyond normal investigation of airflow and temperature profiles in an oven.

ACKNOWLEDGMENTS

The author would like to acknowledge his collaborators and students for their contributions to some of the research reviewed in this manuscript, particularly Dr N. Therdthai, Mr T. Adamczak, Ms S.Y. Wong, Dr J. Hua, Dr P. Lin and Dr H. Huang. He is also grateful to Dr J. Hua for reviewing this manuscript.

REFERENCES

1. Piperno, D.R. et al. Processing of wild cereal grains in the Upper Palaeolithic revealed by starch grain analysis. *Nature*, 430, 670, 2004.
2. Therdthai, N., and Zhou, W. Recent advances in the studies of bread baking process and their impacts on the bread baking technology. *Food Science and Technology Research*, 9, 219, 2003.
3. Zhou, W. and Therdthai, N. Bread manufacture. In: *Bakery Products: Science and Technology*, Hui, Y.H., Eds. Blackwell, Oxford, UK, 2006, 301.
4. De Vries, U., Sluimer, P., and Bloksma, A.H. A quantitative model for heat transport in dough and crumb during baking. In: *Cereal Science and Technology in Sweden*, Asp, N.-G., Eds. STU Lund University, Lund, Sweden, 1989, 174.
5. Tong, C.H., and Lund, D.B. Microwave heating of baked dough products with simultaneous heat and moisture transfer. *Journal of Food Engineering*, 19, 319, 1993.
6. Thorvaldsson, K., and Janestad, H. A model for simultaneous heat, water and vapor diffusion. *Journal of Food Engineering*, 40, 167, 1999.
7. Thorvaldsson, K., and Skjoldebrand, C. Water diffusion in bread during baking. *Lebensm.-Wiss. u.-Technol.*, 31, 658, 1998.
8. Zanoni, B., Peri, C., and Pierucci, S. A study of the bread-baking process I: a phenomenological model. *Journal of Food Engineering*, 19, 389, 1993.
9. Zanoni, B., Pierucci, S., and Peri, C. Study of bread baking process-II. Mathematical modeling. *Journal of Food Engineering*, 23, 321, 1994.
10. Zhou, W. Application of FDM and FEM to solving the simultaneous heat and moisture transfer inside bread during baking. *International Journal of Computational Fluid Dynamics*, 19, 73, 2005.
11. Huang, H., Lin, P., and Zhou, W. Moisture transport and diffusive instability during bread baking. *SIAM Journal of Applied Mathematics*, 68, 222, 2007.
12. Matz, S. *Equipment for Bakers*. Elsevier Science Publishers, New York, 1989, 475.
13. Velthuis, H., Dalhuijsen, A., and De Vries, U. Baking ovens and product. *Food Technology International Europe,* 993, 61, 1993.
14. Carvalho, M.G., and Nogueira, M. Improvement of energy efficiency in glass-melting furnaces, cement kilns and baking ovens. *Applied Thermal Engineering,* 17, 921, 1997.
15. Gupta, T.R. Individual heat transfer modes during contact baking of Indian unleavened flat bread (*chapati*) in a continuous oven. *Journal of Food Engineering*, 47, 313, 2001.
16. Carvalho, M.G., and Mertins, N. Mathematical modelling of heat and mass transfer phenomena in baking ovens. In: *Computational Methods and Experiment Measurements V*, Sousa, A., Brebbia, C.A., and Carlomagno, G.M., Eds. Computational Mechanics Publications, Southampton, UK, 359, 1991.
17. De Vries, U., Velthuis, H., and Koster, K. Baking ovens and product qualityóa computer model. *Food Science and Technology Today*, 9, 232, 1995.
18. Noel, J.Y., Ovenden, N.A., and Pochini, I., Prediction of flow and temperature distribution in a domestic forced convection electric oven, in *Proceedings of ACoFoP IV*, Goteborg, Sweden, 491, 1998.
19. Therdthai, N., Zhou, W., and Adamczak, T. Optimisation of temperature profile in bread baking. *Journal of Food Engineering*, 55, 41, 2002.
20. Mathioulakis, E., Karathanos, V.T., and Belessiotis, V.G. Simulation of air movement in a dryer by computational fluid dynamics: application for the drying of fruits. *Journal of Food Engineering*, 36, 183, 1998.
21. Scott, G., and Richardson, P. The application of computational fluid dynamics in the food industry. *Trends in Food Science & Technology,* 8, 119, 1997.
22. Wong, S.Y., Zhou, W., and Hua, J., Improving the efficiency of food processing ovens by CFD techniques. *Food Manufacturing Efficiency*, 1, 35, 2006.
23. Therdthai, N., Zhou, W., and Adamczak, T. Two-dimensional CFD modelling and simulation of an industrial continuous bread baking oven. *Journal of Food Engineering*, 60, 211, 2003.
24. Therdthai, N., Zhou, W., and Adamczak, T. Three-dimensional CFD modeling and simulation of the temperature profiles and airflow patterns during a continuous industrial baking process. *Journal of Food Engineering*, 65, 599, 2004.
25. Zhou, W., and Therdthai, N. Three-dimensional CFD modeling of a continuous industrial baking process. In: *Computational Fluid Dynamics in Food Processing*, Sun, D-W., Eds. Taylor and Francis, NY, 2007, 287.
26. Therdthai, N., Zhou, W., and Adamczak, T. The development of an anemometer for industrial bread baking. *Journal of Food Engineering*, 63, 329, 2004.

27. Wong, S.Y., Zhou, W., and Hua, J. CFD modeling of an industrial continuous bread-baking process involving U-movement. *Journal of Food Engineering*, 78, 888, 2007.
28. Fluent. *FLUENT User's Guide*. Fluent Inc, Lebanon, NH, 2002.
29. Wong, S.Y., Zhou, W., and Hua, J. Designing process controller for a continuous bread baking process based on CFD modeling. *Journal of Food Engineering*, 81, 523, 2007.
30. Therdthai, N., Zhou, W., and Adamczak, T. Simulation of starch gelatinization during baking in a travelling-tray oven by integrating a three-dimensional CFD model with a kinetic model. *Journal of Food Engineering*, 65, 543, 2004.
31. Wong, S.Y., Zhou, W., and Hua, J. Robustness analysis of a CFD model to the uncertainties in its physical properties for a bread baking process. *Journal of Food Engineering*, 77, 784, 2006.

Section IV

Low Temperature Thermal Processing

14 Freezing, Thawing, and Chilling of Foods

Jianfeng Wang
Massey University

Quang Tuan Pham
University of New South Wales

Donald J. Cleland
Massey University

CONTENTS

14.1   INTRODUCTION

Refrigerated food is popular worldwide because refrigeration is the technique that best preserves the quality characteristics of fresh food. The main industrial food refrigeration processes include freezing, thawing and chilling. In order to minimize the energy consumption of refrigeration equipment and maintain food quality, it is essential to understand and quantify the mechanisms of heat and mass transfer between the processed food and the heat transfer medium during those operations [1–3]. The objective of this chapter is to summarize the major methodologies and the key issues related to mathematical modeling of freezing, thawing and chilling processes for foods.

14.2   FOOD CHARACTERISTICS AND REFRIGERATION PROCESSES

Food components include water, minerals, carbohydrates, proteins and fats. In addition, for some foods there can be a significant amount of air voids (e.g., air gaps between the items in a package or natural voidage between cells in a horticultural product). While many foods have a cellular structure, at a macroscopic level the composition of most foods is relatively uniform. Important exceptions are surface layers such as fat cover on meat carcasses and skins on fruits and vegetables. Further, many foods are refrigerated after being packaged. Packaging generally comprises plastic films and cardboard layers. The packaging volume and mass are generally small relative to the food mass so the main influences are their effect on surface heat and mass transfer both directly due to the heat and mass transfer properties of the packaging and indirectly due to air layers trapped between the packaging and the food itself.

Generally foods are frozen, thawed, or chilled by contact of individual product items with the external cooling medium. Occasionally, product items are consolidated together onto pallets or bins prior to freezing, thawing or chilling but this is usually avoided because the large physical size leads to excessively long process times unless the heat transfer medium can penetrate the product stack. The most common cooling/heating medium is air (air blast freezers and chillers), but water, brines or other liquids are also often used where the product or packaging will not be damaged by liquid contact. Cooling or heating by direct contact with a cold or hot surface is occasionally used (e.g., in plate freezers).

During chilling, there is seldom any significant change in the composition and physical structure of the food except for some depletion in moisture content near the surface of the food. However, freezing and thawing result in the phase change of water to ice or vice versa which has significant effect on the composition, structure and the heat and mass transfer characteristics. In particular, ice has lower density so the food volume and/or the air voidage will change. Also, ice has significantly different thermal conductivity and heat capacity from water and the latent heat of freezing/melting must be removed or added.

Unfrozen foods can be considered as a mixture of water, dissolved solutes, insoluble matter and air. The solutes in the water result in freezing point depression so the initial freezing point and final thawing point (θ_f) of most foods is below 0°C. During freezing, when the surface temperature reaches the freezing point, ice crystals will start to nucleate near the surface and grow toward the center of the product. Nucleation may not be instantaneous, so the food may supercool below θ_f. Once nucleation has occurred then generally the food quickly equilibrates to the freezing temperature and further ice crystal growth is controlled by heat transfer rather than by mass transfer. The ice that forms largely excludes any solutes so the remaining solution becomes more concentrated and freezing point depression increases. Therefore, further ice formation must occur at progressively lower temperatures and the full latent heat of freezing is released over a range of temperatures below θ_f. Ultimately, the remaining water is so tightly bound to food macro-molecules that it is not available to freeze even at very low temperatures. Once this point is reached, the food can be considered completely frozen. For most high moisture foods, most of the water that will freeze has done so when the temperature is below about –10 to –15°C. During thawing, nucleation does not exist so the thawing rate is constrained by heat conduction to the ice crystals.

14.3   MATHEMATICAL FORMULATION OF FREEZING, THAWING, AND CHILLING

Heat and mass transfer during freezing, thawing, and chilling generally comprises heat conduction and diffusional mass (moisture) transfer within the solid food, and convection, radiation, and/or evaporation/sublimation heat and mass transfer between the food surface and the external medium.

14.3.1   UNSTEADY STATE HEAT CONDUCTION

In many processes, changes in moisture content within the food are small, mass transfer by moisture diffusion can be ignored and the effect of ice nucleation can be ignored so only heat transfer by conduction needs to be modeled. Conduction involves a temperature change as a function of time t and displacement in the x, y, and z coordinates. The Fourier equation for heat conduction within the food volume (V) is [4]:

$$\rho c \frac{\partial \theta}{\partial t} = \frac{\partial}{\partial x}\left(k \frac{\partial \theta}{\partial x}\right) + \frac{\partial}{\partial y}\left(k \frac{\partial \theta}{\partial y}\right) + \frac{\partial}{\partial z}\left(k \frac{\partial \theta}{\partial z}\right) + \rho Q \quad x, y, z \in V \tag{14.1}$$

where c = specific heat capacity (J/kgK)
   k = thermal conductivity (W/mK)
   Q = internal heat generation e.g., heat of respiration (W/kg)
   t = time (s)
   V = volume of the food item (m^3)
   x, y, z = space coordinates (m)
   θ = temperature (°C)
   ρ = density (kg/m^3)
If thermal conductivity k is constant with temperature, Equation 14.1 can be written as:

$$\frac{\partial \theta}{\partial t} = \psi\left(\frac{\partial^2 \theta}{\partial x^2} + \frac{\partial^2 \theta}{\partial y^2} + \frac{\partial^2 \theta}{\partial z^2}\right) + \frac{Q}{c} \quad x, y, z \in V \tag{14.2}$$

and

$$\psi = \frac{k}{\rho c} \tag{14.3}$$

where ψ = thermal diffusivity (m^2/s)
   To solve Equations 14.1 or 14.2, initial and boundary conditions are required.

14.3.2   BOUNDARY CONDITIONS

The general form of the boundary condition to define the heat transfer from the surface of the solid food (A_s) to the external medium where convection, radiation and evaporative heat transfer occur in parallel is:

$$k\left[l_x \frac{\partial \theta}{\partial x} + l_y \frac{\partial \theta}{\partial y} + l_z \frac{\partial \theta}{\partial z}\right]_s = h(\theta_a - \theta_s) + F\varepsilon\sigma[T_r^4 - T_s^4] + h_m(RH\, p_{wa} - a_w p_{ws})L_e \quad x, y, z \in A \tag{14.4}$$

where A_s = food surface (m²)

  a_w = water activity at product surface
  F = radiation view factor
  h = surface convective heat transfer coefficient (W/m²K)
  h_m = surface mass transfer coefficient based on partial pressure (kg/sm²Pa or s/m)
  L_e = latent heat of evaporation or sublimation (J/kg)
  l_x, l_y, l_z = cosine of outward normal in the x, y, z directions, respectively
  p_{wa} = vapor pressure of water at the air temperature (Pa)
  p_{ws} = vapor pressure of water at the product surface temperature (Pa)
  RH = relative humidity of air as a fraction
  T_r = radiation source/sink temperature (K)
  T_s = food surface temperature (K)
  ε = food surface emissivity
  σ = Stefan–Boltzmann constant (5.67 × 10⁻⁸ W/m²K⁴)
  θ = temperature (°C)
  θ_a = external medium temperature (°C)
  θ_s = food surface temperature (°C)

The symmetry boundary condition is also often used for a perfect insulated surface or an axis of symmetry (A_{ins}) so that the size of the solution domain can be reduced:

$$\left[l_x \frac{\partial \theta}{\partial x} + l_y \frac{\partial \theta}{\partial y} + l_z \frac{\partial \theta}{\partial z} \right]_s = 0 \quad x, y, z \in A_i \tag{14.5}$$

A common simplification of Equation 14.4 is to model radiation and evaporative heat transfer as pseudo-convection so only the first term on the right side of Equation 14.4 is retained and h is replaced by an effective heat transfer coefficient, h_e. For example, if the radiation source and convective air are at the same temperature and evaporative heat transfer is negligible then:

$$h_e = h + F\varepsilon\sigma(T_a + T_s)(T_a^2 + T_s^2) \tag{14.6}$$

where h_e = effective surface heat transfer coefficient (W/m²K).

If h is very high ($h \rightarrow \infty$) and radiation and evaporation/sublimation are low then the surface temperature approximates the external medium temperature and Equation 14.4 simplifies to:

$$T_s = T_a \quad \text{or} \quad \theta_s = \theta_a \tag{14.7}$$

14.3.3 INITIAL CONDITION

The initial condition defines the temperature distribution in the food prior to the start of the process. The most common initial condition is uniform temperature within the food volume:

$$\theta = \theta_{\text{in}} \quad \text{for} \quad t = 0; \quad x, y, z \in V \tag{14.8}$$

where θ_{in} = initial food temperature (°C).

14.3.4 MOVING BOUNDARY CONDITION

A model sometimes used for freezing or thawing is the concept of a moving phase change boundary about which there is a step change in thermal properties from the unfrozen to the frozen values at a unique phase change temperature (θ_f). The moving boundary is always at θ_f and all the latent heat is released at this temperature. The equation defining the moving boundary for the case of one dimensional freezing in the x direction only is:

$$k_u \left[\frac{\partial \theta}{\partial x} \right]_{x_f+} - k_{ff} \left[\frac{\partial \theta}{\partial x} \right]_{x_f-} = -\rho L_f \frac{dx_f}{dt} \quad x = x_f \tag{14.9}$$

where k_u = unfrozen phase thermal conductivity (W/mK)

k_{ff} = fully frozen thermal conductivity (W/mK)

L_f = latent heat of freezing or thawing (J/kg)

x_f = distance from the surface to the phase change front (m)

A model sometimes used for freezing or thawing is the concept of a moving phase change boundary about which there is a step change in thermal properties from the unfrozen values at the unique phase change temperature (θ_f). The moving boundary is always at θ_f and all the latent heat is released at this temperature. Either side of the moving boundary, Equations 14.1 or 14.2 is used to describe the heat conduction. This model is seldom used for foods because phase change occurs over a range of temperature as discussed in Section 14.2.

14.4 THERMAL PROPERTIES

It is essential to know the variation of thermal properties of foods with temperature when modeling freezing, thawing, and chilling [3,5]. The important thermal properties for refrigerated foods are k, ρ, c, enthalpy (H), latent heat of freezing (L_f), initial freezing temperature (θ_f), and heat of respiration (Q). Figure 14.1 shows most of these properties for a typical high moisture frozen food. For chilling temperatures, changes in k, ρ, and c are usually sufficiently small that they can be assumed to be constant. For freezing and thawing, thermal properties must take into account the phase change from water to ice or vice versa.

Recent reviews of thermal property sources for foods are given by ASHRAE [1,2], Rahman [6], and others [3,7–10]. Properties can be either measured or predicted from composition data and knowledge about the food microstructures. Measurements are expensive and time consuming to perform accurately especially for thermal conductivity and enthalpy/specific heat capacity for some materials. Note that the specific heat capacity is normally obtained from measured enthalpy data using:

$$c = \frac{dH}{d\theta} \tag{14.10}$$

where: H = enthalpy (J/kg)

The following equations are often used to curve-fit measured data [11,12]:

$$k = k_f + a(\theta - \theta_f) + b\left(\frac{1}{\theta} - \frac{1}{\theta_f} \right) \quad \text{for } \theta < \theta_f \tag{14.11}$$

$$k = k_f + d(\theta - \theta_f) \quad \text{for } \theta > \theta_f \tag{14.12}$$

$$H = H_f + c_{ff}(\theta - \theta_f) + e\left(\frac{1}{\theta} - \frac{1}{\theta_f} \right) \quad \text{for } \theta < \theta_f \tag{14.13}$$

$$H = H_f + c_u(\theta - \theta_f) \quad \text{for } \theta > \theta_f \tag{14.14}$$

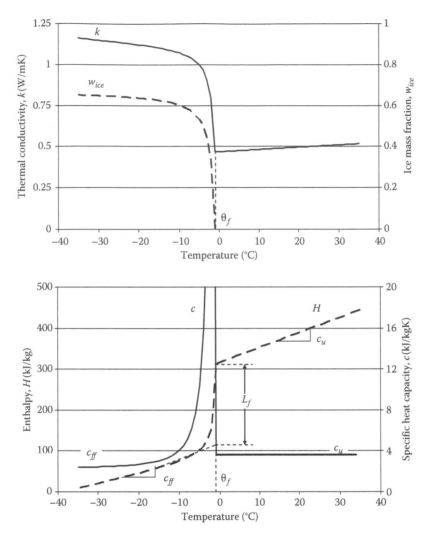

FIGURE 14.1 Thermal properties of a typical high moisture food.

where a, b, d, and e are curve-fit coefficients, and

c_{ff} = fully frozen specific heat capacity (J/kgK)

c_u = unfrozen specific heat capacity (J/kgK)

H_f = enthalpy at θ_f (J/kg)

k_f = thermal conductivity at θ_f (W/mK)

Except for respiring fruits and vegetables, heat generation (Q) is usually zero. Even for chilling of fruits and vegetables, heat of respiration is usually small compared with sensible heat transfer so neglecting it usually does not significantly affect prediction accuracy [3].

While generic thermal property data sets are available for many foods [2,6], the accuracy of these data is often uncertain and the data may not be available for a product sufficiently similar to the food of interest. Therefore, prediction from easily measured composition data is often preferred.

14.4.1 THERMAL PROPERTIES WITHOUT PHASE CHANGE

The density and specific heat of foods can be calculated using weighted averages where the density or specific heat (ρ_i or c_i) and volume fractions (v_i) or mass fraction (w_i) of various components of the foods are known:

$$\rho = \sum_{1}^{N} \rho_i v_i = \frac{1}{\sum_{1}^{N} \dfrac{w_i}{\rho_i}} \tag{14.15}$$

$$c = \frac{\sum_{1}^{N} \rho_i v_i c_i}{\sum_{1}^{N} \rho_i v_i} = \sum_{1}^{N} w_i c_i \tag{14.16}$$

where v_i = volume fraction of component i

w_i = mass fraction of component i

N = number of components

Properties of the various components in foods as a function of temperature [7] and the constant values often used to make predictions for fully frozen foods [13] are shown in Table 14.1.

Unlike density and specific heat capacity, there are no single straightforward models for predicting the thermal conductivity of mixtures of components. This is mainly due to the dependence of k on the microstructures of the food components and that k is a volume rather than a mass based property [8,14]. Table 14.2 shows six fundamental structural models for the effective thermal conductivity, k_e, for a food with two components. Unfortunately, multicomponent foods often have more

TABLE 14.1
Thermal Properties of Major Components of Foods

Thermal Property	Component	Temperature Dependent Models	Constant Value
Density ρ_i (kg/m³)	Water	$\rho = 9.9718 \times 10^2 + 3.1439 \times 10^{-3}\theta - 3.7574 \times 10^{-3}\theta^2$	1000
	Ice	$\rho = 9.1689 \times 10^2 - 1.3071 \times 10^{-1}\theta$	920
	Protein	$\rho = 1.3299 \times 10^3 - 5.1840 \times 10^{-1}\theta$	1300
	Fat	$\rho = 9.2559 \times 10^2 - 4.1757 \times 10^{-1}\theta$	930
	Carbohydrate	$\rho = 1.5991 \times 10^3 - 3.1046 \times 10^{-1}\theta$	1500
	Fiber	$\rho = 1.3115 \times 10^3 - 3.6589 \times 10^{-1}\theta$	1300
	Ash	$\rho = 2.4238 \times 10^3 - 2.8063 \times 10^{-1}\theta$	2400
Specific heat c_i (kJ/ kgK)	Water ($\theta > 0°C$)	$c = 4.1289 - 9.0864 \times 10^{-5}\theta + 5.4731 \times 10^{-6}\theta^2$	4.18
	Water ($\theta < 0°C$)	$c = 4.1289 - 5.3062 \times 10^{-3}\theta + 9.9516 \times 10^{-4}\theta^2$	4.28
	Ice	$c = 2.0623 + 6.0769 \times 10^{-3}\theta$	2.00
	Protein	$c = 2.0082 + 1.2089 \times 10^{-3}\theta - 1.3129 \times 10^{-6}\theta^2$	2.00
	Fat	$c = 1.9842 + 1.47339 \times 10^{-3}\theta - 4.8008 \times 10^{-6}\theta^2$	1.95
	Carbohydrate	$c = 1.5488 + 1.9625 \times 10^{-3}\theta - 5.9399 \times 10^{-6}\theta^2$	1.50
	Fiber	$c = 1.8459 + 1.8306 \times 10^{-3}\theta - 4.6509 \times 10^{-6}\theta^2$	1.80
	Ash	$c = 1.0926 + 1.8896 \times 10^{-3}\theta - 3.6817 \times 10^{-6}\theta^2$	1.05
Thermal conductivity k_i (W/mK)	Water	$k = 5.7109 \times 10^{-1} + 1.7625 \times 10^{-3}\theta - 6.7036 \times 10^{-6}\theta^2$	0.54
	Ice	$k = 2.2196 - 6.2489 \times 10^{-3}\theta + 1.0154 \times 10^{-4}\theta^2$	2.35
	Protein	$k = 1.7881 \times 10^{-1} + 1.1958 \times 10^{-3}\theta - 2.7178 \times 10^{-6}\theta^2$	0.16
	Fat	$k = 1.8071 \times 10^{-1} - 2.7604 \times 10^{-4}\theta - 1.7749 \times 10^{-7}\theta^2$	0.19
	Carbohydrate	$k = 2.0141 \times 10^{-1} + 1.3874 \times 10^{-3}\theta - 4.3312 \times 10^{-6}\theta^2$	0.20
	Fiber	$k = 1.8331 \times 10^{-1} + 1.2497 \times 10^{-3}\theta - 3.1683 \times 10^{-6}\theta^2$	0.18
	Ash	$k = 3.2962 \times 10^{-1} + 1.4011 \times 10^{-3}\theta - 2.9069 \times 10^{-6}\theta^2$	0.30

Source: Adapted from Choi, Y. and Okos, M.R., *Food Engineering and Process Applications*, 1st edn. Elsevier, Amsterdam, 93, 1986.

TABLE 14.2
Six Fundamental Effective Thermal Conductivity Structural Models for Two-Component Foods (Assuming the Heat Flow is in the Vertical Direction)

Model	Structure Schematic	Effective Thermal Conductivity Equation
Parallel (P)		$k_e = v_1 k_1 + v_2 k_2$
Maxwell Eucken 1 (ME1) (k_1 = continuous phase, k_2 = dispersed phase) [15]		$k_e = \dfrac{k_1 v_1 + k_2 v_2 \dfrac{3k_1}{2k_1 + k_2}}{v_1 + v_2 \dfrac{3k_1}{2k_1 + k_2}}$
Effective medium theory (EMT) [16]		$v_1 \dfrac{k_1 - k_e}{k_1 + 2k_e} + v_2 \dfrac{k_2 - k_e}{k_2 + 2k_e} = 0$
Co-continuous (CC) [17]		$k_e = \dfrac{k_{series}}{2}\left(\sqrt{1 + 8k_{parallel}/k_{series}} - 1\right)$
Maxwell Eucken 2 (ME2) (k_1 = dispersed phase, k_2 = continuous phase) [15]		$k_e = \dfrac{k_2 v_2 + k_1 v_1 \dfrac{3k_2}{2k_2 + k_1}}{v_2 + v_1 \dfrac{3k_2}{2k_2 + k_1}}$
Series (S)		$k_e = \dfrac{1}{v_1/k_1 + v_2/k_2}$

complicated structures than these six fundamental model structures. For any food, theoretically its thermal conductivity value is between the parallel and series models; while for isotropic foods, its thermal conductivity value lies within the two Maxwell–Eucken models according to Hashin-Shtrikman bounds theory [18,19]. For isotropic foods, a four-zone theory can be used to further narrow the bounds of the thermal conductivity range [17].

For most high moisture foods at temperatures above θ_f, the difference between component thermal conductivities is small; generally the ratio of k values for different compounds is less than five for foods without air voids and less than 25 for foods with air voids. Therefore, for foods without air voids and at temperatures above θ_f, most models in Table 14.2 except the series model give similar values of k_e. For simplicity and ease of application to foods with more than two components, the parallel model is often applied. However for foods with air voids, it is important to take into account the microstructure and choose an appropriate model to estimate the thermal conductivity [14]. When the air voids are reasonably uniformly dispersed in the food, the recommended approach is to use the parallel model to estimate k_e for the nonair components, and then to use the Maxwell–Eucken model for the combination of the air and nonair components in a sequential manner [13].

14.4.2 THERMAL PROPERTIES WITH PHASE CHANGE

For freezing and thawing an extra difficulty is how to predict the ice fraction of foods during phase change. Several ice fraction models are available as reviewed by Fikiin [20] and Rahman [6]. One of the simplest and most commonly used ice fraction models is that proposed by Pham [21] based on the model postulated by Schwartzberg [22]:

$$w_{ice} = (w_{tw} - w_{bw}) \left(1 - \frac{\theta_f}{\theta} \right) \tag{14.17}$$

$$\frac{w_{bw}}{w_{tw}} = 0.342(1 - w_{tw}) - 4.510 w_{ash} + 0.167 w_{protein} \tag{14.18}$$

where
w_{ash} = mass fraction of ash content
w_{bw} = mass fraction of bound water content
w_{ice} = mass fraction of ice content
$w_{protein}$ = mass fraction of protein content
w_{tw} = mass fraction of total water content

Methods to predict θ_f are summarized by Boonsupthip and Heldman [23]. The best methods use Raoult's law based on the average mass fraction of solute components in the food:

$$\frac{1}{T_f} = \frac{1}{273.15} - \frac{R}{M_w L_w} \ln \left[\frac{(w_{tw} - w_{bw})/M_w}{(w_{tw} - w_{bw})/M_w + \sum_i (w_i/M_i)} \right] \tag{14.19}$$

where
L_w = latent heat of freezing of pure water (J/kg)
M_i = molecular weight of solute component i (g/mol)
M_w = molecular weight of water (18 g/mol)
R = gas constant (8.314 J/molK)
T_f = initial freezing temperature (K)
W_i = mass fraction of solute component i

Equations 14.15 and 14.16 can be used to estimate density and specific heat capacity by treating ice and nonfrozen water as separate components. However, the accurate prediction of effective thermal conductivity (k_e) is difficult because k of ice is significantly higher than those of the other components (Table 14.1). The general approach is to apply two-component models sequentially [24,25]. For example, the parallel model can be used for the solid components excluding ice, then Levy's model [26] for ice as one phase and all other solid components as the other phase, and lastly the Maxwell-Eucken model is used for air voids as the dispersed phase and all other phases as the continuous phase [13,25].

Other k_e modeling approaches often use empirical coefficients that have no physical basis. Examples include the use of a weighted average of the parallel and series model values by Krischer or the geometric weighting approach [8]. Such approaches can not be confidently applied to situations other than those they were derived for. In contrast, recent work has shown that Levy's model combined with the ice fraction calculated using a model similar to Equation 14.17 predicts a range of frozen foods with acceptable accuracy [11,27]. Levy's model is a combinatory model from the two Maxwell–Eucken models, with the physical basis defined by Wang et al [24]. In this model the effective thermal conductivity is given by:

$$k_e = k_1 \frac{2k_1 + k_2 - 2(k_1 - k_2)F}{2k_1 + k_2 + (k_1 - k_2)F} \tag{14.20}$$

$$F = \frac{2/G - 1 + 2v_2 - \sqrt{(2/G - 1 + 2v_2)^2 - 8v_2/G}}{2} \tag{14.21}$$

$$G = \frac{(k_1 - k_2)^2}{(k_1 + k_2)^2 + k_1 k_2 / 2} \tag{14.22}$$

For modeling processed foods with more complicated physical structures, other combinatory models based on the six fundamental models shown in Table 14.2 are worth further investigation [24]. Unfortunately, despite a multitude of models, either theoretical or empirical, there is currently little guidance in the literature on which models should be used in different situations beyond that given above.

By substituting the desired final temperature (freezing) or the initial temperature (thawing) into Equation 14.17, the latent heat of freezing/thawing can be approximated by:

$$L_f = w_{ice} L_w \tag{14.23}$$

Combining Equations 14.10, 14.13, 14.17, and 14.23 it can be shown that for $\theta < \theta_f$ the effective specific heat capacity including latent heat is given as a function of temperature by:

$$c = c_{ff} - \frac{e}{\theta^2} \quad \text{or} \quad c = c_{ff} - \frac{L_f \theta_f}{\theta(\theta - \theta_f)} \tag{14.24}$$

depending on whether measured data or predictions based on compositional data are used, respectively.

14.4.3 Heat and Mass Transfer Coefficients

Knowledge of heat and mass transfer coefficients (h and h_m, respectively) between the air and product surface is also necessary for prediction of process times and weight losses by evaporation or sublimation. Rahman [6] gives detailed information about the techniques for measuring h. The techniques can be divided into steady state, quasi-steady, transient and surface heat flux methods. Since the heat transfer coefficient is one of the most difficult parameters to measure accurately, it can be a major source of uncertainty in process time prediction [6,28]. Uncertainties of at least ±10 to ±20% are typical for h and processing times may be affected in the same proportion if the Biot number is low [29].

Unfortunately only a small amount of convective and effective heat transfer coefficient data are available from the literature [30,31]. For example, for freezing systems values are reported by ASHRAE [2], Rahman [6], Mannapperuma et al. for poultry meat [32,33] and Tocci and Mascheroni for meat ball and hamburgers [34]. By comparison, evaporative mass transfer coefficients (h_m) are even scarcer in literature [34,35]. Often, the analogy between the mass and heat transfer coefficients in air as defined by the Lewis relationship is used to calculate mass transfer coefficients from the heat transfer coefficients [34,36].

In addition, it was found that for the same product and air velocity, experimental values from different researchers differ significantly [30]. In order to avoid confusion in comparing heat transfer coefficients reported by different authors, it is important to distinguish between convective only and effective heat transfer coefficients. The latter generally include the heat transfer by radiation or by moisture transfer at the surface, in addition to convective heat transfer [37]. As h and h_m depend on many factors including thermophysical properties of fluid and product, characteristics of the object, fluid flow and the design of heat transfer equipments, care must be taken when using the reported data for a specific condition in a slightly different situation [10]. It is desirable to develop new measurement and prediction techniques with higher accuracy but they should be applicable to a wide range of situations [30,38].

14.5 HEAT TRANSFER PREDICTION METHODS

Solution techniques available for predicting temperature-time histories during freezing, thawing and chilling processes defined by the equations in Section 14.3 can be classified into analytical, empirical or numerical methods. In general terms, modeling of chilling processes is more straightforward than freezing and thawing because the thermal properties of foods do not change significantly with temperature but evaporation or sublimation can be more important at chilling temperatures.

Most predictive methods assume that the change in product dimensions due to any change in density is negligibly small. Given that density change is less than 5% for most foods even during freezing or thawing, that food item outer dimensions are often constrained, and that air voidage can increase or decrease due to any change in density, this approximation usually does not add significant prediction uncertainty and is commonly used.

14.5.1 ANALYTICAL METHODS

14.5.1.1 Chilling

For chilling of foods of simple geometry with constant thermal properties, constant external conditions, uniform initial conditions, no internal heat generation, and only convection at the boundary, there are exact analytical solutions [39]. For example, for the infinite slab geometry:

$$Y = \frac{\theta - \theta_a}{\theta_{in} - \theta_a} = \sum_{i=1}^{\infty} \frac{2 Bi \cos\left(\alpha_i \frac{r}{R}\right) \sec(\alpha_i)}{Bi(Bi+1) + \alpha_i^2} \exp(-\alpha_i^2 Fo) \qquad (14.25)$$

where α_i are the roots of:

$$\alpha \tan\alpha = Bi \qquad (14.26)$$

$$Fo = \frac{kt}{\rho c R^2} \qquad (14.27)$$

$$Bi = \frac{hR}{k} \qquad (14.28)$$

and R = characteristic half thickness; shortest distance from the product center to the surface (m)
r = distance from the center (m).

Newman [40] showed that the solutions for regular multidimensional objects can be obtained using the product rule. For example, for the three-dimensional rectangular brick shape, it is the product of the slab solutions in the three orthogonal dimensions:

$$Y = Y_x \cdot Y_y \cdot Y_z \qquad (14.29)$$

For high value of *Fo* (e.g., *Fo* > 0.25), Equation 14.25 and the equivalents for infinite cylinders and sphere geometries can be simplified to the first term of the series.

For chilling situations where evaporation and radiation are important surface heat transfer in parallel with convection, Chuntranuluck et al. [41–43] have shown that models such as Equation 14.25 can be used but θ_a should be replaced by an effective external condition defined by:

$$\theta_{ae} = \theta_a - \frac{h_m L_e}{h}(a_w p_{ws} - RH\, p_{wa}) \qquad (14.30)$$

where θ_{ae} = effective cooling medium temperature (°C).

14.5.1.2 Freezing and Thawing

For freezing and thawing, there have been many attempts to develop analytical solutions [9,10]. Most attempts have assumed that the phase change occurs at a unique temperature. For food freezing and thawing, the most commonly used analytical solution of any practical value is Plank's equation. For one-dimensional geometries, the equation to predict the time to freeze or thaw is:

$$t_f = \frac{V}{A_s R}\frac{L_f}{(\theta_f - \theta_a)}\left(\frac{R}{h} + \frac{R^2}{2k_e}\right) \qquad (14.31)$$

where t_f = freezing/thawing time (s) and $k_e = k_u$ for thawing and $k_e = k_{ff}$ for freezing.

The derivation of Plank's equation requires the following simplifications: unique phase change temperature, sensible heat effects are negligible relative to latent heat effects and constant k_e. The net effect is that freezing time predictions are up to 50% too low while thawing predictions are 30% too low to 50% too high. The large range of predicted results is because Plank's equation does not take into account many of the important factors affecting freezing and thawing rates.

14.5.2 Empirical Methods

Empirical approaches are generally developed to keep prediction methods as simple to use as the analytical methods, yet to address the weakness of analytical solutions due to the simplifying assumptions. Empirical methods can either be generally applicable or are effectively dedicated to a specific situation for which they are derived. The focus in this section is on generally applicable methods.

14.5.2.1 Chilling

For chilling, empirical prediction approaches focus on methods to extend the analytical solutions for regular geometries to all irregular geometries, and to simplify the infinite series solutions. Some researchers [44–46] developed curve-fit equations that approximate the transient solutions such as Equation 14.25 to a wide range of conditions (values of Bi and Fo). Many researchers have used the first term approximation to Equation 14.25 but have limited the range of applicability (e.g., to Fo > 0.25 and Y < 0.7). The prediction equation becomes:

$$Y = j \exp(-ft) \qquad (14.32)$$

where j = pre-exponential constant (lag factor)

f = exponential factor (s^{-1})

Many researchers have attempted to develop expressions for A and B as functions of food geometry and heat transfer conditions including Lin et al. [47,48] and Smith [49,50]. The chilling calculation method proposed by Lin et al. [47,48] is currently the most comprehensive. It gives equations and charts to calculate A and B for most geometries across a wide range of chilling conditions for the thermal center, θ_c, or the mass-average temperature, θ_m. For example, for a general three dimensional irregular shape, the equations are:

$$f = \frac{Ek\alpha^2}{3\rho cR^2} \tag{14.33}$$

where α is the first root of:

$$\alpha \cot \alpha + Bi - 1 = 0 \tag{14.34}$$

$$E = \frac{Bi^{4/3} + 1.85}{\dfrac{Bi^{4/3}}{E_\infty} + \dfrac{1.85}{E_0}} \tag{14.35}$$

$$E_0 = \frac{3[\beta_1 + \beta_2 + \beta_1^2(1+\beta_2) + \beta_2^2(1+\beta_1)]}{2\beta_1\beta_2(1+\beta_1+\beta_2)} - \frac{\left[(\beta_1-\beta_2)^2\right]^{0.4}}{15} \tag{14.36}$$

$$E_\infty = 0.75 + 1.01\left[\frac{1}{\beta_1^2} + 0.01P_3\exp\left(\beta_1 - \frac{\beta_1^2}{6}\right)\right] + 1.24\left[\frac{1}{\beta_2^2} + 0.01\exp\left(\beta_2 - \frac{\beta_2^2}{6}\right)\right] \tag{14.37}$$

$$j_c = \frac{Bi^{1.35} + \dfrac{1}{\beta_1}}{\dfrac{Bi^{1.35}}{1.271 + 0.305\exp(0.172\beta_1 - 0.115\beta_1^2) + 0.425\exp(0.09\beta_2 - 0.128\beta_2^2)} + \dfrac{1}{\beta_1}} \tag{14.38}$$

$$j_{ma} = j_c\left[\frac{1.5 + 0.69\,Bi}{1.5 + Bi}\right]^3 \tag{14.39}$$

$$\beta_1 = D_1/2R \tag{14.40}$$

$$\beta_2 = D_2/2R \tag{14.41}$$

where

D_1 = shortest dimension through geometric center of the object taken at right angles to R (m)

D_2 = dimension through geometric center of the object at right angles to both R and D_1 (m)

E = shape factor

E_0 = shape factor for $Bi = 0$

E_∞ = shape factor for $Bi \rightarrow \infty$

j_{ma} = lag factor for mass-average temperature

j_c = lag factor for thermal center temperature

β_1, β_2 = length ratios

For more defined geometries such as rectangular bodies, ellipsoids and finite cylinder more accurate equations are available [48].

14.5.2.2 Freezing

For freezing, there are a number of relatively simple, yet accurate, approximate methods to predict freezing times [9,10]. Most involve empirical methods based on Plank's equation that are restricted to simple one-dimensional geometries. One of the most popular methods is a modified Plank's equation developed by Pham [51,52], which is accurate to within about ±15% for a wide range of freezing conditions and products:

$$t_f = \frac{V}{A_s R}\left[\frac{\Delta H_1}{\Delta \theta_1} + \frac{\Delta H_2}{\Delta \theta_2}\right]\left[\frac{R}{h} + \frac{R^2}{2k_{ff}}\right] \tag{14.42}$$

where

$$\Delta H_1 = \rho\, c_u\, (\theta_{in} - \theta_{fm}) \tag{14.43}$$

$$\Delta H_2 = \rho L_f + \rho\, c_{ff}\, (\theta_{fm} - \theta_c) \tag{14.44}$$

$$\Delta \theta_1 = \frac{(\theta_{in} + \theta_{fm})}{2} - \theta_a \tag{14.45}$$

$$\Delta \theta_2 = \theta_{fm} - \theta_a \tag{14.46}$$

$$\theta_{fm} = 1.8 + 0.263\, \theta_c + 0.105\, \theta_a \tag{14.47}$$

ΔH_1 = heat released in precooling (J/m^3)
ΔH_2 = heat released in freezing (J/m^3)
$\Delta \theta_1$ = temperature driving force for precooling (°C)
$\Delta \theta_2$ = temperature driving force for freezing (°C)
θ_{fm} = mean freezing temperature (°C)
θ_c = centre temperature at end of freezing (°C).

The exact way in which shape affects freezing is complex, so considerable research has considered simple ways to take into account food shape [52–55]. Hossain et al. [56–58] showed that using a simple shape factor to adjust predictions for a one-dimensional slab shape with the same R was sufficiently accurate for a wide range of conditions:

$$t_f = \frac{t_{slab}}{E} \tag{14.48}$$

where t_{slab} = freezing time of slab (s).

This is the same as replacing $A_s R/V$ in Equation 14.42 by E. Hossain et al. [56–58] derived simple expressions to estimate E for a wide range of geometries. The most general equation for a three dimensional irregular shape is:

$$E = 1 + \frac{\left(1 + \dfrac{2}{Bi}\right)}{\left(\beta_1^2 + \dfrac{2\beta_1}{Bi}\right)} + \frac{\left(1 + \dfrac{2}{Bi}\right)}{\left(\beta_2^2 + \dfrac{2\beta_2}{Bi}\right)} \tag{14.49}$$

$$\beta_1 = \frac{A_{XS}}{\pi R^2} \tag{14.50}$$

$$\beta_2 = \frac{3V}{4\pi\beta_1 R^3} \tag{14.51}$$

$$Bi = \frac{hR}{k_{ff}} \tag{14.52}$$

where A_{XS} = smallest cross-sectional area of the food object through the thermal centre (m^2).

14.5.2.3 Thawing

For thawing, less effort has gone into developing greatly applicable prediction methods. Reasons include that microwave heating is often used to create nonuniform internal heat generation and/or products often undergo significant shape and size changes as they thaw (e.g., a large frozen block breaks up into smaller items) which both complicate the development of prediction methods. In addition, thawing is frequently performed in the home under relatively uncontrolled conditions compared with industrial thawing processes, and consequently accurate estimates of thawing rate are less likely to be possible or deemed useful. Cleland et al. [59] gives some generally applicable empirical formulae for thawing by conduction for one-dimensional geometries based on Plank's equation. The approach defined by Equations 14.48 through 14.52 to take account of the shape is equally applicable to thawing.

14.5.3 NUMERICAL METHODS

The analytical and empirical solutions are useful for situations that can be modeled by convection only at the surface, where h and T_a are constant, and where heat generation is negligible. They are most accurate for regular geometries. In other cases, the problems are best handled by numerical techniques. For simplicity of presentation we limit the examples and equations to two-dimensional systems (Figure 14.2). All numerical solutions are based on approximations to Equations 14.1 through 14.8. For foods, the moving boundary condition (Equation 14.9) is seldom used because phase change in foods normally occurs over a range of temperature.

The time derivative in Equation 14.1 is usually approximated by a forward difference finite difference (FD) approximation, Crank–Nicolson approximation or similar:

$$\rho c \frac{\partial \theta}{\partial t} \approx \rho^i_{m,n} c^i_{m,n} \frac{\theta^{i+1}_{m,n} - \theta^i_{m,n}}{\Delta t} \quad \text{or} \quad \rho c \frac{\partial \theta}{\partial t} \approx \rho^{i+(1/2)}_{m,n} c^{i+(1/2)}_{m,n} \frac{\theta^{i+1}_{m,n} - \theta^i_{m,n}}{\Delta t} \tag{14.53}$$

where
$c^i_{m,n}$ = effective specific heat capacity evaluated at $\theta^i_{m,n}$ (J/kg K)
$\rho^i_{m,n}$ = density evaluated at $\theta^i_{m,n}$ (kg/m^3)
$\theta^i_{m,n}$ = temperature for node m in the x direction and n in the y direction at time step i
and similarly for the other properties and temperatures where the superscripts designate the time increment while the subscripts indicate the nodal position. The space discretization is usually based on a FD [4], finite element (FE) [60,61], or finite volume (FV) [62,63] grids of nodes to represent the food object. Temperature between the nodes is usually interpolated linearly although for FE grids higher order approximation can be used. For example, for the two-dimensional FD grid shown in

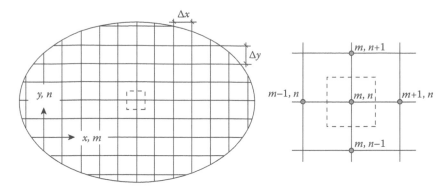

FIGURE 14.2 Finite difference nodal grid for two-dimensional heat conduction.

Figure 14.2, the partial derivative in the x or y direction on the right hand of Equation 14.1 can be approximated at the ith time level by:

$$\frac{\partial}{\partial x}\left(k\frac{\partial \theta}{\partial x}\right) \approx \frac{1}{(\Delta x)^2}\left[k^i_{m+1/2,n}\left(\theta^i_{m+1,n} - \theta^i_{m,n}\right) - k^i_{m-1/2,n}\left(\theta^i_{m,n} - \theta^i_{m-1,n}\right)\right] \tag{14.54}$$

$$\frac{\partial}{\partial y}\left(k\frac{\partial \theta}{\partial y}\right) \approx \frac{1}{(\Delta y)^2}\left[k^i_{m,n+1/2}\left(\theta^i_{m,n+1} - \theta^i_{m,n}\right) - k^i_{m,n-1/2}\left(\theta^i_{m,n} - \theta^i_{m,n-1}\right)\right] \tag{14.55}$$

where $k^i_{m,n+1/2}$ = thermal conductivity evaluated at $\theta^i_{m,n+1/2}$ (W/m K) and similarly for $k^i_{m,n-1/2}$, $k^i_{m+1/2,n}$ and $k^i_{m-1/2,n}$. For this example, $\theta^i_{m,n+1/2}$ is the average of $\theta^i_{m,n+1}$ and $\theta^i_{m,n}$ and similarly for $\theta^i_{m,n-1/2}$, $\theta^i_{m+1/2,n}$, and $\theta^i_{m-1/2,n}$.

In general, it is desirable for the numerical approximations to be centrally balanced in both time and space [3]. Combining such approximations means that the difference equations can be written for each and every node in the space grids. Solving these equations for the $\theta^{i+1}_{m,n}$ values allow a prediction for one time step (Δt) into the future starting from the initial condition. The process can be repeated to predict the full temperature history over time for all nodal positions. FD or FV approximations can be either explicit where each nodal temperature in a time step can be evaluated independently of the calculation for other nodes (e.g., thermal properties all evaluated at the ith time level) or implicit where the future nodal temperatures must be estimated simultaneously because the nodal equations are interdependent (e.g., if thermal properties are evaluated at the $i + 1/2$ time level as for the Crank–Nicolson approximation of the time derivative). FE approaches are inherently implicit because temperature within an element is determined in term of the temperature at each of the nodes and a node may be part of multiple elements, so the nodal temperatures need to be determined simultaneously. In general terms, smaller time steps (Δt) and finer grids (smaller Δx and Δy) give more accurate mathematical approximations and overall predictions, but this must be counterbalanced by the decrease in computation speed and increased computer truncation errors (the high floating point precision of modern computers mean that this later issue is seldom important). The time and space discretization and the type of numerical approaches also affect the stability of the numerical solution. Explicit schemes have strict stability limits. For example, for an explicit one-dimensional slab FD scheme, the stability criteria are:

$$\Delta t < \frac{\rho c \Delta x^2}{2k} \text{ for internal nodes} \tag{14.56}$$

$$\Delta t < \frac{\rho c \Delta x^2}{2(k + h\Delta x)} \text{ for a surface node with convection heat transfer only} \quad (14.57)$$

Generally for the numerical predictions to be convergent and accurate, Δt should be significantly smaller than the above criteria. While implicit schemes are inherently stable, Δt still has significant effect on numerical accuracy.

FD methods are only practical for regular geometries where boundaries are perpendicular to each other and an orthogonal rectangular grid can be used. FE and FV methods are easily applied to irregular geometries and foods with heterogeneous structures and therefore are the basis of most commercial packages designed specially for predictions of conduction and/or diffusional problems. Many commercial packages have limited ability to deal with temperature-variable thermal properties that are necessary to do freezing or thawing predictions so often specialized purpose written software is necessary. Further detail about FD, FE and FV methods and their applications are given in [60–68].

Computational fluid dynamics (CFD) packages have increasingly been used to model both heat conduction and mass transfer within foods and the heat transfer and fluid flow behavior in the heat transfer medium [69,70]. Most commercial CFD software packages adopt FE or FV methods and are able to deal with coupled heat and mass transfer equations where complex food geometries, combinations of boundary conditions, temperature-variable thermal properties and the flow field of the heat transfer medium can be taken into account simultaneously.

14.5.3.1 Chilling

Chilling predictions can be performed by most commercial packages because thermal properties are normally constant so the approximation of Equations 14.1 through 14.8 is relatively straightforward. Some commercial software requires user defined code in order to model evaporation heat transfer at the boundary due to the highly nonlinear relationship between the partial pressures of water vapor and the temperature in many circumstances.

Most FE and FV based packages allow irregular shapes and time-variable boundary conditions (e.g., h, T_a, T_r).

14.5.3.2 Freezing and Thawing

Numerical methods for freezing and thawing must be designed to deal with variation in thermal properties particularly k and c. In particular, as shown in Figure 14.1, the effective specific heat capacity goes through very rapid change in value near θ_f. A direct approximation of Equation 14.1 such as Equations 14.53 through 14.55 runs the risk of partially jumping over the latent heat peak in c as the temperature decreases from just above θ_f to just below. Cleland et al. [71] used a heat balance method to check that the Δt and space grid were fine enough to minimize this problem. The heat balance within the volume (V) and across the boundary surface (A_s) is:

$$\left[\int_V \rho H dV\right]^{t=t} - \left[\int_V \rho H dV\right]^{t=0} = \int_0^t \int_{A_S} \{h(T_a - T_s) + F\varepsilon\sigma[T_r^4 - T_s^4] + h_m(RH\, p_{wa} - a_w p_{ws})L_e\} dA\, dt \quad (14.58)$$

Alternatively, the terms in the integral on the right hand side of Equation 14.58 could be replaced by the terms on the left hand side of Equation 14.4.

The most common approach to avoid the latent heat jumping problem has been to use the enthalpy transformation [9,10]. Equation 14.1 is rewritten as:

$$\rho \frac{\partial H}{\partial t} = \frac{\partial}{\partial x}\left(k \frac{\partial \theta}{\partial x} \right) + \frac{\partial}{\partial y}\left(k \frac{\partial \theta}{\partial y} \right) + \frac{\partial}{\partial z}\left(k \frac{\partial \theta}{\partial z} \right) + \rho Q \qquad (14.59)$$

H is a smoother function of temperature than c with less rapid changes in value. With the enthalpy transformation an explicit numerical scheme is commonly used because temperature must be calculated from the nodal enthalpy at the end of each time step yet the relationship between T and H is not linear. Pham [68] proposed a quasi-enthalpy transformation method with temperature correction. This method was shown to be faster, more accurate, and more easily programmed.

The rapid change in thermal conductivity around the freezing point also contributes to the difficulty in accurately modeling phase change processes. The Kirchhoff transformation, which is equivalent to the enthalpy transformation for k, can be used to smooth the function [72,73]. The transformation is:

$$u = \int_{\theta_{ref}}^{\theta} k \partial \theta \qquad (14.60)$$

Using both transformations the heat conduction equation becomes:

$$\rho \frac{\partial H}{\partial t} = \frac{\partial^2 u}{\partial x^2} + \frac{\partial^2 u}{\partial y^2} + \frac{\partial^2 u}{\partial z^2} + \rho Q \qquad (14.61)$$

where u = transformed temperature (W/m).

It was found that the Kirchhoff transformation leads to a significant reduction in computation time [73,74]. Many packages can be used to solve the enthalpy and Kirchhoff transformed equations. Use of Kirchhoff transformation can be problematic to implement in food that must be modeled as composites rather than homogeneous materials. For example, at the boundary of two materials (e.g., a fat layer over a meat), a simple node may have several possible u values depending on which material is being considered.

For most foods, nucleation and supercooling are quite short term phenomena and have relatively little effect on freezing rate. However, for some circumstances it can be important. A number of researchers have attempted to numerically model freezing where nucleation and ice crystal growth rates are rate limiting (not instantaneous) and the process is partially mass transfer controlled [9,75–79]. For example, Nahid et al. [75] found that to model freezing of butter it was necessary to use Equation 14.1 with c values excluding latent heat effects but to replace Q by a term related to release of latent heat of ice crystallization as a function of supercooling extent and duration.

Some numerical methods included changes in dimensions with time due to density variations [80]. The key issue is to predict the change in dimension and then to decide how to map the temperature profiles prior to the displacement onto the new spatial grid at the end of each time step.

14.5.3.3 Pressure Shift Freezing and Thawing

Pressure shift freezing has been gaining attention as a freezing method for high quality or freeze-sensitive foods [81–85]. In pressure shift freezing, the food is cooled under very high pressure to sub-zero temperatures. Because the freezing point decreases with the increase in pressure, phase change does not take place. Once the product temperature has equilibrated at low temperature, the

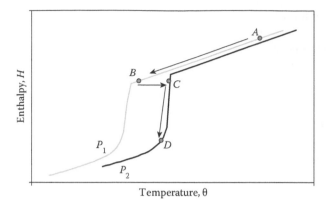

FIGURE 14.3 Pressure shift freezing from high pressure (P_1) to low pressure (P_2) shown on an enthalpy-temperature diagram.

pressure is released suddenly to atmospheric. The food becomes supercooled and nucleation takes place spontaneously throughout the product, causing an almost instantaneous rise to the temperature of a partially frozen food with the same enthalpy at atmospheric pressure. Figure 14.3 shows a pressure shift process on the enthalpy-temperature diagram where the food is cooled at high pressure from A to B, pressure is released along BC causing partial freezing, and then freezing is completed along CD. The benefits of the method are uniform nucleation that ensures evenly small crystal size and minimal textural damage.

High pressure thawing has also been investigated as a fast thawing method. Because of the lowering of the freezing point, the difference between the product and ambient temperatures is increased so faster thawing can be achieved, without the risk of microbial growth if external medium temperature is increased. To model high pressure thawing and freezing with numerical methods, the effect of pressure on thermal properties (enthalpy, freezing point, and thermal conductivity) must be taken into account [81–85]. To handle any temperature and pressure regimes, Pham [9,68] recommended a simple and efficient method using the enthalpy or quasi-enthalpy formulation at every time step where the relationship between nodal enthalpy and the nodal temperature varies with pressure so that a step change in pressure will result in a commensurate change in temperature.

14.6 CONCLUSION

This chapter has summarized the major methodologies and the key issues related to mathematical modeling of freezing, thawing and chilling processes for solid foods. If the assumptions or simplifications inherent in the derivation of the calculation methods are physically justified, the prediction's accuracy is usually limited more by the accuracy of input data such as thermal properties and surface heat transfer conditions, than by the inherent uncertainty in the prediction method itself.

NOMENCLATURE

a_w	water activity at product surface
A	food surface (m²)
Bi	Biot number
c	specific heat capacity (J/kgK)
D_1	shortest dimension through geometric centre of the object taken at right angles to R (m)
D_2	dimension through geometric centre of the object at right angles to both R and D_1 (m)
E	shape factor

E_∞ Shape factor for $Bi \to \infty$
E_0 shape factor for $Bi = 0$
F radiation view factor
f exponential factor (s^{-1})
Fo Fourier number
h surface convective heat transfer coefficient (W/m^{2}K)
h_m surface mass transfer coefficient based on partial pressure (kg/sm^{2}Pa or s/m)
H enthalpy (J/kg)
ΔH_1 heat released in precooling (J/m^3)
ΔH_2 heat released in freezing (J/m^3)
j lag factor
k thermal conductivity (W/mK)
L_e latent heat of evaporation or sublimation (J/kg)
L_f latent heat of freezing or thawing (J/kg)
L_w latent heat of freezing of pure water (J/kg)
l_x, l_y, l_z cosine of outward normal in the x, y, z directions
M molecular weight (g/mol)
N number of components
p_w vapor pressure of water (Pa)
Q internal heat generation, e.g. heat of respiration (W/kg)
r distance from the centre (m)
R characteristic half thickness; shortest distance from the product centre to the surface (m)
R   gas constant (8.314 J/mol K)
RH relative humidity of air as a fraction
t time (s)
T temperature (K)
u transformed temperature (W/m)
v volume fraction
V volume of the food item (m^3)
w mass fraction
x, y, z space coordinates (m)
Y dimensionless temperature

Greek
α root of transcend equation
β length ratio
ε food surface emissivity
θ temperature (°C)
$\Delta\theta_1$ temperature driving force for precooling (°C)
$\Delta\theta_2$ Temperature driving force for freezing (°C)
ρ density (kg/m^3)
σ Stefan-Boltzmann constant (5.67 × 10^{-8} W/m^{2}K^4)
ψ thermal diffusivity (m^2/s)

Subscript
a external cooling medium
bw   bound water
c thermal centre
e effective
f initial freezing state

ff	fully frozen state
fm	mean freezing state
i	food component
in	initial state
ins	perfect insulated or symmetric
ma	mass-average
m, n	for node m in the x direction and n in the y direction
r	radiation source/sink
s	surface
tw	total water
u	unfrozen state
w	water
XS	smallest cross-sectional area through the thermal centre

Superscript

i	time step

REFERENCES

1. ASHRAE. *ASHRAE Handbook: Fundamentals*. American Society of Heating Refrigeration and Air-conditioning Engineers, Atlanta, GA, 2005.
2. ASHRAE. *ASHRAE Handbook of Refrigeration*. American Society of Heating Refrigeration and Air-conditioning Engineers, Atlanta, GA, 2006.
3. Cleland, A.C. *Food Refrigeration Processes Analysis, Design and Simulation*, Elsevier, London, 1990.
4. Incropera, F.P. et al. *Fundamentals of Heat and Mass Transfer*, 6th ed. John Wiley & Son, Hoboken, NJ, 201, 2007.
5. Tocci, A.M. and Mascheroni, R.H. Freezing times of meat balls in belt freezers: experimental determination and prediction by different methods. *Int. J. Refrig.*, 17, 455, 1994.
6. Rahman, S. *Food Property Handbook*, CRC Press, Boca Raton, FL, 1995.
7. Choi, Y. and Okos, M.R. Effect of temperature and composition on the thermal properties of foods. In: Maguer, L. and Jelen, M. (ed.), *Food Engineering and Process Applications*, 1st edn. Elsevier, Amsterdam, 93, 1986.
8. Carson, J.K. Review of effective thermal conductivity models for foods. *Int. J. Refrig.*, 29, 958, 2006.
9. Pham, Q.T. Modeling heat and mass transfer in frozen foods: a review. *Int. J. Refrig*, 29, 876, 2006.
10. Delgado, A.E. and Sun D.W. Heat and mass transfer models for predicting freezing processes-a review. *J. Food Eng.*, 47, 157, 2001.
11. Pham, Q.T. and Willix, J. Thermal conductivity of fresh lamb meat, offal and fat in the range 40 to + 30°C: measurements and correlations. *J. Food Sci.*, 54, 508, 1989.
12. Willix, J., Lovatt, S.J. and Amos, N.D. Additional thermal conductivity values of foods measured by a guarded hot plate. *J. Food Eng.*, 37, 159, 1998.
13. Wang, J.F. et al. A thermal conductivity prediction method for frozen foods. *ICEF 10*, 20–24 April 2008, Vina del Mar, Chile.
14. Carson, J.K. et al. Predicting the effective thermal conductivity of unfrozen, porous foods. *J. Food Eng.*, 75, 297, 2006.
15. Maxwell, J.C. *A Treatise on Electricity and Magnetism*, 3rd edn. Dover Publications Inc., New York, reprinted 1954, Chapter 9.
16. Landauer, R. The electrical resistance of binary metallic mixtures. *J. Appl. Phys.*, 23, 779, 1952.
17. Wang, J.F. et al. A new structural model of effective thermal conductivity for heterogeneous materials with co-continuous phases. *Int. J. Heat Mass Transfer*, 51, 2389, 2008.
18. Hashin, Z. and Shtrikman, S. A variational approach to the theory of the effective magnetic permeability of multiphase materials. *J. Appl. Phys.*, 33, 3125, 1962.
19. Carson, J.K. et al. Thermal conductivity bounds for isotropic, porous materials. *Int. J. Heat Mass Transfer* 48, 2150, 2005.
20. Fikiin, K.A. Ice fraction prediction methods during food freezing: a survey of the eastern European literature. *J. Food Eng.*, 38, 331, 1998.

21. Pham, Q.T. Prediction of calorimetric properties and freezing time of foods from composition data. *J. Food Eng.*, 30, 95, 1996.
22. Schwatzberg, H. Effective heat capacities for the freezing and thawing of foods. *J. Food Sci.*, 41, 152, 1976.
23. Boonsupthip, W. and Heldman, D.R. Prediction of frozen food properties during freezing using product composition. *J. Food Science*, 72, 254, 2007.
24. Wang, J.F. et al. A new approach to modeling the effective thermal conductivity of heterogeneous materials. *Int. J. Heat Mass Transfer*, 49, 3075, 2006.
25. Cogne, C. et al. Experimental data and modeling of thermal properties of ice creams. *J. Food Eng.*, 58, 331, 2003.
26. Levy, L. A modified Maxwell-Eucken equation for calculating the thermal conductivity of two-component solutions or mixtures. *Int. J. Refrig.*, 4, 223, 1981.
27. Fricke, B.A. and Becker, B.R. Evaluation of thermophysical property models for foods. *HVAC&R Res.*, 7, 311, 2001.
28. Hung, Y.C. Prediction of cooling and freezing times. *Food Tech.*, 44, 241, 1990.
29. Mallikarjunan, P. and Mittal, G.S. Effects of process parameters on beef carcass chilling time and mass loss predictions using a finite element heat and mass transfer model. *J. Muscle Food*, 9, 75, 1998.
30. Willix, J., Harris, M.B. and Carson, J.K. Local surface heat transfer coefficients on a model beef side. *J. Food Eng.*, 74, 561, 2006.
31. Kondjoyan, A. and Daudin, J.D. Heat and mass transfer coefficients at the surface of a pork hindquarter. *J. Food Eng.*, 32, 225, 1997.
32. Mannapperuma, J.D., Singh, R.P. and Reid, D.S. Effective surface heat transfer coefficients encountered in air blast freezing of whole chicken and chicken parts individually and in packages. *Int. J. Refrig.*, 17, 263, 1994.
33. Mannapperuma, J.D., Singh, R.P. and Reid, D.S. Effective surface heat transfer coefficients encountered in air blast freezing of single plastic wrapped whole turkey. *Int. J. Refrig.*, 17, 273, 1994.
34. Tocci, A.M. and Mascheroni, R.H. Heat and mass transfer coefficients during refrigeration, freezing and storage of meats, meat products and analogues. *J. Food Eng.*, 26, 147, 1995.
35. Daudin, J.D. and Swain, S.V.L. Heat and mass transfer in chilling and storage meat. *J. Food Eng.*, 12, 95, 1990.
36. Ozisik, N.M. *Heat Transfer—A Basic Approach*. McGraw-Hill, New York, 1985.
37. Kondjoyan, A. et al. Heat and mass transfer coefficients at the surface of elliptical cylinders placed in a turbulent air flow. *J. Food Eng.*, 20, 339, 1993.
38. Harris, M.B., Lovatt, S.J. and Willix, J. Development of a sensor for measuring local heat transfer coefficients on carcass-shaped objects. *Proc. 20th Int. Congress Refrig.*, Sydney, 1999.
39. Carslaw, H.S. and Jaeger, J.C. *Conduction of Heat in Solids*, 2nd edn. Clarendon Press, Oxford, 1959.
40. Newman, A.B. Heating and cooling rectangular and cylindrical solids. *Ind. Eng. Chem.*, 28, 545, 1936.
41. Chuntranuluck, S., Wells, C.M. and Cleland, A.C. Prediction of chilling times of foods in situations where evaporative cooling is significant—part 1. model development. *J. Food. Eng.*, 37, 111, 1998.
42. Chuntranuluck, S., Wells, C.M. and Cleland, A.C. Prediction of chilling times of foods in situations where evaporative cooling is significant—part 2. experimental testing. *J. Food. Eng.*, 37, 127, 1998.
43. Chuntranuluck, S., Wells, C.M. and Cleland, A.C. Prediction of chilling times of foods in situations where evaporative cooling is significant—part 3. applications. *J. Food. Eng.*, 37, 143, 1998.
44. Fikiin, A.G. and Fikiina, I.K. Calculation de la duree de refrigeration des produits alimentaires et des corps solides. *Proc 13th Int. Cong. Refrig.*, Washington, DC, 2411, 1971.
45. Baehr, H. D. Die berechnung der kuhldauer bei ein- und mehrdimensionalen warmefluss. *Kaltetechnik*, 5, 255, 1953.
46. Rutov, D.G. Calculation of the time of cooling of food products. *Refrig. Sci. Technol.*, 415, 1958.
47. Lin, Z. et al. A simple method for prediction of chilling times for objects of two dimensional irregular shape. *Int. J. Refrig.*, 19, 95, 1996.
48. Lin, Z. et al., A simple method for prediction of chilling times: extension to three dimensional irregular shapes. *Int. J. Refrig.*, 19, 107, 1996.
49. Smith, R.E. and Nelson, G.L. Transient heat transfer in solids: theory versus experimental. *Trans ASAE*, 12, 833, 1969.
50. Smith, R.E., Nelson, G.L. and Henrichson, R.L. Analyses on transient heat transfer from anomalous shapes. *Trans ASAE*, 10, 236, 1967.

51. Pham, Q.T. An extension to Plank's equation for predicting freezing times for food stuffs of simple shape. *Int. J. Refrig.*, 7, 377, 1984.
52. Pham, Q.T. Simplified equation for predicting the freezing time of foodstuffs. *J. Food Tech.*, 21, 209, 1986.
53. Ilicali, C., Cetin, M. and Cetin, S. Methods for the freezing time of ellipses. *J. Food Eng.*, 28, 361, 1996.
54. Becker, B.R. and Fricke, B.A. Evaluation of semi-analytical/empirical freezing time estimation methods part II: Irregularly shaped food items. *HVAC&R Res.*, 5, 171, 1999.
55. Mittal, G.S., Hanenian, R. and Mallikarjunano P. Evaluation of freezing time prediction models for meat patties. *Can. Agril. Eng.*, 35, 75, 1993.
56. Hossain, M.M., Cleland, D.J. and Cleland A.C. Prediction of freezing and thawing times for foods of regular multidimensional shape by using an analytical derived geometric factor. *Int. J. Refrig.*, 15, 227, 1992.
57. Hossain, M.M., Cleland, D.J. and Cleland A.C. Prediction of freezing and thawing times for foods of two dimensional irregular shape by using a semi-analytical derived geometric factor. *Int. J. Refrig.*, 15, 235, 1992.
58. Hossain, M.M., Cleland, D.J. and Cleland A.C. Prediction of freezing and thawing times for foods of three dimensional irregular shape by using a semi-analytical derived geometric factor. *Int. J. Refrig.*, 15, 241, 1992.
59. Cleland, D.J., et al. Prediction of thawing times for food of simple shape. *Int. J. Refrig.*, 9, 220, 1986.
60. Puri, V.M. and Anantheswaran R.C. The finite element method in food processing: a review. *J. Food Eng.*, 19, 247, 1993.
61. Zienkiewicz, O.C. and Taylor R.L. *The Finite Element Method*, 2nd edn. McGraw-Hill, London, 1991.
62. Voller, V.R. and Swaminathan C.R. Treatment of discontinuous thermal conductivity in control volume solution of phase change problems. *Num. Heat Transfer Part B-Fundamentals*, 19, 175, 1991.
63. Voller, V.R. An overview of numeric methods for solving phase change problems. *Adv. Numer. Heat Transfer*, 1, 341, 1996.
64. Patankar, S.V. *Numerical Heat Transfer and Fluid Flow*, 1st edn, Hemisphere, London, 1980.
65. Pham, Q.T. Comparison of general-purpose finite-element methods for the Sefan problem. *Num. Heat Transfer Part B-Fundamentals*, 27, 417, 1995.
66. Cleland, D.J. et al. Prediction of freezing and thawing times for multi-dimensional shapes by numeric methods. *Int. J. Refrig.*, 10, 32, 1987.
67. Cleland, D.J. et al. Prediction of freezing and thawing times for multidimensional shapes by using numerical methods. *Int. J. Refrig.*, 10, 32–39, 1992.
68. Pham, Q.T. A fast unconditionally stable finite-difference method for heat conduction with phase change. *Int. J. Heat Mass Transfer*, 28, 2079, 1985.
69. Wang, L. and Sun, D.W. Recent developments in numerical modeling of heating and cooling processes in the food industry—a review. *Trends Food Science Tech.*, 14, 408, 2003.
70. Norton, T. et al. Applications of computational fluid dynamics in the modeling and design of ventilation systems in the agricultural industry—a review. *Bioresource Tech.*, 98, 2386, 2007.
71. Cleland, A.C. and Earle, R.L. Assessment of freezing time prediction methods. *J. Food Sci.*, 49, 1034, 1984.
72. Kirchhoff, G. Vorlesungen uber die theorie der warme, 1894. In: Carslaw, H.S. and Jaeger, J.C. (ed.), *Conduction of Heat in Solids*, 2nd edn. Clarendon Press, Oxford, 1986.
73. Fikiin, K.A. Generalized numerical modelling of unsteady heat transfer during cooling and freezing using an improved enthalpy method and quasi-one-dimensional formulation. *Int. J. Refrig.*, 19, 132, 1996.
74. Scheerlink, N. et al. Finite element computation of unsteady phase change heat transfer during freezing or thawing of food using a combined enthalpy and Kirchhoff method. *ASAE Trans.*, 44, 429, 2001.
75. Nahid, A. et al. Modeling the freezing of butter. *Int. J. Refrig.*, 31, 152, 2008.
76. Pham, Q.T. Effect of supercooling on freezing time due to dendritic growth of ice crystals. *Int. J. Refrig.*, 12, 295, 1989.
77. Menegalli, F.C. and Calvelo, A. Dendritic growth of ice crystals during freezing of beef. *Meat Science*, 3, 179, 1978.
78. Miyawaki, O., Abe, T. and Yano, T. A numerical model to describe freezing of foods when supercooling occurs. *J. Food Eng.*, 9, 143, 1989.
79. Devireddy, R.V. Smith, D.J. and Bischof, J.C. Effect of microscale mass transport and phase change on numerical prediction of freezing in biological tissues. *J. Heat Transfer ASME*, 124, 365, 2002.

80. Kim, C.J. et al. Two-dimensional freezing of water-filled between vertical concentric tubes involving density anomaly and volume expansion. *Int. J. Heat Mass Transfer*, 36, 2647, 1993.
81. LeBail, A. et al. High pressure freezing and thawing of foods: a review. *Int. J. Refrig.*, 25, 504, 2002.
82. Chourot, J.M. et al. Numerical modelling of high pressure thawing: application to water thawing. *J. Food Eng.*, 34, 63, 1997.
83. Oterol, L. and Sanz, P. High-pressure shift freezing. Part 1. Amount of ice instantaneously formed in the process. *Biotech. Process*, 16, 1030, 2000.
84. Denys, S., Vanloey, M.E. and Hendrickx, M.E. Modelling heat transfer during high-pressure freezing and thawing. *Biotech. Progress*, 13, 416, 1997.
85. Denys, S., Vanloey, M.E. and Hendrickx, M.E. Modeling conductive heat transfer during high-pressure thawing process: determination of latent heat as a function of pressure. *Biotech. Progress*, 16, 447, 2000.

15 Food Refrigeration Aspects

Ibrahim Dincer
University of Ontario Institute of Technology (UOIT)

CONTENTS

15.1 INTRODUCTION

A wide variety of food processing and preservation methods are now available to maintain and enhance the appearance and taste of food. These methods also create products that are convenient for consumers, such as products that are ready to eat or require minimal preparation and cooking. Although food preservation techniques such as freezing and canning are the results of relatively new food science technology, the preservation of food products has been practiced since the beginning of mankind by many methods such as sun drying, salting, smoking, and food fermentation. Many years ago, our ancestors used chilling and freezing techniques for food preservation, for example, the use of water/ice taken from frozen lakes for chilling fish and meat.

The technology of food preservation advanced dramatically with the discovery of microorganisms late in the eighteenth century and the appreciation of their role in food deterioration that followed. Due to the increasing energetic, environmental and technical problems, research on food preservation processes and technologies now focuses the following key targets: (i) better product quality, (ii) better process efficiency, (iii) better cost effectiveness, (iv) better processing conditions, and (v) better environment.

The means of preservation by drying, salting, smoking, cooking, etc. are often nutritionally less good and alter the taste of the produce; since they are generally less expensive, they are useful in feeding the population. However, applying refrigeration results in foods of higher value.

The cooling of fruits and vegetables implies removal of the field heat before processing, transporting or storing. Cooling inhibits growth of decay-producing microorganisms and restricts enzymatic and respiratory activity during the postharvest holding period, inhibits water loss, reduces ethylene production and reduces the sensitivity of products to ethylene. The holding period may be the relatively short time required to transport and sell or process the product, or it may include a long-term storage period as well. It is important to mention that slowing down metabolism can give rise to physiological disorders which are called *cold storage injuries*. For this reason, the cooling temperature and exposing period must be suitable for the produce held. The uses of refrigeration include the storage and preservation of foodstuffs, the maintenance of comfortable living conditions and industrial applications. The economic impact of refrigeration technology throughout the world is already very impressive and more important than is generally believed. It is therefore that the refrigeration industry is called the *invisible industry*. The importance of refrigeration is bound to increase since it will be an essential factor in solving two major problems of the future through key targets as listed above. Refrigeration technology will play a key role in countless ways to improve living conditions [1].

It is evident that the world population has increased rapidly during the past century. The time taken for an increase of one billion in population has gone down dramatically in the 1900s. Despite a diminution in the rate of increase, the world population in the year 2025 is expected to be somewhat above eight billion. During the 1960s and 1970s, it was indicated that the world will be facing a food shortage during the next few decades. However, during the 1980s, the main issue was the surplus in industrialized countries despite the fact that 15 million people die every year of starvation in some less developed countries. It is estimated that today we have enough food in the world to feed six billion people, whereas the population is about five billion, and that while we have about 3.2 million hectares of cultivable land, we are using only 1.4 billion hectares for food production. There are still large margins for both increasing the land presently utilized and improving the output per hectare in many developing countries [2].

The main goal of this book chapter is to discuss a large number of aspects and issues related to food preservation, food refrigeration, food precooling methods and applications, cool and cold storage techniques, control atmosphere and modified atmosphere (MA) storage techniques, refrigerated transport applications, cooling process parameters, and cooling heat transfer parameters along with illustrative examples to highlight the importance of the topics and show how to conduct design analysis and thermal calculations.

15.2 FOOD DETERIORATION

Cooling and freezing are probably the most popular forms of food preservation in use today. In refrigeration, the idea is to slow bacterial action to a crawl so that it takes food much longer (perhaps a week or two, rather than half a day) to spoil. In freezing, the idea is to stop bacterial action altogether. Frozen bacteria are completely inactive with the exception of a few which can survive freezing temperatures. Since foods are of plant and animal origin, it is worthwhile to consider the intrinsic and extrinsic parameters of foods that affect microbial growth. The parameters of plant and animal tissues that are an inherent part of the tissues are referred to as some key intrinsic parameters:

- pH
- Moisture content
- Oxidation-reduction potential
- Nutrient content
- Antimicrobial constituents
- Biological structures

The extrinsic parameters of foods are the properties of the storage environment that affect both the foods and their microorganisms. They are of greatest importance to the welfare of foodborne organisms as follows:

- Storage temperature
- Relative humidity of environment
- Presence of gases in the environment
- Concentration of gases in the environment

15.3 FOOD PRESERVATION

Food preservation is considered one of the most crucial food processing techniques and aims to prolong the shelf life of food products and maintain their quality in terms of color, texture and flavor. The term *food preservation* covers a wide range of methods from short-term techniques (e.g., cooking, cold storage) to long-term techniques (e.g., canning, freezing, drying). From a good food preservation technique, one can expect to fulfil the following:

- Keeping the food as fresh as possible
- Keeping the food as safe as possible
- Retaining the nutrients
- Preventing the spoilage of the food
- Maintaining the quality of the food
- Being technically feasible
- Being inexpensive and simple
- Requiring less amount of labor

- Requiring less processing time
- etc.

In practice, numerous food preservation methods are available and can be classified into three main categories (e.g., physical methods, chemical methods and biological methods). These methods can be used individually or together in order to achieve numerous goals including:

- Prevention or delay of microbial decomposition
- Prevention or delay of self-decomposition of the food product
- Prevention of damage caused by insects, animals, mechanical processing
- etc.

15.4   FOOD QUALITY

The most significant factors in attaining and maintaining good quality are harvesting at the fully-ripe stage, avoiding physical injuries during all handling steps, enforcing strict quality control procedures, prompt precooling, and providing proper temperature and relative humidity during transport and handling at destination.

Cooling of food products as quickly as possible after harvest has become a widely used method for maintaining the quality, preventing spoilage and maximizing postharvest life. This thermal processing is very common in the food industry and the design of equipment for such processing, for example, cooling, depends heavily upon knowledge of the thermal properties of the foods to be processed. This knowledge is also important for economic reasons when energy balances are being considered in the design of process equipment.

In order to provide optimum product quality, there is a need for proper harvesting, handling, grading, packaging and storing of fruits and vegetables. At the marketplace, the following factors are considered in the evaluation of food quality such as crispness and freshness, taste, appearance and condition, nutritive value and price. In this regard, producers who are able to produce and package their produce in such a way to enhance these variables are most successful in the market place. Due to the perishable nature of fruits and vegetables, harvesting and handling speed is of utmost importance as soon as harvest maturity has occurred, before precooling. After harvesting, we aim to extend postharvest shelf life of the perishable products by:

- Reducing respiration by lowering temperature
- Slowing respiration by maintaining optimal gaseous environment
- Slowing water loss by maintaining optimal relative humidity

Removal of field heat by the process of cooling to a recommended storage temperature and relative humidity is absolutely necessary to maintain the quality of fruits and vegetables. The quality of most products will rapidly deteriorate if field heat is not removed before loading into transportation equipment. The rate of respiration and ripening increases two to three times for every 10°C above the recommended storage temperature.

Refrigerated transportation equipment is designed to maintain temperature and should not be used to remove field heat from products packed in shipping containers. The refrigeration units also are not capable of raising or controlling the relative humidity. A high temperature difference between the refrigeration unit evaporation coil and the product will increase the loss of product moisture. This will cause the evaporator to frost and the products to shrivel or wilt and weigh much less. Most fruits and vegetables have a water content between 80 and 95%, respectively.

Postharvest handling includes all steps involved in moving a commodity from the producer to the consumer including harvesting, handling, cooling, curing, ripening, packing, packaging, storing, shipping, wholesaling, retailing and any other procedure that the product is subjected to. Because vegetables can change hands so many times in the postharvest sector, a high level of management is necessary to ensure that quality is maintained. Each time someone fails to be conscientious in carrying out his or her assigned responsibility, quality is irreversibly sacrificed.

Maintaining produce quality from the farm to the buyer is a major prerequisite of successful marketing. The initial step required to insure successful marketing is to harvest the crop at the optimum stage of maturity. Full red, vine-ripened tomatoes may be ideal to meet the needs of a roadside stand, but totally wrong if the fruit is destined for long distance shipment. Factors such as size, color, content of sugar, starch, acid, juice or oil, firmness, tenderness, heat unit accumulation, days from bloom and specific gravity can be used to schedule harvest. Vegetable producers should gather as much information as possible on maturity indices for their particular commodities. The result of harvesting at an inappropriate stage of development can be a reduction in quality and yield [3].

15.5 FOOD PRECOOLING (OR CHILLING)

Precooling, *chilling*, is a cooling process in which the field temperature of the fruits and vegetables is reduced to approximately optimum storage and/or transportation temperature (i.e., chilling temperature) in the shortest time possible after harvest in the field prior to the storage and/or transportation. The main objective of this treatment is to reduce the rates of biochemical and microbiological changes in order to prevent spoilage of produce, maintain its quality (all possible preharvest freshness and flavor) and extend its storage life [1,2].

Harvesting should be conducted in early morning hours to minimize field heat and the refrigeration load requirement on precooling equipment. Harvested products should be protected from the sun with a covering until they are placed in the precooling facility.

Many products are field or shed packed and then precooled. Wirebound wood or nailed crates or wax impregnated fiberboard boxes are used for packed products that are precooled with water or ice after packing. Precooling of products packed in shipping containers and stacked in unitized pallet loads is especially important as air circulation around and through the packaging may be limited during transportation and storage.

Precooling is particularly important for products which produce a lot of heat. The following are examples of products which have high respiration rates and short transit and storage lives: artichokes, brussels sprouts, onions (green), asparagus, carrots (bunched), okra, beans (lima), corn (sweet), parsley, beans (snap), endive, bean sprouts, raspberries, blackberries, lettuce, spinach, broccoli, mushrooms, strawberries, watercress, etc. [4].

Fresh produce (or any perishable item) continues to breathe or respire after it has been harvested. As it respirates, it breaks down. Fresh produce is precooled to reduce the temperature of the produce, slow the respiration rate and slow down product degradation. In practice, a proper temperature management through the distribution chain allows consumers to buy the freshest possible produce anywhere. Factors such as maturity, ripeness, harvesting date, variety and origin influence the optimum storage temperature and maximum storage lives of fruits and vegetables. The optimum storage temperature and storage life are influenced by, for example,

- The ripeness of bananas, lemons, pineapples, tomatoes, etc.
- The variety of apples, pears, etc.
- The harvesting date of potatoes, etc.
- The origin of tomatoes, grapefruit, oranges, cucumbers, etc.

Precooling is the first postharvest operation in the cold chain. Produces are living organisms that even after harvesting have an active metabolism. In addition, respiration, heat and respiratory gases and certain metabolic products such as carbonic acid, ethylene gases are released during the storage with undesirable consequences to the quality. It is found that a delay of one hour to start precooling of grapes after the harvest reduces the shelf life of grapes by a few days. In general, precooling is necessary due to the following key reasons:

- Products are perishable in nature and more so at high ambient temperature.
- Products are seasonal but demand is continuous.
- Food quality becomes essential and consumers display an increased awareness for it. It is now the international standard for quality of the products to be exported.
- The high ambient temperature causes the following effects.
- Higher metabolic activity results in product ripening or ageing.
- Increased moisture loss results in product drying and shrivelling.
- Increased growth of microorganisms causes decay of the products.

All these undesirable effects are temperature dependent, the higher the temperature, the larger the damage and culminate in product quality degradation and shortening of shelf life. Therefore, by quick and uniform cooling after harvesting and subsequent maintaining of the products at optimum temperature the metabolic process is largely inactivated and the produce is brought to a dormant condition. Hence, produce can be stored for a longer period of time without losing its freshness. It is important to indicate that precooling extends product life by reducing [e.g., 4]:

- The field heat
- The rate of respiration (heat generated by the product per unit mass or unit volume)
- The rate of ripening
- The loss of moisture (shrivelling and wilting)
- The production of ethylene (ripening gas generated by the product)
- The spread of decay

Understanding cooling requirements of perishable products requires adequate knowledge of their biological responses. Fresh fruits and vegetables are living organisms, carrying on many biological processes essential to the maintenance of life. They must remain alive and healthy until processed or consumed. Energy that is needed for these life processes comes from the food reserves that accumulated while the commodities were still attached to the plant.

The success of a food precooling process depends normally upon [e.g., 4]:

- Time duration between harvesting and precooling
- Type of shipping container if product is packed beforehand
- Initial product temperature
- Velocity or amount of cold air, water or ice provided
- Final product temperature
- Sanitation of the precooling air or water to reduce decay organisms
- Maintenance of the recommended temperature after precooling

During precooling, the sensible heat (or field heat) from the product is transferred to the ambient cooling medium. The rate of heat transfer, or cooling rate, is critical for the efficient removal of field heat and is dependent upon three factors: time, temperature and contact [1,2]. In order to achieve maximum cooling, the product must remain in the precooler for sufficient time to remove the heat. This is particularly important during busy periods when it may be tempting to push product through the precooler. A correctly sized precooler should have sufficient capacity so as to provide adequate resident time for precooling, while at the same time not slowing subsequent packing and/or handling operations. The cooling medium (e.g., air) must be maintained at a constant temperature throughout the cooling period. If the refrigeration system is undersized for the capacity of product requiring precooling, the temperature of the medium will increase over time. Inappropriately-designed containers can markedly reduce flow of the cooling medium.

The cooling rate is not only dependent upon time, temperature and contact with the commodity; it is also dependent on the cooling method employed. The rate of fresh produce cools depending on various parameters including [e.g., 5]:

- The rate of heat transfer from the produce to the air or water used to cool it. (i.e., the faster cold air moves past the product the quicker the product cools).
- The difference in temperature between the produce and the cooling air or water (i.e., the greater the difference between the two the faster the product cools).
- The nature of the cooling medium. (i.e., cold water has a greater capacity to absorb heat than cold air).
- The nature of the produce which influences the rate heat is lost (i.e., leafy vegetables have a greater thermal conductivity than potatoes and so cool faster).

15.6   FOOD PRECOOLING SYSTEMS

Precooling, the rapid removal of heat from freshly harvested vegetables, allows the grower to harvest produce at optimum maturity with greater assurance that it will reach the consumer at maximum quality. The major types of precooling methods, namely hydrocooling, contact icing, air cooling, hydraircooling (wet air) and vacuum cooling, are widely used in practice. Each of the above methods has its own advantages, disadvantages, suitability or otherwise to a particular product. It is found that invariably forced air cooling or wet air cooling can be used for precooling of all types of products as it offers the following advantages. The following is a brief summary of these precooling methods [e.g., 2]:

- Room cooling: stacking containers of products in a refrigerated room. Some products are misted or sprayed with water during room cooling.
- Forced air cooling or wet pressure cooling: drawing air through stacks of containers of products in a refrigerated room. For some products, water is added to the air. Forced air cooling can take one or two hours depending on the amount of packaging, while room cooling may take 24–72 hours. Packaging must allow ventilation of heat for these methods to be successful.
- Hydrocooling: flushing product in bulk tanks, bins or shipping containers with a large quantity of ice water. In hydrocooling, fruits and vegetables are cooled by direct contact with cold water flowing through the packed containers and absorbing heat directly from the produce. Products and packaging must be able to withstand direct water contact in hydrocooling.
- Vacuum cooling: removing heat from products packed in shipping containers by drawing a vacuum in a chamber. The vacuum cooling process produces rapid evaporation of a small quantity of water, lowering the temperature of the product to the desired level. In vacuum cooling, it is necessary that the products have a large surface area, low density and high moisture content. The boxes and wrapping must allow ventilation of heat.
- Hydrovacuum cooling: adding moisture to products packed in shipping containers before or during the vacuum process, to speed the removal of heat.
- Package-icing or contact icing: injecting slush or crushed ice into each shipping container of product. Some operations use bulk containers. In contact icing, crushed ice is placed in the package or spread over a stack of packages to precool the contents. Package-icing provides effective cooling and a high relative humidity for products and packaging that can withstand direct contact with ice.

One of the above precooling methods may be selected by considering a number of criteria, including:

- The cooling rate
- The type of commodity and subsequent storage
- The nature, value and quantity of the product
- Shipping conditions
- Equipment and operating costs
- Cost of labor
- Convenience
- Effectiveness
- Applicability
- Efficient energy use
- Operating conditions
- Personal preference
- Product requirements

Product physiology in terms of harvest maturity and ambient temperature during harvesting has a large effect on cooling requirements and cooling methods. For example, some products, primarily vegetables such as asparagus, snap beans, broccoli, cauliflower, cabbage, and ripened tomatoes must be cooled in the shortest time possible. On the other hand, for some products, especially vegetables such as white potatoes, winter squash, pumpkins, and green tomatoes, there is no urgent need to cool them quickly but they need some time to be cured or ripened.

The preferred method of precooling varies according to the physical characteristics of the vegetable. Hydrocooling is recommended for asparagus, beet, broccoli, carrot, cauliflower, celery, muskmelon, pea, radish, summer squash and sweet corn (maize); cabbage, lettuce and spinach are suited to vacuum cooling; air cooling is preferred for bean, cucumber, eggplant, pepper and tomato. After the produce is precooled, it is desirable to maintain low temperature by shipping in refrigerator cars or trucks, by storing in cold-storage rooms, and by refrigeration in retail stores.

Here, we present in detail four key most common methods of precooling such as hydrocooling, forced air cooling, hydraircooling and vacuum cooling with some technical details as follows.

15.6.1 Hydrocooling

It is an effective method of quickly removing field heat from produce by cascading chilled water, is a commonly used cooling method, particularly for tree fruits and other low-moisture produce. Bulky products such as asparagus, peas, sweet corn, radishes, carrots, cantaloupes, peaches and tart cherries are successfully precooled by this method. Some of their advantages are simplicity, effectiveness, rapidity and cost. Despite their advantages, some disadvantages may be considered as follows:

- Packing and/or handling difficulties (because the products are taken out from the system in a very wet state)
- Heavy contamination with soil, plant, sap, plant debris, and disease organisms (because the cooling water is recirculated to keep the cooling effect)
- Some effluent problems (because of the disposal of used cooling water)

The type of hydrocooling system selected for fruits and vegetables depends on various criteria including [7,8]:

- The product, whether it is in bulk or shipping containers
- The type of container, whether it is handled individually or in unit loads
- Personal preference

Hydrocooling of products is accomplished by flooding, spraying or by immersing the product or product load in a chilled-water pool. Flooding refers to the showering of products with a liberal supply of cold water flowing under a gravity head from overhead flood pans. Spraying is accomplished by overhead nozzles. The flood pan system uses much less pump horsepower than does the spray nozzle system because of the pressure developed at the nozzles. The product to be cooled is immersed in, flooded with, or sprayed with cool water (especially near 0°C) containing a mild disinfectant such as chlorine or an approved phenol compound. There are two basic hydrocooling systems: the flow-through and the batch system. In the flow-through system, the product is conveyed in continuous flow through the hydrocooling tunnel either in bulk lots or in packages. Cooling in the hydrocooling tunnel may be accomplished by flooding, spraying, immersion (or parts of each), depending on the particular type of hydrocooler.

Hydrocooling is accomplished most effectively when 0°C water is made to sweep across each product such that the temperature of the entire surface promptly becomes essentially equal to the temperature of the water. Tightly packed containers reduce this cooling effectiveness. In order to increase system efficiency, it is necessary to keep the hydrocooler away from the heat-gain sources (e.g., from solar insolation).

Figure 15.1 shows an immersion type hydrocooling system and its components. As shown in the figure, the primary elements of the apparatus are a conventional mechanical refrigerating unit and a cooling pool (test section with inner dimensions of 2.0 × 0.8 × 0.4 m) that consists of a lidded pool,

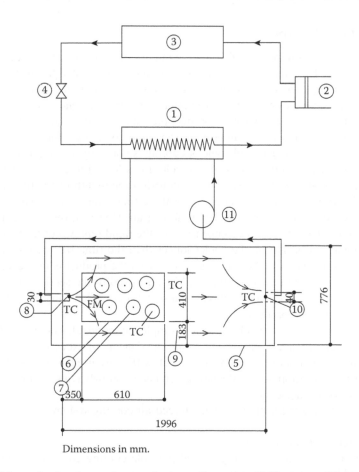

Dimensions in mm.

FIGURE 15.1 Schematic of an immersion type hydrocooling system (1) Evaporator, (2) compressor. (3) condenser, (4) expansion valve, (5) cooling tank, (6) polyethylene crate, (7) product, (8) water inlet, (9) water flow, (10) water exit, (11) water circulation pump. FM, flow meter; TC, thermocouple.

insulated with glass wool, through which cooled water was pumped. The pool-water level was held at two thirds full. The operational details and experimental procedure used in the hydrocooling is outlined in Dincer [1,2,9,10].

Hydrocooling is one of the most efficient of all methods for precooling. Produce is drenched with cold water, either on a moving conveyor or in a stationary setting. In some cases, commodities may be forced through a tank of cold water. Good water sanitation practices must be observed and once cooled, the produce should be kept cold. The cold water must come in direct contact with the product, so it is essential the containers be designed and filled in such a way that the water does not simply channel through without making contact. In practical applications, hydrocoolers are not adequately insulated and as much as half of the energy for refrigeration is wasted. Because of this, some efficient energy utilization techniques have received highest attention.

Some other potential hydrocooling techniques are available for precooling of fruits and vegetables as follows [e.g., 2]:

- Hydrocooling system using ice or ice-slush cooling
- Hydrocooling system using artificial ice
- Hydrocooling system using natural ice
- Hydrocooling system using natural snow
- Hydrocooling system using compacted snow

15.6.2 FORCED-AIR COOLING

It is by far the most universal method. The cooling rate is determined by the available refrigeration capacity, the heat transfer capacity of cooling air, and the heat transfer parameters of the product and its packaging. In large bulk bins or tight carton stows, the apparent conduction resistance may limit the cooling rate severely or even lead to an internal temperature increase as a result of respiratory heat, but in all normal cases, when the product presents a large surface to the air flow, the first two influences mentioned are the most important. Depending on design, the cooling rate varies over a wide range for different products. This type of cooling is normally accomplished through the use of fans and strategically placed barriers so that cold air is forced to pass through the containers of produce. This method usually takes from one-fourth to one-tenth of the time required to cool produce by passive room cooling, but takes two or three times longer than hydro or vacuum cooling. Room coolers are relatively easy to adapt to forced-air. However, the refrigeration capacity of the room may need to be increased to compensate for the rapid removal of heat from the produce. Commodities should not be left on the cooler longer than the time required to reach about one-eighth of their initial temperature because water loss from the product is increased. The humidity control is another significant parameter that will be discussed in the section on cold storage. The speed of food cooling depends on product type and size, refrigeration capacity of the room and the air flow over and around the produce.

Forced-air cooling is widely used because it is simple, economical, sanitary and relatively non-corrosive to equipment. Its major disadvantages are the dangers of excessive dehydration and possibility of freezing the product if air temperatures below 0°C are used. Some products cooled with air are meat, citrus fruits, grapes, cantaloupes, green beans, strawberries, plums, nectarines, sweet cherries, cauliflowers and apricots.

In practice, food products are precooled in a forced-air cooling system:

- With air circulated in cold rooms
- In rail cars or highway vans using special portable cooling equipment which cools the product load before it is transported
- With air forced through the voids of bulk products moving through a cooling tunnel on continuous conveyors or air-cooled refrigerated rail cars or highway vans

- On continuous conveyors in wind tunnels
- By the forced-air method of passing air through the containers by pressure differential

Each of these techniques is employed commercially and each is suitable for certain commodities when properly applied.

One should note that the rate at which heat can be transferred from the product to an air stream depends primarily on:

- The temperature difference between the product and the cooling air
- The velocity of air passing through the products

Besides, the size and shape of product packages and the method of stacking within the cooling unit definitely have a considerable effect on the cooling times of products.

The following products are more predominantly air cooled in a cold room: strawberries, blueberries, raspberries, blackberries, pears, leaf lettuce, celery, brussels sprouts, cauliflower, grapes, spinach, beans, peas, squash, cantaloupe, and other melons.

As mentioned earlier, there are several methods of forced air cooling. Precooling can be obtained in an insulated room or cold storage, in specialized containers or shipping trailers. In these systems, heat exchangers are very common and there are generally two types of heat exchangers used:

- Dry-air is cooled by passing it through a refrigerated coil.
- Wet-air is cooled by passing it through a fill or packing material and cooled by direct contact with chilled water that is sprayed over the surface media

Figure 15.2 presents a schematic diagram of a small scale forced-air cooling system which consists of two main parts namely a conventional refrigeration system and a cooling cabinet where the products are precooled accordingly. Further information is available elsewhere [e.g., 2,11].

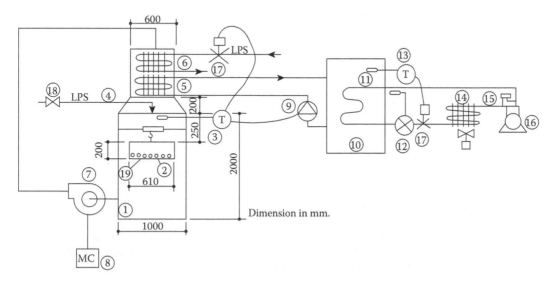

FIGURE 15.2 Schematic of a forced-air cooling system. (1) Cooling chamber, (2) product, (3) thermostat, (4) low pressure steam input, (5) cold water heat exchanger, (6) steam heat exchanger, (7) radial fan, (8) fan speed controller, (9) water pump, (10) water tank, (11) evaporator, (12) thermostatic expansion valve, (13) thermostat, (14) air cooled condenser, (15) presostat, (16) compressor, (17) solenoid valve, (18) valve, (19) Polyethylene crate.

In selecting cooling methods, engineers must be guided by the ultimate condition of the cooled product, which often requires recommendations from agricultural scientists or food technologists. The marketplace may often dictate some preferences of the cooling method. While deluge or submersion hydrocooling has long been used for items such as melons, more and more shippers are now leaning toward air cooling.

The industry trend toward field packing (where the product is picked and placed in its shipping carton) is the major reason for the change to air cooling. First, the containers holding product must allow as much air to come in contact with the product as possible, consistent with acceptable container strength and durability. Venting areas equal to a minimum of 6% of the total face area of a carton on the incoming air side are considered acceptable. It will be well worth the time spent with carton suppliers to develop container ventilation that gives good cooling results. Although some salespeople may prefer a container because It keeps the product cool longer outside a cooler, there is concern that the container may not allow very good ventilation for cooling [12].

In practice, there are various forced air precooling techniques available which are employed. All use the same principle: to force cold air around and through the product to cool it quickly. The actual technique to be used will depend on existing infrastructure, containers to be cooled and product type. Forced-air cooling provides for cold air movement through, rather than around, containers and results in a slight pressure gradient to cause air to flow through container vents, achieves rapid cooling as a result of direct contact between cold air and warm product. With proper design, fast, uniform cooling can be achieved through unitized pallet loads of containers. Various cooler designs can be used, depending on the specific needs. Converting existing cooling facilities to forced-air cooling is often simple and inexpensive, provided sufficient refrigeration capacity and cooling surfaces (evaporator coils) are available. Some forced-air cooler designs are described elsewhere [13].

In practice, various modified techniques of air cooling are employed to precool fruits and vegetables including:

- Tunnel forced air cooling using a free standing fan
- Cold-wall-type tunnel forced air cooling
- Serpentine cooling
- Single pallet forced air cooling
- Room cooling (with storage and shipping)
- Ice bank forced-air cooling system
- Forced air cooling with winter coldness

15.6.3 HYDRAIRCOOLING

It is a combined version of both water and air cooling. In practice, most vegetables before shipment are hydrocooled using large amounts of water in flow-through or batch systems. For flow-through systems, the handling of vegetables in pallet loads necessitated the building of flood or spray-type tunnel hydrocooling systems that could accommodate the pallet loads. These cooling systems are reasonably effective but leave "hot spots" throughout the load, especially loads of sweet corn and celery that are packed in wire-bound crates. With the batch system of hydrocooling, vegetables are placed in rooms where water from nozzles is sprayed onto the vegetables. Slower cooling rates are achieved. For example, when the water used was at 3.3°C, the time required to reduce the temperature of sweet corn from 32.2°C to 7.2°C was over 3 hours [14]. Also, because of the large amount of water required for effective commercial hydrocooling, the maintenance of water cleanliness is impractical and therefore soilborne, decay-producing microorganisms in the water present a problem. Existing hydrocooling systems are not desirable for the cooling of vegetables stacked in fiberboard. Vacuum cooling can be adaptable, but it is limited particularly to leafy vegetables and is more expensive due to its high level of energy requirement. Because of such disadvantages, Henry et al. [14] first developed a more effective cooling technique (so called: hydraircooling) for the vegetables in pallet loads. They conducted a number of experimental tests, using various methods of applying chilled water to the product and varying water flow rates, and also used circulating cold air mixed with the chilled water over the product, as shown in Figure 15.3, to determine the effect of the addition of air on the cooling response. Their results showed that this is a more effective method of cooling vegetables in pallet loads and is available for other vegetables such as cabbage, bell peppers and cucumbers that are not normally hydrocooled.

In general, these hydraircooling systems use wet type forced air precooler air handler. In the operation, water from a chiller located outside the precooling room is pumped and sprayed over surface media inside the air unit. Warm air drawn in by a fan is cooled by direct contact with the chilled water.

In the past, research has shown that relative humidity levels are very important to produce quality and that relative humidities as high as 98% preserve produce weight and quality significantly better than humidity levels below 90%. Recently, new interest has stimulated the development of more efficient wet air coolers combined with ice thermal storage systems. Initially, this cooling technique was employed for high-cost flowers and plants, and later it has been used successfully for the pre-cooling and/or storage of the following food products, such as mushrooms, cauliflowers, white and red cabbage, brussels sprouts, lettuce, celery, leeks, cucumbers, gherkins, strawberries, spinach, chicory, chicory roots, carrots, tomatoes, potatoes, cheese, etc. The main advantage of such system is that it provides higher product cooling rates, especially in the beginning of the cooling process, and that the products therefore, can be cooled more quickly compared with a conventional cooling system. Wet air cooling offers more advantages, namely: simplicity, effectiveness, less cost, shorter cooling time, minimum moisture loss, better product quality, reduced cooling capacity, etc.

15.6.4 VACUUM COOLING

It is an alternative method for the rapid removal of field heat from produce to bring its temperature to the storage temperature. This method is extremely effective for products possessing the following two properties, namely (i) a large surface to mass ratio (leafy vegetables), and (ii) an ability to release internal water readily.

Most fruits and vegetables have large water content. When the produce is subjected to a suitable vacuum, some of this water is evaporated, taking its heat of vaporization from the produce, thereby cooling it. In fact, to cool produce by 10°C requires on average evaporation of water equivalent to 1.8% of the weight of the produce. Vacuum cooling is most successfully applied to flowers, mushrooms and thin leafy produce, e.g., lettuce. It is not suitable for fruits, except for some berries, particularly strawberries.

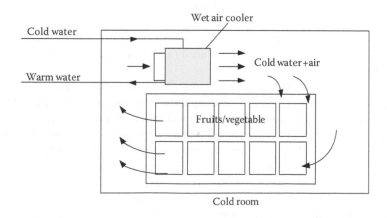

FIGURE 15.3 Schematic representation of a hydraircooling system.

Commodities may be enclosed in a sealed container from which air and water vapor are rapidly pumped out. As the air pressure is reduced, the boiling point of water is lowered, so the product is cooled by surface water evaporation. Vacuum cooling works best with products that have high surface to volume ratio, such as lettuce or leafy greens. The method is effective on produce that is already packaged providing there is a means for water vapor to escape. Moisture loss from the commodity is generally within the range of 1.5–5.0% (about 1.8% of the weight for each 10°C). This can be reduced by wetting the product before cooling. Vacuum chambers vary in size from very large, about rail car size to smaller portable units that may be taken to the field.

This method involves exposing the product to a pressure of about 4–5 mmHg until evaporative cooling yields the desired temperature. Cooling times for products such as lettuce are less than one hour. The extension of cooling is proportional to the water evaporated. Regardless of the product, the temperature is lowered about 5°C for each 1% reduction in water content. Prewetting is useful for many products, especially those that have high initial temperatures and those with properties such as retention of substantial amounts of the added water on their surfaces until the vacuum is applied. For suitable products, vacuum cooling is advantageous because of its rapidity and economy (especially reductions in the cost of labor and packaging, and in product damage). The main disadvantage is the necessity for high capital investment. Equipment need are a large and strong chamber, a pressure-reduction device and a water condensation device. Large amounts of lettuce and moderate but increasing amounts of celery, cauliflower, green peas and sweet corn are vacuum cooled. For example, in British Columbia, Canada, field lettuce is vacuum cooled after harvest to remove field heat, maximizing its crispness and moisture retention. Crisper lettuce varieties such as iceberg and romaine are less perishable than softer types of lettuce such as red leaf, green leaf and butter lettuce.

Vacuum cooling is a rapid evaporative cooling technique which can be applied to specific foods and in particular vegetables. Increased competitiveness together with greater concerns about product safety and quality has encouraged some food manufacturers to use vacuum cooling technology. The advantages of vacuum cooling include shorter processing times, consequently energy savings, improved product shelf life, quality and safety. However, the cooling technique has limited range of application. Traditionally, products such as lettuce and mushrooms have been cooled under vacuum. Recent research has highlighted the possible applications of vacuum cooling for cooling meat and bakery products, fruits and vegetables.

In vacuum cooling, the following factors affect the final product temperature:

- Initial product temperature
- Cooling period in the tank
- Pressure

Also, there are some other factors that influence the final temperature of the product such as prewetting and packaging.

Vacuum cooling is used primarily with head lettuce. During the process, produce is placed in a vacuum retort. The atmospheric pressure is reduced to a point where water boils and evaporates at temperatures close to 0°C.

There are four types of vacuum cooling systems which use water as refrigerant are *steam ejector, centrifugal, rotary,* and *reciprocating.* Among these systems, the steam ejector type of vacuum cooler is best suited for displacing extremely high volumes of water vapor encountered at the low pressures needed and has few moving parts, without a compressor to condense the water vapor. The steam ejector vacuum cooler has advantages such as portability, usability of high-volume pumping, and easy adaptability to water vapor refrigeration. However, it is limited due to inherent mechanical difficulties at the high rotative speeds required to generate the low pressures required. Rotary and reciprocal vacuum coolers make it possible to generate the low pressures required for vacuum cooling and have the advantage of portability. However, they have low volumetric capacity and separate refrigeration systems to condense much of the water vapor that evaporates off the product.

In industry, vacuum coolers are now custom built to meet the exact requirements with either round, oval or rectangular cross section. Vacuum chambers can accommodate one or two rows of pallets or any special size. Hydraulic, pneumatic or electric-powered doors may open by swinging up and over the top (most common), straight up, slide to the side or swing to the side. Doors at both ends and conveyors provide automatic loading. Another loading option is the pallet shuttle platform with door attached. All vacuum coolers are fully automatic using industrial microprocessor based controls.

Table 15.1 summarizes some advantages some disadvantages of vacuum cooling to different sectors of the food processing industry. Most vacuum cooling apparatus are operated in a batch wise process. That is, foods are placed in a vacuum chamber, the chamber is evacuated to predetermined level, foodstuff are cooled and then removed. However, this production method is time consuming and inefficient. In some incidences, it may be necessary to hold products temporarily until they are cooled using vacuum cooler equipment. In this case, the holding time can vary greatly. For example, some cooked product batches may have to be held at high temperature for longer periods than other batches, which can have negative effects on both product safety and quality. However, a recent development aims to make vacuum cooling a more continuous process. Products to be cooled are placed in a plurality of containers designed to hold products for cooling. The containers are inserted successively into a hollow cylinder from one end. The level of vacuum in the containers is then increased through a plurality through holes formed in the cylinder in an axial direction, and

TABLE 15.1
Advantages and Disadvantages of Vacuum Cooling for Some Food Commodities

Applications	Advantages	Disadvantages
Fruits	• Increased shelf life • Rapid cooling times resulting in quicker distribution[a] • Low running costs[a] • Accurate temperature control	• Applicable mostly to large leafy produce • Moisture loss resulting in product weight loss[b] • High capital investment[a]
Meat products	• Increased hygiene and product safety[a] • Reduced microbial counts • Very rapid cooling resulting in significant financial savings • Less maintenance due to the compact cooling units[a,d]	• Narrow product range, applicable to those products which can freely lose water[a] • High loss of product yield due to moisture loss[c] • Some loss of product quality due to moisture loss
Bakery products	• Very rapid cooling of delicate confectionery items • Smaller weight losses than in other vacuum cooled product • Extending crust life of breads and improving product shape • Increased shelf life of many products due to absence of moulds during cooling • Increased productivity due to shorter cooling times[a]	• Specialized modulated vacuum cooling technology required for satisfactory results • Some loss of product aroma due to vacuum cooling

Source: Adapted from McDonald, K. and Sun, D.-W., *J. Food Eng.*, 45, 55–65, 2000.

[a] Generally applicable to all food areas.

[b] Losses can be controlled by prespraying of water onto the produce prior to cooling.

[c] Losses can be controlled or reduced by using modulated vacuum cooling technology and increased brine injection levels in some products.

[d] Units will vary in size depending on the application.

through holes formed in the containers, thereby cooling the food in the containers. The containers are then removed from the cylinder from the other end one after another after being released from evacuation. In essence, the major advantage of vacuum cooling over other techniques of cooling is the short time required to cool a suitable product to a given temperature.

There are other types of modified cooling methods used for food precooling such as (i) hydrovac cooling (a combined version of vacuum and hydrocooling techniques); (ii) evaporative cooling (using wet pads and fans to cool the air in greenhouses); and (iii) direct ice cooling (either by package icing or by bulk application to the top of a load).

In summary, Table 15.2 gives the precooling methods which are recommended for various fruits and vegetables.

TABLE 15.2
Recommended Precooling Methods for Fruits and Vegetables

Produce	Precooling Method
Vegetables	
Herbs	Room-cooling
Lettuce	Hydro-cooling, package icing
Melons	Hydro-cooling, package icing, forced-air cooling
Okra	Room-cooling, forced-air cooling
Onions	No precooling needed
Onions, green	Hydro-cooling, package icing
Oriental vegetables	Package icing
Peas	Forced-air cooling, hydro-cooling
Peppers	Room-cooling, forced-air cooling
Potato	Room-cooling, forced-air cooling
Pumpkin	No precooling needed
Radish	Package icing
Rhubarb	Room cooling, forced-air cooling
Rutabagas	Room cooling
Spinach	Hydro-cooling, package icing
Squash, summer	Forced-air cooling, room cooling
Squash, winter	No precooling needed
Sweet potato	No precooling needed
Tomato room	Room cooling, forced-air cooling
Turnip	Room cooling, hydrocooling, vacuum, package icing
Watermelon	No precooling needed
Fruits	
Apples	Room cooling, Forced-air cooling, hydro-cooling
Apricots	Room cooling, hydro-cooling
Berries	Room cooling, forced-air cooling
Cherries	Hydro-cooling, forced-air cooling
Grapes	Forced-air cooling
Nectarines	Forced-air cooling, hydro-cooling
Peaches	Forced-air cooling, hydro-cooling
Pears	Forced-air cooling, room cooling, hydro-cooling
Plums	Forced-air cooling, hydro-cooling

Source: Adapted from Gast, K.L.B. and Flores, R.A., *Precooling Produce*, MF-1002, Cooperative Extension Service, Kansas State University, Kansas, KS, 1991.

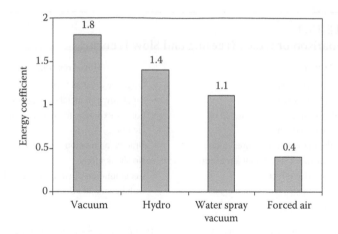

FIGURE 15.4 Energy coefficient data for four types of precooling systems. (Data from Thompson, J.F. and Chen, Y.L., *ASHRAE Trans.*, 94(1), 1427–1433, 1988.)

15.6.5 ENERGY ASPECTS

Here, we look at some energy aspects of precooling systems through energy coefficient as introduced by Thompson and Chen [6]. They conducted a study on energy use of commercial scale cooling systems such as vacuum, hydro, water spray vacuum, and forced-air systems and expressed energy efficiency data as an energy coefficient as follows:

$$EC = SH/EN \tag{15.1}$$

where *SH* is the sensible heat removed from the product, assuming a specific heat of 3.8 kJ/kg°C for fruits and 4.0 kJ/kg°C for vegetables, and *EN* is the electrical energy consumed in operating the cooler (kW).

They found that there are differences in energy efficiency between cooler types and the average energy coefficient, based upon the above equation, was 1.8 for vacuum cooling, 1.4 for hydrocooling, 1.1 for water spray vacuum cooling and 0.4 for forced-air cooling. Also, it is mentioned that there are large differences between coolers of the same type for vacuum coolers and hydrocoolers (a range of 0.8 and 1.3, respectively). Their results are compared in Figure 15.4. Energy use efficiency of cooling systems varies with the type of cooler used. Vacuum coolers are the most efficient, followed by hydrocoolers, water spray vacuum coolers and forced-air coolers. Levels of nonproduct heat input and operational practices have been identified as reasons for the differences. Energy use efficiency varies significantly within coolers of the same type. Level of product throughout, commodity type and operational procedures are also identified as major reasons for this.

15.7 FOOD FREEZING

Freezing, which is a refrigeration process, is used to reduce the temperature of all parts of the produce below freezing point. For most food products, the final quality is better if freezing is done rapidly, especially if the 0–5°C zone is passed through rapidly and if the temperature is reduced to, and maintained at, a sufficiently low level. In fact, food products have high water content (55–90%), and the water exists in different forms in the tissues. Freezing a product consists of converting major part of the water contained in tissues into ice. The freezing point depends on the concentrations of dissolved substances and not on water content. To the degree to which the product is cooled below its freezing point, an increasing quantity of water is transformed into ice crystals. The crystals

TABLE 15.3
Comparison of Quick Freezing and Slow Freezing

Quick Freezing	Slow Freezing
• Small ice crystals formed	• Large ice crystals formed
• Blocks or suppresses	• Breakdown of metabolic rapport
• Brief exposure to concentration of adverse constituents	• Longer exposure to adverse or injurious factors
• No adaptation to low temperatures	• Gradual adaptation
• Thermal shock (too brutal a transition)	• No shock effect
• No protective effect	• Accumulation of concentrated solutes with beneficial effects
• Microorganisms frozen into crystals	
• Avoiding internal metabolic imbalance	

Source: Adapted from Jay, J.M., *Modern Food Microbiology*, 5th Edn., Chapman & Hall, New York, NY, 1996.

are smaller and cause less injury to the tissues if the freezing is rapid enough. The concentration of the residual solutions increases during this operation; this is why a great degree of lowering of temperature is necessary to freeze the maximum amount of water. Below −30°C the content of ice hardly increases at all, and there remains a fraction of water which cannot be frozen. Frozen fruits and vegetables have no further metabolic activity. Freezing permits most perishable foods to be kept for several months. The freezing storage period depends on the kind of product and the level of the temperature.

The rate of freezing of foods is dependent primarily on several factors such as the freezing method, the temperature, the circulation of air or refrigerant (cryogenic fluid), the size and shape of package and the kind of product. There are two basic ways to achieve food freezing such as quick or slow freezing. Apparently, quick freezing is the process by which the temperature of foods is lowered to −20°C within 30 minutes [17], and it may be achieved by direct immersion or indirect contact of foods with the refrigerants (or cryogenics) and the use of air blast freezing. Slow freezing is the process whereby the desired temperature is achieved within 3–72 hours and is commonly used in domestic freezers. Quick freezing has some advantages over slow freezing from the standpoint of overall product quality. These two methods are compared in Table 15.3. Although freezing is an excellent preservation method against further deterioration, it also causes some undesirable changes in the product. Living tissues are destroyed in the process, and fruits and vegetables have completely changed texture and much reduced quality in the thawed state. Despite these facts, freezing is considered certainly the most universally applicable and satisfactory of all long-term food preservation methods available.

15.8 COOL AND COLD STORAGE

Fresh vegetables are living organisms and there is a continuation of life processes in vegetable after harvesting. Changes that occur in the harvested, nonprocessed vegetable include water loss, conversion of starches to sugars, conversion of sugars to starches, flavor changes, color changes, toughening, vitamin gain or loss, sprouting, rooting, softening and decay. Some changes result in quality deterioration; others improve quality in those vegetables that complete ripening after harvest. Postharvest changes are influenced by such factors as kind of crop, air temperature and circulation, oxygen, and carbon dioxide contents and relative humidity of the atmosphere, and disease-incitant organisms. To maintain fresh vegetable in the living state, it is usually necessary to slow the life processes, though avoiding death of the tissues, which produces gross deterioration and drastic differences in flavour, texture and appearance.

Cold storage slows produce respiration and breakdown by enzymes, slows water loss and wilting, slows or stops growth of decay-producing microorganisms, slows the production of ethylene, the natural ripening agent, and "buys time" for proper marketing. Metabolic activity of fruits and vegetables produces heat. Produce also stores and absorbs heat. The objective of optimum storage conditions is to limit the production, storage and absorption of heat by produce.

In the past, the differentiation between cold and cool storages was in terms of the storage of food products with ice or without ice. In broad terms, cool stores operate at air temperatures above −2°C (typically −2°C to 10°C) and the cold store generally operates below −2°C (e.g., −15°C to −30°C). Storage under refrigerated conditions is used to slow deterioration of quality in many food products, especially fruits and vegetables. Maintenance of the ideal product temperature and minimization of water loss are both of importance. These are achieved by maintaining uniform conditions (i.e., temperature, relative humidity and air velocity) in the storage environment. Design and operational factors such as layout of the store, insulation levels, door protection devices, frequency of door use, air cooling coil and fan designs, associated control system design, air flow patterns, and product stacking arrangements can all influence the uniformity of environmental conditions and therefore the rate of change of product quality [18].

In a storage operation, the following should be kept in mind:

- Minimizing the exposure of products to ambient temperatures
- Laying out methods of handling and routes
- Never leaving the doors open when personnel or products are passing through them

Cold storage of produce is refrigerated storage above freezing. Controlled atmospheres (CAs) such as CO_2 or other inert gas around the product and appropriate humidity control in combination with cold storage gives a longer life of a stored product.

15.8.1 CHILLING INJURY

Chilling injury is caused by low, nonfreezing temperatures and generally affects fruits and vegetables with a higher recommended storage temperature. Symptoms of this disorder include decay; failure to ripen properly or uniformly; pitting of the surface; discoloration (russetting on the surface or darkening of the flesh); and watery consistency. Not all symptoms are obvious and several vegetables have specific symptoms that are not described. The injury incurred is dependent on time and temperature. The lower the temperature below recommendation, the quicker and more severe the injury. Chilling injury is particularly troublesome because symptoms do not appear until after the injury has occurred. Produce that looks healthy in the cooler may develop symptoms during transit or marketing that were caused by chilling before it left the farm. Here are some examples for the lowest safe temperatures of some products: asparagus 0–2.2°C, cucumbers 7.2°C, eggplants 7.2°C, watermelon 4.4°C, sweet peppers 7.2°C, potatoes 3.3°C, pumpkins 10°C, ripe tomatoes 7.2–10°C, respectively [2].

15.8.2 OPTIMUM STORAGE CONDITIONS ASPECTS

In cold storage of perishable foods, the following optimum conditions are of considerable significance:

- Optimum temperature
- Optimum relative humidity
- Condensation of water vapor on the product
- Optimum air movement

- Optimum stacking
- Sanitation

These conditions are also of particular importance in transit cooling of perishable foods. Here, we detail these as follows:

Optimum temperature: as harvested fruit ages, it is particularly important to manage the temperatures under which it is stored. For example, respiration largely involves enzymatic processes, which are significantly controlled by ambient temperature. The rate of chemical change in fruit generally doubles for every increase of 10°C (at room temperature). Changes that take place during storage as fruit begins to over-ripen may include extreme color formation, development of strong off-flavors with intense aroma, softening of the flesh, onset of physiological disorders and manifestations of disease. In addition, fruit can be injured by overcooling. Chilling injury may be manifested by pitting and browning of the surface and by pitting and darkening of the flesh. In the refrigeration process, as mentioned earlier, there is an optimum storage temperature for each food product. The product must be stored at this temperature. In many cases, the lower limit will be the product freezing point. For fruits and vegetables, it will be an absolute limit.

Optimum relative humidity: relative humidity of the air around the produce is important in both short- and long-term storage to avoid decay of the product due to the associated microorganisms. The relative humidity of air affects the achievement of cooling and product quality. Dry air may cause desiccation of the product, which can affect the appearance and certainly reduces the saleable weight. Very damp air with high relative humidity causes growth of molds and bacteria on chilled carcass meat and also the development of various fungal disorders on many fruits and vegetables. The relative humidity must be high enough to avoid excessive moisture loss from the product. As known, the water content of fresh produce varies between 55 and 95%. In general, the relative humidity in storage should be 85–90% for fruits, 90–95% for leafy vegetables and root crops, and about 85–90% for other vegetables [19]. Higher than these values will encourage decay and low humidity will increase weight loss. The relative humidity of the air around the produce is dependent on [e.g., 20]: (i) the water activity at the surface of the product, (ii) the rate of fresh air ventilation, (iii) the relative humidity of the fresh air, and (iv) the temperature of the refrigerant coil relative to the dew point of the air in the store.

Humidity control is another significant aspect for cold stores. The relative humidity of store atmosphere is a measure of its drying power associated with the maximum relative humidity. Since the majority of produce that is stored loses moisture and hence quality easily, it is important, particularly for long-term storage, that the store relative humidity be maintained as high a level as practicable. If care is taken at the design stage, it should be possible to avoid the need for any kind of humidifying equipment in the store. At this stage, the choice of a suitable cooling system, with adequate surface area and correct evaporating temperature for the refrigerant, would go a long way towards avoiding the need for humidification. The simplest method of humidification is to spray water into the air stream after the cooling system. If this is done, it is important to avoid liquid water being carried over into the produce and wetting it, possibly resulting in increased rate of rotting. When water is added to an already fairly moist storage atmosphere, little or none of the water may evaporate. Another method is to spray water over the cooling coil of the evaporator. But care must be taken to ensure that water does not freeze on the coil and cause it to become blocked. In some cases, a controlled quantity of steam is used to humidify the store. Its advantage is that it requires no extra heat for evaporation, since it is already a vapor. Careful control of the quantity of steam is required. In addition to these techniques, new humidifiers have been developed for these types of applications. Sometimes, dehumidification is necessary during the storage period of some foodstuff. For long-term storage of some produce, e.g., onions it may be necessary to remove some moisture from them. For this reason, low relative humidity should be used in the store. Normally, this can be done by proper control of the cooling systems. If not, some dehumidifiers are available for such applications.

Condensation of water vapor on the product surface: this is one of the main concerns and is most marked when the relative humidity is highest and when the difference in temperature between the air and product surface is greatest. It is necessary to minimize condensation, which causes growth of microorganisms, and to take steps to evaporate the condensed moisture immediately.

Optimum air movement: in storage, it is very important to provide good air circulation between the boxes and around the products. Adequate air distribution is required to maintain a uniform temperature throughout the storage to prevent stratification of heat and moisture and, for CAs, to provide a uniform atmosphere around the products. Ventilation by bringing in outside air is held to a minimum to maintain optimum temperature and relative humidity. However, ventilation is sometimes recommended to decrease mold or other microorganisms such as scald in storage.

Optimum stacking (stowage): stowage is one of the important factors in all types of storage and transport and is particularly affected by the packaging of the commodity, whether it be by carton, pallet, net bag or hanging meat [20]. In cold storage, according to the produce to be stored, the stacking density of fruits and vegetables in boxes is 200–300 kg per useful m^3. For this reason, the stowage must be stable to avoid damage during handling, in storage or in transit, yet it must permit air to circulate freely through and around the commodity. Storage boxes must be used and stacked with regard to direction of air flow from the cooling unit. The tight stores of cartons on pallets are prone to slow cooling if the product is loaded warm.

Sanitation: certain foods such as fish and citrus give off characteristic odors which can be readily absorbed by other products such as meat, butter and eggs. For this reason, different commodities must be stored separately in different spaces. If they are stored successively, the space must be thoroughly cleaned and deodorized. Further, care must be taken so that undesirable odors are not transmitted to the stored products from the materials used to insulate the cold rooms or those used as renderings or as wood preservatives (e.g., creosote, bitumen). Rodents must be prevented from coming into the storage through proper use of metal, screen, etc.

15.8.3 THERMAL ASPECTS

During storage to maintain the quality of chilled or frozen foods, it is essential to apply the correct storage temperature. In addition, the following must be avoided in storage (including loading and unloading processes):

- Low relative humidity
- Retention beyond the expected storage life
- Fluctuations in storage temperature
- Physical damage to the products and packaging
- Contamination of the products by hazards

In the design of a store, an exact heat transfer analysis is required and the following cases should be considered:

- Heat infiltration through walls, floors, etc.
- Heat input by fans, lights, open doors, people, etc.
- Heat transfer to/from the product
- Heat transfer to/from room fittings and structures

Some sources of water vapors are largely:

- Through open doors
- Water loss from product
- People, hot water sources, etc. in the store

Typical operation of a store during a 24-hour day will show two distinct phases: the part of the day when the store is effectively shut and the so-called working period. In the latter, doors are opened as required for product loading and unloading, lights are on, and people and machinery are present. During this period the heat load on the system can vary enormously, depending primarily on the door opening patterns, resulting in the rapid rise in air temperature over short periods of time. Further, the external temperature during the full 24 hours may change as a function of both weather conditions and unassociated, but nearby, activities in the processing plant. The irregularities are caused by door openings, movement of people and intermittent operation of machinery [21].

The location of the storage is generally dependent on the type of marketing operation and location of the orchard or field. On the one hand, it is desirable to have the storage as close to the production area as possible, while on the other hand, an attractive storage would be of considerable value in making roadside fresh retail sales. It may be desirable to locate the storage so that the product could be sold separately from the farmland [19].

The general design of a cold store is determined by the requirements for effective and safe handling of the products and a suitable storage climate for the products. The normal arrangement is that rooms are built side by side between road and railway loading banks. Thus, all rooms can communicate directly with loading banks and traffic yards. Frequently, the cool and cold stores are built with prefabricated concrete or steel structures. The insulation can be placed on the outside or inside of the structure. In the design, the following considerations must be taken into account:

- Frost heave. Frost heave under stores is prevented by a special under-floor heating system or a ventilated space under the floor.
- Insulation. This represents a large percentage of the total cost for a normal store. It is therefore of great importance that it be designed from an economical point of view. The insulation system must be chosen carefully and supply the optimum insulation required.
- Refrigeration system. The refrigeration system must be designed with regard to the requirements of the climatic conditions for the stored products and be adequate to allow for sufficient safety on peak days and summer conditions. In order to improve safety and make control easier and cheaper, most modern refrigeration plants are automated. The degree of automation may vary, but normally room temperature, compressor capacity, lubrication, cooling water, defrosting, pumps, fans, current and voltage of the main supplies etc. are controlled and supervised by a central control panel in the engine room.
- Lighting. Stores are working places for forklift drivers and others concerned with handling of the products. For this reason, the lighting in the store must be sufficient, but at the same time it must be remembered that lighting is adding to the heat load in the store.
- Layout. The layout of the storage should be such that it reduces the maximum extent possible the ingress of warm air and exposure of product to atmospheric temperatures. The products are conveyed or palletized into stores. The layout of a store is determined by the type of product, packaging, method of palletization, accessibility required and the equipment used for handling. Passageways should be clearly defined and in the interests of safety and quick handling, these should be kept free from obstruction at all times. The floors of large stores are often marked off with a grid and the grid spaces numbered so that the location of goods can be recorded thus enabling quick retrieval. Products stored near doorways will come into frequent contact with warm moist air entering the store when the door is open. Some form of partition may be used to reduce the effect of this warm air on products stacked in this area.
- Jacketed stores. These stores allow storage at nearly 100% relative humidity and at uniform and constant temperatures. These conditions are obtained by circulating

refrigerated air in a jacket around the load space to absorb the heat conducted through the insulation before it can enter the load space. This technique reduces weight losses of unpackaged foods and frost formation inside packaged frozen foods and increases the storage life.

• Equipment. In equipping the store, care should be taken to choose equipment which is suitable to the product being handled and which minimizes the possibility of damage or contamination. For example, timber pallets are suitable for properly packed products but lightly packed semiprocessed stock may need a pallet constructed of metal or some similar washable and less easily damaged material.

15.8.4 REFRIGERATION CAPACITY

The capacity of the refrigeration plant must be based on a thorough heat load calculation for each individual project. Refrigeration load can vary widely for stores of the same capacity depending on design, local conditions, product mix, etc. Therefore, no rule of thumb can be applied. In past practice, a safety margin of some 50% of the theoretical calculation has been used. Today, with more thorough knowledge of practical cold store operation, combined with theoretical knowledge, the safety margin can be reduced to a more realistic level. The following discussion is limited to general considerations serving as guidelines and introduction to more detailed studies of the factors influencing the purchase and installation of refrigeration plants.

Heat leakage or transmission load can be calculated fairly closely using known overall heat transfer coefficient of various portions on the insulated enclosure, the area of each portion and the temperature difference between the cold room temperature and the highest average air temperature likely to be experienced over a few consecutive days.

Heat infiltration load varies greatly with the size of the room, number of door openings, protection of door openings, traffic through the doors, cold and warm air temperatures and humidity. The best basis for this calculation is experience. The type of store has a marked influence on heat load, as has the average storage time. In comparing long-term storage, short-term storage and distribution operation, there is a 15% increase in refrigeration load for short-term storage as compared to long-term storage, whereas refrigeration load in the distribution operations is in the order of 40% higher than that of long-term storage, due mainly to additional air exchanges.

Most large cold stores are equipped with two-stage ammonia refrigeration installations. For smaller plants, usually less than 6000 kcal/h refrigeration capacity, approved refrigerant will probably be used in single stage systems operating with thermostatic expansion valves. Such systems are thermodynamically less efficient, but in areas where only staff with relevant refrigerant experience is available the system may be preferred for service reasons.

The refrigeration system should be designed for high reliability, and easy and proper maintenance. Once a cold store plant has been pulled down in temperature, it is expected to maintain this temperature, literally, forever. Even maintenance jobs that need carrying out only every 5–10 years must be taken into consideration.

Calculation of cold store refrigeration loads: a good deal of experience is required to make accurate calculation of a cold store's refrigeration requirement and this should therefore only be done by a qualified person. The following calculation is not complete but it serves two purposes. It allows the reader to make a similar calculation for his/her own store and thereby obtain an approximate refrigeration requirement. It also helps the reader to appreciate the number of factors that have to be taken into account in calculating the heat load and also gives him/her some idea of their relative importance. One important heat load that has been omitted in the calculation is the heat gain due to solar radiation. This factor depends on a number of conditions which are related to both the location of the store and its method of construction. In some cases, solar heat load may not be significant but in other instances, precautions may be necessary to reduce its effect.

Example 15.1

For the following design conditions and dimensions, determine the cold store refrigeration load.

1. Specification
   Dimension=20 m×10 m×5 m=1000 m³
   Insulation thickness=0.25 m
   External store surface area=771.5 m²
   Maximum ambient temperature=35°C
   Store temperature=−30°C

2. Load calculation
   (a) Insulation heat leak through walls, roof and floor
       Conductivity of polystyrene=0.033 kcal/hm°C
       Temperature difference between ambient and store=35°C and −30°C=65°C
       Thickness of polystyrene=0.25 m
       Surface area of store=771.5 m²
       Heat leak→$Q_a=A×U×\Delta T$=771.5×0.033×65=7422 kcal/h

   (b) Air changes
       Average air change=2.7 times per 24 h
       Store volume=1000 m³
       Heat gain (35°C and 60% RH air)=40 kcal/m³
       Air change heat gain→$Q_b=V×H×C$=1000×40×2.7/24=4500 kcal/h

   (c) Lights (left on during working day)
       Q_c=1000 W=860 kcal/h

   (d) Staff working
       For one man working at −30°C→Q_d=378 kcal/h
       For n number of people at −30°C→$Q_d=n×378$ kcal/h

   (e) Product load
       Product capacity at average temperature of −20°C=5.5 kcal/kg
       Mass of the product per day=35000 kg/24 h
       Product load→$Q_e=m×P$=3500×5.5/24=8020 kcal/h

   (f) Fan load
       For three fans at 250 W→$Q_f=3×250$ W=644 kcal/h

   (g) Defrost heat
       One defrost of 8440 W for 1 h (recovered over 6 h)→Q_g=1209 kcal/h

   (h) Total refrigeration load (sum of items a–g)→Q_T=23411 kcal/h

   (i) Total refrigeration requirement with allowances→Q_D=23411×24/18=31215 kcal/h

If a pump is used to circulate refrigerant, the heat equivalent must be added to the capacity of the refrigeration condensing unit but not to the capacity of the room cooler.

The minimum refrigeration requirement will be when there is only an insulation heat load and the fans are in operation. In this example, the minimum load corresponds to only about 25% of the capacity of the installed refrigeration plant. This minimum load factor will vary considerably with the type of store and mode of operation but some account may have to be taken of this difference between the maximum and minimum refrigeration requirements. Large cold stores should be operated with a number of compressors, which can be switched on and off as required. Large compressors may be fitted with off-loading equipment which allows them to work efficiently on partial loads. The reliance on one large compressor for a large cold store could be catastrophic in the event of its failure. In the case of smaller stores, it may be that only one compressor is viable. Other arrangements can be made to cater for variation in refrigeration demand. What must not happen is that a large compressor should operate with a low load and hence operate with a very low suction pressure or stop and start too frequently. The first condition is bad for the compressor and the second for the electrical equipment.

15.9   CONTROLLED ATMOSPHERE STORAGE (CAS)

Since the 1920s a controlled atmosphere (CA) has been utilized throughout the world for land-based storage of apples, but its utilization for other fruits and vegetables was limited until the 1980s. The Controlled Atmosphere Storage (CAS) is used to extend the storage life of seasonal perishable produce when refrigeration alone is not sufficient. This technique can be used for many fruits and vegetables and, historically, has been the principle storage method for the world's apple crop. Apples still remain the pre-eminent produce stored under CA conditions but, in recent years, it has become an important storage technique for many other commodities.

To store fruit successfully for long periods, the natural ripening of the produce has to be delayed without affecting the eating quality. This is achieved firstly by reducing the temperature of the fruit to the lowest level possible without causing damage through freezing or chilling. To delay the ripening even further, the atmosphere in the storage room is altered by reducing oxygen and allowing CO_2 to increase. The precise level of temperature, oxygen and CO_2 required to max-imise storage life and minimise storage disorders is extremely variable. This will depend on the type of produce, cultivators, growing conditions, maturity and postharvest treatments. Storage behavior can even vary from farm to farm and from season to season. Recommendations are pub-lished regularly by various national research bodies and preservation consultants who offer the best advice on the difficult compromise between extending life and minimizing storage disorders in their locality.

The term *controlled atmosphere* is derived from the fact that the composition of the atmospheric gases in contact with the products is controlled at precise levels during storage or transportation in order to prolong and extend the storage and market life of fresh fruits and vegetables. The amount of fruits and vegetables shipped internationally has increased drastically under a CA.

The CA technology is a standard technique for the postharvest handling of fruits, vegetables and stored grains and is one of the main reasons that high quality produce is available throughout the year. CA technology is a simple and environmentally safe way to manage postharvest insect and disease problems in cold storage. The use of low oxygen (anoxia) and high carbon dioxide and nitrogen environments simply smother insects. Anoxic environments could be a practical, cost-effective and safe way to kill all life stages of insects and mites that infest greenhouse flower crops. Unfortunately, nearly all information on effects of anoxic environments on insects concerns pests of fruits, vegetables and grains in storage.

Note that CA storage does not improve fruit quality, but it can slow down the loss of quality after harvest. Successful CA storage begins by harvesting fruit at its proper maturity. Apples should be cooled rapidly and recommended atmospheric conditions achieved shortly after field heat is removed. The longer it takes to adjust carbon dioxide and oxygen levels, the less effective the dura-tion of storage.

CA is one of the most advanced methods for keeping fruits and vegetables fresh and the timing of harvest is critical to good storage results. In a CA storage, temperature, moisture, oxygen and carbon dioxide are controlled, in order to keep fruits and vegetables under controlled conditions. The CA method is better than any other method of preservation, both in terms of preserving period and keeping the fresh quality. Note that the common specifications of a CA store (installed with plates) are as follows:

- Temperature: −2 to −3°C
- Oxygen (O_2): 3–5%
- Carbon dioxide (CO_2): 1–3%
- Moisture content: 80–90%

It is important to point out that the CAS is a nonchemical process. O_2 levels in the sealed rooms are reduced, usually by the infusion of nitrogen (N_2) gas, from the approximate 21% in the air we breathe to 1 or 2%. Temperatures are kept at a constant 0–2.2°C. Humidity is maintained at 95% and carbon dioxide levels are also controlled. Exact conditions in the rooms are set according to the variety of the product. Researchers develop specific regimens for each variety to achieve the best possible quality.

In CAS, O_2 level is reduced and CO_2 increased. Lower respiration occurs, thus extending the life of some products in storage (Figure 15.5 and Table 15.4) and providing additional distribution and marketing possibilities for fresh products. It is important to maintain the proper relationship of O_2 and CO_2 depending on the product and temperatures of storage. In most instances, O_2 and CO_2 are

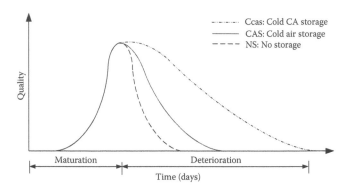

FIGURE 15.5 Illustration of the quality of food product as a function of time. (Adapted from Malcolm, G.L. and Beaver, E.R., *Proceedings of the International Conference on Technical Innovations in Freezing and Refrigeration of Fruits and Vegetables*, 9–12 July, Davis, CA, 51–57, 1989.)

TABLE 15.4
Typical Storage Life of some Fruits and Vegetables in CAS with PRISM Alpha System

Produce	Life (Days)	Temperature (°C)	O_2 (%)	CO_2 (%)
Apples	200+	−0.5–3.5	1–3	1–5
Artichokes	10–16	0.5–1.5	2–3	3–5
Asparagus	14–28	0.0–3.5	8–21	5–10
Bananas	100+	13.0–14.0	2–5	2–5
Blueberries	14–28	0.0–1.5	1–3	1–5
Broccoli	10–14	0.5–3.5	1–2	5–10
Cabbage	90–180	0.0–3.5	3–5	5–7
Cherries	14–21	1.0–3.5	3–10	10–12
Corn	4–6	0.5–3.0	2–4	10–20
Kiwi	100+	−0.5–0.0	1–2	3–5
Lemons	30–50	13.0–15.0	5–8	<10
Lettuce	20–40	0.0–4.5	1–3	0–2
Mangoes	14–25	12.0–13.5	5	5
Nuts	700+	0.0–1.0	<1	Varies
Papaya	10–20	12.0–13.5	2–5	<3
Peppers	12–18	7.5–11.5	3–5	0–3
Pineapples	14–36	7.0–10.0	5	10
Tomatoes	20–70	11.5–20.0	3–5	2–3

Source: Malcolm, G.L. and Beaver, E.R., *Proceedings of the International Conference on Technical Innovations in Freezing and Refrigeration of Fruits and Vegetables*, 9–12 July, Davis, CA, 51–57, 1989.

maintained between 2% and 5% by volume [22], with N_2 comprising the remainder of the mixture. The optimum gas composition varies for each commodity and cultivar. Dewey [23] stated that the composition of an ordinary cold storage room atmosphere varies within the range 19–21% O_2 and 0–2% CO_2 depending on the temperature, kind and quantity of food products stored, and the air-tightness of the structure. Dewey [23] indicated that possibly 10% of all commercial apples grown for fresh market purposes are stored in a CA, and that optimum operating conditions are 0–8% CO_2, 1–7% O_2, and from −1 to 4.5°C.

The CAS offers some remarkable advantages particularly for fruits and vegetables including:

- Longer storage period than regular storage
- Higher preserved quality (near to fresh fruits and vegetables)
- Longer shelf life
- Having bacteriostatis
- Better storage for what is difficult to preserve in common high-temperature cold storage
- Nonpollutant
- Environmentally benign

Recently, considerable research has devoted to application of the CA technology to different fruits and vegetables. During the past decade, significant advances have been made in this technology, e.g., the development of new CA equipment. The equipment is a module composed primarily of hollow polymeric fiber. The properties of the fiber are specifically tailored to remove preferentially O_2 and CO_2 from a feed stream of compressed air. The resultant high N_2 stream is used to generate the CA in any airtight enclosure. Malcolm and Beaver [22] summarized its benefits for both consumers and the food industry as follows:

- Extending the fresh life of produce and increasing the marketing flexibility available to food distributors
- Using only natural atmospheric gases and eliminating the use of chemical growth regulators and pesticides
- Shipping large quantities of fresh produce and eliminating air cargo capacity constraints
- Improving product quality
- Consumer accessibility to an increasing number of fresh fruit and vegetable varieties
- Availability of seasonal fruits and vegetables during the whole year

15.9.1 Controlled Atmosphere Storage (CAS) and Ripening, and Waxing

This is a procedure for storing fruits and vegetables, particularly apples, under an atmosphere that differs from air. Its aim is to increase the storage life of the foods. The most important dietary component of apples is dietary fibre, which is unlikely to change appreciably during CA storage. Significant nutritional changes in other fruits and vegetables would not be expected. For uniform ripening of some fruits, most notably tomatoes and bananas, brief storage under a "ripening gas" is used. This can initiate ripening or speed up the process. Fruit produced for market in this way is unlikely to be significantly different in nutrient composition compared with fruit that has matured normally, although it may taste differently. Without CA storage many seasonal fruits would not be available throughout the year. Many fruits and vegetables have a natural coating of wax, which is removed when these foods are cleaned before appearing on the supermarket shelf. To make them shiny and attractive and promote their sale, some fruits and vegetables are artificially waxed. The waxes are dispersed in water and coated over the food to provide a thin film of wax, which gives a glossy appearance. Apples coated this way are likely to sell more readily. In addition to this cosmetic effect, the wax coating for a short time slows the loss of moisture, which causes weight loss and wilting. The nutritional advantage of waxing, if any, would be expected to be very small.

At present, there is no reason to believe that the use of waxes approved for this purpose is hazardous to health.

15.9.2 Container Controlled Atmospheres (CAs)

Practical systems presently offered for container CAs may be divided into three main categories as follows:

- Controlled-MA systems
- Pressure swing absorption systems
- Membrane separation systems

Controlled-modified atmosphere (MA) systems: these systems are designed to maintain an atmosphere which is first modified by injecting gas mixtures after loading the produce into the container. In these systems, air leakage is the main concern and affects the success of the operation. For this reason, part of the preparatory service consists of leak testing and sealing the container against air leakage. This often results in extensive utilization of sealants throughout the interior of the container. Cargo respiration consumes O_2 and generates CO_2. Controlled admission of air from the outside replaces the lost O_2 while excess CO_2 is removed by a scrubber located inside the container. Good quality service and well-trained staff are required in order to perform pretrip testing, sealing, initial gas injection and controller set-up. The principles of atmosphere control have been applied successfully to long-term storage in cold stores, especially for apples and pears. It is now applied to transport and packaging, not as a replacement but as an enhancement of good temperature control. CA and MA with reduced O_2 content and increased CO_2 content, with appropriate temperature control, can retard deterioration and maintain quality while increasing storage life of various fruits and vegetables (Table 15.5). The beneficial effects of CA and MA are as follows:

- Retarding fruit ripening and leaf senescence (aging)
- Control fungal and bacterial spoilage and insects
- Control of physiological disorders, e.g., spotting in leaf crops and bitter pit in apples
- Reduction of ethylene production
- Reduction of sensitivity to ethylene

TABLE 15.5
Comparison of Storage Life in Air and CA Storages. What Temperature?

	Storage Life	
Produce	Air	CA
Apples	5 months	10 months
Mango	2 weeks	4 weeks
Avocado	1 month	2 months
Strawberry	4 days	7 days

Source: Adapted from Frith, J., *The Transport of Perishable Foodstuffs*, Shipowners Refrigerated Cargo Research Association, Cambridge Refrigeration Technology (CRT), Cambridge, UK, 1991.

Produce in sealed polyethylene film generates the atmospheres within the package. The initial composition depends on the rate of respiration of the produce in the pack and gas permeability of the film and its surface area. The relative humidity inside the package is dependent on the water vapor permeability of the film and external conditions. The packaging film is selected and tailored to the requirements of the produce and to the intended temperature of use. Packages may be individual retail packs, or systems for pallets or cartons. In transit cooling, modified atmospheres (MAs) are limited to use in short journeys (up to 10 days) and both tomatoes and strawberries have benefited from this method.

Pressure swing absorption systems: these systems use the selective absorption characteristics of certain minerals under pressure. By using more than one absorbent, they cannot only separate oxygen and nitrogen but also carbon dioxide, as required, and ethylene continuously. Instead of purging it, the gas within the container envelope is processed, thus humidity control and raised levels of CO_2 are possible giving a fully CAS. Systems tend to be more complicated and contain more components than membranes however the absorbents have to all intents an infinite life. There is a small selection of equipment currently available offering a variety of performance and operational features, for both CA and MA. These systems filter the atmosphere through beds of molecular sieve materials, selectively removing the unwanted gases. This main advantage is the controlled ability to move gases, but they have some disadvantages such as higher cost, complexity, and lost cargo space, as well as the requirement for complicated piping and wiring to control gas flow to several beds.

Membrane separation systems: these systems offer the least complexity of any active atmosphere control system, have minimal weight and power draw, and can be completely integrated within the reefer unit so that cargo space is not compromised. Rushing [24] reported the results of a series of five land-based CA experiments with peaches, plums and nectarines using a commercial CA shipping container, and found the following results:

- In peaches, plums and nectarines, firmness and ground color are two quality characteristics which are most consistently affected by CA. Fruits held under CA are generally firmer and develop their characteristics ripe ground color more slowly than fruits in normal atmosphere.
- Internal browning, bleeding and mealiness are reduced in some cultivars, but the effects are generally more pronounced when fruits are subjected to a CA for a longer time, 21 days compared to 14 days in his study.
- Identification of the most responsive cultivars is essential for companies that wish to use CA in transit.

15.9.3 PACKAGING

Packaging is a fundamental element in the transport and storage of temperature controlled products. It is essential to protect cargoes from damage and contamination. The correct design and highest quality of materials need to be used to ensure it can withstand the refrigeration process and transit. Where appropriate, packaging materials must be able to [2]:

- Protect products from damage as a result of "crushing,"
- Be able to withstand "shocks" occurring in intermodal transport
- Be shaped to fit on pallets or directly into the container for stowage
- Prevent dehydration or reduce the water vapor transmission rate
- Act as an oxygen barrier preventing oxidation
- Withstand condensation and maintain its wet strength
- Prevent odor transfer
- Withstand temperatures of $-30°C$ or colder

15.10   REFRIGERATED TRANSPORT

Refrigerated transport is considered as an essential part of food refrigeration chain, and its primary goal is to supply the consumer with safe, high-quality products. Of course, refrigerated transportation can also be used for nonfood goods such as flowers and plants, pharmaceuticals and medicines, human organs, photographic material and a variety of industrial commodities. There are three basic types of transport: sea transport (conventional ships, container ships), land transport (road, rail) and air transport. Intermodal transport combines more than one of these types of transport. Prompt utilization of this technology has allowed shipment from greater distances and made sea transportation a viable alternative to air freight. Frozen goods are transported at a temperature of −18°C or lower, chilled goods at a temperature above freezing point.

After precooling, the products must be properly loaded and transported at or near the recommended storage temperature and relative humidity to maintain quality. The design and condition of the transport equipment, and the loading method used are critical to maintaining product quality. The mode of transportation and carrier should be chosen carefully. The mode of transportation and type of equipment used should be based on destination, transportation duration, transportation type, product value, degree of product perishability, amount of product to be transported, storage conditions, outside conditions, etc.

The modern food distribution system is expected:

- To deliver foods requiring different storage and transportation temperatures to food stores
- To control temperature within a narrow range
- To have a rapid temperature recovery after loading and door openings during delivery
- To have a safe, quiet system to minimize noise levels and allow night deliveries to food stores
- To have a transport refrigeration system with environmentally benign refrigerants

The requirements for the product environment during transportation and distribution are generally the same as for cold storage. But the time of exposure is normally much shorter and technical difficulties in maintaining low temperatures are greater. Also, there is a need for better temperature control in transportation.

In transport by sea, some problems (e.g., mainly temperature rise) in maintaining satisfactory conditions are relatively slight and the situation has been improved considerably over the past few years. The types of insulation and refrigeration unit used are much the same as in any stationary cold store. Suitable ships are now available to satisfy every requirement for any type of refrigerated food.

In transport by land, more difficult problems from a technical and economic standpoint arise. The small individual units for road or rail trucks set narrow limits on the complexity and cost of the cooling equipment. A high degree of reliability is required.

Dincer [2] gives a list of transport temperatures and conditions for fruits, vegetables, meat, dairy products and fish, and summarizes that fruits are essentially tolerant of low temperatures, are carried at temperatures in the range from −0.5 to 0.5, most vegetables are carried close to 0°C, and chilled meat and dairy products are carried at temperatures in the range from −1.5°C to+5°C.

The refrigeration unit fans cause temperature-controlled air to circulate around the inside of the vehicle floor, walls, doors and roof to remove heat which is conducted from the outside. Some of the air should also flow through and between the cargos, particularly when carrying fruits and vegetables, in which case heat of respiration may be a significant proportion of the heat load.

On refrigerated semitrailers, the compressor is usually driven by a two-speed diesel engine, which also drives the condenser and evaporator fans. Some units also include an electric motor drive as a standby for use in ferries and where noise regulations are in force. A thermostat is mounted on the front casing of the unit with the sensor placed in the return air stream. A dial thermometer is usually fitted in a prominent position where it can be seen by the driver, with the sensor placed adjacent to the thermostat sensor.

For land transport, the refrigerated semitrailer is the most popular vehicle. Rigid vehicles are used for local deliveries and short-distance journeys. For marine transport, two types of container are more common: one having an inbuilt refrigeration unit similar to that of the refrigerated semi-trailer and the other having two circular apertures on the front wall through which refrigerated air may be ducted. The second one is known as a porthole unit and may be refrigerated in land or road transport by a detachable refrigeration unit using a conventional vapor compression system, liquid nitrogen or dry ice.

15.10.1 REEFER TECHNOLOGY

In practice, "reefer" is the generic name for a temperature controlled container. The containers, which are insulated, allow temperature controlled air to circulate within the container. A reefer is not only a refrigerator and can provide heat as well as remove it.

It is expected that CA containers maintain the ideal atmosphere for sensitive produce. CA technology can prolong the shelf life of your goods. CA reefer containers maintain a constant atmosphere by replacing consumed oxygen through a unique air exchange system, maintaining an ideal atmosphere in equilibrium with the product's deterioration rate with a number of advantages, including:

- Delaying aging, ripening and associated changes in product
- Reducing water loss and weight shrinkage
- Allowing longer transit time for destinations and/or to new markets
- Providing better quality control
- Improving control of insects in some commodities

15.10.2 FOOD QUALITY ASPECTS

Under the best circumstances, the quality of fruits and vegetables can only be maintained, not improved, during transportation. Most of these products are high-value and very perishable. Therefore, product quality should be the highest possible and of course, the best conditions help products to [4]:

- Have a longer shelf life
- Allow more time for transportation, storage and marketing
- Satisfy importers, brokers and consumers
- Increase repeat sales and profits
- Help expand markets

During transportation, storage and marketing, products may be exposed to [4]:

- Rough handling during loading and unloading,
- Compression from the overhead weight of other containers of products
- Impact and vibration during transportation
- Loss of moisture to the surrounding air
- Higher than recommended temperatures
- Lower than recommended temperatures
- Ethylene gas from vehicle exhaust or product ripening
- Odors from other products or residues
- etc.

By selecting and packing only top quality products, shippers can help ensure good arrival condition of fruits and vegetables transported over long distances. Grading, good packaging, precooling and proper transportation equipment are essential to maintaining product quality from the field to the consumer in cold chain.

15.10.3   PACKAGING

Proper packaging of fruits and vegetables is essential to maintaining product quality during transportation and marketing. In addition to protection, packaging in the form of shipping containers, serves to enclose the product and provide a means of handling. It makes no sense to ship high quality, high value, perishable products in poor quality packaging which will lead to damage, decay, low prices, or outright rejection of the products by the customer. Packaging is expected to be strong, reliable, flexible, humidity friendly, etc.

Packaging materials are chosen on the basis of needs of the product, packing method, precooling method, strength, cost, availability, buyer specifications and freight rates. Importers, buyers and packaging manufacturers provide valuable recommendations. Materials used include: fiberboard bins, boxes (glued, stapled, interlocking), lugs, trays, flats, dividers or partitions and slipsheets [4]:

- Wood bins, crates (wirebound, nailed), baskets, trays, lugs, pallets
- Paper bags, sleeves, wraps, liners, pads, excelsior and labels
- Plastic bins, boxes, trays, bags (mesh, solid), containers, sleeves, film wraps, liners, dividers, and slipsheets
- Foam boxes, trays, lugs, sleeves, liners, dividers and pads

Bins, boxes, crates, trays, lugs, baskets, and bags are considered shipping containers. Baskets however, are difficult to handle in mixed loads of rectangular boxes. Bags only provide limited product protection. Fiberboard box is the most widely used container.

Fiberboard boxes for products which are packed wet or with ice must be wax-impregnated or coated with water resistant material. The compression strength of untreated fiberboard can be reduced more than one half in conditions of 90% relative humidity. In addition to maintaining box strength, wax helps reduce the loss of moisture from the product to the fiberboard. All glued boxes should be made with water resistant adhesive. Holes are provided in most fiberboard boxes to provide ventilation of product heat (respiration) and allow circulation of cold air to the product. Handholds provide a means of handling boxes during loading and unloading. All holes must be designed and placed in a manner that does not substantially weaken the box.

Wood crates are still popular with some shippers due to material strength and resistance to high humidity during precooling, transit and storage. Wood crates are constructed in a manner that allows a lot of air circulation around the packed product.

The majority of fiberboard boxes and wood crates are designed to be stacked top to bottom. Compression strength and product protection are sacrificed when boxes or crates are stacked on their ends or sides. Misaligned boxes can lose up to 30% of their strength, while cross-stacked boxes can lose up to 50% of their top to bottom compression strength.

Various materials are added to shipping containers to provide additional strength and product protection. Dividers or partitions, and double or triple thickness sides and ends in fiberboard boxes provide additional compression strength and reduce product damage.

Note that pads, wraps, sleeves and excelsior also reduce bruising. Pads also are used to provide moisture as with asparagus; provide chemical treatment to reduce decay as with sulfur dioxide pads for grapes; and absorb ethylene as with potassium permanganate pads in boxes of bananas and flowers.

Plastic film liners or bags are used to retain moisture. Perforated plastic is used for most products to allow exchange of gases and avoid excessive humidity. Solid plastic is used to seal the products and provide for a MA by reducing the amount of oxygen available for respiration and ripening. This is done for bananas, strawberries and tomatoes.

Paper and polystyrene foam liners help to insulate the product from hot or cold temperatures when they are shipped in unrefrigerated air cargo holds. Wet newsprint is used to provide moisture to fresh cut herbs and flowers.

15.10.4 TRANSPORT STORAGE

After harvesting, fruits and vegetables continue to "live and breathe" consuming oxygen and producing carbon dioxide. This respiration means loss of sweetness and a change in texture in most produce, however, the major effect of respiration is heat generation, which speeds up all other forms of deterioration. Proper control of the following is important to reduce deterioration of perishables and provide better operation of the refrigeration system [e.g., 2,25]:

Cooling: as mention earlier, precooling of fresh produce is the simplest and most powerful technique for minimizing deterioration and should applied as quickly as possible after harvesting. Reefer containers should not be used to reduce the temperature of produce they are designed to maintain, rather than to lower the temperature of the cargo.

Temperature: the temperature of the air which is delivered from the reefer unit either into the storage facility or the container and the temperature of the air leaving the interior of the cold store or container before being returned to the reefer unit should also be controlled accordingly.

Relative humidity: the relative humidity of the air around a product is important in determining their storage conditions. A low relative humidity will cause moisture loss and hence weight loss of the product and in certain circumstances even packaging. However, a high relative humidity can also produce undesirable affects in the product such as encouraging the growth of bacteria and mold as well as producing physiological disorders.

Moisture loss: moisture (weight) loss during both storage and transportation is prevented by maintaining the correct temperature and relative humidity.

Air circulation: if the air cannot circulate correctly in the cold store or container, then problems can arise affecting the shelf life of the product. Poor circulation can affect the commodity's temperature, relative humidity and weight loss. If a cargo is "hot" i.e., it has not been adequately precooled, it is essential that there is an adequate air flow to cool the produce as quickly as possible.

Ethylene gas removal: another way to minimize deterioration is to remove ethylene gas, a natural by-product of most fruits and vegetables, accelerating the ripening process. Although ethylene release cannot be avoided fully, adequate ventilation can minimize and control its accumulation.

Loading: the loading pattern is very important in the handling of reefer cargo. The correct loading pattern is based on the commodity's characteristics, individual packaging and the type of airflow system used. For frozen or less temperature-sensitive cargo, a tight block loading pattern can be adopted while in temperature sensitive cargo a palletized loading pattern with ample ventilation should be used to allow an even flow of cool air throughout the load.

Packaging: the most common cause of damage during transit is the shaking, bumping and over-packing of perishables. These problems can be avoided if suitable packaging materials are used. For example, reefer cargo can be palletized by bringing together a number of small product items such as carton boxes to make up a single larger unit.

Stowage: during storage and transportation within the cold chain, the stow has to be stable to ensure that the consignment is not physically damaged and that there are no air flow restrictions or short circuiting of the airflow.

Storage: whether the cargo is stored in a large cold store, a trailer, container or a retail temperature controlled display, storage of products within the cold chain should be controlled and maintained at each stage to ensure the integrity of the cold chain.

Product mix: it is very important that when mixing fresh produce in a single reefer that the mix is the correct balance. The carriage temperature if too low for one commodity could cause chill damage and if the temperature is too high for another commodity could cause increased senescence. Shippers must also be aware of the respiration levels and gases produced by the commodities at certain temperatures and maturity levels and their degree of tolerance or susceptibility to those gases.

15.10.5 Recommended Transit and Storage Procedures

Harvesting and packaging of most products should be closely coordinated with transportation to minimize time in transit and storage and maximize product freshness in the hands of consumers. Some products however, can be consolidated in storage before or after transportation to obtain lower freight rates or higher prices.

During transportation and storage of loads of one product, the temperature and relative humidity should be as close as possible to the recommended levels to achieve maximum product life. While transport refrigeration unit thermostats are sometimes set higher to avoid freezing injury, storage facilities are better able to control temperature and can provide conditions at the recommended level without damaging the products.

During transportation of refrigerated loads in trailers and van containers, the operation of the refrigeration unit and temperature of the load compartment should be checked regularly by the carrier. Gauges are provided for this purpose on most equipment. Many van containers also are provided with an exterior electronic or mechanical temperature recorder.

To improve transport of refrigeration units, their use and their control in refrigerated transport, and the following can strongly be recommended:

- Use of the truck's power takeoff as an alternative, or even primary source of energy for the refrigeration unit
- Ways to maintain the desired humidity level, both high and low in refrigerated transport vehicles, particularly at internal temperatures close to 0°C or slightly lower
- Continuous fan speed at preset levels or the ability to change fan speeds remotely as required
- Instrumentation to monitor and record temperature, air circulation rate, humidity and atmospheric composition (O_2, CO_2, and C_2H_4).

15.11 RESPIRATION (HEAT GENERATION)

Living fruits and vegetables produce heat owing to respiration. This tends to increase the product's surface temperature thereby driving force for moisture transfer. The effect is generally small for moderate vapor pressure differences. On the other hand, it may become a dominant factor at humidities close to saturation. In addition to heat generation, respiration produces additional weight loss due to CO_2 evolution. The net loss is that of carbon which is different from transpiration moisture loss.

Respiration is the process by which food reserves are converted to energy. Through a complex sequence of steps, stored food reserves (sugars and starches) are converted to organic acids and subsequently to simple carbon compounds. Oxygen from the surrounding air is used in the process while CO_2 is released. Some of the energy is used to maintain life processes while excess energy is released in the form of heat, called "vital heat." This heat must be considered in the temperature management program.

The respiration rate varies with commodity, in addition to variety, maturity or stage of ripeness, injuries, temperature and other stress-related factors. Strawberries have a high respiration rate, 12–18 mg CO_2/kg h at 0°C. The major determinate of respiration activity is product temperature. Since the final result of respiration activity is product deterioration and senescence, achieving as low a respiration rate as possible is desirable. For each 10°C temperature increase, respiration activity increases by a factor of two to four [5]. For example, the respiration of strawberries at 10°C is 49–95 mg CO_2/kg h, four to five times greater than at 0°C [5]. Therefore, strawberries must be rapidly pre-cooled to slow their metabolism (physiological deterioration) in order to provide maximum quality and storage life for shipping and handling operations.

The cooling process for freshly harvested fruits and vegetables is complicated by the heat generation from their respirational activity. The relationship between heat generation and temperature is assumed to involve breakdown of glucose as follows:

$$C_6H_{12}O_6 + 6O_2 \rightarrow 6CO_2 + 6H_2O + 2817 \text{ kJ} \qquad (15.2)$$

As with most chemical reactions, the heat generation of fruits and vegetables (see Table 15.6) is an exponential function of absolute temperature as [26]:

$$Q = A\exp(BT) \qquad (15.3)$$

where A and B are the heat generation delated constants and may take different values as listed in Table 15.7.

Maximum heat generation values of some fruits, vegetables and flowers due to respiration in air are tabulated in Table 15.6.

TABLE 15.6
Maximum Heat Generation Values of Some Products

Temperature (°C)	Heat Generation (W/kg)				
	0	5	10	15	20
Fruits					
Apples	0.010	0.019	0.030	0.039	0.046
Blackberries	0.063	0.094	0.177	0.214	0.444
Black currants	0.045	0.077	0.111	0.257	0.372
Cherries, sweet	0.016	0.047	0.091	0.130	0.160
Grapefruit	–	0.019	0.030	0.048	0.078
Oranges	0.014	0.023	0.038	0.063	0.103
Pears	0.011	0.039	0.073	0.110	0.156
Plums	0.018	0.036	0.063	0.105	0.165
Raspberries	0.069	0.158	0.092	0.389	0.576
Rhubarb, forced	0.040	0.060	0.100	0.126	0.155
Strawberries	0.043	0.080	0.147	0.245	0.374
Tomatoes	0.017	0.026	0.043	0.066	0.086
Vegetables					
Artichokes	0.100	0.140	0.212	0.330	0.533
Asparagus	0.080	0.126	0.180	0.300	0.363
Beans, broad	0.104	0.155	0.259	0.357	0.432
Brussels sprouts	0.051	0.089	0.149	0.223	0.268
Cabbage, winter white	0.009	0.021	0.024	0.039	0.060
Carrot (without tops)	0.039	0.051	0.057	0.071	0.098
Cauliflower	0.060	0.101	0.134	0.199	0.375
Celery	0.021	0.027	0.036	0.044	0.098
Cucumber	0.017	0.022	0.037	0.040	0.042
Green peppers	0.023	0.031	0.057	0.063	0.100

(Continued)

TABLE 15.6 (Continued)

Temperature (°C)	Heat Generation (W/kg)				
	0	5	10	15	20
Leeks	0.060	0.083	0.149	0.223	0.328
Lettuce, cabbage	0.048	0.071	0.092	0.149	0.238
Mushrooms	0.130	0.210	0.348	0.570	0.930
Onion, dry bulb	0.009	0.015	0.021	0.021	0.024
Parsnip	0.021	0.033	0.077	0.098	0.146
Peas (in pod)	0.140	0.164	0.357	0.506	0.744
Peas (shelled)	0.217	0.290	0.460	1.070	1.600
Potato, early	0.030	0.045	0.060	0.089	0.119
Radish (without tops)	0.027	0.040	0.065	0.109	0.183
Red beet (without tops)	0.012	0.021	0.033	0.051	0.057
Spinach	0.149	0.208	0.238	0.357	0.447
Sweet corn	0.089	0.158	0.259	0.409	0.605
Turnip (without leaves)	0.019		0.057	0.065	0.069
Flowers					
Carnation	0.028	0.045	0.086	0.190	0.690
Narcissus	0.074	0.140	0.246	0.410	0.620

Source: Adapted from Anon, *Refrigerated Storage of Fruits and Vegetables*, Ministry of Agriculture, Fisheries and Food, Her Majesty's Stationary Office, London, UK, 1978.

Table 15.7 gives the values of constants A and B for different fruits and vegetables. Gogus et al. [26] obtained these values using least mean square fitting of a straight line to plots of T, ln Q representations of the heat generation functions of these types of produce. They found that the coefficients of the exponent B are in the range of 0.07–0.14 1/K except for potatoes, for which 0.034 1/K was found.

15.11.1 MEASUREMENT OF RESPIRATORY HEAT GENERATION

The rate of heat generation is one of the most important parameters that affect postharvest cooling of fresh fruits and vegetables. Most researchers have determined heat generation rate by measuring CO_2 gas evolution from the product. According to a chemical equation for the complete combustion of glucose, 10.8 J of thermal energy is released per 1 g of CO_2 gas evolved. However, this thermal equivalence is not universally applicable when there are abnormal metabolic processes in the product. The common procedure for determining respiratory heat generation was by using data on the thermal response of the food sample placed in calorimeters. In order to monitor accurately this thermal response, heat loss through the calorimeter wall should be eliminated by regulating the surrounding air temperature using a control system.

15.12 TRANSPIRATION (MOISTURE LOSS)

Transpiration (so-called: *moisture loss* or *water loss*) of fresh fruits and vegetables is a mass transfer operation that involves the transport of water vapor from the surface of the product to the air medium. The main elements of transpiration are the transpiration coefficient and transpiration rate as follows:

TABLE 15.7
Constants for Heat Generation of Fruits and Vegetables in the Range of 0–20°C, Based on Equation 15.3

Fruits	A (W/kg)	B (1/K)	Vegetables	A (W/kg)	B (1/K)
Apple, early	3.39×10^{-14}	0.098	Asparagus	8.82×10^{-17}	0.088
Apple, late	4.00×10^{-14}	0.096	Bean, green	1.64×10^{-18}	0.104
Apricot	2.95×10^{-16}	0.117	Bean, broad	5.05×10^{-22}	0.130
Blackberry	5.90×10^{-18}	0.133	Brussels sprout	6.08×10^{-19}	0.107
Blackcurrant	8.40×10^{-18}	0.129	Cabbage, red	1.05×10^{-17}	0.094
Cherry, sweet	8.00×10^{-17}	0.121	Cabbage, white	4.92×10^{-17}	0.087
Cherry, sour	1.46×10^{-16}	0.119	Carrot, bunch	3.78×10^{-17}	0.088
Gooseberry	1.56×10^{-14}	0.127	Cauliflower	2.84×10^{-19}	0.108
Grape	4.02×10^{-14}	0.109	Cucumber	4.24×10^{-20}	0.112
Grapefruit	1.94×10^{-14}	0.098	Leek	6.86×10^{-21}	0.124
Lemon	3.87×10^{-13}	0.087	Garlic	1.74×10^{-14}	0.097
Nut	2.06×10^{-14}	0.094	Lettuce	2.18×10^{-19}	0.108
Orange	4.44×10^{-16}	0.112	Melon	2.11×10^{-17}	0.089
Peach	8.84×10^{-17}	0.121	Mushroom	1.30×10^{-16}	0.089
Pear, early	5.93×10^{-19}	0.138	Onion	3.46×10^{-14}	0.061
Pear, late	7.88×10^{-19}	0.136	Paprika	3.13×10^{-15}	0.074
Plum	6.09×10^{-16}	0.114	Potato	5.74×10^{-11}	0.034
Plum, yellow	2.79×10^{-15}	0.114	Radish, red	3.69×10^{-16}	0.083
Raspberry	1.81×10^{-14}	0.106	Spinach	5.63×10^{-20}	0.116
Strawberry	5.75×10^{-13}	0.092	Tomato, ripe	5.66×10^{-18}	0.093

Source: Malcolm, G.L. and Beaver, E.R., *Proceedings of the International Conference on Technical Innovations in Freezing and Refrigeration of Fruits and Vegetables*, 9–12 July, Davis, CA, 51–57, 1989.

- *The transpiration coefficient* is expressed as the mass of moisture transpired per unit area of product, per unit environmental water vapor pressure deficit per unit area, kg/m²skPa. Sometimes it is expressed per unit mass of product.
- *The transpiration rate* is the mass of moisture transpired per unit area of product per unit time. Also, this is sometimes expressed per unit mass of product.

A better understanding of the mechanism of transpiration from fruits and vegetables to the surrounding air should help in developing new systems for handling, transport, storage and improving existing systems so that moisture loss from fresh produce can be reduced and the initial quality at the time of harvest can be kept for a longer period of time [28]. Knowledge of the transpiration rate of the stored fruits and vegetables is essential for a rational analysis of desired medium conditions in their cold storage.

In transpiration rate measurements, care must be taken in:

- Controlling experimental conditions very well
- Keeping test produce at the same temperature as the environment especially in the initial part of the experiment
- Minimizing radiative heat transfer effects
- Providing sufficient air movement around the produce

Also, it is important for an experimental method for determining transpiration rates to consider all the factors given above.

Several factors influence moisture loss from stored fruits and vegetables, but these can be easily visualized by regarding transpiration as the interaction between a driving force and a resistance, as described by Sastry [29] as follows:

$$M = K_m F (P_s - P_a) \tag{15.4}$$

where

$$K_m = \frac{1}{1/k_s + 1/k_a} \tag{15.5}$$

where the resistance term k_m can be divided into two contributing terms as given in Equation 15.5: one due to the effect of skin resistance, and the other due to the boundary layer resistance. Here k_s is also known as the transpiration coefficient.

In Table 15.8, transpiration coefficients of some fruits and vegetables and their ranges as reported in literature are listed. The rate of transpiration is affected by a number of factors that particularly influence the driving force and resistance terms. Discussions of all these factors and phenomena and their inter-relationships with transpiration are given elsewhere [28,29]. A summary of these factors includes the following:

- Water vapor pressure deficit
- Condition of water
- Air flow velocity
- Respiratory heat generation
- Product shape, size, product surface area and structure
- Product maturity
- Product seeds
- Product skin and tissue permeability
- Dissolved substances in water
- Evaporative cooling
- Package microenvironments
- Physical and physiological conditions
- Packaging

15.12.1 SHRINKAGE

Shrinkage of stored perishable foods because of moisture loss involves not only a loss of saleable weight but also losses because of quality deterioration. Loss of turgidity in surface cells of fruits and vegetables can frequently render an entire product unsalable. Proper design and operation of cold storages must involve minimization of losses as a criterion. There are a number of factors affecting shrinkage of the product, as follows [e.g., 29]:

- Optimum storage temperatures and relative humidities
- Transpiration rates
- Effects of packaging on water loss
- Effects of heat sources adding heat to the facility
- Performance of evaporators in removing sensible and latent heat
- The effectiveness of the control system in handling various loads
- The nature of load variations

TABLE 15.8
Transpiration Coefficients of Some Fruits and Vegetables

Produce and Variety	Transpiration Coefficient (mg/kg-s MPa)	Common Range of Coefficients as given in Literature
Apples		
Johathan	35	16–38
Golden delicious	58	29–250
Average for all varieties	42	16–100
Brussels sprouts		
Average for all varieties	6150	3250–9770
Cabbage		
Average for all varieties	223	16–667
Carrots		
Nantes	1648	106–1896
Average for all varieties	1207	106–3250
Celery		
Average for all varieties	1760	104–3313
Grapefruit		
Average for all varieties	81	29–167
Grapes		
Thompson	204	180–223
Average for all varicties	123	21–254
Leeks		
Average for all varieties	790	530–1042
Lemons		
Average for all varieties	186	139–229
Lettuce		
Average for all varieties	7400	680–8750
Onions		
Average for all varieties	60	13–123
Oranges		
Average for all varieties	117	25–227
Peaches		
Average for all varieties	572	142–2089
Pears		
Average for all varieties	69	10–144
Plums		
Average for all varieties	136	110–221
Potatoes		
Average for all varieties	44	20–171
Tomatoes		
Average for all varieties	140	

Source: Sastry, S.K., Baird, C.D. and Buffington, D.E., *ASHRAE Trans.,* 84, 237–255, 1978, including literature references for data used in the table.

Note that minimization of shrinkage in a cold storage facility requires the use of evaporator temperatures very close to the desired dry-bulb temperature in the facility. Under conditions of high load, this involves removal of large amounts of heat with minimum temperature differences. Therefore, it is necessary to maximize the following contact factor, which is a measure of how closely the dry-bulb and dew point temperatures in storage approach one another [29]:

$$Cf = \frac{(T_{a1} - T_{a2})}{T_{a1} - T_s} \tag{15.6}$$

To maximize contact factor, a high heat transfer coefficient and a large contact area between air and the evaporator-coil surface is required.

15.13   COOLING PROCESS PARAMETERS

In food processing, there are many situations in which the temperature at any point in the product is a function of time and coordinate. This situation is called *transient heat transfer*. The most notable food processing examples of transient heat transfer are heating, cooling, precooling, freezing, drying, blanching, etc. In transient heat transfer during food cooling, the temperature at a given point within the food depends on the cooling time and position (coordinate). For practical food cooling applications, factors that influence temperature change and cooling rate are the following:

- Temperature and flow rate of cooling medium (coolant)
- Thermal properties of food product
- Physical dimensions and shape of food product

Regardless of the type of cooling technique, knowledge and determination of the cooling process parameters are essential to provide efficient and effective food cooling at the micro and macro scales. Some major design process factors for a food cooling process are given below:

- Cooling process conditions in terms of temperature, flow rate and relative humidity
- Arrangement of the individual products and/or product batches
- Depth of the product load in the cooling medium
- Initial and final product temperatures

The parameters in terms of cooling coefficient, lag factor, half cooling time and seven-eighths cooling time are the most important and meaningful variables in the food cooling process. They can be used to evaluate and present cooling rate data and cooling behavior of food products. In the literature, these cooling process parameters are also called *precooling process parameters*. A number of experimental and modeling studies to determine these parameters for various food products, particularly for fruits and vegetables, have been undertaken [e.g., 1,2].

15.13.1   COOLING COEFFICIENT

The cooling coefficient, which is an indication of the cooling capability of a food product subject to cooling, denotes the change in the product temperature per unit time of cooling for each degree temperature difference between the product and its surroundings.

15.13.2   LAG FACTOR

The lag factor, which is a function of the size and shape and the thermal properties of the product, such as the effective heat transfer coefficient, thermal conductivity, and thermal diffusivity, quantifies the resistance to heat transfer within the product to its surroundings. With this definition, it is very clear that the lag factor is directly related to the Biot number.

15.13.3 HALF COOLING TIME

In practical food cooling applications, the cooling rate data are formed in the cooling times. The most common cooling times are half cooling times and seven-eighths cooling times (Figure 15.6). The cooling times of food products are mainly influenced by the following:

- Heat transfer characteristics of the food products
- Physical dimensions and properties of the food products
- Heat transfer characteristics of the cooling medium
- Geometric details and heat transfer characteristics of the packaging materials (or containers), when used.

The half cooling time is the time required to reduce the product temperature by one-half of the difference in temperature between the product and the cooling medium.

15.13.4 SEVEN-EIGHTHS COOLING TIME

The seven-eighths cooling time, which is one of most meaningful parameters, describes the cooling rate in terms of the time required to reduce the product temperature by seven-eighths of the difference in temperature between the product and the cooling medium.

15.14 ANALYSIS COOLING PROCESS PARAMETERS

One of the main objectives of a cooling study is to produce useable cooling data and technical information that will help improve the existing cooling systems and processes, and provide optimum operation conditions. In this regard, the following become beneficial to the industry:

- A procedure for analyzing cooling process parameters
- A procedure for using cooling data to design cooling systems for efficient food cooling applications
- A basic data documentation

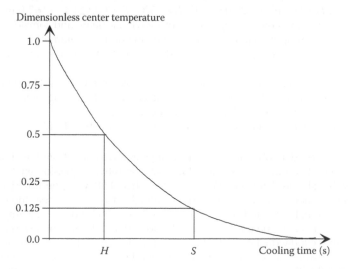

FIGURE 15.6 Cooling curve to show the cooling times.

Below we introduce the semiexperimental method for the cooling process parameters. In this method, it is considered that the thermal and physical properties of the product and the cooling medium are those occurring during operation under unsteady-state conditions. The dimensionless temperature in terms of the product and medium temperatures can be written as

$$\theta = \frac{(T - T_a)}{(T_i - T_a)} \tag{15.7}$$

The dimensionless temperature is generally expressed in the form of an exponential equation, including the cooling parameters in terms of the cooling coefficient (C) and lag factor (G) as

$$\theta = G\exp(-Ct) \tag{15.8}$$

From the definition of the half cooling time, by substituting $\theta=0.5$ into Equation 15.8, the half cooling time

$$H = \frac{\ln 2G}{C} \tag{15.9}$$

Also, by substituting $\theta=0.125$ into Equation 15.8, the seven-eighths cooling time becomes

$$S = \frac{\ln 2G}{C} \tag{15.10}$$

Example 15.2

In this example, we explain how to measure the center temperature distributions of individual spherical products, namely tomato and pear subjected to forced air cooling (the system given in Figure 15.2) and determine its cooling process parameters. A detailed description of the experimental apparatus, instrumentation and procedure is given in Dincer and Genceli [30,31]. Here are the diameters, initial temperatures of the products, cooling air temperatures and velocities:

- For pear: $D=0.06$ m, $T_i=22.5\pm0.5°C$, $T_f=5.0°C$, $T_a=4.0\pm0.1°C$, $U=1.25$ m/s.
- For tomato: $D=0.07$ m, $T_i=21.0\pm0.5°C$, $T_f=5.0°C$, $T_a=4.0\pm0.1°C$, $U=2.0$ m/s.
- After measuring the center temperatures of individual pears and tomatoes, the following methodology was applied to determine the cooling process parameters:
- Nondimensionalization of the measured center temperatures of the products by Equation 15.7
- Application of a curve-fitting technique to each of the dimensionless temperature data set in the form of Equation 15.8 for obtaining cooling coefficients and lag factors
- Determination of the half cooling times via Equation 15.9
- Determination of the seven-eighths cooling times via Equation 15.10.

The cooling process data obtained for individual pear and tomato from the above analysis are given in Table 15.9. It can be seen in Table 15.9 that a sensitive regression analysis in the exponential form (with high correlation coefficients of more than 0.98) was performed, and the cooling coefficients and lag factors were determined accordingly. It is obvious that the cooling process parameters were strongly affected by an increase in flow velocity of air. It is important to mention that the lag factors became greater than 1, due to certain internal and external resistances to heat transfer from the individual pear and tomato to the air flow. In light of the results, the cooling process parameters were found to be very much dependent upon the experimental conditions in terms of air flow velocities.

TABLE 15.9
Experimental Cooling Process Data for the Individual Products

Product	U (m/s)	r^2	G	C (1/s)	H (s)	S (s)
Pear	1.25	0.994	1.051475	0.0003039	2246.0	7007.7
Tomato	2.00	0.982	1.012360	0.0004720	1494.5	4431.6

15.15 THE FOURIER–REYNOLDS CORRELATIONS

Transient heat transfer is essential in maintaining fruit and vegetable quality during postharvest handling operations. For many fruits and vegetables, cooling is the recommended procedure to extend storage life sufficiently for shipping and retailing. There are two main objectives for a practical food cooling application:

- To obtain the relevant information, knowledge and data concerning recommended storage conditions and cooling techniques
- To develop simple but accurate models, correlations and methods for cooling process and cooling thermal parameters

In conjunction with this, refrigeration engineers and/or people who work in the food cooling industry, in practice prefer over complex techniques or processes, a body of information that can easily and effectively be used in the workplace to design cooling systems for their particular applications. In this respect, determination of the half cooling time and the seven-eighths cooling time is of great importance for providing optimum heat transfer and cooling processing conditions. There is a need to estimate product cooling times by using simple models or graphs in system design and improvements in the process, without experimental studies. In spite of extensive efforts to analyze transient heat transfer during cooling of food products, the literature on experiments conducted to determine the cooling process and thermal parameters for food cooling applications is limited [1]. Nevertheless, there are no simple models or correlations to estimate cooling times for fruits and vegetables without referring to experimental measurements. Here, we introduce the development of new Fourier–Reynolds correlations for such purposes.

In light of our background work and the illustrative example given earlier, it can be pointed out that the cooling medium conditions, particularly flow properties, considerably influence the cooling process parameters and hence cooling times. This influence brought to the forefront the Reynolds number, which reflects the coolant flow properties. On the other hand, it is well known that the Fourier number is also identified as dimensionless time and is affected by flow properties. This became our starting point to introduce the development of a Fourier–Reynolds correlation system [32]. Despite the simplicity of the methodology, there is still a need to use experimental temperature data in modeling of cooling process parameters. For practical applications, ready data or simple correlations and/or graphs are preferred over experimental methods or measurements. Experimental work means money and time and hence the following correlation maybe a good alternative.

The Reynolds number is:

$$\text{Re} = \frac{U_a(2Y)}{v} \tag{15.11}$$

where Y is l (i.e., half thickness) for slab products and R (i.e., radius) for spherical or long cylindrical products.

From the Fourier number (Fo$=at/Y^2$), we define new Fourier numbers for food cooling applications, which are the Fourier number for half cooling time and the Fourier number for seven-eighths cooling time, as follows [32]:

$$\text{Fo}_H = \frac{aH}{Y^2} \tag{15.12}$$

$$\text{Fo}_S = \frac{aS}{Y^2} \tag{15.13}$$

The thermal diffusivity of fruits and vegetables can be estimated, depending on their water content by the following Riedel correlation in [33]:

$$a = 0.088 \cdot 10^{-6} + (a_w - 0.088 \cdot 10^{-6})W \tag{15.14}$$

where a_w is the thermal diffusivity of water at the product temperature (e.g., 0.148×10^{-6} m²/s at 25°C).

Using the experimental half cooling time and seven-eighths cooling time data in Equations 15.12 and 15.13, we determined the experimental Fourier numbers. The Reynolds numbers were calculated from Equation 15.11 by using product dimensions, flow velocities and the kinematic viscosity of air. Then, experimental Fourier numbers were regressed against the Reynolds numbers by using the least-squares curve-fitting technique. The following Fourier–Reynolds correlations were developed to estimate half cooling times and seven-eighths cooling times of fruits and vegetables cooled with air for 100 < Re < 100000 [32]:

$$\text{Fo}_H = 42.465 \text{Re}^{-0.54426} \tag{15.15}$$

$$\text{Fo}_S = 125.21 \text{Re}^{-0.53913} \tag{15.16}$$

In addition to the above Fourier–Reynolds correlations, the following Fourier–Reynolds correlations for estimating the half cooling times and seven-eights cooling times of fruits and vegetables cooled with water were developed for 1000 < Re < 70000 in previous work performed by Dincer [32]:

$$\text{Fo}_H = 0.3693 \text{Re}^{-0.11871} \tag{15.17}$$

$$\text{Fo}_S = 1.2951 \text{Re}^{-0.1600} \tag{15.18}$$

Consequently, if we know the dimensions of the product to be cooled and the flow velocity of the coolant (air), the Reynolds number can be calculated by employing Equation 15.11. Then, we can calculate the Fourier number for the half cooling time from Equation 15.15 and that for the seven-eighths cooling time from Equation 15.16. After inserting thermal diffusivity into Equations 15.12 and 15.13 (if the thermal diffusivity is not known, Equation 15.14 can be used for its calculation), we can estimate the half cooling time and seven-eighths cooling time from these equations, without making any experimental temperature measurement.

Example 15.3

This example is given to provide a better understanding and verification of these correlations we provide this example. This example consists of two different applications. We processed the data for carrot taken from Ansari and Afaq [34] and for strawberry taken from Guemes et al. [35].

1. *Application I*. Product, carrot; shape, cylinder; fluid, air; $R=0.0165$ m; $U=6.6$ m/s; $W=0.88$; experimental $H=517$ s; experimental $S=1447$ s.
2. *Application II*. Product, strawberry; shape, sphere; fluid, air; $R=0.0132$ m; $U=3$ m/s; $W=0.89$; experimental $H=409$ s; experimental $S=1173$ s.

First, the values of thermal diffusivity were estimated from Equation 15.14 by using their water contents as $a=1.408\times10^{-7}$ m²/s for carrot and $a=1.414\times10^{-7}$ m²/s for strawberry. Then, the Reynolds numbers were calculated from Equation 15.11 and found to be Re=15557.2 and Re=5657.1, using an average value of $v=14\times10^{-6}$ m²/s for air. The Fourier number for half cooling time and Fourier number for seven-eighths cooling time were determined by the present correlations, i.e., Equations 15.15 and 15.16: $Fo_H=0.222$ and $Fo_S=0.688$ for carrot, and $Fo_H=0.385$ and $Fo_S=1.187$ for strawberry. The half cooling times and seven-eighths cooling times were found, from Equations 15.9 and 15.10)to be $H=492$ s and $S=1330$ s for carrot, and $H=474$ s and $S=1462$ s for strawberry. The maximum difference between the experimental and calculated values is within ±15%. This shows that the agreement is considered high for these types of thermal applications. The results of this study indicated that these new, simple but accurate Fourier–Reynolds correlations were developed to estimate the half and seven-eighths cooling times for single products to be exposed to cooling in an air flow. It is believed that these correlations could be of benefit in the food cooling industry.

15.16   COOLING HEAT TRANSFER COEFFICIENT CORRELATIONS

Cooling perishable commodities as quickly as possible after harvest has become a widely used method of maximizing postharvest life, preventing spoilage and maintaining quality. Significant factors in the design of a refrigerated food chain are the cooling heat transfer parameters (e.g., specific heat, thermal conductivity, thermal diffusivity and heat transfer coefficient) for food products. It is essential that design engineers know the quantity of heat to be released and the time taken to remove it; consequently cooling heat loads are calculated based on the effects of refrigeration on the product batches or the quality changes during storage. During the past three decades, there has been continuing interest in analyzing transient heat transfer that takes place during food cooling applications in order to provide the optimum processing conditions.

Thermal processing is widely used in the food industry, and the design of equipment for such processing, for example, cooling, depends strongly on a knowledge of the cooling heat transfer parameters of the foods to be cooled. It is also important especially for economic reasons, that the energy balances be considered in the design of cooling equipment. Since a number of processes such as cooling are of a transient nature, the cooling heat transfer parameters provide the most important design and process magnitudes. Therefore, the determination of such parameters has become a primary concern and has received the most attention in food cooling applications.

As defined by Newton's law of cooling, the effective heat transfer coefficient is a proportionality constant relating the heat flux from a surface to the temperature difference between the surface and the fluid stream moving past the surface. Actually, the heat transfer coefficient is not a direct property of the food product being cooled, however, it defines the rate of heat convection from the surface of a food product. It is mainly dependent upon the flow velocity of coolant, but also depends on several variables such as fluid properties of the coolant and the shape, size and surface texture of the product, as well as the temperature difference between the surface and the coolant.

The analysis of transient heat transfer to determine the effective heat transfer coefficients for food-cooling applications has been reported by a large number of different investigators. Majority of these studies focus on the development of mathematical models, correlations or approximations for the determination of effective heat transfer coefficients. These details are available elsewhere [1,2].

Dincer and Dost [36] introduced the development of simpler effective heat transfer coefficient correlations for food products, especially fruits and vegetables, cooled in both water and air, in terms of the cooling coefficient. In this study, they used the same modeling technique, but went one step further for correlating the experimental heat transfer coefficient values obtained from earlier works using a least squares curve-fitting method. These correlations were found with correlation coefficients of 0.88 and 0.79 and are given below:

For water cooling applications:

$$h = 27.356\exp(1381.836C) \tag{15.19}$$

For air cooling applications:

$$h = 213.5497\exp(256.9278C) \tag{15.20}$$

where C is the cooling coefficient.

In order to determine the effective heat transfer coefficients, it is necessary only to obtain the cooling coefficients' values using the methodology presented earlier.

15.17 EFFECTIVE NUSSELT–REYNOLDS CORRELATIONS

In the previous section, new models for determining the effective heat transfer coefficients for food products subject to cooling were obtained. In practice, working people prefer quite simple correlations, graphs, and/or tools, without having to make any measurements. In this section, we are going one step further to provide newer and simpler correlations, so-called *effective Nusselt–Reynolds correlations*. In various heat transfer books, many Nusselt–Reynolds correlations are proposed to estimate the heat transfer coefficients for solid objects being cooled or heated in any fluid flow, but these correlations lead to steady-state heat transfer. However, during cooling of food products, unsteady-state heat transfer occurs. Using these existing Nusselt–Reynolds correlations for the unsteady-state case may cause discrepancies of up to ±50%. For this reason, the development of effective Nusselt–Reynolds correlations is required to estimate more accurate effective heat transfer coefficients for food products, which will be helpful in system design and process optimization.

Two past publications [37,38] deal with effective Nusselt–Reynolds correlations developed for spherical and cylindrical fruits and vegetables, which are the most common shapes. In the first publication, Dincer [37] did both experimental and theoretical studies. In the experimental work, the center temperature distributions of several fruits and vegetables being cooled with air flow were measured and used as process data for the modeling. In the theoretical study, the modeling technique (as given in the previous section) was used. Also, some experimental cooling data were taken from the literature. Then, the effective heat transfer coefficients were employed to obtain Nusselt numbers via $Nu = hY/k$. The Reynolds numbers were found using Equation 15.11. Thus, $Nu/Pr^{1/3}$ data were correlated against the Reynolds number. Such diagrams for spherical and cylindrical fruits and vegetables and detailed information on the effective heat transfer coefficients, data reduction methods and experimental conditions can be found in Dincer [37]. The Nusselt–Reynolds correlations obtained are given below:

For spherical fruits and vegetables with a correlation coefficient of 0.77:

$$\frac{Nu}{Pr^{1/3}} = 1.560\,Re^{0.426} \quad \text{for } 100 < Re < 100{,}000 \tag{15.21}$$

where $Nu = h(2Y)/k_a$ and $Re = Ua(2Y)/v_a$.
and for cylindrical fruits and vegetables with a correlation coefficient of 0.99:

$$\frac{Nu}{Pr^{1/3}} = 0.291\,Re^{0.592} \quad \text{for } 100 < Re < 100{,}000 \tag{15.22}$$

In the second paper, Dincer [38] also provided a combination of experimental and theoretical studies and covered water and air cooling applications of fruits and vegetables, and developed the Nusselt–Reynolds diagram for these applications.

Methodology was the same but the aim was slightly different and more general correlations were obtained as follows:

For spherical and/or cylindrical products cooled in water with a correlation coefficient of 0.88:

$$\frac{Nu}{Pr^{1/3}} = 0.2672 Re^{0.4324} \quad \text{for} \quad 100 < Re < 100,000 \tag{15.23}$$

and for individual spherical or cylindrical products being cooled in air with a correlation coefficient of 0.82:

$$\frac{Nu}{Pr^{1/3}} = 0.264 Re^{0.4493} \quad \text{for} \quad 100 < Re < 100,000 \tag{15.24}$$

The results presented here show that new Nusselt–Reynolds correlations are very useful sources for practical cooling applications. However, further research needs to be carried out to develop various Nusselt–Reynolds correlations for different food commodities.

Example 15.4

For a better understanding and verification of the present Nusselt–Reynolds correlations, we provide this example. This example consists of three different applications. The data for a carrot taken from Ansari and Afaq [34], for an apple were from Ansari et al. [39], and for a cucumber were from Dincer and Genceli [31].

- *Application I.* Product, carrot; shape, cylinder; fluid, air; R, 0.0165 m; U, 6.6 m/s; experimental h, 61.12 W/m²°C (for details, see [34]).
- *Application II.* Product, apple; shape, sphere; fluid, air; R, 0.0388 m; U, 6.6m/s; experimental h, 33.59 W/m²°C (for details, see [39]).
- *Application III.* Product, cucumber; shape, cylinder; fluid, water; R, 0.019 m; U, 0.05 m/s; experimental h, 182.5 W/m²°C (for details, see [31]).

As the initial step, for an air temperature of 1.5°C, the Prandtl number, kinematic viscosity and thermal conductivity of air were taken as Pr=0.71, v_f=14.0×10^{-6} m²/s, k_f=0.0239 W/m°C. Also, for a water temperature of 1.0°C, we took Pr=13.1, v_f=1.75×10^{-6} m²/s, and k_f=0.568 W/m°C. The Reynolds numbers were calculated as 15557.15 for carrot, 36582.85 for apple, and 1085.7 for cucumber through the Fourier number equation. After inserting the values of the thermophysical properties of the cooling fluids with the corresponding Reynolds numbers into Equation 15.22 for carrot, Equation 15.21 for apple and Equation 15.23 for cucumber, the effective heat transfer coefficients were extracted as 57.00 W/m²°C for carrot, 37.66 W/m²°C for apple, and 193.42 W/m²°C for cucumber. Therefore, the differences between the actual and correlation results are found to be 6.7% for carrot, 10.8% for apple and 5.6% for cucumber. It has been shown that these Nusselt–Reynolds correlations provide reliable effective heat transfer coefficient results. In this respect, we can conclude that for practical cooling applications, the utilization of these correlations saves money and time for refrigeration engineers and technicians and provides effective and efficient operations.

15.18   THE DINCER NUMBER

Cooling is one of the most important thermal processes in a wide range of engineering applications, from the cooling of food products to hot processing of solid metals. Transient heat transfer takes place during cooling of food products and is obviously of significant practical interest due to the vast amount of cooling applications.

Many studies on the determination of temperature distributions and heat transfer rates within solid objects of regular or irregular shapes cooled in a fluid flow have been undertaken. Limited studies have been carried out to determine cooling process parameters and cooling heat transfer parameters, as mentioned earlier. A detailed literature survey on these subjects has been published [1]. In a recent publication [40], a new dimensionless number (the so-called Dincer number) for forced-convection heat transfer was introduced. The Dincer number defines the relationship between the cooling or heating medium and the cooling or heating rate of the object.

Here, it will be used for food cooling applications. The Dincer number expresses the effect of the flow velocity of the cooling fluid on the cooling coefficient (i.e., cooling process parameter) for food products with regular or irregular shapes. It is defined as follows:

$$\text{Di} = \frac{U}{CY} \tag{15.25}$$

where Y denotes the characteristic length in meters, such as radius (i.e., R) for spherical and cylindrical products and half thickness (i.e., l) for slab products.

In Equation 15.25, the ratio of these quantities defines the relative magnitude of the cooling capability of the food product being cooled in a fluid flow and is the connecting link between the flow velocity and the cooling rate.

As indicated above, there is a strong relationship between the properties of the cooling fluid and the cooling process parameters. It is known that the Nusselt number is a dimensionless parameter that provides a measure of the convection heat transfer occurring at the product surface. In light of existing Nusselt–Reynolds correlations, after examining Equation 15.25 more carefully, we can note a strong similarity. The Nusselt number is then a function of the Dincer number [40] as follows:

$$\text{Nu} = \frac{h(2Y)}{k_f} = f(\text{Di}) \tag{15.26}$$

where k_f denotes the average thermal conductivity of the surrounding fluid in W/m°C.

The existing Nusselt–Reynolds correlations are based on steady-state conditions but Equations 15.21 through 15.24 are the result of comprehensive unsteady-state heat transfer analyses as the commonly encountered in food cooling applications. The Nusselt–Dincer correlations are extremely useful from the standpoint of suggesting how fluid properties and cooling process parameters affect each other. It is well known that the Nusselt–Reynolds correlations are used particularly for determining heat transfer coefficients for the corresponding objects. In this case, if we go one step further, we can say that the Nusselt–Dincer correlations can be used for determining effective heat transfer coefficients for food products.

Therefore, the experimental Nusselt number and Dincer number data were correlated to obtain the following Nusselt–Dincer correlation (with a correlation coefficient of 0.78) was found for the range $10^4 < \text{Di} < 10^6$ [40]:

$$\text{Nu} = 2.2893 \cdot 10^{-4} \text{Di}^{1.0047} \tag{15.27}$$

Example 15.5

Let's apply this procedure to the experimental data published earlier. The experimental cooling coefficients and Nusselt numbers for products, namely tomatoes, pears, cucumbers, figs and grapes, cooled in a forced-air cooling system were taken from Dincer [40]. Some additional experimental data for bananas and carrots were taken from Ansari and Afaq [34], and the cooling coefficients were determined according to the methodology described. These experimental data are given in Table 15.10. As can be seen in Table 15.10, the cooling coefficients and effective heat transfer

TABLE 15.10
The Values of U, C, h, Nu, and Di for Some Products (k_f=0.0239 W/m°C)

U (m/s)	C (1/s)	h (W/m²°C)	Nu	Di
	**Tomatoes (Y=0.035 m, T_a=4°C)[a]**			
1	0.0001980	10.89	30.98	144,300.14
1.25	0.0002302	13.08	37.22	155,144.59
1.5	0.0002371	13.56	38.58	180,755.55
1.75	0.0002555	14.90	42.39	195,694.71
2	0.0002861	17.24	49.05	199,730.36
	Pears (Y=0.030 m, T_a=4°C)[a]			
1	0.0002763	12.62	30.78	120,641.81
1.25	0.0003039	14.18	34.58	137,106.50
1.5	0.0003315	15.82	38.58	150,829.56
1.75	0.0003361	16.10	39.26	173,559.45
2	0.0003897	19.51	47.58	191,071.76
	Cucumbers (Y=0.019 m, T_a=4°C)[a]			
1	0.0003957	18.22	28.14	133,008.79
1.25	0.0004251	19.86	30.67	154,762.34
1.5	0.0004504	21.31	32.91	175,282.78
1.75	0.0004800	23.06	35.62	191,885.96
2	0.0005367	26.56	41.02	196,130.34
	Grapes (Y=0.0055 m, T_a=4°C)[a]			
1	0.0026019	23.72	10.91	69,879.00
1.25	0.0028321	26.05	11.99	80,248.83
1.5	0.0031315	29.18	13.43	87,091.57
1.75	0.0033387	31.43	14.46	95,301.11
2	0.0034538	32.72	15.06	105,285.87
	Figs (Y=0.0235 m, T_a=4°C)[a]			
1.1	0.0006217	23.77	46.74	75,291.15
1.5	0.0006677	26.16	51.44	95,596.50
1.75	0.0006907	27.41	53.90	107,815.38
2.5	0.0007828	32.71	64.32	135,900.58
	Banana (Y=0.015 m, T_a=5.9°C)[b]			
6.6	0.0018500	63.44	79.63	237,837.83
	Carrot (Y=0.0165 m, T_a=1.5°C)[b]			
6.6	0.0013200	61.12	84.39	303030.30

Source:
[a] Dincer, I., *Determination of the Effective Process Parameters and Heat Transfer Characteristics of Several Food Products during Cooling*, Dincer, I., *Energy* 18(4), 335–340, 1993.
[b] Ansari, F.A. and Afaq, A., *Int. J. Refrigeration* 9, 161–163, 1986.

coefficients are dependent on experimental conditions (i.e., flow velocity) in addition to product-related properties. It is apparent that the Nusselt number increases with Dincer number because of the increment in cooling coefficient. It is obvious that an increasing effective heat transfer coefficient increased the cooling coefficient and hence the Dincer number and vice versa. This strong linkage between the cooling coefficient and effective heat transfer coefficient as defined earlier.

The Dincer number is highly sensitive to the flow velocity, cooling coefficient and size of the product. In this problem, increasing flow velocity increased the cooling coefficient and hence increased the Dincer number. For instance, increasing the air-flow velocity from 1 to 2 m/s (i.e., by 100%) for tomatoes with $R=0.035$ m increased the Dincer number from 144,300.14 to 199,730.36 (i.e., by 38.4%).

As engineers, our interest in cooling heat transfer parameters is directed principally toward the parameters U, C, and Y. From the knowledge of these parameters, we can compute the Dincer number and hence the effective heat transfer coefficient using the present Nusselt–Dincer correlation for products exposed to air cooling. It is therefore understandable that expressions that relate U, C and Y to one another reflect the cooling behavior of product.

As a result, a new Dincer number has been developed for food products subject to cooling in a fluid flow. Also, a Nusselt–Dincer correlation was obtained for products cooled with air flow. These are useful tools for food cooling applications. The development of more Nusselt–Dincer correlations for various cooling applications will be beneficial in solving problems and making the phenomena more understandable.

15.19   CONCLUSIONS

This chapter has dealt with various aspects of food refrigeration in terms of food cooling, food freezing, food storage, cooling process parameters, cooling heat transfer parameters, precooling systems. Other aspects such as their applications, advantages and disadvantages, refrigerated transport and their technical and operational issues and optimum processing criteria, CA storage and relevant issues, cooling process parameters and their determination, cooling heat transfer parameters are also discussed. In order to highlight the importance of these topics, some practical examples and applications are presented. Also, some illustrative examples are given to show how to use the models, correlations, charts and tables in food refrigeration applications.

NOMENCLATURE

a	thermal diffusivity, m²/s
A	Parameter in Equation (3), W/kg
B	paramater in Equation (3), 1/K; constant rate of temperature increase
C	cooling coefficient, 1/s
Cf	contact factor in Equation (6)
D	diameter, m
Di	Dincer number
EC	energy coefficient
EN	electrical energy, kW
F	product surface area, m²
Fo	Fourier number
G	lag factor or intercept
h	effective heat transfer coefficient, W/m²°C
H	half cooling time, s
k	thermal conductivity, W/m°C
k_m	area-based overall moisture transfer coefficient of the product, kg/m² s kPa
l	half thickness, m
L	thickness, m
L_c	characteristic length, m
m	mass, kg
M	rate of moisture loss from product, kg/s; volume fraction of dispersed component

Nu	Nusselt number
P	water vapor pressure (kPa); parameter; mass fraction of protein
Pr	Prandtl number
q	alternative dimensional ratio
Q	specific heat respiration capacity, W/kg; heat transfer rate, W; heat flux, W/m
r^2	correlation coefficient
R	radius or half-thickness, m
Re	Reynolds number
S	seven-eighths cooling time, s; mass fraction of solid; the ratio of volume to area, m
SH	sensible heat removed from the product, kW
t	cooling time, s
T	temperature, °C
U	flow velocity, m/s
V	volume, m³
W	water content by weight, in decimal units
Y	diameter for sphere and cylinder and thickness for slab, m (i.e. R for spherical and cylindrical products and l for slab products), m
θ	dimensionless temperature
ρ	density, kg/m³
ν	kinematic viscosity, m²/s

Subscripts

a	surrounding medium; ambient; air film, air
a1	air entering evaporator
a2	air leaving evaporator
c	center; continuous
d	dispersed or discontinuous
f	final, surrounding fluid
H	half cooling time
i	initial
m	mass average
s	average effective coil surface; evaporating surface on the product; skin; surface
S	seven-eighths cooling time
w	water

REFERENCES

1. Dincer, I. 1997. *Heat Transfer in Food Cooling Applications.* Taylor & Francis, Washington, DC.
2. Dincer, I. 2003. *Refrigeration Systems and Applications.* Wiley, London, UK.
3. Wagner, A.B., Dainello, F.J. and Parsons, J.M. 1999. *Harvesting and Handling.* Center for Food Safety and Applied Nutrition, US Food and Drug Administration, Washington, DC.
4. USDA. 2001. *Tropical Products Transport Handbook.* The US Department of Agriculture (USDA), Washington, DC.
5. Sargent, S.A., Talbot, M.T. and Brecht. J.Y. 1988. Evaluating precooling methods for vegetable packing-house operations. *Proc. Fla. State Hort. Soc.,* 101, 175–182.
6. Thompson, J.F. and Chen, Y.L. 1988. Comparative energy use of vacuum, hydro, and forced air coolers for fruits and vegetables. *ASHRAE Trans.,* 94(1), 1427–1433.
7. ASHRAE. 2001. *Handbook of Fundamentals.* American Society of Heating, Refrigerating and Air-Conditioning Engineers, Inc., Atlanta, GA.

8. ASHRAE. 2002. *Handbook of Refrigeration*. American Society of Heating, Refrigerating and Air-Conditioning Engineers, Inc., Atlanta, GA.

9. Dincer, I. 1993. *Determination of the Effective Process Parameters and Heat Transfer Characteristics of Several Food Products during Cooling*. PhD Thesis, Faculty of Mechanical Engineering, Istanbul Technical University, Istanbul, Turkey.

10. Dincer, I., Yildiz, M., Loker, M. and Gun, H. 1992. Process parameters for hydrocooling apricots, plums and peaches. *Int. J. Food Sci. Technol.*, 27, 347–352.

11. Dincer, I. and O.F. Genceli, 1994. Cooling process and heat transfer parameters of cylindrical products cooled both in water and in air. *Int. J. Heat Mass Transfer*, 37, 625–633.

12. Ohling, R.S. 1990. Rapid air precooling of fruits and vegetables. *ASHRAE J.*, 32(3), 60–65.

13. SARDI. 2000. *Forced Air Cooling, Information Kit No. 11* (compiled by Matthew Palmer). South Australian Research and Development Institute, Adelaide, Australia.

14. Henry, F.E., Bennett, A.H. and Segall, R.H. 1976. Hydraircooling: A new concept for precooling pallet loads of vegetables. *ASHRAE Trans.*, 82(2), 541–547.

15. McDonald, K. and Sun, D.-W. 2000. Vacuum cooling technology for the food processing industry: a review. *J. Food Eng.*, 45, 55–65.

16. Gast, K.L.B. and Flores, R.A. 1991. *Precooling Produce*, MF-1002. Cooperative Extension Service, Kansas State University, Kansas, KS.

17. Jay, J.M. 1996. *Modern Food Microbiology*, 5th Edn. Chapman & Hall, New York, NY.

18. Amos, N.D., Cleland, D.J., Banks, N.H. and Cleland, A.C. 1993. A survey of air velocity, temperature and relative humidity in a large horticultural coolstore. *Proceedings of the International Meeting on Cold Chain Refrigeration Equipment by Design*, 15–18 November, Palmerston, New Zealand, 414–422.

19. Hall, C.W. 1972. *Processing Equipment for Agricultural Products*. Avi Publishing Co., Westport, CT.

20. Frith, J. 1991. *The Transport of Perishable Foodstuffs*. Shipowners Refrigerated Cargo Research Association, Cambridge Refrigeration Technology (CRT), Cambridge, UK.

21. Cleland, A.C. 1990. *Food Refrigeration Processes: Analysis, Design and Simulation*. Elsevier Applied Science, London, UK.

22. Malcolm, G.L. and Beaver, E.R. 1989. Utilization of advanced technology CA equipment for storage, transportation, and marketing of perishable commodities. *Proceedings of the International Conference on Technical Innovations in Freezing and Refrigeration of Fruits and Vegetables*, 9–12 July, Davis, CA, 51–57.

23. Dewey, D.H. 1983. Controlled atmosphere storage of fruits and vegetables. In: *Developments in Food Preservation-2*, S. Thorne (Ed.). Applied Science Publishers, London, UK, 1–24.

24. Rushing, J.W. 1993. Simulated shipment of peaches, plums, and nectarines under controlled atmosphere. *Proceedings of the International Meeting on Cold Chain Refrigeration Equipment by Design*, 15–18 November, Palmerston, New Zealand, 425–435.

25. OOCL. 2001. *Conditions for Storage Commodities*. The Orient Overseas Container Line Limited, http://www.oocl.com/reefer/technology.htm.

26. Gogus, A.Y., Akyurt, M. and Yavuzkurt, S. 1972. Unsteady cooling of unit loads with exponential heat generation, *Int. Inst. Refrigeration-Annexe*, 1, 227–239.

27. Anon. 1978. *Refrigerated Storage of Fruits and Vegetables*. Ministry of Agriculture, Fisheries and Food, Her Majesty's Stationary Office, London, UK.

28. Sastry, S.K., Baird, C.D. and Buffington, D.E. 1978. Transpiration rates of certain fruits and vegetables. *ASHRAE Trans.*, 84, 237–255.

29. Sastry, S.K. 1985. Factors affecting shrinkage of foods in refrigerated storage. *ASHRAE Trans.*, 91, 683–689.

30. Dincer, I. and Genceli, O.F. 1995. Cooling of spherical products: Part I Effective process parameters. *Int. J. Energy Res.*, 19, 205–218.

31. Dincer, I. and Genceli, O.F. 1995. Cooling of spherical products: Part II Heat transfer parameters. *Int. J. Energy Res.*, 19(3), 219–225.

32. Dincer, I. 1995. Development of Fourier-Reynolds correlations for cooling parameters. *Appl. Energy*, 51(2), 125–138.

33. Sweat, V.E. 1986. Thermal properties of foods. In: *Engineering Properties of Foods*, M.A. Rao, S.S.H. Rizvi (Eds). Marcel Dekker, New York, NY, 49–87.

34. Ansari, F.A. and Afaq, A. 1986. Precooling of the cylindrical food products. *Int. J. Refrigeration*, 9, 161–163 (1986).

35. Guemes, D.R., Pirovani, M.E. and di Pentima, J.H. 1988. Heat transfer characteristics during air precooling of strawberries. *Int. J. Refrigeration* 12, 169–173.
36. Dincer, I. and Dost, S. 1996. New correlations for heat transfer coefficients during direct cooling of products. *Int. J. Energy Res.,* 20, 587–594.
37. Dincer, I. 1994. Development of new effective Nusselt–Reynolds correlations air-cooling of spherical and cylindrical products. *Int. J. Heat Mass Transfer,* 37(17), 2781–2787.
38. Dincer, I. 1997. New effective Nusselt–Reynolds correlations for food-cooling applications. *J. Food Eng.,* 31(1), 59–67.
39. Ansari, F.A., Charan, V. and Varma, V.K. 1984. Heat and mass transfer in fruits and vegetables and measurement of thermal diffusivity. *Int. Commun. Heat and Mass Transfer,* 11, 583–590.
40. Dincer, I. 1996. Development of a new number (the Dincer number) for forced-convection heat transfer in heating and cooling applications. *Int. J. Energy Res.,* 20(5), 419–422.
41. Dincer, I. 1993. Heat-transfer coefficients in hydrocooling of spherical and cylindrical food products. *Energy,* 18(4), 335–340.

16 Heat Transfer and Air Flow in a Domestic Refrigerator

Onrawee Laguerre
UMR Génie Industriel Alimentaire Cemagref-ENSIA-INAPG-INRA

CONTENTS

16.1   INTRODUCTION

Domestic refrigerators are widely used in industrialized countries. There are approximately 1 billion domestic refrigerators worldwide[1] and the demand in 2004, was 71.44 million units (including 11.2 in China, 10.7 in the United States, 4.43 in Japan, 3.36 in India, 3.14 in Brazil[2]). In developing countries, the production is rising steadily: total production rose 30% in 2000.[3] In France, there are 1.7 refrigerators per household.[4]

Epidemiological data from Europe, North America, Australia and New Zealand indicate that a substantial proportion of foodborne disease is attributed to improper food preparation practices in consumers' homes.[5] Data also illustrate that a large proportion of consumers lack knowledge of adequate refrigeration temperatures. Surveys carried out in various countries on the temperature and the microbial contamination in the refrigerating compartment under real use conditions show an alarming situation.[6–13] Compared with surveys on the refrigerating compartment, a few surveys have been carried out on temperatures in domestic freezers.[14] Product temperature is a quality and safety-determining factor. It is therefore necessary to fully understand the mechanism of heat transfer and airflow.

Three types of domestic refrigerators are available in the market: static, brewed, and no-frost. The static type (Figure 16.1a) is widely used in Europe. In this case, heat is transferred principally by natural convection and airflow is due to variations in air density. These variations are related principally to the temperature and humidity gradients. The vertical force which results from air weight and buoyancy is ascendant if air is locally lighter than the average and descendant where the opposite is true (hot/humid air is lighter than cold/dry air). There is a combination of heat transfer and airflow inside refrigerators i.e., heat transfer from air to the evaporator (vertical plate), to the other walls (cavity) and to products (various forms). Due to the principal of heat transfer, temperature heterogeneity is often observed in this type of refrigerator. The position of the evaporator

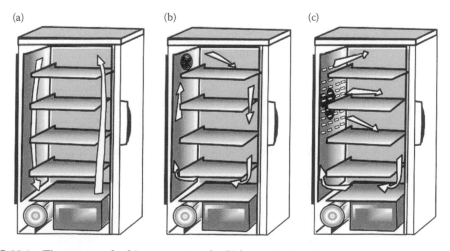

FIGURE 16.1   Three types of refrigerator: (a) static, (b) brewed, (c) no-frost.

(horizontal/vertical, top/bottom of the compartment) determines the location of cold and warm zones. The brewed type is a static refrigerator equipped with a fan (Figure 16.1b). It allows air circulation and the temperature decreases rapidly after door opening. Air temperature is more homogeneous in this case than in the static type but the energy consumption is higher due to the fan. In a no-frost refrigerator (Figure 16.1c), a fan (embedded in the back wall) pushes air to flow over the evaporator before entering into the refrigerating compartment. Air temperature is more homogeneous compared to the two other refrigerator types. Disadvantages of no-frost type are noise, energy consumption, drying on food surface and high price.

It should be remembered that the refrigerator design in the United States and in Europe is quite different: particularly, in the United States, the size of appliances is bigger and there are more refrigerators equipped with fan.

Generally, the external dimensions of commercialized refrigerator are 60×60 cm (width $\times$ depth) and the height varies between 90 cm and over 2 m. The wall thickness is approximately 4 cm and the refrigerator is generally made of polystyrene (inner liner), polyurethane (insulating material) and metal sheet (outer liner).

Knowledge of air temperature and velocity profiles in a refrigerator is important for food quality control. In fact, if the consumer knows the position of warm and cold zones in the refrigerator, the product can be placed correctly. Knowledge of thermal and hydrodynamic boundary layers near the evaporator and the other walls is also important. If the product is too close to the evaporator wall, freezing can occur, and if it is too close to the other walls, there may be health risks.

The objective of this chapter is to present the state of the art of knowledge on heat transfer and airflow by natural convection in domestic refrigerators. Several subjects are dealt with: literature review of natural convection in closed cavity, cold production system of domestic refrigerators, temperatures and different heat transfer modes in appliances. Finally, an example of heat transfer analysis and numerical simulation in a typical refrigerator will be shown.

16.2 LITERATURE REVIEW OF NATURAL CONVECTION IN CLOSED CAVITY

A literature review on natural convection in domestic refrigerators, near a vertical plate, in empty closed cavities and in cavities filled with porous media will be presented. Some limits of the application of these studies to our case (refrigerator loaded with a food product) will also be given.

16.2.1 STUDIES IN DOMESTIC REFRIGERATORS

Several experimental studies were carried out on empty and loaded refrigerators.[7,15] The objective was to analyze the effects of several parameters on the temperature in the refrigerating compartment (thermostat setting, frequency of door openings, filled volume, temperature and humidity of ambient air). However, few studies were carried out on airflow measurement due to the complexity of metrology techniques compared to temperature. Airflow measurement in a freezer compartment under real operating conditions was carried out by Lacerda et al.[16] using PIV (particles imagery velocimetry). It was observed that the flow field was strongly influenced by the temperature variations due to the "on" and "off" operation cycles of compressor. This behavior was attributed to natural convection and strong temperature dependency of air viscosity. Another study on airflow in a ventilated domestic freezing compartment was carried out by Lee et al.[17] In this study a comparison of velocity field obtained by CFD simulation and by experiments (PIV measurements) was undertaken. These authors observed that the flow was very complex: jet-like flow around entrance ports, impinging and stagnation flow on the walls and a large recirculation flow in cavity. To our knowledge, no study was carried out on air velocity measurement in a refrigerating compartment. Moreover, airflow being strongly influenced by the aspect ratio (height/width) of the cavity; the flow in a freezer is therefore different from the one in a refrigerating compartment.

To obtain useful information on natural convection in a domestic refrigerator, airflow in some well known configurations will be presented: near a warm (or cold) vertical plate, empty cavity and cavity filled with product. The temperature of the cold wall is constant for these three configurations in spite that this temperature fluctuates due to the "on" and "off" compressor working cycles in a real refrigerator.

16.2.2 Heat Transfer and Airflow Near a Vertical Plate

For a first approach, literature on flow adjacent to a cold vertical plate placed in a warm environment (without other limiting walls) can be applied for a good understanding on how airflow by natural convection nears the refrigerators evaporator. If a tracer (e.g., smoke) is injected at one end of the plate to visualize the flow, laminar flow is firstly observed near the wall and then turbulence appears (Figure 16.2). The air velocity (u) is zero at the plate surface, then, it increases rapidly at locations away from the plate to attain a maximum value (u_m). Air velocity then decreases and approaches zero, which is the velocity far from the plate (Figure 16.3). The zone of nonzero velocity ($u > u_m/100$) is called the hydrodynamic boundary layer and its thickness (δ) increases in the flow direction (x).

The air temperature (T) increases from the wall temperature (T_w) to the ambient temperature (T_∞) (Figure 16.4). The zone where the temperature differs from ambient (($T - T_\infty$) > ($T_w - T_\infty$)) is called the thermal boundary layer and its thickness (δ_T) increases in the flow direction (x).

The equivalent boundary thermal layer thickness ($\delta_{T,eq}$) is also frequently used in practice; it is defined as $\delta_{T,eq} = \lambda/h\delta$.

When Prandtl number (Pr) is near 1 such as in the case of air, the equivalent boundary layer thickness ($\delta_{T,eq}$) is of the same order of magnitude as that of the thermal boundary layer (δ_T). For example, for laminar forced convection, $\delta_{T,eq} = 2/3 \cdot \delta_T$.

The flow regime in natural convection is characterized by the Rayleigh number (Ra) defined as:

$$\text{Ra} = \frac{g\beta\Delta T L^3}{\alpha\nu} \tag{16.1}$$

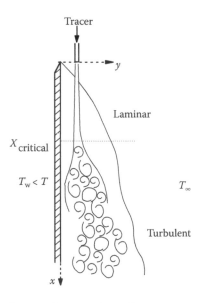

FIGURE. 16.2   Air flow by natural convection near the wall.

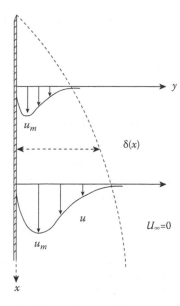

FIGURE. 16.3 Hydrodynamic boundary layer and velocity profile in natural convection flow.

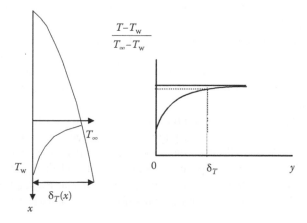

FIGURE 16.4 Thermal boundary layer, temperature profile and dimensionless profile in natural convection.

In general, the critical Rayleigh number, which distinguishes the transition from laminar to turbulent flows, is approximately 10^9 (depending on the geometry and boundary conditions.[18]

Heat transfer phenomena depend on the flow regimes (laminar or turbulent). Khalifa[19] presents a literature review of natural convection heat transfer correlations for vertical or horizontal plates. More than 40 articles are presented in this review. The experimental conditions are summarized: dimension of the tested surface, fluid type, temperature difference between the plate and the fluid, and Rayleigh number range. A strong variation in the values of heat transfer coefficient was found from using these different correlations.

In general, the correlations are presented in the following form:

$$Nu = a.Ra^n \qquad (16.2)$$

"*a*" and "*n*" are coefficients whose value depends on the flow regime. For example, for a vertical plate, Incropera and Dewitt[18] proposed:

$a = 0.59$ and $n = 1/4$ for laminar flow,
$a = 0.10$ and $n = 1/3$ for turbulent flow.

16.2.3 Heat Transfer and Airflow in Empty Closed Cavity

Several experimental studies were carried out to measure air temperature and/or velocity in closed cavities.[20–24] Ostrach[25], Catton[26], and Yang[27] carried out a literature review on this subject, which included both the experimental and modeling results (2D and 3D). These authors emphasize the importance of the aspect ratio of the cavity and the temperature difference between walls on the flow regimes.

When the bottom horizontal wall is cold, stable temperature stratification is observed in the cavity (cold zone at the bottom and warm zone at the top), and there is no airflow. When the upper horizontal wall is cold, unstable flow is observed[25] due to gravity. The state of unstable equilibrium occurs until a critical density gradient is exceeded. A spontaneous flow then results that eventually becomes steady and cellular-like. When a vertical wall is cold, circular flow is observed along walls and the air is almost stagnant at the center of the cavity; thermal stratification is also observed. This case is similar to that of a domestic refrigerator, since the evaporator is often inserted in the vertical back wall.

There are fewer experimental studies on natural convection than on forced convection due to experimental difficulties in terms of metrology for low velocity and design of experimental devices maintaining given wall conditions. In fact, measurement is very sensitive to experimental and boundary conditions. Henks and Hoogendoorn[28] compared some experimental results obtained with a standard case (Ra = 5×10^{10}, cavity aspect ration $H/L = 1$ in 3D, adiabatic horizontal walls). Good agreement between results was found, particularly about the temperature and velocity profiles within the boundary layers.

Ramesh and Venkateshan[29] used a differential interferometer to visualize conditions in the boundary layer along the wall ($10^5 <$ Ra $< 10^6$). They found that it is generally stable except in the corner. Mergui and Penot[23] carried out a visualization of flow in an empty cavity using a laser tomography (Ra = 1.7×10^9); they observed the same phenomena as Ramesh and Venkateshan[29].

Deschamps et al.[30] reported that in a domestic refrigerator, the Rayleigh number varies from 10^8 and 10^9, and that flow is therefore, in the transition regime between laminar and turbulent flow.

Heat exchange by radiation between internal walls of the cavity is as important as that achieved with natural convection and this should be taken into account. Several authors[31–34] showed by experimental and numerical approaches that these two heat transfer modes occur simultaneously. Ramesh and Venkateshan[32] showed experimentally that for a square enclosure (vertical walls maintained at 35 and 65°C, adiabatic horizontal walls, Ra = 5×10^5), the heat transfer by convection and radiation between high emissive vertical walls ($\varepsilon = 0.85$) is twice of that of polished ones ($\varepsilon = 0.05$). Balaji and Venkateshan[31] proposed correlations established from numerical simulations to express the convection and radiation in a square cavity as a function of ε, Ra, T_c/T_h and a radiation convection interaction parameter

$$N_{RC} = \frac{\sigma T_{wh}^4 H}{\lambda(T_{wh} - T_{wc})}.$$

These correlations show that the radiation effect increases when the wall emmisivity and/or wall temperatures increase. Moreover, Li and Li[34] reported that the radiation relative to convection increases as the size of enclosure increases. An estimation of convection and radiation heat transfer in a refrigerator was carried out in our previous study[35], which confirms the importance of radiation.

16.2.4 HEAT TRANSFER AND AIRFLOW IN CAVITY COMPLETELY OR PARTIALLY FILLED WITH POROUS MEDIA

Several reviews on heat transfer by natural convection in a cavity filled with porous media have been carried out.[36–39] In the case of porous media, the Rayleigh number is defined as:

$$\text{Ra}_p = \frac{g\beta\Delta T \cdot H \cdot K}{\alpha_p \nu} \tag{16.3}$$

When Rayleigh number is less than a critical value (Ra_c), the heat transfer is dominated by conduction. When $\text{Ra}_p > \text{Ra}_c$, airflow is observed, which leads to a heat transfer dominated by convection. Oosthuizen[39] reported a value of 40 for Ra_c in a rectangular cavity heated from below.

Airflow in a cavity filled with porous media is generally laminar. Circular flow, similar to that of an empty cavity, is observed in the boundary layer along the walls (Figure 16.5). Velocity is much smaller at the center of the cavity.

Literature concerning heat transfers in porous media and in packed beds[40–42] presents several approaches taking into account heat transfer by conduction, convection and radiation. Moreover, these studies distinguish the one-temperature models, in which local equilibrium between product and air is assumed, from the two-temperature models, in which different temperatures represent product and air statement not clear.

The literature on cavity filled with porous media cannot be applied directly to the case of loaded domestic refrigerator principally due to the large variation in products dimension. For refrigerators, the ratio between the dimension of product and cavity is about 0.10 (~5 cm product width and ~50 cm cavity width) while this ratio is ≤ 0.02 for porous media. There is notably a great influence of product position on heat transfer compared to the case of porous media. This was shown in our previous studies[43] that demonstrate the influence of these parameters on the heat transfer at low air velocity (< 0.2 m/s) in a stack of spheres.

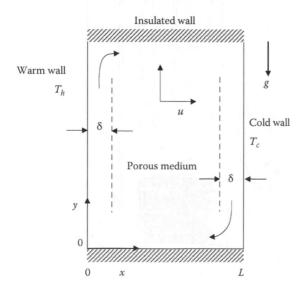

FIGURE 16.5 Two-dimensional rectangular porous layer held between differentially heated side walls.

16.3   COLD PRODUCTION SYSTEM IN DOMESTIC REFRIGERATORS/FREEZERS

The most common refrigerators and freezers have four major parts in their refrigeration system—a compressor, a condenser, an expansion valve and an evaporator (Figure 16.6). In the evaporator section, a refrigerant (commonly R600a and R134; still in use R12 and ammonia) is vaporized to absorb heat added into the refrigerator due to heat transfer across the refrigerators walls and infiltration through the door and seals and during door opening. The refrigerant boils at −18 to −20°C when pressurized at 0.9–1 bar, so the evaporator temperature is maintained at or near that temperature if the appliance is working correctly. In the next stage, an electric motor runs a small piston compressor and the refrigerant is pressurized. This raises the temperature of the refrigerant and the resulting superheated, high-pressure gas (it is still a gas at this point) is then condensed to a liquid in an air-cooled condenser. In most refrigerators and freezers, the compressor is in the base and the condenser coils are at the rear of the appliance. From the condenser, the liquid refrigerant flows through an expansion valve (almost always a capillary tube), in which its pressure and temperature are reduced and these conditions are maintained in the evaporator. The whole process operates continuously, by transferring heat from the evaporator section (inside the refrigerator) to the condenser section (outside the refrigerator), by pumping refrigerant continuously through the system described above. When the desired temperature is reached, the pump stops and so does heat transfer.

The refrigerator/freezers may be equipped with one or two compressors. In the case of one compressor, the operating cycle is both controlled by the air temperature in the refrigerating compartment and in freezer. In the case of two compressors, each operating cycle is independently controlled

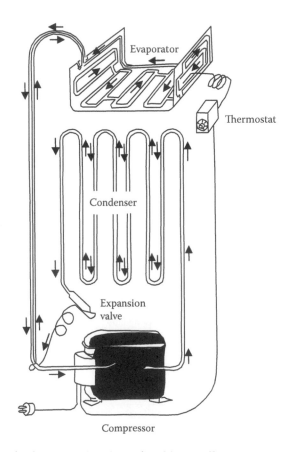

FIGURE 16.6   Cold production system in a domestic refrigerator/freezer.

by the air temperature in the refrigerating compartment and in the freezer. The temperature in each compartment is, therefore, better regulated but the price is higher.

16.4 TEMPERATURES AND HEAT TRANSFER MODES IN DOMESTIC REFRIGERATORS

The "on" and "off" operating cycle of the compressor leads to temperatures fluctuations in the refrigerator. The air temperature inside the appliance is regulated by a thermostat. When this temperature, measured at a given position, is higher than the maximum setting value, the compressor is "on" until the minimum setting value is reached, and then it is switched "off." The difference between maximum and minimum settings is fixed by the manufacturer.

16.4.1 TEMPERATURE IN REFRIGERATING COMPARTMENT

The temperatures in a refrigerating compartment of an empty one-door refrigerator equipped with one compressor, measured using calibrated T-type thermocouples are presented here. The internal dimensions of this compartment were 50×50 cm (width $\times$ depth) and the height was 90 cm. The vertical walls exchanged heat with the external ambience. The thickness of these walls was 4 cm and the overall thermal conductivity was 0.027 W/(m°C). The evaporator was fitted inside the vertical back wall and it was 50 cm wide and 30 cm high.

The temperature variations are shown in Figure 16.7 for a thermostat setting at 6°C and the average ambient temperature at 20°C. The "on" and "off" compressor work cycles lead to variations in the evaporator wall temperature (T_{evap}) (Figure 16.7a). It can be seen that the temperature varies within a range of $+ 7$°C to -12°C (average temperature -1.2°C).

The air temperature was measured at the top, middle, and bottom levels at the center of the refrigerating compartment (Figure 16.7b). It can be seen that the temperature is heterogeneous in the cavity: air at the bottom is cooler than that at the top. The mean air temperature calculated from 25 measurements is also shown in Figure 16.7b and the average value of this temperature over 24 h (T_{ai}) is 6.3°C (minimum value 3.8°C and maximum value 8.3°C).

The wall temperature variations are shown in Figure 16.7c, the mean value at the top level being 9.1°C, at the middle level 5.4°C and at the bottom level 5.7°C. The average value of these three temperatures (T_{wi}) is 6.7°C.

16.4.2 TEMPERATURE IN FREEZER

The air temperature fluctuations in a domestic freezer equipped with one compressor are generally more significant than those equipped with two compressors. An example of these fluctuations is presented in Figure 16.8. Both refrigerators are two-door models, with a refrigerating compartment on the top and a freezing compartment on the bottom. Air-temperature stratification can be observed in the compartment in both cases. The air on the top shelf was slightly higher than that on the middle one (cold air is heavier). The characteristics of these two refrigerators are presented in Table 16.1. The wall of the freezing compartment of these refrigerators was composed of an inner liner (1 mm of polystyrene, $\lambda = 0.15$ W.m^{-1}.K^{-1}), foam (polyurethane, $\lambda = 0.02$ W.m^{-1}.K^{-1}, 5.8 cm for the one compressor refrigerator and 6.3 cm for the two compressor refrigerator) and a metal outer sheet (0.7 mm, $\lambda = 50$ W.m^{-1}.K^{-1}). It was clearly shown, in this example, that the temperature was less stable in the freezing compartment of the one compressor refrigerator.

16.4.3 HEAT TRANSFER MODES IN DOMESTIC REFRIGERATOR

In an empty refrigerator, cold air near the evaporator flows downward and warm air near the door and the other side walls flows upward (Figure 16.9). The heat exchanges inside the cavity

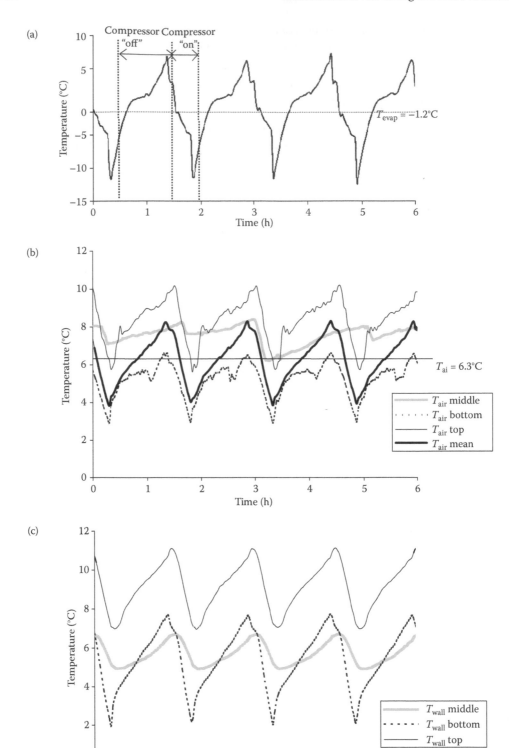

FIGURE 16.7 Example of temperature variations in a refrigerating compartment of one compressor appliance for a thermostat setting at 6°C: (a) evaporator wall, (b) air temperature, (c) wall temperature.

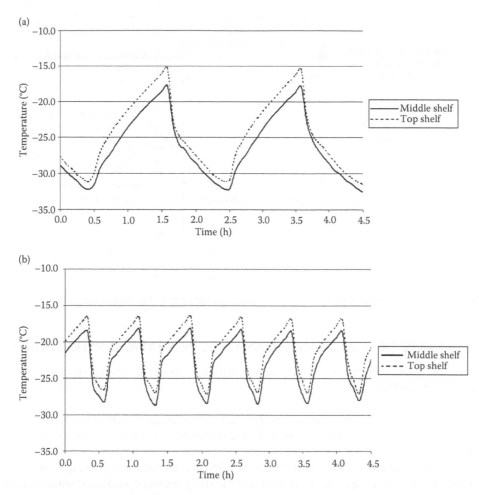

FIGURE 16.8 Example of air-temperature variations in the freezing compartment: (a) one compressor refrigerator, and (b) two compressor refrigerator.

TABLE 16.1
Characteristics of Refrigerators

	One-compressor Refrigerator	Two-compressor Refrigerator
External dimensions (height×width×depth)	185 cm×60 cm×60 cm	195 cm×60 cm×60 cm
Internal dimensions of the freezing compartment	62 cm×48 cm×38 cm	64 cm×47 cm×42 cm
Thermostat setting	+4°C (impossible to set the temperature of freezing compartment)	+4°C (refrigerating compartment) and −18°C (freezing compartment)
Power of compressor	120 W	160 W

are governed by natural convection between internal walls and air, radiation between evaporator and the other walls and conduction within the walls.[35] In the case of a refrigerator filled with products, the products are cooled by natural convection, radiation between the surface of the products and the internal walls of the refrigerator, and through conduction and radiation between products.

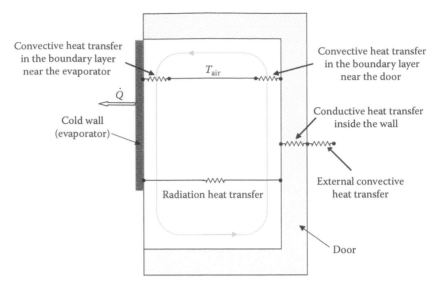

FIGURE 16.9   Various heat exchange modes and airflow inside a domestic refrigerator.

16.5   EXAMPLE OF HEAT TRANSFER ANALYSIS IN A TYPICAL REFRIGERATOR

In order to study heat transfer inside a refrigerator, natural convection theories covering the following cases can be applied:

- Rectangular closed cavity representing heat transfer inside the refrigerating compartment
- Cold vertical plate placed in a warm ambience representing exchanges between the evaporator and air
- Rectangular closed cavity partially filled with porous media representing a loaded refrigerator. This case is more complex and the study requires numerical simulations as presented in Section 16.6.

16.5.1   EMPTY REFRIGERATOR CONSIDERED AS BEING A CLOSED RECTANGULAR CAVITY

The simplest approach that can be used in order to approximate the transfers inside an empty refrigerator is to consider it as a rectangular cavity.[35] Circular air circulation is established, cool air close to the evaporator moves downward and hot air in contact with the door moves upward.

In order to simplify the study, the exchanges inside the cavity are initially, considered as a two dimensional problem of heat transfer between a cold vertical wall (T_{evap}) and a hot vertical wall (T_{wi}). The vertical walls are assumed to have a homogeneous temperature and the horizontal walls are adiabatic. This is a rough approximation because, in fact, only part of the vertical wall of a refrigerator is taken up by the evaporator and heat losses occur through at least three vertical walls.

Despite this, the order of magnitude of the Rayleigh number (Ra) can give some qualitative information regarding the hydrodynamic and thermal boundary layers. The refrigerator described in Section 16.4.1 will be used for the analysis. The aspect ratio (H/L) of this typical refrigerator is equal to 1.8. The Rayleigh number can be based on height or width, but since $H/L = 1.8$, the values are of the same order of magnitude. For simplicity's sake, the walls and air temperatures are assumed to be constant. The difference between the inner wall temperature and the evaporator temperature ($\Delta T = T_{wi} - T_{evap}$) is equal to 7.9°C.

The physical properties of air, such as conductivity (λ), thermal expansion coefficient (β), diffusivity (α) and kinetic viscosity (ν), are calculated at the reference temperature (T_f) defined as the average temperature of the evaporator and the other walls ($T_f = 275.8$ K). The Prandtl number for air is taken constant and equal to 0.72.

The Rayleigh number based on the width of the refrigerating compartment ($L = 0.5$ m) and the temperature difference between the evaporator and the other walls (ΔT) is equal to 1.43×10^8.

Thus, the air flow inside the refrigerator is laminar (Ra $< 10^9$). This result is in agreement with that of Deschamps et al.[30] who showed that Ra varies between 10^8 and 10^9 in refrigerators.

As the Rayleigh number is higher than 10^3, a stationary core region can be expected, indicating very low air velocities in the area where food is stored.

For this range of Rayleigh number and aspect ratio, a vertical thermal gradient of approximately $(T_{wi} - T_{evap})/2$ is expected,[44] that is 4°C in our case.

These expectations are qualitatively confirmed by wall temperature measurement inside the refrigerator (Figure 16.7c). In particular, a wall temperature difference between the top and bottom levels of 3.4°C was observed.

In order to estimate the heat transfer coefficient in our refrigerator, the correlation proposed by Catton,[26] valid for $1 < H/L < 2$, $10^{-3} < Pr < 10^5$ and $10^3 < (Ra_L Pr)/(0.2 + Pr)$ is used:

$$\text{Nu}_L = \frac{h_{gl} \cdot L}{\lambda} = 0.18 \left(\frac{Pr}{0.2 + Pr} Ra_L \right)^{0.29} = 38.9 \tag{16.4}$$

Thus, the overall heat transfer coefficient (h_{gl}) between the hot and cold walls and the refrigerating capacity ($\dot{Q}$) can be estimated as:

$$h_{gl} = 1.94 \ W/(m^2 \cdot °C)$$

$$\dot{Q} = h_{gl} \cdot A_w \cdot (T_{wi} - T_{evap}) = 25.3 \ W$$

The experimental refrigerating capacity was also measured using a fluxmeter (measuring dimension 4×4 cm) attached to the surface of the evaporator. The average value of measurements taken every 2 min over 24 h is 10.4 W, which is significantly lower than the calculated value.

Furthermore, with the 2D approximation and due to the symmetry of the cavity, the average air temperature inside the refrigerator is estimated as $T_{ai} = (T_{wi} + T_{evap})/2 = 2.8$°C while the experimental average air temperature is 6.3°C. The differences are essentially due to the simplifying hypothesis of 2D and steady state heat transfer in the cavity.

This approximate calculation of heat transfer in an air-filled cavity applied to an empty refrigerator gives quantitative results quite different from those observed experimentally. Thus, another approach is proposed using correlations available in the literature results for natural convection between vertical plates and air while taking also into account the effect of radiation.

16.5.2 Empty Refrigerator Considered as a Combination of Vertical Plates

Heat transfer in the refrigerator is examined on the basis of the theory of natural convection between air and vertical plates which are the evaporator and the side walls in our case.[35]

The application of these correlations is now presented for heat transfer between the evaporator or side walls and air inside the refrigerator.

16.5.2.1 Estimation of the Convective Heat Transfer Coefficient by Natural Convection between the Evaporator and Air (h_{evap})

In order to estimate the heat transfer coefficient, which depends on the Rayleigh number, the measured values of the evaporator and the air temperatures inside the refrigerator are used: −1.2°C and 6.3°C, respectively.

The Rayleigh number based on the height of evaporator (H_{evap} = 0.3 m) and the difference between the temperature of the evaporator and that of air is: $Ra_{evap} = 2.0 \times 10^7$.

Since Ra < 10^9, the correlation proposed by Incropera and Dewitt[18] for laminar flow is used:

$$Nu = 0.59 \cdot Ra^{1/4} = 39.4, \quad thus \quad h_{evap} = 3.28 \text{ W/(m}^2.°C).$$

16.5.2.2 Estimation of the Radiative Heat Transfer Coefficient between the Evaporator and the other Walls (h_r)

Heat transfer by radiation occurs between the evaporator and the other walls. An equivalent radiative heat transfer coefficient h_r can be defined. For parallel walls of emissivity near 1:

$$h_r = \sigma \cdot \varepsilon_1 \cdot \varepsilon_2 \cdot (T_{evap}^2 + T_{wi}^2)(T_{evap} + T_{wi}) \qquad (16.5)$$

$$\sigma = \text{Boltzmann constant} = 5.67 \times 10^{-8} \vdots \text{W/(m}^2.K^4)$$

In the case of the refrigerator, the emissivity of the internal surfaces is approximately $\varepsilon_1 = \varepsilon_2 = 0.9$ and we obtain h_r = 3.85W/(m^2.K).

One can note that the value of the radiative heat transfer coefficient is of the same order of magnitude as that of natural convection. Therefore, it should be taken into account when investigating heat transfer in refrigerators.

16.5.2.3 Estimation of the Convective Heat Transfer Coefficient between Air and the Internal Walls of the Refrigerator (h_{wi})

The same approach is used for the side walls as for the evaporator. The Rayleigh number, based on the temperature difference between that of inner side wall and inner air and on the height of the walls is 4.2×10^7 (Ra < 10^9).

The correlation proposed by Incropera and Dewitt[18] for laminar flow is used:

$$Nu_w = 0.59 \cdot Ra^{1/4} = 47.4, \quad thus \quad h_{wi} = 1.3 \text{ W/(m}^2.°C)$$

16.5.2.4 Estimation of the Overall Heat Transfer Coefficient and Refrigerating Capacity

The thermal resistances of the refrigerator are represented in Figure 16.10.

While considering that the surface of the evaporator, A_{evap} is $0.5 \times 0.30 = 0.15$ m^2 and the total vertical wall surface, A_w is $4 \times (0.5 \times 0.9) - 0.15 = 1.65$ m^2, the values of the heat transfer resistances are:

Heat resistance by natural convection between internal air and the evaporator, $R_{evap} = 1/(h_{evap} \cdot A_{evap}) = 2.032°C/W$.

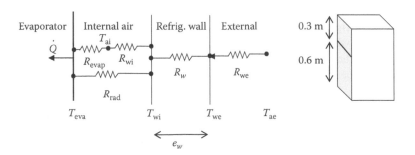

FIGURE 16.10 Thermal resistance between the evaporator and the external ambience.

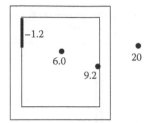

FIGURE 16.11 Estimated internal temperatures (°C) of air and walls of the refrigerator (evaporator and ambient temperatures were determined experimentally).

Heat resistance by natural convection between internal air and the walls, $R_{wi} = 1/(h_{wi} \cdot A_w) = 0.466°C/W$.

Heat resistance by radiation between evaporator and the walls, $R_{rad} = 1/(h_r \cdot A_{evap}) = 1.732°C/W$.

Heat resistance by conduction in the walls, $R_w = e_w/(A_w \cdot \lambda_w) = 0.90°C/W$.

The thermal resistance between the external walls and the ambient air (R_{we}) is assumed to be constant:

$$R_{we} = 1/(h_{ext} \cdot A_w) = 0.060°C/W \ (h_{ext} = 10 \ W/m^2/°C).$$

By using the value of the different heat transfer resistances cited previously, it was found that the overall heat transfer resistance between the evaporator and the external ambience is $R_{g1} = 1.98°C/W$ and the refrigerating capacity, $Q = (T_{ae} - T_{evap})/R_{g1} = 10.7/W$. This value is slightly higher than the measured value (10.4 W).

The different temperatures can also be evaluated as shown in Figure 16.11.

The predicted internal air temperature is close to measured values, while the predicted wall temperature is higher. The estimations are much more accurate than those of the first approach (2D cavity), but the heat transfer coefficient between air and the walls appears to be underestimated.

16.5.2.5 Estimation of the Thickness of the Thermal Boundary Layer Near the Evaporator

According to the Rayleigh number (Ra_{evap}) $< 10^9$, one can consider that air flow in the thermal boundary layer is laminar. The mean thickness of the equivalent thermal boundary layer can be estimated as follows:

$$\delta_{T,eq} = \frac{\lambda}{h_{evap}} = \frac{0.025}{3.3} = 7.6 \times 10^{-3} m \sim 8 \ mm$$

According to Figure 16.12, it can be observed that there is a zone, inside this boundary layer, where the temperature is below 0°C. The thickness of this zone is about 2 mm. This result makes it possible to estimate the minimum distance from the evaporator where foods should be placed in order to avoid the freezing.

It should be borne in mind that this calculation is only an estimation of the order of magnitude since three important assumptions are used:

- The temperature of the evaporator is constant at −1.2°C, while in reality, it varies within a range of + 7 to −12°C.
- The temperature of air in the refrigerating compartment is constant and homogeneous at 6.3°C, while in reality, it varies within a range of 3.8–8.3°C.
- The evaporator is equivalent to a fine plate, while in reality it is embedded in a wall.

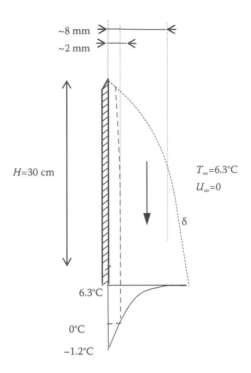

FIGURE 16.12 Thermal boundary layer and temperature profile near a vertical plate representing the evaporator.

It was observed by experiment and by numerical simulation that the heat transfer and airflow in domestic refrigerator are 3D.[45,46] To study this more complex 3D configuration, numerical simulation is presented in the next section.

16.6 NUMERICAL SIMULATION IN DOMESTIC REFRIGERATOR

The simple mathematical approach shown in Section 16.5 has limitations. For more thorough study, CFD simulation was carried out within the refrigerating compartment of a domestic refrigerator without a fan.[46]

16.6.1 REFRIGERATOR CHARACTERISTICS

A single-door appliance with only a refrigerating compartment (without a freezer) was considered. Its general characteristics are shown in Table 16.2.

Three cases were studied (Figure 16.13): an empty refrigerator without shelves, empty refrigerator fitted with glass shelves (5 mm thickness, thermal conductivity of glass 0.75 W m^{-1} K^{-1}) and a refrigerator equipped with glass shelves and loaded with a "test product." This product is made of aqueous methylcellulose gel (thermal conductivity 0.5 W m^{-1} K^{-1}) and the dimensions of one package are 10×10×5 cm (length×width×depth). The arrangement of the packages is shown in Figure 16.13c. All experiments were carried out in a temperature-controlled room (20±0.2°C). As shown in Figure 16.13, the evaporator is located in the upper part of the cabinet. The indentation observed in the lower right area of the figures represents the compressor placement. To avoid a too complex geometry, the containers for butter, eggs, and bottles usually attached to the door were removed during our experiments. This facilitates the meshing of the refrigerator and the result interpretation.

TABLE 16.2
Characteristics of the Refrigerator used for Numerical Simulation

External dimensions (height × width × depth)	149 cm × 60 cm × 59 cm
Internal dimensions (height × width × depth)	136 cm × 52 cm × 44 cm
Dimensions of the evaporator	90 cm × 48 cm
Thermostat setting	+5 °C
Number of shelves	4

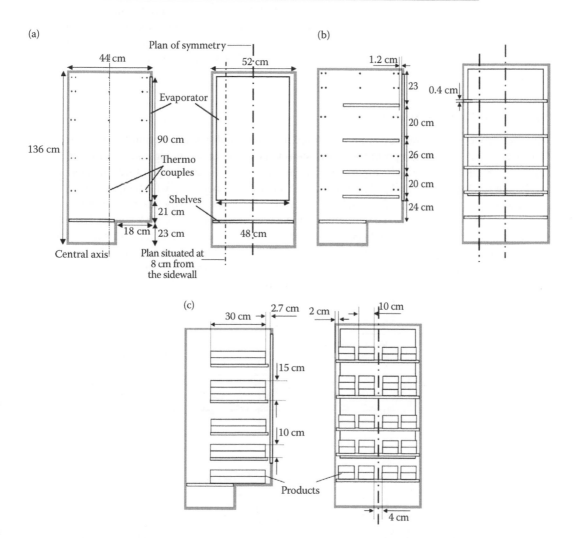

FIGURE 16.13 Domestic refrigerator geometry: (a) empty refrigerator, (b) refrigerator fitted with glass shelves, (c) refrigerator with glass shelves and products.

16.6.2 MEASUREMENT OF THE THERMAL RESISTANCE OF REFRIGERATOR INSULATION

Measurement of the thermal resistance of refrigerator insulation was carried out in a temperature-controlled room (6°C). A heating coil was placed inside the "switch off" refrigerator. The heat supplied to the coil is equal to the heat loss to external air through the walls. The heating power was adjusted in such a manner as to maintain the average internal air temperature at 30°C. In this manner, the average temperature of the insulating walls is almost the same as under real operating conditions. To ensure a homogeneous air temperature inside the refrigerator, a small fan was installed near the heating coil. The internal air temperature (T_{int} controlled at 30°C), external air temperature (T_{ext} controlled at 6°C), power supplied to the heating coil (Q_1) and fan (Q_2) were recorded when the steady state was attained (after 12 h) and the average values were calculated over 3 h. Thus, the thermal resistance of the refrigerator insulation can be calculated knowing $Q_1 + Q_2$ and $T_{int} - T_{ext}$.

The measurement was used afterward for the boundary conditions in the CFD simulation. In fact, this experimental thermal resistance takes into account the thermal resistance between external air and internal walls. Therefore, a correction was undertaken on the measured value by subtracting the thermal resistance between internal air and walls. This correction is weak because the thermal resistance between air and internal wall represents only around 7% of the overall thermal resistance (between external and internal air). In our case, the internal convective heat transfer coefficient was assumed to be about 10 W m^{-2} K^{-1}.

16.6.3 Temperature Measurement

Air and product temperatures were measured experimentally using calibrated thermocouples (T-type) placed in different positions of the symmetry plane of the refrigerator and on the plane situated at 8 cm from side wall (Figure 16.13). On each plane, the air temperature was measured at five height levels (31.0, 61.0, 94.0, 114.5, 134.5 cm) and for each height, five air temperature measurements were recorded (1, 2, 21.5, 42, 43 cm from the evaporator). Firstly, the refrigerator operated over 24 h to ensure stabilization conditions, then the temperatures were recorded every 2 min for 24 h and the average value was calculated at each measurement point. An example of temperature evolution inside the refrigerator is shown in Figure 16.14. It can be seen that the evaporator temperature varies from −16°C to + 7°C, due to the thermal inertia, the air temperature varies less, from + 3.5°C to + 7°C, and the wall temperature varies from 4°C to 9°C.

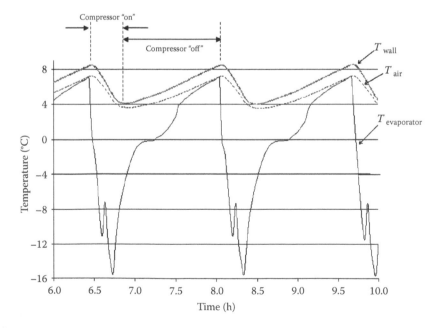

FIGURE 16.14 Air (average value on the symmetry plan), side wall (average value of three measurements: top, middle and bottom levels) and evaporator temperature changes in the empty refrigerator without shelves (thermostat setting at 5°C).

16.6.4 MODELING

16.6.4.1 Main Assumptions and Boundary Conditions

In the present study, the Rayleigh number (Ra) is about 6×10^8 (estimation based on the height of the evaporator and the temperature difference between the internal air and the cold-wall surface). Laminar flow assumption was made for the flow regime in our simulation since $Ra < 10^9$. Furthermore, several numerical studies showed that turbulence does not change the predicted air temperature pattern.[30,47] Boussinesq approximation was used since the air temperature variation is small compared with the mean absolute value.

The thermal boundary conditions are based on experimental data:

- Uniform global heat transfer coefficient between external air and internal wall (0.34 W $m^{-2} K^{-1}$).
- Constant external air temperature (20°C).
- Constant evaporator temperature (−0.5°C) which is the average value during "on" and "off" running cycles of compressor. This constant temperature is used in order to avoid excessive complexity in the calculation and to reduce calculation time.

The simulations were performed with the finite volume method using CFD software Fluent 6.1 with the resolution parameters indicated in Table 16.3.

Transient simulation was performed but only the results obtained after simulation convergence were used for the comparison with the experimental values.

16.6.4.2 Mesh

Structured mesh was used to describe the geometry of the refrigerator. Finer meshes were used near walls, shelves and products. The number of cells used in each case is shown in Table 16.4 and mesh structures are shown in Figure 16.15. To ensure that the results were not influenced by the cell numbers, a sensitivity study was carried out beforehand. Only one half of the refrigerator was meshed because of the symmetry plane.

16.6.4.3 Discrete Ordinate (DO) Method for Radiation

The discrete ordinate (DO) method[48] was successfully used to simulate the coupling of convection and radiation in closed cavity.[49,50]

This model can take into account the participating medium. However, in our case, air is considered as transparent (with neither absorption nor diffusion). The general equation of heat transfer by radiation (in a given $\vec{s}$ direction) is

$$\vec{\nabla} \cdot (I(\vec{r},\vec{s})\vec{s}) = 0 \tag{16.6}$$

$I(\vec{r},\vec{s})$ is radiative intensity in $\vec{s}$ direction (at $\vec{r}$ position) (W m^{-2} per unit solid angle).

For a gray surface of emissivity ε_r, the net radiative flux leaving the surface is

$$\Phi_{rad\_out} = (1-\varepsilon_r) \underbrace{\int_{\vec{s} \cdot \vec{n} > 0} I_{in}\vec{s} \cdot \vec{n} d\Omega}_{\text{incident flux}} + \varepsilon_r \sigma T_s^4 \tag{16.7}$$

The walls are assumed as gray diffuse: $I_{out} = \phi_{rad\_out}/\pi$

I_{in}: intensity of incident radiation in $\vec{s}$ direction (at $\vec{r}$ position)

I_{out}: intensity of radiation leaving the surface (at $\vec{r}$ position)

$\vec{n}$: normal vector

T_s: surface temperature, K

Ω: solid angle

TABLE 16.3
Resolution Parameters used in Simulation

	Relaxation Factor	Type of Discretization
Pressure	0.8	Presto
Density	1	–
Gravity forces	1	–
Momentum	0.2	Second order upwind
Energy	1	Second order upwind
Radiation	1	–
Pressure–velocity	–	Simple

TABLE 16.4
Number of Cells used for the Simulations

Mesh Number	Height (136 cm)	Half width (26 cm)	Depth (44 cm)	Total
Empty refrigerator	138	28	66	255,024
Refrigerator with shelves	222	28	66	410,256
Refrigerator with shelves and products	240	62	74	1,101,120

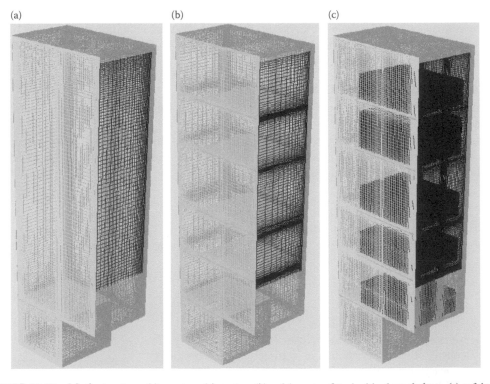

FIGURE 16.15 Mesh structure: (a) empty refrigerator, (b) refrigerator fitted with glass shelves, (c) refrigerator loaded with the "test product."

A sensitivity study of solid angle discretization was carried out beforehand in order to ensure that the simulation results were not influenced by the number of solid angle subdivisions.

16.6.5 Numerical Results (Taking into Account Radiation)

The results presented in this paragraph concern simulation, which takes into account heat transfer by convection between walls and air and by radiation between the internal walls of the refrigerating compartment. Two results will be presented: air temperature and velocity fields. A comparison between numerical and experimental results is carried out only for temperature. In fact, air velocity in a real refrigerator is difficult to measure with precision in practice.

16.6.5.1 Temperature Fields

The temperature fields obtained from simulations for the different cases studied are shown in Figure 16.16. Considering only the main cavity (excluding the vegetable box), for all cases, thermal stratification is observed with the cold zone at the bottom (~2°C) of the refrigerating compartment and the warm zone at the top (8–9°C). In addition, a cold zone is also observed along the back wall. This is related to cold air coming from the evaporator. When the refrigerator is loaded with products, the temperature of the product located near the evaporator is lower than that located near the door. In the top half of the compartment, the temperature is relatively homogeneous at a given height (except in the boundary layers near the walls). The temperature of the vegetable box is almost constant for all cases studied (~8°C).

The temperature field is slightly influenced by the presence of obstacles: shelves and products. A slightly lower temperature is observed at the bottom and a slightly higher one at the top compared with the empty refrigerator case. This is due to the fact that the shelves and/or the products slowed down the air circulation in the central zone of the refrigerator. The presence of shelves and/or products also influenced the main air circulation in the boundary layers situated along the evaporator and the side walls. However, this influence is weak because of the presence of air spaces between the shelves and the vertical walls (1.2 cm between the back wall and the shelves), which facilitates the air flow. In our previous study, it was found that the thickness of the boundary layer was less than 2 cm.[45]

In addition to the overall thermal stratification in the cavity, stratification is also observed in each gap between two shelves or between a shelf and a product. It is to be emphasized that for the refrigerator loaded with the "test product," the symmetry plane is located in the gap between two piles. This explains why the packages are invisible on this plane (Figure 16.16c). On the plane situated at 8 cm from a side wall which cuts the product pile (Figure 16.16d), a cold product zone near the evaporator can be clearly distinguished. This is related to the blockage of cold air by the product.

The average and maximum air temperatures in all cases are reported in Table 16.5. The air temperatures increase with increasing numbers of obstacles.

16.6.5.2 Air Velocity Field

Figure 16.17 presents the air velocity fields on the symmetry plane (Figure 16.17a–c) and on the plane situated at 8 cm from the side wall (Figure 16.17d) for the different cases studied. Considering only the main cavity (excluding the vegetable box), for all cases, the main air circulation is observed near the walls, and constitutes a recirculation loop. Air flows downward along the evaporator while its velocity increases along the course to attain a maximum value at the bottom of the refrigerator ($u_{max} \approx 0.2$ m s^{-1}). Air then flows upward along the door and the side walls of the refrigerator while its velocity decreases progressively and becomes stagnant at the top of the refrigerator. This observation is in agreement with the air temperature field shown in Figure 16.16, with cold air located at the bottom of the cavity and warm air at the top. It can also be observed that there is a weak horizontal

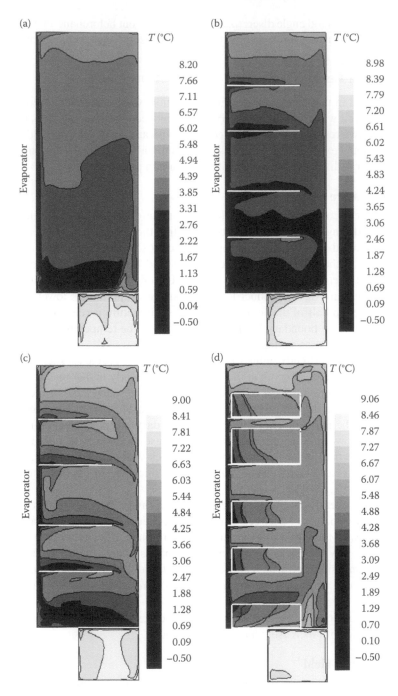

FIGURE 16.16 Predicted temperature fields (°C): (a) on the symmetry plan of empty refrigerator, (b) on the symmetry plan of refrigerator with glass shelves, (c) on the symmetry plan of refrigerator loaded with products, (d) on the plan situated at 8 cm from the side wall of refrigerator loaded with products.

air flow from the door to the evaporator. However, the air velocity at the center of the cavity is very low (< 0.04 m s⁻¹). In the case of the refrigerator fitted with glass shelves, in addition to the main air flow along the walls as mentioned previously, there are also small air loops between the shelves. For the refrigerator loaded with products, air flows in the gaps between the shelves and the products (Figure 16.17d).

TABLE 16.5
Average and Maximum Air Temperatures for the Three Simulations

	Average Temperature in the Main Cavity (°C)	Maximum Temperature in the Main Cavity (°C)	Average Temperature in the Vegetable Box (°C)
Empty refrigerator	3.8	8.2	7.4
Refrigerator with glass shelves	4.0	9.0	8.2
Refrigerator with glass shelves and products	5.1	9.1	8.0

It should be remembered that the containers attached to the door were not represented in our study. In practice these containers are an obstacle to airflow along the door and tend to reduce the air velocity in this area.

Considering the vegetable box, one or two air recirculation loops were observed (Figure 16.17). This is due to the presence of the glass shelf (cold wall), which separates the vegetable box from the main cavity, and the five other walls which are warmer (heat loss through these walls).

16.6.5.3 Comparison with Numerical Simulation without Radiation

Figure 16.18 presents the air temperature field on the symmetry plane obtained by simulation without taking into consideration radiation (between internal walls of the refrigerating compartment, shelves and product surface). It was observed that overall the temperature field is similar to that present when radiation is taken into account (a cold zone at the bottom and a warm zone at the top). However, stratification is more pronounced without radiation, and this leads to a higher temperature at the top of the cavity. In fact, for an empty refrigerator, the maximum temperature rises from 8°C (with radiation) to 15°C (without radiation). This temperature increase can be explained by the fact that, without radiation, there is no heat exchange between the warm top wall and the other colder walls, particularly the evaporator wall. This contributes to a high air temperature at the top position. When radiation is taken into account, the heat exchange between the top wall and the other walls tends to reduce the top wall temperature and consequently reduces air temperature near this wall. From a microbiological point of view, the growth rate is much higher at 15°C than at 8°C. It is therefore necessary to take into consideration radiation in the simulation in order to better describe the phenomena occurring in domestic refrigerators.

16.6.5.4 Comparison between the Measured and Predicted Air Temperature

Figure 16.19 presents a comparison between the measured and predicted air temperature results (with and without taking into account radiation). It can be seen that the simulation results with radiation agreed with the experimental values to a greater extent, while simulation without radiation overestimated the air temperature, particularly at the top of the refrigerator. The peaks observed on the temperature profile in the presence of shelves and/or products can be explained by the higher conductivity of glass compared with air and by the cold air flow along the upper sides of the shelves.

The agreement between the experimental and simulation results is relatively poor in the case of a loaded refrigerator, even though the radiative heat exchange between the product and the walls was taken into account. This may be explained by the geometry complexity. Further grid refinement could lead to a better agreement, but the computing time is already very excessive (about 8 days using a cluster of four processors of 2 Gigaoctets (Go) of RAM).

16.7 CONCLUSIONS

Several studies illustrate that a large proportion of refrigerators operate at too high temperatures. These refrigerators are often static types (without fan) in which heat is transferred by natural

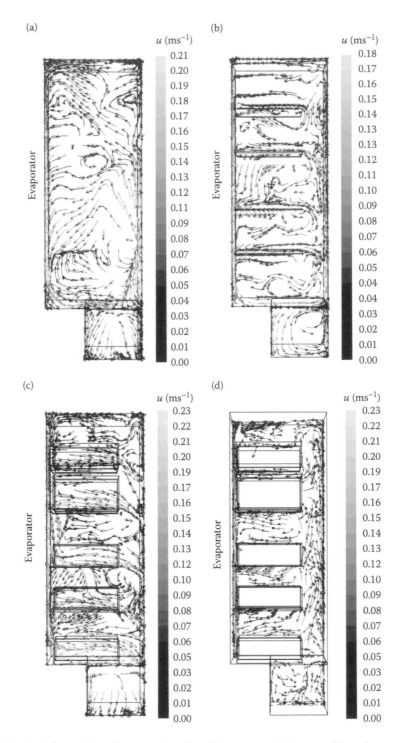

FIGURE 16.17 Path lines: (a) on the symmetry plan of the empty refrigerator, (b) on the symmetry plan of the refrigerator fitted with glass shelves, (c) on the symmetry plan of the refrigerator loaded with products, (d) on the plan situated at 8 cm from the side wall of refrigerator loaded with products.

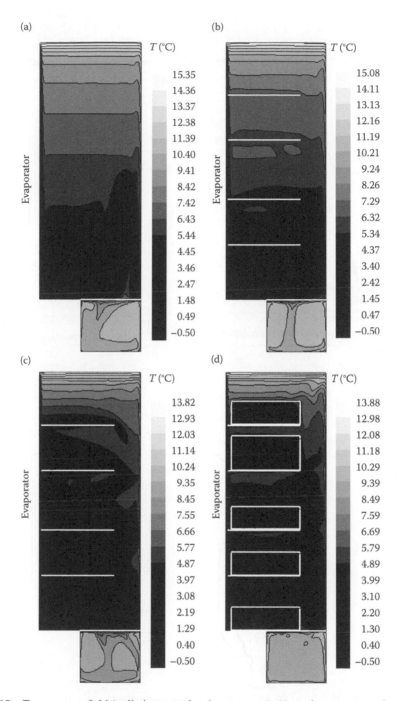

FIGURE 16.18 Temperature field (radiation not taken into account): (a) on the symmetry plan of the empty refrigerator, (b) on the symmetry plan of the refrigerator fitted with glass shelves, (c) on the symmetry plan of the refrigerator loaded with the "test product," (d) on the plan situated at 8 cm from the side wall of the refrigerator loaded with products.

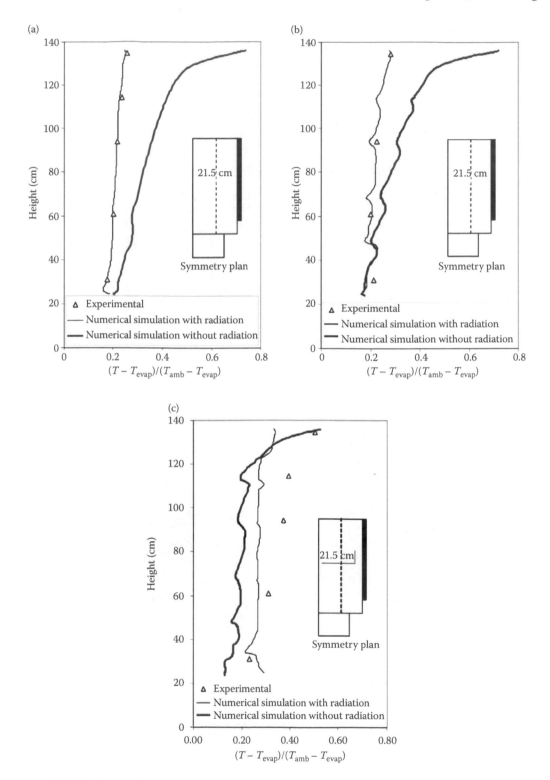

FIGURE 16.19 Comparison between experimental air temperatures and predicted values obtained by simulation with and without radiation: (a) empty refrigerator, (b) refrigerator fitted with glass shelves, (c) refrigerator loaded with products.

convection. From a food quality and safety point of view, it is necessary to fully understand the mechanism of heat transfer and airflow in refrigerators.

A literature review on natural convection near a vertical plate, in empty and filled cavity was presented in this chapter. Cold production systems and temperature evolutions in domestic refrigerators were shown. Several heat transfer modes which may occur in refrigerators are presented: convection, conduction and radiation. It is to be emphasized that radiation may be the same order of magnitude as natural convection.

The heat transfer by natural convection in a typical domestic refrigerator was analyzed using results recorded in the literature concerning:

- Transfer in a rectangular empty cavity
- Transfer between vertical plates and air

In spite of a simplified hypothesis, the simple mathematical approach provided some useful information on the refrigerator:

- The model, which includes natural convection between the evaporator and air and between the walls and air, radiation between the walls, and conduction inside the walls, gives a good approximation of the required refrigerating capacity and air temperature. This model slightly overestimates the wall temperatures.
- Inside the boundary layer near the evaporator, there is a zone a few millimeters wide in which the temperature is below 0°C. Therefore, placing food in this zone can lead to freezing.
- Outside the boundary layers, the air is practically stagnant. This means that air circulation is induced near the evaporator and the side walls. However, in the core region, where food is stored, there are low velocities, which do not ensure marked convective heat transfer between air and products.
- The temperature difference between the top and bottom levels of refrigerator can be estimated as being a half of the temperature gradient between the lateral walls and the evaporator. In this type of refrigerator, sensitive food should not be stored on the top shelf.

CFD numerical simulation seems to be a powerful tool to study the complex 3D heat transfer and airflow in refrigerator. Three configurations were studied: an empty refrigerator, an empty refrigerator fitted with glass shelves and a refrigerator loaded with products. When radiation was taken into consideration in simulation, the predicted air temperatures were in good agreement with the experimental values. However, when radiation was not taken into account, the temperature was over-estimated, particularly at the top of the refrigerator. Radiation allows heat exchange, particularly between the top wall and the cold wall (evaporator); consequently, it limits the stratification phenomena.

The obstacles (shelves and/or products) slow down the air circulation in the central zone of the refrigerator and mildly influence the main air circulation along the walls. This is confirmed by the maximum values of air temperature: 8.2°C for an empty refrigerator without shelves and 9.1°C for a refrigerator loaded with products.

Whatever the configuration studied (empty with/without shelves, loaded with products) for this type of refrigerator, the air temperature at the top of the refrigerator is about 5°C higher than the average air temperature, and therefore it is important to avoid placing sensitive products in this position.

The CFD simulation developed by our work can be further used as a tool to study the influence on the temperature and velocity fields of operating conditions: evaporator temperature (parameter related to the thermostat setting by the consumer), dimensions of the evaporator (parameter related to design) and percentage of product-occupied volume in the refrigerating compartment.

ACKNOWLEDGMENT

The author would like to thank Professor Denis Flick (AgroParisTech, 16 rue Claude Bernard, 75231 Paris Cedex 05, France) for the review of this chapter.

NOMENCLATURE

Cp	Thermal capacity (J/kg/°C)
h	Heat transfer coefficient (W/m²/°C)
H	Height (m)
I	Intensity of radiation (W/m²) per unit solid angle
L	Width or characteristic length (m)
N_{RC}	Radiation-convection interaction parameter
g	Acceleration due to gravity (9.81 m/s²)
r	Radius (m)
R	Thermal resistance (°C/W) or Radius (m)
T	Temperature (°C or K)
ΔT	Temperature difference between cold and warm walls, °C or K
$\dot{Q}$	Refrigerating power (W)
u	Air velocity in flow direction (m/s)

Greek symbol

α	Thermal diffusivity (m²/s)
β	Thermal expansion coefficient (K⁻¹)
λ	Thermal conductivity (W/m/°C)
ν	Kinetic viscosity (m²/s)
δ	Boundary layer thickness (m)
ρ	Density (kg/m³)
ε	Emissivity of the wall
Ω	Solid angle
Φ	Radiative flux (W/m²)

Subscript

a	Air
∞	Far from walls or bulk
ai	Air inside the refrigerator
ae	Air outside refrigerator
ext	External
evap	Evaporator
f	Film
gl	Overall
p	Product
r	Radiation
s	Surface
T	Thermal
w	wall
wi	Internal wall
we	External wall

wc   Cold wall

wh   Hot wall

Dimensionless number

Pr   Prandtl number $= \dfrac{\upsilon}{\alpha}$

Nu   Nusselt number $= \dfrac{hL}{\lambda}$

Ra   Rayleigh number $= \dfrac{g\beta\Delta TL^3}{\alpha\upsilon}$

REFERENCES

1. International Institute of Refrigeration (IIR). *Report on Refrigeration Sector Achievements and Challenges*. International Institute of Refrigeration, Paris, France, 77, 2002.
2. JARN, Japan Air Conditioning, Heating & Refrigeration, News, Tokyo, Japan, 157, 41, 2005.
3. Billiard, F. Refrigerating equipment, energy efficiency and refrigerants. *Bull. Int. Inst. Refrig.*, 85, 12, 2005.
4. Association Française du Froid (AFF). *Livre blanc sur les fluides frigorigènes*. Conseil National du Froid, Paris, France, 51, 2001.
5. Redmond, E.C. and Griffith, C. Consumer food handling in the home: a review of food safety studies. *J. Food Protection*, 66, 130, 2003.
6. Flynn, O.M., Blair, I. and McDowell, D. The efficiency and consumer operation of domestic refrigerators. *Int. J. Refrig.*, 15 (5), 307, (1992)
7. James, S.J. and Evans J. The temperature performance of domestic refrigerators. *Int. J. Refrig.*, 15, 313, 1992.
8. Lezenne Coulander de P.A. Koelkast temperature thuis. Report of the regional Inspectorate for Health Protection, Leeuwarden, The Netherlands, 1994.
9. O'Brien, G.D. Domestic refrigerator air temperatures and the public's awareness of refrigerator use. *Int. J. Envi. Health Res.*, 7, 141, 1997.
10. Sergelidis, D. et al. Temperature distribution and prevalence of *Listeria spp.* in domestic, retail and industrial refrigerators in Greece. *Int. J. Food Microbiol.*, 34, 171, 1997.
11. Laguerre, O., Derens, E. and Palagos B. Study of domestic refrigerator temperature and analysis of factors affecting temperature: a French survey. *Int. J. Refrig.*, 25, 653, 2002.
12. Jackson, V. et al. The incidence of significant food borne pathogens in domestic refrigerators. *Food Control.*, 18, 346, 2007.
13. Azevedo, I. et al. Incidence of *Listeria spp.* in domestic refrigerators in Portugal. *Food Control*, 16, 121, 2005.
14. Olsson, P. and Bengstsson, N. Time temperature conditions in the freezer chain. Report No.309, SIK-The Swedish Food Institute, Gothenberg, Sweden, 1972.
15. Masjuki, H.H. et al. The applicability of ISO household refrigerator-freezer energy test specifications in Malysia. *Energy*, 26, 723, 2001.
16. Lacerda, V.T. et al. Measurements of the air flow field in the freezer compartment of a top-mount no-frost refrigerator: the effect of temperature. *Int. J. Refrig.*, 28, 774, 2005.
17. Lee, I.S. et al. A study of air flow characteristics in the refrigerator using PIV and computational simulation. *J. Flow Visual. Image Proc.*, 6, 333, 1999.
18. Incropera, F.P. and Dewitt D.P. *Fundamentals of Heat and Mass Transfer*, 4th Edition. John Wiley and Sons, New Jersey, USA, 1996.
19. Khalifa, A.J.N. Natural convective heat transfer coefficient - a review I. Isolated vertical and horizontal surfaces. *Energy Conv. Manag.*, 42, 491, 2001.
20. Tian, Y.S. and Karayiannis T.G. Low turbulence natural convection in an air filled square cavity : part I: the thermal and fluid flow fields. *Int. J. Heat Mass Transfer*, 43, 849, 2000.
21. Ampofo, F. and Karayiannis, T.G. Experimental benchmark data for turbulent natural convection in an air filled square cavity. *Int. J. Heat Mass Transfer*, 46, 3551, 2003.

22. Betts, P.L. and Bokhari I.H. Experiments on turbulent natural convection in a closed tall cavity. *Int. J. Heat Mass Transfer*, 21, 675, 2000.
23. Mergui, S. and Penot, F. Convection naturelle en cavité carrée différentiellement chauffée: investigation expérimentale à Ra = 1.69×109. *Int. J. Heat Mass Transfer*, 39, 563, 1996.
24. Armaly, B.F., Li, A. and Nie, J.H. Measurements in three-dimensional laminar separated flow. *Int. J. Heat Mass Transfer*, 46, 3573, 2003.
25. Ostrach, S. Natural convection in enclosures. *J. Heat Transfer*, 110, 1175, 1988.
26. Catton, I. Natural convection in enclosures. *Proceedings 6th International. Heat Transfer Conference*, Toronto, Canada, 6, 13, 1978.
27. Yang, K.T. Natural convection in enclosures. In: *Handbook of Single-Phase Heat Transfer*. Wiley, New York. 1987.
28. Henkes, R.A. and Hoogendoorn, W.M. *Turbulent Natural Convection in Enclosures-A Computational and Experimental Benchmark Study*. Editions Europeenes Thermiques et Industries, Paris, France, 1993.
29. Ramesh, N. and Venkateshan S.P. Effect of surface radiation on natural convection in a square enclosure. *J. Thermophys. Heat Transfer*, 13, 299, 1999.
30. Deschamps, C.J. et al. Heat and fluid flow inside a household refrigerator cabinet. In *20th International Congress of Refrigeration*, Sydney, Australia, 1999.
31. Balaji, C. and Venkateshan, S.P. Correlations for free convection and surface radiation in a square cavity. *Int. J. Heat Fluid Flow*, 15, 249, 1994.
32. Ramesh, N. and Venkateshan S.P. Experimental study of natural convection in a square enclosure using differential interferometer. *Int. J. Heat Mass Transfer*, 44, 1107, 2001.
33. Velusamy, K., Sundarajan, T. and Seetharamu K.N. Interaction effects between surface radiation and turbulent natural convection in square and rectangular enclosures. *Trans. ASME*, 123, 1062, 2001.
34. Li, N. and Li Z.X. Relative importance of natural convection and surface radiation in a square enclosure. *Int. J. Nonlinear Sci. Numerical Simulat.*, 3, 613, 2002.
35. Laguerre, O. and Flick D. Heat transfer by natural convection in domestic refrigerators. *J. Food Eng.*, 62, 79, 2004.
36. Nield, D.A. and Bejan, A. *Convection in Porous Media*. Springer-Verlag Inc., New York, 1992.
37. Cheng, P. Heat transfer in geothermal systems. *Adv. Heat Transfer*, 14, 1, 1979.
38. Kaviany, M. *Principles of Heat Transfer in Porous Media*, 2nd Edition. Springer, New Jersey, USA, 1991.
39. Oosthuizen P.H. Natural convective heat transfer in porous media filled enclosures. In: Vafai Kambiz (Ed.), *Handbook of Porous Media*. Marcel Dekker Inc., New York, USA, 489, 2000.
40. Padet, J. Convection libre. In: *Principes des Transferts Convectifs*. Ed. Polytechnica, Paris, France, 174, 1997.
41. Rohensenow, W.M., Hartnett, J.P. and Cho I.Y. *Handbook of Heat Transfer*, Chapter 3 and 4, 3rd Edition. McGraw-Hill Handbooks, Ohio, USA, 1998.
42. Wakao, N. and Kaguei S. *Heat and Mass Transfer in Packed Beds*. Gordon and Breach Science Publishers, New Jersey, USA, 1982.
43. Ben Amara, S., Laguerre, O. and Flick D. Experimental study of convective heat transfer during cooling with low air velocity in a stack of objects. *Int. J. Thermal Sci.*, 43, 1212, 2004.
44. Raithby, G.D. and Hollands, K.G.T. Natural convection. In: W.M. Rohsenow, J.P. Hartnett, Y. Ohio, USA, Cho (Eds.), *Handbook of Heat Transfer*. McGraw Hill, Ohio, USA, 1998.
45. Laguerre, O., Ben Amara, S. and Flick D. Experimental study of heat transfer by natural convection in a closed cavity: application in a domestic refrigerator. *J. Food Eng.*, 70, 523, 2005.
46. Laguerre, O. et al. Numerical simulation of air flow and heat transfer in domestic refrigerators. *J. Food Eng.*, 81, 144, 2007.
47. Kingston, P., Woolley, N. and Tridimas, Y. Fluid flow & heat transfer calculations in a domestic refrigerator. *FIDAP UK User meeting*, Fluent France SA, 1–11, 1994
48. Chui, E. H. and Raithby, G.D. Computation of radiant heat transfer on a non-orthogonal mesh using the finite-volume method. *Num. Heat Transfer*, Part B, 23, 269, 1993.
49. Colomer, G. et al. Three dimensional numerical simulation of convection and radiation in a differentially heated cavity using the discrete ordinates method. *Int. J. Heat Mass Transfer*, 47, 257, 2004.
50. Sanchez, A. and Smith T.F. Surface radiation exchange for two dimensional rectangular enclosures using the discrete-ordinates method. *J. Heat Transfer*, 114, 465, 1992.

17 Mathematical Analysis of Vacuum Cooling

Lijun Wang
North Carolina Agricultural and Technical State University

Da-Wen Sun
National University of Ireland

CONTENTS

17.1   INTRODUCTION

Vacuum cooling, like vapor-compression refrigeration, is based on liquid evaporation to produce cooling effect. The difference between vacuum cooling and conventional refrigeration methods is that for the vacuum cooling the cooling effect is achieved by evaporating some water from a product directly rather than by blowing cold air or other cold medium over the product.[1-4] High speed and efficiency are two features of vacuum cooling, which are unsurpassed by any conventional cooling method, especially when cooling boxed or palletized products. Cooling time in the order of 30 minutes ensures that tight delivery schedules and strict cooling requirements for the safety and quality of foods can be met.[1-4]

Any product which has free water and whose structure will not be damaged by the removal of such water can be vacuum cooled. The speed and effectiveness of vacuum cooling are mainly related to the ratio between its evaporation surface area and the mass of foods.[3] Vacuum cooling has been satisfactorily used to remove field heat of horticultural products in the United States since the 1950s,[5] and proven to be an effective method for precooling certain type of fresh vegetables such as lettuce and mushroom.[6] It can significantly reduce postharvest deterioration of vegetables, thus, prolonging their storage life. For the same reason, vacuum cooling is also a very effective method for precooling of floricultural products to prolong the vase life of flowers.[7-9] However, if there is a low ratio between the surface area and the mass, or an effective barrier to water loss from the produce surface, vacuum cooling may be very slow. Produce such as tomatoes, apples, and peppers, which have a relatively thick wax cuticle, are not suitable for vacuum cooling.[10] Vacuum cooling has been successfully applied to the processing procedures for some foods such as liquid foods,[11,12] and baked foods[13] to reduce the cooling time for improvement of the efficiency of the processes and reduction of the distribution time to markets. Recently, vacuum cooling has been investigated as an effective cooling method for cooked meats as typical conventional cooling methods cannot achieve a rapid cooling for cooked meats.[14-16]

Although much experimental work has been carried out on vacuum cooling of foods since the first commercial vacuum cooling plant was built in the United States for cooling lettuce, only limited research was published for modeling vacuum cooling process of liquid foods[11,12] and solid foods.[17,18] The cooling load during vacuum cooling is determined by the evaporation rate and the latent heat of water evaporation. During vacuum cooling, the transfer of mass and heat is coupled together and should be considered simultaneously. For modeling of vacuum cooling of solid foods such as cooked meats, a more complex model is needed to describe the transient, simultaneous, coupled mass, and heat transfer through the foods under vacuum.[18] Since the evaporated water, which is the cooling source, comes from the product directly, it is very important to investigate the characteristics of the product such as size, shape, porosity, and pore size on the performance of a vacuum cooler.[19] The operating parameters of a vacuum cooler mainly including chamber free volume, pumping speed of vacuum pumps, condenser temperature also affect the performance of the vacuum cooler.[20] In this chapter, a mathematical model, which was used to describe a vacuum cooling system, and mass and heat transfer in food products under vacuum is introduced. The model was further used to analyze the effects of the characteristics of food products and the opeationg conditions of a vacuum cooler on the performance of the vacuum cooler in terms of the cooling rate and weight loss.

17.2   VACUUM COOLING PRINCIPLE AND EQUIPMENT

17.2.1   Vacuum Cooling Principle

Liquid evaporation is the most popular cooling sink in the refrigeration industry. Whenever any portion of a liquid evaporates to become its vapor state, an amount of heat equal to the latent heat of evaporation must be absorbed by the evaporated portion either from the liquid body or from the surroundings, resulting in reduction of the temperature of the liquid body or surroundings.[1,3] Water boils at 100°C if it is subjected to the atmospheric pressure (1 atm). However, reduction in the imposed pressure on water lowers the boiling temperature of water, that is, water can also boil at as a low temperature as 0°C if the imposed pressure is reduced to 611 Pa. The imposed pressure on water determines the minimum temperature, which can cause water to boil to produce cooling effect.[1,3]

If porous and moist foods are subjected to a vacuum pressure, part of water within the foods can boil out to generate cooling effect for the foods at a temperature as low as the saturation temperature of water at the vacuum pressure. Generally, most foods have two main compositions: water and solid texture. Therefore, the refrigerant used in a vacuum cooler is not pure water but one of the food compositions. Energy balance during vacuum cooling can be expressed as:

$$m_f c \Delta T = \Delta m_w h_{fg} \tag{17.1}$$

The temperature decrease caused by per unit of percentage weight loss is thus determined by

$$\frac{\Delta T}{\Delta m_w / m_f} = \frac{h_{fg}}{c} \tag{17.2}$$

As shown in Equation 17.2, the specific heat of foods determines the temperature decrease per unit of the percentage weight loss. Suppose a vacuum cooler is used to cool 1000 kg vegetables from 30 to 5°C. The average latent heat of evaporation is about 2500 kJ/kg and the average specific heat is 3.5 kJ/kg°C. The evaporated water calculated by Equation 17.1 should be about 35 kg. Each percentage of weight loss will decrease the temperature of the vegetable by 7.1°C as calculated by Equation 17.2.

17.2.2 Typical Vacuum Cooler

Since porous and moisture foods can be cooled directly by boiling part of moisture in the foods at a low temperature under vacuum pressure, a vacuum cooler is really a system to maintain the required vacuum pressure. A typical vacuum cooler is illustrated in Figure 17.1, which consists of two basic components: a vacuum chamber and a vacuum pumping system.[1,3] The vacuum cooling process occurs in two fairly distinct stages: (a) the removal of most of the air in the chamber to the flash point, or the saturation pressure of water at the initial temperature of foods, with relatively little cooling, and (b) the drop of the pressure in the vacuum chamber continuously to final pressure with main cooling phase of the products.

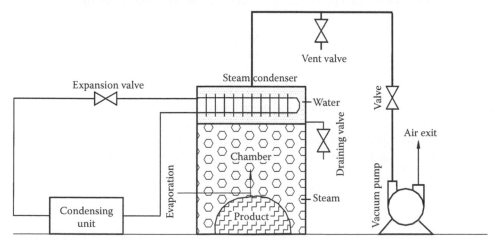

FIGURE 17.1 Schematic diagram of the experimental vacuum cooler. (Adapted from Wang, L.J. and Sun, D.W., *Trans. ASAE*, 46, 108, 2002. With permission.)

The vacuum chamber, which is normally horizontal with cylindrical or rectangular construction, is used to keep the food. During cooling process, the door of the chamber is hermetically sealed and any leakage of air into the vacuum cooler increases the load of the vacuum pumping system. The pumping system may have two elements, which are a vacuum pump and a vapor-condenser. The vacuum pump is usually designed to reduce the pressure in the vacuum chamber from the atmospheric pressure to the saturation pressure at the initial temperature of food such as vegetables and fruits within 3–10 minutes. The rotary oil-sealed vacuum pump is widely chosen for a vacuum cooler. The vapor-condenser in effect acts as a vacuum pump to remove the vapor in the vacuum chamber by condensing the vapor back into water and then draining the water out. However, it should be noted that the cooling effect for food comes from the water evaporation in food, and application of a vapor-condenser in a vacuum cooler is only for practical and economical removal of a large amount of vapor generated.[1,3]

17.3   MATHEMATICAL MODELING OF VACUUM COOLING SYSTEM

17.3.1   Vacuum Chamber and Condensing Unit

The vacuum chamber is the place where a food product is kept during cooling. The volume of the chamber is determined by the requirement of the process. The vacuum pressure of the chamber is the most important factor since the cooling process is controlled by the boiling of water. The maximum pressure in the chamber, which can cause the water to boil away from the food, is the saturation pressure at the temperature of the food. On the other hand, the maximum temperature of the vapor condenser, which can cause the vapor to condense on the cold surface of the condenser, is the saturation temperature at the vapor partial pressure in the chamber. The relationship for the saturation pressure and temperature is determined by:[11,17]

$$P_{\text{sat}} = \exp\left(23.209 - \frac{3816.44}{T_K - 46.44}\right) \tag{17.3}$$

The total pressure in the chamber is the sum of the partial pressures of air and water vapor:

$$P_{\text{vc}} = P_a + P_v \tag{17.4}$$

The decrease rate of the total pressure in the vacuum chamber is thus given by

$$\frac{dP}{dt} = \frac{dP_a}{dt} + \frac{dP_v}{dt} \tag{17.5}$$

17.3.2   Vacuum Pump

During cooling, the water from food will boil when the pressure in the chamber reaches the saturation vapor pressure at the initial temperature of the food. The time at the beginning of boiling is usually called the flash point. The speed of vacuum pumps required for a given vacuum system, S, is determined by the evacuation time to the flash point, t_{fp}, which is given by:[17]

$$S = \eta \frac{V_f}{t_{\text{fp}}} \ln \frac{P_{\text{atm}}}{P_{\text{fp}}} \tag{17.6}$$

Generally, the pumping speed is a function of the vacuum pressure and it decreases with pressure. However, the pumping speed slightly decreases at the range of the pressure chosen by a vacuum cooling cycle.

17.3.3 MASS BALANCE OF AIR IN THE VACUUM CHAMBER

Figure 17.2 gives the flow chart of vapor and air in a vacuum chamber. As shown in Figure 17.2, the vapor evaporated from the foods and the ingress air contribute to the increase of the pressure in the chamber. Meanwhile, the vapor-condenser and the vacuum pumps contribute to the decrease of the pressure in the chamber.

The mass flow rate of air through the vacuum pump can be expressed as:[17]

$$\dot{m}_{a,o} = S\rho_a \tag{17.7}$$

where the density of air in the chamber is given by

$$\rho_a = \frac{P_a M_a}{RT_{K,vc}} \tag{17.8}$$

In practice, air leakage into a vacuum cooler is unavoidable. Air leakage of the vacuum cooler is assumed to be adiabatic compressible flow through a nozzle. The rate of the air leakage is calculated by:[11,17,21]

$$\dot{m}_{a,i} = A_i \left[\left(\frac{2\gamma}{\gamma - 1} \right) \left(\frac{P^2_{atm} M_a}{RT_{K,vc}} \right) \right]^{0.5} \left[\left(\frac{P}{P_{atm}} \right)^{\frac{1+\gamma}{\gamma}} - \left(\frac{P}{P_{atm}} \right)^2 \right]^{0.5} \tag{17.9}$$

In the above equation, A_i is the leakage area of a vacuum system determined by evacuating the system and then monitoring the pressure profiles. If the pressure in the chamber is lower than the critical pressure, the vacuum pressure, P, should be substituted by the critical pressure, which is determined by

$$P_{cr} = P_{atm} \left(\frac{2}{\gamma + 1} \right)^{\gamma/\gamma - 1} \tag{17.10}$$

The mass accumulated rate of air in the vacuum chamber is thus expressed as:

$$\dot{m}_{a,e} = \dot{m}_{a,o} - \dot{m}_{a,i} \tag{17.11}$$

The decrease rate of the air pressure in the chamber is calculated by

$$\frac{dP_a}{dt} = -\frac{\dot{m}_{a,e} R T_{K,vc}}{M_a V_f} \tag{17.12}$$

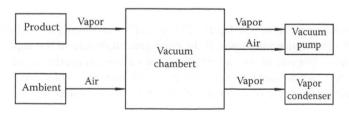

FIGURE 17.2 Flow chart of vapor and air in a vacuum chamber. (Adapted from Wang, L.J. and Sun, D.W., *Internal. J. Refrig.*, 25, 857, 2002. With permission.)

and the air partial pressure after Δt time of cooling is thus given by

$$P_a' = P_a + \frac{dP_a}{dt}\Delta t \qquad (17.13)$$

17.3.4 Mass Balance of Water Vapor in the Vacuum Chamber

The vapor generation rate is different from one food product to another. Normally, for a given food during the vacuum cooling, the rate of vapor generation is a function of the food temperature and the vacuum pressure, which is generally illustrated as:

$$\dot{m}_{v,i} = f(T_f, P_{vc}) \qquad (17.14)$$

For a solid food such as cooked meat, the transfer of mass and heat through the food body during vacuum cooling are coupled together and the temperature of the food is not homogeneous. A mass and heat transfer model must be used to calculate the vapor generation rate and the temperature profiles of the food. The mass and heat transfer model for vacuum cooling of cooked meat is discussed in Section 17.4. For a liquid food, such as soup, Equation 17.14 can be simply expressed as:[11,17]

$$\dot{m}_{v,i} = h_m A_s (a_w P_{f,\text{sat}} - P_{vc}) \qquad (17.15)$$

where $P_{f,\text{sat}}$ is the saturation pressure of water at the temperature of the liquid food, a_w is the water activity ($a_w = 1$ for pure water), and h_m is the mass transfer coefficient of boiling, which can be experimentally determined (e.g., $h_m = 8.4 \times 10^{-7}$ (kg/Pa · m² · s) for the boiling of pure water[17]).

The transient temperature of the liquid food can be simply determined by

$$cm_f \frac{dT}{dt} = \dot{m}_{v,i} h_{\text{fg}} \qquad (17.16)$$

The density of the vapor in the chamber can be calculated by

$$\rho_v = \frac{P_v M_w}{R T_{K,vc}} \qquad (17.17)$$

and the mass flow rate of vapor through the vacuum pump can be determined by

$$\dot{m}_{v,o} = S\rho_v \qquad (17.18)$$

The vapor-condenser as shown in Figure 17.1 is used to economically and practically remove a large volume of generated water vapor. If the mass generation rate of the vapor from the food is smaller than the mass flow rate of the vapor through the vacuum pump, the vacuum pump will cause the reduction of the vapor pressure in the chamber and the condenser thus loses its function. In this case, the mass accumulated rate of vapor in the chamber is expressed as:

$$\dot{m}_{v,e} = \dot{m}_{v,o} - \dot{m}_{v,i} \qquad (17.19)$$

The decrease rate of the vapor pressure in the chamber is then calculated by

$$\frac{dP_v}{dt} = -\frac{\dot{m}_{v,e} R T_{K,vc}}{M_w V_f}$$ (17.20)

and the vapor partial pressure after Δt time period of vacuum cooling is thus given by

$$P_v' = P_v + \frac{dP_v}{dt} \Delta t$$ (17.21)

However, if the mass generation rate of the vapor is not smaller than the mass flow rate of the vapor through the vacuum pump, the vapor-condenser will work and its required condensation ability is determined by:

$$q_{vcd,rq} = h_{fg}(\dot{m}_{v,i} - \dot{m}_{v,o})$$ (17.22)

If the condensation ability of the vapor-condenser is not smaller than the required one, the condenser can efficiently condense all the generated water vapor during the cooling. In this case, the vapor pressure in the chamber is thus maintained at the saturation pressure corresponding to the temperature of the condenser. The mass accumulated rate of vapor in the chamber is zero. Otherwise, if the condensation ability of the condenser cannot meet the requirement, the condenser can only remove part of the generated vapor and the vapor partial pressure in the chamber will increase. In this case, the mass accumulated rate of vapor in the chamber can be calculated by

$$\dot{m}_{v,e} = \left(\dot{m}_{v,o} + \frac{q_{vcd,max}}{h_{fg}} \right) - \dot{m}_{v,i}$$ (17.23)

The increase rate of the vapor pressure and the vapor partial pressure after one time step of cooling can also be calculated by Equations 17.20 and 17.21, respectively.

17.4 MATHEMATICAL MODELING OF MASS AND HEAT TRANSFER THROUGH SOLID FOODS UNDER VACUUM

17.4.1 Mass Transfer

Cooked hams are used as an example of solid foods to be vacuum cooled. The cooked ham is formed by small pieces of boned-out pork legs and there are abundant of macropores among the meat pieces as shown in Figure 17.3. These pores can increase the evaporation surface significantly and thus a rapid cooling can be achieved by vacuum cooling. In the pores, the vapor and water is the major component and air is negligible. The vapor movement through the pore spaces is hydrodynamic as a result of pressure difference shown in Figure 17.4.[18] Since the sample used is a rectangular bricked shape, which is symmetric in length, width, and height directions, the model can be established in one-eighth of the meat joint. The governing equation of the mass transfer, which is used to describe a three-dimensional transient vapor hydrodynamic movement with the inner vapor generation in the meat, is given in the Cartesian coordinate system by:[18]

$$\frac{\partial(\rho_v u_v)}{\partial x} + \frac{\partial(\rho_v u_v)}{\partial y} + \frac{\partial(\rho_v u_v)}{\partial z} + \dot{m}_{pv} - \omega \frac{\partial \rho_v}{\partial t} = 0$$ (17.24)

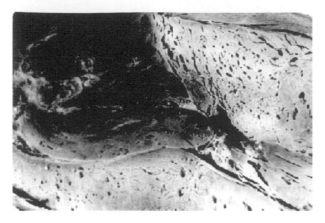

FIGURE 17.3 The visualized cross section of cooked hams by a computer vision system. (Adapted from Wang, L.J. and Sun, D.W., *Internal. J. Refrig.*, 25, 864, 2002. With permission.)

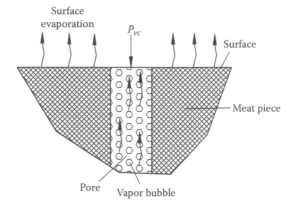

FIGURE 17.4 Vapor movement in the cooked meat during vacuum cooling. (Adapted from Wang, L.J. and Sun, D.W., *Internal. J. Refrig.*, 25, 864, 2002. With permission.)

For laminar flow in a pipe (Reynolds number $du_v \rho_v / \mu$ is below 2,100), the mean velocity can be given by:[18]

$$u_v = \frac{d^2}{32\mu} \frac{\partial P}{\partial n} \tag{17.25}$$

Therefore, Equation 17.24 can be re-written as:

$$\frac{\partial}{\partial x}\left(k_v \frac{\partial P}{\partial x}\right) + \frac{\partial}{\partial y}\left(k_v \frac{\partial P}{\partial y}\right) + \frac{\partial}{\partial z}\left(k_v \frac{\partial P}{\partial z}\right) + \dot{m}_{pv} - \omega \frac{\partial \rho_v}{\partial t} = 0 \tag{17.26}$$

where the mass transfer coefficient is given by

$$k_v = \rho_v \frac{d^2}{32\mu} \tag{17.27}$$

The initial condition is

$$P = P_{\text{sat},0} \tag{17.28}$$

As shown in Figure 17.4, if bubbles can release out of the pores, they should overcome the imposed pressure at the exit of the pores. Therefore, the boundary conditions are

$$\text{on the symmetric surface,} \quad \frac{\partial P}{\partial n_{x,y,z}} = 0;$$

$$\text{on the outer surface,} \quad P = P_{vc}$$

(17.29)

If the water vapor is assumed as an ideal gas, its density is thus given by

$$\rho_v = \frac{PM_w}{RT_K}$$

(17.30)

17.4.2 HEAT TRANSFER

The governing equation of heat transfer, which is used to describe a three-dimensional transient heat conduction with the inner heat generation in the meat, is given in the Cartesian coordinate system by:[18]

$$\frac{\partial}{\partial x}\left(\lambda \frac{\partial T}{\partial x}\right) + \frac{\partial}{\partial y}\left(\lambda \frac{\partial T}{\partial y}\right) + \frac{\partial}{\partial z}\left(\lambda \frac{\partial T}{\partial z}\right) + q_{pv} - \rho c \frac{\partial T}{\partial t} = 0$$

(17.31)

The initial condition for Equation 17.31 is

$$t = 0, \quad T = T_0$$

(17.32)

and the boundary conditions are

$$\text{on the symmetric surface,} \quad \frac{\partial T}{\partial n_{x,y,z}} = 0;$$

$$\text{on the outer surface,} \quad -\lambda \frac{\partial T}{\partial n_{x,y,z}} = h_t(T_{sf} - T_{vc}) + q_{sfv}$$

(17.33)

Under the vacuum, the heat released from the meat body to the cooling medium by convection is negligible. The heat transfer coefficient required by the boundary conditions in Equation 17.33 is the radiative heat transfer coefficients, which is calculated by:[22]

$$h_t = h_r = \sigma\varepsilon(T_{Ksf}^2 + T_{Kvc}^2)(T_{Ksf} + T_{Kvc})$$

(17.34)

17.4.3 WATER EVAPORATION

Water evaporation occurs in the micropores and on the surface of the meat. The water evaporation in the micropores is related to the surface area of the pores, surface mass transfer coefficient, and the pressure difference between the saturation pressure on the wall of the pores and the bulk pressure in the pores. If the pore is assumed to be a tube, the ratio of the wall surface area to the tube volume is thus $4/d$, the evaporation rate per unit volume of the cooked meat can then be expressed as:[18]

$$\dot{m}_{pv} = 4\frac{\omega}{d}h_m(a_w P_{sat} - P) \tag{17.35}$$

On the surface of the meat, the evaporation rate per unit surface area is given by

$$\dot{m}_{sfv} = K_p(a_w P_{sf,sat} - RH \cdot P_{vc}) \tag{17.36}$$

where RH is the ratio of the vapor pressure to the vacuum pressure in the vacuum chamber.

The surface mass transfer coefficient can be determined by pure water since the water environment on the wall of the pores is similar to bulk water. For the boiling of pure water, the experimental result gives $h_m = 8.4 \times 10^{-7} (\text{kg/Pa} \cdot \text{m}^2 \cdot \text{s})$.[17] On the surface of the meat joint, if the boiling does not occur (i.e., at the beginning of the cooling), the surface mass transfer coefficient is set at $K_p = 3.5 \times 10^{-8} (\text{kg/Pa} \cdot \text{m}^2 \cdot \text{s})$, which is the surface evaporation coefficient of the cooked meat during the slow air cooling.[23,24] However, if the boiling occurs on the surface of the meat, $K_p = h_m$. The saturation pressure on the pore surface is a function of the wall temperature, which can be determined by Equation 17.2.

The inner heat per unit volume generated due to water evaporation is thus, given by

$$q_{pv} = \dot{m}_{pv} h_{fg} \tag{17.37}$$

The evaporation heat per unit surface area is given by

$$q_{sfv} = \dot{m}_{sfv} h_{fg} \tag{17.38}$$

The sample is assumed to be composed of water, protein, fat, salt, and vapor. The pores are assumed to be filled with vapor gradually during cooling. The thermal conductivity, density and specific heat of the sample are expressed as functions of compositions.[24] The thermal conductivity, density and specific heat of each composition are given in the literature as functions of temperature, based on regression of experimental data.[24]

17.4.4 Finite Element Analysis

In order to find solutions to the mass and heat transfer models, The governing Equations 17.26 and 17.31 are transformed into their corresponding numeric formula using finite element analysis, which are expressed as:[18]

$$[K_v]\{P\} + [N_v]\frac{\partial\{\rho_v\}}{\partial t} = \{F_v\} \tag{17.39}$$

$$[K_h]\{T\} + [N_h]\frac{\partial\{T\}}{\partial t} = \{F_h\} \tag{17.40}$$

The load vector $\{F\}$, conductance $[K]$, and capacitance $[N]$ matrices are updated at each time step to handle the varying physical properties and operating conditions. The set of transient differential Equations 17.39 and 17.40 can be solved by a finite difference scheme, which is expressed as:[18,24]

$$\left[\frac{[N_v][V]_{n+1}}{\Delta t}+\alpha[K_v]\right]\{P\}_{n+1}=\left[\frac{[N_v][V]_n}{\Delta t}-(1-\alpha)[K_v]\right]\{P\}_n+\{F_v\} \tag{17.41}$$

$$\left[\frac{[N_h]}{\Delta t}+\alpha[K_h]\right]\{T\}_{n+1}=\left[\frac{[N_h]}{\Delta t}-(1-\alpha)[K_h]\right]\{T\}_n+\{F_h\} \tag{17.42}$$

where α is a weighting factor, which must be chosen between 0 and 1. A number of different schemes can be obtained by choosing the value of α such as $\alpha=0$ for forward difference scheme, $\alpha=1/2$ for Crank–Nicholson scheme, and $\alpha=1$ for backward difference scheme.

In Equation 17.39, $\{\rho\}$ is derived by multiplying the $[V]$ matrix and $\{P\}$ vector. The $[V]$ matrix is a diagonal matrix, which is given by[18]

$$[V]=\begin{bmatrix} \dfrac{M_w}{RT_{K1}} & 0 & \cdots & 0 \\ 0 & \dfrac{M_w}{RT_{K2}} & \cdots & 0 \\ \vdots & \vdots & \ddots & \vdots \\ 0 & 0 & \cdots & \dfrac{M_w}{RT_{KN}} \end{bmatrix} \tag{17.43}$$

17.5 SIMULATION APPROACH

A computer program was written in visual C++ language to implement the above mathematical model by an interative procedure. The flow chart of the computional algorithm is illustrated in Figure 17.5. The input information can be divided into three groups: (1) operating parameters such as temperature of the chamber, temperature of the condenser, free volume in the chamber, pumping time from the atmospherical pressure to the flash point of water in the food, and the defined final vacuum pressure; (2) system parameters such as the air leakage area, the maximum condensation load of the condenser and the pump efficiency; and (3) the initial temperature of the food, atmospherical pressure, latent heat of water evaporation and time step.

The pumping speed of a vacuum cooler is practically chosen based on the pumping time to the flash point for a given food. Condensation pressure is determined by the temperature of the condenser. The initial conditions in the vacuum chamber are that the total pressure is equal to atmospheric pressure and the vapor partial pressure is equal to the saturation pressure at the temperature of the vapor condenser. The initial air partial pressure is the difference between the total pressure and the vapor partial pressure.

During the simulation process, the air and vapor partial pressures in the chamber after one time step of computation are initally assumed to be equal to the air and vapor partial pressures before the computation, respectively. The average air partial pressure before and after the computational cycle is used to calculate the air flow rate through the pump and the leakage rate into the chamber by Equations 17.7 and 17.9, respectively. The decreased rate of the air partial pressure and the final air partial pressure after one time step computation are then calculated by Equations 17.12 and 17.13, respectively.

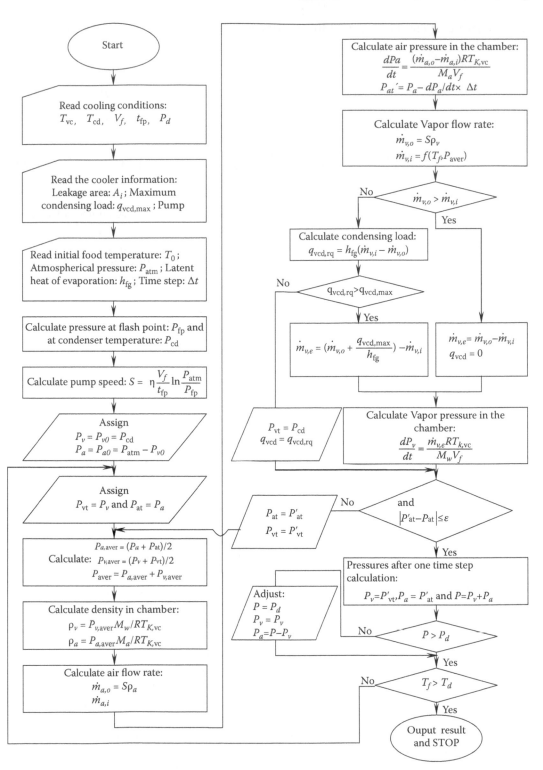

FIGURE 17.5 Flow chart of the computer simulation program. (Adapted from Wang, L.J. and Sun, D.W., *Internal. J. Refrig.*, 25, 859, 2002. With permission.)

If the food is liquid, the average total pressure and vapor partial pressures before and after the computational cycle are used to caculate the vapor generation rate by Equation 17.15. If the food is solid, the vapor generation rate is determined by the mass and heat transfer model presented in Section 17.4 and the vapor release rate through the pump by Equation 17.18. The vapor releasing rate is compared with the vapor generation rate. If the former is bigger than the latter, the vapor condenser will lose its function and the decreased rate of the vapor partial pressure and the final vapor partial pressure are calculated by Equations 17.20 and 17.21, respectively. Otherwise, the condenser will work. In this case, the required condensation load is calculated by Equation 17.22. The maximum condensation ability of the condenser is compared with the required one. If the former is bigger than the latter, the condenser is assumed to work efficiently and the calculated vapor partial pressure is still the saturation pressure at the condenser temperature. However, if the maximum condensation ability of the condenser cannot meet the requirement, the condenser can only remove part of the generated vapor and the vapor partial pressure in the chamber will increase. The increasing rate of the vapor partial pressure and the final vapor partial pressure are also calculated by Equations 17.20 and 17.21, respectively.

The calculated air and vapor partial pressures are then compared with their initialy assumed values, if the error between the calculated and assumed values is within a set value, the computational cycle is finished. Otherwise, the calculated values are reassigned as air and vapor partial pressure after one time step of computation and the computaion is repeated. The calculated final total vacuum pressure is compared with the defined value. If the former is smaller than the latter, the defined vacuum pressure is assigned to the calculated value and the air partial pressure is substituted by the difference between defined vacuum pressure and the calculated vapor partial pressure. The calculations are repeated until the defined temperature of food is achieved.

17.6   MODEL VALIDATION

17.6.1   Validation with Vacuum Cooling of Water

In order to verify the model, water was used in the experiment as the physical properties of water are well-known and the water temperature during cooling can be assumed to be homogeneous. The experiments were carried out to measure the vacuum pressure in the chamber and the temperature histories of water during the cooling. The leakage area of the vacuum cooler was varied by opening the bleeding valve to different levels. For this experiment, there were two stages of cooling achieved by two different leakage areas. Before the cooling, the bleeding valve was opened to a given level and the vacuum chamber was evacuated. After the first stage of cooling process, the bleeding valve was closed to a small level for the second stage of cooling.

The measured and predicted histories of the water temperature and the vacuum pressure during the cooling are given in Figure 17.6. As the digital vacuum pressure indicator attached to the vacuum cooler can only display the vacuum level from 0 Pa to 12,000 Pa and therefore only the pressure below 12,000 was recorded during the experiment and plotted in Figure 17.6. It is clearly indicated in Figure 17.6 that there are two cooling stages. The predicted vacuum pressure profile matches with the measured values very well. The maximum deviation between the predicted and measured pressure is within 110 Pa (for the chamber pressure between 12,000 Pa and 2200 Pa). For vacuum cooling of water with weight of 185 g in a 600-ml beaker from 38.5 to 25.5°C, the predicted temperature is in good agreement with that measured and the maximum deviation between the predicted and measured temperature is within 2°C. The deviations may be caused by the assumption of the homogeneous temperature in the liquid body and by the measurement of the temperature.

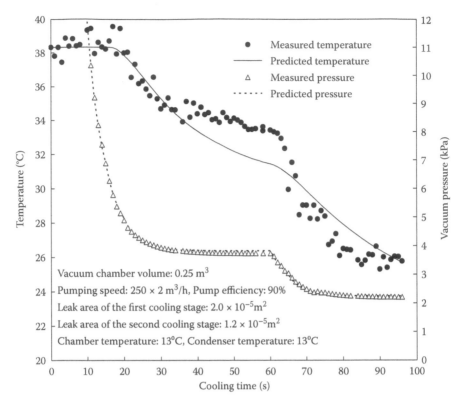

FIGURE 17.6   The measured and predicted temperature and vacuum pressure histories during vacuum cooling of water. (Adapted from Wang, L.J. and Sun, D.W., *Internal. J. Refrig.*, 25, 860, 2002. With permission.)

17.6.2   VALIDATION WITH VACUUM COOLING OF COOKED HAMS

Experiments were also conducted on vacuum cooling of a cooked ham sample to validate the model. In simulation, the initial temperature distribution in the cooked meat was calculated from the measured initial data by an interpolated method. The initial surface temperature was taken from the experimental value measured by two thermocouples at the surface directly. The physical properties of the meat were related to the composition and the temperature of the meat.

The predicted and experimental temperature histories at the core, on the surface and on average are given in Figure 17.7. As shown in Figure 17.7, the predicted results agree with the experimental data. The maximum deviations are within 2.5°C for the core, surface and average temperatures. It is noted in Figure 17.7 that the surface temperature at the end of the cooling procedure is a little lower than 0°C. During the final cooling stage, the vacuum was maintained at around 650 Pa by adjusting the rate of air leakage into the chamber. In this case, water boiling could not occur at the temperature under 0°C as the imposed pressure was above 611 Pa. However, the water evaporation on the meat surface could still occur at the temperature under 0°C since the vapor partial pressure in the chamber might be lower than 611 Pa. The water evaporation during the final cooling stage could reduce the surface temperature to below 0°C.

Weight loss occurs during vacuum cooling. The measured and predicted percentage weight losses are 9.4 and 8.7%, respectively. The deviation of the final total weight loss between the measured and predicted values is about 7.5%. The deviation may be caused by a small amount of liquid water escaping from the meat at the beginning of cooling. The escaped liquid water contributes to the increase in the weight loss and but produces no cooling effect.

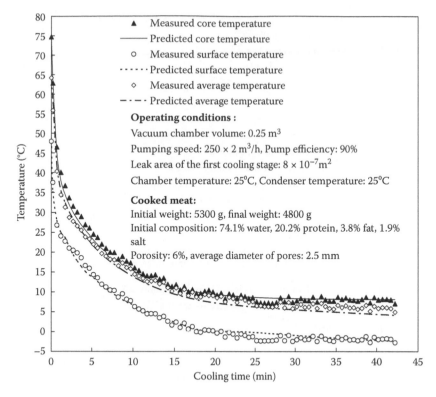

FIGURE 17.7 The predicted and experimental temperature histories at the core, on the surface and on average during vacuum cooling of cooked meats. (Adapted from Wang, L.J. and Sun, D.W., *Internal. J. Refrig.*, 25, 868, 2002. With permission.)

17.7 MATHEMATICAL ANALYSIS OF VACUUM COOLING OF COOKED MEATS

17.7.1 WEIGHT LOSS DURING VACUUM COOLING

During the vacuum cooling, the weight loss comes from water evaporation in the micropores and on the surface of the meat. Figure 17.8 gives the predicted histories of percentage weight loss for the total, inner and surface weight loss during the vacuum cooling, respectively. As shown in Figure 17.8, the weight loss from the inner micropores is 6.9% of the weight before cooling, compared with 1.8% from the surface. Therefore, during the vacuum cooling, most of the cooling effect comes from the water boiling in the micropores. It can also be seen from Figure 17.8 that there is no significant change of the inner weight loss during the final cooling stage, which further confirms that the low evaporation rate during this cooling stage controls the whole cooling rate as indicated in Figure 17.7.

17.7.2 EFFECTS OF OPERATING CONDITIONS ON COOLING RATE

For a given vacuum cooler with a chamber volume of 0.25 m³, if the chamber is used to cool a 5 kg cooked meat joint, the chamber free volume is still almost 0.25 m³. However, if the vacuum cooler is to cool 130 kg cooked meat joints at a bulk density of 1040 kg/m³, the chamber free volume is reduced to about 0.125 m³ and the ratio of the free volume to the chamber volume is 50%. If 195 kg cooked meat joints are placed into the chamber, the chamber has only about 25% free volume, which is 0.0625 m³. A vacuum pump with a pumping speed of 7.56 m³/h can reduce the pressure in the vacuum chamber with 0.25 m³ free volume to the flash point of the meat with an initial temperature

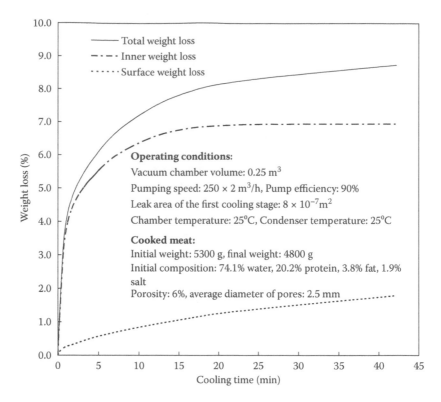

FIGURE 17.8 The predicted histories of the percentage inner, surface and total weight losses during vacuum cooling of cooked meats. (Adapted from Wang, L.J. and Sun, D.W., *Internal. J. Refrig.*, 25, 869, 2002. With permission.)

of 74°C in 2 min. If the same vacuum pump is still used, the flash point can be reached after 1 min and 0.5 min for cooling 130 kg and 195 kg meat joints, respectively.

If a vacuum cooler with a chamber volume of 0.25 m³, pumping speed of 7.56 m³/h and adequate condensing ability to totally remove the evaporated vapor, is used to cool cooked meat joints, the effect of chamber free volume on cooling rates during vacuum cooling is given in Figure 17.9. It can be seen from Figure 17.9 that the smaller the chamber free volume, the higher the cooling rates, as the decrease of chamber free volume can reduce the initial quantity of air in the chamber, thus increasing the dropping rate of air pressure if the same vacuum pump is used.

The vacuum pump is used to remove air in the chamber. The air includes two parts: initial air in free volume of the chamber and leakage air. Therefore, the pumping speed of the vacuum pump should be high enough to efficiently remove the initial air and the leakage air in the chamber in order to cause the pressure reduction in the chamber within a desirable time period. If the vapor condenser is effective to remove the evaporated vapor, the vapor pressure in the chamber is the saturation pressure of water at the temperature of the condenser. In this case, the decreasing rate of air pressure controls the vacuum pressure in the chamber.

Figures 17.10 through 17.12 give the effect of pumping speed on the cooling rates for cooling 5, 130 and 195 kg of cooked meat joints in a vacuum cooler with a chamber volume of 0.25 m³ and sufficient vapor condensation capacity, respectively. As shown in Figure 17.10, an increase in pumping speed can reduce the cooling time significantly. It takes about 5 min for a pump with a pumping speed of 3.02 m³/h to reduce the pressure in the 0.25 m³ chamber with 5 kg meat joints from the atmospheric pressure to the pressure at the flash point of the meat with an initial temperature of 74°C. After cooling for 60 min, the core temperature of meat joint is about 11.5°C and it is difficult to further cool it down to below 10°C. However, if the pumping speed is increased to 4.32 m³/h, the

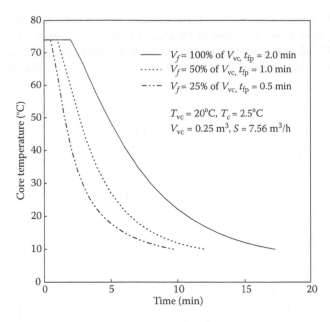

FIGURE 17.9   Effect of free volume in the chamber on the cooling rates during vacuum cooling. (Adapted from Wang, L.J. and Sun, D.W., *J. Food Eng.*, 61, 235, 2004. With permission.)

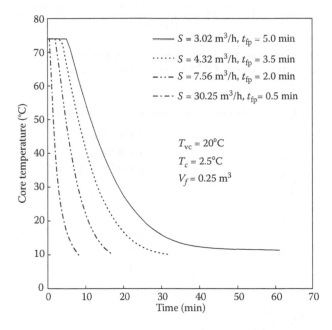

FIGURE 17.10   Effect of pumping speed of vacuum pumps on the cooling rates for cooling 5 kg of cooked meats in a vacuum cooler with a chamber volume of 0.25 m³. (Adapted from Wang, L.J. and Sun, D.W., *J. Food Eng.*, 61, 236, 2004. With permission.)

time to reach the flash point decreases to 3.5 min and the total cooling time for the core temperature of the meat joint from 74 to 10°C is about 32 min. If a vacuum pump with a pumping speed of 7.56 m³/h is used, the flash point can be reached in 2 min and the cooling time is only 17 min. If the pumping speed is further increased to 30.25 m³/h, the time to the flash point is now only 0.5 min and the cooling time is about 8 min.

The cooling rate will be increased with the decreasing free volume of the vaccuum chamber caused by loading more meats as shown in Figures 17.11 and 17.12. However, it can be seen from Figures 17.11 and 17.12 that it is still impossible for a vacuum pump with a pumping speed of 3.02 m³/h to reduce the temperature of cooked meats from 74°C to final temperature of 10°C

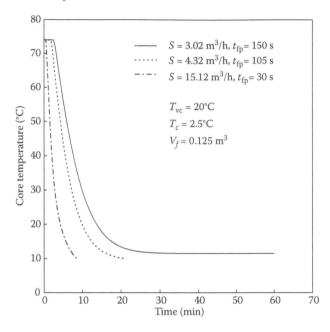

FIGURE 17.11 Effect of volume flow rate of pump on the cooling rates for cooling 130 kg of cooked meats in a vacuum cooler with a chamber volume of 0.25 m³. (Adapted from Wang, L.J. and Sun, D.W., *J. Food Eng.*, 61, 236, 2004. With permission.)

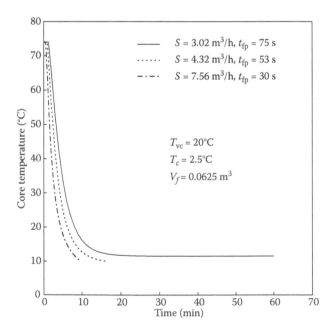

FIGURE 17.12 Effect of volume flow rate of pump on the cooling rates for cooling 195 kg of cooked meats in a vacuum cooler with a chamber volume of 0.25 m³. (Adapted from Wang, L.J. and Sun, D.W., *J. Food Eng.*, 61, 236, 2004. With permission.)

although the free volumes are only 50 and 25% of the total chamber volume after loading 130 and 195 kg meat joints, respectively. This is because the final pressure in the chamber, which can cause water to boil at 10°C, cannot be reached by using a pump with a small pumping speed due to air leakage.[20] Therefore, the selection of a vacuum pump for a vacuum cooler should consider both the desirable cooling rate and final temperature of products.

It can also be seen from Figures 17.11 and 17.12 that the cooling rates are directly determined by the time required to reach the flash point if the vapor condenser can efficiently remove the generated vapor during cooling. For cooling cooked meat joints from 74 to 10°C by using a vacuum cooler with a chamber of 0.25 m³, if the chamber loads 5 kg meat, the chamber free volume is almost 0.25 m³ and a pump with a pumping speed around 30.25 m³/h can be used to reduce the chamber pressure to the flash point in 0.5 min. In this case, the total cooling time is 8.2 min, as shown in Figure 17.10. However, after loading 130 kg meat, the chamber free volume is only 0.125 m³ and a pumping speed of 15.12 m³/h can be used to reduce the chamber pressure to the flash point in 0.5 min and the total cooling time is 8.5 min, as shown in Figure 17.11. If the chamber is loaded with 195 kg meat, the free volume is reduced to only 0.0625 m³ and a pumping speed of 7.56 m³/h is adequate to reduce the chamber pressure to the flash point in 0.5 min and the total cooling time is 9.7 min as shown in Figure 17.12. The small difference between the above total cooling times is affected by air leakage under different free volumes in the chamber.

The vapor condenser, which is an auxiliary vacuum pump, is normally used to remove the large amount of vapor generated by condensing the vapor back to water and draining the water out of the chamber. If the vapor condenser is effective to remove the vapor, the maximum vapor pressure in the chamber is the saturation pressure of water at the temperature of the condenser. The effect of the condenser temperature on cooling rates is shown in Figure 17.13. It can be seen from Figure 17.13 that a decrease of condenser temperature can reduce the vapor pressure in the chamber and thus increase the cooling rate. If the pumping time to the flash point is set at 2 min for vacuum cooling of cooked meats with an initial temperature of 74°C and the condenser temperature of below 5°C is used, the core temperature of meat joints can be reduced from 74 to 10°C within 21 min. However, if the condenser temperature is further reduced from 2.5 to 0°C, the total cooling time cannot be

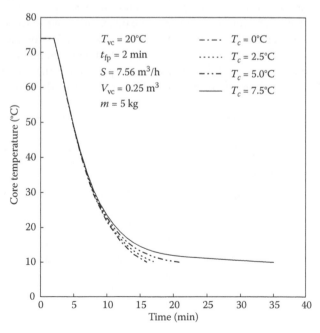

FIGURE 17.13 Effect of condenser temperature on cooling rates during vacuum cooling. (Adapted from Wang, L.J. and Sun, D.W., *J. Food Eng.*, 61, 237, 2004. With permission.)

reduced significantly. In this case, the vapor pressure contributes to a small part of the total pressure in the chamber, compared with air pressure. Therefore, the temperature of the vapor-condenser can be set at around 2.5°C. However, it should be noted that the temperature of the vapor condenser should be above 0°C because water freezes on the outside surface of the condenser when the temperature is below 0°C.

17.7.3 EFFECTS OF OPERATING CONDITIONS ON WEIGHT LOSS

Weight loss occurs during vacuum cooling since cooling effect directly comes from water evaporation (boiling) from food products. The effect of operating conditions on percentage weight losses of cooked meat joints during vacuum cooling are given in Table 17.1. It can be seen from Table 17.1 that operating parameters have no significant effect on percentage weight loss. A slow cooling rate causes a slight decrease in percentage weight loss. The percentage weight loss is closely related to final average temperature if initial temperatures of meat joints are the same. As shown in Table 17.1, there is only small difference in the final average temperatures for the same cooked meat joint under different operating conditions. This means that operating conditions do not have a large effect on the final temperature distribution within the meat joint.

17.7.4 EFFECTS OF CHARACTERISTICS OF SOLID FOOD PRODUCTS ON COOLING RATE

Besides the design and operating parameters of a vacuum cooler, the characteristics of solid food products including their size, porosity, pore size and distribution may also affect the performance of a vacuum cooler. The effects of weight, size, shape, porosity and pore size of cooked meat joints on the cooling rate during vacuum cooling are given in Table 17.2. It can be seen from groups I and II in Table 17.2 that the cooling rate is almost independent of the weight, size and shape of the meat joints, which is different from the traditional cooling processes of cooked meat joints. Therefore, vacuum cooling can be used to cool large food items such as cooked hams, for which traditional cooling methods such as air blast can not rapidly reduce the temperature to a safe level.

Since the water evaporation in the micropores of cooked meat joints causes the cooling effect during vacuum cooling, the porosity, pore size and pore distribution may have effects on the cooling rate. Figure 17.14 gives the effects of the porosity of cooked meat joints on the cooling rate during vacuum cooling. In Figure 17.14, the pores are assumed to be homogeneously distributed within the meat joints and the pore diameter is 2.5 mm based on experimental measurement. It can be seen from Figure 17.14 that the cooling rate increases with the porosity of the cooked meat joints. If the

---

TABLE 17.1
Effect of Cooling Conditions on Weight Losses of Cooked Hams During Vacuum Cooling

	Operating Conditions						Final Average
m (kg)	V_f (%)	S (m³/h)	t_{fp} (s)	T_c (°C)	T_{vc} (°C)	Weight Loss (%)	Temperature (°C)
5	100	7.56	120	2.5	20	8.28	9.4
130	50	7.56	60	2.5	20	8.35	9.3
195	25	7.56	30	2.5	20	8.35	9.3
5	100	4.32	210	2.5	20	8.18	9.8
5	100	30.25	30	2.5	20	8.42	9.0
5	100	7.56	120	0.0	20	8.29	9.1
5	100	7.56	120	5.0	20	8.23	9.7
5	100	7.56	120	7.5	20	8.24	9.9

Source: Adapted from Wang, L.J. and Sun, D.W., *J. Food Eng.*, 61, 237, 2003. With permission.

TABLE 17.2
Total Predicted Cooling Time and Weight Loss of Cooked Hams During Vacuum Cooling

		Cooked Hams						
Group	Mass (kg)	Geometry* (mm)	Porosity (%)	Pore Diameter (mm)	Pore Distribution**	Cooling Time to 10°C (min)	Weight Loss (%)	Final Average Temperature (°C)
I	8	B: 195×195×195	6	2.5	H	17.33	8.25	9.3
	6	B: 177×177×177	6	2.5	H	17.33	8.22	9.3
	4	B: 155×155×155	6	2.5	H	17.33	8.28	9.2
	2	B: 123×123×123	6	2.5	H	17.25	8.27	9.1
II	5	B: 167×167×167	6	2.5	H	17.33	8.29	9.3
		B: 150×176×176	6	2.5	H	17.33	8.27	9.3
		B: 100×216×216	6	2.5	H	17.42	8.31	9.2
		E: 207×207	6	2.5	H	17.25	8.31	9.4
		E: 295.5×173.6	6	2.5	H	17.25	8.30	9.4
III	5	B: 167×167×167	6	1.5	H	15.08	8.25	9.5
		B: 167×167×167	6	1.0	H	14.08	8.23	9.6
		B: 167×167×167	6	0.5	H	13.32	8.22	9.5
IV	5	B: 167×167×167	4	2.5	H	20.67	8.39	9.1
		B: 167×167×167	3	2.5	H	24.42	8.49	8.9
		B: 167×167×167	2	2.5	H	32.25	8.64	8.6
		B: 167×167×167	1	2.5	H	56.83	8.87	8.1
V	5	B: 167×167×167	6	2.5	Block (mm) 30.4×30.4×30.4 No porosity	33.17	8.70	6.6
		B: 167×167×167	6	2.5	Block (mm) 50.1×50.1×50.1 No porosity	51.67	8.90	5.9
		B: 167×167×167	6	2.5	Block (mm) 30.4×30.4×30.4 Porosity: 12%	17.33	8.29	9.3
		B: 167×167×167	2	2.5	Block (mm) 50.1×50.1×50.1 No porosity	71.83	9.16	6.1

Source: Adapted from Wang, L.J. and Sun, D.W., *Trans. ASAE*, 46, 111, 2002. With permission.
* B: brick shape, height/width/length; E: ellipsoid shape, minor/major axis.
** H: homogeneous.

porosity is only 1%, it should take about 57 min for vacuum cooling to reduce the temperature from 74 to 10°C. For vacuum cooling of a cooked meat joint with the porosity of 2%, the total cooling time can shorten by half to 32 min. If the porosity increases from 2 to 4%, the total cooling time decreases by 35% to 21 min. However, if the porosity further increases to be as high as 6%, the cooling time is about 17.5 min with a little decrease compared with that of the 4% porosity meat joints. The increase of the porosity in a cooked meat joint can enhance the mass transfer of water through the meat joint and thus increase the supply of local cooling source. However, during vacuum cooling of cooked meat joints, the coupled mass and heat transfer within the meat joints controls the cooling rate. For vacuum cooling of a cooked meat joint with a very high porosity, the cooling rate is mainly controlled by the local heat conduction. As the porosity of commercial cooked meat joints is between 3 and 7%,[25] large commercial cooked meat joints can be vacuum cooled from an initial

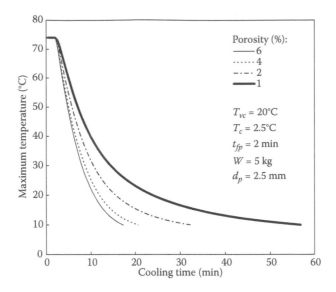

FIGURE 17.14   Effect of the porosity of cooked meat joints on the cooling rate during vacuum cooling. (Adapted from Wang, L.J. and Sun, D.W., *Trans. ASAE*, 46, 111, 2002. With permission.)

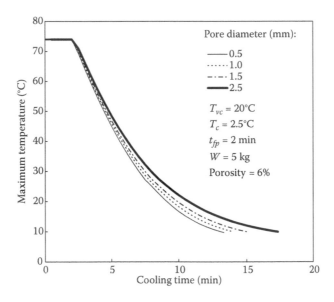

FIGURE 17.15   Effect of the pore diameter of cooked meat joints on the cooling rate during vacuum cooling. (Adapted from Wang, L.J. and Sun, D.W., *Trans. ASAE*, 46, 112, 2002. With permission.)

temperature of 74°C to a final temperature of 10°C within 30 min if the pores are homogeneously distributed in the meat joints.

The effect of pore size on the cooling rate is given in Figure 17.15. As shown in Figure 17.15, there is a slight increase in the cooling rate if the pore diameter decreases from 2.5 to 0.5 mm. On one hand, the decrease of pore size will increase the evaporation surface area per unit volume of cooked meats, resulting in an increase in cooling rate. For a meat joint with the porosity of 6% and the pore diameter of 2.5 mm, the ratio of evaporation surface area to the volume of meat joints is about 96 m²/m³. If the pore diameter decreases to 1 mm and the porosity is still 6%, the ratio will increase to 240 m²/m³. On the other hand, the decrease of pore size will also increase the mass transfer resistance of vapor through the pores and if the pore size is too small, the vapor will move slowly through the pores by diffusion. In this case, the local pressure in the pores will increase and the pressure difference for local water evaporation thus decreases, causing a significant decrease of the cooling rate. Therefore, there is an optimum pore size to maximize the cooling rate. For vacuum cooling of a cooked meat joint, the macropores, which are achieved by piling the meat pieces to form a meat joint and by injecting brine solution with needles, contribute to the high cooling rate. The average diameter of macropores in the commercial cooked meat joints is normally bigger than 1 mm.

17.7.5 EFFECTS OF CHARACTERISTICS OF FOOD PRODUCTS ON WEIGHT LOSS

Weight loss occurs during vacuum cooling since the cooling effect comes from the water evaporation from the cooked meats. The percentage weight losses of cooked meat joints during vacuum cooling are given in Table 17.2. It can be seen from Table 17.2 that the weight loss is about 8–9% of the weight before cooling from the initial temperature of 74°C to the final maximum temperature of 10°C. The percentage weight loss significantly depends on the initial and final average temperatures. As seen from groups I, II, and III in Table 17.2, the weight, size and shape of meat joints, and the diameter of pores have no significant effect on the total percentage weight loss since those factors do not have a large impact on the final average temperature. However, it can be seen from groups IV and V in Table 17.2 that the porosity and pore distribution have a significant effect on the total percentage weight loss. The low total percentage weight loss can be achieved if the porosity is big and the pores are homogeneously distributed within the meat joints.

17.7.6 TEMPERATURE AND VAPOR PRESSURE DISTRIBUTION IN SOLID FOODS DURING VACUUM COOLING

The final temperature distributions within the cooked meat joints with (1) homogeneous pore distribution, (2) a 3.34×3.34×3.34 cm cubic solid block in the centre, and (3) a 3.34×3.34×3.34 cm cubic porous block with double porosity in the centre are given in Figures 17.16, through 17.18, respectively. As shown in Figure 17.16, if the pores are homogeneously distributed within the meat joint, the final temperature distribution is homogeneous and the temperature difference within the meat is within 1°C. However, if there is a cubic solid block in the meat joint, there is a hot region located

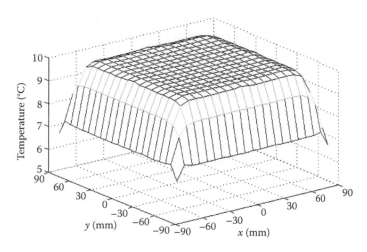

FIGURE 17.16 The final temperature distribution of a cooked meat joint with homogeneous porosity distribution during vacuum cooling (the x-y plan located at z=0). (Adapted from Wang, L.J. and Sun, D.W., *Trans. ASAE*, 46, 113, 2002. With permission).

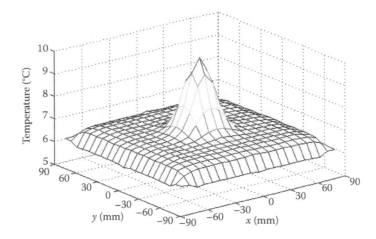

FIGURE 17.17   The final temperature distribution of a cooked meat joint with a 3.04×3.04×3.04 cm solid block during vacuum cooling (the *x-y* plan located at $z=0$). (Adapted from Wang, L.J. and Sun, D.W., *Trans. ASAE,* 46, 113, 2002. With permission.)

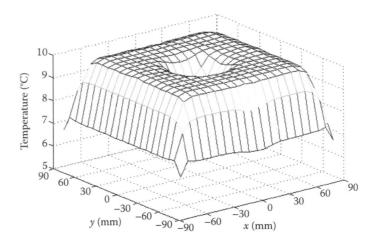

FIGURE 17.18   The final temperature distribution of a cooked meat joint with a 3.04×3.04×3.04 cm double-porosity block during vacuum cooling (the *x-y* plan located at $z=0$). (Adapted from Wang, L.J. and Sun, D.W., *Trans. ASAE*, 46, 113, 2002. With permission.)

in the block as shown in Figure 17.17. The temperature decreases from the core to the surface of the block, which is similar to the temperature distribution for meat joints cooled by traditional cooling methods. If the final core temperature of the block is 10°C, the temperature of the remaining part is only about 6.5°C and therefore the average temperature is 6.6°C as given in Table 17.2. However, if the cubic porous block has double porosity, a cold region locates in the block as shown in Figure 17.18. In this case, if the final maximum temperature of the other inner part is 10°C, the temperature of the block is only 8°C and therefore, the average temperature is 9.3°C. Since the number and locations of the above mentioned blocks are distributed randomly in a meat joint, there is no definite order of the temperature distribution in the meat joints during vacuum cooling. The temperature distribution maps in Figures 17.16 through 17.18 confirm that the block without porosity or with less porosity controls the cooling rate since the safe cooling rate should be in terms of the maximum time permitted to reduce the hottest region to a certain temperature. Meanwhile, some blocks with high porosity may not contribute to the increase of the total cooling rate although those blocks can rapidly reduce the local temperature.

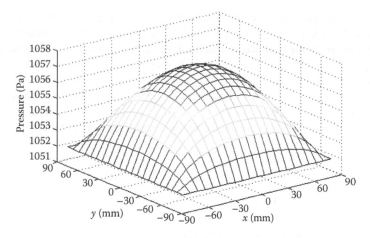

FIGURE 17.19   The final pressure distribution of a ooked meat joint with the pore diameter of 0.5 mm during vacuum cooling (the x-y plan located at $z = 0$). (Adapted from Wang, L.J. and Sun, D.W., *Trans. ASAE*, 46, 114, 2002. With permission.)

The pore size has a significant effect on the pressure distribution in the meat joint. As shown in Figure 17.19, the maximum pressure difference between the pressure in the pore and the vacuum chamber is less than 3 Pa if the pore diameter is bigger than 1 mm. If the pore diameter is 0.5 mm, the maximum pressure difference is about 12 Pa. Figure 17.19 gives the final pressure distribution within the cooked meat joints with an average diameter of 0.5 mm, the porosity of 6% and homogeneous pore distribution. In this case, the pressure distributes from the core to the surface of the meat joints in a decreasing order. It should be noted that if the pore diameter is too small, the vapor movement through the pores should be described by a diffusion equation. However, since the average diameter of pores in the meat joints is normally bigger than 1 mm, the vapor movement through the pores was modeled by the hydrodynamic equation given in this chapter.

17.8   CONCLUSIONS

Vacuum cooling is a rapid cooling technology for porous and moist foods. Heat from food is released by evaporating an amount of water within the food under vacuum directly. Vacuum cooling can provide an effective and efficient cooling technique for a variety of food products such as horticultural and floricultural products, liquid foods, baked products, fishery products, ready-to-eat meals, and cooked meats. A mathematical model was developed to describe the vacuum cooling system, and the simultaneous mass and heat transfer through solid food products under vacuum. The model was validated by experimental data. The validated model was used to analyze the operating conditions including free volume in vacuum chamber, pumping speed of vacuum pumps and condenser temperature, and characteristics of food products including weight, size, shape, porosity, and pore size on the performance of a vacuum cooler in terms of cooling rate and weight loss. The cooling rate increases with the pumping speed and decreases with the increase in free volume in vacuum chamber and condenser temperature. However, it should be noted that a small free space in the chamber is also needed for mass transfer and loading products. The operating conditions have a negligible effect on weight loss during vacuum cooling. The weight, size, and shape of cooked meat joints have negligible effects on the cooling rate during vacuum cooling, which is another advantage of vacuum cooling over the traditional cooling methods beside high cooling rate. The vacuum cooling rate strongly depends on the porosity and pore distribution in foods. An efficient rapid cooling can be achieved for a food item with high porosity and homogeneous pore distribution. The temperature distribution during vacuum cooling is determined by the porosity and pore distribution. If the pores are homogeneously distributed in the foods, the temperature distribution is also homogeneous. For the foods with nonhomogeneous pore distribution, the hot zone locates in the region without or with less porosity, and the cold point is in the part with more porosity. The pressure distribution within the foods is determined by the pore size. However, if the average pore diameter is higher than 1 mm, the maximum pressure difference between the bulk pressure in the pores and the vacuum chamber pressure is less than 3 Pa. The weight loss for vacuum cooling cooked hams from 74 to 10°C is about 8–9% of the weight before cooling. The big weight loss is the main disadvantage of vacuum cooling over the traditional cooling methods. Weight loss during vacuum cooling can be compensated for by injecting a high-level brine solution during the preparation period of cooked hams.

NOMENCLATURE

A	Area (m^2)
a_w	Water activity
c	Specific heat (kJ/kg K)
d	Pore diameter (m)
h_{fg}	Latent heat of evaporation (kJ/kg)
h_m	Evaporation coefficient (kg/Pam^{2}s)
h	Heat transfer coefficient (W/m^2 K)
k	Mass transfer coefficient (kg/msPa)
$\dot{m}$	Mass flow rate (kg/s)
m_f	Mass of food (kg)
$\dot{m}_{pv}$	Inner volumetric vapour generation (kg/m^{3}s)
$\dot{m}_{sfv}$	Water evaporation rate per unit surface area (kg/m^{2}s)
Δm_w	Water loss (kg)
M	Molecular weight (g/mol)
P	Pressure (Pa)
$\dot{q}_{pv}$	Inner volumetric heat generation (W/m^3)
$\dot{q}_{sfv}$	Imposed heat flux on the boundary (W/m^2)
R	Gas constant (8.314 J/molK)
RH	Relative humidity
S	Pumping speed (m^3/s)
T	Temperature (°C)
T_K	Temperature (K)
t	Time (s)
Δ t	Time step (s)
u	velocity (m/s)
v	= Specific volume (m^3/kg)
V	= Volume (m^3)
x, y, z	coordinate (m)

Greek letters

ω	Porosity (%)
γ	Ratio of specific heats at constant pressure and constant volume
σ	Stefan's constant (5.7×10^{-8} W/m^{-2}K^{-4})
λ	Thermal conductivity (W/mK)
ρ	Density (kg/m^{-3})

ε	Emissivity
μ	Viscosity (Pa s)
η	Efficiency of vacuum pump (%)

Subscripts

a	Air
cr	Critical point of gas
cond	Condenser
f	Free or food
fp	Flash point
i	In
o	Out
p	Pore
pv	Vapour in pores
r	Radiation
sf	Surface
sat	Saturation
t	Total
v	Vapour
vc	Vacuum chamber
w	Water
0	Initial

REFERENCES

1. Mellor, J.D. Vacuum techniques in the food industry. *Food Technology Australia,* 32, 397–398 and 400–401, 1980.
2. Sun, D.-W. and Wang, L.J.Vacuum cooling. In *Advances in Food Refrigeration,* Sun D.-W., ed. Leatherhead Publishing, Surrey, UK, 2001.
3. Wang, L.J. and Sun, D.-W. Rapid cooling of porous and moisture foods by vacuum cooling technology. *Trends in Food Science & Technology,* 15, 174–184, 2002.
4. Sun, D-W. and Zheng, L.Y. Vacuum cooling technology for the agri-food industry: past, present and future. *Journal of Food Engineering,* 77, 203–214, 2006.
5. Thompson, J. and Rumsey, T.R. Determining product temperature in a vacuum cooler. In *ASAE Paper No. 84-6543.* American Society of Agricultural Engineering, St. Joseph, MI 49085–9659, 1984.
6. Anon. Vacuum cooling for fruits and vegetables. *Food Processing Industry,* 12, 24–27, 1981.
7. Gao, J.P., Sun, Z.R., Guo, K., and Xu, Y.Y. Methods to accelerate vacuum cooling of cut flowers. *Transactions of the Chinese Society of Agricultural Engineering,* 12, 194–198, 1996.
8. Brosnan, T. and Sun D.-W. Compensation for water loss in vacuum precooled cut lily flowers. *Journal of Agricultural Engineering Research,* 79, 299–305, 2001.
9. Sun, D.-W. and Brosnan, T. Extension of the vase life of cut daffodil flowers by rapid vacuum cooling. *International Journal of Refrigeration,* 22, 472–478, 1999.
10. Longmore, A.P. The pros and cons of vacuum cooling. *Food Industries of South Africa,* May, 6–11, 1973.
11. Houska, M., Podloucky, S., Zitny, R., Gree, R., Sestak, J., Dostal, M., and Burfoot, D. Mathematical model of the vacuum cooling of liquids. *Journal of Food Engineering,* 29, 339–348, 1996.
12. Dostal, M. and Petera, K. Vacuum cooling of liquids: mathematical model. *Journal of Food Engineering,* 61, 533–539, 2004.
13. Anon. Bakery products cooled in minutes instead of hours-Modulated vacuum cooling is the key. *Food Engineering International,* 3, 33–34, 1978.
14. Sun, D.-W. and Wang, L.J. Heat transfer characteristics of cooked meats using different cooling methods. *International Journal of Refrigeration,* 23, 508–516, 2000.

15. Wang, L.J. and Sun, D.W. Experimental evaluation of the performance of vacuum cooling method for large cooked meat joints. *Journal of Food Process Engineering,* 25, 455–472, 2002.

16. Sun, D.-W. and Wang, L.J. Experimental investigation of performance of vacuum cooling for commercial large cooked meat joints. *Journal of Food Engineering,* 61, 527–532, 2003.

17. Wang, L.J. and Sun, D.-W. Modeling vacuum cooling process of cooked meat—Part 1: analysis of vacuum cooling system. *International Journal of Refrigeration,* 25, 852–860, 2002.

18. Wang, L.J. and Sun, D.-W. Modeling vacuum cooling process of cooked meat—Part 2: mass and heat transfer of cooked meat under vacuum pressure. *International Journal of Refrigeration,* 25, 861–872, 2002.

19. Wang, L.J. and Sun, D.-W. Numerical analysis of the three-dimensional mass and heat transfer with inner moisture evaporation in porous cooked meat joints during vacuum cooling process. *Transactions of the ASAE,* 46, 107–115, 2003.

20. Wang, L.J. and Sun, D.-W. Effect of operating conditions of a vacuum cooler on cooling performance for large cooked meat joints. *Journal of Food Engineering,* 61, 231–240, 2003.

21. Moran, M.J. and Shapiro, H.N. *Fundamentals of Engineering Thermodynamics.* John Wiley & Sons, Inc., New York, 1993.

22. Lewis, M.J. *Physical Properties of Foods and Food Processing Systems.* Ellis Horwood Ltd., Chichester, UK, 1987.

23. Daudin, J.D. and Swain, M.V.L. Heat and mass transfer in chilling and storage of meat. *Journal of Food Engineering*, 12, 95–115, 1990.

24. Wang, L.J. and Sun, D.-W. Evaluation of performance of slow air, air blast and water immersion cooling methods in cooked meat industry by finite element method. *Journal of Food Engineering,* 51, 329–340, 2002.

25. McDonald, K. and Sun, D.-W. The formation of pores and their effects in a cooked beef product on the efficiency of vacuum cooling. *Journal of Food Engineering,* 47, 175–183, 2001.

18 Airflow Patterns within a Refrigerated Truck Loaded with Slotted, Filled Boxes

Jean Moureh
Refrigerating Process Engineering Research Unit

CONTENTS

18.1   INTRODUCTION

This work is part of a research activity aiming to improve and optimize air-distribution systems in refrigerated vehicles in order to decrease the temperature differences throughout the palletized cargos. This condition is essential in order to preserve the quality, safety and shelf life of perishable products.

In this chapter, a reduced-scale model and computational fluid dynamics (CFD) predictions are used to investigate experimentally and numerically the airflow patterns within a typical refrigerated truck configuration loaded with slotted pallets filled with spherical objects. The experiments were carried out on a reduced-scale (1:3.3) model of a truck under isothermal conditions. Air velocity measurements were performed by using a Laser Doppler Velocimetry (LDV) and thermal sphere-shaped probes located inside the pallets. The aim was to investigate air velocity characteristics above and within pallets. The performance of ventilation was characterized with and without air duct systems.

Numerical and experimental results made it possible to highlight the confinement effect due to enclosure and the influence of load permeability on the jet penetration, its development and hence the overall heterogeneity of ventilation within the truck.

18.1.1 REFRIGERATED TRUCKS

Within refrigerated vehicle enclosure, air is supplied at relatively high velocities through a small inlet section located adjacent to or near the ceiling. Due to the adherence of the jet on this boundary by the Coanda effect, this design should allow the confined jet to expand while following the room wall surfaces and hence provide a high degree of ventilation throughout the entire enclosure. In the refrigerated enclosure, heat is transferred primarily by convection; therefore, the temperature and its homogeneity are directly governed by the patterns of airflow. Air renewal provided by these air-flows should compensate for the heat exchanged through the insulated walls of refrigerated vehicle or generated by the products.

From an aerodynamic perspective, the key characteristic of transport equipment is the placement of both the air delivery and return on the same face. This configuration is almost universally used as it is practical to place all the refrigeration equipment at one end of the transport unit. The drawback of this asymmetrical design is the presence of a strong pathway between the two sections, implying high velocities in the front of the refrigerated enclosure. In addition, the compactness of the cargo and high resistance to airflow due to narrow air spaces between pallets result in an uneven air distribution in the cargo where stagnant zones with poor ventilation can be observed in the rear part of the vehicle. In these zones, higher temperatures can occur locally within the load [1–5] even though the refrigerating capacity is higher than the heat exchanged by the walls and the products [6].

18.1.2 NUMERICAL MODELING

According to the complexity of direct measurement of local air velocities within a refrigerated truck, CFD has become the methodology of choice for the development of airflow models. Numerical predictions of air velocities and temperature distributions can be obtained by solving sets of differential equations of mass, momentum and energy written in their conservative form using the finite-volume method. To ensure the accuracy and reliability of CFD simulations, predictions need to be validated against reliable measurements obtained in parametric studies where the influence of all pertinent parameters is investigated separately.

In the case of complex 3D systems, comparisons of local velocities and global airflow patterns are required. Nonintrusive techniques for velocity measurements such as LDV and particle image velocimetry (PIV) provide more reliable data for validation. These measurement techniques also provide a means to improve the reliability of the simulations through the imposition of more accurate boundary conditions. Using CFD codes, complex configurations such as refrigerated transport or storage have been studied by many authors [7–19].

The turbulent wall jet (even with an isothermal coflowing or stationary external stream) is known to be a difficult flow to predict [20]. An undeniable difficulty arises through confinement effect induced by wall boundaries, and increased by the compactness of the load, which in turn implied the formation of an adverse pressure gradient along the jet axis. The resulting flow is complicated since it is often includes a combination of free turbulent shear flows, near-wall effect, recirculation areas (including high streamline curvature and probably local separation). In addition, the complexity of the system is increased by the presence of the load which increases the confinement effect and the adverse pressure gradient. Pallets and boxes affect the airflow through surface stresses, porous infiltration, deviations and reattachment and also turbulence generation. They may create secondary recirculating flows, including stagnant zones, and induce high velocities elsewhere.

According to the complexity of the indoor flows underlined above, different authors [21–25] agreed on the inadequacy of the k–ε model to predict airflow patterns and underline its limitation by comparison with experimental data. Wilcox [21] and Menter [22] stated that the k–ε model predicts significantly too high shear-stress levels and thereby delays or completely prevents separation. According to Launder [24], this trend can be more pronounced in the presence of adverse pressure gradient and leads to overprediction of the wall shear stress. Aude [25] pointed out the difficulty in accurately predicting the velocity levels, the turbulent kinetic energy, its dissipation rate and the turbulent viscosity in the stagnant regions with the k–ε model. In this case, improving predictions can be achieved by taking into account the effect of the turbulence anisotropy by using more advanced turbulence models, such as those based on the second-moment closure, or large eddy simulation.

More recently, Tapsoba, Moureh, and Flick [19], Moureh, and Flick [14,15] have tested and compared various turbulence closure models including the standard and the low Reynolds number form of the two-equation k–ε model, and the more advanced Reynolds stress model (RSM). These authors concluded that only RSM was able to accurately predict the separation of the jet from the wall and the general behavior of airflow patterns related to the primary and secondary recirculations. This concerned loaded and unloaded long enclosures where the wall jet was subjected to an adverse pressure gradient and the turbulence anisotropy effect was pronounced.

In this chapter, the numerical model developed was elaborated using the CFD code Fluent and second-moment closure with the RSM as described by Launder [26]. After adequate validation, the model could be used as a design tool to optimize more complex configurations, reducing the need for expensive and time-consuming experiments.

To take into account internal interactions between the fluid and the solid matrix formed by packed products, some authors proposed to treat the porous medium as a continuum by applying the Darcy–Forchheimer equation. The Darcy–Forcheimer–Brinkman equation has been used to describe airflow through vented horticultural packages [27–30]. In a fundamental study, Braga and de Lemos [31] considered that the porous-continuum offers a satisfactory representation of reality when the number of obstacles inside the enclosure is high. However, this point warrants further investigation on a fundamental level especially in the case of highly permeable media. Liu and Masliyah [32] consider that turbulence has little effect on total pressure drop in the porous media. Hence, the Darcy-Forchheimer model can give good predictions of the total pressure drop over the vented porous load.

The validity of the Ergun give further explanation relations was tested by van der Sman [27] using measured pressure drop versus flowrate across beds of potatoes and oranges. It was found to be accurate for these products despite the particle Reynolds number being outside the range for which it is generally assumed to be valid ($Re_p < 300$). Having established the validity of the approach, the author used the finite element solver FIDAP (Fluent Inc., NH) to simulate the airflow through vented cartons of mandarins and tomatoes. Good agreement between predicted and measured overall pressure drops was reported. Zou, Opara, and Mckibbin [29,30] developed CFD models of ventilated boxes undergoing forced-air cooling. The filled boxes were treated as porous media and the Darcy-Forchheimer model was implemented without a turbulence model and was validated against measured product temperatures and a good agreement was reported.

18.2 MATERIALS AND METHODS

18.2.1 DESCRIPTION OF THE SCALE MODEL

The experiments were carried out using a reduced-scale (1:3.3) model of a trailer with respect to the dimensionless Reynolds number (Re = $\rho U_0 D_H/\mu$). Afterward, all the data—dimensions, flow rate

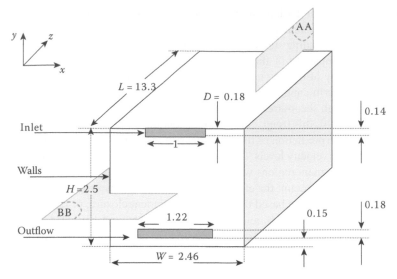

FIGURE 18.1 Dimensions of the enclosure (dimensions are expressed in m).

and results—were expressed in full scale. Airflow inlet and outlet were located in the front face of the parallelepiped as shown in Figure 18.1 in which the full-scale geometry is shown.

The slot-ventilated enclosure was supplied with air from a centrifugal turbine, which can operate at up to 5800 m³/h. The airflow was introduced in the enclosure perpendicular to the front face with a mean velocity of $U_o = 11.5$ m/s (Re = 1.9×10^5) and a relative standard deviation I_{0z} of 10%.

The enclosure was loaded with two rows of 16 polystyrene slotted boxes of size $1.7 \times 0.8 \times 1.2$ m which represent the pallets (Figure 18.2). The slots were spread over the six faces of each pallet and allow air to go through 15% of the surface (Figure 18.3a). The boxes were set down on wooden blocks of dimensions $0.1 \times 0.1 \times 0.1$ m, allowing air circulation under the load. Small gaps of 0.02 m were maintained between the boxes. In order to characterize the influence of air ducts on air distribution within the container, two configurations were investigated:

i. Without air ducts. The entire airflow is blown at the front of the truck (Figure 18.2a).
ii. With air ducts. The airflow was blown at three positions: $z = 0$ (front), $z = L/3$ and $z = 2L/3$ with 40%, 30% and 30% as flow rate distribution, respectively (Figure 18.2c).

18.2.1.1 Velocity Measurements above the Boxes

The walls of the scale model were composed of wood except one side constructed of glass to allow LDV measurements. The mean velocity and its fluctuations were obtained with a one-dimensional anemometer, which comprised of 50 mW laser diode emitting a visible red beam at 690 nm wavelength, a beam splitter, a Bragg (acousto-optic) cell, a focusing and receiving lens to collect scattered light from the measurement point and a photomultiplier. The flow to be measured must contain particles from an atomizer that scatters light as the flow carries them through the measurement volume. The probe was carried on an automatic displacement system.

18.2.1.2 Velocity Measurements within the Boxes

To characterize the ventilation within the filled boxes, heat transfer coefficients were measured with heated spheres of a diameter of 3.38 cm (the same dimension as the products). The probes were made of brass to ensure thermal homogeneity and chromed to limit the radiation effect (Figure 18.3b). The conduction between the neighboring spheres was neglected because they were hollow and made of thin celluloid. The positions of the probes in the boxes are shown in Figure 18.3c.

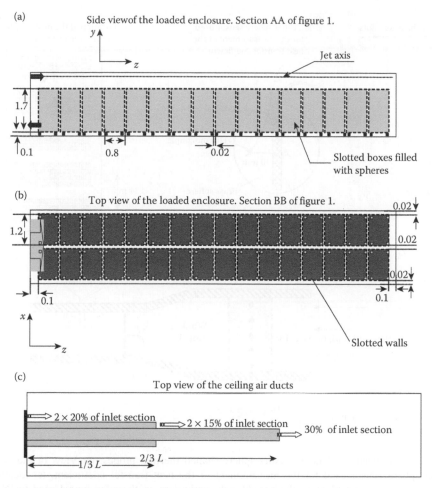

FIGURE 18.2 Configuration of the load in the enclosure. (a) Side view, (b) top view, (c) top view of the ceiling air ducts (dimensions are expressed in m).

The spheres were heated with electrical resistances in order to maintain a difference between the ambient and the sphere temperatures of about 10 K. The convection Nusselt numbers were calculated from:

$$Nu = \frac{hd}{\lambda} \qquad (18.1)$$

The heat transfer coefficient "h" was calculated from temperature measurements conducted at steady state by using:

$$h = \frac{Q_{hs}/S_{hs}}{T_{hs} - T_a} \qquad (18.2)$$

where Q_{hs}, S_{hs}, and T_{hs} are respectively, the heat power, the surface area, and the temperature of the heated sphere.

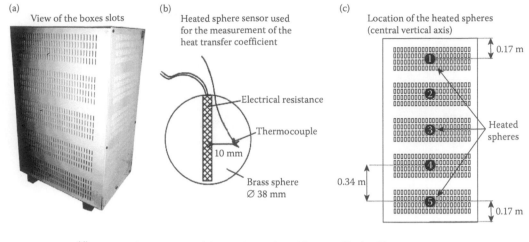

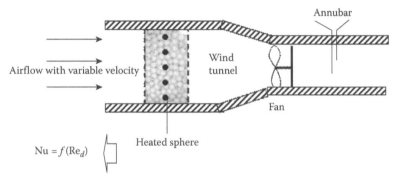

FIGURE 18.3 Experimental device using sphere-shaped thermal probes. (a) View of the boxes slots, (b) heated sphere sensor used for the measurement of the heat transfer coefficient, (c) location of the heated spheres in medium plane of the box, (d) schematic view of the wind tunnel used for $Nu = f(Re_d)$ calibration.

These spheres can be also used as thermal anemometers via a Nusselt-Reynolds relationship [33] obtained by calibration.

$$Nu = \alpha \times Re_d^{\beta} + 2 \tag{18.3}$$

In Equation 18.3 the coefficients α and β are calculated for each probe from the calibration data plotted on Table 18.1. This calibration was obtained in a wind tunnel (Figure 18.3c). After calibration, each heated sphere was used as a thermal anemometer via its corresponding Nusselt–Reynolds relationship: $Nu = f(Re_d)$ (Table 18.1). The differences between the probes were due to the random packaging of spheres. The heat transfer coefficients were influenced by the position of the nearby spheres and closely related to the local airflow behavior. The air temperature just before the bed inlet is measured using a T-type thermocouple (previously calibrated, precision ±0.2°C). Thus, the average precisions of measurements was 2% for Nu and 4% for Re.

The heated spheres were immobilized within the instrumented box in order to avoid disturbing the local arrangement of the porous structure in the vicinity of the heated spheres. The instrumented box was displaced in the enclosure in order to investigate each position.

In order to quantify the influence of load permeability on airflow pattern and ventilation efficiency, some comparisons were performed between the studied case: slotted filled boxes (SFB),

TABLE 18.1
Correlations for Calibrated Heated Spheres: $Nu = \alpha \times Re_d^\beta + 2$

Sphere	1	2	3	4	5
α	2.74	2.28	2.62	2.89	2.72
β	0.48	0.50	0.48	0.48	0.47

slotted empty boxes (SEB), impermeable boxes (IP) and unloaded configuration (without boxes). Unless otherwise statement, the SFB case was considered.

18.3 MODELING APPROACH

Filled boxes were taken into account as a porous medium, which necessitates an additional source term in the standard fluid flow equations and a singular pressure drop was used for slotted boxes walls. Thus, a single domain CFD approach was used in this study since the governing equations in clear regions and porous media are similar. The RSM as described by Launder [31], was used as turbulence model.

18.3.1 GOVERNING EQUATIONS

The time-averaged Navier–Stokes differential equations for steady, high-Reynolds numbers and incompressible flows expressed in their conservative form for mass and momentum conservation were solved by a finite volume method using the Fluent solver. The solved equations can be written as follows:

Mass conservation:

$$\frac{\partial U_j}{\partial x_j} = 0 \tag{18.4}$$

Momentum conservation:

$$\frac{\partial U_j U_i}{\partial x_j} = -\frac{1}{\rho}\frac{\partial P}{\partial x_i} + \frac{\partial}{\partial x_j}\left(\nu\frac{\partial U_i}{\partial x_j} - \overline{u_i u_j}\right) + S_i \tag{18.5}$$

where $\overline{u_i u_j}$, are the unknown Reynolds stresses and $S_i = -((\nu/K)U_i + C_2(1/2)|U_i|U_i)$ represents the source term used to model porous boxes. The first term on the right hand side represents the microscopic viscous drag while the second term accounts for inertial effects due to direction changes and sudden expansion inside the pores and to turbulent dissipation. The K and C_2 coefficients were identified experimentally as described in the following section.

18.3.2 MODELING THE PRESSURE LOSSES WITHIN A FILLED BOX

To determine airflow resistance due to the boxes, the pressure drop (Δp) was measured in a wind tunnel for a slotted wall adjacent to a stack of spheres in different depth within the bed ($l = 0.15$–0.7 m) and for different velocities ranged from 0.3 to 0.8 m/s. The pressure drop was measured with a tilted-tube pressure gauge (micromanometer ± 0.5 Pa). The airflow was measured by means of an

Annubar sensor located in the extraction duct. The analysis of these measurements, described in detail in Zou et al. [30], make it possible to obtain the singular pressure drop caused by the vented wall itself expressed as:

$$\Delta P = \frac{1}{2} C_1 \rho U_\perp^2 \tag{18.6}$$

where ΔP is the singular pressure drop and $U_\perp$ is the mean velocity normal to the wall. The coefficient C_1 was found to be ~130. However, some couples of walls were separated by small gaps. Considering that two slotted walls pressed against each other give about the same pressure drop in the flow as a single one (if the slots are aligned), we considered that the pressure drop coefficient for one slotted wall that is very close to another (C_2) is reduced to half compared with a single one: $C_2 = C_1/2$.

- The evolution of dp/dx as a function of velocity is expressed as:

$$\frac{dp}{dx} = -\frac{\mu}{K} U_D - C_2 \frac{1}{2} \rho U_D^2 \tag{18.7}$$

where $K = 1.39\ 10^{-7}\ \text{m}^2$ and $C_2 = 402\ \text{m}^{-1}$.

The measured pressure drop is slightly higher but follows the same trend as that of Ergun's equation.

18.3.3 BOUNDARY CONDITIONS

The computational domain may be surrounded by inflow and outflow boundaries in addition to symmetry and solid walls. At the inlet, uniform distribution is assumed for velocity components and turbulence. The flow rate was set at 11.5 m/s at the inlet, corresponding to about 70 air-renewals per hour in the enclosure. The inlet turbulence parameters were obtained from the measurements assuming isotropy. The boundary conditions are summarized in Table 18.2 in which the indices (τ, η, λ) represent the local coordinate system of the wall: τ is the tangential coordinate, η is the normal coordinate, and λ is the binormal coordinate. The local Reynolds stresses at the wall-adjacent cells were then computed using the wall kinetic energy.

A computational grid of $52 \times 68 \times 212$ (749,632 cells) was used for this study. The grid sensitivity of the solution was also studied by using different grids ranging from 450,000 to 1,600,000 cells. It was observed that increasing the number of cells above 600,000 does not affect the simulation results. The well-known Simple algorithm was used for coupling pressure and velocity into the continuity equation. The second-order upwind scheme was used in the model for the convection terms of flows and turbulence.

18.3.4 ANALYSIS OF VENTILATION EFFICIENCY

The overall ventilation efficiency is often characterized by the number of times the enclosure's air volume is replaced during one unit time.

$$\tau_0 = \frac{\text{Inlet flow rate (m}^3\text{h}^{-1})}{\text{Enclosure volume (m}^3)} \tag{18.8}$$

TABLE 18.2
Summary of the Boundary Conditions

	Velocity	Pressure	Turbulence
Inlet	$U_x = U_y = 0$ m/s		$k_0 = 3/2(U_0 I_{0z})^2$; $I_{0z} = 10\%$
	$U_z = 11.5$ m/s	–	$\varepsilon_0 = C_\mu^{0.75} k_0^{1.5}/0.07 D_H$; $D_H = 0.$
			$\dfrac{(u_i^2)_0}{k_0} = \dfrac{2}{3}$ and $(u_i u_j)_0 = 0)$
Outflow	–	Constant pressure	–
Walls	$U_x = U_y = U_z = 0$ m/s		$\dfrac{u_\tau^2}{k} = 1.098$; $\dfrac{u_\eta^2}{k} = 1.098$
		–	$\dfrac{u_\lambda^2}{k} = 0.655$; $\dfrac{u_\eta u_\tau}{k} = -0.25$
Symmetry	Zero normal gradient for all variables		

An extension of this concept is proposed here in order to characterize the local ventilation efficiency of the boxes. Thus, we analyze the flow rate $\dot{m}$, flowing in and out of the volume V_b of one box of surface Σ (Figure 18.4a). According to numerical simulations, $\dot{m}$ can be computed by the following integration:

$$\dot{m} = \frac{1}{2} \int_\Sigma \rho |\vec{U}.\vec{n}| d\Sigma \qquad (18.9)$$

where $\vec{n}$ is the unit normal vector of the elementary surface $d\Sigma$.
For each box, a local ventilation efficiency can be written as:

$$\tau = \frac{1}{\rho} \frac{\dot{m}}{V_b} \qquad (18.10)$$

However, the air flowing in the volume of a box is not only composed of fresh injected air but also by recirculating air (Figure 18.4b). If ventilation is used to extract heat which is generated in the enclosure, only the fresh injected air is efficient. In order to characterize the quantity of fresh air entering a box, a fictitious and uniform volumetric heat load per unit volume, q, is applied throughout the pallet domain. This simulates for example heat generation due to respiration of a load of fruit or vegetable. In steady state, the heat balance of a box can be expressed using bulk average air temperatures of the air flowing in and out, named T_{in} and T_{out}, respectively.

$$Q = qV_b = \dot{m} C_p (T_{out} - T_{in}) \qquad (18.11)$$

Moreover, the incoming airflow can be considered as the mixing of a quantity of fresh air named $\dot{m}_{eq}$ at temperature T_0 and a recirculating air flow rate $(\dot{m} - \dot{m}_{eq})$ at temperature T_{out} (Figure 18.4b).
Thus, the heat balance can be rewritten:

$$Q = \dot{m}_{eq} C_p (T_{out} - T_0) \qquad (18.12)$$

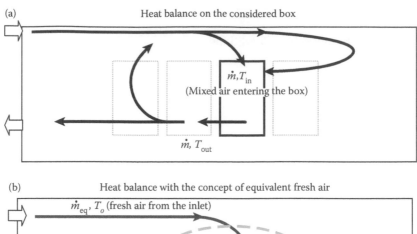

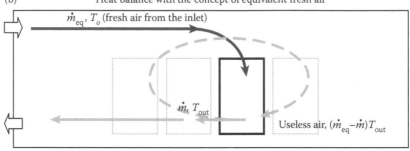

FIGURE 18.4   Introducing the concept of equivalent fresh air renewal for one box.

This means that, in terms of heat extraction capacity, the process gives rise to an equivalent flow rate of injected fresh air $\dot{m}_{eq}$ entering the volume. The additional airflow: $\dot{m} - \dot{m}_{eq}$ flows in a circular manner and does not affect the heat balance.

From the simulations, following quantity A can be computed.

$$A = \int_{\Sigma} \rho |\vec{v} \cdot \vec{n}| T d\Sigma = \int_{\Sigma_{in}} \rho |\vec{v} \cdot \vec{n}| T d\Sigma_{in} + \int_{\Sigma_{out}} \rho |\vec{v} \cdot \vec{n}| T d\Sigma_{out}$$

$$= T_{in} \int_{\Sigma_{in}} \rho |\vec{v} \cdot \vec{n}| d\Sigma_{in} + T_{out} \int_{\Sigma_{out}} \rho |\vec{v} \cdot \vec{n}| d\Sigma_{out} \qquad (18.13)$$

$$= (T_{in} + T_{out}) \dot{m}$$

From Equations 18.12 and 18.14, the temperatures T_{in} and T_{out} can be calculated and finally $\dot{m}_{eq}$ can be obtained from Equation 18.13:

$$\begin{cases} T_{out} = \dfrac{1}{2\dot{m}} \left(A + \dfrac{Q}{C_p} \right) \\[3mm] T_{in} = \dfrac{1}{2\dot{m}} \left(A - \dfrac{Q}{C_p} \right) \\[3mm] \dot{m}_{eq} = \dfrac{Q}{C_p (T_{out} - T_0)} = \dot{m} \dfrac{T_{out} - T_{in}}{T_{out} - T_0} \end{cases} \qquad (18.14)$$

We can then define a local ventilation efficiency based on the equivalent fresh air renewal:

$$\tau_{eq} = \frac{1}{\rho} \frac{\dot{m}_{eq}}{V_b} \tag{18.15}$$

A heat load of 50 W.m^{-3} was applied in the simulations. Adiabatic boundary condition was applied on the enclosure walls and free convection was not taken into account in order to point out only the effect of forced convection inside the enclosure. For these conditions, a dimensional analysis showed that τ and τ_{eq} are independent of the heat load.

18.4 RESULTS AND DISCUSSION

18.4.1 JET BEHAVIOR WITHOUT AIR DUCTS

In order to illustrate the overall behavior of the airflow pattern above the boxes, Figures 18.5 and 18.6 show comparisons between numerical predictions and LDV measurements concerning the evolution of the jet characteristics in the symmetry plane related to the normalized velocity U_z/U_0, and static pressure.

In the front region $z/L \in [0,0.5]$, strong decay of velocity along the jet axis reflects the cumulative effect due to the adverse pressure gradient (Figure 18.6) and to the compactness of the load which in turn enhanced lateral interactions with the return flow in the headspace. Both mechanisms strongly affected jet development and caused it to vanish at $z/L = 0.5$ approximately. In a similar load configuration, but with SEB, Tapsoba et al. [19] found that the limit of the stagnant zone corresponds to the point where the jet leaves the ceiling and dives into the load.

In the rear part region ($z/L \in [0.5,1]$), flow is weak; U_z and velocity magnitude did not exceed 5% of U_0. The uniform static pressure level observed in this area also reflects the presence of a stagnant zone. With regard to the efficiency of air ventilation and its homogeneity within the whole enclosure and throughout the load, this type of airflow is highly undesirable, particularly in the stagnant zone where high levels of temperature and contaminants could be expected due to poor mixing with the inlet jet. Good agreement is obtained between numerical and experimental data concerning the jet decay along the container.

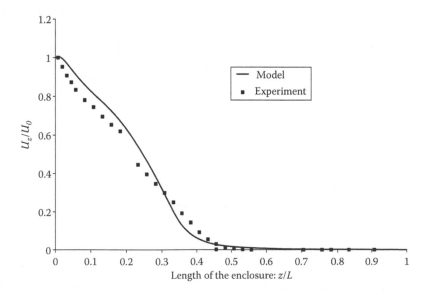

FIGURE 18.5 Evolution of the longitudinal velocity on the jet axis along the enclosure.

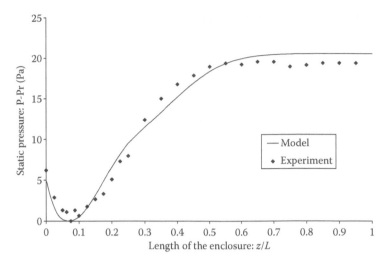

FIGURE 18.6   Evolution of the static pressure along the jet axis: comparison between experiment and numerical results.

18.4.2   OVERALL AIRFLOW DESCRIPTION IN THE HEADSPACE ABOVE BOXES

Figures 18.7 and 18.8 show comparison between numerical and LDV measurements data concerning the contour levels and velocity vectors of longitudinal normalized velocity U_z/U_0 in the symmetry plane and in the inlet-centered horizontal plane, respectively. These contours are obtained from 568 measurement points using the LDV system. To illustrate local jet behavior, Figure 18.9 presents experimental data concerning vertical velocity profiles obtained in the symmetry plane and horizontal profiles obtained in the inlet-centered plane at $z = 1, 2, 4, 6$, and 8 m.

These results show a high degree of heterogeneity in terms of airflow and velocity levels within the enclosure. The high velocities and steep gradients observed in the front part $z/L \in [0,0.5]$, contrasted with the stagnant zone, in the rear part $z/L \in [0.5,1]$. Due to the confinement effect, two lateral vortices structures were induced by jet intrusion into the enclosure as shown in Figure 18.8 (negative values of U_z). These structures dynamically reinforced the confinement effect, controlled the initial growth of the jet and limited its diffusion in the transverse direction. Even with the complexity of the airflow in the transverse direction, where strong aerodynamic interactions occur between the jet and lateral vortices, reasonable agreement is observed between numerical and experimental values concerning airflow patterns.

The limitation of jet transverse diffusion contributes to the enhancement of its wall-normal expansion in the headspace between the ceiling and the top of the load. Experimental results concerning contours and vertical velocity profiles at $z = 1, 2$, and 4 m indicated the presence of return flow with negative values of U_z on the top of the boxes. This takes place between the inlet and $z/L = 0.4$ where the jet attaches to the top of the boxes (contour of 0 m/s in Figure 18.7a). As a consequence, the parabolic profile of the wall jet was progressively transformed into a linear shape at $z/L = 6/13.3$ (Figure 18.9a), which reflects the formation of an overall shear flow in the headspace integrating the jet and the return flow.

Close to the jet load attachment area $z/L \in [0.3,0.5]$, the jet bulk axis experienced a rapid migration from the ceiling to the top of the boxes where higher values of jet velocity were located (Figure 18.10). This clearly indicates the domination of the Coanda effect exerted by the top of the load on the jet ceiling attachment. This effect leads to a Couette flow occupying the whole width of the headspace where the jet attaches simultaneously the ceiling and the top of the load. Further downstream, the jet flows onto the load in the rear as shown by the experimental results obtained in the symmetry plane incomplete sentence. This tendency was not captured by the numerical model since the predicted jet trajectory was still close to the ceiling as shown in Figures 18.10 and 18.11. This clearly indicates that the aerodynamic interaction between the jet and the top of

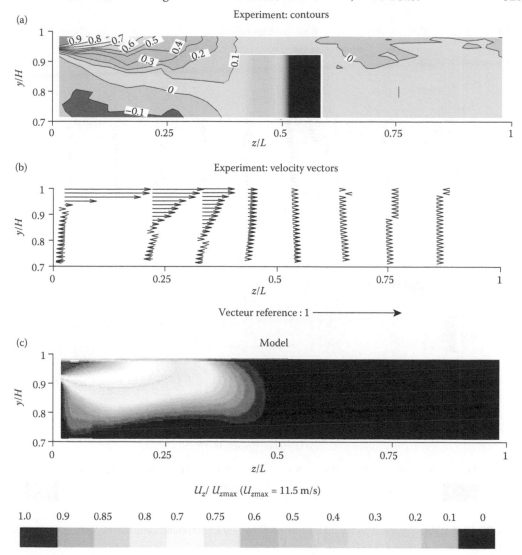

FIGURE 18.7   Velocity levels in the symmetry plane. (a) Experiment: contours of U_z/U_{zmax}, (b) experiment: velocity vectors, (c) model: contours of U_z/U_{zmax}

the porous boxes needs to be improved by taking into account the horizontal frictional resistance exerted by the top of the boxes against the jet. Although this horizontal friction could be neglected in terms of pressure losses compared with the perpendicular effect, it becomes essential for the numerical model to be able to predict the jet attachment on the top of the boxes by the Coanda effect seen experimentally in Figures 18.8d and 18.16a ($z/L = 0.5$). This aspect could also explain the inability of the numerical model to capture the presence of the moderate recirculated zone between the inlet and the jet attachment at $z/L = 0.4$ (Figure 18.7a).

18.4.3   Velocity Levels Characteristics within Filled Boxes

Figure 18.12 shows an overall comparison between numerical and experimental data concerning the evolution within the enclosure of the mean box velocities represented by the average value of the five sensor positions (Figure 18.3c). The average velocity decreases continuously with the model, instead a bimodal maximum was experimentally observed. This lack of agreement with experiment especially in the middle

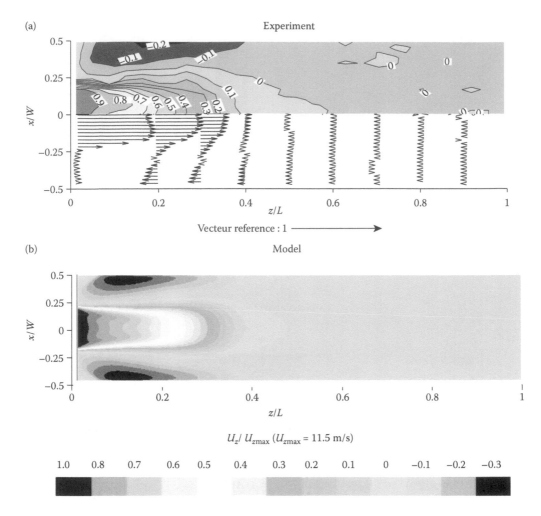

FIGURE 18.8 Velocity levels in the inlet-centered plane. (a) Experiment: contours of U_z/U_{zmax} and velocity vectors, (b) model: contours of U_z/U_{zmax}.

part (boxes 4–6) where the higher differences can reach 33%, 50%, and 40% for the fourth, fifth and sixth boxes, respectively. This could be explained by the inability of the model to predict the circulation flow in the front above the load due to the no attachment of the jet at the top of the load in this area. In addition, the model underpredicts velocities within all boxes. This tendency could be partially explained by the turbulence effect on the velocity measurement with the heated spheres, especially for lower velocities where the fluctuating part is of the same order of magnitude as the mean values (data not shown).

18.4.4 Influence of the Load Permeability on Airflow

Figure 18.13 presents the evolution of the static pressure through the inlet section along the enclosure for filled and empty slotted boxes. Both curves indicated the presence of a high positive (adverse) pressure gradient between the inlet zone and the reattachment of the jet on the top of the load (z/L 0.4). The decrease of the load permeability increased the adverse pressure gradient, which in turn reduced the jet penetration distance. Figure 18.14 shows the jet decay for different studied configurations. This figure clearly shows that jet decay was stronger with lower load permeability. The jet penetration distance varies from $z/L = 0.7$ for empty enclosure to

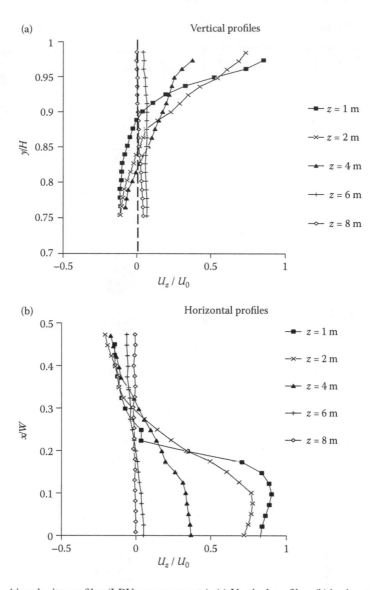

FIGURE 18.9 Air velocity profiles (LDV measurements). (a) Vertical profiles, (b) horizontal profiles.

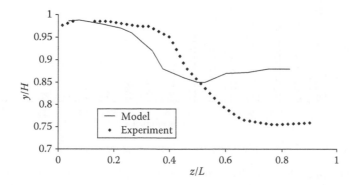

FIGURE 18.10 Jet trajectory in the symmetry plane (y_{max} vs. z).

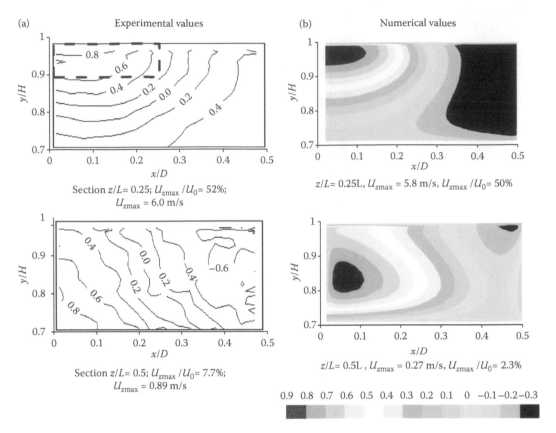

FIGURE 18.11 Contours of longitudinal normalized velocity U_z/U_{max} in transversal sections along the enclosure at $z/L = 0.25$ and 0.5: comparison between numerical and experimental values.

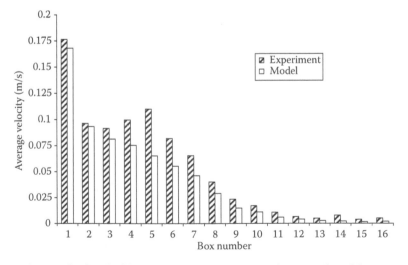

FIGURE 18.12 Average box's velocities: comparison between experiments and model.

$z/L = 0.4$ for an impermeable load. The impermeability of the load (impermeable box case) tends to confine the return flow principally to above the boxes and also limits the wall jet development. Jet penetration was reduced as the load permeability decreased. Numerical results concerning

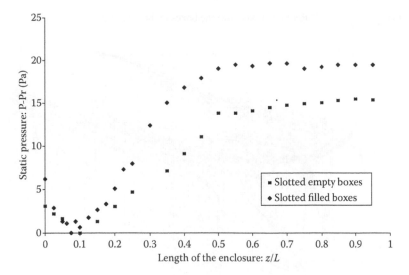

FIGURE 18.13 Evolution of the static pressure along the jet axis: comparison between empty and porous boxes (experimental data).

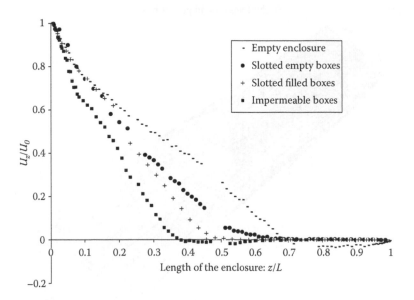

FIGURE 18.14 Influence of the load permeability on the jet decay: experimental data.

streamlines (Figure 18.15) confirm the importance of the short-circuit and recirculation above the boxes especially in the inlet area.

These conclusions clearly show the importance of load permeability on the jet behavior. Figure 18.16 shows the approximate outlines of air patterns for the different configurations.

18.4.5 Influence of Air Ducts

18.4.5.1 Influence of Air Ducts on Airflow above the Load

Figure 18.17 compares the contours of velocity magnitude in the inlet horizontal plane obtained numerically with and without air ducts. Without air ducts, strong aerodynamic interactions were

(a)   Recirculation above the boxes in the inlet are

(b)   Streamlines around porous boxes

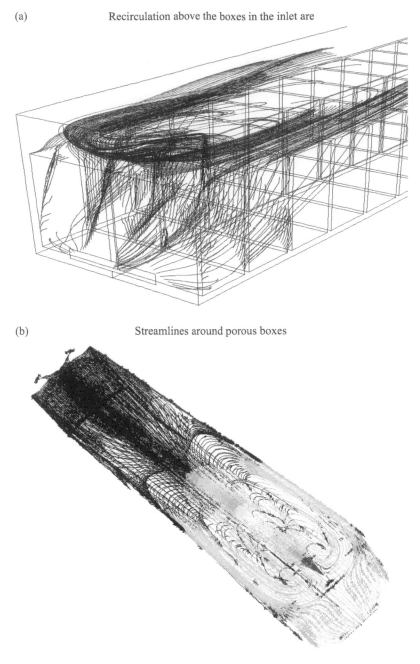

FIGURE 18.15 Streamlines in SFB: numerical results (a) recirculation above the boxes in the inlet area (b) streamlines around boxes.

observed in the transverse direction between the two opposing streams supplying inlet and outlet sections. This limits the wall-jet development and leads to uneven air distribution within the container. The use of air ducts implies a better diffusion of injected air along the container. As it can be seen, the recirculating structure located in the inlet area, in which strong aerodynamic interactions occur between the inlet jet and the return flow was split into three distinct structures with less intensity. This implies a better aerodynamic development for each inlet jet, hence, more homogeneity of ventilation within the whole container.

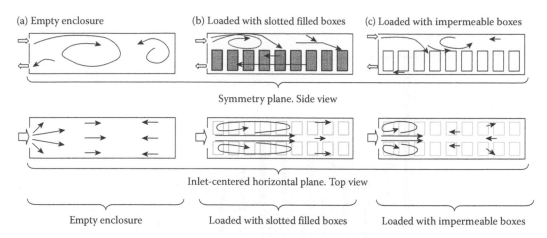

(a) Empty enclosure   (b) Loaded with slotted filled boxes   (c) Loaded with impermeable boxes

Symmetry plane. Side view

Inlet-centered horizontal plane. Top view

Empty enclosure   Loaded with slotted filled boxes   Loaded with impermeable boxes

FIGURE 18.16 Approximate outlines of air patterns, comparison between the empty enclosure and the loaded cases.

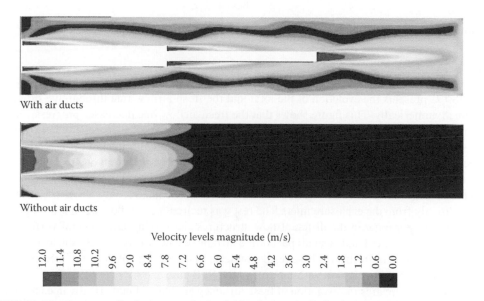

With air ducts

Without air ducts

Velocity levels magnitude (m/s)

12.0 11.4 10.8 10.2 9.6 9.0 8.4 7.8 7.2 6.6 6.0 5.4 4.8 4.2 3.6 3.0 2.4 1.8 1.2 0.6 0.0

FIGURE 18.17 Contours of velocity magnitude obtained numerically in the inlet horizontal plane.

18.4.5.2 Influence of Air Ducts on Airflow within the Load

Figure 18.18 presents an experimental comparison of the box's air velocities obtained with heated sphere probes within boxes with and without an air duct. The box's air velocities were taken as the average value of the five sensor positions. This figure clearly shows that air velocity magnitudes within the boxes were governed by the behavior of the airflow characteristics of the external flow developed by the wall jet in the headspace. Without air ducts, the velocity level remained lower from box 9 to the rear part ($V < 0.05$ m/s). This region was localized under the stagnant zone. The last boxes (11–16) were 20 times less ventilated than the box 5. With the exception of the box 1, it was observed that without ducts, the maximum mean velocity was found on the pallet where the jet penetrates the load (box 5). When using ducts, one can notice three local maxima on the boxes 4, 8 and 16. In addition, the use of ducts tended to improve the ventilation of the last boxes: 14–16. The mean velocity is multiplied by ten compared with the case of no ducts. However, one can observe a reduction in the velocity levels in boxes 2–6, which are of the same order of magnitude as that in boxes 14–16. Finally, the use of air ducts does not affect the intensity of the short-circuit which dominates the first box.

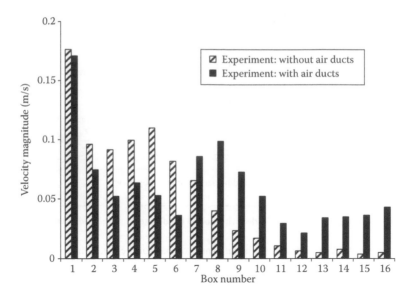

FIGURE 18.18 Influence of air ducts on air velocities within boxes obtained with heated spheres.

18.4.6 VENTILATION EFFICIENCY

Figure 18.19a presents the evolution of the total and the fresh airflow rate through the filled boxes as obtained numerically. This figure shows that the fresh airflow rate decreases progressively from the front to the rear whereas the total airflow rate displayed a higher heterogeneity between boxes with a strong short circuit through the first box. The total flow rate throughout the last box is about 55 times lower than for the first box. The same ratio of fresh air varied from one to ten.

For the first box, the equivalent fresh air flow rate is about five times lower than the total air flow rate. This means that the air flowing into one of these boxes contains only about 1/5 of fresh air (coming directly from the enclosure inlet). The rest was recirculated by flowing first throughout the other boxes or by mixing in the different flow structures (jet mixing layer, lateral vortices above the load…). On the other hand, a small part of the airflow reaches the rear of the enclosure but was essentially composed of fresh air. These results indicate that only 50% of the total injected fresh air contributes to the bulk ventilation within boxes in the SFB case and 67% for SEB.

Figure 18.19b compares the evolution of fresh air renewal obtained numerically in each of the 16 boxes between the SEB and SFB. Excepting the case of box 1, higher values are obtained with SEB than with SFB. For the SEP case, the local ventilation efficiency trend (τ_{eq}) shows some uniformity for the first five boxes at around 120 h^{-1} followed by a rapid decrease, whereas in the SFB case, a uniform decrease along the whole enclosure was observed.

For the SFB case, the value of τ_{eq} obtained for the last box is only ten times lower than that of the highest value obtained for the first box. This figure also indicates that the fresh air renewal for the last boxes is three/four times lower than the overall value. This ratio indicates the deficit of ventilation in the rear enclosure area with respect to the overall value τ_0 (Figure 18.19b).

18.5 CONCLUSIONS

Experiments on a reduced-scale model and CFD simulation were performed to study an enclosure supplied by a ceiling-jet and loaded with slotted filled boxes. In this study, an original approach was developed using an LDV and thermal sphere-shaped probes located inside the boxes. The aim was to investigate air velocity characteristics above and within boxes. This allows to characterize airflow patterns and to quantify the performance of box ventilation in a typical refrigerated truck

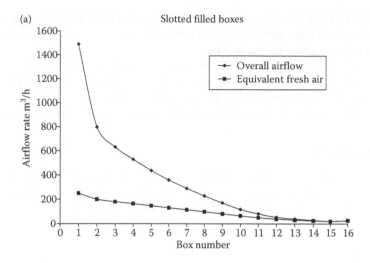

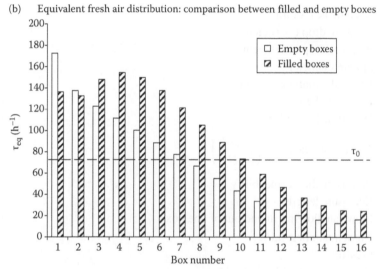

FIGURE 18.19 Ventilation efficiency within boxes: numerical values. (a) Overall airflow rate and equivalent fresh air in SFB, (b) equivalent fresh air distribution: comparison between filled and empty boxes.

configuration without and with an air duct system which provides air release at the inlet and at two additional points along the container.

The results highlight the importance of the load permeability and confinement effect in reducing the reach of the jet within the truck. This leads to a high degree of ventilation heterogeneity between the front and the rear where stagnant zones and low velocities are present. Numerical predictions concerning the evolution of the jet along the enclosure and velocity contours show rather good agreement with experiments.

Experimental results confirm the deviation of the jet from the ceiling and its attachment to the top of the load at $z/L = 0.4$ approximately, where it flows along it to the rear of the enclosure. This leads to the formation of a moderate circulation bubble in the headspace upstream of the jet attachment. This mechanism was not captured by the numerical model and requires taking into account of wall shear stress at the interface between the jet and the slotted walls.

Concerning ventilation efficiency, numerical values indicate that the amount of fresh air decreases progressively from the front to the rear, whereas the total airflow rate displays a higher heterogeneity between boxes. The total flow rate throughout the last box is about 55 times lower than that for the first box. The results also indicate that the fresh air renewal for the last boxes is ten times lower than that of the highest value obtained for the first box.

The use of air ducts contributes significantly to a more even distribution throughout the container by improving air supply toward the rear, whilst reducing air flow intensity at the front. Velocity levels were lowest at the rear, and the maximum near the fifth box. The last box (at the rear) was about 20 times less ventilated than box 5 without air ducts. This ratio drops to four when using air ducts.

ACKNOWLEDGMENTS

The author would like to thanks the Ile de France region and the French Ministry of Agriculture, Food, Fisheries and Rural Affairs for their support and funding this work through an AQS project. He would also thank Dr. O. Laguerre for her help in reviewing this paper.

NOMENCLATURE

C_p	Specific heat of air	[J kg^{-1} K^{-1}]
C_1	Pressure drop coefficient	
C_2	Pressure drop coefficient	
d	Sphere diameter	[m]
D_H	Hydraulic diameter of inlet	[m]
H	Enclosure height	[m]
h	Heat transfer coefficient	[W.m^{-2}.K^{-1}]
I	Turbulence intensity $\left(=\dfrac{\sqrt{\overline{u_i^2}}}{U_i}\right)$	(%)
K	Porous media permeability	[m^2]
L	Enclosure length	[m]
$\dot{m}$	Flow rate flowing in and out of one box	[kg.s^{-1}]
Nu	Nusselt number: hd/λ	
p	Pressure	
Q_{hs}	Heating power of heated sphere	[W]
Re$_d$	Reynolds number based on the sphere diameter: Re$_d$ = ρU_Dd/μ	
S	Surface area	[m^2]
T_{hs}	Temperature of the heated sphere	[K]
$\overline{u_i u_j}$	Reynolds stresses component	[m^2.s^{-2}]
U_i, u_i	Mean and fluctuating velocity component in x_i direction	[m.s^{-1}]
V	Velocity magnitude	[m.s^{-1}]
V_b	Volume of a box	[m^3]
$\dot{V}_b$	Airflow rate of a box	[m^3.s^{-1}]
x, y, z	Lateral, vertical and longitudinal coordinates	[m]
W	Enclosure width	[m]

Greek symbols

Σ	Box surface	[m^2]
α, β, γ	Empirical constants	
ε	Turbulent dissipation rate.	[m^2.s^{-3}]
δ_{ij}	Kronecker symbol	

μ	Laminar dynamic viscosity	[Pa.s]
λ	Thermal conductivity	[W.m^{-1}.K^{-1}]
μ_t	Turbulent viscosity	[Pa.s]
ν	Kinematic viscosity	[m^2.s^{-1}]
ρ	Density	[kg.m^{-3}]
τ	Ventilation efficiency	[h^{-1}]

Subscript

⊥	Normal
0	Relative to inlet boundary condition
a	Air
b	Box
d	Related to sphere diameter
eq	Equivalent
in	Flowing in the considered box
out	Flowing out the considered box
hs	Heated sphere
i, j, k	Relative to coordinate system
ref	Reference value
t	Turbulent
x, y, z	Relative to coordinate system

REFERENCES

1. Lenker, D.H., Wooddruff, D.W., Kindya, W.G., Carson, E.A., Kasmire, R.F., and Hinsch, R.T. Design criteria for the air distribution systems of refrigerated vans. *ASAE Paper*, 28, 2089, 1985.
2. Gögus, A.Y. and Yavuzkurt, S. Temperature pull-down and distribution in refrigerated trailers. In: *Proceedings I.I.F–I.I.R Commissions D2*, Wageningen, The Netherlands, 189, 1974.
3. LeBlanc, D., Beaulieu, C., Lawrence, R., and Stark, R. Evaluation of temperature variation of frozen foods during transportation. *The Refrigeration Research Foundation Information Bulletin (Bethesda, MD)*, December, 1994.
4. Bennahmias, R. and Labonne, R.G. Etude de la distribution de l'air et de la dispersion des températures dans une semi-remorque frigorifique. *Réunion des commissions C2, D1 et D2/3 de l'IIF, Fez (Morocco)*, 241, 1993.
5. Meffert, H.F.Th. and Van Nieuwenhuizen, G. Temperature distribution in refrigerated vehicles. In: *Proceedings I.I.F.–I.I.R. Commissions D1, D2 and D3*, Barcelona, Spain, 131, 1973.
6. Billiard, F., Bennahmias, R., and Nol, P. Nouveaux développements dans les transports à température dirigée routiers. In: *Proceedings I.I.F.-I.I.R. Commissions B2, C2, D1, D2/3*, Dresden, Germany, 793, 1990.
7. Lindqvist, R. Air distribution design for controlled atmosphere in reefer cargo holds. In: *20th International Congress of Refrigeration, IIR/IIF, Sydney*, Australia, 1999.
8. Wang, H. and Touber, S. Simple non-steady state modelling of a refrigerated room accounting for air flow and temperature distributions. In: *Proceedings I.I.F–I.I.R. Commissions B1, B2, C2, D1, D2/3*, Wageningen, The Netherlands, 211, 1988.
9. Meffert, H.F.Th. and Van Beek, G. Basic elements of a physical refrigerated vehicles, air circulation and distribution. In: *16th International Congress of Refrigeration, I.I.F.–I.I.R.*, Paris, France, 466, 1983.
10. Zertal-Ménia, N. Etude numérique et expérimentale de l'aéraulique dans un véhicule frigorifique. *Thèse INA-PG*, 2001.
11. Wang, H. and Touber, S. Distributed dynamic modelling of a refrigerated room. *International Journal of Refrigeration*, 13, 214, 1990.
12. Van Gerwen, R.J.M. and Van Oort, H. Optimization of cold store using fluid dynamics models. In: *Proceedings I.I.F.-I.I.R. Commissions B2, C2, D1, D2/3*, Dresden, Germany, 4, 473, 1990.

13. Hoang, M.L., Verboven, P., De Baermaeker, J., and Nicolaï, B. M. Analysis of air flow in a cold store by means of computational fluid dynamics. *International Journal of Refrigeration*, 23, 127, 2000.

14. Moureh J. and Flick D. Wall air-jet characteristics and Airflow patterns within a slot ventilated enclosure. *International Journal of Thermal Sciences*, 42, 703, 2003.

15. Moureh, J. and Flick, D. Airflow characteristics within a slot-ventilated enclosure. *International Journal of Heat and Fluid Flow*, 26, 12, 2005.

16. Moureh, J., Zertal-Menia, N., and Flick, D. Numerical and experimental study of airflow in a typical refrigerated truck configuration loaded with pallets. *Computer and Electronics in Agriculture*, 34, 25, 2002.

17. Moureh, J. and Flick, D. Airflow pattern and temperature distribution in a typical refrigerated truck configuration loaded with pallets. *International Journal of Refrigeration*, 27, 464, 2004.

18. Nordtvedt, T., Cold air distribution in refrigerated trailers used for frozen fish transport. In: *I.I.F.–I.I.R. Commissions B1, B2, D1, D2/3,* Palmerston North, New Zealand, 2, 539, 1993.

19. Tapsoba, M., Moureh, J. and Flick, D. Airflow pattern in an enclosure loaded with pallets: the use of air ducts. In: *Eurotherm Seminar 77, Heat and Mass Transfer in Food Processing,* June 20–22 Parma. Italy, 2005.

20. Craft, T.J. and Launder, B.E. On the spreading mechanism of the three-dimensional turbulent wall jet. *Journal of Fluid Mechanics*, 435, 305, 2001.

21. Wilcox, D.C. *Turbulence Modeling for C.F.D.* DCW Industries, Inc., La Cañada, CA, 1994.

22. Menter, F.R. Eddy viscosity transport equations and their relation to the k-ε model. *ASME Journal of Fluids Engineering*, 119, 876, 1997.

23. Nallasamy, M. Turbulence models and their applications to the prediction of internal flows: a review. *Computers and Fluids*, 151, 1987.

24. Launder, B.E. On the modeling of turbulent industrial flows. In: *Proceedings of Computational Methods in Applied Sciences,* Hirsch et al. Ed. Elsevier, Amsterdam, 91, 1992.

25. Aude, P., Béghein, C., Depecker, P., and Inard, C. Perturbation of the input data of models used for the prediction of turbulent air flow in an enclosure. *Numerical Heat Transfer, Part B*, 34, 139, 1998.

26. Launder, B.E., Reece, G.J., and Rodi, W. Progress in the development of a Reynolds-stress turbulence closure. *Journal of Fluid Mechanics*, 68, 537, 1975.

27. Van der Sman, R.G.M. Solving the vent hole design problem for seed potato packages with the Lattice Boltzmann scheme. *International Journal of Computational Fluid Dynamics*, 11, 237, 1999.

28. Mirade, P.S., Rougier, T., Daudin, J.D., Picque, D., and Corrieu G. Effect of design of blowing duct on ventilation homogeneity around cheeses in a ripening room. *Journal of Food Engineering*, 75, 59, 2006.

29. Zou, Q., Opara, L.U., and Mckibbin, R. A CFD modeling system for airflow and heat transfer in ventilated packaging for fresh foods: I. Initial analysis and development of mathematical models. *Journal of Food Engineering*, 77, 1037, 2006.

30. Zou, Q., Opara, L.U., and Mckibbin, R. A CFD modeling system for airflow and heat transfer in ventilated packaging for fresh foods: II. Computational solution, software development, and model testing. *Journal of Food Engineering*, 77, 1048, 2006.

31. Braga, E.J. and de Lemos, M.J.S. Heat transfer in enclosures having a fixed amount of solid material simulated with heterogeneous and homogeneous models. *International Journal of Heat and Mass Transfer,* 48, 4748, 2005.

32. Liu, S.J. and Masliyah, J.H. Single fluid flow in porous media. *Chemical Engineering and Communication*, 150, 653, 1996.

33. Wakao, N. and Kaguei, S. *Heat and Mass Transfer in Packed Bed.* Gordon and Breach Science, New York, NY, 408, 1982.

Section V

Non-Thermal Processing of Food

19 Computational Fluid Dynamics Analysis of High Pressure Processing of Food

A.G. Abdul Ghani
Software Design Ltd.

Mohammed M. Farid
The University of Auckland

CONTENTS

19.1   INTRODUCTION

Nonthermal processing of food using high pressure can be applied to a large number of food products (juices, milk, meat, and other solid foods) using batch or continuous treatment units. Some of the high-pressure treated food products such as juices, jams, jellies, yogurts, meat, and oysters are already available in the market in the United States, Europe, and Japan.

19.1.1   What is High Pressure Processing (HPP) of Food?

High pressure processing (HPP) is a nonthermal treatment of food by the application of high pressure of the order of thousands of atmospheres. The process retains food quality such as freshness and maintains its high nutrients content such as vitamins. HPP of foods is an emerging technology of considerable interest because it permits microbial inactivation at low or moderate temperatures.

19.1.2   What are the Advantages of Using High Pressure Processing (HPP)?

Some of the features of HPP are:

1. HPP can be applied without causing significant heating that can damage taste, texture and nutritional value of the food. Since spoilage organisms can be destroyed, foods will stay fresher.
2. HPP is based on "hydrostatic pressure" which is applied to liquid or solid food uniformly in all directions. It does not create shear force to distort food particles. Thus, any moist food such as a whole grape can be exposed to these very high pressures without being crushed.
3. Pressure transmission in HPP is instantaneous and uniform. Pressure transmission is not controlled by product size and hence, it is effective throughout the food items, from the surface through to the center.
4. The "mechanism" of HPP does not promote the formation of unwanted new chemical compounds, "radiolytic" by-products or free-radicals. Vitamins, texture and flavor are basically unchanged during treatment. For example, enzymes can remain active in high pressure treated orange juice.
5. The amount of energy needed to compress food is relatively low. HPP is more energy efficient than many other food production methods that require heat.
6. HPP can also be used for modifying the physical and functional properties of foods.

The main disadvantage of using HPP is the high cost of the equipment needed for processing. Also, food with limited moisture content cannot be treated by HPP as they will shrink and collapse during compression.

19.1.3   How Much Work has been Done to Date?

Extensive experimental works have been conducted on this process during the last decade. However, limited theoretical analysis of the HPP process is available in the literature.

    The temperature in the treatment chamber is a function of position and time and is influenced by both the heat generated due to compression and the external heat transfer rate. The development of a computational model to calculate temperature distribution is very important as experimental testing of every single food at different operating conditions can be very time consuming.

In the past decade, it has become clear that HPP may offer major advantages to the food preservation and processing industry [1,2]. Next to inactivation of microorganisms and spoilage enzymes [3,4], promising results have been obtained in the application of food proteins gelation [5,6] and in improving the digestibility of proteins and tenderization of meat products [7,8].

There is growing interest in the combined effect of temperature and pressure as an effective means of inactivation of microorganisms. Bacterial spores are more resistant to temperature and pressure than vegetative bacteria, and the combined pressure–temperature effect is very efficient for the treatment of spores. Also, it is known that food undergoes minimum nutrient destruction at temperatures below 100°C. Hence, the application of HPP at moderate temperatures can be applied to a large number of food products, especially those contaminated with spores which are usually sterilized thermally at 121°C. The temperature distribution in the high pressure vessel may result in a nonuniform distribution of microbial inactivation and quality degradation. Without fully understanding the combined effects of pressure and temperature on microbial inactivation, the application of such process remains limited and expensive.

19.1.4 Analysis of High Pressure Processing (HPP) Using Computational Fluid Dynamics (CFD)

Computational Fluid Dynamics (CFD) models have been applied in a number of applications, such as aerospace, automotive, nuclear and more recently in food processing. Scott and Richardson [9] discussed mathematical modeling techniques of CFD, which can be used to predict flow behavior of fluid food in food processing equipment. Advances in computing speed and memory allow ever more accurate and rapid calculations to be performed. A number of commercial software packages are now available to carry out these calculations such as FIDAP, FLUENT, FLOW 3-D and PHOENICS, which are used in the simulations discussed in this chapter. CFD models can be of great benefit in a variety of food engineering applications; however, its use has been only recent in the application of HPP of foods. There is growing interest toward the use of mathematical models to predict food temperature during nonthermal treatment of foods such as HPP, but the literature still shows limited attempts.

19.2 MATHEMATICAL MODELING OF HIGH PRESSURE PROCESSING (HPP)

19.2.1 Mathematical Modeling of Conductive Heat Transfer during High Pressure Processing (HPP)

Pressure is transmitted uniformly and immediately through the pressure transferring medium according to the Pascal principal, and thus the effect of pressure is independent of product size and geometry [10]. In fact, heat transfer characterizes every process, accompanied by a period of pressure increase or decrease, because an increase or decrease in pressure is associated with proportional temperature change of the vessel's contents due to adiabatic heating [11]. Heat transfer is caused by the resulting temperature gradients and can lead to large temperature differences especially in large-volume industrial vessels. These limitations should be taken into account in the analysis of HPP processing especially for industrial sized units. By taking into account the nonuniform temperature distribution during the process, it can be assured that the objective of the process has been accomplished everywhere within the food product. For this purpose, the heat conduction equation describing the time-temperature-pressure history of a product must be coupled with the parameters describing the reaction kinetics for inactivation of microorganisms, enzymes and nutrients.

A numerical model for predicting conductive heat transfer during HPP of foods was simulated by Denys et al. [11]. Agar gel was used for testing as a food simulator. HPP processes with a gradual, step by step pressure build-up, pressure release, and pressure cycling HPP processes were included. The model provides a tool to evaluate batch high pressure processes in terms of temperature and/or pressure uniformity. *Bacillus subtilis* α-amylase (BSA) was used as an example to study enzyme inactivation.

A model combining numerical heat transfer and enzyme inactivation kinetics was also developed by Denys et al. [12]. In their work, a numerical conductive heat transfer model for calculating temperature evolution during HPP of foods was tested for two food systems: apple sauce and tomato paste. The uniformity of inactivation of BSA and soybean lipoxygenase during batch HPP was also evaluated. It was found that the residual enzyme activity distribution appeared to be dependant on the inactivation kinetics of the enzyme under consideration and the pressure temperature combination used.

A mathematical model describing the variation of the inactivation rate constant of soybean LOX as a function of pressure and temperature was studied by Ludikhuyze et al. [2]. Temperature dependence of the inactivation rate constants of LOX cannot be described by Arrhenius equation over the entire temperature range, therefore development of another kinetic model was attempted. Hence, Eyring equation which was valid over the entire temperature domain was used. The temperature dependant parameters (k_{refP}, V_a) in the Eyring equation below were replaced by mathematical expressions reflecting temperature dependence of the latter parameters.

$$\ln k = Lnk_{refP} - \left[\frac{V_a}{R(T+273)} (P - P_{ref}) \right] \tag{19.1}$$

Temperature dependence of the activation volume V_a and inactivation rate constant k_{refP} were described by the following equations, respectively.

$$V_a = a_1 T \exp(-b_1 T) \tag{19.2}$$

$$\ln k_{refP} = a_2 T^2 + b_2 T + C_2 \tag{19.3}$$

Subsequently, the proposed model structure was verified to predict likewise the extent of inactivation under different pressure and temperature. It was observed that the multiple application of high pressure enhanced the inactivation of soybean LOX and the effect was becoming more pronounced at low temperature.

19.2.2 MATHEMATICAL MODELING OF CONVECTIVE HEAT TRANSFER DURING HIGH PRESSURE PROCESSING (HPP)

During compression of liquid food, the liquid will be pushed initially by forced convection into the treatment chamber by the action of a high pressure pump/ intensifier. This is because liquids at extreme high pressure are compressible. The increase in temperature due to adiabatic compression induces heat transfer within the liquid and the walls of the pressure chamber. As a consequence, density differences occur leading to a free convection motion of the fluid. The fluid motion generated by forced and free convection strongly influences the temporal and spatial distribution of the temperature which was already observed experimentally [13].

Thermodynamic and fluid-dynamic effects of high pressure treatment are analyzed by means of numerical simulation [13]. Pure water is compressed into a 4-ml chamber at different compression rates and up to 500 MPa. The spatial and temporal evolution of temperature and fluid velocity fields are analyzed. It was found that fluid motion is dominated by forced convection at the beginning of pressurization. Due to density differences, free convection sets in and dominates the fluid motion for a few seconds following initial compression. Also, it is found that the temperature differences occurring in the high-pressure volume depend strongly on the pressure ramp during pressurization.

The influence of heat and mass transfer on the uniformity of high pressure induced inactivation was also investigated by Hartman et al. [14]. The inactivation of *E. coli* suspended in pouched UHT milk was simulated using water as a pressure medium. The result of the simulations showed nonuniformities of more than one log cycle in the residual surviving cell concentration. This non-uniformity is found to depend on the package material and the position and arrangement of the package in the vessel.

Hartman et al. [14] studied the thermofluid dynamics and process uniformity of HPP in a laboratory scale autoclave using experimental and numerical simulation techniques. Treatments at pressure levels of 500 MPa and 300 MPa were analyzed. It was found that a maximum temperature differences after the end of the pressure holding phase (820 s) amount up to 6 K for large compression rate and 500 MPa target pressure. It was also noted that the use of more viscous liquids leads to a substantial increase in temperature nonuniformity (23.4% of the average value). Thermal insulation of the inner wall of the high-pressure chamber was found to be key to both high degree of uniformity and efficient process cycle.

A simulation study of heat transfer during HPP of food using CFD was presented and analyzed by Ghani et al. [15]. In this simulation, modeling the effects of compression with time and the effects of natural and forced convection heating within a three-dimensional cylinder filled with liquid (water) during HPP was analyzed and studied. Temperature, fluid velocity and pressure profiles within the model liquid were computed. The convection currents of liquid food in a nonadiabatic HPP were also studied. The simulation for liquid food shows the effect of forced and free convection flow on temperature distribution in the liquid at the early stages of compression. This is due to the difference between the velocity of the pumping fluid as it enters the cylinder inlet hole (10^{-2}–10^{-3}) ms^{-1} and the much lower velocity in the treatment chamber (10^{-8}–10^{-9}) ms^{-1}. At later processing time, the simulations show that heat transfer is controlled by conduction. The insignificant effect of natural convection in all other stages is due to the very low liquid velocity in the treatment chamber as observed from the results of the simulation.

Numerical simulation of solid–liquid food mixture in a HPP unit was conducted using CFD [16]. In this simulation, temperature distribution, velocity and pressure profiles during high pressure compression (500 MPa) of solid–liquid food mixture (beef fat and water), within a three-dimensional cylinder basket was studied for the first time. The simulation for the solid-liquid mixture shows that the solid pieces were more heated than the liquid, which is due to the difference in their compression heating coefficient. The computed temperature was found to be in agreement with those measured experimentally and reported in the literature.

19.2.3 MATHEMATICAL MODELING OF HEAT TRANSFER DURING HIGH PRESSURE FREEZING AND THAWING

In the field of high pressure freezing and thawing, most studies deal with the impact of high pressure freezing and thawing on quality aspects of a particular food. However, a theoretical based heat transfer model that predicts temperature history within a product undergoing such a process would be very useful [17]. A number of mathematical models have been proposed for atmospheric pressure freezing and thawing over the past three decades. These models allow optimization of the design of industrial freezing and thawing equipment as well as the quality of the end product [18,19]. An existing theoretical model for predicting product temperature profiles during freezing and thawing processes was extended by Denys et al. [17] to the more complex situation of high pressure freezing and thawing processes. This requires knowledge of the product's thermal properties as functions of temperature and pressure. To take into account the influence of elevated pressure, a simplified approach was suggested, consisting of shifting the known thermophysical properties at atmospheric pressure on the temperature scale depending on the prevalent pressure. A numerical solution for two-dimensional heat transfers (finite cylinder) was chosen and a computer program was written for solving the heat transfer equations using an explicit finite difference scheme. The method used in the work also took into account the temperature increase of the high pressure medium during adiabatic compression and the temperature decrease during adiabatic expansion.

19.2.4 Current and Future Works on High Pressure Processing (HPP)

A number of investigations have used first order kinetics for the mechanism of microbial inactivation as usually applied in thermal treatment. The decimal reduction time is taken as a function of both temperature and pressure. The adiabatic heating caused by fluid compression can lead to significant temperature distribution throughout the treated food. Ignoring this temperature variation would lead to incorrect scale up of the small (laboratory) size HPP units to the large industrial size units. To date, a heat transfer model for HPP that includes the above mentioned effects is very limited. The development of a model suitable for the prediction of temperature distribution is important as experimental testing of every single food at different operating conditions can be very time consuming.

The following are the steps in the HPP model presented in this chapter:

1. The modified unsteady state heat conduction equation is used to describe heat transfer in the treatment chamber. Heat generation due to compression is included in the equation as a heat generation term, together with any heat generated or absorbed due to latent heat of solidification of lipid in food.
2. The physical properties of treated food are taken as pressure and temperature dependant.
3. The effect of transient pressure is incorporated in the heat equation as unsteady state heat generation term.
4. Based on the analysis followed, the pressure was assumed to be uniform in the treatment chamber but will vary with time during the pressure build-up and release. The temperature is a function of position and time and is influenced by both heat generation due to compression and external heat transfer.

Hence, one suggestion for future work on HPP is to include the effects of transient pressure and temperature distribution in the treatment chamber on the degree of sterility of food. The development of future HPP model must take into consideration the following proposed steps:

1. The effect of phase change of fat. However, such effect is complicated by the fact that the melting temperature of lipids increases with pressure (10°C/100 MPa). Thus, lipids present in a liquid state will crystallize under pressure at room temperature.
2. The microbial destruction rate should be calculated as a function of temperature, pressure and time, by introducing the kinetics of microbial inactivation, which would allow the calculation of an integral value for the inactivation rate within the treatment chamber at any time.
3. The destruction rate of nutrients such as vitamins due to the effect of pressure and temperature should be calculated as a function of pressure and position in the treatment chamber. The computation would also allow prediction of any important changes in the physical and functional properties of the food systems.

The output from the suggested research work, which can be conducted using any software such as CFX, Fluent, Flow 3D, PHOENICS, can also be used by the industry for optimizing high pressure food processing. This output will provide a comprehensive analysis of some of the most important parameters in food processing applications. The experimental work and strong theoretical analysis will lead to a big leap toward a deeper understanding of the mechanism of heat and mass transfer in one of the most important emerging technologies in food processing.

19.3 NUMERICAL SIMULATION OF LIQUID FOOD IN HIGH PRESSURE PROCESSING (HPP) UNIT USING COMPUTATIONAL FLUID DYNAMICS (CFD)

19.3.1 NUMERICAL APPROXIMATIONS AND MODEL PARAMETERS

In order to analyze the temperature distribution in a liquid food in high pressure treatment, the following process is considered: a high-pressure 300-ml chamber is filled with liquid food (pure water). The liquid food is compressed from ambient pressure to a maximum pressure of 500 MPa. The inflow of water is stopped when the maximum pressure is reached. The pressure level is held at 500 MPa for up to 1000 s. During the holding phase, the pressure remains constant. The initial temperature of the fluid and the wall temperature of the treatment chamber were assumed to be at 20°C during the entire period.

19.3.2 COMPUTATIONAL GRID

A grid system in three-dimensional (angular, radial and vertical) directions is used in the simulation. The whole domain was divided into 100,000 cells: 20 in angular direction, 50 in radial direction and 100 in the vertical direction distributed equally in each direction. The computations domain was performed for a cylinder with a diameter (*D*) of 38 mm and height (*H*) of 290 mm (Figure 19.1). The geometry of the high pressure chamber is constructed with the inlet at its top. Water was used as a model liquid food, compressed to 500 MPa in the 300 ml chamber for a period of 1000 s.

Refinement of different computational grids is used for the numerical solution of the governing equation to provide a mesh independent solution. Different inlet velocities are prescribed from the subroutine available in the software used for the simulation (PHOENICS). The velocity ranges from 10^{-2} to 10^{-3} m/s at the early stages of compression and decrease to very low values at the end of compression.

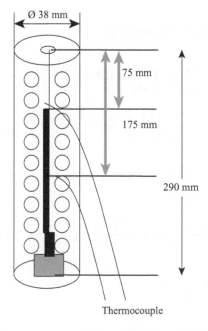

FIGURE 19.1 Geometry of high pressure vessel. (Reprinted from Abdul Ghani, A. G., and Farid, M. M., *Journal of Food Engineering*, 80(4), 1031–1042, 2007. With permission from Elsevier.)

19.3.3 GOVERNING EQUATIONS AND BOUNDARY CONDITIONS

The partial differential equations governing natural convection of the fluid (water) being compressed and heated in a cylinder are Navier-Stokes equations given below [20].

Energy conservation:

$$
\frac{\partial T}{\partial t} + v_r \frac{\partial T}{\partial r} + \frac{v_\theta}{r} \frac{\partial T}{\partial \theta} + v_z \frac{\partial T}{\partial z}
$$
$$
= \frac{k}{\rho C_p} \left[\frac{1}{r} \frac{\partial}{\partial r} \left(r \frac{\partial T}{\partial r} \right) + \frac{1}{r^2} \frac{\partial^2 T}{\partial \theta^2} + \frac{\partial^2 T}{\partial z^2} \right] + \frac{Q}{\rho C_p}
\tag{19.4}
$$

where Q is the volumetric heat generation term due to adiabatic heating (i.e., source term) in Wm^{-3}.

Momentum equation in the radial direction (r):

$$
\rho \left(\frac{\partial v_r}{\partial t} + v_r \frac{\partial v_r}{\partial r} + \frac{v_\theta}{r} \frac{\partial v_r}{\partial \theta} - \frac{v_\theta^2}{r} + v_z \frac{\partial v_r}{\partial z} \right) = -\frac{\partial p}{\partial r} + \mu \left[\frac{\partial}{\partial r} \left(\frac{1}{r} \frac{\partial}{\partial r} (r v_r) \right) + \frac{1}{r^2} \frac{\partial^2 v_r}{\partial \theta^2} - \frac{2}{r^2} \frac{\partial v_\theta}{\partial \theta} + \frac{\partial^2 v_r}{\partial z^2} \right]
\tag{19.5}
$$

Momentum equation in the vertical direction (z):

$$
\rho \left(\frac{\partial v_z}{\partial t} + v_r \frac{\partial v_z}{\partial r} + \frac{v_\theta}{r} \frac{\partial v_z}{\partial \theta} + v_z \frac{\partial v_z}{\partial z} \right)
$$
$$
= -\frac{\partial p}{\partial z} + \mu \left[\frac{1}{r} \frac{\partial}{\partial r} \left(r \frac{\partial v_z}{\partial r} \right) + \frac{1}{r^2} \frac{\partial^2 v_z}{\partial \theta^2} \frac{\partial^2 v_z}{\partial z^2} \right] + \rho_{ref} g (1 - \beta (T - T_{ref}))
\tag{19.6}
$$

Momentum equation in the angular direction (θ):

$$
\rho \left(\frac{\partial v_\theta}{\partial t} + v_r \frac{\partial v_\theta}{\partial r} + \frac{v_\theta}{r} \frac{\partial v_\theta}{\partial \theta} + \frac{v_r v_\theta}{r} + v_z \frac{\partial v_\theta}{\partial z} \right)
$$
$$
= -\frac{1}{r} \frac{\partial p}{\partial \theta} + \mu \left[\frac{\partial}{\partial r} \left(\frac{1}{r} \frac{\partial}{\partial r} r v_\theta \right) + \frac{1}{r^2} \frac{\partial^2 v_\theta}{\partial \theta^2} + \frac{2}{r^2} \frac{\partial v_r}{\partial \theta} + \frac{\partial^2 v_\theta}{\partial z^2} \right]
\tag{19.7}
$$

These equations are coupled with the following equation:

Continuity equation:

$$
\frac{\partial \rho}{\partial t} + \frac{1}{r} \frac{\partial}{\partial r} (r \rho v_r) + \frac{1}{r} \frac{\partial}{\partial \theta} (\rho v_\theta) + \frac{\partial}{\partial z} (\rho v_z) = 0
\tag{19.8}
$$

The boundary Conditions used were:

At the cylinder boundary, $r = R$

$$
v_r = 0, \quad v_\theta = 0, \quad v_z = 0, \quad \frac{\partial T}{\partial r} = 0 \quad \text{or} \quad T = T_{wall} \quad \text{for} \quad 0 \le z \le H
\tag{19.9}
$$

At the bottom and top of the cylinder, $z = 0$ and $z = H$

$$v_r = 0, \quad v_\theta = 0, \quad v_z = 0, \quad \frac{\partial T}{\partial z} = 0 \quad \text{for} \quad 0 \leq r \leq R \tag{19.10}$$

$$\text{At } t = 0, \quad P = 0.1\,\text{MPa}, \quad T = 25^\text{Y}\text{C} \quad \text{for} \quad 0 \leq r \leq R \quad \text{and} \quad 0 \leq z \leq H \tag{19.11}$$

$$\text{At } 30 > t \geq 0, \quad P = 16.663t + 0.1\,\text{MPa} \quad \text{for} \quad 0 \leq r \leq R \quad \text{and} \quad 0 \leq z \leq H \tag{19.12}$$

$$\text{At } t > 30\,\text{s}, \quad P = 500\,\text{MPa} \tag{19.13}$$

At the walls of the pressure chamber, the kinematic boundary condition requires zero fluid velocity which implies a no slip condition. This assumption is valid except at the inlet.

19.3.4 Compression Steps

The pressure generated by the high pressure pump and intensifier of the HPP machine model S-FL-850-9-W used in this work was found to increase with time linearly as shown in Equation 19.12. After compression time of 30 s, the pressure remains constant at 500 MPa.

Simulations were conducted for time up to 1000 s and the results were presented when pressure reached 100, 200, 300, 400, and 500 MPa after 6, 12, 18, 24, and 30 s, respectively. This was based on the performance of the HPP unit, which was provided with a powerful pump. The pressure level of 500 MPa is reached within 30 s only. When the pressure reaches 500 MPa, it was maintained at that pressure until decompression starts. Reynolds number of the water flowing through the inlet hole is found to be small, therefore, the flow can be assumed laminar even at the beginning of the compression.

19.3.5 Physical Properties

An equation of state accounting for the compressibility of pure water under high pressure is implemented in the program to describe the variation of the density with pressure and temperature. This equation was taken from the study of Saul and Wagner [21].

The properties of the model liquid food (water) at atmospheric pressure and ambient temperature of 20°C are: $\rho = 998.23\,\text{kgm}^{-3}$, $C_p = 4181\,\text{Jkg}^{-1}\text{K}^{-1}$, $k = 0.597\,\text{W m}^{-1}\text{K}^{-1}$ [22,23]. As the pressure and temperature changes during compression, the program calculates the new values of physical properties with the aid of written FORTRAN statements. The properties of water are calculated based on updating values at every time step, done using the subroutines in PHOENICS, named CHEMKIN. A call is made to the CHEMKIN routine CKHMS to calculate these properties using the appropriate formula.

The CHEMKIN system used for the liquid is supplied by Sandia National Laboratories and consists of:

- A thermodynamics DATabase.
- A library of FORTRAN subroutines which the user may call from his application programs to supply thermodynamic data.
- A "stand-alone" interpreter program that reads a "plain language" file that specifies the thermodynamic data for the thermo-chemical system under investigation.

Associated with CHEMKIN, and also supplied by Sandia National Laboratories, is a further system that supplies transport data. The transport properties system consists of:

- A transport database.
- A library of FORTRAN subroutines which the user may call from his application programs to supply viscosities, thermal conductivities, diffusion coefficients and thermal diffusion ratios or coefficients calculated according to two approximations.
- A fitting program that generates polynomial fits to the detailed transport properties in order to make the calculations performed by the subroutine library more efficient.

Another built in subroutine used in the simulation is named PRESS0, which is the parameter representing the reference pressure, to be added to the pressure computed by PHOENICS in order to give the physical pressure needed for calculating density and other physical properties. The use of this variable is strongly recommended in cases in which the static component of the pressure is much greater than the dynamic head. The reason is that the static component can be absorbed in PRESS0 leaving the stored pressure field, to represent the dynamic variations which otherwise may be lost in the round off, according to the machine precision and the ratio of dynamic pressure to the static head.

Several built in subroutines are used in this simulation, such as

- DVO1DT, used to calculate the volumetric coefficient of thermal expansion of phase 1 of the material used (water). It is useful for the prediction of natural-convection heat transfer.
- DRH1DP, which is used to calculate the compressibility:

$$\left(\frac{d\rho 1/dP}{\rho 1}\right) \quad \left(\text{i.e. } \frac{d(\ln\rho 1)}{dP}\right) \tag{19.14}$$

If DRH1DP gives a positive value, that value is used for the dependence of the first-phase density on pressure. Recourse to GROUND is necessary when density is a nonlinear function of pressure or a function of other variables.

19.3.6 RESULTS AND DISCUSSIONS

During HPP of food, an increase in temperature of food due to compression is observed as a result of partial conversion of mechanical work into internal energy. This is known as adiabatic heating. In reality, the situation is much more complex. In the simulation presented here, heat exchange between the treated fluid and the cylinder wall as well as the cooling effect caused by the entering fluid during compression are included in the analysis. Temperature distribution, location of the hottest zone (HZ), velocity profile and pressure profile during the process are compared and analyzed based on the simulations conducted. Experimental measurements are used to validate these simulation results.

The calculated pressure field (Figure 19.2) primarily shows the hydrostatic gradient (2.9 kPa at the base). The kinetic head is negligible due to the extremely low liquid velocity (3×10^{-8}–4×10^{-7} ms^{-1}) in the HP cylinder. In order to minimize numerical errors, the pressure field solution calculates the pressure relative to that fixed at the inlet, which was increasing from atmospheric pressure to 500 MPa.

Changes in temperature profile at early stages of compression is due to compression heating and also cooling caused by the pumped water through the hole at the top of the cylinder (inlet). The velocity of the cold water as it enters the inlet is much higher than the velocity inside the cylinder. At this stage, the buoyancy force starts to be effective due to temperature variation of the liquid the cylinder.

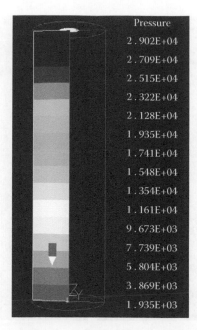

FIGURE 19.2 Pressure variation in the fluid along the cylinder height (1000 s, P = 500 MPa). (Reprinted from Abdul Ghani, A. G., and Farid, M. M., *Journal of Food Engineering*, 80(4), 1031–1042, 2007. With permission from Elsevier.)

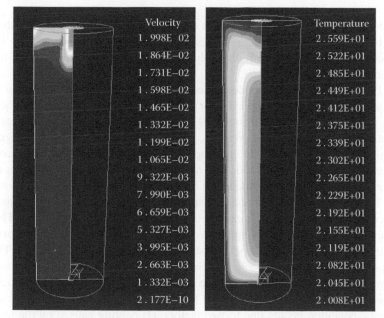

FIGURE 19.3 Temperature and velocity profiles of the fluid at the early stage of compression (t = 12 s, P = 200 MPa). (Reprinted from Abdul Ghani, A. G., and Farid, M. M., *Journal of Food Engineering*, 80(4), 1031–1042, 2007. With permission from Elsevier.)

Figures 19.3 and 19.4 show the temperature and velocity profiles at the early stages of compression after periods of 12 and 24 s at nonadiabatic compression pressures of 200 and 400 MPa, respectively. In Figure 19.3, the fluid velocity at the top (location of the inlet) of the cylinder is large. It is in the range of 10^{-2}–10^{-3} m/s at the top (at location close to the inlet) of the cylinder while it is very small (10^{-7}–10^{-9} m/s) in the rest of the cylinder as shown in Figure 19.7. This is due

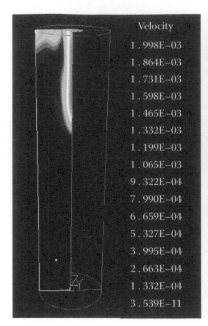 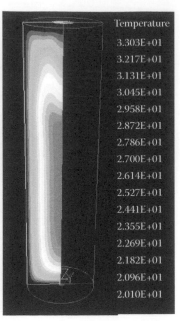

FIGURE 19.4 Temperature and velocity profiles of the fluid at the early stage of compression (t = 24 s, P = 400 MPa). (Reprinted from Abdul Ghani, A. G., and Farid, M. M., *Journal of Food Engineering*, 80(4), 1031–1042, 2007. With permission from Elsevier.)

to the liquid being pumped through the inlet at the top of cylinder in order to fill the shortage of water caused by the effects of compression. This will push the HZ more toward the bottom of the cylinder as clearly seen in the figure. Without this effect, the HZ will move toward the top due to buoyancy effect.

In Figure 19.4, the fluid velocity is decreased to 10^{-3}–10^{-4} m/s at the top of the cylinder and remained very small in the rest of the cylinder while the HZ is pushed even further toward the bottom of the cylinder. Figures 19.3 through 19.5 show the effects of forced and natural convection current at the early stages of compression. At later stages of compression (Figure 19.6), the fluid flow diminishes and the liquid velocity drops to very low values. Under such a condition, heat transfer is more dominated by conduction and to a lesser extent by free convection in the liquid. Figure 19.6 shows the temperature profile in the fluid due to nonadiabatic compression of 500 MPa after 1000 s. Due to the high hydrostatic pressure, the velocities are very small and hence, the effect of free convection heat transfer is expected to be very small in this situation. The increase in the temperature of the HZ after 30 s of compression (500 MPa) is 15.5°C and it occurred almost at the centre of the cylinder, with only very little shift toward the top of the cylinder. This is similar to the increase in temperature due to adiabatic heating usually reported in the literature (3°C per 100 MPa). However, this figure shows that the increase in the fluid average temperature in the cylinder is significantly less than 3°C per 100 MPa, usually reported in adiabatic compression. The nonadiabatic condition occurs due to two factors: (1) cooling caused by the lower temperature of the wall of the cylinder; (2) cooling caused by the incoming fluid through the inlet of the cylinder.

Figure 19.7 shows the velocity profile of the compressed fluid in the cylinder. Due to the external cooling at the wall, fluid adjacent to it will have higher density causing it to flow downward. At all other locations in the bulk of the cylinder, liquid will flow upward as shown in the figure. The velocity is in the range of 10^{-7}–10^{-9} m/s at the end of the process. However, at early stages of compression, the calculated liquid velocity was as high as 10^{-2}–10^{-4} m/s as shown in Figures 19.3 and 19.4 after compression periods of 12 and 24 s, respectively.

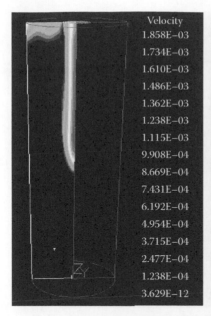

Velocity
1.858E−03
1.734E−03
1.610E−03
1.486E−03
1.362E−03
1.238E−03
1.115E−03
9.908E−04
8.669E−04
7.431E−04
6.192E−04
4.954E−04
3.715E−04
2.477E−04
1.238E−04
3.629E−12

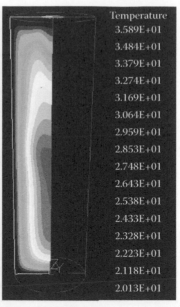

Temperature
3.589E+01
3.484E+01
3.379E+01
3.274E+01
3.169E+01
3.064E+01
2.959E+01
2.853E+01
2.748E+01
2.643E+01
2.538E+01
2.433E+01
2.328E+01
2.223E+01
2.118E+01
2.013E+01

FIGURE 19.5 Temperature and velocity profiles of the fluid at the early stage of compression (t = 30 s, P = 500 MPa). (Reprinted from Abdul Ghani, A. G., and Farid, M. M., *Journal of Food Engineering*, 80(4), 1031–1042, 2007. With permission from Elsevier.)

Temperature
3.555E+01
3.454E+01
3.352E+01
3.250E+01
3.149E+01
3.047E+01
2.946E+01
2.844E+01
2.743E+01
2.641E+01
2.540E+01
2.438E+01
2.337E+01
2.235E+01
2.134E+01
2.032E+01

FIGURE 19.6 Temperature profile of the fluid due to nonadiabatic compression (t = 1000 s, P = 500 MPa). (Reprinted from Abdul Ghani, A. G., and Farid, M. M., *Journal of Food Engineering*, 80(4), 1031–1042, 2007. With permission from Elsevier.)

19.3.6.1 Experimental Validation

Experimental measurement was used to validate the results of the simulation presented. The experimental measurements are conducted at the Innovative Food Processing Laboratory at the Chemical Engineering Department, University of Auckland using model S-FL-850-9-W HPP unit. The maximum operating pressure of the unit is 900 MPa and the operating temperature range is −20°C

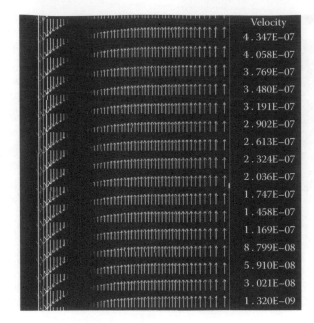

	Velocity
	4.347E−07
	4.058E−07
	3.769E−07
	3.480E−07
	3.191E−07
	2.902E−07
	2.613E−07
	2.324E−07
	2.036E−07
	1.747E−07
	1.458E−07
	1.169E−07
	8.799E−08
	5.910E−08
	3.021E−08
	1.320E−09

FIGURE 19.7 Enlarged section of velocity profile of the fluid due to non-adiabatic compression (t = 1000 s, P = 500 MPa). (Reprinted from Abdul Ghani, A. G., and Farid, M. M., *Journal of Food Engineering*, 80(4), 1031–1042, 2007. With permission from Elsevier.)

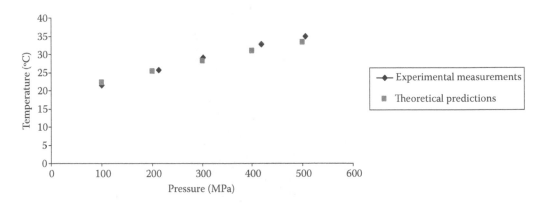

FIGURE 19.8 Experimental measurements and theoretical predictions of temperature of a liquid food (water) at the center of the pressure vessel, 75 mm from its top. (Reprinted from Abdul Ghani, A. G., and Farid, M. M., *Journal of Food Engineering*, 80(4), 1031–1042, 2007. With permission from Elsevier.)

to + 90°C. The pressure vessel is fitted with "T" type thermocouple at the axis of the vessel and can be adjusted to any position along the axis of the vessel. Pressure is controlled by electronic control module with integral digital display. The size of the HPP treatment cylinder is the same size as that used in the simulations, with a diameter of 38 mm and inner height of 290 mm (Figure 19.1). Direct processing of water at pressure level of 500 MPa and within a pressure holding phase ending at 1000 s is considered.

Figures 19.8 and 19.9 show the measured and predicted temperature during the early stages of compression (t < 30 s) at two different locations at the axis of the cylinder. The agreement is reasonable but that was possible only after the effect of cooling from the wall of the cylinder and by the entering fluid at the early stages of cooling. The effect of cooling by the entering fluid during compression is shown more clearly in Figure 19.10. The temperature at the higher location is about

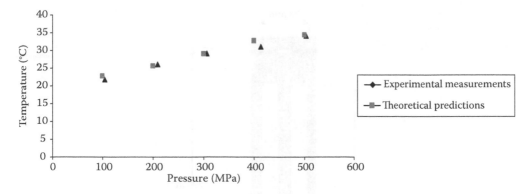

FIGURE 19.9 Experimental measurements and theoretical predictions of temperature of a liquid food (water) at the center of the pressure vessel, 175 mm from its top. (Reprinted from Abdul Ghani, A. G., and Farid, M. M., *Journal of Food Engineering*, 80(4), 1031–1042, 2007. With permission from Elsevier.)

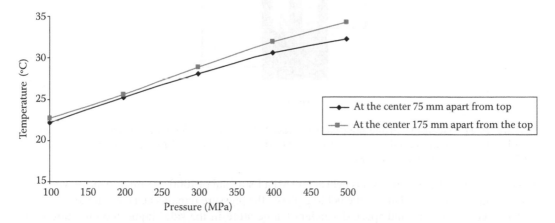

FIGURE 19.10 Comparison between two theoretically predicted points at different heights; top point located at the center of the pressure vessel 75 mm from top. (Reprinted from Abdul Ghani, A. G., and Farid, M. M., *Journal of Food Engineering*, 80(4), 1031–1042, 2007. With permission from Elsevier.)

2°C lower than the center of the cylinder at the end of compression. If the cooling effect is ignored, the temperature at the higher location would experience higher temperature due to buoyancy.

In reality, a true isothermal operation is difficult to achieve. For HPP experiment at elevated temperatures, it is important to recognize that an externally delivered pressure-transmitting fluid may enter the system at a much lower temperature than the pressure-transmitting fluid already in the pressure vessel. The temperature of this fluid should be monitored and reported where possible [24].

19.4 NUMERICAL SIMULATION OF SOLID–LIQUID FOOD MIXTURE IN HIGH PRESSURE PROCESSING (HPP) UNIT USING COMPUTATIONAL FLUID DYNAMICS (CFD)

19.4.1 NUMERICAL APPROXIMATIONS AND MODEL PARAMETERS

In order to analyze the temperature distribution in a solid–liquid food in high pressure treatment, the following process is considered: a high-pressure 300-ml chamber is used and it is filled with pure water and two pieces of solid beef as shown in Figure 19.11. The solid–liquid is compressed from ambient pressure to a maximum pressure of 500 MPa. The inflow of water is stopped when maximum

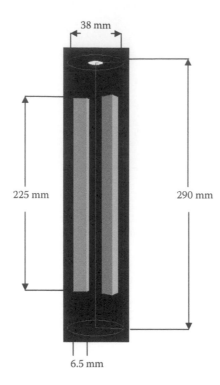

FIGURE 19.11 Geometry of the high pressure vessel shows the configuration of the solid food pieces. (Reprinted from Abdul Ghani, A. G., and Farid, M. M., *Journal of Food Engineering*, 80(4), 1031–1042, 2007. With permission from Elsevier.)

pressure is reached. Pressure level is held at 500 MPa for a total of 1000 s, similar to the previous case of using liquid food only. During the holding phase, the pressure remains at a constant level.

The analysis of time and space dependant temperature in the solid-liquid food mixture under high pressure compression was the principle aim of this investigation. The temperature distribution and the shape of the HZ during compression will be addressed in this section.

19.4.2 COMPUTATIONAL GRID

The three-dimensional grid system (angular, radial and vertical directions) and the computations domain are the same as those used in the simulation of liquid food. The height and thickness of the solid beef slices used are 225 mm and 6.5 mm, respectively. It was taken as a solid material within the same mesh pattern (Figure 19.12) in the domain to simplify numerical analysis. Different computational grid refinement has been done during numerical solution of the governing equation to provide a mesh independent solution.

19.4.3 GOVERNING EQUATIONS FOR THE SOLID

The beef fat pieces are assumed as an impermeable solid and heat is transferred through them by conduction only. In this case, the three convection terms in the left-hand side of Equation 19.4 can be omitted, reducing the governing equation to the well-known diffusion equation.

$$\frac{\partial T}{\partial t} = \alpha_m \left[\frac{1}{r} \frac{\partial}{\partial r} \left(r \frac{\partial T}{\partial r} \right) + \frac{1}{r^2} \frac{\partial^2 T}{\partial \theta^2} + \frac{\partial^2 T}{\partial z^2} \right] + \frac{Q}{\rho C_p} \tag{19.15}$$

Compression steps are similar to those used in Section 19.3.

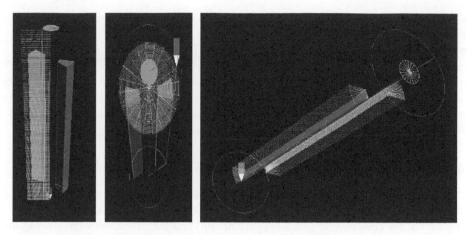

FIGURE 19.12 The three-dimensional grid meshes of the high pressure vessel. The arrow shown in the figure indicates a point in the computation process. (Reprinted from Abdul Ghani, A. G., and Farid, M. M., *Journal of Food Engineering*, 80(4), 1031–1042, 2007. With permission from Elsevier.)

19.4.4 PHYSICAL PROPERTIES

The properties of the solid pieces (beef fat) at atmospheric pressure are: $\rho = 900$ kgm^{-3}, $C_p = 3220$ Jkg^{-1} K^{-1}, $k = 0.43$ Wm^{-1} K^{-1} [22,23]. As the pressure and temperature changes during compression, the program calculates the new values of physical properties with the aid of FORTRAN statements written as discussed below.

The properties of water are calculated based on updating values at every time step, which is done using the subroutines in PHOENICS named CHEMKIN (Section 19.3.5). A call is made to the CHEMKIN routine CKHMS to calculate these properties using the appropriate formula.

Both solid (beef fat) and liquid (water) are compressible under high pressure. Appropriate equations of state describing the density as a function of pressure and temperature have to be added to the equations. For water, these equations are already available in the PHONICS subroutines. For the solid, there are no thermophysical data available at high pressure; therefore the following equations have been incorporated in the software to take into account the pressure dependency of the physical properties following the same approach adopted by Hartman et al. [14]:

$$\rho_{\text{Beef}}(P,T) = \left(\frac{\rho_{\text{Beef}}(T)}{\rho_{\text{Water}}(T)} \right)_{\text{atm}.P} \cdot \rho_{\text{Water}}(P,T) \tag{19.16}$$

$$Cp_{\text{Beef}}(P,T) = \left(\frac{Cp_{\text{Beef}}(T)}{Cp_{\text{Water}}(T)} \right)_{\text{atm}.P} \cdot Cp_{\text{Water}}(P,T) \tag{19.17}$$

$$k_{\text{Beef}}(P,T) = \left(\frac{k_{\text{Beef}}(T)}{k_{\text{Water}}(T)} \right)_{\text{atm}.P} \cdot k_{\text{Water}}(P,T) \tag{19.18}$$

The CHEMKIN system used for the liquid is supplied by Sandia National Laboratories (Section 19.3.5).

19.4.5 RESULTS AND DISCUSSIONS

In the simulation presented here, heat transport between the solid (beef fat) and the liquid (water) leads to transient spatially nonuniform temperature distribution. The degree of nonuniformity depends on the geometrical shape and size of solid and liquid used as well as on their physical properties.

The temperature distribution and the location of the HZ during the process are analyzed for the compression of mixture of a liquid (water) and solid (beef fat).

In order to minimize numerical errors, the pressure is calculated relative to that fixed at the inlet as discussed in Section 19.3. Hence, the calculated pressure field primarily shows the hydrostatic gradient. The kinetic head is negligible due to the extremely low liquid velocity (3×10^{-8}–4×10^{-7} ms^{-1}) in the HP cylinder after an extended time of 1000 s.

19.4.6   COMPRESSION OF MIXTURE OF A LIQUID AND SOLID

Temperature distribution during heating of solid and liquid food is presented in the form of isotherms in Figure 19.13 for different periods of compression (12, 24, and 30 s). The isotherms shown in the figure are for compression pressures of 200, 400 and 500 MPa, respectively. Figure 19.13 shows that the solid (beef fat) was heated more than the fluid (water), which is due to its larger heat of compression of fat. This figure shows clearly how cold fluid enters the cylinder through the hole in the top then deviate when it hits the two pieces of solid.

The HZ in the cylinder was found lying in the middle of the solid food (beef fat) due to the higher heat of compression of beef fat and due to the fact that heat is transferred by conduction only. This is shown more clearly in Figure 19.14. Due to the higher temperature of the solid, heat is transferred from solid to water by free and forced convection heat transfer.

As the compression pressure increased (at t = 30 s and P = 500 MPa), temperature profiles (Figure 19.13) become different from those observed at the beginning of the compression. The figure shows that temperature profile is influenced by cooling caused by the pumped water through the hole at the top of the cylinder (inlet). The velocity of the cold water entering the inlet hole is

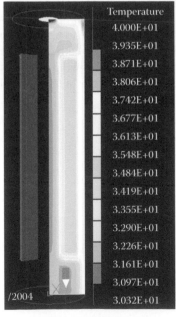
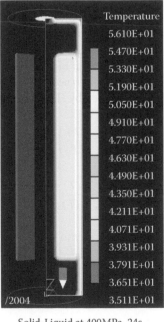
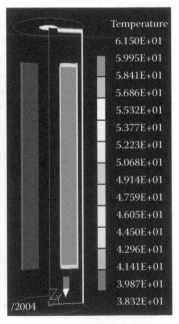

Solid-Liquid at 200MPa, 12s   Solid-Liquid at 400MPa, 24s   Solid-Liquid at 500MPa, 30s

FIGURE 19.13   Radial-vertical temperature profile of the solid–liquid food mixture (beef fat and water) at compression rates of 200, 400, and 500 MPa, respectively. The arrow shown in the figure indicates a point in the computation process. (Reprinted from Abdul Ghani, A. G., and Farid, M. M., *Journal of Food Engineering*, 80(4), 1031–1042, 2007. With permission from Elsevier.)

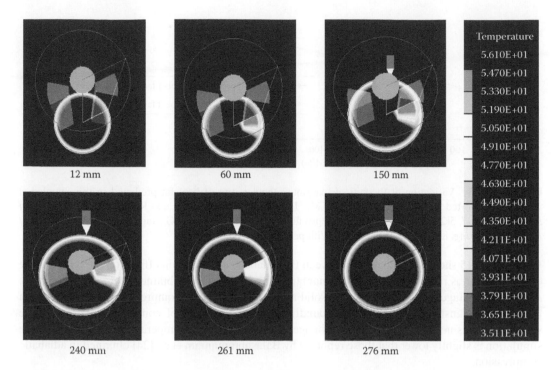

FIGURE 19.14 Radial-angular temperature profile of the solid–liquid food mixture due to compression rate of 400 MPa at different heights of 12, 60, 150, 240, 261, and 276 mm from the bottom, respectively. The arrow shown in the figure indicates a point in the computation process. (Reprinted from Abdul Ghani, A. G., and Farid, M. M., *Journal of Food Engineering*, 80(4), 1031–1042, 2007. With permission from Elsevier.)

10^{-2}–10^{-3} ms^{-1} as calculated is much higher compared to the velocity inside the cylinder (10^{-7}–10^{-9} ms^{-1}) shown in Figure 19.7. At this stage, the buoyancy force also starts to play role due to temperature variation of the liquid.

Under adiabatic compression heating, the sample temperature at process pressure is dictated by the sample's thermodynamic properties and heat transfer between the sample and the pressure-transmitting fluid. Also, the fluid as it enters the pressure vessel during the come-up time would likely influence sample temperature [24]. Some of the commonly used pressure-transmitting fluids are water, food-grade glycol-water solutions, silicone oil, sodium benzoate solutions, ethanol solutions and castor oil. In the simulation used in this work, water was assumed as the pressure-transmitting fluid used in the vessel to transmit pressure uniformly and instantaneously to the food sample.

In reality, a true isothermal test is difficult to achieve. For HPP experiment at elevated temperatures, it is important to recognize that an externally delivered pressure-transmitting fluid may enter the system at a much lower temperature than the pressure-transmitting fluid already in the pressure vessel. The temperature of this fluid should be monitored and reported where possible [24]. This has been taken into account in the simulation reported in this chapter.

19.4.7 SOLID FOOD VALIDATION

In the work reported by Balasubramaniam et al. [24], the theoretically predicted compression heating factors (°C per 100 MPa) of water and various selected food substances were presented and studied. For beef fat, this value ranged from 6.2 to 8.3°C per 100 MPa which was determined at initial product temperature of 25°C. In the simulation presented in this chapter, the compression heating factors was assumed to be 7°C per 100 MPa.

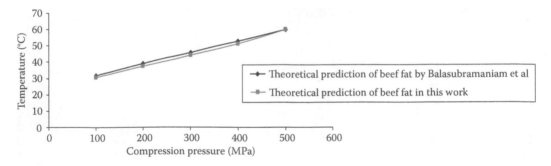

FIGURE 19.15 Validation of theoretical maximum temperature of beef fat predicted in this simulation with those reported by (Balasubramaniam V.M., Ting, E.Y., Stewart, C.M., and Robbins, J. A., *Journal of Innovative Food,* 5(2004), 299–306, 2004; Reprinted from Abdul Ghani, A. G., and Farid, M. M., *Journal of Food Engineering,* 80(4), 1031–1042, 2007. With permission from Elsevier.)

Figure 19.15 shows a comparison between the adiabatic compression effects on the temperature of the beef fat as reported by Balasubramaniam et al. [24] and that obtained from our simulation. At such early stages of compression, the solid maybe assumed to be compressed adiabatically with limited heat transfer with the fluid surrounding it for the purpose of comparison. However, the simulation does not ignore such heat exchange and this is why the temperature obtained from the simulation is slightly lower that those reported by Balasubramaniam et al. [24] obtained in adiabatic compression.

ACKNOWLEDGMENT

This work was supported by the New Zealand Foundation of Research Science and Technology within the project 9071-3502152.

NOMENCLATURE

C_p specific heat of liquid food, $Jkg^{-1}K^{-1}$
D diameter of cylinder, mm
g acceleration due to gravity, ms^{-2}
H height of the cylinder, mm
k thermal conductivity of liquid food, $Wm^{-1}K^{-1}$
P   pressure, Pa
Q volumetric heat generation, Wm^{-2}
r radius of the cylinder, mm
r,θ,z radial, angular and vertical direction of the cylinder
t   compressing time, s
T temperature, °C
T_{ref} reference temperature, °C
T_{wall} wall temperature, °C
v_r velocity in radial direction, ms^{-1}
v_θ velocity in angular direction, ms^{-1}
v_z velocity in vertical direction, ms^{-1}
z distance in vertical direction from the bottom, m
β thermal expansion coefficient, K^{-1}
μ apparent viscosity, Pas
ρ density, kgm^{-3}
ρ_{ref} reference density, kgm^{-3}

REFERENCES

1. Barbosa-Canovas, G.V., Pothakamury, U.R., Paulo, E., and Swanson, B.G. 1997. *Non Thermal Preservation of Foods*. Dekker, New York, NY, 9–52.
2. Ludikhuyze, L., Indrawati, I., Van der Broeck, C., Weemens, C., and Hendrickx, M.E. 1998. Effect of combined pressure and temperature on soybean lipoxygenase. 2. Modeling inactivation kinetics under static and dynamic conditions. *Journal of Agricultural Food Chemistry*, 46, 4081–4086.
3. Seyderhelm, I., Bogulawiski, S., Michaelis, G., and Knorr, D. 1996. Pressure induced inactivation of selected food enzymes. *Journal of Food Science,* 61(2), 308–310.
4. Yen, G.C., and Lin, H.T. 1996. Comparison of high pressure treatment and thermal pasteurization on the quality and shelf life of guava puree. *International Journal of Food Science and Technology*, 31, 205–213.
5. Richwin, A., Roosch, A., Teichgraber, D., and Knorr, D. 1992. Effect of combined pressure and temperature on the functionality of egg-white proteins. *Journal of European Food Science*, 43(7/8), 27–31.
6. Ohshima, T., Ushio, H., and Koizumi, C. 1993. High pressure processing of fish and fish products. *Journal of Trends in Food Science and Technology*, 4, 370–375.
7. Bouton, P.E., Ford, A.L., Harris, P.V., Macfarlane, J.J., and O'Shea, J.M. 1997. Pressure treatment of post rigor muscle: effects on tenderness. *Journal of Food Science*, 42, 132–135.
8. Ohmori, T., Shigehisa, T., Taji, S., and Hayashi, R. 1991. Effect of high pressure on the protease activities in meat. *Journal of BioChemistry*, 55(2), 357–361.
9. Scott, G., and Richardson, P. 1997. The applications of computational fluid dynamics in the food industry. *Journal of Trends in Food Science and Technology*, 8, 119–124.
10. Knor, D. 1993. Effects of high hydrostatic pressure processes on food microorganism. *Journal of Trends in Food Science and Technology*, 4, 370–375.
11. Denys, S., Ludikhuyze, L.R., Van Loey, A.M. and Hendrickx, M.E. 2000. Modeling conductive heat transfer and process uniformity during batch high pressure processing of foods. *Journal of Biotechnology Prog*ress, 16, 92–101.
12. Denys, S., Van Loey, A.M., and Hendrickx, M.E. 2000. A modeling approach for evaluating process uniformity during batch high htdrostatic prcssure processing: combination of a numerical heat transfer model and enzyme inactivation kinetics. *Journal of Innovative Food Science and Immerging Technologies*, 1, 5–19.
13. Hartmann, C. 2002. Numerical simulation of thermodynamic and fluid-dynamic processes during the high-pressure treatment of fluid food systems. *Journal of Innovative Food Science and Emerging Technologies*, 3, 11–18.
14. Hartmann, C., Delgado, A., and Szymczyk, J. 2003. Convective and diffusive transport effects in a high pressure induced inactivation process of packed food. *Journal of Food Engineering*, 59, 33–44.
15. Abdul Ghani, A.G., and Farid, M.M. 2007. Modeling of high pressure food processing using computational fluid dynamics. In *Computational Fluid Dynamics in Food Processing*, Da-Wen Sun (Ed.). Taylor and Francis Group, LLC, Boca Raton, FL.
16. Abdul Ghani, A.G., and Farid, M.M. 2007. Numerical simulation of solid–liquid food mixture in a high pressure processing unit using computational fluid dynamics. *Journal of Food Engineering*, 80(4), 1031–1042.
17. Denys, S., Van Loey, A.M., Hendrickx, M.E., and Tobback P. P. 1997. Modeling heat transfer during high pressure freezing and thawing. *Journal of Biotechnology Progress*, 13, 416–423.
18. Bakal, A., and Hayakawa, K.I. 1973. Heat transfer during freezing and thawing of foods. *Journal of Advanced Food Research*, 20, 217–256.
19. Ramaswamy, H.S., and Tung, M.A.A. 1984. A review on predicting freezing times of foods. *Journal of Food Processing Engineering*, 7, 196–203.
20. Bird, R.B., Stewart, W.E., and Lightfoot, E.N. 1976. *Transport Phenomena*. John Wiley and Sons, New York, NY.
21. Saul, A., and Wagner, W. 1989. A fundamental equation for water covering the range from the melting line to 1273 K at pressures up to 25000 MPa. *Journal of Physical and Chemical Reference Data*, 9, 1212–1255.
22. Rahman, R. 1995. *Food Properties Handbook*. CRC Press, Inc., Boca Raton, FL.
23. Hayes, G.D. 1987. *Food Engineering Data Handbook*. John Wiley and Sons Inc., New York, NY.
24. Balasubramaniam, V.M., Ting, E. Y., Stewart, C.M., and Robbins, J.A. 2004. Recommended laboratory practise for conducting high pressure microbial inactivation experiments. *Journal of Innovative Food Science and Emerging Technologies*, 5(2004) 299–306.

20 Modeling Microbial Inactivation by a Pulsed Electric Field

Michael Ngadi and Jalal Dehghannya
McGill University

CONTENTS

20.1 INTRODUCTION

Pulsed electric field (PEF) technology is a novel nonthermal food preservation method that can provide consumers with microbiologically safe, minimally processed, nutritious, and fresh-like products. The technology has caught the attention of food processors as a technique that can be used to meet consumers' demands.

PEF has been studied extensively for inactivation of different microorganisms in several liquid foods such as juices, cream soups, milk, and egg products [1–11]. Apart from microbial inactivation, different studies have also shown that PEF treatment of biological tissue enhances extraction [12–15], drying [16–19] as well as inactivation or evolution of different enzymes [20–23].

20.2 PULSED ELECTRIC FIELD (PEF) PROCESSING SYSTEM

The major components of a PEF processing system include a voltage power supply, an energy storage capacitor, treatment chamber, discharge switch, and charging resistance. Energy from the high-voltage power supply is stored in the capacitor and is discharged through the food material to generate the necessary electric field across a food product in the treatment chamber. The stored energy can be discharged nearly instantaneously using appropriate high-voltage switches. The short burst of high voltage applied across the two electrodes generates electric current flow for few microseconds through the food. Recirculation of cooling water around the electrodes can be used to control temperature of the sample being treated. When electrical energy is applied in the form of short pulses, bacterial cell membranes are destroyed without significant heating of the food. The pulses are monitored online using an oscilloscope and the electrical parameters such as voltage, pulse duration and current waveforms can be recorded using digital data acquisition system [1].

20.3 MECHANISM OF PULSED ELECTRIC FIELD (PEF) TREATMENT

The mechanism of PEF application in food products is not well known. It is believed that the process induces electroporation or electroplasmolysis of cell membranes thereby degrading cell tissue integrity at the cellular and subcellular levels [24,25]. The antimicrobial effect of PEF is also believed to be due to dielectric rupture resulting from the applied electric field [26]. When an electric field is applied across a medium where a viable cell is suspended, it induces an electric potential across the cell membrane. This electric potential causes an electrostatic charge separation in the cell membrane based on the dipole nature of the membrane's constituents [27]. When the electromechanical stress or trans-membrane potential (TMP) exceeds a critical value, repulsions between charge-carrying molecules occur and stimulates the formation of pores, subsequently weakening the membrane and eventually damaging the cell [8,28].

20.4 FACTORS AFFECTING PULSED ELECTRIC FIELD (PEF) MICROBIAL INACTIVATION

Many factors determine the effectiveness of PEF treatment. These factors can be categorized under three main groups namely process parameters, microbial characteristics and product parameters. The process parameters include electric field strength, number of pulses, treatment time, treatment temperature, pulse shape, pulse length, pulse polarization, frequency, specific energy, configuration of the treatment chamber as well as flow rate and residence time in the continuous processes. Microbial characteristics consist of cell concentration, microorganism resistance to environmental factors, type and species of the microorganism as well as growth conditions such as medium composition, temperature, and oxygen concentration. Product parameters are composed of product composition (presence of particles, sugars, salt, and thickeners), conductivity, ionic strength, pH, water activity, and prior stress conditions.

20.5 MODELS OF MICROBIAL INACTIVATION

The application of PEF treatment in food preservation for the production of nutritionally-stable food products needs reliable models that can explain different trends of various factors involved in the process. The models can then be used to clarify possible mechanisms of action or to design the intensity of PEF application required to achieve stable products. Modeling microbial inactivation by PEF is complicated since a large number of parameters are involved in the process and it is not easy to separate the effect of different parameters from each other. Different models have been used to describe the impact of various factors on PEF processing. These models could be classified into two main categories namely kinetics-based and probability-based models. The aim of this chapter is to review the common approaches which have already been applied or have the potential to be applied to microbial inactivation by PEF.

20.5.1 Kinetics-Based Models

20.5.1.1 Bigelow Model

The first approach used to model PEF process involved the classical and widely practiced first-order kinetic model of Bigelow [29] to describe inactivation of microorganisms and enzymes. It is commonly known that the death of microorganisms is caused by inactivation of some critical enzyme systems. Since enzyme inactivation obeys first-order kinetics [30], thus, the microbial inactivation model results in the following equation:

$$\frac{dN}{dt} = -kN \tag{20.1}$$

where the microorganism population (N) varies with processing time (t) at a constant rate (k) depending on its size. Integration of this expression yields:

$$\frac{dN}{dt} = -kN \tag{20.2}$$

where N_0 is the initial number of microorganisms. Equation 20.2 can be rearranged as follows:

$$\log\frac{N}{N_0} = \log S(t) = -\frac{t}{D} \tag{20.3}$$

where $S(t)$ is the survival fraction after time t (μs^{-1}) and D is the familiar decimal reduction time ($D = 2.303/k$) corresponding to the reciprocal of the first-order rate constant. The resulting semi-logarithmic curve when log $S(t)$ is plotted versus time is frequently referred to as the survival curve. On the other hand, the effect of temperature on the kinetic constant is described using the Arrhenius equation [8,9]:

$$k = A\exp\left(-\frac{E_a}{R\,T}\right) \tag{20.4}$$

where A is an empirical constant (μs^{-1}), E_a is the activation energy (J/mol), R is the gas constant (8.31 J/K/mol), and T is the treatment temperature (K).

Considerable reports can be found in the literature that are based on the foregoing first-order kinetic approach [8–10,20,31]. For example, Amiali et al. [8,9] modeled the reduction of bacterial survival fraction as a function of treatment time at each electric field treatment for inactivation of *Escherichia coli O157:H7* and *Salmonella enteritidis* in liquid egg white and yolk. For liquid egg white, the determination coefficient (R^2) for the first-order kinetic model varied from 0.88 to 0.99 for *E. coli O157:H7* and from 0.96 to 0.99 for *S. enteritidis*. The authors reported R^2 values for liquid egg yolk to vary from 0.95 to 0.99. Change in kinetic rate constants with respect to temperature in the range from 20 to 40°C and electric field of 20 and 30 kV/cm, for *E. coli O157:H7* inactivation in liquid egg yolk yielded the following equation:

$$k = 2.46\times10^6 \exp\left[-\frac{47}{8.31\times10^{-3}T}\right] \tag{20.5}$$

whereas, Equation 20.6 was obtained for *S. enteritidis*:

$$k = 9.24\times10^6 \exp\left[-\frac{48.3}{8.31\times10^{-3}T}\right] \tag{20.6}$$

A similar approach was also reported by Bazhal et al. [10] for inactivation of *E. coli O157:H7* in liquid whole egg using a combination of thermal and PEF treatments. The authors considered the survival fraction as a function of the number of pulses (n) instead of treatment time at each electric field treatment:

$$S = S_0 \exp(-k_{TE}n) \tag{20.7}$$

where S_0 is the survival fraction after thermal treatment alone, n is the number of pulses and k_{TE} is kinetic constant obtained using Equation 20.8:

$$k_{TE} = A \exp\left(-\frac{B}{E^2}\right) \tag{20.8}$$

where A and B are model constants and E is the electric field strength. Figure 20.1 shows typical regression equations obtained from Equation 20.8 with $R^2 = 0.99$. The figure shows that the kinetics of *E. coli O157:H7* inactivation estimated by the k_{TE} value, depend on the treatment temperature and electric field strength. Maximum k_{TE} value was achieved at 60°C and 15 kV/cm. The influence of treatment temperature on the combined thermal and PEF inactivation kinetics was estimated by an Arrhenius' plot shown in Figure 20.2 using an electric field of 15 kV/cm. The figure indicates a threshold temperature of 50°C after which bactericidal effect of the combined treatment was significantly intensified.

20.5.1.2 Hulsheger Model

Hulsheger et al. [32] developed a first-order kinetic model to describe microbial inactivation by PEF, based on the empirically known relationship between the survival fraction and electric field strength:

$$\ln\frac{N}{N_0} = -b_E(E - E_c) \tag{20.9}$$

where b_E is a regression coefficient dependent on different experimental conditions such as treatment medium, treatment time, temperature and target microorganisms, E_c is the extrapolated critical value of E for 100% survival. The equation covers the range of 8–20 kV/cm. In order to apply the Equation 20.9, treatment time has to be kept constant for different electric fields. Hulsheger et al. [32] used two different treatment times of 100 and 360 μs to obtain b_E values in the range of 0.21–0.51 using different electrolytes.

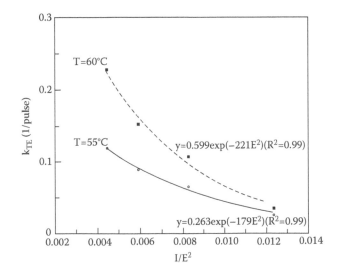

FIGURE 20.1 Influence of electric field strength and temperatures on kinetic constant for inactivation of *E. coli O157:H7*. (Adapted from Bazhal, M.I., Ngadi, M.O., Raghavan, G.S.V., and Smith, J.P., *LWT-Food Sci. Technol.*, 39 (4), 420–426, 2006. With permission.)

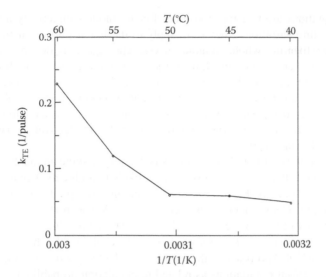

FIGURE 20.2 Arrhenius' plot for combined thermal and PEF inactivation of *E. coli* in liquid whole egg at different temperatures and number of pulses ($E = 15$ kV/cm in all experiments) with 4 min thermal treatment. (Adapted from Bazhal, M.I., Ngadi, M.O., Raghavan, G.S.V., and Smith, J.P., *LWT-Food Sci. Technol.*, 39 (4), 420–426, 2006. With permission.)

On the other hand, the effect of treatment time on the survival fraction can be estimated by [32]:

$$\ln \frac{N}{N_0} = -b_t \ln \left(\frac{t}{t_c} \right) \tag{20.10}$$

where the empirically known relationship between survival fraction and treatment time is correlated by a regression coefficient (b_t) and an extrapolated critical value of t for 100% survival (t_c). To obtain the regression coefficient, Hulsheger et al. [32] applied constant electric field strength of 12 kV/cm.

The treatment time t can be taken as a common parameter which is in addition to the field strength responsible for the degree of survival [32]. The dependency of the survival rate on treatment time is that of a double logarithmic function, whereas the field strength has a linear relation to the logarithmic survival rate. Peleg [33] and Schoenbach et al. [34] also noted that microbial inactivation is exponentially increased by increasing electric field intensity and linearly increased by increasing treatment time.

None of the above-mentioned models directly incorporate the effect of time and electric field strength in a single equation. The following model incorporates all these variables:

$$\frac{N}{N_o} = \left(\frac{t}{t_c} \right)^{-(E-E_c/k)} \tag{20.11}$$

where k is an independent constant factor. Hulsheger et al. [32] acknowledged that the model is influenced by variable parameters such as electric field strength, pulse number, electrolyte concentration, suspension temperature, bacterial cell concentration, and certain kinds of electrolytes. Additionally, a variation of the physiological conditions of type of microorganism may also have remarkable effects. However, several researchers have modeled survival curves under different conditions using this model with a reasonable success ($R^2 > 0.9$) [31,35,36].

Although the application of the kinetic-based models is widely practiced, it must be noted that many survival curves do not really follow the first-order kinetics. For example, although the

Arrhenius-type equations used in the first-order kinetic models effectively account for changes induced by temperature variations, they only apply to the linear portion of the microbial inactivation curves rather than the whole sigmoid curves. The equations do not consider the regions of maximum (shoulder) and minimum (tail) in the inactivation curves thus, limiting their practical utility. Not only are the so-called shoulders and/or tails in the survival curves observed from experimental data [37], downward and upward concavity phenomena are also frequently observed. This is because the microbial population has several subpopulations, each with its own inactivation kinetics. Accordingly, the survival curve is the result of different inactivation patterns, creating the nonlinear nature of the curves [30].

Another drawback of the kinetic-based models is that they assume an identical sensitivity for all microorganisms to the lethal agent. However, it is unlikely that all cells behave the same way and that the death of a single cell is due to one single event, an assumption which is considered in the first-order approach [38]. Consequently, it is necessary to seek new modeling approaches that would be able to better explain the nonlinearity of the survival curves. Curves with upward or downward concavities, or apparent lag time or shoulders can be explained by the probabilistic models of microbial inactivation [30]. Additionally, the models consider the possibility of individual resistance variability within microbial populations as related to population probabilistic distribution [37,39]. Among these are the models based on Weibull, Fermi and Gompertz distributions.

20.5.2 Probability-Based Models

20.5.2.1 Weibull Model

The Weibull distribution is an empirical model which does not link microbial death to mechanistic theories. The model essentially gives a statistical account of a failure time distribution. This is an advantage, because most likely, there is not a single cause of death and therefore, it is not very realistic to apply basic kinetic theories [30]. The Weibull distribution is a flexible model to describe microbial inactivation [39,40] and it has been used successfully in describing the microbial inactivation by PEF [33], radiation [41] and high pressure inactivation [42]. In terms of a survival curve, the Weibull distribution follows that:

$$\log S(t) = -\frac{1}{2.303}\left(\frac{t}{\alpha}\right)^{\beta}$$ (20.12)

where t is a PEF control parameter such as treatment time, total specific energy, pulse frequency, pulse width, etc. The two parameters of the distribution namely α and β are the scale parameter and the shape parameter, respectively. The Weibull distribution corresponds to a concave upward survival curve if $\beta < 1$ and concave downward curve if $\beta > 1$. Interestingly, the Weibull distribution reduces to an exponential distribution when $\beta = 1$. In which case, the model will reduce to the same familiar first-order equation:

$$\log S(t) = -\frac{t}{2.303\alpha}$$ (20.13)

Although Equation 20.13 has the same form as the first-order Equation 20.3, the meaning of the parameters D and α is different. D is the reciprocal of a first-order rate constant, whereas α represents the mean of the distribution describing death times of the microbial population, and thus has a probabilistic interpretation.

It is important to understand the influences of different α and β values and their interpretations on inactivation curves. The scale parameter α is a characteristic time that increases with increasing time if $\beta < 1$ and decreases when $\beta > 1$. If $\beta < 1$, the remaining cells have less probability of dying however if β > 1, the remaining cells become increasingly susceptible to stress. When β = 1, each cell is equally susceptible, no matter how long the treatment lasts; that is there is no biological variation [30]. Figure 20.3 shows three typical examples of inactivation curves using the Weibull distribution when β > 1, β < 1, and β = 1. Superficial inspection reveals that the obtained fits are good; it is also obvious that first-order kinetics do not apply to the data in Figures 20.3A and 20.4B. The classical test for validity of the Weibull model is the so-called hazard plot [46], a double logarithmic plot of ln(−lnS) against ln(t). A straight line should be obtained if the Weibull model applies. Weibull plots for the data given in Figure 20.3 and displayed in Figure 20.4 demonstrate reasonable plots for the model validation.

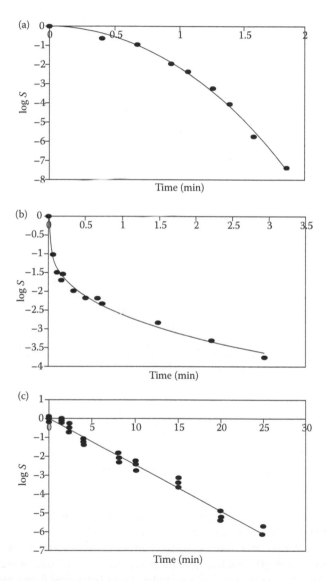

FIGURE 20.3   (A) Example of downward concavity: inactivation of *Listeria monocytogenes* in chicken meat at 70°C [43]. Weibull parameters are α = 0.46 and β = 2.1. (B) Example of upward concavity: inactivation of *Salmonella enteritidis* in egg yolk + 5% NaCl + 5% sucrose at 64°C [44]. Weibull parameters are α = 0.0025 and β = 0.3. (C) Example of linear behavior: inactivation of *Salmonella typhimurium* in ground beef at 57.2°C [45]. Weibull parameters are α = 1.78 and β = 1.0. (Adapted from van Boekel, M., *Int. J. Food Microbiol.*, 74 (1–2), 139–159, 2002. With permission.)

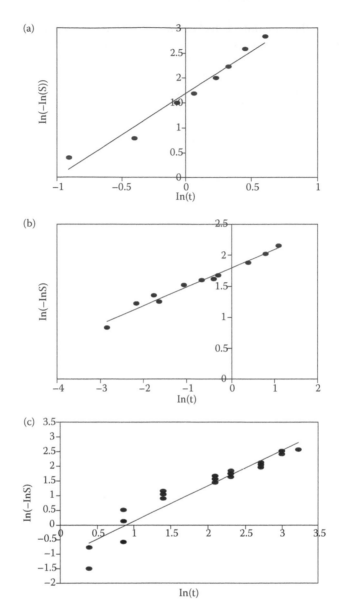

FIGURE 20.4 Weibull plot of the same data displayed in Figure 20.3A, B and C. (Adapted from van Boekel, M., *Int. J. Food Microbiol.*, 74 (1–2), 139–159, 2002. With permission.)

Weibull distribution as a regression model takes into account how the two parameters, α and β, depend on other conditions such as PEF-related parameters and treatment medium characteristics. The dependency of the Weibull parameters on such conditions (for example: pH, temperature, water activity, pressure, ionic strength, electric field intensity, etc.), can be determined by experimental data. van Boekel [30], after analyzing 55 case studies from literature, found that the logarithm of the scale parameter α, is linearly related to temperature in a way that is analogous to the classical D value. However, no clear dependence of the shape parameter β was found with temperature.

Alvarez et al. [3] used the Weibull distribution to model PEF inactivation of *Listeria monocytogenes* with electric field strength in the range from 15 to 28 kV/cm. Alvarez et al. [2] also applied the Weibull distribution to model PEF inactivation of *Yersinia enterocolitica* considering treatment time and total specific energy as two control parameters. For different electric field strengths (5.5–28 kV/cm), the parameters α and β were estimated by fitting Equation 20.12 to the inactivation data shown in Figures 20.5a and b. The determination coefficients (R^2) ranged from 0.98 to 0.99. Both α and β decreased when higher field strengths were applied. There are diverse relationships and values of the Weibull parameters (α and β) regarding different experimental factors such as electric field strength, pH, temperature and microbial cell diameter [4,5,22,40,47–49]. Table 20.1 summarizes some of the values of the Weibull parameters (α and β) for different microorganisms and pH with two different electric field strengths and media considering treatment time and specific energy as control parameters. The different α and β values in the table might be related to the different biological variations of different microorganisms, along with other experimental conditions during PEF treatment. Therefore, due to the diversity of the parameters involved in the PEF process, independent test of the available models using literature data is difficult.

20.5.2.2 Fermi Model

Microorganisms are not practically affected by electric fields of less than about 4–8 kV/cm [33]. Under stronger fields, the number of microorganisms decreases exponentially at a rate that is significantly dependent on the number of pulses. In this lethal region, a plot of the log number of (or fraction or percent) survivors versus the electric field intensity is more or less linear [32,52] and could be described by the model presented by Equation 20.9. However, an alternative approach to model the process is to consider it as continuous in its entire field strength range. According to this approach, there is no abrupt change in the destruction kinetic but a gradual transition from no or minor effect at weak electric fields to effective lethality under strong ones. Peleg [33] demonstrated that this approach is not only compatible with the model of Hulsheger et al. [32] at the pertinent field intensity range but also enables the incorporation of the effects of the number of pulses within its mathematical format. Based on this approach, Peleg [33] developed a mathematical expression capable of modeling the sigmoidal shape of survival curves using Fermi distribution:

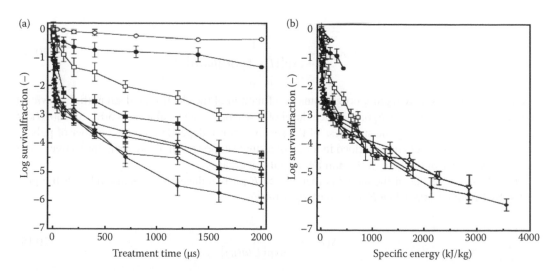

FIGURE 20.5 Influence of the electric field strength and the treatment time (a) or the electric field strength and the total specific energy (b) on the inactivation of *Y. enterocolitica* by PEF treatments: 5.5 kV/kg/pulse (O), 9 kV/cm, 0.42 kJ/kg/pulse (●), 12 kV/cm, 0.71 kJ/kg/pulse (□), 15 kV/cm, 1.08 kJ/kg/pulse (■), 19 kV/cm, 1.69 kJ/kg/pulse (△), 22 kV/cm, 2.2 kJ/kg/pulse (▲), 25 kV/cm, 2.86 kJ/kg/pulse (◇), and 28 kV/cm, 3.56 kJ/kg/pulse (◆). Treatment conditions: McIlvaine buffer pH 7.0; 2 mS/cm; 2 ms; 1 Hz. (Adapted from Alvarez, I., Raso, J., Sala, F.J., and Condon, S., *Food Microbiol.*, 20 (6), 691–700, 2003. With permission.)

TABLE 20.1
Parameters of Weibull Model using Pulsed Electric Field

Control Parameter	Product or Medium	Microorganism	pH	E (kv/cm)	α	β	Reference
Treatment time	Citrate-phosphate McIlvaine buffer	Yersinia enterocolitica	7.0	15	1.67	0.33	[2]
				25	0.02	0.22	
		Lactobacillus plantarum	7.0	25	234.90	0.99	[4]
			6.5	25	41.98	0.98	
			5.0	25	14.54	0.99	
			3.5	25	12.81	0.98	
		Listeria monocytogenes	7.0	15	2101.0	0.85	[5]
				25	111.40	0.62	
			6.5	15	2514.0	0.57	
				25	57.24	0.61	
			5.0	15	108.20	0.36	
				25	4.31	0.39	
			3.5	15	17.64	0.34	
				25	0.95	0.43	
	Orange juice–milk beverage	Escherichia coli CECT 516 (ATCC 8739)	4.05	15	0.68	0.31	[50]
				25	0.17	0.27	
		Lactobacillus plantarum	4.05	15	29.07	0.49	[51]
				25	0.03	0.16	
Specific energy	Citrate-phosphate McIlvaine buffer	Yersinia enterocolitica	7.0	15	0.71	0.32	[2]
				25	0.03	0.22	
		Listeria monocytogenes		15	1135.0	0.85	[3]
				25	158.2	0.62	

$$S(E,n) = \frac{1}{1 + \exp[(E - E_C(n))/a(n)]} \tag{20.14}$$

where $S(E,n)$ is the surviving microorganisms as a function of the electric field strength, E, (kV/cm) and the number of pulses (n), $E_C(n)$ a critical level of E where the survival level is 50%, and a(n) (kV/cm) a parameter indicating the steepness of the survival curve dependent on the number of pulses. Both $E_C(n)$ and $a(n)$ could be described in terms of a single exponential decay term, indicative of the increased lethality of the field as the number of pulse increases.

One of the properties of the model is that at $E \gg E_c$, it is reduced to the same relationship presented by Equation 20.9 when $E \gg E_c$, with $b_E = 1/a(n)$:

$$S(E,n) = \frac{1}{\exp[E/a(n)]} \tag{20.15}$$

or

$$\ln S(E,n) = -\frac{E}{a(n)} \tag{20.16}$$

The applicability of the model has been tested to the individual survival curves of different microorganisms such as *Lactobacillus brevis, Saccharomyces Serevisiae, Staphylococcus aureus, Candida albicans, Listeria monocytogenes, Pseudomonas aeruginosa* [33], *Lactobacillus plantarum* [53], and *Listeria innocua* [54]. Experimental data have shown good agreement with predicted values using the Fermi distribution with R^2 between 0.91 and 0.99.

20.5.2.3 Modified Gompertz Model

The Gompertz model is a double-exponential function that describes asymmetrical sigmoidal curves [55,56]. The model relates the three phases of the microbial growth curve with the logarithm of the fraction survival. The modification of the Gompertz equation can be written as [57]:

$$\ln \frac{N}{N_0} = A \exp\left[-\exp\left(\frac{2.718\mu_{max}}{A}(\lambda - t) + 1 \right) \right] \tag{20.17}$$

where μ_{max} is the maximum specific growth rate defined as the tangent at the inflection point of the microbial growth curve; λ is the lag time defined as the *x*-axis intercepts this tangent; and the asymptote, A, is the maximal value reached, defined by:

$$A = \ln \frac{N_\infty}{N_0} \tag{20.18}$$

Zwietering et al. [57], after comparing several sigmoidal models that describe microbial growth curve, concluded that growth curves are better fitted with the Gompertz model compared to the logistic, linear, quadratic and exponential models. The authors showed that in almost all cases, the Gompertz model can be regarded as the best model to describe growth data.

Gompertz model has also been used to describe the relationships between different parameters involved in inactivation of microorganisms by PEF. The relationship between the shape parameter (β) of the Weibull model and pH of the treatment medium has been modeled using a modified Gompertz equation [4] as following:

$$\beta = A + C\exp[-\exp(-B(pH - M))] \tag{20.19}$$

where A, B, C, and M are model constants.

The model was used to fit the experimental data for PEF inactivation of *Lactobacillus plantarum* in a buffer system with a corresponding R^2 value of 0.99. The Gompertz equation also described the relationship between the shape parameter value and the pH of the treatment medium for *Listeria monocytogenes*. To validate the model in food to assess its predictive capability, Gomez et al. [4] also performed experiments for apple and orange juices. The model validation indicated that the degree of correlation for the inactivation observed in apple and orange juices were not as high as the correlation observed in a buffer system. Inactivation of *L. plantarum* in apple and orange juices were underestimated when the inactivation was higher than 3 $\log_{10}$ cycles. Although such an underestimation in inactivation level could be considered more acceptable than an overestimation, research is needed to find more accurate and competent models.

Additionally, a similar approach has been used to describe the relationship between the scale parameter (α) of the Weibull model and the pH of the treatment medium using a modified Gompertz model [5]:

$$\alpha = A + C\exp[-\exp(-B(pH + M))] \tag{20.20}$$

where A, B, C, and M are model constants (different from the constants of Equation 20.19) that can be obtained by regression analysis.

Gomez et al. [5] reported a good model prediction with an accuracy factor of 1.14 in PEF inactivation kinetics of *Listeria monocytogenes* in apple juice. The accuracy factor was defined as the average of the distance between each point and the line of equivalence as a measure of the closeness of estimations to observations.

20.5.2.4 Other Models

There have been several other attempts on modeling inactivation kinetics based on probabilistic distributions. These include the Baranyi model [58], logistic model [39] and the modified Richards model [59]. These models have not been applied specifically to PEF processing but there are indications that they could be used to describe the inactivation kinetics.

Baranyi model:

$$\ln \frac{N}{N_0} = \mu_{max} A(t) - \frac{1}{m} \ln \left(1 + \frac{\exp(m\mu_{max} A(t)) - 1}{\exp(m(\ln N_{max} - \ln N_0))} \right) \tag{20.21}$$

where μ_{max}, the maximum growth rate; m, a curvature parameter (the special case of $m = 1$ is called the logistic or Pearl–Verlhurst growth model); N_{max}, the maximum population density and $A(t)$ is a function defined as:

$$A(t) = t + \frac{1}{v} \ln \left(\frac{\exp(-vt) + q_0}{1 + q_0} \right) \tag{20.22}$$

where v is a constant specific rate, determined by the quickness of the transition from the lag to the exponential phase and q_0 is a parameter expressing the physiological state of the inoculum.

Baranyi and Roberts [58] stated that with standard subculturing procedure, the physiological state of the inoculum is relatively constant and independent of subsequent growth conditions. Assuming that the specific growth rate follows the environmental changes instantaneously, the model can describe microbial growth in an environment where different factors such as temperature, pH and a_w, change with time.

Logistic model:

$$\log \frac{N}{N_0} = \alpha + \frac{\omega - \alpha}{1 + \exp\left[\frac{4\sigma(\tau - \log t)}{\omega - \alpha} \right]} \tag{20.23}$$

where α is the upper asymptote, ω is the lower asymptote, τ is the position of maximum slope (the logarithm of the time at which the maximum slope is reached) and σ is the maximum slope of the inactivation curve.

Modified Richards model:

$$\ln \frac{N}{N_0} = A \left[1 + v \cdot \exp(1 + v) \cdot \exp\left[\frac{\mu_{max}}{A} \cdot (1 + v) \left(1 + \frac{1}{v} \right) \cdot (\lambda - t) \right] \right]^{(-1/v)} \tag{20.24}$$

where v is a shape parameter and the other parameters were already defined in the modified Gompertz model.

20.6 CONCLUDING REMARKS

In recent years, several models have been proposed to describe microbial inactivation by PEF but none of them has been fully successful in describing the complicated governing process. Research is still ongoing to discover suitable models for PEF microbial inactivation. It should be emphasized that due to the sigmoidal shape of the microbial survival curves, the probability-based models are more realistic than the kinetic-based models. But, due to the high diversity of the PEF process, independent test of the available probability-based models using literature data is difficult; if not impossible. Therefore, to achieve satisfactory results, standardization of research protocols should be sought by researchers in order to reach an agreement on the important factors that need to be monitored to create reliable models.

REFERENCES

1. Rastogi, N. K. Application of high-intensity pulsed electrical fields in food processing. *Food Rev. Int.,* 19 (3), 229–251, 2003.
2. Alvarez, I., Raso, J., Sala, F. J., and Condon, S. Inactivation of Yersinia enterocolitica by pulsed electric fields. *Food Microbiol.,* 20 (6), 691–700, 2003.
3. Alvarez, I., Pagan, R., Condon, S., and Raso, J. The influence of process parameters for the inactivation of Listeria monocytogenes by pulsed electric fields. *Int. J. Food Microbiol.,* 87 (1–2), 87–95, 2003.
4. Gomez, N., Garcia, D., Alvarez, I., Raso, J., and Condon, R. S. A model describing the kinetics of inactivation of Lactobacillus plantarum in a buffer system of different pH and in orange and apple juice. *J. Food Eng.,* 70 (1), 7–14, 2005.
5. Gomez, N., Garcia, D., Alvarez, I., Condon, S., and Raso, J. Modelling inactivation of Listeria monocytogenes by pulsed electric fields in media of different pH. *Int. J. Food Microbiol.,* 103 (2), 199–206, 2005.
6. Sepulveda, D. R., Guerrero, J. A., and Barbosa-Canovas, G. V. Influence of electric current density on the bactericidal effectiveness of pulsed electric field treatments. *J. Food Eng.,* 76 (4), 656–663, 2006.
7. Amiali, M., Ngadi, M. O., Raghavan, V. G. S., and Smith, J. P. Inactivation of Eschericha coli O157 : H7 in liquid dialyzed egg using pulsed electric fields. *Food Bioproducts Process.,* 82 (C2), 151–156, 2004.
8. Amiali, M., Ngadi, M. O., Smith, J. P., and Raghavan, V. G. S. Inactivation of Escherichia coli O157 : H7 and Salmonella enteritidis in liquid egg white using pulsed electric field. *J. Food Sci.,* 71 (3), M88–M94, 2006.
9. Amiali, A., Ngadi, M. O., Smith, J. P., and Raghavan, G. S. V. Synergistic effect of temperature and pulsed electric field on inactivation of Escherichia coli O157 : H7 and Salmonella enteritidis in liquid egg yolk. *J. Food Eng.,* 79 (2), 689–694, 2007.
10. Bazhal, M. I., Ngadi, M. O., Raghavan, G. S. V., and Smith, J. P. Inactivation of Escherichia coli O157 : H7 in liquid whole egg using combined pulsed electric field and thermal treatments. *LWT-Food Sci. Technol.,* 39 (4), 420–426, 2006.
11. Zulueta, A., Esteve, M. J., Frasquet, I., and Frigola, A. Fatty acid profile changes during orange juice-milk beverage processing by high-pulsed electric field. *Eur. J. Lipid Sci. Technol.,* 109 (1), 25–31, 2007.
12. McLellan, M. R., Kime, R. L., and Lind, L. R. Electroplasmolysis and other treatments to improve apple juice yield. *J. Sci. Food Agric.,* 57 (2), 303–306, 1991.
13. Bazhal, M. I., Ngadi, M. O., and Raghavan, V. G. S. Influence of pulsed electroplasmolysis on the porous structure of apple tissue. *Biosystems Eng.,* 86 (1), 51–57, 2003.
14. Bazhal, M. I., Ngadi, M. O., Raghavan, G. S. V., and Nguyen, D. H. Textural changes in apple tissue during pulsed electric field treatment. *J. Food Sci.,* 68 (1), 249–253, 2003.
15. Chalermchat, Y., Fincan, M., and Dejmek, P. Pulsed electric field treatment for solid-liquid extraction of red beetroot pigment: mathematical modelling of mass transfer. *J. Food Eng.,* 64 (2), 229–236, 2004.
16. Arevalo, P., Ngadi, M. O., Bazhal, M. I., and Raghavan, G. S. V. Impact of pulsed electric fields on the dehydration and physical properties of apple and potato slices. *Drying Technol.,* 22 (5), 1233–1246, 2004.
17. Amami, E., Vorobiev, E., and Kechaou, N. Effect of pulsed electric field on the osmotic dehydration and mass transfer kinetics of apple tissue. *Drying Technol.,* 23 (3), 581–595, 2005.
18. Lebovka, N. I., Praporscic, I., Ghnimi, S., and Vorobiev, E. Temperature enhanced electroporation under the pulsed electric field treatment of food tissue. *J. Food Eng.,* 69 (2), 177–184, 2005.

19. Lebovka, N. I., Shynkaryk, N. V., and Vorobiev, E. Pulsed electric field enhanced drying of potato tissue. *J. Food Eng.,* 78 (2), 606–613, 2007.
20. Espachs-Barroso, A., Van Loey, A., Hendrickx, M., and Martin-Belloso, O. Inactivation of plant pectin methylesterase by thermal or high intensity pulsed electric field treatments. *Innovative Food Sci. Emerging Technol.,* 7 (1–2), 40–48, 2006.
21. Giner-Segui, J., Bailo-Ballarin, E., Gorinstein, S., and Martin-Belloso, O. New kinetic approach to the evolution of polygalacturonase (EC 3.2.1.15) activity in a commercial enzyme preparation under pulsed electric fields. *J. Food Sci.,* 71 (6), E262–E269, 2006.
22. Soliva-Fortuny, R., Bendicho-Porta, S., and Martin-Belloso, O. Modeling high-intensity pulsed electric field inactivation of a lipase from Pseudomonas fluorescens. *J. Dairy Sci.,* 89 (11), 4096–4104, 2006.
23. Elez-Martinez, P., Suarez-Recio, M., and Martin-Belloso, O. Modeling the reduction of pectin methyl esterase activity in orange juice by high intensity pulsed electric fields. *J. Food Eng.,* 78 (1), 184–193, 2007.
24. Harrison, S. L., BarbosaCanovas, G. V., and Swanson, B. G. Saccharomyces cerevisiae structural changes induced by pulsed electric field treatment. *LWT-Food Sci. Technol.,* 30 (3), 236–240, 1997.
25. Barbosa-Cánovas, G. V. and Sepulveda, D. R. Present status and the future of PEF technology. In *Novel Food Processing Technologies,* Barbosa-Cánovas, G. V., Tapia M. S., and Cano M. P., Eds. CRC Press, Boca Raton, FL, 2005.
26. Zimmermann, U. Electrical breakdown, electropermeabilization and electrofusion. *Rev. Physiol. Biochem. Pharm.,* 105, 175–256, 1986.
27. Bryant, G. and Wolfe, J. Electromechanical stresses produced in the plasma-membranes of suspended cells by applied electric-fields. *J. Membr. Biol.,* 96 (2), 129–139, 1987.
28. Weaver, J. C. and Chizmadzhev, Y. A. Theory of electroporation: a review. *Bioelectrochem. Bioenerg.,* 41 (2), 135–160, 1996.
29. Bigelow, W. D. The logarithmic nature of thermal death time curves. *J. Infectious Diseases,* 29, 528–536, 1921.
30. van Boekel, M. On the use of the Weibull model to describe thermal inactivation of microbial vegetative cells. *Int. J. Food Microbiol.,* 74 (1–2), 139–159, 2002.
31. Sensoy, I., Zhang, Q. H., and Sastry, S. K. Inactivation kinetics of Salmonella dublin by pulsed electric field. *J. Food Process Eng.,* 20 (5), 367–381, 1997.
32. Hulsheger, H., Potel, J., and Niemann, E. G. Killing of bacteria with electric pulses of high-field strength. *Radiat. Environ. Biophys.,* 20 (1), 53–65, 1981.
33. Peleg, M. A model of microbial survival after exposure to pulsed electric-fields. *J. Sci. Food Agric.,* 67 (1), 93–99, 1995.
34. Schoenbach, K. H., Peterkin, F. E., Alden, R. W., and Beebe, S. J. The effect of pulsed electric fields on biological cells: Experiments and applications. *IEEE Trans. Plasma Sci.,* 25 (2), 284–292, 1997.
35. Zhang, Q. H., Chang, F. J., Barbosacanovas, G. V., and Swanson, B. G. Inactivation of microorganisms in a semisolid model food using high-voltage pulsed electric-fields. *LWT-Food Sci. Technol.,* 27 (6), 538–543, 1994.
36. MartinBelloso, O., VegaMercado, H., Qin, B. L., Chang, F. J., BarbosaCanovas, G. V., and Swanson, B. G. Inactivation of Escherichia coli suspended in liquid egg using pulsed electric fields. *J. Food Process. Preservation,* 21 (3), 193–208, 1997.
37. Cerf, O. Tailing of survival curves of bacterial-spores. *J. Appl. Bacteriol.,* 42 (1), 1–19, 1977.
38. Reichart, O. Reaction kinetic interpretation of heat destruction influenced by environmental factors. *Int. J. Food Microbiol.,* 64 (3), 289–294, 2001.
39. Peleg, M. and Cole, M. B. Reinterpretation of microbial survival curves. *Crit. Rev. Food Sci. Nutrition,* 38 (5), 353–380, 1998.
40. Peleg, M. and Penchina, C. M. Modeling microbial survival during exposure to a lethal agent with varying intensity. *Crit. Rev. Food Sci. Nutrition,* 40 (2), 159–172, 2000.
41. Anellis, A. and Werkowsk, S. Estimation of radiation resistance values of microorganisms in food products. *Appl. Microbiol.,* 16 (9), 1300–1308, 1968.
42. Heinz, V. and Knorr, D. High pressure inactivation kinetics of Bacillus subtilis cells by a three-state-model considering distributed resistance mechanisms. *Food Biotechnol.,* 10 (2), 149–161, 1996.
43. Murphy, R. Y., Marks, B. P., Johnson, E. R., and Johnson, M. G. Thermal inactivation kinetics of Salmonella and Listeria in ground chicken breast meat and liquid medium. *J. Food Sci.,* 65 (4), 706–710, 2000.

44. Michalski, C. B., Brackett, R. E., Hung, Y. C., and Ezeike, G. O. I. Use of capillary tubes and plate heat exchanger to validate US Department of Agriculture pasteurization protocols for elimination of Salmonella enteritidis from liquid egg products. *J. Food Prot.,* 62 (2), 112–117, 1999.
45. Goodfellow, S. J. and Brown, W. L. Fate of Salmonella inoculated into beef for cooking. *J. Food Prot.,* 41 (8), 598–605, 1978.
46. Nelson, W. Theory and applications of hazard plotting for censored failure data. *Technometrics,* 14 (4), 945–966, 1972.
47. Rodrigo, D., Barbosa-Canovas, G. V., Martinez, A., and Rodrigo, M. Weibull distribution function based on an empirical mathematical model for inactivation of Escherichia coli by pulsed electric fields. *J. Food Prot.,* 66 (6), 1007–1012, 2003.
48. Rodrigo, D., Ruiz, P., Barbosa-Canovas, G. V., Martinez, A., and Rodrigo, M. Kinetic model for the inactivation of *Lactobacillus plantarum* by pulsed electric fields. *Int. J. Food Microbiol.,* 81 (3), 223–229, 2003.
49. Lebovka, N. I. and Vorobiev, E. On the origin of the deviation from the first-order kinetics in inactivation of microbial cells by pulsed electric fields. *Int. J. Food Microbiol.,* 91 (1), 83–89, 2004.
50. Rivas, A., Sampedro, F., Rodrigo, D., Martinez, A., and Rodrigo, M. Nature of the inactivation of Escherichia coli suspended in an orange juice and milk beverage. *Eur. Food Res. Technol.,* 223 (4), 541–545, 2006.
51. Sampedro, F., Rivas, A., Rodrigo, D., Martinez, A., and Rodrigo, M. Effect of temperature and substrate on PEF inactivation of *Lactobacillus plantarum* in an orange juice-milk beverage. *Eur. Food Res. Technol.,* 223 (1), 30–34, 2006.
52. Castro, A. J., Barbosacanovas, G. V., and Swanson, B. G. Microbial inactivation of foods by pulsed electric-fields. *J. Food Process. Preservation,* 17 (1), 47–73, 1993.
53. Rodrigo, D., Martinez, A., Harte, F., Barbosa-Canovas, G. V., and Rodrigo, M. Study of inactivation of Lactobacillus plantarum in orange-carrot juice by means of pulsed electric fields: Comparison of inactivation kinetics models. *J. Food Prot.,* 64 (2), 259–263, 2001.
54. Martin, M. F. S., Sepulveda, D. R., Altunakar, B., Gongora-Nieto, M. M., Swanson, B. G., and Barbosa-Canovas, G. V. Evaluation of selected mathematical models to predict the inactivation of *Listeria innocua* by pulsed electric fields. *LWT-Food Sci. Technol.,* 40 (7), 1271–1279, 2007.
55. Buchanan, R. L. Predictive food microbiology. *Trends Food Sci. Technol.,* 4 (1), 6–11, 1993.
56. Geeraerd, A. H., Herremans, C. H., and Van Impe, J. F. Structural model requirements to describe microbial inactivation during a mild heat treatment. *Int. J. Food Microbiol.,* 59 (3), 185–209, 2000.
57. Zwietering, M. H., Jongenburger, I., Rombouts, F. M., and Vantriet, K. Modeling of the bacterial-growth curve. *Appl. Environ. Microbiol.,* 56 (6), 1875–1881, 1990.
58. Baranyi, J. and Roberts, T. A. A dynamic approach to predicting bacterial-growth in food. *Int. J. Food Microbiol.,* 23 (3–4), 277–294, 1994.
59. Richards, F. J. A flexible growth function for empirical use. *J. Exp. Bot.,* 10 (29), 290–300, 1959.

21 Mathematical Modeling and Design of Ultraviolet Light Processes for Liquid Foods and Beverages

Zhengcai Ye
Georgia Institute of Technology

Tatiana Koutchma
Guelph Food Research Center

CONTENTS

21.1 INTRODUCTION

21.1.1 Advantages of Ultraviolet (UV) Light Process for Liquid Foods and Beverages

The intensive exploration of ultraviolet (UV) light treatment for liquid foods has been driven by the search for a mild pasteurization alternative to thermal processing. The outbreaks of foodborne illnesses associated with the consumption of unpasteurized juices and apple cider in 1997–1999 resulted in a rule published by the U.S. Food and Drug Administration (FDA) in order to improve the safety of juice products. The rule[1] required manufacturers of juice products to develop a Hazard Analysis and Critical Control Point (HACCP) plan and to achieve a 5-log reduction in the numbers of the most resistant pathogens.[2] The FDA and US Department of Agriculture (USDA) have concluded that the use of UV irradiation is safe. In 2000, the FDA approved the use of UV-light as an alternative treatment to thermal pasteurization of fresh juice products.[1] In addition, the definition of "pasteurization" was recently revised and now includes any process, treatment, or combination thereof, which is applied to food to reduce most microorganism(s) of public health significance.[3] The processes and technologies examined in the above mentioned report include UV-light irradiation as an alternative to heat for pasteurization purposes.

The key advantage of UV-light treatment over traditional thermal pasteurization is that it is a nonthermal method that does not affect the flavor of foods. Most nutritional components, which are sensitive to heat, are not destroyed by the UV process or potentially have less destruction compared to heat treatment. Moreover, heat pasteurization is cost-prohibitive for small juice producers.[4] Because of these advantages and the growing negative public reaction over chemicals added to foods, UV-light processing can provide an opportunity to reduce the levels of microbial contamination for a wide range of liquid foods and beverages.

Compared to the water and waste water treatments, where UV light was developed into an advanced technology for disinfection and oxidation of organic matters, the application of UV-light to processing liquid foods and beverages is a relatively new area. UV-light is effective in treating water and many clear beverages such as clarified juices, but is less effective in treating turbid liquids (e.g., apple cider and orange juice) where UV-light is strongly absorbed or reflected. The absorption coefficients of juices are significantly higher than that of water and typically vary from 10 to 40 cm^{-1}, depending on the brand of juice and the processing conditions used in its manufacturing. However, the correct choice of the UV-reactor design and UV-light source will make UV light processing of fresh juices, beverages and other absorptive food liquids more effective and will serve to further commercialization of this technology for food applications.

21.1.2 MATHEMATICAL MODELING OF ULTRAVIOLET (UV) PROSSESSING

Mathematical modeling is an essential tool in the design of a UV process for liquid foods and beverages. Modeling can be used for a number of purposes. Firstly, mathematical modeling can be used to predict the efficiency of microbial inactivation in a UV-reactor for a specific application based on inactivation kinetics and transport phenomena as shown schematically in Figure 21.1. Figure 21.1 is a representation of how the elements of a unit operation such as UV-light processing manifest themselves in the microbial reduction kinetics that is measured and/or modeled for the targeted performance objective. The main elements include the transport phenomenon relevant in treating the liquid, the physical or chemical characteristics of the food, and the resistance of the microorganisms to UV treatment. However, the achievement of the performance objective may result in the destruction of nutritional components and the formation of undesired compounds. Secondly, mathematical modeling of UV fluence in the reactor can assist in understanding the fluence distribution and identify the location of the least treated liquid and the dead spot in the reactor. The critical process and product parameters affecting UV fluence distribution in the reactor are shown schematically in Figure 21.2. Since commercial UV reactors are of a flow-through type, they are expected to have a distribution of exposure time or residence time distribution (RTD) and fluence rate distribution (FRD) resulting from UV light attenuation in a medium with high absorptive properties. It can be seen that the emitting characteristics of the UV-light source and absorptive properties of the treated medium, the RTD in the annulus, the annulus size and geometry will determine the UV fluence distribution in the reactor. Computing the UV fluence is another way to evaluate the performance of UV processing reactors, since software is capable of modeling UV fluence and predicting particles and fluid velocities, particle mixing, and RTD.

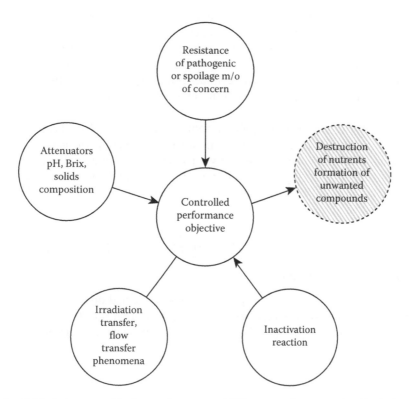

FIGURE 21.1 Critical elements affecting performance of UV reactors for processing liquid foods.

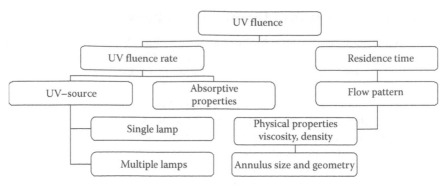

FIGURE 21.2 Critical process and product parameters affecting UV fluence distribution in the reactor.

A further application of the mathematical modeling of UV processing is to calculate the optimal dimensions and geometry of the UV- reactor for maximum inactivation performance taking into account the specific physical properties of food and the requirements of the process.

21.1.3 OBJECTIVES

There are a few basic laminar and turbulent flows UV reactor designs currently being investigated for microbial inactivation in liquid foods. If an open-channel, modular design[5] UV system, used extensively for water and wastewater processing, is applied to pasteurize a fresh juice with a high absorption coefficient, the gap between two neighboring UV lamps should be optimized. Another type of UV reactor design with an annular thin film between two concentric tubes is more suitable for fresh juice treatment. The juice flows through the gap formed by two concentric cylinders. UV-light is irradiated from the inner and/or outer cylinder (quartz sleeve) so that pathogens in the liquid are assured exposure to UV light. When the two concentric cylinders are fixed, the flow pattern in the gap can be laminar Poiseuille flow or turbulent flow depending on flow rates and rheological properties of liquid. If the inner cylinder is rotating, and the rotating speed of the inner cylinder exceeds a certain value, the flow pattern can be either laminar or turbulent Taylor–Couette flow.[6,7]

The objective of this chapter is to apply fundamental principles of mathematical modeling of UV inactivation in liquid foods and beverages in order to compare performance of typical single and multiple lamps UV reactor designs currently used in the food industry. The computational fluid dynamics (CFD) modeling of UV inactivation in a laminar Poiseuille flow, turbulent flow and laminar Taylor–Couette flow will be investigated and recommendations for the optimal reactor designs for fresh juices will be given.

21.2 MODELING OF ULTRAVIOLET (UV) INACTIVATION KINETICS

21.2.1 FIRST ORDER INACTIVATION MODEL

Various modeling approaches have been proposed to describe and predict UV inactivation kinetics[8–10] for microorganisms. Among these, the first order inactivation model is the simplest. It assumes that the inactivation rate changes linearly with respect to pathogen concentration, N, and fluence rate, I, such that:

$$\frac{dN}{dt} = -k_1 I N \qquad (21.1)$$

where k_1 is the first order inactivation constant (cm^2/mJ). The first order inactivation reaction has also been defined by a pseudo-first order model or mixed second order model.[9,11] In the first order model if k_1 and I are constant, then by integration,

$$\frac{N}{N_0} = \exp(-k_1 I t) \qquad (21.2)$$

The first order model was able to reasonably predict microbial inactivation when the fluence was within certain limits. However, the predicted data did not agree well with experimental values at low UV fluence levels.[12] Sigmoidal shaped inactivation curves have often been reported.[13] The first order inactivation constant is small at relatively low UV fluence. This phenomenon is also referred to as the shouldered survival curve.[14] The first order inactivation constant increased with fluence and remained constant within a certain fluence range. Finally, when the fluence was larger than a certain value, the first order inactivation constant decreased with the increase of UV fluence. This phenomenon is referred to as tailing phenomena.

A lag in microbial inactivation at low levels of UV fluence (the shouldered survival curve) arises because microorganisms exposed to sublethal UV fluence may repair their injuries and continue multiplying. The sublethal UV fluence had a slight adverse effect on the analytical procedure used to quantify microbial viability. In contrast, the tailing phenomena may be attributable to the result of heterogeneity among a population of microorganisms. Some organisms may be relatively less resistant to UV radiation while other organisms in the same population may be more resistant. Another reason for the tailing effect is the presence of particles such as pulp in fresh juices. Particles may serve as a hiding space for viable organisms or an opaque surface to shade microorganisms from UV radiation.

Because the tailing phenomenon is mostly observed when log reductions are larger than five, it is not very important in commercial applications of UV disinfection. Other models, for example, multitarget,[9,10] series-event,[9] the Collins-Selleck model,[8] were developed to account for deviations from the first order model at low UV fluence.

21.2.2 Series-Event Inactivation Model

As stated previously, a lag in microbial inactivation at low UV fluence can be observed for various kinds of microorganisms. The series-event inactivation model was proposed by Severin et al.[9] to account for the lag at low fluence. It assumes that inactivation of microorganisms takes place in a stepwise fashion,

$$M_0 \xrightarrow{k_{SE}I} M_1 \xrightarrow{k_{SE}I} \cdots M_i \xrightarrow{k_{SE}I} \cdots M_{n-1} \xrightarrow{k_{SE}I} M_n \xrightarrow{k_{SE}I} \cdots \qquad (21.3)$$

The inactivation rate at each step is first order with respect to the fluence rate I,

$$\frac{dN_i}{dt} = k_{SE}I(N_{i-1} - N_i) \qquad (21.4)$$

where subscript i is event level and k_{SE} is the inactivation constant in the series-event inactivation model. k_{SE} is assumed to be the same for different event levels. When n elements (a threshold) of microorganisms have been inactivated, the microorganisms will become nonviable. If k_{SE} and I are constants, the concentration of surviving microorganisms N is determined by

$$\frac{N}{N_0} = \exp(-k_{SE}It)\sum_{i=0}^{n-1}\frac{(k_{SE}It)^i}{i!} \qquad (21.5)$$

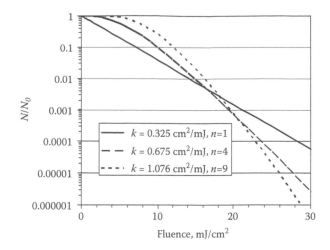

FIGURE 21.3 Microbial UV inactivation curves with different thresholds.

where n is the threshold. It is obvious that if $n = 1$, the above equation will be reduced to the first order model. The physical meaning behind Equation 21.5 is that more than one hit is required for UV inactivation of an individual microorganism. At the beginning of UV inactivation process (low fluence), the probability of an individual microorganism to obtain n hits (where $n > 1$) is rather low. As the UV inactivation continues, more surviving microorganisms have accumulated $n-1$ hits and require only one additional hit to be inactivated completely[14] so that the inactivation curve becomes steeper with increasing fluence. The larger threshold values of n represent microorganisms that are more resistant at low UV fluence. Figure 21.3 illustrates this trend. The inactivation rate constants in Figure 21.3 were obtained by fitting the same experimental data of *Escherichia coli* K12 (ATCC 25253) with different thresholds. At low fluence, the first order model cannot account for the shouldered survival curve and overestimates log reductions. At high fluence, the series event model predicts higher log reductions than the first order model. Moreover, the difference in the microbial log reductions between the first order model and the series event model increases with increase in the threshold value at high fluence. The series event and the first order models, however, predict similar log reductions with intermediate fluence values which are often between 14 and 20 mJ/cm². The shouldered survival curve can be observed in UV inactivation of many types of microorganisms. The first order model ($n = 1$) was valid only for some viruses whose UV sensitive material is single-stranded DNA or single-stranded RNA.[14] For example, it was reported that $n = 1$ for f2 bacterial virus.[12]

Both the first order inactivation model and series-event inactivation model[12] were used in numerical simulations.

21.3 ULTRAVIOLET (UV) INACTIVATION IN LAMINAR FLOW REACTORS

Currently, different continuous flow UV reactor designs are being evaluated for use in fresh juice pasteurization. The first design approach uses an extremely thin film UV reactor to decrease the path length for light and thus minimize problems associated with poor light penetration. Thin film reactors are characterized by laminar flow with a parabolic velocity profile. Maximum velocity of the liquid is observed in the center and this velocity is about 1.5 times as fast as the average velocity of the liquid and results in nonuniform processing conditions.[15] The schematic diagram of thin film annular single lamp reactor is shown in Figure 21.4. The UV annular reactor "UltraDynamics" model TF-1535 (Severn Trent Services Inc., Colmar, PA) is an industrial thin film, high intensity contact purifier and represents one of the simplest and practical means of UV disinfecting liquids

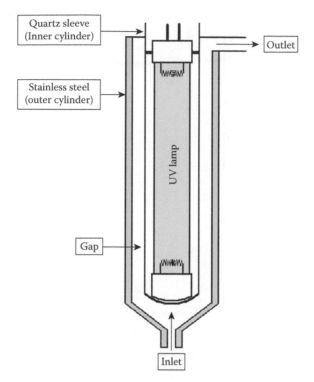

FIGURE 21.4 Schematic diagram of a thin film annular reactor. (From Ye, Z., Koutchma, T., Parisi, B., Larkin, J., and Forney, L. J., *Journal of Food Science*, 75, E271–E278, 2007. With permission.)

with poor UV transmission. The UV unit consists of a chamber, high intensity 254 nm UV lamp, pure fused quartz sleeve and remote plug-in power box with LED lamp indicator. The whole system included four chambers of 80 cm (2 units), 40 cm, and 20 cm length.

21.3.1 NUMERICAL MODELING

The fluid laminar flow in the radiation section of the gap between two cylinders can be approximated as annular Poiseuille flow (Figure 21.5). After the Navier–Stokes equations of laminar flow for Newtonian fluids are simplified and solved, a velocity profile can be obtained.[16]

$$u(r) = C_1 \left[1 - \frac{r^2}{R_2^2} + \frac{1-\kappa^2}{\ln(1/\kappa)} \ln\left(\frac{r}{R_2}\right) \right] U_{av} \qquad (21.6)$$

where

$u(r)$, axial velocity at radius r (cm/s);

U_{av}, average axial velocity, $U_{av} = \dfrac{Q}{\pi(R_2^2 - R_1^2)}$ (cm/s);

κ, ratio of the radius of the inner cylinder to that of the outer cylinder, $\kappa = R_1/R_2$;

C_1, constant determined by κ, $C_1 = \dfrac{2}{1+\kappa^2 - \dfrac{1-\kappa^2}{\ln(1/\kappa)}}$.

The equation describing steady state heat transfer by radiation in a homogeneous medium may be written as Equation 21.7.[17]

$$\frac{dI(s,\Omega)}{ds} = -\alpha I(s,\Omega) \qquad (21.7)$$

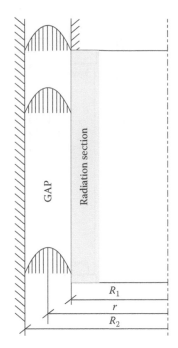

FIGURE 21.5 Schematic diagram of annular Poiseuille flow. (From Ye, Z., Koutchma, T., Parisi, B., Larkin, J., and Forney, L. J., *Journal of Food Science*, 75, E271–E278, 2007. With permission.)

If UV fluence rate is assumed to vary only in radial direction,

$$\frac{d(Ir)}{dr} = -\alpha Ir \tag{21.8}$$

After Equation 21.8 is solved at the boundary condition $I = I_0$ at $r = R_1$, the fluence rate in annular gap can be approximated by Equation 21.9.

$$I(r) = I_0 \frac{R_1}{r} \exp(-\alpha(r - R_1)) \tag{21.9}$$

Because stream lines of the annular Poiseuille flow are parallel, and axial dispersion and diffusion between neighboring layers are negligible, UV fluence It (the product of fluence rate I and exposure time t) can be described by the following equation (Equation 21.10),

$$It(r) = I(r)L/u(r) \tag{21.10}$$

where L is the length of radiation section.

If UV inactivation is known as a function of fluence It,

$$\frac{N}{N_0} = f(It) \tag{21.11}$$

where N and N_0 are concentrations of viable microorganisms after and before exposure respectively, the average concentration of viable microorganisms at the outlet of the reactor, N_{av}, can be obtained by integrating the following equation (Equation 21.12)

$$\frac{N_{av}}{N_0} = \frac{\int_0^{2\pi} \int_{R_1}^{R_2} f(It(r))u(r)rd\theta dr}{\int_0^{2\pi} \int_{R_1}^{R_2} u(r)rd\theta dr} \tag{21.12}$$

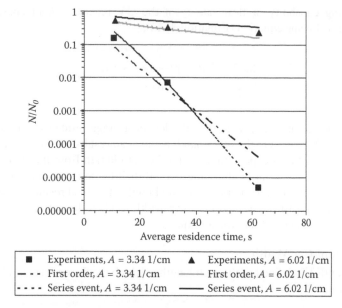

FIGURE 21.6 Comparison *of E. coli* K12 inactivation between experiments and theoretical prediction for $L = 77.9$ cm reactor.

21.3.2 COMPARISON BETWEEN EXPERIMENTS AND NUMERICAL SIMULATIONS

The UV reactor used for experiments was single UV-lamp thin-film annular reactor of the UltraDynamics model TF-1535 (Severn Trent Services Inc., Colmar, PA) shown in Figure 21.4. The outer diameter of the quartz sleeve was 2.45 cm and the inner diameter of the stainless steel tube was 3.48 cm. Correspondingly, the gap formed by the two cylinders was 0.515 cm. The single low pressure, germicidal UV lamp was 77.9 cm long.

Figure 21.6 shows the comparison of experimental data and theoretical predictions of inactivation of *Escherichia coli* K12 (*E. coli* ATCC 25253) in model solutions. The first order model with $k_1 = 0.325$ cm²/mJ and series-event model with $k_{SE} = 0.675$ cm²/mJ and $n = 4$ were used for theoretical calculations.[18] As expected, survival curve did not follow linear relationship when average residence times were used due to nonuniform UV fluence distribution.

According to Figure 21.6, the first order model overestimated inactivation of *E. coli* in regions with low fluence compared with experimental data. On the other hand, the first order model underestimated inactivation in regions with high fluence. This kind of error was avoided when the series-event model was used. Moreover, it can be seen in Figure 21.6, when the absorption coefficients of model solutions increased about two times (from 3 cm⁻¹ to 6 cm⁻¹), the log reduction of *E. coli* in model solution at $A \approx 6$ cm⁻¹ was much less than half the log reductions in a model solutions with $A \approx 3$ cm⁻¹. This was due to fluence rates decreasing exponentially with the path length from the radiation source and slight increase in the absorption coefficient resulting in a large increase of under-irradiated volume.

21.3.3 CALCULATION OF OPTIMUM GAP WIDTH

21.3.3.1 Theoretical Analysis

In design of thin film annular UV reactors, one of the most important parameters is a gap width. The gap width has different effects in the overall inactivation efficiency. In a narrow gap the velocity distribution is not broad (the maximum velocity is about 1.5 times the average velocity), the RTD does not significantly affect the overall disinfection efficiency.

If the processing capability or flow rate Q of the UV reactor is kept constant, N_{av} can be determined by the following equation (Equation 21.13),

$$\frac{N_{av}}{N_0} = \frac{2\pi}{Q} \int_{R_1}^{R_2} f\left(\frac{I_0 R_1 \exp(-\alpha(r-R_1))L}{ru(r)}\right) u(r) r dr \qquad (21.13)$$

In general, when the gap width increases, the longer average residence time may increase the overall disinfection efficiency. On the other hand, a wide gap results in broader fluence distribution because fluence rate decreases exponentially with the path length from the radiation source. This means that in a case of large gap widths, the overall disinfection efficiency may become worse because of larger under-irradiated volumes of treated liquid. The final results can be determined by the factor which is dominating. The optimum gap width needs to be determined for each specific value of absorption coefficient of the treated liquid.

21.3.3.2 Optimum Gap Width

The penetration depth λ is defined as the path length at which the fluence rate is 10% of the incident radiation fluence rate,

$$\frac{I_\lambda}{I_0} = 10\% \qquad (21.14)$$

Therefore, the penetration depth can be determined as the reciprocal of the absorption coefficient (10 base) or $\lambda = 1/A$. The ratio of the penetration depth to gap width (λ/d) shows how far radiation can penetrate in the treated fluid. If $\lambda/d = 1$, the liquid in the gap can be considered illuminated.

Figures 21.7 and 21.8 show the effect of the ratio of the penetration depth to gap width (λ/d) on log reductions of *E. coli* K12. The series-event model with $k_{SE} = 0.675$ cm^2/mJ and $n = 4$ was used for theoretical predictions. The radius of the outer cylinder was changed to create different gap widths while the radius of inner cylinder was kept constant at 1.225 cm. UV radiation was exposed from the inner cylinder at the fluence rate of $I_0 = 12$ mW/cm^2 and UV lamp length of $L = 77.9$ cm.

In Figure 21.7, the absorption coefficients were varied from 3 cm^{-1} to 20 cm^{-1} and flow rate was equal to 12.5 mL/s. In Figure 21.8, the absorption coefficients were varied from 30 cm^{-1} to 60 cm^{-1}

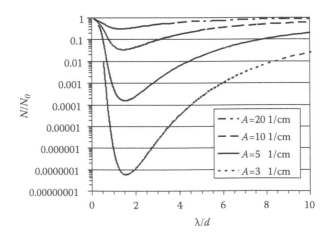

FIGURE 21.7   Effect of varying the ratio λ/d on log reductions at $Q = 12.5$ mL/s.

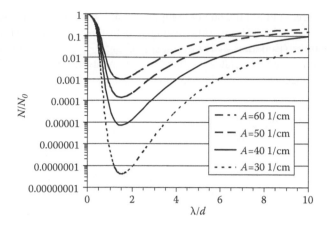

FIGURE 21.8 Effect of varying the ratio of λ/d on log reductions at $Q = 1.25$ mL/s.

TABLE 21.1
Optimum Ratio of λ/d at Different Absorption Coefficients

A (cm^{-1})	3	5	10	20	30	40	50	60
Optimum λ/d	1.55	1.51	1.43	1.33	1.49	1.53	1.51	1.42

and flow rate was decreased to 1.25 mL/s in order to achieve significant log reductions. It is proven in Figures 21.7 and 21.8 that for each value of absorption coefficient there was a specific optimum ratio of λ/d where maximum microbial inactivation was achieved. This specific ratio can be used for determination of the optimum gap width of the annulus. The optimum ratio of λ/d for a range of absorption coefficients is given in Table 21.1.

According to Table 21.1, the range of optimum ratios of λ/d was $1.33 \leq \lambda/d \leq 1.55$. Most of the optimum λ/d values oscillated around 1.5. When the ratio of λ/d = 1.5, the minimum fluence rate was 22% of the incident radiation fluence rate. This means that when λ/d < 1.5, pathogens are partially over-exposed to the UV radiation. However, inactivation of the overexposed fraction cannot compensates for more surviving pathogens in the underexposed fraction resulting from the wider gap. According to Figures 21.7 and 21.8, inactivation levels decreased greatly when λ/d < 1.0. For example, when λ/d = 0.2, log reduction was about 5% of the optimum log reduction. When λ/d further decreased to 0.1, the log reduction was only 1–2% of the maximum log reduction.

Based on the above calculations, it can be concluded that for the range of absorption coefficients of juices between 10 and 40 cm^{-1} the optimum gap width shall be in the range of 0.17–0.67 mm. Therefore the gap width in laminar UV reactors used for fresh juice pasteurization should not be more than 1 mm.

21.4 ULTRAVIOLET (UV) INACTIVATION IN TURBULENT FLOW REACTORS

21.4.1 NUMERICAL MODELING

The conservation equations of microbial concentrations in axisymmetrical coordinates can be written as,[16]

$$u_r \frac{\partial N_i}{\partial r} + u_z \frac{\partial N_i}{\partial z} = \frac{\mu_t}{\rho Sc_t}\left(\frac{1}{r}\frac{\partial}{\partial r}\left(r\frac{\partial N_i}{\partial r}\right) + \frac{\partial^2 N_i}{\partial z^2}\right) + k_{SE}I(N_{i-1} - N_i) \tag{21.15}$$

where u_r and u_z are radial and axial velocity components and μ_t is turbulent viscosity. These three variables can be obtained by solving time-averaged continuity equation (Equation 21.16) and *Navier–Stokes* equations (Equation 21.17) by means of the k–ε two-equation model.[19] Sc_t is the turbulent Schmidt number, which relates the turbulent momentum transport to the turbulent transport of the pathogen concentration N_i. The recommended Sc_t for most cases is 0.8.[20]

$$\frac{1}{r}\frac{\partial(ru_r)}{\partial r}+\frac{\partial u_z}{\partial z}=0 \tag{21.16}$$

$$\frac{1}{r}\frac{\partial(r\rho u_r \phi)}{\partial r}+\frac{\partial(\rho u_z \phi)}{\partial z}=\frac{1}{r}\frac{\partial}{\partial r}\left(r(\mu+\mu_t)\frac{\partial \phi}{\partial r}\right)+\frac{\partial}{\partial z}\left((\mu+\mu_t)\frac{\partial \phi}{\partial z}\right)+S_\phi \tag{21.17}$$

Here, ρ is density and ϕ in Equation 21.17 can be either u_r, u_z, k, and ε. S_ϕ varies depending on the definition of ϕ.

21.4.2 Turbulent Flow Modeling

The FLUENT software from Fluent, Inc (Lebanon, NH), one of the most popular commercial CFD software packages, was used to solve Equations 21.15 through 21.17. In the FLUENT software, three k–ε models are provided: the standard k–ε model,[19] the RNG k–ε model[21] and the realizable k–ε model.[22] All three k–ε models were investigated in numerical modeling of UV inactivation in turbulent flow. No big differences were found among these models since both the realizable and RNG k–ε models were developed to account for streamline curvature and swirling flows, and the streamlines in turbulent UV reactors are almost parallel. Therefore, the standard k–ε model can be used for modeling turbulence.

In the FLUENT software, user-defined scalars (UDS) are provided to solve the conservation equations of microbial concentrations (Equation 21.15). However, since the source terms in Equation 21.15 are not in standard forms, user-defined functions (UDFs) were utilized to express the source terms in a special format.

21.4.3 Comparison between Experiments and Numerical Simulations

21.4.3.1 Ultraviolet (UV)-treatment Experimental Set-up Configurations

The turbulent flow UV-treatment system is shown in Figure 21.9. The entire system consisted of four chambers with various lengths of 20 cm, 40 cm, 80 cm, and 80 cm. The individual UV reactor used for experiments was the UltraDynamics model TF-1535 (Severn Trent Services Inc., Colmar, PA). Reactors connected in series were used in order to increase the short residence time due to high flow rates. The outer diameter of the quartz sleeve was 2.45 cm and the inner diameter of the stainless steel tube was 3.48 cm. Correspondingly, the gap formed by the two cylinders was 0.515cm wide. For the reactors with lengths of 80 cm, 40 cm and 20 cm, the average fluence rate on the surface of the quartz sleeves was measured to be 12, 16 and 15.4 mW/cm^2 respectively, and the effective radiation lengths were 77.9 cm, 29.2 cm, and 11.2 cm, respectively.

The inactivation experiments using *Yersinia pseudotuberculosis* (*Y. pseudotuberculosis*) were conducted in drinking water and three brands of commercial clear apple juices. The samples from the same batch were run three times and the average log reductions were obtained.

21.4.3.2 Ultraviolet (UV) Inactivation of *Y. pseudotuberculosis* in Water

The experimental data of inactivation of *Y. pseudotuberculosis* in clear water in turbulent flow are shown in Table 21.2. UV inactivation of *Y. pseudotuberculosis* was calculated using $k_1 = 0.557$

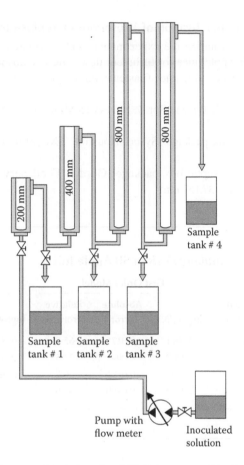

FIGURE 21.9 Schematic diagram of turbulent flow UV-light system. (Ye, Z., Forney, L. J., Koutchma, T., Giorges, A. T., and Peirson, J. A., *Industrial & Engineering Chemical Research*, 47, 3445–3450, 2008. With permission.)

TABLE 21.2
UV Inactivation of *Y. Pseudotuberculosis* in Water

Combination of Reactors	Experimental $\log(N_0/N)$	First Order Model			Series-Event Model		
		$\log(N_0/N)$	Absolute Error	Relative Error (%)	$\log(N_0/N)$	Absolute Error	Relative Error (%)
20 cm	1.8500	1.8344	−0.0156	−0.8	1.6082	−0.2418	−13.1
20 + 40 cm	6.1800*	4.6148	−1.5652	−25.3	5.6101	−0.5699	−9.2
20 + 40 + 80 cm	N/A**						

* As initial concentration was $10^{6.18}$ and no detectable colony was seen on the plates, log (N_0/N) was assumed to be 6.18.
** No detectable colonies were seen on the plates.

cm²/mJ and $k_{SE} = 0.984$ cm²/mJ with threshold $n = 3$.[18] Since drinking water is transparent to UV light and the outer cylinder of the reactors is made of stainless steel, it was assumed in the numerical simulation that 19.5% of UV-light at the surface of the stainless steel was reflected back into the reactors.[23] The inactivation of 1.85-log to 6.18-log of *Y. pseudotuberculosis* was achieved in turbulent flow of water (Table 21.2)

21.4.3.3   Ultraviolet (UV) Inactivation of *Y. pseudotuberculosis* in Apple Juice

Tables 21.3 through 21.5 summarize the experimental and calculated results of UV inactivation of *Y. pseudotuberculosis* in apple juice for turbulent flow. The UV absorption coefficients of three brands of commercial packaged apple juice (pasteurized, no preservatives) were as follows:

- Ocean Spray, plastic bottle (Ocean Spray, Lakeville-Middleboro, MA, abbreviated as OS), $A = 7.155$ cm^{-1},
- Sahara Burst, aseptic box package (Sysco, Houston, TX, abbreviated as SB), $A = 39.093$ cm^{-1},
- Gordon Food Service, aseptic box package (Gordon Food Service, Grand Rapids, MI, abbreviated as GFS), $A = 37.157$ cm^{-1}

TABLE 21.3
UV Inactivation of *Y. Pseudotuberculosis* in SB Apple Juice

Combination of Reactors	Experimental $\log(N_0/N)$	First Order Model			Series-Event Model		
		$\log(N_0/N)$	Absolute Error	Relative Error (%)	$\log(N_0/N)$	Absolute Error	Relative Error (%)
20 cm	0.0267	0.0091	−0.0175	−65.8	0.0039	−0.0227	−85.2
20 + 40 cm	0.0167	0.0207	0.0041	24.5	0.0084	−0.0083	−49.9
20 + 40 + 80 cm	0.0500	0.0399	−0.0101	−20.2	0.0132	−0.0368	−73.7
20 + 40 + 80 + 80 cm	0.0400	0.0590	0.0190	47.6	0.0182	−0.0218	−54.4

TABLE 21.4
UV Inactivation of *Y. Pseudotuberculosis* in GFS Apple Juice

Combination of Reactors	Experimental $\log(N_0/N)$	First Order Model			Series-Event Model		
		$\log(N_0/N)$	Absolute Error	Relative Error (%)	$\log(N_0/N)$	Absolute Error	Relative Error (%)
20 cm	0.0067	0.0100	0.0034	50.5	0.0041	−0.0026	−38.3
20 + 40 cm	0.0367	0.0229	−0.0137	−37.5	0.0087	−0.0280	−76.4
20 + 40 + 80 cm	0.0433	0.0447	0.0013	3.1	0.0137	−0.0296	−68.4
20 + 40 + 80 + 80 cm	0.0700	0.0664	−0.0036	−5.1	0.0191	−0.0509	−72.7

TABLE 21.5
UV Inactivation of *Y. Pseudotuberculosis* in OS Apple Juice

Combination of Reactors	Experimental $\log(N_0/N)$	First Order Model			Series-Event Model		
		$\log(N_0/N)$	Absolute Error	Relative Error (%)	$\log(N_0/N)$	Absolute Error	Relative Error (%)
20 cm	0.1300	0.1385	0.0085	6.6	0.0487	−0.0813	−62.6
20 + 40 cm	0.1567	0.3265	0.1698	108.4	0.1536	−0.0031	−2.0
20 + 40 + 80 cm	0.2267	0.7312	0.5045	222.6	0.4418	0.2151	94.9
20 + 40 + 80 + 80 cm	0.5867	1.1358	0.5491	93.6	0.8018	0.2151	36.7

Sahara Burst and Gordon Food Service brands were enriched with vitamin C. Since vitamin C strongly absorbs UV light at 254 nm[24] and the contents of vitamin C increases the absorption coefficients of juices, the reduction in *Y. pseudotuberculosis* was severely limited when juices enriched with vitamin C were treated. For one complete pass (combination of reactors 20 + 40 + 80 + 80 cm), the vitamin C-free juice yielded approximately 0.6-log reduction, while juices enriched with vitamin C resulted only in 0.1-log reductions. Comparing experimental data with expected values, it was found that, though the absolute error was small, the relative error was up to 90% or more. Besides normal experimental errors and deviations between experiments and models, the possible reason for large relative errors was that log reduction levels were low at less than 0.1. Another possible reason was that vitamin C, which strongly absorbs radiation at 254 nm, degrades during UV treatment. Concentration of vitamin C in SB apple juice decreased from 0.3 mg/mL to 0.12 mg/mL after one full pass. It was estimated that a 0.18 mg/mL loss of vitamin C resulted in decrease of the absorption coefficient of treated juice by 7.47 cm^{-1}. The absorption coefficient of vitamin C at 254 nm is about 41.5 cm^{-1} mL/ mg.[24] However, in the numerical simulation the absorption coefficient of the juice was assumed to be constant. Overall, the observed experimental inactivation data of *Y. pseudotuberculosis* in apple juices enriched with vitamin C were not very reliable because of low log reduction levels (less than 1-log). Other approaches such as increasing the incident UV fluence rate are needed to increase the log reduction levels.

The possible reasons of the low levels of microbial reduction in the turbulent flow were the short residence time and high absorption coefficients of fresh clear apple juices containing vitamin C used in the experimental runs. Although the flow rate of water (180 mL/s) was 1.5 times as large as the flow rate used for inactivation of juices (120 mL/s), the achieved reduction level in water was higher than 6-log when the combination of UV reactors (20 + 40 cm) with one pass in series was used. In contrast, the reduction level in the SB apple juice only reached less than 0.1-log after one complete pass. Moreover, the gap width of 0.515 cm in the rector was too large compared with the penetration depth of ~0.025 cm for juices with absorption coefficients of approximately 40 cm^{-1}.

21.4.4 Calculation of Optimum Gap Width

As in the case of laminar flow, there was an optimum gap width for each value of the absorption coefficient in turbulent flow. Figure 21.10 presents the relationship between microbial log reductions and gap width when the radius of the inner cylinder was kept constant at 1.225 cm. UV light radiation was exposed from the inner cylinder with fluence rate $I_0 = 800$ mW/cm^2 and UV lamp length $L = 29.2$ cm. The radius of the outer cylinder was varied to create different gap widths. The series event model ($k = 0.984$ cm^2/mJ and $n = 3$) was used in the calculations in Figure 21.10.

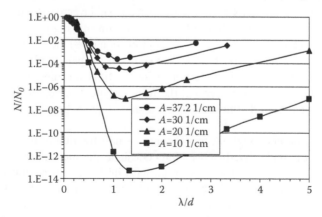

FIGURE 21.10 Relationship between microbial log reductions and ratio of λ/d in turbulent flow as a function of absorption coefficient. (Ye, Z., Forney, L. J., Koutchma, T., Giorges, A. T., and Peirson, J. A., *Industrial & Engineering Chemical Research*, 47, 3445–3450, 2008. With permission.)

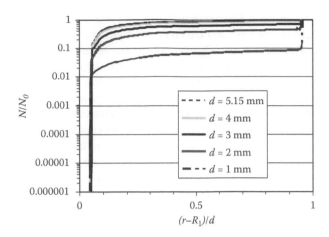

FIGURE 21.11 Concentration profiles of surviving *Y. pseudotuberculo*sis at the outlets of UV reactors with different gap widths.

Figure 21.10 shows that the optimum λ/d did not change dramatically, varying from 1.1 to 1.3 for the range of absorption coefficients tested. Moreover, because log reductions at the ratio $\lambda/d = 1$ are similar to those of the optimum ratio of λ/d, the optimum λ/d can be considered to be equal to 1 in turbulent flow (the penetration depth equals the gap width). In other words, turbulent flow still requires that the radiation has to penetrate the whole gap to obtain maximum inactivation. This requirement is caused by the viscous sublayer on the opposite side of the radiation source (under-irradiated region).

Figure 21.11 presents the concentration profiles of surviving *Y. pseudotuberculosis* at the outlets of UV reactors with different gap widths when the absorption coefficient equals 37.2 cm^{-1}. According to the turbulent flow theories and experimental observations, the near-wall region can be subdivided into three layers:[25] (1) the viscous sublaycr, (2) the fully turbulent region, and (3) the buffer layer between the viscous sublayer and the fully turbulent region. Correspondingly, the gap of the turbulent reactor can be divided into three regions according to the concentration profiles of surviving *Y. pseudotuberculosis*:

1. Over-irradiated region. This region is the viscous sublayer near the radiation source. Most of *Y. pseudotuberculosis* within this region was inactivated because of high fluence rate and long residence time.
2. Under-irradiated region. This region is the viscous sublayer on the opposite side of the radiation source. Most of *Y. pseudotuberculosis* within this region survived because UV was unable to penetrate the juices with high absorption coefficients and reach the viscous sublayer on the opposite side of the radiation source if the gap width is too large.
3. Uniform-irradiated region. This region includes the fully turbulent region and the buffer layer between the viscous sublayer and the fully turbulent region. *Y. pseudotuberculosis* within this region received almost uniform radiation fluence because of turbulent mixing.

Though the under-irradiated region occupies only a small fraction of the gap, about 5% or less, surviving bacteria in the under-irradiated region contribute the most to the surviving bacteria in the whole volume of the gap. Because most of bacteria in the under-irradiated region survives if $\lambda/d \ll 1$, the average concentration of the surviving bacteria at the outlet will be about 5% of its concentration at the inlet. Then, the log reduction is still less than 2 ($-\log_{10}(5\%) = 1.3$) even if most of the bacteria in the fully turbulent region and the viscous sublayer near the radiation source are inactivated. To release the restriction of gap width from the viscous sublayer, one method is to irradiate liquid from both the outer and inner cylinders. Another method is to increase length by connecting the UV reactors in series. So microbes that were present in the under-irradiated region in the previous UV reactor may have a chance of entering over-irradiated region or uniform-irradiated region in the next reactor and being inactivated. Moreover, the short residence time at high flow rates in turbulent flow conditions makes connecting reactors in series necessary.

21.5 ULTRAVIOLET (UV) INACTIVATION IN TAYLOR–COUETTE FLOW

As stated in Section 21.3.1, the flow pattern in the annual gap between two concentric cylinders can be either laminar or turbulent Taylor–Couette flow[6,7] when the outer cylinder is fixed, the inner cylinder is rotating, and the rotating speed of the inner cylinder exceeds a certain value. Since it was proven that turbulent Taylor–Couette flow was not as effective for UV inactivation as laminar Taylor–Couette flow due to strong turbulent axial mixing,[18] UV inactivation of laminar Taylor–Couette flow will be discussed.

21.5.1 Numerical Modeling

For laminar Taylor–Couette flow, the governing equations (Equations 21.18 through 21.20) for fluid flow and microorganism concentrations in cylindrical coordinate can be written as.[16]

$$\frac{1}{r}\frac{\partial(ru_r)}{\partial r} + \frac{1}{r}\frac{\partial u_\theta}{\partial \theta} + \frac{\partial u_z}{\partial z} = 0 \tag{21.18}$$

$$\frac{\partial \phi}{\partial t} + \frac{1}{r}\frac{\partial(ru_r\phi)}{\partial r} + \frac{1}{r}\frac{\partial(u_\theta\phi)}{\partial \theta} + \frac{\partial(u_z\phi)}{\partial z} = \frac{1}{r}\frac{\partial}{\partial r}\left(r\nu\frac{\partial \phi}{\partial r}\right) + \frac{1}{r}\frac{\partial}{\partial \theta}\left(\frac{\nu}{r}\frac{\partial \phi}{\partial \theta}\right) + \nu\frac{\partial^2 \phi}{\partial z^2} + S_\phi \tag{21.19}$$

$$\frac{\partial N_i}{\partial t} + u_r\frac{\partial N_i}{\partial r} + \frac{u_\theta}{r}\frac{\partial N_i}{\partial \theta} + u_z\frac{\partial N_i}{\partial z} = D\left(\frac{1}{r}\frac{\partial}{\partial r}\left(r\frac{\partial N_i}{\partial r}\right) + \frac{1}{r^2}\frac{\partial^2 N_i}{\partial \theta^2} + \frac{\partial^2 N_i}{\partial z^2}\right) + kI(N_{i-1} - N_i) \tag{21.20}$$

where ϕ in Equation 21.19 can be either u_r, u_ϕ, and u_z. S_ϕ varies depending on definition of ϕ. Since the molecular diffusion coefficient D is small, diffusion term in Equation 21.20 can be neglected.

21.5.2 Comparison between Experiments and Numerical Simulations

21.5.2.1 Taylor Vortex Reactor Configuration

A schematic figure of a Taylor vortex reactor is shown in Figure 21.12. The largest difference between the Taylor vortex reactor and a laminar or turbulent reactor is that the inner cylinder in the Taylor vortex reactor is rotating while the inner cylinder in the laminar or turbulent reactor is fixed.

A dimensionless group, Taylor number, is used to describe Taylor–Couette flow (Equation 21.21),

$$\mathrm{Ta} = \frac{R_1\Omega_1 d}{\nu}\left(\frac{d}{R_1}\right)^{1/2} \tag{21.21}$$

where R_1 is the radius of the inner cylinder, Ω_1 is the angular velocity of the inner cylinder, d is the gap width, and ν is the kinematic viscosity. If Ta exceeds a certain value, the flow pattern can change from laminar flow to Taylor–Couette flow.[6,7]

The stator of the tested Taylor vortex UV reactor was constructed of 4.58 cm internal diameter, fused quartz (Vycor, Corning Inc., Corning NY) with a Teflon rotor of 3.43 cm outer diameter corresponding to a gap width of 5.75 mm. The radiation source consisted of six medium-pressure, mercury

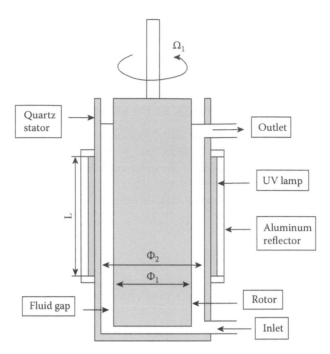

FIGURE 21.12   Schematic representation of a Taylor vortex UV reactor. (Ye, Z., Forney, L. J., Koutchma, T., Giorges, A. T., and Peirson, J. A., *Industrial & Engineering Chemical Research*, 47, 3445–3450, 2008. With permission.)

UVC lamps with diameters of 0.95 cm and effective lengths of 5.34 cm (Pen-Ray Lamps, UVP, Upland, CA), which were distributed evenly around the quartz stator. At the same time the UV lamps were surrounded with an aluminum reflector as shown in Figure 21.12. The radiation fluence rate on the surface of the quartz was assumed to be uniform and equal to 25 mW/cm^2.

21.5.2.2   Effect of Taylor Numbers on Ultraviolet (UV) Inactivation

Figure 21.13 shows the comparison between experimental results and numerical simulations of the inactivation of *E. coli* K12 (ATCC 25253) in laminar Taylor–Couette flow with different Taylor numbers when $Q = 40$ mL/min. and $A = 11$ cm^{-1}. The number of log reductions of *E. coli* K12 increased with the increase in the Taylor number, which was caused by a decrease in the boundary layer thickness for large Taylor numbers.

The results of Figure 21.13 show that the series-event model generally predicted inactivation that agreed with experimental results. The first order model over-estimated the log reductions compared to experimental results.

The motion of fluids in Taylor vortex reactors consists of two parts: one is uniform axial movement downstream and the other is revolution around a vortex center. It has been proven experimentally and numerically that the vortex displacement velocity is about 1.17 times the average axial velocity.[27–30] Figure 21.14 demonstrates the stream function for different Taylor numbers in a moving frame, namely, the original velocity was subtracted by 1.17 times the average axial velocity. In Figure 21.14, the upper and lower boundaries are the outer and inner cylinders, respectively, and the vortices move from left to right. Pathlines of fluids in Taylor–Couette flow can be divided into two parts.[26] One part is within the vortices and advances downstream with a uniform velocity (the vortex part). Another part is winding around the vortices and alternately flowing near the inner and outer cylinder (the winding part). Figure 21.14 shows that the winding part occupies a small percentage of the flow in the gap. However, because the downstream velocity of the vortex part is faster than average axial velocity, the winding part has to move with the lower velocity than the

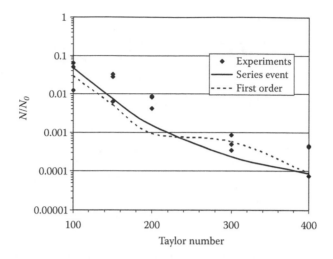

FIGURE 21.13 Comparison of *E. coli* K12 log reductions between experimental results and numerical simulations with different Taylor numbers at $Q = 40$ mL/min and $A = 11$ cm^{-1}.

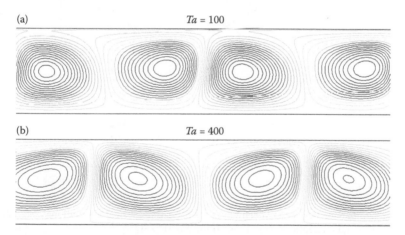

FIGURE 21.14 Stream function for different Taylor numbers in the moving frame (subtracted by 1.17 times the average axial velocity) showing the four vortices produced within the stream.

average axial velocity. As a result, the winding part will be exposed to a higher UV fluence than the average since the winding part flows alternately near the inner and outer cylinder (radiation source). Correspondingly, the vortex part will be exposed to a lower UV fluence than the average. Therefore, it is desirable that the winding part is as small as possible in order to achieve a narrow fluence distribution and better inactivation efficiency.

21.5.2.3 Effect of Flow Rates on Ultraviolet (UV) Inactivation

Figure 21.15 shows the comparison between experimental results and numerical simulations of inactivation for *E. coli* K12 (ATCC 25253) at Ta = 300 at various flow rates (or average residence times) when the Taylor number was fixed and $A = 11$ cm^{-1}. It can be seen from Figure 21.15 that the results of *E. coli* inactivation predicted with the series-event model agreed better with experimental results than the first model especially in a case when the average residence time was shorter than 58 s (or flow rate larger than 40 mL/min). Moreover, the microbial log reduction was nearly proportional to the average residence time. The results show that performance of the Taylor vortex reactors approach plug flow reactors (PFR).[31–33]

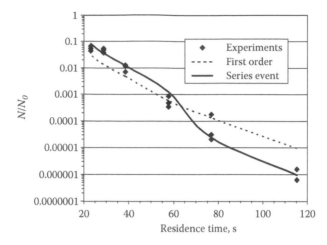

FIGURE 21.15 Experimental results and numerical simulation for inactivation *of E. coli* K12 *at* Ta = 300 and $A = 11$ cm^{-1} for first order and series event models.

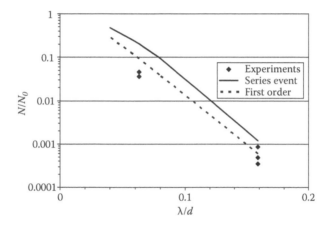

FIGURE 21.16 Comparison between experimental results and numerical simulations *of E. coli* K12 inactivation at different penetration depths at $Q = 40$ mL/min and Ta = 300.

21.5.2.4 Effect of Penetration Depths on Ultraviolet (UV) Inactivation

Figure 21.16 shows a comparison between experimental results and numerical simulations for inactivation of *E. coli* K12 (ATCC 25253) at different penetration depths when Ta = 300 and $Q = 40$ mL/min. The ratio of penetration depth (the reciprocal of the absorption coefficient) to the gap width in the reactor was varied in the range from 0.04 to 0.16. It is observed in Figure 21.16 that the log reduction was almost proportional to the penetration depth.

21.5.3 Calculation of Optimum Gap Width

As in the case of laminar and turbulent flow, there is an optimum gap width for each value of the absorption coefficients of treated fluids in Taylor–Couette flow. Figure 21.17 shows how microbial inactivation expressed in log reductions varies with the change of the gap width when length of the UV-light lamp is $L = 11.2$ cm, the radius of the inner cylinder is 1.225 cm, and the radius of the outer cylinder was varied to create different gap widths. The UV radiation comes from the inner cylinder. In order to keep the log reductions within a comparable range for different values of absorption coefficients, the incident fluence rates were varied, namely, $I_0 = 4$ mW/cm^2 for $A = 10$ cm^{-1}, $I_0 = 8$ mW/cm^2 for $A = 20$ cm^{-1}, $I_0 = 12$ mW/cm^2 for $A = 30$ cm^{-1} and $I_0 = 16$ mW/cm^2

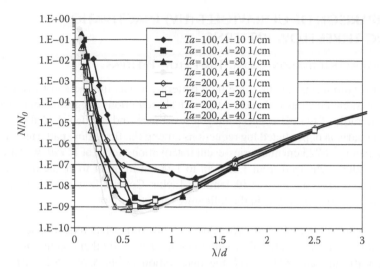

FIGURE 21.17 The impact of λ/d on microbial inactivation for laminar Taylor–Couette flow. (Ye, Z., Forney, L. J., Koutchma, T., Giorges, A. T., and Peirson, J. A., *Industrial & Engineering Chemical Research*, 47, 3445–3450, 2008. With permission.)

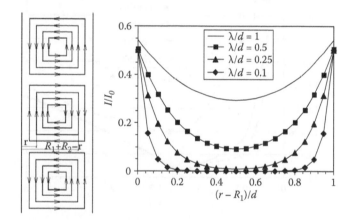

FIGURE 21.18 Profile of UV fluence rate along the gap in laminar Taylor–Couette flow reactor. (Ye, Z., Forney, L. J., Koutchma, T., Giorges, A. T., and Peirson, J. A., *Industrial & Engineering Chemical Research*, 47, 3445–3450, 2008. With permission.)

for $A = 40$ cm^{-1}. From Figure 21.17, it can be seen that the optimum ratio of λ/d was similar across the tested range of absorption coefficients and Taylor numbers. Compared with the optimum values of λ/d equal to 1.5 for the laminar flow and 1.0 for the turbulent flow, the optimum λ/d for Taylor–Couette flow decreased to approximately 0.5. In other words, Taylor–Couette flow requires that the UV radiation reaches the middle of the gap for maximum inactivation. Moreover, if one requires an inactivation level of $N/N_0 = 10^{-5}$, the radiation penetration depth must vary over the range of $0.15 < \lambda/d < 0.3$ for $Ta = 200$. For example, if $Ta = 200$ and $A = 40$ cm^{-1} ($\lambda = 0.25$ mm), then the optimum gap width required is $d \approx 1.7$ mm.

It can be observed in Figure 21.14 that Taylor vortices in the gap is nearly square. If mixing resulting from nearly squared Taylor vortices can be approximated as the mixing between the fluid at r ($R_1 \leq r \leq R_2$) and the fluid at $R_1 + R_2 - r$, the profile of the fluence rate distribution along the gap is shown in Figure 21.18. When $\lambda/d < 0.5$, UV inactivation cannot benefit from mixing in the middle of vortices because mixing only happens among regions exposed to low fluence rates instead of between regions exposed to both low and high fluence rates.

21.6 COMPARISON OF ULTRAVIOLET (UV) INACTIVATION USING THREE FLOW PATTERNS

21.6.1 OPTIMUM INACTIVATION AMONG THREE FLOW PATTERNS

Since different requirements for the three flow patterns, it was impractical to compare microbial reductions at the same flow rates. However, theoretical predictions suggest similar inactivation levels for the same radii of the cylinders, pathogen resistance and incident UV fluence $I_0\tau_{av}$. Figure 21.19 shows comparison of predicted log reductions among the three flow patterns, where the radius of the inner cylinder is 1.225 cm and series-event inactivation model of *E. coli* was used. The radius of the outer cylinder is changed from 1.235 cm to 1.74 cm in order to create different gap widths (Figure 21.19). According to Figure 21.19, laminar Taylor–Couette flow reactor is superior to either turbulent or laminar Poiseuille flow in the following aspects:

1. Laminar Taylor–Couette flow always achieves higher microbial reduction levels than either laminar Poiseuille or turbulent flow with the same $I_0\tau_{av}$ and other conditions. For example, when $A = 40$ cm^{-1} and $\lambda/d = 0.417$, the inactivation levels $N/N_0 = 8.2 \times 10^{-10}$, 8.1×10^{-3}

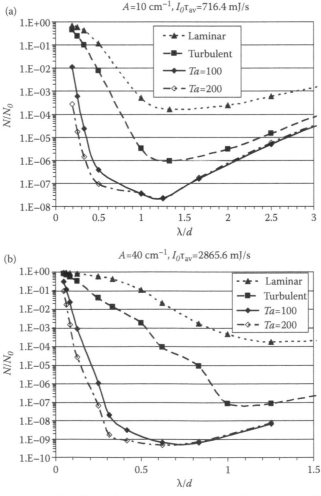

FIGURE 21.19 Comparison of predicted inactivation among three flow patterns. (Ye, Z., Forney, L. J., Koutchma, T., Giorges, A. T., and Peirson, J. A., *Industrial & Engineering Chemical Research*, 47, 3445–3450, 2008. With permission.)

and 0.27 for laminar Taylor–Couette flow (Ta = 200), turbulent and laminar Poiseuille flow, respectively. Very low inactivation levels are achieved in both turbulent and laminar Poiseuille flows when the absorption coefficients of juices are high and λ/d is small. For example, when $A = 40$ cm^{-1} and $\lambda/d = 0.049$, the inactivation levels N/N_0 are 0.81 and 0.95 for turbulent flow and laminar Poiseuille flow, respectively. However, the inactivation level of $N/N_0 = 0.097$ for laminar Taylor–Couette flow (Ta = 200).

2. As stated above, the optimum ratio of λ/d for laminar Taylor–Couette flow decreased to 0.5 compared with the optimum value of 1.5 and 1.0 for laminar Poiseuille flow and turbulent flow reactors, respectively. Thus, laminar Taylor–Couette flow UV reactor has two advantages: (1) the pressure drop of laminar Taylor–Couette flow is smaller than that of other two flow patterns; (2) laminar Taylor–Couette flow reactor is suitable for UV treatment of juices with high absorption coefficients.

21.6.2 Ultraviolet (UV) Fluence Distribution among Reactors with Three Flow Patterns

The definition of fluence distribution function is similar to that of the age distribution function.[34] The age distribution function is defined as the fraction of fluids leaving the reactor that has residence time of $(t, t + dt)$. Because the UV fluence distribution is broad for photochemical reactors, the fluence distribution function has to be defined as the fraction of fluids leaving the reactor that has fluence of $(\log It, \log It + d(\log It))$ instead of $(It, It + d(It))$,

$$E(\log It)d(\log It) = \text{Fraction of fluids with fluence of } (\log It, \log It + d(\log It)) \qquad (21.22)$$

where the UV fluence It is made dimensionless by the average fluence of ideal plug flow reactors (PFR), which is defined by Equation 21.23.

$$It_{av} = I_{av} \times \tau_{av} = \frac{\displaystyle\int_0^{2\pi} \int_{R_1}^{R_2} I_0 R_1 \exp(-\alpha(r - R_1))d\theta dr}{\displaystyle\int_0^{2\pi} \int_{R_1}^{R_2} r d\theta dr} \frac{\pi(R_2^2 - R_1^2)L}{Q} \qquad (21.23)$$

Figure 21.20 shows the comparison of UV fluence distribution function of the three flow patterns when ratio of $\lambda/d = 0.25$ and the absorption coefficient of the fluid equals 10 cm^{-1}.

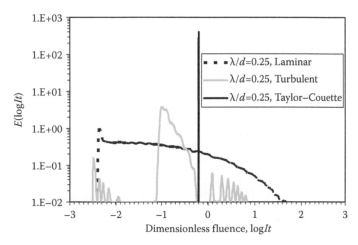

FIGURE 21.20 Fluence distribution function of the three flow patterns when $\lambda/d = 0.25$ and absorption coefficient equals 10 cm^{-1}. (Ye, Z., Forney, L. J., Koutchma, T., Giorges, A. T., and Peirson, J. A., *Industrial & Engineering Chemical Research*, 47, 3445–3450, 2008. With permission.)

According to Figure 21.20, Taylor vortex reactors approach superior characteristics of a PFR. Moreover, for turbulent flow, three peak domains can be observed. The left peak corresponds to the effect of the viscous sublayer on the opposite side of the UV radiation source while the right peak corresponds to the effect of the viscous sublayer near the radiation source. Finally, the middle peak corresponds to the effect of the fully turbulent region and buffer layer.

21.7 MODELING ULTRAVIOLET (UV) FLUENCE IN MULTIPLE LAMPS ULTRAVIOLET (UV) REACTORS

In the above work, it was assumed that the UV fluence rate on the surface of the inner or outer cylinder was uniform and UV fluence rate varied only in the radial direction. However, if multiple UV lamps are used, the steady state radiative transfer equation (RTE) can be solved to estimate nonuniform UV fluence rates.

21.7.1 Multiple Lamps Ultraviolet (UV) Reactors Configuration

Multiple lamps UV reactors such as laminar flow 8-lamp "CiderSure" (CiderSure Model 1500, FPE Inc., Macedon, NY) and turbulent flow 12-lamp "Aquionics" (Hanovia Ltd, Slough, England) are shown in Figures 21.21 and 21.22. The technical parameters of the reactors are given in Table 21.6.

CiderSure UV reactor (Figure 21.21) incorporates three individual chambers connected in tandem with outside tubing. Eight low-pressure mercury arc lamps are mounted within the quartz inside cylinder running centrally through all three chambers. The manufacturer declared that each lamp emits UV rays at a minimum fluence of 60 mJ/cm^2. A stainless steel outside cylinder covers all three chambers and lamps. Apple juice is pumped through a 0.08-cm annular gap between the inner surface of each chamber and the outer surface of the quartz sleeve. The CiderSure model 1500 allows three flow rate settings to regulate the UV fluence.

In the "Aquionics" UV reactor (Figure 21.22), the treatment was achieved by passing apple juice or apple cider through a stainless steel chamber containing 12 UV emitting low-pressure mercury arc-tubes. Each single arc-tube is mounted in a quartz sleeve and fitted within the chamber allowing the liquid to pass the sleeve on all sides.[4] A sealed UV monitor, fitted to the chamber, measures the intensity of UV light being emitted from the arc-tube. A temperature sensor is fitted on top of the chamber. The flow rate of 75 L/min was used in the experiments.

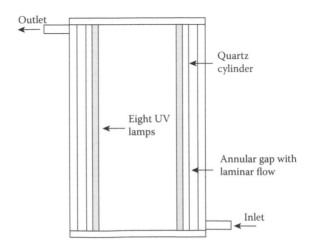

FIGURE 21.21   Schematics representation of a laminar thin film UV reactor (Cider Sure).

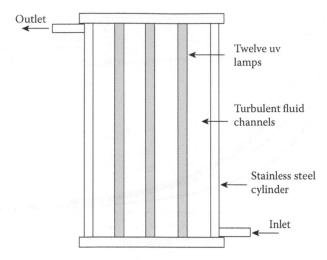

Outlet

Twelve uv lamps

Turbulent fluid channels

Stainless steel cylinder

Inlet

FIGURE 21.22 Schematics representation of turbulent channel reactor (Aquionics).

TABLE 21.6
Parameters of Multiple Lamps UV Reactors

UV reactors	Cider Sure	Aquionics
Flow pattern	Laminar	Turbulent
Reactor volume (L)	0.2172	14.72
Flow rate (mL/s)	56.8	1250
Mean residence time (s)	3.82	14.9
UV lamps output power (W)	8 lamps × 39 W = 312 W	12 lamps × 42 W = 504 W
Manufacturer's declared UV fluence (mJ/cm²)	60 for each lamp	N/A

21.7.2 Flow Dynamics

Flow dynamics was first evaluated for water and apple cider at the entrance section of the UV reactors. The average velocity U_{av} was calculated as ($U_{av} = Q/A_{inlet}$), where Q is a volumetric flow rate and A is the inlet cross section area. Reynolds numbers (Re) were calculated next as (Re = $U_{av}d_{inlet}\rho/\mu$) based on the measured flow rate in each reactor, where d is characteristic dimension, ρ is density of fluid and μ is dynamic viscosity. The magnitude of the Reynolds number (less than 2,000) indicated that at the selected flow rate of 56.8 mL/s, the hydraulic regime was laminar in "CiderSure" UV reactors. However, in "Aquioncs" UV reactor at flow rate of 75 L/min, the flow dynamics can be characterized as a fully developed turbulent flow with Re > 10,000 (Table 21.7).

21.7.3 Simulation Results

21.7.3.1 Ultraviolet (UV) Fluence Rate Distribution

Laminar flow reactor. Multi-point source summation (MPSS) was employed to simulate fluence rate and fluence distribution in the CiderSure UV reactor.[35] Equation 21.24 estimates the total UV light energy received at any point of the receptor site at the reactor I_λ

$$I_\lambda(r,z) = \sum_{i=1}^{n} \frac{\Phi/m}{4\pi l_i^2} \exp[-(a_q t_q + a(r - r_q))]\frac{l_i}{r} \qquad (21.24)$$

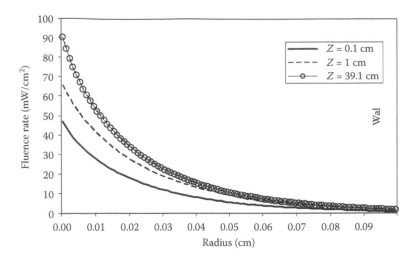

FIGURE 21.23   Radial UV fluence rate profiles in the thin film reactor (CiderSure 1500).

TABLE 21.7
UV Reactors Flow Conditions

		Water			Apple Cider		
Flow Rate (mL/s)	Average Velocity (m/s)	Density (m^3/kg)	Viscosity (Pa.s)	Reynolds Number	Density, (m^3/kg)	Viscosity (Pa.s)	Reynolds Number
Cider Sure UV reactor (inlet i.d = 1 cm)							
56.8	0.07	998	0.001	722	1035	0.005	149
Aquionics (inlet i.d = 2 cm)							
1250	0.99	998	0.001	86863	1035	0.005	16477

where Φ is power of the UV lamp (W), l is the distance (cm) from a point source to a receptor site; m is the number of point sources; r is radial distance from UV lamp axis to receptor site (cm); r_q the outside radius of the UV lamp (cm); a is absorption coefficient of the medium (cm^{-1}); a_q is absorption coefficient of the quartz tube (cm^{-1}); t_q is thickness of the quartz tube (cm). The necessary inputs used for calculation were as follows. The lamps are 39 W, 6″ (15.24 cm) length low-pressure mercury arc with a 2.5-cm diameter quartz sleeve that emits primarily at 253.7 nm. The manufacturer stated that each lamp emits UV rays at a minimum fluence of 60 mJ/cm^2. The cider was pumped as a thin film at the rate of 56.8 mL/s. The number of computational fluid cells used to create a computational domain for the given system was 443,514. The detailed analysis of the modeling approach and the results obtained in this reactor were reported by Unturluk et al.[35] These modeling results of UV fluence rate in apple cider with absorption coefficient of 30 cm^{-1} were used in the current study to calculate UV fluence in the CiderSure reactor. Examples of radial UV fluence rate profiles at the selected vertical distances $z = 0$, 1 and 39.1 cm were calculated based on the MPSS model (Equation 21.24) that is shown in Figure 21.23. The computation results showed variation of UV fluence rate along the annular gap from 90 mW/cm^2 near the UV source down to 1.81 mW/cm^2 near the wall. The resulting average fluence rate was 19.7 mW/cm^2 in apple cider with an absorption coefficient of 3 mm^{-1}. Based on average theoretical residence time in this reactor (3.82 seconds at flow rate of 56.7 mL/s), an average value of UV fluence of 75.25 mJ/cm^2 was obtained in the CiderSure UV reactor.

TABLE 21.8
Fluence Rate in Quadrants of the "Aquionics" UV Reactor

	Average Fluence Rate (mW/cm²)		
	Uncorrected	Corrected	Average Fluence (mJ/cm²)
1 Quadrant	1.907	1.793	26.71
2 Quadrant	1.274	1.197	17.84
3 Quadrant	2.141	2.013	29.99
4 Quadrant	2.141	2.013	29.99

Turbulent flow reactor. UV Calc 2, a software program for multiple lamps ultraviolet reactors (courtesy of Bolton Photoscience Inc, Edmonton, AB, Canada), was used to compute three dimensional fluence rate (irradiance) distribution, average fluence rate and hence the fluence in the "Aquionics" UV reactor. The program is based on the multiple point source summation method with full reflection and refraction accommodation at the air/quartz/water interface.[36]

In order to calculate UV irradiance in multiple lamps reactors using UV Calc 2, the reactor was divided into four quadrants with the center as the origin. Since the quadrants were not symmetrical, the UV fluence rate was computed for each quadrant and the results were averaged. First, the rate was computed only in the central plane through the lamp centers giving the so called "uncorrected" average fluence rate. The product of the "correction factor" and the "uncorrected" average fluence rate resulted in the "corrected" average fluence rate. The use of the correction factor is based on the fact that at a fixed distance r from the center of one lamp, the ratio of the average fluence rate from $x = 0$ to the bottom (or top) of the reactor to that at $x = r$ (r is the "longitudinal" coordinate parallel to the lamp axis) is virtually independent of r. This ratio is called the longitudinal "correction factor."

The input required was as follows: absorption coefficient of apple cider of 5.7 mm^{-1} or transmittance $T = 0.0001\%$; lamp power of 42 W; efficiency of 35%; lamp length of 94 cm; lamp sleeve radius of 1.4 cm and maximum cylinder radius of 9.1 cm.

The calculation of UV fluence rate was made for each of 12 lamps in the quadrants. UV fluence rate "uncorrected" and "corrected" were calculated for the four quadrants and then the results averaged for all four quadrants (Table 21.8).

The variation of the average fluence rate in the quadrants was explained by nonsymmetrical positions of the lamps in each quadrant. The product of the "corrected" fluence rate and the hydraulic residence time resulted in average UV fluence or dose. It can be seen that the UV fluence in apple cider varied from 17.8 to 29.99 mJ/cm² in each of the four quadrants of the reactor. The three dimensional fluence rate (irradiance) distribution in apple cider in one quarter of the reactor with the center as the origin is illustrated in Figure 21.24. UV-fluence rate gradient was observed along the lamps length ranging from 2 to 15 mW/cm².

21.7.3.2 Ultraviolet (UV) Decimal Reduction Fluence

Inactivation of *E. coli* K12 was tested next using apple cider in each reactor. The data used for evaluation of decimal UV fluence and the results obtained are summarized in Table 21.9.

In the "Aquionics" reactor, the destruction of *E. coli* K12 bacteria in apple cider was about 0.2-log per pass through the reactor at flow rate of 75 L/min and mean residence time of 15 s. Thus, the estimated value of UV fluence required for 90% reduction of *E. coli* in apple cider ranged from 90 to 150 mJ/cm².

In the "CiderSure" reactor, exposure of *E. coli* to UV fluence of 75.25 mJ/cm² provided a 3–4-reduction of *E. coli* K12 per pass. Thus, the UV decimal reduction fluence required for inactivation of *E. coli* in apple cider ranged from 25.1 to 18.8 mJ/cm².

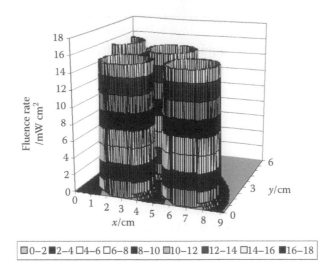

0–2 ■2–4 □4–6 □6–8 ■8–10 ■10–12 ■12–14 □14–16 ■16–18

FIGURE 21.24   UV fluence rate distribution in the central plane of the "Aquionics" UV reactor.

TABLE 21.9
UV Decimal Reduction Fluence for *E. coli* K12 in UV Reactors

Type of UV reactor	Output power (W)	Residence time (s)	Media	Absorption coefficient (cm⁻¹)	UV fluence rate (mW/cm²)	UV fluence (mJ/cm²)	Log reduction per pass	D_{10} fluence (mJ/cm2)
CiderSure	312	3.8	Apple cider	30	12.96	65.2	3–4	21.8
					16.33	90		25.1
Aquionics	504	14.9	Apple cider	57	1.9	17.8	0.22	90
					2.1	30		150

From the preceding data, variation of UV decimal reduction fluence was observed in the reactors. The highest D_{10} value within the range of 90–150 mJ/cm² was observed in a turbulent flow UV reactor. The lower UV fluence of 7.3–7.8 mJ/cm² was required for 1-log inactivation of *E. coli* K12 in the malate buffer and apple juice in the annular single lamp reactor.[15] Nevertheless, when the apple cider with absorption coefficient of 30 cm⁻¹ was pumped through this reactor, the decimal reduction UV fluence was significantly higher with a magnitude about 20.4 mJ/cm². Similar value of the decimal reduction UV fluence of *E. coli* was observed in the thin film UV reactor "Cider Sure" in apple cider that varied from 18.8 to 25.1.mJ/cm². The observed variations in the magnitudes of decimal UV fluence can be attributed to a number of reasons. The increase in decimal reduction fluence in apple cider in comparison to model buffer solution and apple juice may be due to an overestimation of UV fluence rate for fluids with a high particle concentration and smaller soluble absorbance component such as cider.

The uncertainties in measuring UV fluence in static and flow-through systems and treated medium should also be taken into consideration. Significant variations in reported value of D_{10} of *E. coli* can be found in published studies. As an example, Hanovia[37] reported a value for decimal reduction fluence of *E. coli* of 5.4 mJ/cm². Hoyer[38] reported a lower value of UV fluence for 90% reduction of *E. coli* ATCC 11229 and *E. coli* ATCC 23958 of 2.5 and 1.25 mJ/cm², respectively. Wright et al.[39] obtained a value of 1.5 mJ/cm² for a 1-log reduction of *E. coli* O157:H7 in drinking water. Wright et al.[39] reported a significantly higher value of UV fluence of 29,076 mJ/cm² required to achieve 3.6 logs inactivation of *E. coli* O157:H7 in apple cider.

As stated above, the distribution of RTD and LID in flow-through UV reactors results in a variation of UV fluence (product of LID and RTD) that any given microorganism exposed during the disinfection process. Consequently, this variation altered the performance of the reactors and reflected in the changing magnitude of the decimal reduction fluence depending on flow pattern in the reactor.

21.8 CONCLUSIONS

UV inactivation between concentric cylinders in three flow patterns: laminar Poiseuille flow, turbulent flow and laminar Taylor–Couette flow was modeled numerically and compared with experimental results. The principal results obtained can be summarized as:

1. Mathematical modeling can provide an accurate prediction of UV inactivation in thin film annular reactors with three flow patterns: laminar Poiseuille flow, turbulent flow and laminar Taylor–Couette flow. Compared with first order inactivation model, the series-event inactivation model agrees better with experimental results of UV inactivation of *E. coli* K12 and *Y. pseudotuberculosis*.
2. Laminar Poiseuille flow provides inferior (small) inactivation levels while laminar Taylor–Couette flow provides superior (large) inactivation levels. The relative inactivation levels are: Laminar Poiseuille flow < turbulent flow < laminar Taylor–Couette flow.
3. There is an optimum value of gap width for each value of absorption coefficient. The optimum ratios of λ/d for laminar flow reactors, turbulent flow reactors and laminar Taylor vortex reactors are 1.5, 1 and 0.5, respectively. These optimum ratios of λ/d are determined by their characteristic flow patterns.
4. The results of the mathematical simulations of UV fluence rate in the multiple lamps laminar and turbulent flow reactors demonstrated a nonuniform distribution due to significant UV light attenuation in apple cider. Within UV reactors, the absorbance consistently affected the efficacy of UV light inactivation of *E. coli* K12. The differences in inactivation efficiency for laminar and turbulent flow reactors can be explained by the effect of flow dynamics on accumulated UV fluence in liquids with high absorptive properties.
5. It is essential that in order to achieve the required performance standard of 5-log reduction, flow hydrodynamics and mixing behavior along with reactor design should be taken into consideration when a juice UV-processing system is being developed.
6. The development and application of mathematical modeling can improve the design and evaluation of UV fluence delivery and distribution within the flow-through reactors for liquid foods and beverages.

NOMENCLATURE

A	absorption coefficient with 10 base, cm^{-1}, $I_x = I_0 \times 10^{(-Ax)}$
A_{inlet}	inlet cross section area, cm^2
C_1	constant determined by κ, $C_1 = \dfrac{2}{1 + \kappa^2 - \dfrac{1 - \kappa^2}{\ln(1/\kappa)}}$
d	gap width, cm
d_{inlet}	diameter of inlet, cm
D	molecular diffusion coefficient, cm^2/s
i	event level
I	fluence rate, mW/cm^2

I_0 incident fluence rate, mW/cm^2
I_{av} average fluence rate, mW/cm^2
$I(r)$ fluence rate at the position of the gap r, mW/cm^2
I_x fluence rate at the path length x, mW/cm^2
It UV fluence, mJ/cm^2
k turbulent kinetic energy, m^2/s^2
k_1 the first order inactivation constant, cm^2/mJ
k_{SE} inactivation constant in the series-event inactivation model, cm^2/mJ
l   distance from a point source to a receptor site, cm
L   length of radiation section, cm
n threshold for inactivation in the series-event model
m number of point sources
N concentration of viable microorganisms, CFU/mL
N_0 concentration of viable microorganisms before exposure, CFU/mL
N_{av} average concentration of viable microorganisms at the outlet of the reactor, CFU/mL
N_i concentration of viable microorganisms at event level i, CFU/mL
Q   volumetric flow rate, mL/s
r radius, cm
r_q outside radius of UV lamp, cm
R_1 radius of inner cylinder, cm
R_2 radius of outer cylinder, cm
Re   Reynolds number
s linear coordinate along the direction Ω, cm
s_ϕ source term depending on variable ϕ
Sc$_t$ turbulent Schmidt number
t   time, s
Ta   Taylor number
t_q thickness of the quartz tube, cm
u velocity, cm/s or m/s
u_r radial velocity components, cm/s
u$_z$ axial velocity components, cm/s
U_{av} average velocity, $U_{av} = \dfrac{Q}{\pi(R_2^2 - R_1^2)}$, cm/s

x the path length or x coordinate, cm
z axial coordinate, cm
α absorption coefficient with e base, cm^{-1}, $I_x = I_0 \exp^{(-\alpha x)}$
α_q absorption coefficient of the quartz tube, cm^{-1}
ε turbulent dissipation rate, m^2/s^3
ϕ variable
Φ power of the UV lamp, W
θ angle around the cylinder, degree, $0 \le \partial \le 2\pi$
κ ratio of the radius of the inner cylinder to that of outer cylinder, $\kappa = R_1/R_2$
λ penetration depth, cm
μ dynamic viscosity, Pa·s
μ_t turbulent viscosity, Pa·s
ν kinematic viscosity, m^2/s
ρ density, g/cm^3
τ_{av} average residence time, s
Ω unit vector in the direction of propagation
Ω_1 angular velocity of inner cylinder, s^{-1}

REFERENCES

1. US FDA 2000. 21 CFR Part 179. Irradiation in the production, processing and handling of food. *Federal Register,* 65, 71056–71058.
2. US FDA. 2001. Hazard analysis and critical control point (HACCP): Final rule. *Federal Register,* 66(13), 6137–6202.
3. Sugarman, C. 2004. Pasteurization redefined by USDA committee. *Food Chemical News,* 46(30), 21–22.
4. Koutchma, T., Keller, S., Chirtel, S., and Parisi, B. 2004. Ultraviolet disinfection of juice products in laminar and turbulent flow reactors. *Innovative Food Science and Emerging Technologies,* 5, 179–189.
5. Water Environment Federation. 1996. *Wastewater Disinfection Manual of Practice FD-10.* Water Environment Federation, Alexandria, VA.
6. Taylor, G. I. 1923. Stability of a viscous liquid contained between two rotating cylinders. *Proceeding of the Royal Society of London. Series A, Mathematical and Physical Sciences,* 223, 289–343.
7. Lueptow, R. M., Dotter, A., and Min, K. 1992. Stability of axial flow in an annulus with a rotating inner cylinder. *Physics of Fluids,* 4, 2446–2455.
8. Collins, H. F. and Selleck, R. E. 1972. Process kinetics of wastewater chlorination. *SERL Report.* University of California, Berkeley, CA, 72–75.
9. Severin, B. F., Suidan, M. T., and Engelbrecht, R. S. 1983. Kinetic modeling of U. V. disinfection of water. *Water Research,* 17, 1669–1678.
10. Kowalski, W. J. 2001. Design and optimization of UVGI air disinfection system. PhD Thesis, Pennsylvania State University, PA.
11. Chiu, K., Lyn, D. A., Savoye, P., and Blatchley, E. R. 1999. Effect of UV system modification on disinfection performance. *Journal of Environmental Engineering,* 125, 7–16.
12. Severin, B. F., Suidan M. T., Rittmann B. E., and Engelbrecht, R. S. 1984. Inactivation kinetics in a flow-through UV reactor. *Journal WPCF,* 56, 164–169.
13. Harris, G. D., Adams, V.D., Sorensen, D. L., and Curtis, M. S. 1987. Ultraviolet inactivation of selected bacteria and viruses with photorcactivation of the bacteria. *Water Research,* 21, 687–692.
14. Harm, W. 1980. *Biological Effects of Ultraviolet Radiation.* Cambridge University Press, Cambridge, UK.
15. Koutchma, T. and Parisi, B. 2004. Biodosimetry of *Escherichia coli* UV inactivation in model juices with regard to dose distribution in annular UV reactors. *Journal of Food Science,* 69, E14–E22.
16. Bird, R. B., Stewart, W. E., and Lightfoot, E. N. 2002. *Transport Phenomena.* Wiley, New York, NY.
17. Cassano, A. E., Martin, C. A., Brandi, R. J., and Alfano O. M. 1995. Photoreactor analysis and design: Fundamentals and applications. *Industrial and Engineering Chemistry Research,* 34, 2155–2201.
18. Ye, Z., Koutchma, T., Parisi, B., Larkin, J., and Forney, L. J. 2007. Ultraviolet inactivation kinetics of *E. coli and Y. pseudotuberculosis* in annular reactors. *Journal of Food Science,* 75, E271–E278.
19. Launder, B. E. and Spalding, D. B. 1972. *Lectures in Mathematical Models of Turbulence.* Academic Press, London, UK.
20. Fox, R. O. (2003) *Computational Models for Turbulent Reacting Flows.* Cambridge University Press, Cambridge, UK.
21. Yakhot, V. and Orszag, S. A. 1986. Renormalization group analysis of turbulence: I. Basic theory. *Journal of Scientific Computing,* 1, 1–51.
22. Shih, T.-H., Liou, W. W., Shabbir, A., Yang, Z., and Zhu. J. 1995. A new k-ε eddy-viscosity model for high Reynolds number turbulent flows—Model development and validation. *Computers Fluids,* 24, 227–238.
23. Blatchley, E. R. 1997. Numerical modelling of UV intensity: application to collimated-beam reactors and continuous-flow systems. *Water Research,* 31, 2205–2218.
24. Mellon, M. G. 1950. *Analytical Absorption Spectroscopy.* John Wiley & Sons, Inc., New York, NY.
25. Davidson, P. A. 2004. *Turbulence: An Introduction for Scientists and Engineers.* Oxford University Press, Oxford, UK.
26. Wereley, S. T. and Lueptow, R. M. 1999. Velocity field for Taylor–Couette flow with an axial flow. *Physics of Fluids,* 11, 3637–3649.
27. Simmers, D. A. and Coney, J. E. R. 1980. Velocity distributions in Taylor vortex flow with imposed laminar axial flow and isothermal surface heat transfer. *International Journal of Heat and Fluid,* 85–91.
28. Gu, Z. H. and Fahidy, T. Z. 1986. The effect of geometric parameters on the structure of combined axial and taylor-vortex flow. *The Canadian Journal of Chemical Engineering,* 64, 185–189.

29. Haim, D. and Pismen, L. M. 1994. Performance of a photochemical reactor in the regime of Taylor-Gortler vortical flow. *Chemical Engineering Science,* 49, 1119–1129.

30. Howes, T. and Rudman, M. 1998. Flow and axial dispersion simulation for traveling axisyrnrnetric Taylor vortices. *AIChE Journal,* 44, 255–262.

31. Kataoka, K., Doi, H., Hongo, T., and Futagawa, M. 1975. Ideal plug-flow properties of Taylor vortex flow. *Journal of Chemical Engineering of Japan,* 8, 472–476.

32. Pudjiono, P. I. and Tavare, N. S. 1993. Residence time distribution analysis from a continuous Couette flow device around critical Taylor number. *Canadian Journal of Chemical Engineering,* 71, 312–318.

33. Pudjiono, P. I., Tavare, N. S., Garside, J., and Nigam, K. D. P. 1992. Residence time distribution from a continuous Couette flow device. *Chemical Engineering Journal,* 48, 101–110.

34. Froment, G. F. and Bischoff, K. B. 1990. *Chemical Reactor Analysis and Design.* John Wiley & Sons, New York, NY.

35. Unluturk, S. K., Koutchma, T., and Arastoopour, H. 2004. Modeling of UV dose distribution in a thin film reactor for processing of apple cider. *Journal of Food Processing,* 65, 125–136.

36. Bolton, J. R. 2000. Calculation of ultraviolet fluence rate distributions in an annular reactor: significance of refraction and reflection, *Water Research,* 34, 3315–3324.

37. Hanovia. 2007. [Online] http://www.hanovia.com

38. Hoyer, O. 1998. Testing performance and monitoring of UV systems for drinking water disinfection. *Water Supply,* 16, 424–429.

39. Wright, J. R., Summer, S. S., Hackney, C. R., Pierson, M. D., and Zoecklein, B. W. 2000. Efficacy of ultraviolet light for reducing *Escherichia coli* O157:H7 in unpasteurized apple cider. *Journal of Food Protection,* 63, 563–567.

40. Ye, Z., Forney, L.J., Koutchma, T., Giorges, A.T., and Peirson, J.A. 2008. Optimum UV Disinfection Between Concentric Cylinders. *Industrial & Engineering Chemical Research,* 47, 3445–3450.

22 Ozone Processing

Kasiviswanathan Muthukumarappan
South Dakota State University and University College Dublin

Colm P. O'Donnell
University College Dublin

Patrick J. Cullen
Dublin Institute of Technology

CONTENTS

22.1 INTRODUCTION

Minimizing the occurrence of deadly microorganisms in fruits, vegetables, juices, meats and other foods is a primary food-safety concern. Consumer preference for minimally processed foods, foods free of chemical preservatives, recent outbreaks of foodborne pathogens, identification of new food pathogens, and the passage of new legislations such as the Food Quality Protection Act in the United States have created demand for novel food processing and preservation systems. At the same time, sanitizers—such as the chlorine used both to wash produce as well as disinfect processing equipment—may potentially harm the environment. Some consumers also prefer these materials not be present as residues in the foods they eat. Bacterial pathogens in food cause an estimated

TABLE 22.1

Estimated Annual Food Borne Illnesses, Hospitalization, and Deaths Due to Selected Pathogens

Disease/Agent	Illness	Hospitalization	Deaths
Bacterial			
Campylobacter spp.	1,963,141	10,539	99
Clostridium perfringens	248,520	41	7
Escherichia coli O157:H7	62,458	1843	52
Listeria monocytogenes	2493	2298	499
Salmonella, nontyphoidal	1,341,873	15,608	553
Staphylococcus	185,060	1753	2
Vibrio cholerae, toxigenic	49	17	0
Vibrio vulnificus	47	43	18
Parasitic			
Toxoplasma gondii	112,500	2500	375

Source: Mead, P.S., Slutsker, L., Dietz, V., McCaig, L.F., Bresee, J.S., Shapiro, C., Griffin, P.M., and Tauxe, R.V., *Emerging Infectious Diseases,* 11, 607, 2005.

80 million cases of human illness, 325,000 cases of hospitalization and up to 5,000 deaths annually in the United States alone, coupled with significant economic losses [1] (Table 22.1). The Center for Disease Control and Prevention estimates the yearly cost of foodborne diseases in the United States as $7–$8 billion [2]. One way the food industry can address food safety and the negative perception of some sanitizing agents is through the use of ozone processing. There are several processing methods available for inactivation of microorganisms in foods namely thermal, high pressure, pulsed electric field, oscillating magnetic field, irradiation, and ozonation. Ozonation processing of solid and liquid food materials for microbial safety and mathematical modeling in liquid food are emphasized in this chapter.

22.1.1 Regulatory Catch-Up

Although use of ozone is relatively recent in the United States, ozone and its oxidizing properties were first discovered as early as 1840. By the early 1900s, France was using ozone to disinfect drinking water, and this application soon spread to the rest of Europe. Ozone continues to be used in water treatment in both small and large applications. Ozone is generally recognized as safe status (GRAS) in the United States for use in treatment of bottled water and as a sanitizer for process trains in bottled water plants [3]. In June 1997, ozone received the GRAS status as a disinfectant for foods by an independent panel of experts, sponsored by the Electric Power Research Institute.

On June 26, 2001, the FDA granted this petition and published its final rule in the *Federal Register.* The amendment to the food additive regulations (Title 21 of the Code of Federal Regulations, part 173) allows the use of ozone when used as a gas or dissolved in water as an antimicrobial agent on food, including meat and poultry.

In the regulations, ozone's uses include the reduction of microorganisms on raw agricultural commodities in the course of commercial processing. The regulations also state, however, that ozone's use may have the potential to fall under the guidelines of the Federal Insecticide, Fungicide, and Rodenticide Act (FIFRA), thus, entering the jurisdiction of the Environmental Protection Agency (EPA). Food processors should check with ozone equipment suppliers and/or directly with the EPA to determine if a particular use of ozone will require a special pesticide registration under FIFRA. In 2001, the Food and Drug Administration (FDA) allowed the use of ozone as a direct-contact food-sanitizing agent [4]. This action eventually cleared the way for the use of ozone in the $430 billion food processing industry [4,5].

Ozone has only recently gained increasing attention of the food and agricultural industries though it has been used effectively as a primary disinfectant for the treatment of municipal and bottled drinking waters for 100 years at scales from a few gallons per minute to millions of gallons per day. Currently, there are more than 3000 ozone-based water treatment installations all over the world and more than 300 potable water treatment plants in the United States [6]. In Europe, ozonation in food processing began taking place shortly after it was first used for water treatment. Only with recent regulatory rulings has the stage been set for ozonation to make inroads into the U.S. food industry, where adoption of the technology has been slower. Ozone offers the food industry another tool in the ongoing food-safety quest. At the same time, it does so in a more environmentally friendly way than many other sanitizing agents. With the new regulations in place, perhaps more U.S. processors will give ozonation a try.

22.2 WHAT IS OZONE?

The passage of new legislations such as the Food Quality Protection Act in the United States has created renewed demand for novel food processing and preservation systems. Also, the accumulation of toxic chemicals in our environment has increased the focus on the safe use of sanitizers, bleaching agents, pesticides, and other chemicals in industrial processing [7]. Hence, there is a demand for safe and judicious usage of these chemicals and preservatives in food processing. Ozone is a gas made up of oxygen. Unlike the more familiar diatomic form of oxygen, ozone has three oxygen atoms on its molecule and is formed from oxygen in the presence of heat and light. Lightning and ultraviolet (UV) rays both form naturally occurring ozone and the gas also makes up the familiar UV-shielding "ozone layer" in the Earth's upper atmosphere. Besides its function as planetary protector, ozone also is a strong oxidizer. This makes it effective in killing microorganisms because it oxidizes their cell membranes. In fact, ozone is more effective at killing a wider variety of potential pathogens than chlorine. Unlike many other sanitizing agents, ozone contributes no negative environmental impact because it quickly and easily degrades into diatomic oxygen.

Ozone is a naturally occurring substance found in our atmosphere and it can also be produced synthetically. The characteristic fresh, clean smell of air following a thunderstorm represents freshly generated ozone in nature. Structurally, the three atoms of oxygen are in the form of an isoscales triangle with an angle of 116.8 degree between the two O–O bonds. The distance between the bond oxygen atoms is 1.27 angstroms. The name "Ozone" is derived from the Greek word "Ozein" which means "to smell." Ozone as a gas is blue; both liquid (−111.9°C at 1 atmosphere) and solid ozone (−192.7°C) are an opaque blue-black in color [8]. It is a relatively unstable gas at normal temperatures and pressures, is partially soluble in water, has a characteristic pungent odor, and is the strongest disinfectant currently available for use with foods [9–11]. The relatively high (+2.075 V) electrochemical potential (E^0, Volt) indicates that ozone is a very favorable oxidizing agent (Equation 22.1). The various physical properties of ozone are summarized in Table 22.2.

$$O_3(g) + 2H^+ + 2e^- \Leftrightarrow O_2(g) + H_2O \ \{E^0 = 2.075 \ V\} \tag{22.1}$$

22.2.1 PRODUCTION OF OZONE

In addition to being effective over a broader spectrum of microorganisms, ozone is of particular interest to the food industry because it also is 52% stronger than chlorine. Unlike typical sanitizers, however, ozone's short half-life means it cannot be delivered, handled or stored in the usual way.

TABLE 22.2
Physical Properties of Ozone

Physical Properties	Value
Boiling point, °C	−111.9
Density, kg/m³	2.14
Heat of formation, kJ/mole	144.7
Melting point, °C	−192.7
Molecular weight, g/mole	47.9982
Oxidation strength, V	2.075
Solubility in water, ppm (at 20°C)	3
Specific gravity	1.658

Fortunately, ozone is easily generated with devices that create an electrical discharge across a flow of either pure oxygen or air. Ozone generators are compact and can be installed right where they're needed on the processing line. This is particularly handy because ozone may, depending on the product, actually be applied in more than one way. Ozone is generated by the exposure of air or another gas containing normal oxygen to a high-energy source. High-energy sources such as a high voltage electrical discharge or UV radiation convert molecules of oxygen to molecules of ozone. Ozone must be manufactured on site for immediate use since it is unstable and quickly decomposes to normal oxygen. The half-life of ozone in distilled water at 20°C is about 20–30 minutes [12]. Ozone production is predominantly achieved by one of three methods: electrical discharge method, electrochemical method, and UV radiation method. Electrical discharge method, which is the most widely used commercial method, have relatively high efficiencies (20–30%). The other two methods (electrochemical and UV) are less cost effective.

22.2.1.1   Electrical (Corona) Discharge Method

In this method, adequately dried air or O_2 is passed between two high-voltage electrodes separated by a dielectric material, which is usually glass. Air or concentrated O_2 passing through an ozonator must be free from particulate matter and dried to a dew point of at least −60°C to properly protect the corona discharge device. The ozone/gas mixture discharged from the ozonator normally contains from 1 to 3% ozone when using dry air, and 3–6% ozone when using high purity oxygen as the feed gas [10,11].

The electrodes are typically either concentric metallic tubes or flat, plate-like electrodes. When a voltage is supplied to the electrodes, a corona discharge forms between the two electrodes, and the O_2 in the discharge gap is converted to ozone (Figure 22.1). A corona discharge is a physical phenomenon characterized by a low-current electrical discharge across a gas-containing gap at a voltage gradient, which exceeds a certain critical value [13]. First, oxygen molecules (O_2) are split into oxygen atoms (O), and then the individual oxygen atoms combine with remaining oxygen molecules to form ozone (O_3).

Considerable electrical energy at high voltage (5000 V) is required for the ozone producing electrical discharge field to be formed. In excess of 80% of the applied energy is converted to heat that, if not rapidly removed, causes the O_3 to decompose into oxygen atoms and molecules, particularly above 35°C. In order to prevent this decomposition, ozone generators utilizing the corona discharge method, must be equipped with a means of cooling the electrodes. The temperature of the gas inside the discharge chamber must be maintained between the temperature necessary for formation of O_3 to occur and the temperature at which spontaneous decomposition of O_3 occurs [14]. The cooling is usually accomplished by circulating a coolant such as water or air over one surface of the electrodes so that the heat given off by the discharge is absorbed by the coolant.

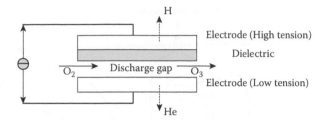

FIGURE 22.1   Ozone generation by corona discharge method.

22.2.1.2   Electrochemical (Cold Plasma) Method

Usually, in the electrochemical method of ozone production, an electrical current is applied between an anode and cathode in electrolytic solution containing water and a solution of highly electronegative anions. A mixture of oxygen and ozone is produced at the anode. The advantages associated with this method are use of low-voltage DC current, no feed gas preparation, reduced equipment size, possible generation of ozone at high concentration, and generation in water. However, this method is less cost-effective compared to electrical discharge method previously described.

22.2.1.3   Ultraviolet (UV) Method

In the UV method of O_3 generation, the ozone is formed when O_2 is exposed to UV light of 140–190 mm wavelength, which splits the oxygen molecules into oxygen atoms, which then combine with other oxygen molecules to form O_3 [10,11]. The method has been reviewed thoroughly by Langlais, Reckhow, and Brink [15]. However, due to poor yields, this method has limited uses.

22.3   MODELING OZONE IN FOOD MATERIALS

Predicting the ozone profile in a bubble column and contact chambers is important for determination of the log reduction and bromate formation in any ozone application in liquid such as water treatment, microbial inactivation in fruit juices, etc. For improvement of operational management of ozonation by model control, the model must be able to predict the ozone profile for changes in different control parameters and water/fruit juice quality parameters.

22.3.1   Modeling Ozone Bubble Columns

Bubble columns are utilized as multiphase contactors and reactors in various food, chemical, petro-chemical, biochemical and metallurgical industries [16]. Processes include oxidation, chlorination, alkylation, polymerization, hydrogenation, and various other chemical and biochemical processes such as fermentation and biological wastewater treatment [17,18]. A bubble column reactor is a cylindrical vessel with a gas diffuser to sparge a gas (ozone, oxygen, carbon dioxide, etc.) into either a liquid phase or liquid–solid dispersions (Figure 22.2). The design of a bubble column is limited by the gas-liquid mass transfer [19], which is controlled by the gas hold-up, specific interfacial area, and bubble size distribution [20]. Available literature shows that the design and modeling of ozone bubble columns are based on determination of overall mass transfer coefficient (K_{La}), gas hold up (ε_G), and Sauter mean diameter (SMD, SD) defined as the diameter of a bubble that has the same volume/surface area which can be determined as follow:

$$d_s = \sqrt{\frac{A_b}{\pi}}$$
(22.2)

$$d_v = \sqrt[3]{\left(\frac{6V_b}{\pi}\right)}$$
(22.3)

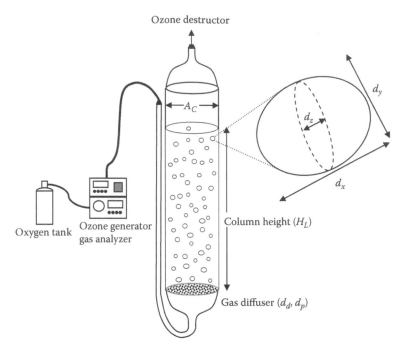

FIGURE 22.2   A bubble column reactor.

where A_b and V_b are the surface area and volume of the bubble, respectively. d_s and d_v are usually measured directly using image analysis. Individual bubble diameter may be determined by Equation 22.5 assuming that the bubble is an ellipsoid (Figure 22.2). This three dimensional technique may be simplified by assuming that shortest length of the bubble, d_x, and width of the bubble, d_z, are of equal length, thus reducing this measurement to a two dimensional approach [20]:

$$d_b = \sqrt[3]{d_x d_y d_z} = \sqrt[3]{d_x d_y^2} \tag{22.4}$$

The SD for a bubble is:

$$SD = \frac{d_v^3}{d_s^2} \tag{22.5}$$

The overall mass transfer coefficient (K_{La}) is used to describe the absorbance or consumption of ozone gas into the liquid or solid-liquid phase, defined as follow:

$$\psi = K_{La}(C_L^e - C_L) \tag{22.6}$$

where ψ is ozone consumption or absorption rate (mgL^{-1}s^{-1}), C_L^e is dissolved ozone concentration in equilibrium with the ozone gas (mgL^{-1}), and C_L is dissolved ozone concentration (mgL^{-1}) in the liquid sample under investigation. Dissolved ozone concentration can be determined using an ozone analyzer or through chemical tests. K_{La} can be determined which usually depends upon operating conditions and liquid characteristics such as surface tension, viscosity, density, surface tension, etc.

Although the determination of K_{La} is useful for design purposes, it is important in many cases to determine the local mass transfer coefficient (K_L, m.s^{-1}) in order to evaluate the factor for ozone absorption with chemical reaction or gas-liquid reaction. Therefore, one must know the value of gas bubbles' specific interfacial area (a, m^{-1}), which is equal to the ratio between the bubbles' surface area (A, m²) and the volume of the dispersed phases (V, m³). However, due to the difficulties associated with determining A, the value of a can be calculated using the following relationship:

$$a = \frac{6\varepsilon_G}{SD} \tag{22.7}$$

SD can be determined as described above, whereas gas hold up (ε_G) can be determined by measuring pressure change in bubble column using a pressure transducer and the following relationship

$$\varepsilon_G = 1 - \frac{\Delta P}{\rho_L g \Delta x} \tag{22.8}$$

where, ΔP is the pressure difference (Pa), ρ_L is the density of the liquid under investigation (kg.m⁻³), g is acceleration due to gravity (9.81 m.s⁻²), and Δx is the distance between two measuring points (m).

Within bubble columns ozone is not supplied in a pure form since, ozone gas supply passing through the diffuser is a mixture of ozone and oxygen. Moreover, oxygen mass transfer is usually neglected in the ozone dissolution models. However, this information is desirable for proper operation of ozone dissolution in a bubble column. Thus, the mass transfer of ozone (A) and oxygen (O) from the gas to liquid phase can be described by the two-film model [21]. According to the theory of surface renewal oxygen based K_{La} has to be converted to ozone based K_{La}. As the ozone is dissolved in water, it may be consumed via the self-decomposition ($2O_3 \rightarrow 3O_2$) and oxidation with the organic matter (suspended soluble fraction of juice) (B) present in the liquid. Regarding the spontaneous ozone decomposition and reaction with the suspended solids in liquid, e.g., pulp fraction and soluble solid in fruit juice. Chang et al. [22] proposed the following pseudo-first-order and second-order reaction rate expressions:

$$\frac{dC_{ALb}}{dt} = K_d C_{ALb} - \alpha_{AB} K_{AB} C_{ALb} C_{BLb} \tag{22.9}$$

$$\frac{dC_{BLb}}{dt} = -K_{AB} C_{ALb} C_{BLb} \tag{22.10}$$

$$\frac{dC_{OLb}}{dt} = -\frac{3K_d C_{ALb}}{2} \tag{22.11}$$

where

C_{ALb}: dissolved ozone concentration in liquid (mgL⁻¹)
C_{BLb}: concentration of organic matter in liquid (mgL⁻¹)
K_{ab} : ozonation rate constant of organic mater in liquid
K_d : self decomposition rate constant of ozone (s⁻¹)
α_{AB} : stoichiometric yield ratio (ml O_3 consumed/mol organic matter)

With the ozone consumption and oxygen formation, the mass transfer rates of ozone and oxygen may be enhanced or retarded [22]. The ratios of the mass transfer rates of ozone and oxygen with the ozone consumption and the oxygen formation to those without may be designated by the enhancement factor of ozone consumption (E_A) and the retarding factor of oxygen formation (R_{FO}), respectively. E_A is defined by Danckwerts [21] as follow:

$$E_A = \frac{\text{Rate of gas absorption with chemical reactions}}{\text{Rate of maximum pure physical gas absorbtion}} \tag{22.12}$$

The enhancement factor is a function of the reactivity of organic matter present in test sample for instance pulp in juice toward ozone, the ozone diffusivity in the liquid phase, and the local mass transfer coefficient, K_{La}. Danckwerts's surface theory [21] is as follows:

$$\frac{K_{La}O_3}{K_{La}O_2} = \sqrt{\frac{DO_3}{DO_2}} \qquad (22.13)$$

where, DO_3 and DO_2 are the molecular diffusivities of ozone and oxygen gas in water. The above equation can be used to correct K_{La} from oxygen to ozone base, the use of this equation is reported and has been validated earlier [23,24].

22.4 APPLICATIONS OF OZONE IN FOOD PROCESSING

22.4.1 Mechanism of Microbial Inactivation

When a cell becomes stressed by viral, bacterial or fungal attack, its energy level is reduced by the outflow of electrons, and becomes electropositive. Ozone possesses the third atom of oxygen which is electrophlic i.e., ozone has a small free radical electrical charge in the third atom of oxygen which seeks to balance itself electrically with other material with a corresponding unbalanced charge. Diseased cells, viruses, harmful bacteria and other pathogens carry such a charge and so attract ozone and its by-products. Normal healthy cells cannot react with ozone or its by-products, as they possess a balanced electrical charge and a strong enzyme system.

Because of its very high oxidation reduction potential, ozone acts as an oxidant of the constituent elements of cell walls before penetrating inside microorganisms and oxidizing certain essential components e.g., unsaturated lipids, enzymes, proteins, nucleic acids, etc. When a large part of the membrane barrier is destroyed causing leakage of cell contents, the bacterial or protozoan cells lyse (unbind) resulting in gradual or immediate destruction of the cell. Most pathogenic and foodborne microbes are susceptible to this oxidizing effect.

22.4.2 Application of Ozone in Solid Food Materials

Ozone is one of the most potent sanitizers known and is effective against a wide spectrum of microorganisms at relatively low concentrations [12]. Sensitivity of microorganisms to ozone depends largely on the medium, the method of application, and the species. Susceptibility varies with the physiological state of the culture, pH of the medium, temperature, humidity, and presence of additives, such as, acids, surfactants, and sugars [25]. The antimicrobial spectrum and sanitary applications of ozone in food industry arc summarized in Table 22.3.

Currently, researchers are looking into ozone's potential for directly cleaning the surface of animal carcasses. Sheldon and Brown [26] investigated the efficacy of ozone as a disinfectant for poultry carcasses. The microbial counts of ozone treated carcasses stored at 4°C were significantly lower than carcasses chilled under nonozonated conditions. Gorman, Sofos, Morgan, Schmidt and Smith [27] evaluated the effect of various sanitizing agents (5% hydrogen peroxide, 0.5% ozone, 12% trisodium phosphate, 2% acetic acid, and 0.3% commercial sanitizer), and water (16–74°C) spray-washing interventions for their ability to reduce bacterial contamination of beef samples in a model spray-washing cabinet. Hydrogen peroxide and ozonated water were found to be more effective than the other sanitizing agents. In another study, the effect of different treatments (74°C hot-water washing, 5% hydrogen peroxide, and 0.5% ozone) in reducing bacterial populations on beef carcasses was studied and the researchers have found that water at 74°C caused higher bacterial reduction than those achieved by the other sanitizing agents [28]. Ozone and hydrogen peroxide treatments had minor effects and were equivalent to conventional washing in reducing bacterial populations on beef.

TABLE 22.3
Antimicrobial Spectrum and Sanitary Applications of Ozone in Food Industry

Sanitation	Dosage	Susceptible Microorganisms
Animal	>100 ppm	HVJ/TME/Reo type 3/murine hepatitis virus
Black berries	0.3 ppm	*Botrytis cinerea*
Cabbage	7–13 mg/m³	Shelf-life extension
Carrot	5–15 mg/m³	Shelf-life extension
	60 μl/L	*Botrytis cinerea/Scerotinia sclerotiorum*
Dairy	5 ppm	*Alcaligens faecalus/P. fluorescens*
Fish	0.27 mg/L	*P. putida/B. thermospacta/L. plantarum/Shewanella putrefaciens/ Enterobacter* sp.
	0.111 mg/L	*Enterococcus seriolicida*
	0.064 mg/L	*Pasteurella piscicida/Vibrio anguillarum*
Media	3–18 ppm	*E. coli* O157:H7
Peppercorn	6.7 mg/L	3–6 log reduction of microbial load
Potatoes	20–25 mg/m³	Shelf-life extension
Poultry	0.2–0.4 ppm	*Salmonella* sp./*Enterobacteriaceae*
Shrimp	1.4 ml/L	*E. coli/Salmonella typhimurium*
Water	0.35 mg/L	*A. hydrophila/B. subtilis/E.coli/V. cholerae/ P.aeruginosa/L. monocytogenes/Salm. typhi/Staph. aureus/Y. enterocolitica*

Source: From Muthukumarappan, K. Halaweish, F., and Naidu, A.S., *Natural Food Anti-microbial Systems*, CRC Press, Boca Raton, FL, 783, 2000. With permission.

Da Silva, Gibbs, and Kirby [29] investigated the bacterial activity of gaseous ozone on five species of fish bacteria and reported that ozone in relatively low concentration ($< 0.27 \times 10^{-3}$ g/l) was an effective bactericide of vegetative cells. Kaothien, Jhala, Henning, Julson, and Muthukumarappan [30] evaluated the effectiveness of ozone in controlling *Listeria monocytogenes* in cured ham. There was a significant ($p > 0.05$) reduction (about 90%) in bacterial population, with ozone concentration in range of 0.5–1.0 ppm, with exposure time of 1–15 minute at an exposure temperature of 20°C.

On fresh, raw meat, for example, levels as low as 0.04 ppm can retard and control microbial growth. Increasing the levels to approximately 0.10 ppm can help cure or age beef to make it tender. If the meat starts with a low bacterial count, ozone storage may even increase shelf-life by up to 40%. For fish, ozone has benefits that begin even closer to the source. By freezing ozonated water, for example, it can be stored on fishing boats and used at sea. The melted water can be used for washing and in processing, while the remaining frozen ice achieves atmospheric benefits in storage areas. Using such techniques, fishing boats can stay at sea for up to 14 days. Closer to consumers, ozonated ice in retail display cases may help extend the shelf-life by one to three days.

Within the food industry, ozone has been used routinely for washing and storage of fruits and vegetables [31,32]. Ozone can be bubbled through water into which it will partially dissolve. This ozonated water then can be used for washing and/or in transfer flumes to reduce the microbial loads of berries and other fruits and vegetables. Controlled studies report that ozonated water may actually provide greater than 90% reduction of total bacterial counts for some vegetables. Such treatments also have been shown to reduce fungi populations and, subsequently, reduce fungal decay. During such processes, ozone is consumed, so wash water must be ozonated continually. The environment for ozonation also should have at least a 50% humidity level, with optimum effectiveness on fruits and vegetables between 90 and 95% humidity. Below 50% humidity, ozone is less effective because microorganisms must be in a swelled state in order to be attacked. As humidity approaches 100% (or in the presence of steam), ozone's effectiveness also decreases somewhat. Besides reducing microbial loads in wash water, ozone can extend the shelf-life of produce in storage. During storage, fruits and vegetables ripen more quickly as they absorb respiration and decay gasses emitted by other fruits and vegetables. As more and more pieces ripen, this effect accelerates in a cascade effect. Ozone in the storage environment can oxidize the metabolic products of decomposition and help to slow this cascade of accelerated ripening and decay. In some cases, ozone treatment in storage may nearly double the shelf-life of fresh produce. For storage applications, the ozone is simply emitted periodically into the storage area.

Ozone can be used during the washing of produce before it is packaged and shipped to supermarkets, grocery stores and restaurants. With 99.9% kill rate, it is far more effective than current sanitizing methods, such as commercial fruit and vegetable washes. Also, processors who chill fruits or vegetables after harvest using water held at approximately 1°C can ozonate the water to prevent product contamination. Cooling fruits and vegetables helps slow product respiration, and preserve freshness and quality. Fruit and vegetable processing systems that incorporate ozone-generating technology will be able to produce cleaner food while using substantially less water. It will destroy bacteria that can cause premature spoilage of fruits and vegetables while also ensuring a safer product for consumers without any toxic residues. The ozone dissipates within minutes following the washing process.

Recent investigations involving the use of ozone for dried foods have shown that gaseous ozone reduced Bacillus spp. and Micrococcus counts in cereal grains, peas, beans and whole spices were reduced by up to 3 log units, depending on ozone concentration, temperature and relative humidity conditions [33,34]. Zhao and Cranston [35] used gaseous ozone as a disinfectant in reducing microbial populations in ground black pepper, observing a 3–6 log reduction depending on the moisture content with samples ozonized at 6.7 ppm for 6 h. Furthermore, ozonated water has been applied to fresh-cut vegetables for sanitation purposes reducing microbial populations and extending the shelf-life [36,37]. Treatment of apples with ozone resulted in lower weight loss and spoilage. An increase in the shelf-life of apples and oranges by ozone has been attributed to the oxidation of ethylene. Fungal deterioration of blackberries and grapes was decreased by ozonation processing [38]. Ozonated water was found to reduce bacterial content in shredded lettuce, blackberries, grapes, black pepper, shrimp, beef, broccoli, carrots, tomatoes, and milk [25,35,39–41].

In the 1980s, studies showed that cheese stored with periodic ozonation experienced no mold growth for four months while the control cheese began growing mold in as little as one month of storage. Later in that same decade, ozone treatment was used to reduce airborne microorganisms in a confectionery processing facility. Not only did this reduce the incidence of airborne microorganisms over a year and a half, it actually extended the shelf-life of the facility's products by seven days due to inhibited bacterial growth. In the 1990s, Japanese researchers discovered that exposing grains, flour and raw noodles to ozone yielded significant reductions in microbial growth. Some of the recent applications of ozone are freshly caught fish [42], poultry products [26,43], meat and milk products [44,45], to purify and artificially age wine and spirits [46], to reduce aflatoxin in peanut and cottonseed meals [47], to sterilize bacon, beef, bananas, eggs, mushrooms, cheese, and fruit [48,49], to preserve lettuce [25], strawberries [50], green peppers [51] and sprouts [52].

22.4.3 Application of Ozone in Liquid Food Materials

Most contemporary applications of ozone include treatment of drinking water [53] and municipal wastewater [6,54]. Effectiveness of ozone against microorganisms depends not only on the amount applied, but also on the residual ozone in the medium and various environmental factors such as medium pH, temperature, humidity, additives (surfactants, sugars, etc.), and the amount of organic matter surrounding the cells [25,55]. It is difficult to predict ozone behavior under such conditions and in the presence of specific compounds. Residual ozone is the concentration of ozone that can be detected in the medium after application to the target medium. Both the instability of ozone under certain conditions and the presence of ozone-consuming materials affect the level of residual ozone present in the medium. Therefore, it is important, to distinguish between the concentration of applied ozone and residual ozone necessary for effective disinfection. It is advisable to monitor ozone availability during treatment [56]. Efficacy of ozone is demonstrated more readily when targeted microorganisms are suspended and treated in pure water or simple buffers (with low ozone demand) than in complex food systems where it is difficult to predict how ozone will react in the presence of organic matter [57]. Food components are reported to interfere with bactericidal properties of ozone against microbes [58].

In apple cider, Dock [59] determined that the mandatory 5-log reduction could be achieved without harming essential quality attributes. Inactivation of *Escherichia coli O157:H7* and *Salmonella* in apple cider and orange juice treated with ozone in combination with antimicrobials such as dimethyl dicarbonate (DMDC; 250 or 500 ppm) or hydrogen peroxide (300 or 600 ppm) was evaluated by Williams, Summer and Golden [60]. In their first study they found that the treatments resulted in a 5-log colony-forming units (CFU)/mL reduction of either pathogen. However, in their second study they found that all combinations of antimicrobials plus ozone treatments, followed by refrigerated storage, caused greater than a 5-log CFU/mL reduction, except ozone/DMDC (250 ppm) treatment in orange juice. They have concluded that the ozone treatment in combination with DMDC or hydrogen peroxide followed by refrigerated storage may provide an alternative to thermal pasteurization to meet the five-log reduction standard in cider and orange juice. Recently a number of commercial fruit juice processors in the United States began employing this ozone process for pasteurization resulting in industry guidelines being issued by the FDA [61].

22.4.4 Disinfection of Food Processing Equipment and Environment

Within the food industry much attention is given to the cleaning and sanitizing operations of food-processing equipment both in preventing contamination of products and in maintaining the functionality of equipment [62]. Since ozone is a strong oxidant, it can be used for the disinfection of processing equipment and environments. It has been reported that ozone decreased surface flora by 3 $\log_{10}$ units when tested in wineries for barrel cleaning, tank sanitation, and clean-in-place processes [63]. In 2000, researchers at the Department of Food Science and Technology at The Ohio State University, Columbus, studied the potential effectiveness of ozonated water in decontaminating the surfaces of both stainless steel and laminated aseptic food-packaging material [64]. They treated both types of surfaces and confirmed that sterility of naturally contaminated packaging material could be achieved when treated with ozone in water for as little as one minute. Dried films of spores could be eliminated by higher concentrations of ozone in water for both the packaging material and stainless steel. The researchers concluded that ozone is an effective sanitizer for both potential applications.

Water containing low concentrations of ozone can be sprayed onto processing equipment, walls or floors to both remove and kill bacteria or other organic matter that may be present. Ozone has been shown to be more effective than chlorine, the most commonly used disinfectant, in killing bacteria, fungi and viruses, and it does this at one–tenth (2–10 ppm) of the chlorine concentration. Ozone can react up to 3000 times faster than chlorine with organic materials and does not leave any residual toxic by-products. Currently, ozone is the most likely alternative to chlorine in food applications.

22.4.5 Effects of Ozone on Product Quality

Applying ozone at doses that are large enough for effective decontamination may change the sensory qualities of food and food products. The effect of ozone treatment on quality and physiology of various kinds of food products have been evaluated by various researchers. Ozone is not universally beneficial and in some cases may promote oxidative spoilage in foods [65]. Surface oxidation, discoloration or development of undesirable odors may occur in substrates such as meat, from excessive use of ozone [12,66].

Richardson [67] reported that ozone helps to control odor, flavor and color while disinfecting wastewater. Dock [59] reported no detrimental change in quality attributes of apple cider when it was treated with ozone. However, much research still needs to be conducted before it can effectively be applied to fruit juice. No change in chemical composition and sensory quality of onion was reported by Song, Fan, Hilderbrand, and Forney [53]. Ozonated water treatment resulted in no significant difference in total sugar content of celery and strawberries [68] during storage periods. Ozone is expected to cause the loss of antioxidant constituents, because of its strong oxidizing activity. However, ozone washing treatment was reported to have no effect on the final phenolic content of fresh-cut iceberg lettuce [37]. Contradictory reports are found in the literature regarding ascorbic acid, with decomposition of ascorbic acid in broccoli florets reported after ozone treatment by Lewis, Zhuang, Payne, and Barth [69]. Conversely Zhang, Lu, Yu, and Gao [68] reported no significant difference between ascorbic acid contents for treated and non-treated celery samples. Moreover, increases in ascorbic acid levels in spinach [70], pumpkin leaves [71], and strawberries [72] were reported in response to ozone exposure.

Slight decreases in vitamin C contents were reported in lettuce [37]. Ozone treatments were reported to have minor effects on anthocyanin contents in strawberries [72] and blackberries [40]. The most notable effect of ozone on sensory quality of fruits was the loss of aroma. Ozone enriched cold storage of strawberries resulted in reversible losses of fruit aroma [72,73]. This behavior is probably due to oxidation of the volatile compounds. In spite of its efficacy against microorganisms both in the vegetative and spore forms, ozone is unlikely to be used directly in foods containing high-ozone-demand materials, such as meat products [74]. Applying ozone at doses that are large enough for effective decontamination may change the sensory qualities of these products. Due to increased concern about the safety of fruit, vegetable and juice products, the FDA has mandated that these must undergo a five-log reduction in pathogens. The effect of ozone treatment on apple cider quality and consumer acceptability was studied over 21 days. Ozone-treated cider had greater sedimentation, lower sucrose content and a decrease in soluble solids by day 21 [75].

Recently researchers in Spain evaluated the effects of continuous and intermittent applications of ozone gas treatments, applied during cold storage to maintain postharvest quality during subsequent shelf-life, on the bioactive phenolic composition of "Autumn Seedless" table grapes after long-term storage and simulated retail display conditions [76]. They found that the sensory quality was preserved with both ozone treatments. Although ozone treatment did not completely inhibit fungal development, its application increased the total flavan-3-ol content at any sampling time. Continuous $0.1\ \mu L\ L^{-1}\ O_3$ application also preserved the total amount of hydroxycinnamates, while both treatments assayed maintained the flavonol content sampled at harvest. Total phenolics increased after the retail period in ozone treated berries. Therefore the improved techniques tested for retaining the quality of "Autumn Seedless" table grapes during long-term storage seem to maintain or even enhance the antioxidant compound content.

22.5 SAFETY REQUIREMENTS

Ozone is a toxic gas and can cause severe illness, and even death if inhaled in high quantity. It is one of the high active oxidants with strong toxicity to animals and plants. Toxicity symptoms such as sharp irritation to the nose and throat could result instantly at 0.1 ppm dose. Loss of vision could arise from 0.1 to 0.5 ppm after exposure for 3–6 h. Ozone toxicity of 1–2 ppm could cause distinct irritation on the upper part of throat, headache, pain in the chest, cough and drying of the throat. Higher levels of ozone (5–10 ppm) could cause increase in pulse, and edema of lungs. Ozone level of 50 ppm or more is potentially fatal [11]. The ozone exposure levels as recommended by the Occupational Safety and Health Administration (OSHA) of the United States are shown in Table 22.4.

TABLE 22.4
Approved Levels of Ozone Application

Exposure	Ozone Level, ppm
Detectable odor	0.01–0.05
OSHA 8 h limit	0.1
OSHA 1.5 minute limit	0.3
Lethal in few minutes	>1700

Source: Muthukumarappan, K. Julson, J.L., and Mahapatra, A.K., *Souvenir 2002 Proceedings of College of Agricultural Engineering Technology Alumni Meeting,* Bhubaneswar, India, 32, 2002.

22.6 LIMITATIONS OF USING OZONE

As discussed earlier applying ozone at doses that are large enough for effective decontamination may result in changes in the sensory or nutritional qualities of some food products including; surface oxidation, discoloration, and the development of undesirable odors. Additionally, microorganisms embedded in product surfaces are more resistant to ozone than those readily exposed to the sanitizer. Hence, suitable application methods have to be used to assure direct contact of ozone with target microorganisms. The rapid reaction and degradation of ozone diminish the residuals of this sanitizer during processing. The lack of residuals may limit the processor's ability for in-line testing of efficacy.

Also, there are existing restrictions relating to human exposure to ozone, which must be addressed. Plant operators seeking to employ ozone will be faced with system design and process operation challenges. However, ozone monitors and distructors may be employed to overcome such challenges. The initial cost of ozone generators may be of concern to small-scale food processors but as the technology improves the cost of the generators are coming down.

22.7 CONCLUSIONS

The presence of harmful microorganisms in the food products is the main concern for the processors as well as consumers. The present consumer preference is for minimally processed foods and foods that don't contain any residual chemicals used for disinfection of the raw foods and process equipment. Treatment with ozone meets the above demand in totality. In addition to this, ozone extends the shelf-life of the produce during storage by oxidizing the metabolic products produced during storage of the product. In this chapter we presented the potential applications of ozone, current status in the United States on the use of ozone, different methods of production of ozone. Mathematical modeling of ozone applications is presented with special reference to bubble column. Application of ozone for solid as well as liquid foods and its effect on product quality is presented in this chapter. Lastly safety requirement when working with ozone and limitations of using ozone are dealt with in detail.

ACKNOWLEDGMENTS

This work was partially supported by Agricultural Experiment Station, College of Agricultural and Biological Sciences, South Dakota State University, Brookings, SD, and the Irish Department of Agriculture and Food under Food Institutional Research Measure. The authors thank Chenchaiah Marella for his time and effort in reviewing this chapter.

REFERENCES

1. Mead, P.S., Slutsker, L., Dietz, V., McCaig, L.F., Bresee, J.S., Shapiro, C., Griffin, P.M. and Tauxe, R.V. Food-related illness and death in the United States. *Emerging Infectious Diseases*, 11, 607, 2005.
2. West, P., Kim, J., Huang, T.S., Carter, M., Weese, J.S. and Wei, C.I. Bactericidal activity of electrolyzed oxidizing water against *E. coli*, *L. monocytogenes* and *S. enteritidis* inoculated on beef and chicken. *Presented at the IFT Annual Meeting*, New Orleans, LA, 2001.
3. FDA. Beverages: bottled water; final rule. *Food and Drug Admin.*, *Fed. Reg.* 60, 57075, 1995.
4. Hampson, B.C. Emerging technology—Ozone. *Presented at the IFT Annual Meeting*, New Orleans, LA, 2001.
5. Johannsen, E.J., Muthukumarappan, K., Julson, J.L. and Stout, J.D. Application of ozone technology in beef processing. *Presented at the North Dakota-South Dakota 2nd Biennial Joint EPSCoR Conference on Stimulating Competitive Research*, Fargo, ND, 1999.
6. Rice, R.G., Overbeck, P. and Larson, K.A. Costs of ozone in small drinking water systems. In: *Proceedings Small Drinking Water and Wastewater Systems*. National Science Foundation Int., Ann Arbor, MI, 27, 2000.
7. Julson, J.L. Muthukumarappan, K. and D Henning, D. Effectiveness of ozone for controlling *L. monocytogenes* in ready to eat cured ham. *Report NPPC Project #99-221*, South Dakota State University, Brookings, SD, 2001.
8. Hunter, B. Ozone applications: an in depth discussion. *Health Freedom News*, 14, 1995.
9. Mahapatra, A.K., Muthukumarappan, K. and Julson, J.L. Applications of ozone, bacteriocins and irradiation in food processing: a review. *Critical Reviews in Food Science and Nutrition*, 45, 447, 2005.
10. Muthukumarappan, K. Julson, J.L. and Mahapatra, A.K. Ozone applications in food processing. In: *Souvenir 2002 Proceedings of College of Agricultural Engineering Technology alumni meeting*, S.K. Nanda (Ed.). Bhubaneswar, India, 32, 2002.
11. Muthukumarappan, K. Halaweish, F. and Naidu, A.S. Ozone. In: *Natural Food Anti-microbial Systems*, A.S. Naidu (Ed.). CRC Press, Boca Raton, FL, 783, 2000.
12. Khadre, M.A., Yousef, A.E. and Kim, J.G. Microbial aspects of ozone applications in food: a review. *Journal of Food Science*, 66, 1242, 2001.
13. Taylor, P.A., Futrell, T.O., Dunn Jr., N.M., Michael, P., DuBois, P.C.R and Capehart, J.D. US Patent # 5 547 644, 1996.
14. Miller, A.D., Grow, W.R., Dees, L.A., Mitchell, M.R. and Manning, T.J. A history of patented methods of ozone production from 1897 to 1997. *Laboratory of Physical Environmental Sciences*, Department Chemistry, Valdosta State University, Valdosta, GA, 2002. http://www.valdosta.edu/~tmanning/research/ozone/
15. Langlais, B., Reckhow, D.A. and Brink, D.R. *Ozone in Water Treatment: Application and Engineering.* Lewis Publishers, MI, 225, 1991.
16. Degaleesan, S., Dudukovic, M. and Pan, Y. Experimental study of gas-induced liquid-flow structures in bubble columns. *AIChE Journal*, 47, 1913, 2001.
17. Shah, Y.T., Godbole S.P. and Deckwer, W.D. Design parameters estimations for bubble column reactors, *AIChE Journal*, 28, 353, 1982.
18. Prakash, A., Margaritis A. and Li, H. Hydrodynamics and local heat transfer measurements in a bubble column with suspension of yeast. *Biochemical Engineering Journal*, 9, 155, 2001.
19. Cramers, P., Beenackers, A. and Vandierendonck, L.L. Hydrodynamics and mass-transfer characteristics of a loop-venturi reactor with a downflow liquid jet ejector. *Chemical Engineering Science*, 47(13–14), 3557, 1992.
20. Baawain, M.S., El-Din, M.G., and Smith, D.W. Artificial neural networks modeling of ozone bubble columns: mass transfer coefficient, gas hold-up, and bubble size. *Ozone: Science and Engineering*, 29, 343, 2007.
21. Danckwerts, P.V. *Gas–Liquid Reactions.* McGraw-Hill, Inc., New York, NY, 1970.
22. Chen, Y.H., Chang, C.Y. Chiu, C.Y. Yu, Y.H. Chiang, P.C., KU, Y., Chen, J.N. Dynamic behavior of ozonation with pollutant in a countercurrent bubble column with oxygen mass transfer. *Water Research*, 37, 2583, 2003.
23. Sherwood, T.K., Robert, L.T., and Charles, R.W. *Mass Transfer.* McGraw-Hill, Inc., NY, 1975.
24. El-Din, M.G. and Smith, D.W. Ozonation of kraft pulp mill effluents: process Dynamics. *Journal of Environmental Engineering Science*, 1, 45, 2002.
25. Kim, G.J., Yousef, A.E. and Dave, S. Application of ozone for enhancing safety and quality of foods: a review. *Journal of Food Protection*, 62, 1071, 1999.
26. Sheldon, B.M and Brown, A.L. Efficacy of ozone as a disinfectant for poultry carcasses and chill water. *Journal of Food Science*, 51, 305, 1986.

27. Gorman, B.M., Sofos, J.N., Morgan, J.B., Schmidt, G.R. and Smith, G.C. Evaluation of hand-trimming, various sanitizing agents, and hot water spray-washing as decontamination interventions for beef brisket adipose tissue. *Journal of Food Protection*, 58, 899, 1995.

28. Reagan, J.O., Acuff, G.R., Bueye, D.R., Buyck, M.J., Dickson, J.S., Kastner, C.L. Marsden, J.L., Morgan, J.B., Nickelson, R., Smith, G.C. and Sofos, J.N. Trimming and washing of beef carcasses as a method of improving the microbiological quality of meat. *Journal of Food Protection*, 59, 751, 1996.

29. Silva da, M.V., Gibbs, P.A. and Kirby, R.M. Sensorial and microbial effects of gaseous ozone on fresh scad (Trachurus trachurus). *Journal of Applied Microbiology*, 84, 802, 1998.

30. Kaothien, P., Jhala, R., Henning, D., Julson, J.L., Muthukumarappan, K. and Dave, R.I. Effectiveness of ozone for controlling *L. monocytogenes* in cured ham. *Presented at the IFT Annual Meeting*, New Orleans, LA, 2001.

31. Hampson, B.C., Montevalco, J. and Williams, D.W. Regulation of ozone as a food sanitizing agent: application of ozonation in sanitizing vegetable process wash waters. *Presented at the IFT Annual Meeting*, Book of Abstracts, Chicago, IL, 140, 1996.

32. Liangji, X. Use of ozone to improve the safety of fresh fruits and vegetables. *Food Technology*, 53, 58, 1999.

33. Naitoh, S., Okada, Y. And Sakai, T. Studies on utilization of ozone in food preservation: III. Microbicidal properties of ozone on cereal grains, cereal grain powders, peas, beans and whole spices. *Journal of Japanese Society of Food Science and Technology*, 34, 788, 1987.

34. Naitoh, S., Okada, Y. and Sakai, T. Studies on utilization of ozone in food preservation: V. Changes in microflora of ozone treated cereals, grains, peas, beans and spices during storage. *Journal of Japanese Society of Food Science and Technology*, 35, 69, 1988.

35. Zhao, J. and Cranston, P.M. Microbial decontamination of black pepper by ozone and effects of treatment on volatile oil constituents of the spice. *Journal of the Science of Food and Agriculture*, 68, 11, 1995.

36. Beltra´n, D., Selma, M.V., Tudela, J.A. and Gil, M.I. Effect of different sanitizers on microbial and sensory quality of fresh-cut potato strips stored under modified atmosphere or vacuum packaging. *Postharvest Biology and Technology*, 37, 37, 2005.

37. Beltra´n, D., Selma, M.V., Mari´n, A. and Gil, .M.I. Ozonated water extends the shelf life of fresh-cut lettuce, *Journal of Agricultural and Food Chemistry*, 53, 5654, 2005.

38. Beuchat, L.R. Surface disinfection of raw produce. *Dairy Food Environmental Sanitation*, 12, 6, 1992.

39. Chen, J.S., Balaban, M.O., Wei, C.I., Marshall, M.R. and Hsu, W.Y. Inactivation of polyphenol oxidase by high pressure CO2. *Journal of Agricultural and Food Chemistry*, 40, 2345, 1992.

40. Barth, M.M., Zhou, C., Mercier, J. and Payne, F.A. Ozone storage effects on antocyanin content and fungal growth in blackberries. *Journal of Food Science*, 60, 1286, 1995.

41. Sarig, P., Zahavi, T., Zutkhi, Y., Yannai, S., Lisher, N. and Ben-Arie, R. Ozone for control and post-harvest decay of table grapes caused by Rhizopus stolonifer. *Physiology Molecular Plant Pathology*, 48, 403, 1996.

42. Goche, L. and Cox, B. Ozone treatment of fresh H&G Alaska salmon. *Report to Alaska Science and Technology Foundation and Alaska Department of Environmental Conservation*, November, Seattle, WA, Surefish, 1999.

43. Dave, S.A. Effect of ozone against Salmonella enteritidis in aqueous suspensions and on poultry meat. *MSc Thesis*. Ohio State University, Columbus, OH, 26, 1999.

44. Dondo, A., Nachman, C., Doglione, L., Rosso, A and Genetti, A. Foods: their preservation by combined use of refrigeration and ozone. *Ingenieria Alimentaria y Conserve Animale*, 8, 16, 1992.

45. Gorman, B.M., Kuchevar, S.L., Sofos, L.W., Morgan, J.B., Schmidt, G.R. and Smith, G.C. Changes on beef adipose tissue following decontamination with chemical solutions or water 351C or 741C. *Journal of Muscle Foods*, 8, 185, 1997.

46. Hill, D.G. and Rice, R.G. Historical background properties and applications. In: *Handbook of Ozone Technology and Applications*, Vol 1. Rice, R.G. and Netzer, A. (Eds). Ann Arbor Science, Ann Arbor, MI, 1982.

47. Dwankanath, C.I., Rayner, E.T., Mann, G.E and Dollar, F.G. Reduction of aflatoxin levels in cotton seed and peanut meals by ozonization. *Journal of American Oil Chemical Society*, 45, 93, 1968.

48. Kaess, G. and Weidemann, J.F. Ozone treatment of chilled beef. Effect of low concentrations of ozone on microbial spoilage and surface color of beef. *Journal of Food Technology*, 3, 325, 1968.

49. Gammon, R. and Karelak, A. Gaseous sterilization of foods. *American Institute of Chemical Engineering Symposium Series*, 69, 91, 1973.

50. Lyons-Magnus. Ozone use survey data. Ozone treatment of fresh strawberries, Data submitted to EPRI Agriculture and Food Alliance, Lyons-Magnus, Fresno, CA, September 28, 1999.

51. Han, Y., Floros, J.D., Linton, R.H., Nielsen, S.S. and Nelson, P.E. Response surface modeling for the inactivation of E. coli O157: H7 on green peppers by ozone gas treatment. *Journal of Food Science*, 67, 3188, 2002.

52. Singha, N., Singh, R.K. and Bhuniab, A.K. Sequential disinfection of E. coli O157: H7 inoculated alfa alfa seeds before and during sprouting using aqueous chloride dioxide, ozonated water and thyme essential oil. *Lebensmittel Wissenschaft und Technologie*, 36, 235, 2003.

53. Bryant, E.A., Fulton, G.P. and Budd, G.L. *Disinfection Alternatives for Safe Drinking Water.* Van Nostrand Reinhold, New York, NY, 1992.

54. Stover, E.L. and Jarnis, R.W. Obtaining high level wastewater disinfection with ozone. *Journal of Water Pollution Control Federation*, 53, 1637, 1981.

55. Restaino, L., Frampton, E., Hemphill, J. and Palnikar, P. Efficacy of ozonated water against various foods related micro-organisms. *Applied and Environmental Microbiology*, 61, 3471, 1995.

56. Pascual, A., Llorca, I. and Canut, A. Use of ozone in food industries for reducing the environmental impact of cleaning and disinfection activities. *Trends in Food Science and Technology*, 18, S29, 2007.

57. Cho, M., Chung, H. and Yoon, J. Disinfection of water containing natural organic matter by using ozone-initiated radical reactions. *Applied and Environmental Microbiology*, 69, 2284, 2003.

58. Guzel-Seydim, Z.B., Grene, A.K. and Seydim, A.C. Use of ozone in food industry. *Lebensmittel Wissenschaft und Technologie*, 37, 453, 2004.

59. Dock, L.L. Development of thermal and non-thermal preservation methods for producing microbially safe apple cider. Thesis, Purdue University, West Lafayette, IN, 1995.

60. Williams, R.C., Sumner, S.S. and Golden, D.A. Inactivation of Escherichia coli O157:H7 and Salmonella in apple cider and orange juice treated with combinations of ozone, dimethyl dicarbonate, and hydrogen peroxide. *Journal of Food Science*, 70, M197, 2005.

61. FDA. *Guidance for Industry: Recommendations to Processors of Apple Juice or Cider on the Use of Ozone for Pathogen Reduction Purposes.* http://www.cfsan.fda.gov/~dms/juicgu13.html. October 2004.

62. Urano, H. and Fukuzaki, S. Facilitation of alumina surfaces fouled with heat-treated bovine serum albumin by ozone treatment. *Journal of Food Protection*, 64, 108, 2001.

63. Hampson, B.C. Use of ozone for winery and environmental sanitation. *Practical Winery and Vineyard*, 27, 2000.

64. Khadre, M.A. and Yousef, A.E. Decontamination of multi-laminated aseptic food packaging material and stainless steel by ozone. *Journal of Food Safety*, 21, 1, 2001.

65. Rice, R.G., Farguhar, J.W. and Bollyky, L.J. Review of the applications of ozone for increasing storage times of perishable foods. *Ozone Science and Engineering*, 4, 147, 1982.

66. Fournaud, J. and Lauret, R. Influence of ozone on the surface microbial flora of frozen boot and during thawing, *Industrial Alimentaria y Agriculture*, 89, 585, 1972.

67. Richardson, S.D. Drinking water disinfection by-products. In *The Encyclopedia of Environmental Analysis and Remediation*, Vol 3. R.A. Meyers (Ed.). John Wiley & Sons, New York, NY, 1994.

68. Zhang, L., Lu, Z., Yu, Z. and Gao, X. Preservation fresh-cut celery by treatment of ozonated water. *Food Control*, 16, 279, 2005.

69. Lewis, L., Zhuang, H., Payne, F.A. and Barth, M.M. Beta-carotene content and color assessment in ozone-treated broccoli florets during modified atmosphere packaging. In *1996 IFT Annual Meeting Book of Abstracts*, Institute of Food Technologists, Chicago, IL, 1996.

70. Luwe, M.W.F., Takahama, U. and Heber, U. Role of ascorbate in detoxifying ozone in the apoplast of spinach (Spinacia oleracea L.) leaves. *Plant Physiology*, 101, 969, 1993.

71. Ranieri, A., Urso, G.D., Nali, C., Lorenzini, G. and Soldatini, G. F. Ozone stimulates apoplastic antioxidant systems in pumpkin leaves. *Physiologia Plantarum*, 97, 381, 1996.

72. Perez, A.G., Sanz, C., Rios, J.J., Olias, R. and Olias, J.M. Effects of ozone treatment on postharvest strawberry quality. *Journal of Agricultural and Food Chemistry*, 47, 1652, 1999.

73. Nadas, A., Olmo, M. and Garcia, J.M. Growth of Botrytis cinerea and strawberry quality in ozone-enriched atmospheres. *Journal of Food Science*, 68, 1798, 2003.

74. Kim, J.G., Yousef, A.E. and Khadre, M.A. Microbiological aspects of ozone applications in food: a review. *Journal of Food Science*, 66, 2035, 2001.

75. Choi, L.H. and Nielsen, S.S. The effects of thermal and nonthermal processing methods on apple cider quality and consumer acceptability. *Journal of Food Quality*, 28, 13, 2005.

76. Artes-Hernandez, F., Aguayo, E., Artes, F. and Tomas-Barberan, F.A. Enriched ozone atmosphere enhances bioactive phenolics in seedless table grapes after prolonged shelf life. *Journal of the Science of Food and Agriculture*, 87, 824, 2007.

Section VI

Other Thermal Processing

23 Computer Modeling of Microwave Heating Processes for Food Preservation

Tatiana Koutchma
Agriculture and Agri-Food Canada

Vadim Yakovlev
Worcester Polytechnic Institute

CONTENTS

23.1 INTRODUCTION

Microwave (MW) heating of foods is a technology that, due to its unique features, can provide the food processor with powerful and efficient options in improving the performance characteristics of the preservation processes. Rapid and principally internal heating, possibility of selective heating of materials through differential absorption and self-limiting reactions present opportunities and benefits that are not available from conventional heating and make MW-based preservation technology an attractive alternative for a wide variety of food products. This includes shelf life extension of solid foods (such as prepacked products), sterilization of low acid semi-solid foods (such as mashed potatoes and other cooked vegetables), pasteurization of liquid foods (such as milk, citrus, orange and other fresh juices and beverages), etc. Higher product quality can be achieved due to reduced processing times as compared to conventional thermal processes. In addition, MW processing systems can be more energy efficient; the heating can be instantly turned on and off—this constitutes important preconditions for controllability of operations.

While domestic MW ovens have been in vast household use worldwide, it is less known that considerable investment has also been made over the last 30 years in the development of industrial MW processing systems for a range of applied processes. Examples include meat tempering, cooking of bacon, drying of cookies, pasteurization and sterilization of ready-to-eat meals and others. Typically, these systems were designed by MW companies in joint development arrangements with food companies so that the users get proprietary processes and the designers keep exclusive manufacturing licenses. As a result, most MW heating technologies that have been developed up to date are not commonly known/available.

On the other hand, absorption of energy of high frequency electromagnetic field by the materials with dielectric losses is known to be a complex multiphysical occurrence which, in addition to electromagnetic and thermal processes, may involve evaporation, mass transfer, and other phenomena. Due to exceptional complexity of interaction of the electromagnetic field with food matrix, simply placing a processed sample in a MW heating system and expecting it to be heated efficiently is rarely fruitful. Nonuniformity of heating and unpredictable energy coupling are two major issues which must be accounted. Numerous publications directly or indirectly addressing these challenges discuss the difficulties in making broad observations and generalizations about common trends in the MW heating technology in general and its application for food preservation (i.e., in pasteurization or sterilization processes) in particular. As reported by IFT and US FDA, in 2000 only two commercial systems worldwide that performed MW pasteurization and/or sterilization of food in accordance with their standards were recognized [1].

Historically, the progress in this area has been traditionally associated with laboratory and industrial experiments. At the same time, with the increased use of computers and computational technologies in modern engineering, there has been growing interest in modeling of MW heating. Moreover, the designers of MW applicators were interested in tools not only for analysis, but also for synthesis of processes and systems with desirable performance. There has been an expectation that, for development of microbiologically safe preservation processes based on MW heating, computer-aided design (CAD) may assist in achieving required technological objectives and save significant time and resources. However, modeling approaches which were used in the period of 1980s–1990s relied on dramatic idealizations, e.g., considering 1-D and 2-D approximations (and using partially analytical schemes) or assuming that the product in a MW heating system is exposed to a plane wave. In this context, the related modeling techniques included concepts and characteristics describing electromagnetic processes in free space, or near the boundary of two infinite media, or in the layered media, and they were applied to the processes inside closed MW heating cavities.

The significant progress made in computational electromagnetics in the 1990s has resulted, in addition to a number of keystone publications in technical literature, in the emergence of several commercial software packages that demonstrated remarkable increase in efficiency and quality of modeling, in accelerating the design and improving characteristics of wireless, telecommunication and radar systems and their elements. A common access to the powerful electromagnetic simulators featuring universal capabilities has effectively started a new era of wider modeling activities in MW power engineering. Particularly, they were aiming at practical developments in MW processing of food where, prior to that time, modeling was normally considered, despite a number of well-known milestone papers (e.g., [2–6]), rather intangible academic exercises having not much in common with "real" technologies. The examples of early applications of electromagnetic modeling tools of new generation in MW power engineering were reported in 1998–2002 [7–11].

The extensive literature on this subject that have accumulated since, makes it clear that modeling of MW heating has become a research activity that is undertaken in intricate projects and addresses complex problems (including those that are genuinely multiphysical in nature) and systems (see, e.g., [12–40]). It has been made evident that some modern modeling tools do allow for getting valuable and reliable data about the characteristics of a specific food processing system prior to construction of its physical prototype. For many researchers and engineers, modeling has become a routine operation in studying processes, developing applications and designing microwavable food and MW applicators.

In this chapter, we firstly, briefly outline the effects of microwaves (MWs) on biological characteristics of food products (including microbial destruction and quality degradation) that are critical in MW preservation of foods and, secondly, give a general review of the current state-of-the-art in modeling of MW heating of food and other similar technologies.

23.2 MODELING MICROBIAL PARAMETERS OF A MICROWAVE (MW) PRESERVATION PROCESS

23.2.1 DEVELOPMENT OF MICROWAVE (MW)-BASED PRESERVATION PROCESSES

When development of an MW-based food preservation process relies on mathematical/computer modeling, it is crucial to have reliable information about process and food matrix parameters which are needed as input data to the model. In this case, simulated temperature fields would be invaluable information allowing for computational evaluation of microbial characteristics of the preservation process.

The performance objective of the process (e.g., shelf life extension, pasteurization, or sterilization) is to achieve targeted microbial reduction. This means that MW preservation process cannot be developed without data on kinetic parameters of microbial destruction during MW heating. Two most important characteristics here are decimal reduction time, or D-value, and thermal resistance of microorganisms, or Z-value. In addition to the required performance objective, a higher product quality and superior retention of nutrients can be the targets of the process design. The latter goals can be achieved with shorter come-up time to the process temperature and shorter duration of the MW process in comparison with conventional heating. Similar kinetic parameters of degradation of food quality parameters, enzymes and vitamins are also needed for establishing an efficient and safe MW preservation process. As for the basic MW heating characteristics (i.e., energy coupling, the electric field, fields of dissipated power and temperature), they are primarily conditioned by dielectric and thermal properties of the processed food product and their variations (as well as variations in its composition, geometry, weight and shape) in the course of thermal processing.

Computer modeling can provide important information about various aspects of MW heating processes developed for food preservation and save (possibly, significant) time and resources at the different phases of the process development. Modeling can begin with setting the performance goals which may be associated with the time characteristic of the product temperature $\mathbf{T}(t)$ specifying the heating process at all points of the processed material. The goals may refer to particular values of temperatures which are expected to be reached in the product in a particular time; for a particular process/product, the choice of these values should be motivated experimentally or computationally – through the preliminary/simplified computations of D- and Z-values.

Consequently, computer simulation could be employed for determining characteristics of interaction of MWs with the material processed in an MW applicator of a chosen design, i.e., the level of reflections from the applicator back to the energy source, distribution of the electric field in the cavity and patterns of dissipated power and temperature within the processed material as functions of heating time. The resulting temperature field $\mathbf{T}(t)$ can then be used for computing microbial reduction and quality degradation which should be provided in the considered scenario. If these characteristics seem to be satisfactory, the modeling phase can be considered completed, and corresponding physical prototype could be produced. Otherwise, an alternative applicator configuration/design could be introduced and modeled, and this process can go on. Simulation of MW heating characteristics in an applicator with a given processed material and subsequent computation of microbial characteristics and quality degradation could be put in a loop controlled by an appropriate optimization procedure.

23.2.2 MICROWAVE (MW) MICROBIAL DESTRUCTION

In order to evaluate microbial lethality of an MW-based thermal process, knowledge of inactivation kinetics parameters of target pathogenic or spoilage microorganisms is required. The kinetics parameters of bacterial destruction in food products during MW heating are documented by the FDA [1]. Bacteria shown to be inactivated by MWs include *Bacillus cereus, Clostridium perfringens, Clostridium sporogenes, Campylobacter jejuni, Escherichia coli, Listeria monocytogenes, Staphylococcus aureus, Salmonella, Enterococcus,* and *S. cerevisiae.* There appear to be no obvious MW-resistant foodborne pathogens. As with conventional heating, bacteria are more resistant to thermal inactivation by MW heating than yeasts and molds whereas bacterial spores are more resistant than vegetative cells.

It appears that insufficient knowledge on the kinetics of microwave (MW) bacterial destruction has been one of the reasons preventing MW preservation technology from wider applications. Difficulties in direct temperature control of heating in domestic MW ovens make it difficult to obtain kinetics data. Moreover, most related findings reported in the literature does not seem to accurately represent the kinetics of microbial destruction solely due to MW heating since they also include the effects of the holding period following MW heating. Hence, a fair comparison cannot be made with the destruction kinetics data obtained due to conventional heating.

The destruction of microorganisms and inactivation of enzymes during conventional thermal processing can be generally modeled as an *n*th order chemical reaction which can be represented by the equation

$$\frac{dC}{dt} = -kC^n \tag{23.1}$$

where dC/dt is the rate of change of concentration C, k is the reaction rate constant, n is the order of reaction. For many foods, the following first-order kinetics model adequately describes the microbial destruction:

$$\ln\frac{C}{C_0} = -kt \tag{23.2}$$

where C_0 is the initial microbial population. The traditional approach to describing changes in microbial populations as a function of time uses the survivor curve equation

$$\log\frac{C}{C_0} = -\frac{t}{D} \tag{23.3}$$

where D is the decimal reduction time, or time required for a 1-log cycle reduction in the microbial population. Comparing Equation 23.2 and 23.3, the relationship between the decimal reduction time and the first-order reaction rate constant becomes

$$k = \frac{2.303}{D} \tag{23.4}$$

The influence of temperature on the microbial population inactivation rates is usually expressed in terms of the thermal resistance constant (Z-value) using the following model

$$\log\frac{D}{D_{T_R}} = -(T - T_R)Z \tag{23.5}$$

The thermal resistance $Z(T)$ is the temperature increase needed to accomplish a 1-log cycle reduction in the D-value. The reference decimal reduction time (D_{T_R}) is the magnitude at a reference temperature (T_R) within the range of temperatures used to generate experimental data.

The D_{T_R}-values and Z-values are well documented and fairly extensively reported in the literature and recommended for the use in the development of MW processes [41]. The basic model for process development is built on the survivor curve (Equation 23.3):

$$F = D\log\frac{C}{C_0} \tag{23.6}$$

where F is the total time required to reduce the microbial population by a specified magnitude needed to ensure product safety, under the conditions defined by the D-value. When thermal resistance of a microorganism is known, it is possible to calculate the equivalent time F_R at the reference temperature necessary for thermal treatment by integrating the time-temperature history in the cold spot of the product $T_c(t)$ using the formula:

$$F_R = \int_0^t 10^{\frac{T_c(t)-T_R}{Z}}\, dt \tag{23.7}$$

This approach has been traditionally followed in conventional thermal processes. Similar concept can be applied for determining kinetics parameters during MW heating. Since nonisothermal heating conditions are involved in this case, the resulting D-values can be computed with the following expression:

$$D = \frac{F_R}{\log\dfrac{C_0}{C}} \tag{23.8}$$

where F_R can be obtained using either computationally or experimentally determined time-temperature profiles. While traditionally evaluation of lethality effects of MW heating is carried out without considering thermal history of the product [42,43], the technique employing equation (Equation 23.8) [44–46] appears to provide more adequate description of kinetics in the course of MW heating.

Table 23.1 shows D-values of E. coli K12 estimated from the experimental survivor data at three reference temperatures using computed effective times for five heating conditions listed. The lowest D-values are observed in continuous-flow MW heating scenarios with and without holding section. Furthermore, the values in the continuous systems are considerably lower than those in the batch heating systems.

TABLE 23.1
Experimental D-values of E. Coli K-12 During MW and Conventional Heating

Temperature, °C	D-values, s		
	55	60	65
Thermal batch method	173.0	18	1.99
Continuous-flow, hot water	44.70	26.80	2.00
Continuous-flow, steam	72.71	15.61	2.98
Continuous-flow, MW	12.98	6.31	0.78
Continuous-flow, MW + holding	19.89	8.33	1.98

Source: Le Bail, A., Koutchma, T., and Ramaswamy, H.S., *J. Food Process Eng.*, 23, 1, 2000.

Riva et al. [45] reported different Z-values for *Enterobacter cloacae* and *Streptococcus faecalis* bacteria for batch conventional heating (4.9°C and 5.8°C) and MW heating (3.8°C and 5.2°C). These differences are due to different heating kinetics and nonuniform local temperature distributions during MW heating. Heddleson et al. [47,48] examine chemical composition of five types of food to determine important factors in achieving uniform temperature when heating these samples in a 700-W MW oven. The chemical composition of foods appears to be critical for temperature distribution and destruction of MW-heated *Salmonella*, *Listeria monocytogenes*, and *Staphylococcus aureus*. Thermal resistance is strongly influenced by food components such as lipids, salts and proteins. It is also shown that heating to the end point temperatures is a better method of achieving consistent destruction.

Selected data on MW-based inactivation effects are collected in Table 23.2. It is worth noting that in these studies, MW pasteurization and/or sterilization were performed in systems which were

TABLE 23.2
Inactivation Effects of MW Heating on Selected Microorganisms

Microorganism	MW Treatment Conditions	Heated Product	Observed Inactivation or Kinetic Data	Source
Escherichia coli	2.45 GHz, 55°C, 30 s	Frozen shrimp	3-logs kill	Odani et al. [49]
Escherichia coli K12	2.45 GHz, 600 W, 50 and 60°C	0.9% NaCl Solution	3-logs kill	Woo et al. [50]
Staphilococcus aureus	55°C, 30 s	Frozen shrimp	3-logs kill	Odani et al. [49]
Staphilococcus aureus	2.45 GHz, 800 W, 61.4°C, 110 s	Steel disks	6-logs kill	Yeo et al. [51]
Bacillus cereus spores	100°C, 90 s	Frozen shrimp	3-logs kill	Odani et al. [49]
Bacillus subtilis KM 107 cells	2.45 GHz, 600 W, 60–80°C	0.9% NaCl Solution	2-logs kill	Woo et al. [50]
Aspergillus nidulans, spores	2.45 GHz, 700 W, 85°C, 30 min	Glass beads	3-logs kill	Diaz-Cinco and Martinelli [52]
Listeria monocytogenes	2.45 GHz, 600 W, isothermal 75, 80 and 85°C	Beef frankfurters	8-logs at 75°C after 13.5 min; 8-logs at 80°C after 11–13 min 8-logs at 85°C after 12 min	Huang [53]
Listeria monocytogenes Scott A	2.45 GHz, 750 W	Non fat milk	$D_{60} = 228$ s $D_{82.2} = 0.57$ s 4–5 logs at > 71.1 for 15 s	Galuska et al. [54]
Saccharomyces cerevisae	2.45 GHz, 700 W, continuous flow, 52.5; 55 and 57.5°C	Apple juice	$D_{52.5} = 4.8$ s $D_{55} = 2.1$ s $D_{57.5} = 1.1$ s $z = 7$°C	Tajchakavit et al. [55]
Lactobacillus plantarum	700 W, continuous flow, 52.5; 55 and 57.5°C	Apple juice	$D_{52.5} = 14$ s $D_{55} = 3.8$ s $D_{57.5} = 0.79$ s $z = 4.5$°C	Tajchakavit et al. [55]
Lactobacillus plantarum	650 W, 50°C, 30 min	Culture broth	4-logs reduction	Shin and Pyun [56]

not designed/optimized specifically for this purpose. In this regard, the study by Huang [53] is of special interest because it presented a computer-controlled MW heating system. The infrared sensors monitoring surface temperature of the heated product are used to control input power delivered to the MW oven. Results show that a pulsing regime of MWs is able to maintain surface temperature of frankfurters near the respective set point. The pasteurization process is able to achieve an 8-log reduction of *L. monocytogenes* in inoculated beef frankfurters using a 600-W MW oven within 11–13 min.

23.2.3 MICROWAVE (MW) QUALITY DEGRADATION

The potential of MWs to offer significant reductions in heating time and better food quality retention makes them attractive for the food industry. Reviewing key findings in the field for nearly 30 years, Decareau [57] concluded that MW high temperature short time (HTST) sterilization is no less successful than conventional retorting of the pouches.

Similarly to the conventional thermal scenarios, degradation kinetics of food constituents (such as quality parameters, attributes, enzymes and vitamins) are required for establishing a MW preservation process. A critical review covering the effects of MWs on nutritional components of foods is given by Gross and Fung [58]—the considered components include moisture, animal proteins, non-animal proteins, carbohydrates, lipids, minerals and vitamins of fat and water soluble varieties. However, similar to studies of microbiological effects, most investigations on food quality and nutrients did not attempt to compare equivalent time-temperature treatments of foods during MW heating. A summary of reported MW heating effects on essential vitamins in foods is presented in Table 23.3.

Datta and Hu [71] evaluated the influence of MW heating on quality from the engineering viewpoint and on the basis of time-temperature history. Their experiments show that while heating by MWs can generally be more uniform than conventional process, considerable nonuniformity can result from spatially varying rates of heat generation. This seems to be consistent with the results of other authors' studies suggesting that the ability of MWs to provide a preserved product of better (compared to conventional heating) quality is not universally true. When MW heating is less thermally degrading than equivalent conventional process, it is attributed to faster and more uniform heating by MWs. However, as mentioned above, one can aim to reach uniformity of MW heating only in appropriately designed applicators.

Ohlsson and Bengtsson [72,73] evaluated quality effects of traditional thermal processes of canning and retorting foil pouches as well as of MW sterilization of plastic pouches by integrating time and temperature profiles in the pouches and calculating cooking value or C-value. The results shown in Table 23.4 lead to the conclusion that high degree of uniformity of heating is needed to achieve superior quality in MW sterilization.

The effects of MW processing on water soluble vitamins (such as ascorbic acid, niacin (stable vitamin B), thiamin and riboflavin) using a first order reaction model are described by Okmen and Bayindirli [60]. The reported kinetic parameters such as the reaction rate constant k at 90°C, activation energy E_a, thermal resistance and the ratio of rate constant at temperatures which differ by 10°C, Q_{10} are given in Table 23.5. It is seen that the most sensitive vitamin is ascorbic acid. The authors conclude that MW heating causes considerably less degradation in water soluble vitamins in comparison with conventional thermal processes.

23.2.4 MICROWAVE (MW) HEATING CHARACTERISTICS

Evaluation of heating characteristics of a MW heating process is important for determining the most efficient conditions for treatment. Heating characteristics of MW-induced process include time-temperature curves during transient and steady state periods, heating rates, absorbed power, coupling

TABLE 23.3
Summary of Reported Effects of MW Treatments on Essential Vitamins in Foods

Vitamin	Product	MW Treatment	Reported Effect	Source
Vitamin A	Corn-soy milk (CSM)	915 MHz, 60 kW	No effects	Bookwalter et al. [59]
Vitamin B1	Model	2.45 GHz, 700 W, 60–90°C	MW inactivation parameters	Okmen and Bayindirli [60]
	Model	2.45 GHz, 110, 115 and 120°C	$z = 26.6$°C	Welt and Tong [61]
	Human milk	2.45 GHz at 30% and 100% power, 850 W,	No effects at temperature up to 77°C	Ovesen et al. [62]
	Corn-soy milk (CSM)	915 MHz, 60 kW	No effects	Bookwalter et al. [59]
Vitamin B2	Model	2.45 GHz, 700 W, 60–90°C	MW inactivation parameters	Okmen and Bayindirli [60]
	Model	N/A tubular pasteurization reactor	Microbial reduction 0–0.01 with 60–75% destruction of initial riboflavin	Aktas and Ozligen [63]
Vitamin B12	Raw beef, pork, milk	2.45 GHz, 500 W, 100°C, 20 min	Loss of 30–40%. Conversion of B12 to inactive B12 degradation products	Watanabe et al. [64]
Vitamin C	Orange juice	100–125°C	$E_a = 64.8$ kJ/mK	Vikram et al. [65]
	Apple juice	2.45 GHz, 1.3 kW, 83°C, 50 s	20–28% destruction	Koutchma and Schmalts [66]
	Model	2.45 GHz, 700 W, 60–90°C	MW inactivation parameters; the most sensitive to MW	Okmen and Bayindirli [60]
	Broccoli, green beans, asparagus	700 W, 50–100% power, 1–4min	Concentration increased due to moisture losses, when adjusted moisture losses it decreased	Brewer and Begum [67]
Vitamin E	Egg yolk	2.45 GHz, 1,500 W	Reduced by 50% traditional and MW cooking	Murcia et al. [68]
	Human milk	2.45 GHz at 30% and 100% power, 850 W	No effects at temperature up to 77°C	Ovesen et al. [69]
	Soya bean oil	2.45 GHz, 500 W, 4–20 min	Remained > 80% in soaked beans after 20 min of roasting	Yoshida and Takagi [70]

efficiency (defined as a ratio of absorbed energy to the nominal incident power). Knowledge of spatial temperature distribution is critical for locating the least heated areas in the product and further evaluation of process lethality. In the case of continuous flow process, temperature rise versus flow rate/residence time and input power needs to be known. In addition, a flow regime (laminar or turbulent) and the residence time distribution (RTD) can be characterized by calculating Reynolds and/or Dean numbers.

TABLE 23.4
Effect of Packaging and Process Temperature on C-value of a 225-g Pack of Solid Food

Package	Dimensions, mm	Process	Cook time, min	C-value
Can	⌀73 × 49	Retort 120°C	45	180
Foil pouch	120 × 80 × 20	Retort 125°C	13	65
Plastic pouch	120 × 80 × 18	MW 128°C	3	28

Source: Ohlsson, T. and Bengtsson, N., In: *Advances in Food and Nutrition Research*, Academic Press, New York, NY, 66, 2001.

TABLE 23.5
Kinetics Parameters of Vitamin Microwave Degradation in a 700-W Microwave Oven

Vitamins	E_a (J/mol)	z (°C)	Q_{10}	k at 90°C (min^{-1})
Vitamin C	4.58×10^4	65.8	1.41	2.8×10^{-3}
Thiamin	5.47×10^3	83.3	1.32	2.0×10^{-4}
Riboflavin	4.59×10^3	99.7	1.26	1.32×10^{-3}
Niacin	3.65×10^3	125.0	0.86	6.33×10^{-4}

Source: Okmen, Z. and Bayindirli, A.L., *Int. J. Food Prop.*, 2, 255, 1999.

The report of Datta [74] is an example of a study of MW heating safety through a simplified computation of spatial temperature profiles in two cylindrical containers of 11.2 cm (500 cc load) and 17.0 cm (2,000 cc loads) diameter filled with solid and liquid (tap water) materials placed in a MW oven. The computed axial temperature profiles show larger variations when compared to the radial profiles. Cold spots are found on the bottom of the containers that is consistent with the measurement. The large variation in temperature is attributed to oven design in which MW energy could not penetrate efficiently through the bottom of the containers.

Le Bail et al. [75] evaluate three simple numerical models to predict the time-temperature profiles under continuous flow conditions to compare efficiency of MW and conventional steam heating systems. A suspension of *E. coli* cells is pumped through two helical glass coils (125 ml volume). For MW heating, the coils are placed in two MW ovens (700 W, 2.45 GHz) and for steam heating—in a steam cabinet. Calculations are based on three different assumptions for the flow: perfectly mixed flow (PMF), plug or piston flow (PF) and laminar flow (LF). The different profiles are assumed to be based on the experimentally determined Reynolds numbers (900–1400 normally corresponding to a laminar regime) and coil geometry which causes the presence of the secondary flows and thus of additional mixing of the flow characterized by Dean number (200–300). These models are based on dividing the tube into elements of equal length along the axis of the coil. Models 1 (PMF) and 2 (PF) were considered to be equivalent to "an ideal" flow with uniform temperature and velocity distribution. In Model 2, as fluid was considered a moving plug, a radial temperature spread based on heat transfer across the tube wall and across the concentric layers of flow is accommodated. Model 3 (LF) is based on conventional LF distribution in tubes. The PMF model gives the best predictions of time-temperature history during MW and steam heating (due to the Dean effect caused by the helical coils).

Microbial lethality at the exit of the MW heating system is found to increase during the initial transient period and reaching a characteristic maximum a little after the target exit temperature is achieved (1–2 min after turning the MW ovens on). The survival ratio, however, reaches its minimum value at about 5 min when full steady state heating conditions is established. All subsequent microbial evaluations are made only during the steady state period.

Typical experimental survival curves of *E. coli* K-12 obtained in continuous-flow MW and steam heating systems for steady state exit temperatures in the range 50 and 70°C are shown in Figure 23.1. This figure is the conventional approach to presenting data for comparative purposes. It is obvious that both survival curves had similar characteristics. The figure shows small differences in *E. coli* survivors between MW and steam heating.

The mean residence time for different heating conditions (temperature, heating mode) are also shown in Figure 23.2 for comparison purposes. It can be seen that for the same exit temperature, the residence times in MW and steam heating conditions are different. Obviously, the residence time in each system steadily increases as the exit temperature is elevated, and this gives a compounded destruction effect due to higher temperature and longer residence time combination. The steep drop in survivor ratios is the result of such combination effects. It can also be seen that at each exit temperature, the residence time under steam heating is relatively higher than the residence time under MW heating conditions.

Figure 23.2 shows the results of computation of lethality values (based on kinetics data obtained from the conventional batch heating approach) immediately after the heating coils using the PMF model. This model predicts the exit temperature with better agreement with respect to the experimental ones at the ports for both heating modes. Comparison with experimental values of lethality and microbial survival ratio shows that the PMF model provides good predictions whereas the other two models do not yield good matching. However, it fails to accurately predict the lethality and microbial survival under MW heating alone by demonstrating large

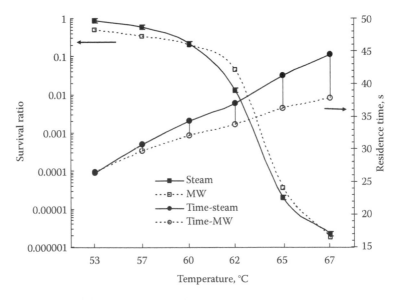

FIGURE 23.1 Survival curves of *E. coli* K-12 and temperature profiles obtained under continuous-flow MW and steam systems in steady state heating. (From Le Bail, A., Koutchma, T., and Ramaswamy, H.S., *J. Food Process Eng.*, 23, 1, 2000.)

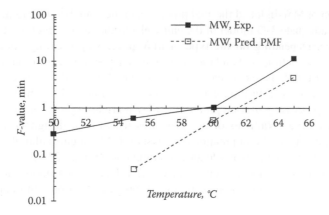

FIGURE 23.2   Numerically predicted lethality (F_o) of *E. coli* in a MW heating system. (From Le Bail, A., Koutchma, T., and Ramaswamy, H.S., *J. Food Process Eng.*, 23, 1, 2000.)

underestimations with only slightly better prediction at the exit Port 2 after accommodating hold time contributions.

As a general conclusion, one can therefore notice that modeling results generally confirm the suitability of the conventional kinetics numerical models for prediction of microbial destruction under steam heating conditions while they tend to underestimate the destruction under MW heating.

Finally, knowledge of dielectric characteristics of food is the most important material parameter in modeling of specific MW heating applications. The complex permittivity of the heated material is described as $\varepsilon = \varepsilon' - j\varepsilon''$, where ε' is dielectric constant and ε'' is the loss factor. The real part of complex permittivity is a measure of an impact made by a dielectric of the electric field. The imaginary part measures the dissipation energy of the electric field into heat. The dielectric properties of food products depend mainly on frequency, temperature and water content. Chemical composition of food with respect to water, salts, ions and sugar is another major factor influencing dielectric properties (and therefore MW heating characteristics). The presence of proteins, starch and gums can further modify these properties. In addition, physical and chemical changes such as gelatinization of starch and denaturation of protein at elevated temperatures can significantly change dielectric properties.

23.3   MODELING MICROWAVE (MW) HEATING SYSTEMS

23.3.1   What's and How's

In MW heating applications, efficient volumetric conversion of MW power into heat within dielectric loads requires much higher levels of electromagnetic energy than in telecommunication applications, so the related processes are to be performed in closed cavities with standing or travelling waves. A processed material occupying a notable part of cavity's volume is usually characterized by high values of dielectric constant and the loss factor. For these reasons, one cannot expect a reasonably accurate computation of electric field distributions, dissipated power and temperature inside the loads if the analysis is done on the basis of approximations involving a free-space field (e.g., a plane wave) or 1- or 2-D idealizations. The problem is essentially about the field in partially filled cavities, and the space to be considered is dictated by the cavity's boundaries enclosing the electromagnetic field, so numerical algorithms capable of performing full-wave 3-D analysis in arbitrarily shaped electrically large structures are the only option here.

Characterization of MW-induced thermal processes in the load therefore requires, as a precondition, a sufficiently adequate 3-D model of the entire electromagnetic structure in which the heating process takes place. Depending on the type of MW processing (heating, thawing, sterilization/pasteurization, quality control, etc.) and the processed material, this electromagnetic model is to be coupled with other algorithms taking care of the associated physical processes in the system, e.g., heat and/or mass transfer within the load, evaporation, etc. Figure 23.3 shows three MW heating devices for thermal processing of food products (a MW household/professional oven, a batch cavity and a conveyor system) as examples of the structures which their designers may wish to model. Another popular application of computational analysis is CAD of particular elements of MW heating systems, e.g., devices such as water loads and circulators (Figure 23.4).

Numerical techniques capable of getting practical valuable information deal with Maxwell's equations and transform the continuous differential/integral equations into approximate discrete

(a)   (b)

(c)

FIGURE 23.3 Typical MW heating systems. (a) Compact GM 1600 EK-a commercial 2.45 GHz microwave oven by Gigatherm AG. (b) MIP 10, a batch 915 MHz MW tempering oven by Ferrite Company, Inc. (c) MIP 9, a booster 915 MHz MW system by Ferrite Company, Inc. (Courtesy of (a) Gigatherm AG, Grub AR, Switzerland and (b,c) Ferrite Company, Inc., Nashua, NH.)

(a)   (b)   (c)

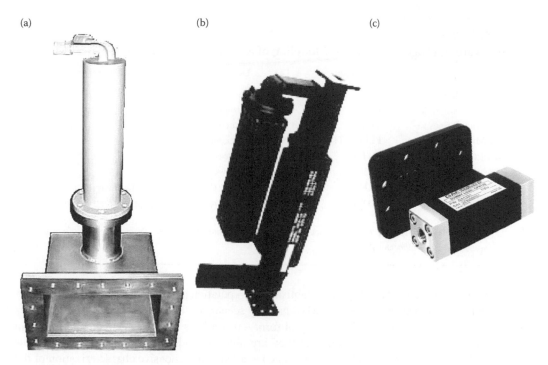

FIGURE 23.4 Typical components of MW heating systems. (a) A 915 MHz high power water load by Ferrite Company, Inc. (b) A 2.45 GHz four-port circulator by Ferrite Company, Inc. (c) A 5.8 GHz water load by Gerling Applied Engineering, Inc. (Courtesy of (a,b) Ferrite Company, Inc., Nashua, NH and (c) Gerling Applied Engineering, Inc., Modesto, CA).

formulations accounting for the system's boundary conditions and all media interfaces. The earlier reviews [11,76] have suggested three selection criteria determining suitability of an electromagnetic simulator for typical MW heating scenarios. Computational output of a full-wave 3-D modeling tool should include at least the following data:

a. Reflections from the system and/or from its particular elements
b. Distribution of the electric field in the system
c. Distribution of dissipated power inside a processed material

These criteria are motivated by critical knowledge which the designer may wish to have in order to succeed in the design of a MW system and a heating process in this system. Information about the level of reflection allows one to calculate energy coupling and thus evaluate system efficiency; awareness of the field structure helps identify favorable (in terms of relative uniformity of the field) and problematic (in terms of singularities in the field behavior which may allow for arching) areas inside the cavity; finally, patterns of the electromagnetic power dissipated in the load reveal heating profiles of the MW-induced heating.

Using these criteria, the reviews [11,76] identified, at the time of publication (2000–2001), 17 modeling tools available on the market and potentially applicable to modeling of MW heating—full-wave 3-D simulators based on fairly diverse numerical techniques. Presently, after several years of exploring the existing computational opportunities, the list of software packages actually associated with this field is much shorter and contains the packages shown in Table 23.6 and based on the transmission line matrix (TLM) method, finite integration technique (FIT), finite element method (FEM) and finite-difference time-domain (FDTD) technique. A conformal FDTD technique (including its version called finite integration technique (FIT) given by Weiland to a matrix

TABLE 23.6
Software Packages Suitable for Modeling of Microwave Heating Systems

Modeling Package	Technique	Developer/vendor (web site)
ANSYS® Multiphysics	FEM	ANSYS, Inc. (www.ansys.com)
COMSOL® Multiphysics	FEM	COMSOL Group (www.comsol.com)
CST MicroStripes	TLM	CST (www.cst.com)
CST Microwave Studio	FIT (FDTD)	CST (www.cst.com)
FIDELITY®	FDTD	Zeland Software (www.zeland.com)
HFSS®	FEM	Ansoft, LLC (www.ansys.com)
MEFiSTo®	TLM	Faustus Scientific Corporation (www.faustcorp.com)
QuickWave-3D*	FDTD	QWED Sp. z.o.o. (www.qwed.com.pl)

* Also distributed within the *Concerto* package by Cobham Technical Services – Vector Fields Software (www.cobham.com/technicalservices/)

representation of the method [77,78]) has established its reputation as the most resourceful, versatile and powerful instrument in analysis of MW heating scenarios. The conformal FDTD schemes start from governing equations in local integral form – this allows including into a cell more than one medium. It has been demonstrated that they are able to handle larger problems, needs less memory and operates faster than FEM algorithms. Detailed comprehensive characterization of the conformal FDTD method in its applications to the problems of MW power engineering (including systematic comparison with FEM) is given by Celuch and Gwarek [38].

According to capabilities of the numerical techniques, some pieces of modeling software have confirmed their potential applicability to MW heating problems whereas others have become principal modeling tools of the field. *QuickWave-3D (QW-3D)* (conformal FDTD) and *CST microwave studio (MWS)* (FIT) have demonstrated the best functionality in handling large and complex scenarios with capabilities of regular PCs (e.g., with 32-bit Windows XP and up to 2 GB RAM). In addition, QW-3D [79] has proved to be particularly useful for MW heating systems designers due to a number of its specific extensions and functions that are beneficial for the field.

The next level of complexity in modeling of MW heating systems, a simulation of coupled phenomena, is represented by computational technologies which are not as well developed as the tools for electromagnetic modeling alone. *COMSOL® multiphysics* and *ANSYS® Multiphysics* claim availability of functions for straightforward combining their electromagnetic solvers with heat and/or mass transfer ones in the same computational environment. In this approach, all differential equations and boundary condi tions are discretized by the same finite-element mesh and the problem is reduced to solving a "united" matrix equation. However, the current literature lacks report of successful use of this technique in solving applied problems, so its actual functionality in solving two-way coupled problems with these tools remains unclear. Moreover, substantial requirements for computer resources (memory and speed) conditioned by the underlying numerical technique basically limit application of these packages to very simple (and thus likely impractical) MW heating scenarios [9,37,80,81]. Nevertheless, a possibility for convenient generation of coupled models under the same interface available in *COMSOL, multiphysics* and *ANSYS,* may be a valuable option for relatively simple scenarios.

It appears that the most advanced technique for solving electromagnetic-thermal coupled problem is offered in *QW-3D* package version 5.0 (released in May 2005) and newer. The core FDTD simulator is able to solve the thermal problem with *QuickWave-3D Basic Heating Module (QW-BHM)*. The electromagnetic and thermal solvers operate in the frameworks of an iterative regime (Figure 23.5) relying on the fundamental physical principle of MW heating processes in

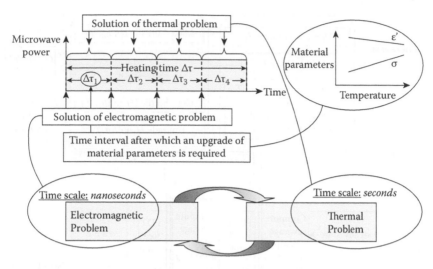

FIGURE 23.5 Concept of iterative solution of the coupled electromagnetic-thermal problem.

which electromagnetic and thermal processes occur on essentially different time scales. The solution starts from the electromagnetic portion; real time associated with heating is not involved in this consideration. Electromagnetic solution is then used to solve the thermal part of the problem, i.e., to obtain the temperature field induced in the load after heating it for time Δ_τ^0. At the next step, temperature (T)-dependent electromagnetic and thermal material parameters, i.e., $\varepsilon'(\tau)$, electric conductivity $\sigma(\tau)$, thermal conductivity $k(\tau)$, density $\rho(\tau)$ and specific heat $c(\tau)$ are upgraded in each FDTD cell in accordance with new values of temperature, and the procedure returns to the electromagnetic problem. The same FDTD mesh used for both numerical solutions simplifies the exchange of data between the solvers and helps avoid numerical diffusion in the interfacing procedures. Kopyt and Celuch [36] have shown that the enthalpy (H)-based mode of operation of the thermal solver allows for full representing history of the heating process (even in the phase change areas, e.g., around thawing or freezing points where temperature may stop changing) and thus ensures high accuracy of computation of temperature and adequacy of numerical exchange between the two simulators.

The structure of the FDTD solution of the electromagnetic-thermal coupled problem implemented in the *QW-3D* package is shown in greater detail in Figure 23.6. First computation of temperature through *QW-BHM* is preceded by electromagnetic simulations involving FDTD steps necessary to reach steady state and a special preparatory phase of computation of time-average dissipated power P_d released throughout the load during the heating time $\Delta\tau_1$. FDTD computation of temperature is performed by the special routine called *QW Heat Flow Module (QW-HFM)* which communicates with *QW-BHM* by getting information about the initial state of the heated load (namely, the enthalpy field and fields of dissipated power and temperature) in an *.hfe file and by generating an *.hfi file containing renewed data on these parameters. After upgrading material parameters in accordance with the new temperature field, the electromagnetic FDTD computations are resumed with the previously obtained steady-state field, but using new data on ε', σ, k, etc.—this allows for reaching new steady state (at the point N_3) much quicker than if the process starts from zero initial electromagnetic conditions.

The *QW-3D* package is also capable of coupling its FDTD electromagnetic solver with *Fluent* [82], the advanced computational fluid dynamics (CFD) software capable of solving complex problems in heat transfer, phase change and radiation; in the context of the FDTD solution scheme in Figure 23.6, *QW-HFM* takes up the role of an "intelligent" interface with *Fluent* solvers.

QW-3D functionalities in modeling MW heating problems as electromagnetic-thermal coupled scenarios with *QW-HFM* kernel have recently found the use in a number of applied projects

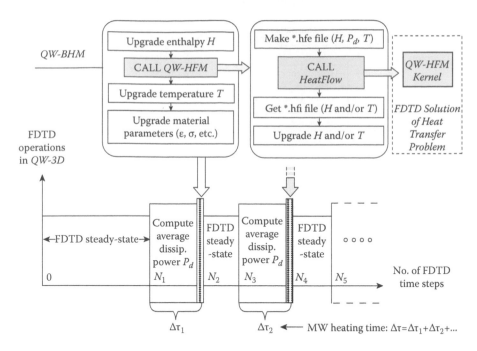

FIGURE 23.6 Flow-chart of the FDTD-based iterative technique of solving electromagnetic-thermal coupled problem in *QW-3D* package.

[20,83–90] including those in which modeling results received excellent experimental verification [13,14,36]. While the known use of the option combining *QW-3D*'s electro magnetic solver with *Fluent* routines is still very limited [36,86], this opportunity appears to be particularly attractive as it opens ways for developing more comprehensive multiphysics models of MW heating involving such phenomena as free or forced convection, radiation, evaporation and other CFD effects.

Known alternative techniques for modeling multiphysics phenomena include the algorithm combining "classical" (i.e., nonconformal) FDTD solvers for electromagnetic and thermal problems [24,25] and the algorithms for solving electromagnetic, convective heat transfer and continuous (non-Newtonian) fluid flow problems also based on classical FDTD technique [30,31]. Two approaches joining and interfacing different numerical techniques and solving one-way coupled electromagnetic-thermal problems are also worth mentioning. In [35], MW heating of food products is modeled by sending data of electromagnetic analysis generated by a "classical" FDTD scheme with tensor product meshes [91] to an algorithm [92] employing the FIT technique for solving heat transfer problem. The conformal FDTD electromagnetic solver from the *QW-3D* package is interfaced with the *COMSOL*'s FEM heat transfer module in [93].

The reviewed computational opportunities in modeling MW heating processes show that the key activities in this field currently lie around exploitation of available solvers in certain combinations (or in more general computational schemes) and further improvement of numerical techniques toward more efficient modeling of particular complex scenarios.

23.3.2 STRATEGY OF FINITE-DIFFERENCE TIME-DOMAIN (FDTD)-BASED COMPUTER-AIDED DESIGN (CAD)

FDTD-based techniques appear to be particularly convenient for modeling MW heating scenarios because they could be naturally involved in CAD of practical systems. All their major characteristics can be addressed in the framework of the following three-stage computational scheme.

1. Modeling starts with pulse excitation of the input ports positioned in the feeds supplying MW energy. Due to the use of Fourier transform, pulse excitation allows for getting parameters of a modeled device in a frequency band after only one simulation. The first stage consists of computation of S-parameters, more specifically, of the magnitudes of the reflection coefficients $|S_{mm}|$ determined in the mth port in the frequency range around the operating frequency f_0. When modeling a system with one source of energy, the problem is reduced to finding the characteristic $|S_{11}|(f)$. Since MW heating takes place in closed cavities with no leaking/radiating electromagnetic field, the energy coupling C, i.e., the part of energy absorbed by the processed material, depends on $|S_{11}|$ and can be found using the formula (Equation 23.9) plotted in Figure 23.7:

$$C \sim (1-|S_{11}|^2). \tag{23.9}$$

Being one of the most important characteristics of MW heating systems, parameter C can be interpreted as a measure of their efficiency. An interest in the values of the reflection coefficient in the range (f_1, f_2) surrounding f_0 is explained by peculiar characteristics of magnetrons traditionally used in MW heating systems as the sources of electromagnetic energy. A magnitude of the magnetron's output signal in the frequency range (f_{m1}, f_{m2}) (such that $f_{m1} > f_1$ and $f_{m2} < f_2$) vary dramatically from one device to another (so that no two sources are identical) and depend on many factors including parameters of the magnetron's power supply, temperature and its other physical conditions, load's parameters, power generated by the magnetron, etc. [6]. A MW heating system is said to be less sensitive to magnetron replacements (and is thus always highly efficient) if it is characterized by a relatively smooth function $|S_{11}|(f)$ with sufficiently low values in the interval (f_1, f_2) [17]. Narrow strong resonances on this curve would suggest that the system's operation may be unstable and hardly predictable whereas the curve displaying quite high values of $|S_{11}|$ in the entire interval (f_1, f_2) would correspond to a fairly inefficient system. As a part of a CAD project, analysis of a system with its pulse excitation may therefore be very informative as it can identify a potential inefficiency and/or instability of the considered system and suggest certain changes in its parameters in order to modify an unfavorable $|S_{11}|(f)$.

At this first stage of the modeling procedure, computation of other S-parameters of the system may be also beneficial. Placing another (supplementary) port in a place of particular interest allows more detailed information about the analyzed system to be obtained [94]. In systems with multiple sources of energy, other ports could also be placed in other feeds so determination of corresponding transmission coefficients would allow for analyzing the effect of magnetron's crosstalk [6,95].

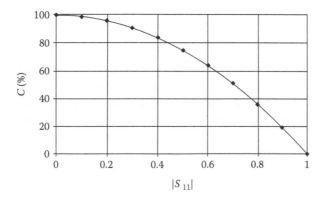

FIGURE 23.7 Energy coupling expressed in percent as function of the reflection coefficient.

2. At the next stage, modeling is continued with sinosoidal excitation of the input ports at f_0 with the goal to compute the electric field in the analyzed system. When performing such simulation with *QW-3D*, has an option to visualize instantaneous fields naturally updated in time by the time-domain solver that gives the user an opportunity to monitor field propagation in the system. A field envelope **E** computed as time-maxima of the field in all FDTD cell provides information about the structure of the electric field throughout the cavity including locations of problematic areas with strong field concentrations raising chances for arching. Distribution of dissipated power in the processed material $\mathbf{P}_d$ is directly computed from the electric field **E** in accordance with the well known formula:

$$\mathbf{P}_d = 2\pi\varepsilon_0\varepsilon'' f |\mathbf{E}|^2 \tag{23.10}$$

where ε_o is free space permittivity.

A thermal solver takes over at this point and computes the temperature field **T** inside the processed material on the basis of information about k, ρ, c, and $\mathbf{P}_d$. While in systems with high energy coupling, poor thermal conductivity of the product and temperature-independent material parameters, $\mathbf{P}_d$ patterns may provide a good idea about the heat intensity within the product, for quantitative characterization of a heating process in terms of temperature computation of a **T** field is required.

3. The two stages of the computational scheme considered above represent the analysis phases and normally do not give the designer direct instructions concerning how the system should be changed in order to improve its performance. Efficient system design guaranteeing the best possible characteristics is usually an ultimate goal of a CAD project.

In addition to the energy coupling specifying system efficiency, another most important characteristic of a MW heating system is distribution of temperature within the processed material. Most applications employing MW thermal processing of food products require uniform temperature field, but generally the problem is about making the process of MW heating controllable. The goal of simultaneous optimization of these two characteristics has been probably the major driving force behind experimental activities of engineers in the field of MW power engineering since the emergence of this discipline nearly half a century ago. However, because of exceptional complexities of the physics of the process, internal/volumetric character of heat release, variation of essential material parameters with temperature, etc., there has not been much progress so far towards obtaining a general solution to this challenge.

It has been recently suggested that the problems of optimization of C and **T** can be addressed with the use of techniques of numerical optimization backed by full-wave 3-D FDTD modeling. The problem of efficiency optimization has been formulated for the first time by in [17,96] for a single-source scenario as follows:

A. Find a configuration of the structure such that $|S_{11}|$ is less than the assigned level (S_0) in the frequency range $f_1 < f_0 < f_2$ (Figure 23.8)

Being a multivariable function of f and n system parameters, $|S_{11}|$ is an objective function of the optimal design, and $S_0, f_1,$ and f_2 are interpreted as the relevant constraints. The goal of optimization process is to get the geometry which would be characterized by the $|S_{11}|$ frequency response passing the interval (f_1, f_2) below the desired level S_0—as the dashed curve in Figure 23.8.

Several approaches to the problem in this formulation are described in the literature. In [17], problem (A) is solved with response surface methodology (RSM) combined with the sequential quadratic programming (SQP) algorithm. Cubical, rectangular, cylindrical cavities are optimized with the use of the Lebenverg–Marquardt (LM) technique and the measured characteristics $|S_{11}|(f)$ in [97] as well as with genetic algorithm (GA) and *MWS* in [98]. The radial basis function (RBF) network technique introduced in [99–101] and backed by data from *QW-3D* demonstrates viable

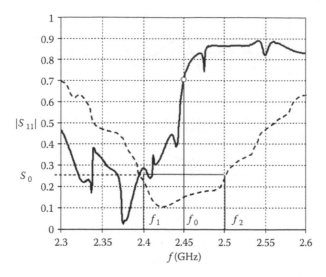

FIGURE 23.8 Conventional frequency characteristics of the reflection coefficient in a nonoptimized (solid curve) and optimized (dashed curve) systems. (From Mechenova, V.A. and Yakovlev, V.V., *J. Microwave Power Electromag. Energy*, 39, 15, 2004. With permission.)

performance in optimizing a practical-size MW oven with a food load on a round table. The RBF network optimization is proved to be more accurate and robust than the RSM-SQP and LM techniques. It also demonstrates the potential to find the "best" local optimum in the specified domain [101] and thus can be considered more efficient (i.e., working with essentially less computational resources, CPU time, and RAM) than global GA optimization.

As for computational tools assisting in CAD of MW heating systems providing uniform (or generally speaking, desirable) temperature field within the processed material, corresponding techniques are still on the initial phase of their development. So far, there is no generally established criterion for measuring uniformity of MW-induced temperature fields. While some researchers believe that temperature can be homogenized by evening out the pattern of the electric field [28,102–105], other authors quantify the level of uniformity of MW heating through the spatial distribution of dissipated power [106,107].

A general modeling-based technique for solving the fundamental problem of intrinsic non-uniformity of the resulting internal heating pattern has been proposed in [88–90]. The problem is formulated as follows:

B. Find the MW heating process which minimizes the time t_f required to raise the minimum temperature of the load, T_m, to a prescribed goal temperature, T_{min}, provided that the maximum temperature, T_M, remains below a preset threshold temperature, T_{max} (Figure 23.9).

Characteristics of the MW system, which, when altered, dramatically affect the resulting temperature field **T**, are identified through the preceding systematic FDTD modeling and are chosen as design variables for the optimization. A pulsing regime (in which periods of thermal relaxation are allowed between periods of MW heating) is always included in the optimization algorithm to let heat diffusion reach the *cold spots* that cannot be accessed by MW heating given the specific set of design variables. The solution technique based on a special FDTD-conditioned technique for measuring the uniformity of the temperature field includes also the synthesis procedure generating a description of the optimal heating process with the resulting uniform 3-D temperature field [89].

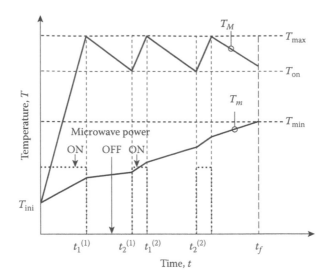

FIGURE 23.9 Conventional interpretation of pulsing MW heating toward uniform temperature field. (From Cordes, B.G., Eves, E.E., and Yakovlev, V.V., *Proc. 11th Conf. Microwave and High Frequency Heating*, Oradea, Romania, 305, 2007. With permission.)

23.3.3 EXAMPLES OF FINITE-DIFFERENCE TIME-DOMAIN (FDTD) MODELING AND OPTIMIZATION

The following examples illustrate functionality of the described FDTD-based computational tools at all three steps of the described computational strategy.

1. The first illustration describes a simplified MW thermal processing scenario shown in Figure 23.10. A rectangular ($A \times B \times C$) resonator imitating a cavity of a MW oven is excited by the dominant TE_{10} mode of a rectangular ($g_a \times g_b \times g_l$) waveguide centrally connected to the resonator and shifted down (by s) from the top edge of one of its walls. The oven contains a rectangular ($a \times b \times c$) block of food lying on a centered roundtable of thickness t elevated above the oven's bottom by the height h. The parameters chosen in the considered computational experiments and fixed in the model correspond to the 700 W MW oven characterized in Table 23.7. Corresponding *QW-3D* model consists of about 443,000 cells and occupies about 42 MB RAM.

A collection of frequency characteristics of $|S_{11}|$ computed with this model and shown in Figure 23.11 provides valuable information concerning the potential of this system to operate with high efficiency. It is seen that a round table makes a significant impact on energy coupling, and there may be an optimal thickness providing, for a processed material with given material parameters and dimensions, minimal reflections at $f_0 = 2.45$ GHz (Figure 23.11e–f). While $|S_{11}|$ is unlikely strongly dependent on the round-table's diameter (Figure 23.11c), significant changes in the position of the roundtable above the oven's bottom may shift the position of the resonance located in the neighborhood of f_0 and thus lead to notable variations of coupling (Figure 23.11d). Although parameter C appears to be dependent on the dimensions of the load (Figure 23.11a), this dependence is not critical. On the other hand, it is seen that a profile of function $|S_{11}|(f)$ is significantly affected by the load's coordinates in the xy-plane (Figure 23.11b), so the energy coupling can be substantially different for different positions of the load on the roundtable. This suggests, for instance, that two MW heating processes involving rotations of the same load on different positions of the roundtable could, in terms of energy coupling, be fairly different.

This brief analysis illustrates that initial characterization of a MW heating system through a series of frequency responses of the reflection coefficient may be practically fruitful by suggesting concrete CAD solutions. For example, in the considered scenario, the function $|S_{11}|(f)$ does not seem to be characterized by narrow deep resonances in the interval $(f_1, f_2) = (2.4, 2.5$ GHz$)$, so no principle alterations in the system design may be necessary. Furthermore, at least ~60% efficiency

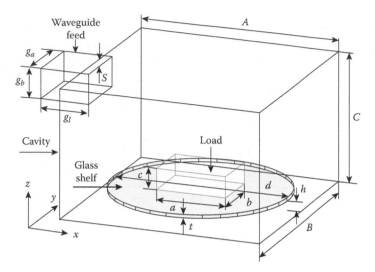

FIGURE 23.10 3-D view and a geometrical layout of a MW oven.

TABLE 23.7
Parameters of the Sanyo *Direct*
***Access* Oven**

Parameter	Value
$A \times B \times C$	$290 \times 300 \times 185$ mm
$g_a \times g_b \times g_l$; s	$86 \times 43 \times 70$ mm; 10 mm
Load	Meat: $\varepsilon = 40.0$; $\sigma = 3.0$ S/m
Shelf	Glass: $\varepsilon = 6.0$, $\sigma = 0$

of the system can be secured by choosing the thickness of the roundtable t such that the resonance minimum is located at f_0.

2. Patterns of the electric field and dissipated power visualized in Figures 23.12 and 23.13 in the coordinate planes through the load are obtained with the sinusoidal excitation of the same applicator at 2.45 GHz. Each pattern shows a relative distribution of the related entity. It is seen that the chosen waveguide excitation is able to generate the field resembling the structure of the deformed resonant TE_{432} mode. The food load inserted in the oven disturbs the electric field, but mostly by affecting the magnitude in the field maxima rather than by breaking the field structure. Distribution of dissipated power inside the load resembles distribution of the electric field there, so the load is expected to be overheated in the corners (following the field singularities around them), in two internal hot spots and in the central area on the bottom surface of the load.

3. An eloquent illustration of efficiency optimization is given by Murphy and Yakovlev [101] for a MW heating scenario similar to the one considered above (Figure 23.10 and Table 23.7) but with a larger cavity size and a different food load—a horizontal cylinder with rounded ends and complex permittivity of cooked beef. The RBF neural network procedure performs five-parameter optimization and determines the values of h and t as well as of the cylinder's dimensions and its position on the roundtable that correspond to the function $|S_{11}|(f)$ characterized by the values of $|S_{11}|$ smaller than $S_0 = 0.3$ (i.e., the efficiency greater than 91%) in the frequency range $(f_1, f_2) = (2.4, 2.5)$ GHz.

While MW heating scenario considered in [101] can be regarded as somewhat artificially constructed to illustrate the capabilities of the optimization procedure, in [15], a similar neural network

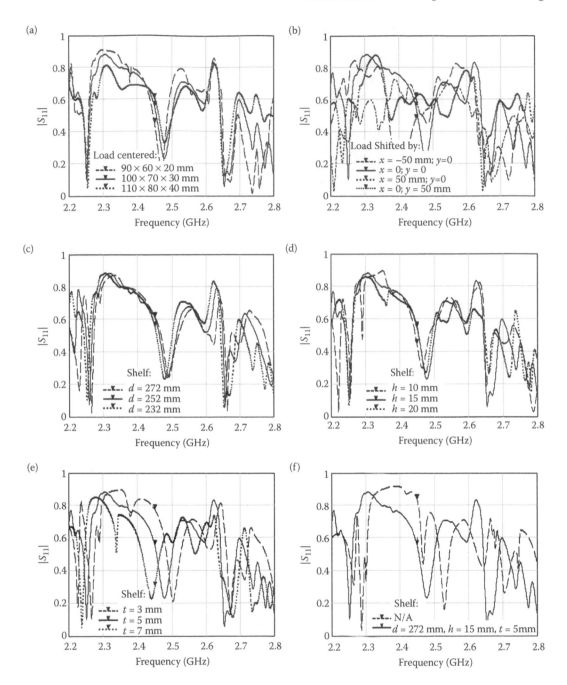

FIGURE 23.11 Frequency characteristics of the reflection coefficient in the MW oven (Figure 23.10, Table 23.7) for (a,b) variable dimensions and position of the load and (c–f) variable dimensions and positions of the shelf: (b–f) $a \times b \times c = 100 \times 70 \times 30$ mm, (a,b,d–f) $d = 272$ mm, (a–c,e,f) $h = 15$ mm, (a–d,f) $t = 5$ mm.

optimizer is applied to a practical system—a twin slotted waveguide applicator for heating aqueous emulsions. In the applicator of an initially chosen (nonoptimized) configuration, the efficiency is very low—around 8% at $f_0 = 2.45$ GHz (thin curve in Figure 23.14). However, an optimal configuration determined by a two-parameter optimization is characterized by much better functions $|S_{11}|(f)$. A few merged resonances produce the curves with low values of the reflection coefficient in the wide range around the operating frequency (thick curves in Figure 23.14). This guarantees

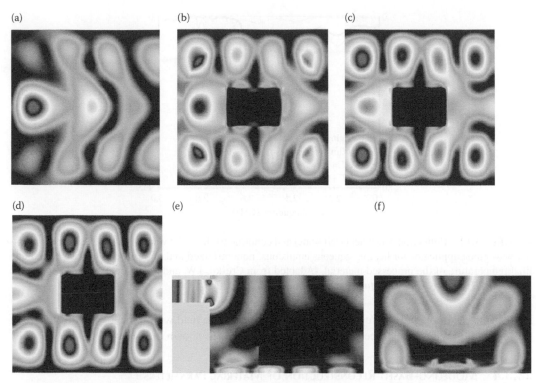

FIGURE 23.12   Electric field envelope in the MW oven (Figure 23.10, Table 23.7, centered load) in the horizontal xy-plane through the (b) bottom, (c) center, and (d) top of the load and in the vertical (e) xz- and (f) yz-planes through the center of the load for $a \times b \times c = 100 \times 70 \times 30$ mm, $d = 272$ mm, $h = 15$ mm, $t = 5$ mm. Pattern (a) shows the field envelope in the same plane as (c), but in the absence of the load.

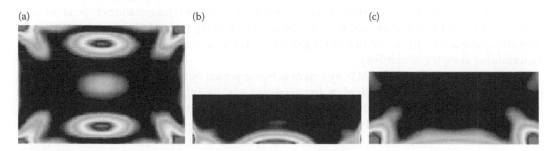

FIGURE 23.13   Patterns of the dissipated power in the MW oven (Figure 23.10, Table 23.7, centered load) in the central (a) xy-, (b) xz-, and (c) yz-planes through the load for $a \times b \times c = 100 \times 70 \times 30$ mm, $d = 272$ mm, $h = 15$ mm, and $t = 5$ mm.

the efficiency of the applicator's performance to be greater than 90% in the operating temperature range (25–80°C).

The examples illustrating a possibility of modeling-based synthesis of an optimized MW heating process resulting in uniform temperature field are given by Cordes et al. [88,90] in support of the corresponding optimization procedure. For instance, computational experiment described in [88] exemplifies the synthesized processes producing homogeneous ($T_m > T_{min} = 60°C$ and $T_M < T_{max} = 70°C$) temperature distribution in the rectangular ($100 \times 76 \times 30$ mm) block with material parameters of raw beef processed in a two-feed rectangular ($400 \times 264 \times 180$ mm) cavity for $t_u \sim 30$ min—this corresponds to about a four times reduction in time-to-uniformity compared to

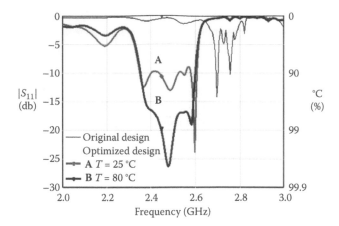

FIGURE 23.14 Reflection coefficient (left scale) and coupling (right scale) versus frequency in the twin slot-ted waveguide applicator for heating aqueous emulsions: nonoptimized and optimized design for initial and final temperature of the processed material. (Adapted from Cresko, J.W. and Yakovlev, V.V., *Proc. 9th Conf. Microwave High Frequency Heating*, Loughborough, UK, 317, 2003.)

results from a simple pulsing regime. Variation of temperature field within the load in the course of this optimal MW heating process is illustrated by the patterns shown in Figure 23.15.

23.3.4 Modeling-Based Reconstruction of Material Parameters

Knowledge of complex permittivity of materials processed in applications involving MW heating is crucial for creating adequate models and thus for successful system design [108,109]. While dielectric measurements is the field well established for electronics and telecommunication applications, reliable data for dielectric constant ε' and the loss factor ε'' characterizing the actual food product are not always available. Measuring complex permittivity at elevated temperatures typical for MW pasteurization and sterilization has been particularly challenging [110,111]. Lack of data regarding realistic materials motivates further development for versatile and robust practical techniques of determining complex permittivity.

Once an efficient design of a MW system may be supported by modeling, it is logical to make a related numerical simulator involved in *determination* of ε' and ε''. Since these parameters are not directly measured but calculated from some measurable characteristics, then more difficult computational tasks may be assigned to a simulator while reducing the experimental part to elementary measurements. This approach is taken in the techniques using the FEM [112–116] and the FDTD method [117–123] for modeling of entire experimental fixtures. For example, in [122], complex permittivity is found with an optimization algorithm designed to match complex S-parameters obtained from measurements and simulations. The method is developed on a two-port (waveguide-type) fixture and deals with complex reflection and transmission characteristics at the frequency of interest. A computational part is constructed as an inverse RBF-neural-network-based procedure that reconstructs ε' and ε'' of the sample from the FDTD modeling data sets and the measured reflection and transmission coefficients. As such, it is applicable to samples and cavities of arbitrary configurations provided that the geometry of the experimental setup is adequately represented by the model.

Modeling-based techniques of complex permittivity reconstruction are currently considered a valuable part of CAD activities in the field of MW power engineering. However, while these techniques may be convenient in supporting particular projects, they are unlikely to be vital in diminishing the deficit of reliable data on media parameters of practical materials. The principal issue here is the access to practical and reliable techniques for measuring ε', ε', and k at elevated

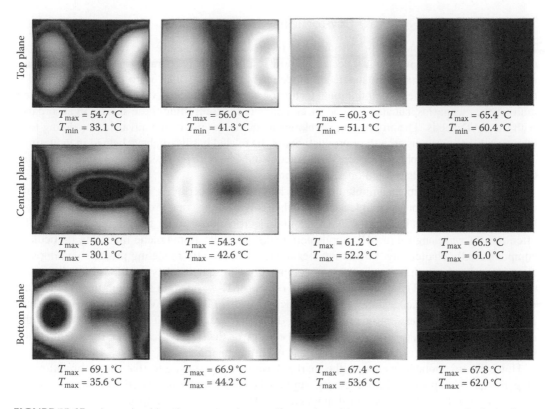

FIGURE 23.15 An optimal heating process: intermediate and resulting temperature patterns in the horizontal planes through a rectangular load in a MW applicator at four particular instances of the process. (From Cordes, B.G., Eves, E.E., and Yakovlev, V.V., *Proc. 11th Conf. Microwave and High Frequency Heating*, Oradea, Romania, 305, 2007. With permission.)

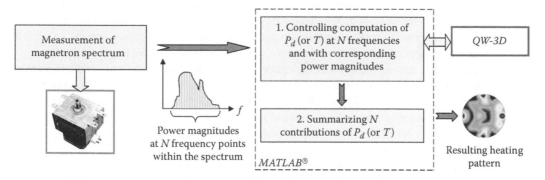

FIGURE 23.16 Computation scheme for finding heating patterns in the system fed by specific magnetrons.

temperatures. On the other hand, the fact that dielectric constant and the loss factor of many materials are temperature-dependent is not a problem for modern modeling tools; e.g., in the *QW-3D* package, a precise upgrade of material parameters on the cell level is an intrinsic part of the iterative solution of the coupled problem (Figure 23.6). However, in order to be given to the simulator, the functions $\varepsilon'(T)$ and $\varepsilon''(T)$ should be known. This implies that success in CAD of a particular MW heating fixture is currently limited not by computational capabilities, but by the absence of data on temperature-dependent electric and thermal properties, namely $\varepsilon'(T)$, $\varepsilon''(T)$, and $k(T)$.

23.3.5 ONGOING AND FUTURE DEVELOPMENTS

The concepts and techniques briefly outlined in previous sections are the result of accomplishments made primarily over the last decade. Nowadays, in terms of accuracy and adequacy, the core numerical solvers substantially supersede modeling approaches which were in use in the field in the preceding era. Modern computational technologies are not limited to 3-D analysis of genuinely multiphysics scenarios, but also include special procedures controlling these solvers in accordance with the CAD needs. The progress in development and in the use of the advanced modeling tools (along with an enormous growth of productivity of the related hardware) have helped in mounting a deeper insight into the physical nature of interaction of MWs with processed materials and the ways this interaction is simulated by numerical techniques.

As such, new generation computational tools have reached a certain level of maturity and sophistication which allows for acceptable accuracy when solving quite complex and large modeling problems on commonly available PCs. A substantial electrical size (more than several wavelengths, often tens of wavelengths) of MW applicators makes their electromagnetic modeling particularly computationally challenging. Still, known examples of successful modeling projects dealing with electrically large scenarios on the edge of computational resources of typical PCs include simulation of a *QW-3D* model of an industrial system of almost 11-wavelength size with the use of nearly 12 million FDTD cells on a standard 32-bit machine with 2 GB RAM. When running it with Intel 2.0 GHz Centrino processor, this project takes ~11–12 h to reach $|S_{11}|(f)$ in steady state. While this may be regarded as a reasonable computational cost, given the size of the system, this example also highlights a fundamental limitation in terms of operating memory.

However, two notable events on the market have recently suggested that these limitations no longer exist . Computational productivity can now be substantially increased with the use of special versions of FDTD simulators running, e.g., on *Acceleware* [124], a special piece of hardware designed for accelerating certain types of computationally intensive simulations by raising productivity of graphics-processing units (GPUs) rather than traditional CPUs as processing tools. A series of comparative tests show that, for example, *QW-AccelSim*, corresponding version of *QW-3D*, provides, depending on the problem, speeds up to factors of five to 16 [79]. Furthermore, with 64-bit operating systems recently become functional on PCs, the principle limit of 3 GB RAM is eliminated allowing for simulations of projects requiring virtually unlimited memory. By allowing for modeling and optimizing more complex and larger systems of MW thermal processing, these newly available options open new horizons in this computationally intensive business; reports on exploiting these options in applied modeling projects are expected to appear in the literature soon.

On the other hand, despite these widening opportunities in increasing accuracy and adequacy of modeling MW heating processes, there is a fundamental issue making precise virtual reproduction of actual performance of practical industrial systems difficult. The problem is caused by uncertainty in characterizing magnetrons, the standard sources of the MWs, connected with applicators by waveguides. In the models built on the considered numerical techniques, a magnetron as an electronic device is rarely modeled, but replaced by a port generating the dominant TE_{10} mode of a rectangular waveguide. While this approach may appear intuitively natural, it has also been advocated by specific studies (e.g., [125,126]) that the standard magnetron launchers indeed form the TE_{10} mode even in a short-length waveguide. However, in the context of the FDTD-based strategy described above, there is an issue. While modeling with pulse excitation leads us to an intrinsic characteristic $|S_{11}|(f)$ providing the values of energy coupling at all frequencies in the interval (f_1, f_2), simulation with sinusoidal excitation produces the electric field, dissipated power and temperature at f_0. Still, magnetrons never generate signals at a single frequency. Rather, they irradiate MWs in certain frequency ranges and profiles of those spectra are known to be dependent on many factors such as type of a magnetron's power supply, magnetron's physical conditions (e.g., its temperature), output power generated by the magnetron, load conditions, etc. [6,14]. This means that performance of the same applicator with different magnetrons may be different.

This implies that for practical MW heating systems experimental validation of modeling results is characterized by a principal difficulty. While $|S_{11}|(f)$ can be measured by replacing a magnetron with a low-power source available in a network analyzer, in order to experimentally capture a heating pattern (e.g., with an infrared camera), an applicator is supposed to operate with the MW power produced by a magnetron. In the first case, parameters of the system are completely deterministic, and the model can surely be adequate for the experimental setup; in the second case, functionality of the system is determined not only by the system geometry and material parameters, but also depends on unknown and "unpredictable" characteristics of the magnetron. It is worth noting that the divergence of simulated and measured results observed in the latter case to no extend can be interpreted as the evidence of inaccuracy of the model.

If preceding measurements of a magnetron is a practical option, a laborious but technically possible resolution of the issue can be achieved through the approach illustrated in Figure 23.16. After getting a magnetron spectrum in its operational mode and incorporating the data on its profile into a special controlling procedure (implemented, e.g., in MATLAB®), computation of the resulting heating patterns may be performed from a series of simulations with sinusoidal excitation at N points of the magnetron spectrum with corresponding magnitudes of the input power. In order to take into account the load conditions, the magnetron spectrum could be measured several times in the presence of the load in question and the computational procedure can then be repeated for each spectrum. Attractive alternative (non-modeling, hardware-related) approach to the same issue [127] is also worth mentioning.

Despite the uncertainty associated with the presence of actual sources of energy, modeling of MW heating systems gradually gains more popularity in applied projects. Design engineers may find it feasible to be limited to experimental validation of only $|S_{11}|(f)$ computed by a presumably accurate model and a subsequent exploitation of the model for "virtual experimentation" with the system (including the sinusoidal excitation) and even its possible CAD and/or optimization. A physical prototype can be produced on the basis of computational results, and a final tune of the system can be made experimentally.

Therefore, it has been shown that high-performance computing systems and suitable efficient numerical techniques play an important role in designing MW heating systems and processes. Beyond the traditional use of mathematical modeling in CAD projects, it can also be useful in virtual prototyping, development of original fixtures and new processes, design of microwavable food and reconstruction of material parameters. Ultimate success in modeling of MW heating processes, however, may be presently dependent on other factors (e.g., availability and accuracy of data on material parameters involved) rather than on the quality of numerical algorithms and computational resources.

23.4 CONCLUSIONS

The remarkable advances in numerical mathematics and computer software and hardware developments have recently led to a significant increase in the use of computer modeling in analysis and CAD of systems and processes of MW heating. Thermal processing of food products is known to be one of the most widespread uses of MW energy. In this chapter, we have shown that availability and applicability of modeling and optimization tools of new generation makes MW-based food preservation, an exceptionally complicated multiphysics process and a highly sophisticated interdisciplinary technology, particularly attractive for further development and exploitation. There is growing interest, both in academia and industry, in using advanced modeling technologies for designing applicators in which MW sterilization or pasteurization is controllable in large-scale industrial applications.

Bringing the technology of MW-induced food preservation on the industrial level requires successful experimental validation of its performance in terms of microbial reduction and quality degradation, and this seems to be a particularly difficult barrier which apparently was responsible for keeping this technology from wider acceptance and acknowledgement. It appears that the modern modeling tools reviewed in this chapter could form the basis for an innovative approach which could be used for addressing this challenge. Indeed, they could dramatically reduce the required experimental studies and potentially decrease the cost of process development. Employing suitable modeling tools in the frameworks of optimization procedures generating optimal processes guaranteeing the desirable microbial parameters—the concept suggested in Section 23.2.1—can be considered one of those inventive techniques.

Finally, it is worth mentioning that this paper exercises an original approach to modeling of MW processes and systems for food preservation. Here, the authors attempt to combined the vision of MW thermal processing typical for food engineers as the end users and computer modelers involved in practical design of MW heating systems. It appears that such an approach could be beneficial and help further develop successful MW processes in food applications.

REFERENCES

1. Anonymous. *Kinetics of Microbial Inactivation for Alternative Food Processing Technologies. Microwave and Radio Frequency Processing.* US FDA, CFSAN, 2000.
2. Dibben, D. and Metaxas, A.C. Finite element time domain analysis of multimode applicators using edge elements. *J. Microwave Power Electromag. Energy,* 29, 242, 1994.
3. Ma, L., Paul, D.L., Pothecary, N., Railton, C., Bows, J., Barratt L., and Simons, D. Experimental validation of a combined electromagnetic and thermal model of a microwave heating process. *IEEE Trans. Microwave Theory Tech.,* 43, 2565, 1995.
4. Sundberg, M., Risman, P.O., Kildal P.-S., and Ohlsson, T. Analysis and design of industrial microwave ovens using the finite-difference time-domain method. *J. Microwave Power Electromag. Energy,* 31, 142, 1996.
5. Torres F. and Jecko, B. Complete FDTD analysis of microwave heating processes in frequency-dependent and temperature-dependent media. *IEEE Trans. Microwave Theory Tech.,* 45, 108, 1997.
6. Chan, T.V.C.T. and Reader, H.C. *Understanding Microwave Heating Cavities,* Artech House, Boston, MA, 2000.
7. Risman, P.O. A microwave oven model—examples of microwave heating computations. *Microwave World,* 19, 20, 1998.
8. Risman, P.O. and Celuch-Marcysiak, M. Electromagnetic modeling for MW power applications. *Proc. 13th Int. Conf. Microwaves, Radar and Wireless Communications (MIKON-2000),* Wrocław, Poland, 3, 167, 2000.
9. Komarov, V.V. and Yakovlev, V.V. Simulations of components of MW heating applicators by FEMLAB, MicroWaveLab and QuickWave-3D. *Proc. 36th Microwave Power Symp.,* San Francisco, CA, 1, 2001.
10. Yakovlev, V.V. Efficient electromagnetic models for systems and processes of MW heating. *Proc. Int. Seminar Heating by Internal Sources (HIS-01),* Padua, Italy, 285, 2001.
11. Yakovlev, V.V. Commercial electromagnetic codes suitable for modeling of MW heating—a comparative review. In *Scientific Computing in Electrical Engineering,* van Reinen, U., Gunther, M., and Hecht, D., Eds. Springer, Berlin, Germany, 87, 2001.
12. Ratanadecho, R., Aoki, K., and Akahori, M. A numerical and experimental investigation of the modeling of MW heating for liquid layers using a rectangular waveguides (effects of natural convection and dielectric properties. *Appl. Math. Model.,* 26, 449, 2002.
13. Kopyt, P. and Celuch-Marcysiak, M. FDTD modeling and experimental verification of electromagnetic power dissipated in domestic microwave oven. *J. Telecomm. Inform. Technol.,* 59, 2003.
14. Kopyt, P., Celuch-Marcysiak M., and Gwarek, W.K. Microwave processing of temperature-dependent and rotating objects: development and experimental verification of FDTD algorithms. In *Microwave and Radio Frequency Applications,* Folz, D.C., Booske, J.H., et al., Eds. The American Ceramic Society, Westerville, OH, 7, 2003.
15. Cresko, J.W. and Yakovlev, V.V. A slotted waveguide applicator design for heating fluids. *Proc. 9th Conf. Microwave and High Frequency Heating,* Loughborough, UK, 317, 2003.
16. Gwarek W.K. and Celuch-Marcysiak, M. A review of microwave power applications in industry and research. *Proc. 15th Int. Conf. Microwaves, Radar and Wireless Communications (MIKON-2004),* Warsaw, Poland, 3, 843, 2004.
17. Mechenova, V.A. and Yakovlev, V.V. Efficiency optimization for systems and components in microwave power engineering. *J. Microwave Power Electromag. Energy,* 39, 15, 2004.

18. Plaza-Gonzalez, P., Monzó-Cabrera, J., Catalá-Civera, J.M., and Sánchez-Hernández, D. Effect of mode-stirrer configurations on dielectric heating performance in multimode microwave applicators. *IEEE Trans. Microwave Theory Tech.*, 53, 1699, 2005.

19. Yakovlev, V.V. Examination of contemporary electromagnetic software capable of modeling problems of microwave heating. In *Advances in Microwave and Radio Frequency Processing*, Willert-Porada, M., Ed. Springer, Berlin, Germany, 178, 2006.

20. Celuch, M., W.K. Gwarek, W.K., and Sypniewski, M. A novel FDTD system for microwave heating and thawing analysis with automatic time-variation of enthalpy-dependent media parameters. In *Advances in Microwave and Radio Frequency Processing*, Willert-Porada, M., Ed. Springer, Berlin, Germany, 199, 2006.

21. Watanabe, S., Kakuta Y., and Hashimoto, O. Analytical and experimental examination of defrosting spots of heated frozen material in a microwave oven. *Digest IEEE MTT-S Int. Microwave Symp.*, Anaheim, CA, 284, 2006.

22. Komarov, V.V. and Yakovlev, V.V. Coupling and power dissipation in a coaxially excited $TM_{01\ell}$ mode cylindrical applicator with a spherical load. *Microwave Opt. Tech. Lett.*, 48, 1104, 2006.

23. Risman, P.O. and Schonning, U. Microwave tunnel ovens using multiple open-ended evanescent mode applicators. *Proc. 40th Microwave Power Symp.*, Boston, MA, 105, 2006.

24. Al-Rizzo, H.M., Tranquilla, J.M., and Feng, M. A finite difference thermal model of a cylindrical MW heating applicator using locally conformal overlapping grids: Part I Theoretical formulation. *J. Microwave Power Electromag. Energy*, 40(1), 17, 2007.

25. Al-Rizzo, H.M., Adada, R., Tranquilla, J.M., Feng, M., and Ionescu, B.C. A finite difference thermal model of a cylindrical microwave heating applicator using locally conformal overlapping grids: Part II Numerical results and experimental evaluation. *J. Microwave Power Electromag. Energy*, 40(2), 17, 2007.

26. Komarov, V.V. and Yakovlev, V.V. CAD of efficient TM_{mn0} single-mode elliptical applicators with coaxial excitation. *J. Microwave Power Electromag. Energy*, 40, 174, 2007.

27. Ehlers, R.A. and Metaxas, A.C. An investigation on the effect of varying the load, mesh and simulation parameters in microwave heating applications. *J. Microwave Power Electromag. Energy*, 40, 251, 2007.

28. Dominguez-Tortajada, E., Monzo-Cabrera J., and Diaz-Morcillo, A. Uniform electric field distribution in microwave heating applicators by means of genetic algorithms optimization of dielectric multilayer structures. *IEEE Trans. Microwave Theory Tech.*, 55, 85, 2007.

29. Pedreño-Molina, J.L., Monzó-Cabrera, J., and Catalá-Civera, J.M. Sample movement optimization for uniform heating in microwave heating ovens. *Int. J. RF Microwave CAE*, 17, 142, 2007.

30. Zhu, J., Kuznetsov, A.V., and Sandeep, K.P. Mathematical modeling of continuous flow microwave heating of liquids (effects of dielectric properties and design parameters). *Int. J. Thermal Sci.*, 46, 328, 2007.

31. Zhu, J., Kuznetsov, A.V., and Sandeep, K.P. Numerical simulation of forced convection in a duct subjected to microwave heating. *Heat Mass Transfer*, 43, 255, 2007.

32. Komarov, V.V., Pchelnikov, Yu.N., and Yakovlev, V.V. Cylindrical double-ridged waveguide as a basic unit of microwave heating applicators. *Microwave Opt. Tech. Lett.*, 49, 1708, 2007.

33. Wappling-Raaholt, B., Janestad. H., and Vinsmo, L.G. Demonstration of the usefulness of electromagnetic modeling in product development work: microwave defrosting, tempering and heating of ready meals. *Proc. 11th Conf. Microwave and High Frequency Heating*, Oradea, Romania, 45, 2007.

34. Leonelli, C., Veronesi, P., and Grisoni, F. Numerical simulation of an industrial microwave-assisted filter dryer: criticality assessment and optimization. *J. Microwave Power Electromag. Energy*, 41(3), 5, 2007.

35. Tilford, T., Baginski, E., Kelder, J., Parrott, A.K., and Pericleous, K.A. Microwave modeling and validation in food thawing applications. *J. Microwave Power Electromag. Energy*, 41, 41.4.29, 2007.

36. Kopyt, P. and Celuch, M. Coupled electromagnetic-thermodynamic simulations of microwave heating problems using the FDTD algorithm. *J. Microwave Power Electromag. Energy*, 41, 41.4.17, 2007.

37. Sabliov, C.M., Salvi, D.A., and Boldor, D. High frequency electromagnetism, heat transfer and fluid flow coupling in *ANSYS Multiphysics*. *J. Microwave Power Electromag. Energy*, 41, 41.4.4, 2007.

38. Celuch, M. and Gwarek, W. Properties of the FDTD method relevant to the analysis of microwave power problem. *J. Microwave Power Electromag. Energy*, 41, 41.4.45, 2007.

39. Soltysiak, M., Erle, U., and Celuch, M. Load curve estimation for microwave ovens: experiments and electromagnetic modeling. *Proc. 17th Int. Conf. MWs, Radar and Wireless Communications (MIKON-2008)*, Wrocław, Poland, 2008.

40. Chen, H., Tang J., and Liu, F. Simulation model for moving food packages in microwave heating processes using conformal FDTD method. *J. Food Eng.*, 88, 294, 2008.

41. Stumbo, C.R. *Thermobacteriology in Food Processing*, 2nd edn. Academic Press, New York, NY, 1973.
42. Hiroshi, F., Hiroshi, U., and Yasuo, K. Kinetics of *Escherichia Coli* destruction by microwave irradiation. *Appl. Environ. Microbiol.*, 58, 920, 1992.
43. Heddleson, R.A., Doores, S., Anantheswaran, R.C., Kuhn, G.D., and Mast, M. Survival of *Salmonella* species heated by microwave energy in a liquid menstruum containing food components. *J. Food Protection*, 54, 637, 1991.
44. Riva, M., Lucisano, M., Galli, M., and Armatori, A. Comparative microbial lethality and thermal damage during microwave and conventional heating in mussels (*Mytilus edulis*). *Ann. Microbiol.*, 41, 147, 1991.
45. Riva, M., Franzetti, L., Mattioli, A., and Galli, A. Microorganisms lethality during microwave cooking of ground meat. 2: Effects of power attenuation. *Ann. Microbiol. Enzymol.*, 43, 297, 1993.
46. Ramaswamy, H.S., Koutchma, T., and Tajchakavit, S. Enhanced thermal effects under microwave heating conditions. In *Engineering and Food for the 21st Century*, Welti-Chanes, J. and Barbosa-Canovas, G.V., Eds. CRC Press, Boca Raton, FL, 2002.
47. Heddleson R.A., Doores S., and Anantheswaran, R.C. Parameters affecting destruction of *Salmonella* spp. by MW heating. *J. Food Sci.*, 59, 447, 1994.
48. Heddleson., R.A., Doores., S., Anantheswaran, R.C., and Kuhn G.D. Viability loss of Salmonella species, *Staphylococcus aureus*, and *Listeria monocytogenes* in complex foods heated by MW energy. *J. Food Protection*, 59, 813, 1996.
49. Odani, S., Abe, T., and Mitsuma, T. Pasteurization of food by microwave irradiation. *J. Food Hygienic Soc. Japan*, 36, 477, 1995.
50. Woo, I.-S., Rhee, I.-K., and Park, H.-D. Differential damage in bacterial cells by microwave radiation on the basis of cell wall structure. *Appl. Environ. Microbiol.*, 66, 2243, 2000.
51. Yeo, C. B., Watson, I.A., Stewart-Tull, D.E., and Koh, V.H. Heat transfer analysis of *Staphylococcus aureus* on stainless steel with microwave radiation. *J. Appl. Microbiol.*, 87, 396, 1999.
52. Diaz-Cinco, M. and Martinelli, S. The use of microwaves in sterilization. *Dairy Food Environ. Sanitation*, 11, 722, 1991.
53. Huang, L. Computer-controlled microwave heating to in-package pasteurize beef frankfurters for elimination of *Listeria monocytogenes*. *J. Food Process Eng.*, 28, 453, 2005.
54. Galuska, P.-J., Kolarik, R.W., Vasavada, P.C., and Marth, E.-H. Inactivation of *Listeria monocytogenes* by microwave treatment. J. Dairy Sci., 72(1), 139, 1989.
55. Tajchakavit, S., Ramaswamy, H.S., and Fustier, P. Enhanced destruction of spoilage microorganisms in apple juice during continuous flow microwave heating. *Food Res. Int.*, 31, 713, 1998.
56. Shin, J.K. and Pyun, Y.R. Inactivation of *lactobacillu plantarum* by pulsed microwave irradiation. *J. Food Sci.*, 62, 163, 1997.
57. Decareau, R.V. The microwave sterilization process. *Microwave World*, 15(2), 12, 1994.
58. Cross, G.A. and Fung, D.Y.C. The effect of microwaves on nutrient value of foods. *Food Sci. Nutr.*, 16, 355, 1982.
59. Bookwalter, G., Shulka, T., and Kwolek, W. Microwave processing to destroy Salmonellae in corn-soy-milk blends and effect on product quality. *J. Food Sci.*, 47, 1683, 1982.
60. Okmen, Z. and Bayindirli, A.L. Effect of microwave processing on water soluble vitamins: kinetic parameters. *Int. J. Food Properties*, 2, 255, 1999.
61. Welt, B. and Tong, C. Effect of microwave radiation on thiamin degradation kinetics. *J. Microwave Power Electromag. Energy*, 28, 187, 1993.
62. Ovesen, L., Jakobsen, T., Leth, T., and Reinholdt, J. The effect of microwave heating on vitamins B1 and E, and linoleic and linolenic acids and immunoglobulins in human milk. *Int. J. Foods Science Nutrition*, 47, 427, 1996.
63. Aktas, N. and Ozligen, M. Injury of *E. coli* and degradation of riboflavin during pasteurization with microwaves in a tubular flow reactor. *Lebensm.-Wiss. u.-Technol.*, 25, 422, 1992.
64. Watanabe, F., Abe, K., Fujita, T., Goto, M., Hiemori, M., and Nakano, Y. Effects of microwave heating on the loss of Vitamin B12 in foods. *J. Agric. Food. Chem.*, 46, 206, 1998.
65. Vikram, V.B., Ramesh, M.N., and Prapulla, S.G. Thermal degradation kinetics of nutrients in orange juice heated by electromagnetic and conventional methods. *J. Food Eng.*, 69, 31, 2005.
66. Koutchma, T. and Schmalts, M. Degradation of Vitamin C after alternative treatments of juices. *Proc. IFT*, Anaheim, CA, 16, 2002.
67. Brewer, M. and Begum, S. Effect of MW power level and time on ascorbic acid content, peroxidase activity and color of selected vegetables. *J. Food Process. Preserv.*, 27, 411, 2003.

68. Murcia, M.A., Martinez-Tome, M., del Cerro, I., Sotillo, F., and Ramirez, A. Proximate composition and vitamin E levels in egg yolk: losses by cooking in a microwave oven. *J. Sci. Food Agric.*, 79, 1550, 1999.
69. Ovesen, L., Jakobsen, T., Leth, T., and Reinholdt, J. The effect of microwave heating on vitamins B1 and E, and linoleic and linolenic acids and immunoglobulins in human milk. *Int. J. Food Sci. Nutrition*, 47, 427, 1996.
70. Yoshida, H. and Takagi, S. Vitamin E and oxidative stability of soya bean oil prepared with beans at various moisture contents roasted in a microwave oven. *J. Sci. Food Agric.*, 72, 111, 1996.
71. Datta, A. and Hu, W. Optimization of quality in microwave heating. *Food Technol.*, 46(12), 53, 1992.
72. Ohlsson, T. Sterilization of foods by microwaves. *Int. Seminar New Trends in Aseptic Processing and Packaging of Foods Stuffs*, Munich, Germany, 22, 1987.
73. Ohlsson, T. and Bengtsson, N. Microwave technology and foods. In *Advances in Food and Nutrition Research*, Taylor, S.L.., Ed. Academic Press, New York, NY, 66, 2001.
74. Datta, A.K. Mathematical modeling of microwave processing as a tool to study safety. *Proc. Meeting American Soc. Agric. Engineers*, 23, 1991.
75. Le Bail, A., Koutchma, T., and Ramaswamy, H.S. Modeling of temperature profiles under continuous tube-flow MW and steam heating conditions. *J. Food Process Eng.*, 23, 1, 2000.
76. Yakovlev, V.V. Comparative analysis of contemporary EM software for microwave power industry. In *Microwaves: Theory and Applications in Material Processing V. Ceramic Transactions, 111*. The American Ceramic Society, Westerville, OH, 551, 2000.
77. Weiland, T. A discretization method for the solution of Maxwell's equations for six-component fields. *Electron. Commun. (AEÜ)*, 31, 116, 1977.
78. Weiland, T. On the unique numerical solution of Maxwellian eigenvalue problems in three dimensions. *Particle Accelerators*, 17, 227, 1985.
79. QuickWave-3D. QWED Sp. z o.o., ul. Nowowiejska 28, lok. 32, 02-010 Warsaw, Poland, 1997–2008, http://www.qwed.com.pl/.
80. Feldman, D.A., Kiley, E.M., Weekes, S.L., and Yakovlev, V.V. Modeling of temperature fields in 1D and 2D heating scenarios with pulsing microwave energy. *Proc. 41ˢᵗ MW Power Symp.*, Vancouver, British Columbia, Canada, 130, 2007.
81. Veronesi, P., Rosa, R., Leonelli, C., and Garuti, M. Numerical simulation of a novel plasma source, *Proc. 10ᵗʰ Seminar Computer Modeling and Microwave Power Eng*, Modena, Italy, 19, 2008.
82. Fluent, Fluent, Inc., 10 Cavendish Court, Lebanon, NH, 1988-2007, http: //www.fluent.com.
83. Kopyt P. and Celuch-Marcysiak, M. Accurate modeling of microwave heating of rotated objects and verification of the results. *Proc. Future Food Conf.*, Gothenburg, Sweden, 75, 2003.
84. Kopyt, P. and Celuch-Marcysiak, M. Coupled electromagnetic and thermal simulation of microwave heating process. *Proc. 2ⁿᵈ Workshop Inform. Technologies and Computing Tech. Agro-Food Sector*, Barcelona, Spain, 51, 2003.
85. Kopyt, P. and Celuch-Marcysiak, M. Coupled FDTD-FEM approach to modeling of microwave heating process. *Proc. 5ᵗʰ IEE Conf. Computation in Electromagnetics*, Stratford-upon-Avon, UK, 171, 2004.
86. Kopyt, P. and Celuch, M. Coupled simulation of microwave heating effect with QuickWave-3D and Fluent simulation tools. *Proc. 10ᵗʰ Conf. Microwave and High Frequency Heating*, Modena, Italy, 440, 2005.
87. Kopyt, P. and Celuch, M. Towards a multiphysics simulation system for microwave power phenomena. *Proc. Asia-Pacific Microwave Conf.*, Suzhou, China, 5, 2877, 2005.
88. Cordes, B.G., Eves, E.E., and Yakovlev, V.V. Modeling-based minimization of time-to uniformity in microwave heating systems. *Proc. 11ᵗʰ Conf. Microwave and High Frequency Heating*, Oradea, Romania, 305, 2007.
89. Cordes, B.G. and Yakovlev, V.V. Computational tools for synthesis of a microwave heating process resulting in the uniform temperature field. *Proc. 11ᵗʰ Conf. Microwave and High Frequency Heating*, Oradea, Romania, 71, 2007.
90. Cordes, B.G., Eves, E.E., and Yakovlev, V.V. Modeling-based synthesis of a microwave heating process producing homogeneous temperature field. *Proc. ACES Conf. Applied Computational Electromagnetics*, Niagara Falls, Ontario, Canada, 542, 2008.
91. Monk, P. and Suli, E. A convergence analysis of Yee's scheme on non-uniform grids. *SIAM J. Num. Anal.*, 31, 393, 1994.
92. PHYSICA, Physica Ltd., 3 Rowan Drive, Witney, Oxon, UK, 1996–2007, http: //www.physica.co.uk.

93. Knoerzer, K., Regier, M., Schubert, H., and Schuchmann, H.P. Simulation of microwave heating processes and validation using magnetic resonance imaging (MRI). *Proc. 39th Microwave Power Symp.*, Seattle, WA, 21, 2005.

94. Eves, E.E. and Yakovlev, V.V. Analysis of operational regimes of a high power water load. *J. Microwave Power Electromag. Energy*, 37, 127, 2002.

95. Meredith, R.J. *Engineers' Handbook of Industrial Microwave Heating*, IEE, London, 1998.

96. Mechenova, V.A. and Yakovlev, V.V. Efficient optimization of S-parameters of systems and components in microwave heating. *Proc. 3rd World Congress Microwave and Radio Frequency Processing*, Sydney, Australia, M4A.24, 2002.

97. Requena-Pérez, M.E., Monzó-Cabrera, J., Pedreo-Molina, J.L., and Díaz-Morcillo, A. Multimode cavity efficiency optimization by optimum load location—experimental approach. *IEEE Trans. Microwave Theory Tech.*, 53, 2114, 2005.

98. Monzó-Cabrera, J., Díaz-Morcillo, A., Domínguez-Tortajada, E., and Lozano-Guerrero, A. Application of genetic algorithms for microwave oven design: power efficiency optimization. In *Nature Inspired Problem-Solving Methods in Knowledge Engineering*, Mira, J. and Álvarez, J.R., Eds. Springer, Berlin, 580, 2007.

99. Murphy, E.K. and Yakovlev, V.V. FDTD-backed RBF network technique for efficiency optimization of microwave structures. *Proc. 9th Conf. Microwave and High Frequency Heating*, Loughborough, UK, 197, 2003.

100. Murphy, E.K. and Yakovlev, V.V. RBF ANN optimization of systems represented by small FDTD data sets, *Proc. 10th Conf. Microwave and High Frequency Heating*, Modena, Italy, 376, 2005.

101. Murphy, E.K. and Yakovlev, V.V. RBF network optimization of complex microwave systems represented by small FDTD modeling data sets. *IEEE Trans. Microwave Theory Tech.*, 54, 3069, 2006.

102. Kolomeytsev, V.A. and Yakovlev, V.V. Family of operating chambers for microwave thermal processing of dielectric materials. *Proc. 29th Microwave Power Symp.*, Chicago, IL, 181, 1993.

103. Bernhard, J.T. and Joines, W. Dielectric slab-loaded resonant cavity for applications requiring enhanced field uniformity. *IEEE Trans. on Microwave Theory Tech.*, 44, 457, 1996.

104. Sundberg, M., Kildal, P.-S., and Ohlsson, T. Moment method analysis of a microwave tunnel oven. *J. Microwave Power Electromag. Energy*, 33, 36, 1998.

105. Lurie, K.A. and Yakovlev, V.V. Method of control and optimization of microwave heating in waveguide systems. *IEEE Trans. Magnetics*, 35, 1777, 1999.

106. Bradshaw, S., Delport, S., and van Wyk, E. Qualitative measurement of heating uniformity in a multimode microwave cavity. *J. Microwave Power Electromag. Energy*, 32, 87, 1997.

107. Wäppling-Raaholt, B., Risman, P.O., and Ohlsson, T. Microwave heating of ready meals—FDTD simulation tools for improving the heating uniformity. In *Advances in MW and Radio Frequency Processing*, Willert-Porada, M., Ed. Springer, Berlin, 243, 2006.

108. Nelson, S.O. and Datta, A.K. Dielectric properties of food materials and electric field interactions. In *Handbook of Microwave Technology for Food Applications*, Datta, A.K. and Anantheswaran, R.C., Eds. Marcel Dekker, Inc., New York, NY, 69, 2001.

109. Tang, J. Dielectric properties of foods. In *The Microwave Processing of Foods*, Schubert, H. and Regier, M., Eds. CRC Press, Boca Raton, FL, 21, 2005.

110. Mudgett, R. Microwave properties and heating characteristics of foods. *Food Technol.*, 40, 84, 1986.

111. Wang, Y., Wig, T., Tang, J., and Hallberg. L. Dielectric properties of foods relevant to RF and microwave pasteurization and sterilization. *J. Food Eng.*, 57, 257, 2003.

112. Deshpande, M.D., Reddy, C.J., Tiemsin, P.I., and Cravey, R. A new approach to estimate complex permittivity of dielectric material at microwave frequency using waveguide measurements. *IEEE Trans. Microwave Theory Tech.*, 45, 359, 1997.

113. Coccioli, R., Pelosi, G., and Selleri, S. Characterization of dielectric materials with the finite-element method. *IEEE Trans. Microwave Theory Tech.*, 47, 1106, 1999.

114. Olmi, R., Pelosi, G., Riminesi, C., and Tedesco, M. A neural network approach to real-time dielectric characterization of materials. *Microwave Opt. Tech. Lett.*, 35, 463, 2002.

115. Thakur, K.P. and Holmes, W.S. An inverse technique to evaluate permittivity of material in a cavity. *IEEE Trans. Microwave Theory Tech.*, 49, 1129, 2001.

116. Santra, M. and Limaye, K.U. Estimation of complex permittivity of arbitrary shape and size dielectric samples using cavity measurement technique at microwave frequencies. *IEEE Trans. Microwave Theory Tech.*, 53, 718, 2005.

117. Wäppling-Raaholt, B. and Risman, P.O. Permittivity determination of inhomogeneous foods by measurement and automated retro-modeling with a degenerate mode cavity. *Proc. 9th Conf. Microwave and HF Heating*, Loughborough, UK, 181, 2003.

118. Eves, E.E., Kopyt, P., and Yakovlev, V.V. Determination of complex permittivity with neural networks and FDTD modeling. *Microwave Opt. Tech. Lett.*, 40, 183, 2004.
119. Yakovlev, V.V., Murphy, E.K., and Eves, E.E. Neural networks for FDTD-backed permittivity reconstruction. *COMPEL*, 33, 291, 2005.
120. Pitarch, J., Contelles-Cervera, M., Peñaranda-Foix, F.L., and Catalá-Civera, J.M. Determination of the permittivity and permeability for waveguides partially loaded with isotropic samples. *Meas. Sci. Technol.*, 17, 145, 2006.
121. Terhzaz, J., Ammor, H., Assir, A., and Mamouni, A. Application of the FDTD method to determine complex permittivity of dielectric materials at microwave frequencies using a rectangular waveguide. *Microwave Opt. Tech. Lett.*, 49, 1964, 2007.
122. Eves, E.E., Murphy, E.K., and Yakovlev, V.V. Practical aspects of complex permittivity reconstruction with neural-network-controlled FDTD modeling of a two-port fixture. *J. Microwave Power Electromag. Energy*, 41, 41.4.81, 2007.
123. Kopyt, P., Soltysiak, M., and Celuch, M. Technique for model calibration in retro-modeling approach to electric permittivity determination. *Proc. 17ᵗʰ Int. Conf. Microwaves, Radar and Wireless Communications (MIKON-2008)*, Wrocław, Poland, 2008.
124. Acceleware, Acceleware Corp., 1600 37th St. SW, Calgary, Alberta, Canada, http: //www.acceleware.com/.
125. Iwabuchi, I., Kubota, T., Kashiwa, T., and Tagshira, H. Analysis of electromagnetics field in a waveguide feed microwave oven by FDTD method. *Japanese Electronic, Inform. Communication Soc. Spring Meeting*, 2, 564, 1994.
126. Sundberg, M., Risman, P.O., and Kildal, P.-S. Quantification of heating uniformity in multi-applicator tunnel ovens, *Proc. 5ᵗʰ Conf. Microwave and High Frequency Heating*, Cambridge, UK, A5.1, 1995.
127. Wojtasiak, W., Gryglewski, D., and Gwarek, W. High-frequency-stability microwave high-power sources, *Proc. 9ᵗʰ Conf. Microwave and High Frequency Heating*, Loughborough, UK, 305, 2003.

24 Ohmic Heating in Food Processing

Fa-De Li
Shandong Agricultural University

Lu Zhang
The University of Auckland

CONTENTS

24.1 DEFINITION AND PRINCIPLE

Ohmic heating of food is defined as a process where (primarily alternating) electric currents are directly passed through an electrical conductive food directly touching the electrodes, to which sufficient power is supplied. At the same time, heat is produced in the form of internal energy generation within food. Hence, ohmic heating is sometimes also referred to as Joule heating, electrical resistance heating, direct electrical resistance heating, electro-heating or electro-conductive heating.

In fact, the principle of ohmic heating is quite simple, as illustrated in Figure 24.1. It is based on Ohm's law and the heat generated by ohmic heating is derived by Joule's law. It is known as the Joule effect, expressing the relationship between heat generated by the current flowing through a

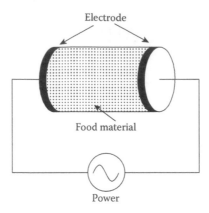

FIGURE 24.1   Schematic diagram of ohmic heating.

conductor. A quantitative form of Joule's law is that the heat evolved per second equals the current squared times the resistance.

24.2   HISTORY

Ohmic heating technology for food processing is not a new concept. It can be traced back to the nineteenth century, when a number of attempts had been made and several processes were patented and described that ohmic heating technology was used for food thawing, blanching, pasteurization, sterilization, and heating of solid foods.[1] In the early twentieth century, ohmic heating was called "electro-pure process".[2] Successful commercial ohmic heating techniques were developed for milk pasteurization,[3] and by the late 1930s, over 50 electric sterilizer units were in operation, serving about 50,000 consumers globally.[4] Unfortunately, such a technology were not widely applied because of technical limitations such as electrode materials, the absence of control equipment, etc. In addition, the decreasing costs of energy from other sources such as oil and gas, gradually made the ohmic heating process uneconomical, resulting in the absence of its use for food processing in succeeding years.[1]

Since the late 1980s, ohmic heating technology has become more attractive again, partially due to the availability of improved electrode materials[5] and designs.[6,7] Moreover, the ohmic heating system can be incorporated with an aseptic filling and packaging system into complete product sterilization or cooking process. A large number of potential applications involving heat transfer exist for ohmic heating including its applications in blanching, evaporation, dehydration, fermentation and extraction[8] as well as pasteurization and sterilization, thawing, grilling and solidification.

Ohmic heating is currently being used in Europe, Asia, and North America to produce a variety of safe and high quality foods, such as low- and high-acid products containing particulates, even whole fruits,[8] liquid egg, and juice.

24.3   FACTORS AFFECTING OHMIC HEATING PERFORMANCE

It is known that a number of factors affect ohmic heating performance for food heating treatment; these factors include the components, physical parameters of food, such as the electrical conductivity, density, specific heat capacity, thermal conductivity and viscosity, and the parameters of the ohmic heating system, such as the features of food (single phase or food mixture), the electrical field intensity, and the configurations of ohmic heater and the electrodes. For the particle-fluid food mixture, the geometry, size (including the size distribution), the orientation and the concentration of the particle in the food mixture play important roles in the heating process.

According to the principle of ohmic heating, the heat generation rate in the center of food materials is proportional to the square of the electric field strength and electrical conductivity of food materials (discussed in Section 24.4.1). Due to the fact that electric field intensity can be easily varied by adjusting electrode gap or the applied voltage, the most important factor is the electrical conductivity of the food and its temperature dependence. Therefore, the following discussion focuses mainly on the effect of some factors on the electrical conductivity of food.

It has been determined that electrical conductivity of food should be in the range 0.01 and 10 Sm⁻¹ at 25°C for ohmic heating process.[9] If electrical conductivity is greater than 10 Sm⁻¹, food does not become hot because current passes through them very easily with a lower resistance. However, when electrical conductivity is too low (under 0.01 Sm⁻¹), it is difficult for current to pass through the food at all.[*] On the other hand, some parameters affecting electrical conductivity of food materials must be carefully considered: firstly, the electrical conductivity of food may be anisotropic, especially for food containing organism tissue; secondly, the value of electrical conductivity increases with increasing temperature. However, these parameters do not apply to food materials during ohmic heating, especially for materials containing raw starch granules.[10,11] This means that the relationship between electrical conductivity and temperature of food is not linear during heating by ohmic heating method.

24.3.1 SOLID FOOD

For most solid food materials of biological origin, when it is heated by conventional method, the electrical conductivity will increase sharply at around 60°C, due to structural changes (e.g., cell wall protopectin and tissue damage) which increases ionic mobility (Figure 24.2). However, when the same cellular tissue is subjected to ohmic heating, the relationship between electrical conductivity and temperature becomes more and more linear as the electric field strength increases (Figure 24.3). It is explained by Palaniappan and Sastry that the electro-osmotic effects could possibly promote the effective conductivity at low temperatures.[5,12]

When an electric field with sufficiently high intensity is supplied to solid food during ohmic heating, the relationship between electrical conductivity and temperature is expressed as follows:[5,12,13]

$$\sigma_T = \sigma_{ref}[1 + m(T - T_{ref})] \tag{24.1}$$

where σ_{ref} is the electrical conductivity (Sm⁻¹) at reference temperature T_{ref} (°C); σ_T is the electrical conductivity (Sm⁻¹) at any temperature T (°C); and m is the temperature coefficient (°C⁻¹).

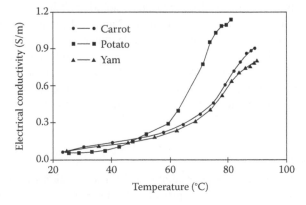

FIGURE 24.2 Electrical conductivity curves for vegetable tissue during convectional heating. (From Palaniappan, S. and Sastry, S.K., *Journal of Food Process Engineering*, 14, 221, 1991. With permission.)

---

* Pitte, G. and Brodeur, C. Ohmic cooking for meat products: the heat is on, http://res2.agr.gc.ca/crda/pubs/art10_e.htm

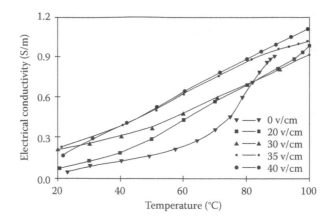

FIGURE 24.3 Electrical conductivity curves for carrot (parallel to stem axis), subjected to various voltage gradients. (From Palaniappan, S. and Sastry, S.K., *Journal of Food Process Engineering*, 14, 221, 1991. With permission.)

In most cases, especially for modeling the ohmic heating process, Equation 24.1 can be written in simple style:

$$\sigma = \sigma_0(1 + mT) \tag{24.2}$$

where σ_0 is the electrical conductivity (Sm^{-1}) when the reference temperature is equal to 0°C; and m is the temperature coefficient (°C^{-1}).

24.3.2 FLUID FOOD CONTAINING ELECTROLYTE OR NONELECTROLYTE

For fluid food with soluble components at constant concentration, electrical conductivity always has a linear relationship with temperature regardless of the heating methods used, which means that Equations 24.1 and 24.2 can be used for describing the relationship between electrical conductivity and temperature of liquid foods. However, it is found that the concentration of solute in liquid food influences electrical conductivity. Hence, the relationship between the electrical conductivity and concentration of soluble components of liquid food is very complex. Some components may influence electrical conductivity of food, based on their electrolytic characteristics.

For aqueous liquid food, such as sodium phosphate solutions,[14] hydrocolloids (containing xanthan, carrageenan, pectin, gelatin or pre-gelled starch),[15] and soybean milk,[16] the electrical conductivities increase with increasing concentration of soluble materials.

Some soluble nonelectrolytic components such as sugar in liquid food or mixed food such as fruit juices[17] and strawberry pulp,[18] have negative effect on electrical conductivities. For example, when concentrated apple and sourcherry juices have different concentrations of soluble solids (20%, 30%, 40%, 50%, and 60%), the electrical conductivities are correlated to concentration and temperature by the equation written as follows:[17]

$$\sigma = aC^n + bT + d \tag{24.3}$$

where a is the empirical concentration constant (Sm^{-1} (mass fraction $\times$ 100)$^{-n}$); b is the empirical temperature constant (Sm^{-1} K^{-1}); d is the empirical constant (Sm^{-1}); n is the constant, being greater than one for all voltage gradients applied; C is total soluble solid concentration of the apple and sourcherry juices (%); T is the temperature (°C). Because the constant a have a negative value, it can be concluded that the drop in electrical conductivity will be at a higher rate as the solids concentration increases.

The reason for the decrease in electrical conductivity with increasing concentration of nonelectrolytes such as fat, oil, sugar, etc., is that nonionic constituents cause an increase in resistance to ions movement.[17,19] This issue is very important in commercial-scale heat treatments especially in concentrated fruit juice processing plants.

24.3.3 Food Mixture Containing Particle(s)

For fluid-particle food mixture the size and concentration of the solid phase (particle) have been confirmed to affect electrical conductivity. In particular, particle content has a significant influence on the electrical property of the food mixture and may have a crucial effect on the performance of the heat treatment process.[19–24] In general, the electrical conductivity of the food mixture increases linearly with the increase in temperature. This result is confirmed by some researchers.[9,11,23,25,26] In addition, it is also noticed that electrical conductivity of fluid-particle mixture decreases with the increase in particle concentration. For fluid containing small particles (such as orange and tomato juices[25]) and fluid with very fine particles (such as raw starch granules[11]), the particles can be dispersed uniformly in the fluid phase the effective electrical conductivity decreases as the concentration of particles separated in the fluid phase increases. Palaniappan and Sastry developed an equation for describing the relationship between electrical conductivity and concentration of particles for tomato and orange juice. They derived the electrical conductivity as a function of temperature as follows:[25]

$$\sigma_T = \sigma_{ref}[1 + K_1(T - T_{ref})] - K_2 C \tag{24.4}$$

where C is the solids content (% w/w); σ_{ref} is the electrical conductivity (Sm^{-1}) of the food mixture at the reference temperature T_{ref} (°C); and K_1 and K_2 are constants.

It has also been confirmed that a special change in electrical conductivity of food containing raw starch granules will occur in the gelatinization temperature range during ohmic heating.[10,11,27] For native starch granule suspension, the relationship between electrical conductivity and temperature of starch suspension during ohmic heating is linear before and after starch gelatinization, but not linear during gelatinization as shown in Figure 24.4. The electrical conductivity of raw starch suspension decreases during gelatinization. The results shown in Figure 24.4 also indicate that the increase in the concentration of starch granules decreases electrical conductivities of the starch suspension.

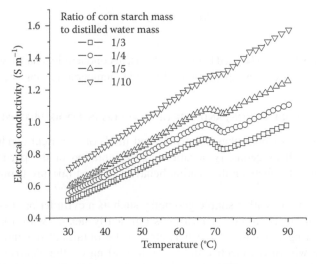

FIGURE 24.4 The electrical conductivity–temperature curves for different concentration starch suspensions. (Reprinted from Li, F.-D. et al., *Journal of Food Engineering*, 62, 113, 2004. With permission from Elsevier.)

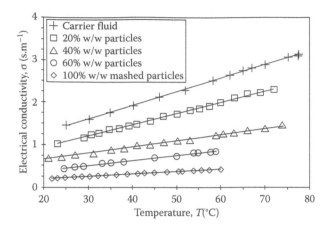

FIGURE 24.5 Electrical conductivity versus temperature and particle concentration (red bean suspension in carrier fluid). (Reprinted from Zareifard, M.R. et al., *Innovative Food Science and Emerging Technologies*, 4, 45, 2003. With permission from Elsevier.)

When a food mixture is a homogeneous suspension composed of large particles, the change in the electrical conductivity with temperature and concentration of the particles are similar to that of food mixture containing fine particles.[23] A typical curve of electrical conductivity versus temperature and particle concentration (red bean suspension in carried fluid) is shown in Figure 24.5. The relationship between particle concentration and electrical conductivity can be described by an exponential function as follows:

$$\sigma = \beta(C) \cdot T + \sigma_{0°C}(C) \tag{24.5}$$

where C is the particle concentration (% w/w); T is the temperature (°C); while $\sigma_{0°C}(C)$ and $\beta(C)$ are functions of particle concentration[9]

$$\sigma_{0°C}(C) = 0.6381 \cdot e^{-0.021C} \tag{24.6}$$

$$\beta(C) = 0.336 \cdot e^{-0.018C} \tag{24.7}$$

Although Equation 24.5 can be used to predict electrical conductivity change with temperature and concentration of solid particles, its application is limited to temperature range between 20 and 80°C.

24.3.4 GEOMETRY, ORIENTATION, AND SIZE OF PARTICLES IN FOOD MIXTURE

Electrical conductivities of both phases (particle and fluid) or electrical conductivity ratio of the particle to fluid, as well as geometry, location, and orientation of the particle within the food mixture have significant effects on the heating behavior of food mixture using the ohmic heating technique.[28–30]

When a single particle, with a simple geometry such as a cylinder or slice dispersed in fluid phase with different electrical conductivities is ohmically heated with a static ohmic heater, the differential heating behavior of the particle can be obtained, depending on the orientation of the particle within the electric field and depending on the electrical conductivities of both phases. When a particle with lower conductivity than the fluid is heated with its longer axis parallel to the electric field, the temperature of the particle would lag behind the fluid; in contrast, when the longer axis of this particle is perpendicular to the electric field, the particle heats faster than the liquid. On the other hand, if the particle is of higher electrical conductivity than the liquid, the results will be reversed: when the axis of the particle is parallel to the electric field, the heating rate of the particle is higher than that of the liquid; if the axis of the particle is perpendicular to the electric field, the temperature of the particle would lag behind the fluid.[28] Some researchers have confirmed these results using parallel circuit or series circuit approximation.[24,29,31]

For a mass of particles with a relatively small dimension existing in a fluid phase, its thermal behavior is affected by its distribution in the fluid (or in the ohmic heating chamber). In practice, it is expected that the particle distribution in the fluid phase is always uniform during ohmic heating. However, in some cases, the mass of particles would sink to the bottom or gather on one side of the static ohmic heating chamber. The former case can be considered as a parallel condition, i.e., the fluid phase and particles form a parallel-circuit in ohmic heating chamber; the latter can be regarded as a series-circuit. For example, in the parallel condition, the fluid phase (4.0% w/w thermo-flo starch solution with 0.5% w/w salt) heats faster than the solid phase (6 mm blanched carrot cubes) while in the series condition, the reverse was observed.[23] The reason for this phenomenon can be explained by using an electrical analogy approach; when the longer axis of the solid phase (particle or mass particles) is parallel to the electric field, the solid phase is in parallel with the liquid phase. In contrast, when the longer axis of the solid phase is perpendicular to the electric field, the solid phase is in parallel with the liquid phase. According to electrical principle, if resistance is lesser in a parallel circuit or greater in a series circuit, it will generate more heat. Thus, the heating rate of the corresponding resistance would be higher.

If the ratio of the length and breadth to the thickness of the particle is greater, the geometrical factors rather than anisotropy have major influence in determining the heating rates of both phases,[28] and if the particles are spheres, cylinders with length equal to diameter, and cubes, the effect of orientation on overall electrical conductivity is small.[14]

In addition, the effects of ratio of particle-to-heater cross-sectional area as well as particle length on heating characteristics of mixtures during ohmic heating[24] cannot be ignored. For the same particle content, the electrical conductivity of food mixture will decrease with increasing particle size, which means that smaller particles may reduce the resistance for ionic movement.[23] In fact, the particles cause an added resistance to the current passage, thus lowering the overall electrical conductivity.[18]

24.3.5 COMPONENTS OF FOOD AND PRETREATMENT

For food mixture, it is expected that the electrical conductivities of respective multicomponent or phase are largely matching each other, which can make a uniform temperature distribution in food during ohmic heating process. Most foods, especially pumpable foods with water content exceeding 30%, can be subjected to ohmic heating because they have enough free water with dissolved ionic electrolytes (such as various salts) to conduct electricity sufficiently well. However, some food materials, for example, covalent, nonionized fluids (such as fats, oils, alcohols and sugar syrups), and nonmetallic solids (such as bone, cellulose and crystalline structures including ice and raw starch granule) cannot be directly heated with ohmic heating technology. In addition, if the food materials have lower electrical conductivity, they also cannot be subjected to ohmic heating process. In these cases, some pretreatments such as adding salt,[11,32,33] salt infusion[34] or blanching[12,35] can be employed.

Although Equations 24.1 or 24.2 can be used for describing the relationship between the electrical conductivity and temperature, especially for solid or gelatinous food, under normal ohmic heating conditions, they do not suit foods containing fat and protein such as minced beef during ohmic heating. It has been discussed in Section 24.3.1 that the curves of the electrical conductivity versus temperature of the solid food (vegetables such as carrot, potato, and yam) containing biological tissue, which is heated with conventional method, exhibited a sharp transition at about 60°C.[12] It is also found that minced beef (mainly containing protein and fat) yields a transition in electrical conductivity at a critical temperature of 45–50°C during ohmic heating, but the trend in electrical conductivity is opposite to that of plant tissue. The electrical conductivity of minced beef increases up to the transition temperature and then the rate decrease as temperature increases. In this case, the relationship between electrical conductivity and temperature is not linear in the overall temperature range (30–60°C), but a typical band having an increasing trend with increasing temperature.[36] The reasons for this phenomenon may be that, firstly, the denaturation of proteins may cause changes in the ionic movements, and secondly, some water containing soluble ionic components may also be removed from the food during ohmic heating.

In comparison with the electrical conductivity of leaner meat (1–3.5 Sm^{-1}), the animal fat has a characteristically lower electrical conductivity (between 0.001 and 0.1 Sm^{-1} at 20°C.[37] Therefore, for minced meat, as fat content increases, the electrical conductivity decreases.[30,33,36] It is important to mention that the fat particles may remain colder than the surrounding leaner portions when food mixture consisting of fat particles and lean meat is cooked with ohmic heating.

24.3.6 ANISOTROPIC TISSUE OF FOOD

It is well known that solid foods such as some vegetables, meat, etc., are anisotropy. The physical properties of these kinds of foods are dependent on the direction of cellular tissue especially for electrical conductivity. As mentioned in Section 24.3.3, the geometrical factors rather than anisotropy primarily determines the heating rates of both phases during ohmic heating when the particles are spheres, cylinders with length equal to diameter, and cubes. However, the structural differences may influence the electrical conductivities of organisms such as muscle and vegetables, resulting from that the constitutive cells in organ tissues are usually neither spherical in shape nor symmetrical in structure. Therefore, the anatomical location of the organisms should be considered in designing processes for ohmic heating of organisms. Mitchell and Alwis[38] and Wang et al.[39] proved that the electrical conductivity of some food materials in different directions is not the same. Some results obtained by them are listed in Table 24.1. According to the datum listed in Table 24.1, the marked differences of electrical conductivities appear when materials such as carrot, leek, bamboo shoot, sugarcane, lettuce stem and mustard stem has distinct growth anisotropy. In addition, the skin of the vegetable also influences electrical conductivity, for example, the electrical conductivity of courgettes with skin is 1.7 Sm^{-1} but the electrical conductivity of it without skin is 3.9 Sm^{-1}.[38]

TABLE 24.1
Comparison of Measured of Value Conductivity for Some Vegetables

Vegetable	Measured Conductivity (S m^{-1})	
	Across Stem	Parallel to Stem Axis
Leeks stem	0.7	3.2
Carrot	2.5	4.2
Bamboo shoots	0.16	0.19
Sugarcane	0.09	0.22
Lettuce stem	0.44	0.23
Mustard stem	0.72	0.46

Source: Adapted from Mitchell, F.R.G. and de Alwis, A.A.P., *Journal of Physics, E: Science Instruments*, 22, 554, 1989 and Wang, C.S. et al., *Journal of Food Science*, 66, 284, 2001.

The electrical conductivity of meat is significantly influenced by the direction of the muscle fibers during ohmic heating. The value of the electrical conductivity of meat with muscle fibers parallel to the electric field is considerably higher than that with muscle fibers perpendicular to the electrical field, and the electrical conductivity of the minced meat is intervenient.[40]

24.3.7 Viscosity of Liquid or Food Mixture

Because a number of hydrocolloids are often used as stabilizer as well as emulsifying agents, homogeneous thick liquids are commonly available in a variety of food products such as puddings, sauces, etc., and carrier fluids for processing particle foods.[41] The effect of fluid viscosity on ohmic heating behavior of fluid-particle mixtures should be seriously considered in the design of ohmic heating processing.

When a food mixture is heated within a static ohmic heater, which consists of only one particle with cylinder or rectangular shape and a liquid with constant electrical conductivity but different, the temperature differences of the liquid phase between different locations, adjacent to the particle, especially around the sides and the front of the particle, increase with the increase in solution viscosity especially if the electrical conductivity of the fluid is not comparable with that of the particle. The reason for this phenomenon is firstly, the distortion of electric field that occurs around the particle due to differences of electrical conductivities causes different rates of heat generation around the particle and secondly, a lack of convection in a highly viscous fluid. However, if electrical conductivities of both the liquid and particle are similar to each other, the temperature difference will be much reduced.[26]

When a food mixture consisting of the same amount of potato cubes (0.7 cm side and particle volume fractions of 0.45) and the liquids (solution of sodium carboxymethyl cellulose (CMC) with sodium chloride), which is with higher electrical conductivity than potato cubes, is heated with a static ohmic heater, the heating rates of fluid and particle are not significantly affected by fluid viscosities. However, if the mixture is agitated in a static ohmic heater or processed with a continuous ohmic heating system, the heating rates of fluid and particles increase with the increase in viscosity of the liquid phase, which is impossible for conventional heating method because the higher viscosity of fluid retards heat transfer between the particle and fluid phases. In contrast, in ohmic heating, the effect of low heat transfer coefficient is compensated with the temperature difference between fluid and particles in the food mixture due to greater heat generating in the fluid with higher viscosity and the higher electrical conductivity. However, further investigation is needed to understand the physics of ohmic heating when electrical conductivity of the particle is greater than that of the fluid.[42]

24.3.8 Frequency and Waveform of Power

If a solid or particle food contains cell tissues, the electrical conductivity and heating rate of food will be influenced by the frequency and waveform of power. In general, the frequency of power supplied to food materials by electrodes is 50 Hz (standard frequency for alternating current in the United Kingdom) or 60 Hz (standard frequency for alternating current in the United States). In most practices, the typical frequency of an alternating current applied to ohmic heating system is the general frequency, 50 Hz or 60 Hz. It is agreed that the current density should not exceed 3,500 Am^{-2} when the lower frequency (50 Hz or 60 Hz) is used during food processed with ohmic heating method.[32,43] In addition, electrochemical processes at the electrode/solution interfaces must be considered. High frequency alternating currents can significantly inhibit electrochemical reaction resulting from minimal charging of electrical double layers. Some researchers performed their research with different waveform[44] power supply at multifrequency[43-45] or pulsed waveform[46,47] for preventing corrosion of electrode or electrochemical reaction.

It is known that biological tissues consist of cells, liquid, membranes, intracellular and extracellular fluids. From an electrical point of view, biological tissue components can be regarded as passive

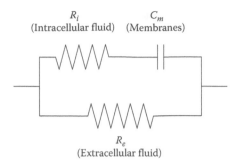

R_i C_m
(Intracellular fluid)   (Membranes)

R_e
(Extracellular fluid)

FIGURE 24.6   An equivalent circuit model for describing impedance behavior of biological tissue.

electrical elements (resistor, capacitor) connected in series and parallel. An electrical equivalent circuit model (Figure 24.6) for describing impedance behavior of biological tissue subjected to electrical fields had been proposed by Fricke.[48,49] The model is composed of a resistive element (R_e) representing extracellular fluids placed in parallel with a capacitive element (C_m) representing insulating membranes in series and a resistive element (R_i) representing intracellular fluids. Due to the existence of the capacitive resistance (C_m) in the equivalent circuit model, the frequency has a significant effect on the total effective resistance of it. According to electrical principle, when power with lower frequency is applied to the model, current flows through the resistor (R_e) without penetrating the capacitor (C_m) and resistor (R_i). However, if power with higher frequency is applied to the model, the capacitor (the membrane) will lose its insulating properties and current can flow through both the resistors and capacitor. Therefore, the frequency of alternating current may be a key parameter in ohmic heating.

For example, when Japanese white radish tissue is heated with a batch ohmic heating device at 40 Vcm^{-1} at different frequencies (50 Hz–10 kHz), the heating rate of the sample increases with decreasing frequency until the temperature reaches about 50°C. However, it becomes almost independent of frequency when temperature is above 50°C, although the impedance of the sample decreases with increasing frequency.[50] Lima et al. also found similar results when turnip tissue was ohmically heated with different frequencies and wave shapes.[44] The reason for this phenomenon is that the influence of frequency and waveform on electrical conductivity is increasingly pronounced as frequency decreases.

However, for some fish foods such as surimi paste and stabilized mince (made from Pacific whiting) as well as Alaska pollock (*Theragra chalcogramma*) surimi-starch paste, the impedance of the sample decreases slightly with increasing frequency at frequencies from 55 Hz to 200 kHz[43] and the electrical conductivity of surimi-starch paste increases when frequency increases.[51]

On the other hand, biological tissues especially muscle are anisotropic. The orientation of vascular fibers to electric field strongly influences the electrical conductivity or impedance of biological tissue especially at low frequency. Damez et al. demonstrated that the impedance in the direction transverse to the grain of beef muscle fibers is roughly $\pi/2$ times that observed in the longitudinal direction at low frequency, and that the anisotropy disappears at high frequencies, since the cellular membranes no longer act as electrical barriers.[48]

24.4   MODELING OF OHMIC HEATING

Heating rate is always predicted for designing and optimizing ohmic heater. A number of studies have attempted either to model the ohmic heating process or to characterize the electrical conductivities of selected food materials. However, multiphase foods are generally complex materials of irregular geometry and composed of components with different electrical conductivities. For example, as mentioned in Section 24.3, there are many factors influencing electrical conductivity, heating rates and temperature distribution of food mixture during ohmic heating process. On the other hand, in addition to involving the electric field and temperature field for modeling the ohmic heating process, the motion of food must be considered especially in a continuous ohmic heating system. Generally, the more complex the model, the more accurate the results will be. However, it is not possible to consider all the phenomena involved in the process to be modeled. For simplifying a complex phenomenon, some assumptions must be made, and in order to simulate a special ohmic heating process or to solve a model, some initial and boundary conditions for food materials, electric field and temperature field must be specified.

Before discussing a model for an ohmic heating system, some basic equations must be discussed.

24.4.1 JOULE'S LAW

According to Joule's law, heat generated from food during ohmic heating is proportional to the square of the current. In general, Joule's law is expressed as:

$$Q = I^2 \cdot R \cdot t \tag{24.8}$$

where Q is the electrical heat (J); I is the current (A); R is the electrical resistance (Ω), and t is the time (s).

If the materials obey Ohm's law, the electric field intensity at any point in the material is determined from:

$$J = \sigma E \tag{24.9}$$

where J is the current density (Am^{-2}); σ is the electrical conductivity (Sm^{-1}) of the medium; and E is the electric field intensity (Vm^{-1}). At each point in the medium, the electric potential V, can be found from:

$$E = -\nabla V \tag{24.10}$$

Combining Ohm's law, for the case of constant voltage, the differential equation of Joule's law is expressed as:

$$\dot{Q} = \sigma |\nabla V|^2 \tag{24.11}$$

where $\dot{Q}$ is the volumetric internal heat generation rate (Wm^{-3}); ∇V is the voltage gradient (Vm^{-1}).

According to Equation 24.11, both the electrical conductivity and voltage gradient of the local point in food affect the local rate of heat generation. In addition, the electrical conductivity not only influences the local rate of heat generation directly, but also governs the electric field distribution and thus, the local rate of heat generation.

24.4.2 VOLTAGE DISTRIBUTION

The voltage distribution within a medium can be developed from Maxwell's equations, or by combining Ohm's law and the continuity equation for current:[52]

$$\nabla \cdot (\sigma \nabla V) + \frac{\partial \rho_c}{\partial t} = 0 \tag{24.12}$$

where ρ_c is the density of charge (Cm^{-3}); t is the time (s).

For a steady-state case, the density of charge is constant; therefore, the voltage is given by:

$$\nabla \cdot (\sigma \nabla V) = 0 \qquad (24.13)$$

Because the voltage field in the material is a function of some variables, such as: (1) electrode and material geometries, (2) electrical conductivity of the material and (3) the voltage applied to the electrode,[53] appropriate electrical boundary conditions depending on these factors must be known for solving Equation 24.13. They can be given in general as follows:[46]

$$V(x,y,z,t)\big|_{\text{wall}} = 0 \text{ V} \qquad (\text{ground})$$
$$V(x,y,z,t)\big|_{\text{wall}} = V_0 \text{ V} \qquad \forall t,\, x,\, y,\, z \in [\text{electrodes}] \qquad (24.14)$$

$$\nabla V(x,y,z,t) \cdot \vec{n}\big|_{\text{wall}} = 0, \qquad \forall t, x, y,\, z \in [\text{vessel or heater surface except electrodes}] \quad (24.15)$$

where V_0 is the voltage (V) supplied to the electrodes; $\vec{n}$ is the normal vector.

Because the geometry of electrode is important parameters during ohmic heating, the type of electrode geometry should be discussed.

As an ohmic heating system for food processing, the main modules are a heating chamber (or an ohmic column), manufactured with nonconductive material for containing food, in which a pair (or a series) of electrodes are mounted, an alternating power supply, and a control panel for controlling the voltage or current as well as the temperature of the food. For a continuous ohmic heating system, another main accessorial system is an appropriate pump system for conveying food.

There are many types of ohmic heating systems, depending upon difference in the operation mode and it can be classified into two types: batch processing system or static ohmic heater and continuous processing system. Irrespective of type, electrodes are considered the key part of the ohmic heating system.

For static ohmic heater, the most typical configurations is of a horizontal or an acclivitous insulated cylinder (Figure 24.7a) as well as a cubic insulated box (Figure 24.7b) with a pair of electrodes mounted in both ends of the tube (for cylinder) or one of wall (for cubic box) opposite to each other. A new typical batch ohmic heating system had been developed by Jun and Sastry.[46,47] They designed a flexible package (Figure 24.7c) for food reheating and sterilization. The package is manufactured using flexible pouch materials and incorporates a pair of foil electrodes with V-shaped configuration to permit ohmic heating of food materials. The electrode is placed between a folded laminate with the electrodes extending out and heat-sealed on the edges. In addition, Farid patented a new apparatus for cooking hamburger patties, which consists of two metallic plates acting not only as a heating plate for thawing the frozen patty and cooking it but also as an electrode. Each plate is connected to a source of alternating current so that the interior of the patties can be heated by an electric current passing through the patty as the patty is also heated by the plates (Figure 24.7d).[54,55]

For continuous ohmic heating process system, there are also many kinds of designs of ohmic heating column. However, according to the relationship between the directions of the electric field and the liquid movement, they can be classified into cross-field and in-field configurations.[56] If the stream of the food flows in the direction perpendicular to the electric field, the configuration is often termed cross-field. If the stream of the food flows in the direction parallel to the electric field, the configuration is termed in-field. For cross-field configurations, the types of heating columns or chambers include a simple duct (Figure 24.8a) with pairs of opposing

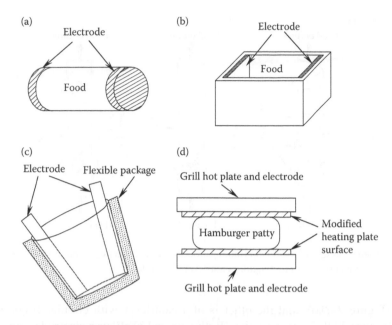

FIGURE 24.7 Basic configurations for batch processing ohmic heaters. (a) Cylinder, (b) cubic box, (c) flexible package, (d) grill hot plate. ((d) Jun, S. and Sastry, S., *Journal of Food Process Engineering*, 28, 417, 2005. Reproduced with permission from Wiley-Blackwell.)

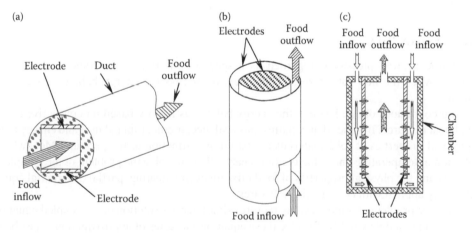

FIGURE 24.8 Some configurations for continuous process ohmic heaters termed cross-field. (a) Flat electrode, (b) coaxial tube electrode, (c) perforated flat electrode.

electrodes mounted on the duct wall opposite each other, a set of coaxial tubes (Figure 24.8b) acting as electrodes with the food flowing between them and a chamber with a pair of perforated flat plates (Figure 24.8c), which is based on displacement of the overheated boundary layer by the cold liquid flowing from the side channels through perforated electrodes into the heating zone.[*] For in-field configurations, there are two kinds of typical electrodes mounted in the ohmic heating columns, which is a vertical or gradient tube; one characteristic shape of the electrode

---

[*] Žitný, R., Šesták, J., and Zajíček, M. Continuous direct ohmic heating of liquids. (http://www.fsid.cvut.cz/en/u218/pe0ples/Zitny/Chisa98a.pdf)

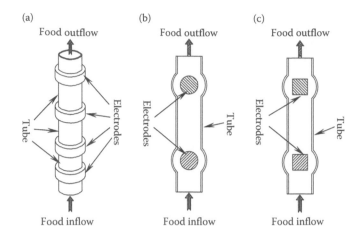

FIGURE 24.9 Some configurations for continuous process ohmic heaters termed in-field. (a) Ring electrode, (b) circular cantilever electrode, (c) rectangular cantilever electrode.

is of a ring (Figure 24.9a)[57] and the other is of a cantilever with circular (Figure 24.9b)[58] or rectangular (Figure 24.9c)[59] cross-section. Both ring and cantilever electrodes are embodied at regular intervals along the moving direction of food. The choice of the best configuration will obviously depend on the food being processed and the objectives of the process (e.g., cooking, pasteurization, and sterilization).

24.4.3 ENERGY BALANCE

Static ohmic heating process is regarded as a special condition of continuous ohmic heating processing when the flow rate is zero. Thus, we will discuss the energy balances of continuous treatment.

During continuous ohmic heating, the energy balance is mainly based on convective heat with internal heat resource. In general, it contains mass balance, momentum balance and energy balance. The physical properties of food materials are based on a function of temperature. Meanwhile, the distribution of temperature and velocity affects each other. To solve the problem of the complex heat convection and the physical properties of food mixture during heating, partial differential equations are normally applied with commercial softwares.

To solve the problem of ohmic heating requires simultaneous solution of the coupled equations of electric, thermal and flow fields. To solve these equations, a series of approximations must be made according to the specific situations, to simplify the problems to the point where it can be solved computationally.

24.4.3.1 Models for Simulating the Behavior of Food in Continuous Ohmic Heating System

24.4.3.1.1 Model for Liquid Food

A complete description of flow field requires solutions of mass and momentum conservation equations that govern the fluid motion. The differential equation of mass conservation in a incompressible fluid with a constant density can be written as:[60]

$$\nabla \cdot \vec{V} = \frac{\partial u}{\partial x} + \frac{\partial v}{\partial y} + \frac{\partial w}{\partial z} = 0 \tag{24.16}$$

For a fluid with constant density and viscosity, the momentum equation, known as the Navier–Stokes equation in vector form, is

$$\rho\frac{D\vec{V}}{Dt} = -\nabla p + \mu\nabla^2\vec{V} + \rho\vec{g} \tag{24.17}$$

In Equations 24.16 and 24.17, the quantity $\nabla\cdot\vec{V}$ denotes the divergence of the velocity vector; D/Dt is total derivative.

A general form of the energy equation for a fluid with constant thermal properties is given by:

$$\rho c_p\left(\frac{\partial T}{\partial t} + u\frac{\partial T}{\partial x} + v\frac{\partial T}{\partial y} + w\frac{\partial T}{\partial z}\right) = \lambda\left(\frac{\partial^2 T}{\partial x^2} + \frac{\partial^2 T}{\partial y^2} + \frac{\partial^2 T}{\partial z^2}\right) + \dot{Q}$$
$$+ 2\mu\left\{\left(\frac{\partial u}{\partial x}\right)^2 + \left(\frac{\partial v}{\partial y}\right)^2 + \left(\frac{\partial w}{\partial z}\right)^2\right\} + \mu\left\{\left(\frac{\partial u}{\partial y} + \frac{\partial v}{\partial x}\right)^2 + \left(\frac{\partial u}{\partial z} + \frac{\partial w}{\partial x}\right)^2 + \left(\frac{\partial v}{\partial z} + \frac{\partial w}{\partial y}\right)^2\right\} \tag{24.18}$$

where T is temperature (°C) at position (x, y, z); u, v and w are the fluid velocities (ms^{-1}) in the x, y and z directions, respectively; $\dot{Q}$ is the volumetric heat generation rate (Wm^{-3}); the λ, ρ, c_p are the thermal conductivity (Wm^{-1}K^{-1}), density (kgm^{-3}), and specific heat (Jkg^{-1}K^{-1}), respectively. The velocities u, v, and w can be obtained from solving the Navier–Stokes equation (Equation 24.17). Equation 24.18 is given here in Cartesian coordinates and the same equation in other coordinate system can be found in other textbooks. A common simplification is to drop the last two terms on the right hand side of Equation 24.18, representing viscous dissipation of heat. Dropping this term leads to the more familiar and simple heat equation:

$$\rho c_p\left(\frac{\partial T}{\partial t} + u\frac{\partial T}{\partial x} + v\frac{\partial T}{\partial y} + w\frac{\partial T}{\partial z}\right) = \lambda\left(\frac{\partial^2 T}{\partial x^2} + \frac{\partial^2 T}{\partial y^2} + \frac{\partial^2 T}{\partial z^2}\right) + \dot{Q} \tag{24.19}$$

For solving Equation 24.18, some initial and boundary conditions must be known. Due to limited knowledge about some phenomena and the physical properties of food materials, a number of assumptions must be made for special cases.

When a fluid food flows in an in-field continuous ohmic heating column, a vertical pipe is considered as the physical model, and the engineering complexities of getting electrodes into and out of the pipe are neglected (Figure 24.10).[61] In addition, some assumptions must be made for simplifying the problem in analysis: (1) the liquid food is uncompressible and its flow is a fully developed laminar flow, (2) the physical properties of fluid food are independent of temperature, except for density and electrical conductivity, and (3) the temperature or heat flux at the pipe wall is given. On the basis of these assumptions, it can be known that a nonuniform temperature distribution at any cross-section in the pipe exists as a result of variations in residence times of food flowing near the pipe wall and near the centre of the pipe resulting from the velocity gradient in the radius direction.

According to the above assumptions, the momentum Equation 24.17 for the vertical velocity (v) in the z direction, in cylindrical coordinates, can be simplified and reduced to Equation 24.20:

$$0 = -\frac{\partial p}{\partial z} + g(\rho - \rho_0) + \frac{1}{r}\frac{\partial}{\partial r}\left(\mu\, r\frac{\partial v}{\partial r}\right) \tag{24.20}$$

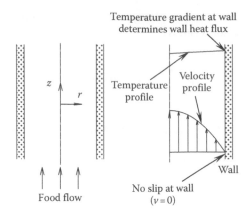

FIGURE 24.10 Ohmic heating column. (Reprinted from Quarini, G.L. *Journal of Food Engineering*, 24, 561, 1995. With permission from Elsevier.)

where ρ is the density of fluid food (kgm^{-3}), in this case, the density will change with temperature; ρ_0 is the initial fluid density (kgm^{-3}); v is the fluid velocity (ms^{-1}) in the z direction; p is the pressure of fluid food (Pa); μ is the viscosity of fluid food (Pas).

In comparison with convective heat, the conductive heat in the direction z is so small that it can be neglected. Therefore, the energy Equation 24.19 can be simplified and reduced to Equation 24.21 in cylindrical coordinate form:

$$\rho c_p v \frac{\partial T}{\partial z} = \dot{Q} + \frac{1}{r}\frac{\partial}{\partial r}\left(\lambda r \frac{\partial T}{\partial r}\right) \tag{24.21}$$

where λ is the thermal conductivity (Wm^{-1}K^{-1}) of the fluid; $\dot{Q}$ is the volumetric heat generation rate (Wm^{-3}).

Because the density of the liquid is not regarded as a constant, from Boussinesq approximation, can be written as follows:[62]

$$\rho = \rho_0[1 - \beta(T - T_0)] \tag{24.22}$$

where β is the thermal expansion coefficient of fluid (K^{-1}); T_0 and ρ_0 are the temperature (K) and density (kgm^{-3}) of fluid at the initial condition respectively; T is the temperature of the liquid food (K).

In order to solve Equation 24.21, $\dot{Q}$ must be known. It can be derived from solving Equations 24.11 and 24.13 with the corresponding electrical boundary conditions, Equations 24.14 and 24.15.

Therefore, Equations 24.11, 24.13, 24.20, 24.21 and 24.22 are the coupled equations for the electric, flow and thermal fields in this case; they can be solved with appropriate boundary conditions. For the flow and thermal fields, combining with the assumptions made above, the common boundary conditions can be written as follows:

$$\frac{\partial v}{\partial r} = 0; \quad \frac{\partial T}{\partial r} = 0 \text{ at } r = 0 \tag{24.23}$$

$$v = 0; \quad \lambda_f \frac{\partial T}{\partial r} = \lambda_w \frac{\partial T}{\partial r} \text{ at } r = R \tag{24.24}$$

where λ_f and λ_w are the thermal conductivities (Wm⁻¹K⁻¹) of the fluid and the pipe wall, respectively; R is the radius of the pipe (m).

If the pipe wall is adiabatic, Equation 24.24 for describing the boundary condition will be reduced to:

$$v = 0; \qquad \frac{\partial T}{\partial r} = 0 \text{ at } r=R \tag{24.25}$$

As mentioned in Section 24.3.1, the relationship between the electrical conductivity of the fluid food and the temperature is commonly expressed by Equation 24.2. Therefore, Equation 24.21 can be written as follows:

$$\rho c_p v \frac{\partial T}{\partial z} = |\nabla V|^2 \, \sigma_0(1+mT) + \frac{1}{r}\frac{\partial}{\partial r}\left(\lambda r \frac{\partial T}{\partial r}\right) \tag{24.26}$$

Thus, a set of coupling equations, such as Equations 24.13, 24.20, and 24.26 as well as Equation 24.22, can be used for describing the electric, flow and thermal fields when the density of fluid food is influenced by temperature.

Under the condition of constant density of fluid food during ohmic heating, if the conductive heat transfer is slow in comparison to the heat generation rate, and the tube wall is assumed adiabatic, Equation 24.26 can be simplified as follows:

$$\rho c_p v \frac{\partial T}{\partial z} = |\nabla V|^2 \sigma_0(1+mT) \tag{24.27}$$

Integration of Equation 24.27 with the boundary condition $T=T_0$ at $z = 0$ gives

$$\ln\left(\frac{1+mT}{1+mT_0}\right) = \frac{m|\nabla V|^2 \sigma_0 z}{\rho c_p v(r)} \tag{24.28}$$

If fluid food is a Newtonian fluid with constant density when its flow is a fully established laminar flow in a circular channel, the velocity distribution has the shape of parabola with the apex at the centerline of the pipe. Equation 24.20 can be integrated as follows:

$$v(r) = \frac{1}{4\mu}\frac{\partial p}{\partial z}(R^2 - r^2) \tag{24.29}$$

For the flow of a non-Newtonian fluid in a tube, the velocity profile, $v(r)$, may be a commonly used expression. For example, for a power law fluid of consistency coefficient K and flow behavior index n, the following expression may be used.[5]

$$v(r) = v_m\left(\frac{3n+1}{n+1}\right)\left[1-\left(\frac{r}{R}\right)^{n+1/n}\right] \tag{24.30}$$

where v_m is the mean fluid velocity (ms⁻¹) over the tube cross-section; and R is the radius of the pipe (m).

24.4.3.1.2 Models for Food Mixture

One of the advantages of ohmic heating process is that ohmic heating methods offer a way of processing food mixture with much better quality at the rate of high-temperature short-time (HTST) processes because both solid particles and liquid phase of the food materials have electrical conductivity which can result in a volumetric heating and uniform temperature distribution during ohmic heating. However, due to the presence of particles especially some large particles with diameter in the same order of magnitude as the ohmic heating column in the food mixture, the flow of the food mixture becomes a hydrodynamic perturbation, which is caused by irregular geometry, size and distribution of the particles. On the other hand, higher concentration of particles (the particle volume fraction is usually more than 50%) will increase the interparticle and particle–wall interactions during food flows through the ohmic heating column. From Section 24.3, it is known that the heating rate of particles is significantly different from the fluid because the electrical field is strongly modified when electrical conductivity of the particle does not match that of the fluid. In addition, the thermal fields of food mixture are directly influenced by the particle–fluid, particle–wall, and fluid–wall heat exchanges. The dependency of physical properties on temperature imposes a tight coupling between the dynamic and thermal fields as well as the electric field. An accurate simulation of such systems is of unresolved scientific difficulty and of very high cost.[63]

In order to simulate the temperature distribution in food mixture, a physical model is still necessary. Here, we discuss that food mixture is heated with a continuous ohmic heating system (as shown in Figure 24.9) during it flows through the heating column, which is a cylindrical tube surrounded by air at constant temperature. Regardless of the geometry and size of electrode, and ignoring the engineering complexities of getting electrodes into and out of the pipe, the only section of the process to be studied is the heating tube. Therefore, the electrical field is established longitudinally in ohmic heating column.

The voltage distributions for the particle and fluid are also obtained by solving Equation 24.13 holding with the electrical boundary conditions, Equations 24.14 and 24.15. In addition, the other special electrical boundary conditions on the axis ($r=0$) and at the wall of the tube ($r=R$) can be written as follows:

$$\frac{\partial V}{\partial r} = 0 \qquad (24.31)$$

Electrical conductivity has been described in Equation 24.2. The heat generation in particle and fluid during ohmic heating can also be derived from solving Equation 24.11.

For particle, the energy equation can be derived from Equation 24.19. In this case, where there is no movement in particle, Equation 24.19 can be reduced to Equation 24.32, for governing the temperature behavior of particle during ohmic heating process:

$$\rho_p c_{pp} \frac{\partial T_p}{\partial t} = \nabla \cdot (\lambda_p \nabla T_p) + \dot{Q}_p \qquad (24.32)$$

where ρ_p, c_{pp}, and λ_p represent the density (kgm^{-3}), specific heat (Jkg^{-1}K^{-1}), and thermal conductivity (Wm^{-1}K^{-1}) of the particle, respectively. The subscript p notes particle (similarly hereinafter).

Applying Equations 24.2 and 24.11 into Equation 24.32 produces:

$$\rho_p c_{pp} \frac{\partial T_p}{\partial t} = \nabla \cdot (\lambda_p \nabla T_p) + |\nabla V|^2 \sigma_{0p}(1 + m_p T_p) \qquad (24.33)$$

where σ_{0p} is the electrical conductivity (Sm^{-1}) of the particle at reference temperature (0°C); m_p is the temperature coefficient (°C^{-1}).

Equation 24.33 is subjected to the following initial condition and boundary condition, respectively:

$$T_p = T_0 \quad \text{at} \quad t = 0 \tag{24.34}$$

$$-\lambda_p \nabla T_p \cdot \vec{n}|_s = h_{fp}(T_{ps} - T_f) \tag{24.35}$$

where h_{fp} is the fluid–particle heat transfer coefficient (Wm^{-2}K^{-1}); T_f is the temperature of the fluid (K); T_{ps} is the temperature at the surface of the particle (K).

For the fluid phase, the temperature distribution can be obtained from:[58]

$$\rho_f c_{pf} \bar{v}_z v_{ff} \frac{\partial T_f}{\partial z} = \beta(v_{ff}) \nabla \cdot (\lambda_f \nabla T_f) - n_p A_p h_{fp}(T_f - T_{ps}) + \dot{Q}_f v_{ff} \tag{24.36}$$

where ρ_f, c_{pf}, and λ_f are the density (kgm^{-3}), specific heat (Jkg^{-1}K^{-1}) and thermal conductivity (Wm^{-1}K^{-1}) of the fluid, respectively. A_p is the surface area of a particle (m^2); n_p is the number of solid particles per unit volume of fluid mixture (m^{-3}); h_{fp} is the fluid–particle heat transfer coefficient (Wm^{-2}K^{-1}); T_f is the temperature of the fluid (K); T_{ps} is the temperature at the surface of the solid particle (K); v_{ff} is the fluid volume fraction in the ohmic heating column (–); $\bar{v}_z$ is the mean velocity of mixture in the axial direction. Assuming plug flow, it can be concluded that there is no slip velocity between the particle and fluid. Therefore, we can solve for $\bar{v}_z$. The subscript f notes fluid (similarly hereinafter).

In Equation 24.36, the first term in the right hand side indicates the conductive heat transfer through the mixture in the fluid phase. $\beta(v_{ff})$ represents the fraction of conductive heat transfer. Although the exact form of $\beta(v_{ff})$ is unknown, an expression may be derived based on the fact that conduction occurs across an elemental surface. Hence $\beta(v_{ff})$ should represent the area fraction of the fluid phase. Based on the Kopelman model, the area fraction of fluid is given by:[58]

$$\beta(v_{ff}) = 1 - (1 - v_{ff})^{2/3} \tag{24.37}$$

Orangi et al. pointed out that the particular form of this expression did not have a measurable effect on the results because of the very small conduction heat within fluid in compassion with heat generation and heat transfer terms in Equation 24.36. Compared with the radial condition term, the axial conduction term ($\partial^2/\partial z^2$) included in the first term on the right hand side of Equation 24.36 is also small (this situation corresponds to a large Peclet number) and can be neglected.[58]

The heat generation term in the fluid phase, on the right hand side of Equation 24.36 can be gained by applying Equation 24.2 into Equation 24.11. Therefore, Equation 24.36 can be written as follows:

$$\rho_f c_{pf} \bar{v}_z v_{ff} \frac{\partial T_f}{\partial z} = \beta(v_{ff}) \nabla \cdot (\lambda_f \nabla T_f) - n_p A_p h_{fp}(T_f - T_{ps}) + |\nabla V|^2 \sigma_{0f}(1 + m_f T_f) v_{ff} \tag{24.38}$$

where σ_{0f} is the electrical conductivity (Sm^{-1}) of the fluid at reference temperature (0°C); m_f is the temperature coefficient (°C^{-1}).

In order to obtain the solution of Equation 24.38, the initial and appropriate boundary conditions must be known. For the initial condition, the following applies:

$$T_f = T_0, \quad \text{at} \quad z = 0 \tag{24.39}$$

For the boundary condition, the balance between convective and conductive heat transfer on the outer surface of the tube is given by:

$$-\lambda_f \nabla T_f \cdot \vec{n}\big|_w = h_w (T_f - T_a)\big|_w \qquad (24.40)$$

where $\vec{n}$ is the unit vector normal to the surface of the tube's wall (namely w), h_w is the overall heat transfer coefficient (Wm^{-2}K^{-1}); T_a is the temperature of the air surrounding the walls of the heater (K); T_f is the temperature of the fluid (K). For obtaining the boundary condition at $r = R$, Equation 24.40 must be applied to Equation 24.38. Another boundary condition on the axis ($r = 0$) can be gained by applying Equation 24.38 in the limit as $r \rightarrow 0$.

In Equation 24.38, assuming that flow of the particle-fluid mixture is a homogeneous plug flow, the value of $\bar{v}_z$ can be easily obtained. However, in most general cases of flow with a velocity gradient (assuming a power law fluid behavior, or the food mixture is regarded as a composite non-Newtonian fluid), $\bar{v}_z$ can be described by Equation 24.29. In this case, $v(r)$ will replace $\bar{v}_z$ in Equation 24.38.[56,58]

Therefore, a set of coupled equations, such as Equations 24.13, 24.33, and 24.38 and their respective initial and boundary conditions, can be solved iteratively for this problem, utilizing the Crank–Nicholson scheme for Equation 24.38, a central-difference formula for Equation 24.13, Galerkin's method and the Crank–Nicholson scheme for Equation 24.33.

In the above discussion, it is assumed that both particle and fluid phases possess the same velocities during flow longitudinally through the ohmic heating column. However, even though the velocities of both particle and fluid phases on average are essentially equal, it has been tested that the slip velocity between the fluid and particles may exist although the fluid is of sufficiently high viscosity to ensure entrainment of particles and prevent phase separation.[64] On the basis of the principle of a mean slip velocity between particle and fluid phases in plug flow, Benabderrahmane and Pain established a model for simulating the thermal behavior of a particle–fluid mixture flowing in an ohmic heating sterilizer. In their model, the thermal diffusion in particles and the particle–fluid slip velocity are taken into account.[63] For details, please refer to the respective publications.

Another mathematical model for describing thermal behavior of food mixture during heating with a continuous ohmic heating column is developed by Sastry[19] on the basis of a circuit analogy to approximate electrical conductivity. For food mixture, it is difficult to solve Equation 24.13 for every particle within the continuous ohmic heating column since the location and properties of every particle at all points in time must be known. In this model, some assumptions are included that: (1) the concentration of particles in the mixture is high, and both particle and fluid phases are uniformly mixed. In addition, the flow of the mixture in the ohmic heating column is of a plug flow in steady state, and (2) the particles are of uniform size and geometry with cubic geometry. Thus, the orientation effects on the thermal behavior of the particle can be neglected. The electric field intensity for each particle is assumed to be identical within each incremental section.

On the basis of these assumptions, the ohmic heater column can be considered as a set of incremental sections in series, as shown in Figure 24.11, the equivalent electrical circuit for describing each incremental section i of thickness Δx_i lying between n and $n + 1$ can be illustrated in Figure 24.12, which is that of parallel fluid (R_{Pfi}) and particle (R_{Ppi}) resistances in series with a fluid (R_{Sfi}) resistance. The effective resistance (R_i) of the incremental section i is calculated as:

$$R_i = R_{Sfi} + \frac{R_{Pfi} \cdot R_{Ppi}}{R_{Pfi} + R_{Ppi}} \qquad (24.41)$$

where:

$$R_{Sfi} = \frac{\Delta x_{Sfi}}{A_{Sf}\sigma_{fi}} \qquad (24.42)$$

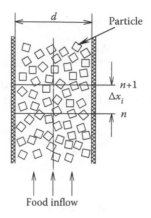

FIGURE 24.11 Ohmic heating column.

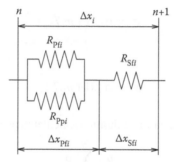

FIGURE 24.12 The equivalent circuit for incremental section.

$$R_{Ppi} = \frac{\Delta x_{Ppi}}{A_{Pp}\sigma_{pi}} \tag{24.43}$$

$$R_{Pfi} = \frac{\Delta x_{Pfi}}{A_{Pf}\sigma_{fi}} \tag{24.44}$$

where subscript Pp and Pf represent the particle and fluid in the parallel circuit section in Figure 24.12 respectively, and subscript Sf represents the fluid phase in the series circuit section in Figure 24.12; subscript p and f represent particle and fluid, respectively; σ is the electrical conductivity (Sm^{-1}); A is the cross-sectional area (m^2).

In the above set equations, A_{Sf} is equal to the cross-section area of the heater and:

$$A_{Sf} = \frac{\pi}{4}d^2 = A_{Pf} + A_{Pp} \tag{24.45}$$

The length or thickness of the incremental heater section (Δx_i) is related to the lengths of each phase:

$$\Delta x_i = \Delta x_{Sfi} + \Delta x_{Pfi} \tag{24.46}$$

and

$$\Delta x_{Ppi} = \Delta x_{Pfi} \tag{24.47}$$

In a similar way to Kopelman, it was assumed that the cross-sectional area and length of the discontinuous phase (particles) could be estimated from the volume fraction of particle phase (v_{fp}) by[19]

$$A_{Ppi} = \frac{\pi}{4} d^2 v_{fp}^{2/3} \tag{24.48}$$

and

$$\Delta x_{pi} = \Delta x_i v_{fp}^{1/3} \tag{24.49}$$

According to Equation 24.2, the electrical conductivities of fluid (σ_f) and particle (σ_p) are calculated, respectively as functions of temperature in the incremental section i as:

$$\sigma_{fi} = \sigma_{0f}(1 + m_f T_{fmi})$$

and

$$\sigma_{pi} = \sigma_{0p}(1 + m_p T_{pmi})$$

The total resistance of the ohmic heating column is calculated as follows:

$$R = \sum_{i=1}^{N} R_i \tag{24.50}$$

According to Ohm's law, the total current flowing through the system can be gained from Equation 24.51:

$$I = \frac{\Delta V}{R} \tag{24.51}$$

The voltage distribution is calculated assuming that all equipotential zones are approximately planar and perpendicular to tube walls (a reasonable approximation when the phases are uniformly mixed). Thus, the voltage drops over increment i can be calculated by Equation 24.52:

$$\Delta V_i = IR_i \tag{24.52}$$

On the basis of the results of voltage drop, the voltage gradient and energy generation within each incremental section can be easily calculated.

For the fluid phase within any incremental section of thickness Δx (as shown in Figure 24.11), between locations n and $n+1$, the equation for an energy balance is written as follows.

$$\dot{m}_f c_{pf}(T_f^{n+1} - T_f^n) = \dot{Q}_f v_f + n_p h_{fp} A_p (T_{psm} - T_{fm}) - h_w A_w (T_{fm} - T_a) \tag{24.53}$$

where $\dot{m}_f$ is the mass flow rate of the fluid (kgs^{-1}); c_{pf} is the specific heat of the fluid (Jkg^{-1}K^{-1}); v_f is the volume of fluid in the incremental element (m^3); n_p is the number of the particles in the mixture; h_{fp} is the convective heat transfer coefficient at the fluid–particle interface (Wm^{-2}K^{-1}); h_w is the overall heat-transfer coefficient of the heater wall (Wm^{-2}K^{-1}); A_p is the surface area of one particle (m^2); A_w is the interface area between the volume element and the heater wall (m^2), being equal around side surface area of the volume element (Δx); T_f is the temperature of fluid (K); T_a is the temperature of ambient air (K); T_{ps} is the temperature of the particle surface (K); T_{psm}, T_{fm}, and v_f are defined as follows:

$$T_{psm} = \frac{T_{ps}^{n+1} + T_{ps}^n}{2} \tag{24.54}$$

$$T_{fm} = \frac{T_f^{n+1} + T_f^n}{2} \tag{24.55}$$

$$v_f = \frac{\pi}{4} d^2 \Delta x v_{ff} \tag{24.56}$$

In Equation 24.56, v_f is the volume fraction of the fluid in mixture (–); d is the diameter (m) of the heating column.

Applying Equation 24.2 into Equation 24.11, we obtain Equation 24.57 as follows:

$$\dot{Q}_f = |\nabla V|^2 \sigma_{0f}(1 + m_f T_{fm}) \tag{24.57}$$

The fluid temperature at each successive incremental location ($n+1$) can be determined from Equation 24.53 if the voltage field (V) and mean particle surface temperature (T_{pm}) is known.

The heat transfer problem for particles is the conduction heat transfer equation with temperature-dependent internal energy generation. The temperature distribution of the particle is also governed by Equation 24.33, being subjected to the initial condition, Equation 24.34, and boundary condition, Equation 24.35. For convenience, Equation 24.33 is rewritten as Equation 24.58 here:

$$\rho_p c_{pp} \frac{\partial T_p}{\partial t} = \nabla \cdot (\lambda_p \nabla T_p) + |\nabla V|^2 \sigma_{0p}(1 + m_p T_p) \tag{24.58}$$

The above problem is solved using the three-dimensional finite element method in space and Crank–Nicolson finite differencing in time, as recommended by Sastry.[19]

This model simplifies the approach for solving electrical field distribution equation. However, if long thin particles are used, errors between the model prediction and the experimental results will occur since this model is based on the assumptions that for cubic particles, orientation effects are considered small and that the electric field strength for each particle is the same within each incremental section.

24.4.3.2 Model for Simulating the Thermal Behavior of Food in Static Ohmic Heater

24.4.3.2.1 Model for Single Phase Food

In this case, the electric field distribution in static ohmic heater containing single phase food, either liquid or solid foods, can also be derived by solving Equation 24.13 with the electrical boundary conditions, Equations 24.14 and 24.15.

It is known that the natural convection mode is an important component of heat transfer in liquid food under gravity conditions with in static state. However, if little convective heat transfer occurs or the convective effect can be ignored in some situations, for example, in a microgravity environment, the temperature differences between different regions of a food system will be more pronounced resulting from conduction of heat existing in food, which is the worst case senario.[47] One of the objectives for creating a mathematical model for ohmic heating process is to simulate and find the worse case scenario during ohmic heating. Therefore, in a static ohmic heater, when convective heat transfer is ignored or density of food is considered a constant, the Navier-Stokes equation (Equation 24.17) can be neglected. Thus, the temperature distribution in food can also be determined by Equation 24.19, but due to the static state without natural convection, Equation 24.19 can be reduced to Equation 24.59.

$$\rho c_p \frac{\partial T}{\partial t} = \lambda \left(\frac{\partial^2 T}{\partial x^2} + \frac{\partial^2 T}{\partial y^2} + \frac{\partial^2 T}{\partial z^2} \right) + \dot{Q} \tag{24.59}$$

where the heat generation rate ($\dot{Q}$) in food can also be calculated by solving Equation 24.57.

For solving Equation 24.59, the corresponding initial and boundary conditions must be known at first on the basis of the fact that the configuration of the static ohmic heater is confirmed. In general, the thermal initial and boundary conditions for static ohmic heater can be generally written as Equations 24.60 and 24.61, respectively.

$$T(x,y,z,t) = T_0, \qquad \forall x, \forall y, \forall z, t = 0 \tag{24.60}$$

$$-\lambda \nabla T(x,y,z,t) \cdot \vec{n}\big|_{\text{wall}} = h_w [T(x,y,z,t) - T_a]$$

$$\forall t, x, y, z \in [\text{vessel or heater surfaces}] \tag{24.61}$$

where T_0 is the initial temperature of food (°C); h is the overall heat transfer coefficient (Wm^{-2}K^{-1}); $\vec{n}$ is the normal vector. T_a is the temperature of ambient (°C). If heat losses to the environment are negligible, the thermal boundary condition can be reduced to Equation 24.62

$$-\lambda \nabla T(x,y,z,t) \cdot \vec{n}\big|_{\text{wall}} = 0$$

$$\forall t, x, y, z \in [\text{vessel or heater surfaces}] \tag{24.62}$$

For food reheating and sterilization by ohmic heating to minimize equivalent system mass during long-duration space missions, Jun and Sastry developed a new typical flexible package, equipped with a pair of V-shaped electrodes (shown in Figure 24.7c), for food reheating and sterilization. They solved the governing equation, such as Equations 24.13, 24.59 and 24.57 with computational fluid dynamics (CFD) software Fluent (version 6.1, Fluent, Inc., Lebanon, NH) in the 2D and 3D environments.[46,47]

24.4.3.2.2   Model for Food Mixture

When food mixture consisting of particles and fluid with different electrical conductivities is processed in an ohmic heating system, both particle and fluid phases may exhibit different thermal behavior, which results from different heat generation rates occurring at distinct localized regions in food mixture. Therefore, the cold zone temperature in food mixture must be predicted with an appropriate mathematical model.

For simulating thermal behavior of food mixture in a static ohmic heater, two different modeling approaches had originally been developed by de Alwis and Fryer[53] and Sastry and Palaniappan,[14] respectively.

In de Alwis and Fryer's model, the voltage distributions in fluid and particle are obtained by solving Equation 24.13 with the electrical boundary conditions, Equations 24.14 and 24.15. On the basis of the supposition that food mixture is stationary with nonconvective effects between fluid and particle, the temperature distributions in fluid and particle can simply be determined according to the following equations:

$$\rho_i c_{pi} \frac{\partial T_i}{\partial t} = \nabla(\lambda_i \nabla T_i) + \dot{Q}_i \qquad (24.63)$$

where the subscript i represents the phase, ρ_i, c_{pi}, and λ_i are the density (kgm^{-3}), specific heat (Jkg^{-1}K^{-1}) and thermal conductivity (Wm^{-1}K^{-1}) of phase i, respectively. The heat generation rate term in Equation 24.63 can also be obtained from solving Equation 24.11, applying Equation 24.2, it is given by:

$$\dot{Q}_i = |\nabla V|^2 \sigma_{0i}(1 + m_i T_i) \qquad (24.64)$$

where i still represents the phase.

Equation 24.63 is subjected to an initial thermal condition (Equation 24.65) and an external time-dependent boundary condition, which is one of convection to the surroundings, for the fluid surface (s) in contact with the vessel wall (w) (Equation 24.66)

$$T_i(x,y,z,t) = T_0, \qquad \forall x, \ \forall y, \ \forall z, \ t = 0 \qquad (24.65)$$

$$-\lambda_f \nabla T_f \cdot \vec{n} \mid_w = h_w(T_f - T_a) \qquad (24.66)$$

where h_w is the overall heat-transfer coefficient at the heater wall; T_a is the temperature of surroundings; $\vec{n}$ is the normal vector.

The system of Equations 24.13, 24.63, and 24.64 with their respective corresponding initial and boundary conditions can be solved iteratively using the Galerkin–Crank–Nicholson algorithm, a hybrid-spatially finite element, temporally finite difference scheme. However, if there are many particles existing in the mixture, the most accurate solution has to be obtained that considers the existence of particle size distribution and its position, and having the heat transfer problem solved for each particle. Such a solution is unlikely to need extreme computational effort. In order to simplify the problem about multiple particles in food mixture, on the basis of de Alwis and Fryer's model for simulating thermal behavior of food mixture within a static ohmic heater, Zhang and Fryer simplified the food mixture containing multiple particles with spherical shape uniform distribution flowing through the ohmic heating tube into a "unit cell," in which the particles lie on a cubic lattice or body-centered pattern. In this assumption, the ohmic heating tube can be divided into a number of "unit cells," just like a special static heater, thus the solutions of each "unit cell" can be used to describe the entire tube.[20]

In Sastry and Palaniappan's model, a circuit analogy approach is used to approximate electrical conductivity and thus, heat generation for a static heater with particles immersed in a well-mixed fluid. In this method, the equipotential lines along the heater are assumed parallel. Thus, the static ohmic heater can be regarded as an equivalent electrical circuit (as shown in Figure 24.12). The effective resistance in each incremental section i along the axis of the heater and corresponding voltage gradients across each increment can be gained from solving Equation 24.41. The effective resistance for the entire static ohmic heater can be calculated by Equation 24.50. Therefore, for each increment section of the static ohmic heater, the current and voltage gradient through the increment section can also be solved with Equations 24.51 and 24.52, respectively.

For fluid phase in the well-mixed food mixture processed with a static ohmic heating system, the energy balance can be written as follows:

$$m_f c_{pf} \frac{(T_f^{n+1} - T_f^n)}{\Delta t} = \dot{Q}_f v_f + n_p h_{fp} A_p (T_{pm} - T_{fm}) - h_w A_w (T_{fm} - T_a) \tag{24.67}$$

where:

$$T_{pm} = \frac{T_p^{n+1} + T_p^n}{2}$$

$$T_{fm} = \frac{T_f^{n+1} + T_f^n}{2}$$

where v_f is the volume of fluid in the heater (m³); n_p is the particle number (–); h_{fp} is the convective heat transfer coefficient (Wm⁻²K⁻¹) at the fluid–particle interface; h_w is the overall heat transfer coefficient (Wm⁻²K⁻¹) at the heater wall, A_p and A_w are the surface area (m²) of the particle and heater wall, respectively; T_a is the surroundings temperature of the heater.

The energy balance for the particle can also be calculated using Equations 24.63 and 24.64 with the initial thermal condition, Equation 24.65, and the thermal boundary condition, Equation 24.35.

The equation governing fluid temperature can be solved using a forward difference scheme, while that governing the particle temperature can be solved using the Galerkin–Crank–Nicolson algorithm.[14,19,24]

By comparing heat generation expressions in both models discussed above, Zhang and Fryer[65] indicated that the results of both approaches were consistent with each other (with 10%) when particle content was more than 30% in the mixture. However errors occurred if particle content was lower than this point when the circuit theory was used, which was caused by the difference of electrical conductivities of the particle and fluid in a low particle concentration situation. However, under worst-case conditions, for example, there is only one particle with lower electrical conductivity compared to the fluid, within the fluid in the static ohmic heater. The model developed by Sastry and Palaniappan and extended by Sastry is shown to be more conservative than the model developed by de Alwis and Fryer and extended by Zhang and Fryer, due to ignoring the convection effects in de Alwis and Fryer's model. In contrast, when the particle is more conductive than the fluid, de Alwis and Fryer's model is more conservative. Considering computational cost, independence of particle positions and a means of predicting large-scale variations in electric field intensity when voltage is applied along the flow, the circuit analogy approach is better than de Alwis and Fryer's model.[22,24] Due to reducing the computation time involved in the whole-field simulation when the heater sections are large in comparison with the particles, the circuit analogy is useful in determining voltage drop. Thus, it is noted as a "forest" level model, while the de Alwis and Fryer's model is called as a "tree" level model for resolving local nonuniformities.[24]

For a model simulating thermal behavior of food in a static ohmic heater, an important parameter called the ohmic heating system performance coefficient (SPC) by Icier and Ilicali,[19,66] or as the electricity-to-heat conversion efficiency (η) by Ye et al.[67] needs to be considered.

In Icier and Ilicali's definition, the ohmic heating SPC is defined by using the energies given to the system and taken up by the food. It can be calculated from Equation 24.68:

$$SPC = \frac{\sum (\Delta V I t) - E_{loss}}{\sum (\Delta V I t)} \tag{24.68}$$

where ΔV is the voltage (V) supplied to the electrodes of the static ohmic heater; I is the current (A) passing through the electrodes; t is the time (s); and E_{loss} is the rate of energy loss term. E_{loss} mainly includes the heat transfer rate to heat the ohmic heating vessel and the electrodes, etc., the rate of heat loss to the surroundings from the external surface of the heating vessel by convection and the portion of the heat generation rate used for purposes other than heating the liquid, i.e., electrochemical reaction, phase change. It is found that E_{loss} increases with the increase in electric field intensity. The rate of energy loss can be calculated from Equation 24.69:

$$E_{loss} = \sum (\Delta VIt) - mc_p(T_2 - T_1) \qquad (24.69)$$

where m is the mass (kg) of food; c_p is the average specific heat of food (Jkg^{-1}K^{-1}); T_1 is the initial temperature of food (K); T_2 is the final temperature of food (K).

If the electric field intensity is supposed to be uniform and constant in the ohmic heater, and uniform heat generation and no temperature gradient in the food are assumed, on the basis of the definition of SPC, the energy balance on food can be written as follows:[17,66]

$$mc_p \frac{dT}{dt} = \text{SPC} \cdot \Delta V^2 \sigma / K_c \qquad (24.70)$$

where σ is the electrical conductivity of food (Sm^{-1}); ΔV is the voltage (V); K_c is the ohmic heater constant (m^{-1}), it can be calculated from Equation 24.71:

$$K_c = \frac{L}{A_e} \qquad (24.71)$$

where L is the distance between the electrodes (m); A_e is the cross-sectional area of the electrodes (m^2).

For simulating temperature distribution of food mixture in a static ohmic heating system, Ye et al.[67] established a mathematical model on the basis of de Alwis and Fryer's model. The governing equation of the electric field intensity for both particle and fluid are still Equation 24.13 with the electrical boundary conditions, Equations 24.14 and 24.15, the temperature distributions for particle and fluid are also Equation 24.63, respectively, with Equation 24.65 as the initial conditions. For fluid in contact with the vessel wall, the boundary condition is written as Equation 24.66. For particle in contact with the fluid, the boundary condition is written as Equation 24.72:

$$-\lambda_p \nabla T_p \cdot \vec{n} \mid_s = h_{fp}(T_{ps} - T_f) \qquad (24.72)$$

where the subscript p and s represents particle and surface of the particle, respectively; h_{fp} is the convective heat transfer coefficient (Wm^{-2}K^{-1}) of the fluid–particle interface.

According to Equation 24.72, the convective effect between the particle and fluid is considered for simulating the thermal behavior of food mixture. In addition, for solving Equation 24.63, they put forward a new parameter for modifying heat generation rate within food during ohmic heating. This new parameter is named as the electricity-to-heat conversion efficiency (η), being used for modifying the error between the value of simulating with de Alwis and Fryer's model and that of the experiment with a static ohmic heater when the convective effect is ignored. When conversion efficiency is ignored, it was found that the temperature at the potato center could be over-predicted by about 25°C.[67] Salengke and Sastry[24,31] also found this phenomenon occurring during sterilization of particle-fluid mixture by a static ohmic heating system. Thus, the electricity-to-heat conversion efficiency (η) should not be one; it means that the electrical energy cannot be transformed into heat completely because some of the electrical energy, for example, is used to drive the movement of ions or to produce electrolytic gas resulting from electrochemical process occurring at the electrode/solution interfaces. Therefore, Equation 24.64 should be changed into Equation 24.73, considering the electricity-to-heat conversion efficiency (η):

$$\dot{Q}_i = \eta \, |\nabla V|^2 \, \sigma_{0i} (1 + m_i T_i) \tag{24.73}$$

For estimating the value of the electricity-to-heat conversion efficiency (η), an equation is defined as by Ye et al. as follows:[67]

$$\eta = \frac{m_1 c_{p1} \Delta T_1 + m_2 c_{p2} \Delta T_2}{\int_0^t VI(t)dt} \tag{24.74}$$

where subscripts 1 and 2 represent the sample and ohmic heating vessel, respectively; m is the mass (kg); c_p is the specific heat ($Jkg^{-1}K^{-1}$); ΔT is the temperature change (K); V and $I(t)$ are the voltage supplied to the sample (V) and the current (A), respectively.

In comparison with Equation 24.68, both Equations 24.68 and 24.74 have no essential differences; they are all defined as a ratio of the energy taken up by the food to the energy given to the static ohmic heating system. However, the energy taken by the ohmic heating vessel is regarded as the effective energy by Ye et al.[67]

Using special experiments performed with some fruit juices at different concentrations, such as apple, sourcherry, peach and apricot purees, it was determined that the ohmic heating SPC is in the range of 0.47–1.00. In addition, the SPC depends strongly on electric field intensity, electrical conductivity and the components of food. In general, the SPC decreases with the increase in electric field intensity and electrical conductivity of the food, respectively. However, according to the results reported by Ye et al.[67] the values of η ranges from 0.78 to 0.85 when both food samples, one being a salt solution containing CMC, the other cut potatoes (cylindrical and rectangular), were heated with a static ohmic heating system, respectively.

24.5 CONCLUSION

The advantages of ohmic heating technology for food processing have been addressed[6,26,56,68,69] in the literature as volumetric heating, better quality of products, easy process control, higher energy efficiency and environmentally-friendly. Furthermore, a high solids loading capacity by the ohmic heating process is expected when the capital investment can be reduced and product safety issues are addressed. By combining an aseptic filling and packaging system, ohmic pasteurization can produce food products with a much longer shelf life, i.e. ambient-temperature storage and distribution. The comprehensive mathematical modeling can benefit the industrial ohmic heater design configuration, with accurately predicted heat, mass and velocity profiles of the food materials during processing.

However, some of the disadvantages of ohmic heating technology for food processing are the higher initial operational costs, lack of information or validation procedures,[56] and lack of applicable temperature monitoring techniques for locating cold/hot spots within food mixture,[6] especially those cooked with continuous ohmic heating systems. Although reducing risks of fouling and burning of food products is one of the advantages of ohmic heating technology over conventional heating methods, fouling of the electrode surface during ohmic processing of food materials, especially liquid food containing protein such as milk[70–73] and soybean milk[74,75] is a serious problem. Deposition on electrode surfaces causes additional electrical resistance, an increase in electricity consumption, and consequently leads to an increase in the temperature of electrode surfaces and consequent sparking. More research is required to further develop this technique for food processing.

REFERENCES

1. de Alwise, A.A.P. and Fryer, P.J. The use of direct resistance heating in the food industry. *Journal of Food Engineering*, 11, 3, 1990.
2. Anderson, A.K. and Finklestein, R. A study of the electropure process of treating milk. *J. Dairy Sci.*, 2, 374, 1919.
3. Getchel, B.E. Electric pasteurization of milk. *Agriculture Engineering*, 16, 408, 1935.
4. Moses, B.D. Electric pasteurization of milk. *Agriculture Engineering*, 19, 525, 1938
5. Sastry, S.K. Ohmic heating. In *Minimal Processing of Foods and Process Optimization: an Interface*, Singh, R.P. and Oliveria, F.A.R., Eds. CRC Press, Inc., Boca Raton, FL, 17, 994.
6. Ruan, R. et al., Ohmic heating. In *Thermal Technologies in Food Processing*, Richadson, P., Ed. CRC Press, Cambridge, UK, 2001.
7. Morrissey, M.T. and Almonacid, S. Rethinking technology transfer. *Journal of Food Engineering*, 67, 135, 2005.
8. Sastry, S.K., and Barach, J.T. Ohmic and inductive heating. *Journal of Food Science (Supplement)*, 65, 42, 2000.
9. Legrand, A. et al. Physical, mechanical, thermal and electrical properties of cooked red bean (*Phaseolus vulgaris* L.) for continuous ohmic heating process. *Journal of Food Engineering*, 81, 447, 2007.
10. Wang, W.-C. and Sastry, S.K. Starch gelatinization in ohmic heating. *Journal of Food Engineering*, 34, 225, 1997.
11. Li, F.-D. et al. Determination of starch gelatinization temperature by ohmic heating. *Journal of Food Engineering*, 62, 113, 2004.
12. Palaniappan, S. and Sastry, S.K. Electrical conductivity of selected solid foods during ohmic heating. *Journal of Food Process Engineering*, 14, 221, 1991.
13. Sastry, S.K. and Palaniappan, S. Ohmic heating of liquid-particle mixtures. *Food Technology*, 4, 364, 1993.
14. Sastry, S.K. and Palaniappan, S. Mathematical modeling and experimental studies on ohmic heating of liquid-particle mixtures in a static heater. *Journal of Food Process Engineering*, 15, 241, 1992.
15. Marcotte, M., Piette, J.P.G., and Ramaswamy, H.S. Electrical conductivities of hydrocolloid solution. *Journal of Food Processing Engineering*, 21, 503, 1998.
16. Li, F.-D. Application study on ohmic heating and energy-efficient electrostatic drying of foods. PhD Thesis, China Agricultural University, Beijing, China, 2002.
17. Icier, F. and Ilicali, C. Electrical conductivity of apple and sourcherry juice concentrates during ohmic heating. *Journal of Food Process Engineering*, 27, 159, 2004.
18. Castro, I. et al. The influence of field strength, sugar and solid content on electrical conductivity of strawberry products. *Journal of Food Process Engineering*, 26, 17, 2003.
19. Sastry, S.K. A model for heating of liquid-particle mixtures in a continuous flow ohmic heater. *Journal of Food Process Engineering*, 15, 263, 1992.
20. Zhang, L. and Fryer, P.J. Models for the electrical heating of solid–liquid food mixtures. *Chemical Engineering Science*, 48, 633, 1993.
21. Sastry, S.K. and Li, Q. Modeling the ohmic heating of foods. *Food Technology*, 50, 246, 1996.
22. Sastry, S.K. and Salengke, S. Ohmic heating of solid–liquid mixtures: a comparison of mathematical models under worst-case heating conditions. *Journal of Food Process Engineering*, 21, 441, 1998.
23. Zareifard, M.R. et al. Ohmic heating behaviour and electrical conductivity of two-phase food. *Innovative Food Science and Emerging Technologies*, 4, 45, 2003.
24. Salengke, S. and Sastry, S.K. Models for ohmic heating of solid–liquid mixtures under worst-case heating scenarios. *Journal of Food Engineering*, 83, 337, 2007.
25. Palaniappan, S. and Sastry, S.K. Electrical conductivities of selected juices: influences of temperature, solid content, applied voltage, and particle size. *Journal of Food Engineering*, 14, 247, 1991.
26. Fryer, P.J. et al. Ohmic processing of solid-liquid mixtures: heat generation and convection effects. *Journal of Food Engineering*, 18, 101, 1993.
27. Haden, K., de Alwis, A.A.P., and Fryer, P.J. Changes in the electrical conductivity of foods during ohmic heating. *International Journal of Food Science and Technology*, 25, 9, 1990.

28. de Alwis, A.A.P., Halden, K., and Fryer, P.J. Shape and conductivity effects in the ohmic heating of foods. *Chemical Engineering Research Design*, 67, 159, 1989.

29. Davies, L.J., Kemp, M.R., and Fryer, P.J. The geometry of shadows: effects of inhomogeneities in electrical field processing. *Journal of Food Engineering*, 40, 245, 1999.

30. Sirsat, N. et al. Ohmic processing: electrical conductivities of pork cuts. *Meat Science*, 67, 507, 2004.

31. Salengke, S. and Sastry, S.K. Experimental investigation of ohmic heating of solid–liquid mixtures under worst-case heating scenarios. *Journal of Food Engineering*, 83, 324, 2007.

32. Yongsawatdigul, J., Park, J.W., and Kolbe, E. Electrical conductivity of pacific whiting surimi paste during ohmic heating. *Journal of Food Science*, 60, 922, 935, 1995.

33. Piette, G. et al. Ohmic cooking of processed meats and its effects on product quality. *Journal of Food Science*, 69, 71, 2004.

34. Wang, W.-C. and Sastry, S.K. Salt diffusion in the vegetable tissue as a pretreatment for ohmic heating: electrical conductivity profile and vacuum infusion studies. *Journal of Food Engineering*, 20, 299, 1993.

35. Sarang, S. et al. Product formulation for ohmic heating: blanching as a pretreatment method to improve uniformity in heating of solid–liquid food mixtures. *Journal of Food Science*, 72, E227, 2007.

36. Icier, F. and Ilicali, C. The use of tylose as a food analog in ohmic heating studies. *Journal of Food Engineering*, 69, 67, 2005.

37. Ede A.J. and Haddow R.R. The electrical properties of food at high frequencies. *Food Manufacturing*, 26, 156, 1951.

38. Mitchell, F.R.G. and de Alwis, A.A.P. Electrical conductivity meter for food samples, *Journal of Physics, E: Science Instruments*, 22, 554, 1989.

39. Wang, C.S. et al. Effect of tissue infrastructure on electric conductance of vegetable stems. *Journal of Food Science*, 66, 284, 2001.

40. Li, X. Study on electrical properties of foods and application. PhD Thesis, China Agricultural University, Beijing, China, 1999.

41. Kim, H.-J. et al. Validation of ohmic heating for quality enhancement of food products. *Food Technology*, 50, 253, 1996.

42. Khalaf, W.G. and Sastry, S.K. Effect of fluid viscosity on the ohmic heating rate of solid-liquid mixtures. *Journal of Food Engineering*, 21, 145, 1996.

43. Wu, H. et al. Electrical properties of fish mince during multi-frequency ohmic heating. *Journal of Food Science*, 63, 1028, 1998.

44. Lima, M., Heskitt, B., and Sastry, S.K. The effect of frequency and wave form on the electrical conductivity-temperature profiles of turnip tissue. *Journal of Food Process Engineering*, 22, 41, 1999.

45. Zhao, Y., Kolbe, E., and Flugstad, B. A method to characterize electrode corrosion during ohmic heating. *Journal of Food Processing Engineering*, 22, 81, 1999.

46. Jun, S. and Sastry, S. Model and optimization of ohmic heating of foods inside a flexible package. *Journal of Food Process Engineering*, 28, 417, 2005.

47. Jun, S. and Sastry, S. Reusable pouch development for long term space missions: a 3D ohmic model for verification of sterilization efficacy. *Journal of Food Engineering*, 80, 1199, 2007.

48. Damez, J.-L. et al. Dielectric behavior of beef meat in the 1–1500 kHz range: simulation with the Fricke/ Cole-Cole model. *Meat Science*, 77, 512, 2007.

49. Damez, J.-L. et al. Electrical impedance probing of the muscle food anisotropy for meat ageing control. *Food Control*, 19, 931, 2008.

50. Imai, T. et al. Ohmic heating of Japanese white radish *Rhaphanus sativus L. International Journal of Food Science and Technology*, 30, 461, 1995.

51. Pongviratchai, P. and Park, J.W. Electrical conductivity and physical properties of surimi–potato starch under ohmic heating. *Journal of Food Science*, 72, E503, 2007.

52. Hayt, W.H. and Buck, H.A. *Engineering Electromagnetics*, 6th edn. McGraw-Hill Co., New York, NY, 2001.

53. de Alwis, A.A.P. and Fryer, P.J. A finite-element analysis of heat generation and transfer during ohmic heating of food. *Chemical Engineering Science*, 45, 1547, 1990.

54. Farid, M. New methods and apparatus for cooking. International Patent Application, #PCT/NZ02/00108, 2001.

55. Özkan, N., Ho, I., and Farid, M. Combined ohmic and plate heating of hamburger patties: quality of cooked patties. *Journal of Food Engineering*, 63, 141, 2004.

56. Vicente, A.A., de Castro, I., and Teixeira, J.A. Ohmic heating for food processing. In *Thermal Food Processing*, Sun, D.-W., Eds. CRC Talyor and Francis Group, New York, NY, 2006.

57. Uemura, K. Isobe, S., and Noguchi, A. Estimation of temperature profile at sequence ohmic heating by finite element method. *Nippon Shokuhin Kagaku Kaishi*, 43, 1190, 1996.
58. Orangi, S., Sastry, S., and Li, Q. A numerical investigation of electroconductive heating in solid-liquid mixtures. *International Journal of Heat and Mass Transfer*, 41, 2211, 1998.
59. Eliot-Godéreaux, S.C. et al. Passage time distributions of cubes and spherical particles in an ohmic heating pilot plant. *Journal of Food Engineering*, 47, 11, 2001.
60. McCabe,W.L., Smith, J.C., and Harriott, P. *Unit Operations of Chemical Engineering*, 6th edn. Chemical Industry Press, Beijing, China, 2003, chapter 4.
61. Quarini, G.L. Thermalhydraulic aspects of the ohmic heating process. *Journal of Food Engineering*, 24, 561, 1995.
62. Ghani, A.G. and Farid, M.M. Using the computational fluid dynamics to analyze the thermal sterilization of solid-liquid food mixture in cans. *Innovative Food Science and Emerging Technologies*, 7, 55, 2006.
63. Benabderrahmane, Y. and Pain, J.-P. Thermal behaviour of a solid/liquid mixture in an ohmic heating sterilizer-slip phase model. *Chemical Engineering Science*, 55, 1371, 2000.
64. Zitoun, K.B. Continuous flow of solid-liquid mixtures during ohmic heating: fluid interstiational velocities, solid area fraction, orientation and rotation. PhD Thesis, The Ohio State University, Columbus, OH, 1996.
65. Zhang, L. and Fryer, P.J. A comparison of alternative formulations for the prediction of electrical heating rates of solid-liquid food materials. *Journal of Food Process Engineering*, 18, 85, 1995.
66. Icier, F. and Ilicali, C. Temperature dependent electrical conductivities of fruit purees during ohmic heating. *Food Research International*, 38, 1135, 2005.
67. Ye, X. et al. Simulation and verification of ohmic heating in static heater using MRI temperature mapping. **Lebensmittel Wissenschaft und Techonologie*, 37, 49, 2004.
68. Frampton, R. Ohmic potential. *Food Manufacture,* May, 1988.
69. Allen, K., Eidman, V., and Kinsey, J. An economic-engineering study of ohmic food processing. *Food Technology*, 50, 269, 1996.
70. Ayadi, M.A. et al. Continuous ohmic heating unit under whey protein fouling. *Innovative Food Science and Emerging Technologies*, 5, 465, 2004.
71. Ayadi, M.A. et al. Experimental study of hydrodynamics in a flat ohmic cell-impact on fouling by dairy products. *Journal of Food Engineering*, 70, 489, 2005.
72. Ayadi, M.A. et al. Thermal performance of a flat ohmic cell under non-fouling and whey protein fouling conditions. *LWT Food Science and Technology,* doi:10.1016/j.lwt.2007.06.022, 2007.
73. Bansal, B., Chen, X.D., and Lin, S.X.Q. Skim milk fouling during ohmic heating, ECI Symposium Series. *Volume RP2: Proceedings of 6th International Conference on Heat Exchanger Fouling and cleaning-Chanllenges and Opportunities*, Iler-Steinhagen, H.M., Malayeri, M.R., and Watkinson, A. P., Eds. Engineering Conferences International, Kloster Irsee, Germany, 133, 2005.
74. Sun, Y.-L., Zuo, D.-W., and Li, F.-D. Experimental study on the continuous ohmic heating system with vertical compartments for liquor. *Journal of Sichuan University (Engineering Science Edition)*, 38, 148, 2006 (in Chinese).
75. Li L.-X. Experimental study on temperature distribution and electrode fouling in continuous ohmic heating system. Master's Degree Thesis, Shandong Agricultural University, Tai'an, China, 2007.

25 Radio Frequency Heating of Foods

Lu Zhang
The University of Auckland

Francesco Marra
Università di Salerno, via Ponte Don Melillo

CONTENTS

25.1 INTRODUCTION

Innovative research on food heating processes (such as cooking, pasteurization/sterilization, defrosting, thawing, and drying) often uses modeling as a virtual tool to investigate areas including assessment of processing time, evaluation of heating uniformity, studying the impact on quality attributes of the final product as well as considering the energy efficiency of these heating processes. In the area of innovative heating, electro-heating accounts for a considerable portion of both the scientific literature and commercial applications, which can be subdivided into either direct electro-heating where electrical current is applied directly to the food (e.g., ohmic heating) or indirect electro-heating (e.g., microwave and capacitive dielectric heating) where the electrical energy is firstly converted to electromagnetic radiation which subsequently generates heat within a product. Some of those electro-heat processes (e.g., ohmic and capacitive dielectric heating), are used only in industrial situations while microwaves can be applied commercially but are also very commonly used domestically. Of these forms of electro-heating in recent years there has been an increased interest in the area of capacitive dielectric heating, also known as radio frequency (RF) or macrowave heating, as evidenced by the increasing number of publications in this area.

    The first attempt of evaluating RF heating on foods was made earlier than that of microwave heating after the Second World War: Sherman[1] described "electric heat," how it is produced, and suggested possible applications in the food industry. These early efforts employed RF energy for applications such as the cooking of processed meat products, heating of bread, dehydration and blanching of vegetables. However, the work did not result in any commercial installations, predominately due to the high overall operating costs of RF energy at that stage. By the 1960s, studies on the application of RF energy to foods focused on the defrosting of frozen products, which resulted in several commercial production lines.[2,3] The next generation of commercial applications for RF energy in the food industry was postbake drying of cookies and snack foods which started in the late 1980s.[4,5] In the 1990s, the area of RF pasteurization was studied with attempts made to improve energy efficiency and solve technical problems such as run-away heating.[6,7] This in turn has led to recent investigations on RF applicator modifications and dielectric properties of food at RF ranges.[8-11]

Zhao et al.[7] reviewed research work in the area of RF heating, with considerations for major technological aspects and applications of RF heating and indicated major engineering challenges in the use of RF technology for food processing and preservation. They also discussed the potential use of mathematical modeling for the design of heating systems. Another review by Piyasena et al.[12] reported main industrial applications for RF heating in food processing and discussed dielectric properties for a range of food products suitable for RF heating. The most recent and comprehensive review about RF heating in food processing has been recently published by Marra et al.[13]

The mathematical modeling of RF heating of foods is receiving more attention in the last 3 years;[14-17] in the past, Neophytou and Metaxas[18-21] published a series of work on electrical field modeling of industrial scale RF heating systems but not related to food processing.

The aim of the current chapter is to include a comprehensive description of the mechanism of RF heating and fundamentals of mathematical modeling of both electromagnetic fields and heat transfer during RF heating of foodstuffs.

25.2   MECHANISM OF RADIO FREQUENCY (RF) HEATING

25.2.1   PRINCIPLES OF RADIO FREQUENCY (RF) HEATING

The RF portion of the electromagnetic spectrum (Figure 25.1) occupies a region from 1 to 300 MHz. Domestic, industrial, scientific and medical applications are permitted at the frequencies of 13.56 ± 0.00678, 27.12 ± 0.16272 and 40.68 ± 0.02034 MHz.[22]

The most simple and basic RF heating design employs a high voltage AC signal applied to a set of parallel electrodes. One of the electrodes is grounded which sets up a capacitor to store electric energy. The target material to be heated is placed between but not touching these electrodes. The alternating electrical field applied to the electrodes drives the positive ions in the material toward the negative portion of the field, while the negative ions move toward the positive portion one.[22] Collisions and friction within the food caused by this movement of ions in the electric field lead to the conversion of kinetic energy to heat. Ultimately, heat manifests itself as a temperature rise within the food.[23] More heat generation in the material is due by movements and frictions of dipolar molecules (such as water) that always attempting to align themselves appropriately with the

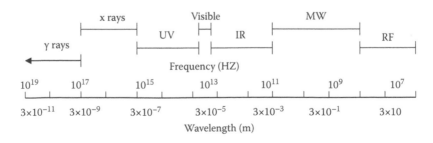

FIGURE 25.1   The electromagnetic spectrum.

changing polarity in the AC electric field. Of these two mechanisms, ionic rotation tends to be the dominant heating mechanism at lower frequencies such as those encountered in the RF range and therefore, at these frequencies dissolved ions are more important for heat generation than water dipoles in which they are dissolved.[24]

In order to include the above mentioned phenomena in equations describing heat transfer, the mathematical description of power generated by the RF applicator within a certain product is needed. The following part of this section is aimed to conduct the reader to a simple formulation of power generated in a dielectric material placed in a RF applicator.

When dielectric materials are placed in a RF applicator, a complex electrical impedance introduced into the RF electrical field[25] is established, according with the following equation:

$$Z_c = \frac{1}{2\pi f C_0} \frac{\varepsilon'' - j\varepsilon'}{\varepsilon''^2 + \varepsilon'^2} \tag{25.1}$$

where Z_c is the capacitance of the material, f is the frequency of the electric field, ε' is the dielectric constant, and ε'' is the dielectric loss factor of the material, respectively. C_0 is the capacitance of free space and $j = \sqrt{-1}$. From Equation 25.1, a finite resistance,

$$R = \frac{1}{2\pi f C_0 \varepsilon''} \tag{25.2}$$

is defined across the capacitor.

Taking the power, P, dissipated in an electrical resistance to be equal to V^2/R, then for a capacitor containing a dielectric material,

$$P = 2\pi f C_0 \varepsilon'' V^2 \tag{25.3}$$

For a parallel plate capacitor, $C_0 = \varepsilon_0 A_p/d$ where A_p is the plate area, d is the plate separation and ε_0 is the permittivity of the free space. As the voltage V is equal to the electric field strength E multiplied by the distance between the two electrodes d, Equation 25.3 can be rewritten as,

$$P = 2\pi f \varepsilon_0 \varepsilon'' E^2 (A_p d) \tag{25.4}$$

Being the term $(A_p d)$ the volume, the power dissipation per unit volume or power density, Q_v, is given by the following expression

$$Q_v = \omega \varepsilon_0 \varepsilon_r'' |E|^2 = 2\pi f \varepsilon'' E_{rms}^2 \tag{25.5}$$

where E_{rms} is the root mean square value of the electric field.

Equation 25.5 thus, states that the power density is proportional to the frequency of the applied electric field and the dielectric loss factor, and is also proportional to the square of the local electric field, which plays a key role in determining how a material will absorb energy in the AC electric field.

25.2.2 RADIO FREQUENCY (RF) HEATING EQUIPMENTS

There are two distinct methods for producing and transmitting RF power to materials: "conventional" equipment (Figure 25.2), and the more recently introduced "50 ohm" (or "50Ω"[22]) equipment (Figure 25.3). Although conventional RF equipment has been used successfully for certain

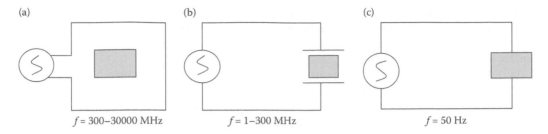

(a)   (b)   (c)

f = 300–30000 MHz f = 1–300 MHz f = 50 Hz

FIGURE 25.2   Schematic arrangement for (a) microwave, (b) RF and (c) ohmic heating.

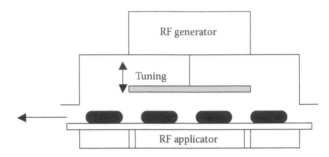

FIGURE 25.3   Components of a conventional RF heating system. (Adapted from Jones, P.L. and Rowley, A.T. In *Industrial Drying of Foods*, Baker C.G.J (Ed). Blackie Academic and Professional, London, UK, 1997.)

applications over many years, the tightening of electromagnetic compatibility (EMC) regulations and the desire for improved process control, are leading to the increase in the usage of 50Ω systems. In a conventional RF system, the applicator is considered as part of the secondary circuit of a transformer, which has the output circuit of the RF generator as its primary circuit.[22] It is often used to control the amount of RF power within set limits. The amount of RF power being delivered is only indicated by the direct current flowing through the high power valve within the generator. For 50Ω RF equipment, the generator is physically separated from the RF applicator by a high power coaxial cable.[22] A crystal oscillator is used to control/fix the operational frequency at exactly 13.56 MHz or 27.12 MHz. After the frequency is fixed, a convenient value such as 50Ω is set as the output impedance for the RF generator, which requires an impedance-matching network to transform the impedance of the RF load (applicator) to 50Ω so that power is transferred efficiently.[22] The advantages of 50Ω systems are: (1) easier to meet the EMC regulation due to crystal control of the frequency; (2) due to the coaxial cable, the RF generator can be sited at a convenient location away from the RF applicator; (3) the applicator can be designed for optimum performance.[22]

25.2.3   ELECTRO-HEATING METHODS

While in conventional heating, once heat reaches the outer surfaces of a foodstuff, it is transferred to the product interior by either conduction (e.g., in solids such as meat), convection (e.g., in liquids such as milk) or in products which display broken heating curves (e.g., some starch containing soups) convection and conduction can alternately dominate at different stages during the heating process[13], in electro-heating applications (e.g., ohmic, microwave and RF (Figure 25.4) heat is generated volumetrically within the material by the passage through, and its interaction with, either alternating electrical current (as in ohmic heating) or electromagnetic radiation (formed by the conversion of electrical energy to electromagnetic radiation at microwave (300–3000 MHz) or radio (1–300 MHz) frequencies). This common characteristic apart, electro-heating technologies are very different in terms of their methods of application.

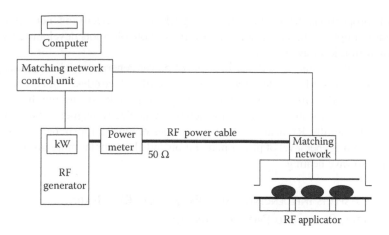

FIGURE 25.4 Components of 50 Ω RF heating system. (Adapted from Jones, P.L. and Rowley, A.T. In *Industrial Drying of Foods*, Baker C.G.J (Ed). Blackie Academic and Professional, London, UK, 1997.)

In ohmic heating, the product is placed in direct contact with a pair of electrodes through which generally a low frequency (50 Hz in Europe or 60 Hz in the United States) alternating current (used because the cyclic change in current direction helps to prevent electrolysis) is passed into the food product. The product needs to be either unpackaged and in direct contact with the electrodes and subsequently packaged, or alternatively be in a sealed pack which has conductive regions which allow electrical current into the product.

In microwave heating, there is no contact between the product and the source of electromagnetic field (the magnetron). In fact, once the microwaves are generated by the magnetron, they pass via a waveguide into an oven cavity in which they essentially bounce off the metal walls of the cavity interior impinging on the product from many directions. A better distribution of the electromagnetic field is promoted by metal stirrers and/or on purposed designed waveguides and/or by rotating the plate where the product is placed.

In RF heating, as in microwave heating, there are no requirements for direct contact between the product and electrodes as RF waves will penetrate through conventional cardboard or plastic packaging. In contrast to microwave heating, no stirrers or other design precautions are used to enhance the distribution of the electromagnetic field.

While microwave and RF heating are both classed as dielectric heating methods, at the lower frequencies encountered in the RF range ionic depolarization is recognized to be the dominant heating mechanism. At frequencies relevant to microwave heating (i.e., 400–3000 MHz) both ionic depolarization and dipole rotation can be dominant loss mechanisms: this also depends upon the moisture and salt content within a product.[13] Under these conditions, positive ions in the product move toward negative regions of the electrical field and negative ions move toward positive regions of the field. In the RF range, dissolved ions are more important for heat generation than the water dipoles in which they are dissolved. Heating occurs because this field is not static with polarity continually changing at high frequencies (e.g., 27.12 MHz). Soon after ions have started to move, the polarity of the electrodes swaps and ions have to move again. The net effect of all of this is that like microwave heating, heat is generated internally by friction (thereby avoiding the lag between the surface and the centre of the product).

Microwave and RF heating also differ in a number of other aspects. As frequency and wavelength are inversely proportional, RF (lower frequency) wavelengths (i.e., 11 m at 27.12 MHz in free space) are much longer than microwave (higher frequency) wavelengths (i.e., 0.12 m at 2450 MHz in free space). As electrical waves penetrate materials with attenuation, the depth in a material where the energy of the wave propagating perpendicular to the surface decreases exponentially. Penetration depth (d_p) is defined as the value of wave energy decreasing to $1/e$ (1/2.72) of the surface

value, which is proportional to wavelength. During RF heating, electromagnetic power can penetrate deeper into samples without surface over heating or hot/cold spots developing which are more
likely to occur with microwave heating.

Microwave and RF heating also differ in terms of power generation. RF energy is generated by
a triode valve and is applied to material via a pair of parallel electrodes.[22] In contrast, microwaves
are generated by special oscillator tubes known as magnetrons or klystrons and then transmitted via
a waveguide to a metal chamber or cavity in which resonant electromagnetic standing wave modes
are established.[12] Generally, RF heating offers advantages of more uniform heating over the sample
geometry due to both deeper level of power penetration and also simpler more uniform field patterns
compared to microwave heating.

25.2.4 OTHER ASPECTS INFLUENCING RADIO FREQUENCY (RF) HEATING: GEOMETRY, SHAPE, AND PRODUCT POSITION

RF heating of foods can be strongly influenced by geometry, shape and product position. Only
recently[13] this aspect has been emphasized, despite results published in recent years[26–28] claimed a
more deep investigation about the role played by geometry and shape (both sample and applicator)
and also by position and orientation of product with respect to electrodes in RF heating.

In an RF applicator, the air gap between the electrodes and load and the air space between
lateral surfaces of the load and electrode edges work as preferential pathways for the electric field
vectors to reach from one electrode to the other due to the low permittivity of air. Furthermore, the
incidence of the electric field on exposed surfaces of the load can be responsible for a differential
heating rate according to the shape of the load itself.

Processing layers of carrot sticks between parallel plate electrodes (8 cm from each other) RF
system, Orsat et al.[26] found different heating rate according with different thickness of treated sample. Particularly, they performed initial experiments on a 1–2 cm layer of carrot sticks: this produced
poor RF coupling and a very slow temperature increase within the product (5 min to reach 40°C
from a starting temperature of 5–6°C). In a second series of experiments, the layer thickness was
increased from 1–2 cm to 4–5 cm, without moving the electrodes: authors reported that internal
temperature increased from 6°C to 60°C in a variable time between two and seven minutes, according with the position. Finally, they performed a third series of experiments with a 6.5 cm layer of
carrots: in this case, reaching an internal temperature of 60°C took between 80 and 140 seconds.
What appeared to be an increase of heating rate due by a better RF coupling, resulting from an
increase in the mass of the load, corresponded to an increase of volume, thus in a change in the
shape of the load (from a thin layer to a thicker box shape) and to a reduction in the gap between the
top surface of the sample and the upper electrode.

Aspects connected to position (location and orientation) of samples during RF heating were
discussed by Wang et al.[27,28] In a first work,[27] they observed that for in-shell walnuts to successfully
undergo RF heating, intermittent stirrings was required to avoid nonuniform heating encountered
when walnuts were heated in a static state, due to differences in orientation and location. In subsequent research,[28] considerations in the design of commercial RF treatments for postharvest pest
control in in-shell walnuts were discussed. Particularly, considering that in-shell walnuts were noni-
sotropic materials due to the irregular shape of the shell and kernel and in such products, authors
concluded that variations in walnut temperature after RF heating were caused not only by the different properties but also by the shape of individual walnuts and different positions of the walnuts
in the applicator.

Birla et al.[29] observed nonuniform heating in oranges (a typical spherically shaped sample) and
apples, when water assisted RF treatment were used. Water assisted RF system was used in order
to prevent overheating and to promote a more uniform temperature distribution within the oranges.
However, in this case, the authors found consistent hot spots at the naval end of the oranges and at the
stem and bottom sides of apples. Therefore, in this case the axis was the shortest route for RF energy

to pass. Failure of water assisted system in promoting more even temperature distribution is due to the shape of the sample: in fact, the authors stated that the consistent hot spots on those two ends suggest that the shape of a fruit highly influences temperature uniformity within a fruit. In order to obtain a more uniform heating within the sample, these authors successfully proposed a fruit-mover to continuously rotate and move the fruit in a water bath placed between two electrodes.

25.3 THE MATHEMATICAL MODELING OF RADIO FREQUENCY (RF) HEATING IN FOOD PROCESSING

Mathematical modeling and computer simulation of RF assisted food processes have developed from the mid 1990s, when modern calculators and new software contributed to support the heavy computational duties required. Modeling RF assisted food processes involves two main problems: the first being the simulation of heat transfer within the product (load) between the electrodes and has focused mainly on the description of transport phenomena inside the foodstuff; the second area has been modeling RF heating in terms of its electric and magnetic fields. As the reader will observe, the two problems are linked, and a complete mathematical description of RF heating of foods must consider both problems together. In fact, As reported in Section 25.2, the power density absorbed by the load is a function of loss factor ε'' and electric field strength. The prediction of electric field distribution in a RF heating system is essential to determine the power density and hence, the temperature of the material.

25.3.1 MATHEMATICS OF RADIO FREQUENCY (RF) HEATING

Electric field strength is part of a more complex electromagnetic field that can be described in terms of electric field intensity $\underline{E}$, magnetic field intensity $\underline{H}$ and electric flux density $\underline{D}$, by the following Maxwell equations in differential form, when the involved media are isotropic, linear and homogenous:[30]

$$\underline{\nabla} \times \underline{E} = -\mu \frac{\partial}{\partial t}(\underline{H}) \tag{25.6}$$

$$\underline{\nabla} \times \underline{H} = \left(\varepsilon_0 \varepsilon_r \frac{\partial}{\partial t} + \sigma_c \right)\underline{E} \tag{25.7}$$

$$\underline{\nabla} \cdot \underline{D} = \rho_c \tag{25.8}$$

$$\underline{\nabla} \cdot \underline{H} = 0 \tag{25.9}$$

where σ_c and ρ_c are the effective electrical conductivity and charge density, respectively. In classical electromagnetism, Equation 25.6 is known as Faraday's Law, Equation 25.7 is known as Ampere Law, Equation 25.8 is known as Gauss' Law for the electric field and Equation 25.9 is known as Gauss' Law for the magnetic field.

In order to be solved, Maxwell's equations need some assumptions on the dependence of fields with respect to time. A usual assumption is to consider the fields being time-harmonic:[20]

$$\underline{E} = \underline{E}_0 e^{j\omega t} \tag{25.10}$$

$$\underline{H} = \underline{H}_0 e^{j\omega t} \tag{25.11}$$

When the time-harmonic assumption is made, Equations 25.6 and 25.7 become

$$\underline{\nabla} \times \underline{E} = -j\mu\omega\underline{H} \tag{25.12}$$

$$\underline{\nabla} \times \underline{H} = j\varepsilon_0\varepsilon_r\omega\underline{E} \tag{25.13}$$

Combining the last two equations, one obtains the following wave equation that is used for the calculation of the frequency domain

$$\underline{\nabla} \times \frac{1}{\mu_r}\underline{\nabla} \times \underline{E} - \omega^2\mu_0\varepsilon_0\varepsilon_r\underline{E} = 0 \tag{25.14}$$

In order to get a unique and proper solution it is necessary to set only the boundary conditions (i.e., time-derivative terms have disappeared). In most RF applications in food processing, external boundaries of the cavity are considered as perfect conductors and the following boundary condition is then used:[20]

$$\underline{n} \times \underline{E} = 0 \tag{25.15}$$

being $\underline{n}$ the unit vector normal to the boundary surfaces.

When a quasi-static approach is considered, Maxwell's equations collapse to the following one:

$$\underline{\nabla} \cdot [(\sigma + j\omega\varepsilon)\underline{\nabla}V] = 0 \tag{25.16}$$

where V is the electrical potential, related to the electric field by

$$\underline{E} = -\underline{\nabla}V \tag{25.17}$$

In this case, as boundary conditions, a source electric potential V_0 is applied to the upper electrode of the capacitor while at the bottom electrical ground conditions is considered

$$V = 0 \tag{25.18}$$

RF applicator shells are electrically insulated, so last boundary conditions are

$$\underline{\nabla} \cdot \underline{E} = 0 \tag{25.19}$$

The mathematical description of heat transfer within the food product placed between the electrodes is given by unsteady heat-conduction equation (assuming that a solid-like foodstuff is processed in the RF applicator) with a generation term represented by the power density already seen in the Equation 25.5:

$$\rho C_p \frac{\partial T}{\partial t} = \nabla k \nabla T + Q_{abs} \tag{25.20}$$

where T is the temperature within the sample, t is the process time, α is the thermal diffusivity, ρ is the density, C_p is the heat capacity.

The heat transport equation to be solved needs initial and boundary conditions: as initial condition, a uniform temperature T_0 can be assumed within the food sample; on boundaries, as general conditions, convective heat transfer from the external surfaces, in accordance with the Newton law, can be set:

$$-k\nabla T = U\left(T - T_{air}\right) \qquad (25.21)$$

where U is the overall convective heat transfer coefficient and T_{air}, is the temperature inside the oven.

25.3.2 RELEVANT RESULTS OF MODELING THERMAL FOOD PROCESSES ASSISTED BY RADIO FREQUENCY (RF)

Modeling RF applications in food processing led to relevant results both for the electromagnetic characterisation of the fields established within the RF applicators and for the heating effects in foods.

Contributions to the general understanding of RF applicators performances started with 2D and 3D FEM models presented by Neophytou and Metaxas[20] in order to establish the validity of electrostatic conditions, that were often assumed to be valid in general RF applicators.[31,32] They solved both Laplace (for electrostatic conditions) and wave equations and compared solutions, both in 2D and 3D, in terms of a ratio between magnetic and electric energy. According to the authors, if electrostatic conditions hold, then magnetic energy should be much less than the electric energy, while at resonance they are equal. Furthermore, they proposed another comparison in terms of power loss inside the processed material predicted by the Laplace and wave equations. The above criteria were analyzed for different type of applicators loaded with paper blocks at different frequencies. The authors concluded that electrostatic conditions (and then the Laplace equation) were valid only in small experimental size applicators. Furthermore, they recommended the use of a 3D model with wave equations, since the 2D model was found to be inappropriate.

A further step toward improving the understanding of RF applicators by means of computer simulation was proposed by Chan et al.:[30] on the basis of the method proposed by Neophytou and Metaxas,[21] the authors added a means of excitation to the tank oscillatory circuit with an external, properly positioned, coaxial source, with the purpose of presenting a way to visualize heating pattern inside the load. These authors did not incorporate heat transfer in their model but, in order to evaluate the goodness of mathematical model, they qualitatively compared the simulated electric field maps over a 1% carboxy-methyl-cellulose (CMC) solution with temperature maps taken with an infrared camera.

Effects of RF heating in foods were analyzed by Yang et al.,[14] who proposed the assessment of a program simulating heating performance of radish and alfalfa sprout seeds packed inside rectangular seed boxes during RF heating. They evaluated the distribution of electric field in a time stepping procedure by means of the transmission line method (TLM), whereas they used a standard explicit finite difference time domain method (FDTD) for the solution of the heat equation in their food samples. Simulated temperature distributions were compared against experimental data, and time-temperature profiles of seeds were then validated using experiments conducted with seeds in a RF heating system. They reported discrepancies between simulated and experimental results especially at the edges of the box.

Mathematical modeling approach was used by Marra et al.[15] to solve and validate coupled EM and heat transfer equations applying the electro-quasi-static hypothesis for a small cavity. Simulations were carried on by means of a commercial FEM-based software—FEMLAB. In their experiments, samples, constituted by cylindrical-shaped meat batters were placed between the electrodes of a 600 W RF system. For the meat batters chosen as sample foods, dielectric and physical property data were available as a function of temperature. Temperature profiles were experimentally measured along the sample axes, for RF output powers of 100, 200, 300, and 400 W (Figure 25.5). The authors evaluated the goodness of fit of the model by comparing numerical results with measured temperature profiles. Results confirmed that the electro-quasi-static hypothesis can be applied for simulating RF heating of a food sample in a small cavity. Both simulations and experiments showed different heating rates within the samples and therefore, uneven temperature distribution.

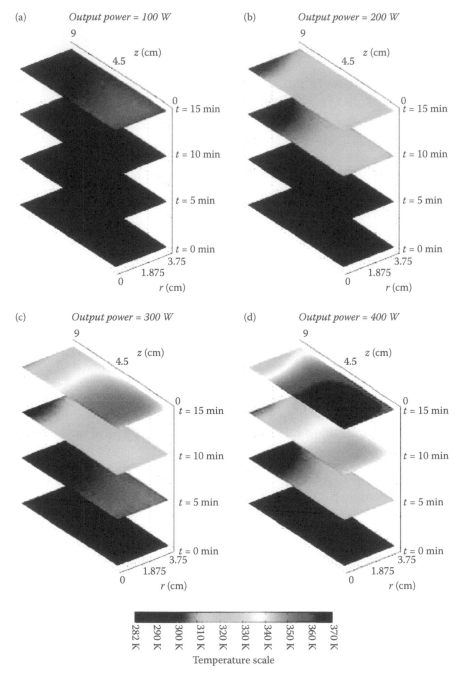

FIGURE 25.5 Simulation of RF heating in luncheon roll emulsion Sample. Temparature maps on the plane (r, z) after 0, 5, 10, 15 minutes of processing at: (a) 100 W, (b) 200 W, (c) 300 W, (d) 400 W of oven power.

Particularly, higher heating rates were detected closer to the sample bottom. Unevenness of temperature distribution was emphasized by the applied RF output power: the higher the applied power, the more uneven was the temperature distribution.

On the basis of the work by Marra et al.,[15] Romano and Marra[16] discussed the numerical analysis of multiphysics phenomena during RF heating of food, shaped as cube, cylinder, or sphere. With a simple modeling approach, they analyzed the effects of sample shape and orientation on heating rate and temperature distribution, once some parameters (such as distance between electrodes, electrode to product air gap, volume occupied by the sample with respect to cavity volume) were defined. For this purpose, they built and solved a 3D multiphysic mathematical model, based on the heat-conduction equation plus a power generation term (Equation 25.20) and on the Gauss Law, stated for electro-quasi-static conduction (Equation 25.16) in inhomogeneous materials (meat batters, with dielectric and physical properties as functions of temperature) with nonuniform permittivity for obtaining electric field distribution. Samples were considered regularly shaped as cube, cylinder, and sphere. Heating rate and temperature distribution were greatly influenced by the sample shape. Among the shapes investigated by the authors, regular cubes were more suitable for RF treatment since cubic-shaped products exhibited a fast and more even heating, with a good absorption of power. In the case of cylindrically shaped products, authors recommended a vertical orientation during treatment, since horizontally oriented cylinders showed slower heating, characterized by uneven temperature distribution. Spherical shapes were found to be the less favored to RF heating. As a practical consequence of such observations, Farag et al.[33] designed experiments for studying thawing of meat, assisted by RF. Experimental results confirmed the conclusions of Romano and Marra,[19] showing that if a regular slab-shaped meat sample was thawed by means of RF heating, relatively uniform temperature distribution occurred. Furthermore, Farag et al.[20] used a simplified mathematical approach in order to quantify the amount of power absorbed by the sample with respect to the total power consumed by the RF apparatus, showing how modeling could be used for practical consideration in analysis and design of new food processes.

The approach used by Marra et al.[15] was also on the basis of the model proposed by Birla et al.,[17] who simulated RF heating of foods taking into account the fluid dynamics of the medium surrounding the sample (a spherical fruit) to be heated by means of Navier–Stoke equations. The model fruit was prepared from 1% gellan gel for experimental validation of the simulation results. Authors showed that spherically shaped samples surrounded with air between RF electrodes and placed in the proximity of electrodes would not heat uniformly. The uneven heating was reduced by means of immersing the model fruit in water, but it created a new problem since different horizontal positions of the model fruits lead to uneven heating. Thus, authors observed that horizontal and vertical model fruit positions with respect to electrodes significantly influenced heating patterns inside the model fruit. Nonuniform heating was attributed to shape, dielectric properties and relative fruit position in the container (Figure 25.6). The study suggested that movement and rotation of the spherical object was the only plausible solution for improving heating uniformity. The authors proposed the prediction of temperature profile by means of computer simulation as a useful tool for designing an RF heating process for disinfestations.

25.4 MATHEMATICAL DESCRIPTIONS OF DIELECTRIC PROPERTIES

The dielectric properties of food materials can be divided into two parts known as the permeability and permittivity (ε). Permeability values for foodstuffs are generally similar to that of free space and as a result are not believed to contribute to heating.[34] Great influence on RF heating is due by the permittivity, which is expressed in terms of the dielectric constant (ε') and the loss factor (ε''), as in the following equation

$$\varepsilon = \varepsilon' - j\varepsilon'' \tag{25.22}$$

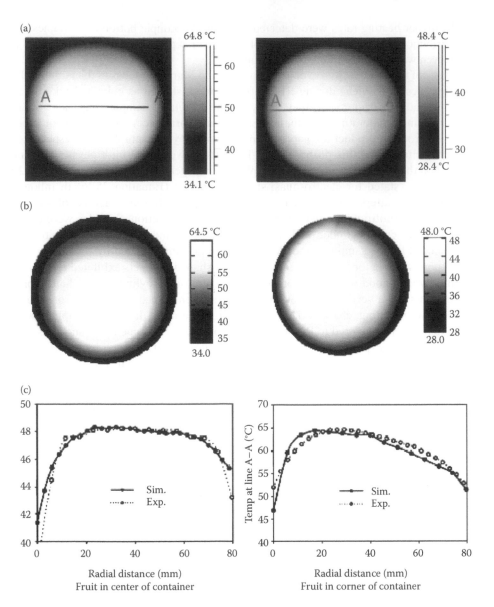

FIGURE 25.6 (a) Experimental and (b) Simulated temperature distribution inside a model fruit (U80 mm) and (c) at horizontal line A-A after 7 minutes of RF heating in 195 mm electrode gap.

The dielectric constant (ε'), the real part of permittivity, is a characteristic of any material and is a measure of the capacity of a material to absorb, transmit and reflect energy from the electric portion of the electrical field[35] and is a constant for a material at a given frequency under constant conditions. The dielectric constant is a measure of the polarizing effect from applied electric field (i.e., how easily the medium is polarized). The loss factor (ε''), the imaginary part of permittivity, measures the amount of energy that is lost from the electrical field, which is related to how the energy from a field is absorbed and converted to heat by a material passing through it.[35] A material with a low ε'' will absorb less energy and could be expected to heat poorly in an electrical field due to its greater transparency to electromagnetic energy.[36] However, it is important to emphasize that thermo-physical properties (especially specific heat) will also have an influence on the magnitude of the temperature rise obtained, as described by Equation 25.20.

The ratio between the dielectric constant and the loss factor determines the tangent of dielectric loss angle ($\tan\delta$), often called the loss tangent or the dissipation (power) factor of the material. For a given material this is equivalent to

$$\tan\delta = \frac{\varepsilon''}{\varepsilon'} \tag{25.23}$$

The loss tangent appears in the expression of the penetration depth (d_p), defined as the depth in a material where the energy of a plane wave propagating perpendicular to the surface has decreased to $^1/_e$ ($^1/_{2.72}$) of the surface value[31]

$$d_p = \frac{C}{2\pi f \sqrt{2\varepsilon'}\sqrt{\sqrt{1+(\tan\delta)^2}-1}} \tag{25.24}$$

where C is the speed of propagation of waves in a vacuum (3×10^8 m s^{-1}) and d_p is expressed in meters. When $\tan\delta$ is low (i.e., far less than 1) Equation 25.24 can be simplified to

$$d_p = \frac{C}{2\pi f \sqrt{\varepsilon'}\tan\delta} = \frac{4.47\times10^7}{f\sqrt{\varepsilon'}\tan\delta} \tag{25.25}$$

Equations 25.24 and 25.25 illustrate the effect of f, ε', and ε'' on d_p. Bengtsson and Risman[37] found that the greatest d_p was experienced when both ε' and ε'' were low.

Among dielectrical properties needed in the model of RF heating, electrical conductivity (σ) indicates the ability of a material to conduct an electric current. In a dielectric food system, σ is related to ionic rotation. It contributes to ε'' and in RF ranges can be calculated from the following equation:[12]

$$\sigma = 2\pi f\varepsilon'' \tag{25.26}$$

As it appears from main equations needed for mathematical description of RF heating, the dielectric properties of food play an important role.[12] The dielectric properties of most materials are influenced by a variety of factors. The content of moisture is generally a critical factor, but the frequency of the applied alternating field, the temperature of the material and also the density, chemical composition (i.e., fat, protein, carbohydrate, and salt) and structure of the material all have an influence.[12] In terms of bulk density samples of an air-particle mixture with higher density generally have higher ε' and ε'' values because of less air incorporation within the samples.[38] In relation to composition, Nelson and Datta[38] stated that the dielectric properties of materials are dependent on chemical composition and especially on the presence of mobile ions and the permanent dipole moments associated with water. While a considerable amount of work on dielectric properties has been published at microwave frequencies a more limited number of studies have examined the dielectric properties of food and agricultural products at RF frequencies. Work completed to the year 2000 was summarized by Piyasena et al.[12]

Most recent works have been analyzed by Marra et al.,[13] including main research on meat products conducted at University College Dublin, Ireland,[39–47] and on a variety of foods (fruits, mashed potatoes, whey protein gel, macaroni, cheese, egg, and salmon) conducted at Washington State University,[48–52] Pullman, WA.

25.5 CONCLUSION

The availability of new software and better performing computers allowed expanding the modeling approach to complex problems, such as RF heating of foods.

The intrinsic multiphysics nature of this process, involving and coupling classical electromagnetic Maxwell's equations and heat transfer, required an interdisciplinary approach and the cooperation among professionals (such as food scientists, food technologists, food engineers and electromagnetic engineers) in order to set up models capable of representing the phenomena occurring during RF heating.

Up to now, mathematical modeling has improved the understanding of RF heating of food and it was essential to the continued development of this technology. More efforts are needed in order to develop computer aided engineering of processes and plants on an industrial scale.

In fact, electro-quasi-static approach has been proven to be reliable and to provide valid information for small scale RF facilities: in any case, the modeling approach for large-scale equipment needs the solution of more complex set of Maxwell's equations, requiring even more powerful computers and better performing software. The computer aided design and scale up of industrial RF facilities requires also scientists to build up databases with dialectical and thermo-physical properties of foods potentially undergoing RF heating to be used in modeling.

REFERENCES

1. Sherman, V.W. 1946. *Food Industry*, 18, 506–509,628.
2. Jason, A.C., and Sanders, H.R. 1962a. Dielectric thawing of fish. I. Experiments with frozen herrings. *Food Technology*, 16(6), 101–106.
3. Jason, A.C., and Sanders, H.R. (1962b). Dielectric thawing of fish. I. Experiments with frozen white fish. *Food Technology*, 16(6), 107–112.
4. Rice, J. 1993. RF technology sharpens bakery's competitive edge. *Food Processing*, 6, 18–24.
5. Mermelstein, N.H. 1998. Microwave and radio frequency drying. *Food Technology*, 52(11), 84–86.
6. Houben, J., Schoenmakers, L., van Putten, E., van Roon, P., and Krol, B. 1991. Radio frequency pasteurisation of sausage emulsions as a continuous process. *Journal of Microwave Power and Electromagnetic Energy*, 26(4), 202–205.
7. Zhao, Y., Flugstad, B., Kolbe, E., Park, J.E., and Wells, J.H. 2000. Using capacitive (radio frequency) dielectric heating in food processing and preservation—a review. *Journal of Food Process Engineering*, 23, 25–55.
8. Laycock, L., Piyasena, P., and Mittal, G.S. 2003. Radio frequency cooking of ground, comminuted and muscle meat products. *Meat Science*, 65(3), 959–965.
9. Zhang, L., Lyng, J.G., and Brunton, N.P. 2004. Effect of radio frequency cooking on texture, colour and sensory properties of a large diameter comminuted meat product. *Meat Science,* 68(2), 257–268.
10. Zhang, L., Lyng, J.G., and Brunton, N.P. 2006. Quality of radio frequency heated pork leg and shoulder ham. *Journal of Food Engineering,* 75(2), 275–287.
11. Birla, S.L., Wang, S., Tang, J., Fellman, J.K., Mattinson, D.S., and Lurie, S. 2005. Quality of oranges as influenced by potential radio frequency heat treatments against Mediterranean fruit flies. *Postharvest Biology and Technology*, 38(1), 66–79.
12. Piyasena, P., Dussault, C., Koutchma, T., Ramaswamy, H.S., and Awuah, G.B. 2003. Radio frequency heating of foods: Principles, applications and related properties. A review. *Critical reviews in food science and nutrition,* 43(6), 587–606.
13. Marra, F., Zhang, L., and Lyng, J. G. Radio frequency treatment of foods: review of recent advances. *Journal of Food Engineering*, 91(4), 497–508.
14. Yang, J., Zhao, Y., and Wells, J.H. 2003. Computer simulation of capacitive radio frequency (RF) dielectric heating on vegetable sprout seeds. *Journal of Food Process Engineering,* 26, 239–263.
15. Marra, F., Lyng, J., Romano, V., and McKenna, B. 2007. Radio-frequency heating of foodstuff: solution and validation of a mathematical model. *Journal of Food Engineering*, 79(3), 998–1006.
16. Romano, V., and Marra, F. 2008. A numerical analysis of radio frequency heating of regular shaped foodstuff. *Journal of Food Engineering,* 84, 449–457.
17. Birla, S.L., Wang, S., and Tang, J. 2008. Computer simulation of radio frequency heating of model fruit immersed in water. *Journal of Food Engineering,* 84, 270–280.
18. Neophytou, R.I., and Metaxas, A.C. 1996. Computer simulation of a radio frequency industrial system. *Journal of Microwave Power and Electromagnetic Energy,* 31(4), 251–259.
19. Neophytou, R.I., and Metaxas, A.C. 1997. Characterisation of radio frequency heating systems in industry using a network analyser. *IEE Proceedings: Science, Measurement and Technology,* 144(5), 215–222.

20. Neophytou, R.I., and Metaxas, A.C. 1998. Combined 3D FE and circuit modeling of radio frequency heating systems. *Journal of Microwave Power and Electromagnetic Energy*, 33(4), 243–262.
21. Neophytou, R.I., and Metaxas, A.C. 1999. Investigation of the harmonic generation in conventional radio frequency heating systems. *Journal of Microwave Power and Electromagnetic Energy*, 34(2), 84–96.
22. Rowley, A.T. 2001. Radio frequency heating. In Richardson, P. (Ed.) *Thermal Technologies in Food Processing* (pp. 162–177). Woodhead Publishing, Cambridge, UK.
23. Buffler, C.R. 1993. Dielectric properties of foods and microwave materials. In Buffler, C.R. (Ed.) *Microwave Cooking and Processing* (pp. 46–69). Van Nostrand Reinhold, New York, NY.
24. Ohlsson, T. 1983. Fundamentals of microwave cooking. *Microwave World*, 4, 4–9.
25. Jones, P.L. and Rowley, A.T. 1997. In Baker, C.G.J (Ed.) *Industrial Drying of Foods*. Blackie Academic and Professional, London, UK.
26. Orsat, V., Bai, L., Raghavan, G.S.V., and Smith, J.P. 2004. Radio-frequency heating of ham to enhance shelf-life in vacuum packaging. *Journal of Food Process Engineering*, 27, 267–283.
27. Wang, S., Yue, J., Tang, J., and Chen, B., 2005. Mathematical modelling of heating uniformity for in-shell walnuts subjected to radio frequency treatments with intermittent stirrings. *Postharvest Biology and Technology*, 35(1), 97–107.
28. Wang, S., Yue, J., Chen, B., and Tang, J. 2008. Treatment design of radio frequency heating based on insect control and product quality. *Postharvest Biology and Technology*, 49(3), 417–423.
29. Birla, S.L., Wang, S., Tang, J., and Hallman, G. 2004. Improving heating uniformity of fresh fruit in radio frequency treatments for pest control. *Postharvest Biology and Technology*, 33(2), 205–217.
30. Chan, T.V.C.T., Tang, J., and Younce, F. 2004. 3-Dimensional numerical modeling of an industrial radio frequency heating system using finite elements. *International Microwave Power Institute*, 39(2), 87–106.
31. Choi, C., and Konrad, A. 1991. Finite-element modelling of RF heating process. *IEEE Transactions on Magnetics*, 27(5), 4227–4230.
32. Stefens, P., and Van Dommelen, D. 1993. CAD software package for high-frequency dielectric applications. *Journal of Microwave Power and Electromagnetic Energy*, 28(1), 3–10.
33. Farag, K.W., Marra, F., Lyng, J.G., Morgan, D.J., and Cronin, D.A. Temperature changes and power consumption during radio frequency tempering of beef lean/fat formulations. *Food and Bioprocess Technology*, doi:10.1007/s11947-008-0131-5.
34. Zhang, H., and Datta, A.K. 2001. Electromagnetics of microwave heating: magnitude and uniformity of energy absorption in an oven. In Datta, A.K and Anantheswaran, R.C. (Eds) *Handbook of Microwave Technology for Food Applications* (pp. 1–28). Marcel Dekker, New York, NY.
35. Engelder, D.S., and Buffler, C.R. 1991. Measuring dielectric properties of food products at microwave frequencies. *Microwave World*, 12(2), 2–11.
36. Decareau, R.V. 1985. *Food Industry and Trade—Microwave Heating—Industrial Applications* (pp. 1–10). Academic Press, Orlando, FL.
37. Bengtsson, N.E., and Risman, P.O. 1971. Dielectric properties of foods at 3 GHz as determined by cavity perturbation technique. *Journal of Microwave Power*, 6(2), 107–123.
38. Nelson, S.O., and Datta, A.K. 2001. Dielectric Properties of food materials and electric field interactions. In Datta, A.K. and Anantheswaran, R.C. (Eds.) *Handbook of Microwave Technology for Food Applications*. Marcel Dekker, Inc., New York, NY.
39. Farag, K. Lyng, J.G., Morgan, D.J., and Cronin, D.A. 2008. Dielectric and thermophysical properties of different beef meat blends over a temperature range of −18 to+10°C. *Meat Science*, 79(4), 740–747.
40. Zhang, L., Lyng, J.G., and Brunton, N.P. 2007. The effect of fat, water and salt on the thermal and dielectric properties of meat batter and its temperature following microwave or radio frequency heating. *Journal of Food Engineering*, 80(1), 142–151.
41. Zhang, L., Lyng, J.G., Brunton, N., Morgan, D., and McKenna, B. 2004. Dielectric and thermophysical properties of meat batters over a temperature range of 5–85°C. *Meat Science*, 68, 173–184.
42. Tang, X., Cronin, D.A., and Brunton, N.P. 2005. The effect of radio frequency heating on chemical, physical and sensory aspects of quality in turkey breast rolls. *Food Chemistry*, 93(1), 1–7.
43. Tang, X., Lyng, J.G., Cronin, D.A., and Durand, C. 2006. Radio frequency heating of beef rolls from *biceps femoris* muscle. *Meat Science*, 72(3), 467–474.
44. Brunton, N.P., Lyng, J.G., Li, W., Cronin, D.A., Morgan, D., and McKenna, B. 2005. Effect of radio frequency (RF) heating on the texture, colour and sensory properties of a comminuted pork meat product. *Food Research International*, 38(3), 337–344.
45. Brunton, N.P., Lyng, J.G., Zhang, L., and Jacquier, J.C. 2006. The use of dielectric properties and other physical analyses for assessing protein denaturation in beef *biceps femoris* muscle during cooking from 5 to 85°C. *Meat Science*, 72(2), 236–244.

46. Lyng, J.G., Cronin, D.A., Brunton, N.P., Li, W., and Gu, X. 2006. An examination of factors affecting radio frequency heating of an encased meat emulsion. *Meat Science*, 75(3), 470–479.

47. Lyng, J.G., Zhang, L., and Brunton, N.P. 2005. A survey of dielectric properties of meats and ingredients used in meat product manufacture. *Meat Science*, 69(4), 589–602.

48. Al-Holy, M., Wang, Y., Tang, J., and Rasco, B. 2005. Dielectric properties of salmon (*Oncorhybchus keta*) and sturgeon (*Acipenser transmontanus*) caviar at radio frequency (RF) and microwave (MW) pasteurization frequencies. *Journal of Food Engineering*, 70(4), 564–570.

49. Guan, D., Cheng, M., Wang, Y., and Tang, J. 2004. Dielectric properties of mashed potatoes relevant to microwave and radio-frequency pasteurilization and sterilization processes. *Journal of Food Science*, 69(1), 30–37.

50. Ragni, L., Al-Shami, A., Mikhaylenko, G., and Tang, J. 2007. Dielectric characterization of hen eggs during storage. *Journal of Food Engineering*, 82(4), 450–459.

51. Wang, S., Tang, J., Johnson, J.A., Mitcham, E., Hansen, J.D., Hallman, G., et al. 2003. Dielectric properties of fruits, nuts, and insect pests as related to radio frequency and microwave treatments. *Biosystems Engineering*, 85(2), 201–212.

52. Wang, S., Tang, J., Johnson, J.A., Mitcham, E., Hansen, J.D., Hallman, G., Drake, S.R., and Wang, Y. 2003. Dielectric properties of fruits and insect pests as related to radio frequency and microwave treatments. *Biosystems Engineering*, 85(2), 201–212.

26 Infrared Heating

Daisuke Hamanaka and Fumihiko Tanaka
Kyushu University

CONTENTS

26.1 INTRODUCTION

Thermal processing of foods is very important in extending the shelf life of various food products. Postharvest heat treatments have become increasingly popular to control insect pests, prevent fungal spoilage and accelerate the ripening of fruits and vegetables. Infrared radiation (IR) heating technology is useful for these purposes because it produces rapid and simple heating. In this chapter, we focus on the microbicidal action of infrared heating and its mechanism of action by means of computational fluid dynamics (CFD) and Monte Carlo (MC) simulation which are based on the fundamentals of quantum physics of radiation. We also simulate the behavior of microorganisms under nonisothermal conditions by coupling predictive microbiology with heat transfer modeling. The combination of microbial modeling and the CFD approach is an emerging and important new field with broad applications in the food and other processing industries. The developments described in this chapter can help in understanding and improving the far-infrared (FIR) heating processes applied to foods.

26.2 MICROBIAL ACTIONS OF INFRARED (IR) HEATING

Infrared rays are used in various industrial fields such as sensing, measuring, analysis, communication and heat treatment [1]. In the food industry in particular, the heating effect of infrared rays has been widely applied to maturing, drying and surface baking, etc [2]. Infrared rays can rapidly and effectively heat the surface of any substance with the exception of very efficient infrared reflectors and little energy dissipation because infrared rays efficiently transfer thermal energy to the target substance without heating the surrounding media such as gas or liquid, a property which is indispensable for heat transfer by convection and conduction [3]. This particular property means that an infrared ray can target a microorganism with large quantities of thermal energy in a very short time. Consequently, IR heat treatment will minimize any deterioration in the internal quality of food due to the penetration of thermal energy. The microbicidal efficiency of IR heating under various conditions is described in the following sections.

26.2.1 INFRARED HEATING FOR BACTERIAL INACTIVATION IN LIQUID MEDIA

Several researchers have investigated the ability of IR heating to inactivate microorganisms in liquid suspension (water saline solution, phosphate buffered saline, etc.). Shimada [4] reported that the effect of irradiation with FIR rays at 55°C on the inactivation of *Staphylococcus aureus* and *Escherichia coli* suspended in physiological saline solution was equal to that of heat treatment in a water bath at 70°C. Hashimoto et al. [5–7] and Sawai et al. [8–13] have reported the pasteurization effects of FIR irradiation on the inactivation of bacteria suspended in phosphate buffered saline solution. Hashimoto et al. [5,6] reported that *S. aureus* and *E. coli* were damaged and inactivated by FIR irradiation at temperatures below those which are lethal to bacteria. They suggested that these bacteria were damaged by FIR irradiation in the area close to the surface of the suspension. Hashimoto et al. [7] also showed that FIR irradiation was more effective for pasteurization than near infrared radiation (NIR). They concluded that the superiority of FIR to NIR pasteurization was a result of the very high absorption coefficient of the bacterial suspension in the FIR region. Sawai et al. [8–11] attempted to evaluate the degree of damage inflicted on cells of *E. coli* by FIR irradiation by monitoring the changes in the sensitivity to selected reagents (penicillin G, chloramphenicol, riphampicin, and nalidixic acid). They found that *E. coli* cells irradiated with FIR became more sensitive to chloramphenicol and rifampicin, thus, indicating that FIR irradiation had damaged RNA polymerase and ribosomes in *E. coli* cells. In addition, these authors reported that FIR irradiation caused the heat activation or death of bacterial spores at a range of temperatures where conductive heating had no effect on bacterial spores [12]. Moreover, they suggested that pasteurization using FIR irradiation may be achieved at a lower temperature than that required to produce the same effect using conductive heating and that enzyme (lipase and α-amylase) activity levels would therefore be maintained. At the lower treatment temperature, the death rate constants for *E. coli* increase while the rate constant for enzyme inactivation decrease at the same bulk temperature of the suspension [13].

26.2.2 INFRARED HEATING FOR BACTERIAL INACTIVATION IN DRY CONDITIONS

Bacterial spores such as those of *Bacillus* and *Clostridium* have high heat resistance under dry conditions. However, it was considered that these spores could be inactivated by subjecting them to a high concentration of thermal energy for a short period of time. Based on this supposition, the use of IR heating to inactivate spores has previously been reported by some researchers [14–17]. Heat resistance of bacterial spores under dry conditions is affected by their water activity (a_w) [18–21]. In addition, spore inactivation by IR heat treatment was highly dependent on varying a_w values of different spores, and we suggest that the resistance of bacterial spores to IR heat treatment might be affected by the infrared absorption efficiency of the bacterial spores [22]. It is well known that infrared rays are characteristically absorbed by any organic matter [23]. Therefore, effective bactericidal processing could be achieved by the use of IR irradiation which causes rapid temperature rise in the target bacterial spores. All living things are composed of water molecules and biopolymers such as proteins, lipids, and carbohydrates. In particular, the characteristics of infrared ray absorption by water molecules inside spores could be the most important factor for bactericidal activity, since infrared rays are easily absorbed by water. Water is a triatomic molecule and there are three vibration modes for OH bands, namely symmetric stretching, asymmetric stretching and deformation vibrations (Figure 26.1). Water molecules inside bacterial cells are bound to polar groups such as -NH$_2$, -COOH, and -COO- contained in biopolymers. The infrared ray absorption characteristics of bacterial spores are mainly determined by the location, vibration state and number of water

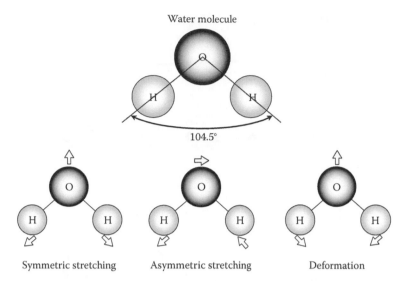

FIGURE 26.1 Vibration modes of water molecules.

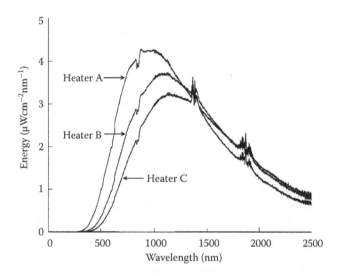

FIGURE 26.2 Distribution of spectral energy in three types of halogen heater (A, B, and C). The peak wavelength (nm) and radiant energy (μ Wcm^{-2}nm^{-1}) of heaters A, B, and C were approximately 950 and 4.2, 1100 and 3.7, and 1150 and 3.2, respectively. (From Hamanaka, D. et al., *Int. J. Food Microbiol.*, 108, 281, 2006. With permission.)

molecules inside the bacterial spores; in other words, the resistance of bacterial spores to IR heat treatment will depend on the binding conditions of water molecules within the spores. We investigated the effect of using infrared heaters of different wavelengths on the inactivation of bacterial spores with various water activities [24]. Three types of infrared heaters with different spectral properties were used as shown in Figure 26.2. The decimal reduction times (D values: time required to decrease the population by 90%), which were calculated using the linear portion of survival curves, were affected by both the a_w values of bacterial cells and the spectra of the infrared heaters (Figure 26.3). As the peak wavelength of the infrared heater was short, the a_w values increased leading to maximum D values for bacterial spores. Generally, an increase in heat resistance of the bacterial spores during heat processing was obtained at a_w levels ranging from 0.2 to 0.4 [21,25].

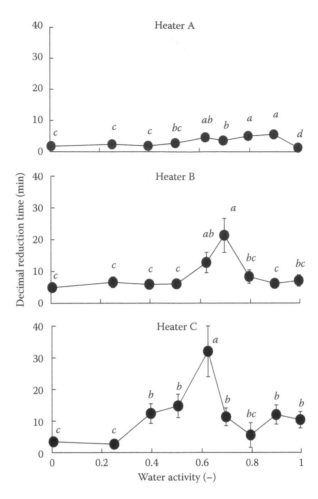

FIGURE 26.3 Relationship between water activity and decimal reduction times of *B. subtilis* spores during IR heating treatment. Decimal reduction times were calculated from the linear decreasing portion of the survival curves. (From Hamanaka, D. et al., *Int. J. Food Microbiol.*, 108, 281, 2006. With permission.)

Hoffman et al. [26] suggested that an important factor regulating heat resistance of bacterial spores was particular conditions of water molecules in the cell. Grecz et al. [27] indicated that molecular masking or caging in the spore by a protective cement consisting of chelators of divalent metal ions was responsible for low water-binding levels. Therefore, electrical and chemical inertness, biological dormancy, and high heat resistance of bacterial spores were induced. Infrared rays are peculiarly absorbed by organic matter (protein, lipid, and carbohydrate) and water molecules, which may or may not be bound to polar groups contained in some biopolymers. It is assumed that the magnitude of molecular motion of organic matter and water molecules within bacterial spores could be affected by qualitative and quantitative changes in infrared ray absorption characteristics. Consequently, the maximum resistance of spores to IR heat treatment can be altered by the characteristic level of absorption of infrared radiation, which varies according to the arrangement of water molecules within spores. The maximum heat resistance of bacterial spores to IR irradiation may depend on the infrared ray absorption characteristics of water molecules within the spore coat and/or cortex rather than those dispersed throughout the entire spore since the core of spores, which has the greatest heat resistance, is extremely dehydrated as a result of contraction of the cortex layer which gives rise to a repulsive electrical force [28,29].

26.3 MODELING OF INFRARED RADIATION HEATING

The FIR heating model is based on the combination of three mechanisms of heat transfer: radiation, convection and conduction. The schematic diagram of heat balance during FIR heat treatment is presented in Figure 26.4.

26.3.1 Radiation Heat Transfer Model

Mathematical modeling techniques are one of the most useful tools for the appropriate design and control of a FIR heating system. Studies on FIR heating of foods have used either a nonlinear radiation boundary condition based on the Stefan–Boltzman law for pairs of diffuse grey surfaces [30], or simply specified the value of the incident radiative heat flux [31]. In complex geometrical configurations, these approaches are not trivial because it is difficult to determine the view factors of the contributing surfaces [32]. MC modeling of FIR heat transfer is appropriate to simulate radiation heat transfer in complex configurations and has been widely described [33–38]. In MC simulations of FIR heat transfer problems, the energy emitted from a surface is simulated by the propagation of a large number of photons. The photon is followed as it proceeds from one interaction to another which is described as a random event. This continues until the photon is absorbed or leaves the computational domain. A large number of trajectories are required to obtain comparably accurate simulation results. The results are used to determine the fraction of energy absorbed at the surface.

Figure 26.5 shows the geometry of the FIR heating cell for heat treatment of a strawberry. The surfaces in the geometry of the heating cell and the strawberry were subdivided into sets of non-overlapping primitive surfaces. These surfaces are boundary faces of volumetric zones that cover the interior computation domain. At the interface of zones there are surfaces that do not interfere with the radiation field. Each primitive surface j is treated as a separate uniform photon source with temperature T_j and an emissivity ε_j. The total number of sources is indicated by N. The total number of histories N_p to be calculated is then divided amongst the sources, according to their emission $S_j \varepsilon_j \sigma T_j^4$ compared to the total emission of all surfaces. Thus the number of photons n_j emitted by surface j is:

$$n_j = N_p \frac{S_j \varepsilon_j \sigma T_j^4}{\sum_{x=1}^{N} S_x \varepsilon_x \sigma T_x^4} \tag{26.1}$$

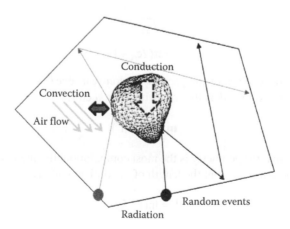

FIGURE 26.4 Schematic diagram of heat balance during FIR heating.

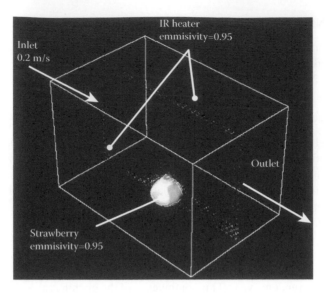

FIGURE 26.5 Configuration of the FIR heater showing the calculated air velocity field and surface temperature of the strawberry.

The random sampling of the source location on a simple rectangular region with local coordinates ($r_1 \leq r \leq r_2, s_1 \leq s \leq s_2, t_1 \leq t \leq t_2$) consists of three random choices of pseudo-random variables $0 \leq \xi_1 \leq 1$, $0 \leq \xi_2 \leq 1$ and $0 \leq \xi_3 \leq 1$.

$$r = r_1 + \xi_1(r_2 - r_1)$$
$$s = s_1 + \xi_2(s_2 - s_1) \tag{26.2}$$
$$t = t_1 + \xi_3(t_2 - t_1)$$

The birth of a photon requires not only a location but also the initial direction of travel. We generate direction cosines from the following expression:

$$r = \sqrt{1 - t'^2} \cos\varphi$$
$$s = \sqrt{1 - t'^2} \sin\varphi \tag{26.3}$$

where

$$t' = \sqrt{\xi_4}$$
$$\varphi = \pi(2\xi_5 - 1) \tag{26.4}$$

where ξ_4 and ξ_5 are two choices of the pseudo-random variable $0 \leq \xi_4 \leq 1, 0 \leq \xi_5 \leq 1$. The initial direction of travel can be expressed as the following unit vector:

$$\mathbf{m} = (r', s', t') \tag{26.5}$$

Tracking photons across the geometry is the most computationally intensive task. The absorption or scattering event is determined from the length of the optical path, L.

$$L = \ln(1 - \xi_6) \tag{26.6}$$

where the pseudo-random variable is $0 \leq \xi_6 \leq 1$.

A photon is absorbed or reflected at a physical surface:

$$\xi_7 \leq \varepsilon \rightarrow \text{absorption}$$

$$\xi_7 > \varepsilon \rightarrow \text{reflection}$$

(26.7)

where ξ_7 is the pseudo-random variable $0 \leq \xi_7 \leq 1$. In the case of reflection on a diffuse surface, a new direction of travel must be sampled according to the Equations 26.3 and 26.4. The number of photons incident on a surface is added.

The irradiation on surface i is:

$$G_i = \frac{1}{S_i} \sum_{j=1}^{N} \frac{n_j^i}{n_j} S_j \varepsilon_j \sigma T_j^4$$

(26.8)

with n_j^i is the number of photons that are emitted by surface j and incident on surface i. Introducing Equation 26.1 into Equation 26.8 leads to the following expression:

$$G_i = \frac{1}{S_i} \frac{\sum_{j=1}^{N} n_j^i}{N_p} \sum_{x=1}^{N} S_x \varepsilon_x \sigma T_x^4 = \frac{1}{S_i} \frac{n^i}{N_p} \sum_{x=1}^{N} S_x \varepsilon_x \sigma T_x^4$$

(26.9)

Therefore, it suffices to add all incident photons on surface i to the photon current n_i and multiply the relative photon current with the available emissive power in the system to obtain the irradiation. The net radiation flux $q_{\text{rad},i}$ leaving the surface i is then equal to:

$$q_{\text{rad},i} = \varepsilon_i \sigma T_i^4 - \varepsilon_i G_i$$

(26.10)

To obtain good estimates of the quantities of interest, many histories of photons need to be generated.

26.3.2 CONVECTION HEAT TRANSFER MODEL

CFD modeling has been applied to a number of food processing activities because it is a powerful tool, based on physical principles which provides detailed information on flow variables and related quantities in complex geometries. A good approximation of three-dimensional momentum and energy transfer in a forced convection oven was determined using this technique [38,39].

The set of governing equations used to describe convection heat transfer are as follows:

$$\frac{\partial u_i}{\partial x_i} = 0$$

(26.11)

$$\rho_a \frac{\partial u_j}{\partial \theta} + \rho_a \frac{\partial}{\partial x_i}(u_i u_j) = \frac{\partial}{\partial x_i} \mu \frac{\partial u_j}{\partial x_i} + \frac{\partial}{\partial x_i} \mu \frac{\partial u_i}{\partial x_j} - \nabla p + \rho_a g \beta(T - T_{\text{ref}})$$

(26.12)

$$\rho_a \frac{\partial H}{\partial \theta} + \rho_a \frac{\partial}{\partial x_i}(u_i H) = \lambda_a \nabla^2 H$$

(26.13)

where $u, x, \rho_a, \theta, \mu, p, g, \beta, H$, and λ_a are air flow rate, direction, density, time, dynamic viscosity, pressure, gravity force, thermal expansion coefficient, enthalpy, and thermal conductivity of air, respectively. These governing equations with appropriate boundary and initial conditions can be solved numerically. At the strawberry surface, a conservative interface heat flux condition was applied (convection + radiation = conduction). The governing equations can be solved by means of a CFD code based on a finite volume method.

26.3.3 CONDUCTIVE HEAT TRANSFER MODEL

It is assumed that conductive heat transfer occurs inside the strawberry, and that convective and radiation heat transfer takes place at the boundaries. The Fourier equation used to describe conductive heat transport through the strawberry is as follows:

$$\rho_s C_{ps} \frac{\partial T}{\partial \theta} = \lambda_s \nabla^2 T \tag{26.14}$$

where ρ_s, C_{ps}, and λ_s are density, heat capacity and heat conductive coefficient for conductive solid, respectively. The thermal properties of the fruit are kept constant in the temperature range used for this simulation and a uniform temperature distribution is chosen as the initial condition for Equation 26.14. The boundary condition is described as follows:

$$\lambda_s \frac{\partial T(x, y, z)}{\partial n} = h(x, y, z)(T_\infty(x, y, z) - T(x, y, z)) + q_{rad}(x, y, z) \tag{26.15}$$

$$(x, y, z) \in \Gamma_s$$

where n is the outward normal to the food surface Γ_s.

26.4 TEMPERATURE DISTRIBUTION DURING HEATING WITH FAR INFRARED (FIR) IRRADIATION

By using a MC FIR radiation model combined with convection–diffusion air flow and heat transfer described above, heating simulations of strawberry fruit were carried out. The predicted time progression of temperature contours on the surface (top view) in a cross-section of the strawberry during 450 s with FIR heater temperature of 200°C is shown in Figure 26.6. The lower part of the figure also shows the measured time progression of temperature contours on the surface (top view) using an IR thermographic camera. The same profiles can be recognized in both simulation and

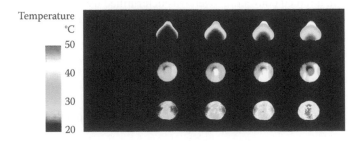

FIGURE 26.6 Predicted (top, inside; middle, top view) and measured (bottom, top of view by means of IR thermographic camera) temperature distributions as a function of heating time (0–450 s, FIR heater temperature 200°C). (From Tanaka, F. et al., *J. Food Eng.*, 79, 445, 2007. With permission.)

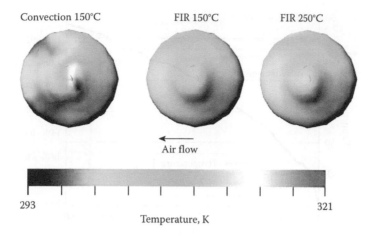

FIGURE 26.7 Temperature of the strawberry surface (top view) under convection (left) and FIR (middle and right) heating (the average surface temperatures are 30°C for each treatments).

measurement. Figure 26.7 compares the surface temperature distribution of the strawberry during three different types of heating: air convection heating at 150°C, FIR heating at 150°C and FIR heating at 250°C. All heating methods used an air velocity of 0.2 ms⁻¹. Convection heating achieved a maximum surface temperature of 45°C, close to the critical limit above which the tissue may be damaged, after 90 s of heating. The average convection coefficient was 9 Wm⁻²°C⁻¹ and this is comparable to the value calculated for a sphere of the same volume as the strawberry. The results in Figure 26.7 show that FIR heating achieved more uniform surface heating than air convection heating with a maximum temperature well below the critical limit of 50°C at the same average temperature. The resulting rate of surface heating was however, smaller or only equal to air convection heating (at 0.2 ms⁻¹) depending on the heater temperature used.

26.4.1 MICROBIAL INACTIVATION MODEL

To model the kinetics of microbial inactivation, the primary model can be used if it is assumed that inactivation follows first-order kinetics:

$$\frac{dN}{d\theta} = -k(T)N \tag{26.16}$$

where N and $k(T)$ are the colony forming unit and inactivation rate constant, respectively. The relationship between the rate constant and temperature is described by the Arrhenius equation:

$$k(T) = k_{\text{ref}} \exp\left(\frac{E_a}{R}\left(\frac{1}{T_{\text{ref}}} - \frac{1}{T}\right)\right) \tag{26.17}$$

The inactivation model has to be coupled to the heat transfer model to elucidate the behavior of the microorganisms. The inactivation of conidia of the important spoilage fungi, *Monilia fructigena* was simulated by using the coupled model in a nonisothermal process. The parameters for describing microbial inactivation are $E_a = 425(\pm28.9)$ kJ/mol, $T_{\text{ref}} = 316$ K and $k_{\text{ref}} = 0.00598(\pm0.000280)$ s⁻¹ [40]. The results in Figure 26.8 show the predicted inactivation curve of *Monilia fructigena* near the top point of the surface of the strawberry during FIR heating using a heater temperature of 200°C (see Figure 26.5). The modeling approach presented here can be used to improve FIR thermal inactivation of microorganisms and to extend the shelf life of food products. It would have been very interesting to include the thermal death for convection heating in Figure 26.8.

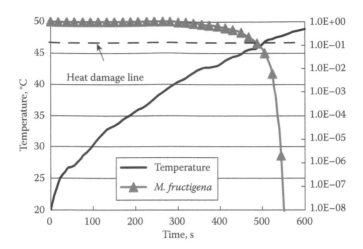

FIGURE 26.8 Predicted inactivation curve for *Monilia fructigena* during FIR heating. The broken line indicates the critical quality damage temperature for the strawberry.

26.5 CONCLUSION

The microbicidal actions of IR heating and modeling of a FIR heating system for postharvest treatment have been outlined in this chapter. A MC FIR radiation simulation combined with convection-diffusion air flow and heat transfer simulations were carried out in a CFD code based on a finite volume method. It has been shown that the proposed method is a powerful tool to describe complex heating configurations which includes radiation, convection and conduction.

ACKNOWLEDGMENT

The authors would like to express our gratitude to Dr. Yen Con Hung from the University of Georgia for reviewing this manuscript.

REFERENCES

1. Takagi, T. *Applications of Infrared rays (in Japanese)*. Zenkoku Shuppan, Tokyo, Japan, 1980.
2. Takano, M. and Yokoyama, M. *Inactivation of Food-borne Microorganisms—Science and Technology (in Japanese)*. Saiwai Shobou, Tokyo, Japan, 1998.
3. Toison, M.L.A. *Infrared and Its Thermal Applications*. Tokyo Electrical Engineering College Press, Tokyo, Japan, 1966.
4. Shimada, Y. Effect of far infrared ray on sterilizing of bacteria (in Japanese with English abstract). *J. Kyushu Den. Soc.*, 36, 307, 1982.
5. Hashimoto, A., Shimizu, M., and Igarashi, H. Effect of far infrared radiation on pasteurization of bacteria suspended in phosphate-buffered saline (in Japanese with English abstract). *Kagaku Kougaku Ronbunshu*, 17, 627, 1991.
6. Hashimoto, A. et al. Effect of far-infrared irradiation on pasteurization of bacteria suspended in liquid medium below lethal temperature. *J. Chem. Eng. Jpn.*, 25, 275, 1992.
7. Hashimoto, A. et al. Irradiation power effect on IR pasteurization below lethal temperature of bacteria. *J. Chem. Eng. Jpn.*, 26, 331, 1993.
8. Sawai, J. et al. Injury of Escherichia coli in physiological phosphate-buffered saline induced by far-infrared irradiation. *J. Chem. Eng. Jpn.*, 28, 294, 1995.
9. Sawai, J. et al. Novel utilization of antibiotics—A finding method of damaged parts in bacteria (in Japanese). *Jpn. J. Bacteriol.*, 51, 589, 1996.
10. Sawai, J. and Shimizu, M. Application of far-infrared heating to food processing part 2, Characteristics of far-infrared ray sterilization (in Japanese). *Food Industry*, 42, 25, 1999.
11. Sawai, J. et al. Far-infrared irradiation-induced injuries to *Eschelichia coli* at below the lethal temperature. *J. Ind. Microbiol. Biotechnol.*, 24 19, 2000.

12. Sawai, J. et al. Pasteurization of bacterial spores in liquid medium by far-infrared irradiation. *J. Chem. Eng. Jpn.*, 30, 170, 1997.
13. Sawai, J. et al. Inactivation characteristics shown by enzymes and bacteria treated with far-infrared radiative heating. *Int. J. Food Sci. Technol.*, 38, 661, 2003.
14. Molin, G. and Östlund, K. Dry-heat inactivation of *Bacillus subtilis* var. *niger* spores with special reference to spore density. *Can. J. Microbiol.*, 22, 359, 1975.
15. Molin, G. and Östlund, K. Dry-heat inactivation of *Bacillus subtilis* spores by means of infra-red heating. *Antonie van Leeuwenhoek*, 41, 329, 1975.
16. Mata-Portuguez, V.H., Sanchez, L., and Acosta-Gio, E. Sterilization of heat-resistant instruments with infrared radiation. *Infec. Control Hosp. Epidemiol.*, 23, 393, 2002.
17. Hamanaka, D. et al. Effects of infrared radiation on inactivation and injury of *B. subtilis* and *B. pumilus* spores (in Japanese with English abstract). *Nippon Shokuhin Kagaku Kogaku Kaishi*, 50, 51, 2003.
18. Murrell, W.G. and Scott, W.J. Heat resistance of bacterial spores at various water activities. *Nature*, 179, 481, 1957.
19. Murrell, W.G. and Scott, W.J. The heat resistance of bacterial spores at various water activities. *J. Gen. Microbiol.*, 43, 411, 1966.
20. Peeler, J.T. et al. Thermal resistance of *Bacillus subtilis* var. *niger* in a closed system. *Appl. Environ. Microbiol.*, 33, 52, 1977.
21. Pfeiffer, J. and Kessler, H.G. Effect of relative humidity of hot air on the heat resistance of *Bacillus cereus* spores. *J. Appl. Bacteriol.*, 77, 121, 1994.
22. Hamanaka, D. et al. Inactivation effect of infrared radiation heating on bacterial spores pretreated at various water activities. *Biocontrol Sci.*, 10, 61, 2005.
23. Ishida, H. et al. Applied techniques of infrared rays. In *Foundation and Application of Infrared Technology (Sekigaisen Kogaku)*, Mitsuishi, A. et al., Eds. Ohmsha, Tokyo, Japan, 1991.
24. Hamanaka, D. et al. Effect of the wavelength of infrared heaters on the inactivation of bacterial spores at various water activities. *Int. J. Food Microbiol.*, 108, 281, 2006.
25. Alderton, G., Chen, J.K., and Ito, K.A. Heat resistance of the chemical resistance forms of *Clostridium botulinum* 62A spores over the water activity range 0 to 0.9. *Appl. Environ. Microbiol.*, 40, 511, 1980.
26. Hoffman, R.K., Gambill, V.M., and Buchanan, L.M. Effect of cell moisture on the thermal inactivation rate of bacterial spores. *Appl. Microbiol.*, 16, 1240, 1968.
27. Grecz, N., Smith, R.F., and Hoffman, C.C. Sorption of water by spores, heat-killed spores, and vegetative cells. *Can. J. Microbiol.*, 16, 573, 1970.
28. Lewis, J.C., Snell, N.S., and Burr, H.K. Water permeability of bacterial spores and the concept of a contractile cortex. *Science*, 132, 544, 1960.
29. Gould, G.W. and Dring, G.J. Heat resistance of bacterial endospores and concept of an expanded osmoregulatory cortex. *Nature*, 258, 402, 1975.
30. Shilton, N., Mallikarjunan, P., and Sheridan, P. Modeling of heat transfer and evaporative mass losses during the cooking of beef patties using far-infrared radiation. *J. Food Eng.*, 55, 217, 2002.
31. Datta, A.K. and Ni, H. Infrared and hot-air-assisted microwave heating of foods for control of surface moisture. *J. Food Eng.*, 51, 355, 2002.
32. Incropera, F.P. and De Witt, D.P. *Fundamentals of Heat and Mass Transfer*, third edition. John Wiley & Sons, Inc., New York, NY, 1990.
33. Guilbert, P. Computer program RAD3D for modeling thermal radiation: technical report. In *AERE-R 13534, Computer Sciences and Systems Division*. AEA, Harwell, UK, 1989.
34. Howell, J.R. Thermal-radiation in participating media, the past, the present, and some possible futures. *J. Heat Transfer—Trans. ASME*, 110, 1220, 1988.
35. Howell, J.R. The Monte Carlo method in radiative heat transfer. *J. Heat Transfer—Trans. ASME*, 120, 547, 1998.
36. Maltby, J.D. and Burns, P.J. Performance, accuracy, and convergence in 3-dimensional Monte Carlo radiative heat-transfer simulation. *Num. Heat Transfer Part B—Fluid Amentals,* 19, 191, 1991.
37. Tanaka, F. et al. Monte Carlo simulation of far infrared radiation heat transfer. *J. Food Proc. Eng.*, 29, 349, 2006.
38. Tanaka, F. et al. Investigation of far infrared radiation heating as an alternative technique for surface decontamination of strawberry. *J. Food Eng.*, 79, 445, 2007.
39. Verboven, P. Investigation of the uniformity of the heat transfer to food products in ovens by means of computational fluid dynamics. PhD Thesis, Katholieke Universiteit Leuven, Belgium, 1999.
40. Marquenie, D. Evaluation of physical techniques for surface disinfections of strawberry and sweet cherry PhD Thesis, Katholieke Universiteit Leuven, Belgium, 2002.

27 Impingement Thermal Processing

Ferruh Erdoğdu
University of Mersin

Brent A. Anderson
Mars, Inc.

CONTENTS

27.1 INTRODUCTION

A promising development in the food processing area to reduce the processing time in freezing, thawing, baking and drying of foods is impingement heat transfer. Impingement (or so called jet impingement) is accomplished by directing a jet or jets of fluid at a solid surface to cause a change. Jet velocities range from 10 to 50 m/s and temperatures range from –50 to 400°C. This process produces a high and spatially variable convective heat (or mass) transfer coefficient that might approach values similar to those found in frying processes [1]. In traditional heating/cooling methods using still air or slowly agitated air, a thermal boundary layer develops around the product which slows the heat transfer rate. A significant effect of impingement is on reducing the thickness of this boundary layer insulation effect between the product and heat transfer medium as a result of higher fluid velocity and induced turbulence [2]. Sarkar and Singh [3] also stated that the boundary layer over the surface does not change significantly beyond a limiting velocity. Hence, there would be no effective increase in the rate of heat transfer beyond a certain velocity value.

The heat flux (q'') at a solid surface can be written as:

$$\lim_{y \to 0} q'' = -k_f \cdot \frac{\partial T}{\partial y} \tag{27.1}$$

where k_f is thermal conductivity of impinging fluid, and $\partial T/\partial y$ is temperature gradient at the surface. With induced turbulence due to higher jet velocities and interactions among the jets, temperature gradient becomes more steep leading to higher heat fluxes and therefore, higher heat transfer coefficient values. Considerable reduction in processing time and improvement in product quality can be obtained by using impingement processing [4]. A point of concern when using air impingement systems is the variation of heat transfer coefficient over the surface. This may lead to undesirable variation in certain quality attributes. Previous studies, for example, have indicated that air impingement systems could result in localized hot and cold spots on the surface [3,5] depending upon the objective of use.

Impingement systems typically have arrays of nozzles (round or slot shaped) directed over products on a moving belt. A schematic diagram of such an impingement system is shown in Figure 27.1. A basic unit consists of a system directing air through a plenum to equilibrate the pressure before it is forced through a nozzle or array of nozzles. The nozzles are directed at the product surface which may be stationary or moving perpendicular to the flow direction. Impingement nozzles come in many different shapes and sizes. Probably the most common shapes are round (circular) or slot (rectangular) nozzles. A rectangular nozzle is called slot jet when its length is at least 10 times higher than its width. Impingement systems are typically characterized by their nozzle length to hydraulic diameter ratio (L/d) and ratio of nozzle-to-surface distance from nozzle exit to the surface where the air impinges (Z/d). Hydraulic diameter is defined as nominal diameter for a round nozzle or twice the width of a slot nozzle. Recently, there have been also studies on the heat transfer characteristics of precessing nozzle jets [6–8]. Precessing nozzles consist of a cylindrical chamber with a small axi-symmetric inlet at one end where the flow separates at this abrupt expansion due to entrainment of ambient air and reattaches nonsymmetrically at the chamber wall [8].

As stated by Sarkar and Singh [2], impingement systems in industrial applications are complicated to analyze due to the interaction of airflow with additional variations in food products. Therefore, understanding the physical phenomena behind the impingement is important for further design and analysis purposes. In this chapter, a brief review for impingement studies in food processing will be given and the analysis of heat transfer and fluid flow phenomena during an impingement process will be covered.

27.2 IMPINGEMENT IN FOOD PROCESSING

Even though air impingement technology has been used in the food industry for several decades, studies that have been published regarding impingement processing for food products is comparatively low. Ovadia and Walker [1] stated that there are known applications of impingement freezers being used in the food industry. Soto and Borquez [9] identified impingement freezers as an alternative to conventional freezing methods indicating the significant reduction in freezing times. Salvadori and Mascheroni [10] carried out an extensive review on correlations that can adequately predict heat transfer coefficients during impingement. They applied a previously developed numerical model to analyze the simultaneous heat and mass transfer during freezing in impingement freezers. Erdoğdu et al. [11] investigated the spatial variation of heat transfer coefficient during air impingement cooling of slab shaped geometries where 1.5 cm in diameter nozzle (with a hydraulic diameter-to-length ratio of 0.2) was used with exit velocities of 14 and 28 m/s. Dirita et al. [12]

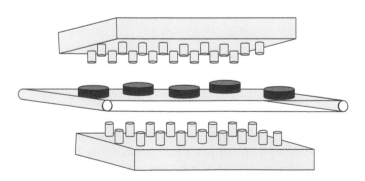

FIGURE 27.1   Schematic of a double-sided air impingement system with round nozzles.

analyzed the air impingement cooling of cylindrical shape foods numerically at the initial stages of cooling/chilling operations reporting the effect of localized forced convection and non-uniform heat flux along the cylindrical surface. Erdoğdu et al. [13] analyzed the flow visualization and heat transfer during air-impingement cooling of hard boiled eggs. They showed the potential of air-impingement systems for faster cooling purposes due to resulting higher local heat flux values. Anderson and Singh [14] predicted thawing of a frozen food product using impingement jets where tylose as a model food system was used for model validation purposes.

Midden [15] reported that the breakfast cereal industry has been using impingement ovens for over 40 years, and Ovadia and Walker [1] that air impingement ovens have been used for baking since 1987. Walker [16] described drying and toasting of breakfast cereals using air impingement ovens while Li and Walker [17] compared cake baking in conventional and impingement ovens. Dogan and Walker [18,19] investigated how various impingement parameters, such as temperature and velocity, affected the baking properties of both cakes and cookies. Walhby et al. [4] compared cooking of yeast buns and meat pieces in a traditional forced convection and an air impingement oven. They found that yeast buns could be cooked to the same quality in an air impingement oven at a lower temperature. They also reported 15–45% reduction in processing time with lower weight losses for cooking meats to the same temperature depending on the size. Walhby and Skjoldebrand [20] investigated heating of buns in an impingement oven with air velocities varied from 2 to 12 m/s. Xue and Walker [21] studied how temperature and product loading affected humidity inside an impingement oven.

In addition, combined air impingement with, such as, microwave processing were reported to be a practical solution to improve the heating uniformity and to better control the moisture transport within the product in aiding the surface crust formation [22].

Comparatively fewer studies have been reported on mass transfer from impingement jets. Francis and Wepfer [23] presented a comprehensive model for impingement drying where mass transfer coefficients were determined theoretically. Predicted moisture contents were then compared to experimentally measured moisture contents in an industrial drier showing a good agreement. Lujan-Acosta et al. [24] used air-impingement to dehydrate tortilla chips. They reported that the chips dried slightly faster due to the effect of increased heat transfer coefficient. Moreira [25] investigated the impingement drying of tortilla and potato chips using hot air and superheated steam indicating the apparent effect of increased heat transfer coefficient on drying rate of potato chips. Braud et al. [26] modeled the impingement drying of corn tortillas using governing equations of simultaneous heat and mass transfer. Bonis and Ruocco [27] modeled the local heat and mass transfer in food slabs during air jet impingement drying where they integrated the time-dependent governing equations to predict local moisture, temperature and velocity distributions.

As seen in the literature review, impingement processing was noted to have its high potential in food processing. The significant effect of impingement was the resulting higher convective heat or mass transfer due to the higher turbulence at higher air velocities. Therefore, analysis of heat transfer and fluid flow should be accomplished accordingly for design and analysis purposes.

27.3 FLUID FLOW AND HEAT TRANSFER IN IMPINGEMENT SYSTEMS

Transport phenomena (heat-mass transfer and fluid flow) from impingement jets have been studied by many authors in different areas of engineering for over 40 years, and general reviews have been published [28–33]. The major advantage of impingement technology, as indicated above, is the higher achievable heat and mass transfer rates. Heat and mass transfer is implicitly a function of fluid flow in impingement systems, and therefore it is important to understand the characteristics of the fluid flow of impinging jets.

As shown in Figure 27.2, air flow from an impingement nozzle can be divided into three characteristic regions. *Potential core* is the region where axial velocity is equal to (or sometimes defined as greater than or equal to 95% of) the velocity at the nozzle exit. Length of the potential core depends

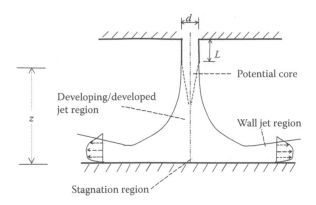

FIGURE 27.2   Flow field of an impinging jet.

on several factors including L/d ratio and degree of turbulence, but in general it has been found to be equal to approximately six to eight times the jet diameter. Outside of and past the potential core region is the area where the forced impingement air mixes with stagnant, ambient air, which slows the axial velocity and increases the jet diameter. According to Schlichting, [34] drop in centerline velocity (u_{center}) and increase in jet width (defined as the radius where $u_r = u_{center}/2$) are both directly proportional to the distance from end of the potential core. This region is called *developing or developed jet region*.

The area where the air impinges on the food product is termed *stagnation region*. In this region, axial velocity dramatically decreases while static pressure and therefore, radial velocity increases. Stagnation is characterized by very high heat transfer coefficient [34] due to a very steep temperature gradient ($\partial T/\partial y$) and very small temperature difference at the solid fluid interface where the surface temperature rapidly comes to the air temperature [33].

According to Polat et al. [29], effect of stagnation region cannot be seen on the jet flow above a distance of 0.25 z (where z is the distance between impingement surface and nozzle exit) from impingement surface. The *wall jet region* is formed in radial direction around the stagnation region. Here the pressure gradient is essentially zero, and boundary layer on the surface increases in size at distances further from stagnation.

Convective heat transfer coefficient (h) for impingement flow varies greatly depending on properties and arrangement of the flow. Heat transfer coefficients for impingement have been found to be dependent on many factors, including Reynolds number (N_{Re}, Equation 27.2), nozzle-to-surface distance (Z/d), nozzle geometry (L/d), Prandtl number (N_{Pr}, Equation 27.3), and turbulence intensity (also often referred as turbulence level, defined to be the ratio of the root-mean-square of the turbulent velocity fluctuations to the mean velocity).

$$N_{Re} = \frac{\rho \cdot d \cdot u}{\mu} \tag{27.2}$$

$$N_{Pr} = \frac{c_p \cdot \mu}{k} \tag{27.3}$$

$$N_{Nu} = \frac{h \cdot d}{k} \tag{27.4}$$

Empirical correlations have been used to correlate the stagnation point Nusselt number (N_{Nu}, Equation 27.4) to the Reynolds number; $N_{Nu} = f(N_{Re}^{n})$ where n varies from 0.5 to 0.8 [28,35,36].

Researchers including Gardon and Akfirat [37], Baughn and Shimizu [38], Lee et al. [39], and Choi et al. [40] have reported that maximum N_{Nu} at the stagnation point occurs when the jet is at a distance of six to eight diameters ($Z/d = 6-8$) away from the impingement surface corresponding to the end of potential core. Dependence of N_{Nu} and therefore the heat transfer coefficient on L/d is less clear. Obviously nozzle design can affect both N_{Re} and turbulence intensity. Hardisty and Can [41] found that more narrow nozzles produced higher heat transfer coefficients compared to geometrically similar ones. Degree of turbulence can also greatly influence the heat transfer from impinging jets [37]. Higher free stream turbulence appears to increase the heat transfer coefficients; however, if the jet becomes turbulent far before it reaches the impingement surface, significant energy can be lost and therefore heat transfer rates might decrease.

Another characteristic of heat transfer from impinging jets is that heat transfer coefficients vary at distances away from stagnation. This distance from stagnation is generally described by the ratio of distance from stagnation to the nozzle diameter (r/d). When the distance from nozzle to the impingement surface is small ($Z/d < 6$), secondary maxima, or even tertiary maxima (in addition to the maxima obtained at the stagnation point) in N_{Nu} could occur. Secondary maxima, for example, has been reported to occur at a radial distance of 0.5–2.0 nozzle diameters ($r/d = 0.5-2.0$). This is sometimes attributed to the transition from laminar to turbulent boundary layer flow [30]. For larger nozzle-to-surface distances ($Z/d > 6$), N_{Nu} generally decreases in radial direction. Martin [28] gives empirical correlations for spatial variation of N_{Nu} on a flat plate. Many correlations for average N_{Nu} are available, but the area over which they are averaged is not always given, which makes them difficult to use [36]. Obviously, if the heat transfer coefficient decreases in radial direction, then the larger the area over which N_{Nu} is averaged, the lower the average value would be. If the product is on a moving impingement surface, spatial variation in heat transfer may be comparatively less important. However, Polat [42] reports that local heat transfer coefficient profile does not change significantly when the ratio of surface velocity to jet velocity is less than 0.3.

Use of multiple nozzles also affects the variation of convective heat transfer coefficient and state of boundary layer. Due to the distance between nozzles, their flow fields would interfere with each other as shown in Figure 27.3.

There can be three possible interactions of jets due to multiple jet configurations [33]: (1) interaction of jets in the mixing domain before impingement depending upon the distance between the nozzles; (2) interaction of two adjacent jets laterally after impinging on the surface leading to strong upward jet fountains [43]; (3) interaction due to presence of axi-symmetrical jets and staggered distribution of exhaust ports. Downs and James [44] stated that the rate of heat transfer from an array of jets might even be lower due to boundary layer separation and flow eddies. Further design and optimization of the jet to jet spacing and exhaust arrangements are required to minimize the jet interaction effects. It was reported by Polat [42] that when the ratio of the distance between multiple nozzles to the distance from the nozzle-to-surface was greater than three, the jets would behave as noninteracting single jets.

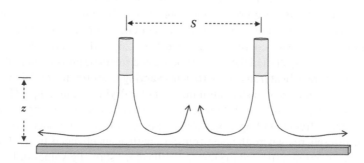

FIGURE 27.3 Interaction of impingement jets.

Entrainment of ambient air into the impinging jet, when the jet and surrounding air are at different temperatures, is also an important parameter to consider in the flow field and heat transfer analysis. Impingement heat transfer, for this case, would actually include the temperatures of the air in the jet, the ambient air and the product surface. When ambient air is at a warmer temperature than that of the impingement jet, ambient air becomes entrained in the jet flow and lowers the temperature of the stream thereby reducing the rate of heat transfer due to the decrease in temperature gradient. If the heat transfer coefficient is defined as a function of difference between surface temperature and air temperature at the nozzle exit, then variation in temperature throughout the jet is not taken into account [45]. To remove this variation from the definition of heat transfer coefficient, the coefficient can be defined using temperature difference between the surface and local adiabatic surface temperature [46]. The local adiabatic surface temperature is the air temperature along the surface when there is no heat transfer between the surface and the air (i.e., at extremely long processing times). For example, if a heated jet passed through air at a cooler temperature, the jet would decrease in temperature along its length as more and more ambient air becomes entrained. When the jet impinged on a surface and flows in radial direction along the surface, it would continue to cool as it incorporated more ambient air and the boundary layer grew. Defining the heat transfer coefficient in this manner eliminates its dependence on temperature difference between the jet and ambient air. Experiments were conducted by Goldstein et al. [46] and Baughn et al. [47] confirming this relation using different nozzle designs and Reynolds numbers.

Confinement (confining walls parallel to the impinging surface) also affects the heat transfer from impingement jets. Mujumdar and Huang [32] report that heat transfer coefficient can be reduced by 10–20% at low wall spacing as compared to free jet impingement. Designing a proper exhaust system can help alleviate this issue [48]. Surface properties, such as curvature and roughness are additional significant parameters in impingement heat transfer. Lee et al. [39] and Tawfek [49] both investigated impingement on convex surfaces and found that stagnation heat transfer coefficient was higher for smaller diameter cylinders due to increased acceleration away from stagnation. Lee et al. [50] found similar results with concave surfaces, with higher stagnation N_{Nu} occurring on surfaces with more curvature. Beitelmal et al. [51] studied the effect of surface roughness on average heat transfer coefficient. They found that heat transfer coefficients were up to 6% higher with rough walls due to increased turbulence on the surface.

Higher heat transfer coefficient and its spatial variation have been noted in each study of impingement thermal processing. Therefore, in analyzing the heat transfer during impingement, the heat transfer coefficient becomes a major issue to experimentally determine.

27.3.1 Heat Transfer Coefficient

Determination of heat transfer coefficient has been generally the first step in analyzing impingement thermal processing. For this objective, either experimental approaches or use of empirical equations are preferred, and quite few studies have been reported in this area.

Nitin and Karwe [52] determined the average heat transfer coefficients of 100 and 225 W/m²K for air velocities of 18 and 38 m/s for cookie shaped objects in a hot air jet impingement oven. Kocer and Karwe [53] reported the variation of the heat transfer coefficient in a multiple jet impingement oven. The values of heat transfer coefficient changed from 26 to 41 for air velocities of 2.5 and 10 m/s, respectively. Li et al. [54] studied the use of superheated steam impingement to dry tortilla chips. They determined the average heat transfer coefficients varying from 100 to 160 W/m²K. Caixeta et al. [55] investigated air and superheated steam impingement for drying potato chips. Heat transfer coefficients were reported to range up to 160 W/m²K. In the frozen temperature range, Soto and Borquez [9] studied impingement freezing of agar particles in a mixing bed as a model food system for vegetable pieces and fish spheres where the heat transfer coefficients changed from 70 to 250 W/m²K. Lujan-Acosta et al. [24,56] investigated drying of tortilla chips showing values of heat transfer coefficient from 60 to 180 W/m²K determined by the empirical correlations reported by Martin [28].

Lumped system methodology has also been applied to determine the heat transfer coefficient. However, lumped system methodology gives only an average value for the heat transfer coefficient [3], and provides no information on its spatial variation over the surface. Based on this methodology, Sarkar and Singh [3] applied the method of microcalorimeters developed by Donaldson et al. [35] to determine the spatial variation of heat transfer coefficient for impingement processes. In these experiments, the microcalorimeter apparatus consisted of a 0.08 cm thick cooper plate (25.4×25.4 cm) with 0.635 cm holes punched in 5.08×5.08 cm square grid spacing. The copper microcalorimeter disks (0.508 cm in diameter and 0.08 cm in thickness) were placed in each hole and soldered by 36 gauge type-T thermocouples. Then, annular space around the copper disks was filled with low thermal conductivity epoxy after insulating the apparatus with 4 cm thick polyurethane. Using the time temperature data recorded for each copper disk, the values of the heat transfer coefficient were determined applying the lumped system analysis providing variation over the given flat surface.

Erdoğdu et al. [11] used this methodology to determine the spatial variation of heat transfer coefficient over a flat surface at air nozzle exit velocities of 14 and 28 m/s for using in a numerical simulation of impingement cooling process. Characteristics of the customized air impingement system were as follows: air blow capacity was 4.72×10^{-3} m³/s at 2.54 cm water column pressure drop; diameter and length of the plenum was 30.5 cm and 1 m, respectively; the nozzle had a hydraulic diameter of 1.5 cm with a hydraulic diameter-to-length ratio of 0.2; and jet exit distance to flat surface was four. The equations obtained for the spatial variation of heat transfer coefficient were as follows:

$v = 14$ m/s:

$$h(y) = 51.78 - 0.56 \cdot y^{1.5} + 19.82 \cdot e^{-y} \tag{27.5}$$

$v = 28$ m/s:

$$h(y) = \left(10349.3 - 735.2 \cdot y\right)^{1/2} \tag{27.6}$$

where y was the axial distance from the center of the given flat surface (cm). Maximum values of heat transfer coefficient at the stagnation point ($y = 0$) were then obtained to be 51.78 and 101.73 W/m²K where the heat transfer coefficient for stagnant air at the given experimental conditions was reported to be 5.5 W/m²K. Sarkar and Singh [3], in the same customized air-impingement system, also reported the variation in heat transfer coefficient values for a wide range of under freeze-thaw and cooling impingement conditions.

Anderson and Singh [57] measured effective heat transfer coefficients during air impingement thawing using an inverse heat transfer method. They found that the effective heat transfer coefficients decreased in radial direction showing a secondary maximum at a distance from stagnation approximately equal to the nozzle diameter. They also stated that the effective heat transfer coefficients tended to increase with time due to the frost formation over the product surface.

With increased abilities of the computational fluid dynamics (CFD) packages, it is also possible to determine the heat transfer coefficient using the temperature variation at the surface boundary. Erdoğdu et al. [13] numerically analyzed the heat transfer and fluid flow for air-impingement cooling of hard boiled eggs where the heat transfer coefficient values and variation over the surface were determined through the applied CFD analysis. For this purpose, solution of thermal boundary layer equation leading to the temperature was used with the temperature gradient at the surface [58]:

$$h = \frac{-k_f \cdot \left.\dfrac{\partial T}{\partial y}\right|_{y=0}}{T_s - T_\infty} \tag{27.7}$$

The highest heat transfer coefficient value was obtained where the jet impinged on the surface (stagnation point). The maximum surface-averaged heat transfer coefficient was 62 while the local maximum was 118. Air velocity at the jet exit, in this study, was reported to be 17.3 m/s (resulting in an N_{Re} of 7000) with 5.5% turbulence intensity. These conditions were obtained from the particle image velocimetry (PIV) flow field measurement results.

There have been also other methodologies introduced to determine the spatial variation of heat transfer coefficient during impingement processing. For example, Goldstein and Timmers [59] applied liquid crystals where the reversible color changes with temperature were used together with measured heat flux values to determine the spatial distribution of heat transfer coefficient. Pan et al. [60] and Pan and Webb [61] studied local heat transfer using a two-dimensional infra-red radiometer. Naphthalene ablation has been used as an indirect approach where heat/mass transfer analogy was applied to measure the spatial variation [62]. In this study, local heat transfer was inferred from local ablation rate of naphthalene. Nitin and Karwe [63] used heat flux gauges to measure local heat flux and surface temperature from which the value of surface heat transfer coefficient was calculated. Accuracy of the heat flux gage measurements is affected by conductivity difference between the gauge and food itself. Deo and Karwe [64] reported that there might be an error of up to 20% in the measurements due to the differences in the conductivity.

For the case of mass transfer, Sherwood number (N_{Sh}) can be related to Nusselt number (N_{Nu}) using the heat and mass transfer analogy theory [65]. This theory relates the dimensionless numbers for heat and mass transfer, N_{Nu} and N_{Sh}:

$$\frac{N_{Nu}}{N_{Sh}} = \left(\frac{N_{Pr}}{N_{Sc}}\right)^n = \left(\frac{\rho \cdot c_p \cdot D_{AB}}{k}\right)^n \tag{27.8}$$

where n is 1/3 in most cases. Li and Tao [66] determined the spatial variation of mass transfer coefficients using sublimation of naphthalene for experimental investigation of slot jet impingement in a rectangular cavity. They determined N_{Sh} and compared results with those obtained from the heat and mass transfer analogy. They found N_{Sh} to vary from 30 to 35 at stagnation point for N_{Re} of 942, which was in a good agreement with their calculated results.

Additional to the difficulties in determining the heat transfer coefficient, solution of fluid flow and heat transfer in impingement thermal processing is a quite challenging problem since the flow and heat transfer equations are required to be solved simultaneously.

27.3.2 Solution of Fluid Flow and Heat Transfer

In the mathematical analysis of impingement, fluid flow and heat transfer are governed by equations of continuity, conservation of momentum and conservation of energy (Navier–Stoke's).

Continuity equation:

$$\frac{\partial U_j}{\partial x_j} = 0 \tag{27.9}$$

Conservation of momentum:

$$\frac{\partial U_i}{\partial t} + \frac{\partial U_i U_j}{\partial x_j} = -\frac{1}{\rho} \cdot \frac{\partial P}{\partial x_i} + \frac{\partial}{\partial x_j}\left[\nu \cdot \left(\frac{\partial U_i}{\partial x_j} + \frac{\partial U_j}{\partial x_i} - \langle u_i' u_j' \rangle\right)\right] \tag{27.10}$$

Conservation of energy:

$$\rho \cdot c_v \cdot \frac{\partial T}{\partial t} + \rho \cdot U_j \cdot c_p \cdot \frac{\partial T}{\partial x_j} = \frac{\partial}{\partial x_j}\left[k \cdot \frac{\partial T}{\partial x_i} - \rho \cdot c_p \cdot \langle u'_j T' \rangle\right] \tag{27.11}$$

where, U is average velocity (m/s), u' is turbulent component of velocity (m/s), $\langle u'_i u'_j \rangle$ is average velocity of fluctuating component, T is average temperature (K), T' is temperature of fluctuating component (K), P is pressure (Pa), ρ is density (kg/m³), v is kinematic viscosity (m²/s), and c_p and c_v are heat capacity (J/kgK) at constant pressure and volume, respectively, and

$$\frac{\partial}{\partial x_i} = \frac{\partial}{\partial x}, \frac{\partial}{\partial y}, \frac{\partial}{\partial z}$$

$$\frac{\partial}{\partial x_j} = \frac{\partial}{\partial x} + \frac{\partial}{\partial y} + \frac{\partial}{\partial z}$$

$$U_j \cdot \frac{\partial}{\partial x_j} = U_x \cdot \frac{\partial}{\partial x} + U_y \cdot \frac{\partial}{\partial y} + U_z \cdot \frac{\partial}{\partial z}$$

For turbulent flow case, velocity magnitude fluctuates with time, and these fluctuations are known as the turbulence where velocity in the turbulent flow can be divided into average and turbulent components. Decomposition of flow field into the average and turbulent (fluctuating) components isolates the fluctuation effects on the average flow. However, the addition of the turbulence in the Navier-Stokes equations, as seen in Equations 27.10 and 27.11, results in additional terms, known as Reynolds stresses, leading to a closure problem increasing the number of unknowns to be solved.

In order to solve this problem, a mathematical path for calculation of turbulence quantities has to be provided. There are special turbulence models to solve this issue, e.g., κ–ε,κ–ω (where κ is the turbulence kinetic energy, ε is the turbulence energy dissipation rate, and ω is the turbulence frequency) and Reynolds stress models. Olsson et al. [67] gives information for comparisons of these models with the experimental data available in the literature. The renormalization group theory (RNG) with κ–ε model is reported to give reasonably good results for impingement applications in the case of rough walls with its certain limitations with smooth wall situations [2].

Solution of Navier-Stokes equations is difficult to solve in their complete forms even for laminar flow regimes, and turbulence adds more complications to the solutions. Therefore, use of CFD models is generally preferred for solution of these equations [68–70].

Flow properties, especially air velocity at the jet exit and turbulence definitely become required parameters for CFD calculations. These are used in specifying boundary conditions and turbulence models and can be obtained through a quantitative experimental determination of the flow field. Actually, understanding and measurement of flow field can directly be used to validate and improve upon the available mathematical models that are used to simulate transport phenomena and also evaluate and/or optimize equipment performance [71]. For experimental flow determination, different approaches have been reported in the literature. Gardon and Akfirat [37] used hot wire anemometry to determine the role of turbulence in determining heat transfer characteristics of impinging jets. Marcroft and Karwe [5] and Markroft et al. [72] applied laser doppler anemometry to determine the axial velocity profiles of single and multiple jets to demonstrate characteristics of multiple turbulent jets in their core region. Kocer and Karwe [53] used Pitot tube fluid velocity measurement technique to determine the jet exit velocities of a multiple jet impingement oven. Erdoğdu et al. [13] determined nozzle exit velocity and turbulence intensity of a single impinging jet using a PIV system, and these values were applied in their CFD simulation model.

27.4 CONCLUSIONS

In the modern food industry, there is an increasing demand for shorter processing of food products (freezing, thawing, drying, baking, etc.) without causing major changes in quality. One of the latest and most promising systems, for this purpose, is the use of air impingement systems. Modeling and simulation of air impingement systems are required for better design of impingement systems. A significant parameter for modeling of different air impingement systems is to know the value and spatial variation of heat transfer coefficient. Use of CFD packages bring a lot of advantages for further modeling of impingement systems where fluid flow and heat transfer can be solved simultaneously with the known flow properties. Therefore, the complex interaction of different factors in impingement systems requires experimental fluid flow studies for better modeling purposes. In addition, even though the studies accomplished with model systems provide insights to the use of this technology, it is obvious that studies involving food applications are required to better model the effects of object geometry with resulting mass transfer phenomena.

NOMENCLATURE

C_p	Specific heat at constant pressure	J/kg-K
c_v	Specific heat at constant volume	J/kg-K
d	Nozzle diameter	m
D_{AB}	Molecular diffusivity	m²/s
h	Convective heat transfer coefficient	W/m²-K
k	Thermal conductivity	W/m-K
k_m	Mass transfer coefficient	m/s
L	Nozzle length	m
N_{Nu}	Nusselt number	
N_{Pr}	Prandtl number	
N_{Re}	Reynold number	
N_{Sc}	Schmidt number	
N_{Sh}	Sherwood number	
P	Pressure	Pa
q''	Heat flux	W/m²
r	Radial distance from stagnation	m
Re	Reynolds Number	
S	Distance between multiple nozzles	m
t	Time	s
T	Average temperature	K
T'	Temperature of fluctuating component	K
u'	Turbulent component of velocity	m/s
$\langle u_i' u_j' \rangle$	Average velocity of fluctuating component	
U	Average velocity	m/s
Z	Nozzle-to-surface distance	m

Greek Symbols

μ	Dynamic viscosity	Pa-s
ρ	Density	kg/m³
ν	Kinematic viscosity	m²/s

Subscripts

f	Fluid
s	Surface

REFERENCES

1. Ovadia, D.Z. and Walker, C.E. Impingement in food processing. *Food Tech.*, 52(4), 46, 1998.
2. Sarkar, A. and Singh, R.P. Air impingement heating In *Improving the Thermal Processing of Foods.* Richardson, P. Ed. CRC Press, Boca Raton, FL, 2004.
3. Sarkar, A. and Singh, R.P. Spatial variation of convective heat transfer coefficient in air impingement applications. *J. Food Sci.*, 68, 910, 2003.
4. Wahlby, U., Skjoldebrand, C., and Junker, E. Impact of impingement on cooking time and food quality. *J. Food Eng.*, 43, 179, 2000.
5. Marcroft, H.E. and Karwe, M.V. Flow field in a hot air jet impingement oven-part I: a single impinging jet. *J. Food Proc. Preserv.*, 23, 217, 1999.
6. Wong, C.Y., Lanspeary, P.V., and Nathan, G.J. Phase-averaged velocity in a fluidic precessing jet nozzle and in its near external field. *J. Exp. Thermal Fluid Sci.*, 27, 515, 2003.
7. Göppert, S., Gürtler, T., Mocikat, H., and Herwig, H. Heat transfer under a precessing jet: effects of unsteady jet impingement. *Int. J. Heat Mass Transfer*, 47, 2795, 2004.
8. Zhou, J.W. and Herwig, H. Heat transfer characteristics of precessing jets impinging on a flat plate: further investigations. *Int. J. Heat Mass Transfer*, 50, 4488, 2007.
9. Soto, V. and Borquez, R. Impingement jet freezing of biomaterials. *Food Control*, 12, 515, 2001.
10. Salvadori, V.O. and Mascheroni, R.H. Analysis of impingement freezers performance. *J. Food Eng.*, 54, 133, 2002.
11. Erdoğdu, F., Sarkar, A., and Singh, R.P. Mathematical modeling of air-impingement cooling of finite slab shaped objects and effect of spatial variation of heat transfer coefficient. *J. Food Eng.*, 71, 287, 2005.
12. Dirita, C., De Bonis, M.V., and Ruocco, G. Analysis of food cooling by jet impingement, including inherent conduction. *J. Food Eng.*, 81, 12, 2007.
13. Erdoğdu, F. Air-impingement cooling of boiled eggs: analysis of flow visualization and heat transfer. *J. Food Eng.*, 79, 920, 2007.
14. Anderson, B.A. and Singh, R.P. Modeling the thawing of frozen foods using air impingement technology. *Int. J. Ref.*, 29, 294, 2006.
15. Midden, T.M. Impingement air baking for snack foods, *Cereal Foods World*, 40, 532, 1995.
16. Walker, C.E. Air impingement drying and toasting of ready-to-eat cereal. *Cereal Foods World*, 36, 871, 1991.
17. Li, A. and Walker, C.E. Cake baking in conventional, impingement and hybrid ovens. *J. Food Sci.*, 61, 188, 1996.
18. Dogan, I.S. and Walker, C.E. Effects of impingement oven parameters and formula variation on sugar snap cookies. *Cereal Foods World*, 44, 597, 1999.
19. Dogan, I.S. and Walker, C.E. Effects of impingement oven parameters on high ratio cake baking. *Cereal Foods World*, 44, 710, 1999.
20. Wahlby, U. and Skjoldebrand, C. Reheating characteristics of crust formed on buns, and crust formation. *J. Food Eng.*, 53, 177, 2002.
21. Xue, J. and Walker, C.E. Humidity change and its effects on baking in an electrically heated air jet impingement oven. *Food Res. Int.*, 36, 561, 2003.
22. Datta, A.K., Geedipalli, S.R., and Almeida, M.F. Microwave combination heating. *Food Tech.*, 59(1), 36, 2005.
23. Francis, N.D. and Wepfer, W.J. Jet impingement drying of a moist porous solid. *Int. J. Heat Mass Transfer*, 39, 1911, 1996.
24. Lujan-Acosta, J., Moreira, R.G., and Seyed-Yagoobi, J. Air impingement drying of tortilla chips. *Drying Tech.*, 15, 881, 1997.
25. Moreira, R.G. Impingement drying of foods using hot air and superheated steam. *J. Food Eng.*, 49, 291, 2001.
26. Braud, L.M., Moreira, R.G., and Castell-Perez, M.E. Mathematical modeling of impingement drying of corn tortillas. *J. Food Eng.*, 50, 121, 2001.
27. Bonis, M.V. and Ruocco, G. Modeling local heat and mass transfer in food slabs due to air jet impingement. *J. Food Eng.*, 78, 230, 2007.
28. Martin, H. Heat and mass transfer between impinging gas jets and solid surfaces. *Advances in Heat Transfer*, 13, 1, 1977.
29. Polat, S., Huang, B., Mujumdar, A.S., and Douglas, W.J.M. Numerical flow and heat transfer under impinging jets: a review. In *Annual Review of Numerical Fluid Mechanics and Heat Transfer*, Vol. 2. Tien, C.L. and Chawla, T.C., Eds. Hemisphere Publishing Corp., New York, NY, 1989.

30. Jambunathan, K., Lai, E., Moss, M.A., and Button, B.L. A review of heat transfer data for single circular jet impingement. *Int. J. Heat Fluid Flow*, 13, 106, 1992.
31. Viskanta, R. Heat transfer to impinging isothermal gas and flame jets. *Exp. Thermal Fluid Sci.*, 6, 111, 1993.
32. Mujumdar, A.S. and Huang, B. Impingement drying. In *Handbook of Industrial Drying*, Vol. 1. Mujumdar, A.S. Ed., Marcel Dekker, Inc, New York, NY, 1995.
33. Sarkar, A., Nitin, N., Karwe, M.V., and Singh, R.P. Fluid flow and heat transfer in air jet impingement in food processing. *J. Food Sci.*, 69, 113, 2004.
34. Schlichting, H. *Boundary-layer Theory*. McGraw-Hill, New York, NY, 1979.
35. Donaldson, C.D., Snedeker, R.S., and Margolis, D.P. A study of free jet impingement. Part 2. Free jet turbulent structure and impingement heat transfer. *J. Fluid Mech.*, 45, 477, 1971.
36. Mohanty, A.K. and Tawfek, A.A. Heat transfer due to a round jet impinging normal to a flat surface. *Int. J. Heat Mass Transfer*, 36, 1639, 1993.
37. Gardon, R. and Akfirat, J.C. The role of turbulence in determining the heat-transfer characteristics of impinging jets. *Int. J. Heat Mass Transfer*, 8, 1261, 1965.
38. Baughn, J.W. and Shimizu, S. Heat transfer measurements from a surface with uniform heat flux and an impinging jet. *J. Heat Transfer*, 111, 1096, 1989.
39. Lee, D.H., Chung, Y.S., and Kim, D.S. Turbulent flow and heat transfer measurements on a curved surface with a fully developed round impinging jet. *Int. J. Heat Fluid Flow*, 18, 160, 1997.
40. Choi, M., Yoo, H.S., Yang, G., Lee, J.S., and Sohn D.K. Measurements of impinging jet flow and heat transfer on a semi-circular concave surface. *Int. J. Heat Mass Transfer*, 43, 1811, 2000.
41. Hardisty, H. and Can, M. Experimental investigation into the effect of changes in the geometry of a slot nozzle on the heat transfer characteristics of an impinging air jet. *Proc. Ind. Mech. Eng.*, *Part C: Mech. Eng. Sci.*, 197, 7, 1983.
42. Polat, S. Heat and mass transfer in impingement drying, *Drying Tech.*, 11, 1147, 1993.
43. Slayzak, S.J., Viscanta, R., and Incropera, F.P. Effects of interaction between adjacent free-surface planar jets on local heat transfer from the impingement surface. *Int. J. Heat Mass Transfer*, 37, 269, 1994.
44. Downs, S.J. and James, E.H. Jet impingement heat transfer—a literature survey. *National Heat Transfer Conference, ASME*, Pittsburgh, PA, 87-HT-35, 1, 1987.
45. Hollworth, B.R. and Wilson, S.I. Entrainment effects on impingement heat transfer: part I - Measurement of heated jet velocity and temperature distributions and recovery temperatures on target surface. *J. Heat Transfer*, 106, 797, 1984.
46. Goldstein, R.J., Sobolik, K.A., and Seol, W.S. Effect of entrainment on heat transfer to a heated circular air jet impinging on a flat surface. *J. Heat Transfer*, 112, 608, 1990.
47. Baughn, J.W., Hechanova, A.E., and Yan, X. An experimental study of entrainment effects on the heat transfer from a flat surface to a heated circular impinging jet. *J. Heat Transfer*, 113, 1023, 1991.
48. Tzeng, P.Y., Soong, C.Y., and Hsieh, C.D. Numerical investigation of heat transfer under confined impinging turbulent slot jets. *Numerical Heat Transfer, Part A*, 35, 903, 1999.
49. Tawfek, A.A. Heat transfer due to a round jet impinging normal to a circular cylinder. *Heat Mass Transfer*, 35, 327, 1999.
50. Lee, D.H., Chung, Y.S., and Won, S.Y. The effect of concave surface curvature on heat transfer from a fully developed round impinging jet. *Int. J. Heat Mass Transfer*, 42, 2489, 1999.
51. Beitelmal A.H., Saad M.A., and Patel C.D. Effects of surface roughness on the average heat transfer of an impinging air jet. *Int. Comm. Heat Mass Transfer*, 27, 1, 2000.
52. Nitin, N. and Karwe, M.V. Heat transfer coefficient for cookie shaped objects in a hot air jet impingement oven. *J. Food Proc. Eng.*, 24, 51, 2001.
53. Kocer, D. and Karwe, M.V. Thermal transport in a multiple jet impingement oven, *J. Food Proc. Eng.*, 28, 378, 2005.
54. Li, Y.B., Seyed-Yagoobi, J., Moreira, R.G., and Yamsaengsung, R. Superheated steam impingement drying of tortilla chips. *Drying Tech.*, 17, 191, 1999.
55. Caixeta, A.T., Moreira, R.G., and Castell-Perez, M.E. Impingement drying of potato chips. *J. Food Proc. Eng.*, 25, 63, 2002.
56. Lujan-Acosta, J. and Moreira, R.G. Reduction of oil in tortilla chips using impingement drying. *Lebensmittel Wiss. Tech.*, 30, 834, 1997.
57. Anderson, B.A. and Singh, R.P. Effective heat transfer coefficient measurement during air impingement thawing using an inverse method. *Int. J. Ref.* 29, 281, 2006.

58. Butler, R.J. and Baughn, J.W. The effect of the thermal boundary condition on transient method heat transfer measurements on a flat plate with a laminar boundary. *J. Heat Transfer*, 118, 831, 1996.
59. Goldstein, R.J. and Timmers, J.F. Visualization of heat transfer from arrays of impinging jets. *Int. J. Heat Mass Transfer*, 25, 1857, 1982.
60. Pan, Y., Stevens, J., and Webb, B.W. Effect of nozzle configuration on transport in the stagnation zone of axisymmetric, impinging free surface liquid jets. Part 2: Local heat transfer. *J. Heat Transfer*, 114, 880, 1992.
61. Pan, Y. and Webb, B.W. Heat transfer characteristic of arrays of free-surface liquid jets. *J. Heat Transfer*, 117, 878, 1995.
62. Angioletti, M., Di Tommaso, R.M., Nino, E., and Ruocco, G. Simultaneous visualization of flow field and evaluation of local heat transfer by transitional impinging jets. *Int. J. Heat Mass Transfer*, 46, 1703, 2003.
63. Nitin, N. and Karwe, M.V. Numerical simulation and experimental investigation of conjugate heat transfer between a turbulent hot air jet impinging on a cookie-shaped object. *J. Food Proc. Sci.*, 69, 59, 2004.
64. Deo, I.S. and Karwe, M.V. Effect of dimensions and thermal properties of heat flux gage on local heat flux. *Proc. ASME Heat Transfer*, 5, 361, 1998.
65. Incropera, F.P. and DeWitt, D.P. *Fundamentals of Heat and Mass Transfer*. John Wiley & Sons, New York, NY, 1996.
66. Li, P.W. and Tao, W.Q. Numerical and experimental investigations on heat/mass transfer of slot-jet impingement in a rectangular cavity. *Int. J. Heat Fluid Flow*, 14, 246, 1993.
67. Olsson, E.E.M., Ahrne, L.M., and Tragardh, A.C. Heat transfer from a slot air jet impinging on a circular cylinder. *J. Food Eng.* 63, 393, 2004.
68. Kumar, A. and Dilber, I. Fluid flow and its modeling using computational fluid dynamics. In *Handbook of Food and Bioprocess Modeling Techniques*. Sablani, S.S., Rahman, M.S., Datta, A.K. and Mujumdar, A.A., Eds. CRC Press, Boca Raton, FL, 2006.
69. Verboven, P., Scheerlinck, N., De Baerdemaeker, J., and Nicolai, B.M. Possibilities and limitations of computational fluid dynamics for thermal process optimization. In *Processing Foods:Quality Optimization and Process Assessment*. Oliveira, F.A.R., Oliveira, J.C., Hendrickx, M.E. and Korr, D., Eds. CRC press, Boca Raton, FL, 1999.
70. Verboven, P., De Baerdemaeker, J. and Nicolai, B.M. Using computational fluid dynamics to optimize thermal processes. In *Improving the Thermal Processing of Foods*. Richardson, P., Ed. CRC Press, Boca Raton, FL, 2004.
71. Chandrasekaran, M., Marcroft, H., Bakalis, S., and Karwe, M.V. Applications of laser Doppler anemometry in understanding food processing operations. *Trends Food Sci. Tech.*, 8, 369, 1997.
72. Marcroft, H.E., Chandrasekaran, M., and Karwe, M.V. Flow field in a hot air jet impingement oven-part II: multiple impinging jets. *J. Food Proc. Preserv.*, 23, 235, 1999.

Section VII

Mass and Momentum Transfer in Food Processing

28 Membrane Processing

Kasiviswanathan Muthukumarappan
South Dakota State University and University College Dublin

Chenchaiah Marella
South Dakota State University

CONTENTS

28.1 INTRODUCTION

Separation of a particular component from a mixture of several components is a requirement in several industrial operations. The target component might be the desired product or an unwanted component, separated to increase the purity of the original mixture. Separations take advantage of differences in physical or chemical properties of the mixture of components. In industrial separations, differences in molecular/particle size, shape, density, color, solubility, and electrical charge and other properties are taken advantage of in order to separate a particular component from a mixture of different components. Some separation processes based on physical properties of the materials are listed in Table 28.1.

Evaporation, drying, crystallization, centrifugation, ion exchange, electrodialysis (ED), extraction, leaching, mechanical separation, sedimentation, and settling are widely used conventional separation processes. These processes involve addition or removal of heat or some form of pretreatments that alters chemical or physical characteristics of the target component in the separation process. Membrane technology is a novel and emerging separation technology. Membrane-based separation processes have the following features:

- Continuous separation processes
- Energy consumption is generally low
- Easily combined with other separation processes
- Operated at room temperature
- Up-scaling is easy
- No additives are required
- No physical or chemical changes required to feed streams

TABLE 28.1
Physical Properties of Material and Separation Processes

Physical Property	Separation Technique
Size, shape	Screening, filtration, dialysis, gas separation, gel permeation chromatography, size exclusion chromatography
Density	Sorting, membrane separation, centrifugation, settling, decanting
Viscosity	Liquid extraction
Vapor pressure	Distillation, membrane distillation
Thermal properties	Evaporation, drying
Surface properties	Froth floatation, foam fractionation
Affinity	Extraction, adsorption, absorption, hyperfiltration, gas separation, pervaporation, affinity chromatography
Charge	Ion exchange, electrodialysis, electrophoresis
Diffusional	Extraction, membrane separation
Solubility	Solvent extraction, thermal denaturation
Optical	Color sorting

Sources: Mulder, M., *Basic Principles of Membrane Technology*, Kluwer Academic Publishers, Norwell, MA, 1991 and Gradison, A.S. and Lewis, M.J., In *Separation Processes in Food and Biotechnology Industries: Principles and Applications*, Gradison, A.S. and Lewis, M.J., Eds., Woodlands Publishing Ltd., Cambridge, UK, 1996.

However the following shortcomings are worth noting:

- Concentration polarization/membrane fouling
- Low membrane life time
- Low selectivity

28.2 HISTORY OF MEMBRANE PROCESSING

The technique of osmosis, transport of water or solvent through a semipermeable membrane has been known since 1784. About a century later in 1855, Fick developed the first synthetic membrane. The artificial membranes made during this period were used for quantitative measurement of diffusion phenomenon and osmotic pressure [5]. Membrane filters were first made available commercially by Sortorius Company in Germany. Even though membrane phenomenon was observed and studied as early as the mid eighteenth century, it was primarily aimed at elucidating the barrier properties and related phenomenon, but not for development of technology for industrial and commercial applications. Technological developments in some membrane processes are given in Table 28.2.

The first commercial cellulose nitrate or cellulose nitrate-cellulose acetate membrane used for practical applications were manufactured by Sortorius in Germany after World War I. These membranes were mostly confined to laboratory scale applications. Some what more dense ultrafiltration (UF) membranes developed during the same time were also mostly used for laboratory applications. A major development in industrial application of membranes was achieved with the development of asymmetric membranes by Sourirajan and Loeb. In the 1960s, modifications were made to the cellulose acetate membranes by heating them in water. This heating and annealing process created symmetry in the structure of the membrane where the membrane behaved differently depending on the side that is exposed to process streams. The asymmetry is characterized by a thin, dense top layer of < 0.5 μm supported by a porous bottom layer of 50–200 μm thickness. The top layer

TABLE 28. 2
Technological Developments in Membrane Processes

Membrane Process	Major Developments
Microfiltration	1. Introduced in Germany in 1920
	2. Cold sterilization of beer in 1963
	3. Cross flow filtration in 1971
Ultrafiltration	1. Introduced in Germany in 1930
	2. Introduction of polysulfone and PVDF membranes in 1966
	3. First hollow fiber UF plant in 1967
	4. First commercial tubular UF plant in 1964
	5. First spiral wound UF modules in 1980
Reverse osmosis	1. Introduced in the United States for sea water desalination in 1960
	2. Anisotropic membranes by Loeb-Sourirajan in 1963
	3. First spiral wound module in 1963
	4. First thin film composite membrane in 1978
Pervaporation	1. Lab scale pervaporation studies during 1965–1980
	2. First pilot scale pervaporation plant in 1988
	3. First commercial scale pervaporation plant in 1996

Sources: Mulder, M., *Basic Principles of Membrane Technology,* Kluwer Academic Publishers, Norwell, MA, 1991 and Baker, R.W., *Membrane Technology and Applications,* 2nd edn, John Wiley and Sons, UK, 2004.

determines the rate of transport while the bottom layer acts as support. This development transformed membrane based separations from laboratory use to commercial use. In 1981, Henis and Tripodi developed a composite membrane by placing a thin homogeneous layer of polymer with high gas permeability on top of an asymmetric membrane with pores in the top layer filled, making it leak-proof. This development made industrial membrane-based gas separations economically feasible. By the mid twentieth century, it became clear that reduction in membrane thickness is the only practical route to obtain commercially feasible permeation rates through the membranes.

28.3 COMMON TERMINOLOGY USED IN MEMBRANE PROCESSING

28.3.1 FLUX

Flux through the membrane is the "permeation rate measured as volume per unit membrane area per unit time."

$$\text{Flux} = \frac{\text{Volume of permeate}}{\text{Area of membrane} \times \text{Time}} = \text{LMH} \qquad (28.1)$$

28.3.2 RELATIVE FLUX

It is the ratio of the flux of solution to the flux of the corresponding permeate.

28.3.3 VOLUME REDUCTION RATIO (VRR)

It is the ratio of volume of feed to the volume of retentate obtained.

$$\text{VRR} = \frac{\text{Volume of feed}}{\text{Volume of retentate}} \qquad (28.2)$$

28.3.4 SEMIPERMEABLE MEMBRANE

Semipermeable membranes are those which permit passage of certain components in a solution and retain others based on molecular weight/size.

28.3.5 ASYMMETRIC MEMBRANE

Asymmetric membranes are thin film composite membranes. They have a dense skin layer on top of a spongy support layer underneath.

28.3.6 MOLECULAR WEIGHT CUT OFF (MWCO)

The ability of a membrane to substantially retain defined macromolecules of known molecular weight and generally used to specify the porosity of the membrane. The term used is MWCO, which is the molecular weight of the smallest test macromolecule that is largely rejected by the membrane.

28.3.7 TRANSMEMBRANE PRESSURE (TMP)

This is the difference between mean inlet and outlet pressures of retentate and pressure of permeate. This is the net hydrostatic pressure responsible for filtration.

```
-----------Recycle---------------------
     ↓                                    ↑
Feed------Pi------------→Membrane---------Po-----→Retentate

                         ↓

                        Pp
                      Permeate
```

$$\text{Transmembrane pressure} = \text{TMP} = \frac{Pi + Po}{2} - Pp \qquad (28.3)$$

28.3.8 REJECTION

$$R = 1 - \frac{Cp}{Cr} \qquad (28.4)$$

C is the concentration of a component in permeate (p) or retentate (r).

  100% Rejection: $Cp = 0$ and $R = 1$
  0% Rejection: $Cr = 0$ and $R = 0$

28.3.9 DIAFILTRATION

A process in which the retentate at the end of UF, is diluted with water and UF is done again. The quantity of water added depends on the extent of solute removal desired and may range from 40 to 200% of the original feed volume.

28.4 DIFFERENT MEMBRANE PROCESSES

28.4.1 INTRODUCTION

Membrane processing is based on selective permeability of a porous membrane. The membrane permits passage of certain components known as "permeate" and retains/rejects other components known as "retentate." These processes are operated either in dead end mode or cross flow mode as shown in Figure 28.1. In dead end mode, the feed is pumped toward the membrane surface. There will be one stream entering the unit, i.e., feed and one stream leaving the unit, i.e., permeate. In this mode, the resistance of the cake increases sharply and the flux drops dramatically. In contrast, in cross flow mode of operation, the feed is pumped tangentially or across the membrane surface. The advantage of this type of operation is that it limits the build up of solids on the membrane surface. In fact, high cross flow velocities cause erosion of the solids formed on the surface of the membrane thereby reducing the cake resistance for the permeate flow. This mode of operation is mainly advantageous in processing materials containing high solids concentration and in fact is used in most of the industrial scale membrane-based processes. The feed solution pumped under pressure over the membrane forms two streams—permeate and retentate as shown in Figure 28.2. Some of the membrane processes working on this principle are microfiltration (MF), UF, nanofiltration, reverse osmosis (RO), pervaporation, etc. Alternate terminology and membrane material used in the various pressure driven membrane processes are given in Table 28.3. Each of these processes differs from one another in several features. The essential features of each of these processes are discussed below.

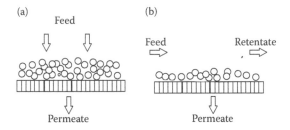

FIGURE 28.1 Modes of operation of membrane separation processes. (a) Dead end and (b) Cross flow.

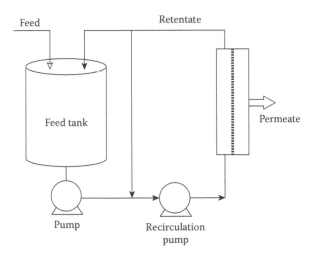

FIGURE 28.2 Schematic of membrane separation process (feed and bleed type).

TABLE 28.3
Pressure Driven Membrane Process

Process	Alternate Terms Used	Membrane Choices
Microfiltration	Cross flow MF, tangential MF	Tortuous path, capillary pore, organic membrane inorganic membrane, alumina, ceramic
Ultrafiltration		Synthetic polymers, (Polysulphone) inorganic zirconium, oxide, alumina)
Nanofiltration	Loose and leaky RO ultraosmosis intermediate RO-UF, RO with solute fractionation	Thin-film composite, e.g., polyamide on mocroporous polysulphones
Reverse osmosis	Hyper-filtration	Single polymer RO, asymmetric thin film composite

28.4.2   COMMONLY USED MEMBRANE PROCESSES

28.4.2.1   Microfiltration (MF)

MF is a pressure driven separation process that uses porous membranes with pores of 100 – 2000 nm. The membrane allows passage of micromolecules while retaining larger macromolecules. The terms cross flow or tangential flow MF are used to signify whether the feed flows tangentially to the membrane surface so as to counter the formation of solid cake on the membrane surface. MF is a clarification process that separates molecules and particles based on size and solubility. MF is also known as loose UF and requires high cross flow velocities. Operative pressures are somewhat lower than UF (0.3–2 bar). Permeation rates are typically higher than those obtained in UF. One main problem in MF is that the colloidal aggregates present in the feed block the pores in the membrane. Moreover, the operational cycle is lower because these membranes foul easily.

MF as a potential commercial process made breakthrough with the development of new ceramic membranes. Highly permeable support, multichannel geometry and high stability to heat, acid and alkali have opened up novel applications for the MF process. The use of ceramic membranes made it possible to develop uniform transmembrane pressure (UTMP) process. In this configuration, the permeate side space is filled with plastic beads of about 2 mm in diameter and an additional circulation pump is used to recirculate the permeate at high speed concurrently with the retentate flow thereby providing UTMP throughout the membrane area. The main advantage of UTMP process is that the formation of secondary filtration layer of solutes on the membrane surface is reduced or minimized. Membrane sieving characteristics are not altered and fouling is limited leading to prolonged/extended operational time. However, the UTMP concept cannot be used with polymeric spiral and composite membranes because this will lead to loosening of the different membrane layers. MF fits between UF and the traditional particle filtration based on perpendicular flow using filters of varying porosity. MF is a pressure driven membrane process usually operating at TMP of the order of 1 bar. It finds its applications in separation of suspended particles in the range of 0.05–10 μm [6]. A wide selection of MF membranes ranging from 0.1 to 1.4 μm is currently available for industrial applications.

28.4.2.2   Ultrafiltration (UF)

It is a method of simultaneously purifying, concentrating, and fractionating macromolecules or fine colloidal suspensions. Molecular weight cut off (MWCO) ranges from 1000 to 200,000 Da. In processing of biological fluids such as milk, UF membranes retain proteins, fats, colloidal salts, suspended particles, etc. It is generally known as the sieving process. Membrane pore size of a typical UF membrane ranges from 1 to 50 nm and operating pressures are between 1 and 15 bar. The first UF plant was installed by Abcor in 1960. Using tubular membrane, module paint was recovered from automobile paint shop rinse water [7]. Application of UF in the cheese whey processing followed in

1970. Tubular as well as plate and frame modules were used in the earlier designs. Romicon developed the first hollow fiber UF unit followed by introduction of spiral wound UF modules by Abcor [8]. UF uses membranes of fairly porous surface layer on a more open/porous support layer. The support layer gives the required strength to the top layer. This also provides flow passages for the permeate.

28.4.2.2.1 Filtration Equation and Resistance in Ultrafiltration (UF)

For batch type UF system, Hagan Poiseuille law for laminar flow gives the flux through the membrane [9] as

$$m = \frac{A\Delta P}{32vp} \times \frac{1}{R_m + R_L} \tag{28.5}$$

where m = mass flow or flux (kg/s), dm/dt; A = area of filter; ΔP = pressure difference across the membrane; vp = kinematic viscosity of permeate (m²/s); R_L = resistance of deposited layer (m⁻¹); R_m = resistance of the membrane.

The resistance of the membrane is directly proportional to the amount of material deposited, i.e., to the thickness of the deposited layer. Resistance can be expressed as

$$R_L = \text{Rsp} \cdot m \cdot x \tag{28.6}$$

where m = mass flow of permeate; x = charge of solute (kg/kg); Rsp = specific resistance which is independent of layer thickness (m/m).

Taking into account the mass flow rate, $m = dm/dt$ the filtration time is obtained as

$$\frac{dm}{dt} = \frac{A\Delta P}{32vp} \times \frac{1}{R_m + \text{Rsp} \cdot m \cdot x} \tag{28.7}$$

$$dt = \frac{32vp}{A\Delta P} \times (R_m + \text{Rsp} \cdot m \cdot x)\, dm \tag{28.8}$$

On integration

$$T = \frac{32vp}{A\Delta P} \times (R_m m + 1/2 \text{Rsp} \cdot m^2) \tag{28.9}$$

For better operational results a continuous flow or cross flow filtration is preferred. Three important factors that control the formation of solute deposition on the membrane surface are (1) The flow of material through the membrane causes solute convection to the membrane surface in a direction which is right angle to the direction of flow of the feed. (2) Due to concentration gradient there will be back diffusion of some of the solutes. (3) Parallel to the membrane, the solutes present in the layer close to the membrane move at velocities which vary according to the increase in axial flow rate. The thickness of the deposited layer depends on the length of the flow path. For effective control of the thickness of this layer, the length of the membrane should be as small as possible and the velocity of flow should be as high as possible. Certain amount of deposit, especially in the boundary layer is unavoidable. If resistance of the membrane and deposit layer is known, the flow for a particular membrane can be estimated.

28.4.2.3 Nanofiltration (NF)

NF falls between UF and reverse osmosis (RO) in terms of membrane pore size/MWCO and permeability characteristics. It removes particles of molecular weight less than 300–1000 Da and retains the rest.

The principal application of NF is in separation of mineral ions in nanometer range. Selective removal of ions based on charge is another feature in NF application and this process is widely used in demineralization of process streams. The newer applications of this process are in the removal of mineral ions that contribute significantly to osmotic pressure and are used prior to the RO process. With recent developments, NF membranes can be used to concentrate liquid whey protein concentrate and whey protein isolate prior to spray drying, thereby saving considerable amount of energy in energy intensive drying operations. This is in addition to the reduced stack losses and improved bulk density of the product. NF uses lower pressures when compared to RO and this is due to the larger pore sizes of these membranes. The mass flow across the membranes can be explained by a combination of pore flow and diffusion. NF is also known as loose RO (or) ultra tight UF process and falls in between both of these. It is a medium pressure process and uses TMPs in the region of 20–30 bar. Its MWCO is 300–1000 Da with a pore size between 1 and 10 nm. In NF of milk/whey, small molecules such as monovalent salts are removed along with water, but lactose, protein and fat are retained.

28.4.2.4 Reverse Osmosis (RO)

RO is essentially a dewatering technique. The membranes used in RO applications have lowest pore sizes which are of the order of 0.5–2 nm and operates at higher pressures. MWCO of the membranes used is 100 Da. RO was originally developed as a process for desalting of water using membranes that are permeable to water and not salts. Large scale industrial applications of RO process was made possible by Loeb-Sourirajan's development of anisotropic cellulose acetate membrane.

28.4.2.4.1 Principle of Reverse Osmosis (RO)

If a semipermeable membrane, selective to a particular component of the mixture separates pure water and salt solution, water will pass through the membrane from pure water side to the salt solution side diluting the salt solution. This is due to the fact that molecules of the dissolved substance exert a pull at the membrane, which causes the molecules of pure solvent to be absorbed through the membrane. The large molecules of solute are too large to pass in the opposite direction through the small pores. This process is called osmosis. Due to thermal motion of the molecules, the membrane is bombarded on one side by the larger molecules which exchange energy with the molecules of the solute. This is the cause of the suction (negative pressure) and of the tendency of the solution to become diluted. If some hydrostatic pressure is applied on the salt solution side, then the flow of water can be retarded and the pressure at which the flow of water ceases is known as osmotic pressure. If the applied pressure is greater than the osmotic pressure, water will flow from the salt solution side to the pure water side. This process is called RO and it is an important method of producing pure water from salt solution and in concentrating several dilute food products. This process is shown in Figure 28.3.

Osmotic pressure of dilute solutions can be calculated using kinetic theory of gases [9].

$$P_{osm} = -H_{osm} \cdot \rho_{sol} \cdot g = C.T.P \tag{28.10}$$

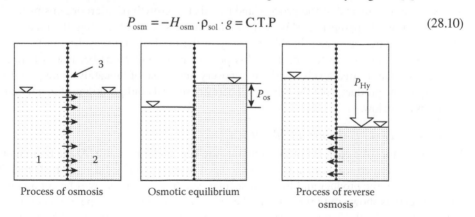

| Process of osmosis | Osmotic equilibrium | Process of reverse osmosis |

FIGURE 28.3 Schematic presentation of reverse osmosis process.

where $C = m/(MV)$ molar concentration (mol/m^3); $R = 8.314$ the universal gas constant (J/mol K); T = the absolute temperature (K).

The mass flow in RO process can be explained as:

$$m = D * \frac{\Delta P - \Delta P_{osm}}{\Delta l} \tag{28.11}$$

where $m = V \times \rho$ is the mass flow rate (kg/s); Δl = thickness of the membrane (m); $\Delta P - \Delta P_{osm}$ = the effective pressure difference; D^* = diffusion coefficient of solvent in the membrane (s).

The smaller the molar mass of a solute, the higher is the osmotic pressure of the solution at the same concentration. The actual basis of separation is still not completely understood in RO. The "preferential sorption capillary flow" mechanism of Sourirajan provides the most logical explanation. In this model, solution containing salts whose surface tension increases with concentration, such as inorganic salts, will have "negative excess" of solute adsorbed on the membrane surface. If the membrane contains pores with diameter twice the thickness of the water layer that is adsorbed on the membrane surface and when that applied pressure is more than the osmotic pressure, the adsorbed water layer will flow through these pores. Thus the control of pore size and providing an appropriate membrane surface is critical to the success of RO.

28.4.3 Applications of Commonly Used Membrane Processes

28.4.3.1 Biotechnology

28.4.3.1.1 Introduction

Remarkable developments in membrane technology drew the attention of R&D groups in biotechnology to take advantage of membrane technology in chemical and biochemical methods to produce organic chemicals, food products, hormones, pharmaceuticals, vitamins and several other biological products. The conventional bioreactors use batch or semibatch process where in recovery and reuse of cell cultures, batch to batch variation in the product, etc. are some of the typical problems associated with these types of reactors. The successful application of membrane technology in bioreactors as an alternative to conventional batch type gave way for design and development of continuous and highly productive reaction and product recovery methods. Membrane bioreactor, characterized by desired reaction process and simultaneous product separation in a single device attracted much attention in commercial circles [5].

28.4.3.1.2 Different Types of Bioreactors

In biotechnological processes, the concentration of the end product, beyond a certain percentage, will be detrimental to the process and retards the growth of microorganisms used in the process. Selection of appropriate MWCO/membrane pore size and recycling the material through the membrane separation system enable recovery of the product and recycling of the biocatalyst and substrate continuously in the reactor. This process is advantageous because it maintains product concentration at desired levels, facilitate recovery and reuse of biocatalyst and permits continuous recovery of the product from the reactors. There are several configurations of membrane bioreactors as discussed below.

28.4.3.1.3 Dead End Stirred Cell (DESC)

In these reactors, the contents of the container are kept under pressure. At the bottom of the reactor, a flat sheet semipermeable membrane is arranged over a porous solid support. Due to TMP across the membrane, the product and lower molecular weight components will be continuously removed from the cell as shown in Figure 28.4a. The main problem with this type of reactor is concentration polarization and cake formation. However, this problem can be minimized by proper agitation of the contents.

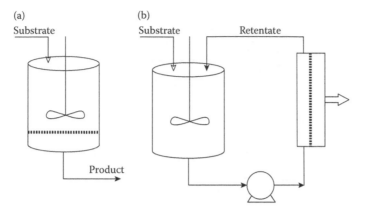

FIGURE 28.4 Different types of membrane bioreactors. (a) Dead end stirred tank, (b) Continuous stirred tank recycle reactor.

28.4.3.1.4 Continuous Stirred Tank Reactor with Recirculation (CSRR)

In this type of reactor, the reaction vessel contains the substrate, biocatalyst/culture and other required nutrients. The contents are continuously agitated for efficient process. A recirculation pump feed the mixture to the membrane unit. The retentate, components that are largely rejected by the membrane, containing the substrate and culture/catalyst is recycled back to the reaction vessel as shown in Figure 28.4b. This reactor is more efficient than the DESC and the concentration polarization and cake formation problem associated with DESC is minimized due to cross flow arrangement. This process has potential for depolymerization of natural macromolecules, reactions requiring pH and temperature control and operations with co-enzyme dependent enzyme systems. Another important feature of this type of reactor is that the product concentration in the reaction vessel and at the outlet will be same, assuming zero rejection for the product components. The kinetics and performance of this type of reactor have been studied [10,11].

28.4.3.1.5 Hollow Fiber Reactor (HFR)

The characteristics of hollow fiber modules, as discussed elsewhere, are also applicable to HFR. Among other things, the main advantage is the large surface area to volume ratio in HFRs. They are well suited for immobilization of soluble enzyme in intact form. In these reactors the catalyst can be located either on the shell side or on the tube side as shown in Figure 28.5.

28.4.3.1.6 Microporous Membrane Reactor (MMR)

In these reactors, the enzymes are loaded on the micropores of the membrane or on the microporous side of anisotropic hollow fiber membranes [12]. Feeding the substrate can be done with several arrangements. A schematic of the MMR is shown in Figure 28.6.

Applications of membrane bioreactors include: production of ethanol [13,14], production of acetone-butanol, production of lactic acid from glucose and from cheese whey, production of citric acid from glucose. Protein hydrolysis has been used to improve the functionality of proteins without loss of nutritive value. Application of membrane reactor reduces the high cost of enzymes and inherent lower efficiency associated with batch processes. A 5–10% increase in productivity was reported with MMR in soy protein hydrolysis [15]. Membrane bioreactors are also used for carbohydrate hydrolysis (starch, cellulose, oligosaccharide lactose and maltose) and in production of high fructose syrup, lipid conversions and L-amino acid production. Ethanol production from wheat flour using integrated membrane bioreactor [16], production of propionic acid from glycerol [17], hydrolysis of k-casein macro peptide to produce bio-active peptides are other examples of application of membrane separation technology.

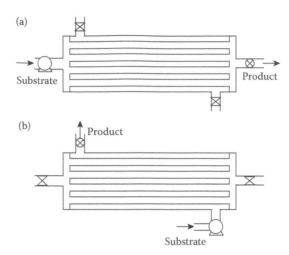

FIGURE 28.5 Different types of membrane bioreactors. Hollow fiber reactor (a) enzyme on shell side and (b) enzyme on tube side.

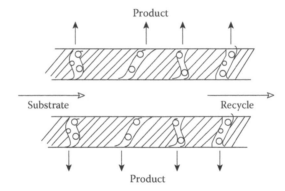

FIGURE 28.6 Different types of membrane bioreactors: microporous membrane reactor.

28.4.3.2 Water Treatment

Quality of water needed for industrial or drinking purposes vary and should be of specific grade. The requirements are specified in water quality standards set for different industries. Raw water contains particulates, colloidal matter, organics, dissolved gases, salts pyrogens, and microorganisms. It cannot be used as process or portable water without purification. Salts present in water gives hardness to water, either of temporary or permanent type. Hard water is not suitable for industrial applications. Historically, purification of raw water for specific applications is done by distillation. MF can remove bacteria, suspended solids and macromolecules without the need for addition of any chemicals. The additional advantage of this type of process is that the quality of water will be consistent.

Processing of water for drinking/portable purposes accounts for the single largest application of membrane technology. Combination of MF and UF is beneficial in removing microorganisms and replaces the need for several conventional processing steps like coagulation, sedimentation, filtration, distillation, etc. The main advantages of membrane based processing are that these processes require less energy and of continuous nature occupying less foot print. Even viruses can be removed from biological and pharmaceutical fluids by selection of appropriate MWCO of the membranes.

28.4.3.3 Waste Water Treatment

Manufacturing and service establishments generate large quantities of waste water daily. Implementation of stringent environmental regulations provides huge scope for application of membrane technology in treatment of waste water. It reduces the total quantity of waste or aid recovery of valuable components from waste water. Removal of biological/organic molecules from waste water also reduces BOD/COD in waste streams going into public sewers.

28.4.3.4 Dairy Processing

Wide spread application of membrane processing is implemented in the dairy industry. The dairy industry accounts for the lion's share of the total membrane area installed in food industries. It is estimated that about 300,000 m² of membrane area is installed in dairy applications world wide [5]. RO is the first process to be used in dairy applications with successful use in cheese whey in 1971. The largest application is in the fractionation and/or preparation of whey protein products and to a lesser extent in preconcentration of milk prior to cheese making, long distance transportation of milk, etc.

Membrane separations have been successfully tried in removal of bacteria, milk standardization and concentration, cheese making, etc. [18]. Application of this technique for whey protein fractionation started in the early 1990s. In an effort to develop a process for obtaining α-lactalbumin enriched product from whey, Roger et al [19] tried hollow fiber membranes with a MWCO of 50 and 2 kDa. α-La/β-Lg ratio of 2.3 and 1.5 were reported in the final product produced from rennet whey and acid whey, respectively. Bottomley [20] used Romicon PM 10 and PM 100 hollow fiber membranes for isolation of an immunoglobulin-rich fraction from whey and reported a 24% recovery of the target protein. In the process of obtaining higher concentrations of α-La from whey, Bottomley [21] reported α-La/β-Lg ratios of 1.75, 2.0 and 3.0 using different processing steps in hollow fiber separation system. Thomas et al. [22] reported 20% enrichment and 90% recovery of immunoglobulins using formed in place membranes on sintered SS tubes. In an effort to develop a process for preparing a fraction of α-La from whey, Uchida et al. [23] heat-treated whey at 120°C followed by membrane filtration using ceramic membranes. α-La/β-Lg ratio of 3.95, 4.07 and 6.32 were reported depending on the process steps and type of whey used. Muller et al. [24] studied the effect of operating conditions on the production of purified α-lactalbumin in a two-step cascade separation process. Ceramic membrane of 150–300 kDa MWCO was used in the study. They reported a 1.76-fold increase in the purity of α-lactalbumin and a yield of 0.53 with ceramic membrane of 300 kDa MWCO. Mehra and Kelley [25] studied a two-step cascade membrane separation process for whey protein fractionation. In step I, MF was used with ceramic membrane and the permeate from this step was used as feed in a second step in which UF membranes of 30, 50 and 100 kDa in hollow fiber configuration were used. Recoveries in terms of α-La/β-Lg achieved were 2.5, 0.9 and 0.55 for 30, 50 and 100 kDa membranes, respectively. These studies highlight the potential of membrane based approach for whey protein fractionation as a viable alternative to the existing processes.

28.4.3.5 Food Processing

Most food products are heat sensitive. As membrane based processing is performed at low temperature and is also gentle in nature, it confers promising advantages in food processing applications. UF is widely used in: (1) Recovery of protein from slaughtered animal blood. When used in place of conventional vacuum evaporation, apart from energy conservation and economic considerations, product quality is preserved. Several slaughter houses in the United States use UF for protein recovery and concentration. (2) Concentration of egg white: the conventional process involves thermal evaporation and a fermentation step to prevent discoloration of the product. In this process, minerals are not removed. Moreover, energy consumption is very high. UF can remove unwanted minerals and glucose in egg white processing.

28.4.3.6 Fruit Juices

Processing of fruit juice and extracts involve removal of suspended material by clarification, concentration of the liquid juice for storage and transportation and deacidification of citrus juices. Membrane technology can be used in all these unit operations. The advantages of membrane processing over the traditional methods are:

1. Traditional methods require refining agents (e.g., bentonite, gelatin, etc.), enzymes, centrifugation, etc. that take several hours of processing. With membrane filtration, clarification and refining are performed in a single step requiring only 2–4 hours of processing time.
2. At product recovery aspect, traditional processes give recovery rates of 80–94% while membrane based separations give recovery rates of 96–98%.
3. The need for filter aids is eliminated with membrane processing.
4. Reliable and consistent removal of suspended solids, colloidal particles, proteins, and polyphenols results in superior quality of the juice.

28.4.3.7 Alcoholic Beverages

Membrane based filtrations have the advantage of preserving natural flavor without exposure to heat and oxygen. It is economically favorable to use MF. MF can be used to remove yeast and haze proteins after fermentation and before storage of the product. MF membrane of 0.45 μm can remove most bacteria. Combination of prefilters, cross flow MF and final filters can preserve the natural characteristics of beer. Another possible application is in the recovery of yeast flocculates which settle at the tank bottom after fermentation. Recovery of beer from these tank bottoms is possible without the use of filter aids and it provides a significant product recovery avenue for brewers.

28.4.4 NOVEL MEMBRANE PROCESSES AND APPLICATIONS

28.4.4.1 Pervaporation

In the pervaporation process, feed liquid flows on one side of the membrane and the permeate is removed as vapor from the other side of the membrane. Pervaporation is the only membrane process where a phase transition occurs with the feed being liquid and permeate being vapor. This is made possible by maintaining partial vacuum on the permeate side of the membrane. The components to be separated from the mixture need to be absorbed by the membrane, should diffuse through it and is expected to easily go into the gaseous phase on the other side of the membrane [26]. The required vapor pressure difference across the membrane can be maintained by a vacuum pump or by condensing the vapor produced which spontaneously creates partial vacuum.

Pervaporation process is known since 1917. However, it got much attention in 1970s but research efforts were mainly conducted on lab scale/pilot scale applications. However, by the end of the 1980s, advances in membrane technology made it possible to develop economically viable pervaporation systems [8].

Pervaporation process can be effectively used for removal of water from liquid organics, water purification and organic/organic separations. Novel application of pervaporation is in purification/ separation of ethanol from fermentation broths. As ethanol forms azeotrope with water at 95% concentration, pervaporation process appears promising because simple distillation will not work under these conditions. Among different membranes tried for this type of application, polyvinylalcohol gives the best outcome. Availability of extremely water selective membranes allows pervaporation systems to produce ethanol with a purity as high as 99.9%. Some plants producing ethanol are using a combination process of distillation-pervaporation. Pervaporation is less capitol and energy intensive operation compared to distillation for small plants treating less than 5000 L/h of feed solution [8]. Economics can favor its use for bigger plants with development of novel membranes. One such development under investigation is poly-n-isopropyacrylamide-co-nisopropyl methacrylamide, grating on a porous polyethelenemembrane by means of plasma filling polymerization technique.

This membrane responds differently under different concentrations of ethanol in the fermentation broth. At higher concentrations, the pores in the membrane expand and open up allowing ethanol to permeate through it. As the concentration of ethanol decreases to less than 8%, the pores of the membrane contract by acting as a mechanical valve responding to the concentration of ethanol. Combining this membrane separation with a zeolite type, pervaporation membrane can concentrate ethanol to 80%. Further concentration can be done using a dehydration membrane. The cost of this process is expected to be one third of the cost involved in distillation.

Pervaporation process is successfully used in production of pure water. A variety of membranes have been tried in these applications. Some of the most commonly used materials are silicon rubber, polytutadiene, natural rubber and polyamide-polyether copolymers. The concentration of volatile organic compounds (VOCs) in permeate will be typically 100 times that of feed water. However, concentration polarization plays an important role in selection of operating variables, membrane materials and design of the pervaporation systems. Other applications of pervaporation for removal of VOCs include purification of contaminated ground water, recovery of volatile flavor and aroma compounds from fruit and vegetable processing streams.

Another novel application and less investigated one of pervaporation is the separation of organic mixtures. The mixtures to which pervaporation can be applied include aromatics/parafins, branched hydrocarbons from n-parafins, olefins/parafins, isometric mixtures and purification of dilute streams [27]. Organic-organic separations will be of considerable interest for future large scale applications of the pervaporation process. However, more research is needed in the membrane materials and process developments before industrial applications are seen.

28.4.4.2 Electrodialysis (ED)

ED is a membrane based demineralization process and uses ion exchange membranes. It is widely used in demineralization of liquid foods such as milk and whey and is used extensively in desalination of sea water. ED is known since 1890 but the first successful installation of ED plant was in 1952. The principle of ED process is based on the fact that when an aqueous solution containing ions of different mobilities is subjected to an electric field, the ionic species migrate to the respective opposite polarities of the field as shown in Figure 28.7 [26]. The ionic mobility is directly proportional to the specific electrical conductivity of the solution and is inversely proportional to the ionic concentration.

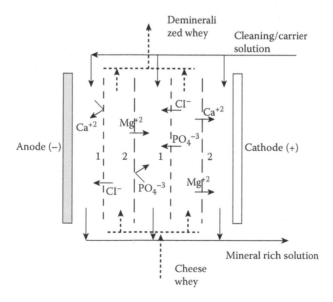

FIGURE 28.7 Schematic of electrodialysis process using anionic and cationic membranes.

In an ED system, anionic and cationic membranes are arranged in a plate and frame configuration (just like the classic plate heat exchanger) and are placed alternately. The feed solution is pumped to the cells of the system and electrical potential is applied. The positively charged ions migrate toward the cathode and negatively charged ions move toward anode. Cations easily pass through the negatively charged cationic exchange membranes but are retained by positively charged anionic exchange membranes. Similarly, anions pass through anion exchange membranes but are retained by the cation exchange membranes. The net result is that one cell (pair of anionic and cationic membrane) becomes enriched/concentrated in ionic species while the adjacent cell becomes depleted of ionic species. The presence of impurities and precipitated materials, as in the case of biological material causes severe concentration polarization of the membranes. The problem is more severe with anionic membranes which are clogged by large organic anions (such as amino acids), precipitated calcium phosphate and denatured proteins [26]. This anionic membrane specific problem can be partially overcome by using neutral membranes in the place of anionic membranes. The advantages of using neutral membranes are that concentration polarization is reduced, easier cleaning cycles and extended process runs. However, the disadvantage includes low degree of separation because only one set of membranes is selective.

28.4.5 Membrane Processes Under Development

28.4.5.1 Solvent Extraction

Solvent extraction is a widely used industrial separation process for extracting flavors, hydrocarbons and food stuffs from different mixtures. Membrane based solvent extraction is a novel process that is currently under development. It has high volumetric mass transfer rate. The disadvantages associated with the conventional extraction process such as need for density differences between the components to be separated, limited scale up nature, high capitol investment, high running, and maintenance costs, etc. are overcome with membrane based solvent extraction. The membranes used in this type of application are microporous hydrophilic, hydrophobic, or composite membranes. When a hydrophobic membrane is used, the organic phase on one side of the membrane will spontaneously wet the membrane and pass through the pores to the other side of the membrane. When an immiscible aqueous phase is maintained on the other side of the membrane, at pressure equal to or higher than that of the organic phase, the organic phase will not pass through to the other side of the membrane. By carefully controlling the pressure on both sides of the membrane, the interface is immobilized and solutes transfer through this interface goes to the other side of the membrane [28]. The membrane based extraction have been studied and used in extraction of gold, copper, phenol, benzene, trichloroethene, tetrachloroethane, and in pharmaceutical applications.

28.4.5.2 Membrane Distillation

Membrane distillation is an evaporation process for separating volatile solvent from one side of a nonwetted microporous membrane. The evaporated solvent is condensed or moved on the permeate side of the membrane. When a hot solution and a cold aqueous solution are separated by a nonwetting membrane, water vapor will diffuse from the hot solution/membrane interface to the cold solution/membrane interface and condense there. So long as the membrane pores are not wetted by both solutions, the pressures on both sides can be different. The microporous membrane in this case acts as liquid phase barrier as water evaporation continues. This arrangement is called the direct contact membrane distillation. Microporous hydrophobic membranes of PTFE, PP, and PVDF have been used in this type of application. The main advantages associated with membrane distillation are:

- No possibility of entrainment
- Possibilities of horizontal configurations
- Low temperature energy sources can be used

- Reduced problem of fouling due to the use of hydrophobic membranes
- Possibility of highly compact designs such as hollow fiber configuration

In addition to direct contact type, other configurations that are being investigated include air gap membrane distillation, vacuum membrane distillation and sweep gas membrane distillation.

28.4.5.3 Separations Using Liquid Membranes

In separation processes using liquid membranes, the solutes diffuse through liquid contained in a porous support. These separations can be either gas or liquid separations. The solute molecules undergo dissolution in the membrane at the feed/membrane interface. The dissolved solutes diffuse through the membrane and are desorbed at the other membrane surface. Applications using liquid membranes includes waste water treatment: removal of phenol [29], removal of thiomersol from vaccine production effluents [30], trace metal treatment from natural waters. Other applications include removal of citric acid, acetic acid from fermentation broths, separation of gas mixtures, toxic heavy metal ions, separation of sugars, etc.

28.5   MEMBRANE MATERIALS

It is very difficult to give a clear cut definition of what a membrane is. It is a selective barrier between two phases. Membrane material is the actual filter medium. It is the heart of the membrane processing systems. All membranes are "asymmetric" in morphology, i.e., they have a dense skin layer (0.1–2 µm thick) on top of a support layer (50–200 µm thick and highly porous) underneath. Membranes are usually homogeneous in material, in that they consist of the same polymer or copolymer throughout their structure.

28.5.1   GENERATIONS IN MEMBRANE MATERIAL DEVELOPMENT

First Generation Membranes:	Cellulose acetate and polymers.
Second Generation Membranes:	"Thin film composite" membranes—non-CA polymers usually a polyamide skin on polysulphones support. Asymmetric nature.
Third Generation Membranes:	Newest membranes derived from mineral or ceramic material (e.g., zirconium oxide, Al_2O_3).
	Tolerable to high temperature up to 400°C, wide pH range (0–14), long shelf-live but brittle.

Good quality membrane material is critical for proper operation of MF, UF, NF, and RO plants. Cellulose acetate was the most common material for UF and RO membranes but these have now been completely replaced by polysulphone membranes especially for UF applications. Numerous other materials have been tested e.g., polyamide, polyviniledene fluoride, etc. Mineral membranes especially zirconium oxide and ceramic membranes are now being used increasingly for UF and MF.

28.5.2   COMMONLY USED MEMBRANE MATERIALS

Different membrane materials used in pressure driven membrane processes are listed in Table 28.4.

28.5.2.1   Cellulose Esters as Membrane Materials

- Cellulose acetate (CA)
  - First commercially developed membranes
  - Good retention capacity
  - A relatively high flux

TABLE 28.4

Membrane Materials Used for Different Pressure Driven Membrane Processes

Process	Materials Used
MF	Regenerated cellulose, cellulose acetate, polysulphones, polycarbonate, polypropylene, polytetrafluoroethylene, polyvinylidenedifluoride, polyamide, polyvinylchloride
UF	Regenerated cellulose, cellulose acetate, polyamide, polyamide hydrazine, polyacrylonitre, polysulphones, polyvinylededifluoride
NF	Modified polyamide, polyamide hydrazine, sulphonated polyvinyl alcohol derivatives, cellulose acetate mixed esters
RO	Cellulose acetate, Cellulose mixed acetates, Polyamide, Polyamide, Polyamide derivatives, Polysulphones/Polyamide composites, Polysulphone/polyurea composites

- Cellulose triacetate
- Cellulose acetate phthalate
- Cellulose acetate butyrate
- Ethyl cellulose
- Regenerated cellulose

28.5.2.2 Polyamide Class of Polymers as Membrane Materials
- Simple aromatic polyamide
- Polyamide hydrazide
- Polyhydrazides
- Polypipperazineamides
- Polybenzamidazol
- Polybenzamidazolones

28.5.2.3 Drawbacks of Cellulose Acetate Membranes
- Limited chemical, thermal and radiation stability and chlorine resistance, maximum temperature limit of 40°C, and pH between 3 and 7
- Susceptible to bacterial attack
- Long term decline in water flux due to pressure-induced compaction—particularly for seawater desalination
- Life is low (< 1 year)
- Single use type—difficulty in cleaning

28.5.2.4 Advantages of Polyamide Class of Membranes
- Better chemical, thermal and radiation stability up to 75°C, pH 2 to 13
- Stable to bacterial attack—but susceptible to bacterial fouling
- Better flux stability in the long term
- Cleaning of these membranes is much easier

28.5.2.5 Disadvantages of Polyamide Class of Membranes
- Highly sensitive to dissolved free chlorine
- Susceptible to microbial fouling
- These are to be synthesized

28.5.3 ESSENTIAL MEMBRANE CHARACTERISTICS

28.5.3.1 Good Selectivity

Good selectivity is an important characteristic of any membrane. The selected membrane for a particular operation should give high water flux and solute separations. Water flux through a membrane is directly proportional to the net pressure gradient and inversely proportional to membrane thickness. Solute separation depends on the concentration across the membrane and membrane properties. Membrane manufacturers characterize the membranes by their salt rejections as in the case for RO and NF membranes. In the case of UF and MF membranes, MWCO and membrane pore size are used to specify the membrane. The rejection of various macromolecules/solutes is the basis for this characterization. However, this single criterion is not adequate for selection of a suitable membrane for a particular application. This criterion is essentially used for initial screening of membranes for specific applications.

Pore size distribution of MF membranes can be determined using a variety of tests viz. bubble pressure break through [5,31–33]. Gas permeability, air flow, permporometry and thermoporometry are also used to obtain pore size distribution. Solute retention is a common method used for characterization of pore size in UF membranes. MWCO, which is the molecular weight above which rejections are expected to be 90%, is commonly used to designate UF membranes. Solutes used in these tests are salts, sugars, purified proteins, dextrans and polyethelene glycols. Construction of a MWCO versus membrane pore size plot [34] will assist estimation of pore size distribution from rejection of various solutes. However, rejection of solute components depends to a large extent on the molecular confrontation of the solutes. Electron microscopy [35] and surface properties [36,37] are also used in characterization of membranes.

28.5.3.2 Chemical Stability

This property helps the membrane withstand harsh chemical cleaning cycles. Different streams can be processed with minimum pretreatment.

28.5.3.3 Mechanical Stability

Membranes should be stable against physical erosion and other dimensional changes during operation.

28.5.3.4 Economical and Availability

Membranes should be easily available to reduce the ultimate cost of separation and to make the technology easily accessible.

28.6 MEMBRANE MODULES

Industrial separations using membrane based processes require large surface area of membranes. The arrangement of this area in the form of definite shape and size is known as module design. Most membrane modules are designed in one of the following configurations. A few tailor-made designs are available that are aimed at specific applications:

- Tubular
- Hollow fiber
- Plate and frame
- Spiral wound

The module selected for a particular application should be of low cost, can accommodate large surface area in unit volume and still provide the required turbulence in the flow of the fluid. The development of membrane modules are proprietary and in-house R&D effort of a module manufacturer.

While cost is always a consideration, membrane fouling and concentration polarization are important issues in designing a module.

28.6.1 Tubular Modules

In tubular modules, a cellulose layer acts as support for the membrane and particularly also acts as a drainage layer for the permeate. The supporting layer and membrane are mounted in stainless steel supporting tubes. Tubes used in these modules typically are 85–600 cm in length and 3–25 mm in diameter. Several tubes can be connected in series or parallel as a bundle and housed in an SS casing. These modules offer good hydrodynamics thereby reducing fouling and concentration polarization. Permeate from each tube is removed to a collection header and sent out. Tubular modules are easy to clean and allow recirculation of feeds with high level of solids and viscosity. However, these modules have the lowest packing density (surface area to volume ratio) thus requiring high feed flow rate leading to high pumping costs.

28.6.2 Hollow Fiber Modules

This design consists of a bundle of very thin and narrow tubes. These fibers are self-supporting and generate high turbulence in the process streams. Operating pressures for these modules are low and are in the range of 170–270 kPa, thus, limiting their use to UF and MF applications. Hollow fiber configurations facilitate back flushing thereby preventing the build up of fouling material on the surface of the fibers. Feed flow arrangement can be on shell side where the permeate passes through the fiber walls and exits through open fiber ends. This design is easy to make and very large area can be accommodated per unit size of the module. As feed is on the outer side of the tube, the fiber experiences high pressure so it is designed for small diameter and thick walls. In contrast, in bore side feed designs feed flows inside the fiber and permeate passes through the walls. In this type of design, the fiber diameter is large.

Hollow fibers operate with velocities of 0.5–2.5 m/s giving laminar flow conditions. Pressure drop depends on flow rate, fiber diameter and fiber length. These modules have higher surface area to volume ratio. Hold up volume of the product is low. Replacement costs are high for this module.

28.6.3 Plate and Frame Modules

Plate and frame design resembles the classic plate heat exchange being used in liquid food pasteurization. These modules were one of the oldest types developed. The module consists of a stack of plates and flat sheet membrane much like plate heat exchanger (PHE) or filter press arrangement. Membranes are placed on both sides of the support plate and a spacer plate separates a number of such plates. The feed flows parallel to the membrane and permeate is channeled out through the support plate. Plate and frame configurations are available both in horizontal and vertical designs. The channel height for feed flow depends on the feed spacer/gasket used in the plate and frame arrangement. The packing density is between that of the hollow fiber and tubular modules. These designs are mostly used in ED and pervaporation applications and in limited number of RO and UF applications.

28.6.4 Spiral Wound Modules

In this type of design, a number of flat sheet membranes secured at one end to a hollow tube are spiraled to form a cylindrical configuration. Each sheet is separated by feed spacers. These spacers provide the required turbulence in the module. This is the most compact and inexpensive design. The feed is pumped length wise along the unit and permeate is collected at the centre into a collection tube. This configuration is widely used in dairy and food industry and is very cost effective, compact and has low liquid holdup. The feed channel height depends on the thickness of the spacer used. Spacers with 0.56–3.1 mm thickness are available. Narrow channel spacers are used for dilute and clear liquids while large spacers are used for viscous and products containing suspended material. As the spacer thickness decreases, the module can accommodate large membrane area and vice versa. Spiral modules operate under turbulent conditions and pressure drops are relatively high because of parasitic drag exerted by the spacer.

28.7 FACTORS AFFECTING PERFORMANCE

28.7.1 CONCENTRATION POLARIZATION

In any membrane processing operation the permeate flux is always lower than pure water flux and permeate flux increases non- linearly with TMP. After exceeding a certain pressure, flux becomes independent of pressure and plateaus to a constant value. With mass transfer conditions remaining constant, feed concentration has a logarithmic relationship with the so called limiting flux. This relationship is due to the influence of concentration polarization. Concentration polarization arises due to rejection of hydrocolloids, macromolecules such as proteins and other relatively large solutes. These tend to form a layer on the surface of the membrane. This offers further resistance which is in addition to the resistance of the membrane and the stagnant boundary layer. Concentration polarization phenomenon is shown in Figure 28.8. It occurs as a consequence of convective movement of solids toward the membrane during processing. The rejected solids accumulate on the surface, causing a steep concentration gradient of solutes within the boundary layer. This causes a back transport of retained solutes back to the bulk solution due to diffusion.

A steady state is reached where the two phenomena—the convective transport toward the membrane and the diffusive transport away from the membrane balances each other. Eventually, the solute concentration in the concentration polarization layer reaches a maximum commonly known as the "gel concentration." Gel concentration thickness is a function of the hydrodynamic conditions in the membrane system and is independent of the physical properties of the membrane [38,39]. It is due to this consolidated gel layer on the membrane that flux becomes independent of pressure. At this point, increasing TMP nearly always results in a thicker layer. After a momentary rise, flux returns to its initial value.

The reduction in flux may be due to an increased solute concentration on the membrane surface leading to higher osmotic pressure, thereby decreasing the driving force and hydrodynamic resistance of the boundary layer. As the solute concentration in the gel layer reaches maximum, this could lead

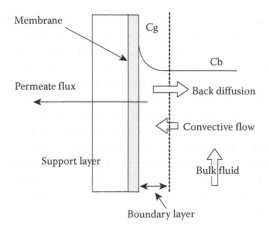

FIGURE 28.8 Concentration polarization on a membrane surface.

to precipitation of the solute and fouling of the membrane. Unlike membrane fouling, concentration polarization is a reversible phenomenon, and it is assumed that the system would reverse back to the pressure-controlled regime upon changing the operating conditions such as lowering the pressure or feed concentration or increasing feed velocity. In practice, this may be difficult to achieve. However, flux in the pressure-independent regime will be controlled by the efficiency of minimizing boundary layer thickness and enhancing the rate of back transport of the polarized molecules [8].

28.7.2 MEMBRANE FOULING

The fouling of membranes is another major limiting factor in membrane processing. Contrary to CP, which is considered a time-independent, reversible process, fouling is an irreversible phenomenon. During fouling, flux drops with time, usually rapidly in the initial stages and slowly at the later stages. Fouling is generally attributed to:

- The accumulation or adsorption of macromolecules or colloidal particles such as proteins, lipids, macroorganisms and/or inorganic salts, on the membrane surface
- The precipitation of permeable solutes such as sugars and salts due to overcrowding within the membrane pores

Fane [40] identified three separate phases of the fouling phenomenon. In the first few minutes, the initial rapid drop in flux is due to concentration polarization and this drop is reversible upon flushing the membrane with water. The second phase is characterized by a continuous drop in flux, initially rapidly due to fouling. This drop cannot be restored without chemical cleaning of the membrane. The third phase, a quasi-steady state period is reached where flux declines slowly, possibly due to further accumulation of deposits or to the consolidation of the fouling layer. The basis for assessing the degree of fouling is clean water flux of a membrane [5]. Clean water permeability of a membrane can be expressed as

$$A_w = \frac{J_w}{P_T} \tag{28.12}$$

Where A_w is in LMH/bar and J_w is the water flux and P_T is the TMP. As viscosity is included in A_w, it is essential to properly consider temperature and quality of water.

28.8 CONCLUSIONS

Separation, concentration, and purification are common processes in food and biomaterial processing. These processes involve separating one or more components from a mixture of several species. The components may vary in their physical or chemical characteristics, such as solubility, electrostatic charges, density, size and shape, etc. Evaporation, drying, crystallization, centrifugation, ion exchange—ED, extraction, leaching, mechanical filtration, sedimentation, and settling etc. are widely used as conventional separation processes. Membrane processing is a novel and green process technology. It is fast gaining widespread application in several fields. In this chapter we presented material covering important aspects of membrane processing. Membrane processes have been conveniently grouped into conventional, novel processes and current investigated processes. Developments in membrane materials have been highlighted. Current and future promising applications of membrane technology are discussed in detail. Finally, important factors determining system performance are presented. The authors are of the view that with current research efforts to develop the thinnest membrane possible and production of membranes with uniform pore sizes, membrane processing will continue to be an area of considerable interests in its application in industrial separations.

ACKNOWLEDGMENTS

We thank and acknowledge financial support provided by the Dairy Management Inc, Chicago, IL, and Agricultural Experiment Station, South Dakota State University, Brookings, SD for carrying out this work.

REFERENCES

1. Mohesenin, N.N. *Physical Properties of Plant and Animal Material.* Gordan and Breach, NY, 1980.
2. Lewis, M.J. *Physical Properties of Foods and Food Processing Systems.* Ellis Horwood, Chichester, UK, 1990.
3. Mulder, M. *Basic Principles of Membrane Technology.* Kluwer Academic Publishers, Norwell, MA, 1991.
4. Gradison, A.S. and Lewis, M.J. Separation process-an over view. In *Separation Processes in Food and Biotechnology Industries: Principles and Applications.* Gradison, A.S. and Lewis, M.J., Eds. Woodlands Publishing Ltd., Cambridge, UK, 1996.
5. Cheryan, M. *Ultrafiltration and Microfiltration Handbook.* Technomic Publishing Co., Inc. PA, 1998.
6. Bird, J. The application of membrane systems in the dairy industry. *J. Soc. Dairy Technol.,* 49, 16, 1996.
7. Goldsmith, R.L., deFilippi, R.P., Hossain, S. and Timmins, R.S. Industrial ultrafiltration. In *Membrane Processes in Industry and Biomedicine.* Bier, M., Ed. Plenum Press, NY, 267, 1971.
8. Baker, R.W. *Membrane Technology and Applications,* 2nd edn. John Wiley and Sons, UK, 2004.
9. Kessler, H.G. *Food Engineering and Dairy Technology.* Verlag A Kessler, Germany, 1981.
10. Deeslie, W.D and Cheryan, M. Continuous enzymatic modification of proteins in an ultrafiltration reactor. *J. Food Sci.,* 46, 1035, 1981.
11. Mannheim, A. Continuous hydrolysis of milk proteins in a membrane reactor. MSc Thesis, University of Illinois, Urbana, IL, 1989.
12. Engasser, J.M., Laumon, J. and Marc, A. Hollow fiber enzyme reactors for maltose and starch hydrolysis. *Chem. Eng. Sci.,* 35, 99, 1980.
13. Watanabe, T., Aoki, T., Honda, H., Taya, M. and Kobayashi, T. Production of ethanol in repeated-batch fermentation with membrane-type bioreactor. *Ferment. Bioeng.,* 69, 33, 1990.
14. Cheryan, M. and Mechaia, M.A. Membrane bioreactors. In *Membrane Separation in Biotechnology.* McGregor, W.C., Ed. Marcel Dekker, NY, 1986.
15. Cheryan, M. and Deeslie, W.D. Soy protein hydrolysis in a membrane reactor. *J. Am. Oil Chem. Soc.,* 60, 1112, 1983.
16. Argeino, V., Canepa, P., Gerbi, V. and Tortia, C. Continuous starch fermentation by integrated membrane bioreactor. A*gro Food Industry HI Tech.,* 2, 33, 1991.
17. Boyaval, P., Corre, C and Madec, M.N. Propionic acid production in a membrane bioreactor. *Enzyme Microbial Technol.,* 16, 883, 1994.
18. Rosenberg, M. Current and future applications for membrane processes in the dairy industry. *Trends Food Sci. Technol.,* 6, 12, 1995.
19. Roger, L., Maubois, J.L., Brule, G. and Piot, M. Process for obtaining a α-lactalbumin enriched product from whey, and uses thereof. U.S. Patent No 4,485,040, 1984.
20. Bottomley, R.C. Isolation of Immunoglobulin rich fraction from whey. U.S. Patent No. 5,194,591, 1993.
21. Bottomley, R.C. Process for obtaining concentrates having a high α-lactalbumin content from whey. U.S. Patent No. 5,008,376, 1991.
22. Thomas, R.L., Cordle, C.T., Criswell, L.G., Westfall, P.H. and Barefoot, S.F. Selective enrichment of proteins using formed in place membranes. *J. Food Sci.,* 57, 1002, 1992.
23. Uchida, Y., Schimatani, M., Mitsuhashi, T. and Koutake, M. Process for preparing a fraction of α-Lactalbumin from whey and nutritional compositions containing such fractions. U.S. Patent No 5,503,864, 1996.

24. Muller, A., Chaufer, B., Merin, U. and Daufin, G. Preparation of α lactalbumin with Ultrafiltration ceramic membranes from acid casein whey: study of operating conditions. *Lait,* 83, 111, 2003.

25. Mehra, R. and Kelley, P.M. Whey protein fractionation using cascade membrane filtration. Bull No. 389. International Dairy Federation, Brussels, Belgium, 2004.

26. Kessler, H.G. *Food and Bioprocess Engineering: Dairy Technology.* Verlag A. Kessler, Germany, 2002.

27. Fleming, H.L. and Slater, C.S. Pervaporation: applications and economics. In *Membrane Handbook.* Winston Ho, W.S. and Sirkar, K.K, Eds. Van Nosrtrand Reinhold, NY, 1992.

28. Prasad, R and Sirkar, K.K. 1992. Membrane based solvent extraction. In *Membrane Handbook,* Winston Ho, W.S. and Sirkar, K.K, Eds. Van Nosrtrand Reinhold, NY, 1992.

29. Cahn, R.P. and Li, N.N. Separation of phenol from waste water by liquid membrane technique. *Separation Sci.,* 9, 505, 1974.

30. Fortunato, R., Afonso, C.A.M., Crespo, J.G. and Reis, M.A. *17th Forum for Applied Biotechnology: Proceedings.* 41, Gent, Belgium, 2003.

31. Kulkarni, S.S., Funk, E.W. and Li, N.N. Membranes. In *Membrane Handbook.* Winston Ho, W.S. and Sirkar, K.K, Eds. Van Nosrtrand Reinhold, NY, 1992.

32. Capannelli, G., Vigo, F and Munari, S. Ultrafiltration membranes-characterization methods. *J. Membr Sci.,* 15, 289, 1983.

33. Munari, S., Bottino, A., Moretti, P., Capannelli, G. and Becci, I. Permoporometric study on ultrafiltration membranes. *J. Membr. Sci.,* 41, 69, 1989.

34. Sarbolouki, M. Properties of asymmetric polimide ultrafiltration membranes: pore size and morphology characterization. *Sep. Sci. Technol.,* 17, 381, 1982.

35. Glaves, C.L. and Smith, D.M. Membrane pore structure analysis via NMR spin-lattice relaxation measurements. *J. Membr. Sci.,* 46, 167, 1989.

36. Fonteny, M., Bijsterboch, B.H. and Van't Reit, K. Chemical characterization of ultrafiltration membranes by spectroscopic techniques. *J. Membr. Sci.,* 36, 14, 1987.

37. Oldani, M and Schock, G. Characterization of ultrafiltration membranes by IR, ELISA and contact angle measurements. *J. Membr. Sci.,* 43, 243, 1989.

38. Howell, J. A. and Velicangli, O., Protein Ultrafiltration: theory of membrane fouling and its treatment with immobilized proteases. In *Ultrafiltration Membrane Applications.* Cooper, A.R., Ed. Plenum Press, NY, 217, 1980.

39. Aimar, P., Howell, J.A., Clifton, M.J. and Sanchez, V. Concentration polarization buildup in hollow fibers: a method of measurement and its modeling in ultrafiltration. *J. Membr. Sci.,* 59, 81, 1991.

40. Fane, A.G. Ultrafiltration: factors influencing flux and rejection. In *Progress in Filtration and Separations.* Wakeman, R.J., Ed. Elsevier Science Publishers, Amsterdam, The Netherlands, 101, 1986.

29 Modeling of Membrane Fouling

Fouling

Alice A. Makardij
Orica Chemnet NZ

Mohammed M. Farid
The University of Auckland

Xiao Dong Chen
Monash University

CONTENTS

29.1 INTRODUCTION

Cross-flow microfiltration (MF) and ultrafiltration (UF) have been considered as two promising technologies to solve many separation problems especially in food and beverage industries. This is due to the simplicity and gentle nature of the processes (i.e., high temperature and phase change are usually not required). The low energy requirements and often low capital and operating costs are also advantageous.

One of the major obstacles that hinder widespread application of membrane separation for food processing is that the permeate flux declines with time—a phenomenon commonly termed as "membrane fouling." Typically understood as the permanent (or irreversible) loss of permeability, whereas flux decline is due to several factors (including both membrane-solute and solute-solute interactions), which are consequences of concentration polarization. "In-pore" fouling reduces processing rate and increases the complexity of membrane filtration operations, as the system has to be halted frequently to restore flux by periodic chemical cleaning. High operational costs make MF and UF economically less attractive compared to other separation methods in some applications.

Due to the complexity, mathematical models have been developed either based on simple empirical approach or based on fundamental principles, which usually lead to complicated formulation. The effects of 'concentration polarization' has been studied using complex models; e.g., a model presented by Clifton et al.[1] and the improved model of Chen et al.[2] In these models, details of the velocity and concentration profiles need to be resolved. More recently, computational fluid dynamics (CFD) packages are increasingly being used to investigate membrane applications especially those involving complex geometry.[3,4]

Despite these developments, membrane process designs are still dealing with relatively simple geometries. A good compromise would be a model with a level of complexity that is in between the empirical and fundamental models. This compromise is logical because the fundamental models still cannot cope with the details of fouling (indeed, these details have not fully emerged in the research so far and cannot really be fully described by mathematical models).

As mentioned earlier, membrane fouling and concentration polarization are two aspects of the same problem, which is the build up of retained species in the boundary layer neighboring a membrane surface. Both phenomena gradually reduce the permeation flux through the membrane and change the selectivity of the process as stated by Zeman and Zydney.[5]

Several models have previously been developed to describe the reduction in flux. The mechanisms responsible for fouling can be classified into three categories: (a) resistance in series, (b) gel polarization, and (c) osmotic pressure. In the resistance-in-series model, flux decline occurs due to the "additive" resistances caused by fouling or solute adsorption and concentration polarization. However, in the gel formation and concentration polarization model according to Blatt et al.,[6] flux decreases due to the hydraulic resistance of the gel layer formed at the membrane surface as well as concentration polarization.

On the other hand, in the osmotic pressure model described by a number of authors,[7–9] flux reduction results from the decrease in the effective transmembrane pressure which occurs as the osmotic pressure of the retentate increases.

Much work has been done to address the problem of fouling over the last 15 years. Although the physical processes which give rise to fouling have been characterized, they are not well understood.[10] So far, some theoretical studies have attempted modeling the dynamic behavior of various membranes. Stamatakis and Tein[11] developed a model based on the concept of particle adhesion. The main argument in their work was that only a fraction of the particles is brought to the membrane surface which could deposit on it. The fraction of deposited particles was estimated from a force balance and a probability analysis. Using this model, they were able to calculate the accumulation of the retained particles on the membrane surface and fit the time dependent permeate flux using their model.

Romero and Davis[12] noted that fouling dynamics in cross-flow filtration could be better described by dividing the entire channel into two distinct regions: the equilibrium and nonequilibrium regions. The movement of the front of the equilibrium was determined with the method of characteristics, which attributes the front movement solely to particle deposition induced by local permeate flux. Bhattacharjee and Datta[13] stated in their research that none of these models could fit experimental data with high accuracy.

While data of time-dependent flux have been collected from numerous MF and UF experiments under various conditions, not much progress has been made in understanding the fundamental mechanisms of membrane fouling. It is not clear how basic parameters such as shear rate, applied pressure and particle size affect fouling processes in cross flow filtration.[14] The delineation of fouling mechanisms and the quantitative presentation of fouling process remain big challenges in the field.

Despite these developments, a simple model or a well defined procedure for quantitative description of the fouling dynamics, which can be used by practicing engineers in process design and operation of cross-flow filtration, is currently not available.

In this chapter, the flux-time patterns and the key aspects of the physics that influence membrane fouling have been explored. Perceiving membrane fouling as a dynamic process has developed a simple model for the time dependent flux in cross-flow filtration. The time scales for pore blocking, cake formation and the time to reach steady state have been investigated. The effects of membrane pore size, membrane type, transmembrane pressure, temperature, solute concentration, and fluid velocity have all incorporated in a single model for the first time.

29.2 MODEL DEVELOPMENT

In practice, flux decline from the initial value to the steady value is observed over time. According to Darcy's law, permeate flux is inversely proportional to the cake and membrane resistances, being time dependent as follow:

$$J = \frac{\Delta P}{\mu(R_m + R_{cp} + R_c)} \qquad (29.1)$$

where J is the permeate flux, ΔP is the transmembrane pressure, μ is the fluid viscosity, and R_m, R_{cp}, and R_c are the membrane resistance, the concentration polarization resistance and the resistance of the cake layer, respectively.

For developing a simple but effective model in this work, it was assumed that the initial flux would drop suddenly upon the start of the membrane operation due to concentration polarization. The flux will then decline gradually due to the net effect of particle deposition on or into the membrane, and deposit removal due to the cross-flow of the retentate. This may be described as follows.

The rate of flux decline is equal to the rate at which solids or solutes are brought to the membrane surface *less* the rate at which deposit is removed from the membrane.

In mathematical form, this may be expressed as follows:

$$-\frac{dJ}{dt} = k_1 C_0 J - k_2 \mathrm{Re}^n \qquad (29.2)$$

where C_0 is the feed concentration (kg.m^{-3}), k_1 is the rate constant for flux decline (m^3.kg^{-1}.s^{-1}), k_2 (m^3.m^{-2}.s^{-2}) is the rate constant for deposit removal from the membrane, and the Reynolds number $\mathrm{Re} = \rho u d/\mu$. Where ρ is the retentate density (kg.m^3), u is the retentate cross-flow velocity (m.s^{-1}), μ is the viscosity of the retentate (Pa.s). If the channel through which the feed fluid flows is not of circular cross section, d is defined as the hydraulic mean diameter, which is calculated by dividing four times the cross sectional area of the flow by the wetted perimeter.

The power n needs to be established experimentally. Equation 29.2 defines the local permeate flux at any position in the membrane. If the permeate flux is small compared to the total flow (as in the small unit used in this work) then both C_0 and Re may be assumed to be equal to their values at the feed entry.

The initial conditions may be specified as follows:

$$t = 0, \quad J = J_0 \quad \text{and} \quad k_1 C_0 J \gg k_2 \mathrm{Re}^n \qquad (29.2a)$$

Hence, k_1 may be calculated from the initial flux by dropping the second term of the right-hand side of Equation 29.2:

$$k_1 = \frac{(dJ/dt)_{\text{initial}}}{C_0 J_0} \tag{29.3}$$

When approaching a steady state, on the other hand, one has

$$\frac{dJ}{dt} \Rightarrow 0, \quad \text{thus} \quad k_2 \approx k_1 \left(\frac{C_o J}{\text{Re}^n} \right)_{\text{equilibrium}} \tag{29.4}$$

Equation 29.3 could have been used to calculate the values of k_1 from the measurements of initial flux decline. However, neither J_0 nor dJ/dt can be measured accurately at zero time; hence it was decided to calculate k_1 from the integral form of Equation 29.2. Firstly Equation 29.2) may be simplified to the following form:

$$-\frac{dJ}{dt} = a(J - b) \tag{29.5}$$

where $\quad a = k_1 C_0 \quad$ and $\quad b = \frac{k_2}{k_1 C_0} \text{Re}^n$

Integrating Equation 29.5 from $t=0$ to $t>0$ gives:

$$J = b + (J_o - b)e^{-at} \tag{29.6}$$

The above equation shows an exponential decay of the permeate flux with both time and concentration. Equation 29.6 shows that as $t \rightarrow \infty$, J approaches the equilibrium flux $b = J_{eq}$. This is in agreement with the measurements available in the literature.[5]

In essence Equation 29.2 is, by nature, similar to the idea proposed for describing the fouling behavior in heat exchangers, i.e., the Kern–Seaton type equation where the net rate of fouling is the result of deposition rate *less* deposit removal rate.[10,15,16]

29.3   EXPERIMENTAL

A high-pressure cell test unit (Environment Products International Ltd., Auckland, New Zealand) was used to produce the data for model validation. A stainless steel plate and frame cross-flow membrane module was used in this study. It consists of two halves bolted to each other. The flat membrane sheet can be placed in the middle of these two stainless steel plates and is supported by a porous metal plate, which sits in the grooved chamber underneath the membrane.

The membrane unit has a channel height of 5.2×10^{-3} m and a filtration area of 7.41×10^{-3} m^2. The cell may be viewed as a semi-cross flow type because flow direction is parallel to the membrane surface, but the feed must travel in a spiral path across some of the membrane to the retentate exit. The permeate is collected in a preweighted plastic jar and the weight was continuously measured by a "Sartorius" digital microbalance. A high-pressure pump (HPP), which is a positive displacement reciprocating diaphragm type, was used to pump the feed to the filtration unit. Two pressure gauges were installed at the inlet and outlet of the membrane unit to measure and monitor the transmembrane pressure. Figure 29.1 shows the schematic diagram of the membrane system.

29.3.1   MEMBRANES

Custom-cut membranes for the plate-and-frame set-up were purchased from EPI Ltd. These were: polyvinylidene di-fluoride (PVDF) MF membrane type JX, with an average pore diameter of 0.3 mm, a polysulphone (PS) UF membrane type G20 with 3,500 molecular weight cut-off (MWCO) on polyethylene glycol (PEG) and finally type G50 with 8000 MWCO on PEG. The maximum operating temperature of these membranes is about 50°C. Both the MF and UF membranes were polymeric porous membranes.

29.3.2 OPERATING PARAMETERS

The transmembrane pressure (P_{TM}) examined was 0.69×10^3 kPa, 1.38×10^3 kPa and 2.07×10^3 kPa. The pressure was controlled via pressure regulators mounted on the feed and retentate sections. The average feed velocities studied was between 0.15 and 0.45 m.s^{-1}.

Temperature effect was studied starting from ambient temperature up to 55°C to avoid protein denaturation. The feed tank was provided with a thermostat controller to provide constant feed temperature. The effect of varying feed concentrations was also studied.

The model was tested against different operating conditions as summarized in Table 29.1.

29.3.3 FILTRATION AND FOULING TRIALS

Three different types of feed solutions were used in this study with three different types of membrane. In the current study, 10 wt% skim milk (pH 6.7, fat content of <0.1 wt%) solutions were prepared freshly using deionized water before each run. High mixing techniques (mechanical mixer adjusted at 200 RPM) were used to achieve maximum solubility. Milk solution was left at room temperature for about two hours before use. Milk with different solid concentrations was also used using the same method. Skim milk permeates were collected for further analysis.

Membrane fouling during filtration of vinegar solutions was also studied. Vinegar was purchased from Bluebird® Foods, New Zealand. The main vinegar solution used consisted of raw vinegar taken from the acetator vat.

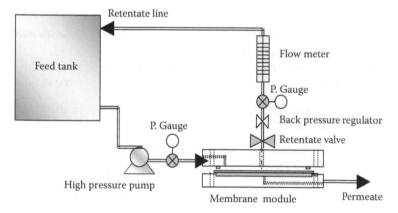

FIGURE 29.1 Schematic diagram of the experimental set-up. (From Makardij, A., Farid, M., and Chen X. D., *Can. J. Chem. Eng.*, 80, 28, 2002. With permission.)

TABLE 29.1
Operating Conditions Employed in this Work

Parameter	Low	Moderate	High
P_{TM} (kPa)	690	1380	2070
Temperature (°C)	23	30	50
Feed velocity (m.s^{-1})	0.17	0.35	0.44
Feed solids content (wt%)	6	8	10

Source: Makardij, A., Farid, M., and Chen X. D., *Can. J. Chem. Eng.*, 80, 28, 2002.

Fouling was also studied using baker's yeast solution. Baker's yeast was reconstituted in saline solution (8.5×10^{-3} g.m^{-3} NaCl). The sodium chloride balances the osmotic pressure across the cell wall, thus preventing cell rupture. A further advantage of using saline solution for yeast suspensions is that it is a simple media, which will enable the study of fouling with yeast cells in the absence other complicating foulants. Table 29.2 summarizes the membrane types and the fluid used in this study. Some of the physical properties of the milk, vinegar, and yeast suspension are shown in Appendix A.

29.4 RESULTS AND DISCUSSION

29.4.1 FLUX-TIME DATA

Filtration experiments were carried out to establish the traditional flux-time data (Figure 29.2). The experimental data were replotted as Ln(J–J_{eq}) versus time (Figure 29.3) to evaluate the values of initial flux, the flux decline coefficient k_1 (m^3.kg^{-1}.s^{-1}) and the cross-flow coefficient k_2 (m^3.m^{-2}.s^{-2}) as defined earlier.

The linear fit shown in Figure 29.3 is the first support to the model suggested. Based on Equation 29.6, $k_1 C_0$ is the slope of the line in Figure 29.3, while the initial flux J_0 was evaluated from the intercept of the line. The value of k_2 was calculated from the equilibrium flux b and slope $k_1 C_0$. Different values for the exponent (n) in Re^n were examined. A plot of k_2 versus Re number is first made to estimate the dependence of the k_2 value on Reynolds number, i.e., to find an approximate value of the exponent n. Values of (n) from 0.2 to 0.6 gave a large variation in the calculated k_2, whereas insignificant variation in the values of k_2 was possible when (n) was greater than 0.7. It was decided then to set (n) equal to 0.8. Where this value is generally used to describe turbulent flow conditions (Re > 2,000) for heat and mass transfer conditions.[15]

TABLE 29.2
Main Filtration Set-Ups

Filtration	Membrane	Feed Solution	Specification
MF JX	Polyvinyldenedifluoride (PVDF)	Yeast	0.3 mm
UF G20	Polysulphone (PS)	Skim milk, vinegar	3500 MWCO
UF G50	Polysulphone (PS)	Skim milk	8000 MWCO

Source: Makardij, A., Farid, M., and Chen X. D., *Can. J. Chem. Eng.*, 80, 28, 2002.

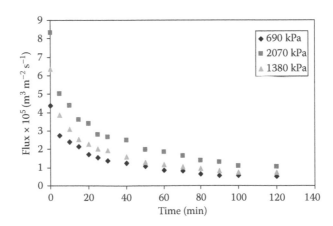

FIGURE 29.2 Flux decline data for 10 wt% skim milk using UF G50 at different pressures, $T = 23$°C, at cross-flow velocity = 0.44 m s^{-1}. (From Makardij, A., Farid, M., and Chen X. D., *Can. J. Chem. Eng.*, 80, 28, 2002. With permission.)

The value of k_1 is the rate constant for flux decline or the rate at which pore blocking is expected to occur. The second term in Equation 29.2 has a negative coefficient $(-k_2)$, and hence k_2 represents the rate constant for deposit removal. If these rate constants k_1 and k_2 are only a function of membrane pore size and the type of feed and are not functions of the operating conditions such as pressure, temperature and cross-flow velocity, the model parameters k_1 and k_2 would perhaps be considered to be an indicator of the specific membrane. This would make the entire exercise much more worthwhile.

29.4.2 Effect of Transmembrane Pressure

Transmembrane pressure (P_{TM}) has often been taken as the arithmetic mean of the module inlet and outlet pressures since the permeate leaves the unit at atmospheric pressure. The pressure difference (ΔP) between the retentate inlet and the outlet was very small in the unit used in this investigation; therefore, P_{TM} was taken as the inlet value. The transmembrane pressures applied were 0.69×10^3, 1.38×10^3, and 2.07×10^3 kPa. Data were collected and values of k_1 and k_2 were calculated as described earlier. The results are shown in Table 29.3.

The average variations in the values of k_1 due to pressure change is $\pm 7\%$ for UF G20/skim milk and vinegar, and is less than 5% for UF G50/skim milk and MF JX/yeast. While the average variations in the values of k_2 due to pressure change is $\pm 7\%$ for UF G20/skim milk and is $\pm 18\%$ for UF G50/skim milk.

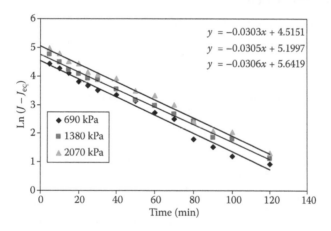

$$y = -0.0303x + 4.5151$$
$$y = -0.0305x + 5.1997$$
$$y = -0.0306x + 5.6419$$

FIGURE 29.3 Plot of $\text{Ln}(J - J_{eq})$ against time, to evaluate J_0, k_1, and k_2. Using G50 UF with 10 wt% skim milk at 23°C, cross-flow velocity = 0.44 m s^{-1}. (From Makardij, A., Farid, M., and Chen X. D., *Can. J. Chem. Eng.*, 80, 28, 2002. With permission.)

TABLE 29.3

The Consistency of the Model for the Systems Studied at 23°C and Different P_{TM} ($n = 8$) (CFV = 0.44 m s^{-1}), k_1 (m^3 kg^{-1} s^{-1}) and k_2 (m^3 m^{-2} s^{-2})

Membrane/Feed	0.69×10^3 kPa		1.38×10^3 kPa		2.07×10^3 kPa		Average Coefficients	
	k_1	k_2	k_1	k_2	k_1	k_2	k_1	k_2
UF G20/skim milk	0.0078	0.00070	0.0071	0.00072	0.0070	0.00082	0.0073 ($\pm 7\%$)*	0.00075 ($\pm 7\%$)
UF G50/skim milk	0.0030	0.00076	0.0031	0.00095	0.0031	0.00108	0.0031 ($\pm 3\%$)	0.00093 ($\pm 18\%$)
UF G20/vinegar	0.0057	0.00089	0.0056	0.00099	0.0053	0.00113	0.0055 ($\pm 4\%$)	0.00100 ($\pm 13\%$)
MF JX/yeast	0.0044	0.00186	0.0045	0.00216	0.0049	0.00238	0.0046 ($\pm 7\%$)	0.00222 ($\pm 16\%$)

Source: Makardij, A., Farid, M., and Chen X. D., *Can. J. Chem. Eng.*, 80, 28, 2002.

These variations in k_1 and k_2 are much smaller than their variation due to process variables (membrane/and type of feed). For example, the values of k_1 of the two cases, UF G20/skim milk and UF G50/skim milk differed by more than 100%. Based on these findings, k_1 and k_2 may be assumed to be independent of P_{TM}.

29.4.3 Effect of Temperature

The effect of temperature on fouling for a fixed P_{TM} (1.38×10^3 kPa) and cross-flow velocity (0.44 m.s^{-1}) was studied. Flux patterns were plotted and the effect of temperature on flux decline was analyzed.

As expected, increasing feed temperature resulted in higher permeate flux. The main effect of increasing temperature is to decrease liquid viscosity, which leads to a higher Re number, and hence more efficient deposit removal. There was a limit to the operating temperature, especially with milk, as high temperature could cause protein denaturation causing more fouling. While in vinegar, high temperature may cause evaporation of the volatile carboxylic groups. Table 29.4 shows the calculated values of k_1 and k_2 using the current model, based on the laboratory flux-time data.

Table 29.4 shows that the average variation in k_1 and k_2 due to temperature is small ($< \pm 8\%$) compared to their variations due to changing membrane or type of fluid used. The value $n = 0.8$ was found to give the least variation in k_2 values.

29.4.4 Effect of Feed Velocity

Table 29.5 illustrates the application of unsteady state model to different cross-flow velocities. The maximum variation in k_1 was $\pm 14\%$ for UF G50/skim milk, while it was ($< \pm 4\%$) for all other

TABLE 29.4
The Consistency of the Unsteady State Model for the Systems Studied at 1.38×10^3 kPa and Different Temperatures, CFV = 0.44 m s^{-1} ($n = 8$), k_1 (m^3 kg^{-1} s^{-1}) and k_2 (m^3 m^{-2} s^{-2})

	35°C		40°C		55°C		Average Coefficients	
Membrane/Feed	k_1	k_2	k_1	k_2	k_1	k_2	k_1	k_2
UF G20/skim milk	0.0077	0.00089	0.0080	0.00084	0.0083	0.00086	0.0080 ($\pm 4\%$)	0.00086 ($\pm 3\%$)
UF G50/skim milk	0.0031	0.00095	0.0034	0.00107	0.0031	0.00109	0.0033 ($\pm 6\%$)	0.00103 ($\pm 8\%$)
UF G20/vinegar	0.0053	0.00110	0.0047	0.00107	0.0049	0.00108	0.0050 ($\pm 6\%$)	0.00108 ($\pm 2\%$)
MF JX/yeast	0.0042	0.00230	0.0046	0.00255	0.0042	0.00229	0.0043 ($\pm 7\%$)	0.00238 ($\pm 7\%$)

Source: Makardij, A., Farid, M., and Chen X. D., *Can. J. Chem. Eng.*, 80, 28, 2002.

TABLE 29.5
The Consistency of the Unsteady State Model for the Systems Studied at 23°C, 1.38×10^3 kPa and Different Cross-flow Velocities ($n = 8$), k_1 (m^3 kg^{-1} s^{-1}) and k_2 (m^3 m^{-2} s^{-2})

	0.17 m.s^{-1}		0.35 m.s^{-1}		0.44 m.s^{-1}		Average Coefficients	
Membrane/Fluid	k_1	k_2	k_1	k_2	k_1	k_2	k_1	k_2
UF G20/skim milk	0.0067	0.00061	0.0070	0.00062	0.0071	0.00062	0.0069 ($\pm 4\%$)	0.00062 ($\pm 2\%$)
UF G50/skim milk	0.0029	0.00090	0.0024	0.00096	0.0031	0.00095	0.0028 ($\pm 14\%$)	0.00094 ($4\pm\%$)
UF G20/vinegar	0.0051	0.00105	0.0053	0.00105	0.0053	0.00109	0.0052 ($\pm 2\%$)	0.00106 ($\pm 2\%$)
MF JX/yeast	0.0046	0.00103	0.0047	0.00105	0.0044	0.00106	0.0045 ($\pm 4\%$)	0.00104 ($\pm 2\%$)

Source: Makardij, A., Farid, M., and Chen X. D., *Can. J. Chem. Eng.*, 80, 28, 2002.

selected systems. The corresponding variation in k_2 was small ($< \pm 5\%$) for the different systems in this study. Hence, it is reasonable to assume k_1 and k_2 are independent of feed velocity.

29.4.5 EFFECT OF FEED CONCENTRATION

The same conclusion can be reached regarding the effect of feed concentration. Values of k_1 and k_2 are calculated and shown in Table 29.6. The maximum variations in k_1 and k_2 are $\pm 10\%$. These variations are much smaller than their variation due to process variables (type of membrane and processed feed). The values of k_1 and k_2 due to the use of different types of membrane is sometimes higher than 100%. Based on these findings, k_1 and k_2 may be assumed to be independent of feed concentration.

29.5 OVERALL VALIDATION OF THE MODEL

The overall average values of k_1 and k_2 for all experiments in which pressure, temperature, cross-flow velocities and feed concentration were varied are shown in Table 29.7. The average variation in k_1 and k_2 was from ± 3 to $\pm 12\%$, which is small compared to their variations when different membrane or process fluid are used. The values of the variance for each system individually are also listed in Table 29.7.

Examining the values of k_1 and k_2 can determine whether the membrane tend to foul easily or be cleaned easily, which makes these parameters physical rather than empirical. High values of k_1 (deposition rate) indicate high fouling tendency for the membrane such as in the case of UF G20/skim milk (k_1 is the highest), while high values of k_2 indicates easiness of cleaning such as in UF G20/vinegar. Therefore, the different membrane/filtering media can be put in this order as shown in Table 29.8.

These conclusions are compared with the flux-time data and it was found acceptable in describing fouling tendency (flux decline pattern) for the four systems.

TABLE 29.6
The Consistency of the Unsteady State Model at 1.38×10^3 kPa, 23°C, 0.44 m.s^{-1} and Different Concentrations, k_1 (m^3 kg^{-1} s^{-1}) and k_2 (m^3 m^{-2} s^{-2})

	6 wt%		8 wt%		10 wt%		Average Coefficient	
Membrane/Feed	k_1	k_2	k_1	k_2	k_1	k_2	k_1	k_2
UF G50/skim milk	0.0028	0.00112	0.0024	0.0011	0.0029	0.00104	0.0027 ($\pm 10\%$)	0.00108 ($\pm 7\%$)
MF JX/yeast	0.0040	0.00273	0.0044	0.00287	0.0045	0.00252	0.0043 ($\pm 7\%$)	0.00271 ($\pm 7\%$)

Source: Makardij, A., Farid, M., and Chen X. D., *Can. J. Chem. Eng.*, 80, 28, 2002.

TABLE 29.7
Values of k_1 (m^3 kg^{-1} s^{-1}) and k_2 (m^3 m^{-2} s^{-2}) for the Selected Systems

Membrane/Filtering Media	k_1	Variance	k_2	Variance
UF G20/skim milk	0.0074 ($\pm 10\%$)	2.83×10^{-4}	0.0008 (± 12)	1.52×10^{-8}
UF G20/vinegar	0.0053 ($\pm 5\%$)	9.51×10^{-4}	0.0011 ($\pm 3\%$)	4.87×10^{-9}
MF JX/yeast solution	0.0044 ($\pm 6\%$)	1.97×10^{-5}	0.0002 ($\pm 8\%$)	9.11×10^{-8}
UF G50/skim milk	0.0031 ($\pm 11\%$)	2.99×10^{-5}	0.0010 ($\pm 11\%$)	2.75×10^{-8}

Source: Makardij, A., Farid, M., and Chen X. D., *Can. J. Chem. Eng.*, 80, 28, 2002.

Theoretical values of permeate flux were then calculated based on average values of k_1 and k_2 for all conditions tested for each type of membrane and type of feed used. The measured and predicted fluxes for UFG50/skim milk were shown in Figures 29.4 through 29.6 for different temperatures, cross-flow velocities and concentrations. Despite average values of k_1 and k_2, the agreement is reasonable between model predictions and experimental data.

TABLE 29.8
The Significance of k_1 and k_2 on Fouling and Cleaning Ability for the Different Systems Studied

Membrane/Filtering Media	Fouling Tendency	Cleaning Capability
UF G20/skim milk	High	Moderate
UF G20/vinegar	High	High
MF JX/yeast solution	Moderate	Low
UF G50/skim milk	low	High

Source: Makardij, A., Farid, M., and Chen X. D., *Can. J. Chem. Eng.*, 80, 28, 2002.

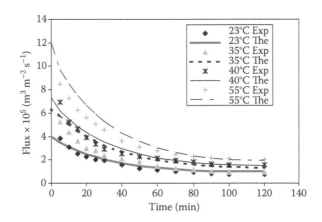

FIGURE 29.4 Unsteady state model predictions versus experimental results for UF G50 of 10 wt% skim milk at different temperatures, feed velocity=0.44 m/s, $P_{TM}=1.38\times 103$ kPa, $k_1=0.0047$ m^3 kg^{-1} s^{-1}, and $k_2=0.0021$ m^3 m^{-2} s^{-2}. (From Makardij, A., Farid, M., and Chen X. D., *Can. J. Chem. Eng.*, 80, 28, 2002. With permission.)

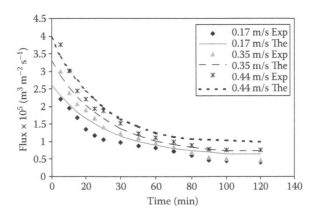

FIGURE 29.5 Unsteady state model predictions vs. exp. results for UF G50 of 10wt% skim milk at different cross-flow velocities. $T=23^{\circ}$C, $P_{TM}=1.38\times 10^4$ kPa. $k_1=0.0047$ m^3 kg^{-1} s^{-1}, and $k_2=0.0021$ m^3 m^{-2} s^{-2}. (From Makardij, A., Farid, M., and Chen X. D., *Can. J. Chem. Eng.*, 80, 28, 2002. With permission.)

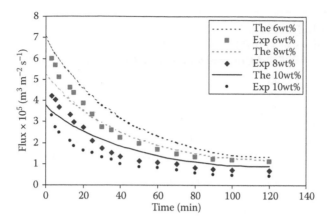

FIGURE 29.6 Unsteady state model predictions versus experimental results for UF G50 skim milk at different concentrations. $P_{TM} = 1.38 \times 10^3$ kPa, $T = 23°C$, feed velocity = 0.44 m.s^{-1} $k_1 = 0.0047$ m^3 kg^{-1} s^{-1}, and $k_2 = 0.0021$ m^3 m^{-2} s^{-2}. (From Makardij, A., Farid, M., and Chen X. D., *Can. J. Chem. Eng.*, 80, 28, 2002. With permission.)

29.6 COMPARISON WITH LITERATURE

In order to validate the kinetic model further, it was necessary to apply it to some experimental data published in the literature by other research groups using different set-ups and operational conditions. Relevant data were selected, one set from Bhattacharjee and Datta[13] and the other set by Frenander and Jonsson.[17] The unit of flux was deliberately left as reported in these publications.

29.6.1 Bhattacharjee and Datta, 1997

These authors developed a mass transfer model for the prediction of flux decline during ultrafiltration of PEG 6000. The author used a cellulose acetate (CA) membrane, operating at 30°C. They studied the effect of different P_{TM} and PEG 6000 bulk concentrations on its flux decline. We have used their raw data to generate the Ln(J–Jeq) time plots. The new model parameters k_1 and k_2 were calculated following the procedure described earlier (see Table 29.9 for the values of k_1 and k_2 at different P_{TM} and different concentrations).

An overall average value for k_1 (0.0044 ± 11%) and k_2 (0.0068 ± 12%) for all conditions tested by Bhattacharjee and Datta were used to predict the theoretical flux values for all experiments reported in their work.

The measured and predicted fluxes were plotted in Figure 29.7 for different P_{TM} and Figure 29.8 for different concentrations. Despite using a single set of values for k_1 and k_2, the agreement is reasonable between the model and the experimental measurements.

29.6.2 Frenander and Jonsson, 1996

In this paper, the performance of the dynamic membrane filtration (DMF) was initially investigated by microfiltration of baker's yeast suspension. The influence of transmembrane pressure was investigated using 25 g L^{-1} baker's yeast suspensions as feed. Yeast was suspended in deionized water without any pretreatment. The feed temperature varied during the trials between 15 and 25°C. Both retentate and permeate were recirculated to the feed tank. The Pall™ microfiltration membrane, Ultipor N66 (with 0.2 µm pore rating), was used in all experiments.

Raw data were extracted from the publication of Frenander and Jonsson and the Ln($J - J_{eq}$) time graphs were plotted and used to calculate the model parameters k_1 and k_2. A single value for

Mathematical Modeling of Food Processing

TABLE 29.9
Values of k_1 (m³ kg⁻¹ s⁻¹) and k_2 (m³ m⁻² s⁻²) at Different P_{TM} and Feed Concentration for CA/PEG 6000 at 30°C

100 Psi		120 Psi		Average Coefficient	
k_1	k_2	k_1	k_2	k_1	k_2
0.0039	0.0059	0.0044	0.0081	0.0042	0.0070

50 kg.m⁻³		70 kg.m⁻³		Average Coefficient	
0.0039	0.0059	0.0052	0.0073	0.0046	0.0066

Source: Makardij, A., Farid, M., and Chen X. D., *Can. J. Chem. Eng.*, 80, 28, 2002.

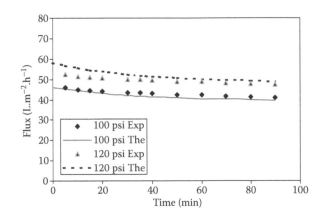

FIGURE 29.7 Unsteady state model validation for UF of PEG at different pressures, $T=30°C$, $C_{feed}=50$ kg.m⁻³, feed velocity$=1.5$ m.s⁻¹. k_1 (0.0044 m³.kg⁻¹.s⁻¹) and k_2 (0.0068 m³.m⁻².s⁻²). (From Makardij, A., Farid, M., and Chen X. D., *Can. J. Chem. Eng.*, 80, 28, 2002. With permission.)

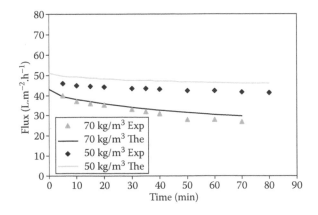

FIGURE 29.8 Unsteady state model validation for UF of PEG at different concentrations. $P_{TM}=100$ psi, $T=30°C$, feed velocity$=1.5$ m.s⁻¹. k_1 (0.0044 m³.kg⁻¹.s⁻¹) and k_2 (0.0068 m³.m⁻².s⁻²). (From Makardij, A., Farid, M., and Chen X. D., *Can. J. Chem. Eng.*, 80, 28, 2002. With permission.)

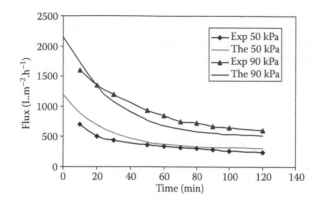

FIGURE 29.9 Unsteady state model validation for MF of baker's yeast at different P_{TM} $C_0=25$ g.L^{-1}, $T=25$°C. k_1 (0.0041 m^3 kg^{-1} s^{-1}) and k_2 (0.0021 m^3 m^{-2} s^{-2}). (From Makardij, A., Farid, M., and Chen X. D., *Can. J. Chem. Eng.*, 80, 28, 2002. With permission.)

k_1 (0.0041 ± 2%) and k_2 (0.0021 ± 7%), as calculated in this chapter, were used to predict the theoretical flux decline in all experiments reported in their work. The measured and predicted fluxes were plotted in Figure 29.9 for two different transmembrane pressures. The agreement is acceptable between the model and experimental data.

The above discussion based on independent measurements provides strong support to the model presented in this chapter. This work was first published in 2001 in *The Canadian Journal of Chemical Engineering*.[18]

29.7 CONCLUSIONS

The simple model developed in this work describes the transient flux decline observed during cross-flow UF/MF operations. The model provides semitheoretical expressions for the rate of flux decline,

$$-\frac{dJ}{dt} = k_1 C_0 J - k_2 Re^n$$

where the rate constants, k_1 (deposition rate constant), k_2 (removal rate constant), and n are physical parameters that are determined from cross-flow filtration experiments. The higher the value of k_1, the higher the tendency of fouling, whilst the higher the value of k_2, the higher the cleaning ability.

The model incorporates the combined effects of cross-flow velocity and temperature through a single parameter, which is the Reynolds number. Using experimental measurements of this work, it was possible to evaluate the model parameters (k_1 and k_2), which were found to be approximately independent of the operating conditions of pressure, temperature, cross-flow velocity and concentration. This simple model, which is supported by experimental measurements performed in this study and also those performed by others, can form the basis for further research and advanced analysis.

ACKNOWLEDGMENTS

Fonterra Whareroa, Blue Bird Foods Ltd., New Zealand, donated the materials used in this study. The technical support provided by Keith Towel, Dragan Ajvaz. We would also like to thank the University of Auckland for the research fund grant no. 9271/3417512.

REFERENCES

1. Clifton, M. J., Abidine, N., Aptel, P., and Sanchez, V. Growth of the polarisation layer in ultrafiltration with hollow fibre membranes. *J. Mem. Sci.*, 21, 233, 1984.

2. Chen, X. D., Ai, G. M., and Chen, J. J. J. Modelling hollow fibre ultrafiltration. *Proc. of Chemeca'99, The 27th Australasian Chem. Eng. Conference*, Newcastle, Australia, 1999.
3. Hillis, P., Padley, M. B., Powell, N. I., and Gallagher, P. M. Effects of backwash conditions on out-to-in membrane Microfiltration. *Desalination*, 118, 197, 1998.
4. Cao, Z., Wiley, D. E., and Fane, A. G. CFD simulations of net-type turbulence promoters in a narrow channel. *J. Mem. Sci.*, 185, 157, 2001.
5. Zeman, L. J. and Zydney, A. L. *Microfiltration and Ultrafiltration Principles and Applications*, 2nd edn. Dekker Inc, NY, 1996.
6. Blatt, W. F., Dravid, A., Michaels, A. S., and Nelson, L. M. Solute polarization and cake formation in membrane ultrafiltration: causes, consequences and control techniques. In *J. Mem. Sci. Tech*. Flinn, J. E., Ed. Plenum Press, New York, 1970.
7. Koziniski, A. A., and Lightfoot, E. N. Protein ultrafiltration: a general example of boundary layer filtration. *A.I.Ch.E. J.*, 18, 1030, 1972.
8. Trettin, D. R., and Doshi, M. R. Pressure independent ultrafiltration - Is it gel limited or osmotic pressure limited? *Synthetic Membranes*, 2. ACS Symposium Series, 154. ACS, Washington DC, 373, 1981.
9. Koutake, M., Matsuno, I., Nabetani, H., Nakajima, M., and Watanabe, A. Osmotic pressure model of membrane fouling applied to the ultrafiltration of whey. *J. Food Eng.*, 18, 313, 1993.
10. Fryer P. J. The uses of fouling models in the design of food process plant. *J. Dairy Tech.*, 42(1), 23, 1989.
11. Stamatakis, K., and Tien, C. A simple model of cross flow filtration based on particle adhesion. *AIChE J.*, 39, 1292, 1993.
12. Romero, C. A., and Davis, R. H. Transient model of cross-flow microfiltration. *Chem. Eng. Sci.*, 45, 13, 1990.
13. Bhattacharjee, C., and Datta, S. A mass transfer model for the prediction of rejection and flux during UF of PEG-6000. *J. Mem. Sci.*, 125, 303, 1997.
14. Song, L. Flux decline in cross-flow microfiltration and ultrafiltration: mechanisms and modelling of membrane fouling. *J. Mem. Sci.*, 139, 183, 1998.
15. Kern, D. Q., and Seaton, R. A. A theoretical analysis of thermal surface fouling, *Chem. Eng.*, 4, 258, 1959.
16. Charma, L. M., and Webb, R. L., Modelling liquid-side particulate fouling in enhanced tubes, *Int. J. Heat Mass Trans.*, 37(4), 571, 1994.
17. Frenander, U., and Jonsson, A. S. Cell harvesting by cross-flow microfiltration using a shear enhanced module. *Biotech. Bioeng.*, 52, 397, 1996.
18. Makardij, A. Farid, M., and Chen X. D. A Simple and effective model for cross flow microfiltration and ultrafiltration, *Can. J. Chem. Eng.*, 80, 28, 2002.

APPENDIX A: IMPORTANT PHYSICAL PROPERTIES

TABLE A1
Physical Properties for Feed and Permeate Materials used in this Study

Process	Density (kg.m⁻³)		Viscosity (Pa.s)	
	Feed	Permeate	Feed	Permeate
UF/skim milk*	1049	1020	1.631×10^{-3}	1.121×10^{-3}
MF/yeast suspension	1038	1000	1.323×10^{-3}	0.894×10^{-3}
UF/raw vinegar	1021	1009	1.191×10^{-3}	0.981×10^{-3}

Note: All specifications are given at standard conditions, 25°C, and 1 atm.
* Permeates for the UF G20 and G50 were assumed to have the same specifications. Yeast permeate specification was taken as pure water properties. All temperature effect on physical properties was taken into account during model calculations.

30 Crystallization in Spray Drying

Timothy A.G. Langrish
University of Sydney

CONTENTS

30.1 INTRODUCTION

It is widely reported that spray-dried powders are amorphous, particularly with respect to lactose and milk powders.[1,2] Chan and Chew[3] have also found the phase of spray-dried powders to be amorphous, as found by Vidgren et al.[4] for sodium cromoglycate and by Chawla et al.[5] for salbutamol sulfate, in both cases of pharmaceutical-scale or very small-scale spray dryers. The physical reason is that the atoms or molecules in the spray-dried substance are formed from a liquid into a solid form within seconds, sometimes fractions of a second, without allowing these atoms or molecules time to crystallize. The subsequent amorphous products are disordered, allowing a considerable amount of moisture to be trapped within such disordered and tangled atomic or molecular arrangements, giving them higher moisture contents than the corresponding crystalline products. However, such amorphous products are known to be unstable, or at best metastable,[6] and tend to transform into crystalline ones. In contrast to the amorphous products typically reported for spray-dried lactose and milk[1,2], it has also been reported[7,8] that sodium chloride solutions, when spray-dried, only give crystals, so the situation (regarding the degree of crystallinity of spray-dried powders) appears to depend on what materials are being processed. Figure 30.1, taken from the work of Wang,[9] shows crystals of spray-dried sodium chloride taken immediately from the dryer solids outlet. Partially-crystallized particles can be seen in a mixture of largely crystallized particles.

In order to define what is meant by crystallization in this context, it is useful to note one method used to detect the transition from amorphous material to crystalline ones. The method utilised by Lai and Schmidt[10] was based on using scanning electron micrographs (SEM) of skim milk powder.

FIGURE 30.1 Scanning electron micrograph of salt particles produced immediately after spray drying, showing the rapid crystallization process inside the spray dryer at a voltage of 15 keV. A Buchi B290 spray dryer was used at 38 m³/hr main air flow rate, 437 L/hr nozzle air flow rate, inlet air temperature 180°C, outlet air temperature 97°C, liquid feed rate 15 g/min, salt concentration 25%. The bar shows a length distance of 10 μm. (From Wang, S., *Crystallization of Amorphous Materials Produced by Spray Dryers*, B.E. (Chemical and Biomolecular) Thesis, The University of Sydney, Sydney, Australia, 41, 2006. With permission.)

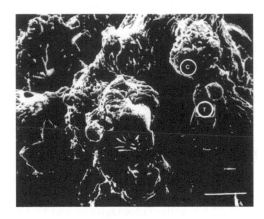

FIGURE 30.2 Scanning electron micrograph of spray-dried skim milk powder showing lactose crystals growing from within the amorphous particles. Magnification = 1650 times. The particles were kept at a water activity of 0.94 (94%) and 20°C for two days. C = cracks, T = tomahawk crystals, circled. The bar shows a length of 25 μm. (From Lai, H.M. and Schmidt, S.J., *J. Food Sci.*, 55, 994, 1990. With permission.)

They reported that lactose crystallization was evident in the milk particles in the form of feathery structures on the surfaces, or tomahawk-shaped crystals (Figure 30.2). The crystallization process could be seen as crystals growing within and out from the surface of amorphous powders, so there were both amorphous and crystalline regions in the same particles. Overall, many authors[11–13] have found that the degree of crystallinity of a mixture can be meaningfully defined as the fraction of all regions in all particles that are crystalline (or well ordered).

There is significant evidence that the local difference between the actual particle temperature and the glass-transition temperature affects the rate at which the glass-transition process takes place.[11–13] Furthermore, this difference in temperatures may then affect other processes, such as the deposition of particles on walls, that occur inside spray dryers.[14] Keeping the particle temperature low, using a freezer, is therefore likely to slow down the rate of glass transition. Powders produced by very small-scale spray dryers can be kept free flowing for a long time by freezing them, because the freezing reduces the particle temperature, reducing the rate of transformation from amorphous to crystalline product. In turn, this reduces the rate at which water is expelled from the solid as it is transformed from an intrinsically highly-moist amorphous solid (a lot of moisture can fit between the tangled solid atoms or molecules) to an intrinsically low-moisture crystal (very little moisture can fit between more dense and reorganised solid atoms or molecules). A low rate of moisture expulsion from the solid at low temperatures means that the surface of the solid is less likely to dissolve. This solids dissolution, which is more likely to occur with the high rate of water expulsion at higher temperatures with higher rates of phase transformation, means that liquid, then solid bridges, are likely to form between the solids, making them more difficult to flow. In other words, freezing may improve flowability by allowing the solid material in particles to become crystalline slowly without allowing solid bridges to form between particles. This explanation suggests why pilot-scale and larger dryers give "different" products to those from pharmaceutical-scale or very small-scale spray dryers. The larger dryers allow significant phase transformation to occur, while pharmaceutical-scale or very small-scale spray dryers produce only dried, but not otherwise transformed, products.

This explanation is also consistent with fluidized beds being placed at the outlets of most industrial spray dryers. The general explanation is that they are there to remove the residual moisture, at low moisture contents, which may take a "long time" to dry. However, even at these low moisture contents, a few seconds in a spray dryer should be enough. Another explanation, in addition to the removal of residual moisture, may be that their function is to allow the phase transformation to occur while breaking up or preventing the formation of liquid or solid bridges.

Amorphous and crystalline powder products possess different properties. Amorphous products are described by Chan and Chew[3] to be "*hygroscopic; more cohesive and difficult to flow and disperse*" compared with crystalline products. It has also been stated by Chidavaenzi et al.[15] that "*different physical forms of lactose can result in drug deposition changes from dry powder inhalers.*" In industry, wall deposition, where particles stick to the walls of spray dryers, is another problem faced due to amorphous products. Though more flowable than amorphous products and thus easier to handle, crystalline products have the disadvantages of low porosity and are also harder to dissolve. Price and Young[16] stated that the propensity of crystallization from amorphous products to crystallize limits the ability of amorphous products to provide high bioavailability for drugs (i.e., to improve solubility of the drug). Since amorphous products are metastable[3] and tend to become sticky, causing caking, a more crystalline product may reduce the amount of wall deposition. There are therefore both advantages and disadvantages in solids being in crystalline and amorphous forms. It is thus also important to investigate the parameters affecting the crystallization process from amorphous products to control the degree of wall deposition. Hence bioavailability, sorption characteristics, stickiness and flowability are all product properties that are fundamentally affected by the degree of crystallinity of the powders. The universal nature of the amorphous to crystalline transformation after spray drying for a wide range of materials, including tea, coffee, plant extracts, and milk powder, has been noted.[17]

30.2   EVIDENCE FOR PHASE TRANSFORMATIONS IN SPRAY DRYERS

Chan and Chew[3] referred to a patent by Chickering et al.,[18] in which the drying time was extended by inserting a secondary drying chamber between the main drying chamber and the gas-particle separation equipment to allow more time for crystallization to occur. The fluidized beds that have been put at the outlets of most industrial spray dryers in the dairy industry for many years prior to 2001 probably have the same effect, and they suggested carrying out the process of inducing crystallization in a closed container or in a fluidized bed dryer at a controlled temperature and humidity. Chan and Chew[3] indicated that the control over the temperature and moisture-content history of particles and droplets is therefore a key issue in spray drying.

Chiou et al.[19] have shown, for a Buchi B290 spray dryer processing of pure lactose, that varying the inlet air temperature from 210 to 134°C changed the size of the sorption peak heights in gravimetric soprtion from 8.3–8.49% (210°C) to 8.76–9.3% (134°C). Harjunen et al.[20] have found that the use of ethanol or water as a solvents in the feed for spray drying can change the degree of crystallinity of the spray-dried lactose product from 0 to 100% crystalline. This is possibly due to different solubilities of lactose in the solvent and to the different glass-transition temperatures for the solvent affecting the glass-transition temperatures in the drying droplets, where the glass-transition temperatures are predicted to affect the crystallization rate, according to the Williams-Landel-Ferry theory.[11] Changing the feed concentration (from 10 to 40% solids) has been shown to affect the degree of product crystallinity for lactose from 82 to 100% is reported by Chidavaenzi et al.[15] The results of Forbes et al.[21] and Chan et al.[22] show that the effects of feed composition (mannitol/lactose,[21] and mannitol/calcitonin[22]) affect the final extent of product crystallinity from nearly all amorphous to nearly all crystalline, through changing the glass transition temperature of the droplets and particles.

Methods to control the rate of crystallization have also been suggested,[3] including conditioning amorphous particles by exposure to a controlled environment at 35–85% relative humidity (RH), or to an organic vapor. This process has been used for the production of antiasthmatic drugs, such as salbutamol sulphate, ipratropium bromide, formaterol fumurate dehydrate, terbutaline sulphate and budesonide.[23] These conditioning techniques have also been used[24] to transform spray-dried lactose, which is used as a carrier for sodium cromoglycate and budesonide, to the crystalline α-monohydrate form. However, on its own, conditioning an albuterol sulphate-lactose formulation at a high humidity was found to be insufficient.[25] Ward and Dwivedi[25] found that it was necessary to remove fine (less than 3 μm diameter) albuterol sulphate particles to maintain formulation stability during storage. They attributed the instability of small particles to their "more energetic" nature. In this context, "more energetic" could refer to the greater ratio of attractive to repulsive forces as particles get smaller, where this ratio is inversely proportional to the diameter squared. The greater attractive forces compared with the repulsive ones may pull the smaller particles into closer contact with other particles, leading to more solid bridges forming with increasing time.

Hence, to produce the most flowable and useful material out of a spray dryer directly, it may be necessary to keep the particle temperature as high as possible for as long as possible (maximum time) within the dryer itself, while the particles are not in contact with each other or with the walls. This procedure is likely to maximize both the rate of transformation or change from amorphous to crystalline material and the amount of change (being proportional to time for a given rate of change). In effect, this process would condition particles within the dryer itself.

This discussion has suggested the possibility that the crystallization reaction, and possibly not the drying step in itself, is the rate-limiting step for many scales of spray-dryer operation, from pilot-scale upwards. What is known about the rates of crystallization within solid materials will now be reviewed, starting with research on lactose and milk powders.

30.3   MEASURING THE DEGREE OF CRYSTALLINITY

Lai and Schmidt[10] found that the crystallization process could be followed by nuclear magnetic resonance measurements, but it was necessary to enrich the skim milk powder with heavy water and deuterium salts, suggesting that this method has only limited practical value. X-ray diffraction (XRD) and differential scanning calorimetry (DSC) are also common methods used to determine the crystallinity of the material. Crystalline products diffract X-ray beams at distinct angles, which produce clear peaks on XRD plots, while amorphous materials produce indistinct plots of scattering intensity as a function of angle. In XRD, a beam of X-rays hits a material and diffracts at angles that are unique to different materials, and the angles are functions of the spacing between successive planes in the crystalline lattice. For lactose, examples of such angles are 20.9° and 16.4°[26,27] for the crystalline α and β forms of lactose.

DSC is another method used to determine the degree of crystallinity. In DSC, sample temperature is increased until the crystallization temperature is reached. When this occurs the amorphous materials transform into crystalline form and energy is released. When the sample is further heated, the melting point is reached and energy is absorbed as latent heat. A graph of energy of the system against temperature is plotted. The heat flow associated with crystallization can be used to assess the degree of crystallinity in solid products. The area under the graph within the positive region is proportional to the amount of amorphous materials present. Both XRD and DSC methods have been shown suitable to quantify the degree of crystallinity in spray-dried solids by Corrigan et al.[28]

For hygroscopic and sorption measurements,[29] the mass of the samples is observed to increase first, due to absorption of moisture by the amorphous material. A peak in mass is then observed before a decrease in mass occurs due to desorption of moisture, where the desorption of moisture is associated with crystallization because there is less space in the crystalline matrix for water to be held compared with the amorphous state.

30.4 CRYSTALLINE TO AMORPHOUS TRANSITIONS IN SPRAY-DRIED MATERIALS

Due to the important role of lactose in the food industry, its crystallization behavior has been more closely studied than other materials, especially regarding the effects of RH. Past work done by Jouppila and Roos,[29] who measured water adsorption curves for skim milk powder, concluded that the rate of crystallization is reduced by increasing the fat content in skim milk powder. They have also found that the crystallization time for lactose is about 40 hours at room temperatures and ambient RH (~25°C and 50% RH). Chidavaenzi et al.[15] have found, using techniques such as XRD and microcalorimetry, that polyethylene glycol (PEG) increases the degree of crystallinity of the product, since PEG forms hydrogen bonds with water, thus decreasing the rate of drying. The molecules thus, have a longer period to rearrange and form crystalline products. Some data on the crystallization times for different materials from literature are shown in Table 30.1. This time is measured when 95% crystallization has occurred. For hygroscopic measurements, this time is measured when 95% of the water desorption process has finished.

As shown in Table 30.1, Haque and Roos[27] have used XRD to measure the crystallization behavior of amorphous lactose from spray drying. Compared with lactose, the crystallization behavior of other materials and the effects of temperature have been investigated much less, although the crystallization behavior of sucrose[30] and tea, coffee, and maltodextrin[17] have also been reported. From these data, it is observed that the crystallization times taken for the same material at similar conditions using XRD and hygroscopic methods are similar. It is also observed that lactose requires an RH of 44% and above for crystallization to occur.

While the crystallization rate of the lactose is increased by using higher RH, the experiments performed on sucrose[30] at 370 K, 390 K and 410 K appear to show an unusual trend of crystallization rate with temperature. These results have been measured using a differential scanning calorimetry (DSC) with freeze dried products and are shown in Figure 30.3.

30.4.1 LACTOSE (ADSORPTION AND DESORPTION)

Lactose in spray-dried milk has been reported to be in a nonequilibrium condition.[34] Drying the milk slowly at 50°C gives nearly all of the lactose as the α-monohydrate form, although King[35] suggests that the β-anhydrate is actually formed first. The physical structure of dried milk has been reviewed by King,[35] who points out that spray-dried milk, which is produced rapidly, is in the amorphous state, with a ratio of α to β-lactose of about 1:1.5. This amorphous state is very hygroscopic, crystallizing into the α-monohydrate form with the uptake of moisture, and this α form has very low solubility. The lack of crystallinity in spray-dried powder has been confirmed by X-ray crystallography and polarized light microscopy.

TABLE 30.1
Data on Crystallization Times for Different Materials, from the Literature

Material	Temperature	Relative Humidity	Times for 95% Crystallization	Method	Source
Lactose	Room temperature	54.5%	50 hours	XRD	Haque and Roos[27]
Lactose	Room temperature	65.5%	35 hours	XRD	
Lactose	Room temperature	76.1%	15 hours	XRD	
Lactose	Room temperature	Ambient	50 hours	Hygroscopic	Bushill et al.[26]
Skim milk powder	Room temperature	Ambient	60 hours	Hygroscopic	
Skim milk powder	Room temperature	76.1%	15 hours	Hygroscopic	
Lactose	Room temperature	55%	> 30 hours	Hygroscopic	
Lactose	Room temperature	44%	None observed	Hygroscopic	
Lactose	Room temperature	33%	None observed	Hygroscopic	
Sucrose	370 K	<10%	60 seconds	DSC	Kedward et al.[30]
Sucrose	390K	<10%	70 seconds	DSC	
Sucrose	410 K	<10%	90 seconds	DSC	
Lactose	Room temperature	Ambient	40 hours	XRD	Jouppila and Roos[31]
Lactose	60–110°C	20–80%	1 minute (100°C and 80%)	Hygroscopic	Ibach and Kind[32]
Whey, whey-permeate	60–110°C	20–80%	5 minutes (100°C and 80%)	Hygroscopic	
Tea	Room temperature	Ambient	>350 hours	Hygroscopic	Chiou and
Coffee	Room temperature	Ambient	115 hours	Hygroscopic	Langrish[17]
Maltodextrin, DE 18	Room temperature	Ambient	72 hours	Hygroscopic	
Lactose	Room temperature	75%	16 hours	Hygroscopic	Wang and
Lactose	40°C	75%	1.6 hours	Hygroscopic	Langrish[33]
Sucrose	40°C	32%	5.1 hours	Hygroscopic	
Sucrose	40°C	51%	1.9 hours	Hygroscopic	
Sucrose	40°C	75%	2.2 hours	Hygroscopic	

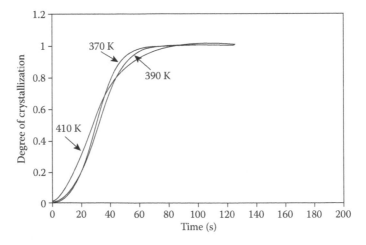

FIGURE 30.3 The crystallization of sucrose from the amorphous solid state at different temperatures. (From Kedward, C.J., MacNaughtan, W., and Mitchell, J.R., *Carbohydrate Res.*, 329, 423, 2000. With permission.)

Bushill et al.[26] have described the process of lactose crystallization in milk powder as being one in which *"spray-dried milk powder when exposed to a humid atmosphere rapidly absorbs moisture up to about 12%. This is immediately followed by the crystallisation of the amorphous lactose present in the milk powder and also the loss, while still in the humid atmosphere, of some of the previously absorbed moisture."* They showed that the lactose crystallized in the α-monohydrate form, and they measured the amount of moisture absorption on a sensitive balance as a function of time.

An initial rapid absorption of moisture was followed by desorption of moisture, all at a virtually constant RH. The initial rapid absorption behavior was attributed by Bushill et al.[26] to the uptake of moisture by the amorphous material. The subsequent desorption behavior was attributed to the loss of moisture as the spray-dried material transformed from a hygroscopic amorphous form to a less hygroscopic crystalline form. The greater degree of order in a crystalline form compared with an amorphous form means that there is less room in the crystalline structure than in a more random amorphous one for moisture, making the crystalline structure less hygroscopic. This sorption and desorption test is a reflection of the relative amount of amorphous material in a powder compared with the total amount of material, both amorphous and crystalline. A purely crystalline powder would show no peak moisture absorption, since this peak absorption is associated with material in an amorphous form in a powder.

Bushill et al.[26] also presented optical properties, namely the refractive indices and the specific rotatory power for polarized light, the XRD patterns and the infra-red spectra for the different crystalline forms of lactose. These forms include the α-monohydrate, the β-anhydride, the unstable α-anhydride, the stable α-anhydride and the amorphous mixture of α and β forms.

Prior to this work, Supplee[36] had shown that the desorption equilibrium curve for skim milk powder was continuous, while the equilibrium curve for adsorption was broken at the point of 50% RH. The equilibrium moisture content from the desorption isotherm decreased as the fat content of the milk increased, with the moisture content at equilibrium approximately halving when the fat content was 55%. The initial rapid adsorption, over the first 33 hours of being exposed to a RH of 50% or more, was followed by much slower desorption, over more than 56 days.

The discontinuity in the adsorption isotherm at a RH between 0.35 and 0.6 has been confirmed by Berlin et al.,[37] and Troy and Sharp.[34] A subsequent investigation by Berlin et al.[38] found that a second adsorption process, after an initial adsorption-desorption process, resulted in no discontinuity in the adsorption isotherm, since the lactose is then already in a crystalline state. It was also suggested[38] that casein is a significant moisture adsorption site in milk powder at low RH, with lactose becoming dominant as the RH approaches 0.5. Above RH of 0.5, when the lactose has converted from a glassy to a crystalline form and moisture is released, it was suggested that the proteins again become the dominant moisture sorption sites.

The reason why only adsorption or desorption in one direction is found for RH below 40%, may be that the moisture content is too low at these RH and the glass-transition temperatures are too high for crystallization to occur at a measureable rate. Lai and Schmidt[10] found no crystallization at ambient temperatures and RH below 43% over a two-week period. Above RH of 85%, unidirectional adsorption is also found, and Lai and Schmidt[10] suggested that this is due to the adsorption of the water released by the crystallizing lactose by the proteins.

Jouppilla and Roos[29] summarize a considerable amount of evidence that the rate of crystallization is dependent on the difference between the particle temperature and its glass-transition temperature. The glass-transition temperature is lowered by increasing the moisture content of the particle, which can be achieved by increasing the RH around the particle. These workers also found that the presence of fat decreased the rate of crystallization, but not the final amount of sorbed water after complete crystallization. As with Bushill et al.,[26] Supplee[36], and Lai and Schmidt,[10] they found that lactose crystallization occurred within 24 hours at RH greater than 40% for pure lactose or greater than 50% for lactose in milk powders. King[35] also noted the evidence that spray-dried skim milk powder at a moisture content of 3–5% could be stored for over 600 days at 37°C without showing significant crystallization. At a moisture content of 7.6%, crystallization can be seen after one day at 37°C, after ten days at 28.5°C, and after 100 days at 20°C.

Jouppilla and Roos[29] also found that the Guggenheim–Anderson–de Boer (GAB) equation best fitted the moisture sorption behavior of milk powders, in preference to the Brunauer–Emmett–Teller (BET) and Kühn equations. They found that the effect of fat on the equilibrium isotherms could be accounted for by assuming that the fat played no part in moisture sorption and carrying out all calculations on a solids-no-fat (SNF) basis.

30.4.2 MATHEMATICAL MODELING OF THE PROCESS OF SOLID-STATE CRYSTALLIZATION

Several equations have been proposed to represent the crystallization process within solids, and these will now be discussed. The implicit assumption in these approaches appears to have been that nucleation is a rapid and easily attained process, so that the equations have been applied to the crystal growth process, in effect.

30.4.2.1 Initial Empirical Approach

A considerable amount of work has been done on the properties of spray-dried lactose and milk powder. Troy and Sharp[34] studied the transformation between the β form of lactose, which crystallises from a concentrated solution of lactose at the boiling point, and the α-monohydrate form. The rate of change for the amount of α form that has transformed to the β form, x, is given by a rate law of the following form:

$$\frac{dx}{dt} = -Kx \tag{30.1}$$

They reported that the rate constant, K, was approximately constant in the range from 0.46 to 0.48 hr^{-1} over a range of pH from 2 to 9, with the rate of change approaching infinity below a pH of 0 and above a pH of 9. The presence of sucrose was reported to have little effect on the rate of change.

30.4.2.2 Growth of Crystals from Solution

Given that water evaporates from solutions in droplets during spray drying, the concentration of the solutions increases, making the formation of crystals likely if the solution concentration becomes greater than the solubility limit. Such a process of crystallization, in the conventional sense of crystals forming from a solution, has been described by conventional crystallization during the drying of thin films of sucrose and lactose by Shastry and Hartel,[39] Ben-Joseph et al.,[40] and Ben-Joseph and Hartel.[41] The explanation is interesting and may be useful to describe crystallization during the period when the droplet is evaporating, but it does not account for the transformation from solid amorphous to solid crystalline products that may subsequently occur.

30.4.2.3 Williams, Landel, and Ferry Equation

Actual measurements of, and correlations for, the rate of crystallization of amorphous lactose were reported by Williams, Landel, and Ferry[11] and Roos and Karel.[42] They found that the ratio (r) of the time for crystallization (θ_{cr}) at any temperature (T) to the time for crystallization (θ_g) at the glass-transition temperature (T_g) could be correlated by the following equation (the WLF equation):

$$\log_{10} r = \log_{10}\left(\frac{\theta_{cr}}{\theta_g}\right) = \frac{-17.44(T - T_g)}{51.6 + (T - T_g)} \tag{30.2}$$

The use of universal constants in Equation 30.2 has been queried by Peleg,[43] on the grounds that these constants are likely to be product specific. Roos and Karel[44] and Senoussi et al.[45] have also noted the finding that the rate of crystallization is dependent on the difference between the particle and glass-transition temperatures. Both Peleg and Hollenbach[46] and Joupilla and Roos[29,47] have pointed out the decrease in the glass-transition temperature from the absorption of water, increasing the temperature difference and the rate of crystallization. They suggested that the release of water during this crystallization process creates liquid bridges between particles, causing agglomeration and caking.

The integrated kinetics in the WLF equation have also been reported and used by Palzer[13] from Nestlé's research organisation in 2005, suggesting that they are still relevant and in use and have not been superseded by more recent work. The WLF equation predicts a decrease in the time required for crystallization by two to three orders of magnitude for every 10°C increase in the temperature above the glass-transition temperature (Figure 30.4). It is possible that the equation may be applied below the glass-transition temperature to predict a finite time for crystallization even under these conditions. However, Williams, Landel and Ferry[11] suggested that the mobility of water molecules below this transition temperature (in the glassy state) is different to that above the transition temperature (in the rubbery state) "*below T_g, log a_T (e.g. the ratio of θ_{cr} to θ_g) actually increases less rapidly with decreasing temperature* (than predicted by Equation 30.2)." Hence the rate of crystallization will be different in magnitude to that predicted by Equation 30.2, even though the rate may still be less than at the glass-transition temperature or above it.

Although the WLF equation does not involve the effects of moisture content and molecular weight explicitly, it includes them implicitly through their effects on the glass-transition temperature. Higher moisture contents lower the glass-transition temperatures, and a high molecular weight in a material is also usually associated with a high glass-transition temperature.[48] Lactose and sucrose both have the same molecular weight of 342.34 g/mol but they have different glass-transition temperatures (62–72°C for sucrose, 101°C for lactose) since their structures are different. Using the WLF equation, this implies that the difference in structures between lactose and sucrose might cause them to have different crystallization rates.

30.4.2.4 Avrami Equation

There is some theoretical basis for the Avrami model as applied to the crystallization of amorphous components in solids.[49,50] Nevertheless, there is often considerable difficulty in applying this theory in practice. The Avrami equation has been used by Haque and Roos[27] and Corrigan et al.[28] to fit solid state crystallization data. The equation assumes "*negligibly small 'droplets' of the stable*

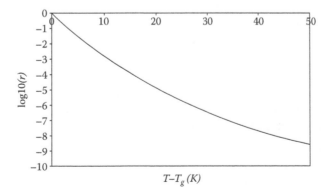

FIGURE 30.4   The logarithm of the ratio of the time for crystallization (θ_{cr}) at any temperature (T) to the time for crystallization (θ_g) at the glass-transition temperature (T_g) as a function of the difference between the temperature and the glass-transition temperature.

phase nucleate from the metastable background; the droplets subsequently grow independently without substantial deformation and the stable phase is pictured as randomly placed, freely overlapping, growing spheres."[42] The Avrami equation is shown in Equation 30.3:

$$\theta = 1 - e^{-kt^n} \tag{30.3}$$

Here θ is the degree of crystallinity, k is the rate constant, t is time and n is the Avrami exponent. The degree of crystallinity can be calculated from XRD data using the equation below. This equation estimates the degree of crystallinity from the extent of completion $(I_f - I_t)$ compared with the total change $(I_f - I_o)$. This approach can also be used for sorption data with the moisture content used instead of the peak intensity in XRD.

$$\theta = 1 - \frac{I_f - I_t}{I_f - I_o} \tag{30.4}$$

Here I_f is the peak intensity at the maximum crystalline product content, I_t is the peak intensity at time t and I_0 is the peak intensity at time zero. The Avrami equation is often expressed in the form given in Equation 30.5 when applied to XRD data:

$$\ln[-\ln(1-\theta)] = \ln k + n \ln t \tag{30.5}$$

From Equation 30.5, it is expected that a plot of $\ln[-\ln(1-\theta)]$ against $\ln t$ should give a straight line with $\ln k$ as the y-intercept and a gradient of n. Values of k and n vary for different materials and conditions. Haque and Roos[27] have reported a series of these values for lactose in different RH at different XRD angles. Some of these values corresponding to 20.9° are shown in Table 30.2.

There has been some debate about the applicability of the Avrami equation. It is used as a correlation for fitting data from isothermal crystallization experiments, and the equation does not explicitly account for the effect of temperature. An attempt by Kedward et al.[30] to include temperature into the equation, assuming a linear relationship between rate of crystallization and temperature, was successful for sucrose but failed for lactose. The failure was attributed to the multiple crystal forms for lactose compared with the single crystal form for sucrose. Only one form of crystal was considered when analyzing the XRD results. To get a true representation, both forms of crystals need to be accounted for. Roos and Karel[51] have also tested the accuracy of the equation by fitting data produced by measuring the crystallization rate of amorphous lactose using XRD for a diffraction angle of 20.9°, which is one of the characteristic angles for both α and β forms. They concluded that the results did not fit well with the Avrami equation. However, in a separate experiment performed by Corrigan et al.[28] measuring the degree of crystallinity for

TABLE 30.2
Values of k and n for the Avrami Equation using XRD Measurements at 20.9° of Lactose at Different Conditions

Relative Humidity (%)	k (h^{-1})	n
54.5	3.2×10^{-5}	3.2
65.6	0.41	0.49
76.1	0.48	0.66

Source: Haque, M.K. and Roos, Y.H., *Carbohydrate Res.*, 340, 293, 2005. With permission.

lactose using XRD at an angle of 16.4°, which represents α lactose monohydrate, it was concluded that the data fitted well into the equation *"with K = 0.239 per day (±0.0127), R^2 = 0.96."* Recently, Ibach and Kind[52] have used this equation to correlate the crystallization kinetics of amorphous lactose, whey-permeate, and whey powders. They found a crystallization time of 1 minute for lactose at 100°C and a RH of 80%, with whey and whey-permeate taking up to five minutes under these conditions, indicating that salts and protein slow down the crystallization process. The mean order, n, for lactose was 1.9.

On closer examination of Equation 30.5, it may also be realised that a regression has been carried out with $\ln(-\ln(1-\theta))$ against $\ln t$. Since taking the logarithm of any number reduces the amount of variability, it is possible that plotting $\ln(-\ln(1-\theta))$ against $\ln t$ will give a straight line, even if no real correlation is present. Hence this may give fitted constants of k and n even if these constants have little or no physical meaning. The parameters in the Avrami equation appear to be much less universal than those in the WLF equation, so the approach using the Avrami equation currently appears to have less generality than that using the WLF equation.

30.4.2.5 Molecular Dynamics

In a different approach to the above empirical methods, the dynamics and kinetics of the crystallisation process can also be modelled computationally. The molecular interactions between individual molecules with moisture, and rearrangements of these molecules with each other that are affected by moisture, may be modeled under different conditions over various periods of time. One of the disadvantages of this method is that it involves complex calculations, which tend to take long computational times. As an example of this, to simulate the dynamics of 1000 atoms over a time scale of 10^{-8} seconds required approximately 10 hour of computation time in 1995.[53] Using this rate of processing, to model the molecular dynamics of 1 g of sucrose for 1 seconds would require 2.26×10^{21} years of processing time. Under Moore's Law of computer processing progress, for every two years, the processing speed of computers doubles. Thus the present computer processing rate would have increased by 32 times since 1995, reducing the time requirement to 7.06×10^{19} years. In the recent molecular dynamics simulations carried out by Panagopoulos et al., the modelling of 30,000 atoms over 3 ns was achieved in a reasonable time.[54] This number of particles is only a modest increase over the 1000 in 1995. Future improvements in computing technology would potentially decrease the processing time and increase the feasibility of this approach. However, at the present this method, though accurate, is time consuming and has high computational requirements.

While existing correlations and formulae, such as the WLF equation and the Avrami equation, are empirical they are simple to use. Computational modeling of the molecular dynamics, on the other hand, requires large computational resources, and long periods of processing time. Unlike the previous methods mentioned, viewing the problem as an activated rate process may allow the principles of solid phase crystallization to be understood and quantified clearly while keeping the experimental and mathematical modeling requirements relatively simple.

30.4.2.6 Reaction Engineering

Another approach, which may be applicable to the quantitative modeling of the crystallization process for amorphous solids, is the reaction engineering approach pioneered by Chen and Xie[55] for the mathematical modeling of the drying kinetics for particles.

This reaction engineering approach, which includes a similar parameter to the activation energy for any activated rate process, is a tempting one for describing the crystallization process in solids in a quantitative manner. However, Williams, Landel and Ferry[11] comment that, when the data for different materials at various temperatures are assessed according to this view of an activated rate process, the apparent activation energy is not a universal or unique function of temperature. They show evidence that Equation 30.2, although it is not expressed in a rate form, gives a more universal correlation of the rate data.

From a qualitative perspective, viewing the crystallization process in solids as an activated rate process allows the concept of a reverse reaction from the crystalline to the amorphous form once there is sufficient activation energy. This reverse reaction may occur in the impact of crystalline materials on walls, where the surfaces of the particles may become heated, transforming the surface material back into an amorphous form that is sticky. Indeed, stickiness in α lactose monohydrate particles on the walls of pneumatic conveying lines has been noted,[56] and such a reverse reaction, forming some amorphous material at the surface, may possibly explain these findings. The reverse reaction, from crystalline to amorphous, has also been noted in the work of Willart et al.[57] on ball milling for trehalose. The reversibility of the reaction is also implied in Figure 30.5, taken from Bhandari and Howes.[58] The interpretation of Figure 30.5 needs some care, and this figure probably needs to be extended, because not only is the process of crystallization from the rubbery state likely to be affected ($f(T,t)$) by temperature (T) and time (t), but the process is also likely to be affected by the material moisture content (X), which in turn is affected by the local temperature and RH. However, the basic ideas in Figure 30.5, particularly the effect of the time scales for transformations forwards and backwards from the crystalline and amorphous states, remain very important.

The time scale for the crystallization process appears to be somewhat longer than that for the drying process. In Figure 30.6, taken directly from Jouppila and Roos,[29] time scales for the crystallization process of the order of 100 hours are shown, while Figure 30.7, taken directly from Roos and Karel,[42] time scales for the crystallization process of the order of 10^3 seconds are shown. By contrast, Figure 30.8, taken directly from Lin and Chen,[59] shows time scales for the drying process of the order of 300 seconds, albeit with larger droplets than are typical for the spray-drying process. However, Ibach and Kind[32] have found crystallization times that are as low as one minute for pure lactose at 100°C and 80% RH, but they also found that the proteins and salts in whey and whey-permeate slowed down the crystallization process under the same conditions to five minutes.

One of the most noteworthy features of Figure 30.9, for the moisture adsorption and desorption behavior of spray-dried citrus fibre and hibiscus extract,[60] is its similarity to Figure 30.6, which shows the adsorption and desorption behavior of milk powders containing lactose. In Figure 30.9, the fibre-only sample shows almost no initial peak, which is consistent with this solid fibre sample staying almost purely crystalline throughout the spray drying process. In contrast, the samples of spray-dried mixtures of both fibre and extract show peaks in the plots of moisture content as a function of time, which is similar to the behavior of the adsorption behavior of milk powders containing lactose. The extract is a liquid, which is likely to form an amorphous solid on drying, so the spray-dried mixtures of fibre and extract are likely to contain both crystalline fibre and amorphous

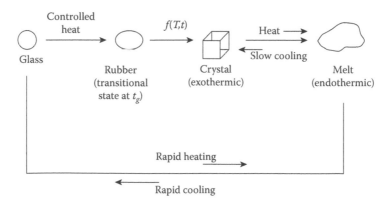

FIGURE 30.5 Change of physical state of an amorphous glass through rubbery (transitional) to crystalline state. t, temperature; t, time. (From Bhandari, B.R. and Howes, T., *J. Food Eng.*, 40, 71, 1999. With permission.)

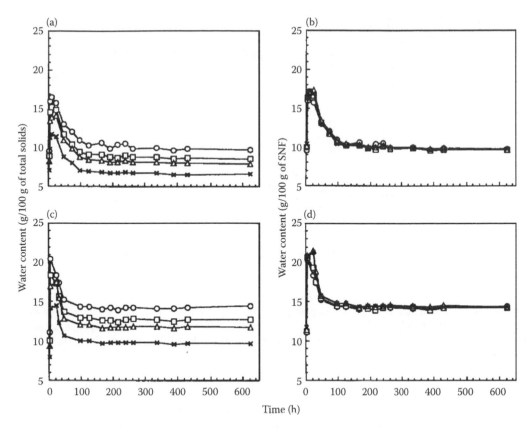

FIGURE 30.6 Water sorption of milk powders containing lactose at relative humidity of 66.2: total solids (A) and SNF basis (B) and 76.4% total solids (C) and SNF basis (D) at 24°C. Experimental water contents are shown for dehydrated skim milk (o), low fat milk with 1% (□) and 1.9% fat (Δ), and whole milk (x). The amount of fat affected the rate of loss of adsorbed water caused by lactose crystallization. (From Jouppila, K. and Roos, Y.H., *J. Dairy Sci.*, 77, 1798, 1994. With permission.)

extract. Hence the amorphous extract may crystallise with increasing time, giving the peaks seen in the plots of moisture content as a function of time in Figure 30.9.

This similarity suggests that the moisture adsorption behavior of spray-dried citrus fibre and hibiscus extract is also showing the crystallization of amorphous material in the spray-dried product. It is therefore suggested that the universal nature of the crystallization process from spray-dried amorphous products may be a serious possibility, and that this process may occur, to some extent, in spray dryers themselves.

Returning to the quantitative aspects of the crystallization reaction within solids and the Williams–Landel–Ferry approach, as stated in Equation 30.2, a limitation of this approach is that the reaction kinetics for the crystallization reaction are expressed in an integrated form. In this integrated form, essentially the time for the crystallization process is expressed as a function of temperature. The time for the process is not expressed as a function of concentration of any reactant, suggesting that the reaction is zero order with respect to concentrations. Expressing the kinetics in a rate form would be very useful for simulation purposes.

The view of this crystallization process within solids as an activated rate process can be illustrated in Figure 30.10. The amorphous state, being more disordered, has a higher energy level than the crystalline one.

Despite the discussion in Williams, Landel, and Ferry,[11] it might still be possible to use the activated rate process view in a quantitative way to express the kinetics in a rate form rather than an

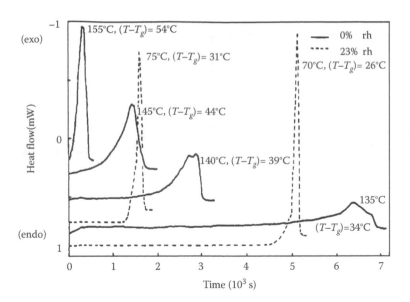

FIGURE 30.7 Isothermal DSC curves for amorphous lactose at various temperatures after equilibration at 0% and 23% relative humidity. The thermograms show exothermal crystallization of amorphous lactose having a peak at θ_{cr} that is specific to each $T - T_g$. (From Roos, Y.H. and Karel, M., *Biotech. Progr.*, 6, 159, 1990. With permission.)

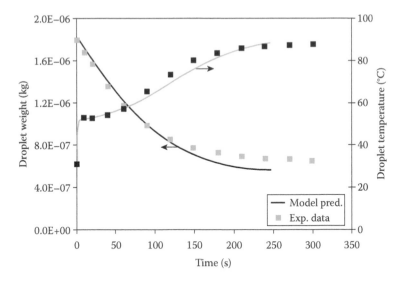

FIGURE 30.8 Comparison of the experimental data and predictions given by the REA (reaction engineering approach) model for 30 wt% whole milk drying at $T_b = 90.3°C$, $v_b = 0.55$ m.s^{-1}, and $H = 0.073$ kg.kg^{-1} (droplet initial weight 1.793×10^{-6} kg). (From Lin, S.X.Q. and Chen, X.D., *Dry. Tech.*, 23, 1395, 2005. With permission.)

integrated form. Chen and Xie[55] create a drying "fingerprint" for each material, which is a unique function that expresses the relationship between the activation energy for drying and the moisture content, for any given material. An example of this type of fingerprint function for drying is shown in Figure 30.11 for silica gel. A similar approach might be taken to create a unique function that expresses the relationship between the activation energy for the crystallization reaction and the temperature, as a type of "fingerprint" for the crystallization kinetics for a material.

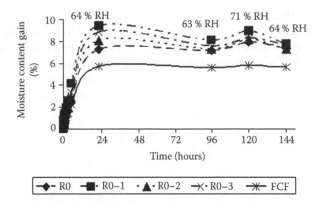

FIGURE 30.9 Moisture gain as a function of time for spray-dried citrus fibre and hibiscus extract with equal weights of the fibre and the solids content of extract, illustrating the adsorption and desorption behavior (From Chiou, D. and Langrish, T.A.G., *J. Food Eng.*, 82, 84, 2007. With permission.)

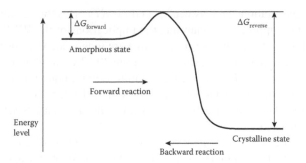

FIGURE 30.10 Schematic diagram showing the transformation between amorphous and crystalline states in a solid, viewed as an activated rate process.

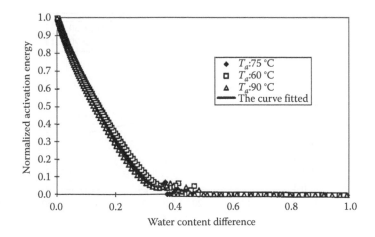

FIGURE 30.11 Fingerprint function for the drying of silica gel: normalized relationship between the apparent activation energy and the water content during drying of silica gel particle layer. (From Chen, X.D. and Xie, G.Z., *Trans. IChemE, FBP Eng.*, 75, 213, 1997. With permission.)

A simple analysis of the WLF equation is enough to demonstrate their comment that, when the data for different materials at various temperatures are assessed as an activated rate process, the apparent activation energy is not a universal or unique function of temperature. In effect, they stated that the activation energy was not a unique function of temperature.

If it may be assumed that the time for crystallization (θ_{cr}) is inversely related to the rate constant (k_{cr}), and that the data used to derive the WLF equation were from isothermal experiments, then the equation may be written in the following form:

$$\log_{10}\left(\frac{k_g}{k_{cr}}\right) = \frac{-17.44(T - T_g)}{51.6 + (T - T_g)} \tag{30.6}$$

A simple analysis might use the following rate function (in an Arrhenius form) for the reaction process:

$$k = q\exp\left(-\frac{E_A}{RT}\right) \tag{30.7}$$

Then, the slope of the relationship between $\ln k$ and $-1/(RT)$ gives the activation energy as a function of temperature. This approach to analysing the implications of the WLF equation can readily be shown to give apparent activation energies, E_A, that are independent of the rate at the glass-transition temperature, k_{cr}, in Equation 30.6. However, Figure 30.12 shows that the relationship between the apparent activation energies, E_A, and the temperature is not unique if two materials have different glass-transition temperatures. This outcome shows that, if two materials have different glass-transition temperatures, the apparent activation energy is not a unique function of temperature if analysed in such a simple way.

However, activated rate processes cannot be analyzed so simplistically, as is well known in solvation and corrosion theory. For example the Butler–Volmer equation[61] for activation polarization in corrosion shows that the rate is affected significantly by the position of the activation energy barrier. The position of this energy barrier is shown by the symbols α_1 and α_2 in Figure 30.13, along the reaction co-ordinate between the amorphous and the crystalline energy states. Translating the Butler–Volmer equation into the form used here gives the following rate equation:

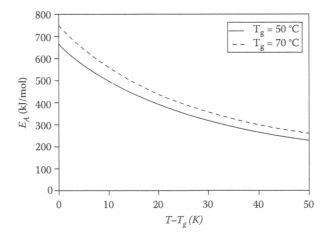

FIGURE 30.12 Illustration of the activation function, as derived from the Williams–Landel–Ferry equation, for two materials with different glass-transition temperatures, illustrating the lack of uniqueness for the activation energy as a function of temperature from the simple analysis.

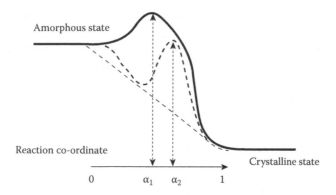

FIGURE 30.13 Schematic diagram showing the effect of the position of the activation energy barrier along the reaction coordinate in activated rate processes

$$k = q \left[\exp\left(\frac{-\alpha E_A}{RT} \right) - \exp\left(\frac{\{1-\alpha\} E_A}{RT} \right) \right] \tag{30.8}$$

There is also the suggestion from Roos and Karel[51] that the kinetics may be more complex than zero order with respect to concentration, in that they reported that *"The increase of crystallinity at constant relative humidity was found to proceed at a fairly constant rate until about 20% of the material had crystallized. This was followed by an exponential increase of crystallinity until completion."* If the crystallization reaction were zero order with respect to concentration, then the rate would not vary with the degree of crystallinity. This observation[51] suggests some concentration effect on the reaction kinetics.

The amorphous and crystalline states appear to be universal packing configurations for chain molecules, in terms of energy levels and structures, for many polymeric materials.[62] The energy levels for crystalline (lowest) and amorphous (second lowest) energy states can be predicted by quantum mechanics when considering the possible energy levels for a particle in a box,[62] so the energy levels are discrete and specific ones.

30.5 INTERACTION BETWEEN PHASE TRANSFORMATIONS AND OTHER PHYSICAL PROCESSES OCCURING WITHIN SPRAY DRYERS

A connection between all of these concepts is that the drying kinetics of the particles and the solids-gas equilibria affect the moisture content and temperature of the solids, which then affect the glass-transition temperature of the particles, since water acts as a depressant for the glass-transition temperature. The solids temperature is the same as the particle temperature, so the difference between the particle temperature and the glass-transition temperature is affected by the drying kinetics of the particles and the solids–gas equilibria. The flow patterns of the gas and the particles are still key aspects of the process, because deposition will still occur with dry crystalline particles, as shown with the well-known data sets of Papavergos and Hedley[63] and McCoy and Hanratty.[64] This significance of the gas and particle flow patterns means that studies using computational fluid dynamics and different types of experimental flow simulations are still very important, but they need to be seen in the context of the overall situation involving a number of complex processes inside spray dryers.

A possible reason why this crystallization phenomenon has not been more widely reviewed as a critical part of the spray-drying process may be that the crystallization process within solids is difficult to follow easily. The process of crystallization of solids from a saturated or supersaturated liquid solution may be followed optically, but the internal solids-phase change involved with the crystallization process within solids is more difficult to follow. The difficulties in analyzing the degree of crystallinity may partially explain the scarcity of research into the crystallization reaction within solids. XRD studies may be applied, but these are not typically standard and readily-available laboratory techniques. DSC and thermo-gravimetric analysis may also be used, but these techniques depend on being able to separate crystallization events clearly from other thermal events within solids. The techniques themselves, involving increasing the temperature of the solids, may affect the degree of crystallinity of the solids, which is the key parameter being measured. Some workers[10] have noted the appearance of crystals on the surface of the solids particles during this crystallization process within solids but these changes in appearance occur at such a fine level that they need to be followed with an SEM. It is possible that manufacturers of spray dryers may already realise implicitly that this crystallization process within solids may be a rate-limiting step for the equipment design.

Not only might this crystallization process explain many aspects of the sizing and operation of spray dryers but it may also explain, at least partially, the apparent stickiness of crystals during solids conveying and in cyclones used to separate particles from gases. With crystals, the kinetic energy of their impact on walls and corners may be sufficient to raise their energy up to a level where part of the outside of the crystal transforms back into an amorphous solid. This process would create a very thin layer of sticky amorphous material, almost undetectable and possibly only a molecular layer thick, on the outside of the part of the crystals. It is also possible that the crystallization process, or its reverse, may complicate the interpretation of results from the recently-proposed stickiness test using a cyclone.[65] Given that energy dissipation within solids may reverse the crystallization reaction, vibration, or agitation of solids may also lead to some amorphous material being formed on the outside of particles within agglomeration processes, such as fluidized beds and rotating drums.

Parameters that are likely to affect the rate of crystallization include the particle temperature, which will be influenced by the surrounding gas temperature, the moisture content, which affects the glass-transition temperature, and time. The composition of the particle is also likely to be important. Hence, the influence of the operating parameters for spray drying on the rate of crystallization is likely to be complex. Increasing the inlet gas temperature is likely to increase the particle temperature and increase the rate of crystallization. At the same time, the moisture content will decrease as the inlet gas temperature increases, increasing the glass-transition temperature and decreasing the rate of crystallization. The design of the dryer will influence the residence time and hence the extent of the crystallization reaction, so the amount of crystalline or amorphous material may depend strongly on the dryer design. As with crystallization of solids from liquids in crystallizers, the use of seed crystals in the feed liquid to the dryers may also be beneficial.

Increasing the molecular weight increases the glass-transition temperature of food polymers[66,67] according to the Fox and Flory[68] relationship. In this relationship, the glass-transition temperature, T_g, is equal to the glass-transition temperature at a very high molecular weight, $T_{g\infty}$, less a factor K/M, where K is a positive constant (around 25,000 K) and M is the average molecular weight. This relationship was found by To and Flink[69] to apply to glucose polymers with more than three glucose chains. The effect of the molecular weight on the glass-transition temperature is another way, through altering the composition to include low or high molecular weight components, in which the rate of crystallization may be controlled. The possibilities for changing the sorption behavior by adjusting the inlet air temperature (an operating condition) during spray drying have been demonstrated by Chiou et al.,[70] as shown in Figure 30.14. The lower amount of sorption (lower sorption peak height) for the powders produced at 210°C compared with those produced at 134°C, with all the other drying and sorption conditions being the same, indicates greater crystallinity in the 210°C samples, since crystalline material shows much less sorption behavior than amorphous material.

30.6 CONCLUSIONS

It has always been somewhat of a mystery why spray dryers appear to be much larger than they need to be. In other words, why the particle residence time in spray dryers is much larger than the time

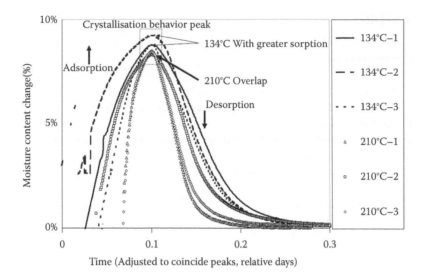

FIGURE 30.14 Sorption data for lactose powders produced with inlet air temperatures of 134°C and 210°C from a Buchi-290 spray dryer at 38 m³/hr main air flow rate, 1190 L/hr nozzle air flow rate, liquid feed rate 6.4 g/min, lactose solution concentration 15%. (From Chiou, D., Langrish, T.A.G., and Braham, R., *J. Food Eng.*, 86, 288, 2008. With permission.)

required to dry the particles has been an unsolved puzzle, suggesting that the size of spray dryers could be reduced. Some might say that previous trial and error approaches to design have been ineffective, but this explanation is unlikely to be enough to justify the difference, frequently an order of magnitude or larger, between the required drying time of a droplet and the actual residence times.

The suggestion here is that spray dryers do produce dry powders, but if that was all that they do, then the powders would be almost useless. It is suggested that spray dryers act as particle transforming reactors, allowing a certain amount of phase transformation to occur between the initially-produced amorphous powders and the final, partially-crystalline products. This phase transformation might be regarded as a reaction process, so by this reasoning, spray dryers could be regarded as reactors in which the "water loss reaction," namely drying, occurs, as well as a one-way reaction from amorphous to crystalline solid phase forms. This reasoning would also explain why, with powders produced by very small-scale spray dryers, the powders can be kept free flowing for a long time by freezing them.

Although definitive evidence is not present to support this hypothesis, there is strong suggestive evidence. Some of this evidence is the common finding that many spray dryers are much larger than they need to be in order to do the required drying duty. Additional evidence is that the rate of crystallization appears to be significantly slower than the rate of drying, and that crystalline materials are less sticky than amorphous ones. Some patents appear, fundamentally, to be suggesting that additional residence time in a spray dryer increases yield, even when the residence time is adequate to do the drying duty. This hypothesis may also partly explain the differences in particle properties from pharmaceutical-scale, pilot-scale and industrial-scale dryers. The transformation of spray-dried amorphous powders into crystalline ones occurs for lactose and for *Hibiscus* extracts, and this transformation may occur in all spray-dried products, even inside the dryers.

ACKNOWLEDGMENTS

Acknowledgments and grateful thanks are due to Dr. M.N. Haque, CSIRO Minerals, for reviewing this chapter and for his helpful and constructive criticism and feedback, which are greatly appreciated. Thanks are due to D. Chiou, S. Wang and I. Islam for their valuable contributions to this work.

NOMENCLATURE

E_A	activation energy (J mol^{-1})
K	rate constant (s^{-1})
k	rate constant (s^{-1})
k_{cr}	rate of crystallization at any temperature (s^{-1})
k_g	rate of crystallization at the glass transition temperature (s^{-1})
I_0	peak intensity at time zero (counts s^{-1})
I_f	peak intensity at the maximum crystalline product content (counts s^{-1})
I_t	peak intensity at time t (counts s^{-1})
M	average molecular weight (kg mol^{-1})
n	Avrami exponent (–)
q	pre-exponential factor (s^{-1})
R	Universal gas constant (J mol^{-1} K^{-1})
r	the ratio of θ_{cr} over θ_g (–)
t	time (s)
T	temperature (K)
T_g	glass-transition temperature (K)
x	ratio of α to β form of lactose (–)

Greek:

θ	degree of crystallinity (–)
θ_{cr}	time for crystallisation at any temperature (T) (s)
θ_g	time for crystallisation at the glass-transition temperature (T_g) (s)

REFERENCES

1. White, G.W., and Cakebread, S.H. The glassy state in certain sugar containing food products. *Journal of Food Technology*, 1, 73, 1966.
2. Vega, C., and Roos, Y.H. Invited review: spray-dried dairy and dairy-like emulsions—compositional considerations. *Journal of Dairy Science*, 89, 383, 2006.
3. Chan, H.-K., and Chew, N.Y.K. Novel alternative methods for delivery of drugs for the treatment of asthma. *Advanced Drug Delivery Review*, 55, 793, 2003.
4. Vidgren, M.T., Vidgren, P.A., and Paronen, T.P. Comparison of physical and inhalation properties of spray-dried and mechanically micronized disodium cromoglycate. *International Journal of Phamacy*, 35, 139, 1987.
5. Chawla, A. et al. Production of spray dried salbutamol sulfate for use in dry powder aerosol fomulation. *International Journal of Phamacy*, 108, 233, 1994.
6. Bhandari, B.R., Datta, N., and Howes, T. Problems associated with spray drying of sugar-rich foods. *Drying Technology*, 15, 671, 1997.
7. Langrish, T.A.G. New engineered particles from spray dryers: research needs in spray drying. *Drying Technology*, 25, 971, 2007.
8. Patil, M.N., Pandit, A.B., and Thorat, B.N. Ultrasonic atomisation assisted spray drying. In *Proc. 5th Asian Drying Conference*, G. Chen, Ed. Hong Kong, 255, 2007.
9. Wang, S. *Crystallization of Amorphous Materials Produced by Spray Dryers*. B.E. (Chemical and Biomolecular) Thesis, The University of Sydney, Sydney, Australia, 41, 2006.
10. Lai, H.M., and Schmidt, S.J. Lactose crystallization in skim milk powder observed by hydrodynamic equilibria, scanning electron microscopy and 2H nuclear magnetic resonance *Journal of Food Science*, 55, 994, 1990.
11. Williams, M.L., Landel, R.F., and Ferry, J.D. The temperature dependence of relaxation mechanisms in amorphous polymers and other glass-forming liquids. *Journal of the American Chemical Society*, 77, 3701, 1955.
12. Roberts, C.J., and Debenedetti, P. Engineering pharmaceutical stability with amorphous solids. *AIChE Journal*, 48, 1140, 2002.
13. Palzer, S. The effect of glass transition on the desired and undesired agglomeration of amorphous food powders. *Chemical Engineering Science*, 60, 3959, 2005.

14. Ozmen, L., and Langrish, T.A.G. An experimental investigation of the wall deposition of milk powder in a pilot-scale spray dryer. *Drying Technology*, 21, 1253, 2003.
15. Chidavaenzi, O.C., Buckton, G., and Koosha, F. The effect of co-spray drying with polyethylene glycol 4000 on the crystallinity and physical form of lactose. *International Journal of Pharmaceutics*, 216, 43, 2001.
16. Price, R., and Young, P.M. Visualisation of the crystallisation of lactose from the amorphous state. *Journal of Pharmaceutical Sciences*, 93, 155, 2004.
17. Chiou, D., and Langrish, T.A.G. Crystallisation of amorphous components in spray-dried powders. *Drying Technology*, 25, 1423, 2007.
18. Chickering, D.E. III et al. Spray drying method. U.S. Patent 6223455, 2001.
19. Chiou, D., Langrish, T.A.G., and Braham, R. Partial crystallisation of materials in spray drying: simulations and experiments. In *Proc. 6th Asian Drying Conference*, G. Chen, Ed. 13–15 August, paper H-013, 2007.
20. Harjunen, P. et al. Effects of ethanol to water ratio in feed solution on the crystallinity of spray-dried lactose. *Drug Development and Industrial Pharmacy*, 28, 949, 2002.
21. Forbes, R.T. et al. Water vapor sorption studies on the physical stability of a series of spray-dried protein/sugar powders for inhalation. *Journal of Pharmaceutical Sciences*, 87, 1316, 1998.
22. Chan, H.-K. et al. Physical stability of salmon calcitonin spray-dried powders for inhalation. *Journal of Pharmaceutical Sciences*, 93, 792, 2004.
23. Briggner, L.-E. et al. Pharmaceutical formulation. U.S. Patent 5874063, 1999.
24. Kussendrager, K.D., and Ellison, M.J.H. Carrier material for dry powder inhalation. WO World IPO 0207705, 2002.
25. Ward, G., and Dwivedi, S. Physically stabilized dry powder formulation. WO World IPO 0176560, 2001.
26. Bushill, J.H. et al. The crystallization of lactose with particular reference to its occurrence in milk powder. *Journal of the Science of Food and Agriculture*, 16, 622, 1965.
27. Haque, M.K., and Roos, Y.H. Crystallization and X-ray diffraction of spray-dried and freeze-dried amorphous lactose. *Carbohydrate Research*, 340, 293, 2005.
28. Corrigan, D.O., Healy, A.M., and Corrigan, O.I. The effect of spray drying solutions of polyethylene glycol (PEG) and lactose/PEG on their physicochemical properties. *International Journal of Pharmaceutics*, 235, 193, 2002.
29. Jouppila, K., and Roos, Y.H. Water sorption and time-dependent phenomena of milk powders. *Journal of Dairy Science*, 77, 1798, 1994.
30. Kedward, C.J., MacNaughtan, W., and Mitchell, J.R. Isothermal and non-isothermal crystallization in amorphous sucrose and lactose at low moisture contents. *Carbohydrate Research*, 329, 423, 2000.
31. Jouppila, K., and Roos, Y.H. Glass transitions and crystallization in milk powders. *Journal of Dairy Science*, **77**, 2907, 1994.
32. Ibach, A., and Kind, M. Crystallization kinetics of amorphous lactose, whey-permeate and whey powders. *Carbohydrate Research*, 342, 1357, 2007.
33. Wang, S., and Langrish, T.A.G. Measurements of the crystallization rates of amorphous sucrose and lactose powders from spray drying. *International Journal of Food Engineering*, 3, 1, 2007.
34. Troy, H.C., and Sharp, P.F. Alpha and beta lactose in some milk powders. *Journal of Dairy Science*, 8, 140, 1930.
35. King, N. The physical structure of dried milk *Dairy Science Abstracts*, 27, 91, 1965.
36. Supplee, G.C. Humidity equilibria of milk powders. *Journal of Dairy Science*, 4, 50, 1926.
37. Berlin, E., Anderson, B.A., and Pallansch, M.J. Water vapor sorption properties of various dried milks and wheys. *Journal of Dairy Science*, 51, 1339, 1968.
38. Berlin, E., Anderson, B.A., and Pallansch, M.J. Effect of temperature on water vapor sorption by dried milk powders. *Journal of Dairy Science*, 53, 146, 1970.
39. Shastry, A.V., and Hartel, R.W. Crystallization during drying of thin sucrose films. *Journal of Food Engineering*, 30, 75, 1996.
40. Ben-Joseph, E., Hartel, R.W., and Howling, D. Three-dimensional model of phase transition of thin sucrose films during drying. *Journal of Food Engineering*, 44, 13, 2000.
41. Ben-Joseph, E., and Hartel, R.W. Computer simulation of sugar crystallization in confectionary products. *Innovative Food Science and Emerging Technologies*, 7, 1294, 2006.
42. Roos, Y.H., and Karel, M. Differential scanning calorimetry study of phase transitions affecting the quality of dehydrated materials. *Biotechnology Progress*, 6, 159, 1990.
43. Peleg, M. On the use of the WLF model in polymers and foods. *Critical Reviews in Food Science and Nutrition*, 32, 59, 1992.

44. Roos, Y.H., and Karel, M. Amorphous state and delayed ice formation in sucrose solutions. *International Journal of Food Science and Technology*, 26, 553, 1991.
45. Senoussi, A., Dumoulin, E.D., and Berk, Z. Retention of diacetyl in milk during spray-drying and storage. *Journal of Food Science*, 60, 894, 1995.
46. Peleg, M., and Hollenbach, A.M. Flow conditioners and anticaking agents. *Food Technology*, 38, 93, 1984.
47. Jouppila, K., and Roos, Y.H. Glass transitions and crystallization in milk powders. *Journal of Dairy Science*, 77, 2907, 1994.
48. Roos, Y.H. Melting and glass transitions of low molecular weight carbohydrates. *Carbohydrate Research*, 39, 238, 1993.
49. Ramos, R.A., Rikvold, P.A., and Novotny, M.A. Test of the Kolmogorov-Johnson-Mehl-Avrami picture of metastable decay in a model with microscopic dynamics. *Physical Review B*, 59, 9053, 1999.
50. Novotny, M.A. et al. Simulations of metastable decay in two- and three-dimensional models with microscopic physics. *Journal of Non-Crystalline Solids*, 274, 356, 2000.
51. Roos, Y.H., and Karel, M. Crystallization of amorphous lactose. *Journal of Food Science*, 57, 775, 1992.
52. Ibach, A., and Kind, M. Crystallization kinetics of amorphous lactose, whey-permeate and whey powders. *Carbohydrate Research*, 342, 1357, 2007.
53. Debenedetti, P.G. *Metastable Liquids: Concepts and Principles.* Princeton University Press, Princeton, NJ, p. 336 and 441, 1996.
54. Panagopoulos, G.N., Bruttini, R., and Liapis, A.I. A molecular dynamics modeling and simulation study on determining the molecular mechanism by which formulations based on trehalose could stabilize biomolecules during freeze drying. In *Proc. 15th International Drying Symposium IDS 2006*, I. Farkas, A.S. Mujumdar, Eds. Budapest, Hungary, August 20–23, Volume A, 114, 2006.
55. Chen, X.D., and Xie, G.Z. Fingerprints of the drying behaviour of particulate or thin layer food materials established using a reaction engineering model. *Trans. IChemE, Part C: Food and Bioproduct Engineering*, 75, 213, 1997.
56. Mcleod, J.S., and Paterson, A.H.J. A preliminary study into the role of sliding contact in the adhesion of alpha lactose monohydrate to a conveying line wall. In *Proc. Chemeca 2005, Brisbane*, 26–29 September, CDROM, paper number 205, 2005.
57. Willart, J.F. et al. Direct crystal to glass transformation of trehalose induced by ball milling. *Solid State Communications*, 119, 501, 2001.
58. Bhandari, B.R., and Howes, T. Implication of glass transition for the drying and stability of dried foods. *Journal of Food Engineering*, 40, 71, 1999.
59. Lin, S.X.Q., and Chen, X.D. Prediction of air-drying of milk droplet under relatively high humidity using the reaction engineering approach. *Drying Technology*, 23, 1395, 2005.
60. Chiou, D., and Langrish, T.A.G. Development and characterisation of novel nutraceuticals with spray drying technology. *Journal of Food Engineering*, 82, 84, 2007.
61. Atkins, P.W. *Physical Chemistry*, 2nd edn. Oxford University Press, Oxford, UK, 1049, 1983.
62. Porter, D. *Group Interaction Modelling of Polymer Properties*. Marcel Dekker, NY, 33, 97, 1995.
63. Papavergos, P.G., and Hedley, A.B. Particle deposition behaviour from turbulent flows. *Trans. I.Chem.E. (Chem. Eng. Res. Design)*, 62, 275, 1984.
64. McCoy, D.D., and Hanratty, T.J. Rate of deposition of droplets in annular two-phase flow. *International Journal of Multiphase Flow*, 3, 319, 1977.
65. Boonyai, P., B.R. Bhandari, and T. Howes Stickiness measurement techniques for food powders: a review. *Powder Technology*, 34, 145(1), 2004.
66. Slade, L., and Levine, H. Water and the glass transition-dependence of the glass transition on composition and chemical structure: special implications for flour functionality in cookie baking. *Journal of Food Engineering*, 22, 143, 1994.
67. Roos, Y.H., and Karel, M. Water and molecular weight effects on glass transitions in amorphous carbohydrates and carbohydrate solutions. *Journal of Food Science*, 56, 1676, 1991.
68. Fox, T.G., and Flory, P.J. Second-order transition temperature and related properties of polystyrene. I. Influence of molecular weight. *Journal of Applied Physics*, 21, 581, 1950.
69. To, E.C., and Flink, J.M. 'Collapse', a structural transition in freeze dried carbohydrates 2. Effect of solute composition. *Journal of Food Technology*, 13, 567, 1978.
70. Chiou, D., Langrish, T.A.G., and Braham, R. The effect of temperature on the crystallinity of lactose powders produced by spray drying. *Journal of Food Engineering*, 86, 288, 2008.

31 Extrusion of Foods

Prabhat Kumar and K.P. Sandeep
North Carolina State University

Sajid Alavi
Kansas State University

CONTENTS

31.1 INTRODUCTION

Food extrusion refers to a process in which a food material is forced to flow, under one or more varieties of conditions of mixing, heating and shear through a die which is designed to form and/or puff-dry the ingredients [1]. The extrusion mechanism can be as simple as a piston contained in a cylinder with a die at one end. Material is loaded into the cylinder and emerges in its defined shape from the die due to the pressure created by the moving piston. This type of extrusion is a batch process. However, an extrusion process can be made continuous by replacing the piston with a helical screw. In this case, material is fed continuously into a hopper and transported toward the die by rotating screws. As the material reaches the die, the pressure increases to the level required to propel the extrudate through the die. The friction between the material and the screw surface results in heating of the material [2]. Thus, extrusion combines several unit operations such as mixing, cooking, kneading, shearing, shaping and forming to make food products of high quality, high throughput and low cost [3].

The first extruded food products were expanded corn snacks which were commercially produced in the mid to late 1940s. The early 1950s saw the first application of extrusion for producing dry expanded pet foods in the United States. Pet foods have grown into the largest single commercial application of extrusion cooking. The 1960s saw the first commercial production of dry expanded ready-to-eat (RTE) breakfast cereals using extrusion. Textured vegetable protein (TVP) production using extrusion became commercial in the 1970s and 1980s brought commercialization of production of feed for aquatic animals using extrusion. These applications represent only a few of the many products that can be made using extrusion [2].

The major advantages of extrusion include [3,4]:

i. Versatility—The extruder can produce a variety of products with different shapes, textures and colors by changing ingredients and the operating conditions.
ii. Reduced cost—Extrusion has a lower processing cost as compared to other cooking and forming processes. Extrusion processing also requires less space per unit of operation than traditional cooking systems.
iii. Automated production—The extruder can provide continuous high throughput and can be fully automated.
iv. High product quality—Extrusion minimizes degradation of nutrients because it is a high temperature-short time heating process.
v. Absence of process effluents—Extrusion produces little or no waste streams and thus it prevents processors to install effluent treatment system.

31.2 RAW MATERIALS FOR EXTRUSION OF FOODS

Raw materials play an important role in forming the structure of the extruded products. Inside the extruder, raw materials provide a continuous phase which binds all other particulate matter of the dispersed phases and retain the gases released during expansion.

31.2.1 STARCH-BASED INGREDIENTS

Starch-based ingredients include cereal grains (wheat, corn, rice and oat) and potato derivatives (potato granules, potato flour, and potato starches). These ingredients are usually the largest component which helps in forming the structure of the extrudate.

Wheat is commonly used in the form of fine flours to manufacture baked products or as semolina for pasta. Starch granules of wheat occur in a bimodal size distribution with two groups comprising of large granules (20–40 μm) and small granules (1–10 μm). There is slight variation in the overall composition of starch granules in wheat. Among cereals, wheat has relatively high protein levels (8–15 wt%). The main proteins present in wheat, glutenin and gliadin, hydrate in water to form a rubbery elastic mass which is highly stretchable [5].

Corn occurs in many varieties, differing in grain morphology and color. Two types of endosperm exist within each grain of corn. The outer endosperm is hard and contains densely packed starch and proteins, leading to polygonal shaped starch granules. The endosperm in the center of the grain is soft and contains loosely bound starch and proteins along with air cavities. Starch granules in the central endosperm are globular with smooth surfaces. Different varieties of corn differ in the amounts of the two types of endosperm. Starch granules within each type of endosperm range in size from 5 to 20 μm. Protein content in corn ranges from 6 to 10 wt% and the major protein is water insoluble zein protein. Zein protein swells in water and behaves in a manner similar to that of wheat gluten [5].

Rice has different varieties based on grain morphology and endosperm texture. Most rice grains have a hard endosperm with small amounts of chalky endosperm penetrating the main structure. The hard endosperm has a strong bonding between starch granules and proteins. The chalky endosperm has weak bonding between starch and proteins along with some air cavities. Starch granules in rice are very small (2–8 μm) and in the shape of polygons. These granules are present in groups of five to eight granules joined together. Protein content in rice ranges from 6 to 8 wt% and the proteins are predominantly of the glutenin and gliadin types [5].

Oats are less widely grown compared to other cereals and are available in different varieties based on grain morphology and endosperm composition. Prior to milling, oat grains are steamed to inactivate enzymes located in the outer layers of the kernel and dried to a moisture content of

6–8 wt%. Starch granules in oat are in the form of small angular shapes with size ranging from 2 to 12μm. Lipid content of oat is high (7–9 wt%) compared to other cereal grains. Fiber content of oats is high and the fiber is in the form of a mixture of glucans and insoluble component of the kernel. The protein content is similar to wheat and the predominant protein is globulin. Oats have the lowest starch levels among cereal grains [5].

Potatoes are used in extrusion in the form of granules, flour and starches. Potato granules are formed from diced potatoes. Diced potatoes are tempered to activate enzymes for softening cell walls and then cooked and dried. The cooked potato is mixed with emulsifier and dried potato granules to reduce moisture content and dehydrate the cellular structure. Potato granules retain their cellular structures except at the cut surfaces. Potato flour is prepared by roller drying of cooked potato slurry with small amounts of emulsifier to prevent adhesion of starch to the metal surface of the extruder. Potato flour is mainly comprised of gelatinized starch with little cell structure. Potato flour makes stiff and adhesive dough on mixing with water at 35–50 wt% moisture content. Potato starches are used to increase expansion in snack products. The size of potato starch varies from 60 to 100 μm [5].

31.2.2 PROTEIN-BASED INGREDIENTS

Protein-based ingredients include oil-seed proteins such as soy flour and proteins from cereals such as wheat. These ingredients form the dispersed phase in the main continuous phase matrix.

Soybeans are processed to form a pressed cake by extracting the oil. Protein rich flours are obtained by further processing the pressed cake. The most basic soy-based raw material for extrusion are pressed flakes which are defatted products containing 50 wt% protein with a variable protein digestibility index of 55–90%. Soy flours and grits are made from heat-treated pressed cake by milling. Preparation of soy protein concentrates involves extraction of soluble carbohydrates from soy flakes with aqueous alcohol, followed by milling to a coarse powder. Protein content of soy protein concentrate varies from 65 to 70 wt%. Soy protein isolates are prepared by isolation of proteins by solubilization and reprecipitation, followed by drying, to form powder with a protein content of 90 wt% [5].

The protein rich extract of wheat flour with a protein content of 65–70 wt% is obtained by washing and spray drying. This protein consists mainly of prolamin with long hydrophobic chains. Wheat proteins are denatured at high temperatures to form a melt phase in the extruder and texturize in the same manner as soy proteins [5].

31.2.3 LIPIDS

Lipids, which include oils and fats, act as lubricants during extrusion and modify the texture of the finished product. Lipids melt inside the extruder and are mixed in the continuous matrix in the form of very small (< 1 μm) droplets. The addition of lipids in the range of 0.5–1 wt% prevents degradation of carbohydrates during low-moisture (< 25 wt% moisture) extrusion. However, if oil content is raised above 2–3 wt%, it becomes difficult to disperse starch, resulting in a less expanded product. Expanded products can be made at high oil (> 5 wt%) and moisture content (30–35 wt%) because expansion is aided by swelling and diffusion mechanisms. Lipids are encapsulated in the glassy structure of starch in the extrudate, but grinding of extrudates might expose lipids to oxygen, leading to lipid oxidation. Emulsifiers are special types of lipids which have higher melting points. They behave as lipids during an extrusion process [5].

31.2.4 NUCLEATING AGENTS

Texture of an extrudate can be changed from coarse to fine with the addition of small amounts (1–2 wt%) of a nucleating agent. Ideal nucleating agents are fine powders which can remain insoluble during extrusion and provide surfaces where bubbles may form during release of water vapor. Examples of nucleating agents include calcium carbonate, sodium bicarbonate, magnesium carbonate, and silicon dioxide [5].

31.2.5 Nutritional Ingredients

Vitamins and minerals are added during extrusion for fortification of products. They do not have significant effects on processing variables because they are added in very small quantities.

31.2.6 Flavoring Agents

Salt, sugar, and spices are main flavoring agents used during extrusion. Salt is added at levels of 1–1.5 wt% to create a flavor profile balancing other ingredients. Salt itself has little effect on processing variables because it dissolves in water during extrusion. Sugar may be added at levels of up to 10 wt% without causing significant changes to processing variables. The flavor of sugar becomes noticeable at levels greater than 5 wt%, but sweetness is perceived only at levels greater than 10 wt% [5].

31.2.7 Coloring Agents

Natural and synthetic colorants can be added during extrusion to produce products of different colors. Reducing sugars such as glucose and fructose along with protein can also be used to produce colors by Maillard browning reactions [5].

31.3 CLASSIFICATION OF EXTRUDERS

Extruders are classified based on the method of operation and type of construction [3].

31.3.1 Method of Operation

31.3.1.1 Cold Extruders

The temperature of the food remains at ambient conditions in cold extruders. They are used to mix and shape foods such as pasta, liquorice, fish pastes, surimi, and pet foods [4].

31.3.1.2 Hot Extruders

Food is heated above 100°C in hot extruders. The food is then passed to the section of the barrel having the smallest clearance where pressure and shearing is further increased. The food is then forced through one or more dies at the discharge end of the barrel. The food expands to the final shape and cools rapidly as moisture is flashed off as steam [4].

31.3.1.3 Isothermal Extruders

Isothermal extruders operate at an essentially constant product temperature throughout the extruder. They are mainly used for forming different shapes.

31.3.1.4 Adiabatic Extruders

Adiabatic extruders generate all heat by friction and no heat is removed from the extruder. They operate at low moisture levels (8–14 wt%).

31.3.2 Type of Construction

31.3.2.1 Single-Screw Extruder

Single-screw extruders consist of a cylindrical screw that rotates in a grooved cylindrical barrel. The length to diameter ratio of the barrel varies from 2:1 to 25:1. The screw is driven by a variable speed electrical motor with power sufficient enough to pump food against pressure generated in the barrel. Compression is achieved in the extruder barrel by back pressure exerted by the die, increasing screw diameter, decreasing screw pitch, using a tapered barrel, and placing restriction in screw flights. Single-screw extruders can further be classified based on the extent of shear on the food product as high shear, medium shear and low shear [3,4].

31.3.2.1.1 High Shear

High shear extruders create high pressures and temperatures by using shallow flights. They are used to make breakfast cereals, candy, crisp breads, expanded snack foods, and expanded pet foods.

31.3.2.1.2 Medium Shear

Medium shear extruders are used for breadings, texturized proteins and semi-moist pet foods.

31.3.2.1.3 Low Shear

Low shear extruders create low pressure by using a smooth barrel, deep flights and low screw speed. They are used to make pasta, meat products, and gums.

31.3.2.2 Twin-Screw Extruder

Twin-screw extruders have several advantages over single-screw extruders. They are [3]:

i. Handling of viscous, oily and sticky materials
ii. Use of wide range of particle sizes
iii. Less wear compared to a single-screw extruder
iv. Easy clean-up because of its self-wiping nature

Twin-screw extruders are classified according to the direction of screw rotation as counter-rotating or corotating.

31.3.2.2.1 Counter-rotating

In counter-rotating twin-screw extruders, the screws rotate in opposite directions. These extruders are most commonly used in the plastic industry for their ability to process nonviscous materials requiring low speed and long residence time. Counter-rotating twin-screw extruders are not widely used in the food industry, but are used to make gum, jelly, and licorice confections. Counter-rotating twin-screw extruders can be further classified into intermeshing and nonintermeshing based on the manner in which the screws engage.

31.3.2.2.1.1 Intermeshing
In an intermeshing extruder, the flight of one screw engages the channels of the other screw. It provides a positive pumping action, efficient mixing and is self-cleaning.

31.3.2.2.1.2 Nonintermeshing
In a nonintermeshing extruder, the flight of one screw does not engage the channels of the other screw. This type of extruder depends on friction for extrusion.

31.3.2.2.2 Corotating

Corotating twin-screw extruders are most commonly used in the food industry. In these extruders, the screws are in constant contact with each other, creating a natural wiping action. These extruders provide high degrees of heat transfer, efficient pumping, controlled residence time distribution (RTD), self-cleaning mechanism and a uniform process. Corotating twin-screw extruders can also be classified into intermeshing and nonintermeshing based on the manner in which the screws engage.

31.4 ELEMENTS OF AN EXTRUDER

31.4.1 METERING OR FEEDING SYSTEM

A variable speed metering and feeding screw is used for continuous and uniform discharge of materials to the extruder. There are two types of feeding systems—volumetric and gravimetric.

Volumetric feeding system controls feed rate by adjusting feeder speed. Thus, the actual feed rate is affected by changes in bulk density of raw materials. It is simple in design and lower in cost. The actual feed rate in a gravimetric feeding system is not affected by bulk density of raw materials because the system controls feed rate automatically. Gravimetric feeding systems have become a standard for most extrusion processing systems because they offer constant feed rate for dry ingredients and liquid additives.

31.4.2 PRECONDITIONER

The feeding system can feed materials directly into the extruder or into a preconditioner. Preconditioning is beneficial for extrusion processes involving higher moisture content. The purpose of a preconditioner is to pre-blend steam and water with dry ingredients and mixing them long enough for temperature and moisture equilibration. The goal of preconditioning is to plasticize or soften the materials. The addition of water lowers the glass transition temperature and steam heats the materials above the glass transition temperature. This glass transition changes the materials from a glassy state to a rubbery (soft) state. Further moisture and heat addition results in melt transition which makes the material flowable. Some of the advantages of preconditioning includes increased throughput, decreased wear and tear of the extruder, and improved flavor and texture.

Preconditioners could be either pressurized preconditioners or atmospheric preconditioners. Pressurized preconditioners operate at elevated pressures resulting in a discharge temperature above 100°C. Atmospheric preconditioners operate at atmospheric pressure and thus the discharge temperature is limited to 100°C. A properly designed preconditioner can achieve mixing, hydration, cooking and pH modifications. The conditioning cylinders used in preconditioning is equipped with single, dual or differential diameter/speed shafts [3].

31.4.3 SCREW

The screw is the central part of an extruder. The screw rotates inside the barrel and conveys the material from the feed end to the discharge end. The screw of the extruder is divided into three sections—feed section, compression section and metering section. The portion of the screw which accepts food material is known as the feed section. The feed section should have sufficient material for conveying it down the screw. This section is typically 10–25% of the total length of the screw. The compression section is the portion of the screw between the feed and metering section. In this section, material is heated, compressed and worked into a continuous mass. The material changes from a granular state to an amorphous state. This section is approximately half the length of the screw. The metering section is the portion of the screw nearest to the discharge of the extruder. This section usually has very shallow flights of the screws to increase the shear rate in the channel to the maximum level.

The screw of an extruder can be built from different types of conveying and mixing elements (forward screw, kneading screw and reverse screw elements) as shown in Figure 31.1. The forward screw elements are characterized by forward flighted screws which convey the material in the forward direction. The kneading elements are mild flow restricting elements which have no conveying effect. However, they can be combined and oriented to cause static mixing, and/or weak forward or backward conveying effect. The reverse screw elements are characterized by a reverse flight conveying the material backward. They act as the back pressure element in the extruder and thus they are placed nearer to the die end of the extruder.

The critical dimensions of a screw are shown in Figures 31.2 and 31.3. The dimensions and their terminology have been described below [6].

Screw diameter (D_s) is determined as:

$$D_s = D - 2\delta \qquad (31.1)$$

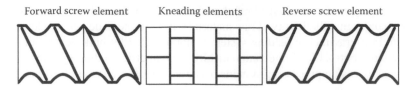

FIGURE 31.1 Different screw elements of an extruder.

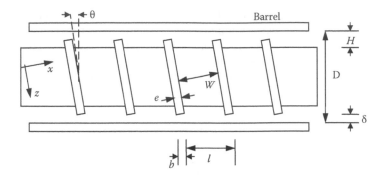

FIGURE 31.2 Geometry of a screw of an extruder.

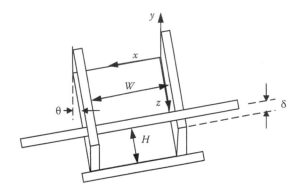

FIGURE 31.3 Geometry of the cross-section of a screw channel.

where D is the internal diameter of the barrel (which is also known as the bore) and δ is the radial clearance between the barrel and the screw.

Flight height (H_s) is given by:

$$H_s = H - \delta \tag{31.2}$$

where H is the distance between the surface of the barrel and root of the screw.

Root diameter (D_r), which is the diameter of the root of the screw, is given by:

$$D_r = D - 2H \tag{31.3}$$

Lead (l) is the axial distance from the leading edge of a flight at its outside diameter to the leading edge of the same flight in front of it. Axial flight width (b) is the width of a flight measured at the diameter of the screw. Flight width ($e = b\cos\theta$) is the flight width measured perpendicular to the face of the flight. Axial channel width ($B=1-b$) is the axial distance from the leading edge of one flight to the trailing edge of the next flight at the diameter of the screw. Channel length ($Z=1/\sin\theta$) is the length of the screw channel in the z direction. The number of flight turns (p) is the total number of single flights in the axial direction. A screw with $p=1$ is known as single flighted screw whereas a screw with $p=2$ is known as double flighted screw. Height to diameter ratio (H/D) is the ratio of flight height to screw diameter.

31.4.4 BARREL

Barrel is the cylindrical casing which fits tightly around the screw of the extruder. Bore (D) is the inside diameter of the barrel and it ranges from 5 cm to 25.4 cm. Length (L) of the extruder is the distance from the rear edge of the feed section to the discharge end of the metering section. Length to diameter ratio ($L:D$) is the ratio of the length to the diameter of the barrel and it usually ranges from 1:1 to 20:1. A barrel is made up of several segments which makes it relatively easy to change the configuration of the barrel and to replace the discharge section which wears out the fastest. Barrel is constructed of special hard alloys such as Xaloy® 306 and stainless steel (SS 431) to withstand the pressure developed in the barrel and to resist wear. The barrel is also provided with a removable sleeve to resist wear. The interior surface of the barrel is grooved to prevent slippage of materials at the walls. The presence of grooves also increases the ability of the extruder to pump materials against high back pressure. The outside of the barrel is covered with hollow cavities for circulation of the heat transfer medium such as water or steam [6].

31.4.5 DIES

The extruder barrel is equipped with a die with one or more openings through which the extrudate flows. The openings shape the final product and provide a resistance against the flow of extrudate. Dies may be designed to be highly restrictive for increasing barrel fill, residence time and energy input. Coextrusion dies are used to make a single product from two different products. One such example is a product with an outer shell and an inner filling material. Sheeting dies are used to make sheets of thickness ranging from 0.8 mm to 1.5 mm.

31.4.6 ROTATING KNIFE

Rotating knife, installed at the exit of the dies, cuts the product, giving it the final shape and length.

31.5 PROCESS PARAMETERS

There are two types of process parameters in an extrusion process—independent and dependent. Independent variables are those variables which can be controlled directly. Dependent variables are those variables which are dependent upon the magnitude of the independent variables. Both independent and dependent variables have been discussed below for clarity [6].

31.5.1 INDEPENDENT VARIABLES

31.5.1.1 Feed Ingredients and Feed Rate

The properties of an extruded product are dependent upon the composition of feed ingredients. Therefore, it is very important to specify, characterize, and control the feed ingredients. Extruders can be operated such that the feed section is not completely full. Under these conditions, feed rate acts as an independent variable and can be changed to different levels during an extrusion process.

31.5.1.2   Preconditioning Variables

The preconditioner preblends steam and water with dry ingredients and mixes them long enough for temperature and moisture equilibration. Preconditioning variables are characterized by the change in temperature and moisture of the material.

31.5.1.3   Screw Speed

Most of extruders have a variable speed drive which enables the speed of the screw to be changed easily.

31.5.1.4   Configuration of Screw and Die

The configuration of the screw can be changed by using different conveying and mixing elements such as forward screw, kneading screw, and reverse screw elements. These elements can be attached together in different ways to achieve variation in the geometric configuration of the extruder. The size, shape, number and location of the dies are also independent variables.

31.5.1.5   Water and Steam Injection

The amount and location of water and steam injection are independent variables because they can be controlled separately in an extruder. Water can be added directly to the feed, injected into the barrel or added in the form of steam to the preconditioner or barrel.

31.5.1.6   Speed of Rotating Knife

The speed of the rotating knife determines the throughput in an extrusion process. It can be altered to adjust the rate of product formation.

31.5.1.7   Set Point Temperature of Barrel

Set point temperature of the barrel can be changed to achieve a particular temperature profile within the barrel in the extruder.

31.5.2   DEPENDENT VARIABLES

31.5.2.1   In-Barrel Moisture

In-barrel moisture is the actual moisture in an extrusion process. Moisture is added in the form of either steam or water in the extruder.

31.5.2.2   Specific Mechanical Energy (SME)

Specific mechanical energy (SME) is the amount of mechanical energy per unit mass dissipated as heat inside an extruder.

31.5.2.3   Specific Thermal Energy

Specific thermal energy is the amount of thermal energy per unit mass added from heat sources or sinks in the extruder.

31.5.2.4   Residence Time

Residence time is a measure of the time a material spends in the extruder. Residence time is used to determine optimal processing conditions for mixing, cooking and shearing reactions during an extrusion process.

31.5.2.5   Pressure

The pressure at the discharge of the extruder regulates the output of the extruder by affecting the flow through the die.

31.5.3   PRODUCT CHARACTERISTICS

Product characteristics are a measure of the quality of the product that results from changes made to independent or dependent variables. Some of the product characteristics include moisture, bulk density, solubility, absorption, texture, color, and flavor [3].

31.5.4   CRITICAL PARAMETERS

The four critical parameters in extrusion are moisture content of the product, mechanical energy input, thermal energy input and residence time. Critical processing parameters are functions of independent and dependent variables which control the quality of the product. Moisture is the actual moisture in an extrusion process. Mechanical energy input can be expressed as either gross mechanical energy (GME) or SME as follows [4]:

$$\text{GME} = \frac{\text{Power } (kW)}{\dot{m}}$$

$$\text{SME} = \frac{\text{Power}_{\text{loaded}} - \text{Power}_{\text{empty}} (kW)}{\dot{m}} \tag{31.4}$$

where $\dot{m}$ is the feed rate (kg.h^{-1}). Thermal energy input is expressed in the same energy units as mechanical energy (kWh.kg^{-1}). Average retention time ($\bar{t}$) is determined as [3]:

$$\bar{t} = \frac{m}{\dot{m}} \tag{31.5}$$

where m is the amount of material in the extruder.

   If all four critical parameters are kept constant, then product with consistent quality can be produced in an extrusion process.

31.6   PERFORMANCE ANALYSIS OF AN EXTRUDER

Extrusion is an energy efficient process because a substantial amount of mechanical energy from the drive motor is converted to heat due to viscous dissipation in the material. Energy consumption and extruder efficiency, which are two important ways to evaluate performance of an extruder, are described below.

31.6.1   ENERGY CONSUMPTION

Energy consumption in an extruder depends on a number of factors such as properties of the material, design of extruder, type of motor drive and processing conditions [7]. SME is a commonly used measure of the energy consumption in an extruder. Total SME (SME$_t$) for a twin-screw extruder can be calculated as [7]:

$$\text{SME}_t = \frac{\%T_t}{100} \times \frac{P_r}{N_r} \times \frac{N}{\dot{m}} \tag{31.6}$$

where $\%T_t$ is the percent total torque generated by the drive motor, N is the screw speed (rpm), and P_r and N_r are rated horsepower (W) and maximum screw speed (rpm) respectively. Total torque (T_t), which is the sum of the torques required for shearing, pumping and turning the screws is given by:

$$T_t = T_s + T_p + T_e \tag{31.7}$$

where T_s and T_p are the torque for shearing and pumping of material respectively, and T_e is the torque required for turning the empty extruder.

SME represents the mechanical energy consumption per unit mass of the product, but it does not indicate the overall extrusion performance. Thus, energy consumption is not a good indicator for the performance analysis of an extruder.

31.6.2 EXTRUDER EFFICIENCY

Extruder efficiency (η_{ext}) is defined as the ratio of the theoretical power required to the actual power consumed in an extruder. Total power (P_t) input to the shaft of an extruder in the metering section can be expressed as [6]:

$$P_t = P_h + P_p + P_k + P_\delta \tag{31.8}$$

where P_h is the power due to viscous dissipation in the screw channel, P_p is the power required to raise the pressure of the melt, P_k is the power required to raise the kinetic energy of the melt, and P_δ is the power due to viscous dissipation in the flight clearance. Assuming P_k is negligible, expression for P_t is given as [6]:

$$P_t = p\frac{(\pi ND)^2 L}{\sin\theta}\left[\mu\frac{W}{H}(\cos^2\theta + 4\sin^2\theta) + \mu_\delta\frac{e}{\delta}\right] + p\frac{\pi NDWH}{2}\Delta P(\cos\theta) \tag{31.9}$$

where p is the number of flight turns, μ is the viscosity of the material in the screw channel, μ_δ is the viscosity of the material in the flight clearance, e is the flight width, and δ is the radial clearance between the barrel and the screw.

Total torque can be calculated by dividing P_t with the screw speed, N. Specific power (P_{sp}) which is the total power per unit mass flow rate (feed rate) is expressed as [6]:

$$P_{sp} = \frac{P_t}{\dot{m}} \tag{31.10}$$

Liang et al. [7] gave another expression to calculate total theoretical power (P_t) in an extruder as:

$$P_t = \dot{m}\int_T^{T_o} c_p dT + \frac{\dot{m}\Delta P}{\rho} + \dot{m}\Delta H \tag{31.11}$$

where c_p is the specific heat capacity of the material (J.kg^{-1}.K^{-1}), T_i and T_o are the inlet and outlet temperatures (K) of the material respectively, ΔP is the discharge pressure (Pa), ρ is the density (m.kg^{-3}), and ΔH is the change in enthalpy (J.kg^{-1}) associated with the reactions (starch gelatinization, protein denaturation, fat melting, etc.) taking place inside the extruder.

The actual power consumption is given by [7]:

$$P_{actual} = \dot{m}(\text{SME}_t + \text{SHE}_t) \tag{31.12}$$

where SME$_t$ is the total SME and SHE$_t$ is the total specific heat energy.

By definition, extruder efficiency (η_{ext}) is given as [7]:

$$\eta_{ext} = \frac{P_t}{P_{actual}} = \frac{P_t}{\dot{m}(\text{SME}_t + \text{SHE}_t)} = \frac{P_{sp}}{(\text{SME}_t + \text{SHE}_t)} \tag{31.13}$$

31.7 RESIDENCE TIME DISTRIBUTION (RTD)

One of the most important parameters in evaluating the performance of an extruder is the residence time distribution (RTD) of the materials during their flow through the extruder. RTD is a plot of residence time against the fraction of flow having that particular residence time. RTD in an extruder is used to determine the optimal processing conditions for mixing, cooking and shearing reactions during the process [8]. RTD is usually determined by a stimulus response technique using a tracer. RTD is generally described by exit age distribution function, $E(t)$, and the cumulative distribution function, $F(t)$. These distributions are calculated as [9]:

$$E(t) = \frac{C(t)}{\int_0^\infty C(t)dt}$$

(31.14)

$$F(t) = \int_0^t E(t)dt$$

where $C(t)$ is the tracer concentration at time, t.

Pinto and Tadmor [10] derived an expression for the RTD of a Newtonian fluid with constant viscosity as:

$$E(t)dt = \frac{3\frac{y}{H}\left\{1 - \frac{y}{H} + \left[1 + \frac{2y}{H} - 3\left(\frac{y}{H}\right)^2\right]^{1/2}\right\}}{\left[1 + 2\frac{y}{H} - 3\left(\frac{y}{H}\right)^2\right]^{1/2}}$$

(31.15)

where y is the position of particle within the channel.

The average or mean residence time $(\bar{t})$ is related to $E(t)$ by the following equation [6]:

$$\bar{t} = \int_0^\infty tE(t)dt$$

(31.16)

Substituting $E(t)dt$ in the above equation and integrating gives the mean residence time (MRT) as [10]:

$$\bar{t} = \frac{2L}{V_b \sin\theta \cos\theta(1-a)}$$

(31.17)

where a is the negative ratio of pressure flow to drag flow. V_b is the tangential velocity of the barrel and is directly proportional to the screw speed (N). Thus, the average residence time is proportional to the reciprocal of N.

The work of Pinto and Tadmor [10] was extended by Bigg and Middleman [11] to determine RTD for isothermal non-Newtonian flow. They used the velocity profiles for a power law fluid and determined $F(t)$ to be a function of both $t/\bar{t}$ and a dimensionless parameter, G_z, given by:

$$G_z = \frac{\partial P}{\partial z}\frac{H^{n+1}\sin\theta}{mV^n}$$

(31.18)

where H is the height of the flight, m is the consistency coefficient, n is the flow behavior or power law index, and V (πDN) is the velocity of barrel relative to the screw. Modeling of RTD and mean residence time has been reviewed in detail by Ganjyal and Hanna [8].

31.7.1 FACTORS AFFECTING RESIDENCE TIME DISTRIBUTION (RTD)

31.7.1.1 Configuration and Speed of Screw

Configuration of the screw is an important process parameter affecting the RTD in an extruder. The type, length and position of the mixing elements have been shown to significantly affect the RTD and mixing [12]. The results showed that the mean residence time of the material in the extruder was significantly increased with the incorporation of mixing elements (kneading and reverse screw elements). The mean residence time increased as the mixing elements were moved farther away from the die, as the elements were made longer, and as the spacing between the elements was increased. Mean residence time and extent of mixing were lower for screw profiles with kneading elements than those with reverse screw elements. RTD in an extruder is also affected by the pitch of the screw. An increase in screw pitch increases the mean residence time in an extruder.

The rotation speed of the screw determines the duration a material remains in the extruder. Thus, it is one of the important process parameters affecting RTD. With other variables remaining constant, an increase in the speed of the screw reduces the MRT in an extruder [8]. Lee and McCarthy [13] showed that the speed of the screw had a strong effect on the $E(t)$ and $F(t)$ diagrams with MRT varying inversely with the screw speed. Singh and Rizvi [14] showed a sharp decrease in average residence time from 119.2 to 80.8 s as the speed was increased from 150 to 200 rpm, but decreased moderately from 80.8 to 74.8 s when the screw speed was further increased to 250 rpm. Gogoi and Yam [15] developed the following empirical equation to predict MRT based on the screw speed:

$$\bar{t} = a_m \left(\frac{1}{N} \right)^{b_m} \tag{31.19}$$

where a_m and b_m are empirical parameters and N is the screw speed (rpm). The value of parameter b_m, ranged from 0.3 to 0.6. Yeh et al. [16] developed an empirical correlation to express the dispersion number as a power law function of feed rate (N_f) and screw speed (N) in a twin-screw extruder as:

$$\frac{D}{VL} = k \left(\frac{N_f}{N} \right)^{-n_m} \tag{31.20}$$

where D is the diffusivity ($m^2.s^{-1}$), V is the axial velocity in the extruder ($m.s^{-1}$), L is the length of the extruder (m), N_f is the speed of feeding auger (rpm), N is the screw speed (rpm), and k and n_m are empirical parameters. Dispersion number is the reciprocal of the Peclet number (P_e) which is used to measure the extent of axial dispersion. When Peclet number approaches infinity, dispersion is negligible and hence we have plug flow. As Peclet number decreases, the dispersion increases leading to a mixed flow.

31.7.1.2 Die Geometry

The size and shape of the die affect the RTD in an extruder. Decrease in the diameter of the die increases die pressure due to increased degree of fill in the extruder. De Ruyck [17] showed that a decrease in diameter of the die below 4 mm increased RTD and MRT significantly. However, there was no significant effect of die diameter on the RTD above 4 mm. Altomare and Ghossi [18] reported that die diameter did not have a significant effect on RTD in an extruder.

31.7.1.3   Moisture Content and Viscosity of the Material

Moisture content and viscosity of a material affect the RTD in an extruder. Gogoi and Yam [15] showed that increasing moisture content slightly affected the mean residence time in general and reduced it for some cases. However, Altomare and Ghossi [18] reported that there was no significant effect of increasing moisture content up to 28 wt% on mean residence time. Seker [19] showed that increasing the moisture content from 28.5 to 41.2 wt% did not significantly affect mean residence time. Moisture content affects the rheology of material in two opposite ways in the extruder. Increase in moisture content of material results in a decrease in material viscosity in the barrel and hence, lower SME is required to pump the material through the die. However, lower viscous dissipation in the die decreases the temperature of material. This decrease in temperature increases viscosity at the die, which tends to restrict flow through the die. The effect of moisture content on mean residence time is expected to be the result of these two opposing effects on the rheology of material in the barrel and die of the extruder [19].

31.7.2   MEASUREMENT TECHNIQUES

31.7.2.1   Stimulus Response Technique

Stimulus response technique is the most common method to determine RTD in an extruder. A stimulus is provided by a tracer (usually a colored dye) and the response is measured as dye concentration in the extrudate either by absorption spectrophotometry or by reflectance colorimetry. Tracers that have been used in this method are yellow dye and carbon black [20], blue tracer [21], FD&C No. 40 red dye [18], rhodamine B [22], and erythrosine [23]. This method can be used to determine RTD accurately. However, a considerable amount of time is needed for sample preparation and determination of tracer concentration [12].

31.7.2.2   Online Techniques

Online measurement of RTD using radioactivity, dielectric properties, optical properties and electrical conductivity of tracer has been reported. Radioactive tracers that have been used include $^{56}MnSO_4 \cdot H_2O$ [24], ^{64}Cu [25], and Indium-113 [26]. Tracer activity in the die is measured by either a scintillation detector or a probe. However, the high cost of radioactive tracers and extensive safety requirements make practical application of this technique difficult. Golba [27] used dielectric properties of carbon black for online measurement of RTD in an extruder. The test cell consisted of a specially designed parallel plate capacitor incorporated in a slit die. However, this method can be applied to only a slit die. Chen et al. [28] used optical property of the tracer for online measurement of RTD in an extruder. The limitations of this method were fluctuation and drift of the photomultiplier output signal and nonuniformity of tracer concentration in the cross-section of the flow at the end of the barrel. A technique based on electrical conductivity of materials in the die for online measurement of RTD in an extruder was developed by Choudhury and Gautam [12]. The method used a series circuit consisting of a 5 mm diameter die, a 10-ohm resistor, and a 15 V dc power supply. The electrical conductivity of the materials was altered by the addition of an electrolyte (sodium nitrate) tracer at the feed inlet. The voltage across the resistor was measured under steady state. Material conductivity in the die was monitored as a proportional voltage response across the 10-ohm resistor. The RTD results from this method correlated well with traditional erythrosine dye method.

31.8   MATHEMATICAL MODELING OF EXTRUSION PROCESSING

Modeling of extrusion processing has mainly focused on the metering section because it controls the rate of extrusion, accounts for most of the power consumption and causes uniform pressure behind the die [6].

31.8.1 FLUID FLOW

The main reason to model fluid flow in an extruder is the insight it provides for the mechanism of mixing, RTD, prediction of flow rates, pressure drop and power consumption. However, the basic problem in modeling fluid flow inside an extruder is the non-Newtonian and nonisothermal nature of flow in the compression and metering sections of an extruder. One of the simplified theories assumes steady-state laminar flow of an incompressible isotropic Newtonian fluid with constant viscosity and neglects any slip at the walls, gravity, inertial forces and any curvature of the channel around the screw axis. It also assumes that the barrel is rotating and the screw is stationary.

For this simplified theory, the momentum transfer equation can be written as [29]:

$$\frac{1}{\mu}\left(\frac{\partial P}{\partial z}\right) = \left[\frac{\partial^2 v_z}{\partial x^2} + \frac{\partial^2 v_z}{\partial y^2}\right] \tag{31.21}$$

where P is the pressure in Pa, μ is the viscosity in Pa.s, and v is the velocity in m.s^{-1}. The flow in the z direction needs to be considered to determine the extrusion output. A solution to the momentum transfer equation can be written as [29]:

$$v_z = \frac{4V_z}{\pi}\sum_{g=1,3,5\ldots}^{\infty}\frac{1}{g}\frac{\sinh\left(\dfrac{g\pi y}{W}\right)}{\sinh\left(\dfrac{g\pi H}{W}\right)}\sin\left(\frac{g\pi x}{W}\right)$$

$$-\frac{1}{\mu}\left(\frac{\partial P}{\partial z}\right)\left[\frac{y^2}{2}-\frac{Hy}{2}+\frac{4H^2}{\pi^3}\sum_{g=1,3,5\ldots}^{\infty}\frac{1}{g^3}\frac{\cosh\left(\dfrac{g\pi}{2H}\right)(2x-W)}{\cosh\left(\dfrac{g\pi W}{H}\right)}\sin\left(\frac{g\pi y}{H}\right)\right] \tag{31.22}$$

Once v_z is known, the following equation can be used to obtain an expression for the flow rate (Q).

$$Q = p\int_0^H\int_0^W v_z\, dy\, dx \tag{31.23}$$

where p is the number of channels in parallel (one for single flighted and two for double flighted screws). Using the boundary conditions of $v_z(x,0)=0$, $v_z(x,H)=V_z$, $v_z(0,y)=0$, and $v_z(W,y)=0$, which indicate a no slip condition at the channel boundaries, the above equation can be integrated to yield [29]:

$$Q = p\frac{V_z WH}{2}F_d + p\frac{WH^3}{12\mu}\left(\frac{\partial P}{\partial z}\right)F_p = Q_d + Q_p \tag{31.24}$$

where $V_z = \pi DN\cos\theta$, $W = \pi D\tan(\theta/p)\cos\theta - e$, N is the speed of screw in revolutions per unit time, F_d is the drag flow shape factor, and F_p is the pressure flow shape factor. The first term in the above equation is called the drag flow (Q_d) which results from viscous drag and is proportional to N. Drag flow is similar to the flow of a liquid between two plates. Movement of one of the plates causes the liquid at the surface of the moving plate to move at the speed of the moving plate. Drag flow results in a forward movement of material in extruder. The second term is the pressure flow (Q_p) which is the flow due to the pressure gradient between the feed zone and the metering zone of the extruder.

The extent of this flow can be controlled by the size of the die opening and screw configuration. This flow is from the die end toward the feed zone under normal operation.

In many food applications of extrusion, drag flow is much larger than pressure flow. Thus, net flow rate increases linearly with N. The ratio of pressure flow to drag flow is an important parameter and is given by [6]:

$$a = -\frac{Q_p}{Q_d} = \frac{H^2}{6V_z\mu}\frac{\partial P}{\partial z}\frac{F_p}{F_d} \tag{31.25}$$

The shape factors F_d and F_p are given by [6]:

$$F_d = \frac{16W}{\pi^3 H}\sum_{i=1,3,5...}^{\infty}\frac{1}{i^3}\tanh\left(\frac{i\pi H}{2W}\right)$$

$$\tag{31.26}$$

$$F_p = 1 - \frac{192H}{\pi^5 W}\sum_{i=1,3,5...}^{\infty}\frac{1}{i^5}\tanh\left(\frac{i\pi W}{2H}\right)$$

These shape factors approach unity as H/W approaches zero. Considering the case when $H/W=0$ and there is no slip at the root of the channel or at the barrel [$v_z(x,0)=0$, $v_z(x,H)=V_z$], the expression for v_z (velocity along the z direction) can be written as [6]:

$$v_z = \frac{V_z y}{H} + \frac{y^2 - Hy}{2\mu}\left(\frac{\partial P}{\partial z}\right) \tag{31.27}$$

which can further be rewritten as [6]:

$$v_z = V_z\left[(1-3a)\frac{y}{H}+3a\left[\frac{y}{H}\right]^2\right] \tag{31.28}$$

where a equals the negative ratio of pressure flow to drag flow.

A similar analysis can be performed for the cross-channel velocity profile with the assumption that $\partial^2 v_x/\partial x^2$ is small. Using the boundary conditions $v_x(z,0)=0$ and $v_x(z,H)=-V_x=-\pi DN\sin\theta$ and integrating the momentum transfer equation twice, the expression for v_x can be written as [6]:

$$v_x = -\frac{V_x y}{H} - \frac{1}{2\mu}\frac{\partial P}{\partial x}(y^2 - yH) \tag{31.29}$$

There is no net flow in the x direction because of the flights on the screw. This can be written as [6]:

$$\int_0^H v_x dy = 0 \tag{31.30}$$

Integrating the above equation yields:

$$\frac{\partial P}{\partial x} = -6\mu\frac{V_x}{H^2} \tag{31.31}$$

Substitution of the above equation in the expression for v_x (velocity in the x direction) yields [6]:

$$v_x = \frac{y}{H}\left(2 - \frac{3y}{H}\right)V_x \tag{31.32}$$

It can be concluded from the above equation that the cross channel velocity is not affected by the down channel pressure gradient.

The flow rate in an extruder can be more conveniently written in terms of common geometric parameters of D, e, L, H, N, and θ as [6]:

$$Q = G_1 N F_{dt} + \frac{G_2}{\mu}\left(\frac{P_1 - P_2}{L}\right)F_{pt} \tag{31.33}$$

where

$$G_1 = \frac{\pi^2}{2}D^2 H\left(1 - \frac{ep}{\pi D \sin\theta}\right)\sin\theta\cos\theta$$

$$G_2 = \frac{\pi}{2}DH^3\left(1 - \frac{ep}{\pi D \sin\theta}\right)\sin^2\theta \tag{31.34}$$

$$F_{dt} = F_d F_{de} F_{dc}$$

$$F_{pt} = F_p F_{pe} F_{pc}$$

F_{de} and F_{pe} are the end correction factors for drag and pressure flow which account for the oblique end of a real screw. F_{dc} and F_{pc} are the curvature correction factors for drag and pressure flow which take into account the curvature of the channel, P_1 is the pressure at the beginning of the metering section, and P_2 is the pressure at the discharge of the metering section. Booy [30,31] developed correlations for curvature and end correction factors for both drag and pressure flow.

Apart from drag and pressure flow, there is a third type of flow in an extruder which arises due to the clearance (δ) between the screw and the barrel. This type of flow, known as the leakage flow (Q_l), is caused by the cross channel pressure gradient. As the screw wears with time, clearance increases which in turn increases leakage flow. Increase in clearance also effectively reduces H, thereby reducing drag flow in the extruder [6].

Li and Hsieh [32] developed an analytical model for an isothermal and Newtonian fluid flow in a single-screw extruder. The analytical solution obtained satisfied the actual boundary conditions in single screw extruders. The model of Li and Hsieh [32] was further extended by Ferretti and Montanari [33] who used the finite-difference method to simulate flow of a Newtonian fluid through a barrel to predict downstream velocity profile in an extrusion process. The use of Microsoft Excel to implement the model made it straightforward and easy to use.

For a non-Newtonian fluid, the above model can lead to errors because drag flow and pressure flow terms become dependent on each other. However, the above model can be used with reasonable accuracy by substituting viscosity (μ) by apparent viscosity (η). Apparent viscosity should be calculated at an effective shear rate ($\dot{\gamma}_H$) given as [6]:

$$\dot{\gamma}_H = \frac{\pi D N}{H} \tag{31.35}$$

The models developed above assume isothermal conditions in the extruder which is not true in practice. Temperatures of the barrel and screw are influenced by the steam jackets surrounding the barrel and the presence of steam in a hollow screw, respectively. Assuming nonisothermal Newtonian flow and a linear viscosity profile across the channel, a solution to the momentum transfer equation in the z direction can be written as [6]:

$$Q = G_1 N F_{\mu d} F_{dt} + \frac{G_2}{\mu_m} \left(\frac{P_1 - P_2}{L} \right) F_{\mu p} F_{pt} \tag{31.36}$$

where μ_m is the mean viscosity in the channel and is given by:

$$\mu_m = \frac{\mu_b + \mu_s}{2} \tag{31.37}$$

where μ_b is the viscosity at the barrel surface and μ_s is the viscosity at the root of screw.

The terms $F_{\mu d}$ and $F_{\mu p}$ are viscosity factors for the drag and pressure flow, respectively. These factors are a function of the parameter γ (μ_b/μ_s) and are given as [29]:

$$F_{\mu d} = 2 \left[\frac{\gamma}{\gamma - 1} - \frac{1}{\ln \gamma} \right]$$
$$F_{\mu p} = \frac{3(1 + \gamma)}{1 - \gamma} \left[1 - 2 \left(\frac{\gamma}{\gamma - 1} - \frac{1}{\ln \gamma} \right) \right] \tag{31.38}$$

When $\gamma = 1$, both $F_{\mu d}$ and $F_{\mu p}$ equal unity and the above flow rate Equation 31.36 reduces to Equation 31.33, previously developed for flow with a constant viscosity in the screw channel. If the screw is cooled, the viscosity at the channel root becomes more than that at the barrel surface so that γ is less than 1. In this case, $F_{\mu d}$ decreases, which reduces the drag flow rate. Heating the screws can either increase or decrease the net flow rate depending on the relative magnitude of drag and pressure flows.

For a real extrusion process, the flow is nonisothermal and non-Newtonian. Development of an analytical model considering nonisothermal and non-Newtonian flow is complex. White and Chen [34] developed a model to simulate nonisothermal flow in the screw and kneading elements in a modular corotating twin screw extruder. The flow was described by the basic equations of lubrication theory which balances pressure gradients and shear stress fields.

Yu and Gunasekaran [35] developed a model to analyze flow for isothermal and nonisothermal conditions for both Newtonian and non-Newtonian fluids (power-law and Bird-Carreau models) inside a deep channel single-screw extruder. The traditional flat plate model for a single-screw extruder was modified to address the effects of flight and channel curvature. Finite difference and finite element methods were used to obtain numerical solutions to the flow and energy equations. Dimensionless numbers, Brinkman number (B_r) and Peclet number (P_e) strongly affected the temperature profile. The results showed that flight effect was more important than curvature effect in terms of flow rate.

The operating characteristics of an extruder are determined by coupling the flow in the barrel of the extruder with that in the die at the discharge of the extruder. For common die cross-sections such as circular, slit or annular, neglecting entrance and exit losses, the Hagen–Poiseuille equation can be used to determine the flow rate of a Newtonian fluid as follows [6]:

$$Q = k \frac{\Delta P}{\mu_d} \tag{31.39}$$

where K is the geometric constant which depends on the type of die opening, ΔP is the pressure drop across the die, and μ_d is the viscosity of the material at the die opening. For some common cross-sections, K is given as [6]:

$$Circular: K = \frac{\pi R^4}{8L_d}$$

$$Slit: K = \frac{2wC^3}{3L_d} \tag{31.40}$$

$$Annular: \ K = \frac{\pi (R_o + R_i)(R_o - R_i)^3}{12L_d}$$

where R is the radius of the die, L_d is the length of the die, w is the width of the slit, and $2C$ is the height of the slit.

The Hagen–Poiseuille equation cannot be applied to a non-Newtonian fluid because of the non-linear relationship between Q and ΔP. Pressure drop across the die for a non-Newtonian fluid is estimated by calculating shear rate ($\dot{\gamma}$) and shear stress (τ).

$$Circular: \dot{\gamma} = \frac{3n+1}{4n}\left(\frac{4Q}{\pi R^3}\right), \qquad \tau = \frac{R\Delta P}{2L_d}$$

$$Slit: \dot{\gamma} = \frac{2n+1}{3n}\left(\frac{3Q}{2C^2W}\right), \quad \tau = \frac{C\Delta P}{L_d} \tag{31.41}$$

where n is the flow behavior index and can be determined from data on a rheological model of the material. The apparent viscosity can also be determined from the rheological model. Since apparent viscosity is the ratio of shear stress to shear rate, the pressure drop across the die can be calculated from the above equations.

The above equations for flow through an extruder die assume that there is no pressure drop associated with end effects on the die. However, pressure drop occurs due to contraction and expansion at the entrance and exit respectively. The end correction for an individual die hole can be written as [6]:

$$\tau = \frac{\Delta P}{2\left(\dfrac{L}{R} + \dfrac{L^*}{R}\right)} \tag{31.42}$$

where τ is the effective shear stress at the wall and L^*:R is the ratio of the equivalent length to radius for a hole of zero length.

For a die of irregular shape, the radius, R, in the above equations is replaced by hydraulic radius, R_H, given by the following equation [6]:

$$R_H = \frac{s}{\psi} \tag{31.43}$$

where s is the cross-sectional area and ψ is the wetted perimeter of the die.

31.8.2 Mixing

Mixing in an extruder is important for the uniformity of the product. To ensure good mixing, the dry and wet ingredients are pre-blended before they enter the extruder. Mixing occurs in an extruder due to laminar shear flow. The extent of mixing during extrusion can be measured by the scale of segregation which is a measure of the actual size of the minor component after mixing [6]. Increase in mixing is related to the total strain received by the material. Shear strain (γ) is defined as [6]:

$$\gamma = \dot{\gamma}t \tag{31.44}$$

where $\dot{\gamma}$ is the shear rate and t is the time of deformation. Pinto and Tadmor [10] defined a weighted average total strain (WATS), $\bar{\gamma}$, which combines shear strain and RTD as:

$$\bar{\gamma} = \int_{0}^{\infty} \gamma E(t)dt \tag{31.45}$$

WATS is a measure of the total deformation the extrudate experiences in the extruder. It gives a quantitative measure of the extent of mixing. A high value of WATS indicates a well mixed extrudate. WATS depends on three parameters—L/H, the helix angle (θ) and the pressure to drag flow ratio (a). WATS is independent of net flow rate which means the same extent of mixing can be achieved at high flow rates as at low flow rates provided the pressure to drag flow ratio remains constant [10]. Their results showed that there is little influence of the helix angle (θ) over the range of 20–75°. However, decrease in the helix angle below 20° increases WATS. This means extruders with screws with lower helix angle will provide better mixing of the material.

WATS can be normalized by using a nominal strain (γ_N) which assumes that shear rate in an extruder is approximately V_z/H and is exerted for an average residence time of $\bar{t}$. Thus, nominal strain can be written as [6]:

$$\gamma_N = \frac{\pi D N (\cos\theta)\bar{t}}{H} \tag{31.46}$$

Normalized strain is defined as:

$$\bar{\gamma}^* = \frac{\bar{\gamma}}{\gamma_N} \tag{31.47}$$

Bigg and Middleman [11] calculated normalized strain for a power law fluid as a function of flow behavior index (n) and flow rate. The results showed that for a fixed flow rate, normalized strain is smaller for non-Newtonian fluids and the difference is greater at lower flow rates.

31.8.3 Heat transfer

Understanding heat transfer is very important for proper control and optimization of the cooking process in an extruder. The material in an extruder is heated due to heat transfer into the barrel and viscous energy dissipation within the barrel. Some of the dimensionless numbers which are used to determine the rate of heat transfer during extrusion are described below [36].

    a. Nusselt number (N_u) is the ratio of total heat transfer to heat transfer by conduction and is given by:

$$N_u = \frac{hL}{k} \tag{31.48}$$

where h is the heat transfer coefficient (W.m^{-2}.K^{-1}), L is the characteristic length of the extruder (m), and k is the thermal conductivity (W.m^{-1}.K^{-1}).

b. Prandtl number (P_r) is the ratio of momentum diffusivity to thermal diffusivity and is given by:

$$P_r = \left(\frac{\mu c_p}{k} \right)$$
(31.49)

where c_p is the specific heat of the material (J.kg^{-1}.K^{-1}).

c. Graetz number (G_z) is the ratio of heat transfer by convection to heat transfer by conduction and is given by:

$$G_z = \frac{\alpha L}{VH^2}$$
(31.50)

where α is the thermal diffusivity (m^2.s^{-1}), V is the velocity of barrel wall with respect to the screw (m.s^{-1}), and H is the characteristic length (m).

d. Brinkman number (B_r) is the ratio of heating due to conversion of mechanical energy by viscous dissipation to heating due to conductive heat transfer and is given by:

$$B_r = \frac{\mu V^2}{k}$$
(31.51)

where μ is the Newtonian viscosity (Pa.s).

e. Peclet number (P_e) is the ratio of heat transfer by convection to heat transfer by conduction and is given by:

$$P_e = \frac{VL}{D}$$
(31.52)

Analysis of heat transfer in an extruder can be done by either of the two approaches described below [37].

The first approach considers extruder as a whole and quantifies mechanical and thermal energy involved in the extruder as a whole. This approach determines the energy consumption of the process and quantifies the energy transferred to the material. The global energy balance equation can be written as [37]:

$$P_{\text{mechanical}} + P_{\text{heating}} = P_{\text{cooling}} + P_{\text{losses}} + P_{\text{material}}$$
(31.53)

where $P_{\text{mechanical}}$ is the mechanical power supplied by the motor (W), P_{heating} is the thermal power supplied by the heating system (W), P_{cooling} is the thermal power supplied by the cooling system (W), P_{losses} is the thermal loss to the environment (W), and P_{material} is the power absorbed by the material (W).

The second approach considers heat transfer in the barrel alone. Thermal changes in the material, as it passes through the extruder are determined by solving a one-dimensional model of heat transfer given by [37]:

$$dH = dq + h_{m/b} \cdot dS_{m/b} \cdot (T_b - T_m) + h_{m/s} \cdot dS_{m/s} \cdot (T_s - T_m)$$
(31.54)

where T_b, T_m, and T_s are temperatures of the barrel, material and screw, respectively. dH is the change in internal energy of the material, dq is the heat added to the material. dq is positive if there is a viscous dissipation or exothermic reaction inside the extruder and is negative when there is an endothermic reaction inside the extruder. $h_{m/b}$ is the convective heat transfer between the material and the barrel (W.m^{-2}.K^{-1}), $h_{m/s}$ is the convective heat transfer between the material and the screw (W.m^{-2}.K^{-1}), $dS_{m/b}$ is the heat transfer area between the material and the barrel (m^2), and $dS_{m/s}$ is the heat transfer area between the material and the screw (m^2). In most heat transfer analyses, heat transfer between the material and the screw is neglected.

Convective heat transfer between the material and barrel depends on convective heat transfer coefficient (h). For materials in powder form, heat transfer coefficient has been found to range from 30 to 2000 W.m^{-2}.K^{-1}. For molten material, h has been found to range from 136 to 768 [37]. Levine and Rockwood [38] developed an expression for the Nusselt number as a function of Peclet number, Brinkman number and the extruder geometry as follows:

$$N_u = \frac{hD}{k} = \left(\frac{P_e H}{L}\right)^{K_1} B_r^{K_2} \tag{31.55}$$

where k_1 and k_2 are empirical parameters. Heat transfer coefficient (h) can be calculated once Nusselt number is known.

Mohamad et al. [39] developed a theoretical model to study the effects of material properties, geometry and operating conditions of the extruder on heat transfer coefficient in single-screw food extruders. Todd [40] developed another expression for the Nusselt number as:

$$N_u = \frac{hD_{ext}}{K} = 0.94\left(\frac{D_{ext}^2 N \rho}{\mu}\right)^{0.28} P_r^{0.33}\left(\frac{\mu}{\mu_w}\right)^{0.14} \tag{31.56}$$

where D_{ext} is the external screw diameter (m) and μ_w is the viscosity of water (Pa.s).

White et al. [41] gave another expression for the Nusselt number for heat transfer between the material and the barrel surface as:

$$N_u = \frac{hD}{K} = 0.807 F\left(\frac{\rho c_p D^2 N}{k}\right)^{1/3}\left(\frac{\pi D^2 (\cos\theta)}{L H}\right)^{1/3} \tag{31.57}$$

Qualitative description of two phase flow in the melting zone of an extruder was described by Maddock [42]. It was proposed that solid particles in contact with the hot surface of the barrel partially melts and forms a film of molten polymer over the surface of the barrel. Tadmor and Klein [43] proposed a model based on Maddock's work for calculating the length of the melting zone. However, the viscous dissipation in the channel, fluid flow and heat transfer in the downstream direction were neglected in this model. Jepson [44] developed the theory of heat penetration to consider the wiping effect. Wiping effect considers the fact that after the flight of the screw has wiped a certain area on the inner barrel surface, a fresh layer of material attaches to the same region and remains there for approximately one revolution. The amount of heat penetrating into this layer by conduction is homogeneously distributed throughout the bulk of the material. Jepson's model is valid only when the penetration depth is small in comparison to the boundary layer thickness.

Yacu [45] described a heat transfer model to simulate three sections (solid conveying, melt pumping and melt shearing) of a twin-screw co-rotating extruder. The model assumed a non-Newtonian and nonisothermal viscosity model, laminar flow, steady state conditions, negligible gravity effects, negligible heat transfer to the screw shaft and flow in the axial direction. In the solid conveying section, energy balance for an element normal to the axial direction was expressed as [45]:

$$\dot{m}_s c_{ps} T + F U_s A (T_b - T) \partial x = \dot{m}_s c_{ps} (T + dT) \tag{31.58}$$

where $\dot{m}_s$ is the feed rate (kg.s^{-1}), c_{ps} is the specific heat capacity of solid material (J.kg^{-1}.K^{-1}), T is the temperature at distance x, F is the degree of fill, U_s is the pseudo heat transfer coefficient (W.m^{-2}.K^{-1}) which considers both convection and conduction, and A is the surface area (m^2). Using the boundary condition, $T(x=0)=T_0$, the above equation yields the temperature profile as:

$$T = T_b - (T_b - T_f)\exp\left(\frac{-FU_s A x}{\dot{m}_s c_{ps}}\right) \tag{31.59}$$

In the melt pumping section, material changes from a solid powder to a fluid melt and the screw becomes completely filled. The energy converted to heat by viscous dissipation per screw channel in one screw was estimated to be a sum of energy converted within the channel, between the flight tip of the screw and the inside surface of the barrel, between flight tip of one screw and the bottom of channel of other screw, and flights of opposite screw parallel to each other. The total viscous dissipation (Z_p) per channel per screw turn was expressed as [45]:

$$Z_p = C_{1p}\mu N^2 \tag{31.60}$$

where C_{1p} is the pumping section screw geometry parameter and is described as [45]:

$$C_{1p} = \frac{\pi^4 D_e^3 D(\tan\theta)}{2H} + m^*\left[\frac{\pi^2 D^2 e C_e}{\delta} + \frac{8\pi^2 I^3 e}{\varepsilon} + \frac{\pi^2 I^2 H \sqrt{(D^2 - I^2)}}{2\sigma}\right] \tag{31.61}$$

where D_e is the equivalent twin screw diameter, m^* is the number of screw flights, e is the screw flight tip width in the axial direction, C_e is the equivalent twin-screw circumference, I is the distance between the screw shafts, ε is the clearance between flight tip and channel bottom of two opposite screws, and σ is the clearance between flights of opposite screws parallel to each other. The energy balance for an element in the melt pumping zone was described as [45]:

$$\frac{\partial T}{\partial x} = C_{2p}\exp(-b_1 T) + C_{3p}(T_b - T) \tag{31.62}$$

where

$$C_{2p} = \frac{C_{1p}\mu N^2}{\pi D(\tan\theta)\dot{m}_s c_{pm}} \tag{31.63}$$

and

$$C_{3p} = \frac{FA}{\dot{m}_s c_{pm}}$$

c_{pm} is the specific heat capacity of the molten material (J.kg^{-1}.K^{-1}), and b$_1$ is the temperature coefficient of viscosity.

In the melt shearing section, shearing is achieved by the reverse screw elements. The mechanism of fluid flow and heat transfer in this section is complex due to the existence of cross-channels normal to the screw channel. The temperature profile in this section was expressed by Yacu as [45]:

$$\frac{\partial T}{\partial x} = (C_{2rs} + C_{4rs})\exp(-b_1 T) + C_{3rs}(T_b - T)) \tag{31.64}$$

where C_{2rs} and C_{3rs} are similar to C_{2p} and C_{3p} described in the melting section. C_{4rs} is calculated as [45]:

$$C_{4rs} = \frac{\dot{m}D\mu\dot{\gamma}^{-n}\left(1 - \dfrac{m_1 BG}{\pi DH}\right)}{c_{pm}H^5(\tan\theta)} \tag{31.65}$$

where m_1 is the number of reverse screw elements, B is the width of reverse cross-channel and G is the depth of reverse cross-channel. Tayeb et al. [46] further improved the model by Yacu [45] for flow through the reverse screw element of a twin-screw extruder. There was an additional term in the energy balance equation to account for the energy generated by the friction of the material in the space between the screw tip and the barrel wall.

Bouvier et al. [47] presented a one-dimensional model of heat transfer during extrusion cooking of defatted soy flour in a single-screw extruder. The model assumed wider and shallow ($H/W \leq 1$) screw channel, steady state, constant thermophysical properties and uniform flow along the z axis. The energy balance equation combining both convective and viscous heating was given as [47]:

$$\rho c_p v_z(y)\frac{\partial T}{\partial z} = k\frac{\partial^2 T}{\partial y^2} + \mu\left(\frac{\partial v_z}{\partial y}\right) \tag{31.66}$$

where ρ is the density of the material, c_p is the specific heat of the material, k is the thermal conductivity of the material, μ is the viscosity of the material, and T is the temperature of the material. Using the boundary conditions, $T=T_i$ at $z=0$, $T=T_b$ at $y=0$ and $y=H$ for temperature and the expression of velocity (v_z) from Section 31.8.1, the above equation was solved using the finite difference to determine the temperature profile in the y direction.

LU model, developed by van Zuilichem et al. [36], was based on the penetration theory of Jepson [44]. This model calculates total heat transferred in the extruder for every position along the screw axis. The model considers heat penetration from the barrel to the material and heat generated by viscous dissipation. The model was incorporated into a computer program which generated a plot of the temperature of the material as a function of axial distance in the extruder [36]. Mohamed and Ofoli [48] developed a model which incorporated viscous dissipation effects and a heat transfer coefficient based on Brinkman and Graetz number to predict temperature profiles of non-Newtonian dough in a twin-screw extruder. The model assumed uniform product temperature in the direction normal to the shaft of the screw.

Chiruvella et al. [49] presented numerical (finite difference) and analytical methods to solve for velocity and temperature profiles in the screw channel of a single-screw extruder. The equations were solved iteratively to satisfy conservation of mass, momentum and energy in the extruder and to match the pressure drop across the die. Sastrohartono and Jaluria [50] presented a numerical simulation of fluid flow and heat transfer in a corotating twin-screw extruder to predict temperature, velocity and pressure along the screw channels. The flow was assumed to be taking place in translational (flow similar to that in single-screw channel) and intermeshing (between two screws) regions. Simulation for both regions was done separately and coupled for each screw section to model the transport process. The finite difference method was employed for modeling in the translational region whereas the modeling in the intermeshing region was done by a finite element method.

White and Chen [34] developed a model to simulate heat transfer in the screw and kneading elements in a modular corotating twin screw extruder. Heating by viscous dissipation and conduction from the solid surface was considered in the model. Mean temperature distribution in the channels was considered rather than detailed temperature profiles. White and Bawiskar [51] extended their previous model [34] to describe melting in a self-wiping corotating twin screw extruder. The melting process in a modular screw configuration occurs in specific sections such as the kneading section. The model assumed formation of two stratified layers during melting—melt layer at the hot barrel surface and solid bed in contact with the screw. The model predicted both the location and length of melting in an extruder. Effect of operating variables (mass flow rate and screw speed) on melting was also studied. White et al. [41] further extended their previous models to predict axial mean temperature and screw temperature profile in a modular self-wiping corotating twin-screw extruder. The screw temperatures were much lower than the mean temperature in the screw channel. The results also showed that under the same conditions, an extruder with larger diameter produces more heating due to viscous dissipation compared to that of a small diameter extruder. The models developed by White [34,41,51] are available as a computer software (AKRO-CO-TWIN-SCREW®) from Temarex Corporation (Akron, OH). Version 1 of this software, released in 1990, was capable of computing pressure and fill factors for a modular screw/barrel system. Version 2 of the software, released in 1994, was able to handle nonisothermal and non-Newtonian flow. Version 3, released in 1998, also included integrated modeling of solids conveying, melting and melt flow in modular screw configurations. Version 3A, released in 2001, included all features of Versions 1, 2, and 3. Additionally, it is more user-friendly, easy to install and easy to operate.

A global computer software to obtain variables such as pressure, mean temperature, residence time and filling ratio along the screw in a twin screw extruder was developed by Vergnes et al. [52]. Flow in the forward screw, kneading screw and reverse screw elements was analyzed using a simplified one dimensional approach. The individual local models were linked together to obtain a global description of the flow field along the extruder. This model is now available as LUDOVIC© software. For predetermined criteria, this software can also be used for scale-up from pilot scale to industrial scale. SME is an important parameter used for scale-up because it governs product quality in extrusion of foods. LUDOVIC can be used to determine processing conditions required for a specified SME.

Wang et al. [53] developed a mathematical model using finite element modeling to simulate nonisothermal plug flow of starch granules in the feed section, melting in transition (compression) section, and non-Newtonian melt flow in metering section and die channel during processing in a single-screw extruder. The model considered variations in rheological and thermal properties of biomaterials. Characteristic curves generated by the model were used to determine the operating point of the extruder. The model was also able to predict bulk product temperature and pressure along the down channel of the screw and die during extrusion.

Apart from the AKRO-CO-TWIN-SCREW® and LUDOVIC© softwares which are commercially available, Fluent CFD (ANSYS Inc., NH) software can be used to visualize the velocity, pressure and temperature distribution in single and twin-screw extruders. The Fluent CFD software is capable of quantifying mixing, residence time and shear rate in the extruder. The software can also be used to evaluate the effects of changes in design and operating conditions on the extent of mixing. The unique mesh superposition technique employed by the software describes the complexity of screw rotation accurately. POLYFLOW is another software from Fluent which is capable of solving the Navier-Stokes equation in combination with the energy equation. This software can be used to determine the behavior of the extrudate as it passes through die of the extruder after considering the flow pattern, local pressure drop, deformation in each section and temperature profile. POLYFLOW can also be used to design the die for a given shape of the finished product.

31.8.4 Scale-Up

Scale-up is an issue which comes up while taking a pilot plant process and designing an industrial scale production process. Scale-up of an extrusion process is difficult because flow rate and power requirements of an extruder are not a linear function of the extruder geometry. When pressure flow is small compared to drag flow in an extruder ($a \approx 0$), flow rate and power requirements follow a cubical relation with the extruder geometry for a given screw speed. Thus, if all dimensions in the extruder were increased by a scale-up factor, Φ (ratio of diameter of large extruder to the diameter of small extruder), the flow rate and power will theoretically increase by a factor of Φ^3 [6].

In most extrusion processes, heat transfer from the barrel and viscous dissipation in material play an important role in the operation of the extruder. Thus, for proper scale-up, D, H, and N should be adjusted such that the heat transfer area, flow rate and viscous dissipation increase at approximately the same rate. Shear rate in the screw channel and clearance over screw flight are other factors that affect scale-up. During scale-up, the shear rate should remain the same in both the extruders [6].

Residence time is another parameter which should remain constant during scale-up of an extrusion process because residence time affects the behavior of a material inside the extruder. A first order approximation of average residence time can be given as [6]:

$$\bar{t} = \frac{V_s}{Q_D} = f\left(\frac{L}{DN}\right) \tag{31.67}$$

where V_s is the volume of screw channel (m³).

The die of the extruder should also be considered during scale-up. The number of die holes should be increased by the extruder scale-up factor (Φ). This will ensure similar pressure drop across the die and shear rate of material through die for both the extruders [6].

In the most common scale-up method, screw flight height (H) is increased by the square root of scale-up factor and screw speed is decreased by the square root of the scale-up factor. This increases flow rate proportionally to the square of scale-up factor. The apparent shear rate in the screw channel and clearance remain constant. However, the average residence time and the peripheral screw speed increase by a factor of square root of the scale-up factor. Increased residence time and peripheral screw speed increase screw power consumption by a factor of $\Phi^{2.5}$, resulting in a higher melt temperature. Analysis of this common scale-up method shows that the pumping capacity and solid conveying capacity increase more than necessary whereas the increase in melting capacity is not sufficient. This could be overcome by increasing the barrel surface area in the melting section and decreasing the barrel surface area in the pumping and solid conveying section [54].

Rauwendaal [55] performed a basic analysis of scale-up in extruders and compared the effects of various existing scale-up methods on performance of the extruder. The performance was analyzed in terms of solids conveying, melting, melt conveying, mixing, residence time, heat transfer, power consumption, and specific energy consumption. Three most critical parameters which should remain constant during scale-up are SME, flow rate (solids conveying, melting and pumping), and the ratio of surface area to throughput. None of the existing scale-up methods resulted in these three critical parameters being constant. They proposed two new scale-up methods that resulted in constant SME and high throughput rates. The first scale-up method, which holds true at high Brinkman number (nearly adiabatic process), keeps SME and specific surface area constant. However, the melting rate is insufficient at low Brinkman number (negligible viscous dissipation). Therefore, the second scale-up method keeps the melting rate at low Brinkman number equal to the pumping rate.

Ganzeveld and Janssen [56] proposed scale-up methods for counter-rotating closely intermeshing twin screw extruders and extended the use of this scale-up method for an extruder working as a chemical reactor. They concluded that the important factors during scale-up are extent of filling of the barrel and leakage flows. For high Brinkman number, the throughput is proportional to the cube of screw diameter. For low Brinkman number, screw speed must be inversely proportional to the square of screw diameter for a throughput proportional to screw diameter.

Bigio and Wang [57] developed scale-up methods both experimentally and theoretically for a nonintermeshing counter-rotating twin-screw extruder. The effects of various extrusion parameters such as percentage of drag flow, screw stagger and screw speed on the extent of mixing were studied. A new scale-up method was developed which increased the diameter (D) of the extruder but kept the ratios L/D and H/D constant. Total shear (γ_{total}) given to the fluid was given as [57]:

$$\gamma_{total} = \frac{L}{H\sin\theta} \tag{31.68}$$

Using this scale-up method, screw speed could be changed because it does not affect total shear. Thus, this method gives the flexibility to change screw speed to balance solids conveying, melting or pumping rate.

31.9 EXTRUSION IN THE FOOD INDUSTRY

The first use of extrusion in the food industry was in the mid-1940s when an expanded cornmeal-based snack was produced using a single-screw extruder. In the food industry, single-screw extruders are more popular than the twin-screw extruder because twin-screw extruders were developed for the food industry only in the early 1980s [58].

31.9.1 DIRECT EXPANDED SNACKS

The majority of extruded snacks belong to the category of direct expanded snacks. Raw material is fed into the extruder where it is exposed to moisture, heat and pressure. Extruders for direct expanded snacks are normally short in length (L:D less than 10:1). The moisture content of direct expanded snacks is between 8 and 10 wt% (on wet basis) and additional drying to 1–2 wt% moisture content is required to produce the desired product crispiness [58].

31.9.2 BREAKFAST CEREALS

Production of RTE breakfast cereals using extrusion offers several advantages over conventional processing methods such as rotary cookers. Some of the advantages include faster processing time, lower processing costs, less space requirements and greater flexibility in making cereals with varied shapes and sizes. The extruder cooks and shapes the final cereal product. After extrusion, the product is further dried and blended with additives before being packed [58].

Extrusion cookers are used extensively in the RTE cereal industry to cook dough for indirect expanded RTE cereals which are cold-formed into various shapes. The process of making indirect expanded RTE cereals involves mixing of flour blends with other minor ingredients, followed by preconditioning and cooking in the extruder, forming and cutting of the cooked dough ball, followed by forming of cooled cooked dough ball in a single-screw extruder. The dense unpuffed pellets are dried to 9–11 wt% moisture and tempered for at least 24 hours before being puffed in gun or tower puffers. Expansion of the pellets in gun or tower puffers occurs in three-axial plane whereas extruders expand products in a double-axial plane. The process of making direct expanded cereal is less complicated compared to the indirect expanded method [58].

Some of the raw materials which can be used in the production of RTE breakfast cereals include corn flours, whole wheat flour, white and brown rice flour, whole and defatted oat flours, sugar, wheat starch, sodium carbonate, colors, barley malt extract, and salt.

During extrusion of RTE breakfast cereals, starch present in the raw materials is gelatinized. The screw profile and configuration of the extruder is specifically designed to produce RTE cereals. Cereals with bran in the formula need preconditioning to maintain the quality of the product. Preconditioning ensures adequate hydration of dry grain-based raw materials before they enter the barrel of the extruder [58].

31.9.3 TEXTURED VEGETABLE PROTEINS (TVPs)

Meat extenders are textured vegetable food proteins which have been hydrated to 50–65 wt% moisture and blended with meat or meat emulsions to replace 20–30 wt% of the meats. Meat extenders represent the largest portion of TVP. Extrusion is also used to transform vegetable proteins into meat analogs. Meat analogs can be flavored and formed into sheets, disks, patties, strips and other shapes. TVP provides health benefits because they are free from cholesterol and contain low amount of fat [58].

Some of the raw materials which can be used for TVP are defatted soy flour, soy flakes, soy grits, soy concentrates, soy isolates, mechanically extracted soy meal, wheat gluten and other legumes/grains. Traditionally, soybean proteins have been used to produce TVP. However, proteins of wheat, pea, peanut, cottonseed, rapeseed, sesame, sunflower and lentils have also been used to produce TVP. The protein dispersibility index (PDI), which measures total protein dispersed in water under controlled conditions of extractions, is generally recognized as being more accurate and reliable. TVP products have been produced with raw materials having protein content of 50–70 wt% and a PDI of 50–70. Fat content in raw materials used to produce TVP vary from 0.5 to 6.5 wt%. As fat content increases, higher processing temperature and shear energy is required to maintain the quality of TVP products. Fiber content in raw materials used to produce TVP varies from 0.5 to 7 wt% and has a negative effect on the quality of TVP. Fiber partially blocks cross-linking of protein molecules and thus affects the structure and texture of TVP. Particle size of raw materials for TVP production varies from 45 to 150 μm. Particles smaller than 45 μm can cause clumping on wetting whereas particles over 180 μm are difficult to precondition [58].

During extrusion, protein molecules are denatured and they develop continuous plastic-like consistency. The extruder barrel, screws and die give a laminar profile to TVP foods by aligning denatured protein molecules in the direction of flow. The unpleasant volatile flavor compounds of vegetable proteins are flashed off with steam during extrusion. During extrusion of TVP, uniformity of minor ingredients throughout the protein matrix is possible due to homogeneous dispersion of all minor ingredients in the protein matrix. The TVP is shaped with the final die at the end of the extrusion process. The rotating cutting knife cuts and sizes it into its final shape [58].

Single-screw extruders are still the preferred method for manufacturing TVP-type products. However, twin-screw extruders are able to produce TVP with raw materials which are outside the specifications for single-screw extruders. Preconditioning and configuration of screws and die play an important role in producing TVP in an extruder. It is difficult to produce TVP without preconditioning because vegetable proteins without preconditioning tend to expand rather than have the laminar profile. Configuration of screws during production of TVP should be such that the hydrated and heated protein molecules are realigned in a laminar and stretched manner. Proper shearing and heating of materials is very important during TVP production because excessive shearing can decrease the strength and water holding ability of the final product. During production of TVP, dies should provide smooth streamlined flow so that cross-linked and laminar structures of the protein molecules are not disrupted [58].

Several minor ingredients are added to improve the quality of finished product during production of TVP. Calcium chloride is added in the range of 0.5–2 wt% to increase textural integrity and smooth the surface of the TVP product. Bleaching agents such as hydrogen peroxide are added in the range of 0.25–0.5 wt%. Pigments such as titanium dioxide are used to lighten the color of the product. pH plays an important role in the texturization of vegetable proteins. An increase in pH increases protein solubility and decreases the textural integrity of the final product. A decrease in pH decreases the solubility, thereby making protein difficult to process. Addition of salt decreases the textural integrity whereas addition of sodium alginate can increase chewiness, water holding capacity and density of the final TVP product. Soy lecithin added up to levels of 0.4 wt% assists in the development of laminar profile in the extruder and die. Sulfur is added in the range of 0.01–0.2 wt% to increase expansion and smooth the surface of TVP product [58].

31.9.4 CONFECTIONERY PRODUCTS

Twin-screw extruders are preferred in the processing of confectionary products because of their ability to convey materials, better temperature control and incorporation of fat, milk solids, nuts, color, and flavors. Twin-screw extruders used for confectionary products have a longer length to diameter ratio for enhanced heat transfer to the products such as licorice and caramel [59].

31.9.5 PASTA PRODUCTS

Pasta products are generally wheat-based products that are formed from dough without leavening. Durum semolina is the best material for making pasta flour because durum wheat is hard wheat having less wheat gluten. Semolina is mixed with water to attain a moisture content of 31 wt% and formed into small dough balls. It is desirable to keep air out of the dough during mixing by using either an air tight mixer or a degasser. Incorporation of air weakens the final dried product and activates lipoxygenase enzymes that can bleach the dough. A single-screw extruder with deep flight channels is used for the extrusion of pasta products. The extruder is also equipped with a jacket filled with cold water to keep the temperature less than 45°C. Low temperature combined with low moisture content results in little or no expansion and is desired in pasta production. After extrusion, pasta production involves a long drying step ranging from 10 to 16 hours [59].

31.9.6 THIRD-GENERATION (3G) SNACKS

Third generation snack products are sometimes referred to as semi- or half- products. Third generation products are cooked and formed as pellets in extruders. The pellets are dried to 6–8 wt% moisture and distributed to a snack processor where the pellets are expanded by hot oil, hot air or microwaves. The expanded products are then seasoned with spices, packaged and sold to consumers as RTE snacks. The pellets can also be sold directly to consumers for home consumption [58].

Raw materials, which are used to produce 3G snacks, should contain relatively high amount of starch (greater than 60 wt%) to maximize expansion of the final product. 30–35 wt% of proteins and protein enrichments can be added to 3G snacks without affecting the quality of final product. Oils or emulsifiers are used to reduce stickiness. Salt assists in uniform migration of moisture throughout the pellet after drying during the moisture equilibration period. Sodium bicarbonate adds flavor and texture to the finished product [58].

During manufacture of 3G snacks, it is very important that the starch is completely gelatinized. Temperatures in the cooking zone of the extruder vary from 80 to 150°C and the temperatures in the forming zone vary from 65 to 90°C. Total energy requirements for producing 3G snacks is lower for tuber starch such as potato as compared to whole cereal grains or high protein wheat flours because more energy is required to fully gelatinize starch granules [58].

31.10 FUTURE TRENDS

Industry and academic research and development on extrusion in the coming years will focus increasingly on (1) hardware modifications for improved product quality and flexibility in processing, (2) improved metallurgy for reduced wear of extruder parts, (3) advanced sensor and process control technologies, and (4) novel uses of extruders as continuous reactor systems for nontraditional applications.

Some of the hardware modifications, which are currently being rolled out by industry include— novel die designs and interchange systems that allow quick product changeover and multishape and multicolor products without the need for postextrusion blending, product density control systems involving a back pressure valve or a postextrusion pressure-control chamber, and special screw designs like conical twin screws that lead to more efficient processing. It is unlikely that there will be a sudden technological leap in extrusion processing in the near future, however hardware innovations such as those described above are going to provide incremental improvement in the coming years. Another aspect of hardware where developments are likely to be seen is the metallurgy of high-wear parts such as screws and barrels. Development of premium metallurgy for screw elements, barrel segments and barrel-liners would increase the life of these parts by several fold, reduce downtime and costs for replacement and repair of parts, and ensure long-term product consistency.

Research and development is also focused on process control. Rapid advances in data processing and sensor technologies will be utilized toward better online monitoring and control of process parameters and product characteristics. Online sensors based on methods such as electrical capacitance and resistance, and near infra-red (NIR) spectroscopy would increasingly be a feature in extrusion systems enabling real-time control of critical parameters such as moisture content of the product, residence time in the extruder barrel, and specific mechanical and thermal energy input.

Increasingly, the high pressure, shear and temperature environment encountered during extrusion is being utilized to facilitate thermo-chemical reactions, and extruders are being explored as high throughput continuous reactors in nontraditional applications such as production of industrial chemicals, modification of starches, encapsulation of ingredients and fabrication of nanocomposite materials. Research is also focused on using extrusion for breakdown of lignocellulosic biomass for more efficient downstream hydrolysis and fermentation to produce ethanol.

With increasing consumer demand for healthy food products, the use of extrusion to produce RTE high nutrition food products, and the effect of extrusion on functional foods and other healthful food components will also likely be the focus of studies in future.

REFERENCES

1. Rossen, J.L., Miller, R.C. 1973. Food extrusion. *Food Technology*, 27(8): 46–53.
2. Hauck, B.W., Huber, G.R. 1989. Single screw vs twin screw extrusion. *Cereals Foods World,* 34(11): 930–939.
3. Riaz, M.N. 2000. Introduction to extruders and their principles. In: Riaz, M.N., editor. *Extruders in Food Applications*. Technomic Publishing Company, Inc., Lancaster, PA. pp. 1–23.
4. Fellows, P. 2000. *Food Processing Technology: Principles and Practice*. Woodhead Publishing, Boca Raton, FL. pp. 294–308.
5. Guy, R.C.E. 1994. Raw materials for extrusion cooking processes. In: Frame, N.D., editor. *The Technology of Extrusion Cooking*. Chapman and Hall, Glasgow, UK. pp. 52–72.
6. Harper, J.M. 1981. *Extrusion of Foods*. Vol. 1. CRC Press, Inc., Boca Raton, FL. pp 1–212.
7. Liang, M., Huff, H.E., Hsieh, F.H. 2002. Evaluating energy consumption and efficiency of a twin-screw extruder. *Journal of Food Science*, 67(5): 1803–1807.
8. Ganjyal, G., Hanna, M. 2002. A review on residence time distribution (RTD) in food extruders and study on the potential of neural networks in RTD modeling. *Journal of Food Science*, 67(6): 1996–2002.
9. Levenspiel, P. 1972. Nonideal flow. In: Levenspiel, O., editor. *Chemical Reaction Engineering*. Wiley, New York, NY. pp. 252–325.
10. Pinto, G., Tadmor, Z. 1970. Mixing and residence time distributions in melt screw extruders. *Polymer Engineering and Science,* 10(5): 279–288.
11. Bigg, D., Middleman, S. 1974. Mixing in a screw extruder. A model for residence time distribution and strain. *Industrial & Engineering Chemistry Fundamentals*, 13(1): 66–71.
12. Choudhury, G.S., Gautam, A., 1998. On-line measurement of residence time distribution in a food extruder. *Journal of Food Science,* 63(3): 529–534.
13. Lee, S.Y., McCarthy, K.L. 1996. Effect of screw configuration and speed on RTD and expansion of rice extrudate. *Journal of Food Engineering*, 19: 153–170.
14. Singh, B., Rizvi, S.S.H. 1998. Residence time distribution (RTD) and goodness of mixing (GM) during CO_2-injection in twin screw extrusion. Part II: GM studies. *Journal of Food Engineering*, 21: 111–126.
15. Gogoi, B.K., Yam, K.L. 1994. Relationships between residence time and process variables in a corotating twin-screw extruder. *Journal of Food Engineering*, 21(2): 177–196.

16. Yeh, A.I., Hwang, S.J., Guo, J.J. 1992. Effects of screw speed and feed rate on residence time distribution and axial mixing of wheat flours in a twin-screw extruder. *Journal of Food Engineering*, 17(1): 1–13.
17. De Ruyck. 1997. Modeling of the residence time distribution in a twin screw extruder. *Journal of Food Engineering*, 32(4): 375–390.
18. Altomare, R.E., Ghossi, P. 1986. An analysis of residence time distribution patterns in a twin screw cooking extruder. *Biotechnology Progress*, 2: 157–163.
19. Seker, M. 2005. Residence time distributions of starch with high moisture content in a single-screw extruder. *Journal of Food Engineering*, 67: 317–324.
20. Kao, S.V., Allison, G.R. 1984. Residence time distribution in a twin-screw extruder. Polymer Engineering and Science, 24(9): 645–651.
21. Colonna, P. Melcion, J.P., Vergnes, B., Mercier, C. 1983. Flow, mixing, and residence time distribution of maize starch within a twin screw extruder with a longitudinal split barrel. *Journal of Cereal Science*, 1: 115–125.
22. Tsao, T.F. 1976. Available lysine retention during extrusion processing. PhD thesis, Colorado State University, Fort Collins, CO.
23. Yeh, A.I., Jaw, Y. 1998. Modeling residence time distribution for single screw extrusion process. *Journal of Food Engineering*, 35: 211–232.
24. Olkku, J., Antila, J., Heikkinen, J., Linko, P. 1980. Residence time distribution in a twin-screw extruder. In: P. Linko, Y. Malkki, J. Olkku, J. Larinkari, editors. *Food Process Engineering*. Vol. 1. Elsevier Applied Science Publishers, London, UK. pp. 791–794.
25. Van Zuilichem, D.J., Jager, T., Stolp, W. 1988. Residence time distribution in extrusioncooking. Part III: Mathematical modeling of axial mixing in a conical, counterrotating, twin-screw extruder processing maize grits. *Journal of Food Engineering*, 8: 109–127.
26. Kiani, A., Burkhardt, U., Heidemeyer, P., Franzheim, O., Rische, T., Stephan, M., Baetz, H., Pallas, R., Sahoub, M., Zeuner, A. 1996. A new online technique for morphology analysis and residence time measurement in a twin screw extruder. *Society of Plastics Engineers, ANTEC*, 54: 427–434.
27. Golba, J.C. 1980. A new method for the on-line determination of residence time distributions in extruders. Society of Plastics Engineers, ANTEC, 26: 83–85.
28. Chen, T., Patterson, W.I., Dealy, J.M. 1995. On-line measurement of residence time distribution in a twin-screw extruder. International Polymer Processing, 10(1): 3–9.
29. Squires, P.H. 1964. Screw extrusion—flow patterns and recent theoretical developments. *SPE Transactions*, 4: 7–16.
30. Booy, M.L. 1963. Influence of channel curvature on flow, pressure, distribution and power requirements of screw pumps and melt extruders. *SPE Transactions*, 3(3): 176–185.
31. Booy, M.L. 1967. Influence of oblique channel ends on screw-pump performance. *Polymer Engineering and Science*, 7(1): 5–16.
32. Li, Y., Hsieh, F. 1996. Modeling of flow in a single screw extruder. *Journal of Food Engineering*, 27: 353–375.
33. Ferretti, G., Montanari, R. 2007. A finite-difference method for the prediction of velocity field in extrusion process. *Journal of Food Engineering*, 83: 84–92.
34. White, J.L., Chen, Z. 1994. Simulation of non-isothermal flow in modular co-rotating twin screw extrusion. *Polymer Engineering and Science*, 34(3): 229–237.
35. Yu, C., Gunasekaran, S. 2004. Modeling of melt conveying in a deep-channel single-screw cheese stretcher. *Journal of Food Engineering*, 61(2): 241–251.
36. Van Zuilichem, D.J., ven der Laan, E., Kuiper, E. 1990. The development of a heat transfer model for twin-screw extruders. *Journal of Food Engineering*, 11: 187–207.
37. Mottaz, J., Bruyas, L. 2001. Optimized thermal performance in extrusion. In: Guy, R., editor. *Extrusion Cooking: Technology and Applications*. CRC Press, Inc., Boca Raton, FL. pp. 51–82.
38. Levine, L., Rockwood, J., 1986. A correlation for heat transfer coefficients in food extruders. *Biotechnology Progress*, 2(2): 105-108.
39. Mohamed, I.O., Morgan, R.G., Ofoli, R.Y. 1988. Average convective heat transfer coefficient in single-screw extrusion of non-Newtonian food materials. *Biotechnology Progress*, 4(2): 68–75.
40. Todd, D., 1988. Heat transfer in twin-screw extruders. *Society of Plastic Engineers, ANTEC*, 54–58.
41. White, J.L., Kim, E.K., Keum, J.M., Jung, H.C. 2001. Modeling heat transfer in screw extrusion with special application to modular self-wiping co-rotating twin-screw extrusion. *Polymer Engineering and Science*, 41(8): 1448–1455.

42. Maddock, B.H. 1959. A visual analysis of flow and mixing in extruder screws. *Society of Plastic Engineers Journal,* 15(5): 383–389.
43. Tadmor, Z., Klein, I. 1970. *Engineering Principles of Plasticizing Extrusion.* Van Nostrand-Reinhold, New York, NY. pp. 185–190.
44. Jepson, C.H. 1953. Future extrusion studies. *Industrial and Engineering Chemistry*, 45(5): 992.
45. Yacu, W.A. 1985. Modeling of a twin-screw co-rotating extruder. *Journal of Food Engineering,* 8: 1–21.
46. Tayeb, J., Vergnes, B., Della Valle, G. 1988. Theoretical computation of the isothermal flow through the reverse screw element of a twin-screw extrusion cooker. *Journal of Food Science,* 53(2): 616–625.
47. Bouvier, J.M., Fayard, G., Clayton, J.T. 1987. Flow rate and heat transfer modeling in extrusion cooking of soy protein. *Journal of Food Engineering*, 6: 123–141.
48. Mohamed, I.O., Ofoli, R.Y. 1990. Prediction of temperature profiles in twin screw extruders. *Journal of Food Engineering*, 12: 145–164.
49. Chiruvella, R.V., Jaluria, Y., Abib, A.H. 1995. Numerical simulation of fluid flow and heat transfer in a single-screw extruder with different dies. *Polymer Engineering and Science*, 35(3): 261–273.
50. Sastrohartono, T., Jaluria, Y. 1995. Numerical simulation of fluid flow and heat transfer in twin-screw extruders for non-Newtonian materials. *Polymer Engineering and Science*, 35(15): 1213–1221.
51. White, J.L., Bawiskar, S. 1998. Melting model for modular self wiping co-rotating twin screw extruders. *Polymer Engineering and Science*, 38(5): 727–740.
52. Vergnes, B., Della Valle, G., Delamare, L. A global computer software for polymer flows in corotating twin screw extruders. *Polymer Engineering and Science,* 38(11): 1781–1792.
53. Wang, L.J., Ganjyal, G.M., Jones, D.D., Weller, C.L., Hanna, M.A. 2004. Finite element modeling of fluid flow, heat transfer, and melting of biomaterials in a single-screw extrusion. *Journal of Food Science*, 69(5): E212–E223.
54. Chung, C.I. 1984. On the scale-up of plasticating extruder screws. *Polymer Engineering and Science,* 24(9): 626–632.
55. Rauwendaal, C. 1987. Scale-up of single screw extruders. *Polymer Engineering and Science*, 27(14): 1059–1068.
56. Ganzeveld, K.J., Janssen, L.P.B.M. 1990. Scale-up of counter-rotating closely intermeshing twin screw extruders without and with reactions. *Polymer Engineering and Science*, 30(23): 1529–1536.
57. Bigio, D., Wang, K. 1996. Scale-up rules for mixing in a non-intermeshing twin-screw extruder. *Polymer Engineering and Science*, 36(23): 2832–2839.
58. Sevatson, E., and Huber, G.R. 2000. Extruders in the food industry. In: Riaz, M.N., editor. *Extruders in Food Applications*. Technomic Publishing Company, Inc., Lancaster, PA. pp. 167–204.
59. Gray, D.R., Chinnaswamy, R. 1995. Role of extrusion in food processing. In: Gaonkar, A.G., editor. *Food Processing: Recent Developments*. Elsevier Science. London, UK. pp. 241–268.

Section VIII

Biofilm and Bioreactors

32 Modeling of Food Biofilms: A Metabolic Engineering Approach

Paul Takhistov
Rutgers, The State University of New Jersey

CONTENTS

32.1 INTRODUCTION

Bacteria on the surface develop a biofilm-associated community with higher resistance to toxic compounds [1,2] than their planktonic counterparts in the bulk. In general, biofilms result from physicochemical conditions and interactions in the bacteria/environment complex [3,4]. A biofilm consists of a living microbial biomass surrounded by an exopolysaccharide (EPS) envelope, proteins and nucleic acids, which the biofilm microorganisms produce. These components help bacteria to attach to surfaces, stabilize local environment, and spatially organize communities that need to collaborate to use the substrate effectively [5]. The process of the microorganism's attachment to a surface is very complex, and the nature of both the microbial cell surface and the supporting surface (substratum) is critical for successful attachment [6]. Surface adherence is an important survival mechanism for microorganisms. Moreover, the adhesion kinetics is the unique characteristic of a specific microorganism, differing even among phenotypes and strains [7]. Several major factors affect attachment and consequently biofilm formation: the nature of the cell surface, the chemistry and texture of the attachment surface, the nature of the surrounding medium and the temporal and spatial distribution of available nutrients [8–10].

32.2 BIOFILMS: MICROBIAL LIFE ON A SURFACE

A biofilm can be defined as a layer of microorganisms immobilized at a substratum held together in a multinature matrix polymer matrix [11]. This matrix consists mainly of water (97%), microbial cells (2–5%), polysaccharides (neutral and polyanionic) (1–2%), proteins, including enzymes (1–2%), DNA and RNA from lysed cells (1–2%) [12]. Usually, a mix consortium of microbes makes up this ecosystem, and they "team up," in order to protect themselves from stress and maximize nutrient uptake. One of the main components of a biofilm is the EPS, which often consists of one or more family of different polysaccharides produced by at least some of the biofilm microorganisms. These components aid the attachment of cells to surfaces, stabilizing the local environment and spatial organization of the microbial communities, which may need to cooperate with each other to effectively use the available substrate [5].

32.2.1 BIOFILM ARCHITECTURE

A biofilm is a multiphase system. It consists of the biofilm itself, the overlying gas and/or liquid layer and the substratum on which it (the biofilm) is immobilized. This system can be classified in terms of phases and compartments. The phases comprises of the solid, liquid and gas components, whereas the compartments comprise of the substratum, the base film, the surface film, the bulk liquid and the gas. Biofilms are heterogenous by nature, however, having stacks of cells scattered in a glycocalyx network with fluid-filled channels [13]. Structurally, a biofilm is approximately 2-D, with its thickness ranging from a few micrometers to millimeters [5]. This structure allows for the diffusion of nutrients and metabolic substances within the matrix.

Biofilm organization. The microbes in a biofilm are typically organized into microcolonies embedded in the EPS polymer matrix [5]. These microcolonies attain distinct 2-D or 3-D structural patterns. Initially, microcolonies are separated by void spaces but ultimately they merge into unique structures forming a mature biofilm. This spatial organization is very important to the biological activity of the biofilm.

EPS. EPS may vary in chemical and physical properties, but it is primarily composed of polysaccharides [11]. Some of these polysaccharides can be neutral or polyanionic. Generally, this polymer can accommodate considerable amount of water into its structure by hydrogen bonding. Overall, the EPS has an important role of holding the biofilm together [14]. As the EPS layer thickens, the biofilm microenvironment changes due to the activities of the bacteria. Therefore, a mature biofilm is a heterogeneous matrix. This heterogeneity concept is descriptive for both mixed and pure culture biofilms common on abiotic surfaces including medical devices [15].

Quorum sensing. Quorum sensing gene expression has been proposed as an essential component of biofilm physiology, since biofilm typically contains high concentration of cells [16]. Generally, the irreversible attachment of bacteria to a substratum triggers alteration to an array of gene expression and phenotypes of these cells [17]. In the quorum sensing process, cell-cell communication is accomplished through the exchange of extracellular signaling molecules [16]. For most gram-negative bacteria, the quorum sensing regulation involves a freely diffusible auto-inducer, acylhomoserine lactone (AHL) signaling molecule. For instance, the quorum-sensing ability in *P. aeruginosa* is dependent upon two distinct but interrelated systems, *las* and *rhl* [18], which directs formation of the AHL. In gram-positive bacteria, structurally diverse peptides act as quorum sensing regulators [19]. The QS system is not necessarily involved in the initial attachment and growth stages of biofilm formation but is very important in the overall biofilm differentiation process [20]. During biofilm formation, QS signaling molecule mutants may develop thicker, more acid resistant [21] or "abnormal" biofilms [22] than the wild type strains.

32.2.2 Biofilm Life Cycle

Once immobilized on a contact surface, microorganisms have the potential to form a biofilm. Attachment to the surface is beneficial to the microbe for a number of reasons. First, the surface represents important microbial habitats since in the microenvironment of a surface; nutrient levels may be much higher than they are in the bulk solution [23]. Secondly, it increases the microbes' resistance to mechanical and chemical stresses. Overall, biofilm formation is a dynamic process, comprised of four main stages [24]: migration of cells to the substratum, adsorption of the cells to the substratum, growth and metabolic processes within the biofilm, and detachment of portions of the biofilm (Figure 32.1). These steps can be divided into three phases: initial events, exponential accumulation and steady state.

Surface conditioning film formation. The conditioning film is created when organic materials (polysaccharides and proteins) settle on the surface [11]. It can be derived from the microbes in the vessel or from the bulk fluid. Adsorption of a conditioning film is relatively quick compared to the other steps. This film has the potential to alter the physicochemical properties of the substratum and thus greatly impacts bacterial attachment.

Cell migration to the surface. Migration of microorganism to the substratum is considered the second step in biofilm formation. This process can be mediated by different mechanisms depending on the system under consideration. Thus, transport can be active, facilitated by flagella [25] or passive, facilitated by Brownian diffusion, convection or sedimentation. In quiescent systems (batch culture), sedimentation rates for bacteria are generally low due to their size and specific gravity [11], and microbes with a diameter 1–4 μm^3 have small Brownian diffusivity. Therefore, motility may be the limiting factor of transport in such systems. In a laminar flow system, although motility affects transports, diffusion remains the controlling factor. In a turbulent flow system, Brownian diffusion

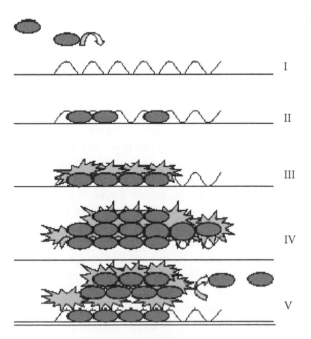

FIGURE 32.1 Schematic representation of biofilm formation. Step I conditioning film formation. Step II bacteria migration to the conditioned surface. The cells start to produce extracellular polysaccharides (EPS), which cause an irreversible attachment Step III. Gradually, the biofilm increases through growth of the irreversibly attached cells and new ones from the solution Step IV. Cells near the outer surface can dislodge from the biofilm and escape to colonize new microenvironments Step V. (Modified from Boom, R.M., et al., *International Journal of Food Microbiology*, 80(2), 117–130, 2003. Reproduced with permission from Elsevier.)

has minute contributions to transport, but forces such as frictional drag force, eddy diffusion, lift force and turbulent bursts are significant.

32.2.3 Factors that Impact Interactions of Bacteria with a Substrate

Microbial adhesion is mediated by specific interactions between cell surface structures and specific molecular groups on the substratum. Moreover, the adhesion process is determined by physicochemical and molecular interactions. It is believed that primary adhesion between bacteria and abiotic surfaces is generally determined by nonspecific (e.g., hydrophobic) interactions, whereas adhesion to living or devitalized tissue is accomplished through specific molecular (lectin, ligand or adhesion) mechanisms [26].

Physicochemical interactions. Generally, two types of physicochemical interactions are used to describe the adhesion of a microorganism to a planar surface. The DLVO approach relates to the interaction energies (attraction and repulsion) primarily to electrostatic and van der Waals forces, but chemical forces can operate [27]. Typically, attraction between microbe and surface occurs at a long range (5–8 nm), a secondary minimum or at a shorter range, the primary minimum. Thus, adhesion can be reversible (at the secondary minimum) or irreversible (towards the primary minimum) [11]. In the DLVO approach, the ionic charge of the medium, the physicochemistry of both the bacteria and substratum surface, and the physicochemistry of biosurfactant determine the extent of adhesion.

Alternatively, in the thermodynamic approach, adhesion is described as the formation of a new interface between the substratum surface and adhering bacteria at the expense of the interfaces between bacteria and the suspending liquid, and the substratum-liquid interface [5]. Each interface contains a specific amount of interfacial energy (or surface tension). The extent of adhesion is determined by the surface properties of all three phases, the surface tension of adhering particles, of the substratum and of the medium [28]. The more hydrophilic a substrate, the higher is its surface tension.

Molecular interactions. Bacterial adherence is also mediated by molecular mechanisms. Bacteria are able to adhere to animal cells [29], such as muscle meats through protein-protein interactions on the surface. These proteins sometimes function as ligands to receptors when the bacteria invade target cells and/or have specific affinity for host components [30]. The colony-opacity-associated (Opa) outer membrane proteins or ligands (often called an adhesion) confer intimate bacterial association with mammalian cells. Two classes of cellular receptors for Opa protein receptors have been identified: adhesio-sulfate proteoglycan (HSPG) receptors and members of the carcinoembryonic antigen (CEA) or CD66 family [31]. *Listeria monocytogenes* surface proteins Internalin A (InlA) and B (InlB) are involved in the attachment of this bacterium to host cells [32].

The magnitude of the cell substratum interaction forces, the chemical heterogeneity and the roughness of the substratum surface greatly affect the extent of microbial adhesion. As food traverses from the farm to the table, it comes in contact with fabricating equipments, utensils, gaskets, conveyor belts, packaging materials, storage containers and chopping boards. These surfaces are usually metallic, plastic, rubber or wood. Food processing equipments are often made of stainless steel, transport crates of high-density polyethylene (HPDE), conveyor belts of rubber, chopping boards of wood, packaging materials of aluminum. Other storage and packaging-type materials include Polypropylene, PVC and Teflon. Sometimes the contact time between foods and surface may be 24–48 hours depending on the processing conditions including design of equipment cleaning and sanitation techniques.

Substratum surface hydrophobicity. The hydrophobicity of the substratum has substantial effect on bacterial adhesion. Typically, hydrophilic surfaces such as stainless steel and glass have a high free surface energy and thus, allow greater bacteria attachment than hydrophobic surfaces such as Teflon [25]. For instance, a general trend of decreasing colonization density was observed for *Staphylococcus epidermis* and *Pseudomonas aeruginosa* with an increase in substratum hydrophobicity [33]. In the above-mentioned study, the packaging materials used were stainless steel, polyvinyl chloride (PVC), polystyrene and glass. Likewise, biofilm formation by *L. monocytogenes* LO28 was faster on hydrophilic (stainless steel) than on hydrophobic (polytetrafluoroethylene, PTFE) [34].

Substratum-surface roughness/topography. Many reports have indicated that metal surfaces with a high degree of roughness serve as a better substrate for bacterial attachment than smooth ones, since the surface area of the former is greater [33,35]. Arnold and others discovered that resistance to bacteria attachment decreased in the order: *electropolished > sanded > blasted > untreated stainless steel*. Bowers and others illustrated that surface topography is extremely important in biofilm formation and resistance [1]. Still studies such as that of Barnes and others [36] claimed that the difference in bacteria attachment due to difference in surface topography is minimal.

Substratum coverage with organic material. The layer of organic substances present on the surface can be favorable or unfavorable to bacteria adhesion. Barnes and others [36] discovered that proteins that adsorbed to a stainless steel surface inhibited bacterial attachment. The dominating mechanism was suspected to be competitive inhibition, since the proteins were able to interact with the hydrophilic surface. The adhesion process begins provided that the conditioning film is favorable to bacterial attachment.

Bulk nutrient composition. Generally, bulk fluid conditions influence surface hydrophobicity, adhesion expressions and other factors that affect adhesiveness [37]. Microbes are usually exposed to a range of nutrients concentrations from as low as 1 ug/L to 500 g/L, and this range has an effect on biofilm growth [5]. At the highest nutrient concentrations, biofilms can appear to be uniform with few or no pores. This is common of biofilms associated with animal and food surfaces. Various reports suggest that the lower the concentration of nutrients, the greater the rate of attachment and biofilm development [38–40]. *Escherichia coli* O157:H7 biofilms developed in minimal salts medium (MSM) developed faster, had thicker extracellular matrix and cells detached much slower compared to those grown in trypticase soy broth (TSB) [38].

Temperature. Temperature effect is particularly important in the food industry, since food experience differentials in temperature (temperature abuse) from farm to fork. This abuse is a consequence of cell wall and attachment factors changes [41]. Stopforth and others showed that a greater number of *L. monocytogenes* cells adhered to stainless steel templates at 59 and 77°F compared to 41 and 95°F [42]. Moreover, Stepanovic and others [43] suggested that a microaerophilic environment support biofilm formation. Presently, there are few conclusive reports that support this claim.

Ionic strength. The atomic ions present in the medium can indirectly affect attachment of the bacteria to other substratum. These ions may act as chelator, forming bridges between protein molecules on the bacteria surface and adsorbed proteins on the substratum surface. For example, with milk-treated steel, ferrous ions in solution increased *L. monocytogenes* attachment [36]. Ions in the solution can also act as shields, shielding the surface charge of the substratum and the bacteria [44] and increasing bacteria attachment [45,46]. The ionic composition in the bulk may affect the composition of the metabolic by products of biofilm cells but not necessarily affect the physical property of the biofilm [47].

Hydrodynamics. The flow velocity in close proximity to the substratum and the liquid boundary (hydrodynamics) has marked influence on the cellular interaction and the biofilm structure. Cells behave as particles in a liquid, and the rate of settling and association with a submerged surface depends greatly on the velocity characteristics of the liquid [48]. After the bacteria has attached, flow rate or shear force of the liquid affects the biofilm structure and content [25]. More compact, stable and denser biofilms were formed at relatively higher hydrodynamic shear force [49].

Atmosphere. The incubation atmosphere also influences biofilm formation. For instance, a microaerophilic and carbon dioxide rich environment provided a relatively high rate of biofilm formation, whereas least biofilm was formed under anaerobic conditions [43]. On the other hand, anaerobic growth favors maintenance of mucoid, alginate (polysaccharide) production by *Pseudomonas* in cystic fibrosis airways [50].

32.2.4 Methods for Studying Bacteria Adhesion and Biofilms' Formation

In order to completely characterize a biofilm, some key procedures must be conducted. The constituent members of the biofilm have to be isolated, the physiology and gene expressions of the cells must be characterized, and the physical and spatial aspects of the biofilm should be quantified. It is important that in the characterization of a biofilm, the model system allows for aseptic and nondestructive removal of the sample. The samples should also be easily identifiable, representative and as reproducible as possible. The choice of a particular system however, is most often determined by the specific biofilm process being investigated, the convenience of sampling, cost and the scientific background of the investigator [11,51]). Table 32.1 highlights the most common experimental designs used to study microbial adhesion and biofilm development. These model systems have either flow or nonflow characteristics which are based on the objective of the research.

Parallel flow chamber. The parallel flow chamber is the most frequently used design for flow systems. It allows for in situ observation of initial adhesion, surface growth and detachment. This device is usually made of noncorrosive material (e.g., nickel-coated brass [52]) to allow sterilization. The bottom plate can be of any material and has a groove in the middle in which the substratum can be placed. The top plate is made of glass allowing for microscopic detection and control of flow and temperature. A container containing different mediums can be attached to the system and an air bubble can be blown in to create stress [52]. The images can then be recorded with a charge coupled device (CCD) camera and processed by an image analyzer.

Robbins device. The Robbins device is widely used to investigate biofilm growth. This device is a modification of the parallel flow chamber with replaceable sample ports. It consists of a cylinder, which can be inserted directly into a pipe or bypass system. The cylinder is usually made of glass, metal or clear plastic, and consists of removal studs [53]. The surface of the removable metal studs forms an integral part of the metal surface available for colonization. Removal of the biofilm is frequently accomplished by scraping or sonication [54].

Flow cell and microfluidic device. Flow cells for biofilm studies are generally of two designs. The most commonly used is one where parallel grooves (about 4 mm²) are pounded into a plexiglass base [55]. A cover glass is used to cover the open side of the grooves, thereby forming a closed channel. Inlets and outlets are made by boring holes at the end of the plexiglass through to the channel. The other design is one where cover slip is used for the top and bottom of a channel formed by silicone gasket. This model has two channels with similar dimensions to a standard microscope slide. Sterilization of this chamber can be difficult. Microfluidic devices are somewhat similar but the parallel groove sizes are in the micrometer range. Cox and others used a microfluidic device to monitor the adhesion of glial cells to coated glass [56].

Fixed-bed reactor. A fluidized bed is a bed of small particles (0.2–2.0 mm in diameter) freely suspended in the upward flow of water or a water and gas mixture [11]. Biomass immobilization is achieved by adhesion to a support media such as particles [57] and fibers [58]. Good stability and robustness is the main advantage of this system while channeling and clogging problems represent a major drawback [59].

Microplate wells. The microplate dish usually used has 96 wells, made of PVC or polystyrene. Bacteria suspensions are transferred to the wells and then incubated. The wells are subsequently stained with a bacteria-interacting dye (e.g., crystal violet), washed with distilled water then air-dried [43,60]. The plates are subsequently washed with ethanol for destaining. A microtiter plate reader determines the optical density of an aliquot of the destaining solution. The absorbance value of the dye present provides an indirect measure of biofilm production.

Slide method. The slide method encompasses all assays in which a microbial suspension remains stationary with respect to an exposed substratum surface [51]. After exposure, the substratum is usually rinsed and adhered microorganisms are enumerated. The exposed substratum is commonly swabbed, vortexed, or scraped to enumerate the attached cells.

TABLE 32.1
Characteristic Properties of Chamber Designs Employed to Cultivate Biofilms and Monitor Cell Adhesion

Chamber Design		Quantification		Measures	
Name	Scheme	Direct*	Destructive	Adhesion	Biofilm
Parallel flow chamber		Y	N	Y	Y
Robbins device		N	Y	Y	Y
Flow cell		Y	N	Y	Y
Microfluidic devices		N	N	Y	N
Fixed-bed Reactor		N	N	N	Y
Microtiter plate		Y	Y/N	Y	Y
Slide		N	Y/N	Y	Y

* Quantification of bacteria cells can be done in real-time (direct) or at a certain time after adhesion or biofilm development (indirect).

32.3 MODELING OF SURFACE CONTAMINATION AND BIOFILM DEVELOPMENT

32.3.1 TEMPORAL AND SPATIAL PATTERNS IN *LISTERIA* BIOFILMS

Figure 32.2 depicts the process of *L. monocytogenes* biofilm development over an aluminum substrate. Both individually isolated bacteria and large areas of the surface covered with distinct clusters

(a) (b) (c)

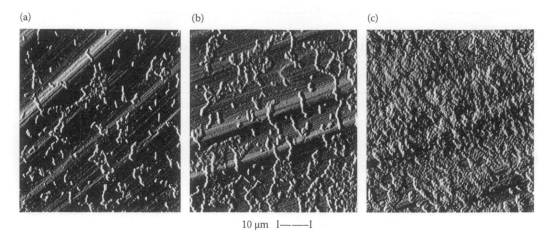

10 μm I———I

FIGURE 32.2 *L. monocytogenes* surface colonization and biofilm development: (a) individual colonies; (b) bacterial web; (c) matured biofilm.

can be observed during the transition from the exponential growth phase of the batch culture to the stationary phase [61].

Some cell clusters (microcolonies) observed on the surface, have high local concentration of cells even at the initial stages of biofilm development. These clusters are created because of the cells' affinity to adhere to preconditioned surface sites with high concentration of adsorbed polysaccharides and proteins [62]. These regions are developed by previously attached cells by production of excessive amounts of extracellular polymers. As can be observed from Figure 32.2, bacterial community consists of separate colonies (clusters) randomly distributed over the substrate.

32.3.2 Diffusion-Limited Control of Bacterial Colony Growth

As demonstrated by experimental data, the spatial organization of microorganisms and biofilm development are often limited by nutrient availability. At low population densities, connected/clustered cells grow towards the regions of higher nutrient concentration and form branched structures [4,63,64] similar to clusters observed as a result of diffusion-limited colloidal aggregation (DLA model, [65,66]). On the contrary, if neighboring cells are present in colony vicinity (i.e., high population density), colony cells tend to grow as close to each other as possible forming continuous dense domains, which correspond to the reaction-limited (Eden-type [67]) growth patterns [68,69]. The extent of these limitations can be examined in terms of nutrient diffusion transport. Nutrient transport toward surface population of growing microorganisms is unsteady-state problem with moving boundaries. Therefore, behavior of the system will be determined by two characteristic times: the time of nutrient diffusion transport ($t_D \approx x^2 D^{-1}$) and the cell doubling time ($t_\mu \sim 1200$ s). Substituting characteristic size of bacterial cell ($x \sim 10^{-6}$m) and diffusivity of nutrient ($D \sim 10^{-6}$cm^{2}s^{-1}), it is clear that $t_\mu \gg t_D$, and therefore, one can consider nutrient diffusion transport and cell growth as independent processes.

Diffusion-controlled nutrient flux toward the colony can be expressed as:

$$J_n = -D\frac{\partial C_n}{\partial r} \tag{32.1}$$

Solving Equation 32.1 in a hemispherical domain for $r = [r_0, \infty]$:

$$J_n = -\frac{DC_{n0}}{r} - \frac{DC_{n0}}{\sqrt{\pi Dt}} \tag{32.2}$$

Single colony growth. Now it is possible to determine the maximum size of a single 2-D colony growing under nonlimited nutrient conditions. The number of microorganisms in a single colony of radius r is:

$$C_c = 4\left(\frac{r}{a}\right)^2 \tag{32.3}$$

The condition of balanced growth:

$$C_c = \gamma C_n \tag{32.4}$$

Therefore, required amount of nutrient for the colony will be:

$$I_n = \frac{4}{\gamma}\left(\frac{r}{a}\right)^2 \tag{32.5}$$

The amount of nutrient delivered to the colony:

$$I_{nD} = \pi r^2 J_n = -\pi r D C_{n0} \tag{32.6}$$

A critical colony size can be obtained from the nutrient mass balance using Equations 32.5 and 32.6:

$$r_{cr} = \frac{1}{4}\pi \gamma a^2 D C_{n0} \tag{32.7}$$

Substituting $t_\mu \sim 1200$ s, γC_{n0} for BHI~10^8cell/ml (γC_{n0} is the maximum carrying capacity of the medium), and characteristic cell size $a \sim 1$ μm, the critical colony size can be estimated as $r_{cr} \sim 10^{-4}$cm, which is in good agreement with our experimental data.

Growth of bacterial population on a surface. A nutrient flux to the surface can be expressed as [70]:

$$J_s = -C_{n0}\sqrt{\frac{D}{\pi t}} \tag{32.8}$$

Using material flux balance, i.e., equating nutrient consumption (Equation 32.2) and delivery (Equation 32.8), the radius of the diffusion zone can be expressed as:

$$r_D = \sqrt{r\sqrt{\pi D t} + r^2} \tag{32.9}$$

Lets consider a dense square lattice configuration of the diffusion zones. A footprint of nutrient depletion zones (surface coverage α) can be estimated from simple geometrical considerations as:

$$\alpha = \frac{S_D}{S} = \frac{\pi}{4} \tag{32.10}$$

On the other hand,

$$\alpha = \frac{\pi r_D}{d^2} = \frac{\pi r \sqrt{\pi Dt} + \pi r^2}{d^2} \qquad (32.11)$$

Using Equations 32.10 and 32.11, one can determine a critical distance between two independently growing colonies (with no overlap of the depletion zones):

$$d \le 2\sqrt{a\sqrt{\pi Dt_\mu} + a^2} \approx 2(\pi Dt_\mu)^{1/4} a^{1/2} \qquad (32.12)$$

A critical distance between the colonies obtained from the expression, Equation 32.12, is $d_{cr} \sim 49$ μm. However, bacteria in the experiments were observed to change their growth rate only after 50% of nutrient depletion. For half-overlapped nutrient depletion zones, the distance between the colonies is ~ 25 μm, which is in good agreement with observed patterns.

The change in the colony organization (circular to dendric) as a function of nutrient availability was experimentally observed for the first time by Matsushita et al. [71,72]. Similar experiments performed by McLandsborough group [73] using growth media with various nutrient contents have confirmed that *L. monocytogenes* also changes the colony shape depending on the amount of nutrient available.

Typically, cell cluster shape replicates the form of an available free surface. Initially, cell colony tries to avoid coalescing/merging with other cell groups, keeping certain distance (about two to five cell lengths) between itself and other cell aggregates. Our experimental observations show that there is no continuous front-to-front approach between two colonies. When merging becomes inevitable, the closest bacteria in each colony start to grow towards another colony, advancing their neighbors and building cell "bridges" thus, connecting two colonies. After that, the colony spreads filling its interstitial space [74]. One possible explanations of the observed "bridge"-forming phenomenon can be made through the presence of communication between cells in different colonies (quorum sensing). This communication possibly leads to coordinated growth of the boundary cells in the colonies towards each other. There are several comprehensive models of intercellular communication and its effect on bacterial population [75–77]. We will consider the simplest linear case only to explain observed phenomenon.

Without any loss of generality, one can assume that cells generate autoinducer at constant rate and release it through their surfaces by diffusion. When autoinducer concentration reaches its critical (threshold) value, it triggers cell metabolism, influences bacterial growth rate, and changes social behavior of the cell population. Lets consider a 1-D system of two semicylindrical colonies depicted in Figure 32.3.

The autoinducer generated inside the cell is released from bacteria surface by diffusion, therefore, autoinducer flux through the cell surface can be estimated as linearized concentration gradient:

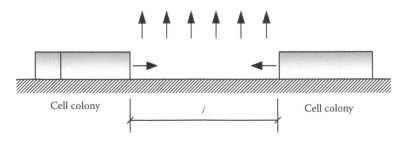

FIGURE 32.3   An interaction between two colonies: diffusion-controlled process of autoinducer release.

$$\frac{dC}{dx} \sim \frac{C}{r}.$$

The reaction-diffusion kinetics can be expressed in terms of characteristic reaction length:

$$L_{ai} = \sqrt{\frac{D}{k}} \tag{32.13}$$

The physical sense of this expression becomes more clear recalling that Equation 32.13 is a function of modified Damköhler number ($Da = r/L_{ai}$), which represents the ratio between diffusion and reaction mass transfer rates.

We will derive the autoinducer accumulation condition from a steady-state mass balance equation:

$$-S_s D \frac{dC}{dx} + S_0 k_s C = 0 \tag{32.14}$$

where $S_s = \pi r l$ is the surface area of an intercolony space, $S_0 = (1/2)\pi r^2$, the sidewall area of a colony that releases autoinducer. The surface reaction can be expressed in terms of bulk reaction substituting $k_s S = kV$ [78]. The condition of autoinducer accumulation is obtained by rearranging Equation 32.14 with expression 32.13 taken into account:

$$4\frac{l}{r}\frac{L_{ai}^2}{r^2} < 1 \tag{32.15}$$

In this model l represents the distance between colonies, therefore growing of the cell colony leads to l decreasing. From the expression (Equation 32.15), continuous cell reproduction and bacteria chain elongation result in the shrinkage of intercolony space and growth of the autoinducer concentration. Physically, the inequality (Equation 32.15) represents current relative distance between two groups of microorganisms, corrected by the Damköhler number.

32.4 BACTERIAL ATTACHMENT AND GROWTH ON ROUGH SURFACES

Our study indicates that cells initially adhere to the substratum randomly, with substantial amount of space between cells (Figure 32.4a). Cell population density increases with time, leaving less space between them (Figure 32.4b). With time, cell groups develop branched structures, corresponding to the diffusion-limited kinetics of microbial growth. Cell colonies grow towards regions of highest nutrient concentration and smallest number of cells. As more cells adhere to the surface, they aggregate forming larger branched colonies and developing local cellular networks. After two hours, clustering of cells becomes more pronounced and a distinct polysaccharide film, which is an extracellular polymeric substance associated with biofilm formation [79,80], is observed to surround each cluster (see Figure 32.4b). At this stage, microbial colonies also appear to be uniting with each other (Figure 32.4c), thereby forming a microbial web structure over the surface.

The author suggests that cell adhesion is virtually not influenced by cell growth during the initial period of cell culture contact with the surface, when the time of contact does not exceed the characteristic time of bacteria reproduction. For L. monocytogenes the characteristic doubling time is ~20 min.

Surface patterns affect microorganism proliferation and influencing nutrient transport to the surface. The effects of size, shape and morphology of surface constraints on Listeria adhesion and growth were studied by direct observations of irreversibly adhered L. monocytogenes cells on

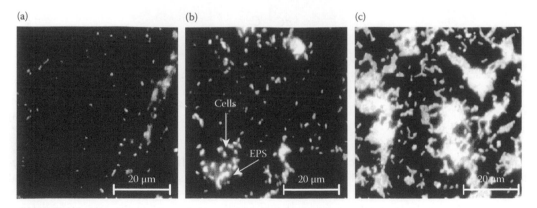

FIGURE 32.4 Bacterial colony formation on an aluminum surface at 1 min (a), 10 min (b), 210 min (c) contact time. (Takhistov, P. and George, B., *Bioprocess and Biosystems Engineering*, 26, 259–270, 2004. With permission from Springer Science & Business Media.)

prepatterned aluminum surfaces by fluorescence and scanning probe microscopy. Two geometric parameters can be used to characterize and distinguish between surface patterns: confine aspect ratio $K = H/W$, where W, H are the width and the depth of the confine respectively, and characteristic bacterium size a. Surface elements can be divided into three groups based on their geometry: plain surface elements ($H \leq a$, $W \gg a$); wide constraints or low-profile obstacles ($W > 50a$, $K \leq 1$); and narrow confines ($W \sim 10\text{–}15a$, $K \gg 1$). The majority of "real-life" surfaces can be represented by a combination of these morphological surface units.

We found that there is a critical size of surface pattern which triggers bacteria behavior on the surface; it can be estimated as $W_{cr} \sim 15a$. Comparing cell adhesion in the narrow ($W < 10a$) and wide ($W > 20a$) grooves, it has been observed that *Listeria* cells preferentially adhere in corners of narrow grooves and to the center of wide ones.

A successful strategy of bacteria growth is to find the balance between maximum security/protection for existing cells and unrestricted nutrient access to them. Narrow confines with high aspect ratio are characterized by diffusion-limited nutrient supply; hence bacteria that settled in these confines will eventually experience starvation stress. Microorganisms were observed to attach in corners of narrow grooves first (Figure 32.5a), maximizing surface area available for subsequent adhesion and developing more compact and protective EPS "umbrella." This highly adaptive surface colonization strategy is likely to exist only in motile microorganisms, which corresponds with the observations of Scheurman [10] that only motile organisms can be found at the bottom of narrow grooves. Furthermore, biofilm development in narrow confines consists of two steps: colony spreading over the confine base and development of 3-D pillar structures in the middle of the groove (Figure 32.5b).

The mechanism of pillar development might be explained as follows. To survive in a deep surface confine, bacteria either have to build a 2-D biofilm over its walls or develop 3-D structure. However, it is difficult for bacteria to adhere to vertical walls of the confine, since the number of cells that settled on its base is not sufficient to produce enough EPS film to cover walls of the constraint. Therefore, newborn cells prefer to grow on top of existing colony, erecting the next biofilm layer and developing 3-D structure. This way, they get better access to the energy source.

On the contrary, wide and relatively shallow constraints allow unrestricted nutrient access to the whole surface. Cells adhere preferably in the center of these grooves, initiating colonies equidistant from the sidewalls (see Figure 32.5a). As cells grow and population density increases, bacteria fill the bottom of the wide surface constraints, spreading towards the walls and eventually growing over the edges, merging with colonies outside the obstacle (Figure 32.5b).

Bacterial attachment determines surface colonization during the initial contact of metal (aluminum) surface with bacterial culture. As biofilm grows and spreads over the surface, planktonic

FIGURE 32.5 Biofilm development in a narrow groove: (a) bacteria attach in the corners; (b) 3-D cell pillars grow in the constraint with a high aspect ratio.

and sessile (surface) cell populations begin competing for nutrient supply. Bacteria on the surface are more stable and stress-resistant [2,81], but their planktonic competitors probably have greater growth rate. It is difficult to predict the result of this competition, but it is clear that surface topography should play a major role in the survival strategies of sessile bacterial population.

As we believe, surface patterns impact microorganism proliferation and biofilm development often restricting free nutrient access to the growing bacterial population. The extent of these limitations can be examined in terms of nutrient diffusion transport. Let us consider a model surface with a cylindrical confine (see Figure 32.6) immersed in solution with an initial nutrient concentration C_{n0}. Nutrient consumption at the confine bottom is performed by the bacterial population, and the metabolic products are diffusing out of the confine. This nutrient consumption results in an external nutrient diffusion flux to the confine from the bulk solution.

To examine possible limits of nutrient transport due to surface topography, we will consider two problems: external nutrient transport to the confine from the bulk medium, and internal diffusion transport of nutrients inside the confine. The analysis of the internal nutrient transport allows estimation for a critical confine depth (H_{cr}) at which bacteria settled on the bottom start to experience nutrient deficiency. A critical size (W_{cr}) of confine opening (footprint) can be obtained from the mass balance between the bulk and the confine.

The governing diffusion equation for the internal transport problem can be written as:

$$J_i = D_n \frac{dC_n}{dx} \tag{32.16}$$

where C_n, D_n are the concentration and diffusivity of the nutrient.

Boundary conditions (BCs) for Equation 32.16 can be obtained by examining the physical limits of an idealized confine. Boundary condition at the confine bottom describes the balance between nutrient concentration at the confine entrance (mouth) and the equilibrium nutrient concentration in the bulk. The second BC reflects the fact that there is a flux of nutrient at the confine bottom due to bacteria metabolic activity:

$$C_n = C_{n0}|_{x=0}$$

$$\frac{dC_n}{dx} = J_c \Big|_{x=H} \tag{32.17}$$

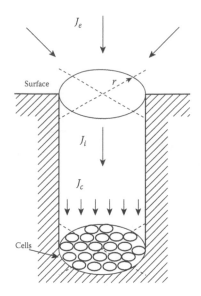

FIGURE 32.6   Model representation of the surface confine.

The number of microorganisms that settle on the confine bottom is $C_C \sim S/a^2$, where S is the footprint of the confine, and a is the average bacteria size. Total amount of nutrient required to support this population is $C_n \sim C_c \gamma$, where γ is the biomass yield coefficient [82]. Finally, nutrient flux due to its consumption by the bacteria can be determined from the amount of nutrient that is required to double cell biomass per division time:

$$J_c = \frac{\mu}{a^2 \gamma} \tag{32.18}$$

where μ is the bacteria specific growth rate.

Linearizing Equation 32.16 as

$$D_n \frac{dC_n}{dx} \approx D_n \frac{C_n|_{x=0} - C_n|_{x=L}}{H}$$

where $C_{n|x=0,H}$ are the nutrient concentration at the confine mouth and its bottom respectively, the maximum nutrient flux in the confine ($C_{n|x=H} \rightarrow 0$) can be estimated as follows:

$$J_{i\,max} = \frac{D_n C_{n0}}{H} \tag{32.19}$$

Combining Equations 32.18 and 32.19, we can estimate the critical depth of the surface constraint, which allows yet unrestricted development of bacterial population:

$$H_{cr} = \frac{D_n C_{n0}}{\gamma \mu} = \frac{D_n C_{c\,max}}{\mu} \tag{32.20}$$

where C_{cmax} is the maximum carrying capacity of a medium with given nutrient concentration (experimentally determined C_{cmax} value for BHI is approx. 10^9 cfu/mL). Performed calculations indicate that the critical depth of the surface constraint is ~10 μm, which corresponds well with our experimental observations.

For the external transport problem, diffusion-controlled nutrient flux toward the confine opening can be expressed as:

$$J_e = -D_n \frac{\partial C_n}{\partial r} \tag{32.21}$$

where r is the confine radius (half width, $W/2$), and the BCs for Equation 32.21 are:
$C_n|_{x=\infty} = C_{n0}; C_n|_{x=r} = 0$. Solving Equation 32.21 over the hemispherical domain $r = [r_0, \infty)$:

$$J_e = -\frac{D_n C_{n0}}{r} - \frac{D_n C_{n0}}{\sqrt{\pi D t}} \tag{32.22}$$

Using nutrient mass balance as suggested earlier, the critical radius of the confine can be estimated with equating expressions 32.19 and 32.22:

$$r_{cr} = \frac{H_{cr}\sqrt{\pi D_n t}}{\sqrt{\pi D_n t} - H_{cr}} \xrightarrow{t=\infty} H_{cr} \tag{32.23}$$

Therefore, in all surface constraints with the critical width (W_{cr}) less than ~20 μm and the confine ratio greater than 1, growing bacterial cells will experience nutrient deficiency. One of possible responses of bacterial population to this starvation stress is to change its spatial organization, e.g., by building 3-D pillar structures as was observed in our experiments (see Figure 32.5).

32.5 MODEL OF BIOFILM DEVELOPMENT IN BATCH REACTOR

All models describing bacterial population dynamics are based on two types of kinetic models: Monod and logistic. Both models are semiempirical and require fitting to get species- and conditions-specific parameters values. Logistic model better represents the behavior of microbial population on various substrata, and fits changes in lag phase of bacteria growth cycle due to changes in the environment. Since nutrient concentration is not a model parameter, this model can be considered as species-oriented and does not reflect differences in metabolic activities of different bacteria with similar growth curves. This is a big advantage when one is investigating the behavior of the same bacteria type in various environments/substrata; consequently this is the most popular approach for predictive food safety microbiology. Monod kinetics is the basic model to investigate the dynamics of microbial population in advance systems, in which not only biomass production, but also nutrient consumption and products yield should be taken into account.

Microbial populations in natural and industrial environments exist in most cases in both sessile and planktonic states. Depending on the process, environment and microorganisms' nature, the ratio between two populations greatly differs. The dynamics of microbial population growth has been intensively investigated during the past years for both industrial applications [83,84], and natural ecosystems [64,85]. However, there is a lack of knowledge in understanding of the processes controlling the interactions between the sessile (biofilm) and planktonic parts of monospecies bacterial population. Both parts compete for the same ecological niche using different life strategies. It is of great scientific interest and big practical importance to understand major trends and obtain valid timescale estimations for this fundamental process.

Two phenomena determine the entire process of biofilm initiation and development: bacterial adhesion (adsorption) to the surface and cell growth. In an ideal case a nonoccupied solid (in our study metal) surface with sites available for microorganism attachment contacts with a liquid medium containing bacteria inoculum. There are two limit cases that can be modeled and investigated.

A system with high initial cell concentration ($>10^5$ cfu/mL). Planktonic bacterial community in the stationary phase experience lack of nutrients, therefore cells cannot grow exponentially. Hence, all cell activities are directed toward finding available sites with sufficient nutrient supply for adhesion. Cell deposition onto the surface occurs by the mechanism similar to that of a diffusion-limited aggregation [67] process. Characteristic time of cell deposition is significantly shorter than the characteristic reproduction time of the bacteria.

A system with low initial cell concentration ($<10^5$ cfu/mL) has a relatively high concentration of nutrients in the media to support cell growth. Therefore, "critical decision" about cell adhesion to the surface involves choosing the survival/life strategy for the cell. It has two choices: (1) a highly competitive environment (bulk medium) with more nutrients and degrees of freedom, and (2) a protective surface environment with lower nutrient availability. In the system with low initial concentration of bacteria cells attach to the surface and at the same time grow in the bulk, changing external conditions for attachment. Cells on the surface and those in planktonic state have enough time and nutrients to reproduce. Biofilm growth patterns reflect the competition occurring between surface biofilm formation and bulk media saturation (nutrients decay). Nutrient-sufficient conditions are the characteristic of the exponential phase during the constant volume (batch-type) cell growth. This phase has been extensively studied and has resulted in the development of a concept of balanced growth [86]. There are two factors from the exponential growth phase that influence the physiology of bacteria in the stationary phase [87]. Firstly, it is the actual growth rate (Monod kinetic constant) which determines the overall cell composition. Secondly, the selection of nutrients supplied basically determines the biochemical potential.

Nonsteady-state biofilm development occurs in nonflow systems that can be described by simplified model of a batch-type chemical reactor. Let us consider a batch system with bacteria inoculated solution with uniform population distribution at the beginning. There are several models developed that allow description of bacterial growth. We have chosen to represent bacterial growth by the Monod equation. This model allows adequate estimation of the influence of various external factors on the efficacy of cell growth. Describing changes of cell population and nutrient consumption by similar equations allows the development of a unified approach that considers changes in microbial concentration and external conditions.

The microbial growth described in terms of Monod kinetics can be described as follows:

$$\begin{cases} \dfrac{\partial C_c}{\partial t} = \dfrac{\mu_0 C_n}{K_0 + C_n} C_c \\[2mm] \dfrac{\partial C_n}{\partial t} = -\dfrac{1}{\gamma_n} \dfrac{\mu_0 C_n}{K_0 + C_n} C_c \end{cases} \tag{32.24}$$

The Monod constant K_0 is the concentration of the growth-limiting nutrient at which the specific growth rate is equal to one half of its maximum value. It represents the affinity that the organism has for a particular nutrient. The values of μ_0 and K_0 depend upon the organism, the nutrient itself, fermentation medium, and environmental factors such as pH and temperature of the growth medium. The system describing bacterial kinetics uses stochiometric yield coefficient γ_n to determine the amount of biomass produced with each unit of substrate (nutrient) utilized. Yield coefficient describes the effectiveness of nutrient conversion into bacterial biomass. Equation 32.24 is only valid when the rate of product formation is proportional to the rate of cell growth, i.e., balanced growth conditions exist [79].

Initial conditions for bulk cell growth are represented by uniform density of the bacterial population, corresponding initial amount of nutrients and absence of adhered bacteria on the surface:

$$C_c|_{t=0} = C_{c0}$$
$$C_n|_{t=0} = C_{n0} \tag{32.25}$$
$$C_s|_{t=0} = 0$$

The boundary no-flux condition at the reactor top ($x = L$) is:

$$\frac{\partial C_c}{\partial t}\bigg|_{x=L} = 0 \tag{32.26}$$

The change of nutrient consumption at the surface (BC at $x = 0$):

$$\frac{\partial C_n}{\partial t}\bigg|_{x=0} = -\frac{1}{\gamma} \frac{\mu C_n|_{x=0}}{K_0 + C_n|_{x=0}} C_s \tag{32.27}$$

The size of bacteria surface population determined by bacteria concentration is the result of two simultaneous processes: cell adsorption from the bulk and division of the adhered microorganisms on the surface. As a result, the equation describing microorganisms population growth can be written as:

$$C_s = C_{s\max} \frac{K_a C_c|_{x=0}}{1 + K_a C_c|_{x=0}} + \int_0^t \frac{\mu_s C_n|_{x=0}}{K_{0s} + C_n|_{x=0}} C_s \, dt \tag{32.28}$$

The first term in Equation 32.28 represents the surface concentration dependence on the microbial adsorption/adhesion (based on the Langmuir isotherm) from the bulk solution. The second integral term is determined by bacteria growth kinetics. Differentiating the first term by t one obtains the rate of bacterial concentration changes at the surface due to their adsorption from the bulk solution and self-reproduction:

$$\frac{\partial C_s}{\partial t} = C_{s\max} \left[\frac{K_a}{1 + K_a C_c|_{x=0}} - \frac{K_a^2 C_c|_{x=0}}{\left(1 + K_a C_c|_{x=0}\right)^2} \right] \frac{\partial C_c}{\partial t}\bigg|_{x=0} + \frac{\mu_s C_n|_{x=0}}{K_{0s} + C_n|_{x=0}} C_s \tag{32.29}$$

Equations 32.24 through 32.29 comprise a closed, strongly nonlinear system that can be solved only numerically. However, for practical applications an analytical solution is more preferable, since it allows investigating the direct impact of process parameters of bacterial growth and biofilm dynamics. To develop realistic estimation of bacteria growth dynamics and nutrient consumption, we will obtain an analytical solution for the simplified (linearized) problem with certain physical assumptions. Our model is based on the main assumption of equilibrium reaction kinetics, neglecting possible diffusion limitations. This assumption can be used here since most of the experimental data were obtained measuring volume-averaged properties of bacterial population. Figure 32.7 shows that according to the Monod model, the bacteria growth rate increases with concentration of a nutrient. When this concentration exceeds a critical value, growth rate reaches its maximum, which is the maximum specific growth rate.

In a typical batch culture, nutrient availability generally becomes limiting towards the end of the growth phase. This is reflected in decreasing of the growth rate as cells enter the stationary phase. End product inhibition can also lead to cessation of growth. During the exponential phase of a batch microbial fermentation, nutrient concentration is considerably higher than that of the Monod constant. As this concentration becomes a growth-limiting factor substrate, the specific growth rate of a batch culture reduces until all growth ceases due to the unavailability of that nutrient.

The two major mechanisms determine growth of cell population on the surface and biofilm development:

Adhesion of planktonic bacteria from the bulk solution onto the surface. This process is described by the adsorption kinetics and characterized by the maximum surface bacteria concentration (C_{smax}) and adsorption constant (K_a). In this chapter, we consider this process as irreversible;

Bacterial cells reproduction and growth. If a bacterium exists on the surface for a period of time which is longer than its characteristic doubling time ($t \gg \mu_s^{-1}$), then an increase in the surface population due to cell division should be taken into account.

32.5.1 Cell Surface Population Growth due to Adsorption

Let us consider the case when nutrient concentration is related to cells concentration as:

$$C_n = \gamma_n C_c \tag{32.30}$$

which is based on the cells "balanced growth" conception, when bacterial population growth is limited only by the amount of nutrient available. Hence, the maximum concentration of bacteria in the bulk medium can be estimated as $C_{cmax} = \gamma_n C_{n0}$.

Characteristic Monod reaction time can be obtained by linearizing Equation 32.24:

$$t_m = \frac{\gamma_n(K_0 + C_{n0})}{\mu_0 C_{c0}} \tag{32.31}$$

Nutrient availability has great influence on bacteria specific growth rate. If a nutrient is available in concentrations that limit the growth of cells, then that nutrient becomes the growth-limiting factor. Nutrient consumption changes in time can be expressed as a function of the growth kinetics, initial bacteria concentration and biomass yield coefficient. Representing all concentration changes as linear functions of time and determining the rate of concentration change as C_{cmax}/t_m, see Figure 32.7 and Equation 32.30, the following expression for the nutrient concentration in the system can be written:

$$C_n = C_{n0} - \frac{\mu_0 C_{n0} C_{c0}}{\gamma_n(K_0 + C_{n0})}t \tag{32.32}$$

$C_c(t)$ is the linearized function that represents temporal change of bacteria concentration in the bulk solution. It can be obtained as:

$$C_c = \frac{\mu_0 C_{n0} C_{c0}}{K_0 + C_{n0}}t \tag{32.33}$$

Adhesion of bacteria to the surface is well described by the adsorption kinetics based on the Langmuir isotherm [88].

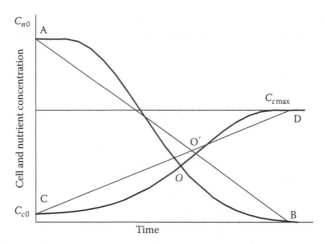

FIGURE 32.7 Actual and linearized kinetics of bacteria growth and nutrient consumption in batch system. (Takhistov, P. and George, B., *Bioprocess and Biosystems Engineering*, 26, 259–270, 2004. With permission from Springer Science & Business Media.)

$$\Gamma_s = C_{s\,max} \frac{K_a C_c}{1 + K_a C_c} \tag{32.34}$$

By the analogy with Equation 32.33, we can determine surface concentration of bacteria using Equation 32.31 and substituting $C_{s0} = K_a C_{c0} C_{s\,max}$ as an initial surface concentration:

$$t_{ms} = \frac{\gamma_n (K_0 + C_{n0})}{\mu_0 K_a C_{c0} C_{s\,max}} \tag{32.35}$$

Combining Equations 32.33 through 32.35 and using the Taylor expansion, the surface concentration of adhered bacteria can now be expressed as:

$$C_{sc} = \frac{\mu_0 K_a C_{s\,max} C_{c0} C_{n0} t}{K_0 + C_{n0} + \mu_0 K_a C_{c0} C_{n0} t} \approx \mu_0 C_{c0} \frac{K_a C_{s\,max} C_{n0}}{K_0 + C_{n0}} t \tag{32.36}$$

Finally, we can obtain the characteristic surface time t_{sc}, which is needed to reach the maximum surface concentration of bacteria due to adsorption from the bulk medium:

$$t_{sc} = \frac{1}{\mu_0} \frac{K_0 + C_{n0}}{K_a C_{c0} C_{n0}} \tag{32.37}$$

The amount of nutrient available at the surface can be determined as the difference between the initial nutrient concentration and the amount of nutrient consumed by bacteria in the bulk:

$$C_n = C_{n0}\left[1 - \frac{\mu_0 C_{c0}}{\gamma_n (K_0 + C_{n0})} t\right] \tag{32.38}$$

Now we are able to obtain the surface time of bacteria population growth due to adsorption from the bulk:

$$t_s = \frac{\gamma_n (K_0 + C_{n0})}{\mu_0 C_{c0}\left(1 + K_a C_{s\,max} \gamma_n^2\right)} \tag{32.39}$$

The presented model is based on the assumption of irreversible adsorption kinetics and considers linearized equilibrium system of equations with no diffusion limitations considered.

32.5.2 BIOFILM DEVELOPMENT: MIXED KINETICS

L. monocytogenes is often considered as a nonbiofilm-forming microorganism, since it cannot directly adhere to the surface and create protein-ligand bonds between cell membranes and the surface [88]. Instead, this bacteria lie on a relatively thick layer of EPS. This layer is homogeneous by chemical content but is not homogeneous in mechanical properties. After attachment, cells require a certain time to adjust to the new environmental conditions. At this stage, their reproduction rate decreases significantly but the metabolic activity remains very high and is directed towards intensive production of EPS matrix components. Development of the polysaccharide protective shell allows bacteria to resist mechanical (high shear stress during washing and brushing) and chemical (detergents and antimicrobial agents usage) treatment procedures.

There are many publications where biofilms formed by *Listeria* have been observed [89,90]. This inconsistency in the interpretation of experimental facts is deeper than just a terminological difference. There is a disagreement in the biofilm research community about categorizing microorganisms into biofilm-forming and non-forming groups. Biofilm is not just the settlement of microorganisms on a surface. Instead, it is a dynamically interacting community of one or multispecies bacterial populations, characterized by temporal (development), spatial (space organization) and functional (variation in metabolic activity) heterogeneities, hierarchical structure and well-determined evolutionary stages. In general, the difference between microorganism settlement and biofilm can be defined using two criteria:

- Significant role of the surface in bacterial population growth
- Existence of cell-to-cell communication within the population

The growth kinetics of planktonic and sessile bacterial communities is described by the same model, and only kinetic parameters used in model reflect the difference in the physiological response and metabolic reaction of both bacterial groups. The kinetics of bacterial growth on a surface, i.e., the surface specific growth rate parameter μ_s is different from the bulk growth kinetics due to extended lag-phase of the bacterial surface population. Numerous literature data show slower specific growth rate of the bacteria in biofilms. On the other hand, the value of the surface Monod constant K_{0s} is close to its bulk value, because nutrient consumption by the sessile bacteria is almost the same as their planktonic analogs. Therefore, for further derivations we can assume that $K_{0s} \sim K_0$.

Using the same approach as before, we can develop the model determining the values of critical parameters describing biofilm growth. Changes in surface concentration of cells are now determined by the two processes: bacteria adsorption from the bulk solution, and their division and growth on the surface:

$$C_{cs} = \frac{\mu_0 \mu_s K_a^2 C_{c\max}{}^2 C_{c0}^2 C_{n0}}{\gamma_n (K_0 + C_{n0})^2} t^2 \tag{32.40}$$

Hence, the characteristic time for cell population proliferation can be expressed as:

$$t_s = \frac{(K_0 + C_{n0})}{C_{c0} K_a} \left(\frac{\gamma_n}{\mu_0 \mu_s C_{n0} C_{s\max}} \right)^{1/2} \tag{32.41}$$

32.5.3 COMPETITION FOR NUTRIENTS BETWEEN SURFACE AND BULK BACTERIA POPULATIONS

Two components (sessile and planktonic) of microbial population in batch reactor systems have different growth rates. As follows from literature data [88] and our observations, the surface population has lower growth rate than the bulk. This is due to the necessity for bacteria on a surface to undergo several physiological processes before they are able to reproduce: adhesion, adjustment of environment and building the EPS protective envelope. Therefore, the consumption of nutrients is primarily made by planktonic bacteria. Microorganisms on a surface, due to their low population density, consume fewer nutrients so there is no conflict with the planktonic cells. However, in the course of growing, surface bacteria require greater energy, and at a certain surface population density their needs for nutrients become vital for their survival.

Nutrient consumption by the bacterial surface population growing due to cell adsorption from the bulk can be determined as:

$$C_{na} = C_{n0} \frac{\mu_0 K_a C_{c0} C_{smax}}{\gamma_n (K_0 + C_{n0})^2} t \tag{32.42}$$

Nutrient consumption by cells that settle on the surface in the mixed kinetics case of simultaneous adsorption and growth can be easily obtained using Equation 32.40:

$$C_{nag} = C_{n0} \frac{\mu_0 \mu_s}{\gamma_n^2} \frac{C_{c0}^2 C_{smax}^2 K_a^2}{(K_0 + C_{n0})^2} t^2 \tag{32.43}$$

Now we can determine the time required to reach the equilibrium between nutrient consumption by the surface bacterial population and amount of nutrient available in the solution:

$$t_n = \frac{\gamma_n (K_0 + C_{n0})}{C_{c0} \mu_0 (K_a C_{smax} + 1)} \tag{32.44}$$

Taking into account cell growth on the surface:

$$t_{nag} = \frac{1}{2} \frac{\sqrt{\mu_0^2 + 4\mu_0 \mu_s C_{smax}^2 K_a^2} - \mu_0}{\mu_0 \mu_s C_{c0} C_{smax}^2 K_a^2} \gamma_n (K_0 + C_{n0}) \tag{32.45}$$

The ratio of the characteristic growth times for the bulk and surface microbial populations determines the condition of sufficient nutrient supply to the biofilm:

$$n = \frac{t_s}{t_n} < 1 \tag{32.46}$$

If this ratio is greater than 1, i.e., $C_n(t) > C_{ns}(t)$, the biofilm microbial community has better nutrient supply than the planktonic bacteria. If it is less than 1, the surface population has smaller amount of nutrient available than bacteria in the bulk. Microorganisms experiencing this starvation stress have only three choices: they can escape, adapt, or die.

Based on Equation 32.46, one can obtain the critical conditions for two cases, when the surface bacterial population is stressed by the limited nutrient supply before it reaches the maximum possible concentration C_{smax}.

In the case of bacterial adsorption to the surface, characterized by short contact time or high concentration of inoculum, condition of nutrient limitation can be interpreted as a ratio of maximum possible surface concentration of the bacteria and maximum concentration of microorganisms in the solution with given initial nutrient content:

$$n_a = \frac{t_s}{t_n} = \frac{C_{s\max}K_a + 1}{C_{n0}K_a\gamma_n} \approx \frac{C_{s\max}}{C_{n0}\gamma_n} < 1 \tag{32.47}$$

In the case of bacterial adsorption to the surface with simultaneous cell growth on the surface, characterized by long cell-surface exposure time or low inoculum concentration, nutrient limitation occur at the same condition as in the previous case (Equation 32.47) corrected by the ratio of normalized reaction rates of microbial growth and adsorption process:

$$n_{ag} = n_a \frac{2\mu_s K_a C_{s\max}}{\sqrt{\mu_0^2 + 4\mu_0\mu_s C_{s\max}^2 K_a^2} - \mu_0} < 1 \tag{32.48}$$

The radicand in Equation 32.48 represents an average growth rate of bacteria in the presence of the surface colonization due to adhesion.

To understand the dynamics of bacteria surface colonization, one has to investigate separate impact of factors responsible for cell adhesion and biofilm proliferation processes. The three main questions should be answered experimentally to obtain information about the underlying mechanisms of surface colonization:

What is the threshold value of the available nutrient at which cells become unable to maintain normal metabolism and are forced to change their life strategy or environment?

How does the adsorption/adhesion kinetics differ from the kinetics of cell proliferation; and are biofilm structure and bacterial physiological patterns different during these processes?

What happens to surface bacterial population when nutrient supply continuously decreases?

We address these questions both theoretically and experimentally, analyzing nutrient access-driven competition between surface and bulk bacteria cultures in a batch reactor. Based on our theoretical considerations, it is possible to conclude that there are two factors that influence surface colonization and the number of adhered bacteria: nutrient access (the amount of available nutrient) and surface preparedness. Previously settled bacteria consume nutrients intensively, therefore the nutrient distribution over the surface is nonuniform. The primary choice of newly arrived cells is surface regions with the lowest population density of microorganisms, hence the largest amount of available nutrient.

However, already adhered cells surround themselves with polysaccharides to adjust the microenvironment and increase binding with the surface. Preliminary adsorption of organic molecules and polysaccharides (surface conditioning) is necessary for successful attachment of bacteria, and bacteria approaching the surface consider previously adhered cells as the preferable sites for attachment. From this point of view, the best place for new bacteria settlement is at the border of an existing colony, where preconditioning polysaccharide film already exists and energy source is not exhausted yet. Depending on the initial cell concentration, surface conditions, bacteria motility and life cycle, the combination of these two competitive factors determines actual colonization strategy of bacteria population.

Two separate experiments realizing different regimes of *Listeria* attachment have been performed. One is an exposure of highly concentrated *Listeria* culture to a metal surface for a short period of time to determine cell adhesion kinetics. This experiment allows the separation of adsorption/desorption processes during bacteria adhesion from bacteria growth and proliferation during the course of biofilm development. The other experiment involves biofilm formation assay including inoculation of a small amount of bacteria into a batch system and analysis of the surface population by fluorescent microscopy.

32.5.4 Adhesion of *L. monocytogenes* to the Aluminum Surface

Immersion of metal samples into the bacteria inoculated medium with high concentration of micro-organisms allows monitoring of cell adhesion process if the time of metal-culture contact is not longer than two to three characteristic bacteria reproduction (doubling) times. Due to the existence of a lag-phase, there is no time for bacterial growth on a surface, hence all observed bacteria have settled on a surface by adsorption. Technically, this experiment is similar to the classical characterization of adsorption process. Acquired data allows us to obtain (a) the maximum bacterial concentration on a surface and (b) kinetic parameters of bacterial adhesion.

Figure 32.8 represents the adhesion kinetics of *L. monocytogenes* to the aluminum surface. After approximately 3–5 seconds of contact, a few cells are observed to randomly adhere to the substratum, with substantial amount of space between the cells. Initial uniform distribution of adhered cells becomes more clustered with time. The plot shows that all available surface area is quickly filled with individual bacteria. For the first 10 seconds, their adhesion rate essentially follows first order kinetics (Henry law), i.e., there is linear increase in the number of cells adhered to the surface per time period.

Adsorption takes place only at specific surface sites, and the saturation coverage corresponds to complete occupancy of these sites. However, beyond this interval the number of newly attached cells decreases (Langmuir kinetics). Obtained data clearly indicate that for a given time and fixed cell concentration in the bulk, there is a limit to the number of cells the surface can support.

Using this data, it is possible to find the adsorption constant K_a for *L. monocytogenes* settling on the aluminum surface. Equation 32.36 describes the changes of surface concentration due to adsorption taking into account bacterial population growth in the bulk. Growth parameters (μ_0, K_0) of the planktonic population have been determined by the Lineweaver-Burke method analyzing the growth curves of *Listeria* in a 96-well microplate. The maximum surface concentration of the bacterial population is still unknown. We can estimate this parameter theoretically as the maximum number of cells that are able to cover the whole surface (defined as the ratio of surface area and single bacterium footprint) corrected by the surface factor, due to the presence of some constraints

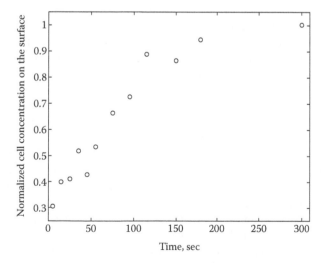

FIGURE 32.8 Kinetics of cell adhesion to the aluminum substrate and progress of bacterial colony formation on an aluminum surface. (Takhistov, P. and George, B., *Bioprocess and Biosystems Engineering*, 26, 259–270, 2004. With permission from Springer Science & Business Media.)

on the surface not the whole surface is available for settlement. From our experimental observations, we have evaluated this factor to be ~0.2. Fitting the experimental data depicted in Figure 32.8 and substituting values of known and theoretically obtained parameters into Equation 32.36, we have found that $K_a = 1.71 \times 10^{-8}$.

32.5.5 Biofilm Dynamics on Metal Surface: Experiments and Modeling

A single adhered bacterial cell has very limited ability to modify its surrounding physico-chemical environment. Bacteria growing on a surface in close associations (biofilms) are able to develop adaptive strategies both at the microscale (individual cells level) and at the scale of biofilm as a whole.

As known from the literature [91], bacteria that adhere to the surface have lower reproduction rate and growth kinetic parameters than the planktonic counterparts. This is especially important during the initial period of biofilm development, when cell population exists on the surface as a consortium of individual cells and isolated colonies. At the initial time of contact between metal and bacterial culture, adhesion is the main process controlling bacteria population on the surface.

Cells adhere to the surface looking for more stable and secure environment. However, the contact with a surface, especially if surface constraints exist, is considered by cells as a significant environmental stress. This stress results in the delay of cell reproduction due to elongation of its lag-phase. Bacteria respond to stress using a multistep adaptation process. It includes analysis of the environment by chemical receptors and following homeostatic regulation of metabolic activity, which is a sequence of the "decision-making" processes on the most effective survival strategy under given conditions. The amount of bacteria in the surface increases with time and the population dynamics depends on the surface growth rate, not on the planktonic population changes. At this stage, adhesion and growth on a surface are competitive processes. As the biofilm spreads over the surface, there is competition between planktonic and sessile cell populations for nutrient supply. Bacteria on a surface are more stable and stress-resistant but their planktonic competitors have greater growth rate [10,92].

Experimental observation of biofilm development on plain aluminum surface reveals increase in the number of singly attached cells, as well as colonies on the surface over time, but only up to a certain point. After 120 hours, cells start to detach from the surface. The amount of polysaccharide film spanning the surface also progressively decreases. The EPS has an important role of keeping cells attached to the surface, regardless of its topography. However, as cells grow they use up carbon from the EPS film as nutrient and consequently the polysaccharide material disintegrates, making it easier for starving cells to detach from the surface. In addition, as cells die they lose body mass and the polysaccharide capsule weakens and shrinks. This phenomenon provides a means by which the entire polysaccharide film and its underlying cells are dislodged.

In fact, macroscopic observation of batch cultures indicates that although the suspension in each of the wells was turbid with a light brown slime, the wells of 120 and 144 hours old appeared to be only slightly turbid on day 7. The extent of slime formation and turbidity could indicate the variation of the physiological state of bacteria over time.

The data depicted in Figure 32.9 were obtained using the image analyzing procedure from experimental results. Using the kinetic parameters obtained from the growth and adsorption kinetics analysis, we are able to estimate the surface population kinetics using Equation 32.43. All parameters of the model are known except the growth rate of bacteria on the surface. Our observations indicate that during surface colonization, *L. monocytogenes* has surface growth rate of almost the same order of magnitude as the bulk population. Theoretically evaluated surface concentration of bacteria is depicted in Figure 32.9. Even with linearized and simplified model, the theoretical result is in good qualitative agreement with the experimental data. It is interesting to evaluate the critical time of the system when the surface bacterial population is stressed by nutrient depletion due to extensive growth of the planktonic part of the population. This characteristic time is the major

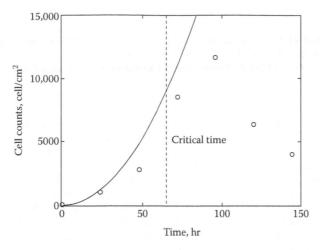

FIGURE 32.9 Dynamics of *L. monocytogenes* surface population in a batch system and theoretical estimation of the growth kinetics and critical time. (Takhistov, P. and George, B., *Bioprocess and Biosystems Engineering*, 26, 259–270, 2004. With permission from Springer Science & Business Media.)

parameter, which determines the population dynamics for the given conditions and can be estimated using Equation 32.45. Substituting variables described above, we determine that the characteristic time for nutrient depletion in case of simultaneous adsorption, surface and bulk bacteria growth is equal to 75.1 hours, which is supported by the experimental data in Figure 32.9. The parabolic growth of surface population at this point slows until it reaches its maximum value at ~90 hours.

It is clear that massive cell detachment from colonized surface occurs in response to stress induced by the lack of nutrients. Cells liberated from the surface are still alive and grow in the bulk. Dead cells do not detach from the surface, as shown by numerous observations of control samples.

32.6 CONCLUSION

Biofilm initiation and development is a complex process, which includes several major stages. In this chapter the dynamics of biofilm's formation on food-relevant (packaging, equipment, etc.) surfaces has been investigated. The theoretical linearized model of the cell adhesion and bacteria growth on a surface has been developed, allowing obtainment of values of important kinetic parameters of the process. The derived set of equations describes the kinetics of surface population growth and characteristic times for adsorption and combined growth processes, including characteristic time for the nutrient supply depletion. All equations contain variables based on the fundamental characteristics of bacterial population and can be easily determined from the experimental data or estimated theoretically.

It has been found that at the beginning bacteria adsorption is a linear function of time. The developed approach allows determining an adsorption constant for the bacteria material/system, which is a very important parameter for biomaterials, packaging and biotechnology applications. The combined effect of bacteria growth on the surface and their adsorption from bulk result in the surface population growth is proportional to t_2. The data obtained through the experimental observations are in good agreement with theoretical results, supporting the applicability of the suggested model to biofilm development studies. A complex process of biofilm formation can be represented as a sum of two separate processes: cell adhesion and colony proliferation. The overall dynamics of surface colonization and biofilm development is the result of these processes. It has been confirmed by the experimental observations that the cell adhesion process can be described in terms of the Langmuir isotherm and characterized by adsorption/desorption kinetics. The maximum surface concentration and kinetic constants can be determined from the experimental data by conventional analysis.

REFERENCES

1. Bower, C.K. and M.A. Daeschel Resistance responses of microorganisms in food environments. *International Journal of Food Microbiology*, 1999, 50(1–2): 33–44.
2. Mah, T.-F.C. and G.A. O'Toole. Mechanisms of biofilm resistance to antimicrobial agents. *Trends in Microbiology*, 2001, 9(1): 34–9.
3. Loosdrecht, M.V., C. Picioreanu, and J. Heijnen. A more unifying hypothesis for biofilm structures. *FEMS Microbiology Ecology*, 1997, 24, 181–83.
4. Wimpenny, J.W.T. and R. Colasanti. A unifying hypothesis for the structure of microbial biofilms based on cellular automaton models. *FEMS Microbiology Ecology*, 1997, 22(1): 1–16.
5. Evans, L.V. *Biofilms: Recent Advances in Their Study and Control*, ed. L.V. Evans. Harwood Academic Publishers, Newark, New Jersey, 2000, 466.
6. Blake, R., W. Norton, and G. Howard. Adherence and growth of a Bacillus species on an insoluble polyester polyurethane. *International Biodeterioration & Biodegradation*, 1988, 42: 63–73.
7. Kalmokoff, M.L., et al. Adsorption, attachment and biofilm formation among isolates of *Listeria monocytogenes* using model conditions. *Journal of Applied Microbiology*, 2001, 91(4): 725–34.
8. Apilanez, I., A. Gutierrez, and M. Diaz. Effect of surface materials on initial biofilm development. *Bioresource Technology*, 1998, 66(3): 225–30.
9. Bower, C.K., J. McGuire, and M.A. Daeschel. The adhesion and detachment of bacteria and spores on food-contact surfaces. *Trends in Food Science & Technology*, 1996, 7(5): 152–57.
10. Scheuerman, T., A. Camper, and M. Hamilton. Effects of substratum topography on bacterial adhesion. *Journal of Colloid and Interface Science*, 1998, 208: 23–33.
11. Characklis, W. *Biofilms*, ed. K.M. Mitchell. New York, NY: Wiley Interscience Publication John Wiley and Sons, 1990.
12. Sutherland, A.J. and I.W. Sutherland. The biofilm matrix—an immobilized but dynamic microbial environment. *Journal of Organic Chemistry*, 2001, 66(26): 9033–37.
13. Morton, L.H.G., et al. Consideration of some implications of the resistance of biofilms to biocides. *International Biodeterioration & Biodegradation*, 1998, 41(3–4 SU): 247–59.
14. Tsuneda, S., et al. Extracellular polymeric substances responsible for bacterial adhesion onto solid surface. *FEMS Microbiology Letters*, 2003, 223(2): 287–92.
15. Donlan, R.M. Biofilms: microbial life on surfaces. *Emerging Infectious Diseases*, 2002, 8(9): 881–90.
16. McLean, R.J., et al. Evidence of autoinducer activity in naturally occurring biofilms. *FEMS Microbiology Letters,* 1997, 154(2): 259–63.
17. Nakagawa, I., S. Hamada, and A. Amano. Studying initial phase of biofilm formation: molecular interaction of host proteins and bacterial surface components. *Methods in Enzymology*, 1999, 310: 501–13.
18. Shirtliff, M.E., J.T. Mader, and A.K. Camper. Molecular interactions in biofilms. *Chemistry & Biology,* 2002, 9(8): 859–71.
19. Lazazzera, B.A. Quorum sensing and starvation: signals for entry into stationary phase. *Current Opinion in Microbiology*, 2000, 3(2): 177–82.
20. Pearson, J.P., et al. Pseudomonas aeruginosa cell-to-cell signaling is required for virulence in a model of acute pulmonary infection. *Infection & Immunity*, 2000, 68(7): 4331–34.
21. Zhu, J. and J.J. Mekalanos. Quorum sensing-dependent biofilms enhance colonization in *Vibrio cholerae*. *Developmental Cell*, 2003, 5(4): 647–56.
22. Parsek, M.R., et al. Regulation of the alginate biosynthesis gene algC in *Pseudomonas aeruginosa* during biofilm development in continuous culture. *Science*, 1998, 280(5361): 295–98.
23. Brock, T.D. *Brock Biology of Microorganisms*, 10th edn, ed. J.M.M. Michael, T. Madigan, and J. Parker. Upper Saddle River, NJ: Prentice Hall, 2002.
24. Boom, R.M., et al. Quantifying recontamination through factory environments--a review. *International Journal of Food Microbiology*, 2003, 80(2): 117–30.
25. Chmielewski, R.A.N. and J.F. Frank. Biofilm formation and control in food processing facilities. *Comprehensive Reviews in Food Science and Safety*, 2003, 2: 22–31.
26. Dunne, W.M., Jr. Bacterial adhesion: seen any good biofilms lately? *Clinical Microbiology Reviews,* 2002, 15(2): 155–66.
27. Ellwood, D.C. *Adhesion of Microorganisms to Surfaces*, ed. D.C. Ellwood. London, UK: Academic Press, 1979, 216.
28. Zingg, W., et al. Surface thermodynamics of bacterial adhesion. *Transactions—American Society for Artificial Internal Organs*, 1983, 29: 146–51.

29. Cossart, P., J. Pizarro-Cerda, and M. Lecuit. Invasion of mammalian cells by *Listeria monocytogenes*: functional mimicry to subvert cellular functions. *Trends in Cell Biology*, 2003, 13(1): 23–31.

30. Shimoji, Y., et al. Adhesive surface proteins of *Erysipelothrix rhusiopathiae* bind to polystyrene, fibronectin, and Type I and IV collagens. *Journal of Bacteriology*, 2003, 185(9): 2739–48.

31. Koomey, M. Implications of molecular contacts and signaling initiated by *Neisseria gonorrhoeae*. *Current Opinion in Microbiology*, 2001, 4(1): 53–57.

32. Roberts, A.J. and M. Wiedmann. Pathogen, host and environmental factors contributing to the pathogenesis of listeriosis. *Cellular & Molecular Life Sciences*, 2003, 60(5): 904–18.

33. Eginton P.J., G.H. Handley, and P.S. Gilbert. The influence of substratum properties on the attachment of bacterial cells. *Colloids and Surfaces B: Biointerfaces*, 1995, 5: 153–59.

34. Chavant, P., et al. *Listeria monocytogenes* LO28: surface physicochemical properties and ability to form biofilms at different temperatures and growth phases. *Applied & Environmental Microbiology*, 2002, 68(2): 728–37.

35. Arnold, J.W. and G.W. Bailey. Surface finishes on stainless steel reduce bacterial attachment and early biofilm formation: scanning electron and atomic force microscopy study. *Poultry Science*, 2000, 79(12): 1839–45.

36. Barnes, L.M., et al. Effect of milk proteins on adhesion of bacteria to stainless steel surfaces. *Applied and Environmental Microbiology*, 1999, 65(10): 4543–48.

37. Reid, G. Adhesion of urogenital organisms to polymers and prosthetic devices. *Methods in Enzymology*, 1995, 253: 514–19.

38. Ratih Dewanti, A.C.L.W. Influence of culture conditions on biofilm formation by *Escherichia coli* O157:H7. *International Journal of food Microbiology*, 1995, 26: 147–64.

39. Boe-Hansen, R., et al. Bulk water phase and biofilm growth in drinking water at low nutrient conditions. *Water Research*, 2002, 36(18): 4477–86.

40. Turnbull, A.G., et al. The role of motility in the in vitro attachment of *Pseudomonas putida* PaW8 to wheat roots. *FEMS Microbiology Ecology*, 2001, 35(1): 57–65.

41. Gorski, L., J.D. Palumbo, and R.E. Mandrell. Attachment of *Listeria monocytogenes* to radish tissue is dependent upon temperature and flagellar motility. *Applied & Environmental Microbiology*, 2003, 69(1): 258–66.

42. Stopforth, J.D., et al. Biofilm formation by acid-adapted and nonadapted *Listeria monocytogenes* in fresh beef decontamination washings and its subsequent inactivation with sanitizers. *Journal of Food Protection*, 2002, 65(11): 1717–27.

43. Stepanovic, S., et al. Influence of the incubation temperature, atmosphere and dynamic conditions on biofilm formation by Salmonella spp. *Food Microbiology*, 2003, 20(3 SU): 339–43.

44. Bouttier, S., et al. Attachment of *Salmonella choleraesuis choleraesuis* to beef muscle and adipose tissues. *Journal of Food Protection*, 1997, 60(1): 16–22.

45. Mercier-Bonin, M., et al. Study of bioadhesion on a flat plate with a yeast/glass model system. *Journal of Colloid and Interface Science*, 2004, 271(2): 342–50.

46. Sadr Ghayeni, S.B., et al. Adhesion of waste water bacteria to reverse osmosis membranes. *Journal of Membrane Science*, 1998, 138(1): 29–42.

47. Batte, M., et al. Biofilm responses to ageing and to a high phosphate load in a bench-scale drinking water system. *Water Research*, 2003, 37(6): 1351–61.

48. Donlan, C.J., F. Courchamp, and B.A. Lazazzera. Quorum sensing and starvation: signals for entry into stationary phase. *Proceedings of the National Academy of Sciences of the United States of America*, 2002, 99(2): 791–96.

49. Liu, Y. and J.-H. Tay. The essential role of hydrodynamic shear force in the formation of biofilm and granular sludge. *Water Research*, 2002, 36(7): 1653–65.

50. Hassett, D.J., et al. Anaerobic metabolism and quorum sensing by *Pseudomonas aeruginosa* biofilms in chronically infected cystic fibrosis airways: rethinking antibiotic treatment strategies and drug targets. *Advanced Drug Delivery Reviews*, 2002, 54(11): 1425–43.

51. Bos, R., H.C. van der Mei, and H.J. Busscher. Physico-chemistry of initial microbial adhesive interactions--its mechanisms and methods for study. *FEMS Microbiology Reviews*, 1999, 23(2): 179–230.

52. Gottenbos, B., H.C. van der Mei, and H.J. Busscher. Models for studying initial adhesion and surface growth in biofilm formation on surfaces. *Methods in Enzymology*, 1999, 310: 523–34.

53. Dibdin, G. and J. Wimpenny. Steady-state biofilm: practical and theoretical models. *Methods in Enzymology*, 1999, 310: 296–322.

54. Bagge, D., et al. Shewanella putrefaciens adhesion and biofilm formation on food processing surfaces. *Applied & Environmental Microbiology*, 2001, 67(5): 2319–25.

55. Palmer, R.J., Jr. Microscopy flowcells: perfusion chambers for real-time study of biofilms. *Methods in Enzymology*, 1999. 310: 160–66.
56. Cox, J.D., et al. Surface passivation of a microfluidic device to glial cell adhesion: a comparison of hydrophobic and hydrophilic SAM coatings. *Biomaterials*, 2002, 23(3): 929–35.
57. Leitao, A. and A. Rodrigues. Dynamic behaviour of a fixed-bed biofilm reactor: analysis of the role of the intraparticle convective flow under biofilm growth. *Biochemical Engineering Journal*, 1998, 2(1): 1–9.
58. Lee, K.C. and B.E. Rittmann. Applying a novel autohydrogenotrophic hollow-fiber membrane biofilm reactor for denitrification of drinking water. *Water Research*, 2002, 36(8): 2040–52.
59. Alves, M.M., et al. Effect of lipids and oleic acid on biomass development in anaerobic fixed-bed reactors. Part I: Biofilm growth and activity. *Water Research*, 2001, 35(1): 255–63.
60. Djordjevic, D., M. Wiedmann, and L.A. McLandsborough. Microtiter plate assay for assessment of Listeria monocytogenes biofilm formation. *Applied & Environmental Microbiology*, 2002, 68(6): 2950–58.
61. Allerberger, F. Listeria: growth, phenotypic differentiation and molecular microbiology. *FEMS Immunology and Medical Microbiology*, 2003, 35(3): 183–89.
62. Oliveira, R., et al. Polysaccharide production and biofilm formation by *Pseudomonas fluorescen*: effects of pH and surface material. *Colloids and Surfaces B: Biointerfaces*, 1994, 2(1–3): 41–46.
63. Soutourina, O.A. and P.N. Bertin. Regulation cascade of flagellar expression in Gram-negative bacteria. *FEMS Microbiology Reviews*, 2003, 27(4): 505–23.
64. Williams, H.T., et al. Two-dimensional growth models. *Physics Letters A*, 1998, 250(1–3): 105–10.
65. Barzykin, A.V., K. Seki, and M. Tachiya. Kinetics of diffusion-assisted reactions in microheterogeneous systems. *Advances in Colloid and Interface Science*, 2001, 89–90: 140–47.
66. Enmon, R., et al. Aggregation kinetics of well and poorly differentiated human prostate cancer cells. *Biotechnology & Bioengineering*, 2002, 80(5): 580–88.
67. Ivanenko, Y.V., N.I. Lebovka, and N.V. Vygornitskii. Eden growth model for aggregation of charged particles. *European Physical Journal B*, 1999, 11(3): 469–80.
68. Mimura, M., H. Sakaguchi, and M. Matsushita. Reaction-diffusion modelling of bacterial colony patterns. *Physica A: Statistical Mechanics and its Applications*, 2000, 282(1–2): 283–303.
69. Crampin, E. Pattern formation in reaction–diffusion models with nonuniform domain growth. *Bulletin of Mathematical Biology*, 2002, 64: 747–69.
70. Crank, J. *The Mathematics of Diffusion*, 2nd edn. Oxford Science Publications. Oxford, UK: Clarendon Press, 414, 1975.
71. Matsushita, M., et al. Interface growth and pattern formation in bacterial colonies. *Physica A: Statistical and Theoretical Physics*, 1998, 249(1–4): 517–24.
72. Matsushita, M., et al. Formation of colony patterns by a bacterial cell population. *Physica A: Statistical Mechanics and its Applications*, 1999, 274(1–2): 190–99.
73. Apostilides, E. and L. McLandsborough. Characterization of a swarming phenotype of *Listeria innocua* on semi-solid surfaces. In *IAFP Annual Meeting*. New Orleans, LA, 2003.
74. Takhistov, P. and B. George. Early events and pattern formation in Listeria monocytogenes. *Biofilms*, 2004, 1: 351–59.
75. Golding, I., et al. Studies of bacterial branching growth using reaction-diffusion models for colonial development. *Physica A: Statistical and Theoretical Physics*, 1998, 260(3–4): 510–54.
76. Dockery, J.D. and J.P. Keener. A mathematical model for quorum sensing in *Pseudomonas aeruginosa*. *Bulletin of Mathematical Biology*, 2001, 63(1): 95–116.
77. Ward, J., et al. Mathematical modelling of quorum sensing in bacteria. *Journal of Mathematics Applied in Medicine and Biology*, 2001, 18(3): 263–92.
78. Levenspiel, O. *Chemical Reaction Engineering*, 2nd edn. New York, NY: John Wiley & Sons, 578, 1972.
79. Characklis, W. *Biofilms*, ed. K.M. William Characklis. New York, NY: John Wiley and Sons, Inc. 796, 1990.
80. Christensen, B.E. The role of extracellular polysaccharides in biofilms. *Journal of Biotechnology*, 1989, 10(3–4): 181–202.
81. Cloete, T.E. Resistance mechanisms of bacteria to antimicrobial compounds. *International Biodeterioration & Biodegradation*, 2003. 51(4): 277–82.
82. Bazin, M. Microbial population dynamics. In *Mathematical Models in Microbiology*, ed. M. Bazin. Boca Raton, FL: CRC Press. 202, 1982.
83. Augustin, J.-C. and V. Carlier. Mathematical modelling of the growth rate and lag time for Listeria monocytogenes. *International Journal of Food Microbiology*, 2000, 56(1 SU): 29–51.

84. Buchanan, R.L., J.L. Smith, and W. Long. Microbial risk assessment: dose-response relations and risk characterization. *International Journal of Food Microbiology,* 2000, 58(3): 159–72.
85. Pham, H., G. Larsson, and S.-O. Enfors. Modelling of aerobic growth of Saccharomyces cerevisiae in a pH-auxostat. *Bioprocess Engineering,* 1999, 20: 537–44.
86. Abee, T. and J.A. Wouters. Microbial stress response in minimal processing. *International Journal of Food Microbiology,* 1999, 50(1–2): 65–91.
87. Kjellerberg, S. *Starvation in Bacteria,* ed. S. Kjellerberg. New York, NY: Plenum Press, 277, 1993.
88. Abdelmalek, F., M. Shadaram, and H. Boushriha. Ellipsometry measurements and impedance spectroscopy on Langmuir-Blodgett membranes on Si/SiO2 for ion sensitive sensor. *Sensors and Actuators B: Chemical,* 2001, 72(3): 208–13.
89. Wong, A.L. Biofilms in food processing environments. *Journal of Dairy Science,* 1998, 81: 2765–70.
90. Chae, M.S. and H. Schraft. Cell viability of Listeria monocytogenes biofilms. *Food Microbiology,* 2001, 18(1): 103–12.
91. Rudney, J. and R. Staikov. Simultaneous measurement of the viability, aggregation, and live and dead adherence of *Streptococcus crista, Streptococcus mutans* and *Actinobacillus actinomycetemcomitans* in human saliva in relation to indices of caries, dental plaque and periodontal disease. *Archives of Oral Biology,* 2002, 47(5): 347–59.
92. Arnold, J. and G. Bailey. Surface finishes on stainless steel reduce bacterial attachment and early biofilm formation: scanning electron and atomic force microscopy study. *Poultry Science,* 2000, 79: 1839–45.
93. Takhistov, P. and B. George. Linearized kinetic model of Listeria monocytogenes biofilm growth. *Bioprocess and Biosystems Engineering,* 2004, 26: 259–70.

33 Kinetics of Biological Reactions

Roy Zhenhu Lee
Johnson Matthey Catalysts, Inc.

Qixin Zhong
University of Tennessee

CONTENTS

33.1   INTRODUCTION

Food systems are complex from compositional, structural and chronicle points of view. Compositionally, each food product differs in the type and concentration of chemical elements it contains. The compositional variation in food products requires that we should analyze case by case for all possible reactions—formation of physical and chemical bonds between molecules, consumption of nutrients by microorganisms, enzymatic or microbial breakdown of molecules, among others. Structurally, molecules in food systems form nanoscopic, microscopic, and macroscopic domains, which may strain the type and rate of reaction. Chronically, reactions in food systems occur during processing and storage and hardly stop before being consumed. All the above aspects pose challenges to develop a comprehensive description of kinetics of biological reactions within real food products.

Nevertheless, with the accumulation of our understanding in food chemistry, microbiology, processing and engineering, food scientists have made significant strides toward analyzing biological reactions during processing and storage of food products. Chemical reaction engineering has been applied to food science to better predict and control food production practices. Besides analyses of chemical reactions where reactants and products follow defined reaction schemes, quality parameters, e.g., L^*, a^*, and b^* values used to measure colors, have been used to describe the kinetics of quality changes.[1]

This chapter does not intent to describe all biological reactions relevant to food processing and storage. Indeed, kinetics of specific reactions relevant to food systems has been recently reviewed.[2] In addition, enzymatic reactions are discussed in a separate chapter. Our objective is to provide a generalized mathematical description of biological chemical reactions, with the expectation that the descriptions may be used for modeling and optimization of reactions, which occur during food processing and storage. Our discussions are within the framework of traditional chemical reaction engineering. However, readers are advised that physical interactions are also critical to understanding the kinetics of biological reactions in foods, which are excluded in this chapter. Readers are referred to Israelachvili's book to understand the physical interactions between molecules and structures.[3]

We start this chapter with an introductory discussion of chemical reactions that are known and well-understood in food science. Generalized reactions and reaction rates are then described, followed by examples of a few important reactions. Kinetics is then incorporated in the context of food processing, where a unit operation is generalized as a reactor, which leads us to discuss homogenous reactions in ideal reactors. Hydrogenation of vegetable oils and glucose/fructose isomerization are finally discussed as examples of heterogeneous reactions. Abundant information is available on the kinetics of biological reactions relevant to food systems; it is our hope that the generalization of these reactions and their kinetics will eventually enable readers to apply the principles to their case-specific reactions and processes.

33.2   BIOLOGICAL REACTIONS IN FOOD SYSTEMS

Water/ice, carbohydrates, lipids, amino acids, and their oligomers (peptides) and polymers (proteins, including enzymes) are major components of food products, while minor food components include vitamins, minerals, colorants, flavors, among others.[4] Synergetic interactions of major and minor components and their reactions during handling, processing and storage determine the quality (texture, flavor, color, etc.), nutritive value and safety of a product. Many reactive functional groups, e.g., aldehydes, amines, ketones, alkenes, and hydroxyls, are present in food molecules—some are available for reactions, while others need molecular structural rearrangements to expose reactive groups (e.g., formation of disulfide bonds after denaturation of beta-lactoglobulins). Some chemical and biochemical reactions, important to food quality and safety, include enzymatic and nonenzymatic browning, oxidation, hydrolysis, metal interactions, lipid isomerization, lipid cyclization, lipid polymerization, protein denaturation, protein cross-linking, polysaccharide synthesis, and glycolytic changes.[5] The occurrences of these reactions and their reaction rates are affected by product factors (physicochemical properties of individual constituents including catalysts, oxygen content, pH, water activity) and environmental factors (thermal history i.e., temperature, time and rate of changes, composition of the atmosphere, the chemical, physical or biological treatments imposed, exposure to light, contamination, physical abuse, etc.).[5]

Due to the complexity of molecular structures and compositions of food systems, a precise description of the kinetics of a specific reaction therefore requires a thorough research of potential reactive species at the corresponding conditions. Further, considerations should be given to chemical and environmental factors that may inhibit a given reaction and those that may limit the applicability of models based on simplified conditions (e.g., heterogeneity of reactions due to mass transfer limitations caused by the structure of food matrices).

33.3 REACTORS, REACTIONS, AND REACTION RATES

33.3.1 REACTORS

Reactors are physically or arbitrarily defined boundaries within which a reaction or reactions take place. For example, we can consider a block of cheese as a reactor to describe a proteolytic reaction during storage or a tubular heat exchanger to describe the degradation of vitamin C during pasteurization of fluidic foods. An obvious challenge to describe reactions and reaction rates in food systems or food processing unit operations is the spatial variations of reactant/product concentrations and possibly also temperature. Nevertheless, numerical simulations may enable us to predict/model reactions in complex systems once fundamentals of reactions are clearly understood. In this chapter, we will discuss kinetics within ideal reactors, as depicted in Figure 33.1.[6] In an ideal batch reactor, there is a uniform distribution of reactants and products whose concentrations change with time. In a continuous stirred tank reactor (CSTR), a feed stream continuously enters a batch reactor that is mixed uniformly, and a stream continuously leaves the reactor with a composition identical to the composition in the reactor at the moment when this stream leaves. While in a continuous plug flow reactor (PFR), we assume that there is a plug flow within the tube and that the composition on a plane perpendicular to the flow direction in the tube is uniform.

33.3.2 HOMOGENEOUS AND HETEROGENEOUS REACTIONS

In chemical reaction engineering, "homogeneous" and "heterogeneous" are two terms used to classify reactions and reactors.[6] If one reaction occurs within one phase only, this reaction is called a homogeneous reaction. Otherwise, a reaction is heterogeneous when at least two phases are involved. Ambiguity does exist for this classification in some cases, including reactions in colloidal systems

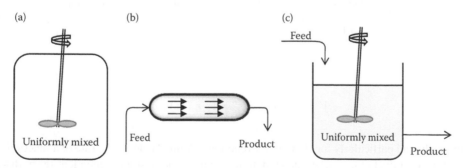

FIGURE 33.1 Three types of ideal reactors: (a) batch reactor, (b) plug flow reactor (PFR) and (c) continuous stirred tank reactor (CSTR). (Adapted from Levenspiel, O., *Chemical Reaction Engineering*, 3rd edn. John Wiley & Sons, New York, NY, 1999.)

TABLE 33.1
Different forms for Defining Reaction Rates

Reaction Systems	Bases of Defining Reaction Rates	Definitions of Reaction Rates	
Reacting fluids	Unit volume	$r_i = \dfrac{1}{V}\dfrac{dN_i}{dt} = \dfrac{\text{moles } i \text{ formed}}{(\text{volume of fluid}) (\text{time})}$	(33.1)
Fluid–solid systems	Unit mass	$r_i' = \dfrac{1}{W}\dfrac{dN_i}{dt} = \dfrac{\text{moles } i \text{ formed}}{(\text{mass of solid}) (\text{time})}$	(33.2)
Two-fluid systems or gas–solid systems	Unit interfacial surface or unit surface of solid	$r_i'' = \dfrac{1}{S}\dfrac{dN_i}{dt} = \dfrac{\text{moles } i \text{ formed}}{(\text{surface}) (\text{time})}$	(33.3)
Gas–solid systems	Unit volume	$r_i''' = \dfrac{1}{V_s}\dfrac{dN_i}{dt} = \dfrac{\text{moles } i \text{ formed}}{(\text{volume of solid}) (\text{time})}$	(33.4)
Overall reactors	Unit volume of reactor	$r_i'''' = \dfrac{1}{V_r}\dfrac{dN_i}{dt} = \dfrac{\text{moles } i \text{ formed}}{(\text{volume of reactor}) (\text{time})}$	(33.5)

Source: Adapted from Levenspiel, O., *Chemical Reaction Engineering*, 3rd edn. John Wiley & Sons, New York, NY, 1999.

that have structures with a dimension between 1 and 1000 nm. For example, enzymes are proteins that can be dissolved in an aqueous solution to catalyze specific reaction(s) (one phase); however, enzymes are colloidal particles where reactions occur on the molecule surface. Classification of an enzymatic reaction then depends on how we define a system: homogeneous if the system is treated as a whole and heterogeneous if we examine reactants that bind onto the molecule surface and products that leave the surface.

33.3.3 Reaction Rates

For a homogeneous reaction, the reaction rate of a component i can be defined in different interexchangeable ways (Table 33.1; Equation 33.6). For heterogeneous reactions, a reaction rate can be defined based on convenience.[6]

$$\begin{pmatrix}\text{volume} \\ \text{of fluid}\end{pmatrix} r_i = \begin{pmatrix}\text{mass of} \\ \text{solid}\end{pmatrix} r_i' = \begin{pmatrix}\text{surface} \\ \text{of solid}\end{pmatrix} r_i'' = \begin{pmatrix}\text{volume} \\ \text{of solid}\end{pmatrix} r_i''' = \begin{pmatrix}\text{volume} \\ \text{of reactor}\end{pmatrix} r_i'''' \quad (33.6)$$

or

$$V\, r_i = W\, r_i' = S\, r_i'' = V_s\, r_i''' = V_r\, r_i''''$$

Commonly and particularly for elementary reactions (defined below), the rate of a reaction can be expressed as a product of an energy-dependent term and a composition-dependent term, discussed individually in the following sections.[6] This energy can be any format that may affect a reaction. Particularly in food processing, the energy terms may include temperature, light, magnetic field, electric field, hydrostatic/dynamic pressure, etc. Temperature is the only energy term discussed in this chapter, which leads to a rate equation of:

$$r_i = f_1(\text{temperature}) \cdot f_2(\text{composition}) = k \cdot f_2(\text{composition}) \tag{33.7}$$

33.3.4 CONCENTRATION-DEPENDENCE OF A RATE EQUATION

33.3.4.1 Stoichiometry

Stoichiometry describes the mass conservation of substances entering a reaction and those produced after the reaction and provides a quantitative relationship between reactants and products. For example, one mole of benzoic acid in humans is eliminated from the body via conjugation with an equal mole of glycine to form one mole of hippuric acid:[7]

$$\tag{33.8}$$

Benzoic acid   Glycine   Hippuric acid

When a single stoichiometric equation and single rate equation represents the progress of a reaction, this reaction is called a single reaction. In contrast, multiple reactions are needed to describe a situation where more than one kinetic expression is needed to follow the progress of all reaction components using multiple stoichiometric equations. These include series and parallel reactions. In series reactions, intermediate products produced from reactants undergo further reaction(s) (Equation 33.9).[8] In parallel reactions, reactants may have more than one reaction path (Equation 33.10). A more complicated situation may be presented in Equation 33.11, which shows series reactions with respect to A, R, and S or parallel reactions with respect to B.[6]

$$A \rightarrow R \rightarrow S \tag{33.9}$$

$$
\begin{array}{ll}
A \rightarrow R & A \rightarrow R \\
\downarrow \qquad \text{and} & \\
S & B \rightarrow S \\
\text{competitive} & \text{side by side}
\end{array}
\tag{33.10}
$$

$$
\begin{array}{l}
A + B \rightarrow R \\
R + B \rightarrow S
\end{array}
\tag{33.11}
$$

33.3.4.2 Elementary and Nonelementary Reactions

A reaction whose rate equation follows a stoichiometric equation is called an elementary reaction. In contrast, nonelementary reactions do not have a direct correspondence between stoichiometry and reaction rate.[6] The reaction in Equation 33.8 is an example where the rate controlling mechanism is the collision or interaction between a single molecule of benzoic acid (A) and a single molecule of glycine (B) and the reaction rate is proportional to the number of collisions. For this elementary reaction, the reaction rate (rate of reactant consumption) can be written as in Equation 33.12.

An exemplary nonelementary reaction is that between hydrogen and bromine (Equation 33.13a), whose rate (Equation 33.13b) does not correspond to stoichiometry.[6]

$$-r_A = kC_A C_B \tag{33.12}$$

$$H_2 + Br_2 \rightarrow 2\ HBr \tag{33.13a}$$

$$-r_{HBr} = \frac{k_1 C_{H_2} \left(C_{Br_2} \right)^{1/2}}{k_2 + C_{HBr}/C_{Br_2}} \tag{33.13b}$$

33.3.4.3   Reaction Rates of Elementary Reactions

33.3.4.3.1   Simple Reactions

Consider an elementary reaction with a stoichiometric equation

$$a\ A + b\ B + c\ C + d\ D + \cdots \rightarrow p\ P + r\ R + s\ S + \cdots \tag{33.14}$$

where a, b, c,... p, r, s... are the stoichiometric coefficients of the corresponding species.

The molecularity, applicable to elementary reactions only, of the above reaction is used to describe the number of molecules that are involved in the reaction. These numbers must be integers to represent the reaction mechanism and are usually one, two and occasionally three.[6]

For a single irreversible reaction, the rate expression is conventionally described as[9]

$$-r_A = kC_A^{m_A} C_B^{m_B} C_C^{m_C} \cdots = k \prod C_j^{m_j} \tag{33.15}$$

where $\prod$ indicates the product of all terms, C_j is the concentration (molar or partial pressure) of the jth species, and m_j is the exponent in the power-law expression used to describe the concentration dependence on the jth species. Note that m_j is not necessarily identical to the stoichiometric coefficient.

The reaction order is the power raised for the concentration of species. Therefore, this reaction is the $m_A{}^{th}$-order with respect to A, the $m_B{}^{th}$-order with respect to B, etc. The overall order (n^{th}) of this reaction can also be defined to be the summation of all powers:

$$n = m_A + m_B + m_C + m_D + \cdots = \sum m_j \tag{33.16}$$

This order n does not need to be an integer. For example, a reaction can be a 1.5-order reaction.

33.3.4.3.2   Reversible and Parallel Reactions

For a reversible reaction with known stoichiometry, the overall reaction rate of a species can be treated by the summation of both forward and back (or reverse) reactions:[9]

$$-r_A = k_f \prod C_j^{m_{fj}} - k_b \prod C_j^{m_{bj}} \tag{33.17}$$

where subscripts of f and b stand for forward and back reactions, respectively.

Similarly, for a system where a reactant undergoes multiple reactions, the reaction rate of this reactant can be treated by considering the mass balance of all reactions:

$$-r_A = \sum_i \left(k_i \Pi c_j^{m_{ij}} \right) \tag{33.18}$$

where the summation is made for all reactions (i) each with a reaction rate constant of k_i.

33.3.4.4 Kinetics of Nonelementary Reactions—An Example of Lipid Autoxidation

The kinetic description of nonelementary reactions becomes more complicated because the kinetics does not match its stoichiometry, as in Equation 33.13. Multiple elementary steps are to be postulated to describe the kinetics of a specific nonelementary reaction. Conventionally, intermediates are proposed in the postulated elementary steps. For example, free radicals and the chain reaction mechanism can be used to describe the kinetics of lipid autoxidation.

A three-step simplified, free radical scheme has been postulated for lipid autoxidation (Equation 33.19),[8] based on the model proposed for the autoxidation of ethyl linoleate. At the first initiation step, free radicals are generated due to the hydroperoxide decomposition, metal catalysis or exposure to light. At the propagation step, more free radical species $R^{\bullet}$ and $ROO^{\bullet}$ are produced. At the termination step, free radicals interact to form more stable, nonradical products.

$$2\,ROOH \rightarrow \text{free radicals } (R^{\bullet},\, ROO^{\bullet}) \qquad \text{Initiation} \tag{33.19a}$$

$$\left.\begin{array}{l} R^{\bullet} + O_2 \rightarrow ROO^{\bullet} \\ ROO^{\bullet} + RH \rightarrow POOH + \rightarrow R^{\bullet} \end{array}\right\} \qquad \text{Propagation} \qquad\qquad \begin{array}{l}(33.19b)\\(33.19c)\end{array}$$

$$\left.\begin{array}{l} R^{\bullet} + R^{\bullet} \rightarrow RR \\ R^{\bullet} + ROO^{\bullet} \rightarrow ROOR \\ ROO^{\bullet} + ROO^{\bullet} \rightarrow ROOR + O_2 \end{array}\right\} \text{Nonradical products} \quad \text{Termination} \quad \begin{array}{l}(33.19d)\\(33.19e)\\(33.19f)\end{array}$$

where RH is the substrate fatty acid and ROOH is hydroperoxide.

At a high oxygen pressure, reactions 33.19d and e are negligible because free radicals $R^{\bullet}$ will be reacted through reaction 33.19b fairly quickly. The kinetics of autoxidation can be simplified based on reactions 33.19a, b, c, and f only. Therefore, the rate of autoxidation with respect to the oxygen consumption can be expressed as:

$$-\frac{d[O_2]}{dt} = -\left\{ -k_2[R^{\bullet}]\,[O_2] + k_6[ROO^{\bullet}]^2 \right\} \tag{33.20}$$

With respect to the change of the free radical $R^{\bullet}$, the initiation step is a second-order reaction, and we can write

$$-\frac{d[R^{\bullet}]}{dt} = -\left\{ k_1[ROOH]^2 - k_2[R^{\bullet}]\,[O_2] + k_3[ROO^{\bullet}]\,[RH] \right\} \tag{33.21}$$

This rate is negligible, i.e.,

$$k_1[ROOH]^2 - k_2[R^{\bullet}]\,[O_2] + k_3[ROO^{\bullet}]\,[RH] = 0 \tag{33.22}$$

After rearranging, we have

$$k_2[R^{\bullet}]\,[O_2] - k_3[ROO^{\bullet}]\,[RH] = k_1[ROOH]^2 \tag{33.23}$$

Similarly for ROO•, we can write

$$-\frac{d[\text{ROO}^\bullet]}{dt} = -\left\{ k_1[\text{ROOH}]^2 + k_2[\text{R}^\bullet]\,[\text{O}_2] - k_3[\text{ROO}^\bullet]\,[\text{RH}] - 2k_6[\text{ROO}^\bullet]^2 \right\} \quad (33.24)$$

This rate is also negligible, giving

$$k_1[\text{ROOH}]^2 + k_2[\text{R}^\bullet]\,[\text{O}_2] - k_3[\text{ROO}^\bullet]\,[\text{RH}] - 2k_6[\text{ROO}^\bullet]^2 = 0 \quad (33.25)$$

After substituting Equation 33.23 for the second and third groups above, we have

$$k_1[\text{ROOH}]^2 + k_1[\text{ROOH}]^2 - 2k_6[\text{ROO}^\bullet]^2 = 0 \quad (33.26)$$

or,

$$[\text{ROO}^\bullet] = \left(\frac{k_1}{k_6}\right)^{1/2}[\text{ROOH}] \quad (33.27)$$

Substituting the above equation into Equation 33.23 and after rearranging, we have:

$$k_1[\text{ROOH}]^2 + k_3\,[\text{RH}]\left(\frac{k_1}{k_6}\right)^{1/2}[\text{ROOH}] = k_2[\text{R}^\bullet]\,[\text{O}_2] \quad (33.28)$$

After substituting Equations 33.27 and 33.28 into Equation 33.20, we finally have an expression for the rate of autoxidation in terms of the substrate fatty acid and hydroperoxide concentrations:

$$-\frac{d[\text{O}_2]}{dt} = -\left\{ -k_1[\text{ROOH}]^2 - k_3\,[\text{RH}]\left(\frac{k_1}{k_6}\right)^{1/2}[\text{ROOH}] + k_1[\text{ROOH}]^2 \right\}$$

$$\tag{33.29}$$

$$= k_3\left(\frac{k_1}{k_6}\right)^{1/2}[\text{RH}]\,[\text{ROOH}]$$

Similarly, at a low oxygen pressure, reactions 33.19e and f are negligible, and the rate of autoxidation can be derived to be:

$$-\frac{d[\text{O}_2]}{dt} = k_2\left(\frac{k_1}{k_4}\right)^{1/2}[\text{O}_2]\,[\text{ROOH}] \quad (33.30)$$

33.3.5 TEMPERATURE-DEPENDENCE OF A RATE EQUATION

The reaction rate constant k in Equation 33.7 is used to present the temperature-dependence of the reaction rate. In most cases, the Arrhenius' law is satisfactory to describe the temperature-dependence:

$$k = k_0 e^{-E/RT} \quad (33.31)$$

where k_0 is the frequency or pre-exponential factor, E is the activation energy of the reaction, R is the universal gas constant, and T is the temperature in degree Kelvin.

33.4 KINETICS OF HOMOGENEOUS, IRREVERSIBLE REACTIONS IN IDEAL REACTORS

In many practices, we are interested in understanding the kinetics in reactors so that the reaction(s) can be completed within a minimum amount of time (or minimum residence time for continuous reactor) to achieve the maximum yields of the desired products. To better design a reactor and model reactions within this reactor, characteristics (stoichiometry, dependence of reaction rates on compositions and temperatures, etc.) of a reaction as well as mass and energy transfers during the reaction are to be considered. We can then proceed to develop prediction models or perform computer simulations for the process development and/or control. Again, we will only discuss here the simplest situations—ideal reactors. Further, this section discusses the simplest case of isothermal conditions; readers are referred to other textbooks for analyses of nonisothermal and adiabatic reactors that involve energy generation or consumption.

The mass balance of all reactors can be dealt with by treating the reactor as a whole (a black box) or by considering a differential element. The accumulation of a certain species can be written as:

$$[\text{Accumulation}]_i = [\text{Flow in}]_i - [\text{Flow out}]_i + [\text{generation by reaction}]_i \qquad (33.32)$$

For an ideal batch reactor with a constant volume, a reaction time is well defined since no mass enters or leaves the reactor. However, in CSTR and PFR, terms of space time (τ) and space velocity (s) need to be defined as follows:[6]

$$\tau = \frac{1}{s} = \left(\begin{array}{l} \text{time required to process one reactor volume} \\ \text{of feed measured at specific conditions} \end{array} \right) \qquad (33.33)$$

$$s = \frac{1}{\tau} = \left(\begin{array}{l} \text{number of reactor volumes of feed at specified} \\ \text{conditions which can be treated in unit time} \end{array} \right) \qquad (33.34)$$

33.4.1 BATCH REACTOR WITH A CONSTANT VOLUME

In an ideal batch reactor with a constant volume, uniform distributions of species are assumed. This simplification (of no spatial gradients of mass and energy) simplifies the mathematical description of reaction kinetics.

33.4.1.1 Kinetics of Reactions with Different Orders

For a simple reaction of

$$A \rightarrow \text{products} \qquad (33.35)$$

the rate equation and the concentration of A after a reaction time t can be expressed in the following equations when this reaction follows a zero-, first-, second-, or n^{th}-order, respectively.[9]

Zero-order

$$-r_A = -\frac{dC_A}{dt} = k \qquad (33.36a)$$

$$C_A = C_{A0} - kt \qquad (33.36b)$$

First-order

$$-r_A = -\frac{dC_A}{dt} = kC_A \tag{33.37a}$$

$$C_A = C_{A0}e^{-kt} \tag{33.37b}$$

Second-order

$$-r_A = -\frac{dC_A}{dt} = kC_A^2 \tag{33.38a}$$

$$C_A = \frac{C_{A0}}{1 + C_{A0}kt} \tag{33.38b}$$

The nth-order

$$-r_A = -\frac{dC_A}{dt} = kC_A^n \tag{33.39a}$$

$$C_A = C_{A0}\left[1 + (n-1)kC_{A0}^{n-1}t\right]^{1/(1-n)} \tag{33.39b}$$

where C_{A0} is the concentration of A at the time zero.

For a bimolecular, second-order reaction,

$$A + B \rightarrow product \tag{33.40}$$

the rate equation can be written as in Equation 33.12. When the ratio of initial concentrations of B and A ($M = C_{B0}/C_{A0}$) is 1, Equation 33.38 can be applied. When M is not equal to 1, concentrations of A and B after a reaction time t can be expressed as:[6]

$$\ln\frac{C_B C_{A0}}{C_{B0}C_A} = \ln\frac{C_B}{MC_A} = C_{A0}(M-1)\,kt = (C_{B0} - C_{A0})\,kt \quad \text{for} \quad M \neq 1 \tag{33.41}$$

However, care should be paid for the case of an overall second-order reaction with the stoichiometry of Equation 33.42 where the reaction is first-order with respect to both A and B. For this reaction, the rate expression and concentrations of reactants during the progress of reaction can be written as in Equation 33.43.[6]

$$A + 2B \rightarrow products \tag{33.42}$$

$$-r_A = -\frac{dC_A}{dt} = kC_A C_B = kC_{A0}^2(1 - X_A)(M - 2X_A) \tag{33.43a}$$

$$\ln \frac{C_B C_{A0}}{C_{B0} C_A} = \ln \frac{M - 2X_A}{M(1 - X_A)} = C_{A0}(M - 2)kt \quad \text{for } M \neq 2$$

$$\frac{1}{C_A} - \frac{1}{C_{A0}} = \frac{1}{C_{A0}} \frac{X_A}{1 - X_A} = 2kt \quad \text{for } M = 2 \tag{33.43b}$$

where X_A is the fraction of A that is consumed by the reaction.

For a trimolecular, third-order reaction, the rate expression and concentrations of reactants during the progress of reaction can be written as:[6]

$$A + B + D \rightarrow \text{products} \tag{33.44}$$

$$-r_A = -\frac{dC_A}{dt} = kC_A C_B C_D \quad \text{or} \quad C_{A0} \frac{dX_A}{dt} = kC_{A0}^3 (1 - X_A) \left(\frac{C_{B0}}{C_{A0}} - X_A \right) \left(\frac{C_{D0}}{C_{A0}} - X_A \right) \tag{33.45a}$$

$$\frac{1}{(C_{A0} - C_{B0})(C_{A0} - C_{D0})} \ln \frac{C_{A0}}{C_A} + \frac{1}{(C_{B0} - C_{D0})(C_{B0} - C_{A0})} \ln \frac{C_{B0}}{C_B}$$

$$+ \frac{1}{(C_{D0} - C_{A0})(C_{D0} - C_{B0})} \ln \frac{C_{D0}}{C_D} = kt \tag{33.45b}$$

Typically, trimolecular reactions take the form of Equation 33.46, whose kinetics are given in the subsequent expressions:[6]

$$A + 2B \rightarrow \text{Products} \tag{33.46}$$

$$-r_A = -\frac{dC_A}{dt} = kC_A C_B^2 \tag{33.47a}$$

$$\frac{(2C_{A0} - C_{B0})(C_{B0} - C_B)}{C_{B0} C_B} + \ln \frac{C_{A0} C_B}{C_A C_{B0}} = (2C_{A0} - C_{B0})^2 kt \quad \text{for } M \neq 2$$

$$\frac{1}{C_A^2} - \frac{1}{C_{A0}^2} = 8kt \quad \text{for } M = 2 \tag{33.47b}$$

For a reaction with a third-order

$$A + B \rightarrow \text{products} \tag{33.48}$$

$$-r_A = -\frac{dC_A}{dt} = kC_A C_B^2 \tag{33.49a}$$

we can write

$$\frac{(C_{A0} - C_{B0})}{C_{B0}C_B} \frac{(C_{B0} - C_B)}{C_B} + \ln\frac{C_{A0}C_B}{C_A C_{B0}} = (C_{A0} - C_{B0})^2 kt \quad \text{for } M \neq 1$$

(33.49b)

$$\frac{1}{C_A^2} - \frac{1}{C_{A0}^2} = 2kt \quad \text{for } M = 1$$

33.4.1.2 Half-Life of an Irreversible Reaction[6]

If the stoichiometric ratios of reactants remain throughout an irreversible reaction with stoichiometry as Equation 33.14, the rate expression in Equation 33.15 can be rewritten as

$$-r_A = kC_A^{m_A}C_B^{m_B}C_C^{m_C}\ldots = kC_A^{m_A}\left(\frac{b}{a}C_A\right)^{m_B}\ldots = k\left(\frac{b}{a}\right)^{m_B}C_A^{m_A+m_B+\cdots}\ldots = \tilde{k}C_A^n$$

(33.50)

$$\text{or} \quad -\frac{dC_A}{dt} = \tilde{k}C_A^n$$

For $n \neq 1$, integrating the above equation gives

$$C_A^{1-n} - C_{A0}^{1-n} = \tilde{k}(n-1)t$$

(33.51)

The half-life of the reaction, $t_{1/2}$, can then be defined as the time needed to reduce the concentration of the reactants to one-half of the original value:

$$t_{1/2} = \frac{(0.5)^{1-n} - 1}{\tilde{k}(n-1)}C_{A0}^{1-n}$$

(33.52)

For a first-order reaction, the half-life is independent on the initial concentration.

Therefore, the half-life of a reaction can be estimated at series of initial reactant concentrations at the same reaction temperature, and the results can be used to obtain the order of this reaction based on a plot of $\log(t_{1/2})$ versus $\log C_{A0}$ (Equation 33.52).

33.4.2 CONTINUOUS STIRRED TANK REACTOR[6] (CSTR)

Under steady-state conditions in a CSTR, a feed stream enters the reactor and a product stream exits the reactor with the same compositions as those in the reactor whose compositional distributions are assumed to be uniform. The notation of a CSTR is given in Figure 33.2, illustrated for reactant A.

The feed stream enters the CSTR at a volumetric flow rate of v_0. The concentration of A in the feed stream is C_{A0} and the amount of A converted by reaction in the feed stream is X_{A0} (=0, assuming no reaction before entering the reactor). The molar feed rate of A is F_{A0} (=$v_0 C_{A0}$). The reaction rate of A, its concentration and conversion in the reactor is r_A, C_A, and X_A, respectively. The exit stream, with a denoted subscript of f, is similarly defined for variables and compositions as those in the feed stream.

After defining the variables and substituting these variables into Equation 33.32 with respect to changes of A in the reactor as a whole, we have Equations 33.53 and 33.54:

Input of A, moles/time $= F_{A0}(1-X_{A0}) = F_{A0}$
Output of A, moles/time $= F_{A0}(1-X_A) = F_A$
Disappearance of A due to reaction, moles/time $= (-r_A)V$
Accumulation of A $= 0$ (at steady state)

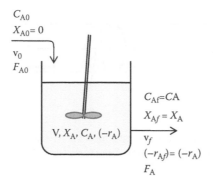

FIGURE 33.2 Notation of a CSTR for reactant A. (Adapted from Levenspiel, O., *Chemical Reaction Engineering,* 3rd edn. John Wiley & Sons, New York, NY, 1999.)

$$F_{A0} X_A = (-r_A)V \tag{33.53}$$

or

$$\frac{V}{F_{A0}} = \frac{\tau}{C_{A0}} = \frac{\Delta X_A}{-r_A} = \frac{X_A}{-r_A}$$

$$\tau = \frac{1}{s} = \frac{V}{v_0} = \frac{VC_{A0}}{F_{A0}} = \frac{C_{A0}X_A}{-r_A} \tag{33.54}$$

For a reaction with a known rate expression, the above equation can be used to estimate the reactor volume and space time needed to achieve a certain conversion rate of reactant A at a defined volumetric feed rate, or vice versa.

33.4.3 PLUG FLOW REACTOR[6] (PFR)

A PFR can be illustrated as in Figure 33.3. The terms in the entrance and exit of the reactor are similarly defined as in a CSTR.

Under plug flow condition, a finite volume along the reactor (dV) can be defined to write a differential equation based on the conservation of for example reactant A:

Input of A, moles/time $= F_A$
Output of A, moles/time $= F_A + dF_A$
Disappearance of A due to reaction, moles/time $= (-r_A)\, dV$
Accumulation of A $= 0$ (at steady state)
Substituting these terms into Equation 33.32, we have

$$F_A = (F_A + dF_A) + (-r_A)dV \tag{33.55}$$

Since $F_A = F_{A0}(1-X_A)$, we have

$$dF_A = d[F_{A0}(1-X_A)] = -F_{A0}\, dX_A \tag{33.56}$$

Substituting Equation 33.56 into Equation 33.55, we have

$$F_{A0}dX_A = (-r_A)dV \tag{33.57}$$

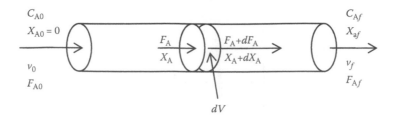

FIGURE 33.3 Notation of a PFR with respect to reactant A. (Adapted from Levenspiel, O., *Chemical Reaction Engineering·* 3rd edn. John Wiley & Sons, New York, NY, 1999.)

Integrating the above equation with respect to the inlet and outlet gives:

$$\int_0^V \frac{dV}{F_{A0}} = \int_0^{X_{Af}} \frac{dX_A}{-r_A}$$

$$\frac{V}{F_{A0}} = \frac{\tau}{C_{A0}} = \int_0^{X_{Af}} \frac{dX_A}{-r_A} \tag{33.58}$$

$$\tau = \frac{V}{v_0} = \frac{VC_{A0}}{F_{A0}} = C_{A0} \int_0^{X_{Af}} \frac{dX_A}{-r_A}$$

In PFR, the reaction rate, $-r_A$, when compared to CSTR, may not be a constant since reactant concentrations may change as the stream flows along the reactor. For a known reaction with defined kinetics, the reactor volume and space time may be solved from the above equation or by computer simulations.

33.5 HETEROGENEOUS REACTIONS

Heterogeneous reactions are rather common in food systems due to the composition and structure of food matrices. Hydrolysis of lipids to fatty acids by lipases is an example. Lipids can be in the continuous or dispersed phase. In order to catalyze the hydrolytic reaction, the water-soluble lipases have to be adsorbed onto an oil/water or water/oil interface. The reaction is thus a heterogeneous one because of the involved multiple phases. In the chemical industries, catalytic reactors are used extensively to produce numerous products or intermediates. In the food industry, both biological and nonbiological catalysts are commonly used. Enzymes are the most common biological catalysts, for example isomerases are used to convert D-glucose to D-fructose in the syrup production. Metal catalysts are used extensively to produce polyols from monosaccharides and to hydrogenate oils to products with desirable functionalities. We illustrate heterogeneous reactions in stirred-tank reactors and packed-bed reactors using examples of hydrogenation of oils and glucose/fructose isomerization, respectively.

33.5.1 Heterogeneous Reactions in Stirred Tank Reactors—An Example of Hydrogenation of Vegetable Oils

We present here a brief description of the hydrogenation of vegetable oils by metal catalysts. One of the predominant factors for us to review this process is that the potential hydrogenation products, *trans*-fatty acids, are of concern to the food industry due to the health issues. Reactions involving metal catalysts are less discussed in food engineering, so we hope this section will shed some light to food engineers when they design or model reactions under analogous conditions.

33.5.1.1 Porous Catalysts in Slurry Reactors—A View from Different Scales

Although continuous reactors are available, batch reactors are commonly used in the hydrogenation of vegetable oils due to low operating costs and a suitability and flexibility for the production of different products with various quantities.[10] A slurry reactor can be viewed from four length scales (Figure 33.4).[9] At the reactor scale, pellets are suspended in oils by an agitator and hydrogen is introduced to mix with the slurry. Therefore, the system is essentially a gas–liquid slurry reactor. At the scale of pellet sizes, reactants, i.e., oils and hydrogen, diffuse from the continuous phase to the pellet surface. At the scale of pores inside catalyst pellets, reactants diffuse into the internal pores. At the scale of molecules, reactants adsorb onto the surface of internal pores where catalytic reactions take place, and products desorb from the surface upon the completion of a reaction. Therefore, major factors to be considered in hydrogenation of oils include (external and internal) mass transfers, catalysts, reaction conditions (temperature, pressure), as discussed individually in the following sections.

33.5.1.2 Physicochemical Properties of Vegetable Oils

Seed oils are triacylglycerols whose individual fatty acids vary in composition and proportion with the resources. For fatty acids with double bonds, the geometric configuration is called a *cis*-structure when the alkyl groups are on the same side of the molecule; otherwise, a *trans*-structure indicates that the alkyl groups are on the opposite sides of the molecules. The *cis*-configuration is the naturally occurring form, while the *trans*-configuration is thermodynamically favored.[8]

Hydrogenation is a process where hydrogen is added onto the double bonds. This is illustrated in Figure 33.5 for simplified schemes where only *cis*-fatty acids are produced when sequentially hydrogenating linolenic acid (Ln) to linoleic acid (L) to oleic acid (O) to stearic acid (S). The isomerization between *cis*- and *trans*-configurations however exists during hydrogenation, which complicates the reaction scheme. A more realistic example of hydrogenation of linoleic acid is illustrated in Figure 33.6 where the isomer of oleic acid is called the elaidic acid (E).

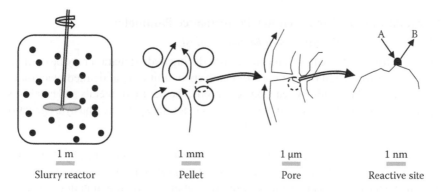

1 m	1 mm	1 μm	1 nm
Slurry reactor	Pellet	Pore	Reactive site

FIGURE 33.4 Illustration of catalytic reactions in a slurry reactor at different length scales. (Adapted from Schmidt, L.D., *The Engineering of Chemical Reactions*, 2nd edn. Oxford University Press, New York, NY, 2005.)

$$Ln + H_2 \xrightarrow{k_3} L + H_2 \xrightarrow{k_2} O + H_2 \xrightarrow{k_1} S$$
$$18:3 \qquad\quad 18:2 \qquad\quad 18:1 \qquad\quad 18:0$$

FIGURE 33.5 Simplified reaction scheme for the sequential hydrogenation of linolenic acid to stearic acid. (Adapted from Veldsink, J.W., Bouma, M.J., Schoon, N.H., and Beenackers, A.A.C.M., *Catal. Rev. Sci. Eng.* 39, 253–318, 1997.)

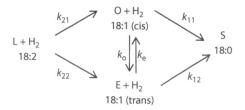

FIGURE 33.6 Reaction scheme for the sequential hydrogenation of linolenic acid to stearic acid considering the isomerization reaction. (Adapted from Veldsink, J.W., Bouma, M.J., Schoon, N.H., and Beenackers, A.A.C.M., *Catal. Rev. Sci. Eng.* 39, 253–318, 1997.)

TABLE 33.2
Melting Points of Different Isomers of Fatty Acids with 18 Carbons and One Double Bond

Systematic Name	Common Name	Melting Point
Octadeca-*cis* 9-enoic	Oleic	16°C
Octadeca-*trans* 9-enoic	Elaidic	42°C
Octadeca-*cis* 6-enoic	Petroselenic	30°C
Octadeca-*trans* 6-enoic	Petroselaidic	51.9°C

Source: Adapted from Patterson, H.B.W., *Hydrogenation of Fats and Oils*, Applied Science Publishers, London, UK, 1983.

The isomerization reaction as in Figure 33.6 may lead to various isomeric products. The positional variations among these isomers result in products with different melting profiles (Table 33.2). In general, fatty acids with *trans*-configurations have a higher melting point than those with *cis*-configurations. The challenge is then to screen reaction conditions to produce a final product with quality parameters matching the desired applications.

33.5.1.3   Metal Catalysts and their Key Performance Parameters

33.5.1.3.1   Performance Parameters of Metal Catalysts

33.5.1.3.1.1   Activity   Activity of a metal catalyst during hydrogenation may be defined as moles of double bonds hydrogenated by unit mass of catalysts in unit time at defined reaction conditions. This activity is affected not only by reaction conditions, the nature of metals and catalyst structures but by impurities that may be present in oils and fats. Generally speaking, the hydrogenation activity of metals follows an order of: $Pd > Pt \gg Ni \gg Cu$.[10]

33.5.1.3.1.2   Selectivity   The selectivity during hydrogenation was defined by Patterson as: "In its most general sense, selectivity in hydrogenation means a preference for hydrogenating one class of unsaturated substance rather than another and in practice an ability to maintain this preference until the concentration of the preferred unsaturated is much decreased."[11] As seen in the reaction schemes illustrated in Figures 33.5 and 33.6, there are more than one selectivity to be defined during the hydrogenation of vegetable oils. In the community of oil chemists, first-order reactions are generally accepted for the hydrogenation of tri-, di-, and mono-unsaturated fatty acids, which leads to definitions of two selectivities (Equation 33.59) based on the three reaction rate constants in Figure 33.5. The S_I is important in the industrial hardening and the hydrogenation is called selective when S_I is greater than 10.[10]

$$S_I = \frac{k_2}{k_1} \quad \text{and} \quad S_{II} = \frac{k_3}{k_2} \tag{33.59}$$

As also seen in Figure 33.6, a selectivity must be defined with respect to the formation of cis- and trans-isomers, where cis-monoenes are usually desirable. This selectivity is called the specific isomerization index, defined as:

$$S_i = \frac{trans \text{ double bond formed}}{\text{double bonds hydrogenated}} \tag{33.60}$$

Because natural fats and oils are triglycerides, triglyceride selectivity (S_T) is also defined to indicate the preference of a certain constituent unsaturated fatty acid of the triglycerides over other fatty acids.[11] A high S_T means constituent unsaturated fatty acids are hydrogenated randomly i.e., no preference.

33.5.1.3.1.3 Catalyst Induction, Fatigue, and Poisoning At the beginning of hydrogenation reactions, the observed reaction rate may be low and this period is called the induction period. This may be caused by partially deactivated sites on the metal surface due to for example oxidization during storage prior to use.[11] The induction period may be avoided by in situ reduction of the catalyst.[10]

With the progression of hydrogenation, the observed reaction rate may decline due to thermal and mechanical attrition of the catalyst surface, mechanical clogging and/or direct chemical poisoning. This is called the catalyst fatigue or fouling. The poisoning of nickel may be caused by carbon monoxide, sulfur, phosphorus, halogens, soap, free fatty acid, oxidized fat, or degradation by contact with excess moisture.[11] Some poisoning mechanisms are reversible, while others are irreversible.

33.5.1.3.1.4 Reusability Form an economic point of view, reusability is a measure of how many times a batch of catalysts can be used without a compromised reaction activity and selectivity. Used catalysts may be regenerated to recover the performance.

33.5.1.3.2 Catalysts Used in Hydrogenation of Vegetable Oils

The most extensively used category of metal catalysts in the hydrogenation of vegetable oils is nickel due to its high activity, tailored linoleic acid and linolenic acid selectivity, low cost and easy removal from oil (by filtration) after reaction.[12] Other active metals, i.e., palladium, copper, ruthenium, cobalt, platinum, and their modified forms have also been researched, especially for the purpose of reducing trans-fatty acids contents. The performance parameters, e.g., activity and selectivity of metals, vary significantly.[10] Selection of a metal catalyst shall be based on the performance parameters, in addition to costs. Palladium has high S_I and S_{II} but tends to produce a product with a high proportion of trans-fatty acids. In regard to reducing trans-fatty acids, platinum may be promising despite its low S_I.[10]

33.5.1.4 Mechanism of Hydrogenation of Vegetable Oils

Chemisorption, adsorption of atoms or molecules on a surface by valence forces, of hydrogen and unsaturated carbon bonds is well known, and their adsorption to metal surfaces is independent to each other.[10] The rate and amount of adsorption depend on the hydrogenation conditions. A simplified half-hydrogenated state mechanism is presented in Figure 33.7. A more detailed, extended reaction scheme, particularly for the hydrogenation of linoleic acid, was proposed by Koritala et al.[13]

As shown in Figure 33.7, the first step of hydrogenating double bonds of fatty acids (in cis-configuration) involves the adsorption of a double bond onto the metal surface, followed by the addition of a hydrogen atom adsorbed on the metal surface—formation of a half-hydrogenated state (configurations b and c). A subsequent addition of a hydrogen atom adsorbed on the metal surface produces a fully saturated carbon-carbon bond (configuration d). Alternatively, losing a hydrogen atom from the half-hydrogenated structures (configurations b and c) reforms double bonds that can now take different carbon positions on the original fatty acids (configurations e, f, and g), giving

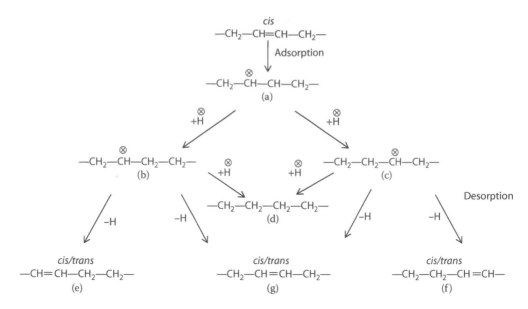

FIGURE 33.7 A half-hydrogenated state mechanism proposed during hydrogenation of a double bond on fatty acids. Symbol ⊗ indicates a linkage to a metal surface. (Adapted from Nawar, W.W., *Food Chemistry*, 3rd edn., Fennema, O.R., Marcel Dekker, Inc., New York, NY, 225–319, 1996.)

TABLE 33.3
Effects of Process Conditions on Hydrogenation Reactions

Increase in the Reaction Parameter*	Response in the Selectivity Ratio	Response in the Ratio of *Trans*-Configuration	Response in the Reaction Rate
Temperature	Increase	Increase	Increase
Pressure	Decrease	Decrease	Increase
Agitation	Decrease	Decrease	Increase
Catalyst concentration	Increase	Increase	Increase

Source: Adapted from Allen, R.R., *J. Am. Oil Chem. Soc.* 55, 792–795, 1978.

* Assuming other reaction conditions are constant. In the commercial production, the operating temperature and pressure in a slurry reactor is usually 393–473 K and 0.1–0.5 MPa, respectively, with a nickel-based catalyst loaded between 0.01 and 0.2 wt%.[11]

either a *cis*- or a *trans*-configuration depending on hydrogenation conditions. Chances do exist for all reaction schemes depicted in Figure 33.7, generating a mixture that determines the quality attributes of a hydrogenated product. The effects of reaction conditions on hydrogenation are generalized in Table 33.3.

33.5.1.5   Intra- and Interparticle Mass Transfers

Mass transfers in a gas-liquid slurry reactor during hydrogenation of vegetable oils were reviewed in detail by Beenackers and van Swaaij.[14] The essential parts of the process are summarized by Veldsink et al.[10] as follows:

The mass transfer paths in a gas-liquid slurry reactor can be presented schematically in Figure 33.8. The flux of hydrogen being transported onto the reactive sites inside porous catalyst pellets (J_{H2}, mol m^{-2} s^{-1}) can be written as the driving force (hydrogen pressure, P_{H2}) divided

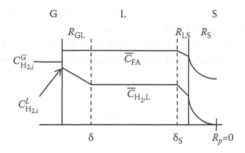

FIGURE 33.8 Schematic of mass transfer processes in a gas-liquid slurry reactor. (Adapted from Veldsink, J.W., Bouma, M.J., Schoon, N.H., and Beenackers, A.A.C.M., *Catal. Rev. -Sci. Eng. 39, 253–318, 1997.*)

by a series of three transport resistances: R_{GL} (gas to liquid), R_{LS} (liquid to pellet surface) and R_S (inside pellets):

$$J_{H_2} A = \frac{He \, P_{H_2}}{R_{GL} + R_{LS} + R_S} \tag{33.61}$$

where A is the gas–liquid surface area (m^2), He is the Henry constant (mol Pa^{-1} m^{-3}), and resistances are defined by

$$R_{GL} = \frac{1}{k_L A}, \quad R_{LS} = \frac{1}{\varepsilon_S \, k_S \, A_S}, \quad R_S = \frac{He \, P_{H_2}}{\varepsilon_S \, \rho_S \, \eta \, |r_{H_2}|} \tag{33.62}$$

where k_L is the liquid side mass transfer coefficient (m s^{-1}), ε_S is the solids holdup (ratio of solids volume to liquid volume), k_S is the liquid–solid mass transfer coefficient (m s^{-1}), A_S is the solid external area (m^2/m$_{cat}^3$), ρ_S is the solids density (kg m^{-3}), η is the effective factor as defined later in Equation 33.66, $|r_{H_2}|$ is the absolute value of the consumption rate of hydrogen during hydrogenation (mol kg$_{cat}^{-1}$ s^{-1}).

The dissolution of hydrogen in oil is the first step required in hydrogenation. A higher hydrogen pressure facilitates the dissolution of hydrogen in oils. In a more realistic gas-liquid reactor at a sparged and agitated condition, the gas-liquid mass transfer may be estimated to be:

$$\frac{k_L A d_s}{D} = 0.06 \left(\frac{N_s d_s^2}{\upsilon} \right)^{1.9} \left(\frac{\upsilon}{D} \right)^{0.5} \left(\frac{N_s^2 d_s}{g} \right)^{0.19} \left(\frac{\sigma_l}{\mu \, u_G} \right)^{-0.6} \left(\frac{N_s d_s}{u_G} \right)^{0.32} \tag{33.63}$$

where d_s is the diameter of the stirrer (m), D is the diffusivity (m^2 s^{-1}), N_s is the stirrer rate (rev s^{-1}), υ is the kinematic viscosity of the oil (m^2 s^{-1}), g is the gravity constant (9.81 m s^{-2}), σ_l is the surface tension of the liquid (N m^{-1}), μ is the dynamic viscosity of liquid (Pa s), and u_G is the superficial gas velocity (m s^{-1}).

Most experimental data have shown negligible external catalyst transfer limitations. The transport coefficient from the bulk liquid to the catalyst surface is expressed by the Sherwood number (Sh) that is a function of the Reynolds number (Re) and Schmidt number (Sc), which may be further estimated by:

$$Sh = \frac{k_S d_p}{D_i} = 2 + f(Re, Sc) = 2 + 0.36 \left(\frac{P_0 d_p^4}{V_L \rho_L \upsilon_L^3} \right)^{1/4} (Sc)^{1/3} \tag{33.64}$$

where d_p is the particle diameter, D_i is the diffusivity of a component i, P_0 is the power input (W), V_L is the liquid volume, ρ_L is the liquid density, and υ_L is the kinematic viscosity of liquid.

Additionally, there is a liquid film through which both hydrogen and triglycerides diffuse from the bulk to the catalyst external surface. The concentration gradient of triglycerides over this liquid film (ΔC_{TAG}) may be written as:

$$\Delta C_{TAG} = \frac{1}{z}\left(\frac{D_H}{D_{TAG}}\right)^{2/3}\Delta C_H \qquad (33.65)$$

where D_H and D_{TAG} are the diffusivity of hydrogen and triglyceride, respectively, and ΔC_H is the hydrogen concentration gradient.

Intraparticle transport limitations are critical in hydrogenation of oils using the pellet form of metal catalysts. As a result, concentration gradients exist in the catalyst for both reactants and products and the available catalyst activity is not fully utilized. The approximate degree of the utilized catalyst activity is expressed by the effectiveness factor (η), which is a function of the Thiele modulus (φ):

$$\eta = \frac{\tanh(\phi)}{\phi} \qquad (33.66)$$

The Thiele modulus may be written as:

$$\phi = L\sqrt{\frac{r}{D_{eff}\ C}} \qquad (33.67)$$

where L is the diffusion distance, D_{eff} is the effective diffusivity inside the catalyst, and C is the concentration of reactants.

The effective diffusivity may be expressed as:

$$D_{eff} = \frac{\varepsilon}{\tau}D^0 \times 10^{-2d_{mol}/d_{pore}} \qquad (33.68)$$

where D^0 is the diffusivity in liquid, ε is the porosity of catalyst (m_{void}^3/m_{cat}^3), τ is the tortuosity factor, d_{mol}, and d_{pore} are the diameters of molecules and pores, respectively.

The intraparticle diffusion limitations of triglycerides and hydrogen have opposite effects on S_I (production of fully saturated triacylglycerols) and the *trans*-fatty acid content, respectively. The selectivity S_I increases with an increase in Thiele modulus when the hydrogen concentration in the liquid phase is treated as a variable, while the opposite is true for triglycerides. Therefore, intraparticle transport limitations of fats should be avoided to receive good selectivities. Additionally, hydrogenation of fats is a mildly exothermic reaction, so temperature gradients in liquid-filled pores are not a big concern.[10]

33.5.1.6   Kinetics of Hydrogenation Reactions

Many kinetic models have been proposed in the literatures and are summarized by Veldsink et al.[10] When utilizing these models, care must be paid for the assumptions used in their derivations as well as in their suitability for the reaction conditions. Special attention shall be made for the mass transfer limitations discussed above.

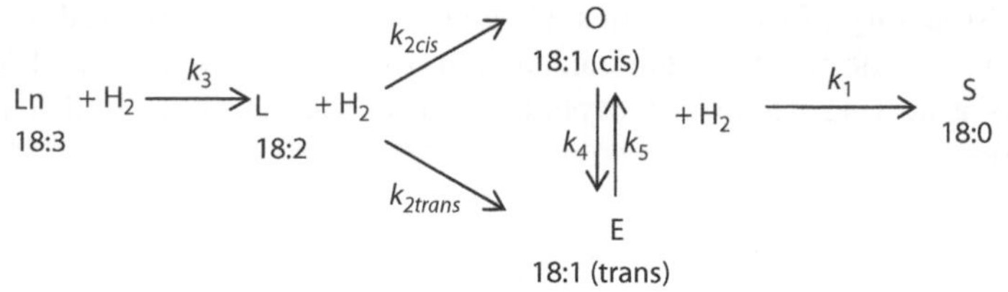

$$Ln \quad +H_2 \xrightarrow{\;k_3\;} L \quad +H_2 \begin{array}{c} \nearrow^{k_{2cis}} \\ \searrow_{k_{2trans}} \end{array}$$

FIGURE 33.9 Reaction scheme during the hydrogenation of soybean oils. (Adapted from Fillion, B., Morsi, B.I., Heier, K.R., and Machado, R.M., *Ind. Eng. Chem. Res.* 41, 697–709, 2002.)

For the simplified reaction scheme illustrated in Figure 33.5, Fillion et al.[15] derived the following rate equations for each species involved during the hydrogenation of soybean oil. The derivations were based on the following assumptions:

1. The adsorptions of hydrogen and double bonds are independent of each other.
2. The reaction rate with respect to each of the triglycerides and hydrogen concentrations is pseudo-first-order.
3. No poisoning of catalysts occurs during hydrogenation and the reaction rates are proportional to the loading of catalysts.
4. The activity of catalysts during the reaction can be described by the following expression:

$$f(t,C_{H_2}) = 1 - \exp(-b_1 C_{H_2}^{b_2} t^{b_3}) \tag{33.69}$$

where b_1, b_2, and b_3 are fit parameters.

$$\frac{\partial C_{Ln}}{\partial t} = \frac{-m_{cat} f(t,C_{H_2})\, k_3 C_{Ln} C_{H_2}}{1 + K_3 C_{Ln} + K_2 C_L + K_1 C_{O+E}} \tag{33.70}$$

$$\frac{\partial C_L}{\partial t} = \frac{m_{cat} f(t,C_{H_2})\,(k_3 C_{Ln} - k_2 C_L)\, C_{H_2}}{1 + K_3 C_{Ln} + K_2 C_L + K_1 C_{O+E}} \tag{33.71}$$

$$\frac{\partial C_{O+E}}{\partial t} = \frac{m_{cat} f(t,C_{H_2})\,(k_2 C_n - k_1 C_{O+E})\, C_{H_2}}{1 + K_3 C_{Ln} + K_2 C_L + K_1 C_{O+E}} \tag{33.72}$$

$$\frac{\partial C_S}{\partial t} = \frac{m_{cat} f(t,C_{H_2})\, k_3 C_{O+E} C_{H_2}}{1 + K_3 C_{Ln} + K_2 C_L + K_1 C_{O+E}} \tag{33.73}$$

$$\frac{\partial C_{H_2}}{\partial t} = k_L a (C_{H_2}^* - C_{H_2}) - 3\frac{m_{cat} f(t,C_{H_2})\,(K_3 C_{Ln} + K_2 C_L + K_1 C_{O+E})\, C_{H_2}}{1 + K_3 C_{Ln} + K_2 C_L + K_1 C_{O+E}} \tag{33.74}$$

where $C_{H_2}^*$ is the solubility of hydrogen in soybean oil that can be estimated by Equation 33.75, k_1, k_2, and k_3 represent the reaction rate constants in Figure 33.5, K_1, K_2, and K_3 are fit parameters, and $k_L a$ is a floating parameter depending on the temperature and hydrogen pressure (Equation 33.76).

$$C_{H_2}^* = 0.113 \exp(-5000/RT) P_{H_2} \tag{33.75}$$

$$k_L a = a_1 + a_2 \, T + a_3 \, T \, P_{H_2} + a_4 \, P_{H_2} + a_5 \, P_{H_2}{}^2 \tag{33.76}$$

Taking into account the *cis-trans* isomerization in Figure 33.6, Fillion et al.[15] also derived the rate Equations 33.77 through 33.82 for the reaction scheme in Figure 33.9:

$$\frac{\partial C_{Ln}}{\partial t} = \frac{-m_{cat} f\left(t, C_{H_2}\right) k_3 C_{Ln} C_{H_2}}{1 + K_3 C_{Ln} + K_2 C_L + K_1 \left(C_O + C_E\right)} \tag{33.77}$$

$$\frac{\partial C_L}{\partial t} = \frac{m_{cat} f\left(t, C_{H_2}\right) \left[k_3 C_{Ln} - \left(k_{2cis} + k_{2trans}\right) C_L\right] C_{H_2}}{1 + K_3 C_{Ln} + K_2 C_L + K_1 \left(C_O + C_E\right)} \tag{33.78}$$

$$\frac{\partial C_O}{\partial t} = \frac{m_{cat} f\left(t, C_{H_2}\right) \left(k_{2cis} C_L C_{H_2} + k_5 C_E C_{H_2}{}^\alpha - k_4 C_O C_{H_2}{}^\beta - k_1 C_O C_{H_2}\right)}{1 + K_3 C_{Ln} + K_2 C_L + K_1 \left(C_O + C_E\right)} \tag{33.79}$$

$$\frac{\partial C_E}{\partial t} = \frac{m_{cat} f\left(t, C_{H_2}\right) \left(k_{2trans} C_L C_{H_2} + k_4 C_O C_{H_2}{}^\beta - k_5 C_E C_{H_2}{}^\alpha - k_1 C_E C_{H_2}\right)}{1 + K_3 C_{Ln} + K_2 C_L + K_1 \left(C_O + C_E\right)} \tag{33.80}$$

$$\frac{\partial C_S}{\partial t} = \frac{m_{cat} f\left(t, C_{H_2}\right) k_1 \left(C_O + C_E\right) C_{H_2}}{1 + K_3 C_{Ln} + K_2 C_L + K_1 \left(C_O + C_E\right)} \tag{33.81}$$

$$\frac{\partial C_{H_2}}{\partial t} = k_L a \left(C_{H_2}^* - C_{H_2}\right) - 3 \frac{m_{cat} f\left(t, C_{H_2}\right) \left[k_3 C_{Ln} + \left(k_{2cis} + k_{2trans}\right) C_L + k1 \left(C_O + C_E\right)\right] C_{H_2}}{1 + K_3 C_{Ln} + K_2 C_L + K_1 \left(C_O + C_E\right)} \tag{33.82}$$

As seen in the above analyses, hydrogenation of vegetable oils is a classical reaction engineering problem that involves the integration of reaction kinetics and transport phenomena. Besides optimizing reaction conditions and efforts in synthesizing novel catalyst and catalytic materials, other technologies studied include: (1) the enhanced transfer of hydrogen to the internal pores of catalysts pellets by supercritical fluids, and (2) electrocatalytic hydrogenation.[12] Interested readers are referred to original publications for details of these developments, as reviewed by Jang et al.

33.5.2 Heterogeneous Reactions Using Biological Catalysts—An Example of Glucose/Fructose Isomerization by Immobilized Glucose Isomerase

Production of high-fructose corn syrup is an important industrial process that is based on the enzymatic isomerization of D-glucose to D-fructose to increase the sweetness. Generally, the enzyme is immobilized on surfaces such as mesopores of porous particles to provide a large surface area for enzymatic reactions.[16] The isomerization reaction can be completed in different types of reactors: CSTR, packed-bed, membrane, or fluidized bed reactors.[17] The analyses of stirred-tank and fluidized bed reactors are analogous to the above slurry reactor for hydrogenation of oils. The analogy is that the substrate glucose has to transport from the bulk to the surface of porous particles, diffuse into the mesopores, and bind with the immobilized enzyme for conversion. The difference is that no gas (hydrogen) phase is involved in the isomerization reaction. Therefore, transport processes (fluid flow, mass and energy transfers) are to be integrated with enzymatic kinetics when designing and analyzing reactors. The rapidly-increasing computational power has enabled the consideration of detailed surface morphologies of porous particles when simulating reactors at various dimensions.[16,18,19] We present here brief descriptions based on traditional continuum equations of mass transfer and reaction that are the bases of mathematical analyses of reactors. We further narrowed our discussion in the context of packed-bed reactors.

33.5.2.1 Reaction Kinetics of Glucose/Fructose Isomerization

At the initial stage of the glucose/fructose isomerization reaction, the classical Michaelis-Menten kinetics based on the two-step elementary reaction mechanism (Equation 33.83) is applicable,[20] with a rate expression in Equation 33.84).[6]

$$S + E \underset{k_{-1}}{\overset{k_1}{\rightleftharpoons}} \times \underset{k_{-2}}{\overset{k_2}{\rightleftharpoons}} E + P \tag{33.83}$$

where S is the substrate (glucose), E is the enzyme, X is the intermediate product, and P is the final product (fructose).

$$-r_S = k \frac{E_o C_S}{K_M + C_S} \tag{33.84}$$

where E_o is the initial active enzyme concentration, C_S is the substrate (glucose) concentration, and K_M is the Michaelis constant.

However, the glucose/fructose isomerization reaction is usually devious from the Michaelis–Menten kinetics and a modified expression—the Briggs–Haldane mechanism is more applicable. The Briggs–Haldane approach adapts an expression similar to Equation 33.84:

$$r_S' = \frac{r_m'(C_S - C_{S,e})}{K_M' + (C_S - C_{S,e})} \tag{33.85}$$

where terms r_m' (maximum reaction rate), $C_{S,e}$ (the glucose concentration in equilibrium with that in the feed, $C_{S,o}$), and K_M' (a constant analogous to the Michaelis constant for the whole isomerization) are defined as:

$$C_{S,e} = \frac{C_{S,o}}{1 + K} \tag{33.86}$$

$$K = \frac{r'_{mf} k_{mr}}{r'_{mr} k_{mf}} \tag{33.87}$$

$$r'_m = [1+(1/K)] \frac{k_{mr} r'_{mf}}{k_{mr} - k_{mf}} \tag{33.88}$$

$$K'_M = \frac{k_{mf} k_{mr}}{k_{mr} - k_{mf}} \left[1 + \left(\frac{1}{k_{mf}} + \frac{1}{k_{mr}} \right) C_{S,e} \right] \tag{33.89}$$

where r'_{mf} and r'_{mr} are the maximum rates of forward and reverse reactions when considering the whole isomerization:

$$k_{mf} = (k_{-1} + k_2)/k_1 \tag{33.90}$$

$$k_{mr} = (k_{-1} + k_2)/k_{-2} \tag{33.91}$$

$$r'_{mf} = k_2 E_t \tag{33.92}$$

$$r'_{mr} = k_{-1} E_t \tag{33.93}$$

where E_t is the actual concentration of the active enzyme.

33.5.2.2　Isomerization in a Packed Bed Reactor

In a realistic reactor, deactivation[21] and substrate protection[22] of enzymes are to be considered. A parameter can be introduced to describe the percentage of active enzymes (ϕ) during the progression of reaction, based on the first-order deactivation kinetics[21]

$$-\frac{d\varphi_t}{dt} = k_d t \tag{33.94}$$

The substrate protection can be incorporated in the above equation by introducing a substrate protection factor, σ, leading to a definition of the residual enzyme activity, a:[23]

$$a = \frac{E_t}{E_o} = e^{-(1-\sigma) k_d t} \tag{33.95}$$

The reaction rate expression can now be written as:[21]

$$r'_S = \frac{a \, r'_m (C_S - C_{S,e})}{K'_M + (C_S - C_{S,e})} \tag{33.96}$$

The above equation can be rewritten in terms of the glucose concentration at the reactor inlet, $C_{S,o}$ and the fraction of glucose converted to fructose, X_s:

$$r_s' = \frac{a\, r_m' C_{S,o}(X_S - X_{S,e})}{K_M' + C_{S,o}(X_S - X_{S,e})} \tag{33.97}$$

The above equation can then be substituted into Equation 33.58 for a PFR:

$$\tau = C_{S,o} \int_0^{X_{S,t}} \frac{dX_S}{\eta_X(-r_s')} \tag{33.98}$$

where the term η is an enzyme effectiveness factor that takes into account the diffusion limitations.[23]

The dependency of η on the substrate conversion can usually be neglected in the industrial production, and an average value $\bar{\eta}$ can be used. Equation 33.98 may then be solved:[23]

$$\tau = \frac{V}{v_0} = \frac{K_M'}{\bar{\eta}\, r_m'\, e^{-(1-\sigma)k_d\, t}}\left[\frac{C_{S,o}}{K_M'}X_S - \ln\frac{X_{S,e} - X_S}{X_{S,e}}\right] \tag{33.99}$$

Further analyses may lead to the estimation of an average residence time that is required to achieve a certain conversion X_s during an operating period, t_p:[23]

$$\bar{\tau} = \frac{(1-\sigma)\, k_d\, t_p\, K_M'}{\bar{\eta}\, r_m'\, \left(1 - e^{-(1-\sigma)k_d\, t_p}\right)}\left[\frac{C_{S,o}}{K_M'}X_S - \ln\frac{X_{S,e} - X_S}{X_{S,e}}\right] \tag{33.100}$$

Lastly, the temperature dependence of reaction rate constants during glucose/fructose isomerization generally follows the Arrhenius equation.[23]

33.6   CONCLUDING REMARKS

What we have tried to illustrate throughout this chapter, understanding and modeling kinetics of biological reactions within food systems and during food processing, is important to improve the quality, safety, and healthfulness of foods. Mathematical analyses of some scenarios have been established by food and chemical engineers, while other scenarios have nonideality, challenging food scientists. With the integration of different disciplines and computational techniques, the extension of established theories may enable us to advance the modeling of biological reactions in complex food systems and processing unit operations in the near future.

REFERENCES

1. Hindra, F. and Baik, O.-D. Kinetics of quality changes during food frying. *Crit. Rev. Food Sci. Nutr.*, 46, 239, 2006.
2. Villota, R. and Hawkes, J. G. Reaction kinetics in food systems. In *Handbook of Food Engineering*, 2nd edn. Heldman, D. R. and Lund, D. B. (eds). CRC Press, Boca Raton, FL, 2007, 125–286.
3. Israelachvili, J. *Intermolecular & Surface Forces*, 2nd edn. Academic Press, San Diego, CA, 1992.
4. Fennema, O. R. *Food Chemistry*, 3rd edn. Marcel Dekker, Inc., New York, NY, 1996.
5. Fennema, O. R. and Tannenbaum, S. R. Introduction to food chemistry. In *Food Chemistry*, 3rd edn. Fennema, O. R. (ed.). Marcel Dekker, Inc., New York, NY, 1996, 1–16.
6. Levenspiel, O. *Chemical Reaction Engineering*, 3rd edn. John Wiley & Sons, New York, NY, 1999.

7. Lindsay, R. C. Food additives. In *Food Chemistry*, 3rd edn. Fennema, O. R. (Ed.). Marcel Dekker, Inc., New York, NY, 1996, 767–823.
8. Nawar, W. W. Lipids. In *Food Chemistry*, 3rd edn. Fennema, O. R. (Ed.). Marcel Dekker, Inc., New York, NY, 1996, 225–319.
9. Schmidt, L. D. *The Engineering of Chemical Reactions*, 2nd edn. Oxford University Press, New York, NY, 2005.
10. Veldsink, J. W., Bouma, M. J., Schoon, N. H., and Beenackers, A. A. C. M. Heterogeneous hydrogenation of vegetable oils: a literature review. *Catal. Rev. Sci. Eng.*, 39, 253–318, 1997.
11. Patterson, H. B. W. *Hydrogenation of Fats and Oils*. Applied Science Publishers, London, UK, 1983.
12. Jang, E. S., Jung, M. Y., and Min, D. B. Hydrogenation for low *trans* and high conjugated fatty acids. *Compr. Rev. Food Sci. Food Safety*, 1, 22–30, 2005.
13. Koritala, S. V., Selke, E., and Dutton, H. J. Deuteration of methyl linoleate with nickel, palladium, platinum and copper-chromite catalysts. *J. Am. Oil Chem. Soc.*, 50, 310–316, 1973.
14. Beenackers, A. A. C. M. and van Swaaij, W. P. M. Mass transfer in gas-liquid slurry reactors. *Chem. Eng. Sci.*, 48, 3109–3139, 1993.
15. Fillion, B., Morsi, B. I., Heier, K. R., and Machado, R. M. Kinetics, gas-liquid mass transfer, and modelling of the soybean oil hydrogenation process. *Ind. Eng. Chem. Res.*, 41, 697–709, 2002.
16. Dadvar, M. and Sahimi, M. Pore network model of deactivation of immobilized glucose isomerase in packed-bed reactors II: Three-dimensional simulation at the particle level. *Chem. Eng. Sci.*, 57, 939–952, 2002.
17. Beck, M., Kiesser, T., Perrier, M., and Bauer, W. Modeling glucose/fructose isomerization with immobilized glucose isomerase in fixed and fluid bed reactors. *Can. J. Chem. Eng.*, 64, 553–560, 1986.
18. Dadvar, M., Sohrabi, M., and Sahimi, M. Pore network model of deactivation of immobilized glucose isomerase in packed-bed reactors I: Two-dimensional simulations at the particle level. *Chem. Eng. Sci.*, 56, 2803–2819, 2001.
19. Dadvar, M. and Sahimi, M. Pore network model of deactivation of immobilized glucose isomerase in packed-bed reactors. Part III: Multiscale modeling. *Chem. Eng. Sci.*, 58, 4935–4951, 2003.
20. Lee, Y. Y., Fratzke, A. R., Wun, K., and Tsao, G. T. Glucose isomerase immobilized on porous glass. *Biotechnol. Bioeng.*, 18, 389–413, 1976.
21. Palazzi, E. and Converti, A. Generalized linearization of kinetics of glucose isomerization to fructose by immobilized glucose isomerase. *Biotechnol. Bioeng.*, 63, 273–284, 1999.
22. Converti, A. and Del Borghi, M. Kinetics of glucose isomerization to fructose by immobilized glucose isomerase in the presence of substrate protection. *Bioproc. Eng.*, 18, 27–33, 1998.
23. Abu-Reesh, I. M. and Faqir, N. M. Simulation of glucose isomerase reactor: optimum operating temperature. *Bioproc. Eng.*, 14, 205–210, 1996.
24. Allen, R. R. Principles and catalysts for hydrogenation of fats and oils. *J. Am. Oil Chem. Soc.* 55, 792–795, 1978.

Section IX

Special Topics in Food Processing

34 Measurement and Control in Food Processing

Brent R. Young and Garth Wilson
The University of Auckland

CONTENTS

34.1 INTRODUCTION

The food industry like most other large processing industries has been faced with growing demand, insufficient supply and increasing consumer awareness. The food industry has turned to automatic measurement and control to improve operating efficiencies and reduce wastage from its plants. This chapter highlights some of the more innovative measurement and control schemes employed by the food industry to manage the imprecise and objective control variables associated with the industry.

Classical measurement and control schemes such as feedback and feedforward control work on a model of true or false. The measured variable is compared to a reference value, if any error exists the control will take action to eliminate the error between reference value and measured value. For the food industry, measured variables are often not so black and white. More advanced techniques are used, such as fuzzy logic and model predictive control (MPC). The merits of each of these classical and more advanced measurement and control schemes will be discussed.

34.2 BACKGROUND

The traditional approach to process control in the food industry was manual operation of valves by individual operators, who were expected to make a scientific assessment of the food quality using human sensors, for example controlling the volume of a tank using a site glass and hand operated valve. As human population has grown the demand for food has also grown, as production plants have gown in size they have required more and more operators. Modern society not only demands more food but also complex and highly processed foods. To keep up with volume and preference, food manufacturers have produced facilities that use large scale machinery and huge flow rates. The large flow rates require precise timing and control, the errors introduced by human operator control caused financial and quality problems. Manufacturers slowly moved toward automated control initially. However automated control has almost totally taken over in the last ten years.[1] Fellows[1] gives three reasons for the rapid automation of plants:

- Increased competition between manufacturers to produce a wide variety of products with little or no down time in between
- The ever increasing cost of labor and raw materials
- Increasing Government regulation and control, from consumer demands for international standard and safe food

As food production plants have grown in size and complexity the need for more and more operators has grown.[1] The small errors made by operators in the timing of valves and motors has caused increasing greater quality issues and financial costs.[1]

34.3 MEASUREMENT

34.3.1 Sensors

34.3.1.1 Real Sensors

A sensor is the first element in the process control loop. The sensor is used to measure a process variable. The measurement is sent to the controller where its deviation from the set point (SP) is determined and appropriate action taken to return the variable to the SP. It is therefore, imperative that the sensor, accurately and precisely measure the process variable. The sensor needs to give accurate information about the process variable to the controller to ensure tight control of the process, to optimize quality and performance. The rate of response of the sensor is also important. If the sensor cannot measure and transmit changes quickly, it may cause the system to be unstable. This is important for systems with small time constants. The selection of the correct type of sensor is essential to optimum operating conditions. Just as important as selection of the sensor, is correct placement of the sensor in the process stream to ensure accurate readings and minimize noise and measurement lags. Some environments may be corrosive or abrasive and damage the sensor if placed directly in process stream, in these cases the sensor may be placed online, at line or off line.[1] Figure 34.1 shows the difference between each of these types of placements. A simple example where placement is crucial is a flow meter used to regulate a control valve, if the sensor is placed too close to the valve significant noise will be generated in the signal. There are two different types of sensors real sensors and soft sensors. Real sensors, measure primary variables like, temperature, flow rate, and pressure. A list of primary sensors is given by Fellows.[1] Mittal,[2] Moreira[3], and Svrcek et al.[8] also give a detailed description of each of the main types of sensor and primary variable. Some of the major types of sensors are shown in Table 34.1.

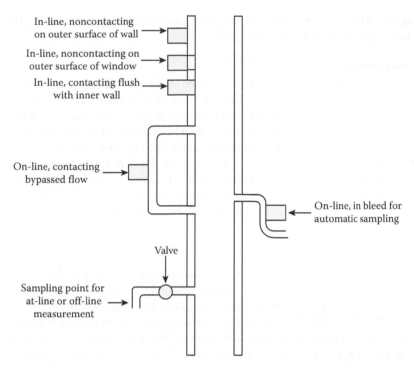

FIGURE 34.1 Different types of sensor placements along a pipeline. (Modified from Fellows, P. *Food processing technology: principles and practices.*, 2nd edn., Boca Raton, FL: CRC Press, 575, 2000.)

TABLE 34.1
Primary Variables and Typical Sensors

Variable	Sensor
Flow	Orifice plate
	Mechanical or electro-magnetic flow meter
Level	Ultrasound
	Dipstick
Pressure	Strain gauge
	Bourdon gauge
Temperature	Thermocouple
	Resistance thermometer
Weight	Strain gauge
	Load cell
Viscosity	Mechanical resonance
	Dipstick

Source: Modified from Fellows, P., *Food Processing Technology: Principles and Practices*, 2nd edn., Boca Raton, FL: CRC Press, 575, 2000.

34.3.1.2 Biosensors

The food industry is also starting to implement biosensors. These sensors combine biomaterial with electrical components.[2] The biomaterial reacts to changes in the process and the electronic components convert this reaction into usable information. These new sensors have the ability to measure variables like sugar, amino acid, fat, protein, drug or alcohol concentration without the need to take samples and test off-line.[2,3] This vastly decreases processing time by eliminating the dead time associated with lab analysis of off-line samples.

34.3.1.3 Soft Sensors

Soft sensors are used where measurement of a desired variable is difficult. As a result the variable is inferred through the measurement of a different but related variable. This is widely used in the food industry as many of the desired parameters cannot be measured directly without destroying the food itself. For example, during the drying of food to extend its shelf life, the moisture content of the dried material is inferred from the humidity of the exhaust gases, calculated from the difference in the wet and dry bulb temperatures.[2,5] Coffee bean flavor is another example. The flavor of coffee beans after roasting is inferred from the amount of volatile organics present, calculated from the light absorbance of the beans. The texture and creaminess of ice cream is a third example. It is inferred from the fat/water content in the mixture and by the freezing rate.[2,4,6]

34.3.2 DATA RECTIFICATION

Most large scale food processing plants will generate process noise and will develop gross errors from uncalculated hold-ups. In any process system, where the flow rates are in the order of tonnes per hour, there will be significant noise in the system. The noise may influence the controllability of the process variable. Pre 1960s the only way of removing noise was to transfer it to a less important stream such as a utility stream. According to Schlat and Hu,[5] Davidson et al. in the 1960s proposed an alternative approach, a system for data reconciliation based on linear process models. The system compared the measured variable to a model to remove noise. The model acted like a moving average that would update itself every time a new measurement was made. Improvements were made on this system by Nogita[6] and Mah et al.[7] who proposed a method for the removal of gross error from the measured variables. Data reconciliation/rectification is now considered good manufacturing practice in large food processing plants.

34.4 CONTROL

34.4.1 PLANT-WIDE CONTROL

For a plant to be as productive as possible, an understanding of the process as a whole must be understood. Plant-wide control is a term used to describe this philosophy (e.g., Svrcek et al.[8]). By gaining an understanding of the chemical and physical conditions of the plant, optimization of the plant becomes easier and boundaries between controllable regions become clearer. From this understanding, energy conversation through recycling and reuse of waste energy can greatly improve the performance of the plant (e.g., Svrcek et al.[8]).

This philosophy has not been widely applied to the food industry, as such there is little or no literature on the application and benefits of plant-wide control in the industry. With this procedure proving itself in other industries, and the rapid uptake of other control technology by the food industry, it will not be long before plant-wide control is applied universally in the food industry.

34.4.2 FEEDBACK CONTROL

Feedback control is the most commonly used control scheme in the food industry. Since the 1960s process plants have been using feedback control schemes to achieve better controllability and accuracy in their products. The original control systems were mechanically driven and were installed directly on the process line. Developments in technology lead to the use of pneumatic and electronically controlled valves with remote installation of the controller. This reduced the amount of equipment located on the processing line and also centralized the controllers in a control room.

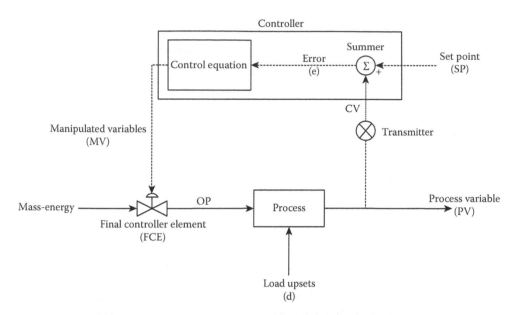

FIGURE 34.2 Block diagram of feedback control system. (Modified from From Svrcek, W.Y., Mahoney, D.P. and Young, B.R. *A real-time approach to process control.*, Chichester, NY: John Wiley & Sons, 307, 2000.)

Feedback control as the name implies alters the manipulated variable (MV) according to distur-bances (d) in the output (PV). The feedback control system is a series of three parts, the sensor and transmitter, the controller and the control valve. The sensor measures a variable in the output, such as pressure, temperature or composition. The sensor and transmitter convert the variable reading into an electronic or pneumatic response and transmit the signal to the controller (CV).

The controller compares the response from the sensor to a reference value, the SP, by subtrac-tion or addition of the two values. The difference is known as the error (e). The controller converts the error into an electronic or pneumatic value (MV), the signal is sent to the control valve (FCE) which adjusts the valve seat position according to the signal received. Figure 34.2 shows this system in graphical form.

There are four main feedback control equations that are used, proportional (P), proportional-integral (PI), proportional-derivative (PD), and proportional-integral-derivative (PID). The control equations differ by the way they respond to an error greater than zero. The generic control equations are shown below:

- P only control equation:

$$MV = K_c e + b \qquad (34.1)$$

where K_c = steady state gain; b-bias, the offset present at zero error and e = error, SP–PV.

- PI control equation:

$$MV = K_c \left(e + \frac{1}{T_i} \int edt \right) = K_c e + K_c \frac{1}{T_i} \int edt \qquad (34.2)$$

where T_i = integral time.

- PD control equation:

$$MV = K_c \left(e + T_d \frac{de}{dt} \right) + b \qquad (34.3)$$

where T_d = derivative time.

- PID control equation:

$$MV = K_c\left(e + \frac{1}{T_i}\int edt - T_d\frac{CV}{dt}\right) \qquad (34.4)$$

For the controller to operate effectively, the controller parameters, K_c, T_i and T_d need to be tuned.

34.4.3.1   Tuning

In order to get the best possible response from the controller the parameters need to be tuned. The tuning methods are based around the response of the controller to an error when the controller parameters are varied. The performance of the controller is evaluated by the time taken for one quarter decay in the response to occur. There are different methods for tuning the controller, "trial and error," process reaction curve and constant cycling methods.

34.4.3.1   Trial and Error

This method relies solely on estimating the parameters and then applying small changes until the best possible response is found. There are several suggested starting points that should allow for quick tuning through fine adjustment. It is recommended to start with K_c, T_i, and T_d all at a minimum and use proportional action as the major control parameter with integral and derivative action used to fine tune the response.

There are also several suggested starting points or rules of thumb described in Svrcek et al.[8] The rules of thumb are displayed in Table 34.2.

34.4.3.2   Process Reaction Curve Methods

The process reaction curve method is carried out in open loop, or no control action. The process response is allowed to reach steady state where the response is a straight line. A small usually step change disturbance is introduced to the process and the response is examined. This method only works on systems where the response curve will return to steady state at some new value. The optimum controller parameters, K_c, T_i and T_d can be determined from the response. A typical response is shown in Figure 34.3.

Ziegler and Nichols,[4,10] developed equations (Table 34.3) to find the controller parameters from the open loop response shown above. The controller parameters should be implemented and then adjusted to achieve one quarter decay between the peaks in the closed loop response.

TABLE 34.2
Table of Trial and Error Suggested Starting Points

Variable	K_c	T_i	T_d	b	SP
Flow	0.4–0.65	0.1 min (or 6 s)			
Level				50%	50%
P	2				
PI	2–20	1–5 min			
Pressure					
Vapor	2–10	2–10 min			
Liquid	0.5–2	0.1–0.25 min			
Temperature	2–10	2–10 min	0–5 min		

Source: Modified from Svrcek, W.Y., Mahoney, D.P. and Young, B.R., *A real-time approach to process control.*, Chichester, NY: John Wiley & Sons, 307, 2000.

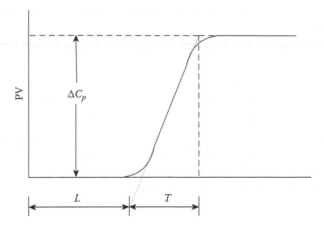

FIGURE 34.3   Modified from Process reaction curve, where ΔCp is the change in process response, L is the lag time and T is the response time. (Modified from Svrcek, W.Y., Mahoney, D.P. and Young, B.R. *A real-time approach to process control.*, Chichester, NY: John Wiley & Sons, 307, 2000.)

TABLE 34.3
Controller Tuning Parameters for Process Reaction Curve According to Ziegler and Nichols

Controller	K_c	T_i	T_d
P-only	$\Delta C_p/NL$		
PI	$0.9\Delta C_p/NL$	$3.33L$	
PID	$1.2\,\Delta C_p/NL$	$2.0L$	$0.5L$

where:

$$N = \frac{\Delta C_p}{T}$$

$$R = \frac{L}{T} = \frac{NL}{\Delta C_p}$$

and ΔC_p is the magnitude to the change made to the process variable, T is the response time and L is the lag time.

Sources: Mittal, G.S., *Computerized control systems in the food industry*, New York, NY: Marcel Dekker, 597, 1997; Ghosh, S. and Coupland, J.N., *Food Hydrocolloids*, 22(1), 105–111, 2008; and Svrcek, W.Y., Mahoney, D.P. and Young, B.R. *A real-time approach to process control*, Chichester, NY: John Wiley & Sons, 307, 2000.

Cohen and Coon[8] expanded on the work of Ziegler and Nichols to remove the sluggish behavior of the controller. The equations they proposed are shown in Table 34.4.

Further work has been carried out on the work done by Ziegler and Nichols. The equations have been developed into the tuning parameters known as the internal model control tuning rules (e.g., Table 34.5). They are covered in detail by Svrcek et al.[8] and Mittal.[2]

34.4.3.3   Constant Cycling Methods

This method is similar to the process reaction curve method described above except that constant cycling is carried out while the loop is closed. The controller parameters are minimized with P-only control used. The gain is increased until the response cycles continuously at constant maximum

TABLE 34.4
Controller Parameters for Reaction Curve According to Cohen and Coon

Controller	K_c	T_i	T_d
P-only	$\Delta C_p/\text{NL}(1 + R/3)$		
PI	$\Delta C_p/\text{NL}(0.9 + R/12)$	$L(30 + 3R)/(9 + 20R)$	
PID	$\Delta C_p/\text{NL}(1.33 + R/4)$	$L(32 + 6R)/(13 + 8R)$	$L4/(11 + 2R)$

Source: Mittal, G.S., *Computerized control systems in the food industry.* New York, NY: Marcel Dekker, 597, 1997; Ghosh, S. and Coupland, J.N., *Food Hydrocolloids*, 22(1), 105–111, 2008; and Svrcek, W.Y., Mahoney, D.P. and Young, B.R. *A real-time approach to process control*, Chichester, NY: John Wiley & Sons, 307, 2000.

TABLE 34.5
Controller Parameters from Constant Cycling Response

Controller	K_c	T_i	T_d
P-only	$K_u/2$		
PI	$K_u/2.2$	$P_u/1.2$	
PID	$K_u/1.7$	$P_u/2$	$P_u/8$

Source: Modified from Svrcek, W.Y., Mahoney, D.P. and Young, B.R. *A real-time approach to process control*, Chichester, NY: John Wiley & Sons, 307, 2000.

values. The two parameters calculated from this oscillation are the ultimate period, P_u, the time between two peaks in the constant oscillation. The second is the K_u, ultimate gain. The ultimate gain is the gain at which the system begins to oscillate. Zielger and Nichols[4,10] suggest equations to find the controller parameters from the oscillations.

34.4.3.4 Multi-Variable Systems

The most simplified control systems contain single input–single output (SISO) loops. However most systems in large scale food processing plants contain multiple inputs–multiple outputs (MIMO). MIMO control schemes arise because most process variables on a piece of equipment can not be assumed to be mutually exclusive, manipulating one has an effect on another. The controller tuning becomes much more difficult as the interaction between variables must be analyzed and compensated for in the control equation. The best combination of measured variable to MV needs to be determined to ensure the system is operating at optimum conditions. In 1966 Bristol proposed the relative gain array (RGA), and this is the most common method used to study variable interaction and pairing (e.g., Gapnepain and Seborg[9]). Gagnepain and Seborg[9] give a full description of how to employ the model. They also suggest other methods such as relative dynamic array, RDA, developed by Witcher and McAvoy. Gagnepain et al.[9] also suggest a new model based on Bristol's RGA method by using average relative gain. Singh and Mulvaney[10] give an example of a cooking extruder where they use Bristol's method to determine the best pairing of the three measured variables and manipulated variables (MVs). The variables considered for analysis were, measured variables, die pressure, motor torque developed and product temperature, the MVs were Barrel temperature SP, moisture content, and screw speed. The feed rate was assumed to be kept constant.[10]

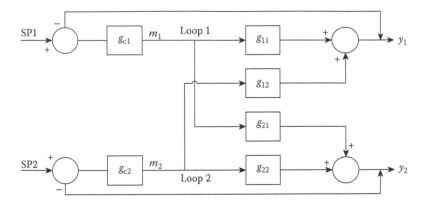

FIGURE 34.4 A 2×2 system with interacting loops. (Modified from Svrcek, W.Y., Mahoney, D.P. and Young, B.R. *A real-time approach to process control*, Chichester, NY: John Wiley & Sons, 307, 2000.)

34.4.3.5 Relative Gain Array (RGA)

RGA compares the interaction of process variables with MVs in MIMO systems. The difficulty of MIMO systems is the complexity of the control equations.[8] The RGA breaks this down into a series of more manageable single inputs single outputs.[8] Svrcek et al.,[8] for example, provide a detailed explanation of the analysis and how to implement it. An example of a generic MIMO loop is shown in Figure 34.4. The system consists of two interacting inputs (m_1, m_2) and outputs (y_1, y_2). First the pairing of m_1 and y_1 is investigated by opening all the loops so no process control is applied. The only factor affecting y_1 is a change in m_1.

The change measured is equal to the steady state gain (g_{11}) between m_1 and y_1.

$$g_{11} = \frac{\delta y_1}{\delta m_1}\bigg|_{m_2=\text{constant}} = \frac{\Delta y_1}{\Delta m_1} = \text{gain}(y_1 - m_1) \text{ Loop with all loops open} \quad (34.5)$$

The next step is to close the second loop and make an identical step change in m_1. It is assumed that perfect control occurs in loop two, m_2 will change to keep y_2 constant [8], resulting in a new gain, g_{11}^*:

$$g_{11}^* = \frac{\delta y_1}{\delta m_1}\bigg|_{y_2=\text{constant}} = \frac{\Delta y_1}{\Delta m_1} = \text{gain}(y_1 - m_1) \text{ Loop with all other loops closed} \quad (34.6)$$

The relative gain (λ_{ij}) is a ratio of the gain of m_j–y_i loop when all other loops are open and closed.[8]

$$\lambda_{ij} = \frac{\left(\dfrac{\delta y_i}{\delta m_j}\right)_{\text{all\_loops\_open}}}{\left(\dfrac{\delta y_i}{\delta m_j}\right)_{\text{all\_loops\_closed\_and\_in\_perfect\_control\_except\_the\_}mj\text{\_loop}}} = \frac{g_{ij}}{g_{ij}^*}$$

$$= \left(\frac{\text{open} - \text{loop gain}}{\text{closed} - \text{loop gain}}\right)_{\text{for\_all\_loop\_}i\text{\_under\_control\_of\_}m_j} \quad (34.7)$$

The relative gains are then placed in a matrix, called the RGA. By evaluating the value of λ_{ij} the stability of the system can be determined.[8]

$\lambda_{ij} = 0$ Manipulated variable m_j has no effect on the output.

$\lambda_{ij} = 1$ Manipulated variable m_j effects y_i without interaction with over control loops.

$\lambda_{ij} < 0$ System is unstable.

$0 < \lambda_{ij} < 1$ The other control loops are interacting with m_j and y_i loop, avoid this paring if $\lambda_{ij} < 0.5$.

$\lambda_{ij} > 1$ The other control loops have an opposite effect on the loop m_j and y_i. This system may become unstable if other loops are open.

For example, Singh and Mulvaney[10] discovered that the cooking extruder problem that they studied could be further simplified into a two by two problem of removing the variable die pressure, because it appears to behave oppositely to open loop response when closed loop control is applied. The remaining problem has the variables of screw speed, barrel SP temperature, moisture content and motor torque developed. Singh and Mulvaney[10] carried out a two by two analysis as above to determine the best possible pairing. They discovered that the pairing of screw speed and moisture content provide the best possible pairing with no negative interaction.

34.4.3.6 Fuzzy Logic

One of the major obstacles to electronic process control in the food industry is that the main parameters of performance and quality are subjective, qualitative measurements. The quality of food is often measured by variables like taste, flavor, appearance, and texture, etc. These variables are difficult to measure using conventional devices. Many food processing plants still rely on skilled operators to make assessments and adjustments according to their sensors. An example of this is the roasting of coffee, the color to the roasting beans is sampled throughout the roasting period by an operator. When the operator observes that the beans are at the correct color for the roast, the batch is removed from the roaster. Classical process control relies on a binary system, of on-spec or off-spec, 1 and 0. Fuzzy logic tries to span the gap between electronic binary control and imprecise operator observation. Rather than two values, 1 or 0, Zadeh (e.g., Mittal[2]) proposed a system where there is a scale from 0 to 1. For example chocolate yogurt is made at a diary processing plant. The diary company has a standard for the color of the yogurt, they have classified the colors as good color, reasonably good color and standard.[3] What is reasonably good color? It is hard to precisely quantify it.[3] The company defines that a color value of above 70 is good color, between 70 and 50 as reasonable color and below 50 as standard.[3] Under classical control any yogurt with a color value of less than 50 or greater than 70 they are not of reasonable color, binary code 0. Similarly any yogurt with a color value of greater than or equal to 50 or less than or equal to 70 they are of reasonable color, binary code 1.

$$\text{IF } x < 50 \text{ OR } x > 70 \quad \text{then “yogurt is or reasonable color”} = 0 \quad \text{(false)}$$

$$\text{IF } x \geq 50 \text{ OR } x \leq 70 \quad \text{“then yogurt is or reasonable color”} = 1 \quad \text{(true)}$$

When fuzzy logic is used the boundary between reasonable color and not becomes less defined. Fuzzy logic ranks yogurts according to a degree of reasonable color, so yogurts with a color value of 50.5 and 70.5 maybe considered of reasonable color with a degree of 0.8.[3] Fuzzy logic allows for the control of a wider range of acceptable SPs than classical control. Moriera et al.[3] provided a detail description of how to implement fuzzy logic control.

34.4.3 ADVANCED CONTROL

34.4.4.1 Feedforward Control

In the feedback control schemes explained above, a disturbance is detected in the output of the system by the sensor and the controller takes action to return the process to the SP. The major short coming of this type of control is that a disturbance must by measured in the output before any control action is taken. Therefore, the system is always slightly deviated from the SP. For most systems this is acceptable however in some cases the SP deviation must be minimized.[8] Feedforward control can be installed to minimize the SP deviation.[8] The feedforward controller ratios input according to the size of the incoming disturbance. The controller basically keeps the mass and energy balance of the system constant. Feedforward is often more effective than feedback control at removing disturbance error, however feedforward control can not measure or control any dynamic error generated inside the process.

Feedforward control is often used in conjunction with feedback control to improve the performance of both. Feedforward control, adjusts for error in the input, while the feedback control adjusts for errors generated within the process measured at the output. Moriera et al.[11] provide an example of feedforward control being used on a cooking extruder. They suggest that feedforward control can help to eliminate variation in the die pressure caused be varying feed rate and feed moisture content. Moriera et al.[11] also suggest that for tight control, feedforward control of the extruder should be used with feedback control.

34.4.4.2   Model Predictive Control (MPC)

Increasingly food processing plants are turning to long range prediction based control.[3] MPC is a computer control scheme that uses a process model to predict the effect of future control actions on the process output, and selects the appropriate action to minimize any disturbance.[3,4] MPC is extensively used in the petrochemical industry.[3] The food industry is starting to implement MPC control. For example Fontera, the biggest dairy processing company in New Zealand, uses MPC control in its milk processing plants. Fontera uses a nonlinear MPC from Pavilion Technology in milk powder production.

Moriera[3] gives an example of a generic MPC control scheme.

The following are known:[3]

1. An appropriate model, $M(\theta)$
2. A vector of past process outputs, $y = [y(k), y(k-1),....]$
3. A vector of future set points, $w = [w(k + 1), w(k + 2),....]$
4. A vector of previous control actions, $u = [u(k-1), u(k-2),....]$
5. A vector of potential future control actions, $u = [u(k), u(k + 1),....]$

The predicted output $y(k + j)$ can be calculated over the prediction horizon. Once the prediction at time k has been made the control action is applied. The model then shifts to time $k + 1$ and the process restarts. The controller acts of the error predictions from the equation shown below.[3]

$$e(k + j) = w(k + j) - y(k + j) \qquad (34.8)$$

At the heart of the MPC system is the process model used to make the predictions from. The more accurate the model the better accuracy is achieved form the control system. Linear discrete models allow for faster control due to less computation power required.[3]

Shaikh and Prabhu[12] have proposed the use of MPC in cryogenic freezing of food. They found that the use of MPC increased the freezing performance by 33%. MPC is the start of intelligent controller design, where the controller learns the variations in the process and begins to predict and adapt its model to compensate for these errors.

34.5   CONCLUSIONS

Automatic process measurement and control is used almost universally by large food processing companies. The shift toward automatic measurement and control has been rapid in the last ten years or so. However many factories still utilize the knowledge of experienced operators to make decisions on the processing line. This is due to two major factors, the cost associated with the installation of new automatic measurement and control, including staff training and ongoing maintenance and also the imprecise and inaccurate behavior of food processing.

Process control of any variable is only as good as its sensors and measurement instruments. This is especially important in the food industry where many of the variables are impossible to measure directly inline without significant delays. Many of the variables must therefore, be inferred from the measurement of other variables that are easier to measure. The development of biosensors has allowed the inline measurement of some variables such as sugar concentration that previously had to be measured offline. This has dramatically reduced delays in the processing time.

Classical control techniques continue to be used extensively in the food industry because of their proven reliability and performance. The down side of classical control is the need to accurately tune and readjust the controller parameters to ensure the optimum settings are being used. Advanced control techniques have been developed to compensate for the inaccuracy and imprecise nature of food processing. Fuzzy logic is a computer model of human subjectiveness. Fuzzy logic compensates for the variability available in product composition and physical appearance. MPC is designed to learn the variability of the variables and control action associated with the variability in order to accurately predict and minimize long range variability.

REFERENCES

1. Fellows, P. *Food processing technology: principles and practices.* 2nd edn. Boca Raton, FL: CRC Press, 575, 2000.
2. Mittal, G.S. *Computerized control systems in the food industry.* New York, NY: Marcel Dekker, 597, 1997.
3. Moreira, R.G. *Automatic control for food processing systems.* Gaithersburg, MD: Aspen Publishers, 333, 2001.
4. Ghosh, S. and Coupland, J.N. Factors affecting the freeze-thaw stability of emulsions. *Food Hydrocolloids*, 2008, 22(1): 105–111.
5. Schladt, M. and Hu, B. Soft sensors based on nonlinear steady-state data reconciliation in the process industry. *Chem. Eng. Process.: Process Intensificat.*, 2007, 46(11): 1107–1115.
6. Nogita, S. Statistical test and adjustment of process data. *Ind. Eng. Chem. Proc. Des. Dev.*, 1972, 11(2): 197–200.
7. Mah, R.S., Stanley, G.M. and Downing, D.M. Reconcillation and rectification of process flow and inventory data. *Ind. Eng. Chem. Proc. Des. Dev.*, 1976, 15(1): 175–183.
8. Svrcek, W.Y., Mahoney, D.P. and Young, B.R. *A real-time approach to process control.* Chichester, NY: John Wiley & Sons, 307, 2000.
9. Gagnepain, J.P. and Seborg, D.E. Analysis of process interactions with applications to multiloop control system design. *Ind. Eng. Chem. Proc. Des. Dev.*, 1982, 21(1): 5–11.
10. Singh, B. and Mulvaney, S.J. Modeling and process control of twin-screw cooking food extruders. *J. Food Eng.*, 1994, 23(4): 403–428.
11. Moreira, R.G., Srivastava, A.K. and Gerrish, J.B. Feedforward control model for a twin-screw food extruder. *Food Control*, 1990, 1(3): 179–184.
12. Shaikh, N.I. and Prabhu, V. Model predictive controller for cryogenic tunnel freezers. *J. Food Eng.*, 2007, 80(2): 711–718.

35 Artificial Neural Networks in Food Processing

Weibiao Zhou
National University of Singapore

Nantawan Therdthai
Kasetsart University

CONTENTS

35.1 INTRODUCTION

Food systems are often very complex due to highly complicated interactions among various components during processing and storage. To establish mathematical models for food processing, an understanding of the underlying mechanisms is necessary. Despite criticism on its lack of physical insight to an underlying relationship, artificial neural networks (ANNs) modeling presents a unique approach with its superb capability of learning to deal with complicated problems.

Artificial neural network (ANN), an adaptive learning system, is able to learn patterns to develop a model from existing observations [1]. The interest in using ANN as a modeling tool for food applications has risen in recent years [2], which may be due to its proven capability in solving complex problems such as those in food processing. In addition, lack of data on food properties under some special conditions in processing such as high pressure may have lead researchers to use black-box approaches including ANN to assist in the modeling and controlling of the process [3].

ANN has been successfully applied to food processing systems including drying, baking, osmotic dehydration, high pressure processing, as well as estimations of a number of food properties and quality indicators. This chapter describes the principles behind ANN and the current status of its applications in food processing.

35.2 FUNDAMENTALS OF ARTIFICIAL NEURAL NETWORKS (ANNs)

ANN is an information processing system imitating biological neural networks in the brain. The brain neural network can learn concepts from data observed over time. With more observations with time, its capability to make decision sand draw conclusions increases. Like the brain neural network, ANN can also increase its ability to recognize and predict patterns by learning and training with observation data. This breaks the limitation of other existing modeling approaches that likely rely on a set of mathematical equations with a few model parameters. Therefore, ANN has become increasingly popular in dealing with complex systems in various fields.

35.2.1 Types of Artificial Neural Network (ANN)

Many types of ANN have been developed, including multilayer perceptron network, Hopfield network, Kohonen network and so on. They deal with different kinds of applications such as classification, data coding, optimization and forecasting. Several of the most popular types of ANN are briefly described in this section.

35.2.1.1 Multilayer Perceptron Network

A multilayer perceptron network, also known as feed-forward network, is a supervised network consisting of three or more major layers: input layer, hidden layer(s) and output layer. Each layer consists of a number of neurons which are connected to those in the next layer through weights and biases. A transfer function is used to generate nonlinear response, mimicking the firing status of a biological neuron. The final output of the network (i.e., from the output layer) is compared with desired output which is derived from prior observations, and the difference between those outputs is used for updating weight and bias values through training. This type of network is very useful for variable prediction and classification.

Learning algorithm used in training is usually backpropagation. More details on backpropagation will be given in Section 35.3. Counter-propagation is also possible for training a network. Counter-propagation was originally used for bidirectional mapping (both forward-mapping and inverse-mapping). However, almost all practical applications consider it only as forward-mapping. This type of network architecture is composed of three layers including an input layer, a hidden layer and an output layer. Learning starts by selecting randomly a training vector pair from the training data set. The input vector is firstly normalized by being divided by the magnitude of the input. The normalized input vector is used for computing weights for neurons in the competitive hidden layer. Then, the connection weights are adjusted by using a small constant term to limit the amount of change in the connection. The process is repeated to make the input associate consistently with the competitive neurons. The connection weights between the wining hidden neurons and the output layer are then adjusted. Learning stops when the difference between the actual output and the weight vector is decreased to an acceptable value.

35.2.1.2 Hopfield Network

A Hopfield network (also called associated memory) contains neurons which are fully connected to each other. The connection link value is dependent on the strength of the connection between two neurons. The overall signal obtained from all other neurons is used to compare with the given threshold, in order to design the activity (−1 or 1). It is useful for image processing where each neuron represents a pixel [4].

35.2.1.3 Kohonen Network

A Kohonen network, also known as self-organizing network or self-organizing map, evaluates a data set for organizing itself to construct a map of relationships among input patterns [5]. Information is brought from an input into each node in the network. The closest node is firstly selected. The link weights for the selected node are adjusted by a small value. After that, each node produces an excitation value in accordance with the data. As the number of training increases, the size of neighborhood is reduced. The training stops when the link weights do not change [4].

35.2.1.4 Adaptive Resonance Network

Adaptive resonance theory (ART) is used to set the network to adapt itself to update newly acquired information, without forgetting previously learned patterns. Unlike backpropagation, previous patterns are memorized while learning proceeds by receiving new data. This network is developed to solve instability of feed-forward systems, particularly the stability-plasticity dilemma [5]. The structure can be either ART1 or ART2. ART1 is simpler but restricted to binary input patterns, while ART2 is more complicated but it can handle continuous input values. This type of network is good at pattern classification and recognition [6].

35.2.1.5 Radial Basis Function (RBF) Network

Similar to multilayer perceptron network, a radial basis function (RBF) network is composed of three layers: input, hidden and output layers. However, transfer function between the hidden layer and the input layer is a RBF. Between the hidden layer and the output layer, linear transfer function was used, thereby neurons in the output layer became summative units. Activity of the hidden neurons is determined as a distance between the input vector and the hidden neuron center. When the activity is passed to the RBF, the hidden output can be calculated and compared with the desired output [7]. Basically, RBF is based on the distance vector from input to weights. When the distance is large, the network might not find the accurate solution. In addition, it needs a large numbers of neurons and training epochs [8].

35.2.2 Learning Paradigms

ANN can also be regarded as a technique to achieve a particular learning task. It is therefore of interest to present a summary of learning paradigms. Learning paradigms are mainly categorized into three types including supervised learning, unsupervised learning and reinforcement learning [5].

The supervised learning paradigm can be comparable to learning with a teacher that provides some supervision and continuous feedback on the quality of solutions. Feed-forward network and RBF network belong to this category. A data set of input-output pairs is given to find matched transfer functions. Errors, calculated from the difference between the output of the network and the desired output, are minimized through a learning algorithm such as backpropagation by adjusting parameters of the network. Applications for this type of learning include pattern recognition and classification.

In contrast, unsupervised learning paradigm only needs input data set. There is no target output. Kohonen network and Hopfield network are examples of unsupervised learning networks. Functions are trial to minimize a function of the input and output of the network. For a successful learning, although unsupervised, the neural network may need some feature-selecting guidelines. Applications for this type of learning include clustering and filtering.

The reinforcement learning paradigm may be taken as a special case of supervised learning where, instead of getting information on how close the actual output is to the desired output, only a single bit of information is available on whether the output is right or wrong. Therefore during training, there is no indication if the output response is moving in the right direction or how close it is to the desired output. Reinforcement learning is sometimes referred as "learning with a critic" as opposed to "learning with a teacher" in typical supervised learning. Applications for this type of learning include economic systems, sequential decision making and games.

Among various ANN types and learning paradigms, the multilayer perceptron network (supervised network) with backpropagation learning algorithm is the most popular and widely used in food applications including sensory analysis, classification, microbial prediction, and process control [2]. RBF network has also been used for food applications, although it has poorer generalization capability than the multilayer perceptron network for some systems, e.g., characterization of product color during and after extrusion as shown in [9].

35.3 FEED-FORWARD ARTIFICIAL NEURAL NETWORK (ANN) WITH BACKPROPAGATION LEARNING ALGORITHM

35.3.1 NETWORK STRUCTURE

ANN is a system composed of interconnected neurons. Each neuron is formed to process data locally in the network. To deal with nonlinear systems in food processing, multilayer perceptron structure (i.e., feed-forward ANN) is often used. Figure 35.1 illustrates a typical multilayer perceptron structure consisting of three layers, i.e., an input layer, a hidden layer and an output layer. Each layer consists of a number of neurons which are connected to form a network. Communication between layers is carried out by transfer functions, to be described in detail in Section 35.3.2. A network could increase its predictability by learning with backpropagation. Various learning algorithms will be described in Section 35.3.3.

The input layer is a layer composed of neurons representing independent parameters that determine the output. Each neuron sends signals to the neurons in next layer. The signals are weighted based on the strength of connection. These weights are ANN model parameters which are to be adjusted to make the network produce desired outputs. Thus, ANN can also be claimed as a parametric model. In an ANN model, there are two sets of weights: the input-hidden layer weights and the hidden-output layer weights. The inputs are transferred to the hidden layer by the input-hidden layer weights. The weighted inputs are then summed, received and transformed by the neurons in

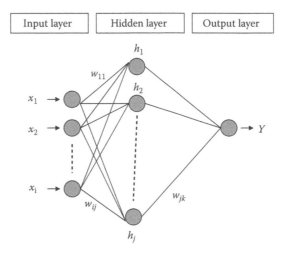

FIGURE 35.1 Structure of ANN with backpropagation.

the hidden layer. The hidden-output weights are used in producing the final outputs of the output layer. These final outputs are compared with the desired values of the output neurons as given in a training data set. The difference is used for the network to adjust and optimize their weights through a learning algorithm, which increases the ANN's capability in data pattern recognition.

35.3.2 TRANSFER FUNCTIONS

For information processing in ANN, a transfer function is required to transform weighted inputs of a neuron to an output. A neuron can be mathematically described by Equation 35.1.

$$O_j = f\left(\sum_{i=1}^{n} w_i x_i + b\right)$$ (35.1)

where O_j is the output of the jth neuron, w_i is the weight associated with the ith input, x is the input, b is a bias input, n is the number of inputs, and f is the transfer function.

To select either linear or nonlinear transfer functions, change of output with regards to input should be considered. When the change of output is bounded within a range, even when f is linear, the actual relationship can be nonlinear. Without any bound, the relationship can be linear, as shown in Figure 35.2.

For a linear transfer function, the output keeps either increasing or decreasing proportionally to the input. The slope of change in the output in response to a change in the input is constant throughout the entire range. The transfer function can be described by the following equation:

$$y = f(x) = ax$$ (35.2)

where f is the transfer function, y is the output, x is the input, and a is the slope.

Among nonlinear transfer functions, sigmoid transfer functions have been most widely used in ANN. A sigmoid transfer function is an S-shape function whose continuous values remain within certain upper and lower limits. Due to continuity, sharp peaks and gaps are not found in a sigmoid function. Therefore, the delta rule can be used to optimize input-hidden layer weights and hidden-output layer weights in accordance with errors. The most popular sigmoid transfer functions are logarithmic sigmoid function and hyperbolic tangent function.

The logarithmic sigmoid transfer function can be described as follows:

$$y = f(x) = \text{logsig}(x) = \frac{1}{1+e^{-x}}$$ (35.3)

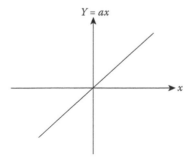

FIGURE 35.2   Linear transfer function.

Output of the logarithmic sigmoid transfer function is in the range of 0–1. Slope of the transfer function indicates rate of change. When the input is at zero, the output is at the middle point, 0.5, and the corresponding slope is at the maximum, as shown in Figure 35.3.

The hyperbolic tangent transfer function can be described as follows:

$$y = f(x) = \tanh(x) = \frac{1 - e^{-x}}{1 + e^{-x}} \tag{35.4}$$

Output of the hyperbolic tangent transfer function is in the range of −1 and 1. When the input is at zero, the output is at zero as well. Similar to the logarithmic function, the greatest slope is found at the middle point of the input range, i.e., 0, as shown in Figure 35.4. However, its slope is steeper than that of the logarithmic function. Therefore, the output of the hyperbolic tangent function approaches the upper and lower bounds more quickly.

Besides sigmoid functions, Gaussian functions are also often used, particularly in RBF networks. The Gaussian functions are in symmetrical bell shape, which include Gaussian standard normal distribution and Gaussian complement distribution.

The Gaussian standard normal distribution can be presented as follows:

$$y = f(x) = e^{-x^2} \tag{35.5}$$

Output of the Gaussian standard normal distribution is in the range of 0 and 1 with its peak value at the input mean, i.e., zero. The output is very sensitive to the input when the input is closer to its mean. In contrast, the output is hardly changed when the input is closer to the tails, as shown in Figure 35.5.

Similarly, the Gaussian complement distribution transfer function has its output in the range of 0–1, and high sensitivity is found at the input mean. However, the maximum outputs are at both tails

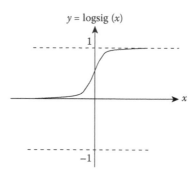

FIGURE 35.3   Logarithmic sigmoid transfer function.

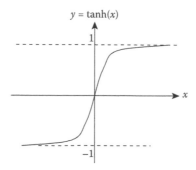

FIGURE 35.4   Hyperbolic tangent transfer function.

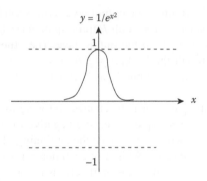

FIGURE 35.5 Gaussian standard normal distribution transfer function.

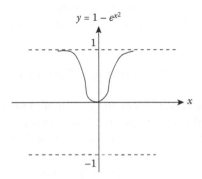

FIGURE 35.6 Gaussian complement distribution transfer function.

rather than at the input mean. That is, the minimum output (zero) is found at the input mean of zero, as shown in Figure 35.6. The Gaussian complement distribution transfer function can be described by the following equation:

$$y = f(x) = 1 - e^{-x^2} \tag{35.6}$$

Other types of transfer function can also be used in ANN, such as arctan (inverse tan function), sine function and so on. To facilitate the learning process in an ANN, values of output variables have to be rescaled into the same range as the output of the chosen transfer function. For input variables, rescaling is not necessary. However, Samarasinghe [1] recommended rescaling the input as well as the output to the same range as that of the transfer function.

35.3.3 LEARNING WITH BACKPROPAGATION

An ANN model needs to be trained using a training data set. This process is known as training or learning. Learning with backpropagation is often used for supervised learning networks. There are many algorithms available for training ANN and a lot of them can be viewed as an application of optimization theory.

Backpropagation is a procedure for changing the weights in an ANN to learn a training data set of input–output pairs. A performance criterion is firstly defined, e.g., sum square error:

$$E = \sum_{p=1}^{P} \sum_{k=1}^{K} \left(e_{pk}^{o} \right)^2 \tag{35.7}$$

where e^o_{pk} is the error of the kth neuron in the output layer corresponding to the pth data point in the training data set, E is the sum square error of all P training patterns (i.e., the number of data points in the training data set) across K units in the output layer (i.e., the number of output variables). Weights are to be updated through the backpropagation procedure so that E is minimized. The backpropagation involves performing computations backward through the ANN layers, using chain rule of calculus.

The simplest form of backpropagation is to change the weights in the direction in which the performance criterion decreases most rapidly, i.e., the negative of the gradient of the sum square error. The procedure can be described as follows. At the beginning, the first pair of input and output data from the training data set is selected. Weight of each neuron is guessed or randomly assigned, a procedure also known as initialization. Transfer functions are assigned to neurons in the hidden layer (denoted as $f(net^h_j)$) and neurons in the output layer (denoted as $g(net^o_k)$). At this stage, error of each neuron across the output layer is computed. After that, error across the hidden layer is computed. The obtained error is used for updating the connection weights of each neuron in the output layer (Equation 35.8) and hidden layer (Equation 35.9) with an objective to reduce the performance criterion.

$$w^o_{kj}(t+1) = w^o_{kj}(t) + \eta e^o_{pk} g'(net^o_k) f(net^h_j) \qquad (35.8)$$

where w^o_{kj} is the weight of connection between the jth neuron in the hidden layer and the kth neuron in the output layer, t is the number of iteration, η is a small value to control the change rate during training cycle (or called learning rate), g' is the differentiation of the function g.

$$w^h_{ji}(t+1) = w^h_{ji}(t) + \eta e^h_{pj} f'(net^h_j) x_i \qquad (35.9)$$

where w^h_{ji} is the weight of connection between the ith node in the input layer and the jth neuron in the hidden layer, η is the learning rate, e^h_{pj} is the error of the jth neuron in the hidden layer corresponding to the pth data point in the training data set, f' is the differentiation of function f, x is the input.

The above procedure is conducted with all other pairs of data points in the training data set. This is called "training epoch." The epoch is repeated for many times to minimize the sum square error E (Equation 35.7). When the ANN can satisfactorily replicate the output of the input patterns, i.e. E is sufficiently small, the learning is stopped [6].

Parameters of a training algorithm need to be optimally chosen to improve accuracy and training speed of the networks [2]. Training algorithms (i.e., error optimization) can be classified into five broad classes including gradient descent, conjugate gradient, Gauss–Newton, Levenberg–Marquardt and Bayesian regularization. The most popular one, which is used as a default in many ANN software, is the gradient descent algorithm.

35.3.3.1   Training Algorithms

35.3.3.1.1   Gradient Descent Algorithm

Gradient descent algorithm is as described in the above in explaining the backpropagation procedure. Weights are updated in the direction of the negative of the gradient of the performance criteria to minimize error, using generalized delta rule.

35.3.3.1.2   Conjugate Gradient Algorithm

Conjugate gradient algorithm is an alternative to the gradient descent algorithm. Instead of updating the weights in the steepest descent direction (i.e., the negative of the gradient), the minimization is performed along conjugate directions, which is proven to converge faster for quadratic functions.

35.3.3.1.3   Quasi-Newton Algorithm

Newton's method is a well-known optimization method. As Newton's method used curvature information, it seems to take a more direct path and faster convergence than the gradient descent does [10]. Quasi-Newton algorithm, also called variable metric method, is a method to solve nonlinear minimization problems. The method is based on Newton's method but with an approximation to the inverse of the second derivative Hessian matrix, which is expensive to calculate. The approximation is done by using only gradients.

35.3.3.1.4   Gauss–Newton Algorithm

Gauss–Newton algorithm is a modification of Newton's method that uses an iterative scheme to solve nonlinear least square problems. The search direction derived from this method is equivalent to the direction by the Newton's method except that the second derivatives are ignored. This makes numerical computation considerably easier.

35.3.3.1.5   Levenberg–Marquardt Algorithm

Levenberg–Marquardt algorithm is also a method for minimizing a nonlinear function without the necessity to compute the second derivative Hessian matrix. The search direction of the Levenberg-Marquardt algorithm is between those of the gradient descent algorithm and Gauss–Newton algorithm. While this method may be slower compared to the Gauss–Newton algorithm, it is more robust, meaning that it is able to find a solution despite poor initial guess.

35.3.3.1.6   Bayesian Regularization

Bayesian regularization aims to improve an ANN's generalization capability to prevent over-fitting of data by the ANN. This is done by modifying the common performance criteria of sum square errors into a linear combination of sum square errors and sum square weights. The combination is minimized often using the Levenberg–Marquardt algorithm. It also modifies the linear combination so that at the end of training the resulting network has good generalization qualities.

35.3.3.2   Factors Affecting Learning Efficiency

Many studies have been carried out to determine the effect of training factors on learning efficiency. These factors include training algorithms, learning rate, number of iteration, momentum and number of hidden layers and hidden neurons.

35.3.3.2.1   Training Algorithms

Training algorithms should be selected specifically for a network. By using a default algorithm, training could be long without significant improvement in the accuracy of the network. As described earlier, the gradient descent algorithm is used as default in many software. However, the gradient descent algorithm may suffer some problems including slow convergence, lack of robustness and too sensitive to learning rate.

Therefore, there are some modifications made to the gradient descent algorithm such as the Levenberg–Marquardt algorithm that is good at nonlinear optimization. In a previous study [2], the Lavenberg–Marquardt algorithm was found to be the best for thermal/pressure food processing, compared to the gradient descent algorithm, the conjugate gradient algorithm, the Newton's method, and the Bayesian regularization. The Levenberg–Marquardt algorithm was also widely used in other applications such as shelf-life prediction of rice snack [11], prediction of physical property [12], prediction of product appearance from instrumental measurement [9] and forecast of material price in time series [13].

Bayesian regularization has been shown being able to reduce over-fitting problem. For faster training speed, the Gauss–Newton algorithm may be used [14].

35.3.3.2.2   Learning Rate

Learning rate is a small value used to control the rate of change for updating weights. Increasing learning rate tends to increase convergence speed. However, a very high learning rate may cause problems of oscillation or divergence to the network. In contrast, a very low learning rate may cause a slow training process without significant improvement in the accuracy of the network.

35.3.3.2.3   Number of Iteration

The performance criterion e.g., sum square error is expected to decrease as training cycles proceed. However, very long training may cause over-fitting; as a result, the network's ability to generalize is reduced.

35.3.3.2.4   Momentum

Momentum is a parameter used to speed up the training rate of the original gradient descent algorithm, by adding a momentum term to the weight adjustment equation. This momentum term presents a contribution to the weight changes from those at the previous time step. With this, the learning rate can be increased without divergent oscillations occurring. As a result, fast convergence can be achieved using momentum.

35.3.3.2.5   Number of Hidden Layers and Hidden Neurons

In general, the more the hidden layers, the lower the error. However, with sufficient number of hidden neurons, large number of hidden layers is not necessary. In practice, most ANN has either one or two hidden layers.

Lower number of hidden neurons may cause under-fitting problem. On the other hand, when there are too many hidden neurons, over-fitting becomes highly possible and the network cannot predict well with unseen data set [9], i.e., poor generalization property. Therefore, the number of hidden neurons should be varied to search for an optimum value thereby the architecture of the network [11,12]. Yu et al. [15] tested the performance of various three-layer neural networks with different number of hidden neurons for predicting the state of *E. coli* O157:H7 in mayonnaise. There were two outputs: survival/death and growth/no-growth. The inputs were incubation temperature, pH, salt content, acetic acid content and sucrose content. It was found that ten to 20 hidden neurons could not result in a network with good performance in terms of small sum square error. At the same time, increasing the number of hidden neurons from 35 to 40 did not significantly increase the accuracy of the network. Therefore, the optimum number of hidden neurons was determined to be 35 for that system.

35.4   MODEL VALIDATION

Model validation is a necessary step for developing any type of models. It is however an especially important issue for ANN models because in comparison to mechanistic models, ANN models are more prone to over-fitting problem and have poorer extrapolation capability.

To determine model performance, data set which has not been seen during training is used for validation. The amount of unseen data points varies from 10 to 30% of those for training the model. The developed neural network processes the unseen inputs to generate output, using the designed structure and transfer functions as well as the derived weights and biases from the training. Then, the network's output is compared with the actual output to evaluate the performance of the network. Methods for performance evaluation include correlation between the network's output and the actual output, mean square error, root mean square error and so on.

If an ANN has a proper structure and has been well trained, the ANN should be robust; therefore the performance from validation should be comparable to that from training. If the performance from validation is significantly worse than that from training, over-fitting during training might be an issue. For such a case, additional data points should be provided during training. Then, new set of weights and biases and transfer function is designed. The model validation procedure should be repeated.

35.5 APPLICATIONS OF ARTIFICIAL NEURAL NETWORK (ANN) IN FOOD PROCESSING

35.5.1 GENERAL APPLICATIONS IN FOOD PROCESSING

Food processing is a complex system containing many parameters. For example, during freezing, process parameters affecting freezing time include product thickness, width, length, convective heat transfer coefficient, thermal conductivity, density, and specific heat of both unfrozen and frozen product, moisture content of the product, initial product temperature and ambient temperature. To include all parameters into a single model, ANN has been shown to be an effective approach, due to unlimited number of input neurons [16]. Excessive input parameters are also found in food smoking processes. Fat–protein ratio, initial moisture content, initial temperature, radius of product, ambient temperature, relative humidity and process time were used as input neurons in a three-layer neural network with backpropagation for predicting average moisture content, core temperature and average temperature of frankfurters during smoking [17].

Besides its superb capability of handling many input parameters in a single model, ANN can also handle many output parameters simultaneously. For a spray drying, feed flow rate, inlet air temperature and atomizer speed of a spray dryer were used as input neurons to predict residual moisture content, particle size, bulk density, average time of wetability, insoluble solids, outlet air temperature and yield [18]. From the study, a four-layer ANN model (with 14 and 10 hidden neurons in the first and second hidden layers, respectively) with backpropagation presented reasonable performance, whereas RBF network could not produce the correct outputs.

Satish and Pydi Setty [19] developed an ANN model with backpropagation (steepest-descent gradient) for a fluidized-bed drying with its input layer consisting of bed temperature, air flow rate, solid flow rate and initial moisture content of solids. The model was developed to predict moisture content of product. In comparison to a tanks-in-series model, improvement in the model accuracy was significant. To predict evaporation rate, heat flux and heat transfer coefficient in a fluidized bed drying, Zbicinski et al. [20] proposed a hybrid ANN model to increase the accuracy of the model. However, only predictability of heat transfer coefficient could be improved over the original ANN model.

Table 35.1 summarizes some recent developments of ANN modeling in food processing, which includes network configuration, training method, validation method, application field and reference detail.

35.5.2 PREDICTION OF FOOD COMPONENTS AND PROPERTIES

For detecting flavor compounds in honey, an electronic nose made of ten different semiconductor gas sensors and one gas flow sensor was used to generate olfactory signals from the product. With ANN, the olfactory signals were used as input information to classify the types of honey efficiently [21]. In another study using ANN with RBF, concentrations of flavor compounds including maltol, ethyl maltol, vanillin and ethyl vanillin were predicted from absorbance spectrum of UV spectrophotometric measurement. Compared to regression methods including partial least square, classical least square and principal component analysis, the RBF network was a better approach [22]. For evaluating final color of extrudate product, a multilayer feed-forward network was used for simulating the relationship between in-line color measurement and final product color [9].

ANN with backpropagation has been applied to predict antioxidant capacity in cruciferous sprouts, using the conjugate gradient descent algorithm. The optimized network consisted of six input neurons, nine hidden neurons and one output neuron [23]. Also, using the conjugated gradient algorithm, ANN with backpropagation was successfully used for predicting the tryptic hydrolysis of pea protein isolate at temperatures of 40, 45 and 50°C [24].

Variation in chemical composition of a product causes changes in product properties that may be predicted by ANN modeling. Therdthai and Zhou [25] developed an ANN model with

TABLE 35.1
Selected Recent ANN Applications in Food Processing

Architecture (Input-Hidden(s)-Output Layers)	Input Neuron	Output Neuron	Transfer Function	Learning Algorithm	Validation Method	Application	Reference
10-40-40-40-1	Thickness, width, length, heat transfer coefficient, conductivity, density and specific heat, moisture content, initial product temperature and ambient temperature	Freezing time	Gaussian function-Gaussian complement-hyperbolic tangent	Back propagation	Relative and absolute error	Prediction of freezing rate and time	[16]
7-40-3	Ratio of fat-protein, moisture content, initial temperature, frankfurter radius, ambient temperature, process time and relative humidity	Average moisture content, average temperature and core temperature	—	Back propagation	Relative and absolute error	Prediction of temperature and moisture content of frankfurter from smoking process	[17]
4-4-4-1	Fat content, protein content, lactose content and temperature	Electrical conductivity of recombined milk	Log-sigmoid	Back propagation	SSE and correlation coefficient	Prediction of electrical conductivity from variation of recombined milk compositions	[25]
ANN-GA	Maximum retort temperature, increasing rate of retort temperature, temperature oscillation and time period of oscillation	Thermal process time, surface cook value and quality retention	—	—	Correlation and relative error	Variable retort temperature thermal process optimization	[31]
11-[]-6	Information from semiconductor gas sensors and gas flow sensor	Different aromatic quality of honey	—	Back propagation	Correctness rate of classifying	Classification of honey using olfactory signal	[21]
6-4-3-2-1	Contents of total phenolic compounds, inositol hexaphosphate, glucosinolates, soluble proteins, ascorbic acid, and tocopherols	Trolox equivalent antioxidant capacity of germinated cruciferous seeds	—	Back propagation (conjugate gradient descent)	RMSE	Prediction of antioxidant capacity of cruciferous sprouts	[23]

—	Orthogonal array data set containing absorption spectra in the 200–350 nm ranges	Content of maltol, ethyl maltol, vanillin and ethyl vanillin	Gaussian function	Radial basis function artificial neural networks	Relative error and % recovery (ratio of predicted content and actual content)	Prediction of flavor enhancer in foods	[22]
2-[4-8]-[4-8]-2	Moisture content, cell wall structure	Shrinkage rehydration	Hyperbolic tangent-hyperbolic tangent-linear	Back propagation (Levenberg–Marquardt algorithm)	Correlation	Prediction of physical property	[12]
5-30-2	Temperature, pH, acetic acid concentration, sucrose concentration, salt concentration	Death/survival No-growth/growth	Hyperbolic tangent-hyperbolic tangent	Back propagation	Number of false classification	Binary classification of microbial growth/survival condition	[15]
20-21-2	Browning degree, hardness, crispness, flavor	Heating time, Heating temperature	Hyperbolic tangent-linear	Back propagation	Prediction error (%)	Sensory quality-based food process control	[38]
2-4-1	Time	Wheat price	Sigmoid	Back propagation (Levenberg–Marquardt)	MAE, MSE, MAPE	Forecasting of material price in time series	[13]
5-5-1	Food characteristics, package properties, storage condition	Shelf-life	Hyperbolic tangent-linear	Back propagation (Levenberg–Marquardt)	MSE correlation	Shelf-life prediction of rice snack	[11]
3-[ ]-3	Contrast, entropy, correlation	Max load, of breaking energy	Sigmoid	Back propagation (Quasi-Newton, Bayesian regularization)	RMSE	Prediction of mechanical properties of fried chicken nuggets	[14]

(Continued)

TABLE 35.1 (Continued)

Architecture (Input-Hidden(s)-Output Layers)	Input Neuron	Output Neuron	Transfer Function	Learning Algorithm	Validation Method	Application	Reference
5-12-2	Pressure applied, compression rate, set temperature, high pressure vessel temperature, ambient temperature	Temperature, time	Sigmoid	Back propagation (Quasi-Newton, Bayesian regularization, Levenberg-Marquardt, gradient descent, conjugate gradient)	Mean prediction error (%), MSE, correlation	Prediction of high pressure process	[2]
3-14-10-7	Feed flow rate, inlet air temperature, atomizer speed	Residual moisture content, particle size, bulk density, average time of wetability, insoluble solids, outlet air temperature, yield	Sigmoid	Back propagation	Correlation	Spray drying of orange juice	[18]

backpropagation to predict changes in electrical conductivity due to variation in fat, protein and water in recombined milk as well as temperature. To overcome ANN's weakness in the lack of providing physical insight to the underlying relationship, a hybrid neural model combining a mechanistic model with an ANN model was further developed as a gray box model. In the hybrid model, the mechanistic model part accounted for the major underlying relationship between the constituents, temperature and electrical conductivity. The ANN model part within the hybrid neural model accounted for the difference between actual electrical conductivity and the output by the mechanistic model [26].

35.5.3 HIGH PRESSURE PROCESSING

For high pressure processing, several process parameters including applied pressure, rate of pressure increase, set-point temperature, high-pressure vessel temperature, and ambient temperature were used for prediction of time needed for re-equilibrate temperature in the sample after pressurization by ANN with backpropagation [27]. The model structure composed of an input layer (five neurons), a hidden layer (three neurons) and an output layer (one neuron). Later, Torrecilla et al. [28] modified the network to predict both time needed for temperature re-equilibrium and product temperature after pressurization. The model structure became 5-7-2 for the input, hidden and output layers respectively, with log-sigmoid transfer function. Torrecilla et al. [2] optimized learning parameters for training the network, including learning rate, momentum and training algorithm. An improvement in the model performance was obtained. The authors stated that the approach was useful as a process design tool for scanning possible high pressure processing conditions and selecting the optimal one for each product.

35.5.4 THERMAL PROCESSING

Sablani and Shayya [29] used ANN as a tool for calculating process time and lethality from g-value (the difference between retort and product core temperatures) and fh/U value (the ratio of heating rate index to sterilizing value) in thermal processing. The calculation could be easily done compared to Stumbo's method. The ANN structure was optimized without optimizing any learning parameters such as momentum and learning rate.

Mittal and Zhang [30] applied a method of shrinking the input and output by natural logarithmic function to improve the accuracy of ANN models. Z-value, cooling lag factor and ratio of heating rate index to sterilizing value were used for predicting the g-value. Moreover, the ratio of heating rate index to sterilizing value could be predicted from z-value, cooling lag-factor and g-value.

Using genetic algorithm with ANN, a variable retort temperature thermal process could be designed based on model predictions [31]. The model could estimate average quality retention, process time and surface cook value. The retort temperature could therefore be adjusted to reduce over-cooked and under-process problems.

35.5.5 BAKING

Baking is a dynamic process where heat and mass transfer phenomena are accompanied by changes in chemical and physical properties of dough and it is the interactions between dough properties and baking parameters that result in different product quality.

Zhou and Fong [32] developed ANN models and hybrid neural models for five quality attributes of bread baked in an industrial bread-baking oven. The five attributes included moisture loss, internal temperature and top, side and average colors that describe browning of bread. Various three-layer feed-forward ANNs with different transfer functions and different number of neurons in the hidden layer were investigated. Tan-sigmoid transfer function was found to be more effective. The optimized ANN models demonstrated significantly improved performance over the earlier developed statistical models. In addition, hybrid neural models were developed by combining mechanistic models and ANN models. Those hybrid neural models provided physical insights to the underlying relationships as well as high accuracy.

On modeling dough properties, Razmi-Rad et al. [33] developed an ANN model to estimate farinographic parameters of Iranian bread dough from chemical composition of wheat flour. The input parameters were protein content, wet gluten, sedimentation value, and falling number, while the output parameters were six farinographic properties including water absorption, dough development time, dough stability time, degree of dough softening after 10 and 20 min, and valorimeteric value. ANN of 4-17-6 structure was found to perform the best. Delta learning rule with momentum was used for training.

Fravolini et al. [34] developed a model-based method for determining the optimal setpoints for the leavening process in a continuous bread-making industrial plant. Modeling of the leavening process was essential to the method. Dough height was predicted by using ANN to identify a NARMA model. Mould temperature, dough temperature and dough height at the current time step were the three input parameters; dough height at the next time step was the only output parameter. An ANN of 3-10-1 structure was trained using the Levenberg–Marquardt algorithm. The model was shown to provide accurate estimation with reasonably small errors.

35.5.6 Prediction of Microbial Growth

To determine microbial growth, microbial enumeration techniques are normally used. These techniques are very time consuming, so statistical models have been developed. However, the statistical models lack accuracy due to nonlinearity in the underlying relationship [35]. ANN has been introduced to deal with the nonlinear relationship between environment factors and microbial growth.

Temperature, time, pH, and water activity are major environment factors for thermal inactivation of *E. coli*. Among these factors, temperature and time play the most important role while pH and water activity are less significant. Compared to ANN, Cerf's model could not explain the influence of water activity on the rate of microbial growth. While the response surface method (RSM) could reveal accurately the influence of water activity, its performance was poor in low pH region. ANN was therefore found to be superior to both Cerf's model and RSM [36]. However, Jeyamkondan et al. [35] found that the performance of ANN from test data set was lower than RSM, although significantly better performance was found from training data set.

Garcia-Gimeno et al. [37] developed a three layer neural network containing the optimized structure of 3-3-1 (input-hidden-output layers) and sigmoid transfer function, to predict kinetic parameters of the growth curve of *Lactobacillus plantarum*. To prevent saturation problems in the sigmoid function, all data were normalized to the range of 0.1 and 0.9. The input parameters included pH, temperature and salt content. By adding 25 data points from the optical density versus time curve to the input layer, accuracy of the ANN model was improved. Compared to RSM, the developed ANN model showed significantly better performance. Therefore, ANN was selected as a tool to predict microbial growth and thereby estimate the product shelf-life.

35.5.7 Fermentation

One of the major challenges in controlling a food fermentation process is lack of on-line sensors for some key variables of the process. ANN could provide an effective approach for generating soft sensors for such variables. Other more readily measured variables can be used as the inputs to the ANN.

For beer fermentation, ANN was used for online estimation of diacetyl concentration during fermentation [39]. A feed-forward ANN of 5-9-1 structure was built with neurons capable of describing dynamic behavior of the process. The five inputs were temperature, time, gravity, turbidity, and pH value. The accuracy of the estimator was shown to be highly satisfactory, within 0.05 ppm diacetyl.

The above approach was further extended in [40] using ANN as a state estimator which was used for optimizing the beer fermentation process. The state variables included gravity, pH, and diacetyl, while temperature was the input variable. For the ANN model, values of the three state variables, temperature and time at the current time step were the inputs and values of the three state variables at future time step were the outputs. It was shown that the predicted trajectories of gravity, pH, and diacetyl were in good agreement with the experimental data measured at an automated pilot fermenter.

In [41], a hybrid neural model was presented for predicting diacetyl concentration from five inputs. A mechanistic model, expressed by an ordinary differential equation, was built in parallel to an ANN model of 5-7-1 structure. The hybrid model was proven to be more effective than an ANN-only model in terms of the number of training sets required for achieving the same accuracy.

35.6 SUMMARY

The paradigm of ANN has been developed by mimicking human brain function, in order to solve complicated problems such as those found in food processing. Most food processes are nonlinear in nature with a large number of interactive parameters and variables. To solve such nonlinear problems with less consistent or noisy data, ANN has been proven as an effective approach from many previous studies.

The most widely used ANN in food processing is the multilayer perception network with back-propagation. However, each system has its own optimum structure and training algorithm, in order to minimize training speed and maximize accuracy of the model for prediction, forecasting, classification, and so on. To do so, parametric factors including number of hidden neurons, number of hidden layers, number of training epoch, training algorithm, learning rate, and momentum have to be optimized. Model validation is a necessary step to ensure that over-fitting does not happen.

It has been documented that ANN have been used for food processing, including prediction of food properties and modeling of food processes such as freezing, drying, baking, high pressure, thermal processing, fermentation, and extrusion. In addition, the impact of food processing on microbial growth has been evaluated by using ANN.

It is forecasted that there will be more applications of ANN to food processing due to its unique learning capability. However, lack of physical insight in ANN to an underlying relationship may lead to more development of hybrid neural models in the future.

ACKNOWLEDGMENT

The authors are grateful to Dr. Jinsong Hua for reviewing the manuscript.

REFERENCES

1. Samarasinghe, S. *Neural Networks for Applied Sciences and Engineering*. Auerbach Publication (Taylor & Francis Group), Boca Raton, Florida, USA, 2007, 570.
2. Torrecilla, J.S., Otero, L., and Sanz, P.D. Optimization of an artificial neural network for thermal/pressure food processing: Evaluation of training algorithms. *Comput. Electron. Agri.*, 56, 101, 2007.
3. Otere, L., et al. A model to design high-pressure processes towards a uniform temperature distribution. *J. Food Eng*, 78, 1463, 2007.
4. Lisboa, P.G.J. *Neural Networks: Current Application*. Chapman and Hall, London, UK, 1992, 279.
5. Kartalopoulos, S.V. *Understanding Neural Networks and Fuzzy Logic*. IEEE Press, New York, NY, 205, 1996.
6. Skapura, D.M. *Building Neural Networks*. ACM Press, New York, NY, 286, 1996.
7. Doganis, P. et al. Time series sales forecasting for short shelf-life food products based on artificial neural networks and evolutionary computing. *J. Food Eng.*, 75, 196, 2006.
8. Demuth, H. and Beale, M. *Neural Networks Version 3*. The Math Works Inc., Natick, Massachusetts, USA, 45, 1998.

9. Valadez-Blanco, R. et al. In-line color monitoring during food extrusion: sensitivity and correlation with product color. *Food Res. Int.*, 40, 1129, 2007.
10. Avriel, M. *Nonlinear Programming: Analysis and Methods.* Dover Publishing, Mineola, New York, USA, 70, 2003.
11. Siripatrawan, U. and Jantawat, P. A novel method for shelf-life prediction of a packaged moisture sensitive snack using multilayer perceptron neural network. *Expert Systems with Applications*, 34, 1562, 2008.
12. Kerdpiboon, S., Kerr, W.L., and Devahastin, S. Neural network prediction of physical property changes of dried carrot as a function of fractal dimension and moisture content. *Food Res. Int.*, 39, 1110, 2006.
13. Zou, H.F. et al. An investigation and comparison of artificial neural network and time series models for Chinese food grain price forecasting. *Neurocomputing,* 70, 2913, 2007.
14. Qiao, J., et al. Predicting mechanical properties of fried chicken nuggets using image processing and neural network techniques. *J. .Food. Eng.*, 79, 1065, 2007.
15. Yu, C., Davidson, V.J., and Yang, S.X. A neural network approach to predict survival/death and growth/no-growth interfaces for *Escherichia coli* O157:H7. *Food Micro.*, 23, 552, 2006.
16. Mittal, G.S. and Zhang, J. Prediction of freezing time for food product using a neural networks. *Food Res. Int.*, 33, 557, 2000.
17. Mittal, G.S. and Zhang, J. Prediction of temperature and moisture content of frankfurters during thermal processing using neural network. *Meat Sci.*, 55, 13, 2000.
18. Chegini, G.R. et al. Prediction of process and product parameters in an orange juice spray dryer using artificial neural networks. *J. Food Eng.*, 84, 534, 2008.
19. Satish, S. and Pydi Setty, Y. Modeling of continuous fluidized bed dryer using artificial neural networks. *Int. Commun. Heat Mass Transfer*, 32, 539, 2005.
20. Zbicinski, I., Strumillo, P., and Kaminski, W. Hybrid neural model of thermal drying in a fluidized bed. *Comput. Chem. Eng.*, 20, S695, 1996.
21. Linder, R. and Poppl, S.J. A new neural network approach classifies olfactory signals with high accuracy. *Food Qual. Preference*, 14, 435, 2003.
22. Ni, Y., Zhang, G., and Kokot, S. Simultaneous spectrophotometric determination of maltol, ethyl maltol and ethyl vanillin in foods by multivariate calibration and artificial neural networks. *Food Chem.*, 89, 465, 2007.
23. Bucinski, A., Zielinski, H., and Kozlowska, H. Artificial neural networks for prediction of antioxidant capacity of cruciferous sprouts. *Trend Food Sci. Technol.*, 15, 161, 2004.
24. Bucinski, A. et al. Modeling the tryptic hydrolysis of pea protein using an artificial neural network. *LWT*, 41, 942, 2008.
25. Therdthai, N. and Zhou, W. Artificial neural network modeling of the electrical conductivity property of recombined milk. *J. Food Eng.,* 50, 107, 2001.
26. Therdthai, N. and Zhou, W. Hybrid neural modeling of the electrical conductivity property of recombined milk. *Int. J. Food Properties*, 5, 49, 2002.
27. Torrectilla, J.S., Otero, L., and Sanz, P.D. A neural network approach for thermal/ pressure food processing. *J. Food Eng.*, 62, 89, 2004.
28. Torrecilla, J.S., Otero, L., and Sanz, P.D. Artificial neural networks: a promising tool to design and optimize high-pressure food processes. *J. Food Eng.*, 69, 299, 2005.
29. Sablani, S.S. and Shayya, W.H. Computerization of Stumbo's method of thermal process calculation using neural networks. *J. Food Eng.*, 53, 209, 2001.
30. Mittal, G.S. and Zhang, J. Prediction of food thermal process calculation parameters using neural networks. *Int. J. Food Micro.*, 79,153, 2002.
31. Chen, C.R. and Ramaswany, H.S. Modeling and optimization of variable retort temperature (VRT) thermal processing using coupled neural networks and genetic algorithm. *J. Food Eng.*, 53, 209, 2002.
32. Zhou, W. and Fong, M. Neural and hybrid neural modeling of a bread baking process. In *Proceedings of the International Conference on Innovations in Food Processing Technology and Engineering.* AIT, Bangkok, Thailand, 35, 2002.
33. Razmi-Rad, E. et al. Prediction of rheological properties of Iranian bread dough from chemical composition of wheat flour by using artificial neural networks. *J. Food Eng.*, 81, 728, 2007.
34. Fravolini, M.L., Ficola, A., and La Cava, M. Optimal operation of the leavening process for a bread-making industrial plant. *J. Food Eng.,* 60, 289, 2003.
35. Jeyamkondan, S., Jayas, D.S., and Holley, R.A. Microbial growth modeling with artificial neural networks. *Int. J. Food Micro.*, 64, 343, 2001.
36. Lou, W. and Nakai, S. Application of artificial neural networks for predicting the thermal inactivation of bacteria: a combined effect of temperature, pH and water activity. *Food Res. Int.*, 34, 573, 2001.

37. Garcia-Gimeno, R.M., Herras-Martinez, C., and de Siloniz, M.I. Improving artificial neural networks with a pruning methodology and genetic algorithms for their application in microbial growth prediction in food. *Int. J. Food Micro.,* 72, 19, 2002.
38. Kupongsak, S. and Tan, J. Application of fuzzy set and neural network techniques in determining food process control set points. *Fuzzy Sets Systems,* 157, 1169, 2006.
39. Kurz, T., Fellner, M., Delgado, A., and Becker, T. Observation and control of the beer fermentation using cognitive methods. *J. Inst. Brewing,* 107, 241, 2001.
40. Becker, T., Enders, T., and Delgado, A. Dynamic neural networks as a tool for the online optimization of industrial fermentation. *Bioprocess Biosyst. Eng.,* 24, 347, 2002.
41. Fellner, M., Delgado, A., and Becker, T. Functional nodes in dynamic neural networks for bioprocess modeling. *Bioprocess Biosyst. Eng.,* 25, 263, 2003.

36 Exergy Analysis of Food Drying Processes and Systems

Ibrahim Dincer
University of Ontario Institute of Technology (UOIT)

CONTENTS

36.1   INTRODUCTION

Drying is widely used in a variety of thermal energy applications ranging from food drying to wood drying. Utilization of high amounts of energy in the drying industry makes drying one of the most energy-intensive operations with great industrial significance. The objective of the dryer is to supply the product with more heat than is available under ambient conditions thus sufficiently increasing the vapor pressure of the moisture held within the product to enhance moisture migration from within the product and significantly decreasing the relative humidity of the drying air to increase its moisture carrying capability and to ensure a sufficiently low equilibrium moisture content.

Exergy, by definition, is the maximum amount of work which can be obtained from a stream of matter, heat or work as it comes to equilibrium with a reference environment, and is a measure of the potential of a stream to cause change, as a consequence of not being completely stable relative to the reference environment. Exergy is not subject to a conservation law but is destroyed due to irreversibilities during any of the thermal processes including drying. It means that reducing the irreversibilities in a drying system will increase the exergy (i.e., availability) and hence the efficiency of the system. Increased efficiency can often contribute in a major way to achieving energy security in an environmentally acceptable way by the direct reduction of irreversibilities that might otherwise have occurred. Increased efficiency also reduces the energy requirement for drying system facilities for the production, transportation, transformation and distribution of the various energy forms; these additional facilities all carry some environmental impacts. This makes exergy one of the most powerful tools that provide optimum drying conditions. Exergy analysis becomes more crucial especially for industrial (large-scale) high-temperature drying applications.

Problems with energy utilization are related not only to global warming, but also to such environmental concerns as air pollution, global climate change, acid rain, and stratospheric ozone depletion. These issues must also be taken into consideration for drying systems simultaneously if humanity has to achieve a bright energy future with minimal environmental impact. Since all energy resources lead to some environmental impact, it is reasonable to suggest that some (not all) of these concerns, at least in part, can overcome the problems through increased energy efficiency, resulting in less energy utilization and less environmental impact, which can be carried out by exergy analysis.

It is important to highlight that exergy of an energy form or a substance is a measure of its usefulness or quality or potential to cause change. A thorough understanding of exergy and the insights it can provide into the efficiency and environmental impact of drying systems are required for engineers or researchers working in the area of drying technology. During the past decade, the need to understand the connections between exergy and energy, and environmental impact has become increasingly significant.

An understanding of the relations between energy, exergy and the environment may reveal the fundamental patterns and forces that affect and underlie changes in the environment, and help researchers to deal better with environmental damage. In this regard, three relationships between energy, exergy, and environmental impact were introduced previously [2,3].

Thus, the primary objective of this book chapter is to discuss both energetic and exergetic aspects of drying processes and systems, introduce a complete analysis of a drying process as an example, and highlight the importance of using exergy as a potential tool for drying processes and systems. This will also be useful in bringing some insights and directions for analyzing and solving environmental problems of varying complexity using the exergy concept.

36.2   EXERGY ANALYSIS VERSUS ENERGY ANALYSIS

Exergy analysis is a thermodynamic analysis technique based on the second law of thermodynamics which provides an alternative and illuminating means of assessing and comparing processes and systems rationally and meaningfully. In particular, exergy analysis yields efficiencies which provide a true measure of how nearly actual performance approaches the ideal, and identifies more clearly than energy analysis the causes and locations of thermodynamic losses. Consequently, exergy analysis can assist in improving and optimizing designs.

Increasing application and recognition of the usefulness of exergy methods by those in industry, government, and academia has been observed in recent years. Exergy has also become increasingly used internationally. The present authors for instance, have examined exergy analysis methodologies and applied them to numerous industrial processes and systems [e.g., 4–11].

Thermodynamics permit the behavior, performance and efficiency to be described for systems for the conversion of energy from one form to another. Conventional thermodynamic analysis is based primarily on the first law of thermodynamics, which states the principle of conservation of energy. An energy analysis of an energy-conversion system is essentially an account of the energies entering and exiting. The exiting energy can be broken down into products and wastes. Efficiencies are often evaluated as ratios of energy quantities and are often used to assess and compare various systems. Power plants, heaters, refrigerators, and thermal storages for example, are often compared based on energy efficiencies or energy-based measures of merit.

However, energy efficiencies are often misleading in that they do not always provide a measure of how the performance of a system nearly approaches ideality. Furthermore, the thermodynamic losses which occur within a system (i.e., those factors which cause performance to deviate from ideality) often are not accurately identified and assessed with energy analysis. The results of energy analysis can indicate the main inefficiencies to be within the wrong sections of the system, and a state of technological efficiency different than actually exists.

Exergy analysis permits many of the shortcomings of energy analysis to be overcome. Exergy analysis is based on the second law of thermodynamics and is useful in identifying the causes, locations and magnitudes of process inefficiencies. The exergy associated with an energy quantity is a quantitative assessment of its usefulness or quality. Exergy analysis acknowledges that although energy cannot be created or destroyed, it can be degraded in quality, eventually reaching a state in which it is in complete equilibrium with the surroundings and hence of no further use for performing tasks.

For industrial drying systems for example, exergy analysis allows one to determine the maximum potential associated with the incoming energy. This maximum is retained and recovered only if the energy undergoes processes in a reversible manner. Losses in the potential for exergy recovery occur in the real world because actual processes are always irreversible.

The exergy flow rate of a flowing commodity is the maximum rate that work may be obtained from it as it passes reversibly to the environmental state, exchanging heat and materials only with the surroundings. In essence, exergy analysis states the theoretical limitations imposed upon a system, clearly pointing out that no real system can conserve exergy and that only a portion of the input exergy can be recovered. Also, exergy analysis quantitatively specifies practical limitations by providing losses in a form in which they are a direct measure of lost exergy.

36.3 ILLUSTRATIVE EXAMPLE

Here, we consider a continuous drying system for food products which is applicable to various small-, mid-, and large-scale operations. A schematic presentation of the systems with input and output terms is shown in Figure 36.1. As clearly seen in this figure, we have four major components to take into consideration as follows:

- State 1: refers to the input of drying air to the drying chamber to dry the products.
- State 2: refers to the input of moist products to be dried in the chamber.
- State 3: refers to the output of the moist air after taking the evaporated moisture from the products.
- State 4: refers to the output of the dried products. In fact, their moisture contents are reduced to a certain level required for each commodity of the product.

36.3.1 ANALYSIS

In this section, we introduce a novel model for energy and exergy analysis of drying processes and initially write the mass, energy, and exergy balance equations for the above system, as a control

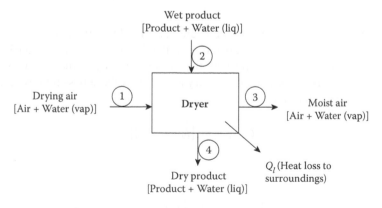

FIGURE 36.1 A general illustration of the drying process.

volume system, shown in Figure 36.1 as necessary for design, analysis, and performance evaluation of such drying systems (for details, see Dincer and Sahin [12]).

- Mass balance equations
  In order to write the mass balance equations for the dryer given above, we considered three materials such as products, air and moisture content in the drying air and products. Therefore, we proceed with the mass balance equations accordingly as follows:

$$\text{Product:} \left(\dot{m}_p \right)_2 = \left(\dot{m}_p \right)_4 = \dot{m}_p \tag{36.1}$$

$$\text{Air:} \left(\dot{m}_a \right)_1 = \left(\dot{m}_a \right)_3 = \dot{m}_a \tag{36.2}$$

$$\text{Moisture content in air:} \; \omega_1 \dot{m}_a + \left(\dot{m}_w \right)_2 = \omega_3 \dot{m}_a + \left(\dot{m}_w \right)_4 \tag{36.3}$$

- Energy balance equation
  The energy balance equation can be written for the entire system in the following manner, by taking input energy terms equal to output energy terms

$$\dot{m}_a h_1 + \dot{m}_p h_{p,2} + \dot{m}_{w,2} h_{w,2} = \dot{m}_a h_3 + \dot{m}_p h_{p,4} + \dot{m}_{w,4} h_{w,4} + \dot{Q}_l \tag{36.4}$$

where

$$h_1 = \left(h_a \right)_1 + \omega_1 \left(h_v \right)_1 \simeq \left(h_a \right)_1 + \omega_1 \left(h_g \right)_1 \tag{36.5}$$

$$h_3 = \left(h_a \right)_3 + \omega_3 \left(h_g \right)_3 \tag{36.6}$$

as can directly be obtained from the psychrometric chart.
The heat transfer rate can be calculated from

$$\dot{Q} = \dot{m}_a \left(h_1 - h_3 \right) - \dot{Q}_l = \dot{m}_p \left(h_{p,4} - h_{p,2} \right) + \dot{m}_{w4} h_{w,4} - \dot{m}_{w2} h_{w,2} \tag{36.7}$$

Note that there are two possibilities where one can obtain the maximum heat transfer rate in the dryer. First, the exit temperature of the drying air would be equal to the inlet temperature of the wet product. Second, the exit temperature of the product would reach the inlet temperature of the hot entering dry air. Each of these possibilities may individually be expressed as follows:

$$\dot{Q}_{a,\max} = \dot{m}_a \left[h_1 - h_3 \left(T_2 \right) \right] \tag{36.8}$$

and

$$\dot{Q}_{p,\max} = \dot{m}_p \left[h_{p,4} \left(T_1 \right) - h_{p,2} \right] + \dot{m}_{w4} h_{w,4} \left(T_1 \right) - \dot{m}_{w2} h_{w,2} \tag{36.9}$$

Hence,

$$\dot{Q}_{max} = \min\left[\dot{Q}_{a,max}, \dot{Q}_{p,max}\right]$$ (36.10)

So, the effectiveness of the dryer can now be shown as

$$\varepsilon = \frac{\dot{Q}}{\dot{Q}_{max}}$$ (36.11)

- Exergy balance equation

In this section, we write the exergy balance equation for the entire system as described above for energy balance equation for the input and output terms. Therefore, it can be written as follows:

$$\dot{m}_a e_1 + \dot{m}_p e_{p,2} + \dot{m}_{w,2} e_{w,2} - (\dot{m}_a e_3 + \dot{m}_p e_{p,4} + \dot{m}_{w,4} e_{w,4}) - \left(1 - \frac{T_0}{T_k}\right)\dot{Q}_l - \dot{E}_d = 0$$

and rearranged as

$$\dot{m}_a (e_1 - e_3) = \dot{m}_p (e_{p,4} - e_{p,2}) + \dot{m}_{w,4} e_{w,4} - \dot{m}_{w,2} e_{w,2} + \left(1 - \frac{T_0}{T_k}\right)\dot{Q}_l + \dot{E}_d = 0$$ (36.12)

where the specific exergy for state 1 can be obtained as

$$e_1 = [(C_P)_a + \omega_1 (C_P)_v](T_1 - T_0) - T_0\left\{[(C_P)_a + \omega_1 (C_P)_v]\ln\left(\frac{T_1}{T_0}\right) - (R_a + \omega_1 R_v)\ln\left(\frac{P_1}{P_0}\right)\right\}$$
$$+ T_0\left\{(R_a + \omega_1 R_v)\ln\left(\frac{1 + 1.6078\omega^0}{1 + 1.6078\omega_1}\right) + 1.6078\omega_1 R_a \ln\left(\frac{\omega_1}{\omega^0}\right)\right\}$$

(36.13)

The specific exergy for state 3 can be obtained as

$$e_3 = [(C_P)_a + \omega_3 (C_P)_v](T_3 - T_0) - T_0\left\{[(C_P)_a + \omega_3 (C_P)_v]\ln\left(\frac{T_3}{T_0}\right) - (R_a + \omega_3 R_v)\ln\left(\frac{P_3}{P_0}\right)\right\}$$
$$+ T_0\left\{(R_a + \omega_3 R_v)\ln\left(\frac{1 + 1.6078\omega^0}{1 + 1.6078\omega_3}\right) + 1.6078\omega_3 R_a \ln\left(\frac{\omega_3}{\omega^0}\right)\right\}$$

(36.14)

The specific exergy for the moist products can be written as

$$e_p = [h_p(T,P) - h_p(T_0,P_0)] - T_0[s_p(T,P) - s_p(T_0,P_0)]$$ (36.15)

and the specific exergy of water content becomes

$$e_w = [h_f(T) - h_g(T_0)] + v_f[P - P_g(T)] - T_0[s_f(T) - s_g(T_0)] + T_0 R_v \ln\left[\frac{P_g(T_0)}{x_v^0 P_0}\right] \quad (36.16)$$

Neglecting pressure drop in dry air stream ($P_1 = P_3$), the exergy difference between states 1 and 3 can be written as follows after some simplifications and mathematical manipulations as follows:

$$e_1 - e_3 = c_{p,m}\left(T_1 - T_3 - T_o \ln\frac{T_1}{T_3}\right) + R_a T_0\left[(1 + 1.608\omega_1)\ln\left(\frac{1 + 1.608\omega_0}{1 + 1.608\omega_1}\right)\right.$$
$$\left. - (1 + 1.608\omega_3)\ln\left(\frac{1 + 1.608\omega_0}{1 + 1.608\omega_3}\right)\right] \quad (36.17)$$

Also, the net change in exergy flow of product and moisture of the product respectively, are:

$$e_{p,4} - e_{p,2} = h_{p,4} - h_{p,2} - T_o(s_{p,4} - s_{p,2}) = c_p\left(T_4 - T_2 - T_o \ln\frac{T_4}{T_2}\right) \quad (36.18)$$

where $c_{p,m} = c_{p,a} + \omega c_{p,v}$.

In addition, the exergy flow due to heat loss can be identified as follows:

$$\dot{E}_q = \dot{m}_a e_q = \dot{m}_a\left(1 - \frac{T_0}{T_{av}}\right)q_l = \left(1 - \frac{T_0}{T_{av}}\right)Q_l \quad (36.19)$$

where T_{av} is the average dryer's outer surface temperature.

Here are some example data for the reference dead state (i.e., the reference environment) for calculations: $T_0 = 32°C$, $P_0 = 1$ atm $= 100$ kPa, $\omega^0 = 0.0153$, and $x_v^0 = 0.024$ (as mole fraction of water vapor in air), respectively.

- Energy and exergy efficiencies
  Here, we define energy and exergy efficiencies of a food drying process as

$$\eta_{en} = \frac{\text{Net energy used for moisture evaporation}}{\text{Energy input by drying air supplied}} \quad \text{and} \quad \eta_{en} = \frac{\dot{m}_p(h_{p,4} - h_{p,2}) + \dot{m}_{w,4} h_{w,4} - \dot{m}_{w,2} h_{w,2}}{\dot{m}_a(h_1 - h_3)} \quad (36.20)$$

and

$$\eta_{ex} = \frac{\text{Net exergy used for moisture evaporation}}{\text{Exergy input by drying air supplied}} \quad \text{and} \quad \eta_{ex} = \frac{\dot{m}_p(e_{p,4} - e_{p,2}) + \dot{m}_{w,4} e_{w,4} - \dot{m}_{w,2} e_{w,2}}{\dot{m}_a(e_1 - e_3)} \quad (36.21)$$

where

$$(\dot{m}_w)_{ev} = (\dot{m}_w)_2 - (\dot{m}_w)_4, \ (e_w)_3 = \left[h(T_3,P_{v3}) - h_g(T_0)\right] - T_0\left[s(T_3,P_{v3}) - s_g(T_0)\right] + T_0 R_v \ln\left[\frac{P_g(T_0)}{x_v^0 P_0}\right]$$

(36.22)

and

$$P_{v3} = (x_v)_3 P_3$$

(36.23)

Example

In this example, we will show how to conduct an exergy analysis of the dryer and investigate the changes in exergy efficiencies versus various system parameters such as mass flow rate of the drying air, temperature of the drying air, the amount of products coming in, the initial moisture content of the product, the final moisture content of the product, specific inlet exergy, humidity ratio, and net exergy use for drying the products.

The following is the procedure to conduct an energy and exergy analysis to determine both energy and exergy efficiencies of the drying process:

    i. Provide $\dot{m}_a$, $\dot{m}_p$, $(\dot{m}_w)_2$, $(\dot{m}_w)_4$, and $\omega_1 \rightarrow$ calculate ω_3

    ii. Provide T_1, P_1, T_2, P_2, T_3, P_3, T_4, and $P_4 \rightarrow$ determine Q_l

    iii. Provide $(C_p)_a$, $(C_p)_v$, R_a, R_v, T_{av}, and $(x_v)_3 \rightarrow$ determine $\dot{E}_d$ and η_{ex}

    iv. Use steam tables, the psychrometric chart or tables and the dead state properties accordingly.

        In the solution, one may consider the following parameters as inputs or known parameters to proceed for the solution:

    v. $\dot{m}_a$, $\dot{m}_p$, $(\dot{m}_w)_2$, $(\dot{m}_w)_4$, ω_1, T_1, and T_2

    vi. In Table 36.1, we list some thermal data related to products and drying air that are used in the calculations to obtain how energy and exergy efficiencies change with process parameters, e.g., mass flow rate of air, temperature of drying air, specific exergy, specific exergy difference, moisture content of the product, and humidity ratio of drying air.

TABLE 36.1
Thermal Data used in the Example

$(C_p)_a$	1.004 kJ/kgK			
$(C_p)_v$	1.872 kJ/kgK			
R_a	0.287 kJ/kgK			
R_v	04615 kJ/kgK			
T_{av}	50°C=323.15 K			
$(x_v)_3$	0.055			
T_0	32°C=305.15 K			
P_0	101.3 kPa			
w_0	0.0153			
$(x_v)_0$	0.024			
$\dot{m}_a$(kg/s)	m_p (kg)	$\dot{m}_p$(kg/s)		
0	1	0.0002778		
1	5	0.0013389		
1.5	10	0.0027778		
2	15	0.0041667		
2.5	20	0.0055556		
Temperature and relative humidity	State 1	State 2	State 3	State 4
Temperature (°C)	55–100	25	25–70	50–95
ϕ (%)	10–35	55–85	60–95	15–30

36.3.2 RESULTS AND DISCUSSION

Here, we study various energetic and exergetic performance of the drying process through energy and exergy efficiencies and heat transfer effectiveness that are mathematically derived and presented above.

The typical resultant thermodynamic efficiencies as energy and exergy efficiencies versus mass flow rate of drying air are shown in Figure 36.2. The energy (first law) efficiency of the process is defined as the ratio of the rate of energy transferred to the stream of products for drying to the total energy transferred from the stream of drying air (including the rate of heat losses). Likewise, the exergy (second law) efficiency of the drying process is described as the ratio of the rate of exergy transferred to the stream of products for drying divided by the total exergy transferred from the stream of drying air (including the rate of thermal exergy losses). As shown in Figure 36.2, higher air mass flow rate results in less energy and exergy efficiencies as the input energy/exergy of the system increases, while the outlet energy/exergy remains constant. Also, results show that the effectiveness, defined in terms of the ratio of heat transfer rate to the maximum possible heat transfer rate within exchanger, does not change with air mass flow rate as it was calculated 59.9%.

Figure 36.3 depicts the distributions of both energy and exergy efficiencies at different inlet air temperatures at given product and air mass flow rates. Both efficiencies show decreasing trend with increasing air inlet temperature since the properties of other points are assumed unchanged. In other words, this figure shows that providing the system with warmer drying air at constant properties of the other points leads to a higher rate of heat loss, thereby decreasing the energy and exergy efficiencies of the dryer.

The influence of inlet air temperature at various air mass flow rates on exergy efficiency is illustrated in Figure 36.4. This figure is in fact another presentation of results obtained for exergy efficiency of the system, previously shown in Figures 36.1 and 36.2. It is obvious that at higher air inlet temperature, while keeping the properties of other points constant, input exergy will increase whereas extracted exergy in the product stream remains unchanged. Therefore, the exergy efficiency of the system decreases. For the same reason, higher mass flow rate with constant properties of all states results in less exergy efficiency due to higher heat loss.

Moreover, the influence of inlet air temperature on heat transfer effectiveness of the dryer is depicted in Figure 36.5. A higher inlet air temperature at constant properties of other states represents a higher amount of heat loss which results in a lower effectiveness of the system.

Figure 36.6 shows the variation of exergy efficiency with inlet-air mass flow rate for various product masses. Increasing inlet-air mass flow rate reduces exergy efficiency. However, increase

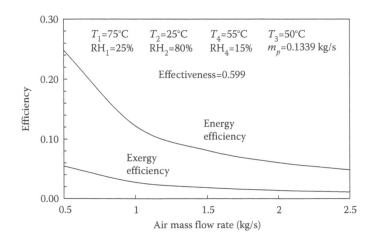

FIGURE 36.2 Effect of air mass flow rate on both energy and exergy efficiencies of the drying process.

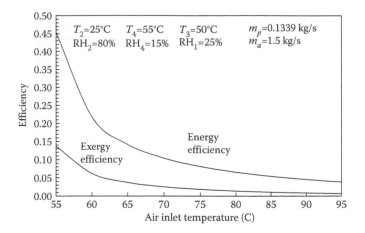

FIGURE 36.3 Variation of energy and exergy efficiencies of the drying process versus air inlet temperature.

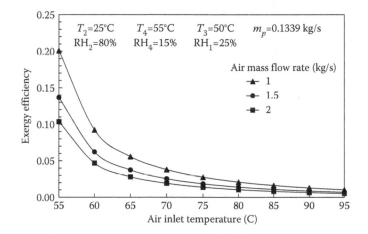

FIGURE 36.4 Variation of exergy efficiency of the drying process versus inlet air temperature at three different air mass flow rates.

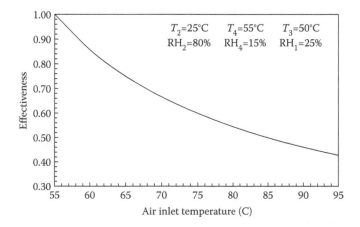

FIGURE 36.5 Variation of heat transfer effectiveness versus inlet air temperature.

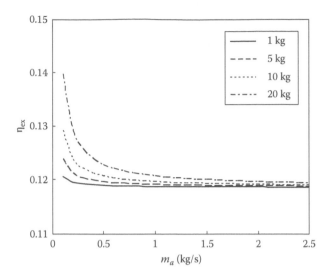

FIGURE 36.6 Variation of process exergy efficiency with mass flow rate of drying air at different product weights.

in inlet-air mass flow rate beyond a certain point do not affect considerably the exergy efficiency. This occurs because increasing inlet-air mass flow rate increases the exergy into the system, which in turn lowers the exergy efficiency based on Equation 36.21. Moreover, increasing product weight (mass) considerably influences the exergy efficiency, i.e., exergy efficiency increases with increasing product mass. In this case, the exergy used to dry the product increases with increasing product mass.

Exergy efficiency also varies with inlet drying air temperature as product mass is varied. Increasing drying air temperature is expected to reduce exergy efficiency, since exergy efficiency is inversely proportional to the exergy rate of drying air. As one may expect, the exergy efficiency changes monotonically as drying air temperature increases. Moreover, the magnitude of exergy efficiency increases considerably with increasing product mass.

Note that the dryer exergy efficiency varies with the difference in specific evaporation exergies of water content at different mass flow rates of drying air in the dryer. So, increasing specific exergy difference results in decreasing exergy efficiency. For the same magnitude of specific exergy difference, a greater mass flow rate of drying air results in a smaller exergy efficiency, due to fact that higher mass flow rates of drying air consume higher energy and hence causing greater exergy losses.

Figure 36.7 illustrates the variation of exergy efficiency with product mass as the mass flow rate of drying air is varied. The exergy efficiency increases linearly with product mass. This increase in exergy efficiency indicates that the mass flow rate of drying air decreases because exergy efficiency is inversely proportional to mass flow rate of drying air. Moreover, this linear increase of exergy efficiency with product mass indicates that the specific exergy difference between the product and the exergy exiting over exergy of the drying air remains constant for the given product mass.

Figure 36.8 depicts the variation of exergy efficiency with moisture content of the incoming products as the mass flow rate of evaporated water is varied. The exergy efficiency increases with increasing moisture content of the products. This variation is more pronounced as evaporation rate increases. In this case, the energy utilized for drying the product increases when moisture content of the products increases. For example, for given air inlet conditions, the energy utilized in the system is reduced, which in turn improves the exergy efficiency of the system. Furthermore, the drying-process exergy efficiency varies with the humidity ratio of the drying air entering the dryer

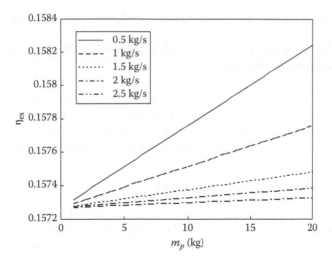

FIGURE 36.7 Variation of process exergy efficiency with product weight at different mass flow rates of air.

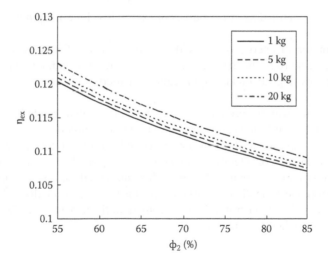

FIGURE 36.8 Variation of process exergy efficiency with moisture content of the product at different moisture evaporation rates.

for various mass flow rates of drying air. A linear relation is observed between exergy efficiency and humidity ratio. Interestingly, we note that exergy efficiency decreases only slightly with increasing humidity ratio of drying air.

The above figures not only demonstrate the usefulness of exergy analysis in thermodynamic assessments of drying systems but also provide insights into their performances and efficiencies. Some advantages of exergy analysis illustrated here for drying are as follows [12]:

- It provides proper accounting of the loss of availability of heat in a drying system using the conservation of mass and energy principles together with the second law of thermodynamics for the goals of design and analysis.
- It gives more meaningful and useful information than energy analysis regarding the efficiency, losses and performance for drying systems.

- It more correctly reflects the thermodynamic and economic values of the operation of drying systems.
- It reveals whether or not and by how much it is possible to design more efficient drying systems by reducing inefficiencies.

The analysis methodology and results presented should be useful to engineers seeking (i) to optimize the designs of drying systems and their components, and (ii) to identify appropriate applications and optimal configurations for drying systems in various applications.

Further information on the above illustrative example can be found in Dincer and Sahin [12].

36.4   FURTHER ILLUSTRATIVE EXAMPLES

Here, we describe the main sources of exergy losses associated with irreversibilities as mentioned earlier, for three different types of industrial drying methods such as air drying, drum drying and freeze drying. Therefore, the ideas discussed in this chapter are demonstrated by practical facts [1].

36.4.1   Air Drying

The following can be identified as three sources of exergy loss for air drying:

- A sizable amount of exergy is lost with exiting air even if it is assumed to reach the wet-bulb temperature in the drying process. At higher wet-bulb temperatures, the water present in the exiting air makes significant contribution to the total exergy loss of the exiting air.
- The exergy exiting with the product is seen to be quite small as might be expected, since it was shown earlier that little exergy was put into the solid products.
- The exergy loss from the walls of the dryer due to heat rejection, is a significant item that needs to be taken into consideration. For example, in spray drying, this amount may reach up to 25% of total exergy input. Of course, this can be reduced by an appropriate insulation of the dryer. Here, another important aspect is the size of the dryer. For example, the jet-type ring dryer has a much smaller loss from its walls than an equivalent-capacity spray dryer due to its smaller dimensions.

36.4.2   Drum Drying

Three major sources of exergy loss can be identified including:

- Some exergy is lost from the drum due to convection of air over the drum surface which is not very large, being of the same order as that loss with solid products. One should realize that a number of small losses can add up to a total value which should be ignored.
- The exergy loss with exhausted vapors is very large when calculated for 1 kg of water. However, it should be recognized that this energy is available at a temperature only slightly above the surrounding temperature (28.9°C) and that it is present in a large volume of air. Therefore, it seems that it would be difficult to develop an efficient means to reclaim this exergy.
- The steam condensate in the drum is another sizable potential loss. The saturated liquid at the drum pressure could be used in a heat exchanger at the same pressure or it could be flashed to atmospheric pressure and then used as a heat exchange medium, though at a lower temperature.

36.4.3 Freeze Drying

Minimizing the exergy losses in freeze drying is more significant compared to others, since the energy requirements are so much higher than for the other drying processes evaluated. Some major sources can be noted accordingly as follows:

- Exergy losses due to radiation heat transfer from the heating plates to the dryer walls and with the exiting products are negligible, being less than 0.1% of the exergy required to remove 1 kg of water.
- Two quite-sizable exergy loss areas which should lend themselves to energy reclamation are heat dissipated in the vacuum pumps and heat rejected to the environment by the refrigeration system condensers. The magnitude of this latter loss is almost equivalent to the exergy required to remove 1 kg of water.
- The largest portion of exergy loss comes from the condenser of the freeze dryer refrigeration system at which 1062.0 kcal/kg of water sublimed must be dissipated, probably either to cooling water or to the ambient air. Under normal refrigeration system operating parameters, most of this heat is available at a temperature of 38°C, a fact which will limit its usefulness. However, refrigeration systems could be designed which would make exergy available at higher temperature, but at a cost of requiring more energy input at the compressor. In this regard, there is a need for an optimum design and implementation study.

Further information on the above examples is available elsewhere [1].

36.5 EXERGY, ENVIRONMENT, AND SUSTAINABLE DEVELOPMENT

Exergy analysis is useful for improving the efficiency of energy-resource use, for it quantifies the locations, types and magnitudes of wastes and losses. In general, more meaningful efficiencies are evaluated with exergy analysis rather than energy analysis, since exergy efficiencies are always a measure of how nearly the efficiency of a process approaches the ideal. Therefore, exergy analysis identifies accurately the margin available to design more efficient energy systems by reducing inefficiencies. Many engineers and researchers agree that thermodynamic performance is best evaluated using exergy analysis because it provides more insights and is more useful in efficiency-improvement efforts than energy analysis [2].

Measures to increase energy efficiency can reduce environmental impact by reducing energy losses. From an exergy viewpoint, such activities lead to increased exergy efficiency and reduced exergy losses (both waste exergy emissions and internal exergy consumption). A deeper understanding of the relations between exergy and the environment may reveal the underlying fundamental patterns and forces affecting changes in the environment and help researchers deal better with environmental damage.

The second law of thermodynamics is instrumental in providing insights into environmental impact. The most appropriate link between the second law and environmental impact has been suggested to be exergy in part, because it is a measure of the departure of the state of a system from that of the environment. The magnitude of the exergy of a system depends on the states of both the system and the environment. This departure is zero only when the system is in equilibrium with its environment.

As stated earlier, exergy analysis is based on the combination of the first and second law of thermodynamics, and can pinpoint the losses of quality or work potential in a system. Exergy analysis is consequently linked to sustainability because in increasing the sustainability of energy use, we must be concerned not only with loss of energy but also loss of energy quality (or exergy). A key advantage of exergy analysis over energy analysis is that the exergy content of a process stream is a better valuation of the stream than the energy content, since the exergy indicates the fraction of energy that is likely useful and thus utilizable. This observation applies equally on the component level, the process level and the life cycle level. Application of exergy analysis to a component, process or sector can lead to insights on how to improve the sustainability of the activities comprising the system by reducing exergy losses.

Sustainable development requires not just that sustainable energy resources be used, but that the resources are used efficiently. The author and others feel that exergy methods can be used to evaluate and improve efficiency and thus, improve sustainability. Since energy can never be "lost" as it is conserved according to the first law of thermodynamics, while exergy can be lost due to internal irreversibilities, this study suggests that exergy losses which represent potential not used, particularly from the use of nonrenewable energy forms, should be minimized when striving for sustainable development. Furthermore, Figure 36.9 clearly summarizes key advantages of exergy as potential for better environment and sustainable development. It is obvious that an understanding of the thermodynamic aspects of sustainable development can help in taking sustainable actions regarding energy. Thermodynamic principles can be used to assess, design and improve energy and other systems, and to better understand environmental impact and sustainability issues. For the broadest understanding, all thermodynamic principles must be used not just those pertaining to energy. Thus, many researchers feel that an understanding and appreciation of exergy, as defined earlier, is essential to discussions of sustainable development.

Furthermore, Figure 36.10 illustratively presents the relation between exergy and sustainability and environmental impact. There, sustainability is seen to increase and environmental impact to decrease as the process exergy efficiency increases. The two limiting efficiency cases are significant. First, as exergy efficiency approaches 100%, environmental impact approaches zero, since exergy is only converted from one form to another without loss, either through internal consumptions or waste emissions. Also, sustainability approaches infinity because the process approaches reversibility. Second, as exergy efficiency approaches 0%, sustainability approaches zero because exergy-containing resources are used but nothing is accomplished. Also, environmental impact approaches infinity because, to provide a fixed service, an ever increasing quantity of resources

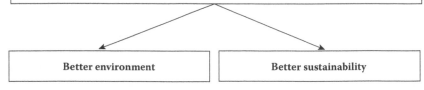

FIGURE 36.9 Illustration of how exergy contributes to better environment and sustainable development.

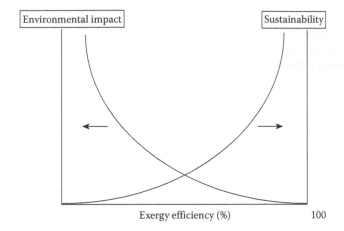

FIGURE 36.10 Qualitative representation of the relationship that exists between environmental impact and sustainability of a process and its exergy efficiency.

must be used and a correspondingly increasing amount of exergy-containing wastes are emitted (see Dincer and Rosen [2] for details).

36.6 CONCLUSIONS

This chapter has discussed thermodynamic aspects of drying systems through energy and exergy analyses. Mass, energy, and exergy balances are written and exergy efficiencies derived as functions of heat and mass transfer parameters. An illustrative example is used to verify the model and illustrate its applicability to actual drying processes at different drying air temperatures, specific exergies of drying air, exergy differences of inlet and outlet products, product weights, moisture contents of drying air and humidity ratios of drying air. The model presented appears to be a significant tool for design and optimization of drying processes for moist solids. The presented illustrative examples demonstrate the relations between energy, exergy and the environment and highlight the importance of utilizing exergy. Exergy appears to be as one of the most powerful tools in addressing and solving environmental problems and providing an effective method using the conservation of mass and conservation of energy principles together with the second law of thermodynamics for the design and analysis of drying systems.

NOMENCLATURE

C_p	Specific heat
e	Specific exergy
$\dot{E}$	Exergy flow rate
h	Specific enthalpy
$\dot{m}$	Mass flow rate
P	Pressure
$\dot{Q}$	heat transfer rate
R	Gas constant
s	Specific entropy
T	Temperature
v	Specific volume
x_v	Mole fraction of vapor
x_v^0	Mole fraction of vapor in air at dead state

Greek Letters

η	Efficiency
ω	Humidity ratio of air
ω^0	Humidity ratio of air at dead state
ε	Effectiveness
φ	Percent relative humidity of air

Subscripts

a	Air
av	Average
d	Destruction
en	Energy
ev	Evaporation
ex	Exergy
f	Saturated liquid state
g	Saturated vapor state
l	Loss
max	Maximum
p	Product
q	Heat transfer related
v	Vapor
w	Water
0	Dead state
1,2,3,4	State points (in Figure 1)

REFERENCES

1. Dincer, I. 2002. On energetic, exergetic and environmental aspects of drying systems. *Int. J. Energy Res.,* 26(8), 717–727.
2. Dincer, I., and Rosen, M.A. 2005. Thermodynamic aspects of renewables and sustainable development. *Renewable Sustainable Energy Rev.,* 9, 169–189.
3. Dincer, I. 2000. Thermodynamics, exergy and environmental impact. *Energy Sources,* 22(8), 723–732.
4. Rosen, M.A., and Dincer, I. 2004. A study of industrial steam process heating through exergy analysis. *Int. J. Energy Res.,* 28(10), 917–930.
5. Szargut, J., Morris, D.R., and Steward, F.R. 1988. *Exergy Analysis of Thermal, Chemical, and Metallurgical Processes.* Hemisphere Publishing Corp., New York, NY.
6. Kotas, T.J. 1995. *The Exergy Method of Thermal Plant Analysis,* Reprint Edition. Krieger, Malabar, FL.
7. Topic, R. 1995. Mathematical model for exergy analysis of drying plants. *Drying Technol.,* 13(1&2), 437–445.
8. Rosen, M.A., and Dincer, I. 1997. On exergy and environmental impact. *Int. J. Energy Res.,* 21, 643–654.
9. Syahrul, S., Hamdullahpur, F., and Dincer, I. 2002. Energy analysis in fluidized bed drying of wet particles. *Int. J. Energy Res.,* 26(6), 507–525.
10. Syahrul, S., Dincer, I., and Hamdullahpur, F. 2003. Thermodynamic modelling of fluidized bed drying of moist particles. *Int. J. Thermal Sci.,* 42(7), 691–701.
11. Dincer, I., and Rosen, M.A., 2007. *Exergy.* Elsevier, London, UK.
12. Dincer, I., and Sahin, A.Z. 2004. A new model for thermodynamic analysis of a drying process. *Int. J. Heat Mass Transfer,* 47, 645–652.

37 Mathematical Modeling of CIP in Food Processing

Konstantia Asteriadou
University of Birmingham

CONTENTS

37.1   INTRODUCTION

Cleaning in place (CIP) technology can clean appropriately designed process equipment and interconnecting piping without disassembly or reconfiguration. CIP methodology and equipment developed in 1950s for dairy plant processes and its implementation greatly reduced manual intervention and time required to clean process equipment, while improving quality and extending product shelf life. The technology has since been applied to many food, beverage, pharmaceutical and biotech processes to remove process soil [1].

The main reasons for CIP are [2]:

1. The equipment is not easily accessible to the operators.
2. Opening the equipment would be harmful to the operators or to the environment.
3. Opening the equipment would be detrimental to product quality.
4. CIP systems can give reproducible results.
5. The cleaning system can be automated.
6. Production down time can be reduced.

After a certain period of running a food production line, cleaning follows. The duration of cleaning depends on the kind of product and the contamination that might occur along with the cleaning conditions selected.

Common CIP steps are:

1. Water prerinse
2. Caustic wash
3. Rinse
4. Acid wash

5. Rinse
6. Disinfection
7. Final rinse

Setting up a production line in a food factory is a major task that involves many aspects. One of them is the need to be reassured that the line is cleanable. Cleaning should be seen as part of the processes and should be taken into account from the start when a new line is designed and commissioned or be validated thoroughly for an already existing line. Usually this is not the case when a line is designed and cleaning has to be faced at a later stage, resulting in more costly and time consuming operations.

Cleaning the line is affected by the following major parameter (Figure 37.1):

1. Equipment has to be hygienic and must be installed to avoid various kinds of contamination and potential damage. The equipment needs to be cleanable at food grade standard. Positioning of the equipment should allow for the best usage of available space and utilities. Thus, it should permit for machinery and technicians to operate and monitor all processes safely and efficiently.
2. The *process design* has to reassure that the product goes through appropriate temperature, pressure etc., for the right amount of time to prevent any contamination from occurring. This will allow a well designed and applied CIP to perform to it its expected standard.
3. *CIP process design* needs to ensure that the product that underwent the well controlled process will be removed and leave a clean line safe for the next batch to take place.
4. Finally, *control* and *monitoring* of all operations of the line is necessary. It is essential to monitor and keep history of key measurements. This helps to assess the performance of CIP while it takes place and between cycles throughout the whole plant operation. Thus, tracking any problems and malfunction will be easier. Automation of the processes should be working properly in order to keep the validated CIP standards.

From the above, it is apparent that hygiene and process design are very much connected and taking into account only one of them, does not lead to desirable results. Food production should take place in equipment and lines that can be cleaned preferably with automated CIP circuit. Unhygienic configurations should be avoided (e.g., dead areas, geometries that are not aligned with the cleaning flow, etc.) and at the same time monitoring of the operation has to be at an adequate level.

In the food industry, contamination of a product might be:

• Physical: e.g., broken pieces from the line or packaging
• Chemical: residuals from cleaning substances
• Microbiological: microorganisms (bacteria, moulds, yeasts) that were allowed to grow due to unsatisfactory pasteurization/sterilization or failed cleaning
• Allergens: due to change-over from an allergen carrying product to an allergen free product

Cleaning is an absolute necessity in the food industry, since deposits are quite likely to form on and adhere to equipment surface. This can have various consequences on the process. Firstly, it

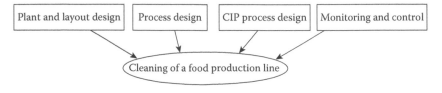

FIGURE 37.1   Factors that affect CIP of a food production line.

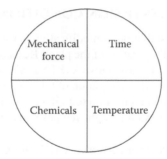

FIGURE 37.2 Factors that affect cleaning process design of a food production line.

potentially permits microbial growth; secondly the thickness of the deposited layer can hinder heat transfer and decrease the overall heat transfer coefficient. This would mean higher energy consumption in order for the product to achieve the required temperature.

There are four factors that affect the overall cleaning process. When designing cleaning procedures, these factors need to be thoroughly considered [3,4] (Figure 37.2):

- Time: the longer a cleaning solution remains in contact with the equipment surface, the greater the amount of food soil that is removed. Increasing time reduces the chemical concentration requirements.
- Temperature: soils are affected to varying degrees by temperature. In the presence of a cleaning solution, most soils become more readily soluble as temperature increases.
- Concentrations of chemicals: they vary depending on the chemical itself, type of food soil and the equipment to be cleaned. Concentration will normally be reduced as time and temperature increases.
- Mechanical force: mechanical force can be simple manual scrubbing with a brush or as complex as turbulent flow and pressure inside a pipeline. Mechanical force aids in soil removal and typically reduces time, temperature, and concentration requirements.

When it comes to cleaning closed equipment, mechanical stress on the wall is of great importance. Such high forces are achieved only with turbulent flows. There should be no dismantling of the line and flow rates should be adequate to perform complete cleaning. Hence, high Reynolds numbers are necessary and turbulent flow conditions should be achieved taking into account product and flow properties when designing the process [5].

Cleaning of closed process equipment aims at disinfection and removal of possible bacterial attachment or any soiling created on the process equipment surfaces. The use of adequate concentration of detergent, at a relatively high temperature is the chemical force that cleans a line. This, in combination with high velocities that achieve intense wall shear stresses, contribute to the detachment of soil and the reassurance of a hygienically designed process. Temperature levels should be adequate to kill pathogens and spoilage bacteria especially when disinfection takes place.

Cleaning should be validated in order to assure its performance and a line that will not lead to cross-contamination between productions. All these can lead to the design of a hygienically manufactured process that will give a safe product. Models and pilot plant experiment can contribute to the above but strong collaboration between food microbiologists, engineers and chemists is needed.

Mathematical modeling of CIP processes has mostly focused on flows and development of cleaning rates on surfaces of difficult to clean equipment, such as heat exchangers. Also when it comes to local phenomena modeling, the foods chosen are usually specific components of a foodstuff. This makes research and validation experiments more straight-forward since food products are complex systems that undergo various changes and reactions when processed.

37.2 MODELING OF PROCESS DESIGN CONDITIONS

As previously mentioned, a CIP installation or modification is usually a retrofit process in an existing line. This impacts on the overall operation of the plant. Researchers [6] used finite state automation representation and equipment state task network (STN), which describes the allowed states and transitions of all plant equipment, for the integration of CIP in an existing multipurpose batch pilot plant. The retrofit problem is formulated as a mixed integer linear program (MILP) and is solved using a branch and bound technique. Adding to the recipes and equipment availability, they handled it as an optimization problem.

Work has been done on heat exchanger networks (HENs) to optimize energy usage and cleaning after fouling has occurred [7]. They used a MILP model for serial and parallel HENs and a combination of the two. In such cases where the state of each piece of equipment affects the overall network performance, analytical methods for the determination of cleaning procedures are inadequate. For continuous networks, an objective function is developed that includes only utility and cleaning costs

$$\min \sum_{p} CU_p \cdot \Theta_p + \sum_{i} \sum_{p} \left(\bar{C}_{ip} \cdot (1 - y_{ip}) + \hat{C}_{ip} \cdot \Delta t_{ip} \right) \tag{37.1}$$

where
   i is the heat exchanger
   p is the time period
   CU_p is the cost coefficient of hot utility
   Θ_p is the temperature of hot utility
   $\bar{C}_{ip}$ is the fixed cleaning cost coefficient over period p
   y_{ip} is the binary variable that equals 1 if exchanger i operates over period p and equals 0 if
        otherwise
   $\hat{C}_{ip}$ is the variable cleaning cost coefficient over period p
   Δt_{ip} is the elapsed time since last cleaning

Equation 37.1 is subject to energy balance, fouling, timing, cleaning, and residence time constraints.

It has been proven that hot utility availability and residence time specifications have significant effect on cleaning and energy management. The models developed cannot be used to networks with split streams. However, the low computational cost of the models encourages their online use [8].

Smaili et al. [9] used a NLP/MINLP (NLP, nonlinear programming) in a HEN used for preheating in a sugar refinery. It was again for continuous operations after having established the fouling behavior through heat transfer data.

37.3 LOCAL PHENOMENA MODELING

37.3.1 HEAT EXCHANGERS DEPOSIT REMOVAL MODELING

Heat exchangers and their cleaning are one of the most crucial pieces of equipment in food processing that causes fouling problems due to its high usage in heat transfer operations. As a consequence, many researchers strive to understand the cleaning and the removal kinetics in heat exchangers. Durr and Grasshoff [10] used a model that was initially developed for the removal of dust from floor carpets by vacuum cleaning. They expressed the relative cleaning rate, i.e., the cleaning rate (v) divided by the remaining soil (r):

$$v = \frac{v}{r} = \left(\frac{R}{T} \right) \left(\frac{t}{T} \right)^{R-1} = \text{relative cleaning rate} \tag{37.2}$$

where

T = time constant; theoretical time to reach 63.2% soil removal
R = slope of cleaning characteristic
t = cleaning time

They compared results with a laboratory-scale CIP test rig. Different chemicals were applied and the model was accepted as an effective method for quick assessment of industrial cleaning processes. Two years later [11], they extended the model and its understanding by testing it on various cleaning applications. Soil removal, s, was expressed by the Weibull function:

$$s = 1 - e^{-(t/T)^R} \qquad (37.3)$$

The slope R is specific to the type of soil, soiled surface and cleaning appliances. The constant T seemed to be an indicator of the tool or the cleaning agent used. R classifies the type of cleaning process [12]: $R = 1$ when processes are of constant cleaning rate, $R < 1$ when it is monotonously decreasing and $R > 1$ when it is monotonously increasing. The history of the process has a positive influence on the relative cleaning rate.

37.3.2 Cleaning in Place (CIP): Removal Mechanisms

The mechanism of soil removal from contact surfaces has been the focal point of studies for many years. Jennings [13] gave the mathematical expression of uniform removal of cleaning solid surfaces as follows (Equation 37.4):

$$\frac{dm}{dt} = -K_R \cdot m \qquad (37.4)$$

where

m = soil mass
t = cleaning time
K_R = cleaning constant

The cleaning constant is influenced by the solubility of the soil in cleaning solution, the nature of the solid surface, temperature, flow rate and chemical concentration in cleaning solution. However, this approach implied that cleaning rate decreases with the amount of soil and the time required to clean the surface totally is infinite [14]. In practice, a limited period of time is available to perform the cleaning operation.

Many food components that are responsible for fouling in the food industry have been studied with regards to their removal from stainless steel surfaces under CIP conditions. Whey protein is one such food component. Its cleaning using NaOH alkaline detergent is described in (Figure 37.3) [15]. At the beginning of the cleaning process, solution swells up the soil as it flows above it. At this stage, removal is slow or might not occur. Following prolonged contact with detergent, the protein deposit softens and breaks up, thus accelerating its removal rate. This is the stage where the removal rate is highest and relatively constant for a certain period of time. It depends on the amount of soil and is also influenced by temperature and concentration of chemicals. Despite removal of the majority of the deposit, small amounts might still remain on the stainless steel surface. This is mostly dependent on the mechanical forces applied by the flow. At this stage, the soil is aged and decaying. (Figure 37.4) shows a typical representation of a cleaning curve where the three above mentioned stages are clearly illustrated: swelling, plateau, and decay.

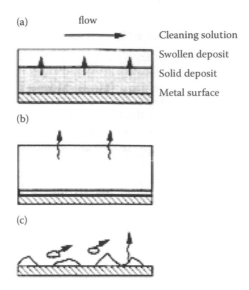

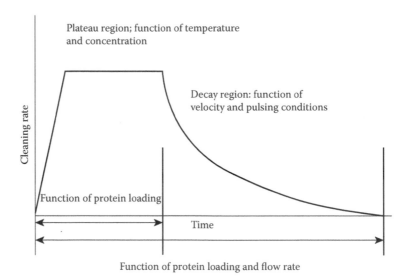

FIGURE 37.3 Stages involved in whey removal mechanism: (a) swelling face, (b) uniform erosion phase, (c) decay/aging. (From Bird, M.R. and Fryer, P.J., Trans. IChemE, *Food Bioprod. Proc.*, 69C, 13–21, 1991. With permission.)

FIGURE 37.4 Schematic representation of a cleaning curve.

Gillham et al. [16] ran experiments to investigate the mechanisms involved in alkali-based cleaning of whey protein deposits on stainless steel surfaces. They used rate deposit surface and heat transfer techniques. The results showed that reaction and diffusive transport processes occurring in the swollen deposit layer determine the cleaning rate under conditions of steady pipe flow. In the plateau/uniform phase, the removal was more sensitive to the deposition/solution interface temperature and less sensitive to hydraulic/external mass transport conditions. In the decay stage, there was a strong dependence on mechanical effects and a threshold temperature above which, there was little effect of temperature. Mass transfer rate and fouling resistance (R_f) was used to model kinetics:

$$R_f = \left| \frac{T_{block} - T_{bulk}}{q} \right| - \frac{1}{U_{clean}} \tag{37.5}$$

where
T_{bulk} = temperature in bulk liquid (K)
T_{block} = temperature of the block experimental assembly (K)
q = heat flux (W m^{-2})
U_{clean} = overall heat transfer coefficient at the end of cleaning experiment (W m^{-2}K^{-1})

Removal of food soils has been studied with a micromanipulation technique [17] that measures the force required to remove food deposits from stainless steel surfaces with a T-shaped probe (Figure 37.5). It has been used for forces on a tomato paste deposit at different heights x from the surface. At small x, the work needed to remove the deposit increases with height and is a function of the nature of the surface, whilst at larger x the work needed decreases with increasing x and is not a function of the surface. Two simple models have been developed which fit the data and observations of cleaning (Figure 37.6); these are based (i) at small x, on the fracture of the deposit and motion across the surface, and (ii) at larger x, on the fracture of deposit–deposit bonds and removal of a uniform layer.

$$F = YA_f + \kappa_s L \text{ at small } x \tag{37.6}$$

where
A_f = the area of fracture equal to $L*x$
L = contact length between the force probe and food deposit at the breakage point and the resisting interfacial tension is κ_s
Y = critical yield stress that must be exceeded to start removal.

Consider a deposit of thickness d, and a probe distance x from the surface. In a case where a layer of deposit is removed along the probe height, the work required is the sum of that required to break

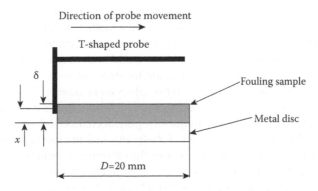

FIGURE 37.5 Schematic of the T-shaped probe, fouling sample and stainless steel disc: the probe cuts a sample of thickness a distance x above the surface of the disc. (From Liu, W., Zhang, Z., and Fryer, P., *Chem. Eng. Sci.*, 61, 7528–7534, 2006. With permission.)

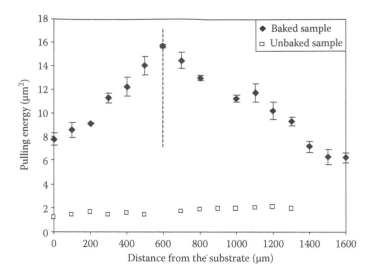

FIGURE 37.6 Data for the pulling energy required to disrupt baked and unbaked tomato deposits, as a function of probe cut height above the surface. (From Liu, W., Zhang, Z., and Fryer, P., *Chem. Eng. Sci.*, 61, 7528–7534, 2006. With permission.)

deposit–deposit bonds over the entire surface, assumed as τ_d (J/m^2), and that required to deform and remove the deposit, assumed as ψ (J/m^3). Thus,

$$W = \tau_d A + \psi A(\delta - x) \tag{37.7}$$

so

$$\sigma = \tau_d + \psi(\delta - x) \tag{37.8}$$

where

σ = apparent adhesive strength (J/m^2)

Xin et al. [18] expressed the constant cleaning rate in the uniform stage of whey protein cleaning as a product of mass transfer coefficient and saturation concentration of disengaged protein molecules. Dissolution and chemical reaction mechanism contributions were investigated for whey protein gel model deposits. Later on [19] they developed a mathematical expression for the cleaning of whey protein concentrate from stainless steel surfaces. They accounted for two of the cleaning stages: the swelling-uniform and decay stages. The model was supported by a number of experimentally determined parameters but did not account for shear stress induced aggregate removal.

On surfaces covered by flowing liquid, reversible adhesion is more probable than irreversible. An additional mechanism that allows microorganisms to remain adhering is the production of a capsule of exocellular polymers (proteins or more frequently, polysaccharides). Polymer threads come close to the surface, where they adhere irreversibly. Firmly attached microorganisms can then multiply and colonize the surface within the polymer matrix thus forming a "slime layer" [20].

With regards to the primary step of adhesion, investigations using a variety of microorganisms confirm thermodynamic provisions: those with high free surface energy adhere to hydrophilic materials, while those with low free surface energy adhere more to hydrophobic materials. If macromolecules, which modify the apparent surface energy of material are excreted by microorganisms, thermodynamic models remain valid provided the new state of the surface is considered.

Wirtanen et al. [21] tested various parameters for estimating cleaning procedures in open systems to eliminate biofilms of *Bacillus subtilus*, *Listeria monocytogenes*, *Pediococcus pentosaceus*, and *Pseudomonas fragi* on stainless steel surfaces. With statistical analysis, they showed that the roughness of the stainless steel surface is the most important factor in cleaning surfaces from biofilms.

Biofilm has been investigated industrially for years [22,23]. Growth of bacteria and their adherence to surfaces has led to the consideration that the hygiene of surfaces, instruments and equipment in the food industry essentially affects the quality of the products processed. In process equipment, open or closed, biofilm has ideal opportunities to develop e.g., bends, seals, crevices, dead ends and grooves. If cleaning and sanitation are inadequate the above sources result in contamination. Microorganisms of food products have varied classes of surface free energies. In general, bacteria have a high-energy surface whereas moulds have a low-energy surface.

Bird and Espig [24] investigated the removal of crude oil films from stainless steel surfaces using nonionic surfactants at different temperatures. The cleaner used was alcohol ethoxylate. Plain water rinsing was not effective and followed Arrhenius kinetics. Surfactant cleaning deviated strongly from Arrhenius kinetics. At high surfactant concentrations, a smooth curve is produced which appears to be physically controlled at high temperatures and a combination of chemically and diffusion controlled at low temperatures. A four-step cleaning mechanism is postulated where the surface modification of the oil surfactant adsorption is the rate-limiting step in the removal process. Bird and Bartlett [25] also elaborated on concentration and temperature optima during the removal of protein, starch and glucose for CIP processes.

The disinfection step kinetics of a CIP process was predicted by applying a multiple linear regression (MLG) method for a brewery [26]. Calibration models were built up by spectra collection data from reaction of peroxyacetic acid and hydrogen peroxide with diphenylamine and determining their residues from a disinfection process in a brewery.

37.3.3 EQUIPMENT AND LINE DESIGN

It is quite common in the production lines to encounter pipework configurations that might be risky and unhygienic. They might be forming areas not aligned with the flow of the product and consequently the cleaning solution flow. These might be due to positioning of sensors such as in line fitted pressure gauge or any other measuring devices (Figure 37.7) [27] They might also be dead ends and T-pieces that are formed when the flow is redirected (e.g., retrofitted installation) (Figure 37.8). Also, opening or closing valves in order to direct the flow through the different paths can create dead spaces in the pipes just before the divergence occurs.

These areas end up trapping the product. Stagnation follows especially for highly viscous products, and, depending on process conditions (temperature, velocity) and product properties (pH, water activity). Bacterial growth could occur in the product due to mass exchange with bulk flow. This is the first difficulty that has to be taken into consideration when there are risky geometries formed across the line. The second is to ensure that cleaning is efficient enough to remove trapped product before new production starts. Any residues should be cleaned off for the line to be considered hygienic.

One could easily suggest detaching these configurations from the lines. This cannot always be implemented for various reasons: primarily due to time constrains, as delay in production costs money to the factory. Interrupting production results in loss of earnings and must be kept to a minimum. Also, changes in pipework or replacing equipment are costly, as it often requires dismantling of the line or buying new components. Cleaning time, on the other hand, should be kept to a minimum to maintain high production time.

In a hygienically designed line, multiplication of microorganisms during processing is limited and with frequent cleaning, is kept to a minimum. A well-designed line can be rapidly cleaned and decontaminated effectively. Reduced frequency of cleaning and shorter cleaning times will reduce production downtime [28].

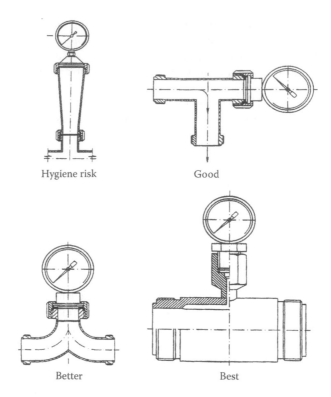

FIGURE 37.7 Locations of a pressure gauge related to hygienic risk. (From Hygiene Manufacturing Guidelines. Confidential Unilever Handbook. With permission.)

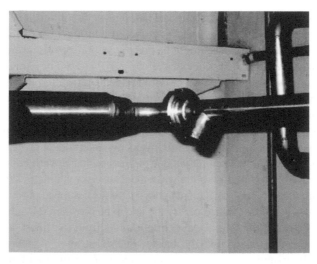

FIGURE 37.8 Example of sealed pipe that might be a source of contamination. (From Unilever R&D Material. With permission.)

There are design guidelines to maintain hygienic aspect in tanks, pumps, and pipework (Joint Technical Committee, FMF/FMA) including basic principles for design and materials for construction that show how and when some geometries are not, or less risky. Although, it is not always easy to anticipate in advance, an investigation or tests should be carried out.

Significant work has been carried out to ensure hygienic design in food processing lines. There are also a number of patents that have suggested fittings or couplings that minimize product trapping between pipes of different dimensions, bends or any other changes in a plant. Applying protective layers onto vessels used in the food industry (such as milk and meat) is another method that has been invented so that microorganisms' adhesion is minimized [29].

The Campden Association have also suggested design rules for liquid handling equipment in the food industry [30] such as use of various couplings, joints, flange assemblies, fittings, drainable arrangements, T-pieces, etc.

Grasshoff [18] has studied the flow behavior of fluids in dead spaces and the influence of fluid movement on cleanability. He found that fluid exchange and local shear stress decrease very rapidly with increasing stagnant space depth. He found that the maximum permissible dead space depth is easily reached. The required rinsing times exceeded the time needed for the individual steps in programed controlled CIP.

Recent studies have investigated the flow in a down stand [31,32] applying a commercial CFD (computational fluid dynamics) code, FLUENT. The studies showed that even for highly turbulent flows of a Newtonian fluid in a horizontal pipe, a dead end forming a stagnant zone could not necessarily reach the desired pasteurization/sterilization temperature (Figure 37.9).

They expressed the temperature along a vertical, down standing T-piece depending on the known inlet flow velocity for flows varying from laminar to turbulent [33]:

$$T = T_{in} \; 0 \leq x \leq \varepsilon \quad \text{and} \tag{37.9}$$

$$T = \left(T_{in} - T_{env} \right) \cdot e^{-B(x-\varepsilon)} + T_{env} \; x \geq \varepsilon \tag{37.10}$$

where
T_{in} = temperature at the inlet of the flow and top of T-piece (K)
T_{env} = air temperature

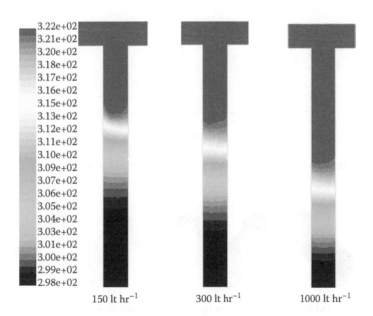

FIGURE 37.9 Contours of temperature distribution along the dead leg for various flows. (From Asteriadou, K., Hasting, T., Bird, M., and Melrose, J., *J. Food Process Eng.*, 30(1), 88–105, 2007. With permission.)

B = pre-exponential factor of the exponential equation (m^{-1})
ε = distance down the pipe where temperature starts to drop (m)

Equation 37.9 is valid for values of x lower than ε and Equation 37.10 is valid for values of x higher that ε. A representing plot is shown in Figure 37.10. B and ε depend on the inlet velocity.

CFD has also been used by Jensen and Friis [34] to model flows in upstands (Figure 37.11) and valves during CIP flows. They used the finite volume method code STAR-CD. Validation was carried out using the standardized cleaning test proposed by the European Hygienic Engineering and Design Group (EHEDG). The controlling factors for cleaning were shear stress and the nature and magnitude of recirculation zones. They concluded that CFD can be a qualitative tool for hygienic design and that complex geometries are not necessarily difficult to clean. Also, three-dimensional modeling is suggested for turbulent flows in that kind of configuration.

They also simulated flow in a mix-proof valve [35]. Flow patterns visualized by laser sheet visualization (LSV) were identified in the same regions using CFD. They compared various mesh and model set-ups and emphasize the importance of resolving flow in the near-wall region for modeling this type of wall-bounded flows. They also worked on simulating the flow through a radial flowcell using STAR-CD [36]. They showed that with appropriate models and mesh configuration, shear stress under transient flow could be estimated. They validated this with the EHEDG test method.

Jensen [37] showed, with various models run with STAR-CD, that by only measuring critical wall shear stress on spherical-shaped valve it was not possible to predict cleanability. Other parameters that should be taken into account are fluid exchange and flow patterns. Generally, he found that quantitative prediction of wall shear stress values for complex flows was quite difficult and the average values predicted by CFD could not justify all areas found cleaned or uncleaned by standard EHEDG test for equipment that contains three-dimensional flow phenomena.

Local wall shear stresses have been investigated by Lelievre et al. [38] in configurations where the radius was suddenly expanding or contracting along the pipe length in relation to bacterial

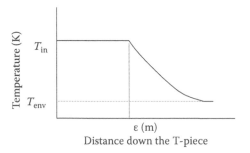

FIGURE 37.10 Shape of temperature drop in the central axis of the T-piece.

FIGURE 37.11 Simulated fluid exchange in an up-stand. The gray fluid was introduced at time 0 s and is gradually replacing the dark fluid. The pictures were taken at time 0.05, 0.2, 0.3, 0.8, and 1.05 s from left to right. (From Friis, A. and Jensen, B.B.B., *Food and Bioproducts Process.*, 80(C4), 281–285, 2002. With permission.)

removal. They found that some low shear stress areas could be very cleanable due to the high tur-bulence observed, which means high shear stress fluctuation rate. Therefore, it is not only necessary to take into account the mean shear stress but also the fluctuation rate.

CIP of a T-piece and a "test-cell" was modeled with CFX by the Campden Association [39] to determine efficiency of the cleaning process. They checked shear stress on the walls and also measured log reduction number of bacteria in biofilms that remained attached on the surfaces after cleaning. It was demonstrated that log reduction increases with shear stress. They also referred to models of butterfly valves and pipe couplings where flow at an angle occurs. They claimed that CFD can be used to assess cleaning quantitatively and by comparing with EHEDG experiments, they suggested that development of a correlation between flow properties and cleaning efficiency would give a very useful tool.

A recent application of CFD is on predicting the displacement of yogurt by water in a pipe [40]. This is a typical operation in the food industry and takes place in product change over and in rinsing. They used Fluent 6.1 and more specifically, the method of volume of fluid (VOF) and the species model. The former was more suitable in predicting the general phase distribution but mixing was not captured. The species model led to an overestimation of mixing.

Celnik et al. [41] presented a green function method that allows direct calculation of wall shear stress and flow rates generated by pulsed flow for use in application of CIP with periodic pressure harmonic gradient.

37.4 CONCLUSIONS

Mathematical modeling in food cleaning processes has not found the merit given in other industries [42,43]. This is the case especially for CIP that is a process where monumental changes have not taken place for the last few years. Research is commonly focused on developing experimental meth-ods to validate its performance. The deposits that have mostly been studied are the ones occurring in the dairy industry, or key components of the dairy industry. New chemicals and detergents are being discovered that are dedicated to specific soils and products. This is because food production is a rather "traditional" area and also because health and safety of the consumers is involved.

Supply chain considerations could save time and energy if production and cleaning are scheduled according to mathematical algorithms.

Product purging techniques are sometimes in place such as pigging, ice pigging [44] air purg-ing [45] that aim to recover product before CIP starts. This can save time and materials. However, it cannot be applicable to all products and equipment. Solid pigs could cope with straight lines but not with vessels. Air recovery can be a rather expensive investment for a food factory. Thus, better understanding of existing CIP methods can already offer great advantages.

Generally, CFD appears to be a good qualitative predictive tool. However, more work needs to be done and more ideas to be investigated with more sophisticated models and correlations or by taking advantage of existing CFD models and combinations thereof.

As previously mentioned, due to the complexity of food systems it is mostly specific components that have been studied to quantify and validate cleaning mechanisms and mathematical models that describe them. Increasing knowledge and understanding of deposit structures under cleaning condi-tions will potentially lead to the development of more accurate models.

REFERENCES

1. Andersen, B.J.. 2007 *Control Design for CIP Systems*. Control Engineering, www.controleng.com, IP1-IP4.Reed Elsevier, Inc. London, UK.
2. Sharp, N.P.B. CIP system design and philosophy. *J. Soc. Dairy Technol.*, 38(1), 17–21, 1985.
3. Jennings, W.G., McKillop, A.A., and Luick, J.R. Circulation cleaning. *J. Dairy Sci.*, 40, 1471–1479, 1957.

4. Bishop, A. Cleaning In The Food Industry. Reprinted by permission of Wesmar Company Inc., Washington, US, *from Basic Principles of Sanitation.*
5. Brodkey, R.S. and Hersey, H.C. *Transport Phenomena.*: A Unified Approach. McGraw-Hill, New York, 1988. Translated in Greek by Lavdakis, K.A., and Tziola, E., Thessaloniki, Greece, 1990.
6. Barbosa, P.F.D and Macchietto, S. Detailed design of multipurpose batch plants. *Comput. Chem. Eng.*, 18(11/12), 1013–1042, 1994.
7. Georgiadis, M.C. and Papageorgiou, L.G. Optimal energy and cleaning management in heat exchanger networks under fouling. *Trans. IChemE*, 78(Part A), 168–179, 2000.
8. Georgiadis, M.C., Papageorgiou, L.G. and Macchietto, S. Optimal cleaning policies in heat exchanger networks under rapid fouling. *Ind. Eng. Chem. Res.*, 39, 441–454, 2000.
9. Smaili, F., Angari D.K., Hatch, C.M, Herbert, O., Vassiliadis, V.S., and Wilson, D.I. Optimisation of scheduling of cleaning in heat exchanger networks subject to fouling: Sugar industry case study. *Trans. IChemE*, 77(Part C), 159–164, 1999.
10. Durr, H. and Grasshoff, A. Milk heat exchanger cleaning: modeling of deposit removal. *Food Bioproducts Process.*, 77 (C2), 114–118, 1999.
11. Durr, H. and Grasshoff, A. A mathematical model for cleaning processes. *Tenside Surfactants Detergents*, 38(3), 148–157, 2001
12. Durr, H. Milk heat exchanger cleaning—Modelling of deposit removal II. *Food and Bioproducts Process.*, 80(C4), 253–259, 2002.
13. Jennings, W.G. Theory and practice of hard surface cleaning. *Adv. Food Res.*, 14, 325–458, 1965.
14. Wilbrett, W. and Sauerer, V. The multi-use test for determining the effectiveness of cleaning in place. *Milchwissenschaft*, 45(12), 763–766, 1990.
15. Bird, M.R. and Fryer, P.J. An experimental study of the cleaning of surface fouled with whey proteins. *Trans. IChemE, Food Bioprod. Proc.*, 69C, 13–21, 1991.
16. Gillham, C.R., Fryer, P.J., Hasting, A.P.M., and Wilson, D.I. Cleaning-in-place of protein fouling deposits: mechanisms controlling cleaning. *Trans. IChemE*, 77(Part C), 127–136, 1999.
17. Liu, W., Zhang, Z., and Fryer, P. Identification and modeling of different removal modes in the cleaning of a model food deposit. *Chem. Eng. Sci.*, 61, 7528–7534, 2006.
18. Xin, H., Chen, X.D., and Ozkan, N. Cleaning rate in the uniform cleaning stage for whey protein gel deposits. *Trans. IChem E*, 80(Part C), 240–246, 2002.
19. Xin, H., Chen, X.D., and Ozkan, N. Removal of a protein foulant from metal surfaces. *AIChE J.*, 50, 1961–1973, 2004.
20. Grasshoff, A. Hygienic design—the basis for computer controlled automation. In ICHEME Symposium Series No. 126, Food Engineering in a Computer Climate, Taylor & Francis/Hemisphere Bristol, PA, USA, 89–109.
21. Wirtanen, G., Ahola, H., and Mattila-Sandholm, T. Evaluation of cleaning procedures for elimination of biofilm from stainless steel surfaces in open process equipment. In Proceedings of a Conference on Fouling and Cleaning in Food Processing. Jesus College Cambridge, UK 23–25 March 1994, ISBN 92-827-4360-8
22. Characklis, W.G. and Marshall, K.C. *Biofilms.* Wiley-Interscience Publication John Wiley & Sons, Inc., New York, 1990.
23. Panikov, N.S. and Nikolaev, Y.A. Growth and adhesion of Pseudomonas fluorescens in a batch culture: a kinetic analysis of the action of extracellular antiadhesins. *Microbiology*, 71(5), 532–540, 2002. Translated from *Mikrobiologiya*, 71(5), 619–628, 2002.
24. Bird, M.R. and Espig, S.W.P. The Removal of crude oil from stainless steel surfaces using non-ionic surfactants. In Proceedings of a Conference on Fouling and Cleaning in Food Processing. Jesus College, Cambridge, UK, 6–8 April 1998, Office of official publications of the European Communities, Luxembourg.
25. Bird, M.R. and Bartlett, M. CIP optimisation for the food industry. *Trans. IChemE*, 73(Part C), 63–70, 1995.
26. Pettas, I.A. and Karayannis, M.I. Simultaneous spectra-kinetic determination of peracetic acid and hydrogen peroxide in a brewery cleaning-in-place disinfection process. *Anal. Chimica Acta*, 522, 275–280, 2004.
27. Hasting, A.P.M. Industrial experience of monitoring fouling and cleaning systems. In Proceedings of a Conference on Fouling and Cleaning in Food Processing. Jesus College, Cambridge, UK, 3–5 April, 2002, ISBN 0-9542483-0-9.
28. EHEDG. Hygienic Equipment Design Criteria. *Trends Food Sci. Technol.*, 4, 225–229, 1993.
29. Van Leeuwen, P.J.M. Method for applying a protective layer to which microorganisms do not adhere to vessels and utensils used in the food industry. Patent 11 of January 1996, WO 96/00505.

30. Thorpe, R.H. and Barker, P.M. Hygienic design of liquid handling equipment for the food industry. Technical Manual No. 17. Food Preservation Research Association, CCFRA, Chipping Campden, UK, 1988.
31. Asteriadou, K., Hasting, A.P.M., Bird, M.R., and Melrose, J. Computational fluid dynamics for the prediction of temperature profiles and hygienic design in the food industry. *Food Bioprod. Process.*, 84 (C2), 1–7, 2006.
32. Asteriadou, K., Hasting, T., Bird, M., and Melrose, J. Predicting cleaning of equipment using computational fluid dynamics. *J. Food Process Eng.*, 30(1), 88–105, 2007.
33. Asteriadou, K. The use of computational fluid dynamics for the microbial assessment of food processing equipment. PhD thesis, University of Bath, UK, May 2005.
34. Friis, A. and Jensen, B.B.B. Prediction of hygiene in food processing equipment using flow modelling. *Food Bioproducts Process.*, 80(C4), 281–285, 2002.
35. Jensen, B.B.B. and Friis, A. prediction of flow in mix-proof valve by use of CFD-validation by LDA. *J. Food Process Eng.*, 27, 65–85, 2003.
36. Jensen, B.B.B. and Friis, A. critical wall shear stress for the EHEDG method. *Chem. Eng. Process.*, 43, 831–840, 2004.
37. Jensen, B.B.B. Hygienic design of closed processing equipment by use of computational fluid dynamics. PhD thesis, Technical University of Denmark, Lungby, September 2002.
38. Lelievre, C., Legentilhomme, P., Gaucher, C., Legrand, J., Faille, C., and Benezech, T. Cleaning in place: effect if local wall shear stress variation on bacterial removal from stainless steel equipment. *Chem. Eng. Sci.*, 57, 1287–1297, 2002.
39. Hall, J.E., Jones, M.R., and Timperley, A.W. Improving the hygienic design of food processing equipment using modelling approaches based on computational fluid dynamics. Confidential R&D Report No. 91 Project No. 29735, Campden and Chorleywood Food Research Association, Chipping Campden, UK.
40. Regner, M., Henningsson, M., Wiklund, J., Ostergren, K., and Tragardh, C. Predicting the displacement of yogurt by water in a pipe using CFD. *Chem. Eng., Technol.*, 30(7), 844–853, 2007.
41. Celnik, M.S., Patel, M.J., Pore, M., Scott, D.M., and Wilson, D.I. Modelling laminar pulsed flow for the enhancement of cleaning, *Chem. Eng. Sci.*, 61, 2079–2084, 2006.
42. Varga, S, Oliveira, J.V., Smout, C., and Hendrickx, M.E. Modelling temperature variability in batch retorts and its impact on lethality distribution. *J. Food Eng.*, 44, 163–174, 2000.
43. Farid, M and Ghani, A.G.A. A new computational technique for the estimation of sterilisation time in canned foods. *Chem. Eng. Process.*, 43, 523–531, 2004.
44. Quarini, J. Ice-pigging to reduce and remove fouling and to achieve clean-in-place. *Appl. Thermal Eng.*, 22, 747–753, 2002.
45. Roscoe, K. www.Whirltech.co.uk personal communication.

Index

Numerical equations
  incomplete solutions, 134–135
  iteration, relaxation and time steps, 135–136
  linearized equations and residuals, 135
Nusselt number, 141–142, 203–204, 815
Nutrient consumption by bacterial surface population, 851–852

O

Oats use in extrusion, 797–798
Ohmic heating
  column, 674, 679
Ohmic heating for food processing, 81, 659–660
  batch processing, basic configurations for, 671
  continuous process, configurations for, 672
  food heating treatment, factors affecting, 660–661
    anisotropic tissue of food, 666–667
    fluid food, electrolytic and nonelectrolytic components, 662–663
    food and pretreatment, components, 665–666
    food mixture particles, 663–664
    geometry, location, and orientation of the particle within food mixture, 664–665
    liquid or food mixture, viscosity of, 667
    power, frequency and waveform, 667–668
    solid food materials, 661–662
  modeling, 668
    energy balances of continuous treatment, 672–686
    Joule's law, 669
    system performance coefficient (SPC), 684–686
    voltage distribution, 669–672
One-dimensional conduction, 73
One-equation effective viscosity models, 17; see also Viscosity
Opa protein receptors, 834
Optimum gap width
  laminar flow reactors, ultraviolet (UV) inactivation
    optimum ratios, range of, 585
    penetration depth, ratio of, 584
    theoretical analysis, 583–584
Optimum storage conditions aspects for cool and cold storage, 417
  air movement, 419
  condensation of water vapor on product surface, 419
  optimum temperature, 418
  relative humidity, 418
  sanitation, 419
  stacking/stowage, 419
Osmosis technique, 737
Osmotic pressure model, 760
Overall heat transfer coefficient and refrigerating capacity, 466–467
Oxygen balances for gas–liquid transfer in bioreactor, 104–105
Ozone
  physical properties of, 610
  production, 609
    corona discharge method, 610–611
    electrochemical and ultraviolet (UV) method, 611
Ozone processing, 607–608
  food processing equipment and environment, disinfection of, 617
  limitations, 620

liquid food materials, application, 616–617
    apple cider, 617
    drinking water, 616–617
  microbial inactivation, mechanism, 614
  modeling ozone in food materials
    bubble columns, 611–614
  product quality, effect, 617–618
  regulatory catch-up
    FDA grant, 608
    Food and Drug Administration (FDA), 609
    water treatment in, 608
  safety requirements, 619
  solid food materials, application, 614–616

P

Package-icing/contact icing, 405
Packed-bed reactors, 883
Palaniappan's model, 683–684
Parallel flow chamber, 836; see also Bacteria
Parallel-flow design spray dryers
  droplet heat balance equations, 305–306
  droplet mass balance equations, 304–305
  droplet trajectory equations, 303–304
  mass and energy balance equations for drying medium, 306
Particle image velocimetry (PIV), 512, 726
Pasta products and extrusion mechanism, 824
Pearl–Verlhurst growth model, 570
Peclet number, 186, 813
Pediococcus pentosaceus biofilms on stainless steel surfaces, 945
PER, see Plug flow reactor (PFR)
Perfectly mixed flow (PMF), 633–634
Pervaporation process, 748–749
  volatile organic compounds (VOCs), concentration of, 749
Pham's method, 79
Phase change/transition
  associated with food, 227
PHOENICS manuals, 185
PHONICS subroutines, 553
Physical aging, 232
Pitot tube fluid velocity measurement technique, 727
PIV, see Particle image velocimetry (PIV)
Planck's equation, 79, 386
Planktonic bacterial community, 846
Plant-wide control, 892; see also Food processing control
Plate and frame modules, 754
  plate heat exchanger (PHE), 754
Plate heat exchangers (PHEs), 208
Plug flow reactor (PFR), 863, 873–874; see also Ideal reactors
Plug/piston flow (PF), 633–634
PMF, see Perfectly mixed flow (PMF)
Poiseuille flow, 578, 581–582
  annular, diagram of, 582
Polyamide membranes, 751
  advantages and disadvantages, 752
POLYFLOW software, 820
Polylactide thermogram, 238
Polymeric materials, glass transition temperature, 235–236
Porous catalysts in slurry reactors, 875